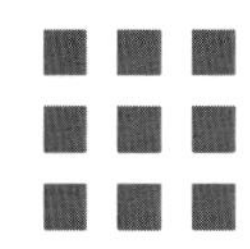

Biopolymeric Nanomaterials

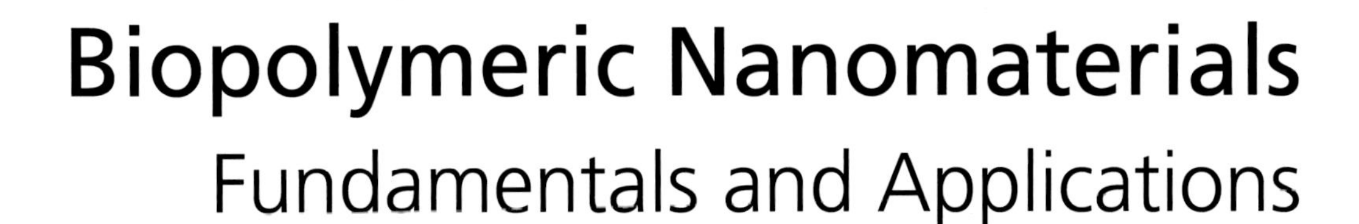

Biopolymeric Nanomaterials
Fundamentals and Applications

Edited by

Shamsher S. Kanwar

Biotechnology and Director of IPRCC, Himachal Pradesh University, Shimla, India

Ashok Kumar

Department of Biotechnology and Bioinformatics, Jaypee University of Information Technology, Waknaghat, India

Tuan Anh Nguyen

Microanalysis Department, Institute for Tropical Technology, Vietnam Academy of Science and Technology, Hanoi, Vietnam

Swati Sharma

University Institute of Biotechnology, Chandigarh University, Mohali, Punjab, India

Yassine Slimani

Department of Biophysics, Institute for Research and Medical Consultations, Imam Abdulrahman Bin Faisal University, Dammam, Saudi Arabia

Elsevier
Radarweg 29, PO Box 211, 1000 AE Amsterdam, Netherlands
The Boulevard, Langford Lane, Kidlington, Oxford OX5 1GB, United Kingdom
50 Hampshire Street, 5th Floor, Cambridge, MA 02139, United States

British Library Cataloguing-in-Publication Data
A catalogue record for this book is available from the British Library

Library of Congress Cataloging-in-Publication Data
A catalog record for this book is available from the Library of Congress

ISBN: 978-0-12-824364-0

For Information on all Elsevier publications visit our website at
https://www.elsevier.com/books-and-journals

Publisher: Matthew Deans
Acquisitions Editor: Simon Holt
Editorial Project Manager: Gabriela Capille
Production Project Manager: Prasanna Kalyanaraman
Cover Designer: Greg Harris

Typeset by Aptara, New Delhi, India

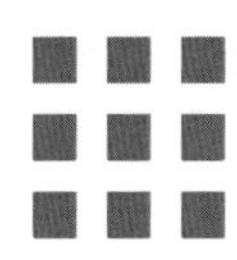

Contents

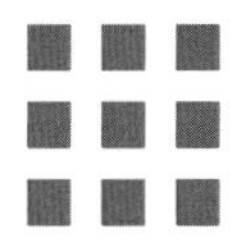

Contributors

Tibor Alpár, Simonyi Károly Faculty of Engineering, University of Sopron, Sopron, Hungary

Maria Jaízia dos Santos Alves, Department of Chemical and Food Engineering, Federal University of Santa Catarina, Florianópolis, SC, Brazil

Daniel López-Ângulo, Laboratory of Food Technology, Faculty of Animal Science and Food Engineering—FZEA/USP, University of São Paulo, Pirassununga, São Paulo, Brazil

Daniel Enrique López Angulo, Faculty of Animal Science and Food Engineering, University of São Paulo, Av. Duque de Caxias Norte, Pirassununga (SP), Brazil

Samin Anjum, Applied Chemistry and Chemical Engineering, University of Dhaka, Dhaka, Bangladesh

Niloofar Arastehnejad, Department of Chemistry, Kansas Polymer Research Center, Pittsburg State University, Pittsburg, United States

P. Marie Arockianathan, PG & Research Department of Biochemistry, St. Joseph's College of Arts & Science, Cuddalore, Chennai, Tamil Nadu, India

Muhammad Bilal Asif, Institute of Environmental Engineering & Nano-Technology, Tsinghua-Shenzhen International Graduate School, Tsinghua University, Shenzhen, Guangdong, China

Rida Badar, Interdisciplinary Research Center in Biomedical Materials, COMSATS University Islamabad, Lahore Campus, Pakistan; Department of Chemistry, COMSATS University Islamabad, Lahore Campus, Pakistan

Ahmad Reza Bagheri, Department of Chemistry, Yasouj University, Yasouj, Iran

Debjyoti Banerjee, Fiber and Polymer Science, Wilson College of Textiles, North Carolina State University, Raleigh, United States

Mattia Bartoli, Department of Applied Science and Technology, Politecnico di Torino, Torino, Italy; Consorzio Interuniversitario Nazionale per la Scienza e Tecnologia dei Materiali (INSTM), Florence, Italy

Muhammad Bilal, School of Life Science and Food Engineering, Huaiyin Institute of Technology, Huai'an, China

Manik Chandra Biswas, Fiber and Polymer Science, Wilson College of Textiles, North Carolina State University, Raleigh, United States

Cătălina Bogdan, Faculty of Pharmacy, Department of Dermopharmacy and Cosmetics, Iuliu Haţieganu, University of Medicine and Pharmacy, Cluj-Napoca, Romania

Jeannine Bonilla, Faculty of Animal Science and Food Engineering, University of São Paulo, Av. Duque de Caxias Norte, Pirassununga (SP), Brazil; Laboratory of Food Technology, Faculty of Animal Science and Food Engineering—FZEA/USP, University of São Paulo, Pirassununga, São Paulo, Brazil

N. Brosse, Université de Lorraine, INRAE, LERMAB, Nancy, France

Federica Catania, Department of Applied Science and Technology, Politecnico di Torino, Torino, Italy

Andreea Cernat, Department of Analytical Chemistry, Faculty of Pharmacy, Iuliu Hatieganu University of Medicine and Pharmacy, Cluj-Napoca, Romania

Wilson Daniel Caicedo Chacon, Department of Chemical and Food Engineering, Federal University of Santa Catarina, Florianópolis, SC, Brazil

Sanjay Chhibber, Panjab University, Chandigarh, India

Cecilia Cristea, Department of Analytical Chemistry, Faculty of Pharmacy, Iuliu Hatieganu University of Medicine and Pharmacy, Cluj-Napoca, Romania

Ankita Dhillon, Department of Chemistry, Banasthali University, Rajasthan, India

Mani Divya, Biomaterials and Biotechnology in Animal Health Lab, Nanobiosciences and Nanopharmacology Division, Department of Animal Health and Management, Alagappa University, Karaikudi, Tamil Nadu, India

Ana-Maria Dragan, Department of Analytical Chemistry, Faculty of Pharmacy, Iuliu Hatieganu University of Medicine and Pharmacy, Cluj-Napoca, Romania

Ali Ehsani, Department of Chemistry, Faculty of Science, University of Qom, Qom, Iran

W. Fatriasari, Research Center for Biomaterials, Indonesian Institute of Sciences (LIPI), Cibinong, Indonesia

Anca Florea, Department of Analytical Chemistry, Faculty of Pharmacy, Iuliu Hatieganu University of Medicine and Pharmacy, Cluj-Napoca, Romania

Talita Ribeiro Gagliardi, Department of Cell Biology, Embryology and Genetics, Federal University of Santa Catarina, Florianópolis, SC, Brazil

Xiaoyan Gao, Key Laboratory for Palygorskite Science and Applied Technology of Jiangsu Province, National & Local Joint Engineering Research Centre for Deep Utilization Technology of Rock-salt Resource, Faculty of Chemical Engineering, Huaiyin Institute of Technology, Huai'an, China

Vijay Singh Gondil, Post Graduate Institute of Medical Education and Research, Chandigarh, India; Panjab University, Chandigarh, India

Valentina Grumezescu, National Institute for Lasers, Plasma, and Radiation Physics, Magurele, Ilfov, Romania

Ram K. Gupta, Department of Chemistry, Kansas Polymer Research Center, Pittsburg State University, Pittsburg, KS, United States

Najam ul Haq, Division of Analytical Chemistry, Institute of Chemical Sciences, Bahauddin Zakariya University, Multan, Pakistan

Kusum Harjai, Panjab University, Chandigarh, India

K.M. Faridul Hasan, Simonyi Károly Faculty of Engineering, University of Sopron, Sopron, Hungary

R. Hashimd, School of Industrial Technology, Universiti Sains Malaysia, Minden, Penang, Malaysia

Kun Hong, Key Laboratory for Palygorskite Science and Applied Technology of Jiangsu Province, National & Local Joint Engineering Research Centre for Deep Utilization Technology of Rock-salt Resource, Faculty of Chemical Engineering, Huaiyin Institute of Technology, Huai'an, China

Seongwoo Hong, Department of Chemistry and Kansas Polymer Research Center, Pittsburg State University, Pittsburg, KS, United States

Péter György Horváth, Simonyi Károly Faculty of Engineering, University of Sopron, Sopron, Hungary

M.H. Hussin, Materials Technology Research Group (MaTReC), School of Chemical Sciences, Universiti Sains Malaysia, Minden, Penang, Malaysia

Sidra Iftekhar, Department of Applied Physics, University of Eastern Finland, Kuopio, Finland

Ewelina Jamróz, Department of Chemistry, Faculty of Food Technology, University of Agriculture, Cracow, Poland

Nisar Alia Adnan Khan, Institute of Chemical Sciences, University of Peshawar, Khyber Pakhtunkhwa, Pakistan

Łukasz Klapiszewski, Faculty of Chemical Technology, Institute of Chemical Technology and Engineering, Poznan University of Technology, Berdychowo, Poznan, Poland

Dinesh Kumar, Department of Chemistry, Banasthali University, Rajasthan, India; School of Chemical Sciences, Central University of Gujarat, Gandhinagar, Gujarat, India

F. Laoutid, Laboratory of Polymeric & Composite Materials, Materia Nova Research Center, Mons, Belgium

N.H. Abdul Latif, Materials Technology Research Group (MaTReC), School of Chemical Sciences, Universiti Sains Malaysia, Minden, Penang, Malaysia

Vesa-Pekka Lehto, Department of Applied Physics, University of Eastern Finland, Kuopio, Finland

Kennya Thayres dos Santos Lima, Department of Chemical and Food Engineering, Federal University of Santa Catarina, Florianópolis, SC, Brazil

Saadat Majeed, Division of Analytical Chemistry, Institute of Chemical Sciences, Bahauddin Zakariya University, Multan, Pakistan

Sumeet Malik, Institute of Chemical Sciences, University of Peshawar, Khyber Pakhtunkhwa, Pakistan

Elena Marras, Department of Applied Science and Technology, Politecnico di Torino, Torino, Italy

Saba Goharshenas Moghadam, Color and Surface Coatings Group, Polymer Processing Department, Iran Polymer and Petrochemical Institute (IPPI), Tehran, Iran

Mirela Moldovan, Faculty of Pharmacy, Department of Dermopharmacy and Cosmetics, Iuliu Haţieganu, University of Medicine and Pharmacy, Cluj-Napoca, Romania

Alcilene Rodrigues Monteiro, Department of Chemical and Food Engineering, Federal University of Santa Catarina, Florianópolis, SC, Brazil

Irina Negut, National Institute for Lasers, Plasma, and Radiation Physics, Magurele, Ilfov, Romania

Tuan Anh Nguyen, Microanalysis Department, Institute for Tropical Technology, Vietnam Academy of Science and Technology, Hanoi, Vietnam

Lingli Ni, Key Laboratory for Palygorskite Science and Applied Technology of Jiangsu Province, National & Local Joint Engineering Research Centre for Deep Utilization Technology of Rock-salt Resource, Faculty of Chemical Engineering, Huaiyin Institute of Technology, Huai'an, China

Hamidreza Parsimehr, Color and Surface Coatings Group, Polymer Processing Department, Iran Polymer and Petrochemical Institute (IPPI), Tehran, Iran; Department of Chemistry, Faculty of Science, University of Qom, Qom, Iran

Vaishali Pawar, Department of Biosciences and Bioengineering, Indian Institute of Technology, Bombay, Maharashtra, India

Tahir Rasheed, School of Chemistry and Chemical Engineering, State Key Laboratory of Metal Matrix Composites, Shanghai Jiao Tong University, Shanghai, China

Sneha Ravi, Department of Biosciences and Bioengineering, Indian Institute of Technology, Bombay, Maharashtra, India

Iulia Rus, Department of Analytical Chemistry, Faculty of Pharmacy, Iuliu Hatieganu University of Medicine and Pharmacy, Cluj-Napoca, Romania

M.R. Saeb, Université de Lorraine, CentraleSupélec, LMOPS, Metz, France

Kowshik Saha, Textile Technology Management, Wilson College of Textiles, North Carolina State University, Raleigh, United States

Sameera Shafi, School of Chemistry and Chemical Engineering, State Key Laboratory of Metal Matrix Composites, Shanghai Jiao Tong University, Shanghai, China

Mika Sillanpää, School of Environment, Tsinghua University, Beijing, China; Institute of Research and Development, Duy Tan University, Da Nang, Vietnam; Faculty of Environment and Chemical Engineering, Duy Tan University, Da Nang, Vietnam; School of Civil Engineering and Surveying, Faculty of Health, Engineering and Sciences, University of Southern Queensland, West Street, Toowoomba, QLD, Australia; Department of Chemical Engineering, School of Mining, Metallurgy and Chemical Engineering, University of Johannesburg, Doornfontein, South Africa

Paulo J.A. Sobral, Faculty of Animal Science and Food Engineering, University of São Paulo, Av. Duque de Caxias Norte, Pirassununga (SP), Brazil; Food Research Center (FoRC), University of São Paulo, São Paulo (SP), Brazil

N.N. Solihat, Research Center for Biomaterials, Indonesian Institute of Sciences (LIPI), Cibinong, Indonesia

Felipe M. de Souza, Department of Chemistry, Kansas Polymer Research Center, Pittsburg State University, Pittsburg, KS, United States

Rohit Srivastava, Department of Biosciences and Bioengineering, Indian Institute of Technology, Bombay, Maharashtra, India

Anna Szarpak, University of Grenoble Alpes, CNRS, CERMAV, Grenoble, France

Alberto Tagliaferro, Department of Applied Science and Technology, Politecnico di Torino, Torino, Italy; Consorzio Interuniversitario Nazionale per la Scienza e Tecnologia dei Materiali (INSTM), Florence, Italy; Faculty of Science, OntarioTech University, Oshawa, Ontario, Canada

Mihaela Tertis, Department of Analytical Chemistry, Faculty of Pharmacy, Iuliu Hatieganu University of Medicine and Pharmacy, Cluj-Napoca, Romania

Florina Truta, Department of Analytical Chemistry, Faculty of Pharmacy, Iuliu Hatieganu University of Medicine and Pharmacy, Cluj-Napoca, Romania

H. Vahabi, Université de Lorraine, CentraleSupélec, LMOPS, Metz, France

Germán Ayala Valencia, Department of Chemical and Food Engineering, Federal University of Santa Catarina, Florianópolis, SC, Brazil

Rachel Auzély-Velty, University of Grenoble Alpes, CNRS, CERMAV, Grenoble, France

Sekar Vijayakumar, Marine College, Shandong University, Weihai, P. R. China

Muhammad Abdul Wasayh, School of Chemical and Materials Engineering, National University of Science and Technology, Islamabad, Pakistan

He-Lin Xu, Department of Pharmaceutics, School of Pharmaceutical Sciences, Wenzhou Medical University, Wenzhou City, Zhejiang Province, China

Muhammad Yar, Interdisciplinary Research Center in Biomedical Materials, COMSATS University Islamabad, Lahore Campus, Pakistan

Alap Ali Zahid, Interdisciplinary Research Center in Biomedical Materials, COMSATS University Islamabad, Lahore Campus, Pakistan; Department of Mechanical and Industrial Engineering, College of Engineering, Qatar University, Qatar; Biomedical Research Center, Qatar University, Qatar

Zhenghua Zhang, Institute of Environmental Engineering and Nano-Technology, Tsinghua-Shenzhen International Graduate School, Tsinghua University, Shenzhen, Guangdong, China

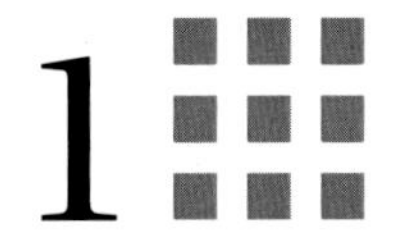

Biopolymeric nanomaterials: design, synthesis, and applications

Mani Divya[a] and Sekar Vijayakumar[b]

[a]BIOMATERIALS AND BIOTECHNOLOGY IN ANIMAL HEALTH LAB, NANOBIOSCIENCES AND NANOPHARMACOLOGY DIVISION, DEPARTMENT OF ANIMAL HEALTH AND MANAGEMENT, ALAGAPPA UNIVERSITY, KARAIKUDI, TAMIL NADU, INDIA.
[b]MARINE COLLEGE, SHANDONG UNIVERSITY, WEIHAI, P. R. CHINA.

Chapter outline

1.1 Introduction

Packaging is a vital constituent of each stage in quality and safety of the material [32]. Presently, most of these materials have environmental issues due to the accretion of plastics and elevate serious problem to global alarms [34]. To compensate this, the utilization of natural biopolymer as packaging materials is mostly biocompatible, easily obtainable, and mainly stable [10]. Some of the bionanocomposites are intended to progress the particular type of plain packaging. Besides, possible nanotechnology has the consideration from scientific researchers to extend superior and quality materials in packaging to many sectors [59]. An emblematic packaging material could supply incredible advantages including quality storage, resistance to physical problems, shelf-life, quality, safety, and controlling the growth of microbes [41,51],. Recently, the utilization of nanomaterials in food industry opens up various promising initiates from nanoadditives for unique properties of polymeric nanomaterials [24]. The expansion of biodegradable packaging materials for packing, even if mounting environmental unease, has obliged the packaging industries to greatly discover new methods and eco-friendly alternatives. The ultimate purpose of packaging intends should be to follow the biomass life series [11]. Packaging materials efficiently

Biopolymeric Nanomaterials: Fundamentals and Applications. DOI: https://doi.org/10.1016/B978-0-12-824364-0.00004-6

manage the objects traditionally. Limitations, such as dampness, oxygen, lightly put forward huge declaration for the main subtle objects [27]. The ecological and bio-based packaging materials have a lot of reflection cause of hygiene and environmental issues [5]. The utilization of pioneer packaging techniques using biopolymeric nanomaterial for packaging leads to improvement in the quality [53].

Biopolymers are produced from natural resources such as chitosan, starch, and various proteins from plants and animals. Biopolymeric films used to solute barriers for recovering superiority packaging. Many researchers' efforts have been decisive on the enhancement of polymer-based nanomaterials in terms of physical, thermal, and mechanical properties. Biopolymeric nanocomposites exhibit better properties with biopolymers or macro-scale composites owing to their elevated aspects proportion and surface region [47]. The chief impediment to utilizing biopolymer as packaging materials can be biodegradability in its matrix [6]. In food packaging, some antimicrobial compounds could take steps as a polymeric matrix to produce active packaging [15]. The commercial-scale manufacture and applications of natural-based packaging films are recent trends in development [19]. For packaging application, the performance properties of nanomaterials are closely related to their structure. Polymers-based nanomaterials grades in better electric conductivity, thermal constancy, etc., nanometric has perked up the obstruction properties in packages and contribution for active packaging system [37]. The advanced nanomaterial enlarged polymers employed assist to increase the remuneration connected to polymers through environmental concerns. Various nanomaterials such as carbon nanotubes [52], graphene [30], chitosan [12], ZnO_2 [18], Ti [31], silver [39] are being extensively used as biopolymeric nanomaterials or packaging materials. The nanotechnology application in polymeric science can provide new awareness for recovering attribute description and outlay proficiency of polymeric materials [14]. In this chapter, the efficacy of advanced biopolymeric nanomaterials used in packaging structure is reported. The content of this chapter, for the most part, concentrates on polymeric-based nanomaterials used as packaging materials for several industrial applications such as food packaging, etc.

1.2 Preparation of biopolymeric nanomaterials

Polymeric nanoparticles (NPs) are classified into two types, nanospheres and nanocapsules. Many synthesis methods have been used for polymeric nanomaterials, the preference depends on their applications and models to be performed [45]. The encapsulation of hydrophobic molecules is used to synthesis polymeric nanomaterials. The drug can be encapsulated or bound to the surface of NPs (Fig. 1.1). In nanocapsules, the solid material environs fluid core. The core is filled with oil-encapsulated lipophilic drug [28].

1.3 Properties of biopolymeric nanomaterials

Biopolymers are flexible in arrangement on the surface of its belongings. This is owing to the construction and exterior possessions of polymers [40]. An outsized quantity of available

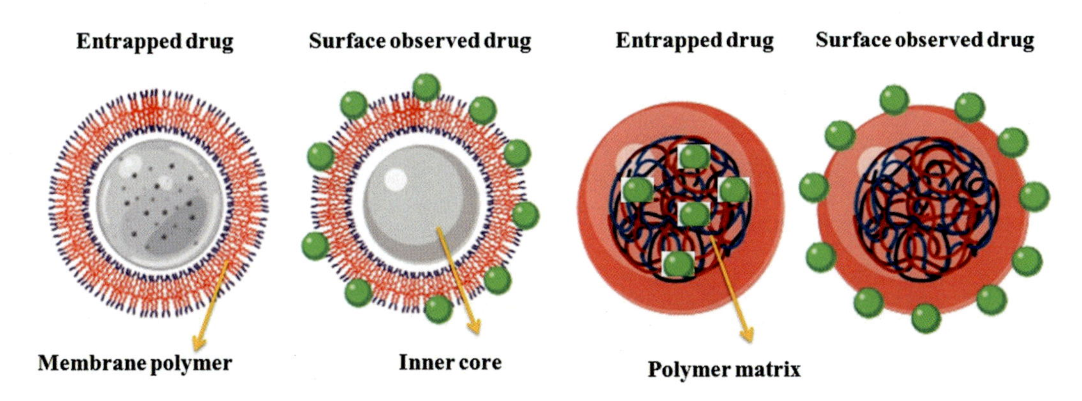

FIGURE 1.1 Polymeric nanomaterials.

information has portrayed the expansion of biopolymer for dropping to environmental property at the same time to raise the assortment of application of biopolymers. The two significant properties construct promising biopolymer.

1. Natural biodegradability
2. Indisputable virtues to the environment.

A recent study fabricated biopolymeric films integrated with quite a few essential oils and plasticizers [3]. The H_2O barrier vapor property for the films is found to get better. Although several modifications in physical and mechanical property are achieved, starch-based films have to discover relevance in packaging industries for the improvement in mechanical properties to be able to rely on form of the particles creature [38]. Consequence advances in biopolymeric nanomaterials with the high quantity of evenness and accurate control of their physiochemical properties [42]. The factor affecting the mechanical properties of polymer depends on the polymer characteristics such as polymer length, porosity properties, etc. [16]. An innovative group of materials, bionanocomposites, has automatic and thermal properties for recovering the property of their biopolymer support with a packaging material [7]. A packaging material with an antimicrobial function documented most capable active packaging system has antimicrobial properties. Thus, the large upgrading properties were found in the biopolymeric nanomaterials namely, mechanical, barrier, biodegradation, antimicrobial, optical, thermal, and other properties (Fig. 1.2).

1.4 Packaging materials

A packaging material emerges to have very great prospect is meant for a wide variety of application in various packaging industries. Degradable packaging materials have gotten a lot of deliberation for the reason that the hygiene and biological concern got with exploit petrochemical

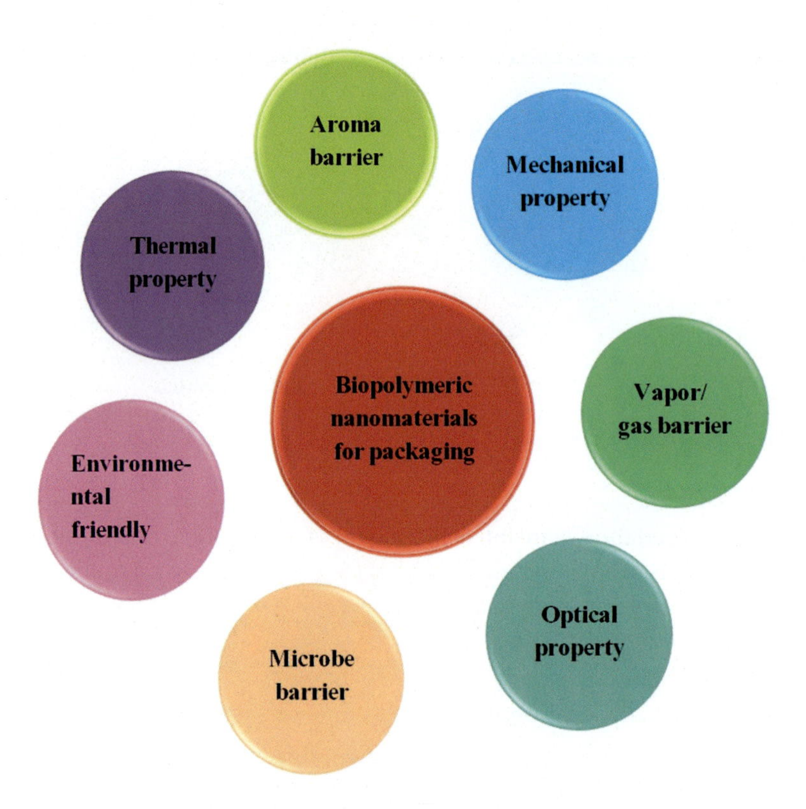

FIGURE 1.2 Different properties of biopolymeric nanomaterials.

pedestal plastics [5]. The eco-friendly biopolymers preserve the following three foremost types on the origin of its cause:

1. Biopolymers originated on or after agricultural properties, as well as polysaccharides, lignocellulosic commodities (cellulose and its derivatives), proteins (whey and collagen), lipids (bee wax), and liberated fatty acids.
2. Biopolymers attained by microbial fermentation, including pollutant and polyhydroxyalkanoates.
3. Biopolymers with monomers synthesized chemically accomplished through ordinary unprocessed materials, for instance poly(lactic acid) [20].

Edible composite films are created by polysaccharides, lipids, proteins, and various biopolymers in disparity to manufacture polymers as packaging materials [59]. Polymeric nanomaterial–based packaging materials diminish the unconstructive ecological collision of packaging material. Edible coating nanomaterial packaging provides a noticeable exclusive class of traditional packaging material such as biodegradable, biocompatible, and edible characteristics [38].

1. *Biopolymeric nanomaterials used as packaging materials.*

Biopolymer-based nanomaterials be able to conquer the removal struggle toward significant enormity. Natural-based biopolymer films should be able to contribute extensive evolution in food packaging industries, pharmaceutical, and treatment center revenue. Red cabbage anthocyanins (KGM/O-ChNCs/RCA) produced films are capable of an active food packaging structure [56]. Also, polylactic acid utilized as a packaging material in the food industry amplifies polymer's thermal stability [35]. Chitosan films have elasticity, low cytotoxic, and exceptional performance in function, simply provided by casting of chitosan integrated with laponite and LAP@AgNPs [58]. Gelatin is used as a packaging material source and in the manufacture of gelatin-based films on a commercial system. It has heat profitability, excellent oxygen barrier. These are all main functional properties for packaging purpose [22]. TiO_2-based starch pectin matrix is used to make starch-pectin blend films. It can improve the prosperity of the edible film for food packaging application [13]. Furthermore, zein-coated oleic acid films provide relevant improvement with optical interaction property in efficient active packaging material [26]. Carboxymethyl cellulose-polyvinylpyrrolidone–based novel hydrogel films provide biodegradability under controlled environment and used as advantageous packaging materials [49].

1.5 Applications of packaging materials in different sectors

Biopolymeric nanomaterials have demonstrated noteworthy development in polymer properties constructive for wrapping applications. The enhanced gas blockade assets of nanocomposites formulate striking and valuable in the food packaging system. Organic modifier was more compatible with structure enhanced desirable properties for short-shelf existence packaging, such as one-way packaging material or medical applications [29]. The polymer clay nanocomposite has attracted extensive attention for their use in various industrial applications such as packaging applications [17]. In most commercial application, direct incorporation of oxygen scavengers into biopolymer films can maintain deficient oxygen level, which is useful for packaging applications [25]. The rational packaging application of nanomaterials provides a sequence to manufactured goods' superiority, integrity, time, history of a product, temperature, etc. [21].

1. *Food packaging materials*
   Recyclable polymers are capable of fabricating sustainable films to stand-in ordinary food packaging as responding to green apprehension and food conservation. More than 40% of the plastics are used for packaging and more or less a part of them are utilized for food packaging in the form of sheets, films, tubs, bottles, tableware, trays, etc. [46]. Food packaging materials after the food is obsessive, they put into the nonbiodegradable community and be inclined to mount up chief ecological apprehension. This composes for recyclable ordinary polymeric nanomaterials used in dynamic food packaging [2]. Numerous researchers have been immobile undertaking an attempt to augment the functional belongings of gelatin films intended for expanding as food packing resources. An assortment as a class of nanofillers to facilitate into FG (Fish gelatin) to fabricate bionanocomposite films, such as polysaccharide, nanoclays, metal ions and metal oxides NPs be too reviewed, make these films a potential bioactive food packaging material. Not long, considering the applications for

edible films/food packing, a great deal notice has been resolutely resting on polysaccharide nanofillers [23]. Wang et al. [55] exposed the interest that carboxymethyl chitosan (CMCS) CMCS/MgO nanocomposites film can be used in particular industries, in particular for food packaging. This makes it dwell in benefit in the purpose of food preservation. These results confirmed a probable application of MgO/CMCS nanocomposite as a capable food packaging. Chitosan-gelatin-based AgNPs films emerge to be capable of a protective packaging material that can enlarge the shelf-life of red grapes to 14 days. This research work presented a narrative composite film that can be functional in antimicrobial food packaging [36]. The food packaging films with antibacterial effects were equipped through integration polyphenolics AgNPs into CA (Cellulose acetate) films green synthesized as secure an ecological mediator. Silver ion discharge side view as of CA nanocomposites films was dependent on top of all nanocomposites films and Ag reductive complexes were the legal restrictions. This swot up displays the prospect of using CA-Ag nanocomposite seeing that energetic food packaging materials [36]. Bionanocomposites film integrating 2% ZnO NPs exhibits the maximum antibacterial activity. Prepared antibacterial bionanocomposites' films prove probable for use as active food packaging. Further examination to appraise latent migration of Zn^{+2} from the film to foods should be demeanor to certify protection [50]. Curcumin-loaded mesoporous silica NP (SBA-15) was built-in into chitosan (CS) film, a bionanocomposites' film with forced liberate purpose, was industrial using CS, SBA-15, and curcumin and these bionanocomposites films could be a capable of active food packaging material [57]. Biopolymeric materials along with chitosan films have been widely utilized as an active food packaging substance in the current period. Chitosan films among further NPs demonstrate not only superior antimicrobial activity next to a mixture of foodborne microorganisms, except beside a betterment in the mechanical properties owing to strengthening consequence of NPs in the polymer medium [44]. Films designed to create dynamic ecological films as a substitute to the adding up of food preservatives alongside this bacterium's spreading out. The shelf-existence conservatory and upholding of the integrity of foods packaged with these films emphasize the prospective function of the corn starch-PVA-films (Polyvinyl alcohol) built-in with maleic anhydride, cellulose nanocrystals, and nisin as a good alternative to replacement of traditional plastic materials [15]. Newly, soluble soybean polysaccharide (SSPS) gelatin blend films direct as aqueous way out were calculated utilizing a vision to extend appropriate contender for food packaging appliances. Fit for human consumption packages as the SSPS, it is imperative to get better the corporal properties of its films like tensile potency, thermal immovability and warmth seal potency. The packaging test points out that SSPS/gelatin eatable blend films are a probable agent for fabricating quick-dissolving packages for crushed foodstuff and brew merchandise [33]. Likewise, poly(ε-caprolactone) PCL one of the thermoplastic polymers that has been worn broadly, is meant for tissue engineering and is also a capable packaging material. It has the quality of biodegradability and nontoxicity. *In-vitro* study showed that PCL with chitosan and Grapefruit seed extract (GFSE)(PCL/chitosan/GFSE) films show signs of outstanding bactericidal and bacteriostatic capability in opposition to diverse foodborne pathogens. Also these films would be deeply valuable for certifying food safety and extending the shelf-life of foodstuff [54]. Nanocomposite packaging supplies might be fabricated with implanting cellulose nanofibers. This dynamic packaging was exceedingly efficient at lipid oxidation, dropping microbial spoilage, and lipolysis. In that way expanded shelf existence for the lamb meat for the duration of frozen storage. Moreover, extremely low stages of TiO_2 NPs wandered as of the packaging into the meat for the period of storage. The nanocomposite films developed maybe apposite for purpose as dynamic packaging materials in the food industry en route for preserving the wellbeing and superiority meat commodities [1]. The applications of nanomaterial assure to develop the biopolymer-based packaging materials. It facilitates to decrease the packaging dissipate connected with processed foodstuff and will prop up the preservation of spanking new foods, lengthen their shelf-existence.

2. *Industrial packaging materials*

   In recent times, the consideration in the wrapping industry on the subject of the use of bioplastics has been variable from ecological materials to bio-based resources. Robinson and Morrison [48], researchers in New Zealand, developed an insulating packaging system based on nanoporous calcium silicate overloaded with a phase-change material (paraffin wax). That could be able to alleviate the special effects of an increase in outside temperature over a tiny period. Likewise another industrial packaging material such as poly(ethylene terephthalate) is one of the polymers for the most part broadly used in the wrapping industry. On the other hand, it is extremely attractive to augment its barrier properties for applications that include gaseous drinks, erstwhile rigid and flexible packaging applications. The nanomaterial offers exclusive possibilities to improve the features of this material, provided that plenty of thermal opposed to and legislation meet the terms of nanoadditives [8]. And also, melt blending uses elevated temperature and elevated cut off forces to scatter nanofillers in a thermoplastic polymer matrix. This is for using conservative apparatus for industrial polymer processing.

   Similarly, starch-based thermoplastic materials have been productively applied on the industrial point for foaming, film blowing, puff molding, inoculation molding, and extrusion applications. Mostly, the industrial applications of nanoclay in multilayer film packaging take place in gaseous drinks, beer bottles and heat formed containers. Packaging materials used by polymer nanocomposites as in electronic packaging, food packaging, and industrial packaging have strained concentration in the scientific area owing to their adaptable properties [4].

3. *Other packaging materials*

   Other packaging applications for bionanomaterials so as to are accepted to be given improved concentration in the outlook such as color-containing films, antioxidant-releasing films, light absorbing scheme, anti-fogging and antisticking films, breathable films, bioactive mediator for inhibited discharge and insect repellent packaging [43]. Plastic films metalized by way of aluminum have been used as chatter barriers and light barriers and as decorative films. Metallic oxide NPs for example TiO_2, MgO, ZnO, and Al_2O_3, and metallic NPs such as silver are extensively used to create nanocoatings on polymeric films, paperboard. A variety of narrative properties of nanocoating materials' self-cleaning smart nanocoatings that act against bacteria, cut off pathogens, or fluoresce under convinced circumstances are under process [9]. Mechanical recycling is a method in which waste materials are recycled into new-fangled raw materials without changing the fundamental structure of the material by reconstituting the innovative polymer belongings. It is a piece of ingrained machinery for the material revival of predictable plastics. Its chief benefit is that division of the assets esteemed for the construction of the plastic materials is not worn out. This expertise of automatic recycling is appropriate to mutually recycling bio-based conservative plastics in addition to general marks of recyclable plastics [4].

1.6 Conclusion

The successful application of nanomaterials with biopolymers has enthused new research on the development of products. Several technologies are already on the market, while many others are at present being studied. All of us have an immense expectation from nanotechnology and its application for scientific relevant as it is one of the majority notable industrial revolutions of the twenty-first century. In a few existences, nanotechnology will permit the packaging industry to make available an additional reasonable retort to require for permanent enhancement of all conditions. Packaging materials by biopolymeric nanomaterial has emerged to have an incredible future for a wide assortment of applications in the packaging industries. In this chapter, the authors tried to integrate all the original aspects and current evolution about the nanomaterials, along with the biopolymers concise development and usage of unique scientific approaches. These resources, not only plentiful on our ground planet other than bio-compatible impending for instances of healthcare relevance. Also, the properties features of all the declared bio-materials and topical encroachment are on form of purposeful in this chapter. Consequently, through technologies imitative from nanoscience we might be able to create required biopolymer stimulated nanomaterials for an extensive series of applications for packaging materials (food, industrial, and other systems).

Acknowledgment

The corresponding author Dr. Sekar Vijayakumar would like to thank Shandong University, Weihai, P. R. China for providing the postdoctoral research fellowship.

References

[1] M. Alizadeh-Sani, E. Mohammadian, DJ. McClements, Eco-friendly active packaging consisting of nanostructured biopolymer matrix reinforced with TiO2 and essential oil: Application for preservation of refrigerated meat, Food Chem. 322 (2020) 126782.

[2] Y.A. Arfat, M. Ejaz, H. Jacob, J. Ahmed, Deciphering the potential of guar gum/Ag-Cu nanocomposite films as an active food packaging material, Carbohydr. Polym. 157 (2017) 65–71.

[3] L. Avérous, E. Pollet, Biodegradable polymers, in: L. Avérous, and E. Pollet (Eds.), Environmental Silicate Nano-Biocomposites, Springer, Cham, 2012, pp. 13–39.

[4] P. Balakrishnan, M.S. Thomas, L.A. Pothen, S. Thomas, M.S. Sreekala, Polymer films for packaging, in: S. Kobayashi, K. Müllen (Eds.), Encyclopedia of Polymeric Nanomaterials, Springer, Berlin, Heidelberg, 2014, pp. 1–8.

[5] F. Bi, X. Zhang, R. Bai, Y. Liu, J. Liu, J. Liu, Preparation and characterization of antioxidant and antimicrobial packaging films based on chitosan and proanthocyanidins, Int. J. Biol. Macromol. 134 (2019) 11–19.

[6] C.M. Bitencourt, C.S. Fávaro-Trindade, P.J.A. Sobral, R.A. Carvalho, Gelatin-based films additivated with curcuma ethanol extract: Antioxidant activity and physical properties of films, Food Hydrocoll. 40 (2014) 145–152.

[7] P. Bordes, E. Pollet, L. Avérous, Nano-biocomposites: biodegradablepolyester/nanoclay systems, Prog. Polym. Sci. 34 (2) (2009) 125–155.

[8] T.G. Bowditch, Penetration of Polyvinyl Chloride and Polypropylene Packaging Films by *Ephestiacautella* (Lepidoptera: Pyralidae) and *Plodiainterpunctella* (Lepidoptera: Pyralidae) Larvae, and *Triboliumconfusum* (Coleoptera: Tenebrionidae), J. Econ. Entomol. 90 (4) (1997) 1028–1101.

[9] J.O. Carneiro, V. Texeira, P. Carvalho, S. Azevedo, Self-cleaning smart nanocoatings, in: A.S.H. Makhlouf, I. Tiginyanu (Eds.), Nanocoatings and Ultra-Thin Films, Woodhead Publishing, Cambridge, U.K., 2011, pp. 397–413.

[10] D.S. Cha, M.S. Chinnan, Biopolymer-based antimicrobial packaging: a review, Crit. Rev. Food Sci. Nutr. 44 (4) (2004) 223–237.

[11] MY. Chan, SC. Koay, Biodegradation and thermal properties of crosslinked chitosan/corn cob biocomposite films by electron beam irradiation, Polym. Eng. Sci. 59 (1) (2019) 59–68.

[12] P.R. Chang, R. Jian, J. Yu, X. Ma, Fabrication and characterisation of chitosan nanoparticles/plasticised-starch composites, Food Chem. 120 (3) (2010) 736–740.

[13] K.K. Dash, N.A. Ali, D. Das, D. Mohanta, Thorough evaluation of sweet potato starch and lemon-waste pectin based-edible films with nano-titania inclusions for food packaging applications, Int. J. Biol. Macromol. 139 (2019) 449–458.

[14] H.M. De Azeredo, L.H. Mattoso, T.H. McHugh, Nanocomposites in food packaging – a review, in: B. Reddy (Ed.), Advances in Diverse Industrial Applications of Nanocomposites, Intechopen, London, UK, 2011, pp. 57–78.

[15] T.V. de Oliveira, P.A. de Freitas, C.C. Pola, J.O. da Silva, L.D. Diaz, S.O. Ferreira, F.F. de, N. Soares, Development and optimization of antimicrobial active films produced with a reinforced and compatibilized biodegradable polymers, Food Packag. Shelf Life 24 (2020) 100459.

[16] B. Dhandayuthapani, Y. Yoshida, T. Maekawa, D.S. Kumar, Polymeric scaffolds in tissue engineering application: a review, Int. J. Polym. Sci. 2011 (2011) 1–19.

[17] T.V. Duncan, Application of nanotechnology in food packaging and food safety: barrier materials, antimicrobials and sensors, J. Coll. Interfaces Sci. 363 (2011) 1–24.

[18] S.K. Esthappan, M.K. Sinha, P. Katiyar, A. Srivastav, R. Joseph, Polypropylene/zinc oxide nanocomposite fibers: morphology and thermal analysis, J. Polym. Mater. 30 (1) (2013) 79.

[19] A. Farhan, N.M. Hani, Active edible films based on semi-refined κ-carrageenan: antioxidant and color properties and application in chicken breast packaging, Food Packag. Shelf Life 24 (2020) 100476.

[20] F. Garavand, M. Rouhi, S.H. Razavi, I. Cacciotti, R. Mohammadi, Improving the integrity of natural biopolymer films used in food packaging by crosslinking approach: A review, Int. J. Biol. Macromol. 104 (2017) 687–707.

[21] A. Garland, Nanotechnology in Plastics Packaging: Commercial Applications in Nanotechnology, Pira International, Leatherhead, Surrey, U.K., 2004, pp. 14–63.

[22] Z.N. Hanani, Y.H. Roos, J.P. Kerry, Use and application of gelatin as potential biodegradable packaging materials for food products, Int. J. Biol. Macromol. 71 (2014) 94–102.

[23] S.F. Hosseini, M.C. Gómez-Guillén, A state-of-the-art review on the elaboration of fish gelatin as bioactive packaging: special emphasis on nanotechnology-based approaches, Trends Food Sci. Technol. 79 (2018) 125–135.

[24] Y. Huang, L. Mei, X. Chen, Q. Wang, Recent developments in food packaging based on nanomaterials, Nanomaterials 8 (10) (2018) 830.

[25] I. Janjarasskul, K. Tananuwong, J.M. Krochta, Whey protein film with oxygen scavenging function by incorporation of ascorbic acid, J. Food Sci. 76 (2011) 561–568.

[26] M. Kashiri, G. López-Carballo, P. Hernández-Muñoz, R. Gavara, Antimicrobial packaging based on a LAE containing zein coating to control foodborne pathogens in chicken soup, Int. J. Food Microbiol. 306 (2019) 108272.

[27] A.M. Khaneghah, S.M.B. Hashemi, S. Limbo, Antimicrobial agents and packaging systems in antimicrobial active food packaging: an overview of approaches and interactions, Food Bioprod. Process. 111 (2018) 1–19.

[28] A.A. Kulkarni, P.S. Rao, Synthesis of polymeric nanomaterials for biomedical applications, in: A.K. Gaharwar, S. Sant, M.J. Hancock, S.A. Hacking (Eds.), Nanomaterials in Tissue Engineering, 2013, p. 2763.

[29] S. Kumar, J.C. Boro, D. Ray, A. Mukherjee, J. Dutta, Bionanocomposite films of agar incorporated with ZnO nanoparticles as an active packaging material for shelf life extension of green grape, Heliyon 5 (6) (2019) s01867.

[30] W. Lee, O. Kahya, C.T. Toh, B. Özyilmaz, J.H. Ahn, Flexible graphene–PZT ferroelectric nonvolatile memory, Nanotechnology 24 (47) (2013) 475202.

[31] B. Li, Q.B. Lin, H. Song, H.J. Wu, X.M. Li, Y. Chen, Determination of titanium, lead, chromium, cadmium in plastics for food packaging by microwave digestion-ICP-AES [J], Chemical Research and Application 2 (2011).

[32] F. Li, Y. Liu, Y. Cao, Y. Zhang, T. Zhe, Z. Guo, X. Sun, Q. Wang, L. Wang, Copper sulfide nanoparticle-carrageenan films for packaging application, Food Hydrocoll. 109 (15) (2020) 106094.

[33] C. Liu, J. Huang, X. Zheng, S. Liu, K. Lu, K. Tang, J. Liu, Heat sealable soluble soybean polysaccharide/gelatin blend edible films for food packaging applications, Food Packag. Shelf Life 24 (2020) 100485.

[34] Y. Liu, Y. Li, L. Deng, L. Zou, F. Feng, H. Zhang, Hydrophobic ethylcellulose/gelatin nanofibers containing zinc oxide nanoparticles for antimicrobial packaging, J. Agric. Food Chem. 66 (36) (2018) 9498–9506.

[35] J. Markarian, Biopolymers present new market opportunities for additives in packaging, Plast. Additives Compound. 10 (3) (2008) 22–25.

[36] D.A. Marrez, A.E. Abdelhamid, O.M. Darwesh, Eco-friendly cellulose acetate green synthesized silver nano-composite as antibacterial packaging system for food safety, Food Packag. Shelf Life 20 (2019) 100302.

[37] S.D. Mihindukulasuriya, L.T. Lim, Nanotechnology development in food packaging: a review, Trends Food Sci. Technol. 40 (2) (2014) 149–167.

[38] R.K. Mishra, S.K. Ha, K. Verma, Recent progress in selected bio-nanomaterials and their engineering applications: an overview, J. Sci. Adv. Mater. Devices 3 (3) (2018) 263–288.

[39] A. Mohammed Fayaz, K. Balaji, M. Girilal, P.T. Kalaichelvan, R. Venkatesan, Mycobased synthesis of silver nanoparticles and their incorporation into sodium alginate films for vegetable and fruit preservation, J. Agric. Food Chem. 57 (14) (2009) 6246–6252.

[40] A. Mokhtarzadeh, A. Alibakhshi, H. Yaghoobi, M. Hashemi, M. Hejazi, M. Ramezani, Recent advances on biocompatible and biodegradable nanoparticles as gene carriers, Exp. Opin. Biol. Ther. 16 (6) (2016) 771–785.

[41] M. Noruzi, Electrospunnanofibres in agriculture and the food industry: a review, J. Sci. Food Agric. 96 (2016) 4663–4678.

[42] R.K. O'Reilly, C.J. Hawker, K.L. Wooley, Cross-linked block copolymer micelles: functional nanostructures of great potential and versatility, Chem. Soc. Rev. 35 (2006) 1068–1083.

[43] M. Ozdemir, J.D. Floros, Active food packaging technology, Crit. Rev. Food Sci. Nutr. 44 (2004) 185–193.

[44] R. Priyadarshi, Y.S. Negi, Effect of varying filler concentration on zinc oxide nanoparticle embedded chitosan films as potential food packaging material, J. Polym. Environ. 25 (4) (2017) 1087–1098.

[45] M.M. Raja, P.Q. Lim, Y.S. Wong, G.M. Xiong, Y. Zhang, S. Venkatraman, Y. Huang, Polymeric Nanomaterials: Methods of Preparation and Characterization, in: S.S. Mohapatra, S. Ranjan, N. Dasgupta, R.K. Mishra, S. Thomas (Eds.), Nanocarriers for Drug Delivery, Springer, Cham, 2019, pp. 557–653.

[46] J.W. Rhim, H.M. Park, C.S. Ha, Bio-nanocomposites for food packaging applications, Prog. Polym. Sci. 38 (2013) 1629–1652.

[47] J.W. Rhim, P.K.W. Ng, Natural biopolymer-based nanocomposite films for packaging applications, Crit. Rev. Food Sci. Nutr. 47 (2007) 411–433.

[48] D.K. Robinson, M.J. Morrison, Nanotechnologies for food packaging: Reporting the science and technology research trends: report for the observatory NANO, Phytother. Res. 15 (2010) 476–480.

[49] N. Roy, N. Saha, T. Kitano, P. Saha, Biodegradation of PVP–CMC hydrogel film: a useful food packaging material, Carbohydr. Polym. 89 (2) (2012) 346–353.

[50] K.P. Satriaji, C.V. Garcia, G.H. Kim, G.H. Shin, J.T. Kim, Antibacterial bionanocomposite films based on $CaSO_4$-crosslinked alginate and zinc oxide nanoparticles, Food Packag. Shelf Life 24 (2020) 100510.

[51] T. Singh, S. Shukla, P. Kumar, V. Wahla, V.K. Bajpai, Application of nanotechnology in food science: perception and overview, Front. Microbiol. 8 (2017) 1501.

[52] S.K. Swain, A.K. Pradhan, H.S. Sahu, Synthesis of gas barrier starch by dispersion of functionalized multi-walled carbon nanotubes, Carbohydr. Polym. 94 (1) (2013) 663–668.

[53] X.G. Tang, P. Kumar, S. Alavi, K.P. Sandeep, Recent advances in biopolymers and biopolymer-based nanocomposites for food packaging materials, Crit. Rev. Food Sci. Nutr. 52 (2012) 426–442.

[54] K. Wang, P.N. Lim, S.Y. Tong, E. San Thian, Development of grapefruit seed extract-loaded poly (ε-caprolactone)/chitosan films for antimicrobial food packaging, Food Packag. Shelf Life 22 (2019) 100396.

[55] Y. Wang, C. Cen, J. Chen, L. Fu, MgO/carboxymethyl chitosan nanocomposite improves thermal stability, waterproof and antibacterial performance for food packaging, Carbohydr. Polym. 236 (2020) 116078.

[56] C. Wu, Y. Li, J. Sun, Y. Lu, C. Tong, L. Wang, Z. Yan, J. Pang, Novel konjac glucomannan films with oxidized chitin nanocrystals immobilized red cabbage anthocyanins for intelligent food packaging, Food Hydrocoll. 98 (2020) 105245.

[57] C. Wu, Y. Zhu, T. Wu, L. Wang, Y. Yuan, J. Chen, Y. Hu, J. Pang, Enhanced functional properties of biopolymer film incorporated with curcurmin-loaded mesoporous silica nanoparticles for food packaging, Food Chem. 288 (2019) 139–145.

[58] Z. Wu, X. Huang, Y.C. Li, H. Xiao, X. Wang, Novel chitosan films with laponite immobilized Ag nanoparticles for active food packaging, Carbohydr. Polym. 199 (2018) 210–218.

[59] A.M. Youssef, S.M. El-Sayed, Bionanocomposites materials for food packaging applications: concepts and future outlook, Carbohydr. Polym. 193 (2018) 19–27.

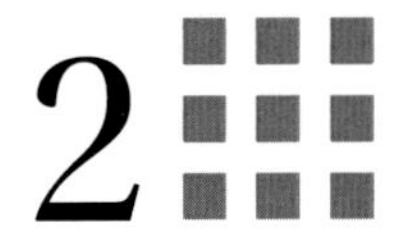

Methods for synthesis of nanobiopolymers

Sidra Iftekhar[a], Muhammad Abdul Wasayh[b], Muhammad Bilal Asif[c,d], Vesa-Pekka Lehto[a] and Mika Sillanpää[e,f,g,h]

[a]DEPARTMENT OF APPLIED PHYSICS, UNIVERSITY OF EASTERN FINLAND, KUOPIO, FINLAND [b]SCHOOL OF CHEMICAL AND MATERIALS ENGINEERING, NATIONAL UNIVERSITY OF SCIENCE AND TECHNOLOGY, ISLAMABAD, PAKISTAN [c]INSTITUTE OF ENVIRONMENTAL ENGINEERING & NANO-TECHNOLOGY, TSINGHUA-SHENZHEN INTERNATIONAL GRADUATE SCHOOL, TSINGHUA UNIVERSITY, SHENZHEN, GUANGDONG, CHINA [d]SCHOOL OF ENVIRONMENT, TSINGHUA UNIVERSITY, BEIJING, CHINA [e]INSTITUTE OF RESEARCH AND DEVELOPMENT, DUY TAN UNIVERSITY, DA NANG, VIETNAM [f]FACULTY OF ENVIRONMENT AND CHEMICAL ENGINEERING, DUY TAN UNIVERSITY, DA NANG, VIETNAM [g]SCHOOL OF CIVIL ENGINEERING AND SURVEYING, FACULTY OF HEALTH, ENGINEERING AND SCIENCES, UNIVERSITY OF SOUTHERN QUEENSLAND, WEST STREET, TOOWOOMBA, QLD, AUSTRALIA [h]DEPARTMENT OF CHEMICAL ENGINEERING, SCHOOL OF MINING, METALLURGY AND CHEMICAL ENGINEERING, UNIVERSITY OF JOHANNESBURG, DOORNFONTEIN, SOUTH AFRICA

Chapter outline

2.1 Introduction

Over the past few decades, materials based on synthetic polymers are extensively employed, which could potentially affect human health and the environment [1]. The issues can not only arise due to the application of polymer-based materials but can also due to the processes

Biopolymeric Nanomaterials: Fundamentals and Applications. DOI: https://doi.org/10.1016/B978-0-12-824364-0.00009-5

involved in the production of synthetic polymers. For instance, the most commonly used raw materials, that is, petrochemicals, will endure severe deficiency in the years to come [2]. Additionally, the nondegradable nature of synthetic polymers contributes to the additional financial and social burden on posttreatment required to minimize their adverse effects on the environment. These challenges related to the use of synthetic polymers encourage the application and development of green polymers, that is, naturally occurring polymers (referred to as biopolymers) for the preparation of new materials [2,3].

With an exceptional development of research in the field of nanotechnology and nanoscience, there is a growing positivity that it can bring about considerable progress in the field. At the same time, the main attraction in nanoparticles is related to their undistinguishable properties such as quantum properties, large surface to mass ratio, and absorption potential for other substances [4]. The combination of nanotechnology with biopolymer science has given a unique perspective to the field of packaging, food preservation, tissue engineering, nanoimplants, drug delivery, and biomedical engineering. In the recent years, biopolymers (such as chitin, cellulose, starch, silk, and zein) have drawn immense attraction because of their easy availability, biocompatibility, biodegradability, and sustainability [5–7]. The direct utilization of natural biopolymers at nanoscale for the fabrication of novel materials with tuneable structure and functionalities is an optimal choice. The benefits related to the structural properties (e.g., nanoconfinement and complex hierarchical structure) of natural material can be efficiently retained by such approaches [1,7].

From 2015 to 2020, a projected growth of 38.6% is predicted for nanobiopolymers, with overall global revenue of around $2 billion in 2020 [8]. In recent years, numerous chemical, mechanical, and other methods have been reported to attain nanobiopolymers successfully from their biological materials. Nanobiopolymers are known to be successfully obtained from agricultural and forestry products, namely cotton, wood, silk fibers, coconut, shells, rice, potato, and wheat, and their utilization has been expanded to various high-tech fields such as energy storage, ultrafiltration membranes, transparent display panels, catalytic supports, drug delivery and devices [1,7,9]. Herein, we are focusing on summarizing the techniques used for the preparation of nanobiopolymers and their respective composites for various applications.

2.2 Approaches for the synthesis of nanobiopolymers

The methods used for the preparation of nanobiopolymers are based on two common approaches, namely top-down and bottom-up approaches (Fig. 2.1) [11]. In the top-down approach, a nanosized structure is formed by breaking off a large structure (bulk material) under harsh and extreme conditions. Several techniques that come under this approach are mechanochemical, ball milling, arc discharge, laser ablation, and electrochemical method [11,12]. The nanostructures produced by this approach possess imperfections such as nonuniform surface structure, a wide size distribution, and crystallographic damage, which can significantly impact the chemical and physical properties of a nanobiopolymer [13]. Alternatively, the bottom-up approach of fabrication depends on the utilization of molecular-level precursors that

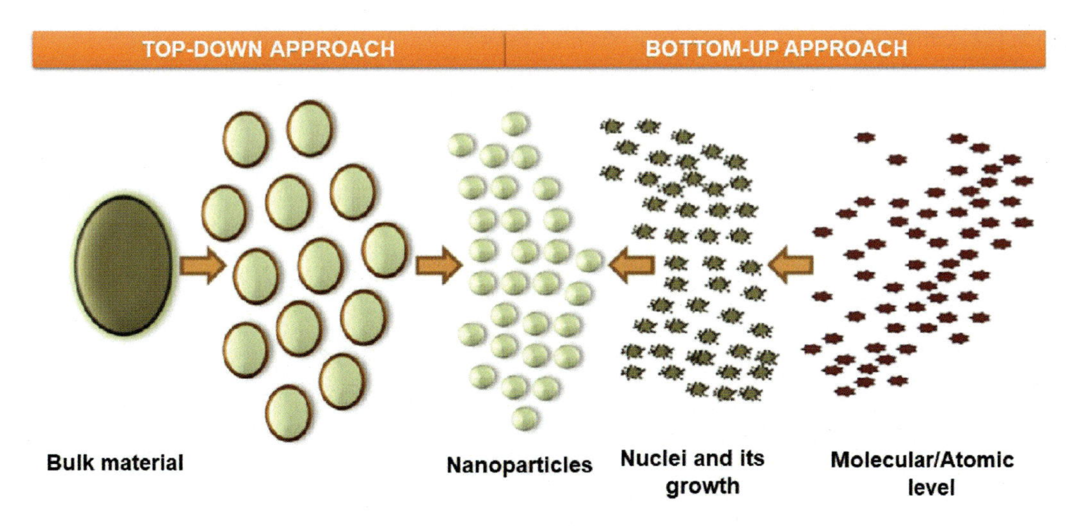

FIGURE 2.1 A schematic illustration of synthesis approaches. Adapted from ref [10].

aggregate through physicochemical methods such as condensation, polymerization, and pyrolysis to produce nanostructures. The method mainly employs decomposition of base material or chemical reduction to form the nanostructures and represents the possibility of preparing the materials of desired features [13]. Various methods under the purview of this approach include solvothermal, hydrothermal, sonochemical synthesis, microwave-assisted pyrolysis, sol-gel method, and coprecipitation. This approach does not require harsh conditions and tuning of nanostructures is simple. Comparing the two approaches, the later seems to be more common than the former approach as it offers large-scale production, as well as inexpensive and expeditious. Additionally, the bottom-up approach presents good control over the shape, size, distribution, and morphology of the nanomaterials [12].

2.3 Preparative methods for synthesis of nanobiopolymers nanocomposites

Various mechanical, chemical, and biological methods have been used for the synthesis of nanobioploymers and their composites. The selection of the method is commonly based on the properties needed for the particular application of nanobiopolymer as explained below.

2.3.1 Mechanical processes

The mechanical methods used to produce nanobiopolymers are mainly dependent on the choice of equipment. For the exfoliation of biopolymers to nanofibers and nanocrystals, various types of mechanical equipment used are grinders, blenders, ultrasonicators, high-pressure homogenizers, and twin-screw extruders.

2.3.1.1 Grinders

Another most common apparatus for the preparation of nanofibrils is grinders. This method employs the passing of a biopolymer pulp between a rotating grindstone and a static grindstone at a speed of around 1500 rpm. The discs create powerful cyclic shearing forces during nanofibrillation to disintegrate them into nanofibrils. The degree of nanofibrillation and structural morphology of fibers mainly depends on the distance between the discs, source of biopolymer, time of repeated processing, and the chemical pretreatments [14,15].

2.3.1.2 Blenders

Along with grinders, nanofibrils from a biopolymer can also be synthesized using high-speed blenders. The blender consists of a bottle with a propeller and a motor. To accelerate the blending operation, the associated tamper is used. During blending, the mixing of suspension is driven by a high-speed propeller generating an immense impact on biopolymer pulp, which leads to nanofibrillation. For example, cellulose wooden pulp treated for 30 min at 37,000 rpm in a blender was reported to achieve a high level of nanofibrillation [16]. Similarly, for nanofibrillation of straw pulp, the balloon-like structure was formed initially, and then the balloon expanded toward the edges and isolates the nanofibrils. The blending process is mainly influenced by the parameters such as agitation time and speed, pulp source, and concentration, which are crucially important in determining the performance, structure, and degree of nanofibrillation [15].

2.3.1.3 Ultrasonication

Ultrasonication is also an intensively used technique for the isolation and generation of cellulosic nanofibers by applying sound energy to physical and chemical systems. For the preparation of nanofibers, the chemical effects of the process are drawn from acoustic cavitation in which the high pressure of water and potential energy of bubbles are transformed into kinetic energy, and at a speed of 100 m/s, the bubbles collide with the surface of biopolymer. As a result of this physical collision, the nanofibers are generated from a biopolymer. This method appears to be nonselective as it removes both amorphous and crystalline cellulose from the environment [17]. However, this method is not only limited to produce only cellulosic nanofibers. In 2007, nanochitin with a size of 25–120 nm is prepared under neutral conditions in 30 min using ultrasonication [18].

2.3.1.4 Twin-screw extrusion

To obtain biopolymer nanofibers, the twin-screw extrusion method can be used. In this technique, substantial forces are generated to separate nanofibrils by combining the processes of kneading and feeding. The extruders can be used to process the pulp with high solid contents, that is from 25 to 40 wt.%, and the resulting nanofibrils may contain high solid contents of up to ~50 wt.%. In addition, the obtained material is in a powder form instead of an aqueous

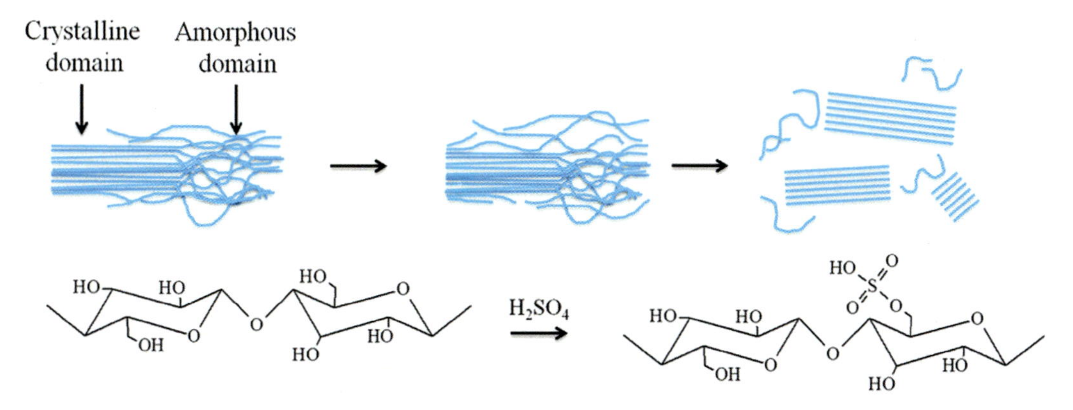

FIGURE 2.2 Acid hydrolysis of cellulose by sulfuric acid. Reprinted with permission from ref [23]. Copyright 2017 Royal Society of Chemistry.

paste or suspension exhibiting benefits for industries in terms of storage, transportation, and production [19].

Other mechanical instruments that can be used are microfluidizers, high-pressure homogenizers, or their combinations to prepare nanobiopolymers. For instance, to obtain nanochitin from lobster shells, high-pressure homogenizers were used, which yielded a nanochitin of a uniform size (<100 nm) [20]. By combining the mechanical nanofibrillation with microfluidization (30,000 psi, 120 mL/min, 10 cycles), grinding (10 passes), and homogenization, nanochitin with a width of 50 nm can be synthesized [21].

2.3.2 Chemical processes

2.3.2.1 Acid hydrolysis

Acid hydrolysis is one of the oldest methods dated back to the late nineteenth century and used to produce nanofiber and nanocrystals from biopolymers. The process includes hydrolysis by both weak and strong acids. The application of concentrated acid hydrolysis was first reported in 1883 for the production of nanocellulose. The main principle of acid hydrolysis is the complete immersion of biopolymer in concentrated acids, for example, 72–65% H_2SO_4, 42% HCl, or 77–83% HNO_3 at low temperature, which leads to homogenous hydrolysis of biopolymer. The acid hydrolysis techniques using concentrated H_2SO_4 is more developed and is widely studied as compared to other acids. Hydrolysis when carried out with 200 g of raw material for 30–120 min at 100 °C with concentrated H_2SO_4 yielded 78–82% of nanomaterials [3,15]. The main factors, which govern the dissolution of biopolymers with strong acids, are the breaking of hydrogen bonds because of the complex formation and depolymerization of large chains. Moreover, the acid hydrolysis considerably lowers the thermal strength of nanocrystals. This method is commonly used for the production of nanofibrous cellulose (Fig. 2.2). Besides, acid hydrolysis requires a longer reaction time; can result in the corrosion of equipment; and can cause environmental pollution, Thus, limiting the applicability of this technique [22].

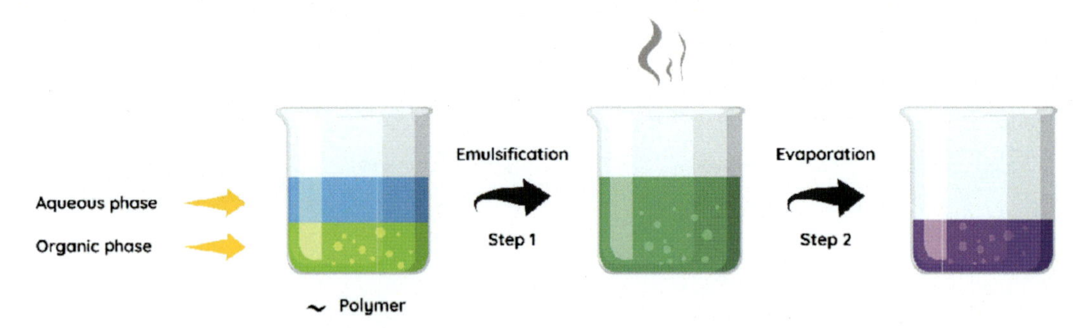

FIGURE 2.3 Typical illustration for the preparation of nanobiopolymers using the solvent evaporation technique. Adapted from ref [24].

2.3.2.2 Solvent evaporation method

Solvent evaporation is employed for the preparation of polymeric nanoparticles via the dispersion of polymers and is now exclusively used in several applications particularly for biopolymers [24]. The volatile solvents such as chloroform, ethyl acetate, and dichloromethane are used for the formulation of polymer solutions. The organic phase is homogenized by mixing with an immiscible nonsolvent in the dispersing phase that comprises an appropriate emulsifying agent. The solvent from the emulsion was then evaporated either by continuous stirring or by reducing the pressure to obtain a nanoparticle suspension (Fig. 2.3). Traditionally two main approaches are used for the preparation of emulsion that is, formulation of a single emulsion (oil in water, o/w) and preparation of double emulsion (water-in-oil-in-water, w/o/w). Normally, the organic solvent in which the polymer is dissolved forms the oil phase, while the aqueous phase consists of a stabilizer. The ultrasonication and high-speed homogenization are required in both approaches followed by solvent evaporation. Later, ultracentrifugation is used for the collection of solidified nanoparticles followed by washing with water to remove the extra additives followed by exposure to lyophilization [25].

From the economic perspective, o/w emulsion is of main interest as water being utilized as nonsolvent, which simplifies the whole process by reducing the requirement for recycling, aiding the washing phase, and reducing agglomeration [24]. Several factors found to have a major impact on the particle size are properties and concentration of stabilizer, the concentration of polymer and speed of homogenization, along with temperature and viscosity of aqueous and organic phase [26]. Notably, high energy in homogenization is required for the scale-up of the process [27].

2.3.2.3 Salting out method

The modified form of the emulsion method was introduced to minimize the use of hazardous surfactants, organic solvents, and other chlorinated solvents (Fig. 2.4). The polymer solution is prepared in a solvent such as acetone, which is miscible in water and is homogenized in the aqueous phase without employing high shear forces, followed by the addition of some

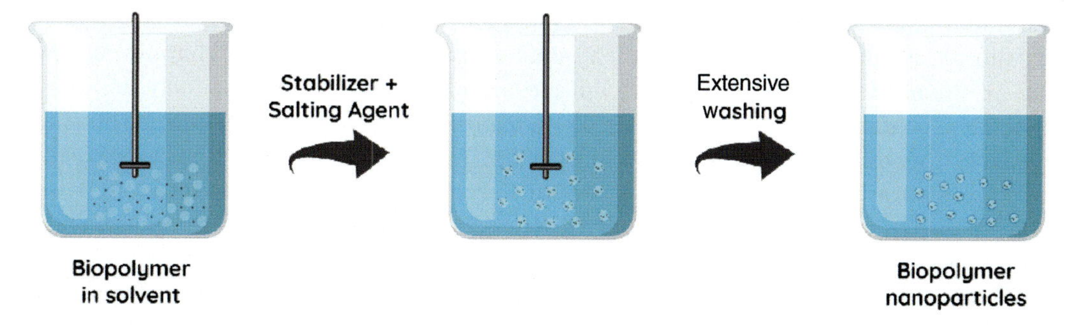

FIGURE 2.4 A generic representation of the salting-out method. Adapted from ref [24].

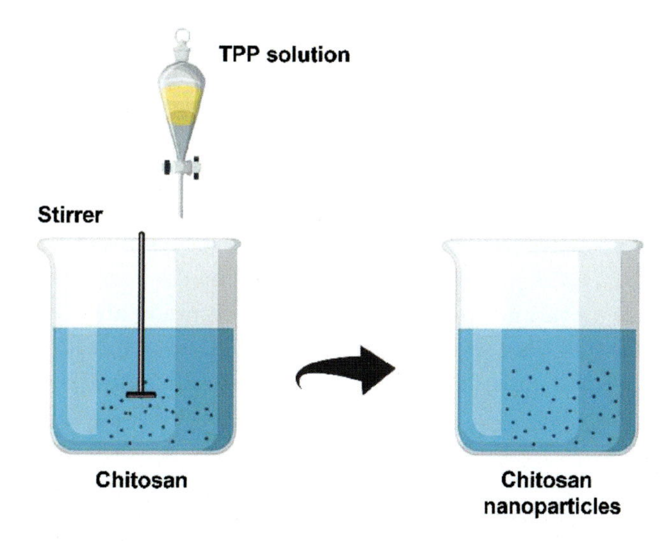

FIGURE 2.5 A schematic for the preparation of chitosan nanoparticles by ionotropic gelation method. Adapted from ref [24].

salting-out agent in the aqueous phase such as a high concentration of sucrose [28]. The miscibility features of solvents with water are tuned according to the electrolyte solubility. The migration of solvent from the emulsion droplets toward the polymer occurs after the reverse process that is, dilution. This is due to a decrease in the concentration of the salting-out agent (sucrose or salt) in the continuous phase, resulting in the precipitation of the dissolved polymer [29–32]. The main advantage of this method is to deal with substances sensitive to heat as it does not need high temperature. However, the nanoparticles produced need extensive washing [24].

2.3.2.4 Ionotropic gelation

The ionotropic gelation methods are often used for the preparation of chitosan and alginate hydrogel nanoparticles. The general illustration of the method is presented in Fig. 2.5. This process is simple in operation; however, optimization is very tricky [77]. For the synthesis of nanoparticles of chitosan, alkaline tripolyphosphate (TPP) solution is added into an acidic chitosan solution. The mixing of both phases leads to the linkages (inter- and intramolecular) between amino moieties of chitosan and TPP phosphates. The tunable chitosan nanoparticles can be obtained by varying the concentration of TPP and chitosan [33]. The parameters that influence the particle size of chitosan are pH of both solutions, chitosan-TPP ratio, mixing speed and procedure, as well as the temperature of the solution, and degree of deacetylation. The electrostatic repulsion takes place between chitosan molecules under acidic conditions due to the protonation of surface amino groups, meanwhile, interchain hydrogen bonding interactions also occur. Therefore, at the same time, two opposite forces play a role in the chitosan molecule, and at some point, both the forces reach equilibrium. Subsequently, after this point with an increase in chitosan concentration, the molecules start to come closer, despite the electrostatic repulsion. This results in the formation of nanoparticles of chitosan. Above the equilibrium concentration, hydrogen bonding can act strongly, thus triggering a huge number of chitosan molecules participating in the cross-linking of a one particle leading to the formation of microparticles. Initially, with the increase in the volume of TPP, the particle size decrease but adding an excess amount of TPP could result in aggregate formation. The most favorable cross-linking of the amino-phosphate group occurs at pH 4.5 and 5.2 [34].

In the case of alginate, the reaction can be initiated using any cationic species such as divalent metal ion (e.g., $CaCl_2$). Like chitosan, for obtaining the desired size particles, alginate optimization depends on various factors including concentration and viscosity of counterions and mixing speed of counterions with alginate solution. All the factors in both cases seem to be interrelated, which makes it difficult to determine the most influential factor/parameter [35].

2.3.2.5 Nanoprecipitation/solvent displacement method

The nanoprecipitation or solvent displacement method is based on the assumption of interfacial polymer deposition, where a semipolar solvent, which is miscible in water, is displaced from a lipophilic solution as shown in Fig. 2.6 [24]. An interfacial deposition is an emulsification/solidification process and is not a polymerization technique [36]. The speedy diffusion of the solvent into the nonsolvent leads to the reduction in the interfacial tension between the phases, resulting in the formulation of small organic solvent droplets with a high surface area. The nonsolvent of the polymer, the polymer-solvent, and the polymer are the three basic components of nanoprecipitation. Commonly, the organic solvent such as acetone or ethanol is selected from polymer solvents as they are miscible with water and removal via evaporation is easier [37]. On the other hand, water is used as a nonsolvent phase in which surfactants are added. The nanoparticles can be formed by the addition of the aqueous phase to the organic phase slowly under moderate stirring and vice versa. The rapid diffusion leads to the formation of narrow distribution with a uniform size of nanoparticles. The main factors that govern the performance of this process successfully are the mixing rate of the aqueous phase, the injection rate of the organic phase, the ratio of aqueous to the organic phase, ways of additions, concentration, and nature of their components. The surfactants are mainly added to avoid the agglomeration

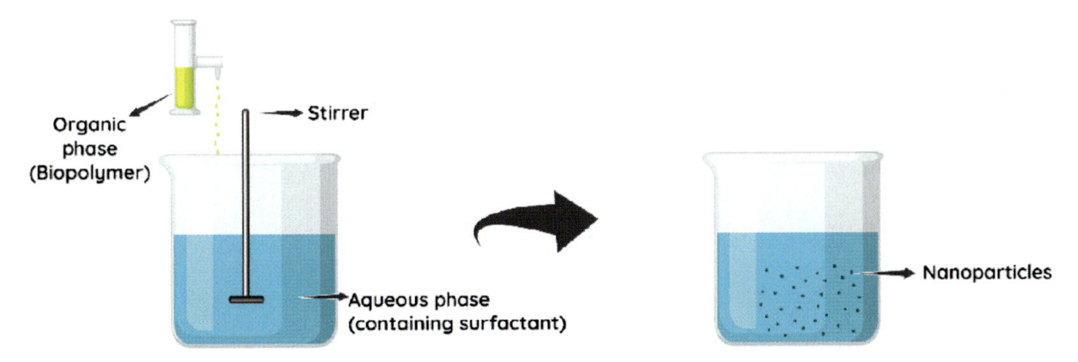

FIGURE 2.6 Representation for the synthesis of nanoparticles by the solvent displacement method. Adapted from ref [41].

of nanoparticles when stored for longer periods, while their addition may affect the particle size [38]. The process mainly consists of three main steps including nucleation, growth, and aggregation [39]. The particle size is determined by the rate at which these steps occur where to get the uniformly distributed particles, the separation between the growth, and the nucleation. Under ideal conditions, high nucleation should be permitted, which depends on lower growth rate and supersaturation. The technique is extensively used in the preparation of nanocapsules and nanospheres due to its simplicity, speed, and reproducibility, while to obtain effective results challenges related to the low concentration of polymer in the organic phase need to be considered [40].

2.3.2.6 Coacervation

Coacervation is mainly used for the preparation of microcapsules and is usually termed as a phase separation technique. This process comprises the separation of the emulsified polymeric solution from the dense polymer-rich phase (known as coacervates) to another layer with low polymer content. The two categories of coacervation, that is, complex and simple are based on the final phase separation and amount of polymers added. In the complex coacervates (Fig. 2.7), strong electrostatic interaction holds the oppositely charged polymers mixture together. On the other hand, a single polymer is used in a single coacervate, and the phase separation is carried out by desolvation through the addition of nonsolvents and inorganic salts along with temperature variations [42,43]. Both coacervates are stabilized and hardened either via chemical cross-linkers such as formaldehyde and glutaraldehyde or by heat treatment. The particle size can be controlled by various factors such as the viscosity of the nonsolvent, quantity of the polymer, and its molecular weight [44,45].

2.3.2.7 Desolvation

The antisolvent thermodynamically governed precipitation process is referred to as desolvation. The formation of nanobiopolymer is the self-assembly method stimulated by adding the

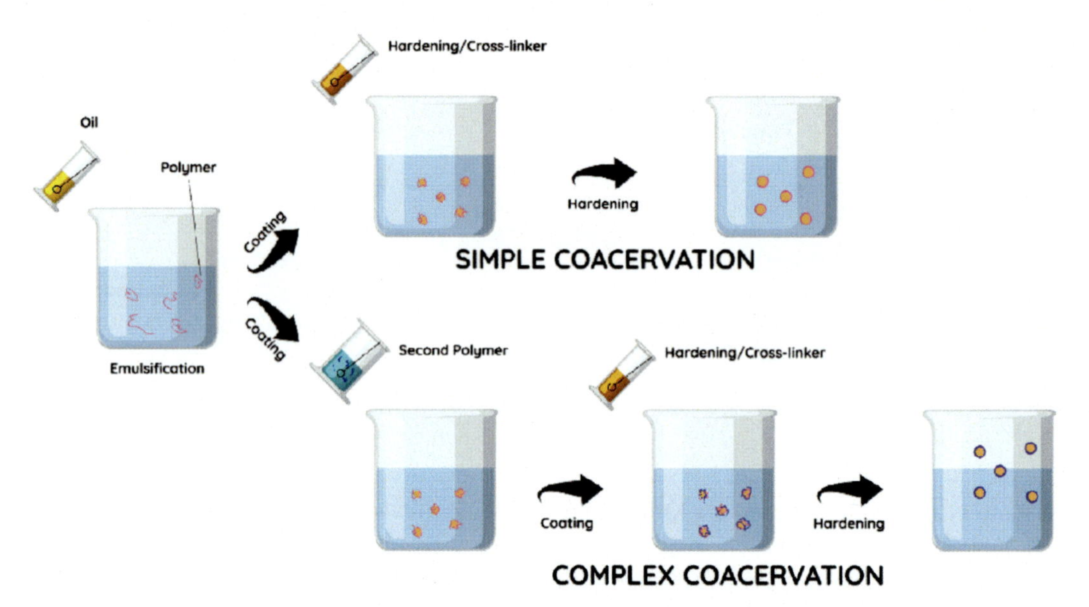

FIGURE 2.7 From emulsification to be encapsulated oily core, the depiction of simple and complex coacervation. (Reprinted with permission from ref [46]).

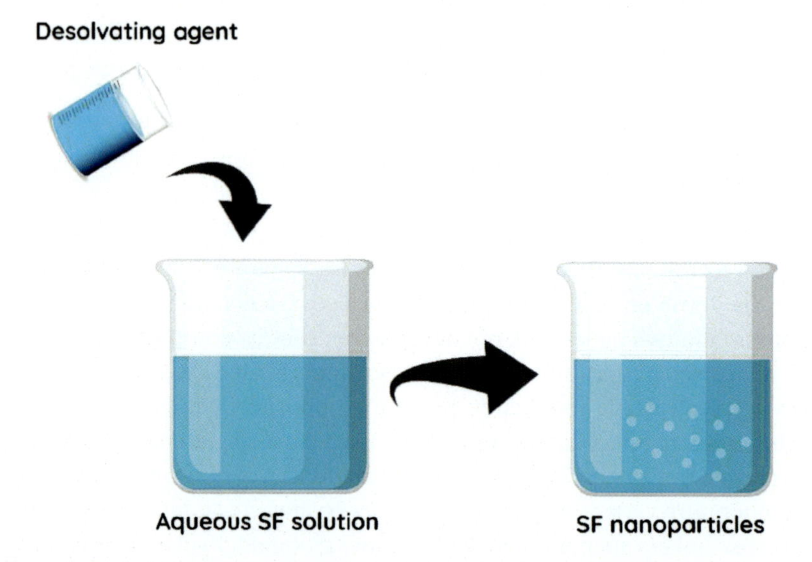

FIGURE 2.8 Preparation of silk fibroin nanoparticles by desolvation method. Adapted from [52].

desolvating agents (Fig. 2.8) [47,48]. Moreover, the processes of nucleation, condensation, and coagulation can increase the particle size. For the slow growth of particles and rapid nucleation, the experimental conditions such as pH, amount of desolvating agent, ionic strength, and concentration of cross-linking agents play an important role [49]. In the case of proteins, the transformation of the 3D stretched to highly coiled conformation forming coacervates happens

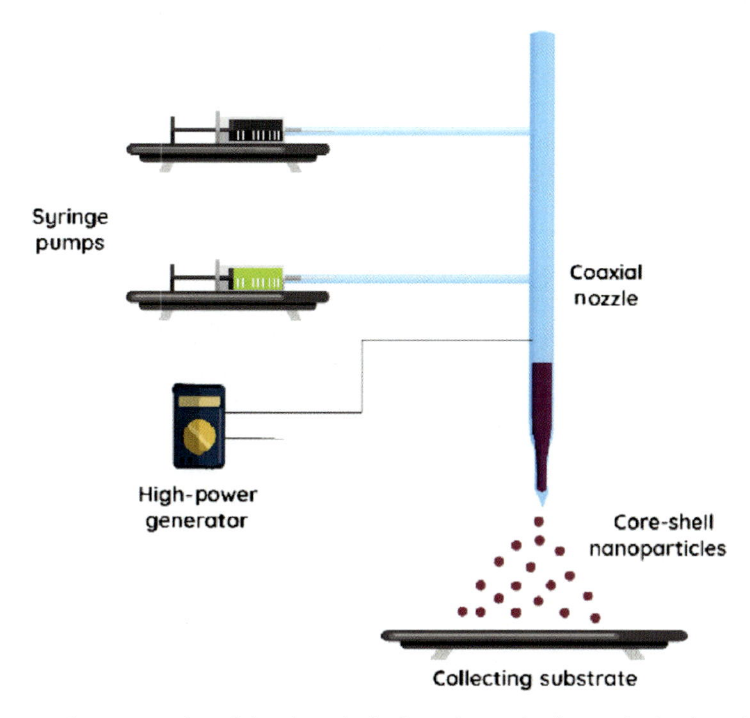

FIGURE 2.9 The general representation of the electrohydrodynamic atomization method. Adapted from ref [56].

due to dehydration, and stabilization is done with the help of enzymes or chemicals. The particles of 190 nm of lactoferrin protein are prepared with a 75% yield using a desolvation procedure under the optimized conditions and are applied as a carrier for curcumin delivery to test its anticancerous activities. Nanoparticles based on human serum albumin and bovine serum albumin are prepared by improved desolvation process for the sustained drug delivery applications, in which keeping the polydispersity below 0.2% the particle size can be regulated up to 100 nm. The particle size can be controlled by the addition of ethanol under controlled desolvating conditions to keep the particle size in a smaller range [50]. Nanogelatin, prepared by the two-step desolvation method, is exclusively used for cancer treatment owing to their promising outcomes in photodynamic therapy. To improve the yield, the desolvation is conducted twice to obtain uniform-sized nanoparticles [51].

2.3.2.8 Electrohydrodynamic atomization

Electrohydrodynamic atomization also termed as electrospray deposition process, is the process in which high-voltage electrical forces are used to generate the cloud of charged droplets through liquid atomization. The charged particles sprayed through the nozzle tip dried while reaching the substrate (Fig. 2.9) [53]. The subsequent nanoparticles are strongly bound to the substrate by electrostatic forces. The main advantage of this method is the absence of any surfactant and solvents. Furthermore, by altering the high voltage of the power supply, the droplet speed, charge, and size of the particle can be easily controlled [54]. The most frequently prepared nanocoating prepared by the electrospray method is the chitosan nanocoating. Zhang et al. [55] reported the one-step method for the preparation of nano- and microparticles where the increase in conductivity of solution and decrease in viscosity were found to alter the particle size. The nanoparticles prepared have an average particle size of 124 nm and showed promising results in pulmonary and oral drug delivery systems [55].

2.3.2.9 Supercritical fluid technology

The preparation of nanoparticles by most of the techniques involves the use of different organic solvents leading to the need for the development of green methods. This feature encourages the utilization of supercritical fluids that are environmentally friendly solvents and have the potential to use for the production of biopolymer nanoparticles with high purity. The method is known to offer an efficient and appealing strategy by preventing most of the shortcomings of the conventional methods for the preparation of nanobiopolymers [24]. Above the critical temperature of the solvent, it remains in single-phase regardless of the pressure (also known as supercritical fluid). The most commonly used supercritical fluid is Supercritical CO_2, ascribing to its undamaging and nonflammable nature and gentle critical conditions (P_c = 73.8 bars, T_c = 31.1 °C). The rapid expansion of critical solution (RESS) and supercritical antisolvent (SAS) are the two common methods that fall under supercritical fluids. In the SAS method, methanol is used as a solvent to dissolve the solute at operating conditions that are fully miscible with CO_2 supercritical fluid. Conversely, in the RESS, solute is dissolved in a supercritical fluid–like supercritical methanol and then sprayed with the help of a nozzle in a low-pressure region. The methods lead to the preparation of nanobiopolymers free of any solvents. Although the techniques are suitable for production at a large scale and environmentally friendly, the initial capital required is considerably high as they need specially designed equipment [57].

2.3.2.10 Emulsification-diffusion method

In this process, the small nano-range droplets are formed by the spontaneous mixing of the organic-aqueous phase (1:2). Generally, the aqueous phase comprises hydrophilic surfactant and water, whereas the organic phase is constituted of hydrophilic solvent, a lipophilic surfactant, and plant oils [58]. The decrease in particle size is attained by increasing the concentration of water-soluble solvent. The higher yield stable nanoparticle dispersion is provided by o/w emulsion through solvent evaporation and the nanoparticles, thus possessing high loading and entrapment efficiencies [59,60]. However, the major setback of this technique is the utilization of a larger volume of water and toxic organic solvents [61,62]. The schematic illustration of the steps involved in the process is shown in Fig. 2.10.

2.3.2.11 Emulsion polymerization

One of the fastest methods available for the preparation of nanobiopolymers is emulsion polymerization, which is subdivided into two groups depending on the use of aqueous and organic

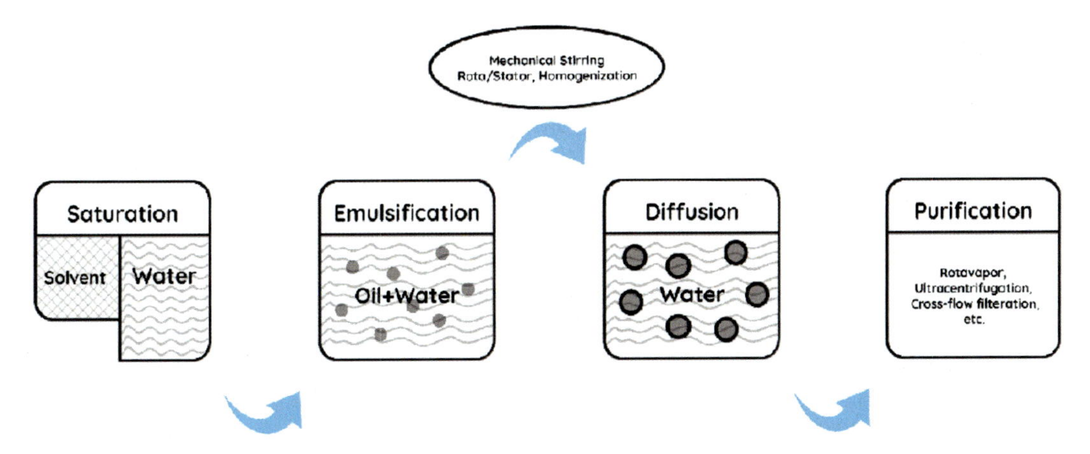

FIGURE 2.10 A schematic illustration of the steps involved in the emulsification-diffusion process. (Adapted from ref [61]).

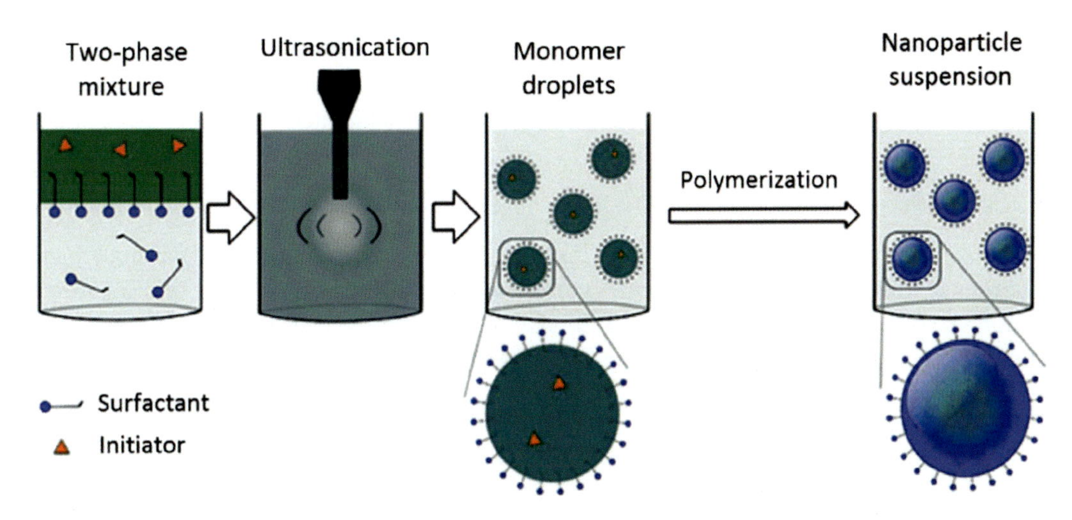

FIGURE 2.11 Preparation of nanoparticles by the emulsification polymerization process. Adapted from ref [24].

phases. As shown in Fig. 2.11, the monomer dispersion is adapted by organic phase methods into an emulsion or any nonsolvent or inverse microemulsion. Alternatively, through applying high energy the monomer can be transformed by radical initiation for the propagation of chain that then joins with other monomer followed by the formation of nanoparticles and phase separation either before or after termination. The surfactants were used in past to avoid clogging; however, due to associated toxicity, the method is not in practice. On the other hand, surfactants and emulsifiers are not needed in the aqueous continuous phase, and the monomer is directly dissolved in an aqueous solution, forming a continuous phase. There are several ways to initiate the polymerization reaction but once propagated, the chain growth occurs rapidly.

2.3.2.12 Microemulsion method

Microemulsion method is another way to produce polymeric nanoparticles that gained considerable attention due to its special feature of producing nanoparticles with higher molar mass, which are completely different when equated kinetically. However, the particle size and number of chains per particle formed by microemulsion is very small. Additionally, in comparison to simple emulsion, which depicts three reaction rate intervals, this method only shows two reaction rate intervals [24]. Commonly, in a thermodynamically stable microemulsion that consists of swollen micelles, the initiators added are water soluble. The polymerization reaction results in the formation of spontaneously formed states that are stable thermodynamically and depend on quantities of surfactants that at the o/w interface have an interfacial tension close to zero. Due to a large number of surfactants present in the system the particles get covered completely with it after some time. At the start, the polymer chains appeared only on the few droplets as the initiation cannot be achieved instantaneously on all microdroplets. After a while, the destabilization of complex microemulsions started because of the elastic and osmotic control of chains increasing the size of particles, growth of empty micelles, and subsequent secondary nucleation. In the final formed material along with empty micelles, minuscule latexes also exist in a size range of 5–50 nm [63]. The main factors, which influence the kinetics of microemulsion polymerization and characteristics of polymeric nanoparticles, include concentration and type of initiators, monomer, surfactant, and temperature of the reaction [64].

2.3.2.13 Mini-emulsion polymerization

A substantial increase in the research work has been seen for the preparation of nanopolymeric materials by this method. The traditional components of this process consist of a mixture of monomer, initiators, water, surfactant, and costabilizer. The main difference between mini-emulsion polymerization and emulsion polymerization is the use of high shear source namely ultrasound and costabilizer that is generally a compound with low molecular mass. To attain the steady-state condition in mini-emulsion, high shear is needed as mini-emulsions are critically stabilized with interfacial tension higher than zero [24]. The nature and morphology of nanopolymeric materials prepared by this method are influenced by the costabilizers and the initiator. The method is suitable for the fabrication of nanoparticles with polymer core shells comprising a biodegradable polymeric shell and magnetic core [65].

2.3.2.14 Interfacial polymerization

The technique is well developed for the preparation of nanobiopolymeric materials [66]. The reaction occurs at the interface of two phases and mainly involves the polymerization of two reactive monomers that are dissolved in dispersed and continuous phases. The particle can be prepared in nano-range by engaging interfacial cross-linking reactions such as radical polymerization, polycondensation, and polyaddition [67]. Oil-containing nanocapsules are formed at the interface of o/w microemulsion by polymerization of monomers. In addition to nanocapsules, the formation of nanospheres was found to be stimulated in the presence of protic solvents such as ethanol, isopropanol, and n-butanol. Consequently, aprotic solvents such as acetonitrile and acetone must be utilized for the formation of nanocapsules. Similarly, water-containing nanocapsules can be formed when the polymerization of the monomer occurs in w/o microemulsions, which is then precipitated to form nanocapsules [24].

2.3.2.15 Controlled or living radical polymerization (C/LRP)

Living radical polymerization technique is a completely different way of polymerization, which opens the new approach compared to the traditional old techniques. The primary contributors to the development of this method are the need for modernizations in the utilization of hydrophilic polymers and the increasing environmental concerns. The technique can be easily adaptable in aqueous dispersed systems. Additionally, the method provides exact particle size and distribution making it valuable to be used at an industrial scale, which is vital for commercialization [68]. Among the numerous existing processes for C/LRP, the three methods namely atom transfer radical polymerization (ATRP), nitroxide-mediated polymerization (NMP), and fragmentation transfer chain polymerization (RAFT) have been studied a lot. Besides temperature, other factors that affect the process in controlling the size of particles are concentration and properties of control agent, initiator, monomer, and type of emulsion. Furthermore, over the years the research community is interested to understand the heterogeneous C/LRP in ATRP, NMP, and RAFT and understands the chemistry behind the method and number of challenges that are still there to overcome. Some of the issues arising are common and some are unique in every system. The common issues are controlling the presence of a residual agent in the final product and comparison of cost with the conventional radical polymerizations. Moreover, other problems, which arise due to the presence of residual control agent is stability, odor, color, and environmental lawmaking. The removal and extraction of these interfering agents seem to be very complex than homogeneous solutions from aqueous dispersions and thus for successful commercialization, the challenges need to be studied deeply [64].

2.3.3 Biological methods

Due to importance in reducing the environmental footprints, biological methods, which use proteins and peptides as green reducing agents for preparation of nanostructure, have gained attraction. The method is subdivided into two categories including microbial synthesis and enzymatic hydrolysis.

2.3.3.1 Microbial synthesis

The preparation of nanostructures by this method involves the utilization of bacteria, viruses, fungi, and yeast as a reducing agent to produce bio-nanostructures. The utilization of microorganisms as reducing agents is instigated by the inherent necessity of chemical decontamination that leads to resistance of microbes to toxic metals. The method provides a straightforward approach for the preparation of nanostructure with biological accessibility. For instance, a metal-accumulating bacterium was used as a nanofactory for the microbial synthesis of Ag-based nanostructure. On the contrary, fungi are known to have significantly greater quantities of proteins and hold substantially easier downstream handling and processing and thus are deemed more biosynthetically productive than bacteria. Likewise, the synthesis of nanostructures mediated by viruses is also worth noticing. By implication of the virus as a biotemplate, nanowire consisting of various functional components could be prepared, which have several applications namely photovoltaic devices, battery electrodes, and supercapacitors. Although using the microbial systems nanostructure could be synthesized; however, the shortcoming of the system is to fully understand the underlying mechanism to control the synthesis of nanostructures makes it a challenge to fully take the advantage of the process [69].

2.3.3.2 Enzymatic hydrolysis

Compared to the chemical methods, rapid hydrolysis is a key feature of the enzymatic process with no associated environmental pollution. The obtained nanobiopolymer possesses high purity, stable chemical and physical properties, and water solubility. The general process mainly involves the dissolution of raw material in 0.5 M acetic acid solution that contained enzymes (such as trypsin, pepsin, and chymotrypsin) followed by stirring at 4 °C for 48 h and filtration [70]. The conditions employed for processing are used to precipitate and dialyzed the filtrate to obtain the collagen nanoparticles. Nanocollagen (type I) was also extracted from animal skins, which is one of the common raw materials. After the skin was separated from the animal, the cuticle layer and hair were removed. Later the skin was added to 2.5 g/L of acetic acid, which contains pepsin as an enzyme, stirred for 48 h followed by low temperature and ultra-speed centrifugation for the harvesting of supernatant [15]. Likewise, to obtain the nanocollagen from the skin of yellow-fin tuna, the pretreatment was conducted with NaOH, and pepsin hydrolysis was carried out with HCl solution [71].

2.4 Green sustainable methods

The idea of preparing the nanomaterials using greener approaches includes the use of natural templates (such as agricultural residues, starch, cellulose), nontoxic solvents, and bio-based reducing agents (such as a microorganism, vitamins, leaf extracts, etc.) [69]. These pathways encompass the utilization of sustainable energy techniques such as microwave irradiation, UV light, and ultrasound. Numerous plant metabolites play a crucial role to produce nanoparticles through bioreduction and thus lessening the need for using toxic chemicals namely organic solvents, strong acids, and bases, and so on. Furthermore, in comparison to microorganism-based biogenic (e.g., algae, bacteria, seaweeds, fungi, diatoms) methods, the plant-based process seems to be more beneficial as it does not involve the use of culture preparation, media, and expensive isolation methods. Moreover, plant-based methods involve only one step, are inexpensive, have short preparation time, and comparatively safe. The extract used can be obtained from various parts of plants such as leaves, roots, shoots, seeds, barks, and stems, and it also consists of secondary metabolites that can be used as stabilizing and reducing agents in the production of nanoparticles [72]. Henceforth, the methods seem to be promising for the synthesis of nanoparticles and a shift from traditional chemical production to biogenic synthesis is possible.

Over the past decade, many researchers reported the biogenic synthesis method for the preparation of nanoparticles. In this regard, chloroauric acid was reduced by fenugreek seed extract, which contains a high concentration of flavonoids and other bioactive compounds such as vitamins, lignin, and saponin, and can also be used as a surfactant [73]. The electrostatic stabilization to nanoparticles was provided by flavonoids, while the functional groups (C=C, C=N, $-COO^-$) act as surfactants. Likewise, Usha et al. [74] reported the biosynthesis of Cu nanoparticles by utilizing tulsi leaf extract, which consists of tannic acids, alkaloids, saponins, glycosides, and aromatic compounds. Aminodextran was used as protective and reducing agents for the fabrication of monodispersed Pt nanoparticles by Yang et al. [75]. Likewise, barberry fruit extract was employed as a stabilizing and reducing agent. A greener process seems to be a viable substitute for traditional solvent-based techniques [76].

2.5 Conclusion and future prospects

For the preparation of nanobiopolymers several methods have been used, which have their pros and cons. Nano-architectures are considered a unique feature of nanobiopolymers and are crucial for the physical properties of the materials. Accordingly, the choice of method and proper optimization of these nanostructures during fabrication and use of nano-building-block to synthesis better material are the features that need due consideration. Although the sustainability of mechanical and chemical processes may be questionable, these processes are the most effective. By contrast, the cost and efficiency of the greener fabrication methods need to improve. Notably, most of these methods are used in combinations, which significantly improve the fineness and uniformity of nanobiopolymers that is a desirable property from a practical point of view. In addition, most of the chemical techniques employ the use of strong acids, surfactants, and organic contaminants, which would induce pollution and become the source of pollution. Likewise, microbial synthesis consumes huge quantities of culture, extraction media, and energy. Hence, to minimize environmental concerns, green approaches need to be considered, which are more environmentally friendly. Importantly, the cost and efficiency of green processes need to be optimized in the future. Therefore, ways to reduce energy consumption and improving production efficiency are of paramount importance for industrial production.

References

[1] N. Yang, W. Zhang, C. Ye, X. Chen, S. Ling, Nanobiopolymers Fabrication and Their Life Cycle Assessments, Biotechnol. J. 14 (2019) 1700754, https://doi.org/10.1002/biot.201700754.

[2] S. Wang, A. Lu, L. Zhang, Recent advances in regenerated cellulose materials, Prog. Polym. Sci. 53 (2016) 169–206, https://doi.org/10.1016/j.progpolymsci.2015.07.003.

[3] S. Ling, D.L. Kaplan, M.J. Buehler, Nanofibrils in nature and materials engineering, Nat. Rev. Mater. 3 (2018) 18016, https://doi.org/10.1038/natrevmats.2018.16.

[4] W.H. De Jong, P.J.A. Borm, Drug delivery and nanoparticles: Applications and hazards, Int. J. Nanomedicine (2008) 133, https://doi.org/10.2147/IJN.S596.

[5] R.J. Moon, A. Martini, J. Nairn, J. Simonsen, J. Youngblood, Cellulose nanomaterials review: structure, properties and nanocomposites, Chem. Soc. Rev. 40 (2011) 3941, https://doi.org/10.1039/c0cs00108b.

[6] P. Balakrishnan, M.S. Sreekala, M. Kunaver, M. Huskić, S. Thomas, Morphology, transport characteristics and viscoelastic polymer chain confinement in nanocomposites based on thermoplastic potato starch and cellulose nanofibers from pineapple leaf, Carbohydr. Polym. 169 (2017) 88–176, https://doi.org/10.1016/j.carbpol.2017.04.017.

[7] H. Kargarzadeh, J. Huang, N. Lin, I. Ahmad, M. Mariano, A. Dufresne, S. Thomas, A. Gałęski, Recent developments in nanocellulose-based biodegradable polymers, thermoplastic polymers, and porous nanocomposites, Prog. Polym. Sci. 87 (2018) 197–227, https://doi.org/10.1016/j.progpolymsci.2018.07.008.

[8] M. Gagliardi, Global markets and technologies for nanofibers, PR Newswire 1 (2016) 1–10.

[9] A. Barhoum, M. Bechelany, A.S.H. Makhlouf (Eds.), Handbook of Nanofibers, Springer International Publishing, Cham, 2019, https://doi.org/10.1007/978-3-319-53655-2.

[10] P. Khanna, A. Kaur, D. Goyal, Algae-based metallic nanoparticles: Synthesis, characterization and applications, J. Microbiol. Methods 163 (2019) 105656, https://doi.org/10.1016/j.mimet.2019.105656.

[11] R. Bayan, N. Karak, Polymer nanocomposites based on two-dimensional nanomaterials, Two-Dimensional Nanostructures Biomed. Technol, Elsevier (2020) 249–279, https://doi.org/10.1016/B978-0-12-817650-4.00008-5.

[12] A. Biswas, I.S. Bayer, A.S. Biris, T. Wang, E. Dervishi, F. Faupel, Advances in top–down and bottom–up surface nanofabrication: Techniques, applications & future prospects, Adv. Colloid Interface Sci. 170 (2012) 2–27, https://doi.org/10.1016/j.cis.2011.11.001.

[13] S. Pramanik, P. Das, Nanomater. Polym. Nanocomposites, Metal-Based Nanomaterials and Their Polymer Nanocomposites, Elsevier, 2019, pp. 91–121.

[14] S. Kalia, S. Boufi, A. Celli, S. Kango, Nanofibrillated cellulose: surface modification and potential applications, Colloid Polym. Sci. 292 (2014) 5–31, https://doi.org/10.1007/s00396-013-3112-9.

[15] S. Ling, W. Chen, Y. Fan, K. Zheng, K. Jin, H. Yu, M.J. Buehler, D.L. Kaplan, Biopolymer nanofibrils: Structure, modeling, preparation, and applications, Prog. Polym. Sci. 85 (2018) 1–56, https://doi.org/10.1016/j.progpolymsci.2018.06.004.

[16] K. Uetani, H. Yano, Nanofibrillation of Wood Pulp Using a High-Speed Blender, Biomacromolecules 12 (2011) 53–348, https://doi.org/10.1021/bm101103p.

[17] W. Chen, Q. Li, Y. Wang, X. Yi, J. Zeng, H. Yu, Y. Liu, J. Li, Comparative Study of Aerogels Obtained from Differently Prepared Nanocellulose Fibers, ChemSusChem 7 (2014) 61–154, https://doi.org/10.1002/cssc.201300950.

[18] H.P. Zhao, X.Q. Feng, H. Gao, Ultrasonic technique for extracting nanofibers from nature materials, Appl. Phys. Lett. 90 (2007) 073112, https://doi.org/10.1063/1.2450666.

[19] T.T.T. Ho, K. Abe, T. Zimmermann, H. Yano, Nanofibrillation of pulp fibers by twin-screw extrusion, Cellulose 22 (2015) 421–433, https://doi.org/10.1007/s10570-014-0518-6.

[20] A.M. Salaberria, S.C.M. Fernandes, R.H. Diaz, J. Labidi, Processing of α-chitin nanofibers by dynamic high pressure homogenization: Characterization and antifungal activity against A. niger, Carbohydr. Polym. 116 (2015) 286–291, https://doi.org/10.1016/j.carbpol.2014.04.047.

[21] D. Liu, Y. Zhu, Z. Li, D. Tian, L. Chen, P. Chen, Chitin nanofibrils for rapid and efficient removal of metal ions from water system, Carbohydr. Polym. 98 (2013) 483–490, https://doi.org/10.1016/j.carbpol.2013.06.015.

[22] A. Karimian, H. Parsian, M. Majidinia, M. Rahimi, S.M. Mir, H. Samadi Kafil, V. Shafiei-Irannejad, M. Kheyrollah, H. Ostadi, B. Yousefi, Nanocrystalline cellulose: Preparation, physicochemical properties, and

applications in drug delivery systems, Int. J. Biol. Macromol. 133 (2019) 850–859, https://doi.org/10.1016/j.ijbiomac.2019.04.117.

[23] H.M.A. Ehmann, T. Mohan, M. Koshanskaya, S. Scheicher, D. Breitwieser, V. Ribitsch, K. Stana-Kleinschek, S. Spirk, Esign of anticoagulant surfaces based on cellulose nanocrystals, Chem. Commun. 50 (2014) 13070–13072, https://doi.org/10.1039/C4CC05254D.

[24] A. George, P.A. Shah, P.S. Shrivastav, Natural biodegradable polymers based nano-formulations for drug delivery: A review, Int. J. Pharm. 561 (2019) 244–264, https://doi.org/10.1016/j.ijpharm.2019.03.011.

[25] N. Anton, J.P. Benoit, P. Saulnier, Design and production of nanoparticles formulated from nano-emulsion templates—A review, J. Control. Release 128 (2008) 185–199, https://doi.org/10.1016/j.jconrel.2008.02.007.

[26] M. Zohri, T. Gazori, S.S. Mirdamadi, A. Asadi, I. Haririan, Polymeric NanoParticles: Production, Applications and Advantage, The Internet Journal of Nanotechnology 3 (1) (2009) 1–14.

[27] K.S. Soppimath, T.M. Aminabhavi, A.R. Kulkarni, W.E. Rudzinski, Biodegradable polymeric nanoparticles as drug delivery devices, J. Control. Release 70 (2001) 1–20, https://doi.org/10.1016/S0168-3659(00)00339-4.

[28] F. Ganachaud, J.L. Katz, Nanoparticles and Nanocapsules Created Using the Ouzo Effect: Spontaneous Emulsification as an Alternative to Ultrasonic and High-Shear Devices, Chem. Phys. Chem. 6 (2005) 209–216, https://doi.org/10.1002/cphc.200400527.

[29] M.L.T. Zweers, G.H.M. Engbers, D.W. Grijpma, J. Feijen, Release of anti-restenosis drugs from poly (ethylene oxide)-poly (DL-lactic-co-glycolic acid) nanoparticles, J. Control. Release 114 (2006) 317–324.

[30] M.L.T. Zweers, G.H.M. Engbers, D.W. Grijpma, J. Feijen, In vitro degradation of nanoparticles prepared from polymers based on DL-lactide, glycolide and poly (ethylene oxide), J. Control. Release 100 (2004) 347–356.

[31] S. Galindo-Rodriguez, E. Allémann, H. Fessi, E. Doelker, Physicochemical Parameters Associated with Nanoparticle Formation in the Salting-Out, Emulsification-Diffusion, and Nanoprecipitation Methods, Pharm. Res. 21 (2004) 1428–1439, https://doi.org/10.1023/B:PHAM.0000036917.75634.be.

[32] S.A. Galindo-Rodríguez, F. Puel, S. Briançon, E. Allémann, E. Doelker, H. Fessi, Comparative scale-up of three methods for producing ibuprofen-loaded nanoparticles, Eur. J. Pharm. Sci. 25 (2005) 357–367, https://doi.org/10.1016/j.ejps.2005.03.013.

[33] K.A. Janes, P. Calvo, M.J. Alonso, Polysaccharide colloidal particles as delivery systems for macromolecules, Adv. Drug Deliv. Rev. 47 (2001) 83–97, https://doi.org/10.1016/S0169-409X(00)00123-X.

[34] W. Fan, W. Yan, Z. Xu, H. Ni, Formation mechanism of monodisperse, low molecular weight chitosan nanoparticles by ionic gelation technique, Colloids Surfaces B Biointerfaces 90 (2012) 21–27, https://doi.org/10.1016/j.colsurfb.2011.09.042.

[35] M. Kaloti, H.B. Bohidar, Kinetics of coacervation transition versus nanoparticle formation in chitosan–sodium tripolyphosphate solutions, Colloids Surfaces B Biointerfaces 81 (2010) 165–173, https://doi.org/10.1016/j.colsurfb.2010.07.006.

[36] C. Pinto Reis, R.J. Neufeld, A.J. Ribeiro, F. Veiga, Nanoencapsulation I. Methods for preparation of drug-loaded polymeric nanoparticles, Nanomedicine Nanotechnology, Biol. Med. 2 (2006) 8–21, https://doi.org/10.1016/j.nano.2005.12.003.

[37] B. Mishra, B.B. Patel, S. Tiwari, Colloidal nanocarriers: a review on formulation technology, types and applications toward targeted drug delivery, Nanomedicine Nanotechnology, Biol. Med. 6 (2010) 9–24, https://doi.org/10.1016/j.nano.2009.04.008.

[38] I. Limayem Blouza, C. Charcosset, S. Sfar, H. Fessi, Preparation and characterization of spironolactone-loaded nanocapsules for paediatric use, Int. J. Pharm. 325 (2006) 124–131, https://doi.org/10.1016/j.ijpharm.2006.06.022.

[39] D. Moinard-Chécot, Y. Chevalier, S. Briançon, L. Beney, H. Fessi, Mechanism of nanocapsules formation by the emulsion–diffusion process, J. Colloid Interface Sci. 317 (2008) 458–468, https://doi.org/10.1016/j.jcis.2007.09.081.

[40] F. Lince, D.L. Marchisio, A.A. Barresi, Strategies to control the particle size distribution of poly-ε-caprolactone nanoparticles for pharmaceutical applications, J. Colloid Interface Sci. 322 (2008) 505–515, https://doi.org/10.1016/j.jcis.2008.03.033.

[41] M. Mudgil, N. Gupta, M. Nagpal, P. Pawar, Nanotechnology: a new approach for ocular drug delivery system, Int. J. Pharm. Pharm. Sci 4 (2012) 105–112.

[42] S. Patra, P. Basak, D.N. Tibarewala, Synthesis of gelatin nano/submicron particles by binary nonsolvent aided coacervation (BNAC) method, Mater. Sci. Eng. C 59 (2016) 310–318, https://doi.org/10.1016/j.msec.2015.10.011.

[43] D. Vecchione, A.M. Grimaldi, E. Forte, P. Bevilacqua, P.A. Netti, E. Torino, Hybrid Core-Shell (HyCoS) Nanoparticles produced by Complex Coacervation for Multimodal Applications, Sci. Rep. 7 (2017) 45121, https://doi.org/10.1038/srep45121.

[44] L. Battaglia, I. D'Addino, E. Peira, M. Trotta, M. Gallarate, Solid lipid nanoparticles prepared by coacervation method as vehicles for ocular cyclosporine, J. Drug Deliv. Sci. Technol. 22 (2012) 125–130, https://doi.org/10.1016/S1773-2247(12)50016-X.

[45] L. Battaglia, M. Gallarate, R. Cavalli, M. Trotta, Solid lipid nanoparticles produced through a coacervation method, J. Microencapsul. 27 (2010) 78–85, https://doi.org/10.3109/02652040903031279.

[46] K. Bruyninckx, M. Dusselier, ACS Sustain, Sustainable Chemistry Considerations for the Encapsulation of Volatile Compounds in Laundry-Type Applications, Chem. Eng. 7 (2019) 8041–8054, https://doi.org/10.1021/acssuschemeng.9b00677.

[47] O.G. Jones, D.J. McClements, Biopolymer Nanoparticles from Heat-Treated Electrostatic Protein-Polysaccharide Complexes: Factors Affecting Particle Characteristics, J. Food Sci. 75 (2010) N36–N43, https://doi.org/10.1111/j.1750-3841.2009.01512.x.

[48] I.J. Joye, V.A. Nelis, D.J. McClements, Gliadin-based nanoparticles: Stabilization by post-production polysaccharide coating, Food Hydrocoll 43 (2015) 236–242, https://doi.org/10.1016/j.foodhyd.2014.05.021.

[49] C. Schmitt, S.L. Turgeon, Protein/polysaccharide complexes and coacervates in food systems, Adv. Colloid Interface Sci. 167 (2011) 63–70, https://doi.org/10.1016/j.cis.2010.10.001.

[50] A.P. Pandey, M.P. More, K.P. Karande, R.V. Chitalkar, P.O. Patil, P.K. Deshmukh, Artif. Cells, Optimization of desolvation process for fabrication of lactoferrin nanoparticles using quality by design approach, Nanomedicine Biotechnol 45 (2017) 1101–1114, https://doi.org/10.1080/21691401.2016.1202259.

[51] J.A. Carvalho, A.S. Abreu, V.T.P. Ferreira, E.P. Gonçalves, A.C. Tedesco, J.G. Pinto, J. Ferreira-Strixino, M. Beltrame Junior, A.R. Simioni, Preparation of gelatin nanoparticles by two step desolvation method for application in photodynamic therapy, J. Biomater. Sci. Polym. Ed. 29 (2018) 1287–1301, https://doi.org/10.1080/09205063.2018.1456027.

[52] Z. Zhao, Y. Li, M.Bin Xie, Silk Fibroin-Based Nanoparticles for Drug Delivery, Int. J. Mol. Sci. 16 (2015) 4880–4903, https://doi.org/10.3390/ijms16034880.

[53] J. Xie, J. Jiang, P. Davoodi, M.P. Srinivasan, C.H. Wang, Electrohydrodynamic atomization: A two-decade effort to produce and process micro-/nanoparticulate materials, Chem. Eng. Sci. 125 (2015) 32–57, https://doi.org/10.1016/j.ces.2014.08.061.

[54] D.H. Paik, S.W. Choi, Entrapment of protein using electrosprayed poly(d,l -lactide-co-glycolide) microspheres with a porous structure for sustained release, Macromol. Rapid Commun. 35 (2014) 1033–1038, https://doi.org/10.1002/marc.201400042.

[55] S. Zhang, K. Kawakami, One-step preparation of chitosan solid nanoparticles by electrospray deposition, Int. J. Pharm. 397 (2010) 211–217, https://doi.org/10.1016/j.ijpharm.2010.07.007.

[56] P. Davoodi, M.P. Srinivasan, C.-H. Wang, Effective co-delivery of nutlin-3a and p53 genes via core-shell microparticles for disruption of MDM2-p53 interaction and reactivation of p53 in hepatocellular carcinoma, J. Mater. Chem. B 5 (2017) 5816–5834, https://doi.org/10.1039/C7TB00481H.

[57] D. Rawtani, N. Khatri, S. Tyagi, G. Pandey, Nanotechnology-based recent approaches for sensing and remediation of pesticides, J. Environ. Manage. 206 (2018) 749–762, https://doi.org/10.1016/j.jenvman.2017.11.037.

[58] Z. Tang, C. He, H. Tian, J. Ding, B.S. Hsiao, B. Chu, X. Chen, Polymeric nanostructured materials for biomedical applications, Prog. Polym. Sci. 60 (2016) 86–128, https://doi.org/10.1016/j.progpolymsci.2016.05.005.

[59] M. Kakran, M.N. Antipina, Emulsion-based techniques for encapsulation in biomedicine, food and personal care, Curr. Opin. Pharmacol. 18 (2014) 47–55, https://doi.org/10.1016/j.coph.2014.09.003.

[60] S. Bohrey, V. Chourasiya, A. Pandey, Polymeric nanoparticles containing diazepam: preparation, optimization, characterization, in-vitro drug release and release kinetic study, Nano Converg 3 (2016) 3, https://doi.org/10.1186/s40580-016-0061-2.

[61] D. Quintanar-Guerrero, M. de la, Luz Zambrano-Zaragoza, E. Gutierrez-Cortez, N. Mendoza-Munoz, Impact of the Emulsification-Diffusion Method on the Development of Pharmaceutical Nanoparticle, Recent Pat. Drug Deliv. Formul. 6 (2012) 184–194, https://doi.org/10.2174/187221112802652642.

[62] C.E. Mora-Huertas, O. Garrigues, H. Fessi, A. Elaissari, Nanocapsules prepared via nanoprecipitation and emulsification–diffusion methods: Comparative study, Eur. J. Pharm. Biopharm. 80 (2012) 235–239, https://doi.org/10.1016/j.ejpb.2011.09.013.

[63] M. Antonietti, K. Landfester, Polyreactions in miniemulsions, Prog. Polym. Sci. 27 (2002) 689–757, https://doi.org/10.1016/S0079-6700(01)00051-X.

[64] J.P. Rao, K.E. Geckeler, Polymer nanoparticles: Preparation techniques and size-control parameters, Prog. Polym. Sci. 36 (2011) 887–913, https://doi.org/10.1016/j.progpolymsci.2011.01.001.

[65] J.L. Arias, V. Gallardo, S.A. Gómez-Lopera, R.C. Plaza, A.V. Delgado, Synthesis and characterization of poly(ethyl-2-cyanoacrylate) nanoparticles with a magnetic core, J. Control. Release 77 (2001) 309–321, https://doi.org/10.1016/S0168-3659(01)00519-3.

[66] K. Landfester, A. Musyanovych, V. Mailänder, From polymeric particles to multifunctional nanocapsules for biomedical applications using the miniemulsion process, J. Polym. Sci. Part A Polym. Chem. 48 (2010) 493–515, https://doi.org/10.1002/pola.23786.

[67] D. Crespy, M. Stark, C. Hoffmann-Richter, U. Ziener, K. Landfester, Polymeric Nanoreactors for Hydrophilic Reagents Synthesized by Interfacial Polycondensation on Miniemulsion Droplets, Macromolecules 40 (2007) 3122–3135, https://doi.org/10.1021/ma0621932.

[68] M.F. Cunningham, Controlled/living radical polymerization in aqueous dispersed systems, Prog. Polym. Sci. 33 (2008) 365–398, https://doi.org/10.1016/j.progpolymsci.2007.11.002.

[69] H. Duan, D. Wang, Y. Li, Green chemistry for nanoparticle synthesis, Chem. Soc. Rev. 44 (2015) 5778–5792, https://doi.org/10.1039/C4CS00363B.

[70] D. Li, C. Mu, S. Cai, W. Lin, Ultrasonic irradiation in the enzymatic extraction of collagen, Ultrason. Sonochem. 16 (2009) 605–609, https://doi.org/10.1016/j.ultsonch.2009.02.004.

[71] J.W. Woo, S.J. Yu, S.M. Cho, Y.B. Lee, S.B. Kim, Extraction optimization and properties of collagen from yellowfin tuna (Thunnus albacares) dorsal skin, Food Hydrocoll 22 (2008) 879–887, https://doi.org/10.1016/j.foodhyd.2007.04.015.

[72] V.V. Makarov, A.J. Love, O.V. Sinitsyna, S.S. Makarova, I.V. Yaminsky, M.E. Taliansky, N.O. Kalinina, "Green" Nanotechnologies: Synthesis of Metal Nanoparticles Using Plants, Acta Naturae 6 (2014) 35–44, https://doi.org/10.32607/20758251-2014-6-1-35-44.

[73] J. Mittal, A. Batra, A. Singh, M.M. Sharma, Phytofabrication of nanoparticles through plant as nanofactories, Adv. Nat. Sci. Nanosci. Nanotechnol. 5 (2014) 043002, https://doi.org/10.1088/2043-6262/5/4/043002.

[74] S. Usha, K.T. Ramappa, S. Hiregoudar, G.D. Vasanthkumar, D.S. Aswathanarayana, Biosynthesis and Characterization of Copper Nanoparticles from Tulasi (Ocimum sanctum L.) Leaves, Int. J. Curr. Microbiol. Appl. Sci. 6 (2017) 2219–2228, https://doi.org/10.20546/ijcmas.2017.611.263.

[75] W. Yang, Y. Ma, J. Tang, X. Yang, "Green synthesis" of monodisperse Pt nanoparticles and their catalytic properties, Colloids Surfaces A Physicochem. Eng. Asp. 302 (2007) 628–633, https://doi.org/10.1016/j.colsurfa.2007.02.028.

[76] M. Nasrollahzadeh, M. Maham, A. Rostami-Vartooni, M. Bagherzadeh, S.M. Sajadi, Barberry fruit extract assisted in situ green synthesis of Cu nanoparticles supported on a reduced graphene oxide-Fe_3O_4 nanocomposite as a magnetically separable and reusable, RSC Adv. 5 (2015) 64769–64780, https://doi.org/10.1039/C5RA10037B.

[77] M. Hamidi, A. Azadi, P. Rafiei, Hydrogel nanoparticles in drug delivery, Adv. Drug Deliv. Rev. 60 (15) (2008) 1638–1649. In this issue. doi:10.1016/j.addr.2008.08.002.

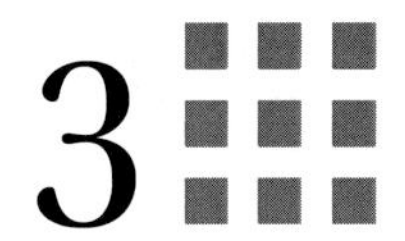

Gelatin-based nanomaterials

Jeannine Bonilla[a], Daniel Enrique López Angulo[a] and Paulo J.A. Sobral[a,b]

[a]FACULTY OF ANIMAL SCIENCE AND FOOD ENGINEERING, UNIVERSITY OF SÃO PAULO, AV. DUQUE DE CAXIAS NORTE, PIRASSUNUNGA (SP), BRAZIL. [b]FOOD RESEARCH CENTER (FORC), UNIVERSITY OF SÃO PAULO, SÃO PAULO (SP), BRAZIL.

Chapter outline

3.1 Introduction

Among the proteins, gelatin is one of the most popular protein biopolymers [42], differentiated from other proteins by the absence of an appreciable odor and the random configuration of polypeptide chains in an aqueous solution [72], such as observed by the authors of this chapter (Fig. 3.1). Moreover, gelatin is considered as having good functional properties, such as water binding capacity, cross-linking possibility, and foaming, emulsifying and film-forming abilities [12].

The gelatin global market demand in 2019 was 620.6 kilotons, and it is projected to expand at a volume-based compound annual growth rate of 5.9% from 2020 to 2027, due to an increased demand for convenience and functional food products, coupled with its use in the pharmaceutical application, over the forecast period [55].

Gelatin can be produced from collagen from different animal sources including pigskin, bovine hides, pig and cattle bones, and other (including marine sources). Pigskin is preferred over other resources for its cheaper price, despite some sociocultural and health concerns associated with pork in Islamic countries [55]. Moreover, depending on the raw material used,

Biopolymeric Nanomaterials: Fundamentals and Applications. DOI: https://doi.org/10.1016/B978-0-12-824364-0.00002-2

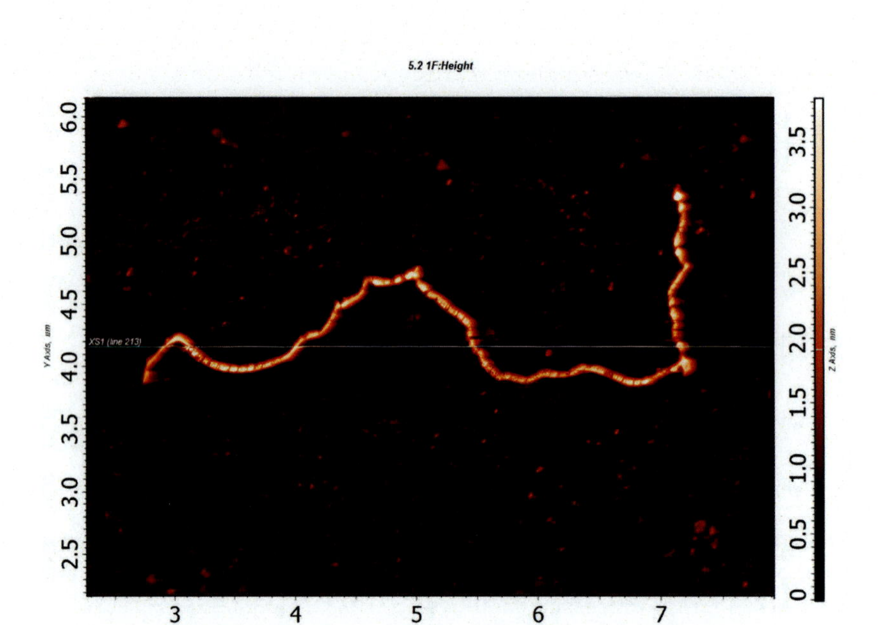

FIGURE 3.1 AFM image of a random coil structure of pig gelatin type A in aqueous solution of 1.5 g/L. Original image of our own work group at FZEA-USP, has not been previously published.

that is source and age of the animal, gelatin does not have exactly the same composition and properties [23]. The most commonly used production process for gelatin manufacture involves the subjection of raw materials to washing, partial hydrolysis with acids (obtaining type A gelatin, with an isoelectric point of 7–9) or alkalis (obtaining type B gelatin, with an isoelectric point of 4–5), neutralization, extraction, filtration, demineralization, concentration, sterilization, and drying [[1,22,72]]. In addition, its physical properties depend to a large extent on two factors: (1) the amino acid composition, represented by proline and hydroxyproline in about 30% of the total amino acids [23], and (2) the molecular weight distribution, which results mainly from processing conditions, for example pH, temperature, and time during both pretreatment and extraction processes [1,36,42].

According to Huang et al., [42], this biopolymer has been recognized as a safe additive in the food industry by the United States Food and Drug Administration, European Commission (EC), Food and Agriculture Organization of the United Nations, and World Health Organization, being extensively studied as an outer covering (i.e., films and/or coating forms) to protect food against drying, light, and oxygen due to its film-forming capacity, biodegradability, biocompatibility, and applicability [13–15,36].

Films composed primarily by proteins as gelatin have suitable mechanical (flexibility, tension) and optical properties (brightness and opacity) but are highly sensitive to moisture and exhibit poor water vapor barrier properties [29,36]. In this sense, different film formulations have been developed to add new functional properties, such as improved gas and moisture barriers, mechanical properties, and/or antimicrobial and antioxidative activities [33,34].

Gelatin is one of the first materials proposed as a carrier of bioactive components [36], becoming an active packaging material, with deliberately incorporated components, which release substances into the food to maintain or improve the condition of the packaged food (EC Regulation No. 450/2009) [62]. These active components can be categorized in antioxidant and antimicrobial agents, and they can be divided according to their origin in organic compounds, including natural biopolymers such as chitosan and chitin, phenols (main compounds of the essential oils and natural extracts), organic acids, and enzymes [40], or inorganic compounds, including metallic and metal oxide nanoparticles (NPs) (e.g., silver (Ag), magnesium (Mg), zinc oxide (ZnO), titanium oxide (TiO), etc.) and clays (e.g., montmorillonite (MMT), laponite, etc.) [76].

A number of organic and inorganic compounds have been shown to produce modifications in the gelatin backbone, thus affecting the strength, solubility, surface hydrophilicity, and morphology of the composite materials obtained [53]. In general, organic compounds present high sensitivity to intense processing conditions (as high temperatures and pressures), but they have the potential to be used as antimicrobial/antioxidant components in food packaging materials [40,42]. On the other hand, the inorganic compounds are most stable under extreme conditions such as high temperature and pressures and some of them are considered nontoxic and even contain mineral elements essential to human health. Hence, all these compounds have been noted as a promising alternative in food packaging market [75].

3.2 Nanocomposite materials

Nanocomposites are a class of hybrid materials composed of polymer matrices charged with NPs [80]. The general idea of nanocomposites is based on the concept of creating a very large interface between the nanosized-building blocks and the polymer matrix [43]. Within the polymer matrix nanocomposites, there are the bionanocomposites, which consist of a biopolymer (polymeric materials obtained from renewable biological resources such as gelatin) matrix reinforced with NPs, sometimes called biocomposites, nanobiocomposites, or green composites [68]. They can be utilized as eco-friendly, cost-effective, and renewable materials for food packaging with improved antimicrobial and antioxidant activities, along with improved dimensional stability and better thermal expansion compared to micro- or macrocomposites ([52,68], Karak, 2019). Their application possibilities and performance are directly related to chemical composition and especially the purity of the NPs [43], which enable these NPs to be charged in various forms such as fillers, fiber (short and continuous), flake, particles, and/or lamina [83].

However, the drawbacks of these NPs can include an increase in viscosity and the sedimentation rate, difficulties in dispersion, and some difficulties in fabrication [75]. According to Karak, (2019), the geometrical shape, surface chemistry, aspect ratio, and size of nanomaterials are the critical parameters in tuning such interactions and hence the properties of the resultant systems.

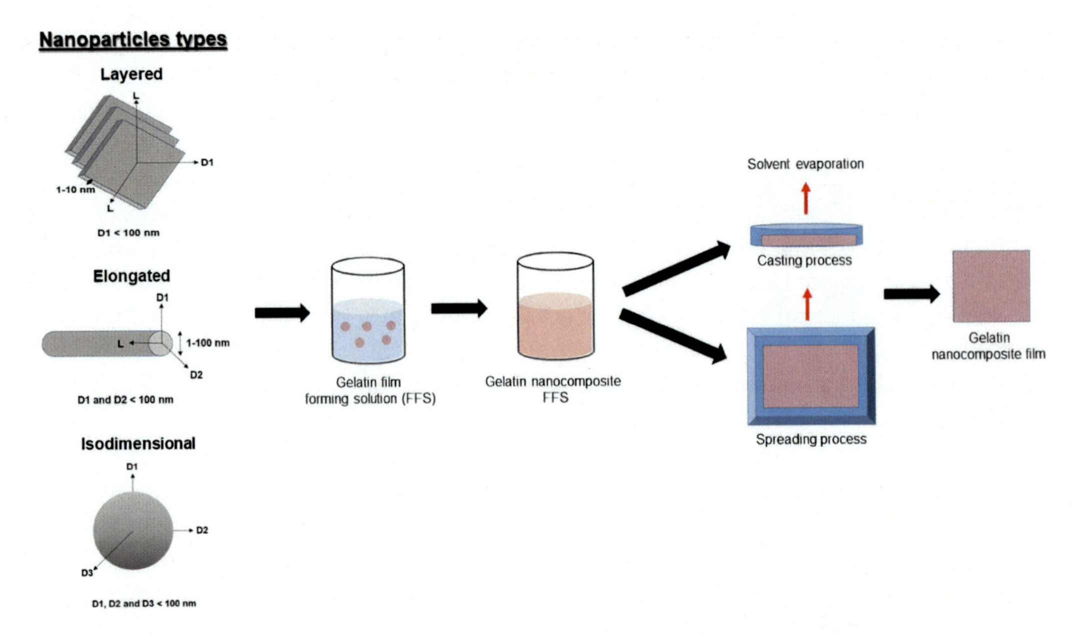

FIGURE 3.2 Processing routes to produce gelatin nanocomposite films. *Source: By the authors.*

Moreover, depending on the nature of the components used and on the preparation method, it is possible to obtain nanocomposites with the three main types of structure [65], namely layered, elongated, and isodimensional NPs ([81,24, Fu et al., 2019; 43]; Zaferani et al., 2018; Karak, 2019).

The layered NPs are characterized by one dimension with less than 100 nm (Fig. 3.2) ([81,24], Fu et al., 2019), and are usually in the form of nanoplatelets, nanosheets, nanodisks, nanoplates, and nanowalls [24]. Common examples of one-dimensional (1D) nanofillers include layered silicates, MMT clay, and nanographene platelets, which are used in applications such as films/coatings fields, among others [24]. Another example of this kind of NP is the Laponite, which is a synthetic crystalline layered clay with structure and composition closely resembling to smectite clays [77]. Among these NP forms, it is widely recognized that the inclusion of impermeable nanoplatelets such as clays in the biopolymer must significantly enhance its barrier properties [81], where the direct mixing of inorganic fillers and polymers may result in strong interactions and then nanometer-level dispersion, showing also unique mechanical and physical properties not shared by microcomposites (Fu et al., 2019).

Moreover, the elongated particles consist in fibrils with two dimensions, a diameter ranging between 1 and 100 nm and length up to several hundred nanometers (Fig. 3.2) [81,24]. These are mostly in the shape of tubes, fibers, or filaments. Carbon nanotubes (CNTs), cellulose nanofibers, gold or silver nanotubes, two-dimensional (2D) graphene, black phosphorus, and clay nanotubes are the most common examples of 2D NPs [81,24]. These particles are in general manufactured by three approaches: laser furnace, arc, and chemical vapor deposition. The first method is not suitable for large-scale production. The problem with the second method is purification, where the removal of a metal catalyst and nonnanotube materials costs more than production itself. And, the latest method is a scalable method whereby the purity can be controlled by careful processing and aligned arrays with a desired diameter and length (Fu et al., 2019).

Finally, isodimensional particles are relatively equiaxed particles having all three dimensions in the nanometer scale, and less than 100 nm (Fig. 3.2) ([81], Fu et al., 2019). They are usually in spherical and cubical shapes, and also commonly called nanospheres, nanogranules, and nanocrystals. Examples of 3D nanofillers include nanosilica, nanoalumina, nanotitanium oxide (TiO_2), nanozinc oxide (ZnO), nanomagnesium hydroxide ($Mg(OH)_2$), carbon black, silicon dioxide (SiO_2), iron oxide (Fe_3O_4), and silver (Ag) [81,24]. These nanofillers are very important in the formulation of polymer nanocomposites due to their inherent properties. Some of these (e.g., TiO_2, Ag, SiO_2, Fe_3O_4, and ZnO) possess good stability, high refractive index, hydrophilicity, ultraviolet (UV) resistance and excellent transparency to visible light, nontoxicity, high photocatalytic activity, and low cost [24]. According to Fu et al. (2019), synthesis of spherical NPs with controllable size is the goal of many research efforts, where the main driving force for this interest is the particle size's effect on their properties.

3.3 Nanocomposites films—processing methods

Nanocomposite film is a blend of a biopolymer and nanofillers to make a solid multiphase material system [72]. So, when producing nanocomposite by the casting method, it is necessary previously to well disperse the NPs in water, because generally it is commercialized in an agglomerate microscopic form (Fig. 3.3) [30]. Then, when the polymer and nanofiller in solution are mixed, the polymer chains intercalate and displace the solvent within the interlayer of the nanocomposite. Upon solvent removal, the intercalated/exfoliated structure remains (Fig. 3.4), resulting in polymer nanocomposite films with improved properties (Zaferani et al., 2018). So, the selection of an appropriate solvent is a primary criterion for obtaining the desired level of intercalated/exfoliated of an NP into the polymer [65]. Flaker et al. [30] observed that the quality of dispersion of NP in water has important effect on its particle size and zeta potential, with evident effect on the properties of nanocomposite films.

The unification of polymer nanocomposite requires good homogeneous dispersion of nanofiller when they come in contact with the matrix, as the latter tend to form agglomerates, stacking of NPs, and/or an incomplete exfoliation [74], causing a poor compatibility among components, and then creating defects that lead to an undesirable effect on the properties of the films.

At an industrial scale, melt processing is generally preferred for obtaining films, due to the difficulty to implement processes requiring the use of huge volumes of solvent [81]. However, studies based on gelatin films, including NPs, and obtained by this process were not found in the literature, due to nonthermoplastic nature of this biopolymer.

Then, different processing routes have been found to produce gelatin nanocomposite films, based on the nature of the raw constituents (polymer source, its molecular weight, solvent to be

FIGURE 3.3 Scanning electronic micrograph of a montmorillonite commercial sample. *Source*: *Flaker (2014).*

used, type and size of nanofiller, etc.) and their field of application [75,81]. At a laboratory scale, solution-mixing process is most commonly used to prepare gelatin-based nanocomposite films using various types of NPs [65,81], which lead to the formation of Van der Waals interactions between their constituents [74]. In this process, the polymer or prepolymer must be solubilized in an appropriate solvent, and the NPs must be dispersed in that solvent too before addition in the film-forming solution (Zaferani et al., 2018). These nanocomposite films can be produced by for either of the two wet techniques, namely casting or spreading technique, as shown in Fig. 3.2 [31,79].

For preparation of biopolymer nanocomposites films, solution casting technique is the most widely used [21,74], which consists in pouring a film-forming solution on small plates (e.g., a petri dish), controlling the average thickness of the resulting films from the mass of suspension poured on the plate, being not suitable for forming films much larger than 30 cm. Most studies report that solution's drying takes place at room temperature or in ovens with forced air circulation at moderate temperatures (30–40 °C), requiring drying times around 24 h [21]. Nevertheless, the temperature of air drying must be critically controlled. Menegalli et al. [58] observe some separation of gelatin and plasticizer components, when the air temperature increased closer to the melting temperature of the film.

On the other hand, spreading technique consists in spreading the film-forming solution on a support with a doctor blade device, which allows controlling the suspension thickness, and after that, the spread suspension is dried at a controlled temperature [21,64]. The heat supply during the drying step can be done by heat conduction, convection, infrared radiation, or by a combination of these mechanisms [21], and the determination of each component

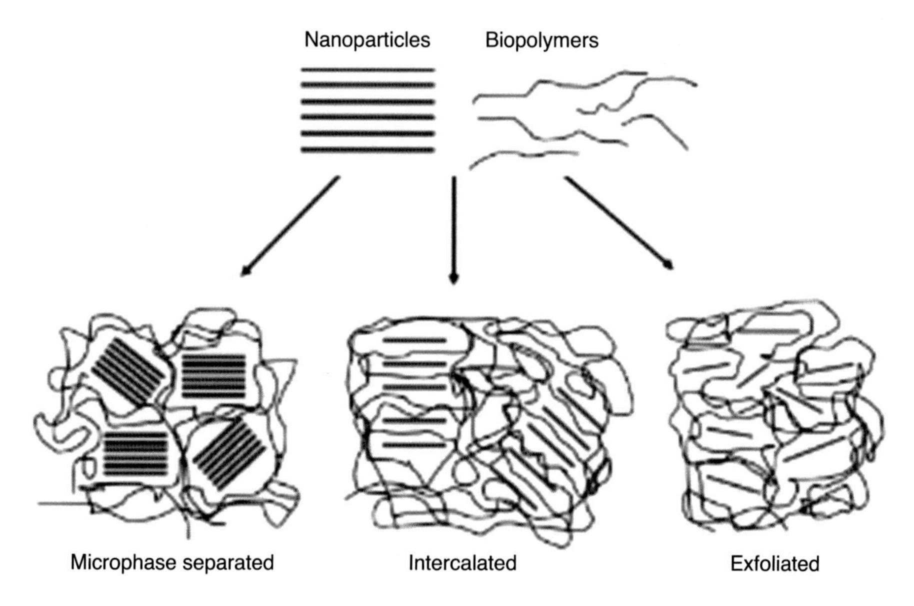

FIGURE 3.4 Principle of the formation of nanocomposites based on a biopolymer charged with NP: quality of NP dispersion on the biopolymer matrix. *Source: Based on Weiss et al. [80].*

concentration is fundamental to achieve the rheological-desired behavior [64]. Interested in developing gelatin-based nanocomposites using a spreading technique, Jorge et al. [46] observed that nanocomposite-forming solutions presented viscoelastic behavior and that their sol–gel and gel–sol transitions temperatures were affected by the MMT concentration.

According to Hári and Pukánszky [43], water-soluble biopolymers can be used with high efficiency for preparing polymer/layered silicate nanocomposites by solution blending. Neat, unmodified layered silicates exfoliate relatively easily in water, or organic solvents, when the polymer does not dissolve in water. Thus, the mixing of the solution made from the biopolymer with the slurry of the silicate is relatively easy, where occasionally homogeneous, stable colloidal distribution of the silicate layers prevails in the mixture and a good quality film can be prepared under mild drying conditions.

3.4 Physicochemical properties of gelatin nanocomposite films

Gelatin nanocomposite films can be devised to acquire some special properties that are not present in pure gelatin films (Zaferani et al., 2018; Karak, 2019). Tables 3.1 and 3.2 show the different improved physicochemical properties, for example, mechanical factors such as tensile strength (TS), modulus, or stiffness (Zaferani et al., 2018), the enhancement of thermal stability [45] and barrier properties, with organic and inorganic compounds added into gelatin nanocomposite film, respectively.

Table 3.1 Gelatin nanocomposite films with organic NPs.

Processing method	Gelatin' source	NP/reinforcing filler	Physical properties	References
Solution/casting	N/S	Bacterial cellulose nanocrystal	Improved the mechanical and thermal properties of gelatin films and reduced the moisture affinity of gelatin.	[35]
	Bovine	Cellulose nanocrystals	Improved mechanical and water vapor barrier properties.	[4]
	Pigskin	Cellulose nanocrystals + montmorillonite	Improved tensile strength and oxygen gas permeability barrier.	[25]
	Bovine	Chitin nanofibers	Improved water vapor permeability, solubility, mechanical properties, and thermal stability.	[69]
	N/S	Chitin nanofibers + zinc oxide NPs	Improved the water barrier properties and thermal stability of the film.	[70]
	Catfish skin	Chitosan NPs + tea polyphenol	Improved tensile strength and oxygen permeability.	[8]
	Bovine skin	Chitosan NPs + tea polyphenol	Improved UV barrier properties.	[51]
	Fish skin	Chitosan NPs + oregano essential oil	Improved elasticity and water vapor permeability.	[39]
	N/S	Chitosan + zinc oxide NPs	Improved thermal stability, elongation at break, and compactness properties.	[49]
	N/S	Guar gum benzoate NP	Improved water vapor permeability and thermal properties.	[50]

* N/S = not specified

3.4.1 Gelatin nanocomposite films charged with organic NPs

Organic NPs have some advantages compared to inorganic ones such as usually coming from food processing byproducts, being compatible to biodegradable polymers, edible, and nontoxic [73]. Table 3.1 shows that chitin, chitosan, and cellulose NPs stand out among the organic compounds charged into gelatin films.

Chitin possesses antimicrobial properties, and it can be achieved from fungi wall and stiff shell of marine animals, being the second abundant substance in the nature, after cellulose [73]. Sahraee et al. [69] studied bovine gelatin films containing chitin at 0%, 3%, 5%, and 10%, and concluded that this incorporation, in any quantity, reduced water vapor permeability and

Table 3.2 Gelatin nanocomposite films with inorganic NPs.

Processing method	Gelatin' source	NP/reinforcing filler	Physical properties	References
Solution/casting	Bovine	Sodium montmorillonite	Improved moisture resistance, water vapor permeability, storage modulus, and tan δ peaks.	[56]
	Bovine hide	Cu (II)-exchanged montmorillonite	Improved tensile strength and water vapor permeability.	[57]
	Fish	Montmorillonite (Cloisite Na^{+})	Improved tensile strength and oxygen and water vapor barrier properties.	[7]
	Tilapia skin	Montmorillonite (Cloisite Na^{+}) + Ethanolic extract (coconut husk)	Improved water vapor barrier properties.	[61]
	N/S	Cloisite 30B + orange peel extract	Improved mechanical properties.	[71]
	N/S	Cloisite 30B + silver	Improved UV barrier, hydrophobicity, and water vapor properties. Increased the tensile strength.	[47]
	Bovine	Zinc oxide nanorod	Improved UV, solubility and water vapor barrier properties.	[60]
	Bovine skin	Zinc oxide nanorods + clove essential oil	Improved mechanical resistance, and oxygen and UV barrier properties.	[26]
	Fish skin	Zinc oxide NPs + basil leaf essential oil	Improved mechanical properties and water vapor permeability. Increased the hydrophobicity, and marked impact on thermal stability.	[6]
	Fish	Zinc oxide	Improved mechanical, UV, crystallinity, and conductivity properties.	[67]
	Olive flounder bone	Zinc oxide	Improved tensile strength, water vapor permeability, and water solubility properties.	[9]
	Bovine	Titanium dioxide NPs	Improve solubility, moisture content, and water vapor permeability.	[32]

(*continued on next page*)

Table 3.2 (*continued*)

Processing method	Gelatin' source	NP/reinforcing filler	Physical properties	References
	Fish skin	Titanium dioxide NPs	Improved tensile strength, elongation at break, water vapor permeability, and UV properties.	[37]
	Fish skin	Bimetallic Ag-Cu	Increased\tensile strength, UV barrier, and thermal stability.	[5]
	Bovine skin	Nano-hydroxyapatite powder	Improved mechanical and thermal properties.	[82]
	Bovine	Carbon nanotubes	Improved water vapor permeability and tensile strength. Decrease in water solubility, water swelling, water uptake.	[48]
Solution/ Spreading	Pigskin	Montmorillonite + ginger essential oil	Improved mechanical properties and slightly reduced their crystallinity. Decreased solubility in water, moisture content and superficial hydrophobicity.	[2]
	Pigskin	Montmorillonite	Improved mechanical properties (tensile strength). Decreased water vapor permeability and superficial hydrophobicity.	[31]
	Pigskin	Montmorillonite	Improved mechanical properties (tensile strength and Young's modulus). Decreased the moisture content.	Jorge et al., 2015

solubility, and higher surface hydrophobicity of the nanocomposite films. Moreover, the use of NPs at 5%, led to the improved thermal stability and mechanical properties of the films. In the same way, Sahraee et al. [70] investigated the effect of chitin nanofibers (0%, 3%, 5%, and 10%) and zinc oxide (0%, 1%, 3%, and 5%) NPs charged in gelatin films. These authors reported an interactive effect, where both NPs added showed an increase in melting point and melting enthalpy, compared to the gelatin film containing each of them. In addition, film containing only chitin nanofibers up to 5% showed an increase in TS.

Moreover, chitosan is a cationic polysaccharide obtained from chitin by deacetylation in the presence of alkali [12]. Their NPs are natural materials with excellent physicochemical properties that are environmentally friendly and bioactive [39]. Liu et al. [51] and Bao et al. [8] studied the influences of the chitosan NPs with free or encapsulated tea polyphenol (TP), in fish or bovine skin gelatin films respectively. The first authors only observed that the light transmittance (250–550 nm) decreased with incorporation of NPs (at 0%, 5%, 1%, and 3%) into the film. The second authors related that NPs (0.3%) decreased the transparency, TS, and oxygen permeability but increased the water vapor permeability of gelatin films.

Likewise, in another research Hosseini et al. [39] studied fish skin gelatin charge with chitosan NPs (6%) and oregano essential oil (0.2%). They indicated that the introduction of NPs into the film matrix induces an increase in the crystallinity and flexibility of the films, with a decrease in their resistance and water vapor permeability. In addition, Kumar et al. [49] studied the effect of gelatin/chitosan nanocomposite films containing zinc oxide (ZnO) NPs, at 1%, 2%, and 4%. These authors concluded that reinforcement with ZnO NPs in the films led to improved thermal stability, elongation at break (EAB), and compactness properties of the films.

On the other hand, cellulose is the most abundant organic compound on the earth and is found in plant cell wall in association with hemicellulose and lignin [4]. From cellulose, the natural nano-reinforcements, called cellulose nanocrystals (CNCs), can be obtained using a controlled sulfuric acid hydrolysis, producing highly crystalline rod-like nanostructures [20], and interacting with the matrix mainly to favor mechanical and barrier properties [4]. In this regard, Alves et al. [4] studied the influence of CNCs (up to 3%) into films based on bovine gelatin or based on blended corn starch/gelatin films. They reported that gelatin films with CNC incorporation displayed a positive effect on the assessed response variables (i.e., improved mechanical properties, reduced water vapor permeability, and increased the maximum degradation temperature). Moreover, increases in gelatin and CNC concentrations in blended films led to increases in film thickness, strength, and EAB parameters of the films.

Echegaray et al. [25] investigated the influence of CNCs (up to 10%) and inorganic MMT (up to 10%) added into pigskin gelatin films. They concluded that gelatin nanocomposites films with MMT improved TS but systems reinforced with nanocellulose showed lower TS than neat gelatin ones. Furthermore, oxygen gas permeability values decreased for all nanocomposites films, especially for MMT systems; however, after the incorporation of reinforcements water vapor permeability increased. In addition, George and Siddaramaiah [35] studied gelatin films with bacterial CNCs (BCNCs, at 1, 2, 3, 4, and 5 wt%) added. They informed that BCNCs resulted in improving the mechanical properties of the films. Also, the addition of these NPs reduced the moisture affinity of gelatin and water vapor permeability, which is very favorable for edible packaging applications. According to Iguchi et al. [44], BCNCs produced in substantial quantities by *Gluconacetobacter xylinus* are well known for its superior properties (distinctly soft texture, high fiber content, and high-water holding capability) compared to plant cellulose.

Finally, Kundu et al. [50] studied gelatin films reinforced with guar gum benzoate (GGB) NPs (2%). Guar gum is a galactomannan obtained directly from *Cyamopsis tetragonoloba* seed pericarp and its NPs were prepared in a solvent displacement technique. The results of this work confirmed an improvement in water vapor permeability (0.75 g/mm kPa h) and thermal conductivity (0.39 W/m K) of the films.

3.4.2 Gelatin nanocomposite films charged with inorganic NPs

The incorporation of inorganic NPs, including nanoclays, nanometals, and metal oxide, into the polymeric matrix to enhance the properties of polymeric materials is well known, where nanometals or metal oxides, such as silver (Ag), copper (Cu), titanium oxide (TiO), copper (II) oxide (CuO), and zinc oxide (ZnO), as reinforcing filler in polymers, have emerged due to their high thermal stability and certain functions such as strong antimicrobial activity [72].

Table 3.2 shows that MMT NPs stand out among the inorganic compounds charged in gelatin films, through a casting or spreading preparation method. According to Jorge et al. [17], MMT is a layered silicate characterized by a moderate negative surface charge that displays a perfect crystalline structure. Its ability to intercalate a huge list of compounds is well known, producing significant improvements in the mechanical and thermal properties of the nanoclay/polymer nanocomposites materials [7].

Different authors have used the solution casting preparation method for obtaining nanocomposite films. Martucci and Ruseckaite [56,57] studied bovine gelatin charged with sodium MMT (Na^{+}MMt) and Cu(II)-exchanged MMT (Cu^{2+}MMt), respectively. These authors concluded that Na^{+}MMt addition (5% w/w) provoked an improvement in storage modulus of its films. In addition, a reduction in polar groups due to hydrogen interactions between gelatin and clay particles was evidenced by the decrease in surface hydrophilicity and water vapor permeability. On the other hand, Cu^{2+}MMt addition (0, 2, 5, 10% w/w) allowed an increase in TS, while the EAB and the water vapor permeability decreased. Moreover, Bae et al. [7] and Nagarajan et al. [61] studied fish gelatin with sodium MMT (Cloisite Na^{+}, at 1, 3, 5, 7, 9% w/w), and Cloisite Na^{+} (1% w/w) with ethanolic coconut husk extract (0%–0.4%) added, respectively. The first authors reported that nanoclay incorporation increased the TS and exhibited improvements in oxygen and water barrier properties. And, the second authors just observed improvements in water vapor barrier with 0.4% of coconut husk extract added.

Shams et al. [71] and Kanmani and Rhim, [47] studied Cloisite 30B (5%) with orange peel extract (at 7, 14, 21%), and Cloisite 30B (5%) and silver (0.5%), charged in whey protein/gelatin and gelatin films, respectively. Both authors informed a decrease in moisture content, water solubility, and water vapor permeability, and an increase in TS in nanocomposite films with organoclay incorporation. And, Valencia et al. [78] developed nanocomposites based on gelatin and charged with Laponite and observed improvement in mechanical properties (elastic modulus, TS, and puncture force) as a consequence of the NP charge. Moreover, the Laponite improved gelatin films thermal stability [19].

Additionally, different authors have used the solution spreading preparation method for obtaining nanocomposite films (Table 3.2). In this sense, Vanin et al. [79], Flaker et al. [31], Jorge et al. [46], and Alexandre et al. [2] studied some properties of nanocomposite films based on gelatin charged with MMT NPs. Vanin et al. [79] and Alexandre et al. [2] also added potassium sorbate and nanoemulsified ginger essential oil (GEO), as well, respectively. All these authors described improvements in mechanical properties. Besides, Jorge et al. [46] informed a decrease in the moisture content, Flaker et al. [31] reported that nanocomposite films were less permeable to water vapor and more hydrophilic than control film, and Alexandre et al. [2] concluded that the incorporation of MMT and nanoemulsified ginger oil decreased the solubility in water, moisture content, superficial hydrophobicity of films (only with MMT), and slightly reduced their crystallinity.

On the other hand, gelatin nanocomposite films charged with different metals or nonmetal NPs have been studied by several researchers (Table 3.2). The zinc oxide nanorods and/or NPs incorporation into gelatin films was investigated. Nafchi et al. [60] studied nanorod-rich zinc oxide (ZnO-nr, up to 0.05 g/g dried gelatin) addition in bovine gelatin films. These authors reported that the introduction of 0.05 ZnO-nr concentrations to gelatin solutions significantly increased the viscosity of the solution and decreased the water vapor permeability (from 8.9×10^{-11} to 1.78×10^{-11} (g/ms Pa), solubility in water (from 30% to 20%), and monolayer water content (from 0.13 to 0.10 g water/g dried solid) of the films. Moreover, the ZnO-nr gelatin films had very low UV transmittance and were able to absorb more than 50% of the near-infrared spectra. Ejaz et al. [26] investigated ZnO-nr (at 2%) with clove essential oil (CEO) (at 25 and 50%, w/w of protein), charged in bovine gelatin films. These authors observed that ZnO-nr addition decreased the flexibility and increased the mechanical resistance of the films, in addition to improving the oxygen and UV barrier properties of nanocomposites films.

Rouhi et al. [67] investigated the ZnO nanorods (up to 5%) into fish gelatin films. The results showed an increase in Young's modulus and TS of 42% and 25% for nanocomposites charged with 5% ZnO-nr, respectively. UV transmission decreased to zero with the addition of a small amount of ZnO NRs, and an increase in the intensity of the crystal facets and the conductivity of the films was observed with the addition of ZnO in the nanocomposite matrix. Moreover, [5] studied the effect of zinc oxide NPs (ZnONP, at 3%) and basil leaf essential oil (BEO, at 50 and 100% w/w, protein) into fish gelatin films. The authors informed that TS decreased, while EAB increased as basil essential oil level increased. However, ZnONP addition resulted in higher TS but lower EAB. The lowest water vapor permeability was observed for the film charged with 100% essential oil and 3% ZnONP. Fourier-transform infrared spectroscopy (FTIR) spectra indicated that films added with essential oil exhibited higher hydrophobicity. Both BEO and ZnONP had a marked impact on thermal stability of the films. Finally, Beak et al. [9] studied olive flounder bone gelatin charged with ZnONPs (, at 1, 3, and 5% w/w protein). These authors observed that TS increased, and water vapor permeability and water solubility decreased by the addition of ZnONP to the films.

On the other hand, He et al. [37] and Fonseca et al. [32] studied the effect of titanium dioxide (TiO_2) NPs (gelatin/TiO_2 ratio of 0.03%, 0.05%, and 0.1% and 0–2 wt%, respectively) in gelatin, from fish or bovine sources, respectively. Both authors reported that NPs addition significantly decreased films water vapor permeability. The first authors also informed an increase in TS and EAB values, as well as an improvement in UV barrier was observed. Moreover, the second authors reported that the NPs incorporation improved solubility and moisture content of the nanocomposite films.

Kavoosi et al. [48] investigated the effect of multiwalled CNTs (at 0.5, 1, 1.5, and 2% w/w gelatin) in bovine gelatin. These authors observed that nanocomposite films containing 1%–1.5% showed the lowest water vapor transmission, and the CNTs incorporation caused a significant increase in TS and decrease in the EAB. Likewise, a significant decrease in water solubility, water swelling, water uptake, and water vapor permeability was reported.

3.5 Applications of active-nanocomposites gelatin films in food packaging

Active films can be used to produce flexible active packaging systems. They contain deliberately incorporated components intended to release (controlled) or absorb substances into or from the packaged food or from the environment surrounding the food with the purpose of the extension of the shelf-life of the food [18]. In this sense, several authors developed active nanocomposite films adding compounds with bioactivity (e.g., antioxidant and/or antimicrobial activities) [2,79].

Moreover, the hydrophilic character of gelatin films can be a drawback in certain applications, due to its tendency to swell or be dissolved when putting in contact with the surface of foodstuffs with high moisture content. Hence, there is interest in developing blends using such components as nanocomposites, oils, or other biopolymers to augment the hydrophobic regions in the films and thus improve the functional properties of gelatin and the shelf-life of food products [66].

Nowadays, thanks to nanotechnology techniques, it is possible to improve the food packaging performances, formulating and inserting more stable antioxidant/antimicrobial ingredients, and introducing active and smart/intelligent functions, among others [3].

Table 3.3 shows different examples of active-nanocomposite gelatin films, with organic or inorganic compounds added, some of them are applied to different food products.

3.5.1 Active-nanocomposites gelatin films charged with organic NPs

Different authors charged chitin or chitosan NPs and/or nanofiber into gelatin films to study their antioxidant and/or antimicrobial activity. Bao et al. [8] reported the effect of catfish skin gelatin films, with TP-loaded chitosan NPs, on retardation of fish oil oxidation. Radical-scavenging activity of the films and the peroxide value (POV) of fish oil samples (2 mL) poured into gelatin bags, and stored at 45 °C during 25 days, were evaluated. These authors reported that nanocomposite films showed a higher level of radical-scavenging activity, and lower POV compared with other samples, especially when TP was unencapsulated in the films, possibly due to antioxidant activity of TP added in the formulation. Likewise, Liu et al. [51] studied the influence of chitosan NPs in various free/encapsulated TP ratios on the antioxidant properties of bovine skin gelatin films and sunflower oil. The authors concluded that the TP-loaded films had ferric reducing and 2,2-diphenyl-1-picrylhydrazyl (DPPH) radical scavenging power that corresponded to the encapsulation efficiencies (EE, ~50%, ~80%, and ~100%). Sunflower oil packaged in bags made of gelatin films embedded with NPs of 80% EE showed the best oxidation inhibitory effect, followed by 100% EE, 50% EE, and free TP, over 6 weeks of storage. However, when the gelatin film was placed over the headspace and was not in contact with the oil, the free TP showed the

Table 3.3 Active gelatin nanocomposite films.

Gelatin nanocomposite films	Antimicrobial or antifungal activity	Antioxidant activity	Application in product	References
Gelatin from catfish skin with chitosan NPs and tea polyphenol		DPPH, peroxide value	Fish oil	[8]
Gelatin from bovine skin with chitosan NPs and tea polyphenol		Total phenol, DPPH, peroxide value, and thiobarbituric acid reactive substances (TBARS).	Sunflower oil	[51]
Bovine gelatin with chitin NPs	*Aspergillus niger*			[69]
Gelatin with chitin nanofibers and zinc oxide NPs	*Aspergillus niger*			[70]
Gelatin from fish skin with chitosan NPs and oregano essential oil	*S. aureus, L. monocytogenes, S. enteritidis,* and *E. coli.*			[39]
Gelatin with guar gum benzoate NP	*E. coli* and *S. aureus*			[50]
Gelatin from bovine skin with zinc oxide nanorods and clove essential oil	*L. monocytogenes* and *S. typhimurium*		Refrigerated peeled shrimp	[26]
Gelatin from olive flounder bone with zinc oxide	*L. monocytogenes*		Fresh spinach	[9]
Bovine gelatin with nanorod-rich zinc oxide	*S. aureus*			[60]
Gelatin with chitosan + zinc oxide NPs	*E. coli* and *S. aureus*			[49]
Gelatin from fish skin with whey protein isolate, zinc oxide NPs and basil leaf essential oil	*L. monocytogenes* and *P. aeruginosa*			[6]
Pigskin gelatin with montmorillonite and ginger essential oil	*S. aureus, P. aeruginosa, E. coli,* and *S. enteritidis*	DPPH		[2]
Gelatin from bovine hide with copper (II)-exchanged montmorillonite	*E. coli* and *L. monocytogenes*			[57]
Gelatin with Cloisite 30B NPs, orange peel extract and tripolyphosphate	*E. coli* and *S. aureus*			[71]
Gelatin with Cloisite 30B and silver NPs	*E. coli* and *L. monocytogenes*			[47]
Gelatin from fish skin with bimetallic Ag-Cu NPs	*L. monocytogenes* and *S. typhimurium*			[5]
Gelatin from fish skin with titanium dioxide NPs	*E. coli* and *S. aureus*			[37]
Bovine gelatin with carbon nanotubes	*P. aeruginosa, E. coli, B. subtilis,* and *S. aureus.*			[48]

best effect. These results indicate that sustained release of TP in the contacting surface can ensure the protective effects, which vary with free/encapsulated mass ratios, thus improving antioxidant activities instead of increasing the dosage.

Sahraee et al. [69,70] investigated the antifungal properties against *Aspergillus niger* of nanocomposite film based on bovine gelatin-chitin NPs (0%, 3%, 5%, and 10%), and gelatin-chitin nanofibers (0%, 3%, 5%, and 10%) with zinc oxide NPs (0%, 1%, 3%, and 5%), respectively. Sahraee et al. [69] informed that the nanocomposite films had antifungal activity in the contact surface zone, where an increment of NPs concentration up to 5% enlarged inhibition zone diameter. According to these authors, the reason of reduced inhibition capacity of nanocomposite films containing 10% NPs could be aggregation and unequal dispersion of NPs which led to less contact surface to microbial cells. Moreover, Sahraee et al. [70] informed that the antifungal activities of the nanocomposite films improved by increasing NPs concentration, where applying each of NPs in gelatin films caused antifungal activity in film, and adding both of them in film formulation created gelatin films with higher antimicrobial activity (inhibition zone diameter $= 31.816 \pm 0.639$ mm). This proved that they have synergistic effect in concentrations of 5% each of them. Additionally, Hosseini et al. [39] studied the antimicrobial properties of fish gelatin/chitosan NPs films, incorporated with oregano essential oil at various concentrations, against *Staphylococcus aureus, Listeria monocytogenes, Salmonella enteritidis*, and *Escherichia coli* Bacteria. They reported that the composite films, with chitosan NPs but without essential oil, showed no antimicrobial activity against the studied microorganisms. Such observation was due to low polarity of these NPs, which makes them diffuse slowly from the polymer matrix into agar medium. Moreover, all active nanocomposite films inhibited the growth of the four pathogens. As the concentration of oregano essential oil (OEO) increased in the films (up to 1.2%, w/v), the zone of inhibition also increased significantly.

On the other hand, Kundu et al. [50] studied GGB NP (0%, 5%, 10%, 20%) reinforced gelatin films for enhanced antimicrobial properties. These authors concluded that inhibition zones were recorded in case of all films with NPs added, with higher activity observed in the films with GGB NPs at 20%. Bacteria growth was terminated and quantitative killing was recorded after 4 h contact time, where strong hydrophobic and the benzoate groups interactions appeared as the responsible factors for these results.

3.5.2 Active-nanocomposites gelatin films charged with inorganic NPs

Different authors charged zinc oxide NPs into gelatin films for studying their antioxidant and/or antimicrobial activity. In this sense, Ejaz et al. [26] studied zinc oxide nanorods (ZnO-nrs, in 2%) and CEO (in 25 and 50% w/w of protein) charged into bovine skin gelatin films, and its applicability for shrimp packaging. The results showed that the highest antimicrobial activity (*in vitro*) against *L. monocytogenes* and Salmonella *typhimurium* pathogens was observed for the film containing 2% ZnO-nrs and 50% CEO, describing a complete inactivation after 7 days incubation, which showed a synergism between the active compound from clove oil and the ZnO-nrs. Additionally, this same active film showed a reduction of 2–3 log for both bacteria inoculated in shrimp at the end of the storage (4 °C), indicating a strong antimicrobial activity and potential as active packaging for peeled shrimp.

Beak et al. [9] characterized an olive flounder bone gelatin film with zinc oxide nanocomposite (1%, 3%, and 5%) added, and its potential application in spinach packaging. They reported that all composite film exhibited antimicrobial activity against *L. monocytogenes*, showing a similar activity for the films charged with 3% and 5% of NP. To investigate its applicability, fresh spinach was wrapped in the film (3% of zinc oxide) and stored for a week. The results indicated this active film showed a strong antimicrobial activity without affecting the quality of spinach, such as vitamin C content and color, probably due to the light barrier property of this packaging film. Nafchi et al. [60] investigated the antimicrobial activity of bovine gelatin films with ZnO-nr added. The result showed that the inhibition zone of the nano-charged films was significantly increased with increasing zinc oxide content (>0.02 g/g dried gelatin), concluding that the nanorods can function as needles that easily penetrate through the cell wall of bacterium.

On the other hand, different authors charged MMT NPs into gelatin films. Alexandre et al. [2] studied the antioxidant and antimicrobial activities of gelatin-based films reinforced with MMT (5 g/100 g gelatin) and activated with nanoemulsion of GEO (2 g/100 g gelatin). These authors reported that the nanocomposite films (gelatin/MMT) presented antioxidant activity (0.09 ± 0.08 μmol Trolox equiv/g of dried film), which was not significantly higher than control sample (0.05 ± 0.05 μmol Trolox equiv/g of dried film), and nanocomposite active films (Gelatin/MMT/GEO) exhibited relatively low antioxidant activities (0.35 ± 0.12 μmol Trolox equiv/g of dried film) but significantly higher than that found for control. This study not showed antimicrobial activity against the four bacteria (*S. aureus, E. coli, Pseudomonas aeruginosa,* and *S. enteritidis*) studied. And, Vanin et al. [79] observed that potassium sorbate, an antimold agent, played a plasticizer effect on nanocomposites based on gelatin and charged with MMT.

Moreover, Martucci and Ruseckaite [57] investigated the antibacterial activity of gelatin/copper (II)-exchanged MMT films (Cu^{2+}MMt, 5% w/w dry gelatin basis). Cu^{2+}MMt showed antibacterial activity *in vitro* against *E. coli* O157:H7 and *L. monocytogenes*, and the gelatin/Cu^{2+}MMt film exhibited antibacterial effectiveness against both pathogens tested under the same conditions, demonstrating a stronger effect on *L. monocytogenes* than on *E. coli* O157:H7, as the cell wall of the latter differs significantly and such difference could influence their vulnerability and response to the active films.

Other studies regarding the antimicrobial activity of gelatin films with Cloisite 30B NPs [47,71], titanium dioxide [37], bimetallic Ag-Cu NPs [5], and CNTs [48] charged have been reported. Kanmani and Rhim [47] studied antimicrobial properties of gelatin-based active nanocomposite films containing silver NPs (AgNPs at 0.5% w/w) and nanoclay (Cloisite 30B, at 5%). The results on inhibitory activities (zone) of the film samples were reported, where composite films included with AgNPs or nanoclay showed clear antibacterial activity. However, the AgNPs-included nanocomposite films showed high antimicrobial activity against both *L. monocytogenes* and *E. coli* bacteria, but the nanoclay-included nanocomposite film showed antimicrobial activity only against to *L. monocytogenes* bacteria. He et al. [37] studied the antibacterial activity of fish skin gelatin–titanium dioxide (TiO_2) nanocomposite films. The authors informed that the antibacterial rates increased gradually for the gelatin film reinforced with increasing amounts of TiO_2 NPs (up to 0.1% w/w). Otherwise, the inhibitory effect of TiO_2-gelatin film for *E. coli* was stronger than that for *S. aureus*. When gelatin/TiO_2 ratio was 10:1, the inhibition rates of nanocomposite film for *E. coli* and *S. aureus* could reach 54.38 ± 1.28 and $44.89 \pm 2.71\%$, respectively. Kavoosi et al. [48] investigated the antimicrobial activity against *P. aeruginosa, E. coli, Bacillus subtilis,* and *S. aureus* bacteria of gelatin/multiwalled CNT (MWCNT, 0.5, 1, 1.5, and 2% w/w gelatin) nanocomposite films. They reported that all gelatin/MWCNT films showed significant antibacterial activities against both Gram-positive and Gram-negative bacteria, through cell membrane damage, cell inactivation, and expression of stress-related gene products.

3.6 Possible migration of NPs in food products

The extent of migration and the potential toxicological risk of the consumption of food containing nanoscale compounds, transferred from the packaging, is the biggest concern associated to the use of nanomaterial-based improved food packaging [3]. According to Huang et al. [41] this migration process involves two stages: (1) the initial migration must be due to those nanomaterials encapsulated within the surface layers, and (2) the subsequent release of nanomaterial from the interior part of the specimen has to pass through voids and other gaps between the polymer molecules, which will depend on polymer properties, and in some cases, the nanomaterials encapsulated well inside the film need to oxidize and migrate out through the polymer matrices.

To improve understanding of risk and benefit, the crucial questions are whether NPs can be released from food contact polymers and under which conditions [27]. Numerous studies showed that the migration can be influenced by several factors, such as the NPs' properties (e.g., particle size, molecular weight, solubility, and diffusivity in the polymer [84]), environmental conditions (temperature, mechanical stress), position of the NPs in the packaging material [16], packaging characteristics (polymer structure and viscosity [11]), interaction and contact time between the NPs and the material [38], food conditions (pH value [59], composition [27]), and the preservation methods of foods [27]. For instance, if the nature of the food is compatible with the type of packaging, the food itself can be absorbed into the polymer matrix, enlarging the gaps between the polymer chains, thereby increasing the migration rate [41].

The EC (No. 10/2011; 2020) [63] has recently published a Union list of 1077 authorized plastic materials and articles (monomers or other starting substances; additives excluding colorants; polymer production aids excluding solvents, and macromolecules obtained from microbial fermentation) intended to come into contact with food, including gelatin, nanoclay (MMT), titanium oxide, zinc oxide, and iron oxide NPs, among others, as shown in Table 3.4. To protect the consumer health and to prevent the foodstuff adulteration, two migration limits have been defined for plastic materials by the European Commission (2020): (1) the overall migration limit, which means the maximum permitted amount of nonvolatile substances released from a material or article into food simulants, and established it as 60 mg of substances/kg of food packaged or food simulants in any case, and (2) the specific migration limit (SML), which means the maximum permitted amount of a given substance released from a material or article into

Table 3.4 List of some authorized additives or polymer production aids intended to come into contact with food. Not all authorized articles are listed here, but examples are included.

Substance name	Restrictions and specifications
Gelatin	
Fats and oils, from animal or vegetable food sources	
Aluminum oxide, zinc oxide, calcium oxide, and iron oxide	
Titanium dioxide and carbon dioxide	
Magnesium hydroxide and potassium hydroxide	
Silicon dioxide	For synthetic amorphous silicon dioxide: primary particles of 1–100 nm that are aggregated to a size of 0.1–1 μm which may form agglomerates within the size distribution of 0.3 μm to the millimeter size.
Titanium nitride, NPs	No migration of titanium nitride NPs. Only to be used in polyethylene terephthalate (PET) up to 20 mg/kg. In the PET, the agglomerates have a diameter of 100–500 nm consisting of primary titanium nitride NPs; primary particles have a diameter of approximately 20 nm.
Titanium dioxide surface-treated with fluoride-modified alumina	Only to be used at up to 25.0% w/w, including in the nanoform.
Zinc oxide, NPs, uncoated	Only to be used in unplasticized polymers.
Montmorillonite clay modified by dimethyl dialkyl (C16-C18) ammonium chloride	Only to be used up to 12% (w/w) in polyolefins in contact with dry foods at room temperature or below. The sum of the specific migration of 1-chlorohexadecane and 1-chlorooctadecane should not exceed 0.05 mg/kg food. Can contain platelets in the nanoform that are only in one dimension thinner than 100 nm. Such platelets should be oriented parallel to the polymer surface and should be fully embedded in the polymer.
Montmorillonite clay modified with hexadecyltrimethylammonium bromide	Only to be used as additive at up to 4.0% w/w in polylactic acid plastics intended for long-term storage of water at ambient temperature or below. Can form platelets in the nanoform that are in one or two dimensions thinner than 100 nm. Such platelets should be oriented parallel to the polymer surface and should be fully embedded in the polymer.

(Source: EC, No. 10/2011; 2020).

food or food simulants, as shown in Table 3.5. Moreover, this legislation provides the simulants and test procedures under which relevant tests such as overall migrations must be performed.

Regarding toxicological studies, there is much controversy among published studies. In some cases, it seems to be really alarming to incorporate NPs into food packaging materials. In this sense, toxicological aspects of different kinds of NPs have been evaluated both *in vitro* and *in vivo*. For example, Enescu et al. [28] evaluated the specific migration of titanium from chitosan

Table 3.5 General list of migration limits for some examples of substances migrating from plastic materials and articles.

Name	Specific migration limit [mg/kg food or food simulant]
Aluminum	1
Arsenic	ND
Calcium	–
Copper	5
Iron	48
Lead	ND
Lithium	0.6
Magnesium	–
Mercury	ND
Nickel	0.02
Potassium	–
Sodium	–
Zinc	5

[a] ND = Not Detectable
(Source: EC, No. 10/2011; 2020).

(2%) films containing TiO_2 (0.5 %) micro- or nanoparticles into different food simulants. The results revealed that titanium can migrate ($<5.44 \times 10^{-4}$ % of the total titanium in the chitosan matrix) from chitosan matrix after incubation in different food simulants over 10 days at 40 °C or 10 days at 5 °C, but only in its ionic form, and no evidence of TiO_2-particle migration was found. Also, cytotoxicity analysis using Caco-2 cells demonstrated that the leachates of chitosan/metal complexes films had no cell toxicity. Becaro et al. [10] evaluated the genotoxic and cytotoxic effects of silver NPs (AgNP, size range between 2 and 8 nm) on root meristematic cells of *Allium cepa*. Tests were carried out in the presence of colloidal solution of AgNP and AgNP mixed with carboxymethylcellulose (CMC), using distinct concentrations of AgNP (1.24 ppm and 12.40 ppm). The results showed that AgNP induced a mitotic index decrease and an increase of chromosomal aberration number for two studied concentrations. When AgNP was in the presence of CMC, no cytotoxic potential was verified, but only the genotoxic potential for AgNP dispersion having concentration of 12.4 ppm. Moreover, Maisanaba et al. [54] studied the cytotoxic effects of different mineral clays (Cloisite Na^+, at 0–125 mg/mL, and Cloisite 30B (0 and 250 mg/mL) in the human cells. The test results of Cloisite Na^+ did not show any cytotoxic or mutagenic effect. Conversely, the Cloisite 30B showed toxic effects in both assays.

3.7 Final remarks

Gelatin is a protein with excellent film-forming ability. Overall, films based on gelatin have good mechanical and optical properties, but are highly sensitive to moisture and exhibit poor water vapor barrier properties. One approach to try to improve gelatin-films properties is to charge this material with reinforcing load of NPs, creating a material the so-called nanocomposite or nanobiocomposite.

Until nowadays, nanocomposites are preferentially produced, in a laboratory scale, by casting or spreading a film-forming solution containing both, gelatin, previously thermally solubilized, and NPs, well dispersed in water. The drying conditions of these solutions must be well controlled to avoid phase separations.

Several kinds of nanocomposites based on gelatin charged with organic or inorganic NPs, and active nanocomposites, produced by the addition of an active component into nanocomposite formulation, have been produced with success. Overall, these materials present better mechanical and thermal properties than their respective gelatin-based films (without charge). Nevertheless, the issue provoked by their sensitivity to water was not yet fixed, probably due to the use of high hygroscopic plasticizer, such as glycerol or sorbitol.

Thus, further studies are necessary to try to avoid (or significantly reduce) the gelatin-based nanocomposite sensitivity to water, and also, to scale up the production of these material. In this sense, studies on extrusion process must be carried out.

Acknowledgments

The authors acknowledge the financial support from the São Paulo Research Foundation (FAPESP) for the Grant (2013/07914-8), and the Brazilian National Council for Scientific and Technological Development (CNPq) for the Research fellowship of Paulo J.A. Sobral (30.0799/2013-6).

References

[1] T. Ahmad, A. Ismail, S.A. Ahmad, K.A. Khalil, Y. Kumar, K.D. Adeyemi, A.Q. Sazili, Recent advances on the role of process variables affecting gelatin yield and characteristics with special reference to enzymatic extraction: a review, Food Hydrocoll. 63 (2017) 85–96. https://doi.org/10.1016/j.foodhyd.2016.08.007.

[2] E.M.C. Alexandre, R.V. Lourenço, A.M.Q.B. Bittante, I.C.F. Moraes, P.J. d. A. Sobral, Gelatin-based films reinforced with montmorillonite and activated with nanoemulsion of ginger essential oil for food packaging applications, Food Packag. Shelf Life. 10 (2016) 87–96. https://doi.org/10.1016/j.fpsl.2016.10.004.

[3] S. Alfei, B. Marengo, G. Zuccari, Nanotechnology application in food packaging: a plethora of opportunities versus pending risks assessment and public concerns, Food Res. Int. 137 (2020). https://doi.org/10.1016/j.foodres.2020.109664.

[4] J.S. Alves, K.C. Dos Reis, E.G.T. Menezes, F.V. Pereira, J. Pereira, Effect of cellulose nanocrystals and gelatin in corn starch plasticized films, Carbohydr. Polym. 115 (2015) 215–222. https://doi.org/10.1016/j.carbpol.2014.08.057.

[5] Y.A. Arfat, J. Ahmed, N. Hiremath, R. Auras, A. Joseph, Thermo-mechanical, rheological, structural and antimicrobial properties of bionanocomposite films based on fish skin gelatin and silver-copper nanoparticles, Food Hydrocoll. 62 (2017) 191–202. https://doi.org/10.1016/j.foodhyd.2016.08.009.

[6] Y.A. Arfat, S. Benjakul, T. Prodpran, P. Sumpavapol, P. Songtipya, Properties and antimicrobial activity of fish protein isolate/fish skin gelatin film containing basil leaf essential oil and zinc oxide nanoparticles, Food Hydrocoll 41 (2014) 265–273. http://dx.doi.org/10.1016/j.foodhyd.2014.04.023.

[7] H.J. Bae, H.J. Park, S.I. Hong, Y.J. Byun, D.O. Darby, R.M. Kimmel, W.S. Whiteside, Effect of clay content, homogenization RPM, pH, and ultrasonication on mechanical and barrier properties of fish gelatin/montmorillonite nanocomposite films, LWT Food Sci. Technol. 42 (2009) 1179–1186. https://doi.org/10.1016/j.lwt.2008.12.016.

[8] S. Bao, S. Xu, Z. Wang, Antioxidant activity and properties of gelatin films incorporated with tea polyphenol-loaded chitosan nanoparticles, J. Sci. Food Agric. 89 (2009) 2692–2700. https://doi.org/10.1002/jsfa.3775.

[9] S. Beak, H. Kim, K.B. Song, Characterization of an olive flounder bone gelatin-zinc oxide nanocomposite film and evaluation of its potential application in spinach packaging, J. Food Sci. 82 (2017) 2643–2649. https://doi.org/10.1111/1750-3841.13949.

[10] A.A. Becaro, M.C. Siqueira, F.C. Puti, M.R. de Moura, D.S. Correa, J.M. Marconcini, L.H.C. Mattoso, M.D. Ferreira, Cytotoxic and genotoxic effects of silver nanoparticle/carboxymethyl cellulose on *Allium cepa*, Environ. Monit. Asses. 189 (2017) 352. https://doi.org/10.1007/s10661-017-6062-8.

[11] K. Bhunia, S.S. Sablani, J. Tang, B. Rasco, Migration of chemical compounds from packaging polymers during microwave, conventional heat treatment, and storage, Compr. Rev. Food Sci. Food Saf. 12 (2013) 523–545. https://doi.org/10.1111/1541-4337.12028.

[12] J. Bonilla, P.J.A. Sobral, Investigation of the physicochemical, antimicrobial and antioxidant properties of gelatin-chitosan edible film mixed with plant ethanolic extracts, Food Biosci. 16 (2016) 17–25. https://doi.org/10.1016/j.fbio.2016.07.003.

[13] J. Bonilla, P.J.A. Sobral, Gelatin-chitosan edible film activated with Boldo extract for improving microbiological and antioxidant stability of sliced Prato cheese, Int. J. Food Sci. Technol. 54 (2018) 1617–1624.

[14] J. Bonilla, T. Poloni, P.J.A. Sobral, Active edible coatings with Boldo extract added and their application on nut products: reducing the oxidative rancidity rate, Int. J. Food Sci. Technol. 53 (2018) 700–708. https://doi.org/10.1111/ijfs.13645.

[15] L.M.J. Bonilla, S.P.J. do Amaral, Application of active films with natural extract for beef hamburger preservation, Ciênc. Rural. 49 (1) (2019). https://doi.org/10.1590/0103-8478cr20180797.

[16] J. Bott, R. Franz, Investigation into the potential migration of nanoparticles from laponite-polymer nanocomposites, Nanomaterials. 8 (2018) 723. https://doi.org/10.3390/nano8090723.

[17] M.F. Coronado Jorge, E.M.C. Alexandre, C.H. Caicedo Flaker, A.M.Q.B. Bittante, P.J.D.A. Sobral, Biodegradable films based on gelatin and montmorillonite produced by spreading, Int. J. Polym. Sci. (2015). https://doi.org/10.1155/2015/806791.

[18] D. Dainelli, N. Gontard, D. Spyropoulos, E. Zondervan-van den Beuken, P. Tobback, Active and intelligent food packaging: legal aspects and safety concerns, Trends Food Sci. Technol. 19 (2008) S103–S112. https://doi.org/10.1016/j.tifs.2008.09.011.

[19] D. López-Angulo, A.M. Q.B.Bittante, C.G. Luciano, G. Ayala-Valencia, C.H.C. Flaker, M. Djabourov, P.J.A. Sobral, Effect of Laponite® on the structure, thermal stability and barrier properties of nanocomposite gelatin films, Food Biosci. (2020) 100596. https://doi.org/10.1016/j.fbio.2020.100596.

[20] J.P. De Mesquita, C.L. Donnici, F.V. Pereira, Biobased nanocomposites from layer-by-layer assembly of cellulose nanowhiskers with chitosan, Biomacromolecules. 11 (2010) 473–480. https://doi.org/10.1021/bm9011985.

[21] J.O. De Moraes, A.S. Scheibe, A. Sereno, J.B. Laurindo, Scale-up of the production of cassava starch based films using tape-casting, J. Food Eng. 119 (2013) 800–808. https://doi.org/10.1016/j.jfoodeng.2013.07.009.

[22] M. Ding, T. Zhang, H. Zhang, N. Tao, X. Wang, J. Zhong, Effect of preparation factors and storage temperature on fish oil-loaded crosslinked gelatin nanoparticle pickering emulsions in liquid forms, Food Hydrocoll. 95 (2019) 326–335. https://doi.org/10.1016/j.foodhyd.2019.04.052.

[23] A. Duconseille, T. Astruc, N. Quintana, F. Meersman, V. Sante-Lhoutellier, Gelatin structure and composition linked to hard capsule dissolution: a review, Food Hydrocoll. 43 (2015) 360–376. https://doi.org/10.1016/j.foodhyd.2014.06.006.

[24] E.I. Akpan, X. Shen, B. Wetzel, K. Friedrich, Design and synthesis of polymer nanocomposites, in: K. Pielichowski, T.M. Majka (Eds.), Polymer Composites with Functionalized Nanoparticles., Elsevier BV, Amsterdam, 2019. https://doi.org/10.1016/b978-0-12-814064-2.00002-0.

[25] M. Echegaray, G. Mondragon, L. Martin, A. González, C. Peña-Rodriguez, A. Arbelaiz, Physicochemical and mechanical properties of gelatin reinforced with nanocellulose and montmorillonite, J. Renew. Mater. 4 (2016) 206–214. https://doi.org/10.7569/JRM.2016.634106.

[26] M. Ejaz, Y.A. Arfat, M. Mulla, J. Ahmed, Zinc oxide nanorods/clove essential oil incorporated type B gelatin composite films and its applicability for shrimp packaging, Food Packag. Shelf Life. 15 (2018) 113–121. https://doi.org/10.1016/j.fpsl.2017.12.004.

[27] D. Enescu, M.A. Cerqueira, P. Fucinos, L.M. Pastrana, Recent advances and challenges on applications of nanotechnology in food packaging. A literature review, Food Chem. Toxicol. 134 (2019). https://doi.org/10.1016/j.fct.2019.110814.

[28] D. Enescu, A. Dehelean, C. Gonçalves, M.A. Cerqueira, D.A. Magdas, P. Fucinos, L.M. Pastrana, Evaluation of the specific migration according to EU standards of titanium from chitosan/metal complexes films containing TiO_2 particles into different food simulants. A comparative study of the nano-sized vs micro-sized particles, Food Packag. Shelf Life. 26 (2020). https://doi.org/10.1016/j.fpsl.2020.100579.

[29] V. Falguera, J.P. Quintero, A. Jiménez, J.A. Muñoz, A. Ibarz, Edible films and coatings: structures, active functions and trends in their use, Trends Food Sci. Technol. 22 (2011) 292–303. https://doi.org/10.1016/j.tifs.2011.02.004.

[30] C.H.C. Flaker, R.V. Lourenço, A.M.Q.B. Bittante, P.J.A. Sobral, Montmorillonite dispersion in water affects some physical properties of gelatin-based nanocomposites films, Biointerface Res. Appl. Chem. 6 (2016) 1093–1098.

[31] C.H.C. Flaker, R.V. Lourenço, A.M.Q.B. Bittante, P.J.A. Sobral, Gelatin-based nanocomposite films: a study on montmorillonite dispersion methods and concentration, J. Food Eng. 167 (2015) 65–70. https://doi.org/10.1016/j.jfoodeng.2014.11.009.

[32] J.d.M. Fonseca, G.A. Valencia, L.S. Soares, M.E.R. Dotto, C.E.M. Campos, R. d. F.P.M. Moreira, A.R.M. Fritz, Hydroxypropyl methylcellulose-TiO_2 and gelatin-TiO_2 nanocomposite films: physicochemical and structural properties, Int. J. Biol. Macromol. 151 (2020) 944–956. https://doi.org/10.1016/j.ijbiomac.2019.11.082.

[33] S. Galus, J. Kadzińska, Food applications of emulsion-based edible films and coatings, Trends Food Sci. Technol. 45 (2015) 273–283. https://doi.org/10.1016/j.tifs.2015.07.011.

[34] S. Ganiari, E. Choulitoudi, V. Oreopoulou, Edible and active films and coatings as carriers of natural antioxidants for lipid food,, Trends Food Sci. Technol. 68 (2017) 70–82. https://doi.org/10.1016/j.tifs.2017.08.009.

[35] J. George, Siddaramaiah, High performance edible nanocomposite films containing bacterial cellulose nanocrystals, Carbohydr. Polym. 87 (2012) 2031–2037. https://doi.org/10.1016/j.carbpol.2011.10.019.

[36] M.C. Gómez-Guillén, M. Pérez-Mateos, J. Gómez-Estaca, E. López-Caballero, B. Giménez, P. Montero, Fish gelatin: a renewable material for developing active biodegradable films, Trends Food Sci. Technol. 20 (2009) 3–16. https://doi.org/10.1016/j.tifs.2008.10.002.

[37] Q. He, Y. Zhang, X. Cai, S. Wang, Fabrication of gelatin-TiO2 nanocomposite film and its structural, antibacterial and physical properties, Int. J. Biol. Macromol. 84 (2016) 153–160. https://doi.org/10.1016/j.ijbiomac.2015.12.012.

[38] M. Heydari-Majd, B. Ghanbarzadeh, M. Shahidi-Noghabi, M.A. Najafi, P. Adun, A. Ostadrahimid, Kinetic release study of zinc from polylactic acid based nanocomposite into food simulants, Polym. Test. 76 (2019) 254–260. https://doi.org/10.1016/j.polymertesting.2019.03.040.

[39] S.F. Hosseini, M. Rezaei, M. Zandi, F. Farahmandghavi, Development of bioactive fish gelatin/chitosan nanoparticles composite films with antimicrobial properties, Food Chem. 194 (2016) 1266–1274. https://doi.org/10.1016/j.foodchem.2015.09.004.

[40] M. Hosseinnejad, S.M. Jafari, Evaluation of different factors affecting antimicrobial properties of chitosan, Int. J. Biol. Macromol. 85 (2016) 467–475. https://doi.org/10.1016/j.ijbiomac.2016.01.022.

[41] J.Y. Huang, X. Li, W. Zhou, Safety assessment of nanocomposite for food packaging application, Trends Food Sci. Technol. 45 (2015) 187–199. https://doi.org/10.1016/j.tifs.2015.07.002.

[42] T. Huang, Z.c. Tu, X. Shangguan, X. Sha, H. Wang, L. Zhang, N. Bansal, Fish gelatin modifications: a comprehensive review, Trends Food Sci. Technol. 86 (2019) 260–269. https://doi.org/10.1016/j.tifs.2019.02.048.

[43] J.A. Hári, B.A. Pukánszky, Nanocomposites: preparation, structure, and properties, in: M. Kutz (Ed.), Applied Plastics Engineering Handbook., Elsevier Inc., Hungary, 2011, pp. 109–142. https://doi.org/10.1016/B978-1-4377-3514-7.10008-X.

[44] M. Iguchi, S. Yamanaka, A. Budhiono, Bacterial cellulose—a masterpiece of nature's arts, J. Mater. Sci. 35 (2000) 261–270. https://doi.org/10.1023/A:1004775229149.

[45] I.Y. Jeon, J.B. Baek, Nanocomposites derived from polymers and inorganic nanoparticles, Materials. 3 (2010) 3654–3674. https://doi.org/10.3390/ma3063654.

[46] M.F.C. Jorge, C.H.C. Flaker, S.F. Nassar, I.C.F. Moraes, A.M.Q.B. Bittante, P.J.A. Sobral, Viscoelastic and rheological properties of nanocomposite-forming solutions based on gelatin and montmorillonite, J. Food Eng. 120 (2014) 81–87. https://doi.org/10.1016/j.jfoodeng.2013.07.007.

[47] P. Kanmani, J.W. Rhim, Physical, mechanical and antimicrobial properties of gelatin based active nanocomposite films containing AgNPs and nanoclay, Food Hydrocoll. 35 (2014) 644–652. https://doi.org/10.1016/j.foodhyd.2013.08.011.

[48] G. Kavoosi, S.M.M. Dadfar, S.M.A. Dadfar, F. Ahmadi, M. Niakosari, Investigation of gelatin/multi-walled carbon nanotube nanocomposite films as packaging materials, Food Sci. Nutr. 2 (2014) 65–73. https://doi.org/10.1002/fsn3.81.

[49] S. Kumar, A. Mudai, B. Roy, I.B. Basumatary, A. Mukherjee, J. Dutta, Biodegradable hybrid nanocomposite of chitosan/gelatin and green synthesized zinc oxide nanoparticles for food packaging, Foods. 9 (2020) 1143. https://doi.org/10.3390/foods9091143.

[50] S. Kundu, A. Das, A. Basu, M.F. Abdullah, A. Mukherjee, Guar gum benzoate nanoparticle reinforced gelatin films for enhanced thermal insulation, mechanical and antimicrobial properties, Carbohydr. Polym. 170 (2017) 89–98. https://doi.org/10.1016/j.carbpol.2017.04.056.

[51] F. Liu, J. Antoniou, Y. Li, J. Yi, W. Yokoyama, J. Ma, F. Zhong, Preparation of gelatin films incorporated with tea polyphenol nanoparticles for enhancing controlled-release antioxidant properties, J. Agric. Food Chem. 63 (2015) 3987–3995. https://doi.org/10.1021/acs.jafc.5b00003.

[52] A. Llorens, E. Lloret, P.A. Picouet, R. Trbojevich, A. Fernandez, Metallic-based micro and nanocomposites in food contact materials and active food packaging, Trends Food Sci. Technol. 24 (2012) 19–29. https://doi.org/10.1016/j.tifs.2011.10.001.

[53] M.C. Gómez-Guillén, B. Giménez, M.E. López-Caballero, M.P. Montero, Functional and bioactive properties of collagen and gelatin from alternative sources: a review, Food Hydrocoll. 25 (2011) 1813–1827. https://doi.org/10.1016/j.foodhyd.2011.02.007.

[54] S. Maisanaba, S. Pichardo, M. Puerto, D. Gutiérrez-Praena, A.M. Cameán, A. Jos, Toxicological evaluation of clay minerals and derived nanocomposites: a review, Environ. Res. 138 (2015) 233–254. https://doi.org/10.1016/j.envres.2014.12.024.

[55] Grand View Research. Market analysis report: Gelatin market size, share & trends analysis report by raw material (pig skin, bovine hides, cattle bones), by function (thickener, stabilizer, gelling agent), by application, by region, and segment forecasts, 2020. Available in: https://www.grandviewresearch.com/industry-analysis/gelatin-market-analysis.

[56] J.F. Martucci, R.A. Ruseckaite, Biodegradable bovine gelatin/Na+-montmorillonite nanocomposite films. structure, barrier and dynamic mechanical properties, Polym. Plast. Technol. Eng. 49 (2010) 581–588. https://doi.org/10.1080/03602551003652730.

[57] J.F. Martucci, R.A. Ruseckaite, Antibacterial activity of gelatin/copper (II)-exchanged montmorillonite films, Food Hydrocoll. 64 (2017) 70–77. https://doi.org/10.1016/j.foodhyd.2016.10.030.

[58] F.C. Menegalli, P.J. Sobral, M.A. Roques, S. Laurent, Characteristics of gelatin biofilms in relation to drying process conditions near melting, Drying Technol. 17 (1999) 1697–1706. https://doi.org/10.1080/07373939908917646.

[59] A.M. Metak, F. Nabhani, S.N. Connolly, Migration of engineered nanoparticles from packaging into food products, LWT Food Sci. Technol. 64 (2015) 781–787. https://doi.org/10.1016/j.lwt.2015.06.001.

[60] A.M. Nafchi, M. Moradpour, M. Saeidi, A.K. Alias, Effects of nanorod-rich ZnO on rheological, sorption isotherm, and physicochemical properties of bovine gelatin films, LWT Food Sci. Technol. 58 (2014) 142–149. https://doi.org/10.1016/j.lwt.2014.03.007.

[61] M. Nagarajan, S. Benjakul, T. Prodpran, P. Songtipya, Properties and characteristics of nanocomposite films from tilapia skin gelatin incorporated with ethanolic extract from coconut husk, J. Food Sci. Technol. 52 (2015) 7669–7682. https://doi.org/10.1007/s13197-015-1905-1.

[62] Commission Regulation, On active and intelligent materials and articles intended to come into contact with food, European Commission Regulation. (2009).

[63] Commission Regulation, On plastic materials and articles intended to come into contact with food, European Commission Regulation (2011).

[64] C.M. Ortiz, J.O. de Moraes, A.R. Vicente, J.B. Laurindo, A.N. Mauri, Scale-up of the production of soy (Glycine max L.) protein films using tape casting: formulation of film-forming suspension and drying conditions, Food Hydrocoll. 66 (2017) 110–117. https://doi.org/10.1016/j.foodhyd.2016.12.029.

[65] F.R. Passador, A. Ruvolo-Filho, L.A. Pessan, Nanocomposites of polymer matrices and lamellar clays, in: O.N. de Oliveira Jr., M. Ferreira, A.L. Da Róz, F. Liete (Eds.), Nanostructures., Elsevier Inc., Brazil, 2017, pp. 187–207. https://doi.org/10.1016/B978-0-323-49782-4.00007-3.

[66] M. Ramos, A. Valdés, A. Beltrán, M.C. Garrigós, Gelatin-based films and coatings for food packaging applications, Coatings 6 (41) (2016) 1–20.

[67] J. Rouhi, S. Mahmud, N. Naderi, C.H. Raymond Ooi, M.R. Mahmood, Physical properties of fish gelatin-based bio-nanocomposite films incorporated with ZnO nanorods, Nanoscale Res. Lett. 8 (2013) 1–6. https://doi.org/10.1186/1556-276X-8-364.

[68] E. Ruiz-Hitzky, M. Darder, F.M. Fernandes, B. Wicklein, A.C.S. Alcântara, P. Aranda, Fibrous clays based bio-nanocomposites, Prog. Polym. Sci. 38 (2013) 1392–1414. https://doi.org/10.1016/j.progpolymsci.2013.05.004.

[69] S. Sahraee, B. Ghanbarzadeh, J.M. Milani, H. Hamishehkar, Development of gelatin bionanocomposite films containing chitin and ZnO nanoparticles,, Food Bioprocess Technol. 10 (2017) 1441–1453a. https://doi.org/10.1007/s11947-017-1907-2.

[70] S. Sahraee, J.M. Milani, B. Ghanbarzadeh, H. Hamishehkar, Physicochemical and antifungal properties of bio-nanocomposite film based on gelatin-chitin nanoparticles, Int. J. Biol. Macromol. 97 (2017) 373–381b. https://doi.org/10.1016/j.ijbiomac.2016.12.066.

[71] B. Shams, N.G. Ebrahimi, F. Khodaiyan, Development of antibacterial nanocomposite: whey protein-gelatin-nanoclay films with orange peel extract and tripolyphosphate as potential food packaging,, Adv. Polym. Technol. 2019 (2019) 1973184. https://doi.org/10.1155/2019/1973184.

[72] S. Shankar, L. Jaiswal, J.W. Rhim, Gelatin-based nanocomposite films: potential use in antimicrobial active packaging, in: J. Barros-Velazquez (Ed.), Antimicrobial Food Packaging., Elsevier Inc., South Korea, 2016, pp. 339–348. https://doi.org/10.1016/B978-0-12-800723-5.00027-9.

[73] S. Shankar, X. Teng, G. Li, J.W. Rhim, Preparation, characterization, and antimicrobial activity of gelatin/ZnO nanocomposite films, Food Hydrocoll. 45 (2015) 264–271. https://doi.org/10.1016/j.foodhyd.2014.12.001.

[74] B. Sharma, P. Malik, P. Jain, Biopolymer reinforced nanocomposites: a comprehensive review, Mater. Today Commun. 16 (2018) 353–363. https://doi.org/10.1016/j.mtcomm.2018.07.004.

[75] R. Sharma, S.M. Jafari, S. Sharma, Antimicrobial bio-nanocomposites and their potential applications in food packaging, Food Control. 112 (2020) 107086. https://doi.org/10.1016/j.foodcont.2020.107086.

[76] M. Tuncer, Effects of chloride ion and the types of oxides on the antibacterial activities of inorganic oxide supported Ag materials, MSc. Thesis, Izmir Institute of Technology, Turkey, 2007.

[77] G.A. Valencia, M. Djabourov, F. Carn, P.J.A. Sobral, Novel insights on swelling and dehydration of laponite, Coll. Interface Sci. Commun. 23 (2018) 1–5. https://doi.org/10.1016/j.colcom.2018.01.001.

[78] G.A. Valencia, R.V. Lourenço, A.M.Q.B. Bittante, P.J. do Amaral Sobral, Physical and morphological properties of nanocomposite films based on gelatin and laponite, Appl. Clay Sci. 124–125 (2016) 260–266. https://doi.org/10.1016/j.clay.2016.02.023.

[79] F.M. Vanin, M.H. Hirano, R.A. Carvalho, I.C.F. Moraes, A.M.Q.B. Bittante, P.J. d. A. Sobral, Development of active gelatin-based nanocomposite films produced in an automatic spreader, Food Res. Int. 63 (2014) 16–24. https://doi.org/10.1016/j.foodres.2014.03.028.

[80] J. Weiss, P. Takhistov, D.J. McClements, Functional materials in food nanotechnology, J. Food Sci. 71 (2006) R107–R116. https://doi.org/10.1111/j.1750-3841.2006.00195.x.

[81] C. Wolf, H. Angellier-Coussy, N. Gontard, F. Doghieri, V. Guillard, How the shape of fillers affects the barrier properties of polymer/non-porous particles nanocomposites: a review, J. Membr. Sci. 556 (2018) 393–418. https://doi.org/10.1016/j.memsci.2018.03.085.

[82] X. Wu, Y. Liu, W. Wang, Y. Han, A. Liu, Improved mechanical and thermal properties of gelatin films using a nano inorganic filler, J. Food Process Eng. 40 (2017) e12469. https://doi.org/10.1111/jfpe.12469.

[83] S.H. Zaferani, Introduction of polymer-based nanocomposites, in: M. Jawaid, M.M. Khan (Eds.), Polymer-Based Nanocomposites for Energy and Environmental Applications: A Volume in Woodhead Publishing Series in Composites Science and Engineering, University of Ottawa Press, Iran, 2018, pp. 1–25. https://doi.org/10.1016/B978-0-08-102262-7.00001-5.

[84] P. Šimon, Q. Chaudhry, D. Bakoš, Migration of engineered nanoparticles from polymer packaging to food—A physicochemical view, J. Food Nutr. Res. 47 (2008) 105–113.

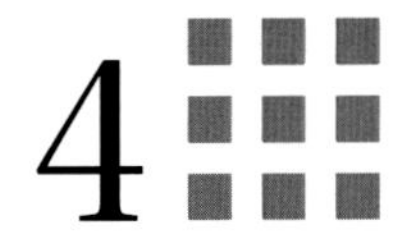

Chitin-based nanomaterials

Marie Arockianathan P

PG & RESEARCH DEPARTMENT OF BIOCHEMISTRY, ST. JOSEPH'S COLLEGE OF ARTS & SCIENCE (AUTONOMOUS), CUDDALORE, TAMIL NADU, INDIA

Chapter outline

4.1 Introduction

Currently researchers are in search of novel materials that promote sustainability and are eco-friendly. This has created renewed interest in bio-based polymers and its composites. Due to this, new and new products are developed by scientists with enhanced properties for various biological applications in different fields. Many biological wastes from forests to marine organisms create various environmental and health issues. The disposals of these wastes are also of great concern. But these wastes provide a great source for the production of biopolymers, which is beneficial not only to the environment but also economically.

Biopolymers such as starch, cellulose, chitin, etc. are naturally occurring polymers obtained from various natural sources that are available in plenty. These polymers are alternative to

Biopolymeric Nanomaterials: Fundamentals and Applications. DOI: https://doi.org/10.1016/B978-0-12-824364-0.00021-6

synthetic polymers because they possess important properties such as renewable, biocompatibility and biodegradability. In spite of these advantages, the biopolymers faced with some pitfalls such as poor mechanical strength, low thermal stability and high gas permeability. To solve these pitfalls, modification of polymeric surface, addition of ionic groups, and functional compounds and fillers are added. So the order of the day is not only to produce new biopolymeric composites with improved properties but also to develop new methods or techniques to create novel products.

Novel nanocomposites prepared from biopolymers form a new alternative to synthetic polymers. These nanocomposite polymers contain nanoparticles of either organic or inorganic origin in the range 10–1000 Å. These particles in polymeric matrix improve the properties significantly. Due to the presence of these particles of nanometer size, the nanocomposites exhibit much improved properties compared to conventional composites or pure polymers. They showed better mechanical strength, improved barrier properties, thermal stability, and much more.

Nowadays polysaccharides are being exploited to produce many nanocomposites as they are available abundantly in nature. Among the polysaccharide, chitin is the second most abundant polysaccharide found in nature. It is present in cell walls of fungi and yeasts, the hard shells of shell fish and marine invertebrates such as crab, shrimp, lobster, oysters, etc. It consists of repeating units of N acetyl-D-glucosamine [172]. It exhibit properties such as nontoxicity, biocompatibility, and biodegradability so that it can be used as reinforcement in polymer composites.

Mostly chitin exists in nature as α-chitin, which is a stable form among the three forms (α, β, and γ) of chitin. These polymeric types depend on the species and molecular orientation. Chitin shows a strong crystalline character owing to its linear structure and the functional groups present on the surface. Due to this, it is insoluble in common solvents. Chitosan, an important natural polymer, is obtained from chitin after deacetylation process. It has repeated units of glucosamine and N-acetyl glucosamine units linked by β-(1-4) glycosidic bonds [18]. It shows a different solubility character in aqueous and dilute acidic solutions, that is insoluble in water. The deacetylation process not only improved its solubility but also its applications. The degree of deacetylation (DD) and the molecular weight have a strong impact on the physiochemical properties of chitosan such as solubility, crystallinity, degradability, etc. It is used extensively in various biological applications and can be processed into different forms such as nanofibers, nanoparticles, sponges, beads, membranes, and films similar to chitin.

4.2 Chitin/chitosan-based nanocomposites

There is a great interest in the development of novel bionanocomposites with remarkable properties. This has paved the way for the researchers to develop new bionanocomposites with superior properties by joining two or more materials. Normally biocomposites are constructed to improve the stability and the functional property of conventional polymers [19]. To improve the properties of conventional polymers, various biopolymers and fillers are combined to tailor-made the desired function.

There are numerous studies that showed the use of chitin and chitosan to form composites of different shapes, sizes, and also for different functions [40,141,173]. This was mainly due to the presence of reactive functional groups on the surface of chitin and chitosan, which makes it possible to combine with other polymers or synthetic substances to form novel composites with superior properties. In spite of low cost, biodegradability, nontoxicity, and biocompatibility, chitin and chitosan inherently have poor mechanical and barrier properties. This limits its applications. So the biopolymer chitin/chitosan should be combined with various polymeric materials or fillers to produce composites of great strength and barrier properties.

Chitosan has more reactive functional groups on its surface than chitin and also show less crystallinity than chitin. Chitosan is soluble in dilute acids, whereas chitin is soluble only in few solvents. This property of chitosan makes it more readily accessible to reagents for modifications compared to chitin. Various chitin/chitosan-based composites with improved properties and applications are shown in Table 4.1.

4.3 Properties of chitin/chitosan-based nanocomposites

4.3.1 Molecular weight

Chitosan, a deacetylation product of chitin, consists of copolymer of β-(1$\rightarrow$4)-2-acetamido-D-glucose and β-(1$\rightarrow$4)-2-amino-D-glucose units. It is nontoxic, biocompatible, and biodegradable. The chemical and physical properties of chitosan depend on various characteristics such as molecular weight, DD, crystallinity, free amino or hydroxyl group, etc.

The molecular weight of chitosan depends on the number of sugar molecules per polymer. The properties of the polymer depend on the molecular weight [231,233,245]. The molecular weight also decides the crystal size and its morphology. The crystallinity of the polymer increases with decrease in molecular weight of the substance. The chitosan molecular weight varies from 50 to 2000 kDa that depends on the DD, source, and the extraction protocol.

4.3.2 Degree of deacetylation (DD)

The DD is one of the key parameters to assess its applicability in the medical, nutritional, and biotechnological fields [107]. This property also determines the crystallinity, hydrophilicity, degradability, and cellular response of the polymer [33,64,76,80,93,164].

Under alkaline conditions, Chitosan undergoes repeated treatment to form more amine groups on the surface and also show higher DD. This DD determines the solubility and crystallinity property of the polymer. The DD also changes the flexibility and charge density of the chitosan molecules. The flexibility of the chitosan chains increases on deacetylation process, which helps them to form intramolecular hydrogen bonding between the chains. This makes the chain weaker in mechanical strength [75]. The DD also plays a major role in cell adhesion and proliferation of chitosan. Lower the DD more is the cell adhesion and proliferation of chitosan [37,99,245,254,256].

Table 4.1 Applications and improved properties of chitin/chitosan-based composites.

Composite	Applications	Improved properties	References
Chitosan–hydroxyapatite (HA) nanocomposite	Bone tissue engineering	Mechanical strength and Antimicrobial	[218]
HA-chitosan-carboxycellulose	Bone tissue engineering	Mechanical strength	[122]
Chitosan-montmorillonite (MMT)-rosemary essential oil (REO) composite	Packaging applications	Tensile strength	[10]
Chitin–bentonite clay-based polyurethane bionanocomposites	Medical applications	Mechanical strength	[262]
Sodium montmorillonite (Na–MMT)-poly(caprolactone) chitosan composites	Medical applications	Tensile strength and Thermal properties	[11]
Organoclay montmorillonite (ORG-MMT)-poly(caprolactone) chitosan composites	Medical applications	Tensile strength and Thermal properties	[11]
Halloysite nanotubes with chitosan composites	Tissue engineering	Tensile strength	[116]
Chitosan-CNT nanocomposites	Membrane separation and electrodes	Mechanical strength	[225]
Unzipped multiwalled carbon nanotube oxides (UMCNO)-chitosan composite	Films	Mechanical strength	[59]
Poly(3-hydroxybutyrate) (PHB)–functionalized MWCNT-Chitosan composite	Membrane separation and electrodes	Mechanical strength	[148]
Chitosan-poly(3,4-ethylenedioxythiophene) (PEDOT)–poly(styrenesulfonate) (PSS)-carbon nanotubes composite films	Membrane	Mechanical strength	[235]
Chitosan-grafted carbon nanotubes	Biosensors	Mechanical strength	[192]
Chitosan-graphene oxide films	Packaging applications	Tensile strength and Thermal property	[247]
α-Chitin whiskers (ChW)-chitosan film	Packaging applications	Mechanical strength	[199]
Poly(vinylalcohol)PVA-chitin whiskers	Packaging applications	Mechanical strength	[213]
Chitosan nanoparticles–plasticized starch composite films	Packaging applications	Tensile strength	[36]

(*continued on next page*)

Table 4.1 (*continued*)

Composite	Applications	Improved properties	References
Chitosan-silica nanocomposite films	Environment remediation	Storage modulus (thermal property) and Glass-transition temperature	[16]
Chitosan/clay nanocomposites	Packaging applications	Thermal property	[202]
Chitosan-alumina-functionalized multiwalled carbon nanotube (fMWCNTs)	Environment remediation	Thermal property and Antimicrobial	[137]
Silver/chitosan-g-PAA/montmorillonite nanocomposites	Packaging applications	Thermal stability and Antimicrobial	[8]
Poly(vinylalcohol) (PVA)-α-chitin composite mat	Packaging applications	Thermal stability and Toughness	[187]
Poly(methyl methacrylate) PMMA/SMCNFs small modified chitin nanofibres nanocomposite	Packaging applications	Thermal stability and Toughness	[14]
Chitosan-cellulose nanofiber	Packaging applications	Thermal stability, Mechanical strength and Improved Barrier properties	[24]
Nanocrystalline cellulose (NCC) reinforced chitosan composite films	Packaging applications	Tensile strength and Improved Barrier properties	[101]
Gelatin –N-chitin composite films	Packaging applications	High hydrophobicity, Reduced WVP and Mechanical property	[181]
Chitosan–MgO nanocomposites	Packaging applications	Improved Barrier properties and UV protection	[53]
Chitin nanofibers (CNFs)-Au NPs-tricyclodecane dimethylol dimethacrylate (TCDDMA) resin	Solar and electronic devices	Transparency, Improved Electrical and optical properties	[257]
Carboxylic acid–imidazole–chitosan	Drug delivery	Transfection and Extended release	[198]
Polyvinyl pyrrolidone (PVP)-chitosan,	Drug delivery	Stability and Targeted delivery	[117]
Genipin-cross-linked chitosan/PEG/zinc oxide (ZnO)/silver (Ag) nanoparticles	Drug delivery	Antimicrobial, Mechanical property and Wound healing	[118]
Cyclodextrin (CD)-chitosan composite	Drug delivery	Absorption and Drug release	[48]
PAMAM(polyamidoamine)-chitosan conjugate based temozolomide nanoformulation (PCT)	Drug delivery	Anticancer and Drug release	[191]

(*continued on next page*)

Table 4.1 *(continued)*

Composite	Applications	Improved properties	References
Chitosan/ellagic acid biocomposite films	Packaging applications	Mechanical and barrier properties	[221]
Hyaluronan with chitin nanowhiskers nanocomposite film	Drug delivery and Tissue engineering	Antimicrobial and mechanical property	[9]
Carboxymethyl chitin/polyvinyl alcohol(PVA) composite	Tissue engineering	Mechanical and compression strength	[190]
Chitosan–pluronic hydrogel	Tissue engineering	Thermal property	[156]
Chitosan-gelatin-bioactive glass ceramic scaffolds	Tissue engineering	Mineralization and Protein absorption	[160]
Gallium-apatite/chitin/pectin (Ga-HA/C/P) nanocomposites	Tissue engineering	Mechanical strength and Swelling ability	[244]
Methotrexate loaded chitin nanogel	Drug delivery	Extended release and Nontoxic	[153]
Chitosan-loaded curcumin nanoparticles	Drug delivery	Entrapment efficiency and Drug release	[142]
Chitosan/sago starch/silver nanoparticles	Wound healing	Antimicrobial, Exchange of gases, Activate platelets and cell repair	[127]
Chitosan-fibrin nanocomposite	Wound healing	Antimicrobial, Exchange of gases, Activate platelets and cell repair	[217]
Chitin/silk fibroin/TiO_2 bionanocomposite	Wound healing	Antimicrobial, Exchange of gases, Activate platelets and cell repair	[66]
pH-sensitive hydroxypropyl chitin/tannic acid/ferric ion hydrogel	Wound healing	Antimicrobial, Exchange of gases, Activate platelets and cell repair	[132]
β-chitin/nanosilver composite	Wound healing	Antimicrobial, Exchange of gases, Activate platelets and cell repair	[106]
Chitosan/polycaprolactone scaffold	Wound repair	Antimicrobial, Exchange of gases, Activate platelets and cell repair	[111]
Chitin–polyaniline composite	Environment remediation	Chelation, Adsorption/absorption and Ion exchange	[98]
Chitin/cellulose membranes	Environment remediation	Chelation, Adsorption/absorption and Ion exchange	[206]
Chitosan/PVA/zeolite composite	Environment remediation	Chelation, Adsorption/absorption and Ion exchange	[74]
Chitin/magnetite/multiwalled carbon nanotubes MWCNTs composite	Environment remediation	Chelation, Adsorption/absorption and Surface area	[180]

4.3.3 Viscosity

The DD and the concentration of chitosan have direct correlation with viscosity of the sample solution. The viscosity increases with DD and concentration, whereas it decreases with temperature. Mostly chitosan produce highly viscous solution due to its high molecular weight and linear structure. Its viscosity decreases with increase in shear stress [1,58,105].

4.3.4 Mechanical properties of chitin/chitosan nanocomposites

The natural polysaccharide chitin and chitosan play a pivotal role in the fabrication of nanocomposites. As it is nontoxic and shows biocompatibility and biodegradability property, it is used as fillers as well as provide matrix for making the composites. But its diverse advantages cannot be utilized optimally for various applications as it lacks mechanical properties. So, various fillers and matrices are employed to boost the mechanical properties of composites.

4.3.4.1 Hydroxyapatite

Hydroxyapatite (HA) is the major component of bone and it possesses properties such nontoxicity, noninflammatory, biocompatible, nonimmunogenic, etc. [218]. The incorporation of a polymer layer on the composite surface can improve the mechanical properties of HA. The chitosan–HA nanocomposite formed by the combination of chitosan (CH) with hydroxyapatite (HA) showed increase in the mechanical strength of the material. This was mainly due to the possible interaction between the amino and hydroxyl group of chitosan with calcium ions of HA. As the mechanical properties of the Ch-HA composites increased, it plays a prominent part in bone tissue engineering applications [128].

The generally used parameter to determine the strength of porous scaffold is compression strength [128,218]. The compression strength of the Ch-HA composites increased with increase in HA concentration. Among the different ratios of Ch-HA composites, 30:70 ratio showed the maximum compression strength of 119.86 MPa [113]. Apart from this, the high-molecular-weight chitosan exhibited higher compression strength compared to low-molecular-weight chitosan [208]. Moreover the increase in temperature also improves the mechanical properties of composites by increasing the interfacial bonding between HA and Ch [242].

Another way to improve the mechanical property is to lower the water content in the composites. This way of enhanced mechanical strength was exhibited by HA-Ch-carboxycellulose composite in a dry state compared to a wet condition [122]. To mimic the properties of natural bones, polylactic acid (PLA) along with Ch was added to HA to form Ch-PLA-HA composites to improve the elastic and mechanical strength of the composites. In case of chitosan–HA composite films, the tensile property is improved significantly by two ways: one by treating with formaldehyde [102] and another by hydrophobic modifications of chitosan [163].

4.3.4.2 Clay particles

Nowadays nanoclay is used widely to prepare polymer-based nanocomposites as it brings significant improvement in the mechanical properties of composites. The use of nanoclay increases the toughness and tensile properties of the composites by transferring the stress to the available reinforcing elements.

To form Ch-clay nanocomposites, nanoclay is well blended with chitosan (Ch) as it interacts and bonds well with amino and hydroxyl groups of chitosan [50,224].

At a low concentration, the addition of nanoclay montmorillonite (MMT) to Ch film increased the tensile property significantly by their strong interaction between them; whereas at a high concentration, it decreases the tensile property due to aggregation of clay particles [239].

The MMT nanoclay and rosemary essential oil (REO) were added to chitosan (Ch) film to form Ch-MMT-REO composite [10]. The addition of up to 3 wt% MMT into chitosan film increased the tensile property but when the concentration increased beyond the 3 wt% there was no effect. At a low concentration of MMT, the tensile property of the film increased due to the strong interaction between MMT and chitosan, whereas at a high concentration there occurs aggregation between the particles that causes decrease in tensile strength [239].

Another clay substance bentonite, obtained by devitrification of volcanic ash, was used to produce chitin–bentonite clay-based polyurethane bionanocomposites [262]. The Delite HPS (Laviosa Chemica Mineraria, Italy) bentonite was used to make this composite. A constant and steady tensile property of the composite was showed by 4% nanoclay content, but it decreases if the content is more than 8%. This decrease in tensile property at 8% is attributed to the larger aggregations of Delite HPS bentonite clay contents. The addition of plasticizers such as glycerol to the composites not only increases the clay loading capacity but also the mechanical property.

The addition of nanoclay-like sodium MMT (Na–MMT) and organoclay-MMT (ORG-MMT) to the poly(caprolactone) chitosan composites [11] and halloysite nanotubes with chitosan composites [116] was studied and showed enhanced tensile property.

4.3.4.3 Carbon nanotubes (CNTs)

Carbon nanotubes (CNT) are tube-like material made up of carbon in a nanometer scale. They possess excellent thermal, mechanical, and electronic properties. It shows Young's modulus as high as 1.2 TPa and tensile strength of 50–200 GPa. So it acts as an ideal material for reinforcement in composites.

Chitosan (Ch) and CNT strongly interact with each other by hydrogen bonds. The interaction of multiple-walled CNT fillers with chitosan matrix produces superior composites with greater mechanical strength. The tensile strength of the Ch-CNT nanocomposites was greatly increased by about 94–99%, when 0.8% of CNTs was added to the chitosan matrix [225]. The tensile strength of the chitosan composite increased even at the addition of a low concentration of nonfunctionalized CNTs, whereas at a high concentration, it did not show relative increase in the strength. But in the case of functionalized CNTs, the tenacity increased relatively. This is due to strong interaction of a carboxyl group of functionalized CNTs with an amino group of chitosan matrix [150].

Unzipped multiwalled CNT oxides (UMCNO) are used as reinforcement materials for chitosan to improve their strength. The addition of a relatively small amount of UMCNO to chitosan matrix has significantly increased the mechanical properties of the composite films that is, 0.2% of UMCNO increased the tensile strength from 69.3 to 142.7 MPa [59]. Likewise many chitin/chitosan nanocomposites were prepared using CNTs to improve their mechanical strength. For example, poly(3-hydroxybutyrate) (PHB)-functionalized multiwalled CNT (MWCNTs) into chitosan matrix [148], chitosan composite films reinforced by poly(3,4-ethylenedioxythiophene) (PEDOT)-poly(styrenesulfonate) (PSS)-treated CNTs [235], and chitosan-grafted CNTs [192].

4.3.4.4 Graphene oxide

Graphene is a crystalline form of carbon in a 2D form with one-atom thickness arranged in a honeycomb-like structure [2,17]. It shows extraordinary properties such as thermal, optical, mechanical, and chemical properties in a single material [168]. The oxidized form of graphene is graphene oxide (GO) which is hydrophilic in nature and acts as a weak cationic exchange resin [60]. The chemical functionalization of graphene was found to improve the dispersion and bonding of graphene with the matrix [219,246].

With the incorporation of 1 wt% GO, Young's modulus and tensile strength of the graphene-based composite materials improved significantly by about 64% and 122%, respectively [247]. With nearly 0.1–0.3 wt% addition of graphene in chitosan matrix, the elastic modulus of chitosan increased over 200%. The chitosan films reinforced and homogenously aligned with GO showed enhanced mechanical properties [152].

4.3.4.5 Chitin/chitosan as nanofiller

Chitin (CH) exists in the form of microfibrils with alternating crystalline and amorphous domains in it. The amorphous part of chitin can be removed by acidolysis to form crystallites in nanoscale called chitin whiskers or chitin nanocrystals. Chitin whiskers used as a new type of nanofillers that act as a reinforcing material in both natural [69,70,88,124,199,230,234] and synthetic polymeric matrices [61,82,85–87,94,95,139,143,144,151,173]. The reinforcing effect of the chitin whiskers depends on the aspect ratio [139].

The α-chitin whiskers (CHW) incorporated in chitosan film form nanocomposite with better mechanical property. The mechanical property increases initially up to 2.96 wt% of chitin whiskers, later it decreases on the increasing the content of chitin whiskers, which is due to the interaction of H-bonding between chitosan and CHW. This interaction makes the composite film more rigid as the concentration of CHW increased and the percentage of elongation at break decreased [199]. The soy protein films with CHW also exhibited a similar type of behavior [124]. Another composite fiber, poly(vinylalcohol) Chitin (PVA-CHW) whiskers, showed excellent mechanical properties that is, tensile strength (1.88 GPa) and toughness (68 J/g) [213]. The concentration of 5% chitin whiskers showed optimal loading in relation to mechanical strength. The orientation of the filler orientation also decides the mechanical property of the composites.

The incorporation of chitosan nanoparticles into plasticized starch composite films showed increased tensile strength as the concentration of chitosan increased but the elongation at break decreased [36]. This is mainly due to the interaction between chitosan nanoparticles and glycerol plasticized starch (GPS) matrix. The agglomeration of chitosan particles decreased the tensile strength, when the concentration of chitosan was added more than 8% to the matrix. Chitosan improved the interaction between the GPS molecules in a GPS matrix by bringing the molecules closer and thereby raising the glass-transition temperature of the composites.

Chitin nanocomposites films showed higher transparency and mechanical properties than cellulose composite films because of high crystallinity and aspect ratio of chitin nanocrystals [79]. Likewise the strength of starch films with chitin whiskers was significantly higher ($P < 0.05$) than pure maize starch films [165].

4.3.5 Thermal properties

Chitin is insoluble in common solvents due to its crystallinity and hydrogen bonding among the functional groups. So it is converted to chitosan to enhance its properties. It possesses excellent properties such as biocompatibility, biodegradability, non-toxic, antimicrobial, etc. Its use is restricted due to its poor mechanical and barrier properties. Due to this, it is blended with various polymers/nanomaterials to enhance the mechanical and barrier properties. Various thermal analyzers are employed to calculate the thermal properties of the composites. The thermal analyzers used are thermogravimetric analyzer, differential scanning calorimetry, and dynamic mechanical analyzer. The thermal properties of the nanobiocomposites are measured as degree of crystallinity, glass transition temperature (T_g), melting temperature (T_m), decomposition temperature (T_d), storage modulus (E^1), loss of modulus (E^{11}), and damping peaks ($\tan\delta$).

Chitosan-silica nanocomposite films were constructed using tetraethoxysilane as a precursor by Al-Sagheer and Muslim [16]. The increase in the loading of silica nanoparticles increases the storage modulus (E^1) of chitosan. The thermal property "storage modulus" is mainly connected with elastic properties of a polymeric material. The storage modulus increased from 3.67 GPa to 6.7 GPa at 50 °C, when silica nanoparticles are added to the composite.

Swain et al. [202] fabricated chitosan/clay nanocomposites using a copper sulfate-glycine complex as a catalyst. The decomposition temperature of this composite is higher than the chitosan because of thermal stability of clay and its strong interaction with chitosan.

The thermal stability of chitosan was also increased by the addition of alumina and functionalized MWCNT (fMWCNT) to form chitosan-alumina-fMWCNTs nanocomposites [137]. Likewise the thermal stability of chitosan increased by the addition of GO [42]. Abbasian and Mahmoodzadeh [8] prepared silver/chitosan-graft polymerization of Acrylic acid (g-PAA)/MMTnanocomposites using polymerization and chemical reduction methods. This composite exhibited higher thermal stability than the chitosan-g-PAA polymer. When 10% cellulose whiskers were added to chitosan, the degree of crystallinity (χc) of chitosan was increased up to 40.6% [174].

SeunghwanChoy et al. [187] developed a tough composite mat with two polymers: PVA and α-chitin. The toughness of this composite mat is 20% higher than PVA. Apart from increased stiffness and extensibility, PVA/α-chitin mat showed high thermal stability with 2.8-fold larger melting enthalpy than PVA.

AklogYihun et al. [14] prepared surface-modified chitin nanofibers (SMCHNFs) dispersed in a poly (methyl methacrylate) (PMMA) matrix to form PMMA/SMCHNFs nanocomposite films. This composite film showed transparency as that of PMMA film. The addition of SMCHNFs to a PMMA matrix increased the thermal stability and toughness of the composite film.

4.3.6 Barrier properties

The introduction of nanoparticles to the biocomposites alters its surface characteristics, which increases the mechanical, biological, and barrier properties. The incorporation of nanoparticles in biocomposites also increases the gas barrier and biosensing properties to maintain the characteristics of the product such as color, texture, and stability [216].

In packaged foods, the presence of oxygen and water vapor causes many adverse reactions such as loss of nutrients, color and flavor change, microbial growth, and also affects respiration rate and ethylene production [171,134]. But for fresh fruits and vegetables whose shelf-life depends on the continual supply of oxygen for its cellular respiration, the barrier to the diffusion of gases is not needed.

In polymeric materials, the permeability of gases depends on the adsorption and diffusion rate of gases in the matrix. The rate of adsorption of gas molecules depends on the formation of free-volume holes in the polymer and also diffusion between the holes. Thus, the free-volume holes in the polymeric matrix, the motions of polymer chains, polymer–polymer interaction, and polymer-gas interaction all affect the permeability of polymeric films. Moreover, the film thicknesses also have direct effect on the diffusion rate of gases [56].

Chitosan, a derived product of chitin, forms films with high tensile and elastic properties. As they provide optimum water vapor permeability and oxygen barrier properties, they are used as a packaging material for moist foods. Azerdo et al. [24] prepared nanocomposite films made up of chitosan-cellulose nanofiber (Ch-CNF) and glycerol of different concentrations to study the mechanical, thermal, and barrier properties of the films. The 15% of Ch-CNF and 18% of glycerol exhibited good mechanical and barrier properties. Khan et al. [101] prepared a nanocrystalline cellulose (NCC) reinforced chitosan composite films with improved tensile strength for the optimum 5% (w/w) NCC concentration. The water vapor permeability (WVP) decreased for this composite film by 27% for the NCC concentration of 5% (w/w).

The WVP and mechanical properties of chitosan films show a linear relationship with storage time and molecular weight and an inverse relation with storage temperature [100].

Samar sahraee et al. [181] studied the physicochemical properties of gelatin composite films containing N-chitin nanoparticles. The incorporation of N-chitin to the composite films significantly reduced the WVP and solubility and also showed higher surface hydrophobicity of the film.

Samar sahraee et al. [182] also studied the effect of corn oil on the gelatin-N-chitin composite film. The addition of N-chitin decreased the water absorption capacity of the films but incorporation of corn oil enhanced the barrier properties compared to chitin nanoparticles.

The incorporation of MgO into the chitosan nanocomposites enhanced the moisture barrier properties due to its dispersion on the matrix [53] and also exhibited UV protection effect.

Ahmed A. Oun and Jong-WhanRhim [5] prepared multifunctional carboxymethyl cellulose (CMC) composite films with the addition of chitin nanocrystals and grapefruit seed extract (GSE).The addition of chitin nanocrystals not only increased the tensile strength by 19.6% but also decreased the WVP by 27% in this composite film.

4.3.7 Electrical and optical properties

The electric properties of the nanobiocomposite depend on the types of bond present in the structure of the composite. The ionic and the covalent bonds formed between the atoms in the structure do not let the electrons leave the atoms, so it will not conduct electricity. When biopolymers are concerned, they are not good conductors of electricity due to the presence of ionic and covalent bonds. Thus, they are used as insulators instead of conductors.

Ömer Bahadır Mergen et al. [147] prepared chitosan/graphene nanoplatelets (Ch/GNPT) and chitosan/MWCNT (Ch/MWCNT) biocomposite films. By increasing the concentration of GNPT or MWCNT in a chitosan composite, the surface conductivity property, light-scattered intensity, and the tensile strength were increased significantly. The electrical and optical properties also match the applied percolation theory.

It was reported that nanocomposite chitin nanofibers (CHNF)-gold nanoparticles (AuNPs) blend with a tricyclodecane dimethylol dimethacrylate (TCDDMA) resin and were polymerized using photoinitiator 2-hydroxy-2-methylpropiophenone to produce a transparent film. The polymerized film exhibit 64% of transmittance to visible light. So it can be exploited as a substrate for a wide range of applications from solar cell to electronic devices [257].

The MgO incorporated to Chitosan (Ch) forms a Ch-MgO nanocomposite film with improvement in the crystallinity. This composite film showed a decrease in the transmission of light in a Ch-MgO film due to the increase in the crystallinity of the sample [46].

Transparency of the polymeric film depends on the polymer and its method of production. Transparency is one of the essential optical properties of packaging material. Several bionanocomposites were prepared and analyzed for its optical properties [71,77,78,188,189]. The addition of nanofillers (ZnO, CNFs, SiO_2) to the composites has varied effect on the optical properties.

4.3.8 Chemical modifications

The biopolymers chitin and chitosan are used for various biological applications, its efficiency or potential can be increased by chemical modifications of the functional groups. Among the two polymers, chitin cannot undergo chemical modification easily because it shows difficulty to undergo selective substitution, insolubility, ambiguity in the structure of product, partial degradation, etc. Due to this, chitosan is used extensively for chemical modifications than chitin. Chitosan can undergo chemical modifications with ease compared to other polysaccharides. The functions of the polymer will be tailor-made based on the groups added.

The functional groups present in the chitosan are two hydroxyl groups and one amine group. These groups undergo chemical modification as per the desired applications [57]. The primary

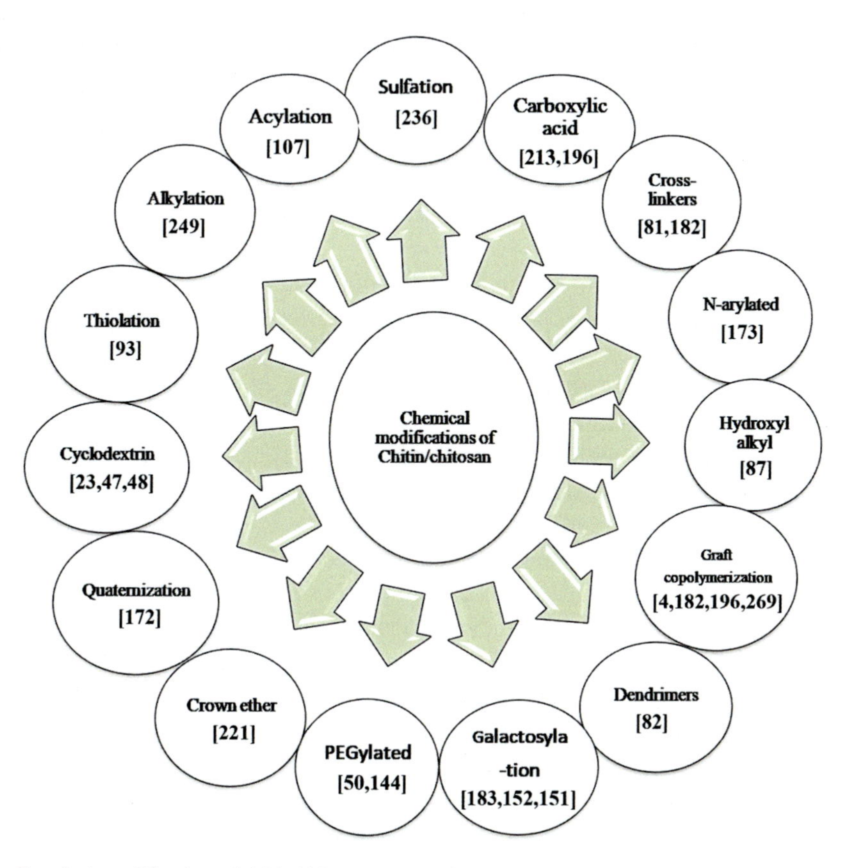

FIGURE 4.1 Chemical modification of chitin/chitosan composite.

amine group is mainly involved in the chemical reactions than hydroxyl groups. The hydroxyl groups of the chitosan can be modified only after masking/protecting the amino group.

Various chemical modifications adopted for chitin/chitosan polymers are shown in Fig. 4.1. The types of chemical modifications are sulfation, phosphorylation, nitration, acylation, Schiff's base formation, alkylation, etc. [91]. The modification of chitin/chitosan not only promotes or enhances the biological function but also increases the mechanical properties [3]. The modified chitosan possesses enhanced properties of unmodified chitosan such as solubility, gelling properties, amphiphilic hydrophobic derivatives, and potential to connect self-assembling nanostructures and conjugates with bioactive molecules; improved biocompatibility; enhanced properties to form DNA complex [31].

The important factor that affects the chitosan to do diverse function is its poor solubility property. This property depends not only on the pH and ionic strength of the medium but also on the concentration and distribution of both acetylated and deacetylated residues on it. Apart from this, the molecular weight and DD form key physicochemical properties of chitin and chitosan. By changing the molecular weight, the properties of the chitin/chitosan composites can be controlled [123].

The polycationic nature of chitosan is important for various activities such as mucoadhesion, absorption enhancement, transfection, antimicrobial, antiinflammatory, etc. So the polycationic derivatives of chitosan, which is non-pH dependent, plays a pivotal role in enhancing these activities [26].

In chitosan, acylation is mostly used to introduce hydrophobic groups on the functional groups than alkylation. This modification increases the biodegradation potential of the composite with enzymes such as lipase. Quaternization is the simplest chemical modification by which chitosan reacts with either methyl or ethyl iodide under basic conditions. Quaternizing chitosan increases the hardness and also shows better hydroxyl radical scavenging property [20]. The mucoadhesive property of chitosan was increased by quaternization, which makes a strong interaction of cationic group of chitosan with anionic group of mucin. Likewise, thermosensitive hydrogel was also formed for a nasal drug delivery [236]. The quaternization of chitosan derivatives improved the solubility, mucoadhesive property, and also managed the cationic character. N,N,N-trimethyl chitosan chloride is the most common quaternization product of chitosan derivative with good mucoadhesive and solubility property [140]. This is used for drug delivery applications [175,198]. To enhance the transfection efficiency and drug delivery process, amphiphilic chitosan derivatives are prepared for example carboxymethyl chitosan [251], hydroxyalkyl chitosans [89], N-arylated chitosan [176], etc. These derivatives self-assemble as micelles and serve as a good drug delivery model.

Carboxylic acid–chitosan derivative obtained by reacting chitosan with carboxylic acid showed improved transfection efficiency compared to unmodified chitosan but their binding capacity with DNA is weak [215]. So, it was modified with cationic polymer chains to form carboxylic acid–imidazole–chitosan to increase its transfection [67,198].

Thiolated chitosan enhances the mucoadhesive property by binding strongly to the mucin proteins [96]. The chemical modification of chitosan with phosphorylcholine showed anticoagulation property. It was reported that sulfated chitosans exhibit anticoagulant property and also show antiviral, antibacterial, antioxidant, and antisclerotic activities [238].

Another modification of chitosan is by grafting copolymerization method. This method of copolymerization is initiated by free radicals, enzymes, and radiation [4,184,198,263]. Acrylic acid is grafted with chitosan to create mucoadhesive hydrophilic derivatives.

The polymer polyethylene glycol (PEG) is used mainly in pharmaceutical products to sustain the release and circulation of drugs in the blood. Chitosan grafted with PEG showed enhanced biocompatibility, good solubility, and sustained circulation in blood. The only issue with PEG is nonbiodegradability. Thus, toxicological studies should be carried out to ascertain its effects for further applications [51,146].

Cross-linkers are used to create chitosan nanocomposites by linking polymer chains. In physical type cross-linking, ionic interactions take place between polymeric chains, whereas in covalent linkage chemical cross-linking takes place [28,184]. The commonly used physical linkers are pentasodium tripolyphosphate (TPP) and sodium alginate and the chemical linkers are glutaraldehyde, genipin, and dextran sulfate. The aldehyde of glutaraldehyde cross-links with functional amino groups of chitosan to form Schiff's base [28]. Many chitosan nanocomposites such as polyvinyl pyrrolidone (PVP)-chitosan, chitosan–PVA composites, etc. were prepared using glutaraldehyde for drug delivery purposes [44,117]. Only problem with glutaraldehyde is its toxicity, so its use is limited in biological applications [184].

To overcome this problem, a water-soluble genipin obtained from gardenia fruit was used as a cross-linker. Several chitosan nanocomposites obtained by cross-linking with genipin are genipin-linked O-CMC-alginate hydrogels [83], chitosan–alginate beads [184], genipin-cross-linked chitosan/PEG/zinc oxide (ZnO)/silver (Ag) nanoparticles [118].

Chitosan is modified with sugar derivatives to form composites such as lactose-modified chitosan, succinyl-sialic acid bound chitosan, galactosylated chitosan, etc. to carry out a specific action on the target cells or molecules such as lectin, cartilage, liver, and so on [138,154,155,185]. The nontoxic cyclic oligosaccharide cyclodextrin (CD) has truncated cone-shaped geometry structure that is useful to carry molecules for drug delivery applications. Many researchers prepared CD derivative molecules grafted with chitosan to make them interactive with the desired hydrophobic substances in the cell [23,48,49]. Dendrimers are nanosized highly symmetrical molecules that act as pathogenic adhesion inhibitors. Ashok Kumar Sharma et al. [191] prepared PAMAM(polyamidoamine)-chitosan conjugate based temozolomide nanoformulation (PCT) against gliomas *in vitro* as well as *in vivo* which showed increased potential of this nanoformulation to treat cancer. Similarly dendrimers with carboxymethyl chitosan containing magnetic nanoparticles were prepared to remove dyes at a specific pH [84]. The molecular structure of crown ether can form a complex with some metal ions such as Pd^{2+}, Au^{3+}, etc. This property can be utilized by binding it with chitosan to absorb selective metal ions [223].

4.3.9 Antimicrobial property

There is a great concern among the researchers to find a suitable drug to fight against multidrug-resistant strains of bacteria keeping in mind the sustainability and eco-friendliness. So their interests turned to natural products from animal, plant, and marine origin. One such natural substance that gained interest nowadays is chitin/chitosan. This biopolymer possesses wonderful properties such as biocompatibility, biodegradability, safe, antimicrobial, etc. which can be used for diverse applications.

The antimicrobial property of chitosan, deacetylation product of chitin, is due to the interaction of polycationic group with lipopolysaccharide and teichoic acid on the surface of Gram-positive and Gram-negative bacteria, respectively. The chelation and agglutination reaction caused by chitosan on the bacterial surface disrupts the cell membrane [32] and leaks the cell contents. The chelating effect of chitosan inhibits the microbial growth and toxin production by coordinating metal ions and also stimulates host defense mechanisms. The inhibitory effect shown by this polymer is dose dependent [166,186].

The physicochemical properties of chitosan such as molecular weight, DD, pH, temperature, solubility, degree of polymerization, derivatization, type of organism, environment, etc. also influence the antimicrobial property [72]. The antimicrobial property of chitosan is enhanced by modifying the functional groups chemically thereby increasing the interaction between its

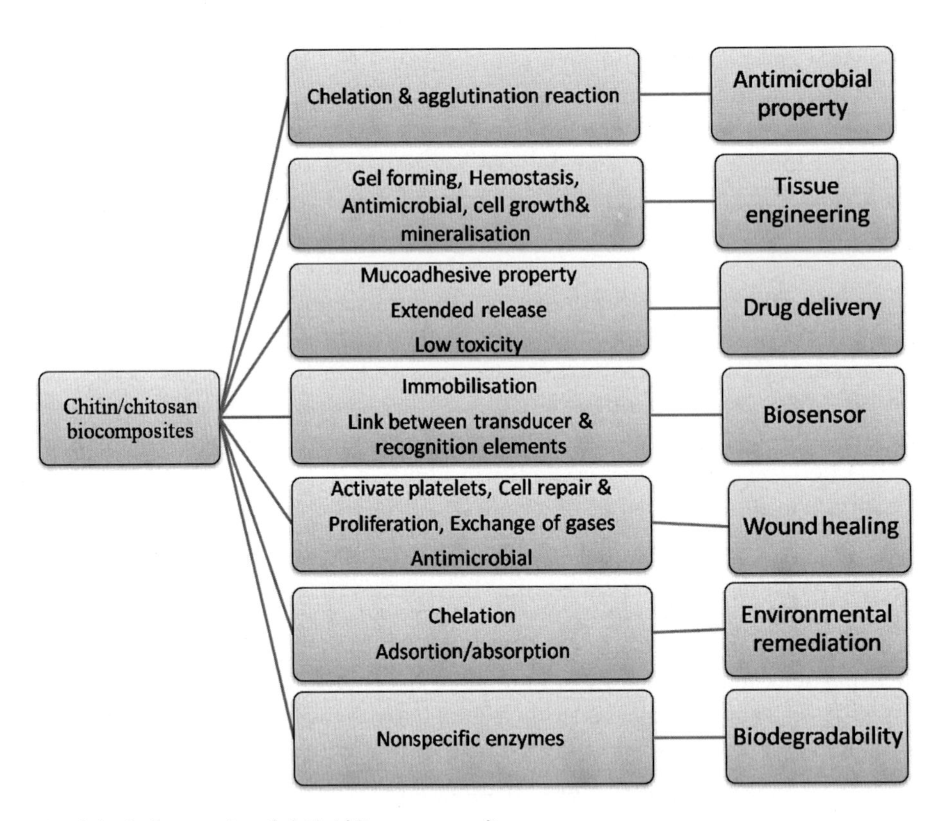

FIGURE 4.2 Biological properties of chitin/chitosan composite.

functional groups and microbes [26]. Not only the modification but the positions of positive charge on the surface play a key role in antimicrobial activity [65,220,240].

Several workers modified chitosan or prepared chitin/chitosan derivatives such as quaternary chitosan salts, alkyl chitosan, aryl chitosan, carboxymethyl chitosan which showed efficacy against Gram-positive and Gram-negative bacteria [177,178,220,240].

Many natural substances such as nisin, essential oils, lysozyme, citric acid, propionic acid, etc. are added to form chitin/chitosan biocomposites [62,226] with enhanced antimicrobial properties. A chitosan-propolis nanoformulation developed by Teik Hwa Ong et al. [149] demonstrated antimicrobial properties against *Enterococcus faecalis*. The chitosan/propolis film prepared by Siripatrawan and Vitchayakitti [196] showed enhanced antimicrobial activity against *Escherichia coli, Staphylococcus aureus, Pseudomonas aeruginosa*, and *Salmonella enteritidis*. Recently many chitin/chitosan nanocomposites developed with enhanced antimicrobial properties were chitosan/ginger essential oil film [169], chitosan/ellagic acid biocomposite films [221], chitin-chitosan nanofiber [179], hyaluronan with chitin nanowhiskers nanocomposite film [9], gelatin-chitin nanocomposite films [181], chitin nanowhiskers/ZnO/Silver nanocomposite film [6], and CMC/chitin nanocrystal/GSE composite films [7]. The chitin/chitosan composites showed a wide range of biological properties and it is depicted in Fig. 4.2.

4.3.10 Tissue engineering

Tissue engineering combines the methods of engineering and life sciences by manipulating/combining the cells to develop suitable substitutes for implantation. It restores, maintains, or improves the function of the tissues.

In tissue engineering applications, chitosan and its derivatives are a preferred material because of their properties such as biodegradability, biocompatibility, antimicrobial activity, porous nature, gel forming capabilities can undergo chemical modification easily and at a low cost and low toxicity. A large number of chitosan derivatives are developed as scaffold to regenerate tissues without any adverse foreign reaction and also maintain hemostasis [63,145]. These chitosan-derived composites are used in a variety of tissue engineering applications such as skin, bone, cartilage, blood, and nerve.

In bone engineering, the addition of chitosan and its derivatives to the composite stimulates the cell growth and also their mineralization. It also eases the inflammation at the site and promotes osteoconduction [51].

The composite chitosan/glycerophosphate with hydroxyethylcellulose was fabricated and used for cartilage regeneration and myocardial-infracted heart [243]. Shalumon et al. prepared and optimized the water-soluble carboxymethyl chitin/PVA composite by varying the concentration of carboxymethyl chitin and PVA [190]. This composite showed the property of cell attachment and proliferation. Similarly poly(L-lactic acid)/chitosan hybrid composite was constructed for bone tissue engineering applications [126]. The other composites used for bone tissue engineering are the chitosan/PVA/HA [81], chitosan–pluronic hydrogel [156,205], amino group–modified nano-HA/chitosan composite scaffolds [22], mPEG-g-chitosan [112], etc.

A high-strength hydrogel made up of HA/GO/chitosan composite was prepared with good biocompatibility and porosity for bone tissue engineering [253]. The composite chitosan-gelatin-bioactive glass ceramic scaffolds were developed with increased mineralization and protein absorption property and also showed a decrease in the swelling and degradable behavior [160]. Another composite chitin hydrogel/chitosan/nanodiopside scaffolds were fabricated for tissue engineering application [135].

Chitosan plays an important role in the scaffolds by changing the surface properties of the materials for the attachment of osteoblasts [109]. Gallium-apatite/chitin/pectin (Ga-HA/C/P) nanocomposites fabricated for bone tissue engineering applications showed increased swelling ability, lower biodegradation rate, and higher mechanical strength [244].

4.3.11 Drug delivery

Nowadays numerous substances obtained from natural source possess very good therapeutic values that can be used to treat various diseases such as diabetes, inflammation, skin diseases, cardiovascular, and microbial infection. The toxicity and biocompatibility are the issues of great concern for them to be used as medicine. Among the natural compounds, biopolymers play a prominent role in medical field. Biopolymeric materials display excellent drug delivery characteristics when the diameter is within 10–1000 nm [30]. Many nanocomposites are fabricated using various synthetic polymers (polyvinyl alcohol, poly-L-lactic acid, polyethylene glycol, and poly(lactic-co-glycolic acid); and natural polymers cellulose, alginate and chitosan) due to their biocompatible and biodegradable properties [34,183,204,229]. The theranostic property of the nanoparticles in the composite can be used to identify the location of the problem, diagnose the disease, identify the stage of the disease, and also give the information about the treatment response.

Chitosan, deacetylation product of chitin, possesses various characteristic properties due to the presence its functional groups [38,203,250]. It is used in the detection and diagnosis of many diseases. Lee et al. [110] loaded iron oxide nanoparticles into oleic acid–conjugated chitosan to examine the tumor cells. Cy5.5 dye was used for imaging these conjugated polymeric nanoparticles. The imaging showed the detectable signal intensity and enhancement in tumor tissues. This nanoparticles' conjugate acts as a valuable probe to detect tumors *in vivo*.

The chitosan biopolymers demonstrate mucoadhesive properties so they are used widely for drug delivery applications mainly for epithelial cells in mouth [162], eye [52], intestine [21], and lungs [15].

Silva et al. [194] evaluated the delivery of the antibiotic ceftazidime to the eye using chitosan/sodium TPP/hyaluronic acid nanoparticles in 0.75% w/w isotonic hydroxypropyl methylcellulose solution. The nanoparticles show good interaction with the mucosal layer of the eye and also enhance the life span of the antibiotic by extending it release.

To release intranasal carbamazepine, carboxymethyl chitosan nanoparticles were prepared to bypass the blood-brain barrier. Due to this, there is a less systemic exposure to drug [119]. Jain and Jain [90] examined the release profile of 5-flurouracil drug from hyaluronic acid–coated chitosan nanoparticles through oral administration. This drug relatively in high concentration even with prolonged exposure causes enhanced efficacy and low toxicity for the treatment of colon cancer.

Wang et al. [227] designed hyaluronic acid–coated chitosan nanoparticles by ion gelation method and increased the efficiency of antitumor effect. In this composite, there occurs a strong interaction between the anionic carboxyl group of hyaluronic acid and the cationic group of chitosan.

Peng et al. [158] prepared a hydrogel by blending PVA solution with CHW for drug delivery applications. In this gel, cross-linking takes place between each constituent so that it provides a large surface area for loading of the drug.

Sundaram et al. [201] prepared a tGNPs-chitin-fibrinogen gel with antibacterial as well as hemostatic property. The antibiotic ciprofloxacin-loaded carboxymethyl chitosan NPs are effective against *E. coli* pathogens compared to antibiotic alone [260].

Panonnummal et al. [153] analyzed the efficacy of methotrexate-loaded chitin nanogel against psoriasis and its distribution compared to methotrexate tablet.

Paula et al. [197] constructed two delivery systems, doxorubicin (DOX)-loaded chitosan and DOX-loaded O-HTCC (ammonium-quaternary derivative of chitosan) nanoparticles, by ionotropic gelation method. O-HTCC nanoparticles release a high amount of DOX than chitosan nanoparticles alone.

Modified carboxymethyl chitosan nanoparticles were developed by Cheng et al. [39] for effective intracellular release of drug and also to facilitate its permeability into the cell membrane. Another nanocomposite made up of trimethyl chitosan-fucoidan was also used for the oral delivery of insulin as it possesses excellent mucoadhesive and absorption-enhancing properties apart from hypoglycemic effect [211]. Nair [142] developed chitosan-loaded curcumin nanoparticles for effective transdermal delivery of drug. These nanoparticles showed entrapment efficiency of more than 80% and also exhibit enhanced permeation, drug release and cell viability. The only problem associated with chitosan in drug delivery process is its poor solubility under physiological conditions [29,261].

A water-soluble glycol chitosan (GC) has functional groups for both encapsulation of drugs and chemical modifications. Recently many GC-based materials were used widely for diagnosing and treating diseases such as cancer and microbial infection apart from cell-imaging applications [114].

4.3.12 Biosensing property

Chitin and chitosan are widely used to make many advanced biomaterials. The potential of these polymers to undergo surface modification helps to extend its application in sensing and diagnostic devices. Nowadays it forms a prominent material in biosensor applications.

Biosensors are the devices that convert biological reactions into detectable signals for applications in clinical diagnosis, food industry, and environment surveillance. The chitosan is mostly sought biopolymer to design biosensors as it has the functional groups to be attached to the transducer molecule after a chemical modification. This thin film on the transducer helps to fix the substances such as enzymes, cells, DNA, organelles, proteins, etc. on its surface. These substances are called biological recognition elements that show high specificity and affinity for their substrates. Moreover, the addition of substances, such as CNTs, graphite, etc., enhances the mechanical strength as well as electron transfers to the transducer.

The detection of glucose is very important for the diagnosis of diabetes. For the detection of glucose molecules, an array of biosensors created with chitosan molecules were reported. A biosensor glucose oxidase immobilized on the chitosan-CNTs composite was prepared to detect glucose levels [252]. Another biosensor constructed for glucose detection was chitosan-gold nanoparticles with glucose oxidase on a platinum electrode [237]. Various other biosensors constructed for glucose detection are Pt electrode-glucose oxidase-Fe_3O_4 nanoparticles on a chitosan-nafion composite [248], chitosan-graphite composite with Fe_3O_4 nanoparticles on a Pt-coated indium tin oxide (ITO) glass electrode [258], glucose oxidase immobilized on chitosan and on which polypyrrole-nafion and MWCNTs were attached [193].

Apart from glucose oxidase, biosensors are constructed for other oxidase enzymes. For example, a lactate biosensor was constructed using lactate oxidase and chitosan-polyvinylimidazole-Os-CNTs composites [45]. Likewise for glutamate and hypoxanthine biosensors, chitosan/GO–polymerized riboflavin using glutamate and xanthine oxidases were created [35]. In addition, xanthine oxidase immobilized on chitosan-polypyrrole-gold nanoparticles was constructed for xanthine biosensor [54].

A simple construct of chitosan-single-walled CNTs with immobilized galactose oxidase was developed for galactose biosensor [210]. The Pt nanoparticles deposited on a multiwalled chitosan-CNTs composite with immobilized cholesterol oxidase was used for detecting cholesterol [212]. A chitosan/titanate nanotubes composite with immobilized choline oxidase was developed to detect choline [47].

The biosensors also constructed for detection of other compounds are immobilized horseradish peroxidase on an alumina nanoparticles-chitosan composite to detect phenolic compounds [120], a chitosan-Prussian blue composite film immobilized with corresponding oxidase to detect glucose, galactose, and glutamate in blood [228], immobilized tyrosinase on a chitosan-nickel nanoparticles film to detect catechol and other phenolic compounds [249], immobilized laccase on a ZnO-chitosan nanocomposite to detect chlorophenol [129], horseradish peroxidase immobilized on a chitin-gelatin nanofiber composite to detect hydrogenperoxide [207], immobilized catalase on chitosan-β-CD to detect hydrogenperoxide [55].

Biosensors are able to detect even trace amounts of metal ions and toxic substances that are of great importance. Various biosensors developed for the detection of metallic ions are chitosan-CNT for the detection of Cd(II) and Hg(II) ions [92], chitosan-α-Fe_3O_4 nanoparticles to detect Ni^{2+}, As^{2+}, and Pb^{2+} ions [13], and organophosphorus hydrolase immobilized on a chitosan-carbon-nanoparticles-HA biosensor to detect pesticides [200].

Chitosan can be readily immobilized with DNA and proteins due to its polycationic nature. A DNA biosensor containing ssDNA of *Salmonella typhi* immobilized on GO/chitosan/ITO nanocomposite to detect typhoid was developed [195]. Likewise DNA biosensor was developed for *E. coli* [241]. Afkhami et al. [12] constructed a gold nanoparticles/chitosan/graphene nanocomposite with immobilized antibodies to measure the bound neurotoxin botulism. Giannetto et al. [68] fabricated an immunosensor to detect the HIV1 capsid protein p24 in human serum by immobilizing p24 on a gold/single-walled carbon nanotubes (SWCNT)/chitosan complex.

4.3.13 Wound healing

Natural polysaccharides, such as alginate, hyaluronic acid, starch, cellulose, chitosan, have been used widely as a wound dressing material due to their inherent biological properties [136]. Normally the wound healing involves four phases: hemostasis, inflammation, cell migration and proliferation, and maturation in any healing process. But sometimes the wound healing is delayed due to diabetes, old age deficiencies, infection, and diseases of blood vessels.

To enhance the wound-healing properties, chitin/chitosan is combined with various natural substance/metals to form composites with excellent healing properties. An ointment of chitosan oligomer with silver nanoparticles was used to treat burn wounds on rats. This showed superior healing properties than pure chitosan alone [97].

The wound dressings' material must possess the characteristic properties such as nontoxicity, biocompatibility, biodegradability, exchange of gases, protection against infection, etc. Chitin/chitosan-based nanocomposites fulfill these properties and behave as excellent biomaterials for wound-healing purpose by activating the platelets and also promote cell repair by adhesion and proliferation [125,159].

Numerous chitosan derivatives or composites used for wound-healing applications are N-carboxybutyl chitosan [73], fluorinated methacrylamide chitosan [232], hydroxybutyl chitosan [131], chitosan/gelatin hydrogel [157], chitosan/sago starch/silver nanoparticles [127], chitosan/heparin/poly(γ-glutamic acid) composite [115], chitosan-fibrin nanocomposite [217], chitin/silk fibroin/TiO_2 bio-nanocomposite [66], pH-sensitive hydroxypropyl chitin/tannic acid/ferric ion hydrogel [132], β-chitin/nanosilver composite [106], chitosan/polycaprolactone scaffolds [111], and chitosan/collagen sponges [209]. As silver nanoparticles exhibit antimicrobial property without showing any résistance, it is mostly incorporated in the chitin/chitosan-based nanocomposite for wound-healing process.

4.3.14 Environment remediation

The treatment of the wastewater treatment is one of the important serious issues of each and every developing country. The common methods employed to remove heavy metals from wastewater are chemical precipitation, coagulation/flocculation and solvent extraction, ion exchange, filtration, adsorption, reverse osmosis, electrochemical treatment, and evaporative recovery [27,170,255]. Moreover the addition of chemicals to treat the wastewater system also creates new problems other than cost. Thus, an alternative method for conventional way of treatment should be explored. In this regard, bionanocomposites obtained from natural sources created a promising avenue for treating wastewater, pollutants, heavy metals, etc. Bionanocomposites are normally formed by combining a biopolymer with inorganic nanoscale minerals [133]. The composites normally show improved properties than its constituent substance.

Various mechanisms involved in the removal of heavy metals by bionanocomposites are chemisorption, metal complex formation, ion exchange, precipitation, and physical adsorption. The factors that affect the adsorption rate of pollutants are chemical nature of pollutants, nature of composite material, external conditions such as pH, temperature, ionic strength, etc. and the structure of the composite.

It was reported that the natural polymer chitin was used to chelate many heavy metal ions from wastewater such as mercury, copper, chromium, lead, cadmium, cobalt, etc. The chitin hydrogel with silicon dioxide composites were prepared and used as absorbents for dyes (Remazol Black B, Erythrosine B, Neutral Red, and Gentian Violet) [43].

Karthick and Meenakshi [98] prepared a novel chitin–polyaniline composite that is used to remove chromium (VI) ions from contaminated water. The electrostatic adsorption and coupled reduction reaction of polyaniline was responsible for the removal of Cr (VI) ions, which is pH dependent. Tang et al. [206] prepared porous chitin/cellulose membranes that can remove heavy metal ions such as mercury, lead, and copper.

The amino and hydroxyl groups of chitosan are responsible for the removal of heavy metals and dyes from contaminated water. The properties of chitosan is improved either by a chemical modification or by cross-linking with formaldehyde, glutaraldehyde, etc., or by incorporation of nanoparticles, clay, etc. [222].

Gandhi et al. [167] prepared chitin/chitosan n-HA composites to remove Cu(II) by an ion exchange mechanism. Habiba et al. [74] fabricated a electrospun chitosan/PVA/zeolite composite for the removal of Cr(VI), Fe(III), and Ni(II) ions.

A chitin/magnetite/MWCNTs composite showed a higher surface area and significant magnetic properties, which helps to remove Cr(VI) ions from the solution [180].

4.3.15 Biodegradability

Biodegradation is the normal process of breaking the high-molecular-weight substance into low-molecular-weight one with the help of microorganisms. Chitosan does not exist in nature but is extracted from chitin after deacetylation process. Only after deacetylation of about 60%, chitosan is obtained. The degradation products of chitin and chitosan are not toxic to humans [25]. They can be degraded by a nonspecific enzymatic method using proteolytic enzyme lysozyme. The degradation of chitin/chitosan-based products depends on various factors such as molecular weight, pH, DD, and the preparation of the sample.

Recently many researchers have attempted to find a suitable alternative to fossil fuels with sustainable natural polymers such as cellulose, chitin, alginate, and gelatin [121] to protect the environment from adverse effects of synthetic chemicals. Chitin/chitosan biopolymer possesses excellent properties that are superior to properties of petrochemical-derived polymers. As its applications are wide spread in the medical field, it is important to understand the property of degradability. Moreover the biocompatibility of chitosan-based composites in physiological medium was proved by many studies [186,214].

The biodegradation of chitosan forms small chains of nontoxic oligosaccharide molecules. These oligosaccharides can either enter metabolic pathways or excreted [108]. There is an interconnection between rate of degradation, molecular weight, DD, and also the amount of N-acetyl D-glucosamine residues as per Zhang et al. [259]. There is also a relationship between DD, degradation rate, and crystallinity. When DD is maximum, crystallinity is minimum and vice versa. Therefore the rate of degradation increases with decrease in crystallinity, when DD of chitosan is 60%.

Many proteases such as leucine aminopeptidases, pectinase isozyme, porcine pancreases, etc. are used to degrade chitosan composites [103,104]. Recently researchers have created a lot of chitin/chitosan-based composites such as PLA/chitin nano-fibrils [41], chitin/HA composite [130], cellulose-chitin polymeric composite [161], etc. with a biodegradable property along with other applications.

4.4 Conclusion

The chitin and chitosan, the main byproduct of the marine waste, have shown an excellent array of properties such as biocompatibility, biodegradability, and nontoxicity. This unique biopolymer has the ability to undergo modifications chemically to form different types of material with a wide range of applications. This versatile polymer can be made into different forms such as films, sponge, scaffold, hydrogel, nanocrystals, fibers, etc. The physicochemical properties showed a significant improvement when combined with either organic or inorganic fillers. Moreover the chitin/chitosan nanocomposites formed was responsible for removing the limitations of pure chitin/chitosan. This chapter highlighted the changes brought about when chitin/chitosan nanocomposites formed.

The formation of chitin/chitosan nanocomposites increased the mechanical, thermal, barrier, and mucoadhesive property. The remarkable enhancement of the properties made them to expand its applications to diverse fields starting from wound healing to biosensors. Due to its vast applications, it has created lot of interest in the mind of researchers to explore various possibilities or combinations with other organic/inorganic fillers to create a promising material that can be used for industrial production.

References

[1] J. Shaji, V. Jain, S. LodhaChitosan, A novel pharmaceutical excipient, Int. J. Pharm. App. Sci. 1 (2010) 11–28.

[2] R. Kumar, R.K. Singh, P.K. Dubey, I.K. Oh, Review on functionalized graphenes and their applications, Smart Nanosyst. Eng. Med. (2012) 18–39.

[3] S.A. Hashmi, A. Chandra, R.K. Singh, A. Chandra, S. Chandra, Electroactive polymers:, Materials and Devices, Allied Publishers (2015).

[4] S.K.H. Gulrez, S. Al-Assaf, G.O. Phillips, Hydrogels: methods of preparation, characterisation and application, In: A. Carpi (Ed.), Progress in Molecular and Environmental Bioengineering—From Analysis and Modeling to Technology Applications, Intechopen, London, 2011, pp. 117–150.

[5] O.A. A., R. Jong-Whan, Preparation of multifunctional carboxymethyl cellulose-based films incorporated with chitin nanocrystal and grapefruit seed extract, Int. J. Biol. Macromol. (2020) 1038–1046. https://doi.org/10.1016/j.ijbiomac.2019.10.191.

[6] O.A. A., R. Jong-Whan, Preparation of multifunctional chitin nanowhiskers/ZnO-Ag NPs and their effect on the properties of carboxymethyl cellulose-based nanocomposite film, Carbohydr. Polym. (2017) 467–479. https://doi.org/10.1016/j.carbpol.2017.04.042.

[7] O.A. A., R. Jong-Whan, Preparation of multifunctional carboxymethyl cellulose-based films incorporated with chitin nanocrystal and grapefruit seed extract, Int. J. Biol. Macromol. (2020) 1038–1046. https://doi.org/10.1016/j.ijbiomac.2019.10.191.

[8] M. Abbasian, F. Mahmoodzadeh, Synthesis of antibacterial silver-chitosan-modified bionanocomposites by RAFT polymerization and chemical reduction methods, J. Elastomers. Plast. 49 (2017) 173–193. https://doi.org/10.1177/0095244316644858.

[9] R.M. Abdelrahman, A.M. Abdel-Mohsen, M. Zboncak, J. Frankova, P. Lepcio, L. Kobera, M. Steinhart, D. Pavlinak, Z. Spotaz, R. Sklenářévá, J. Brus, J. Jancar, Hyaluronan biofilms reinforced with partially deacetylated chitin nanowhiskers: extraction, fabrication, in-vitro and antibacterial properties of advanced nanocomposites, Carbohydr. Polym. 235 (2020). https://doi.org/10.1016/j.carbpol.2020.115951.

[10] M. Abdollahi, M. Rezaei, G. Farzi, A novel active bionanocomposite film incorporating rosemary essential oil and nanoclay into chitosan, J. Food Eng. 111 (2012) 343–350. https://doi.org/10.1016/j.jfoodeng.2012.02.012.

[11] S. Abdolmohammadi, W.M.Z.W. Yunus, M.Z.A. Rahman, N. Azowa Ibrahim, Effect of organoclay on mechanical and thermal properties of polycaprolactone/chitosan/montmorillonite nanocomposites, J. Reinforc. Plast. Compos. 30 (2011) 1045–1054. https://doi.org/10.1177/0731684411410338.

[12] P. Hashemi Afkhami, H. Bagheri, J. Salimian, A. Ahmadi, T. Madrakian, Impedimetric immunosensor for the label-free and direct detection of botulinum neurotoxin serotype A using Au nanoparticles/graphene-chitosan composite, Biosensors Bioelectron 93 (2017) 124–131. https://doi.org/10.1016/j.bios.2016.09.059.

[13] R.A. Ahmed, A.M. Fekry, Preparation and characterization of a nanoparticles modified chitosan sensor and its application for the determination of heavy metals from different aqueous media, Int. J. Electrochem. Sci. 8 (2013) 6692–6708. http://www.electrochemsci.org/papers/vol8/80506692.pdf.

[14] F. Aklog Yihun, H.Saimoto S. Ifuku, H. Izawa, M. Morimoto, Highly transparent and flexible surface modified chitin nanofibers reinforced poly (methyl methacrylate) nanocomposites: mechanical, thermal and optical studies, Polymer. 197 (2020) 122497.

[15] S. Al-Qadi, A. Grenha, D. Carrión-Recio, B. Seijo, C. Remuñán-López, Microencapsulated chitosan nanoparticles for pulmonary protein delivery: in vivo evaluation of insulin-loaded formulations, J. Control. Release. 157 (2012) 383–390. https://doi.org/10.1016/j.jconrel.2011.08.008.

[16] F. Al-Sagheer, S. Muslim, Thermal and mechanical properties of chitosan/SiO_2 hybrid composites, J. Nanomater. 2010 (2010). https://doi.org/10.1155/2010/490679.

[17] M.J. Allen, V.C. Tung, R.B. Kaner, Honeycomb carbon: a review of graphene, Chem. Rev. 110 (2009) 132–145.

[18] P. Anaya, G. Cárdenas, V. Lavayen, A. García, C. O'Dwyer, Chitosan gel film bandages: correlating structure, composition, and antimicrobial properties, J Appl. Polym. Sci. 128 (2013) 3939–3948. https://doi.org/10.1002/app.38621.

[19] M.Mengíbar Aranaz, R. Harris, I. Paños, B. Miralles, N. Acosta, G. Galed, A. Heras, Functional characterization of chitin and chitosan, Curr. Chem. Biol. 3 (2009) 203–230. https://doi.org/10.2174/187231309788166415.

[20] R. Harris Aranaz, A. Heras, Chitosan amphiphilic derivatives. Chemistry and applications, Curr. Organ. Chem. 14 (2010) 308–330. https://doi.org/10.2174/138527210790231919.

[21] P. Artursson, T. Lindmark, S.S. Davis, L. Illum, Effect of chitosan on the permeability of monolayers of intestinal epithelial cells (Caco-2), Pharma. Res. 11 (1994) 1358–1361. https://doi.org/10.1023/A:1018967116988.

[22] B.H. Atak, B. Buyuk, M. Huysal, S. Isik, M. Senel, W. Metzger, G. Cetin, Preparation and characterization of amine functional nano-hydroxyapatite/chitosan bionanocomposite for bone tissue engineering applications, Carbohydr. Polym. 164 (2017) 200–213. https://doi.org/10.1016/j.carbpol.2017.01.100.

[23] R. Auzély, M. Rinaudo, Controlled chemical modifications of chitosan. characterization and investigation of original properties, Macromol. Biosci. 3 (2003) 562–565. https://doi.org/10.1002/mabi.200300018.

[24] L. Attoso Azerdo, R. AvenaBustillos, M. Munford, D. Wood, TH. McHugh, Nanocellulose reinforced chitosan composite films as affected by nanofiller loading and plasticizer content, J. Food Sci. 75 (2010) 1–7.

[25] S. Bagheri-Khoulenjani, S.M. Taghizadeh, H. Mirzadeh, An investigation on the short-term biodegradability of chitosan with various molecular weights and degrees of deacetylation, Carbohydr. Polym. 78 (2009) 773–778. https://doi.org/10.1016/j.carbpol.2009.06.020.

[26] P.S. Bakshi, D. Selvakumar, K. Kadirvelu, N.S. Kumar, Comparative study on antimicrobial activity and biocompatibility of N-selective chitosan derivatives, Reactive Funct. Polym. 124 (2018) 149–155. https://doi.org/10.1016/j.reactfunctpolym.2018.01.016.

[27] Y. Benito, M.L. Ruíz, Reverse osmosis applied to metal finishing wastewater, Desalination 142 (2002) 229–234. https://doi.org/10.1016/S0011-9164(02)00204-7.

[28] J. Berger, M. Reist, J.M. Mayer, O. Felt, N.A. Peppas, R. Gurny, Structure and interactions in covalently and ionically crosslinked chitosan hydrogels for biomedical applications, Eur. J. Pharma. Biopharma 57 (2004) 19–34. https://doi.org/10.1016/S0939-6411(03)00161-9.

[29] N.P. Birch, J.D. Schiffman, Characterization of self-Assembled polyelectrolyte complex nanoparticles formed from chitosan and pectin, Langmuir 30 (2014) 3441–3447. https://doi.org/10.1021/la500491c.

[30] B. Bonifácio, S.Ramos Silva, M. Negri, K. Bauab, T. Chorilli, Nanotechnology-based drug delivery systems and herbal medicines: a review, Int. J. Nanomed. 9 (2014).

[31] S. Bruno, G.F. M., S. Alejandro, das N. José, Chitosan and chitosan derivatives for biological applications: chemistry and functionalization, Int. J. Carbohydr. Chem. (2011) Article ID 802693. https://doi.org/10.1155/2011/802693 .

[32] G.R. C., B.D. de, A.O.B G., A review of the antimicrobial activity of chitosan, Polímeros (2009) 241–247. https://doi.org/10.1590/s0104-14282009000300013.

[33] W. Cao, D. Jing, J. Li, Y. Gong, N. Zhao, X. Zhang, Effects of the degree of deacetylation on the physicochemical properties and Schwann cell affinity of chitosan films, J. Biomater. Appl. 20 (2005) 157–177. https://doi.org/10.1177/0885328205049897.

[34] L. Casettari, L. Illum, Chitosan in nasal delivery systems for therapeutic drugs, J. Control. Release 190 (2014) 189–200. https://doi.org/10.1016/j.jconrel.2014.05.003.

[35] R. Celiesiute, A. Radzevic, A. Zukauskas, S. Vaitekonis, Pauliukaite, A strategy to employ polymerised riboflavin in the development of electrochemical biosensors, Electroanalysis 29 (2017) 2071–2082.

[36] P.R. Chang, R. Jian, J. Yu, X. Ma, Fabrication and characterisation of chitosan nanoparticles/plasticised-starch composites, Food Chem. 120 (2010) 736–740. https://doi.org/10.1016/j.foodchem.2009.11.002.

[37] C. Chatelet, O. Damour, A. Domard, Influence of the degree of acetylation on some biological properties of chitosan films, Biomaterials 22 (2001) 261–268. https://doi.org/10.1016/S0142-9612(00)00183-6.

[38] K. Chen, X. Chen, Design and development of molecular imaging probes, Curr. Top. Med. Chem. 10 (2010) 1227–1236. https://doi.org/10.2174/156802610791384225.

[39] X. Cheng, X. Zeng, Y. Zheng, X. Wang, R. Tang, Surface-fluorinated and pH-sensitive carboxymethyl chitosan nanoparticles to overcome biological barriers for improved drug delivery in vivo, Carbohydr. Polym. 208 (2019) 59–69. https://doi.org/10.1016/j.carbpol.2018.12.063.

[40] D. Ciechańska, J. Wietecha, M. Kucharska, K. Wrzeoeniewska-Tosik, E. Kopania, Biomasa jako Ÿródo funkcjonalnych materiaów polimerowych, Polimery/Polymers 59 (2014) 383–392. https://doi.org/10.14314/polimery.2014.383.

[41] P. Cinellia, M.B. Coltellia, N. Mallegnia, P. Morgantib, A. Lazzeria, Degradability and sustainability of nanocomposites based on polylactic acid and chitin nano fibrils, Chem. Eng. Trans. 60 (2017) 115–120.

[42] M. Cobos, B. González, M. Jesús Fernández, M. Dolores Fernández, Chitosan-graphene oxide nanocomposites: Effect of graphene oxide nanosheets and glycerol plasticizer on thermal and mechanical properties. J. Appl. Polym. Sci. 134 (2017) 1–14.

[43] G.J. Copello, A.M. Mebert, M. Raineri, M.P. Pesenti, L.E. Diaz, Removal of dyes from water using chitosan hydrogel/SiO_2 and chitin hydrogel/SiO_2 hybrid materials obtained by the sol-gel method, J. Hazard. Mater. 186 (2011) 932–939. https://doi.org/10.1016/j.jhazmat.2010.11.097.

[44] E.S. Costa-Júnior, E.F. Barbosa-Stancioli, A.A.P. Mansur, W.L. Vasconcelos, H.S. Mansur, Preparation and characterization of chitosan/poly(vinyl alcohol) chemically crosslinked blends for biomedical applications, Carbohydr. Polym. 76 (2009) 472–481. https://doi.org/10.1016/j.carbpol.2008.11.015.

[45] X. Cui, C.M. Li, J.F. Zang, S. Yu, Highly sensitive lactate biosensor by engineering chitosan/PVI-Os/CNT/LOD network nanocomposite, Biosens. Bioelectron. 22 (12) (2007) 3288–3292. https://doi.org/10.3109/02713683.2016.1145235.

[46] K.K. D., B.A. H., K. Mohammed, Z.A. A., Synthesis, structural, dielectric and optical properties of chitosan-MgO nanocomposite, J. Taibah. Univ. Sci. (2020) 975–983. https://doi.org/10.1080/16583655.2020.1792117.

[47] H. Dai, Y. Chi, X. Wu, Y. Wang, M. Wei, G. Chen, Biocompatible electrochemiluminescent biosensor for choline based on enzyme/titanate nanotubes/chitosan composite modified electrode, Biosensor Bioelectron 25 (2010) 1414–1419. https://doi.org/10.1016/j.bios.2009.10.042.

[48] Y. Daimon, H. Izawa, K. Kawakami, P. Zywicki, H. Sakai, M. Abe, J.P. Hill, K. Ariga, Media-dependent morphology of supramolecular aggregates of β-cyclodextrin-grafted chitosan and insulin through multivalent interactions, J. Mater. Chem. B. 2 (2014) 1802–1812. https://doi.org/10.1039/c3tb21528h.

[49] Y. Daimon, N. Kamei, K. Kawakami, M. Takeda-Morishita, H. Izawa, Y. Takechi-Haraya, H. Saito, H. Sakai, M. Abe, K. Ariga, Dependence of intestinal absorption profile of insulin on carrier morphology composed of β-cyclodextrin-grafted chitosan, Mol. Pharma. 13 (2016) 4034–4042. https://doi.org/10.1021/acs.molpharmaceut.6b00561.

[50] M. Darder, M. Colilla, E. Ruiz-Hitzky, Chitosan-clay nanocomposites: application as electrochemical sensors, Appl. Clay. Sci. 28 (2005) 199–208. https://doi.org/10.1016/j.clay.2004.02.009.

[51] M. Dash, F. Chiellini, R.M. Ottenbrite, E. Chiellini, Chitosan—a versatile semi-synthetic polymer in biomedical applications, Prog. Polym. Sci. 36 (2011) 981–1014. https://doi.org/10.1016/j.progpolymsci.2011.02.001.

[52] A.M. De Campos, A. Sánchez, M.J. Alonso, Chitosan nanoparticles: a new vehicle for the improvement of the delivery of drugs to the ocular surface. Application to cyclosporin A, Int. J. Pharma. 224 (2001) 159–168. https://doi.org/10.1016/S0378-5173(01)00760-8.

[53] R.T. De Silva, M.M.M.G.P.G. Mantilaka, S.P. Ratnayake, G.A.J. Amaratunga, K.M.N. de Silva, Nano-MgO reinforced chitosan nanocomposites for high performance packaging applications with improved mechanical, thermal and barrier properties, Carbohydr. Polym. 157 (2017) 739–747. https://doi.org/10.1016/j.carbpol.2016.10.038.

[54] M. Dervisevic, E. Dervisevic, E. Çevik, M. Şenel, Novel electrochemical xanthine biosensor based on chitosan–polypyrrole–gold nanoparticles hybrid bio-nanocomposite platform, J. Food Drug Anal. 25 (2017) 510–519. https://doi.org/10.1016/j.jfda.2016.12.005.

[55] W. Dong, K. Wang, Y. Chen, W. Li, Y. Ye, S. Jin, Construction and characterization of a chitosan-immobilized-enzyme and β-cyclodextrin-included-ferrocene-based electrochemical biosensor for H_2O_2 detection, Materials 10 (2017). https://doi.org/10.3390/ma10080868.

[56] T.V. Duncan, Applications of nanotechnology in food packaging and food safety: barrier materials, antimicrobials and sensors, J. Coll. Interf. Sci. 363 (2011) 1–24. https://doi.org/10.1016/j.jcis.2011.07.017.

[57] F.M. E., A. Anthony, V.D.M. E., Smart biomaterials design for tissue engineering and regenerative medicine, Biomaterials (2007) 5068–5073. https://doi.org/10.1016/j.biomaterials.2007.07.042.

[58] E.A. El-Hefian, E.S. Elgannoudi, A. Mainal, A.H. Yahaya, Characterization of chitosan in acetic acid: rheological and thermal studies, Turk. J. Chem. 34 (2010) 47–56. https://doi.org/10.3906/kim-0901-38.

[59] J. Fan, Z. Shi, Y. Ge, Y. Wang, J. Wang, J. Yin, Mechanical reinforcement of chitosan using unzipped multiwalled carbon nanotube oxides, Polymer. 53 (2012) 657–664. https://doi.org/10.1016/j.polymer.2011.11.060.

[60] L. Fan, C. Luo, M. Sun, X. Li, F. Lu, H. Qiu, Preparation of novel magnetic chitosan/graphene oxide composite as effective adsorbents toward methylene blue, Bioresour. Technol. 114 (2012) 703–706. https://doi.org/10.1016/j.biortech.2012.02.067.

[61] L. Feng, Z. Zhou, A. Dufresne, J. Huang, M. Wei, L. An, Structure and properties of new thermoforming bionanocomposites based ort chitin whisker-graft-polycaprolactone, J. Appl. Polym. Sci. 112 (2009) 2830–2837. https://doi.org/10.1002/app.29731.

[62] G.C. Feyzioglu, F. Tornuk, Development of chitosan nanoparticles loaded with summer savory (*Satureja hortensis* L.) essential oil for antimicrobial and antioxidant delivery applications, LWT Food Sci. Technol. 70 (2016) 104–110. https://doi.org/10.1016/j.lwt.2016.02.037.

[63] J.K. Francis Suh, H.W.T. Matthew, Application of chitosan-based polysaccharide biomaterials in cartilage tissue engineering: a review, Biomaterials 21 (2000) 2589–2598. https://doi.org/10.1016/S0142-9612(00)00126-5.

[64] T. Freier, H.S. Koh, K. Kazazian, M.S. Shoichet, Controlling cell adhesion and degradation of chitosan films by N-acetylation, Biomaterials 26 (2005) 5872–5878. https://doi.org/10.1016/j.biomaterials.2005.02.033.

[65] X. Fu, Y. Shen, X. Jiang, D. Huang, Y. Yan, Chitosan derivatives with dual-antibacterial functional groups for antimicrobial finishing of cotton fabrics, Carbohydr. Polym. 85 (2011) 221–227. https://doi.org/10.1016/j.carbpol.2011.02.019.

[66] M.M. Ghanbari, K. Ramin, R. Rasul, P. Farzaneh, E. Hosein, F. Vahid, R. Mahdi, S. Roya, K.H. Samadi, Chitin/silk fibroin/TiO_2 bio-nanocomposite as a biocompatible wound dressing bandage with strong antimicrobial activity, Int. J. Biol. Macromol. (2018) 966–976. https://doi.org/10.1016/j.ijbiomac.2018.05.102.

[67] B. Ghosn, S.P. Kasturi, K. Roy, Enhancing polysaccharide-mediated delivery of nucleic acids through functionalization with secondary and tertiary amines, Curr. Top. Med. Chem. 8 (2008) 331–340. https://doi.org/10.2174/156802608783790947.

[68] M. Giannetto, M. Costantini, M. Mattarozzi, M. Careri, Innovative gold-free carbon nanotube/chitosan-based competitive immunosensor for determination of HIV-related p24 capsid protein in serum, RSC Adv. 7 (2017) 39970–39976. https://doi.org/10.1039/c7ra07245g.

[69] K. Gopalan Nair, A. Dufresne, Crab shell chitin whisker reinforced natural rubber nanocomposites. 1. Processing and swelling behavior, Biomacromolecules 4 (2003) 657–665. https://doi.org/10.1021/bm020127b.

[70] K. Gopalan Nair, A. Dufresne, Crab shell chitin whisker reinforced natural rubber nanocomposites. 2. Mechanical behavior, Biomacromolecules 4 (2003) 666–674. https://doi.org/10.1021/bm0201284.

[71] V. Goudarzi, I. Shahabi-Ghahfarrokhi, A. Babaei-Ghazvini, Preparation of ecofriendly UV-protective food packaging material by starch/TiO_2 bio-nanocomposite: characterization, Int. J. Biol. Macromol. 95 (2017) 306–313. https://doi.org/10.1016/j.ijbiomac.2016.11.065.

[72] R. Goy deBritto, S. Campana-Filho, O. Assis, Quaternary salts of chitosan: history, antimicrobial features and prospects, Int. J. Carbohyd. Chem. (2011).

[73] B. Graziella, B. Aldo, M. Riccardo, D. Andrea, D. Giovanni, B. Antonella, R. Giuseppe, Z. Cinzia, R. Carlo, Wound management with N-carboxybutyl chitosan, Biomaterials (1991) 281–286. https://doi.org/10.1016/0142-9612(91)90035-9.

[74] U. Habiba, A.M. Afifi, A. Salleh, B.C. Ang, Chitosan/(polyvinyl alcohol)/zeolite electrospun composite nanofibrous membrane for adsorption of $Cr6^{2+}$, $Fe3^{2+}$ and $Ni2^{2+}$, J. Hazard. Mater. 322 (2017) 182–194. https://doi.org/10.1016/j.jhazmat.2016.06.028.

[75] Y. Habibi, L.A. Lucia, Polysaccharide Building Blocks: A Sustainable Approach to the Development of Renewable Biomaterials, John Wiley and Sons, NJ, United States, 2012. https://doi.org/10.1002/9781118229484.

[76] K.V. Harish Prashanth, F.S. Kittur, R.N. Tharanathan, Solid state structure of chitosan prepared under different N-deacetylating conditions, Carbohydr. Polym. 50 (2002) 27–33. https://doi.org/10.1016/S0144-8617(01)00371-X.

[77] M. Hassannia-Kolaee, F. Khodaiyan, R. Pourahmad, I. Shahabi-Ghahfarrokhi, Development of ecofriendly bionanocomposite: whey protein isolate/pullulan films with nano-SiO_2, Int. J. Biol. Macromol. 86 (2016) 139–144. https://doi.org/10.1016/j.ijbiomac.2016.01.032.

[78] M. Hassannia-Kolaee, F. Khodaiyan, I. Shahabi-Ghahfarrokhi, Modification of functional properties of pullulan–whey protein bionanocomposite films with nanoclay, J. Food Sci. Technol. 53 (2016) 1294–1302. https://doi.org/10.1007/s13197-015-1778-3.

[79] N. Herrera, A.M. Salaberria, A.P. Mathew, K. Oksman, Plasticized polylactic acid nanocomposite films with cellulose and chitin nanocrystals prepared using extrusion and compression molding with two cooling rates: effects on mechanical, thermal and optical properties, Compos. Pt. A. Appl. Sci. Manuf. 83 (2016) 89–97. https://doi.org/10.1016/j.compositesa.2015.05.024.

[80] Y. Hidaka, I. Michio, K. Mori, H. Yagasaki, A.H. Kafrawy, Histopathological and immunohistochemical studies of membranes of deacetylated chitin derivatives implanted over rat calvaria, J. Biomed. Mater. Res. 46 (1999) 418–423. https://doi.org/10.1002/(SICI)1097-4636(19990905)46:3<418::AID-JBM15>3.0.CO;2-T.

[81] Q. Hu, B. Li, M. Wang, J. Shen, Preparation and characterization of biodegradable chitosan/hydroxyapatite nanocomposite rods via in situ hybridization: a potential material as internal fixation of bone fracture, Biomaterials 25 (2004) 779–785. https://doi.org/10.1016/S0142-9612(03)00582-9.

[82] J. Huang, J.W. Zou, P.R. Chang, J.H. Yu, A. Dufresne, New waterborne polyurethane-based nanocomposites reinforced with low loading levels of chitin whisker, Exp. Polym. Lett. 5 (2011) 362–373. https://doi.org/10.3144/expresspolymlett.2011.35.

[83] G.Q. Huang, X.N. Han, J.X. Xiao, L.Y. Cheng, Effects of coacervation acidity on the genipin crosslinking action and intestine-targeted delivery potency of the O-carboxymethyl chitosan–gum arabic coacervates, Int. J. Polym. Mater. Polym. Biomater. 66 (2017) 89–96. https://doi.org/10.1080/00914037.2016.1190924.

[84] K. Hye-Ran, J. Jun-Won, P. Jae-Woo, Carboxymethyl chitosan-modified magnetic-cored dendrimer as an amphoteric adsorbent, J. Hazard. Mater. (2016) 608–616. https://doi.org/10.1016/j.jhazmat.2016.06.025.

[85] S. Ifuku, M. Nogi, K. Abe, M. Yoshioka, M. Morimoto, H. Saimoto, H. Yano, Preparation of chitin nanofibers with a uniform width as α-chitin from crab shells, Biomacromolecules 10 (2009) 1584–1588. https://doi.org/10.1021/bm900163d.

[86] S. Ifuku, M. Nogi, M. Yoshioka, M. Morimoto, H. Yano, H. Saimoto, Fibrillation of dried chitin into 10-20 nm nanofibers by a simple grinding method under acidic conditions, Carbohydr. Polym. 81 (2010) 134–139. https://doi.org/10.1016/j.carbpol.2010.02.006.

[87] S. Ifuku, M. Nogi, K. Abe, M. Yoshioka, M. Morimoto, H. Saimoto, H. Yano, Simple preparation method of chitin nanofibers with a uniform width of 10-20 nm from prawn shell under neutral conditions, Carbohydr. Polym. 84 (2011) 762–764. https://doi.org/10.1016/j.carbpol.2010.04.039.

[88] V. Jacobs, R. Anandjiwala, M. John, A. Mathew, K. Oksman, Studies on Electrospun Chitosan Based Nanofibres Reinforced with Cellulose and Chitin Nanowhiskers, International Conference on Nanotechnology for the Forest Product Industry 2010, TAPPI, Georgia, 2011, p. 1378.

[89] H.P. Jae, S. Kwon, M. Lee, H. Chung, J.H. Kim, Y.S. Kim, R.W. Park, I.S. Kim, B.S. Sang, I.C. Kwon, Y.J. Seo, Self-assembled nanoparticles based on glycol chitosan bearing hydrophobic moieties as carriers for doxorubicin: In vivo biodistribution and anti-tumor activity, Biomaterials 27 (2006) 119–126. https://doi.org/10.1016/j.biomaterials.2005.05.028.

[90] S.K. Jain Jain, Optimization of chitosan nanoparticles for colon tumors using experimental design methodology,, Artif. Cells Nanomed. Biotechnol. 44 (2016) 1917–1926. https://doi.org/10.3109/21691401.2015.1111236.

[91] M. Jalal Zohuriaan Mehr, Advances in chitin and chitosan modification through graft copolymerization: A comprehensive review, Iran Polym J 14 (2005) 235–265.

[92] B.C. Janegitz, L.C.S. Figueiredo-Filho, L.H. Marcolino-Junior, S.P.N. Souza, E.R. Pereira-Filho, O. Fatibello-Filho, Development of a carbon nanotubes paste electrode modified with crosslinked chitosan for cadmium(II) and mercury(II) determination, J. Electroanal. Chem. 660 (2011) 209–216. https://doi.org/10.1016/j.jelechem.2011.07.001.

[93] M. Jaworska, K. Sakurai, P. Gaudon, E. Guibal, Influence of chitosan characteristics on polymer properties. I: crystallographic properties, Polym. Int. 52 (2003) 198–205. https://doi.org/10.1002/pi.1159.

[94] J. Junkasem, R. Rujiravanit, P. Supaphol, Fabrication of α-chitin whisker-reinforced poly(vinyl alcohol) nanocomposite nanofibres by electrospinning, Nanotechnology 17 (2006) 4519–4528. https://doi.org/10.1088/0957-4484/17/17/039.

[95] J.I. Kadokawa, A. Takegawa, S. Mine, K. Prasad, Preparation of chitin nanowhiskers using an ionic liquid and their composite materials with poly(vinyl alcohol), Carbohydr. Polym. 84 (2011) 1408–1412. https://doi.org/10.1016/j.carbpol.2011.01.049.

[96] K. Kafedjiiski, A.H. Krauland, M.H. Hoffer, A. Bernkop-Schnürch, Synthesis and in vitro evaluation of a novel thiolated chitosan, Biomaterials 26 (2005) 819–826. https://doi.org/10.1016/j.biomaterials.2004.03.011.

[97] Y.O. Kang, J.Y. Jung, D. Cho, O.H. Kwon, J.Y. Cheon, W.H. Park, Antimicrobial silver chloride nanoparticles stabilized with chitosan oligomer for the healing of burns, Materials 9 (4) (2016) 215. https://doi.org/10.3390/ma9040215.

[98] R. Karthik, S. Meenakshi, Synthesis, characterization and Cr(VI) uptake study of polyaniline coated chitin, Int. J. Biol. Macromol. 72 (2015) 235–242. https://doi.org/10.1016/j.ijbiomac.2014.08.022.

[99] M.R. Kasaai, Determination of the degree of N-acetylation for chitin and chitosan by various NMR spectroscopy techniques: a review, Carbohydr. Polym. 79 (2010) 801–810. https://doi.org/10.1016/j.carbpol.2009.10.051.

[100] G. Kerch, V. Korkhov, Effect of storage time and temperature on structure, mechanical and barrier properties of chitosan-based films, Eur. Food Res. Technol. 232 (2011) 17–22. https://doi.org/10.1007/s00217-010-1356-x.

[101] R.A.Khan Khan, S. Salmieri, C. Le Tien, B. Riedl, J. Bouchard, G. Chauve, V. Tan, M.R. Kamal, M. Lacroix, Mechanical and barrier properties of nanocrystalline cellulose reinforced chitosan based nanocomposite films, Carbohydr. Polym. 90 (2012) 1601–1608. https://doi.org/10.1016/j.carbpol.2012.07.037.

[102] P. Kithva, L. Grøndahl, D. Martin, M. Trau, Biomimetic synthesis and tensile properties of nanostructured high volume fraction hydroxyapatite and chitosan biocomposite films, J. Mater. Chem. 20 (2010) 381–389. https://doi.org/10.1039/b914798e.

[103] F.S. Kittur, A.B. Vishu Kumar, R.N. Tharanathan, Low molecular weight chitosans—preparation by depolymerization with *Aspergillus niger* pectinase, and characterization, Carbohydr. Res. 338 (2003) 1283–1290. https://doi.org/10.1016/S0008-6215(03)00175-7.

[104] F.S. Kittur, A.B. Vishu Kumar, M.C. Varadaraj, R.N. Tharanathan, Chitooligosaccharides—preparation with the aid of pectinase isozyme from *Aspergillus niger* and their antibacterial activity, Carbohydr, Res. 340 (2005) 1239–1245. https://doi.org/10.1016/j.carres.2005.02.005.

[105] R.R. Klossner, H.A. Queen, A.J. Coughlin, W.E. Krause, Correlation of chitosan's rheological properties and its ability to electrospin, Biomacromolecules 9 (2008) 2947–2953. https://doi.org/10.1021/bm800738u.

[106] P.T.S. Kumar, S. Abhilash, K. Manzoor, S.V. Nair, H. Tamura, R. Jayakumar, Preparation and characterization of novel β-chitin/nanosilver composite scaffolds for wound dressing applications, Carbohydr. Polym. 80 (2010) 761–767. https://doi.org/10.1016/j.carbpol.2009.12.024.

[107] J. Kumirska, M. Czerwicka, Z. Kaczyński, A. Bychowska, K. Brzozowski, J. Thöming, P. Stepnowski, Application of spectroscopic methods for structural analysis of chitin and chitosan, Marine Drugs 8 (2010) 1567–1636. https://doi.org/10.3390/md8051567.

[108] K. Kurita, Y. Kaji, T. Mori, Y. Nishiyama, Enzymatic degradation of β-chitin: susceptibility and the influence of deacetylation, Carbohydr. Polym. 42 (2000) 19–21. https://doi.org/10.1016/S0144-8617(99)00127-7.

[109] B. Lee, Y. Lee, Y. Sohn, SC. Song, A thermosensitive poly(org anophosphazene) gel, Macromolecules 35 (2002) 3876–9229.

[110] C.M. Lee, D. Jang, J. Kim, S.J. Cheong, E.M. Kim, M.H. Jeong, S.H. Kim, D.W. Kim, S.T. Lim, M.H. Sohn, Y.Y. Jeong, H.J. Jeong, Oleyl-chitosan nanoparticles based on a dual probe for optical/MR imaging in vivo, Bioconjug. Chem. 22 (2011) 186–192. https://doi.org/10.1021/bc100241a.

[111] S.L. Levengood, A.E. Erickson, F.C. Chang, M. Zhang, Chitosan-poly(caprolactone) nanofibers for skin repair, J. Mater. Chem. B. 5 (2017) 1822–1833. https://doi.org/10.1039/C6TB03223K.

[112] X Li, X. Kong, S. Shi, Y. Gu, L. Yang, G. Guo, F. Luo, X. Zhao, Y.Q. Wei, Z.Y. Qian, Biodegradable MPEG-g-chitosan and methoxypoly(ethylene glycol)-b-poly([epsilon]-caprolactone) composite films: Part 1. Preparation and characterization, Carbohydr. Polym. 79 (2009) 429–436.

[113] Z. Li, L. Yubao, Y. Aiping, P. Xuelin, W. Xuejiang, Z. Xiang, Preparation and in vitro investigation of chitosan/nano-hydroxyapatite composite used as bone substitute materials, J. Mater. Sci. Mater. Med. 16 (2005) 213–219. https://doi.org/10.1007/s10856-005-6682-3.

[114] F. Lin, F.-G.Jia Hao-Ran, Wu, Glycol chitosan: a water-soluble polymer for cell imaging and drug delivery, Molecules 24 (23) (2019) 4371.

[115] Z. Lin, M. Yina, P. Xiaochen, C. Siyuan, Z. Huahong, W. Shufang, A composite hydrogel of chitosan/heparin/poly (γ-glutamic acid) loaded with superoxide dismutase for wound healing, Carbohydr. Polym. (2018) 168–174. https://doi.org/10.1016/j.carbpol.2017.10.036.

[116] M. Liu, Y. Zhang, C. Wu, S. Xiong, C. Zhou, Chitosan/halloysite nanotubes bionanocomposites: structure, mechanical properties and biocompatibility, Int. J. Biol. Macromol. 51 (2012) 566–575. https://doi.org/10.1016/j.ijbiomac.2012.06.022.

[117] M.V. Risbud, A.A. Hardikar, S.V Bhat, R.R. Bhonde, pH-sensitive freeze-dried chitosan-polyvinyl pyrrolidone hydrogels as controlled release system for antibiotic delivery, J. Control. Release 68 (2000) 23–30.

[118] Y. Liu, H.I. Kim, Characterization and antibacterial properties of genipin-crosslinked chitosan/poly(ethylene glycol)/ZnO/Ag nanocomposites, Carbohydr. Polym. 89 (2012) 111–116. https://doi.org/10.1016/j.carbpol.2012.02.058.

[119] S. Liu, S. Yang, P.C. Ho, Intranasal administration of carbamazepine-loaded carboxymethyl chitosan nanoparticles for drug delivery to the brain, Asian J. Pharma. Sci. 13 (2018) 72–81. https://doi.org/10.1016/j.ajps.2017.09.001.

[120] X. Liu, L. Luo, Y. Ding, Y. Xu, Amperometric biosensors based on alumina nanoparticles-chitosan-horseradish peroxidase nanobiocomposites for the determination of phenolic compounds, Analyst 136 (2011) 696–701. https://doi.org/10.1039/c0an00752h.

[121] X. Liu, L. Ma, Z. Mao, C. Gao, Chitosan-based biomaterials for tissue repair and regeneration, Adv. Polym. Sci. 244 (2011) 81–128. https://doi.org/10.1007/12_2011_118.

[122] J. Liuyun, L. Yubao, X. Chengdong, A novel composite membrane of chitosan-carboxymethyl cellulose polyelectrolyte complex membrane filled with nano-hydroxyapatite I. Preparation and properties, J. Mater. Sci. Mater. Med. 20 (2009) 1645–1652. https://doi.org/10.1007/s10856-009-3720-6.

[123] A. Lizardi-Mendoza, W Monal, FM. G. Valencia, Chemical characteristics and functional properties of chitosan, in: S. Bautista-Baños, G. Romanazzi, A. Jiménez-Aparicio (Eds.), Chitosan in the Preservation of Agricultural Commodities, Elsevier, Amsterdam, 2016, pp. 3–31.

[124] Y. Lu, L. Wen, g, L. Zhang, Morphology and properties of soy protein isolate thermoplastics reinforced with chitin whiskers, Biomacromolecules 5 (2004) 1046–1051. https://doi.org/10.1021/bm034516x.

[125] B. Lu, T. Wang, Z. Li, F. Dai, L. Lv, F. Tang, K. Yu, J. Liu, G. Lan, Healing of skin wounds with a chitosan-gelatin sponge loaded with tannins and platelet-rich plasma, Int. J. Biol. Macromol. 82 (2016) 884–891. https://doi.org/10.1016/j.ijbiomac.2015.11.009.

[126] J.F. Mano, G. Hungerford, J.L. Gómez Ribelles, Bioactive poly(L-lactic acid)-chitosan hybrid scaffolds, Mater. Sci. Eng. C. 28 (2008) 1356–1365. https://doi.org/10.1016/j.msec.2008.03.005.

[127] P. Marie Arockianathan, S. Sekar, B. Kumaran, T.P. Sastry, Preparation, characterization and evaluation of biocomposite films containing chitosan and sago starch impregnated with silver nanoparticles, Int. J. Biol. Macromol. 50 (2012) 939–946. https://doi.org/10.1016/j.ijbiomac.2012.02.022.

[128] K. Masanori, I. Toshiyuki, I. Soichiro, M.H. N, K. Yoshihisa, T. Kazuo, S. Kenichi, T. Junzo, Biomimetic synthesis of bone-like nanocomposites using the self-organization mechanism of hydroxyapatite and collagen, Compos. Sci. Technol. (2004) 819–825. https://doi.org/10.1016/j.compscitech.2003.09.002.

[129] R.K. Mendes, B.S. Arruda, E.F. De Souza, A.B. Nogueira, O. Teschke, L.O. Bonugli, A. Etchegaray, Determination of chlorophenol in environmental samples using a voltammetric biosensor based on hybrid nanocomposite, J. Braz. Chem. Soc. 28 (2017) 1212–1219. https://doi.org/10.21577/0103-5053.20160282.

[130] H. Meng, W. Xiaolan, W. Zhenggang, C. Lingyun, L. Yao, Z. Xinjiang, L. Mei, L. Zhongming, Z. Yu, X. Hong, Z. Lina, Biocompatible and biodegradable bioplastics constructed from chitin via a "green" pathway for bone repair, ACS Sustain. Chem. Eng. (2017) 9126–9135. https://doi.org/10.1021/acssuschemeng.7b02051.

[131] S. Mengjie, W. Ting, P. Jianhui, C. Xiguang, L. Ya, Hydroxybutyl chitosan centered biocomposites for potential curative applications: a critical review, Biomacromolecules (2020) 1351–1367. https://doi.org/10.1021/acs.biomac.0c00071.

[132] M. Mengsi, Z. Yalan, J. Xulin, Thermosensitive and pH-responsive tannin-containing hydroxypropyl chitin hydrogel with long-lasting antibacterial activity for wound healing, Carbohydr. Polym. (2020) 116096. https://doi.org/10.1016/j.carbpol.2020.116096.

[133] M.H. Mhd, C. Yern, A. Luqman, P. Sin, C. Cheng, Review of bionanocomposite coating films and their applications, Polymers (2016) 246. https://doi.org/10.3390/polym8070246.

[134] S.D.F. Mihindukulasuriya, L.T. Lim, Nanotechnology development in food packaging: a review, Trends Food Sci. Technol. 40 (2014) 149–167. https://doi.org/10.1016/j.tifs.2014.09.009.

[135] A. Teimouri Moatary, M. Bagherzadeh, A.N. Chermahini, R. Razavizadeh, Design and fabrication of novel chitin hydrogel/chitosan/nano diopside composite scaffolds for tissue engineering, Ceram. Int. 43 (2017) 1657–1668. https://doi.org/10.1016/j.ceramint.2016.06.068.

[136] G.D. Mogoşanu, A.M. Grumezescu, Natural and synthetic polymers for wounds and burns dressing, Int. J. Pharma 463 (2014) 127–136. https://doi.org/10.1016/j.ijpharm.2013.12.015.

[137] M. Monaheng, N. Lebea, M. Soraya, N. Edward, B. Tobias, M. Sabelo, Antimicrobial properties of chitosan-alumina/f-MWCNT nanocomposites, J. Nanotechnol (2016) 1–8. https://doi.org/10.1155/2016/5404529.

[138] M. Morimoto, H. Saimoto, H. Usui, Y. Okamoto, S. Minami, Y. Shigemasa, Biological activities of carbohydrate-branched chitosan derivatives, Biomacromolecules 2 (2001) 1133–1136. https://doi.org/10.1021/bm010063p.

[139] A. Dufresne Morin, Nanocomposites of chitin whiskers from Riftia tubes and poly(caprolactone), Macromolecules 35 (2002) 2190–2199. https://doi.org/10.1021/ma011493a.

[140] V.K. Mourya, N.N. Inamdar, Chitosan-modifications and applications: opportunities galore, reactive and functional Polymers. 68 (2008) 1013–1051. https://doi.org/10.1016/j.reactfunctpolym.2008.03.002.

[141] R.A.A. Muzzarelli, Biomedical exploitation of chitin and chitosan via mechano-chemical disassembly, electrospinning, dissolution in imidazolium ionic liquids, and supercritical drying, Marine Drugs 9 (2011) 1510–1533. https://doi.org/10.3390/md9091510.

[142] R.S. Nair, A. Morris, N. Billa, C.O. Leong, An evaluation of curcumin-encapsulated chitosan nanoparticles for transdermal delivery, AAPS Pharm Sci. Tech. 20 (2019) Article number: 69. https://doi.org/10.1208/s12249-018-1279-6 .

[143] T.T. Nge, N. Hori, A. Takemura, H. Ono, T. Kimura, Phase behavior of liquid crystalline chitin/acrylic acid liquid mixture, Langmuir 19 (2003) 1390–1395. https://doi.org/10.1021/la020764n.

[144] T.T. Nge, N. Hori, A. Takemura, H. Ono, T. Kimura, Synthesis and orientation study of a magnetically aligned liquid-crystalline chitin/poly(acrylic acid) composite, J. Polym. Sci. B. Polym. Phys. 41 (2003) 711–714. https://doi.org/10.1002/polb.10428.

[145] F. Farshi Azhar Olad, The synergetic effect of bioactive ceramic and nanoclay on the properties of chitosan-gelatin/nanohydroxyapatite-montmorillonite scaffold for bone tissue engineering, Ceram. Int. 40 (2014) 10061–10072. https://doi.org/10.1016/j.ceramint.2014.04.010.

[146] C.E. Olteanu, Applications of functionalized chitosan, Sci. Study Res. VIII (3) (2007) 227–256.

[147] Ömer Bahadır Mergen, Ertan Arda, Gülşen Akın Evingür. Electrical, optical and mechanical properties of chitosan biocomposites, J. Compos. Mater 54 (11) (2019) 1497–1510.

[148] Y.T. Ong, A.L. Ahmad, S.H.S. Zein, K. Sudesh, S.H. Tan, Poly(3-hydroxybutyrate)-functionalised multi-walled carbon nanotubes/chitosan green nanocomposite membranes and their application in pervaporation, Sep. Purif. Technol. 76 (2011) 419–427. https://doi.org/10.1016/j.seppur.2010.11.013.

[149] T.H. Ong, E. Chitra, S. Ramamurthy, R.P. Siddalingam, K.H. Yuen, S.P. Ambu, F. Davamani, Chitosan-propolis nanoparticle formulation demonstrates anti-bacterial activity against *Enterococcus faecalis* biofilms, PLoS ONE 12 (4) (2017) e0176629. https://doi.org/10.1371/journal.pone.0174888.

[150] S. Ozarkar, M. Jassal, A.K. Agrawal, PH and electrical actuation of single walled carbon nanotube/chitosan composite fibers, smart materials and structures. 17 (2008) 055016. https://doi.org/10.1088/0964-1726/17/5/055016.

[151] M. Paillet, A. Dufresne, Chitin whiskers reinforced thermoplastic nanocomposites, Macromolecules 34 (2001) 6527–6529.

[152] Y. Pan, T. Wu, H. Bao, L. Li, Green fabrication of chitosan films reinforced with parallel aligned graphene oxide, Carbohydr. Polym. 83 (2011) 1908–1915. https://doi.org/10.1016/j.carbpol.2010.10.054.

[153] R. Panonnummal, R. Jayakumar, G. Anjaneyan, M. Sabitha, In vivo anti-psoriatic activity, biodistribution, sub-acute and sub-chronic toxicity studies of orally administered methotrexate loaded chitin nanogel in comparison with methotrexate tablet, Int. J. Biol. Macromol. 110 (2018) 259–268. https://doi.org/10.1016/j.ijbiomac.2018.01.036.

[154] Y.K. Park, Y.H. Park, B.A. Shin, E.S. Choi, Y.R. Park, T. Akaike, C.S. Cho, Galactosylated chitosan-graft-dextran as hepatocyte-targeting DNA carrier, J. Control. Release 69 (1) (2003) 97–108.

[155] I.K. Park, J. Yang, H.J. Jeong, H.S. Bom, I. Harada, T. Akaike, S.I. Kim, C.S. Cho, Galactosylated chitosan as a synthetic extracellular matrix for hepatocytes attachment, Biomaterials 24 (2003) 2331–2337. https://doi.org/10.1016/S0142-9612(03)00108-X.

[156] K.M. Park, S.Y. Lee, Y.K. Joung, J.S. Na, M.C. Lee, K.D. Park, Thermosensitive chitosan-Pluronic hydrogel as an injectable cell delivery carrier for cartilage regeneration, Acta Biomater 5 (2009) 1956–1965. https://doi.org/10.1016/j.actbio.2009.01.040.

[157] S. Patel, S. Srivastava, M.R. Singh, D. Singh, Preparation and optimization of chitosan-gelatin films for sustained delivery of lupeol for wound healing, Int. J. Biol. Macromol. 107 (2018) 1888–1897. https://doi.org/10.1016/j.ijbiomac.2017.10.056.

[158] J. Xu Peng, G. Chen, J. Tian, M. He, The preparation of α-chitin nanowhiskers-poly (vinyl alcohol) hydrogels for drug release, Int. J. Biol. Macromol. 131 (2019) 336–342. https://doi.org/10.1016/j.ijbiomac.2019.03.015.

[159] M.H. Periayah, A.S. Halim, A.R. Hussein, A.Z. Mat Saad, A.H. Abdul Rashid, K. Noorsal, In vitro capacity of different grades of chitosan derivatives to induce platelet adhesion and aggregation, Int. J. Biol. Macromol. 52 (2013) 244–249. https://doi.org/10.1016/j.ijbiomac.2012.10.001.

[160] M. Peter, N. Ganesh, N. Selvamurugan, S.V. Nair, T. Furuike, H. Tamura, R. Jayakumar, Preparation and characterization of chitosan-gelatin/nanohydroxyapatite composite scaffolds for tissue engineering applications, Carbohydr. Polym. 80 (2010) 687–694. https://doi.org/10.1016/j.carbpol.2009.11.050.

[161] M.R.S. Poblete, L.J.L. Diaz, Synthesis of biodegradable cellulose-chitin polymer blend from *Portunus pelagicus*, Adv. Mater. Res. 925 (2014) 379–384. https://doi.org/10.4028/www.scientific.net/AMR.925.379.

[162] C. Remuñán-López Portero, M.T. Criado, M.J. Alonso, Reacetylated chitosan micropheres for controlled delivery of anti-microbial agents to the gastric mucosa, J. Microencapsul 19 (2002) 797–809. https://doi.org/10.1080/0265204021000022761.

[163] P. Kithva Pradal, D. Martin, M. Trau, L. Grøndahl, Improvement of the wet tensile properties of nanostructured hydroxyapatite and chitosan biocomposite films through hydrophobic modification, J. Mater. Chem. 21 (2011) 2330–2337. https://doi.org/10.1039/c0jm03080e.

[164] M. Prasitsilp, R. Jenwithisuk, K. Kongsuwan, N. Damrongchai, P. Watts, Cellular responses to chitosan in vitro: the importance of deacetylation, J. Mater. Sci. Mater. Med. 11 (2000) 773–778. https://doi.org/10.1023/A:1008997311364.

[165] Y. Qin, S. Zhang, J. Yu, J. Yang, L. Xiong, Q. Sun, Effects of chitin nano-whiskers on the antibacterial and physicochemical properties of maize starch films, Carbohydr. Polym. 147 (2016) 372–378. https://doi.org/10.1016/j.carbpol.2016.03.095.

[166] K.Von Bargen Raafat, A. Haas, H.G. Sahl, Insights into the mode of action of chitosan as an antibacterial compound, Appl. Environ. Microbiol. 74 (2008) 3764–3773. https://doi.org/10.1128/AEM.00453-08.

[167] M. Rajiv Gandhi, G.N. Kousalya, S. Meenakshi, Removal of copper(II) using chitin/chitosan nano-hydroxyapatite composite, Int. J. Biol. Macromol. 48 (2011) 119–124. https://doi.org/10.1016/j.ijbiomac.2010.10.009.

[168] C. Rao, A. Sood, K. Subrahmanyam, AA. Govindaraj, Graphene: the new two dimensional nanomaterial, Angew. Chem. Int. Ed. Engl. 48 (42) (2009) 7752–7777.

[169] S. Remya, C.O. Mohan, J. Bindu, G.K. Sivaraman, G. Venkateshwarlu, C.N. Ravishankar, Effect of chitosan based active packaging film on the keeping quality of chilled stored barracuda fish, J. Food Sci. Technol. 53 (2016) 685–693. https://doi.org/10.1007/s13197-015-2018-6.

[170] S. Rengaraj, K.H. Yeon, S.H. Moon, Removal of chromium from water and wastewater by ion exchange resins, J. Hazard. Mater. 87 (2001) 273–287. https://doi.org/10.1016/S0304-3894(01)00291-6.

[171] J.W. Rhim, H.M. Park, C.S. Ha, Bio-nanocomposites for food packaging applications, Prog. Polym. Sci. 38 (2013) 1629–1652. https://doi.org/10.1016/j.progpolymsci.2013.05.008.

[172] M. Rinaudo, Chitin and chitosan: properties and applications, Prog. Polym. Sci. (Oxford) 31 (2006) 603–632. https://doi.org/10.1016/j.progpolymsci.2006.06.001.

[173] R. Rizvi, B. Cochrane, H. Naguib, P.C. Lee, Fabrication and characterization of melt-blended polylactide-chitin composites and their foams, J. Cell. Plast. 47 (2011) 283–300. https://doi.org/10.1177/0021955X11402549.

[174] S.Y. Rong, N.M. Mubarak, F.A. Tanjung, Structure-property relationship of cellulose nanowhiskers reinforced chitosan biocomposite films, J. Environ. Chem. Eng. 5 (2017) 6132–6136. https://doi.org/10.1016/j.jece.2017.11.054.

[175] O.V. Rúnarsson, J. Holappa, S. Jónsdóttir, H. Steinsson, M. Másson, N-selective "one pot" synthesis of highly N-substituted trimethyl chitosan (TMC), Carbohydr. Polym. 74 (2008) 740–744. https://doi.org/10.1016/j.carbpol.2008.03.008.

[176] W. Sajomsang, Synthetic methods and applications of chitosan containing pyridylmethyl moiety and its quaternized derivatives: a review, Carbohydr. Polym. 80 (2010) 631–647. https://doi.org/10.1016/j.carbpol.2009.12.037.

[177] W. Sajomsang, P. Gonil, S. Saesoo, Synthesis and antibacterial activity of methylated N-(4-N,N-dimethylaminocinnamyl) chitosan chloride, Eur. Polym. J. 45 (2009) 2319–2328. https://doi.org/10.1016/j.eurpolymj.2009.05.009.

[178] W. Sajomsang, S. Tantayanon, V. Tangpasuthadol, W.H. Daly, Synthesis of methylated chitosan containing aromatic moieties: chemoselectivity and effect on molecular weight, Carbohydr. Polym. 72 (2008) 740–750. https://doi.org/10.1016/j.carbpol.2007.10.023.

[179] A.M. Salaberria, R.H. Diaz, J. Labidi, S.C.M. Fernandes, Role of chitin nanocrystals and nanofibers on physical, mechanical and functional properties in thermoplastic starch films, Food Hydrocoll 46 (2015) 93–102. https://doi.org/10.1016/j.foodhyd.2014.12.016.

[180] M.A. Salam, Preparation and characterization of chitin/magnetite/multiwalled carbon nanotubes magnetic nanocomposite for toxic hexavalent chromium removal from solution, J. Mol. Liq. 233 (2017) 197–202. https://doi.org/10.1016/j.molliq.2017.03.023.

[181] S. Samar, M.J. M., G. Babak, H. Hamed, Physicochemical and antifungal properties of bio-nanocomposite film based on gelatin-chitin nanoparticles, Int.J. Biol. Macromol. (2017) 373–381. https://doi.org/10.1016/j.ijbiomac.2016.12.066.

[182] S. Samar, M.J. M., G. Babak, H. Hamed, Effect of corn oil on physical, thermal, and antifungal properties of gelatin-based nanocomposite films containing nano chitin, LWT- Food Sci. Technol. (2017) 33–39. https://doi.org/10.1016/j.lwt.2016.10.028.

[183] V. Sanna, A.M. Roggio, S. Siliani, M. Piccinini, S. Marceddu, A. Mariani, M. Sechi, Development of novel cationic chitosan-and anionic alginate-coated poly(D,L-lactide-co-glycolide) nanoparticles for controlled release and light protection of resveratrol, Int. J. Nanomed 7 (2012) 5501–5516. http://www.dovepress.com/getfile.php?fileID=14245.

[184] B. Sarmento, J.Das Neves, Chitosan-Based Systems for Biopharmaceuticals: Delivery, Targeting and Polymer Therapeutics, John Wiley and Sons, Portugal, 2012. https://doi.org/10.1002/9781119962977.

[185] H. Sashiwa, J.M. Thompson, S.K. Das, Y. Shigemasa, S. Tripathy, R. Roy, Chemical modification of chitosan: preparation and lectin binding properties of alpha-galactosyl-chitosan conjugates. Potential inhibitors in acute rejection following xenotransplantation, Biomacromolecules 1 (2000) 303–305. https://doi.org/10.1021/bm005536r.

[186] H. Sashiwa, S.I. Aiba, Chemically modified chitin and chitosan as biomaterials, Prog. Polym. Sci. 29 (2004) 887–908. https://doi.org/10.1016/j.progpolymsci.2004.04.001.

[187] H. SeunghwanChoy, Y. YeonjuPark, J. MeeJung, X. Dongyeop MoKoo, D. Oh, SooHwang, Mechanical properties and thermal stability of intermolecular-fitted poly(vinyl alcohol)/α-chitin nanofibrous mat, Carbohydr. Polym. 244 (2020) 116476.

[188] F. Khodaiyan Shahabi-Ghahfarrokhi, M. Mousavi, H. Yousefi, Preparation of UV-protective kefiran/nano-ZnO nanocomposites: physical and mechanical properties, Int. J. Biol. Macromol. 72 (2015) 41–46. https://doi.org/10.1016/j.ijbiomac.2014.07.047.

[189] F. Khodaiyan Shahabi-Ghahfarrokhi, M. Mousavi, H. Yousefi, Green bionanocomposite based on kefiran and cellulose nanocrystals produced from beer industrial residues, Int. J. Biol. Macromol. 77 (2015) 85–91. https://doi.org/10.1016/j.ijbiomac.2015.02.055.

[190] K.T. Shalumon, N.S. Binulal, N. Selvamurugan, S.V. Nair, D. Menon, T. Furuike, H. Tamura, R. Jayakumar, Electrospinning of carboxymethyl chitin/poly(vinyl alcohol) nanofibrous scaffolds for tissue engineering applications, Carbohydr. Polym. 77 (2009) 863–869. https://doi.org/10.1016/j.carbpol.2009.03.009.

[191] A.K. Sharma, L. Gupta, H. Sahu, A. Qayum, S.K. Singh, K.T. Nakhate, U.Gupta Ajazuddin, Chitosan engineered PAMAM dendrimers as nanoconstructs for the enhanced anti-cancer potential and improved in vivo brain pharmacokinetics of temozolomide, Pharma. Res. 35 (2018). https://doi.org/10.1007/s11095-017-2324-y.

[192] Y.T. Shieh, Y.F. Yang, Significant improvements in mechanical property and water stability of chitosan by carbon nanotubes, Eur. Polym. J. 42 (2006) 3162–3170. https://doi.org/10.1016/j.eurpolymj.2006.09.006.

[193] B.K. Shrestha, R. Ahmad, H.M. Mousa, I.G. Kim, J.I. Kim, M.P. Neupane, C.H. Park, C.S. Kim, High-performance glucose biosensor based on chitosan-glucose oxidase immobilized polypyrrole/nafion/functionalized multi-walled carbon nanotubes bio-nanohybrid film, J. Colloid Interf. Sci. 482 (2016) 39–47. https://doi.org/10.1016/j.jcis.2016.07.067.

[194] M.M. Silva, R. Calado, J. Marto, A. Bettencourt, A.J. Almeida, L.M.D. Gonçalves, Chitosan nanoparticles as a mucoadhesive drug delivery system for ocular administration, Marine Drugs 15 (2017). https://doi.org/10.3390/md15120370.

[195] G. Sinsinbar Singh, M. Choudhary, V. Kumar, R. Pasricha, H.N. Verma, S.P. Singh, K. Arora, Graphene oxide-chitosan nanocomposite based electrochemical DNA biosensor for detection of typhoid, Sens. Actuators B. Chem. 185 (2013) 675–684. https://doi.org/10.1016/j.snb.2013.05.014.

[196] U. Siripatrawan, W. Vitchayakitti, Improving functional properties of chitosan films as active food packaging by incorporating with propolis, Food Hydrocoll 61 (2016) 695–702. https://doi.org/10.1016/j.foodhyd.2016.06.001.

[197] P.I.P. Soares, A.I. Sousa, J.C. Silva, I.M.M. Ferreira, C.M.M. Novo, J.P. Borges, Chitosan-based nanoparticles as drug delivery systems for doxorubicin: optimization and modelling, Carbohydr. Polym. 147 (2016) 304–312. https://doi.org/10.1016/j.carbpol.2016.03.028.

[198] T.A. Sonia, C.P. Sharma, Chitosan and its derivatives for drug delivery perspective, Advances in Polymer Science 243 (2011) 23–54. https://doi.org/10.1007/12_2011_117.

[199] P. Supaphol Sriupayo, J. Blackwell, R. Rujiravanit, Preparation and characterization of α-chitin whisker-reinforced chitosan nanocomposite films with or without heat treatment, Carbohydr. Polym. 62 (2005) 130–136. https://doi.org/10.1016/j.carbpol.2005.07.013.

[200] R. Zlatev Stoytcheva, G. Montero, Z. Velkova, V. Gochev, A nanotechnological approach to biosensors sensitivity improvement: application to organophosphorus pesticides determination, Biotechnol. Biotech. Equip. 32 (2018) 213–220. https://doi.org/10.1080/13102818.2017.1389618.

[201] M.N. Sundaram, V. Krishnamoorthi Kaliannagounder, V. Selvaprithiviraj, M. Suresh, R. Biswas, A.K. Vasudevan, P.K. Varma, R. Jayakumar, Bioadhesive, hemostatic and antibacterial in situ chitin-fibrin nanocomposite gel for controlling bleeding and preventing infections at mediastinum, ACS Sustainable Chem. Eng. 6 (2018) 7826–7840. https://doi.org/10.1021/acssuschemeng.8b00915.

[202] S.K. Swain, S.K. Kisku, G. Sahoo, Preparation of thermal resistant gas barrier chitosan nanobiocomposites, Polym. Compos. 35 (2014) 2324–2328. https://doi.org/10.1002/pc.22897.

[203] H.S.Han Swierczewska, K. Kim, J.H. Park, S. Lee, Polysaccharide-based nanoparticles for theranostic nanomedicine, Adv. Drug Deliv. Rev. 99 (2016) 70–84. https://doi.org/10.1016/j.addr.2015.11.015.

[204] Q. Tan, W. Liu, C. Guo, G. Zhai, Preparation and evaluation of quercetin loaded lecithin-chitosan nanoparticles for topical delivery, Int. J. Nanomed 6 (2011) 1621–1630.

[205] Y. Tang, Y. Du, Y. Li, X. Wang, X. Hu, A thermosensitive chitosan/poly(vinyl alcohol) hydrogel containing hydroxyapatite for protein delivery, J. Biomed. Mater. Res. A. 91 (2009) 953–963. https://doi.org/10.1002/jbm.a.32240.

[206] H. Tang, C. Chang, L. Zhang, Efficient adsorption of Hg2+ ions on chitin/cellulose composite membranes prepared via environmentally friendly pathway, Chem. Eng. J. 173 (2011) 689–697. https://doi.org/10.1016/j.cej.2011.07.045.

[207] S. Teepoo, P. Dawan, N. Barnthip, Electrospun chitosan-gelatin biopolymer composite nanofibers for horseradish peroxidase immobilization in a hydrogen peroxide biosensor, Biosensors 7 (2017). https://doi.org/10.3390/bios7040047.

[208] W.W. Thein-Han, R.D.K. Misra, Biomimetic chitosan-nanohydroxyapatite composite scaffolds for bone tissue engineering, Acta Biomater 5 (2009) 1182–1197. https://doi.org/10.1016/j.actbio.2008.11.025.

[209] H.Hao Ti, L. Xia, C. Tong, J. Liu, L. Dong, S. Xu, Y. Zhao, H. Liu, X. Fu, W. Han, Controlled release of thymosin beta 4 using a collagen-chitosan sponge scaffold augments cutaneous wound healing and increases angiogenesis in diabetic rats with hindlimb ischemia, Tissue Eng. A. 21 (2015) 541–549. https://doi.org/10.1089/ten.tea.2013.0750.

[210] J.W.Whittaker Tkac, T. Ruzgas, The use of single walled carbon nanotubes dispersed in a chitosan matrix for preparation of a galactose biosensor, Biosensors Bioelectron 22 (2007) 1820–1824. https://doi.org/10.1016/j.bios.2006.08.014.

[211] L.C. Tsai, C.H. Chen, C.W. Lin, Y.C. Ho, F.L. Mi, Development of mutlifunctional nanoparticles self-assembled from trimethyl chitosan and fucoidan for enhanced oral delivery of insulin, Int. J. Biol. Macromol. 126 (2019) 141–150. https://doi.org/10.1016/j.ijbiomac.2018.12.182.

[212] Y.C. Tsai, S.Y. Chen, C.A. Lee, Amperometric cholesterol biosensors based on carbon nanotube-chitosan-platinum-cholesterol oxidase nanobiocomposite, Sens. Actuators B. Chem. 135 (2008) 96–101. https://doi.org/10.1016/j.snb.2008.07.025.

[213] A.J. Uddin, M. Fujie, S. Sembo, Y. Gotoh, Outstanding reinforcing effect of highly oriented chitin whiskers in PVA nanocomposites, Carbohydr. Polym. 87 (2012) 799–805. https://doi.org/10.1016/j.carbpol.2011.08.071.

[214] P.J. VandeVord, H.W.T. Matthew, S.P. DeSilva, L. Mayton, B. Wu, P.H. Wooley, Evaluation of the biocompatibility of a chitosan scaffold in mice, J. Biomed. Mater. Res. 59 (2002) 585–590. https://doi.org/10.1002/jbm.1270.

[215] A.K. Varkouhi, R.J. Verheul, R.M. Schiffelers, T. Lammers, G. Storm, W.E. Hennink, Gene silencing activity of siRNA polyplexes based on thiolated N, N, N -trimethylated chitosan, Bioconjug. Chem. 21 (2010) 2339–2346. https://doi.org/10.1021/bc1003789.

[216] C. Vasile, Polymeric nanocomposites and nanocoatings for food packaging: a review, Materials 11 (10) (2018) 1834. https://doi.org/10.3390/ma11101834.

[217] W.S. Vedakumari, P. Prabu, T.P. Sastry, Chitosan-fibrin nanocomposites as drug delivering and wound healing materials, J. Biomed. Nanotechnol 11 (2015) 657–667. https://doi.org/10.1166/jbn.2015.1948.

[218] S.K. Kim Venkatesan, Chitosan composites for bone tissue engineering—an overview, Marine Drugs 8 (2010) 2252–2266. https://doi.org/10.3390/md8082252.

[219] R. Verdejo, F. Barroso-Bujans, M.A. Rodriguez-Perez, J.A. De Saja, M.A. Lopez-Manchado, Functionalized graphene sheet filled silicone foam nanocomposites, J. Mater. Chem. 18 (2008) 2221–2226. https://doi.org/10.1039/b718289a.

[220] Vidar Rúnarsson, J. Holappa, C. Malainer, H. Steinsson, M. Hjálmarsdóttir, T. Nevalainen, M. Másson, Antibacterial activity of N-quaternary chitosan derivatives: synthesis, characterization and structure activity relationship (SAR) investigations, Eur. Polym. J. 46 (2010) 1251–1267. https://doi.org/10.1016/j.eurpolymj.2010.03.001.

[221] C. Vilela, R.J.B. Pinto, J. Coelho, M.R.M. Domingues, S. Daina, P. Sadocco, S.A.O. Santos, C.S.R. Freire, Bioactive chitosan/ellagic acid films with UV-light protection for active food packaging, Food Hydrocoll 73 (2017) 120–128. https://doi.org/10.1016/j.foodhyd.2017.06.037.

[222] W.S. Wan Ngah, L.C. Teong, M.A.K.M. Hanafiah, Adsorption of dyes and heavy metal ions by chitosan composites: a review, Carbohydr. Polym. (2011) 1446–1456. https://doi.org/10.1016/j.carbpol.2010.11.004.

[223] Y.Wang Wan, S. Qian, Study on the adsorption properties of novel crown ether crosslinked chitosan for metal ions, J. Appl. Polym. Sci. 84 (2002) 29–34. https://doi.org/10.1002/app.10180.

[224] X. Wang, Y. Tang, Y. Li, Z. Zhu, Y. Du, The rheological behaviour and drug-delivery property of chitosan/rectorite nanocomposites, J. Biomater. Sci., Polymer Edition 21 (2010) 171–184. https://doi.org/10.1163/156856209X410300.

[225] S.F. Wang, L. Shen, W.D. Zhang, Y.J. Tong, Preparation and mechanical properties of chitosan/carbon nanotubes composites, Biomacromolecules 6 (2005) 3067–3072. https://doi.org/10.1021/bm050378v.

[226] H. Wang, R. Zhang, H. Zhang, S. Jiang, H. Liu, M. Sun, S. Jiang, Kinetics and functional effectiveness of nisin loaded antimicrobial packaging film based on chitosan/poly(vinyl alcohol), Carbohydr. Polym. 127 (2015) 64–71. https://doi.org/10.1016/j.carbpol.2015.03.058.

[227] T. Wang, J. Hou, C. Su, L. Zhao, Y. Shi, Hyaluronic acid-coated chitosan nanoparticles induce ROS-mediated tumor cell apoptosis and enhance antitumor efficiency by targeted drug delivery via CD44, J. Nanobiotechnol 15 (2017) Article number: 7. https://doi.org/10.1186/s12951-016-0245-2 .

[228] Y. Wang, J. Zhu, R. Zhu, Z. Zhu, Z. Lai, Z. Chen, Chitosan/prussian blue-based biosensors, Measure. Sci. Technol. 14 (2003) 831–836. https://doi.org/10.1088/0957-0233/14/6/317.

[229] R. Watkins, L. Wu, C. Zhang, R. Davis, B. Xu, Natural product-based nano medicine: recent advances and issues, Int. J. Nanomed 10 (2015) 6055–6074.

[230] P. Supaphol Watthanaphanit, H. Tamura, S. Tokura, R. Rujiravanit, Fabrication, structure, and properties of chitin whisker-reinforced alginate nanocomposite fibers, J. Appl. Polym. Sci. 110 (2008) 890–899. https://doi.org/10.1002/app.28634.

[231] D. Winter Wiegand, U.C. Hipler, Molecular-weight-dependent toxic effects of chitosans on the human keratinocyte cell line HaCaT, Skin Pharmacol. Physiol. 23 (2010) 164–170. https://doi.org/10.1159/000276996.

[232] N. Fountas-Davis Wijekoon, N.D. Leipzig, Fluorinated methacrylamide chitosan hydrogel systems as adaptable oxygen carriers for wound healing, Acta Biomater 9 (2013) 5653–5664. https://doi.org/10.1016/j.actbio.2012.10.034.

[233] Y. Wimardani, D. Suniarti, H. Freisleben, MA.Ikeda Wanandi, Cytotoxic effects of chitosan against oral cancer cell lines is molecular-weight dependent and cell-type-specific, Int. J. Oral Res. 3 (2012) 1–10.

[234] N. Sanchavanakit Wongpanit, P. Pavasant, T. Bunaprasert, Y. Tabata, R. Rujiravanit, Preparation and characterization of chitin whisker-reinforced silk fibroin nanocomposite sponges, Eur. Polym. J. 43 (2007) 4123–4135. https://doi.org/10.1016/j.eurpolymj.2007.07.004.

[235] T. Wu, Y. Pan, H. Bao, L. Li, Preparation and properties of chitosan nanocomposite films reinforced by poly(3,4-ethylenedioxythiophene)-poly(styrenesulfonate) treated carbon nanotubes, Mater. Chem. Phys. 129 (2011) 932–938. https://doi.org/10.1016/j.matchemphys.2011.05.030.

[236] J. Wu, W. Wei, L.Y. Wang, Z.G. Su, G.H. Ma, A thermosensitive hydrogel based on quaternized chitosan and poly(ethylene glycol) for nasal drug delivery system, Biomaterials 28 (2007) 2220–2232. https://doi.org/10.1016/j.biomaterials.2006.12.024.

[237] B.Y. Wu, S.H. Hou, F. Yin, J. Li, Z.X. Zhao, J.D. Huang, Q. Chen, Amperometric glucose biosensor based on layer-by-layer assembly of multilayer films composed of chitosan, gold nanoparticles and glucose oxidase modified Pt electrode, Biosensors Bioelectron 22 (2007) 838–844. https://doi.org/10.1016/j.bios.2006.03.009.

[238] R. Xing, H. Yu, S. Liu, W. Zhang, Q. Zhang, Z. Li, P. Li, Antioxidant activity of differently regioselective chitosan sulfates in vitro, Bioorg. Med. Chem. 13 (2005) 1387–1392. https://doi.org/10.1016/j.bmc.2004.11.002.

[239] Y. Xu, X. Re, n, M.A. Hanna, Chitosan/clay nanocomposite film preparation and characterization, J. Appl. Polym. Sci. 99 (2006) 1684–1691. https://doi.org/10.1002/app.22664.

[240] T. Xu, M. Xin, M. Li, H. Huang, S. Zhou, Synthesis, characteristic and antibacterial activity of N,N,N-trimethyl chitosan and its carboxymethyl derivatives, Carbohydr. Polym. 81 (2010) 931–936. https://doi.org/10.1016/j.carbpol.2010.04.008.

[241] S. Xu, Y. Zhang, K. Dong, J. Wen, C. Zheng, S. Zhao, Electrochemical DNA biosensor based on graphene oxide-chitosan hybrid nanocomposites for detection of *Escherichia coli* O157:H7, Int. J. Electrochem. Sci. 12 (2017) 3443–3458. https://doi.org/10.20964/2017.04.16.

[242] K.Tokuchi Yamaguchi, H. Fukuzaki, Y. Koyama, K. Takakuda, H. Monma, J. Tanaka, Preparation and microstructure analysis of chitosan/hydroxyapatite nanocomposites, J. Biomed. Mater. Res. 55 (2001) 20–27. https://doi.org/10.1002/1097-4636(200104)55:1<20::AID-JBM30>3.0.CO;2-F.

[243] J. Yan, L. Yang, G. Wang, Y. Xiao, B. Zhang, N. Qi, Biocompatibility evaluation of chitosan-based injectable hydrogels for the culturing mice mesenchymal stem cells in vitro, J. Biomater. Appl. 24 (2010) 625–637. https://doi.org/10.1177/0885328208100536.

[244] W.Qiong Yanchao, H. Juan, L. Meng, Z. Zhi, Q. Yusheng, Porous nano-minerals substituted apatite/chitin/pectin nanocomposites scaffolds for bone tissue engineering, Arab. J. Chem. (2020) 7418–7429. https://doi.org/10.1016/j.arabjc.2020.08.018.

[245] J. Yang, F. Tian, Z. Wang, Q. Wang, Y.J. Zeng, S.Q. Chen, Effect of chitosan molecular weight and deacetylation degree on hemostasis, J. Biomed. Mater. Res. B. Appl. Biomater. 84 (2008) 131–137. https://doi.org/10.1002/jbm.b.30853.

[246] Y. Yang, J. Wang, J. Zhang, J. Liu, X. Yang, H. Zhao, Exfoliated graphite oxide decorated by PDMAEMA chains and polymer particles, Langmuir 25 (2009) 11808–11814. https://doi.org/10.1021/la901441p.

[247] X. Yang, Y. Tu, L. Li, S. Shang, X.M. Tao, Well-dispersed chitosan/graphene oxide nanocomposites, ACS Appl. Mater. Interf. 2 (2010) 1707–1713. https://doi.org/10.1021/am100222m.

[248] X.Ren Yang, F. Tang, L. Zhang, A practical glucose biosensor based on Fe_3O_4 nanoparticles and chitosan/nafion composite film, Biosensors Bioelectron. 25 (2009) 889–895. https://doi.org/10.1016/j.bios.2009.09.002.

[249] L. Yang, H. Xiong, X. Zhang, S. Wang, A novel tyrosinase biosensor based on chitosan-carbon-coated nickel nanocomposite film, Bioelectrochemistry 84 (2012) 44–48. https://doi.org/10.1016/j.bioelechem.2011.11.001.

[250] J.Y. Yhee, S. Son, S.H. Kim, K. Park, K. Choi, I.C. Kwon, Self-assembled glycol chitosan nanoparticles for disease-specific theranostics, J. Control. Release 193 (2014) 202–213. https://doi.org/10.1016/j.jconrel.2014.05.009.

[251] L. Yin, L. Fei, F. Cui, C. Tang, C. Yin, Superporous hydrogels containing poly(acrylic acid-co-acrylamide)/O-carboxymethyl chitosan interpenetrating polymer networks, Biomaterials 28 (2007) 1258–1266. https://doi.org/10.1016/j.biomaterials.2006.11.008.

[252] L. Ying, W. Mingkui, Z. Feng, X. Zhiai, D. Shaojun, The direct electron transfer of glucose oxidase and glucose biosensor based on carbon nanotubes/chitosan matrix, Biosensors Bioelectron (2005) 984–988. https://doi.org/10.1016/j.bios.2005.03.003.

[253] Yu, R.Y. Bao, X.J. Shi, W. Yang, M.B. Yang, Self-assembled high-strength hydroxyapatite/graphene oxide/chitosan composite hydrogel for bone tissue engineering, Carbohydr. Polym. 155 (2017) 507–515. https://doi.org/10.1016/j.carbpol.2016.09.001.

[254] Y. Yuan, B.M. Chesnutt, W.O. Haggard, J.D. Bumgardner, Deacetylation of chitosan: material characterization and in vitro evaluation via albumin adsorption and pre-osteoblastic cell cultures, Materials 4 (2011) 1399–1416. https://doi.org/10.3390/ma4081399.

[255] L. Yurlova, A. Kryvoruchko, B. Kornilovich, Removal of Ni(II) ions from wastewater by micellar-enhanced ultrafiltration, Desalination 144 (2002) 255–260. https://doi.org/10.1016/S0011-9164(02)00321-1.

[256] Z. Zakaria, Z. Izzah, M. Jawaid, A. Hassan, Effect of degree of deacetylation of chitosan on thermal stability and compatibility of chitosan-polyamide blend, Bio. Resources 7 (2012) 5568–5580. https://doi.org/10.15376/biores.7.4.5568-5580.

[257] S. Zameer, Chitin-gold nanocomposite film and electro-optical properties, Front. Nanosci. Nanotechnol (2017). https://doi.org/10.15761/FNN.1000158.

[258] W. Zhang, X. Li, R. Zou, H. Wu, H. Shi, S. Yu, Y. Liu, Multifunctional glucose biosensors from Fe_3O_4 nanoparticles modified chitosan/graphene nanocomposites, Sci. Rep. 5 (2015) 11129. https://doi.org/10.1038/srep11129.

[259] H. Zhang, S.H. Neau, In vitro degradation of chitosan by a commercial enzyme preparation: effect of molecular weight and degree of deacetylation, Biomaterials 22 (2001) 1653–1658. https://doi.org/10.1016/S0142-9612(00)00326-4.

[260] B. Zhu Zhao, Y. Jia, W. Hou, et al., Preparation of biocompatible carboxymethyl chitosan nanoparticles for delivery of antibiotic drug, Biomed. Res. Int. 2013 (2013) 236469.

[261] S. Yu Zhao, B. Sun, S. Gao, S. Guo, K. Zhao, Biomedical applications of chitosan and its derivative nanoparticles, Polymers 10 (4) (2018) 462. https://doi.org/10.3390/polym10040462.

[262] K.M. Zia, M. Zuber, M. Barikani, R. Hussain, T. Jamil, S. Anjum, Cytotoxicity and mechanical behavior of chitin-bentonite clay based polyurethane bio-nanocomposites, Int. J. Biol. Macromol. 49 (2011) 1131–1136. https://doi.org/10.1016/j.ijbiomac.2011.09.010.

[263] M. Jalal Zohuriaan-Mehr, Advances in chitin and chitosan modification through graft copolymerization: a comprehensive review, Iran. Polym. J. 14 (2005) 235–265.

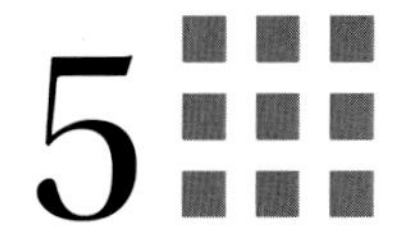

Lignin-based nanomaterials

Łukasz Klapiszewski

FACULTY OF CHEMICAL TECHNOLOGY, INSTITUTE OF CHEMICAL TECHNOLOGY AND ENGINEERING, POZNAN UNIVERSITY OF TECHNOLOGY, BERDYCHOWO, POZNAN, POLAND

Chapter outline

5.1 Lignin—basic information and structure

Plant cells are characterized by a multilayered structure, with cytoplasm with suspended organelles in its center. The entire structure is surrounded by a cytoplasmic membrane, which, in turn, is restricted by a cell wall. A plant cell core is covered by a secondary wall, which consists of three various layers, surrounded by a primary wall [1,2].

The secondary wall is made of properly oriented cellulose and hemicellulose, lignin, and pectins. The outermost part of a plant cell is the middle lamella, the task of which is to separate adjacent plant cells. This outer layer is mainly composed of lignin and polysaccharides [1–4].

A single plant cell, commonly found in nature, is 1–50 mm long and has a diameter of 10–50 μm. The cells are oriented parallel to each other, thus creating the most organized structure, namely, microfibrils. However fibers, which create parallel-packed microfibrils, are characterized by a higher organizational degree. Cellulose molecules present in the secondary wall form crystalline, chain, spirally twisted structures, hence ensuring high mechanical strength of the plant fibers. Hemicellulose found in a plant cell combines with cellulose through hydrogen bonds, thus creating a cellulose-hemicellulose grid. However the task of lignin is to encrust this

Biopolymeric Nanomaterials: Fundamentals and Applications. DOI: https://doi.org/10.1016/B978-0-12-824364-0.00026-5

cellulose and hemicellulose system, giving stiffness to the plant. Hemicellulose and lignin found in the secondary wall of a plant cell form only amorphous areas [1–4].

Lignin, which this chapter focuses on, is a biopolymer responsible for the construction of wooden plant tissues. This substance can be described as certain "cement" binding cellulose fibers [5]. The global annual increase in the amount of lignin due to biosynthesis is estimated at 6×10^{14} tons [6]. The main sources of this material include extraction methods and secondary processing, which specifically define the ultimate physical and mechanical properties of lignin. A biopolymer can be obtained from various materials such as wood, cellulose pulp, paper, bagasse, as well as cereal straw [5–8]. The cellulose and paper industrial segment deals with the production of lignin and its further processing toward lignin-cellulose products. A fairly small amount of lignin is recovered from grasses, branches, leaves, and solid waste in urban and rural areas, where its target content is estimated at approximately 15% [7].

Lignin content in plant cells ranges from 16% to 32%. Its distribution in wood cell walls is not uniform, with its content in the central lamella estimated at approximately 70%, approximately 60% in the primary membrane, and approximately 40%–50% in the external layer of the secondary wall. This is why, based on publicly available data, it can be concluded that the farther away from the center of a single plant cell, the higher lignin content in subsequent secondary wall layers [5,6]. The amount and quality of lignin depends on plant species. In the case of deciduous wood (angiosperm plants), the biopolymer content ranges from 19% to 28%, whereas in softwood (gymnosperm plants), from 24% to 33% [5,7]. In reality, angiosperm plant bark usually contains more hemicellulose and less lignin than the bark of gymnosperm plants [5,7,9,10]. The age of a plant is also important when searching for its individual components, for example young wood has higher lignin content than older wood [7,9,11]. The wooden material delignification process itself causes lignin from the hard bark of a wood to be easier to remove in the course of pulping, at the same time limiting the formation of condensed lignin structures [7,9,11,12]. Fig. 5.1 shows a general plant cell structure, taking into account the locations of its components—biopolymers.

Lignin, which is one of the main plant cell elements, is an amorphous body with a complicated and currently insufficiently researched chemical structure. Although the latest achievements in the field of analytical chemistry and spectroscopy have significantly contributed to extending the knowledge about this biopolymer, there are still many unknowns. Over the years, scientists have confirmed that lignin is a polymer composed of a combination of three basic monomers (monolignols)—alcohols: *p*-coumaryl, coniferyl and sinapyl, which is shown in Fig. 5.1. The main lignin monomer in softwood is coniferyl alcohol, the structure of which contains a methoxy group in the C3 position. The lignin structure in hardwood, besides coniferyl alcohol, also contains sinapyl alcohol molecules, which in the C3 and C5 positions has substituted $-OCH_3$ groups. The third monomer, *p*-coumaryl alcohol, is encountered in grasses and during the so-called wood compression [9,10].

The sequence of reactions aimed at producing a lignin macromolecule in a plant cell is one of the most energy-consuming biosynthetic processes [13]. The pathway of synthesis and enzymatic polymerization of monolignols in lignin is called the phenylpropanoid pathway and has been developed over many years of research involving numerous research groups [14, 15].

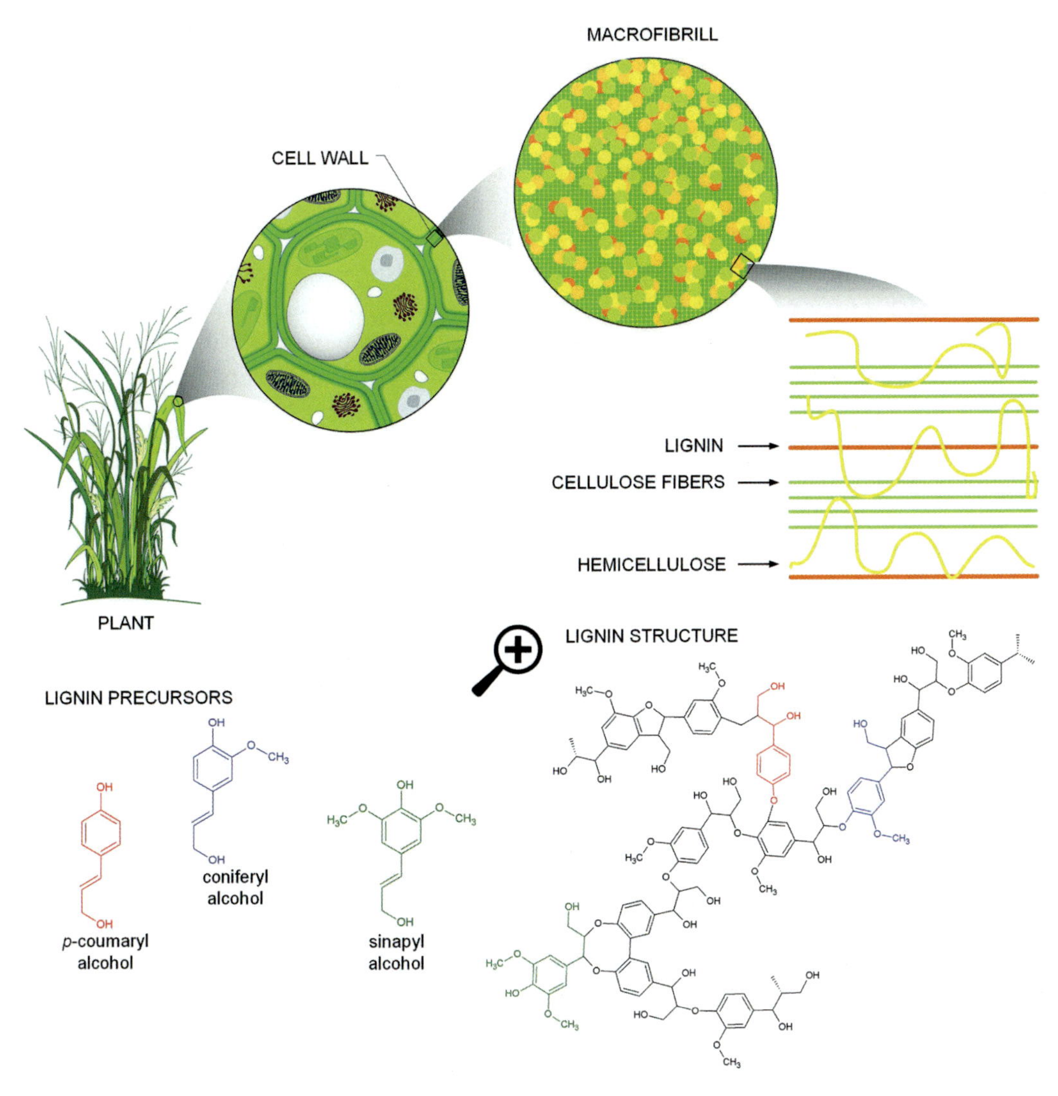

FIGURE 5.1 General plant cell structure, taking into account the locations of its components—biopolymers, based on [7,9,10].

The most important aspect in shaping the lignin structure is the monolignol production process itself, which is basis of lignification. The initial substrate is L-phenylalanine (L-Phe)—a chemical compound of the exogenous amino-acid group, which is the basic construction material for most of naturally occurring proteins. An enzyme, phenylalanine ammonia-lyase, catalyzes the reaction of L-phenylalanine deamination to cinnamic acid—a precursor of coumarates or sinapinates. *p*-Coumaryl acid is formed under the impact of 4-hydroxylase cinnamic acid (C4H), which exposed to phenolic acid-CoA ligase (4CL), leads to the formation of *p*-coumaroyl-CoA. Next, cinnamoyl-CoA reductase (CCR) enables the transition of *p*-coumaroyl-CoA into *p*-coumaryl aldehyde, and further, through cinnamyl alcohol dehydrogenase (CAD), into *p*-coumaryl alcohol, denoted by the H form of lignin.

p-Coumaroyl-CoA can be subjected to the action of CCR or *p*-hydroxycinnamoyl-CoA transferase (HCT). HCT initiates the transition into *p*-coumaroyl-shikimic/quinic acid, so that it is possible to join another hydroxyl group to the aromatic ring under the impact of coumaroyl acid 3-hydroxylase, with caffeoyl-shikimic/quinic acid as the result. After a repeated exposure to the HCT enzyme, we get the caffeoyl-CoA form, which subjected to caffeoyl-CoA O-methyltransferase (CCoA-OMT), enables obtaining feruloyl-CoA. CCR enables transition to coniferyl aldehyde, which is a substrate for obtaining the two other most important alcohols composing lignin. To obtain coniferyl aldehyde from sinapyl alcohol, which forms the so-called S-lignin, three enzymes must take part in the transformations: (1) ferulic acid 5-hydroxylase (F5H) leading to 5-hydroxyconiferyl aldehyde, (2) caffeic acid OMT (COMT) initiating the formation of sinapyl aldehyde, (3) CAD, which provides the end-product. However, to obtain coniferyl alcohol, which form the so-called G-lignin, coniferyl aldehyde should be exposed to CAD [16–18]. Thus, formed monolignols are transported from the cytoplasm, through the cell membrane, near the cell wall, where further reactions take place, aimed at converting precursors into lignin [19,20].

Monolignols formed as a result of the phenylpropanoid pathway are secreted to cell walls, where they undergo polymerization. Lignification is a two-stage process composed of the following phases: (1) initiation and (2) propagation. In the case of lignification, there is no chain termination relative to the radical polymerization process. The polymer culture occurring in this case refers to appropriate structural units, which are monolignol derivatives: (1) *p*-hydroxyphenyl residue (unit H), (2) guaiacyl residue (unit G), and (3) syringyl residue (unit S)—see Fig. 5.2 [16,17].

The research on lignin and its possibilities has been intensified over the recent years. This is directly associated with the method of isolating the biopolymer from the wood material, which is a process hard to implement. All isolating techniques entail the decomposition of covalent bonds, which results in the solubilization of biopolymer fragments. Using an isolation method as a division criterion enables isolating the largest number of various lignin preparations. Therefore, the methods for extracting lignin from plant tissue can be divided into the following groups [21]:

- Methods involving dissolving lignin or its derivatives through extraction, resulting in the so-called soluble lignin. The polysaccharide part of the wood remains undissolved.
- Methods associated with removing polysaccharides (through condensation or oxidation) and introducing them into the solution. In this case, lignin remains in its undissolved form, as the so-called insoluble lignin.

This division differentiates between lignins primarily based on the molecular weight [22]. The first method allows to isolate such lignins such as Brauns, Björkman, dioxane, kraft, as well as lignosulfates. The following lignin types have been distinguished following the second method: Klason, Willstätter, copper, and cellulite enzyme.

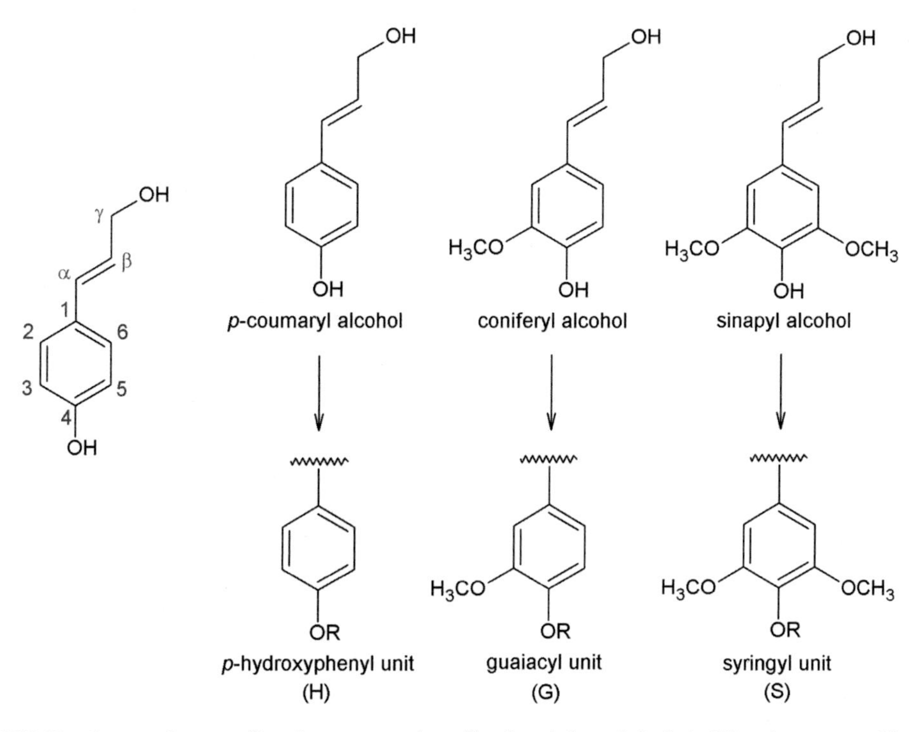

FIGURE 5.2 The three major monolignols, *p*-coumaryl, coniferyl, and sinapyl alcohols. When incorporated into lignin, they are referred to as *p*-hydroxyphenyl (H), guaiacyl (G), and syringyl (S) units, based on [16,17].

The most popular method for lignin acquisition is the kraft method, which is an alkaline process most usually utilized in the paper industry and amounting for approximately 90% of the entire pulp production in the world [23]. In the case of the kraft method, white liquor, which is a combination of sodium hydroxide (NaOH) and sodium sulfide (Na_2S), enables breaking of the lignin-carbohydrate bonds in the wood mass and causes the formation of chemically resistant cellulose. The process undergoes at a temperature of ˜175 °C, in an aqueous solution, and takes 2 h. The end result is obtained black liquor, which is composed of sodium/calcium carbonate, sodium thiosulfate, hemicellulose, lignin, and other ingredients, which are basic components of wood. The kraft method is characterized by high purity, because, compared to other methods, it involves significantly less sulfur, carbohydrate residues, and inorganic compounds [23].

5.2 Lignin modification aimed at creating new, functional nanomaterials

To create advanced nanosystems with unique properties, lignin is subjected to various modifications that can be classified into the following three main categories (Fig. 5.3):

FIGURE 5.3 Three main categories of lignin modifications, based on [24].

- Lignin fragmentation or depolymerization to use the biopolymer as a carbon source or to cleave the biopolymer structure into aromatic monomers.
- Modification through the creation of new chemically active locations.
- Chemical modification of mainly hydroxyl groups [24].

Based on the study results available in the source literature, it was concluded that lignin was an appropriate raw material for the production of low-molecular-weight compounds (e.g., vanillin, simple and hydroxylated aromatic compounds, aldehydes, aliphatic acids, and others). Many thermochemical conversion methods, leading to lignin depolymerization, have been proposed over the last several decades. They include base-catalyzed depolymerization, pyrolysis, gasification, and Lewis' acid-catalyzed solvolysis. Lignin depolymerization has two primary objectives, namely, clarification of the biopolymer structure and composition, and the production of functional, useful products based on a biopolymer separated in the pulp and paper industry, which constitutes postproduction waste. The second objective gains particular importance in the context of depleting and increasingly expensive fossil resources. Therefore, we witness the increasing utilization of various types of lignin (alkaline, sulfite process derivative, or kraft) separated both from coniferous as well as deciduous wood to produce products of high added value through conversion methods, such as thermochemical, biochemical, and/or chemical conversion. The most significant lignin fragmentation methods include hydrogenation (phenols, cresols), hydrolysis, oxidation (vanillin, dimethyl sulfoxide (DMSO), dimethylformamide (DMF)) and/or enzymatic oxidation, gasification (syngas), microbial conversion (ferulic, vanillic, and coumaric acids), and pyrolysis (acetic acid, phenol, aromatic compounds, methane, CO) [24].

Lignin not presubjected to modification processes has usually a limited range of applications. Due to the presence of various functional groups (hydroxy, methoxy, carbonyl, carboxyl groups) in its structure, there are many options for biopolymer modification that is carried out to increase its chemical reactivity, improve its solubility, mainly in organic solvents, and enable enhancing and facilitating product processing. The specificity of its chemical structure, mainly the presence of hydroxy groups and the characteristic position of the aromatic ring bond (mainly in the *ortho* position), gains particular importance in terms of lignin reactivity. Research on the chemical modification methods enabling the introduction of new chemically active substances into the biopolymer structure has been recently developed in this context. Such modifications include nitration, amination, alkylation/dealkylation (grafting of alkyl chains, decaying of methoxy groups), hydroxyalkylation (reaction with formaldehyde), halogenation, or sulfonation [24].

The chemical structure of lignin contains phenolic hydroxy groups and aliphatic hydroxy groups in the C-α and C-γ positions of the lateral chain. Phenolic hydroxy groups are the most reactive functional groups and most strongly impact the chemical reactivity of a biopolymer, thus allowing to conduct a series of reactions, including esterification, etherification, silylation, or oxidation/reduction [24].

Controlled modification of lignin structure enables obtaining functional products of nanometric sizes and for specific applications. Furthermore, a new research trend, which utilizes ionic liquids in "lignin chemistry," has been developed over the recent years [25–28]. Hence, owing to the use of such compounds, it becomes possible to generate advanced, unique nanoproducts, in line with the broadly understood principles of green chemistry.

5.3 The application of lignin and lignin-based materials

Over the recent years lignin has become a very popular biopolymer and its application has been significantly intensified. Research taking into account the modification of lignin and its derivatives enables obtaining newer and newer nanomaterials of unique properties, intended for special applications. Lignin valorization has so far been very promising, owing to the possibility of obtaining highly functional, monomeric or oligomeric products. They can be used as starting materials for various processes within the chemical, pharmaceutical, and other industries [5,6,29]. The crucial issue in the proper engineering of new nanomaterials with lignin is primarily in researching and finally confirming the biopolymer structure, as well as the analysis and correct understanding of the mechanisms ongoing therein.

Lignin and its derivatives are currently deemed as the main source for numerous various aromatic compounds [5,18,30]. Increasing crude oil price and the economic crisis worsening over the recent years have contributed to the constant development of recycling, as well as seeking new methods for utilizing renewable resources [31]. Isolating lignin from black liquor within the kraft process enables obtaining fuels that can be used in the next stages of the conducted process. Converting lignin to liquid fuels is possible through the fragmentation/depolymerization of the complex biopolymer structure. Nowadays, modern biorefineries, which utilize the valorization of the lignocellulosic biomass fragment, produce similar products and chemicals, as in the case of crude oil refining process. Then, the simultaneous utilization of all biomass components within parallel processes takes place [18,31].

Over the recent years we have also seen a very dynamic development of research on lignin as a starting raw material for obtaining a broad base of polymers (e.g., polyethylene (PE), polyether, polyesters, polyurethanes, or polystyrene derivatives). It is possible owing to the presence of hydroxy groups in the aromatic rings [32]. Lignin valorization also indicates potential for the production of biopolymers, that is, polyhydroxyalkanoates or polyhydroxybutyrates [31–33]. Studies involving the conversion of lignin in terms of biopolymer application, mainly on an industrial scale, were conducted by Rajesh Banu et al. [33]. It was concluded that lignin played the role of potential, effective components of biofuels and other biopolymers. However, large-scale lignin-based strategies are a rather difficult task, primarily due to their costly nature. Nonetheless, the significant potential undoubtedly lying in this important strategy will be a source of many intensified research studies in the coming years, resulting in modern, low-waste technologies. In the long run, this will open up the possibility of unlimited engineering of modern and multifunctional composite nanomaterials. The following sections overview examples of industries, where nanomaterials with lignin are used in practice. In addition, Fig. 5.4 summarizes the most important fields of application for lignin and lignin-based materials.

5.3.1 Electrochemistry

The constant technological development, including nanotechnology, makes lignin (or lignin with a modifier), as well as materials with lignin, being used in electrochemistry [28, 34–51].

Milczarek and Inganäs obtained a special biocomposite containing lignin [34]. However, because the aforementioned biopolymer does not conduct electricity, the scientist selected a second component, so that the accumulated energy could be used. This was the function of polypyrrole, which is an organic compound, easy, and cheap to produce from crude oil or plant products. Both substances form a very thin nanolayer, which is a cathode of an ecological battery. The valuable properties of lignin and its specific structure were also used by Milczarek for constructing electrochemical sensors and detectors, which constitute the basis for other publications [35–38]. This topic was expanded in the form of designing silica-lignin

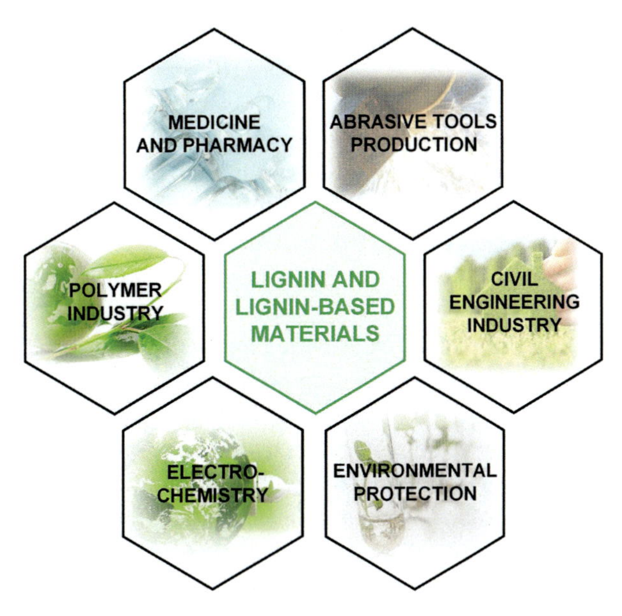

FIGURE 5.4 Application of lignin and lignin-based materials.

hybrid nanomaterials. The use of such systems in electrochemistry enables to demonstrate that a vitreous carbon electrode modified as a result of the electrophoretic deposition of a SiO_2-lignina system exhibited better electrochemical properties than a nonmodified electrode itself. The peak current obtained for a pure electrode was almost threefold lower than the one obtained for a modified electrode [39]. Furthermore, the electrochemical research involved combining the silica-lignin hybrid with carbon nanotubes. The analysis of the results for obtained voltamperograms exhibited the highest electrochemical activity and a decrease in the total capacitive currents for a system containing 20 parts wt. of lignin and SiO_2 obtained in a nonpolar medium. This results from the optimal ionic charge equilibrium between the proton amino groups located on the surface of the modified nanosilica and the negatively charged oxygen-containing groups in the activated kraft lignin. The observed electrochemical activity of silica-lignin systems, especially resulting from the presence of quinone groups, indicates a possibility of using these materials in electrochemical detectors and sensors and/or biosensors [40]. The silica-lignin materials are an example of a hybrid nanoplatform, which is an integrated system consisting of a mixture of inorganic and organic components that, when bonded at the molecular level, enable creating materials with new, often better properties in relation to the starting components. The dynamic technological development and the increasing requirements of the customers make it necessary to design and create increasingly more advanced and stable nanoparticles and analytical devices. An example of devices, which, due to their selectivity and sensitivity, are suitable for a wide range of applications, are the aforementioned biosensors that are a type of chemical sensors containing biological material, such as enzymes, cells, etc., in the receptor layer, which generates an analytical signal, processed by specialized equipment. Designing and creating a platform with desired features enables detecting different substances, depending on the used biological layer. A correctly operating biosensor is distinguished by high selectivity, stability, sensitivity, and accuracy.

Scientists have recently intensified their work on the synthesis and optimization of micro- and nanoplatforms based on natural or nature-inspired materials, mainly magnetite, lignin, polydopamine, dendrimers, and cyclodextrin. The used biopolymers are characterized by unique physicochemical properties that combined form multipurpose utility platforms for biosensors [41–44]. Other studies involved designing a hybrid lignin-polypyrrole material that exhibited very good capacity, in the order of 70 mAh/g. Such a system could be used as a potential source of renewable energy [45].

However Meng et al. described hydrogels with a lignin derivative used to obtain a sensitive biosensor for detecting Cr(VI), thus eliminating pollutants harmful to aqueous environments [46]. Graphene was added to the newly developed sensors to increase material conductivity. In the course of the research, it was found that the hydrogel/graphene/lignosulfonate had twice the conductivity of pure hydrogel/graphene material. This important property was attributed to the reversible charge being transferred by quinone groups in lignin derivatives. Besides hydrogel electrodes with lignin, the scientists also prepared a gel polymer electrolyte by dosing a liquid electrolyte onto a dry lignin membrane. The compatibility of this polymer electrolyte with the lithium anode in the system was found. Compared to the common polymeric gel electrolyte, the formed gel material indicated increased thermal stability, higher conductivity, and increased electrochemical stability. Based on the conducted tests, the scientists demonstrated that the obtained electrolyte had the potential for application in lithium-ion batteries [46].

Another approach to hybrid materials was adopted by a team studying a system based on nickel oxide and lignin derivatives. It carried a potential application for the biopolymer in systems, which could replace traditional lithium-ion batteries. The resulting anode was composed of mesoporous carbon obtained from modified lignin and NiO nanoparticles. The obtained nanospheres had a porous structure (micro- and mesoporous) and a BET (Brunauer-Emmett-Teller) surface area of approximately 852 m^2/g. This material exhibited excellent capacity, which decreased by only 8% after 2000 charging/discharging cycles [47].

Furthermore, other scientists concluded that lignin could increase the energy storage devices, such sodium-ion batteries. In this case, materials with lignin that could be used as precursors in generating inexpensive, highly effective nanomaterials for carbon electrodes to sodium-ion batteries were produced [48].

The authors of other, very interesting studies came to a conclusion that a mixture of silicon micro- and/or nanoparticles and lignin with an addition of carbon black enabled replacing conventional polymer binders. To obtain the desired product, the copper substrate was covered with the appropriate mixture. Next, the whole was subjected to thermal treatment resulting in obtaining a composite electrode with very good electrochemical efficiency. This method enabled producing highly effective and durable negative electrode with silicon micro-/nanoparticles, with an addition of a biopolymer [49].

A very important platform in terms of the original research developed over the recent years was the use of ionic liquids, constituting "green" compounds, for the modification/activation of lignin. A research aspect, extremely important in this context, was also an attempt to develop a method for the recovery of ionic liquid after the biopolymer activation process, hence, indicating the possibility of its repeated use. A properly designed activation process enabled to influence the chemical composition and structure of lignin; Thus, increase its bonding affinity with the inorganic carrier, in the context of engineering new, functional hybrid nanosystems used initially in the preparation of organic materials for electrode cells [28,50,51]

In the perspective of just the last few years, there has been a very significant increase in the interest in lignin and its derivatives in terms of designing new, functional materials/nanomaterials used in electrochemistry. This clearly indicates that, in the coming years, this field will undoubtedly be a source of very compelling, often ground-breaking scientific research and the entailing new solutions and technologies.

5.3.2 Construction industry

Lignin and its derivatives, as well as materials based on this biopolymer, are widely used in the construction industry (Fig. 5.5). The development of research in this aspect recently intensified mainly due to the search for functional nanoadditives to concrete specific properties, for example photocatalytic and antimicrobial properties [52].

Lignin, and even more often its derivatives—lignosulfonates, is used as concrete plasticizers, which are compounds reducing the water-cement ratio. They have been used for over 80 years; however their ability to reduce water is approximately 8%, which is lower than the value obtained for competing products currently used in the industry [53]. To increase the efficiency of the products in question as potential plasticizers, they are subjected to various modifications. For example, a biopolymer is exposed to fractionation in the first stage, and then modified once again as a result of oxidation, exhibits increased ability to reduce water volume [54]. Lignosulfonates can also be obtained with another lignin type, for example alkaline, by subjecting the biopolymer to the sulfomethylation process. Thus, the achieved product enables improving the workability of the concrete mix, owing to lower surface tension and zeta potential value [55].

An attempt to design functional hybrid nanoparticles as concrete admixtures has recently become an innovative approach. The study conducted by Klapiszewski et al. involved engineering two hybrid systems, that is, Al_2O_3-kraft lignin and Al_2O_3-magnesium lignosulfonate [56,57]. Both components were combined using the mechanical grinding method to provide the end product with appropriate homogeneity. In the course of the next stage, these systems were used as concrete admixtures. The application of this biopolymer led to improving the plastic properties of target products, whereas using a mineral ingredient resulted in improved strength properties, compressive strength in particular. Therefore, appropriate production and use of hybrid nanomaterial enables obtaining target cement composites with unique final properties that satisfy the very stringent structural, economic, and operating criteria [56,57]. Another hybrid system tested in terms of application as cement composite admixture comprises silica-lignin nanomaterials. The referenced material positively impacts the rheological and strength properties of cement mortar. Adding hybrid material, with a silica to lignin weight ratio of 1:5, improves the compressive strength of concrete by 40%, relative to the reference mortar. However in the

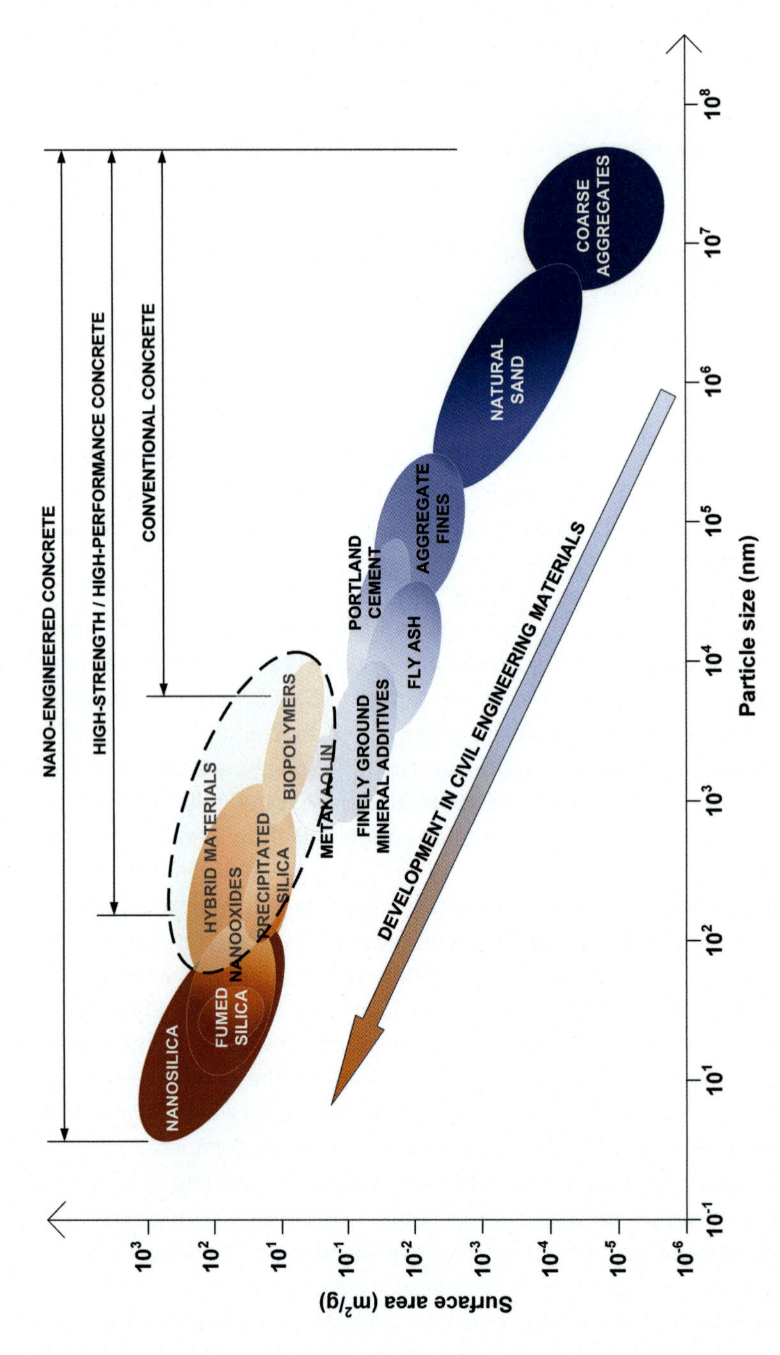

FIGURE 5.5 Development in civil engineering materials, adapted from [52].

case of an admixture with pure silicon dioxide, the authors were able to improve this parameter by only 10%–20% [58].

In addition, Norgbey et al. used lignin as a bitumen filler and modifier [59]. The discussed biopolymer was used to decrease environmental pollution as well as to limit the dependence on bitumen produced from crude oil. Increased lignin content results in higher stiffness and cohesion of the bituminous binder. Furthermore, the obtained binding material exhibited better adhesion with aggregate than pure bitumen. It was also concluded that a binder modified with an addition of waste lignin would return to its original shape faster after removing the force acting on it, and would exhibit lower strain within the tested temperature ranges and at various stress values [59]. Another study found that asphalt mix, where the content of lignin-containing industrial waste was 20% wt., relative to bitumen would exhibit similar resistance to moisture as the test sample and was less susceptible to permanent strain at high temperatures. On the other hand, it is characterized by higher fatigue strength at lower temperatures. Based on the conducted tests, it was concluded that the discussed asphalt mix was suitable for roads with medium and low traffic intensity [60].

The application of lignin as an additive in producing anticorrosion coatings is also important. This includes protecting aluminum and carbon steel, the properties of which make them widely used in numerous industry branches, also in construction. To use biopolymer in this context, it is subjected to various modifications. For example, one method of kraft lignin modification can lead to the formation of silane groups on the macromolecular surface. The main objective of the process in question is to increase the biopolymer adhesion to the material surface, aluminum in this case. Some of the conducted research led to a conclusion that coatings with silanized lignin were better at inhibiting corrosion than a passive layer formed by aluminum oxide [61]. The aforementioned oxide is not stable in alkaline pH, wherein aluminum can be covered by a film of lignin enhanced with carboxy groups, which are good at inhibiting the discussed process of destruction in an alkaline environment [62,63]. On the other hand, Hussin et al. verified three lignin types, namely, kraft, sodium, and organosolv as corrosion inhibitors in mild steel under acidic conditions. Based on the conducted studies, the researchers concluded that the highest corrosion-inhibiting ability was exhibited by a coating with a biopolymer synthesized within the sodium process with its concentration of 500 ppm [64].

The widely understood construction industry has recently evolved significantly, which is associated with intensively developing studies aimed at engineering newer, more functional nanomaterials, also with lignin. These systems can not only lead to improving important strength properties, but also exhibit numerous, often specific features, including antimicrobial activity, which is so important nowadays in terms of care for the environment.

5.3.3 Polymer industry

Lignin, as an aromatic biopolymer, is a potential substitute for polymers obtained from crude oil owing to comparable or improved physicochemical parameters and lower production cost. The presence of numerous functional groups within the aromatic rings, including hydroxy, results in it being used as a starting raw material for the synthesis of many polymer groups (among others, polyethers, polyesters, PE, polyurethanes) [65].

In addition, the source literature indicates the possibility of using lignocellulose materials, including pure lignin or materials with lignin, as fillers for a wide group of polymers, both highly polar (PE terephthalate, PE oxide) [66–69], as well as hydrophobic (polypropylene—PP [70–80], PE [81–87]) in polymer matrices. The studies also involved poly (vinyl chloride) [88–90]. They demonstrated better compatibility of polyvinyl chloride and lignin for a nonplasticized material, relative to plasticized one.

The polymer industry, which has been recently developing very dynamically, is searching for newer solutions in the field of designing functional hybrid nanomaterials used subsequently as polymer fillers. One such example are the pro-ecological silica-lignin hybrid materials, which are distinguished by very good thermal stability properties, obtained at the fundamental research stage [76–78]. However the results of conducted strength tests confirmed that introducing the smallest amount of silica-lignin hybrid material (2.5 wt%) to PP leads to a rather significant increase of elongation at rupture, compared to PP without a filler. When the hybrid filler concentration was increased to 5.0 wt%, the elongation reduced largely, reaching a value, which was much lower than in the case of pure PP. This phenomenon can be associated with reaching the optimal concentration resulting from the presence of inorganic–organic hybrid filler in the composite [76,78].

The research of other scientists involved creating composites as a result of introducing chemical homogenizing agents, causing a reduction in lignin hydrophilicity, and improvement in the mechanical properties of end products. Lu et al. [73] used PP grafted with maleic anhydride, and hyperbranched polymer lubricant as a compatibilizer, in the amount of 3–5 wt%, relative to lignin (PP content was 50 wt%). Adding compatibilizers also increased the impact strength and flexural strength, and scanning electron microscopy (SEM) microscope images also confirmed increased dispersion degree. Similar results were obtained by El-Sabbagh et al. who proved that the homogeneity of created PP-g-MAH (maleic anhydride-grafted polypropylene) resulted in the increased tensile strength and better impact strength of a composite [74].

The effect of compatibilization using the same modifier was used in the research by Morandim-Giannetti et al. [72]. Compatibilization with the use of PP grafted with maleic anhydride named PB 3200, in the amount of 4 wt%, relative to a 10 wt% content of lignin in the composite, resulted in increase of the tensile strength by almost 80%, whereas for a composite without the compatibilizer, the value of this parameter was significantly lower. Comparing the obtained strength test results for the SiO_2-lignin system [76,78] with the composites produced by Morandim-Giannetti et al. (PP with a filler content of 10 wt% and the PP-g-MAH modifier content of 4 wt%) [72], it was observed that better mechanical properties were obtained for silica-lignin materials. Composites characterized in the publication [72] exhibited tensile strength in the order of 20 MPa and elongation at rupture at a level of 2.5%.

In addition, a team led by Klapiszewski and Jesionowski obtained PE-LD (low-density polyethylene) composite films with a silica-lignin hybrid filler addition [85]. This enabled the scientists to conclude that the application of selective extrusion conditions, such as high receiving roller speeds (at least 7.5 m/min) and a 2.5 wt% and 5.0 wt% share of the filler allowed obtaining systems with better barrier and mechanical conditions than in the case of pure PP, without added filler. However, it should be noted that previous homogenization of introduced hybrid fillers in a corotating twin-screw extruder with mixing and kneading zones was of crucial importance in the case of such systems. In other studies, SiO_2-lignin systems were used by the researchers also as proecological poly(vinyl chloride) [88,89] or polylactide [91,92] fillers.

The next step in designing functional hybrid products was the development, by Klapiszewski et al., of other concepts in terms of obtaining ZnO-lignin materials with potential antibacterial properties, which were then used as high-density PE (PE-HD) [84] and PP [79] fillers. Using ZnO allowed to increase the thermal stability and mechanical parameters of the final hybrid material, whereas using the biopolymer enabled reducing its production costs. The extremely important test results, the objective of which was to determine antibacterial properties, confirmed that the ZnO-lignin hybrid material (1:5 wt./wt.) exhibited very good antimicrobial activity, especially toward Gram-positive bacteria (for the tested genuses of *Staphylococcus* and *Bacillus).* These conclusions confirmed the possibility of conducting further research on the application of a hybrid as a PE-HD filler [84]. The next stage of the work involved creating containers with the share of an inorganic-organic hybrid through extrusion with the blow molding process, which contributed to increased compressive strength of the end products, without any deterioration of the material's processing properties, such as the mass flow rate. Furthermore, the modification of the starting material using a hybrid filler in the amount of 5 wt% lead to a twofold increase of the destructive compressive force of the containers made using it, relative to the ones made with pure PE-HD. The favorable change of the critical compression force for containers with the ZnO-lignin material (1:5 wt./wt.) can arise from lignin acting as a plasticizer, which facilitates PE-HD particle stretching and orientation, decreasing material anisotropy. At the same time, a lower composite crystallinity degree does not result in the decline of its mechanical properties. The conducted tests allowed to confirm the possibility to use a hybrid nanofiller with lignin and zinc oxide, as a modifier of PE-HD properties, intended for packaging manufacturing. The originality of these tests is based primarily on the application of a new class of functional nanofillers, which exhibit antimicrobial activity and comprehensive action in a polymer system. An additional advantage is the possibilities of achieving desired features without deteriorating the mechanical and processing properties of end products [84]. Owing to the very interesting results of the tests concerning ZnO-lignin systems, the scientists also attempted to design MgO-lignin hybrid materials, which were successfully used as proecological PP [80] and PE [86, 87] fillers. They were characterized by very good thermal, mechanical, and processing properties, which enabled to draw an initial conclusion regarding the possibility of their practical use in the industry.

Currently, the plastics market is one of the most prospective markets. Therefore, it seems absolutely justify to look for new nanomaterials that will be used as functional, proecological polymer filler with success. It is expected for the activity in this regard to be even more intensified in the nearest future.

5.3.4 Production of abrasives

Owing to the extensive, aromatic chemical structure of lignin and its similarity to phenol-formaldehyde resins in this respect, the search for newer ways to use this biopolymer as a substrate or additive to other polymers are being sought [93].

Phenol-formaldehyde resins are very widespread and used in the manufacturing of, among others, abrasive tools, owing to their high thermal and chemical resistance, good mechanical properties, and low price. An unquestionable disadvantage of phenoplasts is the tendency to release harmful compounds, mainly phenol and formaldehyde, in elevated temperature. This adverse phenomenon caused the developing research work to focus, among others, on the application of lignin in resins, not only as a substitute for phenol at the synthesis stage, but also as alternative phenol resin additives in composites based on them [94,95].

An important research direction in this perspective has become the appropriate design, production, and characterization of innovative abrasives, obtained based on ecological binders with lignin or its derivative—lignosulfonate [96–103]. The primary objective of these studies was to create a binder based on phenol-formaldehyde-lignin resin. These binders are primarily environmentally friendly through limited release of harmful compounds, that is, phenol and formaldehyde. This idea was based on a hypothesis claiming that because lignin or its derivatives were phenol oligomers, they could partially replace phenol in a phenol-formaldehyde resin and be used to obtain a proecological binder with properties similar to the commonly used phenol resin. As part of the research, the scientists demonstrated, among others, that lignin interacted with phenol-formaldehyde resin, acting as a curing (cross-linking) agent. Owing to the appropriate activation of lignin or lignosulfonate, that is, their oxidation (chemical functionalization), the authors changed the structure of the biopolymers, simultaneously causing increased reactivity of these compounds, which is shown in Figs. 5.6 and 5.7 [99].

Alternatively, lignin was combined with selected inorganic components (SiO_2 [97,98], Al_2O_3 [100,103], boron nitride [102]), with the objective of obtaining hybrid bifunctionalized material (a novel binder and a functional abrasive component). Using an inorganic component was also aimed at increasing the thermal stability and mechanical properties, which is crucial in terms of abrasive tools.

The research work [100, 103] involved the creation of Al_2O_3-lignin hybrids of various components weights. For this purpose, the authors proposed mechanical grinding of the precursors with simultaneous mixing using a planetary ball mill. It is an uncomplicated method, which at the same time does not burden the environment with the use of harmful chemicals, often applied in the course of synthesizing advanced materials and nanomaterials. The possibility of using unconventional additives in abrasives has been confirmed by a comprehensive analysis of the test results concerning surface and performance properties, through, that is, reversed gas chromatography, testing the degree of adhesion between the components, testing the thermo-mechanical and rheological properties. By adding a hybrid material to the mixture, thus obtaining the final abrasive, the researchers were able to create products with higher ductility, which was the result of added biopolymer. Furthermore, it was demonstrated that a slight addition of alumina (lignin:Al_2O_3 system equal to 8:1 wt./wt.) can increase the thermal conductivity of

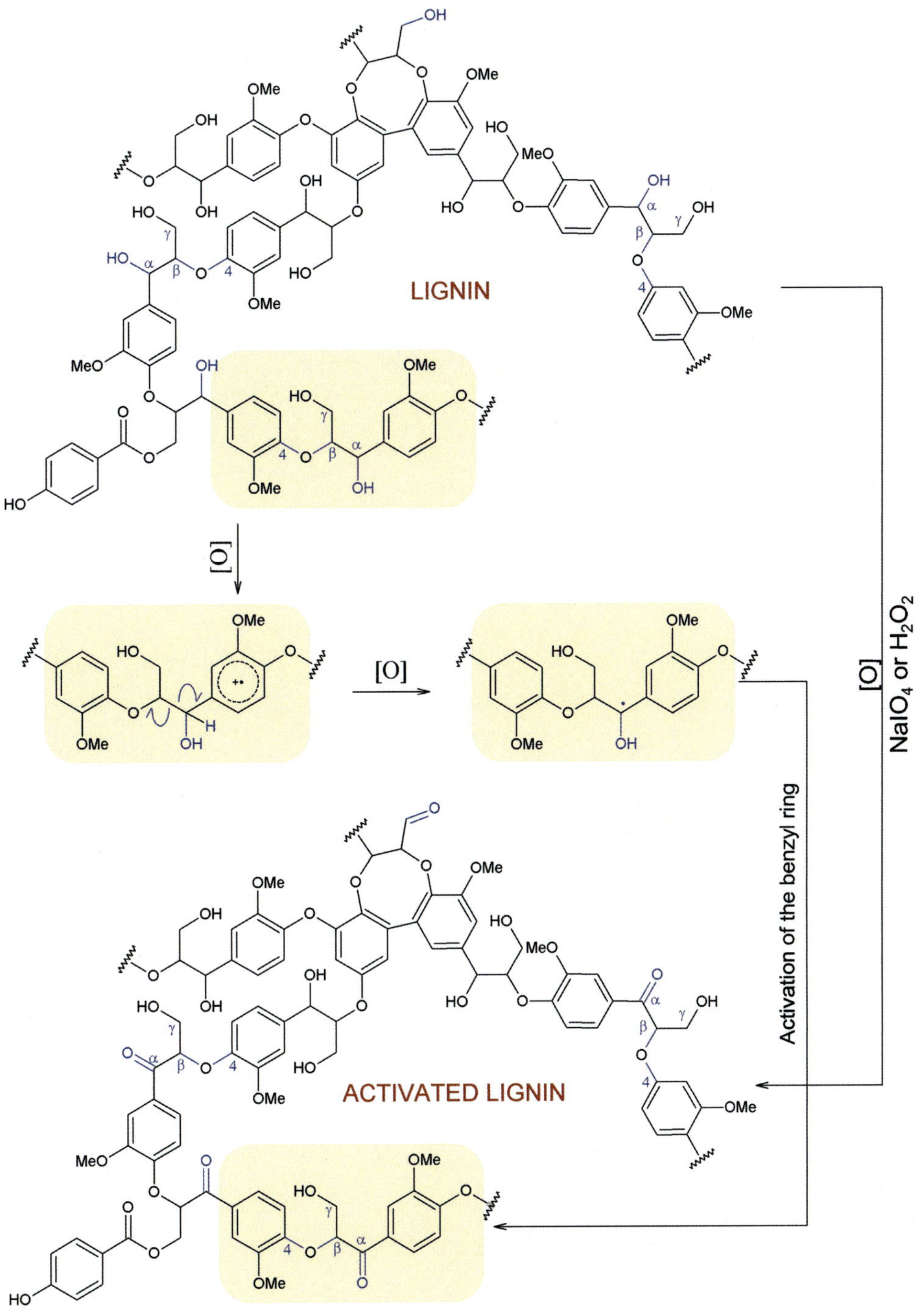

FIGURE 5.6 Proposed mechanism of activation of kraft lignin (chemical functionalization), adapted from [99].

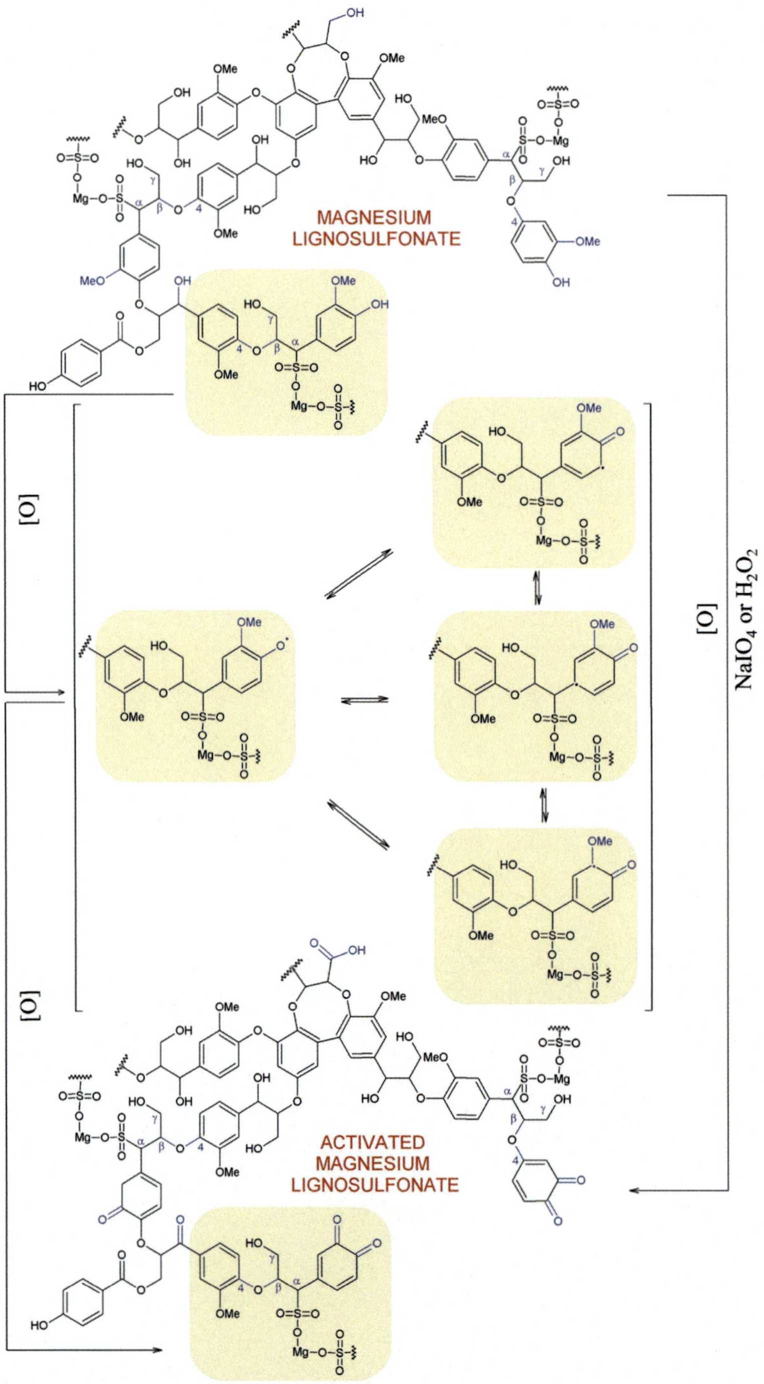

FIGURE 5.7 Proposed mechanism of activation of magnesium lignosulfonate (chemical functionalization), adapted from [99].

lignin; hence, can improve the thermomechanical properties of the end composition. However, the most important conclusion arising from the conducted research is the fact that adding a hybrid material to an abrasive composition significantly reduced phenol emission, as well as slightly restricted formaldehyde emission, compared to the commercially used filler in the form of Micro 20 zeolite, as well as pure kraft lignin [102,103].

A more in-depth step in terms of studying this research topic, associated with designing functional additives to abrasives, was the creation of a composition with boron nitride-lignin hybrid materials [102]. The conducted studies on thermal stability allowed to confirm that boron nitride, owing to its very good thermal stability (mass loss ˜2% in up to 1000 °C), enabled obtaining additives with increased resistance to high temperature. Furthermore, the application of an inorganic component enabled maintaining mechanical properties at an appropriately high level, which does not disqualify using the aforementioned systems in industrial practice. At the same time, the organic part—lignin—plays a very important role in the composition allowing to limit the release of harmful organic compounds (phenol and formaldehyde), generated in the course of operation, but also very often during the storage of abrasive tools. The example of the cBN-lignin hybrid materials clearly shows the contemporary significance of hybrid materials and their precise engineering, so as to maximize the functionality of single components in the perspective of advanced applications.

5.3.5 Application in environmental protection

Lignin and its derivatives can constitute one of the potential, inexpensive, and readily available biosorbents of environmentally harmful compounds [104–110]. This biopolymer is often chemically modified to increase the number of functional groups (i.e., hydroxy, carboxy, phenolic, ether, carbonyl, and ketone) [111,112]. This is fully justified due to the fact that unmodified lignin exhibits poor porous structure parameters, that is, BET surface area (1–5 m^2/g). The mechanism of adsorbing the ions of environmentally harmful metals on the lignin surface is based on the reaction of characteristic functional groups. These groups have the ability to bind environmentally harmful metal ions through returning the electron pair and thus forming complexes within the solution or new chemical bonds with the adsorbate [113].

In terms of engineering functional biosorbents, the attempt to develop new hybrid nanomaterials seems justified. One of the examples of such systems is the SiO_2-lignin hybrid developed by the team led by Klapiszewski and Jesionowski [114, 115]. The conducted adsorption tests clearly indicated a significant sorption capacity of the silica-lignin system relative to nickel (II), cadmium (II), and lead (II) ions [115]. The obtained ion adsorption experimental data finely describe the model of pseudo-second order kinetics model, which is confirmed by the high correlation factor ($r^2 = 0.999$–1.000). In addition, the conducted research also demonstrated that the proposed sorbent was perfect as an effective and economic sorption material, which was confirmed by satisfactory desorption test results. The team of Budnyak et al. also worked on designing SiO_2-lignin nanomaterials [116].

Another nanomaterial designed and obtained by the Klapiszewski team was the TiO_2-lignin system, which was also used in environmental protection to remove inorganic and/or organic pollutants, primarily ions of metals harmful to humans and the environment (lead(II), cadmium(II), copper(II)) [117, 118]. Furthermore, a part of the aforementioned work involved using hybrid TiO_2-SiO_2-lignin [117, 118] and MgO-SiO_2-lignin [118, 119] systems.

Not only the relative high value of the BET surface area of the used material, which in the case of the MgO-SiO_2-lignin system amounts to 312 m^2/g, but also the diversity of functional groups present on the surface of previously activated lignin is of great significance in the adsorption process [118]. It was this feature that was the main reason behind undertaking research on the utilization of the aforementioned systems as effective biosorbents.

A part of the study [117] involved testing the adsorption of Pb^{2+} ions on TiO_2-lignin and TiO_2-SiO_2-lignin carriers. It was concluded that along with the increase of the initial lead(II) ion concentration from 25 to 100 mg/L, over 1–180 min, the maximum adsorption capacity also gradually increased. Adsorption equilibrium was achieved after, 20 and 30 min, respectively. In addition, the detailed tests involving process kinetics enabled a conclusion that the pseudo-second-order model was precise in describing the lead(II) ion adsorption kinetics. This is confirmed by the high values of correlation factors, and the good quality of estimation and matching of experimental data and the determined sorption capacity values. Furthermore, based on the determined experimental data, a conclusion was drawn that the Pb(II) ion adsorption process could be well defined according to the Langmuir isotherm model. Higher sorption capacity for removing the aforementioned ions from aqueous solutions was observed for the TiO_2-SiO_2-lignin hybrid nanomaterial (59.93 mg/g), which indicates that this system is a better sorbent than TiO_2-lignin (35.70 mg/g).

A series of studies addressing the issue of using functional hybrid nanomaterials as effective biosorbents of ion metals harmful to the environment is supplemented by the publications [118, 119]. The studies involved adsorption of copper(II) and cadmium(II) ions on similar systems as in [117], and, in addition, utilized the MgO-SiO_2-lignin hybrid sorbent, which turned out to be the most effective. At the same time, adequate test results associated with the adsorption process and the kinetic aspect parameters were also obtained. It was additionally proven that the adsorption of selected metal ions on the surface of hybrid materials was a heterogeneous, endothermic, and spontaneous reaction, which is confirmed by the calculated values of the thermodynamic parameters, such as free energy ($\Delta G°$), enthalpy ($\Delta H°$), and entropy ($\Delta S°$) for the studied system in various temperatures. To confirm the effective adsorption of selected metal ions, the authors of the publications [117–119] used such research methods such as SEM, low-temperature nitrogen sorption (BET, BJH - Barrett-Joyner-Halenda), X-ray photoelectron spectroscopy, Fourier transform infrared spectroscopy, and the pH-dependent zeta potential analysis. The results confirm the validity of the suggested adsorption process as an effective method for removing environmentally harmful metal ions with the use of functional hybrid systems. The most general presentation of the adsorption process mechanism is shown in Fig. 5.8.

However, as part of other research work, a team led by Wahlström modified the organosolv process product (OLS) using glycidyltrimethylammonium chloride (GTAC) in the presence of sodium hydroxide [120]. The synthesis took 20 h, at a temperature of 60 °C, and the ingredients were constantly mixed using a magnetic stirrer. After completing the reaction, the obtained material was cooled and appropriately cleaned. The thus-obtained substance was then tested

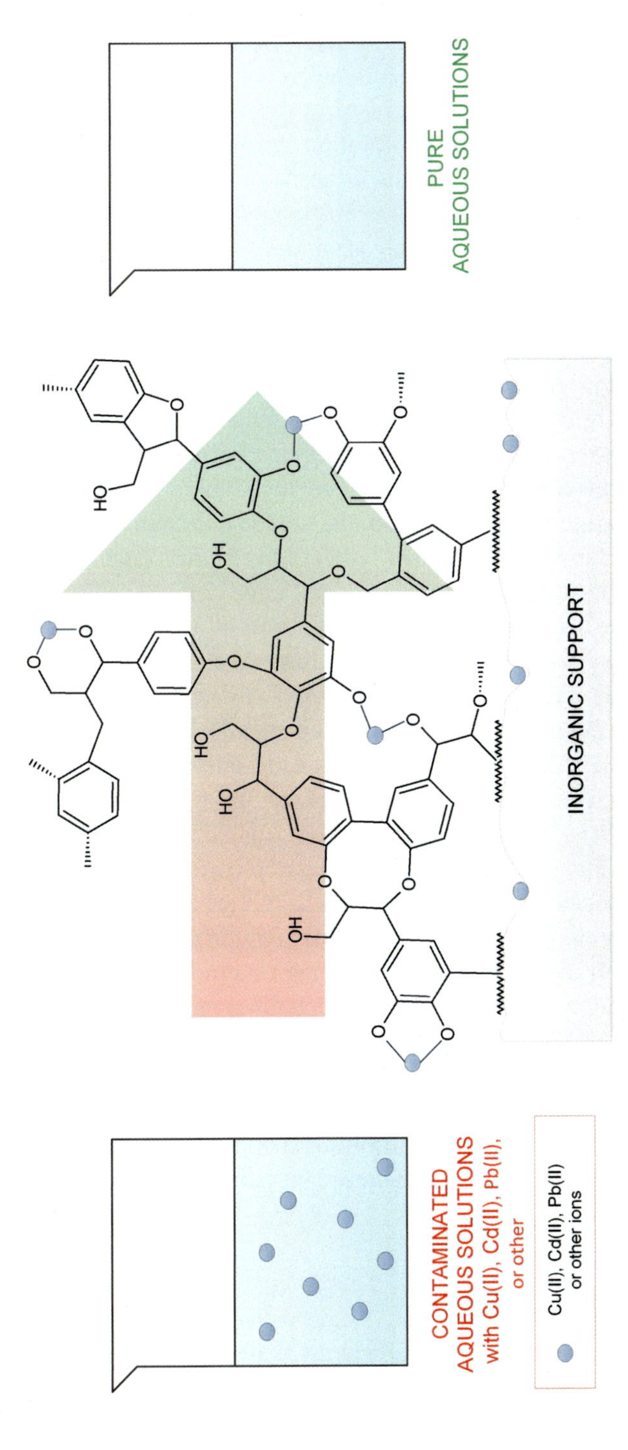

FIGURE 5.8 Diagram of removal of environmentally harmful metal ions using hybrid sorbents, adapted from [118].

in terms of its adsorption and precipitation properties. One of the most popular water pollutants are sulfate ions. The tested product adsorbed sulfates well, yet the effects were not satisfactory. The formed adsorbent–adsorbate system exhibited very good solubility in water, which meant it could not be separated from the solution using simple methods. GTAC-OLS cross-linking would probably enable effective material filtration. Another test studied the ability of modified lignin to precipitate kaolins. The results in terms of the analyzed product are good, with its presence resulting in the studied impurity to be precipitated with the expected outcome, and the postreaction mixture could be easily separated. The Wahlström team also studied the coagulation potential of the overview material. The conducted experiments enable a conclusion that GTAC-OLS, after a few modifications, can be used as a multipurpose material for wastewater treatment [120].

The team lead by Liu developed a material that is able to adsorb ions such as Pd^{2+}, Cu^{2+}, and Cd^{2+} [121]. Lignin obtained from the paper industry was used to create the adsorbent. To fully utilize the selected substance, it was necessary to appropriately prepare it initially. To obtain the adsorbent, the first stage involved mixing together alkaline lignin with N,N'methylenebisacrylamide and ammonium persulfate, and then heating in 60 °C for 30 min. Next, the system was cross-linked with acid radicals. In the last stage, the synthesized product was subjected to swelling and freeze drying. The obtained composite (PAA-g-APL - polyacrylic acid with acid-pretreated alkali lignin) was tested in terms of, among others, the ability to absorb selected water pollutants. The conducted tests indicate that modified lignin causes an increase in the structural strength of the polyacrylic acid (PAA). Owing to the increased durability of the material, it can adsorb greater concentration of heavy metal ions. It was also found that the presence of a biopolymer prevented leaching of adsorbed substances through chelating and blocking them after reaching adsorption equilibrium. The test results allow to conclude that the described material is good for attracting such ions as Pd^{2+}, Cu^{2+}, and Cd^{2+}. An experiment involving mixed wastewater was also conducted. Based on its effect, it can be concluded that the discussed composite was selective in adsorbing lead ions first, copper ions second, and cadmium ions in the end. The presented results are enough to conclude that PAA-g-APL is a very promising material with application potential in the wastewater treatment process. These studies provide also interesting information on the adsorbent, which differentiates between Pb^{2+} and Cu^{2+} ions in polluted waters [121].

The application of lignin and its derivatives as effective biosorbents, primarily owing to the extensive structure of the biopolymer, associated with a significant content of various functional groups, seems to be highly justified. The multitude of engineering possibilities in terms of new hybrid nanomaterials, strongly expanded in the recent years, clearly indicates increasing interest in the aforementioned field.

5.3.6 Medicine and pharmacy

Lignin and its derivatives are still an insufficiently exploited natural carbon source. Despite their availability and structural potential, the interest in using lignin in the course of micro-/nanoparticles development has grown only in recent years. Regardless of such significant

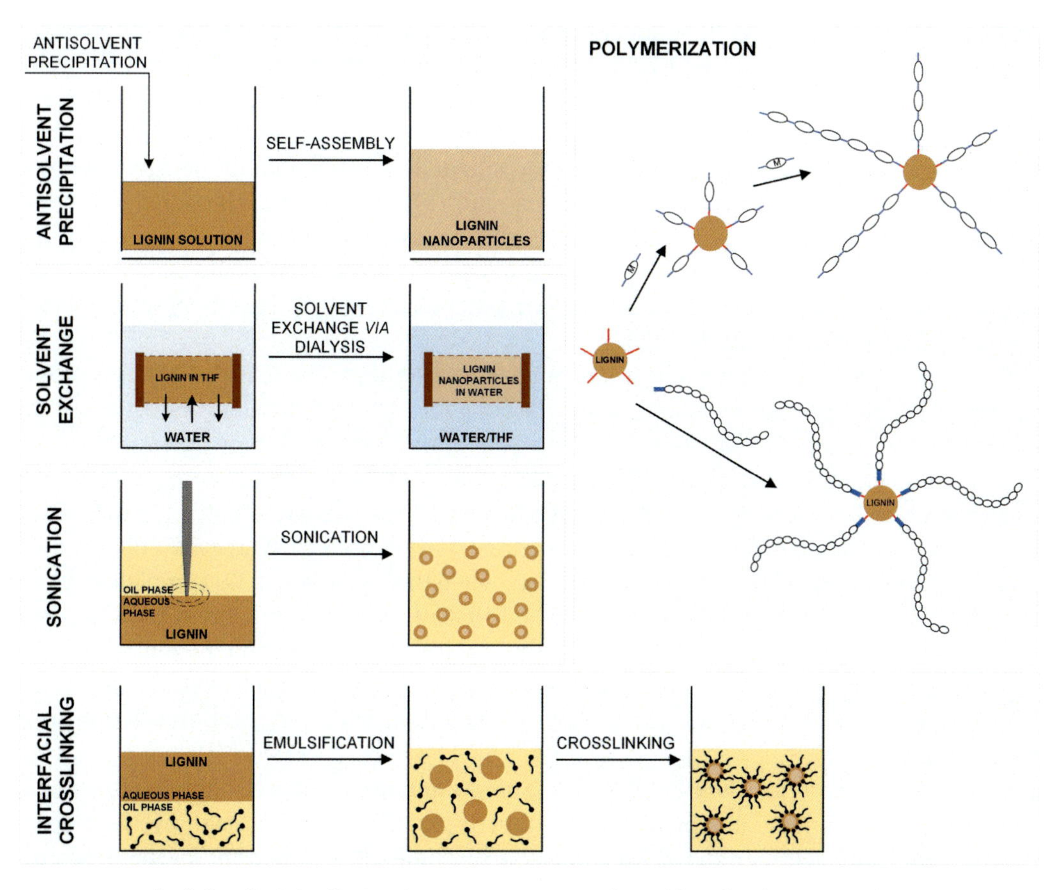

FIGURE 5.9 Methods for obtaining lignin micro-/nanostructures, adapted from [126]. THF - tetrahydrofuran

advantages, its application is limited due to the complex macromolecular structure, which depends on the plant's origin and its processing methods. One of the ways to overcome this restriction is to transform native lignin into micro-/nanoparticles of uniform shape and size [29,122]. Lignin has been successfully used for producing various micro-/nanoparticles, such as micro-/nanomolecules, micro-/nanotubes, micro-/nanofibers, and hydrogels, for different applications in medicine and pharmacy [29,123].

The development of lignin micro-/nanoparticles, also called LNP, has crucial advantages such as improving the polymer mixture properties [124] and higher antioxidant activity of the substances [125]. Furthermore, LNPs contain functional groups, which can be chemically modified, significantly increasing their application-wise potential [126]. Various methods were used to form micro-/nanoparticles, such as antisolvent precipitation, interfacial cross-linking, polymerization, solvent exchange, and sonication [126]. Each of the presented approaches enables producing micro-/nanoparticle of characteristic shapes [22]. Fig. 5.9 shows different methods for obtaining lignin micro-/nanostructures. Micro-/nanoparticles involving the described polymer can be potentially used on the functional surface of the coating (e.g., micro-/nanocomposites), as well as in biomedicine, such as drug/gene delivery or tissue engineering [126].

Anti-solvent precipitation involves introducing another solvent to the lignin solution. The outcome of such an experiment is spontaneously forming micro-/nanoparticles of the studied polymer [127]. Good results are observed for the ethanolic combination of the lignin solution with added water. A lignin solution with a concentration of 2.285 mg/mL is prepared to conduct the experiment. Next, water is added to a 3 L solution using a peristaltic pump, until water amounts for ca. 95% of the solution. The entire mixture has to be constantly agitated using a magnetic stirrer. The solution conducted in this manner enables obtaining empty sphere, which can find a number of applications [127].

Obtaining LNP through polymerization can follow two paths. The first one involves proper preparation of polymer particles, and then subjecting it to selected modifications. The second method enables attaching long chains of selected compounds to fragmented lignin [126].

LNP can also be obtained through the solvent exchange method. They utilize the hydrophobic properties of the biopolymer, which enable the formation of aggregates in an aquatic environment. This micro-/nanostructure formation method can be used to synthesize spheres filled with, for example, drugs. This process can be conducted through dialysis. To complete it, lignin and medicinal substances, dissolved in a selected solvent, should be properly prepared. Then, such a solution is introduced to a dialysis bag to replace the selected diluent with water. This way, anticancer drugs can be administered to be released after being delivered to appropriate cells. This is a very specific application, which can revolutionize the way of administering drugs in the future [128].

One of the methods for obtaining LNP is the sonication of two-phase solutions. To conduct the experiment, the initial solutions should be properly prepared. The next stage involves starting an ultrasonic probe between the phases. The process duration is important for the properties of obtained particles. The longer the sonication process, the smaller particles can be obtained, however, always to a certain limit. To facilitate the formation of micro-/nanoparticles, surfactants, which enable faster dispersion, can be added to the system. Micro-/nanoparticles obtained through this method can be used as carriers for drugs or other substances, which enable more accurate diagnostics of selected diseases [129].

Nanomaterials with lignin, exhibiting antioxidant, radiation-absorbing, and antimicrobial features constitute a good basis for producing agents strengthening nanocomposites, delivering drugs and genes, and for other biomedical purposes [6,29,130]. Lignin can be used to produce a nanoparticle for encapsulating various compounds for pharmaceutical applications, owing to the low costs and eco-friendly properties. Frangville et al. [130] applied the antisolvent precipitation method, where lignin was dissolved in ethylene glycol, thus obtaining biopolymer nanoparticles of various sizes, together with hydrochloric acid. This material did not show any visible cytotoxicity to yeasts and microalgae, which is why it was considered a promising carrier for delivering drugs and stabilizers to cosmetic and pharmaceutical preparations [6,29]. Richter et al. [131] also prepared lignin nanoparticles, using the same precipitation method, and then saturated them with silver ions and coated with poly(diallyldimethylammonium) chloride, the so-called PDAC. This coating promoted the adhesion of LNP saturated with silver ions to bacterial membranes, which aided the bacterial neutralization process [126].

Very2 interesting research results were the achievement of Chinese scientists who suggested a model of empty lignin spheres with controlled porous structure, with no need to remove the matrix or core during synthesis. Such a solution may become an alternative in terms of biomedical applications, that is developing techniques for releasing active substances from carriers in the form of spherical structures [132].

Hydrophobic interaction of lignin in the oil phase has been recently proven. The scientists obtained spherical lignin systems through cross-linking it, using high-intensity ultrasounds and a cross-linking compound addition [133]. Oil-filled capsules were tested in terms of storage and transport of hydrophobic molecules. A fluorescent substance was added to their interior to determine their properties. After placing the capsules in surfactant solutions, the dosed compound was released. This experiment flags the new direction for the applications of modified lignin. Oil-filled capsules constitute a basis for cosmetic applications, where the oil phase will be educed during direct contact with the skin.

As known, the development of contemporary medicine has been recently highly intensified. Therefore, further search for new nanomaterials, which can be used in the field of medicine, is now a priority, providing chances to overcome numerous diseases, unbeatable just up till several years ago. The validity of the research conducted in this respect is indisputable, and the subjects studied by the scientists, indicated in this section, indicate its high significance.

5.4 Conclusions and perspective for further research

Among the technologies currently being implemented, trends related to searching innovative solutions combining low production costs with the use of environmentally friendly nanomaterials are on the rise. Developing such a method can be significantly hindered as the produced material must additionally met predefined requirements and exhibit unique properties. In general, the optimal solution to this issue is using inexpensive and widely available natural raw materials, and their preparation, which would enable obtaining the desired features, while maintaining a number of advantages that determined the selection of this material.

The attractiveness of products of natural origin also arises from the increasing ecological awareness of the society, which is willing to select biodegradable products on a growing number of occasions. Great significance of lignin can be predicted on this basis. A total of 300 M tons of cellulose and paper materials have been manufactured worldwide in the last 10 years. With an average content of lignin in wood cells at 25%, its production scale can be estimated at a level of 75 M tons. Furthermore, its widespread availability in the natural environment can justify the interest in developing the acquisition of valuable everyday products from lignin. The interest in lignin and biomass as an energy source or in "green" chemicals falls in line with the current eco-friendly policy. Alternative lignin application directions, which have been thoroughly presented in individual sections, can also be distinguished.

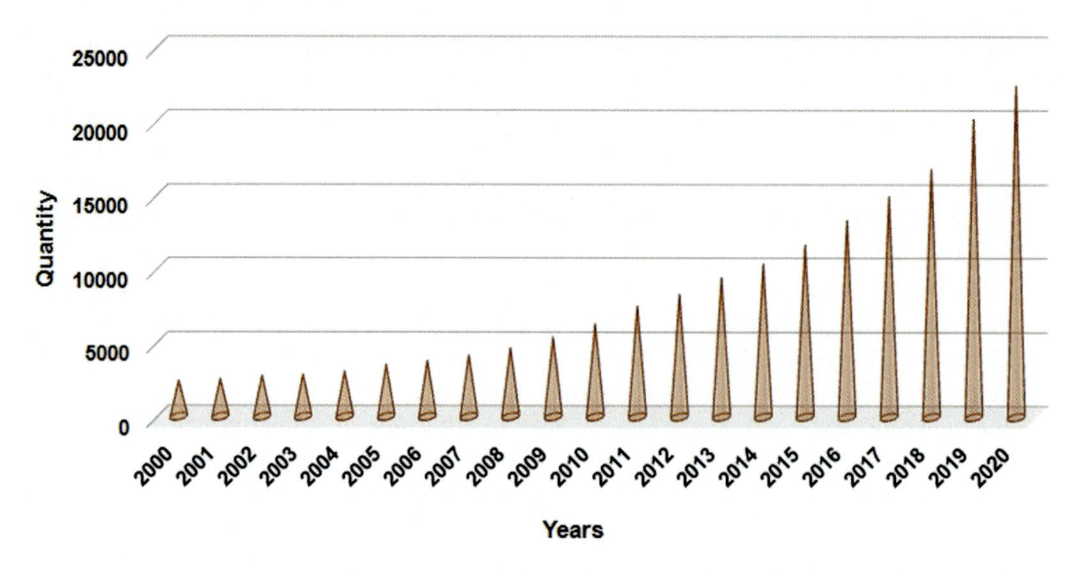

FIGURE 5.10 Search results for 'lignin' in the SCOPUS database from 2000 to 2020.

Increasing the proecological awareness of the society can also be observed based on the lignin-related reports emerging in the source literature. According to the SCOPUS database, over 200 thousand papers with the phrase "lignin" have appeared until 21 November 2020. The increase in the number of citations with the aforementioned term over the recent years has been very intensive, which is demonstrated by the attached Fig. 5.10.

Furthermore, biodegradable materials/biomaterials, hybrid materials, or composites/biocomposites obtained from renewable sources (biomass) of nanometric sizes have recently gained large interest. In the future, such systems can replace synthetic polymers and contribute to reducing the global demand for petroleum products. Organic–inorganic hybrid nanomaterials are gaining importance not only scientifically but also technologically. The systems are obtained from widely available and easily degradable products, in particular. For this reason, materials using biopolymers, mainly lignin, are under intensive development. The selection of this biopolymer is determined primarily by its diverse chemical structure, as well as ease of obtaining, resulting from the production process, which accompanies the acquisition of cellulose fibers. Furthermore, lignin is the second most available substance on the earth, right after cellulose. In addition, this biopolymer stands out among other products owing to such properties as biodegradability, antioxidant and antibacterial activity, good chemical reactivity, ability to reverse electron and proton transport, affinity with inorganic oxides, or the possible sorption of harmful compounds from the environment through the variety of functional groups. The aforementioned characteristics were utilized in the practical chemistry of functional nanomaterials. Moreover, it should be stressed that lignin and its derivatives are renewable resources, which fall in line with the global eco-friendly trends. What is more, their relatively good chemical reactivity makes them very prospective.

The field of research associated with obtaining new materials and/or biomaterials is of huge significance for the development of science, chemical technology, material engineering, and the widely understood environmental protection in particular. The further progress of knowledge in this regard will allow to understand "green technologies" better, above all with care for the natural environment, hence bringing tangible, also social, benefits.

Acknowledgments

The work has been conducted within a project financed from the funds of the National Science Centre Poland No. 2019/35/B/ST8/02535.

References

[1] D. Klemm, B. Heublein, H.P. Fink, A. Bohn, Cellulose: Fascinating biopolymer and sustainable raw material, Angew. Chem. Int. Ed. 44 (2005) 3358–3393.

[2] S. Kalia, B.S. Kaith, I. Kaur, Cellulose Fibers: Bio- and Nano-Polymer Composites, Springer, Berlin, 2011.

[3] A. Rouilly, C. Vaca-Garcia, Bio-Based Materials, in: J. Clark, F. Deswarte (Eds.), Introduction to Chemicals from Biomass: 6., John Wiley & Sons, NJ, 2014, pp. 205–248.

[4] A.J. Ragauskas, G.T. Beckham, M.J. Biddy, R. Chandra, F. Chen, M.F. Davis, B.H. Davison, R.A. Dixon, P. Gilna, M. Keller, P. Langan, A.K. Naskar, J.N. Saddler, T.J. Tschaplinski, G.A. Tuskan, Wyman Ch.E., Lignin valotization: improving lignin processing in the biorefinery, Science 344 (2014) 1246843.

[5] C. Heitner, D. Dimmel, J. Schmidt, Lignin and Lignans – Advances in Chemistry, CRC Press, Boca Raton, 2010.

[6] J. Huang, S. Fu, L. Gan, Lignin Chemistry and Applications, Chemical Industry Press, Amsterdam, 2019.

[7] O. Franuk, M. Sain, Lignin in Polymer Composites, Elsevier, Oxford, 2016.

[8] F. Garcia, C. Floes, J.A. Dobado, Lignin as renewable raw material, ChemSusChem 2 (2010) 1227–1235.

[9] M. Ek, G. Gellerstedt, G. Henriksson, Pulp and Paper Chemistry and Technology Volume 1 Wood Chemistry and Wood Biotechnology, Walter de Gruyter, Berlin, 2009.

[10] L.A. Lucia, O.J. Rojas, The Nanoscience and Technology of Renewable Biomaterials, John Wiley & Sons, NJ, 2009.

[11] A. Björkman, Studies on finely divided wood. Part 1. Extraction of lignin with neutral solvents, Sven. Papp. tidn. Nord. Cellul. 59 (1956) 477–485.

[12] F.E. Brauns, Native lignin I. Its isolation and methylation, J. Am. Chem. Soc. 61 (1939) 2120–2127.

[13] A.M. Boudet, Lignins and lignification: selected issues, Plant Physiol. Biochem. 38 (2000) 81–96.

[14] R. Zhong, Z.H. Ye, Transcriptional regulation of lignin biosynthesis, Plant Signal. Behav. 11 (2009) 1028–1034.

[15] N.D. Bonawitz, C. Chapple, The genetics of lignin biosynthesis: connecting genotype to phenotype, Annu. Rev. Genet. 44 (2010) 337–363.

[16] W. Boerjan, J. Ralph, M. Baucher, Lignin biosynthesis, Annu. Rev. Plant Biol. 54 (2003) 519–546.

[17] R. Vanholme, B. Demedts, K. Morreel, J. Ralph, W. Boerjan, Lignin biosynthesis and structure, Plant Physiol. 153 (2010) 895–905.

[18] H. Lange, S. Decina, C. Crestini, Oxidative upgrade of lignin – recent routes reviewed, Eur. Polym. J. 49 (2013) 1151–1173.

[19] J.M. Whitbred, M.A. Schuler, Molecular characterization of CYP73A9 and CYP82A1 P450 genes involved in plant defense in pea, Plant Physiol. 124 (2000) 47–58.

[20] X. Li, N.D. Bonawitz, J.K. Weng, C. Chapple, The growth reduction associated with repressed lignin biosynthesis in *Arabidopsis thaliana* is independent of flavonoids, Plant Cell 22 (2010) 1620–1632.

[21] T.Q. Hu, Chemical Modification, Properties and Usage of Lignin, Springer, New York, 2002.

[22] M. Hofrichter, A. Steinbüchel, Biopolymers, Lignin, Humic Substances and Coal, 1, Wiley-VCH, Weinheim, 2001.

[23] F.S. Chakar, A.J. Ragauskas, Review of current and future softwood kraft lignin process chemistry, Ind. Crops Prod. 20 (2004) 131–141.

[24] S. Laurichesse, L. Avérous, Chemical modification of lignins: towards biobased polymers, Prog. Polym. Sci. 39 (2014) 1266–1290.

[25] A. Brandt, J. Gräsvik, J.P. Hallett, T. Welton, Deconstruction of lignocellulosic biomass with ionic liquids, Green Chem. 15 (2013) 550–583.

[26] G. Chatel, R.D. Rogers, Review: oxidation of lignin using ionic liquids – an innovative strategy to produce renewable chemicals, ACS Sustain. Chem. Eng. 2 (2014) 322–339.

[27] T.J. Szalaty, Ł. Klapiszewski, T. Jesionowski, Recent developments in modification of lignin using ionic liquids for the fabrication of advanced materials–A review, J. Mol. Liq. 301 (2020) 112417.

[28] T.J. Szalaty, Ł. Klapiszewski, M. Stanisz, D. Moszyński, A. Skrzypczak, T. Jesionowski, Catalyst-free activation of kraft lignin in air using hydrogen sulfate ionic liquids, Int. J. Biol. Macromol. 119 (2018) 431–437.

[29] M. Stanisz, Ł. Klapiszewski, T. Jesionowski, Recent advances in the fabrication and application of biopolymer-based micro- and nanostructures: a comprehensive review, Chem. Eng. J. 397 (2020) 125409.

[30] C. Crestini, H. Lange, M. Sette, D.S. Argyropoulos, On the structure of softwood kraft lignin, Green Chem. 19 (2017) 4104–4121.

[31] J. Hu, Q. Zhang, D.-J. Lee, Kraft lignin biorefinery: a perspective, Bioresour. Technol. 247 (2018) 1181–1183.

[32] A. Abe, K. Dusek, S. Kobayashi, Biopolymers. Advances in Polymer Science, Vol. 232, Springer, Berlin, Heidelberg, 2009.

[33] J. Rajesh Banu, S. Kavitha, R. Yukesh Kannah, T. Poornima Devi, M. Gunasekaran, S.-H. Kim, G. Kumar, A review on biopolymer production via lignin valorization, Bioresour, Technol. 290 (2019) 121790–121806.

[34] G. Milczarek, O. Inganäs, Renewable cathode materials from biopolymer/conjugated polymer interpenetrating networks, Science 335 (2012) 1468–1471.

[35] G. Milczarek, T. Rębiś, Synthesis and electroanalytical performance of a composite material based on poly(3,4-ethylenedioxythiophene) doped with lignosulfonate, Int. J. Electrochem. 130980 (2012) 1–7.

[36] G. Milczarek, Preparation, characterization and electrocatalytic properties of an iodine/lignin modified gold electrode, Electrochim, Acta 54 (2009) 3199–3205.

[37] G. Milczarek, Lignosulfonate-modified electrodes: Electrochemical properties and electrocatalysis of NADH oxidation, Langmuir 25 (2009) 10345–10353.

[38] G. Milczarek, Preparation and characterization of a lignin modified electrode, Electroanalysis 19 (2007) 1411–1414.

[39] T. Jesionowski, Ł. Klapiszewski, G. Milczarek, Kraft lignin and silica as precursors of advanced composite materials and electroactive blends, J. Mater. Sci. 49 (2014) 1376–1385.

[40] T. Jesionowski, Ł. Klapiszewski, G. Milczarek, Structural and electrochemical properties of multifunctional silica/lignin materials, Mater. Chem. Phys. 147 (2014) 1049–1057.

[41] A. Jędrzak, T. Rębiś, Ł. Klapiszewski, J. Zdarta, G. Milczarek, T. Jesionowski, Carbon paste electrode based on functional GOx/silica-lignin system to prepare an amperometric glucose biosensor, Sensor. Actuators B Chem. 256 (2018) 176–185.

[42] A. Jędrzak, T. Rębiś, M. Nowicki, K. Synoradzki, R. Mrówczyński, T. Jesionowski, Polydopamine grafted on received advanced Fe_3O_4/lignin hybrid and their evaluation in biosensing, Appl. Surf. Sci. 455 (2018) 455-464.

[43] A. Jędrzak, T. Rebiś, M. Kuznowicz, T. Jesionowski, Bio-inspired magnetite/lignin/polydopamine-glucose oxidase biosensing nanoplatform. From synthesis, via sensing assays to comparison with others glucose testing techniques, Int. J. Biol. Macromol. 127 (2019) 677–682.

[44] A. Jędrzak, T. Rebiś, M. Kuznowicz, A. Kołodziejczak-Radzimska, J. Zdarta, A. Piasecki, T. Jesionowski, Advanced Ga_2O_3/lignin and ZrO_2/lignin hybrid microplatforms for glucose oxidase immobilization: evaluation of biosensing properties by catalytic glucose oxidation, Catalysts 9 (2019) 1044–1063.

[45] T.Y. Nilsson, M. Wagner, O. Inganäs, Lignin modification for biopolymer/conjugated polymer hybrids as renewable energy storage materials, ChemSusChem 8 (2015) 4081–4085.

[46] Y. Meng, J. Lu, Y. Cheng, Q. Li, H. Wang, Lignin-based hydrogels: a review of preparation, properties and application, Int. J. Biol. Macromol. 135 (2019) 1006–1019.

[47] Z. Zhou, F. Chen, T. Kuang, L. Chang, J. Yang, P. Fan, Z. Zhao, M. Zhong, Lignin derived hierarchical mesoporous carbon and NiO hybrid nanospheres with exceptional Li-ion battery and pseudocapacitive properties, Electrochim. Acta 274 (2018) 288–297.

[48] J. Jin, B.-J. Yu, Z.-Q. Shi, C.-Y. Wang, C.-B. Chong, Lignin-based electrospun carbon mikro-/nanofibrous webs as free-standing and binder-free electrodes for sodium ion batteries, J. Power Sources 272 (2014) 800–807.

[49] T. Chen, Q. Zhang, J. Pan, J. Xu, Y. Liu, M. Al-Shroofy, Y.-T. Cheng, Low-temperature treated lignin as both binder and conductive additive for silicon mikro-/nanoparticle composite electrodes in lithium-ion batteries, ACS Appl. Mater. Interf. 8 (2016) 32341–32348.

[50] Ł. Klapiszewski, T.J. Szalaty, B. Kurc, M. Stanisz, B. Zawadzki, A. Skrzypczak, T. Jesionowski, Development of new acidic imidazolium ionic liquids for activation of kraft lignin by controlled oxidation. Comprehensive evaluation and practical utility, ChemPlusChem 83 (2018) 361–374.

[51] T.J. Szalaty, Ł. Klapiszewski, B. Kurc, A. Skrzypczak, T. Jesionowski, A comparison of protic and aprotic ionic liquids as effective activating agents of kraft lignin. Developing functional MnO_2/lignin hybrid materials, J. Mol. Liq. 261 (2018) 456–467.

[52] F. Sanchez, K. Sobolev, Nanotechnology in concrete - a review, Construct. Build. Mater 24 (2010) 2060–2071.

[53] X. Ouyang, X. Qiu, P. Chen, Physicochemical characterization of calcium lignosulfonate - a potentially useful water reducer, Colloids Surf. A Physiochem. Eng. Asp. 282-283 (2006) 489–497.

[54] S. Li, Z. Li, Y. Zhang, C. Liu, G. Yu, B. Li, X. Mu, H. Peng, Preparation of concrete water reducer via fractionation and modification of lignin extracted from pine wood by formic acid, ACS Sustain. Chem. Eng. 5 (2017) 4214–4222.

[55] C. Huang, J. Ma, W. Zhang, G. Huang, Q. Yong, Preparation of lignosulfonates from biorefinery lignins by sulfomethylation and their application as a water reducer for concrete, Polymers 10 (2018) 841.

[56] I. Klapiszewska, A. Ślosarczyk, Ł. Klapiszewski, T. Jesionowski, Production of cement composites using alumina-lignin hybrid materials admixture, Physicochem. Probl. Miner. Process. 55 (2019) 1401–1412.

[57] Ł. Klapiszewski, I. Klapiszewska, A. Ślosarczyk, T. Jesionowski, Lignin-based hybrid admixtures and their role in cement composites fabrication, Molecules 24 (2019) 3544.

[58] A. Ślosarczyk, I. Klapiszewska, P. Jędrzejczak, Ł. Klapiszewski, T. Jesionowski, Biopolymer-based hybrids as an effective admixtures for cement composites, Polymers 12 (2020) 1180.

[59] E. Norgbey, J. Huang, V. Hirsch, W.J. Liu, M. Wang, O. Ripke, Y. Li, G.E.T. Annan, D. Ewusi-Mensah, X. Wang, G. Treib, A. Rink, A.S. Nwankwegu, P.A. Opoku, P.N. Nkrumah, Unravelling the efficient use of waste lignin as a bitumen modifier for sustainable roads, Construct. Build. Mater. 230 (2020) 116957.

[60] I.P. Pérez, A.M.R. Pasandín, J.C. Pais, P.A.A. Pereira, Use of lignin biopolymer from industrial waste as bitumen extender for asphalt mixtures, J. Clean. Prod. 220 (2019) 87–98.

[61] J.C. de Haro, L. Magagnin, S. Turri, G. Griffini, Lignin-based anticorrosion coatings for the protection of aluminum surfaces, ACS Sustain. Chem. Eng. 7 (2019) 6213–6222.

[62] A.A. Mazhar, S.T. Arab, E.A. Noor, Electrochemical behaviour of Al-Si alloys in acid and alkaline media, Bull. Electrochem. 17 (2001) 449–458.

[63] M.M. El-Deeb, E.N. Ads, J.R. Humaidi, Evaluation of the modified extracted lignin from wheat straw as corrosion inhibitors for aluminum in alkaline solution, Int. J. Electrochem. Sci. 13 (2018) 4123–4138.

[64] M.H. Hussin, A.A. Rahim, M.N.M. Ibrahim, N. Brosse, The capability of ultrafiltrated alkaline and organosolv oil palm (*Elaeis guineensis)* fronds lignin as green corrosion inhibitor for mild steel in 0.5 M HCl solution, Measurement 78 (2016) 90–103.

[65] H. Hatakeyama, T. Hatakeyama, Lignin structure, properties, and application, Adv. Polym. Sci. 232 (2010) 1–63.

[66] J.F. Kadla, S. Kubo, Lignin-based polymer blends: analysis of intermolecular interactions in lignin-synthetic polymer blends, Compos. A Appl. Sci. Manufact. 35 (2004) 395–400.

[67] E. Svinterikos, I. Zuburtikudis, Carbon nanofibers from renewable bioresources (lignin) and a recycled commodity polymer [poly(ethylene terephthalate)], J. Appl. Polym. Sci. 133 (2016) 43936.

[68] A.E. Imel, A.K. Naskar, M.D. Dadmun, Understanding the impact of poly(ethylene oxide) on the assembly of lignin in solution toward improved carbon fiber production, ACS Appl. Mater. Interf. 8 (2016) 3200–3207.

[69] V. Poursorkhabi, A.K. Mohanty, M. Misra, Electrospinning of aqueous lignin/poly(ethylene oxide) complexes, J. Appl. Polym. Sci. 132 (2015) 41260.

[70] A. Gregorova, Z. Cibulkova, B. Kosıkova, P. Simon, Stabilization effect of lignin in polypropylene and recycled polypropylene, Polym. Degrad. Stab. 89 (2005) 553–558.

[71] F. Chen, H. Dai, X. Dong, J. Yang, M. Zhong, Physical properties of lignin-based polypropylene blends, Polym. Compos. 32 (2011) 1019–1025.

[72] A.A. Morandim-Giannetti, J.A. Agnelli, B.Z. Lancas, R. Magnabosco, S.A. Casarin, S.H.P. Bettini, Lignin as additive in polypropylene/coir composites: thermal, mechanical and morphological properties, Carbohydr. Polym. 87 (2012) 2563–2568.

[73] S. Lu, S. Li, J. Yu, D. Guo, R. Ling, B. Huang, The effect of hyperbranched polymer lubricant as a compatibilizer on the structure and properties of lignin/polypropylene composites, Wood Mater. Sci. Eng. 8 (2013) 159–165.

[74] A. El-Sabbagh, Effect of coupling agent on natural fibre in natural fibre/polypropylene composites on mechanical and thermal behavior, Compos. Part B Eng. 57 (2014) 126–135.

[75] B. Bozsódi, V. Romhányi, P. Pataki, D. Kun, K. Renner, B. Pukánszky, Modification of interactions in polypropylene/lignosulfonate blends, Mater. Des. 103 (2016) 32–39.

[76] K. Bula, Ł. Klapiszewski, T. Jesionowski, A novel functional silica/lignin hybrid material as a potential bio-based polypropylene filler, Polym. Compos. 36 (2015) 913–922.

[77] S. Borysiak, Ł. Klapiszewski, K. Bula, T. Jesionowski, Nucleation ability of advanced functional silica/lignin hybrid fillers in polypropylene composites, J. Therm. Anal. Calorim. 126 (2016) 251–262.

[78] Ł. Klapiszewski, K. Bula, M. Sobczak, T. Jesionowski, Influence of processing conditions on the thermal stability and mechanical properties of PP/silica-lignin composites, Int. J. Polym. Sci. 2016 (2016) 1627258.

[79] Ł. Klapiszewski, A. Grząbka-Zasadzińska, S. Borysiak, T. Jesionowski, Preparation and characterization of polypropylene composites reinforced by functional ZnO/lignin hybrid materials, Polym. Test. 79 (2019) 106058.

[80] A. Grząbka-Zasadzińska, Ł. Klapiszewski, T. Jesionowski, S. Borysiak, Functional MgO-lignin hybrids and their application as fillers for polypropylene composites, Molecules 25 (2020) 864.

[81] R.R.N. Sailaja, M.V. Deepthi, Mechanical and thermal properties of compatibilized composites of polyethylene and esterified lignin, Mater. Des. 31 (2010) 4369–4379.

[82] S.K. Samal, E.G. Fernandes, A. Corti, E. Chiellini, Bio-based polyethylene-lignin composites containing a pro-oxidant/pro-degradant additive: preparation and characterization, J. Polym. Environ. 22 (2014) 58–68.

[83] A. Diop, F. Mijiyawa, D. Koffi, B.V. Kokta, D. Montplaisir, Study of lignin dispersion in low-density polyethylene, J. Thermoplast. Compos. Mater. 28 (2015) 1662–1674.

[84] Ł. Klapiszewski, K. Bula, A. Dobrowolska, K. Czaczyk, T. Jesionowski, A high-density polyethylene container based on ZnO/lignin dual fillers with potential antimicrobial activity, Polym. Test. 73 (2019) 51–59.

[85] K. Bula, Ł. Klapiszewski, T. Jesionowski, Effect of processing conditions and functional silica/lignin content on the properties of bio-based composite thin sheet films, Polym. Test. 77 (2019) 105911.

[86] K. Bula, G. Kubicki, T. Jesionowski, Ł. Klapiszewski, MgO-lignin dual phase filler as an effective modifier of polyethylene film properties, Materials 13 (2020) 809.

[87] K. Bula, G. Kubicki, A. Kubiak, T. Jesionowski, Ł. Klapiszewski, Influence of MgO-lignin dual component additives on selected properties of low density polyethylene, Polymers 12 (2020) 1156.

[88] Ł. Klapiszewski, F. Pawlak, J. Tomaszewska, T. Jesionowski, Preparation and characterization of novel PVC/silica-lignin composites, Polymers 7 (2015) 1767–1788.

[89] J. Tomaszewska, Ł. Klapiszewski, K. Skórczewska, T.J. Szalaty, T. Jesionowski, Advanced organic-inorganic hybrid fillers as functional additives for poly(vinyl chloride), Polimery 62 (2017) 52–59.

[90] Ł. Klapiszewski, J. Tomaszewska, K. Skórczewska, T. Jesionowski, Preparation and characterization of eco-friendly $Mg(OH)_2$/lignin hybrid material and its use as a functional filler for poly(vinyl chloride), Polymers 9 (2017) 258–276.

[91] A. Grząbka-Zasadzińska, Ł. Klapiszewski, K. Bula, T. Jesionowski, S. Borysiak, Supermolecular structure and nucleation ability of polylactide-based composites with silica/lignin hybrid fillers, J. Therm. Anal. Calorim. 126 (2016) 263–275.

[92] A. Grząbka-Zasadzińska, Ł. Klapiszewski, S. Borysiak, T. Jesionowski, Thermal and mechanical properties of silica–lignin/polylactide composites subjected to biodegradation, Materials 11 (2018) 2257.

[93] A. Gardziella, L.A. Pilato, A. Knop, Phenolic Resins - Chemistry, Applications, Standardization, Safety and Ecology, Springer, Berlin, 2000.

[94] A.Y. Kharade, D.D. Kale, Effect of lignin on phenolic novolak resins and moulding powder, Eur. Polym. J. 34 (1998) 201–205.

[95] S. Hattali, A. Benaboura, S. Dumarçay, P. Gérardin, Evaluation of alfa grass soda lignin as a filler for novolak molding powder, J. Appl. Polym. Sci. 97 (2005) 1065–1068.

[96] A. Rudawska, Abrasive Technology - Characteristics and Applications (2018).

[97] B. Strzemiecka, Ł. Klapiszewski, A. Voelkel, T. Jesionowski, Functional lignin-SiO_2 hybrids as potential fillers for phenolic binders, J. Adhes. Sci. Technol. 30 (2016) 1031–1048.

[98] B. Strzemiecka, Ł. Klapiszewski, A. Jamrozik, T.J. Szalaty, D. Matykiewicz, T. Sterzyński, A. Voelkel, T. Jesionowski, Physicochemical characterization of functional lignin–silica hybrid fillers for potential application in abrasive tools, Materials 9 (2016) 517.

[99] Ł. Klapiszewski, A. Jamrozik, B. Strzemiecka, D. Matykiewicz, A. Voelkel, T. Jesionowski, Activation of magnesium lignosulfonate and kraft lignin: influence on the properties of phenolic resin-based composites for potential applications in abrasive materials, Int. J. Mol. Sci. 18 (2017) 1224.

[100] Ł. Klapiszewski, A. Jamrozik, B. Strzemiecka, I. Koltsov, B. Borek, D. Matykiewicz, A. Voelkel, T. Jesionowski, Characteristics of multifunctional, eco-friendly lignin-Al_2O_3 hybrid fillers and their influence on the properties of composites for the abrasive tools, Molecules 22 (2017) 1920.

[101] Ł. Klapiszewski, R. Oliwa, M. Oleksy, T. Jesionowski, Calcium lignosulfonate as eco-friendly additive for crosslinking fibrous composites with phenol-formaldehyde resin matrix, Polimery 63 (2018) 23–29.

[102] Ł. Klapiszewski, A. Jamrozik, B. Strzemiecka, P. Jakubowska, T.J. Szalaty, M. Szewczyńska, A. Voelkel, T. Jesionowski, Kraft lignin/cubic boron nitride hybrid materials as functional components for abrasive tools, Int. J. Biol. Macromol. 122 (2019) 88–94.

[103] A. Jamrozik, B. Strzemiecka, P. Jakubowska, I. Koltsov, Ł. Klapiszewski, A. Voelkel, T. Jesionowski, The effect of lignin-alumina hybrid additive on the properties of composition used in abrasive tools, Int. J. Biol. Macromol. 161 (2020) 531–538.

[104] M.B. Šćiban, M.T. Klasnja, M.G. Antov, Study of the biosorption of different heavy metal ions onto kraft lignin, Ecol. Eng. 37 (2011) 2092–2095.

[105] M. Ahmaruzzaman, Industrial wastes as low-cost potential adsorbents for the treatment of wastewater laden with heavy metals, Adv. Coll. Interf.Sci. 166 (2011) 36–59.

[106] M. Betancur, P.R. Bonelli, J.A. Velásquez, A.L. Cukierman, Potentiality of lignin from the Kraft pulping process for removal of trace nickel from wastewater: effect of demineralization, Bioresour. Technol. 100 (2009) 1130–1137.

[107] H. Harmita, K.G. Karthikeyan, X.J. Pan, Copper and cadmium sorption onto kraft and organosolv lignins, Bioresour. Technol. 100 (2009) 6183–6191.

[108] L. Bulgariu, D. Bulgariu, T. Malutan, M. Macoveanu, Adsorption of lead(II) ions from aqueous solution onto lignin, Adsorp. Sci. Technol. 27 (2009) 435–445.

[109] X. Guo, S. Zhang, X. Shan, Adsorption of metal ions on lignin, J. Hazard. Mater. 151 (2008) 134–142.

[110] A. Demirbas, Adsorption of lead and cadmium ions in aqueous solutions onto modified lignin from alkali glycerol delignification, J. Hazard. Mater. 109 (2004) 221–226.

[111] Y. Ge, Z. Li, Y. Kong, Q. Song, K. Wang, Heavy metal ions retention by bi-functionalized lignin: synthesis, applications, and adsorption mechanisms, J. Ind. Eng. Chem. 20 (2014) 4429–4436.

[112] Y. Lei, Y. Huizhen, Modification of reed alkali lignin to adsorption of heavy metals, Adv. Mater. Res. 622 (2013) 1646–1650.

[113] F. Pagnanelli, S. Mainelli, F. Veglio, L. Toro, Heavy metal removal by olive pomace: biosorbent characterization and equilibrium modeling, Chem. Eng. Sci. 58 (2003) 4709–4717.

[114] Ł. Klapiszewski, M. Nowacka, G. Milczarek, T. Jesionowski, Physicochemical and electrokinetic properties of silica/lignin biocomposites, Carbohydr. Polym. 94 (2013) 345–355.

[115] Ł. Klapiszewski, P. Bartczak, M. Wysokowski, M. Jankowska, K. Kabat, T. Jesionowski, Silica conjugated with kraft lignin and its use as a novel 'green' sorbent for hazardous metal ions removal, Chem. Eng. J. 260 (2015) 684–693.

[116] T.M. Budnyak, S. Aminzadeh, I.V. Pylypchuk, A.V. Riazanova, V.A. Tertykh, M.E. Lindström, O. Sevastyanova, Peculiarities of synthesis and properties of lignin-silica nanocomposites prepared by sol-gel method, Nanomaterials 8 (2018) 950.

[117] Ł. Klapiszewski, K. Siwińska-Stefańska, D. Kołodyńska, Preparation and characterization of novel TiO_2/lignin and TiO_2-SiO_2/lignin hybrids and their use as functional biosorbents for Pb(II), Chem. Eng. J. 314 (2017) 169–181.

[118] Ł. Klapiszewski, K. Siwińska-Stefańska, D. Kołodyńska, Development of lignin based multifunctional hybrid materials for Cu(II) and Cd(II) removal from the aqueous system, Chem. Eng. J. 330 (2017) 518–530.

[119] F. Ciesielczyk, P. Bartczak, Ł. Klapiszewski, T. Jesionowski, Treatment of model and galvanic waste solutions of copper(II) ions using a lignin/inorganic oxide hybrid as an effective sorbent, J. Hazard. Mater. 328 (2017) 150–159.

[120] R. Wahlström, A. Kalliola, J. Heikkinen, H. Kyllönen, T. Tamminen, Lignin cationization with glycidyltrimethylammonium chloride aiming at water purification applications, Ind. Crops Prod. 104 (2017) 188–194.

[121] M. Liu, Y. Liu, J. Shen, S. Zhang, X. Liu, X. Chen, Y. Ma, S. Ren, G. Fang, S. Li, C.T. Li, T. Sun, Simultaneous removal of Pb^{2+}, Cu^{2+} and Cd^{2+} ions from wastewater using hierarchical porous polyacrylic acid grafted with lignin, J. Hazard. Mater. 392 (2020) 122208–122245.

[122] D.S. Bajwa, G. Pourhashem, A.H. Ullah, S.G. Bajwa, A concise review of current lignin production, applications, products and their environment impact, Ind. Crop Prod. 139 (2019) 111526–111537.

[123] C. Thulluri, S.R. Pinnamaneni, P.R. Shetty, U. Addepally, Synthesis of lignin-based nanomaterials/nanocomposites: recent trends and future perspectives, Ind. Biotechnol. 12 (2016) 153–160.

[124] S.S. Nair, S. Sharma, Y. Pu, Q. Sun, S. Pan, J.Y. Zhu, Y. Deng, A.J. Ragauskas, High shear homogenization of lignin to mikro-/nanolignin and thermal stability of mikro-/nanolignin-polyvinyl alcohol blends, ChemSusChem 7 (2014) 3513–3520.

[125] T. Chen, Q. Zhang, J. Pan, J. Xu, Y. Liu, M. Al-Shroofy, Y.T. Cheng, Low-temperature treated lignin as both binder and conductive additive for silicon nanoparticle composite electrodes in lithium-ion batteries, ACS Appl. Mater. Interf. 8 (2016) 32341–32348.

[126] P. Figueiredo, K. Lintinen, J.T. Hirvonen, M.A. Kostiainen, H.A. Santos, Properties and chemical modifications of lignin: towards lignin-based nanomaterials for biomedical applications, Prog. Mater. Sci. 93 (2018) 233–269.

[127] H. Li, Y. Deng, B. Liu, Y. Ren, J. Liang, Y. Qian, X. Qiu, C. Li, D. Zheng, Preparation of nanocapsules via the self-assembly of kraft lignin: a totally green process with renewable resources, ACS Sustain. Chem. Eng. 4 (2016) 1946–1953.

[128] P. Figueiredo, K. Lintinen, A. Kiriazis, V. Hynninen, Z. Liu, T. Bauleth-Ramos, A. Rahikkala, A. Correia, T. Kohout, B. Sarmento, J. Yli-Kauhaluoma, J. Hirvonen, O. Ikkala, M.A. Kostiainen, H.A. Santos, In vitro evaluation of biodegradable lignin-based nanoparticles for drug delivery and enhanced antiproliferation effect in cancer cells, Biomaterials 121 (2017) 97–108.

[129] S. Kim, M.M. Fernandes, T. Matamá, A. Loureiro, A.C. Gomes, A. Cavaco-Paulo, Chitosan-lignosulfonates sono-chemically prepared nanoparticles: characterisation and potential applications, Coll. Surf. B Biointerf. 103 (2013) 1–8.

[130] C. Frangville, M. Rutkevičius, A.P. Richter, O.D. Velev, S.D. Stoyanov, V.M. Paunov, Fabrication of environmentally biodegradable lignin nanoparticles, ChemPhysChem 13 (2012) 4235–4243.

[131] A.P. Richter, J.S. Brown, B. Bharti, A. Wang, S. Gangwal, K. Houck, An environmentally benign antimicrobial nanoparticle based on a silver-infused lignin core, Nat. Nanotechnol. 10 (2015) 817–823.

[132] J. Huang, M. Wang, P. Song, Y. Li, F. Xu, X. Zhang, Directed 2D nanosheet assemblies of amphiphilic lignin derivatives: formation of hollow spheres with tunable porous structure, Ind. Crops Prod. 127 (2019) 16–25.

[133] M. Tortora, F. Cavalieri, P. Mosesso, F. Ciaffardini, F. Melone, C. Crestini, Ultrasound driven assembly of lignin into microcapsules for storage and delivery of hydrophobic molecules, Biomacromolecules 15 (2014) 1634–1643.

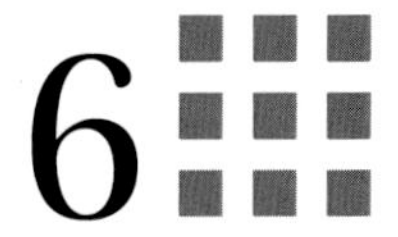

6 Preparation and therapeutic applications of chitosan nanoparticles

Vijay Singh Gondil[a,b], Kusum Harjai[b] and Sanjay Chhibber[b]

[a]DEPARTMENT OF NEPHROLOGY, POST GRADUATE INSTITUTE OF MEDICAL EDUCATION AND RESEARCH, CHANDIGARH, INDIA. [b]DEPARTMENT OF MICROBIOLOGY, PANJAB UNIVERSITY, CHANDIGARH, INDIA.

Chapter outline

Biopolymeric Nanomaterials: Fundamentals and Applications. DOI: https://doi.org/10.1016/B978-0-12-824364-0.00006-X

6.1 Chitosan: a versatile polymer

Chitosan is a deacetylated form of chitin and an abundant polymer on the earth after cellulose. Chitin is mainly found in the cell walls of fungi, insect cuticles, shrimp shells, and crabs [1]. Chitosan is less abundant and found in cell walls of limited fungi [2]. Chitin chemically is a polymer of repeated units of β 1-4 N-acetyl D-glucosamine residues. Chitin is divided into three main classes based on its structure α, β, and γ. α chitin contains antiparallel chains whereas β chitin is composed of intrasheet hydrogen bonds in parallel chains. γ chitin is blend of α chitin and β chitin, which contains both parallel as well as antiparallel chains [3]. Chitosan is formed from deacetylation of chitin and contains β 1-4 N-acetyl D-glucosamine and 2-amino 2-deoxy-β-D-glucopyranose [4]. Chitosan contains two hydroxyl groups and one amino group in hexosamine residue [5]. The chitosan molecules are attached to the side groups by reactive hydroxyl groups without any alteration in its biophysical characteristics (Fig. 6.1). Chitosan molecules contain more than 60% of 2-amino 2-deoxy-β-D-glucopyranose units, which are expressed in terms of degree of deacetylation of chitosan [6]. The degree of deacetylation can be determined by a number of methods such as infrared radiation, UV-visible spectroscopy, potentiometric titrations, ^{1}H-NMR, and ^{13}C-NMR [7–9]. The free amino group of chitosan makes it polyelectrolyte soluble in an acidic solution unlike its parent compound chitin. The biochemical and pharmacological application of chitosan is dependent on its molecular weight, degree of deacetylation, and viscosity in aqueous solution. Some other important parameters that can affect its application range are crystallinity, moisture, heavy metal, and ash content [10]. The density of chitosan solution depends on the solubility and degree of deacetylation. The crystalline structure of chitosan is attributed to the intrahydrogen and interhydrogen bonds between amino and hydroxyl group. The slightly hydrophobic nature of chitosan is due to the presence of acetyl groups in its structure [11]. Chitosan is treated as Generally Recognized as Safe (GRAS) and approved by Food and Drug Administration in food and medical applications in a number of countries, which makes it a more fascinating polymer ([12].

6.2 Preparation of chitosan nanoparticles

Chitosan possesses the ability to form beads and gel on contact with negatively charged ions. The high size of beads limits its applications in drug delivery [13]. Chitosan nanoparticles were first

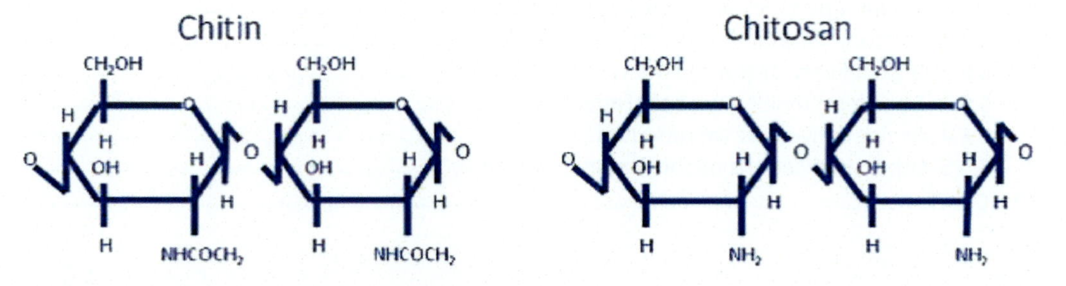

FIGURE 6.1 Chemical structures of chitin and chitosan.

prepared by cross-linking and emulsification by Ohya and co-workers in 1994 [14]. From then onwards, a number of methods have been described for preparation of chitosan nanoparticles that include ionic or ionotropic gelation, microemulsion cross-linking, emulsification solvent diffusion, spray drying, reverse micellar, and complex coacervation method [15]. Ionotropic gelation is most widely used method especially for biomolecules as they do not employ use of organic, harsh solvents, and high shear force [16].

6.2.1 Ionotropic gelation method

Ionotropic gelation method is widely used method for preparing chitosan nanoparticle and was developed by Calvo and coworkers [17]. The present method involves the electrostatic interaction between positively charged amine group of chitosan and negatively charged group of anions (Tripolyphosphate). Chitosan solution is usually dissolved in the acetic acid with or without stabilizer such as polyethylene glycol (PEG) 1500, PEG 6000, and Tween 80 [18]. Polyanion solution is added to the chitosan solution while stirring at room temperature to form nanoparticles. Chitosan solution undergoes ionic gelation by complexation between polymer chitosan and polyanion polytripolyphosphate [19,20]. Size variations can be observed with varying concentrations of polymer and polyanions. The addition of sodium chloride is reported to decrease the size of nanoparticles as sodium chloride decreases the electrostatic repulsion in positively charged amine group of chitosan molecules. It may increase the flexibility of polymer molecules which in turn increases its stability in solution [21]. This method shows various advantages over the present method such as mild processing conditions, aqueous environment, nontoxicity, and does not affect the chemistry of loaded molecule [22].

A modified method of ionic gelation with radical polymerization has also been reported in which acrylic acid solution with or without polyether or polyethylene glycol is added to the chitosan solution. Anion of acrylic acid interacts with cation of chitosan and polymerization sets in following the addition of potassium persulfate at 60–70 °C with a stream of nitrogen. After the polymerization, nanoparticles are allowed to settle down overnight and unreacted monomers are removed by dialysis [23].

6.2.2 Microemulsion cross-linking

Microemulsion cross-linking method was reported by Grenha [24]. In this method, the chitosan and glutaraldehyde are mixed with surfactant in hexane and allowed to stir at room temperature overnight. This process allows the cross-linking of free amine groups of chitosan and cross-linker glutaraldehyde [14]. Organic solvent and surfactants can be removed by evaporation and precipitation with calcium chloride, respectively. A major limitation of microemulsion cross-linking method is the use of glutaraldehyde which can serve as an antigen and also damage the proteins or peptides by covalent cross-linking [17].

6.2.3 Emulsification solvent diffusion

Emulsification solvent diffusion method was proposed by Niwa and coworkers [25]. Emulsion of chitosan solution is prepared by injecting organic phase (acetone and methylene chloride) and stabilizer (lecithin and poloxamer) into chitosan solution along with magnetic stirring. The high-pressure homogenization is applied to oil in water emulsion after magnetic stirring. At reduced pressure and room temperature, methylene chloride is removed from emulsion. Diffusion of acetone into water initiates polymerization of polymer, thus forming chitosan nanoparticles. Generally, water is added in excess to permit the high diffusion of acetone for proper polymer precipitation. The use of high shear pressure and organic acids in preparation, limits its applications in delivery of protein and other biomolecules [26].

6.2.4 Spray drying

Spray drying preparation of chitosan nanoparticles employs the use of nanospray dryer. The basic principle of this method is hot stream–mediated drying of atomized chitosan droplets. Chitosan solution in acetic acid is stored overnight, leading to the formation of small droplets. The solvent from solution is evaporated by using nanospray dryer, resulting in formation of nanoparticles [27,28]. The process is rapid and short drying time lowers the degradation of temperature-sensitive molecules [29,30]. The operational conditions such as temperature, flow rate, polymer concentration, and feed rate influence the morphology and size of particles. One of the major limitations of this method is the formation of amorphous and heterogeneous powder samples [31].

6.2.5 Reverse micellar

Reverse micellar method was proposed by Brunel and coworkers [32]. Water in oil emulsion is prepared by adding a lipophilic surfactant such as CTAB (cetyl trimethyl ammonium bromide) and sodium bisethylhexyl sulfosuccinate into organic solvent. Chitosan solution is added to the water in oil emulsion under magnetic stirring to form reverse micelles. This method avoids the use of toxic organic solvents as well as cross-linkers to form nanoparticles with a narrow size range [33]. The major drawback of reverse micellar method is the use of organic acids such as hexane, which can alter the secondary structure of protein antigen to alter its immunogenicity [34].

6.2.6 Complex coacervation

Coacervation method involves the formation of spherical nanoparticles by the addition of electrostatically interacting liquids. DNA-loaded chitosan nanoparticles are formed by interaction between negative phosphate backbone of DNA and positive amine groups of chitosan [35]. The advantage of using this method is the coacervation process that is performed at a low temperature in an aqueous solution. Mild conditions help in retaining biological activity of encapsulated molecules. The method also shows number of limitations such as low drug loading, low solubility of nanoparticles, and use of toxic chemical linkers such as glutaraldehyde [36,37]. In polyelectrolyte complex method, a negatively charged solution such as dextran sulfate, DNA is added to positively charged chitosan solution under magnetic stirring followed by neutralization of charge. The method is simple, forms nanoparticles instantaneously, and avoids harsh conditions [38]. Low-molecular-weight chitosan containing insulin nanoparticles, prepared with a polyelectrolyte complex method, showed a mean diameter of 200 nm with a sustained release profile [37].

6.3 Cross-linkers for chitosan nanoparticles

6.3.1 Tripolyphosphate

Tripolyphosphate (TPP) is a polyanionic cross-linker that contains multivalent anions in its chemical structure [39]. TPP is nontoxic in nature and can undergo reversible cross-linking with positive amino groups of chitosan through its negatively charged phosphate groups [40,41]. TPP contains five ionizable negative groups with different pK_a and its overall charge is pH-dependent [42]. A number of advantages have been reported for TPP, which include the reversible electrostatic strong interactions with chitosan, plausibly limiting its toxicity [43].

6.3.2 Glutaraldehyde

Glutaraldehyde is considered as a versatile cross-linker due to its high fixation capacity, low cost, and capacity to form a cationic emulsion [44]. Aldehyde group of glutaraldehyde covalently cross-links with amino groups of chitosan to form nanoparticles [45]. The amount of glutaraldehyde added to the chitosan solution affects the size of nanoparticles [46]. Glutaraldehyde confers toxicity and balanced drug integrity, which limits its applications. Cinnamaldehyde is a naturally occurring aldehyde used as a cross-linker for preparations of nanoparticles [47]. A phenolic aldehyde vanillin is also used for preparation of chitosan nanoparticles [48]. The aldehyde group of vanillin and cinnamaldehyde forms cross-linkage with amino group of chitosan to form nanoparticles [49].

6.3.3 Genipin

Genipin is another natural cross-linker obtained from the gardenia fruit and has been explored for applications such as food dyes and herbal medicines. Genipin is considered as 10,000 times less cytotoxic than glutaraldehyde [50]. Genipin has been used in cross-linking reactions for proteins and ciprofloxacin-loaded chitosan-heparin nanoparticles by ionotropic gelation method [51]. Genipin cross-linked nanoparticles also showed a potent antibacterial activity against *Escherichia coli* 443 [52]. However, genipin presents a number of disadvantages such as high cost, difficult extraction, and limited availability, which limits its application as cross-linker in chitosan nanoparticle preparation [53].

6.3.4 Citric acid

Citric acid is a commonly used cross-linker in the process of thermal cross-linking. A defined molar ratio of citric acid and chitosan solution are mixed to form chitosan nanoparticles [54].

In a recent study, chitosan citric acid nanoparticles were prepared by functionalizing the amino group of chitosan which in turn is cross-linked by citric acid. Chitosan citric acid nanoparticles showed the potential as a cleansing agent for chromium-contaminated wastewaters [55].

6.3.5 Other cross-linkers

A number of other cross-linkers, which include tartaric acid, malic acid, and succinic acid, are used for the preparation of nanoparticles. These di-carboxylic and tri-carboxylic acid cross-linkers react with amine groups of chitosan to form nanoparticles [56]. Another important cross-linker used for the nanoparticle formation is sodium sulfate in which sulfates anions cross-link with positively charged amine groups of chitosan to form chitosan nanoparticles [57].

6.4 Stability of chitosan nanoparticles

Stability of nanoparticles is an imperative factor for any successful delivery and efficacy of a therapeutic agent over the time [58]. Some physical factors such as particle agglomeration, coagulation, and flocculation provide instability to the nanoparticles. Other chemical factors such as pH, temperature, composition of formulation, molecular weight, and polymer types also influence the stability of nanoparticles. Some drugs or biomolecules are sensitive to pH and temperature alterations; hence these factors should be optimized for the preparation of stable nanoparticle formulation [59]. The change in pH can lead to deprotonation or protonation of glucosamine groups of chitosan, which can influence the stability of nanoparticle structures [60]. It is reported that the TPP and glycidoxy propyl trimethoxysilane cross-linked chitosan nanoparticles are highly sensitive to alterations in pH [61]. Camptothecin-encapsulated chitosan nanoparticles also showed sensitivity to pH of tumor with change in ratio of a cross-linker to a polymer [62].

6.5 Cellular interactions

Cell viability determination for chitosan nanoparticles has been carried out by cytotoxicity assays. Cytotoxicity analysis commonly involves the use of tetrazolium salts such as XTT (2,3-bis-(2-methoxy-4-nitro-5-sulfophenyl)-2H-tetrazolium-5-carboxanilide), MTT (2,3-Bis-(2-Methoxy-4-nitro-5-sulfophenyl)-2H-tetrazolium-5-carboxanilide and 3-(4,5-dimethylthiazol-2-yl)-2,5-diphenyltetrazolium bromide), and fluorescent dyes such as rhodamine 123 and acridine orange [63,64]. Chitosan nanoparticles have been studied for cytotoxicity against human liver cell line HepG2 and results revealed no morphological alteration after 24 and 48 h of treatment. Sulforhodamine B assay revealed only 12% of cell death after 48 h of exposure with 100 μg/mL of chitosan nanoparticles [65]. Chitosan nanoparticles showed compatible behavior with liver cells at 0.5% of chitosan nanoparticle concentrations up to 4 h. Uptake of chitosan nanoparticles was also observed evidently with enzyme leakage and cell membrane damage [66]. Chitosan DNA nanoparticles mediated cell toxicity was also observed by MTT assay in human mesenchymal stem cells (MSCs) and HEK293 carcinoma cell lines [67]. The cytotoxicity of chitosan nanoparticles was found to be highly influenced by particle size compared to molecular weight of chitosan in mouse hematopoietic stem cells [68]. In another study, chitosan nanoparticles and copper-loaded chitosan nanoparticles were studied against BGC823, BEL7402, COLO320, and L-02 cell lines. L-02 cell line exhibits low cytotoxicity for chitosan nanoparticles and copper-loaded chitosan nanoparticles as compared to other tested cell lines [69]. Chitosan nanoparticles also show higher cellular uptake which was visualized by confocal microscopy after 4 h of treatment with 1% of chitosan nanoparticles [70]. However, cellular killing to some extent was observed, which possibly resulted from membrane damage during the process of electron microscopy [66]. A recent study revealed antiproliferative effect of chitosan gold nanoparticles and showed it to be dependent on cell type, concentration, and size of nanoparticles [71]. In a recent study, Gondil et al. [72] explored chitosan nanoparticles as an endolysin delivery system and results revealed negligible cellular toxicity on lung epithelial cell lines. These nanoparticles also exhibited protective efficcay against pneumococcal pneumonia in animal model as compared to native endolysin.

6.6 Chitosan nanoparticles as a delivery system

6.6.1 Pulmonary/nasal delivery of therapeutics

The efficiency of drugs or other therapeutic agents such as peptides and proteins can be enhanced by using chitosan microparticles or nanoparticles for pulmonary drug delivery. The encapsulation leads to higher bioavailability, increased half-life, and lower toxicity of the drug [73]. Mucoadhesiveness of chitosan nanoparticles prepared by using cross-linkers such as TPP or glutaraldehyde is considered as an important factor for lung drug delivery [74]. Insulin-containing chitosan nanoparticles and terbutaline sulfate nanoparticles showed sustained pulmonary delivery for loaded therapeutics [75]. Chitosan nanoparticles are known to increase the stability of drug by dry powder inhaler formulations of chitosan nanoparticles [76].

Delivery of antibacterial drugs through chitosan nanoparticles has been investigated [77]. Glutaraldehyde cross-linked chitosan nanoparticles containing levofloxacin showed a high antibacterial activity against *Pseudomonas aeruginosa* infection in cases of cystic fibrosis compared to levofloxacin. Dry powder inhaler of levofloxacin was found to be more effective than solution formulation of levofloxacin-containing chitosan nanoparticles [74]. A high antibacterial activity of cefoxitin-loaded chitosan nanoparticles was observed against biofilms of potent pathogens such as Staphylococcus *aureus*, Klebsiella *pneumoniae, E. coli,* and *P. aeruginosa* [151]. Chhibber et al. [78] also demonstrated the antibiofilm activity of boswellia-loaded chitosan hydrogen in treatment of *S. aureus*-mediated murine burn wound infection. Anti-tuberculin drugs, moxifloxacin, rifampicin, ethambutol dihydrochloride, and ofloxacin, have also been explored for chitosan nanoparticles-based pulmonary delivery [76,79]. Cellular uptake and prolonged release of ofloxacin were observed in chitosan-loaded microspheres. Dry powder inhaler showed better uptake by alveolar macrophages and reduced treatment regime as compared to orally delivered treatment [79]. Other chitosan nanoparticle formulations showed better efficacy of rifampicin and rifabutin in the treatment of tuberculosis with improved and targeted delivery [76]. Clindamycin-loaded chitosan microspheres prepared by spray drying showed high flow and prolonged release properties [80]. Chitosan-entrapped gentamicin also showed high antibacterial as well as antioxidant activity for lung delivery applications as compared to gentamicin alone [152]. In a number of studies, efficient delivery of antibacterial drugs by chitosan nanoparticles has been demonstrated [81,153]. Briefly, chitosan nanoparticle–based drug formulations showed (1) high stability, high encapsulation efficiency, enhanced activity against bacterial pathogens; (2) reduced dose frequency; (3) higher permeation; (4) better efficiency and safety; (5) high entrapment efficacy and prolonged release; (6) high alveolar uptake and low cytotoxicity; (7) reduced dose and increased initialization; (8) selective cellular uptake; (9) higher mucoadhesion and swelling index; and (10) increased intracellular delivery.

Proteins are susceptible to enzymatic degradation and are the second class of molecules after antibacterial drugs, investigated for chitosan nanoparticle–based pulmonary delivery [77]. Insulin-loaded chitosan nanoparticles were used to study the effect of chitosan having different molecular weights on the adsorption of insulin in lungs of rats [82]. Positively charged and small-sized chitosan nanoparticles showed high pulmonary adsorption of proteins and peptides. Positively charged chitosan nanoparticles showed high retention and penetration into mucous layers, which is a major contributing factor in the successful delivery of proteins. Insulin-loaded chitosan nanoparticles showed pronounced hypoglycemic effects in rats compared to exclusive insulin administration [83]. Delivery of heparin was also investigated by employing glycol-modified and surface-modified chitosan nanoparticles in a mouse model [84]. The heparin-loaded chitosan nanoparticles showed increased systemic coagulation time in mice during pulmonary delivery. Elcatonin-loaded liposomal chitosan nanoparticles showed increased cellular interaction and therapeutic efficacy in rats up to 48 h after pulmonary administration [85]. Elcatonin-loaded liposomal chitosan nanoparticles facilitated the opening of tight junctions and adherence to lung tissue, supporting the applicability of chitosan nanoparticles in drug delivery. Delivery of medically important enzymes has also been explored through chitosan-based delivery strategies by pulmonary and other routes. Horseradish peroxidase–loaded chitosan nanoparticles showed increased stability in the presence of urea at 37 °C. These nanoparticles showed high binding to the Bcap7, a human breast cell line and high cell death up to 80% in conjunction with prodrug [86]. Endostatin-loaded chitosan nanoparticles led to inhibition of proliferation of human umbilical vascular endothelial cells and antitumor activity in the Lewis lung cancer model compared to free endostatin administration for 14 days [87]. Prolidase-loaded chitosan nanoparticles also showed restoration of prolidase activity in prolidase-deficient cells, postulating the beneficial effects of chitosan nanoparticles in enzyme replacement therapy [88]. In another study, lysozyme-loaded chitosan nanoparticles were prepared and investigated for their antibacterial potential. Antibacterial assay revealed the high antibacterial activity of lysozyme-loaded chitosan nanoparticles against *S. aureus, P. aeruginosa, Bacillus subtilis,* and *E. coli* as compared to free lysozyme [154]. These studies showed that chitosan nanoparticles have potential applications in controlled delivery of alternative therapeutic agents, which include phages and phage products [11,89–91,150,155–157], antibacterial nanoparticles [72,92,93,158], phytochemicals [94,95], pigments [96], and other antibacterial chemical compounds [97,98] to counter multidrug-resistant pathogens and can be explored further in preclinical and clinical studies.

6.6.2 Oral delivery of therapeutics

Chitosan and chitosan nanoparticles have been well recognized for oral site-specific delivery of therapeutic agents [99]. Oral drug delivery is considered as an ideal route for drug administration because of its ease. Oral administration of drugs also faces serious limitations leading to poor absorption such as highly acidic environment in stomach, intestinal barriers, and the presence of degradative enzymes. These conditions limit the systemic absorption of drugs and lower the bioavailability of drug [100]. Nanoparticles-based drug delivery strategies overcome these limitations due to their small size, high surface to area ratio, and surface modifications [101]. Apart from these advantages nanoparticles provide stability to acid labile drugs in gastrointestinal tract compared to lipid-based delivery systems [102]. The number of chitosan formulations has been developed extensively for delivery of various oral drugs. Flavonoids such as catechin and epigallotocatechin are antioxidants in nature but are poorly absorbed by intestinal membranes and show degradation in intestinal environment. Catechin- and epigallotocatechin-loaded chitosan nanoparticles showed increased intestinal absorption of catechin and epigallotocatechin across intestinal barriers [103]. Tamoxifen is a water-soluble drug and considered as a good therapeutic option in cancer treatment. The efficacy of tamoxifen was increased by loading it into chitosan lecithin nanoparticles and this led to increased transport across intestinal membranes by a paracellular pathway [104]. Chitosan and carboxymethyl chitosan nanoparticles with doxorubicin hydrochloride have also been found to increase the absorption of drug by small intestine compared to doxorubicin alone [105]. Chitosan TPP nanoparticle formulation was prepared for alendronate sodium by ionic gelation method and the drug release was found to be pH dependent. Low pH showed increased release up to 80%, whereas slow sustained release was found at relatively higher pH. A tyrosine kinase inhibitor, sunitinib, was loaded into chitosan nanoparticles, and results showed high encapsulation efficiency of 98% and prolonged release up to 72 h [106].

Proteins get degraded and denature readily in acidic environment of intestinal tract and nanoparticles can aid in conferring stability to these oral therapeutics. Oral administration of insulin is a highly needed technical advancement in large section of population. Insulin-loaded chitosan nanoparticles were prepared by ionotropic gelation method using TPP. Reduced nanoparticle size and high stability of insulin were observed in insulin-loaded nanoparticles prepared by cross-linking. High intestinal epithelium uptake was observed along with sensitivity to lower gastric pH [107]. Hepatitis B virus inhibitor, Bay41-4109 (a heteroaryldihydropyrimidine), was loaded into chitosan nanoparticles to increase its solubility and systemic bioavailability. Results showed that Bay41-4109-loaded chitosan nanoparticles were cytotoxic and showed increased drug uptake of Bay41-4109 due to positively charged chitosan nanoparticles [108]. In another study, chitosan carboxymethyl chitosan nanoparticles were evaluated for oral antigen delivery. Results showed a higher level of antibody generation, complement activation, and high lysozymal activity than free antigen [109]. Albumin is a therapeutically important small globular protein and is widely used as a plasma expander in critical patients [110]. The delivery of albumin is mainly intravenous and attempts have been made for its oral administration. In this course, albumin-loaded chitosan nanoparticles were prepared and nanoparticles reflected high stability, high enzymatic activity, and sustained release profile. In *in-vivo* studies, albumin-loaded chitosan nanoparticles administrated to rats showed higher concentrations of albumin in serum compared to free albumin administrated to rats [111]. Chitosan nanoparticles have also been considered as efficient carriers of vaccines by oral delivery. Vaccine-loaded chitosan nanoparticles showed a sustained release and targeted delivery, which is a critical aspect in terms of immunogenicity of a protein or peptide in a vaccine preparation [109].

6.6.3 Transdermal delivery of therapeutics

Transdermal route is considered as a patient-friendly route as it obviates the gastrointestinal complications associated with oral delivery. A transdermal route not only avoids the acidic environment of intestinal tract, intestinal barriers, and metabolic processes but also provides prolonged constant therapeutically effective dosage to the host. These factors lower the chances of side effects of drug and treatment failure [112]. Skin acts as an inherent barrier for delivery of drugs, their penetration, and diffusion [113]. Nanoparticle delivery systems have been explored for the efficient delivery of drugs across the skin layer [114]. Chitosan nanoparticles have been exploited extensively because of the number of desirable properties such as mucoadhesive nature, ability to move across tight junctions, and ability to depolymerize F-actin [115]. Propanolol-loaded chitosan nanoparticles gel prepared by ionotropic gelation showed prolonged permeation of drug in pig ear skin compared to propanolol [115]. Lecithin chitosan nanoparticles were loaded with melatonin and showed enhanced drug flux across the skin. Melatonin-loaded lecithin chitosan nanoparticles did not show any effect neither on plasma membrane integrity nor on cell viability of human skin keratinocytes [116].

A transdermal delivery of proteins has also been attempted by using chitosan nanoparticle–based formulations. Insulin transdermal delivery was attempted by chitosan nanoinsulin formulation. A polyelectrolyte complex method was employed for chitosan nanoinsulin formulation which showed a decreased level of glucose concentration in rat plasma. The bioavailability of insulin formulation was compared with subcutaneous insulin injection, pharmacokinetics and pharmacodynamic data suggested the high bioavailability by a transdermal patch [117]. The effect of polypropylene electret was also studied in delivery of proteins by N-trimethyl chitosan nanoparticles prepared by ionotropic gelation. Polypropylene electret significantly increased the transdermal delivery of proteins as postulated by permeation assays and confocal scanning laser microscopy. Superoxide dismutase-loaded chitosan nanoparticles along with electret showed best inhibition on ear edema in a mouse model [118]. In a recent study, nanogel formulation containing transforming growth factor β_3-loaded chitosan nanoparticles was prepared by ionotropic gelation. The formulation showed layer-by-layer encapsulation and a sustained release of a transforming growth factor β_3, which can be further exploited in designing fiber mat implants [119].

6.6.4 Ocular delivery of therapeutics

Chitosan is widely used in ocular drug delivery as it enhances precorneal retention, stability, and interaction with mucosa of eye [15]. Chitosan, a natural hydrophilic biodegradable polymer, is nontoxic, low irritant, mucoadhesive, permeation, and transfection enhancer and provides a sustained release. These properties make it a better candidate for ocular delivery than other polymers [120]. Chitosan has the ability to form disulfide bonds with glycoproteins within a mucous gel layer [159]. Cyclosporin A–loaded chitosan nanoparticles were prepared and investigated for their therapeutic potential as well. Cyclosporin A–loaded chitosan nanoparticles were able to achieve therapeutic concentrations in cornea and conjunctiva. Cyclosporin A levels were found to be much higher than during instillation of free cyclosporin A and cyclosporin A with chitosan solution [121]. In another study, cholesterol-modified chitosan nanoparticles showed enhanced delivery of cyclosporin A on ocular surfaces [122]. Acyclovir-loaded chitosan nanoparticles also showed sustained delivery of antiviral acyclovir over 24 h in an *in-vitro* release model [123]. Bacterial endophthalmitis was targeted by daptomycin-loaded chitosan nanoparticles, and bacterial strains showed high susceptibility to drug released by daptomycin-loaded chitosan nanoparticles [124]. Thus, with chitosan nanoparticles it is possible to increase the therapeutic potential of drugs [121], delivery of drug into the anterior part of eye [125], a prolonged release of drug [126,127], and enhanced drug levels in the anterior part and vitreous as compared to commercial preparations [128]. The different delivery routes for therapeutic application of chitosan nanoparticles is shown in Fig. 6.2.

6.6.5 Other applications of chitosan nanoparticles

Chitosan nanoparticles are also involved in a number of other applications along with delivery of therapeutic molecules. Chitosan nanoparticles are used in tissue engineering as these can enhance membrane and transmucosal permeability which further leads to structural changes in proteins associated with tight junctions [129]. Besides the therapeutic potential, chitosan nanoparticles also carry ample diagnostic applications. Quantum dot conjugated folic acid carboxycellulose chitosan nanoparticles were used in drug delivery, targeting, and imaging of cancer cells. These nanostructures lower the toxicity of quantum dots, which are promising probes in a number of biomedical applications [130]. In cancer MRI, various MR contrast agents have been used along with chitosan nanoparticles [131,132]. Nowadays, enzymes immobilization on some solid matrices is attaining much consideration because immobilized enzymes generally enhance the stability and reusability of the enzymes [133,134]. Nanoparticles of various materials are generally used as immobilization matrices to fulfill this purpose [134]. Chitosan is also one of the apt materials for immobilization of enzymes on its surface because of the presence of free amine groups and its ability to inhibit chemical degradation [135,136]. Linolenic

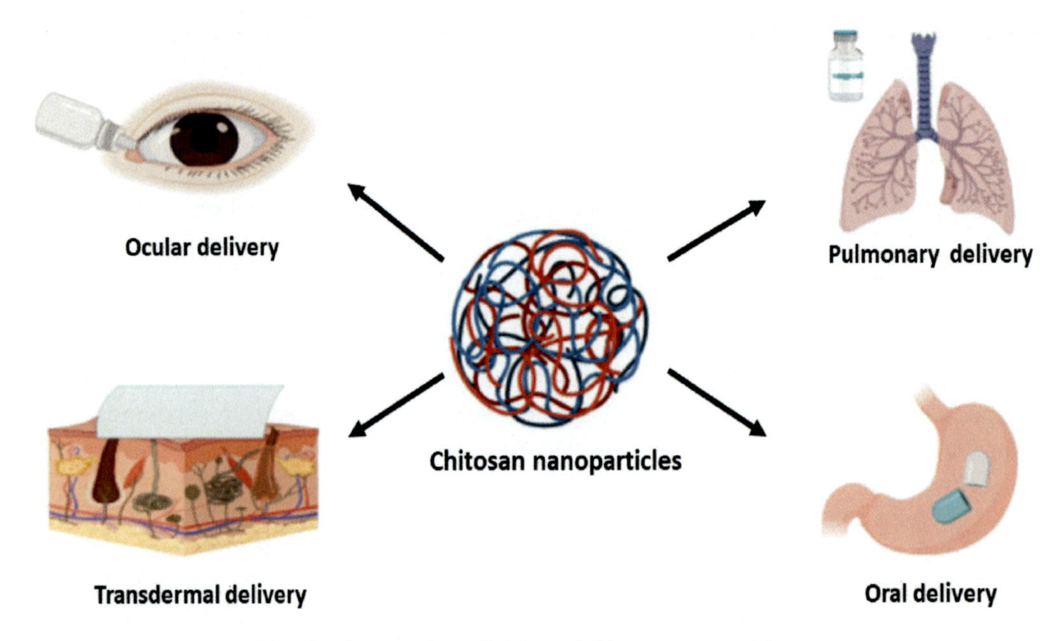

FIGURE 6.2 Image summarizing the therapeutic applications of chitosan nanoparticles.

acid-fabricated chitosan nanoparticles were immobilized with trypsin using glutaraldehyde and this led to enhanced thermal stability as well as achievable optimum temperature for trypsin after immobilization [137]. Lipase immobilization on chitosan magnetic core shell nanoparticles showed increased enzyme loading, adsorption, and activity [136]. Antioxidant activity of chitosan nanoparticles was also studied and chitosan fucoidan nanoparticles showed antioxidant behavior in terms of reactive oxygen species (ROS) and 2,2-diphenyl-1-picrylhydrazyl (DPHH) scavenging activity [138]. Iron chelating and hydroxyl radical scavenging activity of chitosan nanoparticles has been also reported in the literature [139].

In environment management also, chitosan nanoparticles have been explored as free functional groups of chitosan can bind to metal pollutants and pesticides [140]. Chitosan nanostructures have been used for adsorption of Cr, Pb, Cd, and acid green 27 dye [141,142]. Textile effluent was treated with chitosan nanoparticles to remove color and results showed 99% of color removal from effluent [143]. In another study, magnetic chitosan nanoparticles showed high adsorbing ability and recovery rate from effluent because of their magnetic properties [144]. Chitosan nanoparticles have also gained popularity in agricultural applications. Chitosan nanoparticles induce defense genes such as polyphenol oxidase, phenylalanine ammonia lyase, peroxide, and gluconase [145]. Herbicide-loaded chitosan nanoparticles are reported to have reduced herbicide toxicity [146]. Iron chitosan nanoparticle-based biosensors were reported to determine and scavenge heavy metals from plant environment [147]. In other applications chitosan nanoparticles have been reported to lower the bacterial count from fish fingers compared to untreated fish fingers [148]. Chitosan nanoparticles have also been reported to increase the washability and strength of textiles [149].

6.7 Conclusion

Chitosan nanoparticles have established their immense potential in various fields of nanobiotechnology. Among various fields, chitosan nanoparticles are mostly being exploited in medical research and applications. The easy nanoparticle preparation methods and variety of cross-linkers provide a wide range of optimization conditions for nanoparticles design and preparation. Different methods of nanoparticle preparation can be exploited to modulate nanoparticles properties as well as stability of encapsulated drugs. Chitosan nanoparticles are nontoxic in nature; thus they are employed by various delivery routes for delivery of therapeutic drugs in biological systems. These nanoparticles are not limited to drug delivery but are also fascinating nanocarriers in industrial, agricultural and environmental applications. With advent of time, nanobiotechnological advancements can further increase the spectrum as well as applicability of chitosan nanoparticles in various fields of modern life.

References

[1] A. Einbu, K.M. Vårum, Characterization of chitin and its hydrolysis to GlcNAc and GlcN, Biomacromolecules 9 (7) (2008) 1870–1875.

[2] N. Nwe, T. Furuike, H. Tamura, Production, properties and applications of fungal cell wall polysaccharides: chitosan and glucan, in: R. Jayakumar, M. Prabaharan, M. Muzzarelli, A.A. Riccardo (Eds.), Chitosan for Biomaterials II, Springer, Berlin, Heidelberg, 2011, pp. 187–207.

[3] E.F. Franca, R.D. Lins, L.C. Freitas, T.P. Straatsma, Characterization of chitin and chitosan molecular structure in aqueous solution, J. Chem. Theory Comp. 4 (12) (2008) 2141–2149.

[4] J.K. Park, M.J. Chung, H.N. Choi, Y.I. Park, Effects of the molecular weight and the degree of deacetylation of chitosan oligosaccharides on antitumor activity, Int. J. Mol. Sci. 12 (1) (2011) 266–277.

[5] P. Agrawal, G.J. Strijkers, K. Nicolay, Chitosan-based systems for molecular imaging, Adv. Drug Deliv. Rev. 62 (1) (2010) 42–58.

[6] J. Kumirska, M.X. Weinhold, J. Thöming, P. Stepnowski, Biomedical activity of chitin/chitosan based materials—influence of physicochemical properties apart from molecular weight and degree of N-acetylation, Polymers 3 (4) (2011) 1875–1901.

[7] J. Kumirska, M. Czerwicka, Z. Kaczyński, A. Bychowska, K. Brzozowski, J. Thöming, P. Stepnowski, Application of spectroscopic methods for structural analysis of chitin and chitosan, Marine Drugs 8 (5) (2010) 1567–1636.

[8] Y. Zhang, X. Zhang, R. Ding, J. Zhang, J. Liu, Determination of the degree of deacetylation of chitosan by potentiometric titration preceded by enzymatic pretreatment, Carbohydr. Polym. 83 (2) (2011) 813–817.

[9] K. Divya, M.S. Jisha, Chitosan nanoparticles preparation and applications, Environ. Chem. Lett. 16 (1) (2018) 101–112.

[10] M. Rinaudo, Chitin and chitosan: properties and applications, Prog. Polym. Sci. 31 (7) (2006) 603–632.

[11] Y. Yang, S. Wang, Y. Wang, X. Wang, Q. Wang, M. Chen, Advances in self-assembled chitosan nanomaterials for drug delivery, Biotechnol. Adv. 32 (7) (2014) 1301–1316.

[12] T. Kean, M. Thanou, Biodegradation, biodistribution and toxicity of chitosan, Adv. Drug Deliv. Rev. 62 (1) (2010) 3–11.

[13] S. Shiraishi, T. Imai, M. Otagiri, Controlled release of indomethacin by chitosan-polyelectrolyte complex: optimization and in vivo/in vitro evaluation, J. Control. Release 25 (3) (1993) 217–225.

[14] Y. Ohya, M. Shiratani, H. Kobayashi, T. Ouchi, Release behaviour of 5-fluorouracil from chitosan-gel nanospheres immobilizing 5-fluorouracil coated with polysaccharides and their cell specific cytotoxicity., Pure Appl. Chem. A31 (1994) 629–642.

[15] S. Naskar, K. Koutsu, S. Sharma, Chitosan-based nanoparticles as drug delivery systems: a review on two decades of research, J. Drug Target. 5 (2018) 379–393.

[16] S. Kunjachan, S. Jose, T. Lammers, Understanding the mechanism of ionic gelation for synthesis of chitosan nanoparticles using qualitative techniques, Asian J. Pharma 4 (2) (2014) 148–152.

[17] P. Calvo, C. Remunan Lopez, J.L. Vila Jato, M.J. Alonso, Novel hydrophilic chitosan polyethylene oxide nanoparticles as protein carriers, J. Appl. Polym. Sci. 63 (1) (1997a) 125–132.

[18] O. Masalova, V. Kulikouskaya, T. Shutava, V. Agabekov, Alginate and chitosan gel nanoparticles for efficient protein entrapment, Phys. Procedia 40 (2013) 69–75.

[19] R. Fernández-Urrusuno, P. Calvo, C. Remuñán-López, J.L. Vila-Jato, M.J. Alonso, Enhancement of nasal absorption of insulin using chitosan nanoparticles, Pharma. Res. 16 (10) (1999) 1576–1581.

[20] R. Aydin, M. Pulat, 5-Fluorouracil encapsulated chitosan nanoparticles for pH-stimulated drug delivery: evaluation of controlled release kinetics, J. Nanomater. 2012 (2012) 42.

[21] L. Ilium, Chitosan and its use as a pharmaceutical excipient, Pharma. Res. 15 (9) (1998) 1326–1331.

[22] I.C. Gonçalves, P.C. Henriques, C.L. Seabra, M.C. Martins, The potential utility of chitosan micro/nanoparticles in the treatment of gastric infection, Exp. Rev. Anti Infect. Ther. 12 (8) (2014) 981–992.

[23] Y. Hu, X. Jiang, Y. Ding, H. Ge, Y. Yuan, C. Yang, Synthesis and characterization of chitosan–poly (acrylic acid) nanoparticles, Biomaterials 23 (15) (2002) 3193–3201.

[24] A. Grenha, Chitosan nanoparticles: a survey of preparation methods, J. Drug Target. 20 (4) (2012) 291–300.

[25] T. Niwa, H. Takeuchi, T. Hino, N. Kunou, Y. Kawashima, Preparations of biodegradable nanospheres of water-soluble and insoluble drugs with D, L-lactide/glycolide copolymer by a novel spontaneous emulsification solvent diffusion method, and the drug release behavior, J. Control. Release 25 (1–2) (1993) 89–98.

[26] M.H. El-Shabouri, Positively charged nanoparticles for improving the oral bioavailability of cyclosporin-A, Int. J. Pharma 249 (1-2) (2002) 101–108.

[27] A. Mehrotra, R.C. Nagarwal, J.K. Pandit, Fabrication of lomustine loaded chitosan nanoparticles by spray drying and in vitro cytostatic activity on human lung cancer cell line L132, J. Nanomed. Nanotechnol 1 (2010) 103.

[28] S.L. Wang, Đ.M. Hiep, P.M. Luong, N.T. Vui, T.M. Đinh, N.A. Dzung, Preparation of chitosan nanoparticles by spray drying, and their antibacterial activity, Res. Chem. Intermed. 40 (6) (2014) 2165–2175.

[29] P. Schuck, A. Dolivet, S. Méjean, P. Zhu, E. Blanchard, R. Jeantet, Drying by desorption: a tool to determine spray drying parameters, J. Food Eng. 94 (2) (2009) 199–204.

[30] R. Vehring, Pharmaceutical particle engineering via spray drying, Pharma. Res. 25 (5) (2008) 999–1022.

[31] X. Li, N. Anton, C. Arpagaus, F. Belleteix, T.F. Vandamme, Nanoparticles by spray drying using innovative new technology: The Büchi Nano Spray Dryer B-90, J. Control. Release 147 (2) (2010) 304–310.

[32] F. Brunel, L. Véron, L. David, A. Domard, T. Delair, A novel synthesis of chitosan nanoparticles in reverse emulsion, Langmuir 24 (20) (2008) 11370–11377.

[33] M. Mohammed, J. Syeda, K. Wasan, E. Wasan, An overview of chitosan nanoparticles and its application in non-parenteral drug delivery, Pharmaceutics 9 (4) (2017) E53.

[34] I.M. Vanderlubben, J.C. Verhoef, G. Borchard, H.E. Junginger, Chitosan for mucosal vaccination, Adv. Drug Deliv. Rev. 52 (2) (2001) 139–144.

[35] Y. Zhuo, J. Han, L. Tang, N. Liao, G.F. Gui, Y.Q. Chai, R. Yuan, Quenching of the emission of peroxydisulfate system by ferrocene functionalized chitosan nanoparticles: a sensitive "signal off" electrochemiluminescence immunosensor, Sensor Actuators B. Chem. 192 (2014) 791–795.

[36] M. Huang, C.W. Fong, E. Khor, L.Y. Lim, Transfection efficiency of chitosan vectors: effect of polymer molecular weight and degree of deacetylation, J. Control. Release 106 (3) (2005) 391–406.

[37] K.C. Hembram, S. Prabha, R. Chandra, B. Ahmed, S. Nimesh, Advances in preparation and characterization of chitosan nanoparticles for therapeutics, Artific. Cells Nanomed. Biotechnol. 44 (1) (2016) 305–314.

[38] K. Nagpal, S.K. Singh, D.N. Mishra, Chitosan nanoparticles: a promising system in novel drug delivery, Chem. Pharma. Bull. 58 (11) (2010) 1423–1430.

[39] Z. Liu, Y. Jiao, Y. Wang, C. Zhou, Z. Zhang, Polysaccharides-based nanoparticles as drug delivery systems, Adv. Drug Deliv. Rev. 60 (15) (2008) 1650–1662.

[40] K. Konecsni, N.H. Low, M.T. Nickerson, Chitosan–tripolyphosphate submicron particles as the carrier of entrapped rutin, Food Chem. 134 (4) (2012) 1775–1779.

[41] L. Servat-Medina, A. González-Gómez, F. Reyes-Ortega, I.M. Sousa, N.D. Queiroz, P.M. Zago, M.P. Jorge, K.M. Monteiro, J.E. de Carvalho, J. San Román, M.A. Foglio, Chitosan–tripolyphosphate nanoparticles as *Arrabidaeachica* standardized extract carrier: synthesis, characterization, biocompatibility, and antiulcerogenic activity, Int. J. Nanomed. 10 (2015) 3897–3909.

[42] W. Liu, S. Sun, Z. Cao, X. Zhang, K. Yao, W.W. Lu, K.D. Luk, An investigation on the physicochemical properties of chitosan/DNA polyelectrolyte complexes, Biomaterials 26 (15) (2005a) 2705–2711.

[43] B. Hu, C. Pan, Y. Sun, Z. Hou, H. Ye, B. Hu, X. Zeng, Optimization of fabrication parameters to produce chitosan tripolyphosphate nanoparticles for delivery of tea catechins, J. Agric. Food Chem. 56 (16) (2008) 7451–7458.

[44] L.M. Zhao, L.E. Shi, Z.L. Zhang, J.M. Chen, D.D. Shi, J. Yang, Z.X. Tang, Preparation and application of chitosan nanoparticles and nanofibers, Braz. J. Chem. Eng. 28 (3) (2011) 353–362.

[45] T. Banerjee, S. Mitra, A.K. Singh, R.K. Sharma, A. Maitra, Preparation, characterization and biodistribution of ultrafine chitosan nanoparticles, Int. J. Pharma 243 (1-2) (2002) 93–105.

[46] K.A. Janes, P. Calvo, M.J. Alonso, Polysaccharide colloidal particles as delivery systems for macromolecules, Adv. Drug Deliv. Rev. 47 (1) (2001) 83–97.

[47] M.P. Balaguer, J. Gómez-Estaca, R. Gavara, P. Hernandez-Munoz, Functional properties of bioplastics made from wheat gliadins modified with cinnamaldehyde, J. Agric. Food Chem. 59 (12) (2011) 6689–6695.

[48] P.W. Li, G. Wang, Z.M. Yang, W. Duan, Z. Peng, L.X. Kong, Q.H. Wang, Development of drug-loaded chitosan–vanillin nanoparticles and its cytotoxicity against HT-29 cells, Drug Deliv. 23 (1) (2016) 30–35.

[49] J.J. Wang, Z.W. Zeng, R.Z. Xiao, T. Xie, G.L. Zhou, X.R. Zhan, S.L. Wang, Physical adsorption of lipase onto mesoporous silicacent advances of chitosan nanoparticles as drug carriers, Int. J. Nanomed. 6 (2011) 765–774.

[50] A.O. Elzoghby, W.M. Samy, N.A. Elgindy, Novel spray-dried genipin-crosslinked casein nanoparticles for prolonged release of alfuzosin hydrochloride, Pharma. Res. 30 (2) (2013) 512–522.

[51] F. Song, L.M. Zhang, C. Yang, L. Yan, Genipin-crosslinked casein hydrogels for controlled drug delivery, Int. J. Pharma 373 (1-2) (2009) 41–47.

[52] G.V. Kumar, C.H. Su, P. Velusamy, Ciprofloxacin loaded genipin cross-linked chitosan/heparin nanoparticles for drug delivery application, Mater. Lett. 180 (2016) 119–122.

[53] Y. Zhao, Z. Sun, Effects of gelatin-polyphenol and gelatin–genipin cross-linking on the structure of gelatin hydrogels, Int. J. Food Prop. 20 (Suppl 3) (2017) S2822–S2832.

[54] T.A. Ahmed, B.M. Aljaeid, Preparation, characterization, and potential application of chitosan, chitosan derivatives, and chitosan metal nanoparticles in pharmaceutical drug delivery. Drug Des., Dev. Ther. 10 (2016) 483–507.

[55] M. Bagheri, H. Younesi, S. Hajati, S.M. Borghei, Application of chitosan-citric acid nanoparticles for removal of chromium (VI), Int. J. Biol. Macromol. 80 (2015) 431–444.

[56] Z. Keresztessy, M. Bodnár, E. Ber, I. Hajdu, M. Zhang, J.F. Hartmann, T. Minko, J. Borbély, Self-assembling chitosan/poly-γ-glutamic acid nanoparticles for targeted drug delivery, Coll. Polym. Sci. 287 (7) (2009) 759–765.

[57] M.M. Al-Remawi, Properties of chitosan nanoparticles formed using sulfate anions as crosslinking bridges, Am. J. Appl. Sci. 9 (7) (2012) 1091–1100.

[58] W. Abdelwahed, G. Degobert, S. Stainmesse, H. Fessi, Freeze-drying of nanoparticles: formulation, process and storage considerations, Adv. Drug Deliv. Rev. 58 (15) (2006) 1688–1713.

[59] M.J. AlonsoS. Cohen, H. Bernstein (Eds.), Nanoparticulate drug carrier technology, Microparticulate Systems for the Delivery of Proteins and Vaccines (1996) 203–242.

[60] T. López-León, E.L. Carvalho, B. Seijo, J.L. Ortega-Vinuesa, D. Bastos-González, Physicochemical characterization of chitosan nanoparticles: electrokinetic and stability behavior, J. Coll. Interf. Sci. 283 (2) (2005) 344–351.

[61] A.W. Pan, B.B. Wu, J.M. Wu, Chitosan nanoparticles crosslinked by glycidoxypropyltrimethoxysilane for pH triggered release of protein, Chin. Chem. Lett. 20 (1) (2009) 79–83.

[62] L. Fan, H. Wu, H. Zhang, F. Li, T.H. Yang, C.H. Gu, Q. Yang, Novel super pH-sensitive nanoparticles responsive to tumor extracellular pH, Carbohydr. Polym. 73 (3) (2008) 390–400.

[63] Q. Yao, W. Liu, X.J. Gou, X.Q. Guo, J. Yan, Q. Song, F.Z. Chen, Q. Zhao, C. Chen, T. Chen, Preparation, characterization, and cytotoxicity of various chitosan nanoparticles, J. Nanomater. 7 (3) (2013) 035008.

[64] N. Kumar, R.K. Salar, M. Prasad, K. Ranjan, Synthesis, characterization and anticancer activity of vincristine loaded folic acid-chitosan conjugated nanoparticles on NCI-H460 non-small cell lung cancer cell line, Egypt. J. Basic Appl. Sci. 5 (1) (2018) 87–99.

[65] S.A. Loutfy, H.M. El-Din, M.H. Elberry, N.G. Allam, M.T. Hasanin, A.M. Abdellah, Synthesis, characterization and cytotoxic evaluation of chitosan nanoparticles: in vitro liver cancer model, Adv. Nat. Sci. Nanosci. Nanotechnol 7 (3) (2016) 035008.

[66] J.W. Loh, G. Yeoh, M. Saunders, L.Y. Lim, Uptake and cytotoxicity of chitosan nanoparticles in human liver cells, Toxicol. Appl. Pharmacol. 249 (2) (2010) 148–157.

[67] K. Corsi, F. Chellat, L.H. Yahia, J.C. Fernandes, Mesenchymal stem cells, MG63 and HEK293 transfection using chitosan-DNA nanoparticles, Biomaterials 24 (7) (2003) 1255–1264.

[68] O. Zaki, S. Sarah, M.N. Ibrahim, H. Katas, Particle size affects concentration-dependent cytotoxicity of chitosan nanoparticles towards mouse hematopoietic stem cells, J. Nanotechnol. (2015) 919658.

[69] L. Qi, Z. Xu, X. Jiang, Y. Li, M. Wang, Cytotoxic activities of chitosan nanoparticles and copper-loaded nanoparticles, Bioorgan. Med. Chem. Lett. 15 (5) (2005) 1397–1399.

[70] X. Hu, Y. Wang, B. Peng, Chitosan capped mesoporous silica nanoparticles as pH responsive nanocarriers for controlled drug release, Chem. Asian J. 9 (1) (2014) 319–327.

[71] M. Zăhan, A.m. Muţoiu, L. Olenic, I. Miclea, A. Criste, V. Miclea, Cytotoxic effect of chitosan-gold nanoparticles on two cell lines in culture, Bull. UASVM Anim. Sci. Biotechnol. 74 (2017) 139–143.

[72] V.S. Gondil, T. Kalaiyarasan, V.K. Bharti, S. Chhibber, Antibiofilm potential of Seabuckthorn silver nanoparticles (SBT@ AgNPs) against *Pseudomonas aeruginosa*, 3 Biotechnology 9 (11) (2019) 402.

[73] M. Ariful-Islam, T.E. Park, E. Reesor, K. Cherukula, A. Hasan, J. Firdous, B. Singh, S.K. Kang, Y.J. Choi, I.K. Park, C.S. Cho, Mucoadhesive chitosan derivatives as novel drug carriers, Curr. Pharma. Des. 21 (29) (2015) 4285–4309.

[74] M.C. Gaspar, J.J. Sousa, A.A. Pais, O. Cardoso, D. Murtinho, M.E. Serra, F. Tewes, J.C. Olivier, Optimization of levofloxacin-loaded crosslinked chitosan microspheres for inhaled aerosol therapy, Eur. J. Pharma. Biopharma 96 (2015) 65–75.

[75] T.P. Learoyd, J.L. Burrows, E. French, P.C. Seville, Modified release of beclometasone dipropionate from chitosan-based spray-dried respirable powders, Powder Technol. 187 (3) (2008) 231–238.

[76] R.V. Pai, R.R. Jain, A.S. Bannalikar, M.D. Menon, Development and evaluation of chitosan microparticles based dry powder inhalation formulations of rifampicin and rifabutin, J. Aerosol Med. Pulmonary Drug Deliv. 29 (2) (2016) 179–195.

[77] N. Islam, V. Ferro, Recent advances in chitosan-based nanoparticulate pulmonary drug delivery, Nanoscale 8 (30) (2016) 14341–14358.

[78] T. Chhibber, V.S. Gondil, V.R. Sinha, Development of chitosan-based hydrogel containing antibiofilm agents for the treatment of *Staphylococcus aureus*-infected burn wound in mice, AAPS PharmSciTech. 21 (2) (2020;) 1–2.

[79] J.H. Park, H.E. Jin, D.D. Kim, S.J. Chung, W.S. Shim, C.K. Shim, Chitosan microspheres as an alveolar macrophage delivery system of ofloxacin via pulmonary inhalation, Int. J. Pharma 441 (1-2) (2013) 562–569.

[80] M.S. Kamble, O.R. Mane, V.G. Borwandkar, S.S. Mane, P.D. Chaudhari, Formulation and evaluation of clindamycin Hcl-chitosan microspheres for dry powder inhaler formulation, Drug Invent. Today 4 (10) (2012) 527–530.

[81] M. Sandhya, V. Aparna, B. Raja, R. Jayakumar, S. Sathianarayanan, Amphotericin B loaded sulfonated chitosan nanoparticles for targeting macrophages to treat intracellular *Candida glabrata* infections, Int. J. Biol. Macromol. 110 (2018) 133–139.

[82] C. Yan, J. Wang, J. Gu, D. Hou, L. Lei, H. Jing, H. Katsumi, T. Sakane, A. Yamamoto, The influence of molecular parameters of chitosan on pulmonary absorption of insulin loaded chitosan nanoparticles, Latin Am. J. Pharma 32 (6) (2013) 860–868.

[83] S. Al-Qadi, A. Grenha, D. Carrión-Recio, B. Seijo, C. Remuñán-López, Microencapsulated chitosan nanoparticles for pulmonary protein delivery: in vivo evaluation of insulin-loaded formulations, J. Control. Release 157 (3) (2012) 383–390.

[84] A. Trapani, S. Di Gioia, N. Ditaranto, N. Cioffi, F.M. Goycoolea, A. Carbone, M. Garcia-Fuentes, M. Conese, M.J. Alonso, Systemic heparin delivery by the pulmonary route using chitosan and glycol chitosan nanoparticles, Int. J. Pharma 447 (1-2) (2013) 115–123.

[85] M. Murata, K. Nakano, K. Tahara, Y. Tozuka, H. Takeuchi, Pulmonary delivery of elcatonin using surface-modified liposomes to improve systemic absorption: polyvinyl alcohol with a hydrophobic anchor and chitosan oligosaccharide as effective surface modifiers, Eur. J. Pharma. Biopharma 80 (2) (2012) 340–346.

[86] X. Cao, C. Chen, H. Yu, P. Wang, Horseradish peroxidase-encapsulated chitosan nanoparticles for enzyme-prodrug cancer therapy, Biotechnol. Lett. 37 (1) (2015) 81–88.

[87] R.L. Ding, F. Xie, Y. Hu, S.Z. Fu, J.B. Wu, J. Fan, W.F. He, Y. He, L.L. Yang, S. Lin, Q.L. Wen, Preparation of endostatin-loaded chitosan nanoparticles and evaluation of the antitumor effect of such nanoparticles on the Lewis lung cancer model, Drug Deliv. 24 (1) (2017) 300–308.

[88] C. Colonna, B. Conti, P. Perugini, F. Pavanetto, T. Modena, R. Dorati, P. Iadarola, I. Genta, Ex vivo evaluation of prolidase loaded chitosan nanoparticles for the enzyme replacement therapy, Eur. J. Pharma. Biopharma 70 (1) (2008) 58–65.

[89] V.S. Gondil, S. Chhibber, Evading antibody mediated inactivation of bacteriophages using delivery systems, J. Virol. Curr. Res 1 (2017) 55574.

[90] V.S. Gondil, S. Chhibber, Exploring potential of phage therapy for tuberculosis using model organism, Biomed. Biotechnol. Res. J 2 (1) (2018) 9.

[91] V.S. Gondil, T. Dube, J.J. Panda, R.M. Yennamalli, K. Harjai, S. Chhibber, Comprehensive evaluation of chitosan nanoparticle based phage lysin delivery system; a novel approach to counter *S. pneumoniae* infections, Int. J. Pharma 573 (2020) 118850.

[92] M. Kumar, R. Bala, V.S. Gondil, D.V. Jain, S. Chhibber, R.K. Sharma, N. Wangoo, Efficient, green and one pot synthesis of sodium acetate functionalized silver nanoparticles and their potential application as food preservative, BioNanoScience 7 (3) (2017) 521–529.

[93] S. Chhibber, V.S. Gondil, L. Singla, M. Kumar, T. Chhibber, G. Sharma, R.K. Sharma, N. Wangoo, O.P. Katare, Effective topical delivery of H-AgNPs for eradication of *Klebsiella pneumoniae*-induced burn wound infection, AAPS PharmSciTech. 20 (5) (2019) 169.

[94] I.W. Kusuma, H. Kuspradini, E.T. Arung, F. Aryani, Y.H. Min, J.S. Kim, Y.U. Kim, Biological activity and phytochemical analysis of three Indonesian medicinal plants, *Murrayak oenigii, Syzygium polyanthum* and *Zingiber purpurea*, J. Acupunct. Meridian Stud 4 (1) (2011) 75–79.

[95] P. Kalia, N.R. Kumar, K. Harjai, Phytochemical screening and antibacterial activity of different extracts of propolis, Int. J. Pharma. Biol. Res 3 (6) (2013) 219–222.

[96] V.S. Gondil, M. Asif, T.C. Bhalla, Optimization of physicochemical parameters influencing the production of prodigiosin from *Serratia nematodiphila* RL_2 and exploring its antibacterial activity, 3 Biotech 7 (5) (2017) 338.

[97] R.P. Sharma, A. Saini, S. Kumar, J. Kumar, P. Venugopalan, V.S. Gondil, S. Chhibber, T. Aree, Diaquabis (ethylenediamine) copper (II) vs. monoaquabis (ethylenediamine) copper (II): Synthesis, characterization, single crystal X-ray structure determination, theoretical calculations and antimicrobial activities of [Cu (en) 2 (H_2O) 2](2-phenoxybenzoate) 2• H_2O and [Cu (en) 2 (H_2O)](diphenylacetate) 2• $3H_2O$, Polyhedron 123 (2017) 430–440.

[98] S. Kumar, R.P. Sharma, P. Venugopalan, V.S. Gondil, S. Chhibber, V. Ferretti, Synthesis and characterization of new silver (I) naphthalenedisulfonate complexes with heterocyclic N-donor ligands: packing analyses and antibacterial studies, Polyhedron 159 (2019) 275–283.

[99] O. Cota-Arriola, M. Onofre Cortez-Rocha, A. Burgos-Hernández, J. Marina Ezquerra-Brauer, M. Plascencia Jatomea, Controlled release matrices and micro/nanoparticles of chitosan with antimicrobial potential: development of new strategies for microbial control in agriculture, J. Sci. Food Agric 93 (7) (2013) 1525–1536.

[100] K. Bowman, K.W. Leong, Chitosan nanoparticles for oral drug and gene delivery, Int. J. Nanomed 1 (2) (2006) 117.

[101] S.K. Shukla, A.K. Mishra, O.A. Arotiba, B.B. Mamba, Chitosan-based nanomaterials: a state-of-the-art review, Int. J. Biol. Macromol. 59 (2013) 46–58.

[102] J. Palacio, N.A. Agudelo, B.L. Lopez, PEGylation of PLA nanoparticles to improve mucus-penetration and colloidal stability for oral delivery systems, Curr. Opin. Chem. Eng 11 (2016) 14–19.

[103] A. Dube, J.A. Nicolazzo, I. Larson, Chitosan nanoparticles enhance the intestinal absorption of the green tea catechins (+)-catechin and (−)-epigallocatechin gallate, Eur. J. Pharma. Sci. 41 (2) (2010) 219–225.

[104] S. Barbieri, F. Buttini, A. Rossi, R. Bettini, P. Colombo, G. Ponchel, F. Sonvico, G. Colombo, Ex vivo permeation of tamoxifen and its 4-OH metabolite through rat intestine from lecithin/chitosan nanoparticles, Int. J. Pharma 491 (1-2) (2015) 99–104.

[105] C. Feng, Z. Wang, C. Jiang, M. Kong, X. Zhou, Y. Li, X. Cheng, X. Chen, Chitosan/o-carboxymethyl chitosan nanoparticles for efficient and safe oral anticancer drug delivery: in vitro and in vivo evaluation, Int. J. Pharma 457 (1) (2013) 158–167.

[106] J.J. Joseph, D. Sangeetha, T. Gomathi, Sunitinib loaded chitosan nanoparticles formulation and its evaluation, Int. J. Biol. Macromol 82 (2016) 952–998.

[107] M. Diop, N. Auberval, A. Viciglio, A. Langlois, W. Bietiger, C. Mura, C. Peronet, A. Bekel, D.J. David, M. Zhao, M. Pinget, Design, characterisation, and bioefficiency of insulin–chitosan nanoparticles after stabilisation by freeze-drying or cross-linking, Int. J. Pharma 491 (1-2) (2015) 402–408.

[108] M. Xue, S. Hu, Y. Lu, Y. Zhang, X. Jiang, S. An, Y. Guo, X. Zhou, H. Hou, C. Jiang, Development of chitosan nanoparticles as drug delivery system for a prototype capsid inhibitor, Int. J. Pharma 495 (2) (2015) 771-782.

[109] P. Gao, G. Xia, Z. Bao, C. Feng, X. Cheng, M. Kong, Y. Liu, X. Chen, Chitosan based nanoparticles as protein carriers for efficient oral antigen delivery, Int. J. Biol. Macromol 91 (2016) 716-723.

[110] P. Caraceni, M. Tufoni, M.E. Bonavita, Clinical use of albumin, Blood Transfusion 11 (Suppl. 4) (2013) s18.

[111] D. Nashaat, M. Elsabahy, T. El-Sherif, M.A. Hamad, G.A. El-Gindy, E.H. Ibrahim, Development and in vivo evaluation of chitosan nanoparticles for the oral delivery of albumin, Pharma. Dev. Technol 32 (2018) 329-337.

[112] T. Tanner, R. Marks, Delivering drugs by the transdermal route: review and comment, Skin Res. Technol 14 (3) (2008) 249-260.

[113] G. Cevc, U. Vierl, Nanotechnology and the transdermal route: a state of the art review and critical appraisal, J. Control. Release 141 (3) (2010) 277-299.

[114] T.W. Prow, J.E. Grice, L.L. Lin, R. Faye, M. Butler, W. Becker, E.M. Wurm, C. Yoong, T.A. Robertson, H.P. Soyer, M.S. Roberts, Nanoparticles and microparticles for skin drug delivery, Adv. Drug Deliv. Rev 63 (6) (2011) 470-491.

[115] R. Al-Kassas, J. Wen, A.E. Cheng, A.M. Kim, S.S. Liu, J. Yu, Transdermal delivery of propranolol hydrochloride through chitosan nanoparticles dispersed in mucoadhesive gel, Carbohydr. Polym 153 (2016) 176-186.

[116] A. Hafner, J. Lovrić, I. Pepić, J. Filipović-Grčić, Lecithin/chitosan nanoparticles for transdermal delivery of melatonin, J. Microencapsul. 28 (8) (2011) 807-815.

[117] V. Venugopal, Transdermal delivery of insulin by biodegradable chitosan nanoparticles: exvivo and in vivo studies, Iran. J. Pharma. Sci. 8 (1) (2012) 315-321.

[118] Y. Tu, X. Wang, Y. Lu, H. Zhang, Y. Yu, Y. Chen, J. Liu, Z. Sun, L. Cui, J. Gao, Y. Zhong, Promotion of the transdermal delivery of protein drugs by N-trimethyl chitosan nanoparticles combined with polypropylene electret, Int. J. Nanomed 11 (2016) 5549-5561.

[119] S. Sydow, D. de Cassan, R. Hänsch, T.R. Gengenbach, C.D. Easton, H. Thissen, H. Menzel, Layer-by-layer deposition of chitosan nanoparticles as drug-release coatings for PCL nanofibers, Biomater. Sci. 7 (1) (2019) 233-246.

[120] U.G. Kapanigowda, S.H. Nagaraja, B. Ramaiah, P.R. Boggarapu, Improved intraocular bioavailability of ganciclovir by mucoadhesive polymer based ocular microspheres: development and simulation process in Wistar rats, DARU J. Pharma. Sci. 23 (1) (2015) 49.

[121] A.M. De-Campos, A. Sánchez, M.J. Alonso, Chitosan nanoparticles: a new vehicle for the improvement of the delivery of drugs to the ocular surface. Application to cyclosporin A, Int. J. Pharma. 224 (1-2) (2001) 159-168.

[122] X.B. Yuan, H. Li, Y.B. Yuan, Preparation of cholesterol-modified chitosan self-aggregated nanoparticles for delivery of drugs to ocular surface, Carbohydr. Polym 65 (3) (2006) 337-345.

[123] N.N. Rajendran, R. Natrajan, S. Kumar, S. Selvaraj, Acyclovir-loaded chitosan nanoparticles for ocular delivery, Asian J. Pharma 4 (4) (2014) 220-226.

[124] N.C. Silva, S. Silva, B. Sarmento, M. Pintado, Chitosan nanoparticles for daptomycin delivery in ocular treatment of bacterial endophthalmitis, Drug Deliv. 22 (7) (2015) 885-893.

[125] P. Calvo, J.L. Vila-Jato, M.J. Alonso, Evaluation of cationic polymer-coated nanocapsules as ocular drug carriers, Int. J. Pharma. 153 (1) (1997) 41-50 b.

[126] H.J. Kao, Y.L. Lo, H.R. Lin, S.P. Yu, Characterization of pilocarpine loaded chitosan/carbopol nanoparticles, J. Pharma. Pharmacol 58 (2) (2006) 179-186.

[127] H.R. Lin, S.P. Yu, C.J. Kuo, H.J. Kao, Y.L. Lo, Y.J. Lin, Pilocarpine-loaded chitosan-PAA nanosuspension for ophthalmic delivery, J. Biomater. Sci., Polymer Edition. 18 (2) (2007) 205-221.

[128] X. Qu, V.V. Khutoryanskiy, A. Stewart, et al., Carbohydrate-based micelle clusters which enhance hydrophobic drug bioavailability by up to 1 order of magnitude, Biomacromolecules 7 (2006) 3452–3459.

[129] N.A. Peppas, Y. Huang, Nanoscale technology of mucoadhesive interactions, Adv. Drug Deliv. Rev 56 (11) (2004) 1675–1687.

[130] M.E. Mathew, J.C. Mohan, K. Manzoor, S.V. Nair, H. Tamura, R. Jayakumar, Folate conjugated carboxymethyl chitosan–manganese doped zinc sulphide nanoparticles for targeted drug delivery and imaging of cancer cells, Carbohydr. Polym 80 (2) (2010) 442–448.

[131] K. Nwe, C.H. Huang, A. Tsourkas, Gd-labeled glycol chitosan as a pH-responsive magnetic resonance imaging agent for detecting acidic tumor microenvironments, J. Med. Chem 56 (20) (2013) 7862–7869.

[132] N.D. Thorat, S.V. Otari, R.M. Patil, R.A. Bohara, H.M. Yadav, V.B. Koli, A.K. Chaurasia, R.S. Ningthoujam, Synthesis, characterization and biocompatibility of chitosan functionalized superparamagnetic nanoparticles for heat activated curing of cancer cells, Dalton Trans. 43 (46) (2014) 17343–17351.

[133] A. Sharma, Shadiya, T. Sharma, K.R. Meena, R. Kumar, S.S. Kanwar, Biodiesel and the potential role of microbial lipases in its production, in: P.K. Arora (Ed.), Microbial Technology for the Welfare of Society, Springer, 2019, pp. 83–99. doi:10.1007/978-981-13-8844-6_4.

[134] A. Sharma, T. Sharma, K.R. Meena, A. Kumar, S.S. Kanwar, High throughput synthesis of ethyl pyruvate by employing superparamagnetic iron nanoparticles-bound esterase, Process Biochem. (2018). doi:10.1016/j.procbio.2018.05.004.

[135] K. Yang, N.S. Xu, W.W. Su, Co-immobilized enzymes in magnetic chitosan beads for improved hydrolysis of macromolecular substrates under a time-varying magnetic field, J. Biotechnol 148 (2-3) (2010) 119–127.

[136] A. Ghadi, F. Tabandeh, S. Mahjoub, A. Mohsenifar, F.T. Roshan, R.S. Alavije, Fabrication and characterization of core-shell magnetic chitosan nanoparticles as a novel carrier for immobilization of *Burkholderiacepacia* lipase, J. Oleo Sci. 64 (4) (2015) 423–430.

[137] C.G. Liu, K.G. Desai, X.G. Chen, H.J. Park, Preparation and characterization of nanoparticles containing trypsin based on hydrophobically modified chitosan, J. Agric. Food Chem 53 (5) (2005) 1728–1733 b.

[138] Y.C. Huang, R.Y. Li, Preparation and characterization of antioxidant nanoparticles composed of chitosan and fucoidan for antibiotics delivery, Marine Drugs 12 (8) (2014) 4379–4398.

[139] M.T. Yen, J.H. Yang, J.L. Mau, Antioxidant properties of chitosan from crab shells, Carbohydr. Polym 74 (4) (2008) 840–844.

[140] S.M. Dehaghi, B. Rahmanifar, A.M. Moradi, P.A. Azar, Removal of permethrin pesticide from water by chitosan–zinc oxide nanoparticles composite as an adsorbent, J. Saudi Chem. Soc 18 (4) (2014) 348–355.

[141] S.M. Seyedi, B. Anvaripour, M. Motavassel, N. Jadidi, Comparative cadmium adsorption from water by nanochitosan and chitosan, Int. J. Eng. Innov. Technol 2 (9) (2013) 145–148.

[142] M.S. Sivakami, T. Gomathi, J. Venkatesan, H.S. Jeong, S.K. Kim, P.N. Sudha, Preparation and characterization of nano chitosan for treatment wastewaters, Int. J. Biol. Macromol 57 (2013) 204–212.

[143] A. Abul, S. Samad, D. Huq, M. Moniruzzaman, M. Masum, Textile dye removal from wastewater effluents using Chitosan-Zno nanocomposite, J. Textile Sci. Eng 5 (200) (2015) 1–14.

[144] F. Hosseini, S. Sadighian, H. Hosseini-Monfared, N.M. Mahmoodi, Dye removal and kinetics of adsorption by magnetic chitosan nanoparticles, Desalin. Water Treat 57 (51) (2016) 24378–24386.

[145] S. Chandra, N. Chakraborty, A. Dasgupta, J. Sarkar, K. Panda, K. Acharya, Chitosan nanoparticles: a positive modulator of innate immune responses in plants, Sci. Rep 5 (2015) 15195.

[146] R. Grillo, A.E. Pereira, C.S. Nishisaka, R. de Lima, K. Oehlke, R. Greiner, L.F. Fraceto, Chitosan/tripolyphosphate nanoparticles loaded with paraquat herbicide: an environmentally safer alternative for weed control, J. Hazard. Mater 278 (2014) 163–171.

[147] R.A. Ahmed, A.M. Fekry, Preparation and characterization of a nanoparticles modified chitosan sensor and its application for the determination of heavy metals from different aqueous media, Int. J. Electrochem. Sci. 8 (3) (2013) 6692–6708.

[148] E.S. Abdou, A.S. Osheba, M.A. Sorour, Effect of chitosan and chitosan-nanoparticles as active coating on microbiological characteristics of fish fingers, Int. J. Appl. Sci. Technol 2 (7) (2012) 158–169.

[149] J. Panyam, V. Labhasetwar, Biodegradable nanoparticles for drug and gene delivery to cells and tissue, Adv. Drug Deliv. Rev 55 (3) (2003) 329–347.

[150] V.S. Gondil, K. Harjai, S. Chhibber, Endolysins as emerging alternative therapeutic agents to counter drug-resistant infections, Int. J. Antimicrob. Agents 55 (2) (2020) 105844 1.

[151] B. Jamil, H. Habib, S.A. Abbasi, A. Ihsan, H. Nasir, M. Imran, Development of cefotaxime impregnated chitosan as nano-antibiotics: De novo strategy to combat biofilm forming multi-drug resistant pathogens. Front. Microbiol. 2016;7:330.

[152] Y.C. Huang, R.Y. Li, J.Y. Chen, J.K. Chen, Biphasic release of gentamicin from chitosan/fucoidan nanoparticles for pulmonary delivery. Carbohydrate Polymers. 2016;138:114–22.

[153] Z. Sobhani, S.M. Samani, H. Montaseri, E. Khezri, Nanoparticles of chitosan loaded ciprofloxacin: fabrication and antimicrobial activity. Adv. Pharm. Bull. 2017;7(3):427–32.

[154] Y. Liu, Y. Sun, Y. Xu, H. Feng, S. Fu, J. Tang, W. Liu, D. Sun, H. Jiang, S. Xu, Preparation and evaluation of lysozyme-loaded nanoparticles coated with poly-γ-glutamic acid and chitosan. Int. J. Biol. Macromol. 2013;59:201-07.

[155] F.M. Khan, V.S. Gondil, C. Li, M. Jiang, J. Li, J. Yu, H. Wei, H. Yang, A novel Acinetobacter baumannii bacteriophage endolysin LysAB54 with high antibacterial activity against multiple Gram-negative microbes. Front. Cell. Infect. Microbiol. 2021;11:70.

[156] V.S. Gondil, S. Chhibber, Bacteriophage and endolysin encapsulation systems: A promising strategy to improve therapeutic outcomes. Front. Pharmacol. 2021;12:1113.

[157] D. Luo, L. Huang, V.S. Gondil, W. Zhou, W. Yang, M. Jia, S. Hu, J. He, H. Yang, H. Wei, A choline-recognizing monomeric lysin, ClyJ-3m, shows elevated activity against Streptococcus pneumoniae. Antimicrob. Agents Chem. 2020;64(12):e00311-20.

[158] M. Kumar, N. Wangoo, V.S. Gondil, S.K. Pandey, A. Lalhall, R.K. Sharma, S. Chhibber, Glycolic acid functionalized silver nanoparticles: A novel approach towards generation of effective antibacterial agent against skin infections. J. Drug Del. Sci. Technol. 2020;60:102074.

[159] M. Werle, A. Bernkop Schnürch, Thiolated chitosans: useful excipients for oral drug delivery. J. Pharma. Pharmacol. 2008;60(3):273–81.

7 Hyaluronic acid nanoparticles

Irina Negut and Valentina Grumezescu

NATIONAL INSTITUTE FOR LASERS, PLASMA, AND RADIATION PHYSICS, MAGURELE, ILFOV, ROMANIA

Chapter outline

7.1 Introduction

Recently, nanotechnology is considered to be the most promising machinery, extensively used in the fields of medicine [1], energy [2], electronics, and environment [3]. Over the past years, research efforts have been dedicated to maximizing the therapeutic index of drugs and other alleviating substances and reducing the undesirable side effects. Drugs can be unsafe to the entire human body if they are not distributed to the affected site, at an ideal concentration and within the therapeutic window. Due to the fact that the most used treatment strategies dispense drugs to the entire body and cannot be manipulated to maintain a therapeutic dose, innumerable transporters have been advanced to deliver drugs directly to the affected site. In the modern therapeutic field, the nanotechnology tools, nanoparticles (NPs), have attracted a high interest as transporters for the detection, diagnosis, and treatment agents in the view of managing countless diseases [72]. Although NPs are extremely beneficial as drug delivery system, their clinical triumph is contingent with the proper transporter molecules to have an enhanced bioavailability, biocompatibility, surface functionalization, high encapsulation ability, and biodegradability. In clinical applications, the selection of the carrier is highly important, as it has a substantial influence on the pharmacokinetics and pharmacodynamics of the transported drug. Usually, the drug is dissolved, entrapped, encapsulated, or attached to the carrier and then is targeted to the ill tissue or organ with specific recognition function of the carrier; the drug dispersal is dependent on the carrier's features [4]. Polymeric NPs are now investigated for their sizes (between 1 and 1000 nm) to regulate the biodistribution and cellular uptake of these NPs [74]. The NPs consist of a biodegradable macromolecular material from natural, synthetic, or semisynthetic origin which can augment the solubility and bioavailability of drugs, prolong a drug's *in vivo* half-life, protect the drug from the human environment severe conditions, and

Biopolymeric Nanomaterials: Fundamentals and Applications. DOI: https://doi.org/10.1016/B978-0-12-824364-0.00015-0

release the drug in a sustained and/or triggered manner [4]. Furthermore, polymeric NPs can gather in specific tissues such as tumors, through enhanced permeation and retention effect (EPR) [5]. By developing a diversity of natural and synthetic polymers, researchers have designed multipurpose polymeric NPs for the effective delivery of imaging and therapeutic agents [4, 5]. Furthermore, drug/imaging agent carriers have to be biocompatible, biodegradable, and be easily eliminated from the body. By using natural polymeric NPs, as opposed to metallic ones, or other synthetic polymers, it minimizes the concerns regarding toxicity, biodegradability, and physiological stability. Some advantages of polymeric NPs can be noted: (1) increase in the firmness of any volatile therapeutic agents; (2) can be facile and inexpensively fabricated in high amounts by an assortment of techniques; (3) they are an important upgrade over conventional oral and intravenous drug administration pathways, in terms of effectiveness; (4) polymeric NPs can deliver a higher concentration of therapeutic agent to the desired site; (5) polymeric NPs can be combined with other constructs associated to drug delivery, such as tissue engineering [6]. Nowadays, polysaccharides are intensively studied and applied in various divisions of basic and applied sciences, as they combine the useful features of natural biopolymers. Moreover, polysaccharides are "sustainable" and "bio-based" materials that are of increasing prominence to the scientific community. One of the universally and naturally occurring polysaccharides is hyaluronic acid (HA), which is widely distributed in many organs and tissues. HA has countless useful benefits such as biocompatibility, the absence of immunogenicity, chemical adaptableness, nontoxicity, degradability, and high hydrophobicity [7]. Studies show that polysaccharide NPs may reduce the uptake by the mononuclear phagocyte system compared to other types of NPs. Residence time of polysaccharide NPs is extended due to the bioadhesion properties of polysaccharides. Furthermore, polysaccharide NPs own an assortment of derivable groups, which can be exploited to conjugate different agents [8]. Within this chapter, we put in evidence the key physicochemical and physiological functions of HA. Furthermore, we highlight details of recent works on nanostructured HA carrier systems for various drug delivery applications.

7.2 Polysaccharides

Polysaccharides are abundant biopolymers composed from repeated mono- or disaccharides linked via glycosidic bonds. Natural polysaccharides are obtainable from plants (e.g., starch, cellulose, cyclodextrins, pectin), algae (e.g., dextran, xanthan gum), animals (e.g., chitosan, HA), and microbial sources [9]. As natural biomaterials, polysaccharides are very stable, harmless, nontoxic, hydrophilic, and biodegradable. Polysaccharides possess a high number of reactive groups, an extensive assortment of molecular weights (MW), and chemical compositions. Polysaccharides can be categorized into polyelectrolytes and nonpolyelectrolytes; nonpolyelectrolytes can be further classified into positively (such as chitosan) and negatively charged (e.g., alginate, heparin, HA, pectin, etc.) [10]. Mainly, most of natural polysaccharides own hydrophilic groups such as hydroxyl, carboxyl, and amino groups, which could form noncovalent bonds with biological tissues that facilitate the bonding to the mucus layer that shields epithelial surfaces

[11]. Consequently, polysaccharide-transporter systems own prolonged *in-vivo* residence times in the gastrointestinal tract, in that way augmenting the bioavailability of drugs [12]. Positively charged polysaccharides such as chitosan are used for opening the tight junctions between epithelial cells, thus boosting the paracellular penetrability of hydrophilic drugs through the mucosal epithelia [13]. As mentioned above, mucoadhesive polysaccharides have been widely applied for the delivery of vaccines and hydrophilic macromolecular therapeutics via pulmonary or nasal pathways [14]. As most of polysaccharides hold an assortment of reactive functional groups such as hydroxyls, amines, and carboxyls on their backbones, they can be chemically adjusted to obtain derivatives with distinctive properties for biomedical applications. For instance, the sulfonation of pullulan initiates anticoagulant action to the subsequent sulfated-pullulan, while thiolation of chitosan increases mucoadhesive properties of the resultant polysaccharide owing to strong covalent bonds with subdomains of mucus glycoproteins [15]. For example, chitosan, starch, alginate are good bioadhesive materials [16]. The functional groups also tolerate the conjugation of drugs to the main chain by means of functional linkers. The functional groups can also be of use for attaching hydrophobic moieties; in this manner one can prepare amphiphilic derivatives that can self-assemble in biological fluids. Moreover, some polysaccharides have the distinctive ability to identify specific receptors that are overexpressed on surfaces of unhealthy cells. A characteristic example is the case of HA, which binds specifically to the including cluster determinant 44 (CD44) receptor, overexpressed on many tumor cells [17]. Likewise, it has been observed that pullulan has a high affinity for asialoglycoprotein receptors expressed on the surfaces of hepatocytes [18]. The specific linking of polysaccharides to receptors permits the design of (nano)carriers that can distribute, with high selectively, active substances by receptor-mediated endocytosis. Owing to the existence of derivable groups on their molecular chains, polysaccharides can be chemically/biochemically adjusted, resulting in countless varieties of derivatives. In recent years, a great number of studies have been dedicated to polysaccharides and derivatives for their potential application as NPs drug delivery systems [12]. A special accent was set on designing HA NPs for various biomedical applications.

7.2.1 Physicochemical and physiological functions of HA

HA, also termed hyaluronan, represents a naturally occurring nonsulfated glycosaminoglycan nonprotein compound with a linear chain of alternating d-glucuronic acid and N-acetyl-d-glucosamine units linked by β-1,4 and β-1,3 glycosidic linkages [19]. HA is accessible from various sources such as microbial (e.g., Streptococci strains) and animals (e.g., bovine vitreous humor, rooster combs, and human umbilical cord) [20]. It is made in the plasma membrane and signifies a vital component of the extracellular matrix, having an influence on cell movement and proliferation [21]. As HA is present in abundance in many parts of the human body, it naturally possesses biocompatibility and is degraded by native enzymes [22]. The degradation products interrelate with receptors found on macrophages and dendritic cells, resulting in "distinctive immunity" activation [23]. The highest HA quantity can be found in the skin, other important quantities in the synovial fluid, vitreous body, and in the umbilical cord [24]. HA presents

Table 7.1 Examples of some registered HA-based formulations

Commercial product	Applications
Connettivina®	Treatment of localized skin irritations triggered by physical agents (sun, cold, wind, diaper rash, irritation).
Monovisc	A single injection, lightly cross-linked high molecular weight HA for osteoarthritis
Ial-system™	To recover skin elasticity
iAluRil™	Painful bladder syndrome, interstitial or radiotherapy-induced cystitis, cystitis, repeated urinary tract infections
Yabro® and hysan®	Recurrent upper/lower respiratory tract infections, sustained cough
111 Skin® meso infusion overnight micromask and beauty pie's™	A mask for antiaging applications
Hylase wound gel®	Wound care dressing and management
Juvéderm® VOLUMA® with lidocaine	Face filler

outstanding viscoelasticity, moisture retention capacity, biocompatibility, and hygroscopic properties. Owing to its properties, HA acts as a lubricant, shock absorber, joint structure stabilizer, and water balance regulator, and flow resistance regulator [25]. HA has structural functions through the extracellular matrix (ECM) by means of specific and nonspecific interactions. It is beneficial to proteins, and crucial for cellular communication transduction with specific molecules and receptors. There are three main classes of cell surface receptors that HA can bind to: CD44 (a membrane glycoprotein), receptor for hyaluronate-mediated motility (RHAMM), and Intercellular Adhesion Molecule 1 (ICAM-1) [24]. Moreover, other receptors have been recognized for HA binding: lymphatic vessel endothelial hyaluronan receptor (LYVE-1), an HA receptor for endocytosis (HARE), also named Stabilin-2 [24]. Furthermore, a low relative molecular mass HA is supportive for the growth of epithelial tissue cells and macrophages (only molecules ranging from 20 to 450 kDa). Similarly, HA is essential for the healing and scar formation; however, in this case, higher relative molecular mass molecules are the most helpful ones [26] As mentioned earlier, HA is degraded by native enzymes (commonly referred to as hyaluronidases) into oligosaccharides. In the human body, there are a minimum of seven types of hyaluronidase-like enzymes, most of them being tumor suppressors. The primary structure of HA is degraded under pH, temperature, mechanical, free radical, ultrasonic, and enzymatic stresses conditions [27]. Besides, HA can be degraded by nonenzymatic reactions such as acidic, alkaline hydrolysis, and oxidant decomposition. Thus, chemically modified HA have been advanced, with superior mechanical and biological properties, for specific medical and pharmaceutical applications. The most common chemical adjustments of HA are based on the existent carboxylic and/or hydroxyl groups of the main chain. By using the carboxylic group, one can achieve esterification and carbodiimide-mediated reactions. In the case of the hydroxyl group, sulfation, esterification, isourea coupling, and periodate oxidations may take place [28]. Owing to its ideal properties, there are now many commercially available products containing HA. Some of these products are presented in Table 7.1.

7.2.2 HA nanoparticles

7.2.2.1 Methods for obtaining HA NPs

There are some well-described and reproducible procedures to obtain HA NPs, with variable grades of stability. HA NPs have been stabilized by numerous cross-linking agents, comprising particular functional groups of the polysaccharide (e.g., carboxyl, hydroxyl, N-acetyl groups), which are adjusted by chemical reactions. By changing the coupling reactions, it is possible to improve surfaces with a specific action for cell adhesion and metabolic activation [29]. One of most used techniques for obtaining HA NPs is represented by the *conjugate formation*. This technique is utilized for the preparation of HA-drug nanoconjugates by creating a covalent bond between the drug and HA. This preparation technique offers the advantage of improved solubility, pharmacokinetic profile, and *in-vivo* plasma half-life of conjugated drugs. The basic principle of this technique is the "conversion" of the drug into a prodrug derivative. The bond between the HA and the drug molecule should accomplish certain criteria. The most imperative is that the connection must be extracellularly stable to provide the required *in-vivo* half-life. The chemical modification of HA and conjugation of drugs to HA can be made by hydroxyl groups and carboxylic acid functionalities, whereas the aldehyde group can only be used to formulate terminally modified low molecular mass HA-ligand conjugates and to graft HA oligomers to other polymer carrying amino groups [30]. Generally, the conjugation is aimed to be cleaved after reaching the target site; conjugates usually have higher accumulation at tumor sites owing to the EPR. Due to the presence of multiple functional groups on the backbone of HA, it was possible to get conjugated to various compounds and macromolecules such as paclitaxel [31]. HA NPs can also be obtained via the *self-assemblies formation*, where HA is associated with hydrophobic polymers with the resulting conjugates able to self-aggregate in the form of hydrophobic micelles. Polymeric micelles are prepared in an aqueous solution and have a spherical structure with hydrophilic heads at the shell (provides longer circulation times by decreasing the undesirable protein adsorption) and hydrophobic tails at the core (for the encapsulation of therapeutic agents). Many amphiphilic self-assembling HA nanoderivates have been obtained by coupling carboxylic groups from hydrophilic HA through carbodiimide chemistry to different hydrophobic moieties [32]. As HA is a polyanionic polysaccharide, it can react with many cationic compounds and form ionic complexes. *The ionic nanocomplexes formation* can be made by the direct interaction of HA with positively charged molecules, such as DNA and plasmids, or in the presence of positively charged polymers (e.g., chitosan) [33]. Other types of useful HA nanostructures are represented by nanogels. Nanogels are usually obtained by physical or chemical cross-linking of HA to offer colloidal stable particles. The physical cross-linking is usually done via noncovalent attractive forces between the polymer chains (such as hydrophobic interactions, hydrogen bonding, and ionic interactions). Chemical cross-linking would result in the formation of particles with higher stability and consequently with longer half-lives. Even though diverse approaches have been employed for obtaining HA chemically cross-linked nanogels (such as precipitation or suspension), the most widely used technique is represented by the reverse water in oil (w/o) microemulsion cross-linking. In this technique, w/o micelles perform as nanoreactors for HA cross-linking reactions, helping the control of the NPs sizes [30]. A commonly described method for the preparation of HA microemulsions is forming by water as an aqueous phase for HA solution, isooctane as an organic phase, and sodium bis(2-ethylhexyl) sulfosuccinate as a surfactant [34, 35]. A common cosurfactant is represented by 1-heptanol. From the cross-linking agents, divinyl sulfone [36] and 1,4-butanediol diglycidyl ether [37] are to be noted. Regardless of their toxicity displayed in high concentrations of their unreacted form [38], HA has been cross-linked with various materials containing low concentrations of divinyl sulfone or 1,4-butanediol diglycidyl ether; they maintained the biological features of the unlinked HA and are approved by the Food and Drug Administration [36]. Nevertheless, the toxicity of cross-linking agents is a negligible concern in the design of nanogels due their preparation in diluted dispersions. Moreover, cross-linking agents have a critical part on nanogels synthesis as they can affect the particle size and/or properties of NPs. Conversely, microemulsion chemical cross-linking techniques necessitate vigorous mechanical stirring or ultrasonication and the use of organic solvents; these are not considered as advantageous settings for proteins and nucleic acids. In respect to the above mentioned, physical cross-linking is gentler in terms of disturbing structures of drug molecules [30].

7.2.2.2 Pharmacotherapeutics of HA NPs

Cancer therapy

One of the greatest trials of modern medical practice is the successful treatment of cancer, a disease that was estimated to have $\sim$1.7 million new diagnoses in the United States in 2016 alone [39]. Despite the evolution in this area, cancer still remains a challenging disease with lower than desirable survival rates. Several challenges in the case of cancer chemotherapy are the poor solubility of anticancer drugs and the limited selectivity of drug delivery within the human body. The poor solubility of anticancer drugs stops chemotherapeutics from circulating in the blood to reach the tumor site; the latter intensifies adverse effects (such as bone marrow toxicity, immune system impairment, and cardiotoxicity), as drugs provoke cytotoxic effects in unaffected areas [40]. Consequently, novel drug delivery systems are being verified to address these challenges. The nano-sized HA is selectively transferred to cancer cells by the EPR effect [41]. To enhance drug targeting and reduce toxic side effects, the EPR effect is combined with active targeting design. EPR is a complex phenomenon by which NPs are transported to, and accumulated at the tumor site in high concentrations; however, there are tumors that exist with little or no evidence of EPR [42]. Passive targeting by the EPR is a promising strategy for overcoming the specificity of chemotherapeutics. In addition, the selective delivery of HA is enhanced via reaction of HA with a multitude of receptors, CD44, the RHAMM, and LYVE-1. CD44, a cell adhesion receptor of HA, is abundantly overexpressed (relative to normal cells) in the tumor extracellular matrix, at a low level on the surface of epithelial cells, hematopoietic cells, and neuronal cells [17]. The overexpression of CD44 contributes to important cancer progressions comprising tumor invasion, metastasis, recurrence, and chemoresistance [43]. The CD44 targeting is a unique property of HA-based nanomaterials that makes HA a vital biomaterial for CD44-overexpressing

Table 7.2 Summary of HA NPs recent research for the treatment of cancer

Drug loading	Cell lines and *in-vivo* tests	Applications	Reference
IR780 and Dox	MCF-7 (CD44 receptor overexpressing cells) NHDF (do not overexpress the CD44 receptor)	Breast cancer chemo-phototherapy	[47]
IR-780	MB-49 bladder cancer cells on C57BL/6 mice	Orthotropic bladder cancer	[48]
Paclitaxel and Flamma- 552 and Flamma-774	SCC7 cancer cells athymic (nu/nu-ncr, Balb/c mice)	Head and neck cancer	[49]
Methotrexate, octadecylamine and curcumin	HeLa and MCF-7 cells HeLa tumor-bearing mice	Cervical carcinoma and breast cancer	[50]
Paclitaxel, serum albumin, Imidazoacridinone C-1375	A2780 and SKOV3 cells	Ovarian carcinoma	[43]
Raltitrexed	CD44+ murine CRC cell line, CT26 BALB/c mice	Colorectal cancer	[51]
Polyethyleneimine-HA, siRNA	MDA-MB 468, A549, B16F10 (CD44 expressed cancer)	Lung cancer	[73]

cancer cells. Moreover, studies demonstrated that tumor cells produce a large quantity of hyaluronidase, a complex that can initiatively disintegrate HA NPs [44]. Hyaluronidase catalyzes the degradation of HA, forming different fragments of different MW; the low-molecular-weight form, catabolite, is known to have procancerous activity, while the high-molecular-weight one has anticancer activity [45]. In addition, scientific studies have shown the role of HA in tumor metastasis, with HA variably expressed in diverse tumor environments [46]. Consequently, HA NPs can specially target tumor tissues due to both passive accumulation by the EPR effect and active targeting by the receptor-binding of HA to CD44. As a drug delivery system in the detection and treatment of cancer, HA NPs have been functionalized or encapsulated with various chemical drugs, such as paclitaxel and doxorubicin (DOX), and other biopharmaceuticals. In Table 7.2 one can find more information about recent research on HA NPs useful in the treatment of cancer diseases.

A novel HA-based amphiphilic polymer was used for the preparation of polymeric NPs (hyaluronic polymeric NPs—HPN) encapsulating IR780 and DOX aimed to be functional for breast cancer therapy. The results revealed that HPN were able to successfully encapsulate IR780 (IR-HPN) and the IR780-DOX combination (IR/DOX-HPN). The 2D *in-vitro* cell uptake studies proved that the prepared NPs had a higher internalization by breast cancer cells than by normal ones. Furthermore, analyses made in 3D *in-vitro* models of breast cancer showed that HPN can penetrate into spheroids and induce cytotoxicity. The IR/DOX-HPN reduced spheroids cells' viability, and their combination with near infrared (NIR) light encouraged an even stronger therapeutic effect [47]. In another study, a self-assembled tumor-targeting HA-IR-780 NPs for photothermal ablation in overexpressing CD44 in bladder cancer were developed. The NPs demonstrated a stable spherical nanostructure in aqueous conditions with good monodispersity, and their average size was 171.3 ± 9.14 nm. The *in-vitro* cell viability study exhibited that the combination of HA-IR-780 NPs with 808 nm laser irradiation can efficiently ablate MB-49 cells. The *in-vivo* biodistribution showed the HA-IR-780 NPs are targeted for accumulation in bladder cancer cells but have insignificant buildup in a normal bladder wall. The photothermal therapeutic efficacy of HA-IR-780 NPs in the orthotopic bladder cancer model indicated that none of tumors in the HA-IR-780 NPs plus laser irradiation-treated group were visible to the naked eye. The toxicity study demonstrated that the designed HA-IR-780 NPs (2.5–20 mg/kg, intravenously injected) were nontoxic and safe for *in-vivo* applications [48]. In a work by Thomas et al., paclitaxel cancer drug was loaded into HA nanostructures with an optimal encapsulation efficiency (EE) of 77.3% (a factor that determines the drug entrapment capability of the polymer). The results from the 3-(4,5-dimethylthiazol-2-yl)-5-(3-carboxymethoxyphenyl)-2-(4-sulfophenyl)-2H-tetrazolium (MTS) assay revealed that paclitaxel-loaded HA nanostructures had significantly higher cell toxicity profile (1–100 μg/mL) for drug encapsulating micelles compared to the free drug. The *in-vivo* tumor study conducted with 5 mg/kg of paclitaxel in HA-micelles displayed a noteworthy inhibition of tumor growth compared to control, on days 6 and 8 [49]. Other groups synthesized dual-targeting methotrexate-conjugated HA-octadecylamine, in which methotrexate (MTX) not only performed as a specific anticancer drug but also as a tumor-targeting ligand for folate receptors. The advantages of this nanosystem were that MTX was used as a tumor-targeting ligand toward folate receptors due to structural similarity with folic acid and HA served as another tumor-targeting ligand toward CD44 receptors. On the basis of the specific cell recognition by both MTX and HA moiety, and a combination therapy of MTX and curcumin (CUR), the MTX-HA-OCA/CUR NPs increased the cellular uptake and anticancer activity and enhanced the *in-vivo* tumor accumulation [50]. Bovine serum albumin–HA (BSA–HA) conjugates stabilized paclitaxel and prevented its aggregation and crystallization. The diameter of prepared NPs was $<$ 15 nm, therefore permitting CD44 receptor-mediated endocytosis. BSA-HA conjugates were selectively internalized by human ovarian cancer SKOV3 cells overexpressing the CD44, but not by cognate cells (A2780) lacking CD44 expression. Free HA stopped the endocytosis of paclitaxel-loaded BSA-HA conjugates. Paclitaxel-loaded NPs were noticeably more cytotoxic to cancer cells overexpressing CD44 than to cells lacking CD44, due to selective internalization [43]. Combined modality therapy incorporating raltitrexed (RTX), a thymidylate synthase inhibitor, and radiation can lead to improved outcome for rectal cancer patients. To increase delivery and treatment efficacy, Rosch et al. formulated HA NPs encapsulating RTX (HARPs) through layer-by-layer assembly. The cell uptake in CT26 cells determined by flow cytometry exhibited an approximate fivefold growth between untargeted and HA-coated particles. Through viability and DNA damage assays, the potency of the free RTX and HARPs was assessed; an increased DNA harm in cells treated with the RTX-loaded NPs administered concomitantly with radiation was observed. *In-vivo* effectiveness through tumor growth inhibition was explored in a syngeneic murine colorectal cancer model. NPs treatment showed no acute toxicity *in vivo*. HARPs alone slowed tumor growth. Notably, the combination treatment significantly hindered tumor progression relative to the HARPs [51]. In another work, engineered and screened series of CD44 targeting HA self-assembled nanosystems for targeted siRNA delivery were assessed. The HA was functionalized with lipids of varying carbon chain lengths/nitrogen content, as well as polyamines for evaluating siRNA entrapment. Many of obtained HA nanosystems were capable to transfect siRNAs into cancer cells overexpressing CD44. siRNA encapsulated in HA-polyethyleneimine (PEI)/polyethylene glycol (PEG) nanosystems demonstrated dose-dependent and target-specific gene knockdown in both sensitive and resistant A549 lung cancer cells overexpressing CD44 receptors. These siRNA-encapsulated nanosystems showed tumor selective uptake and target specific gene knock down *in vivo,* in solid tumors, along with metastatic tumors [36] methotrexate.

Skin conditions

Skin diseases have been tagged as the fourth leading causes of disability globally, with dermatitis being the highest burden in terms of treatment costs and patients responses to available treatments [52]. To treat skin ailments, there are several benefits of topical delivery systems as paralleled with oral or parenteral ones, for instance escaping hepatic first-pass metabolism, the improvement of patient compliance, ease of access to the absorbing skin membrane, and minimizing adverse effects associated with systemic toxicity [53]. For example, HA has been shaped into microparticles for a sustained release of caffeine on skin to treat cellulite [54], hydrogels for topical delivery of antiinflammatory drug diclofenac to treat actinic keratosis [55], liposomes for topical distribution of HA to treat dermal and subcutaneous wounds [56]. As a result of its exceptional properties, such as viscoelasticity, biocompatibility, biodegradability, nonimmunogenicity, HA and its nanoderivates have been widely applied on skin for its biomedical benefits, such as antiaging, antiwrinkle, tissue regeneration, rejuvenating, and hydration [57]. The CD44 receptor is highly expressed in psoriatic skin condition. The concentration of CD44 is negatively associated with HA spreading [58]. Moreover, scientific tests have displayed that activated Th1 and Th17 T cells (CD4+ T cells), CD8+ T cells, along with increased levels of cytokines (e.g., IL-17, IL-23, TNF-α and IL-27) are implicated in psoriasis immunopathogenesis [59]. Presently, HA is extensively used as an active part of transdermal delivery systems applied to enhance the drug penetration in psoriasis treatment. For the treatment of moderate-to-severe psoriasis, many drugs have been incorporated into HA delivery systems, including methotrexate [60] and tacrolimus (also known as FK506; is a macrolide immunosuppressive drug that obstructs the triggering of T-lymphocyte through inhibition of IL-2 [61], and corticosteroids). An innovative nanotopical drug delivery system based on the incorporation of CUR in HA-modified ethosomes targeted CD44 in the inflamed epidermis of psoriatic skin [58]. Another hybrid nanosystem constructed by combining amphiphilic conjugations of HA-cholesterol-self-assembled NPs and hydrotropic nicotinamide was developed to enhance the permeation of FK506 in the treatment of psoriasis [62]. As in the case of psoriasis, several drug delivery systems comprising HA NPs and pluronic F-127 dual responsive (pH/temperature) hydrogels have been designed to boost the drug infiltration through/into the skin for the treatment of atopic dermatitis. For example, HA-coated tacrolimus-loaded NPs (HA-TCS-CS-NPs) were studied. The HA-TCS-CS-NPs were tested *in vivo* on mice models performed for drug release, drug permeation and retention, and anti-atopic dermatitis; it was found that the obtained nanosystem had superior properties in comparison with the sole tacrolimus solution [63]. In another study, betamethasone valerate—a medium topical corticosteroid was incorporated into HA-NPs. The studies performed proved favorable physicochemical characteristics, with high entrapment effectiveness and loading capability together with a high drug permeation and retention [64].

Treatment of infections

Infections caused by pathogenic microorganisms are still a threat to the human health [65]. However, antibiotics have shaped a new pathway of human resistance to infectious microorganisms. Most of available antibiotics are operative only in high dosages due to the absence of selective targeting toward diseased sites; this results in multiple drug resistance of microbes and side effects to healthy tissues and organs [66]. To efficiently fight against infectious diseases, there is an imperative necessity for antimicrobial substances, either chemical or natural, to realize targeted and controlled drug delivery and release at infected sites. Recently, the attention of the scientific community was distributed toward HA as a drug carrier, as it can bind to CD44 on the surface of some cells [67]. CD44 represents a type I transmembrane glycoprotein that is expressed on endothelial and mesenchymal cells, and especially in inflamed tissues [68]. Inflammatory reactions happen as a defense response of the organism, through the process of pathogenic attack. In this manner, HA is expected to succeed the targeted drug delivery as a transporter for antibiotics. To realize targeted delivery and controlled release of antibiotics, nitric oxide (NO)-sensitive HA-based nanomicelles were developed by modification with NO-reactive moieties. Levofloxacin (LF), a third-generation fluoroquinolone antibiotic (effective against both Gram-positive and Gram-negative bacteria) was carefully chosen as a model antibiotic. The NO-triggered nanomicelles (HA-NO-LF) were prepared by linking a hydrophilic HA chain and a lipophilic LF chain via a functionalized intermediate bearing o-phenylenediamine groups, followed by self-assembly in aqueous condition. *In-vitro* tests showed that the obtained nanostructures were cleaved to gradually release LF and exhibited antibacterial activity against *Staphylococcus aureus* when exposed to NO. At the same time, the NO-triggered nanomicelles displayed biosafety as well as biocompatibility. Moreover, the competition inhabitation assay proved that nanomicelles could enter macrophages via a CD44-mediated endocytosis. In addition, *in-vivo* pharmacodynamic assessment acknowledged their ability to fight against bacteria and reduce inflammatory levels [69]. In another study, nanohydrogels for LF delivery to treat intracellular bacterial infections were designed. The results indicated that LF was captured within the nanoconstructs by nanoprecipitation. The nanohydrogels were obtained by self-assembling of the HA-cholesterol amphiphilic chains in aqueous solutions. The minimum inhibitory concentration values of LF hydrogels were determined for *S. aureus* and *Pseudomonas aeruginosa* strains. Similarly, the research group obtained a very promising intracellular antimicrobial activity of the nanoconstructs on HeLa epithelial cells infected with *S. aureus* and *P. aeruginosa* bacterial strains [70]. An injectable antibacterial strategy based on an HA hydrogel and chlorhexidine (CHX) for cardiovascular implantable electronic device infection treatment was developed. To balance stability and moldability, the HA scaffold was pre-cross-linked by 1,4-butanediol diglycidyl ether and then ground to form an HA microgel. The obtained cross-linking degree of 8.25 %, ensured that the formed microgel preserved a stable structure for ~15 days but remained injectable and moldable when applied. CHX was further cross-linked in the microgel to obtain hybrid cross-linked hydrogels. These hydrogels exhibited shear-thinning/self-recovery behavior, allowing easy injection into the cardiovascular device pocket and good matching with the pocket shape without extra space requirements. Both *in-vitro* and *in-vivo* tests confirmed the biocompatibility and antibacterial properties of the hydrogel [71].

7.3 Concluding remarks and future perspectives

Many natural polysaccharides are abundant, cheap, and suitable to be applied in the design of NPs for the encapsulation of diverse types of bioactive compounds for their delivery in the human body. HA, as a biodegradable polysaccharide, has been extensively used in the controlled-release and targeted drug delivery systems. Many technological issues linked to the development of novel HA-bioactive compounds suggest that there is still an extensive pathway from the laboratory experiments to the clinic, the global market of HA-based products, is rapidly growing. According to current previsions, the global HA market is anticipated to extent to a value of ~USD 15.4 billion by the year 2025 [19]. This market is presently ruled by HA-based dermal fillers, and is developing at a high rate in many countries due to the aging of the population and the growing attentiveness about the use of antiaging chemicals. However, it is said that the prospect of HA as a fully operative drug carrier will be even wider with the discovery of novel materials and the expansion of innovative technologies. HA upgrades the use of nanomedicines by introducing the following properties: high biocompatibility, blood circulation time, cellular uptake, and colloidal stability of inorganic NPs. In particular, various drugs can be loaded into/onto HA nanoconstructs, enabling a diversity of therapeutical applications, such as the treatment of infections, skin conditions, and cancers. Numerous processes were advanced for the preparation of HA NPs that comprise chemical reactions and/or physical treatments, but the option depends on countless aspects such as safety, economic, environmental considerations, etc. In conclusion, the progress of HA-nanosystems seems to be a fundamental goal to achieve in the future. At the same time, the novel bioactive paths introduced by HA, together with a full understanding of the efficacy of HA-nanosystems, fascinate and impulse the biomedical and pharmaceutical research.

Acknowledgments

All authors recognize the financial support by the Romanian National Authority for Scientific Research, UEFISCDI, project no. PN-III-P2-2.1-PED2019-4926.

References

[1] S.K. Saxena, R. Nyodu, S. Kumar, V.K. Maurya, Current advances in nanotechnology and medicine, in: K. Saxena, S.M. Paul Khurana (Eds.), NanoBioMedicine., Springer, Singapore, India, 2020, pp. 3–16. https://doi.org/10.1007/978-981-32-9898-9_1.

[2] R. Neelu, S. Preeti, S.P. Singh, B. Deepali, P.A. Kumar, Efficiency Enhancement of Renewable Energy Systems Using Nanotechnology, Springer Science and Business Media LLC, Berlin/Heidelberg, Germany, 2020. https://doi.org/10.1007/978-3-030-34544-0_15.

[3] B.T. Belete, A.E. Gebrie, A review on nanotechnology: analytical techniques use and applications, Int. Res. J. Pure Appl. Chem. (2019) 1–10. https://doi.org/10.9734/irjpac/2019/v19i430117.

[4] S. Sur, A. Rathore, V. Dave, K.R. Reddy, R.S. Chouhan, V. Sadhu, Recent developments in functionalized polymer nanoparticles for efficient drug delivery system, Nano-Struct. Nano-Objects. 20 (2019) 100397. https://doi.org/10.1016/j.nanoso.2019.100397.

[5] H. Kang, S. Rho, W.R. Stiles, S. Hu, Y. Baek, D.W. Hwang, S. Kashiwagi, M.S. Kim, H.S. Choi, Size-dependent EPR effect of polymeric nanoparticles on tumor targeting, Adv. Healthc. Mater. 9 (2020) 1901223. https://doi.org/10.1002/adhm.201901223.

[6] N. Kamaly, B. Yameen, J. Wu, O.C. Farokhzad, Degradable controlled-release polymers and polymeric nanoparticles: Mechanisms of controlling drug release, Chem. Rev. 116 (2016) 2602–2663. https://doi.org/10.1021/acs.chemrev.5b00346.

[7] A. Lierova, J. Kasparova, J. Pejchal, K. Kubelkova, M. Jelicova, J. Palarcik, L. Korecka, Z. Bilkova, Z. Sinkorova, Attenuation of radiation-induced lung injury by hyaluronic acid nanoparticles, Front. Pharmacol. 11 (2020) 1199. https://doi.org/10.3389/fphar.2020.01199.

[8] M. Swierczewska, H.S. Han, K. Kim, J.H. Park, S. Lee, Polysaccharide-based nanoparticles for theranostic nanomedicine, Adv. Drug Deliv. Rev. 99 (2016) 70–84. https://doi.org/10.1016/j.addr.2015.11.015.

[9] V.R. Sinha, R. Kumria, Polysaccharides in colon-specific drug delivery, Int. J. Pharma. 224 (2001) 19–38. https://doi.org/10.1016/S0378-5173(01)00720-7.

[10] Z. Liu, Y. Jiao, Y. Wang, C. Zhou, Z. Zhang, Polysaccharides-based nanoparticles as drug delivery systems, Adv. Drug Deliv. Rev. 60 (2008) 1650–1662. https://doi.org/10.1016/j.addr.2008.09.001.

[11] N. Dubashynskaya, D. Poshina, S. Raik, A. Urtti, Y.A. Skorik, Polysaccharides in ocular drug delivery, Pharmaceutics. 12 (2020) 22. https://doi.org/10.3390/pharmaceutics12010022.

[12] B.M. Shah, S.S. Palakurthi, T. Khare, S. Khare, S. Palakurthi, Natural proteins and polysaccharides in the development of micro/nano delivery systems for the treatment of inflammatory bowel disease, Int. J. Biol. Macromol. 165 (2020) 722–737. https://doi.org/10.1016/j.ijbiomac.2020.09.214.

[13] X. Lang, T. Wang, M. Sun, X. Chen, Y. Liu, Advances and applications of chitosan-based nanomaterials as oral delivery carriers: a review, Int. J. Biol. Macromol. 154 (2020) 433–445. https://doi.org/10.1016/j.ijbiomac.2020.03.148.

[14] Y. Yan, Y. Sun, P. Wang, R. Zhang, C. Huo, T. Gao, C. Song, J. Xing, Y. Dong, Mucoadhesive nanoparticles-based oral drug delivery systems enhance ameliorative effects of low molecular weight heparin on experimental colitis, Carbohydr. Polym. 246 (2020) 116660. https://doi.org/10.1016/j.carbpol.2020.116660.

[15] V. Puri, A. Sharma, P. Kumar, I. Singh, Thiolation of biopolymers for developing drug delivery systems with enhanced mechanical and mucoadhesive properties: a review, Polymers 12 (2020) 1803. https://doi.org/10.3390/polym12081803.

[16] U.K. Chinedu, Bioadhesive polymers for drug delivery applications, in: K.L Mittal, I.S. Bakshi, J.K. Narang (Eds.), Bioadhesives in Drug Delivery, Wiley Online Library, 2020. https://doi.org/10.1002/9781119640240.ch2.

[17] K. Xiao, Y. Li, J. Luo, J.S. Lee, W. Xiao, A.M. Gonik, R.G. Agarwal, K.S. Lam, The effect of surface charge on in vivo biodistribution of PEG-oligocholic acid based micellar nanoparticles, Biomaterials. 32 (2011) 3435–3446. https://doi.org/10.1016/j.biomaterials.2011.01.021.

[18] Y. Kaneo, T. Tanaka, T. Nakano, Y. Yamaguchi, Evidence for receptor-mediated hepatic uptake of pullulan in rats, J. Control. Release. 70 (2001) 365–373. https://doi.org/10.1016/S0168-3659(00)00368-0.

[19] A. Fallacara, E. Baldini, S. Manfredini, S. Vertuani, Hyaluronic acid in the third millennium, Polymers. 10 (2018) 701. https://doi.org/10.3390/polym10070701.

[20] S.Y. Vafaei, M. Esmaeili, M. Amini, F. Atyabi, S.N. Ostad, R. Dinarvand, Self assembled hyaluronic acid nanoparticles as a potential carrier for targeting the inflamed intestinal mucosa, Carbohydr. Polym. 144 (2016) 371–381. https://doi.org/10.1016/j.carbpol.2016.01.026.

[21] G. Tripodo, A. Trapani, M.L. Torre, G. Giammona, G. Trapani, D. Mandracchia, Hyaluronic acid and its derivatives in drug delivery and imaging: recent advances and challenges, Eur. J. Pharma. Biopharma. 97 (2015) 400–416. https://doi.org/10.1016/j.ejpb.2015.03.032.

[22] E. Hachet, H. Van Den Berghe, E. Bayma, M.R. Block, R. Auzély-Velty, Design of biomimetic cell-interactive substrates using hyaluronic acid hydrogels with tunable mechanical properties, Biomacromolecules. 13 (2012) 1818–1827. https://doi.org/10.1021/bm300324m.

[23] Y. Cheng, Polysaccharide-based Nanoparticles for Gene Delivery, in: Topics in Current Chemistry Collections, Springer International Publishing, Cham, 2018. https://doi.org/10.1007/978-3-319-77866-2_3.

[24] R.C. Gupta, R. Lall, A. Srivastava, A. Sinha, Hyaluronic acid: Molecular mechanisms and therapeutic trajectory, Frontiers in Veterinary Science. 6 (2019) 192. https://doi.org/10.3389/fvets.2019.00192.

[25] A. Ström, A. Larsson, O. Okay, Preparation and physical properties of hyaluronic acid-based cryogels, J. Appl. Polym. Sci. (2015) 132. https://doi.org/10.1002/app.42194.

[26] S. Vasvani, P. Kulkarni, D. Rawtani, Hyaluronic acid: a review on its biology, aspects of drug delivery, route of administrations and a special emphasis on its approved marketed products and recent clinical studies, Int. J. Biol. Macromol. 151 (2020) 1012–1029. https://doi.org/10.1016/j.ijbiomac.2019.11.066.

[27] R. Stern, G. Kogan, M.J. Jedrzejas, L. Šoltés, The many ways to cleave hyaluronan, Biotechnol. Adv. 25 (2007) 537–557. https://doi.org/10.1016/j.biotechadv.2007.07.001.

[28] R.C.S. Bicudo, M.H.A. Santana, Effects of organic solvents on hyaluronic acid nanoparticles obtained by precipitation and chemical crosslinking, J. Nanosci. Nanotechnol. 12 (2012) 2849–2857. https://doi.org/10.1166/jnn.2012.5814.

[29] M. Morra, Engineering of biomaterials surfaces by hyaluronan, Biomacromolecules. 6 (2005) 1205–1223. https://doi.org/10.1021/bm049346i.

[30] D.A. Ossipov, Nanostructured hyaluronic acid-based materials for active delivery to cancer, Exp. Opin. Drug Deliv. 7 (2010) 681–703. https://doi.org/10.1517/17425241003730399.

[31] H.J. Cho, Recent progresses in the development of hyaluronic acid-based nanosystems for tumor-targeted drug delivery and cancer imaging, J. Pharma. Investig. 50 (2020) 115–129. https://doi.org/10.1007/s40005-019-00448-w.

[32] D. Sousa, D. Ferreira, J.L. Rodrigues, L.R. Rodrigues, Nanotechnology in targeted drug delivery and therapeutics, in: S.S. Mohapatra, S. Ranjan, N. Dasgupta, R.K. Mishra, S. Thomas (Eds.), Micro and Nano Technologies, Elsevier, Amsterdam, 2019. https://doi.org/10.1016/B978-0-12-814029-1.00014-4.

[33] M. Kamat, K. El-Boubbou, D.C. Zhu, T. Lansdell, X. Lu, W. Li, X. Huang, Hyaluronic acid immobilized magnetic nanoparticles for active targeting and imaging of macrophages, Bioconj. Chem. 21 (2010) 2128–2135. https://doi.org/10.1021/bc100354m.

[34] A.K. Jha, M.S. Malik, M.C. Farach-Carson, R.L. Duncan, X. Jia, Hierarchically structured, hyaluronic acid-based hydrogel matrices via the covalent integration of microgels into macroscopic networks, Soft Matter. 6 (2010) 5045–5055. https://doi.org/10.1039/c0sm00101e.

[35] M. Kong, X.G. Chen, D.K. Kweon, H.J. Park, Investigations on skin permeation of hyaluronic acid based nanoemulsion as transdermal carrier, Carbohydr. Polym. 86 (2011) 837–843. https://doi.org/10.1016/j.carbpol.2011.05.027.

[36] M.N. Collins, C. Birkinshaw, Hyaluronic acid based scaffolds for tissue engineering—A review, Carbohydr. Polym. 92 (2013) 1262–1279. https://doi.org/10.1016/j.carbpol.2012.10.028.

[37] F.M. Ghorbani, B. Kaffashi, P. Shokrollahi, E. Seyedjafari, A. Ardeshirylajimi, PCL/chitosan/Zn-doped nHA electrospun nanocomposite scaffold promotes adipose derived stem cells adhesion and proliferation, Carbohydr. Polym. 118 (2015) 133–142. https://doi.org/10.1016/j.carbpol.2014.10.071.

[38] J. Fidalgo, P.A. Deglesne, R. Arroyo, L. Sepúlveda, E. Ranneva, P. Deprez, Detection of a new reaction by-product in BDDE cross-linked autoclaved hyaluronic acid hydrogels by LC–MS analysis, Med. Devices. 11 (2018) 367–376. https://doi.org/10.2147/MDER.S166999.

[39] R.L. Siegel, K.D. Miller, A. Jemal, Cancer statistics, 2016,, CA: Cancer J. Clin. 66 (2016) 7–30. https://doi.org/10.3322/caac.21332.

[40] M.E. Fox, F.C. Szoka, J.M.J. Fréchet, Soluble polymer carriers for the treatment of cancer: the importance of molecular architecture, Acc. Chem. Res. 42 (2009) 1141–1151. https://doi.org/10.1021/ar900035f.

[41] G. Huang, J. Chen, Preparation and applications of hyaluronic acid and its derivatives, Int. J. Biol. Macromol. 125 (2019) 478–484. https://doi.org/10.1016/j.ijbiomac.2018.12.074.

[42] K. Greish, Enhanced permeability and retention effect for selective targeting of anticancer nanomedicine: are we there yet?, Drug Discov. Today Technol. 9 (2012) e161–e166. https://doi.org/10.1016/j.ddtec.2011.11.010.

[43] R. Edelman, Y.G. Assaraf, I. Levitzky, T. Shahar, Y.D. Livney, Hyaluronic acid-serum albumin conjugate-based nanoparticles for targeted cancer therapy, Oncotarget. 8 (2017) 24337–24353. https://doi.org/10.18632/oncotarget.15363.

[44] Q. Zhao, J. Liu, W. Zhu, C. Sun, D. Di, Y. Zhang, P. Wang, Z. Wang, S. Wang, Dual-stimuli responsive hyaluronic acid-conjugated mesoporous silica for targeted delivery to CD44-overexpressing cancer cells, Acta. Biomater. 23 (2015) 147–156. https://doi.org/10.1016/j.actbio.2015.05.010.

[45] D. Naor, Editorial: Interaction between hyaluronic acid and its receptors (CD44, RHAMM) regulates the activity of inflammation and cancer, Front. Immunol. 7 (2016) 39. https://doi.org/10.3389/fimmu.2016.00039.

[46] R.H. Tammi, A. Kultti, V.M. Kosma, R. Pirinen, P. Auvinen, M.I. Tammi, Hyaluronan in human tumors: pathobiological and prognostic messages from cell-associated and stromal hyaluronan, Semin. Cancer. Biol. 18 (2008) 288–295. https://doi.org/10.1016/j.semcancer.2008.03.005.

[47] C.G. Alves, D. de Melo-Diogo, R. Lima-Sousa, E.C. Costa, I.J. Correia, Hyaluronic acid functionalized nanoparticles loaded with IR780 and DOX for cancer chemo-photothermal therapy, Eur. J. Pharm. Biopharm. 137 (2019) 86–94. https://doi.org/10.1016/j.ejpb.2019.02.016.

[48] T. Lin, A. Yuan, X. Zhao, H. Lian, J. Zhuang, W. Chen, Q. Zhang, G. Liu, S. Zhang, W. Chen, W. Cao, C. Zhang, J. Wu, Y. Hu, H. Guo, Self-assembled tumor-targeting hyaluronic acid nanoparticles for photothermal ablation in orthotopic bladder cancer, Acta. Biomater. 53 (2017) 427–438. https://doi.org/10.1016/j.actbio.2017.02.021.

[49] R.G. Thomas, M.J. Moon, S.J. Lee, Y.Y. Jeong, Paclitaxel loaded hyaluronic acid nanoparticles for targeted cancer therapy: in vitro and in vivo analysis, Int. J. Biol. Macromol. 72 (2014) 510–518. https://doi.org/10.1016/j.ijbiomac.2014.08.054.

[50] L. Song, Z. Pan, H. Zhang, Y. Li, Y. Zhang, J. Lin, G. Su, S. Ye, L. Xie, Y. Li, Z. Hou, Dually folate/CD44 receptor-targeted self-assembled hyaluronic acid nanoparticles for dual-drug delivery and combination cancer therapy, J. Mater. Chem. B. 5 (2017) 6835–6846. https://doi.org/10.1039/c7tb01548h.

[51] J.G. Rosch, M.R. Landry, C.R. Thomas, C. Sun, Enhancing chemoradiation of colorectal cancer through targeted delivery of raltitrexed by hyaluronic acid coated nanoparticles, Nanoscale. 11 (2019) 13947–13960. https://doi.org/10.1039/c9nr04320a.

[52] M. Wataya-Kaneda, A. Nakamura, M. Tanaka, M. Hayashi, S. Matsumoto, K. Yamamoto, I. Katayama, Efficacy and safety of topical sirolimus therapy for facial angiofibromas in the tuberous sclerosis complex a randomized clinical trial, JAMA Dermatol. 153 (2017) 39–48. https://doi.org/10.1001/jamadermatol.2016.3545.

[53] M.B. Brown, S.A. Jones, Hyaluronic acid: a unique topical vehicle for the localized delivery of drugs to the skin, J. Eur. Acad. Dermatol. Venereol. 19 (2005) 308–318. https://doi.org/10.1111/j.1468-3083.2004.01180.x.

[54] E.E. Simsolo, İ. Eroğlu, S.T. Tanrıverdi, Ö. Özer, Formulation and evaluation of organogels containing hyaluronan microparticles for topical delivery of caffeine, AAPS PharmSciTech. 19 (2018) 1367–1376. https://doi.org/10.1208/s12249-018-0955-x.

[55] M.B. Brown, C. Marriott, G.P. Martin, The effect of hyaluronan on the in vitro deposition of diclofenac within the skin, Int. J. Tissue. React. 17 (1995) 133–140.

[56] M.L. Vázquez-González, A.C. Calpena, Ò. Domènech, M.T. Montero, J.H. Borrell, Enhanced topical delivery of hyaluronic acid encapsulated in liposomes: a surface-dependent phenomenon, Coll. Surf. B: Biointerfaces. 134 (2015) 31–39. https://doi.org/10.1016/j.colsurfb.2015.06.029.

[57] J. Zhu, X. Tang, Y. Jia, C.T. Ho, Q. Huang, Applications and delivery mechanisms of hyaluronic acid used for topical/transdermal delivery – A review, Int. J. Pharma. 578 (2020) 119127. https://doi.org/10.1016/j.ijpharm.2020.119127.

[58] Y. Zhang, Q. Xia, Y. Li, Z. He, Z. Li, T. Guo, Z. Wu, N. Feng, CD44 assists the topical anti-psoriatic efficacy of curcumin-loaded hyaluronan-modified ethosomes: a new strategy for clustering drug in inflammatory skin, Theranostics. 9 (2019) 48–64. https://doi.org/10.7150/thno.29715.

[59] T.A. Luger, K. Loser, Novel insights into the pathogenesis of psoriasis, Clin. Immunol. 186 (2018) 43–45. https://doi.org/10.1016/j.clim.2017.07.014.

[60] H. Du, P. Liu, J. Zhu, J. Lan, Y. Li, L. Zhang, J. Zhu, J. Tao, Hyaluronic acid-based dissolving microneedle patch loaded with methotrexate for improved treatment of psoriasis, ACS Appl. Mater. Interfaces. 11 (2019) 43588–43598. https://doi.org/10.1021/acsami.9b15668.

[61] A.S.B. Goebel, U. Knie, C. Abels, J. Wohlrab, R.H.H. Neubert, Dermal targeting using colloidal carrier systems with linoleic acid, Eur. J. Pharma. Biopharm. 75 (2010) 162–172. https://doi.org/10.1016/j.ejpb.2010.02.001.

[62] T. Wan, W. Pan, Y. Long, K. Yu, S. Liu, W. Ruan, J. Pan, M. Qin, C. Wu, Y. Xu, Effects of nanoparticles with hydrotropic nicotinamide on tacrolimus: permeability through psoriatic skin and antipsoriatic and antiproliferative activities, Int. J. Nanomed. 12 (2017) 1485–1497. https://doi.org/10.2147/IJN.S126210.

[63] F. Zhuo, M.A.S. Abourehab, Z. Hussain, Hyaluronic acid decorated tacrolimus-loaded nanoparticles: efficient approach to maximize dermal targeting and anti-dermatitis efficacy, Carbohydr. Polym. 197 (2018) 478–489. https://doi.org/10.1016/j.carbpol.2018.06.023.

[64] M. Pandey, H. Choudhury, T.A.P. Gunasegaran, S.S. Nathan, S. Md, B. Gorain, M. Tripathy, Z. Hussain, Hyaluronic acid-modified betamethasone encapsulated polymeric nanoparticles: fabrication, characterisation, in vitro release kinetics, and dermal targeting, Drug. Deliv. Transl. Res. 9 (2019) 520–533. https://doi.org/10.1007/s13346-018-0480-1.

[65] N.Y. Lee, W.C. Ko, P.R. Hsueh, Nanoparticles in the treatment of infections caused by multidrug-resistant organisms, Front. Pharmacol. 10 (2019) 1153. https://doi.org/10.3389/fphar.2019.01153.

[66] M. Chen, S. Xie, J. Wei, X. Song, Z. Ding, X. Li, Antibacterial micelles with vancomycin-mediated targeting and pH/lipase-triggered release of antibiotics, ACS Appl. Mater. Interfaces. 10 (2018) 36814–36823. https://doi.org/10.1021/acsami.8b16092.

[67] G. Huang, H. Huang, Hyaluronic acid-based biopharmaceutical delivery and tumor-targeted drug delivery system, J. Control. Release. 278 (2018) 122–126. https://doi.org/10.1016/j.jconrel.2018.04.015.

[68] B. McDonald, P. Kubes, Interactions between CD44 and hyaluronan in leukocyte trafficking, Front. Immunol. 6 (2015) 68. https://doi.org/10.3389/fimmu.2015.00068.

[69] C. Lu, Y. Xiao, Y. Liu, F. Sun, Y. Qiu, H. Mu, J. Duan, Hyaluronic acid-based levofloxacin nanomicelles for nitric oxide-triggered drug delivery to treat bacterial infections, Carbohydr. Polym. 229 (2020) 115479. https://doi.org/10.1016/j.carbpol.2019.115479.

[70] E. Montanari, G. D'Arrigo, C. Di Meo, A. Virga, T. Coviello, C. Passariello, P. Matricardi, Chasing bacteria within the cells using levofloxacin-loaded hyaluronic acid nanohydrogels, Eur. J. Pharm. Biopharm. 87 (2014) 518–523. https://doi.org/10.1016/j.ejpb.2014.03.003.

[71] Q. Dong, X. Zhong, Y. Zhang, B. Bao, L. Liu, H. Bao, C. Bao, X. Cheng, L. Zhu, Q. Lin, Hyaluronic acid-based antibacterial hydrogels constructed by a hybrid crosslinking strategy for pacemaker pocket infection prevention, Carbohydr. Polym. 245 (2020) 116525. https://doi.org/10.1016/j.carbpol.2020.116525.

[72] D. Lombardo, M. Kiselev A., C. Maria Teresa, Smart nanoparticles for drug delivery application: development of versatile nanocarrier platforms in biotechnology and nanomedicine, J. Nanomater. (2019) 3702518. https://doi.org/10.1155/2019/3702518.

[73] S. Ganesh, A.K. Iyer, D.V. Morrissey, M.M. Amiji, Combination of siRNA-directed gene silencing with cisplatin reverses drug resistance in human non-small cell lung cancer, Mol. Therapy Nucleic Acids 2 (2013) e110. https://doi.org/10.1038/mtna.2013.29.

[74] R.A. Meyer, J.J. Green, Shaping the future of nanomedicine: anisotropy in polymeric nanoparticle design, Wiley Interdisc. Rev. Nanomed. Nanobiotech. 8 (2) (2015) 191–207. https://doi.org/10.1002/wnan.1348.

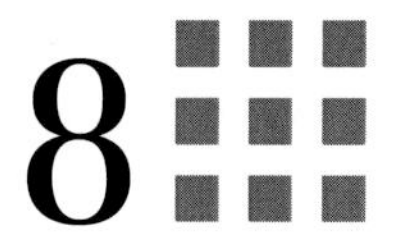

Silk fibroin nanomaterials

He-Lin Xu and De-Li ZhuGe

DEPARTMENT OF PHARMACEUTICS, SCHOOL OF PHARMACEUTICAL SCIENCES, WENZHOU MEDICAL UNIVERSITY, WENZHOU CITY, ZHEJIANG PROVINCE, CHINA.

Chapter outline

8.1 Introduction

Modern biomaterials play an indispensable role in medicine, especially in drug delivery and tissue engineering. The biomaterials for *in-vivo* applications must meet essential properties, such as good biocompatibility, suitable mechanical strength, biodegradable, safety metabolic byproducts. The biological polymer materials can be divided into two categories: synthetic and natural materials. Synthetic polymers such as poly (lactic-co-glycolic acid), polyvinyl alcohol (PVA) are easily obtained and commonly used. However, the limitations such as the interbatches

Biopolymeric Nanomaterials: Fundamentals and Applications. DOI: https://doi.org/10.1016/B978-0-12-824364-0.00016-2

variability, toxicity, and limited half-life in circulation still exist. Despite a variety of synthetic polymers have been investigated for medical applications, natural polymers, including collagen, gelatin, chitosan remain attractive due to their superior advantages, such as biocompatibility and biodegradability. Silk fibroin (SF), a natural polymer that derives from the silkworm cocoons, has been used in textile industry for many centuries. During last 100 years, the surgeons use silk as a suture material for wound ligation considering its biocompatibility and strong mechanical strength [1]. In 1993, SF was approved by the US FDA as a biomaterial [2]. SF applications have been transformed from traditional textile into high-tech medical field. Recently, SF has been developed as substitution of human bodies for tissue regeneration and engineering, including vascular, neural, skin, bone, ocular, and cardiac tissues [2–4]. Moreover, the SF-based materials are also attractive for targeting drug delivery and controllable release [5,6]. Other than these superiorities, the ease to obtain and possibility of mass production also broaden the prospects for the application of SF in biomedical fields.

In addition to composition of polymers, the structure and size of polymers also need full investigation before applying to clinical trials. To date, many biomaterials with different morphologies and sizes have been designed. Within them, the nanomaterials are most popular field and can be used for many applications, such as drug delivery, gene therapy, tissue engineering, biosensing, cancer vaccines [7–10]. In the beginning, the biocompatibility and toxicity of nanomaterials are the main concerns for the researchers, whereas the second generation focused more on the immune escaping abilities, targeting capabilities, and stability. For example, the NPs take advantages of its nanosize and can be used for carrying chemotherapy drugs through intravenous administration. In tumor nanotherapy, travelling through the blood vessels, the NPs can accumulate into tumor due to the enhancing penetration and retention effect (EPR) effects [11]. Moreover, the surface modifications endow NPs more stability and targeting ability in the blood stream, avoiding phagocytes recognition. The NPs technology used in the recent years has great significance in improving the efficacy of the drugs. However, there are still many limitations that need to be solved before applying SF NPs (SFNPs) into clinical practice. For instance, after publishing thousands of papers, the EPR effects in clinical patients are not significant as expected [12]. Moreover, most of NPs are trapped into the liver by Kupffer cells, thus decreasing the effective concentration at the targeted sites [13]. Novel nanomaterials with high biocompatibility and functional effectiveness are essential for medical therapy.

So far, SF has been designed as NPs, nanofibers, microparticles, films, tubes, microneedles, 3D scaffolds and hydrogels, depending on its applications [14]. Therein, due to the high surface to volume ratio, the SF nanomaterials have been widely studied in many fields, such as oncology and wound healing. However, the application obstacles of SF nanomaterials still exist, including instability, fast degradation rate, uncontrollable size as well as toxic solvent involved in materials preparation. Moreover, parameters, such as preparation methods, administration routes, and stability *in vitro* and *in vivo*, greatly vary across studies. More *in-vivo* studies are needed for the application of SF in clinical practice. In this chapter, the different types of silk cocoons as well as general preparation principle of regenerated SF, mainly including steps of degumming, dissolving, and purification are introduced. Moreover, we will summarize the properties of SF, especially the secondary structure of β-sheet. Based on the structure, the SF showed good biocompatibility, controllable degradation rate, and have ability to encapsulation drugs. Then, we will discuss SF-based nanomaterials, including NPs, nanofibers, nanoshells, and nanogels. Finally, the biological application of SF, such as vehicles for delivery drug, tissue engineering, stabilizing other nanoparticles (NPs), signaling pathways activation, and SF composite materials are introduced (Fig. 8.1).

8.2 Preparation of regenerated SF solution

8.2.1 Type of silk cocoons

Silkworm cocoons play protective roles in the process of silkworm pupation, preventing heat or natural enemy. In general, silkworms are classified as either mulberry or nonmulberry. *Bombyx mori* belong to mulberry and is the most extensively used for SF production, while the non-mulberry silkworm, such as *Antheraea pernyi,* is the wide-type silkworm. *Bombyx mori* silk-worm has been cultivated by humans for thousands of years to produce textile fibers, while the *A. pernyi* live in wild. The cocoons produced by these two silkworms are totally different and are often subjected to comparative experiments. From the appearance point of view, the cocoons of the *B. mori* silkworm showed white color with soft texture, while the *A. pernyi* is yellowish brown and hard (Fig. 8.2A). Moreover, the cocoon peduncle is the typical structure in *A. pernyi.* The differences in macroscopic property mainly depend on the component of the cocoons and the arrangement of microscopic network structures. In general component, *B. mori* cocoon contained about 71.8% SF and 28.2% sericin, whereas *A. pernyi* cocoon contained a substantial 84.3% of SF, about 8.7% sericin, and 7.0% mineral [15] (Fig. 8.2B). *Antheraea pernyi* possessed less sericin than *B. mori* and had unique mineral ingredients. The sericin is an amorphous protein with less molecular weight, which showed lower strength than SF. For the microstructure, *B. mori* has thin fibers and shorter intersectional fiber distance, whereas *A. pernyi* has thick fibers and longer intersectional distance (Fig. 8.2C). The high strength networks require thicker fibers and shorter intersectional distance. The combination effects of these two parameters result in five times higher strength in *A. pernyi* than *B. mori* [15]. Therefore, the high mechanical strength was achieved by *A. pernyi* due to difference in components and microstructure. Besides mechanical properties, the amino acids in *B. mori* and *A. pernyi* are also quite different and are shown in Table 8.1. Study found that nonmulberry silkworms, such as *A. pernyi* and *Antheraea mylitta,* possess arginyl-glycyl-aspartic acid (RGD) sequence and were proved to effectively enhance the adhesion of cells to biomaterials in a variety of extracellular matrix as the RGD peptides are the binding sites for cell integrin receptors [16,17] (Fig. 8.2D). However, considering that most studies still use *B. mori* silkworm cocoons, most of our subsequent content will discuss *B. mori* silkworm cocoons.

8.2.2 General preparation principle for regenerated SF solution

The silk from silkworms primarily contains fibroin (70%) and sericin (30% glue-like protein) [20]. Two fibroin fibers occupy the corn of silk wire, while the sericin is covered around the cores

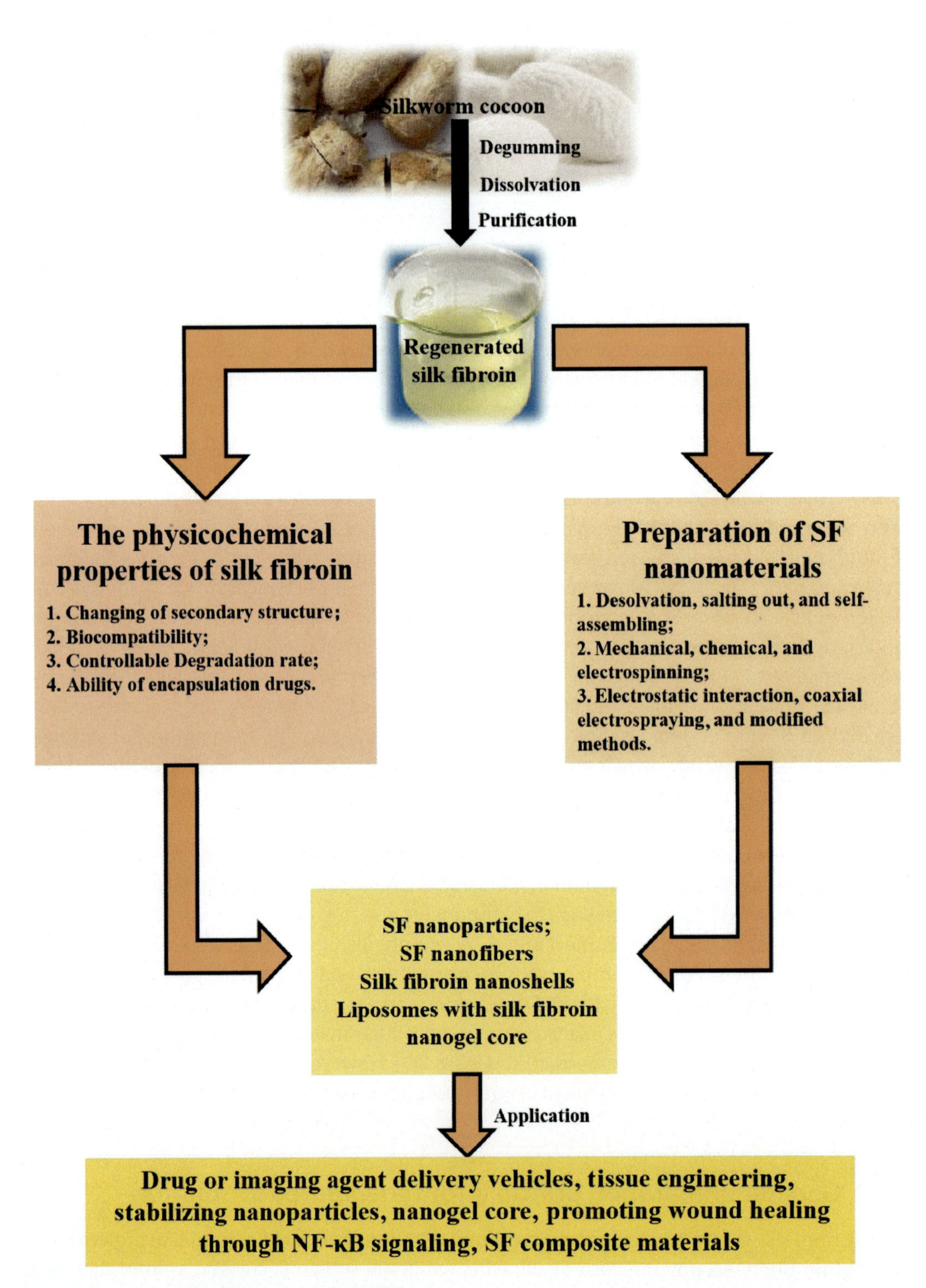

FIGURE 8.1 Summary of silk fibroin nanomaterials.

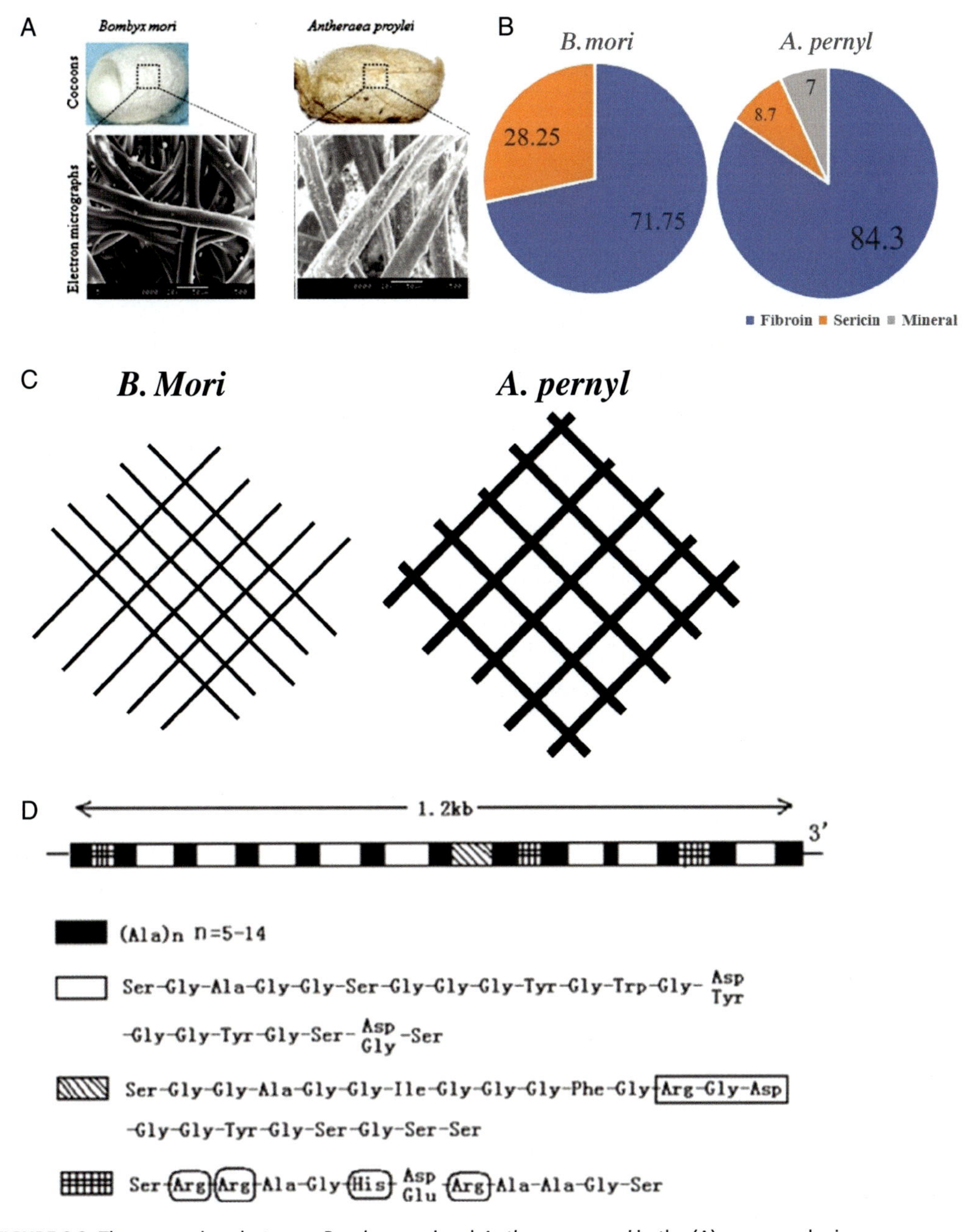

FIGURE 8.2 The comparison between *Bombyx mori* and *Antheraea pernyi* in the (A) macro and micro appearance, (B) component, (C) networks microstructure, (D) amino acid sequence of *A. pernyi*.

Table 8.1 Molar percentage (%) of amino acids in different silkworm cocoons [18,19].

Amino acids	*Bombyx mori*	*Antheraea pernyi*
Glycine	14.48	38.85
Alanine	4.05	24.25
Valine	3.22	0.40
Leucine	0.99	0.50
Isoleucine	0.76	0.22
Serine	35.63	13.00
Threonine	8.14	0.38
Aspartic acid	15.65	6.69
Glutamic acid	4.74	1.22
Lysine	2.72	1.15
Arginine	3.10	3.17
Histidine	1.55	1.11
Tyrosine	3.39	0.89
Phenylalanine	0.56	7.51
Proline	0.57	0.52
Methionine	0.14	0.00
Cysteine	0.29	0.12

(Fig. 8.2). Sericin acts as glue to maintain the morphology of silk. Sericin protein was proved to be useful because of its properties, such as antioxidation, antibacterial, and UV resistant [21]. However, sericin was reported to cause severe allergy and needed to be removed before using fibroin *in vivo* [20]. The regenerated SF refers to the SF after removing sericin by effective methods, which is essential for avoiding immune reaction. The major three steps are used for removing sericin and preparation of regenerated SF solution: (1) Degumming under alkaline solution, (2) dissolve in LiBr, and (3) purification [14]. Generally, in the standard method, cocoons were cut into small pieces and boiled under Na_2CO_3 for degumming [22]. After washing and dying, the degummed fibers were dissolved in LiBr aqueous solution or Ajisawa's reagent ($CaCl_2/H_2O/EtOH$). Then, the solution was added into a dialysis bag for dialysis against deionized water or a low concentration ionic solution to fully remove the LiBr from the SF solution. Finally, the regenerated SF solution was filtered and stored at 4 °C for further use. Due to the high temperature and multiple steps involved in these processing, the chemical reagents using in every step are critical to the quality of regenerated SF.

8.2.3 Degumming

Degumming is the primary step in preparation of regenerated SF. Reagents, such as acid solution, alkaline solution, proteases, and surfactant are used for degumming (Table 8.2). A study compared the effectiveness of different degumming reagents, such as tris-SO_4, β-mercaptoethanol, sodium chloride, sodium carbonate, sodium hydroxide. The study revealed that a sodium carbonate salt-boiling system was the most effective sericin extraction procedure for mulberry silk cocoons [23]. However, due to the long process time and high temperature,

Table 8.2 Summary of reported degumming methods for preparation regenerated SF.

Degumming methods	Reagents	Degumming time (min)	Degumming temperature (°C)	Results	Ref
Alkaline	Na_2CO_3	30, 60, 120	90	Degumming time is an important parameter to influence drug release from SF-based drug delivery systems	[24]
Infrared	/	60, 90, 120	80, 100	This process is much cleaner as recovery of sericin can be done by spray drying the liquor directly.	[27]
Microwave	/	1–15	60, 100	The microwave irradiation reduces the time needed to reach the same degree of degumming by the conventional method.	[28]
Proteases	Alkaline, neutral and acid	5–240	50–65	Alkaline and neutral proteases effectively degummed sericin.	[31]
Silk protein surfactant	Silk fibroin amino acids and lauroyl chloride	30–150	60–100	The sericin of silk fibers can be completely removed after boiling three times for 30 min and using a bath ratio 1:80 (g/mL) and a 0.2% silk protein surfactant aqueous solution.	[32]
Acid	Citric acid	30	98	The silk sericin removal percentage was almost 100% after degumming with 30% citric acid which resulted in a total weight loss of 25.4% in the silk fibers.	[33]
Surfactant	Alkyl polyglycoside (APG)	30	100	Degumming in APG did not induce an evident breakage of the silk fibroin peptide chains, including the light chain and P25 protein.	[34]

the degumming process was likely to cause the structural change of SF, leading to breakage of peptide chain of SF, which would have effects on the qualities of SF, such as molecular weight distribution. Shorter degumming times resulted in a narrow distribution of molecular weights, while longer degumming times caused more degradation by a broad molecular weight distribution and a shift toward low-molecular-weight fragments [24]. Moreover, the quality of SF-based nanomaterials also will be affected by the degumming process. Another study showed that there was a positive correlation between SF degradation caused by long time degumming and a reduction in the mean size and size distribution of SF-based NPs [25]. Sodium dodecyl sulfate (SDS) - polyacrylamide gelelectrophoresis (PAGE) showed that the SF degummed with urea at 80 °C was similar to the natural SF *in vivo*, but boiling in Na_2CO_3 solution leads to serious breakage of the SF peptide chains [26]. Except for chemical reagents-based degumming, physical methods such as infrared heating, microwave served as a greener process for degumming [27,28]. Degumming ratio was frequently used for measuring efficiency of degumming processing. However, due to the different types of cocoons, there is no exact degumming ratio. In addition, although many studies showed no sericin residue after degumming under the electron microscopy, there are still no standard methods for detection and quantification of sericin specifically [23]. Therefore, the standard methods for preparation regenerated SF should be established. Degumming ratio = (Initial weight - degummed weight)/initial weight $\times$100%

8.2.4 Dissolution

In the subsequent dissolution step, regenerated SF was dissolved into high ionic solution to breaking the hydrogen bond within the β-sheet (Table 8.3). Studies compared the dissolution reagents such as lithium bromide, $Ca(NO_3)_2$-MeOH, lithium thiocyanate and found that lithium bromide is the most effective fibroin dissolution system [23]. Moreover, SF produced by dissolving in LiBr aqueous solution for 6 h, showed the highest molecular weight level [29]. SF protein was produced by dissolving SF in Ajisawa's reagent for dissolution times ranging from 3 to 180 min, the molecular weight of the SF decreased with increasing dissolution time. Aqueous solubility of lyophilized SF was higher when the native SF was dissolved in LiBr than when Ajisawa's reagent was used due to the presence of some β-sheet crystals in the resultant protein. Moreover, the molecular weight of SF peptides was found to play an important role in the fiber sizes of the electrospun mats generating thinner fibers when LiBr is used for the dissolution of the native SF than fibers obtained when Ajisawa's reagent or $CaCl_2$ is utilized [22]. In recent year, some novel reagents such as tetrabutylammonium hydroxide and $CaCl_2$—formic acid were also used for dissolving SF due to the mild temperature in processing (Table 8.3). These dissolution reagents dissolve SF under milder conditions, avoiding the breakage of amino acid chains caused by high temperature.

8.2.5 Purification

In the last step of purification, the ionic solution in the secondary step was removed by dialysis against low ionic solution. However, the aggregation of SF in this step was often reported due

Table 8.3 Summary of reported dissolution methods for preparation regenerated SF.

Dissolution reagents	Dissolution time (min)	Dissolution temperature (°C)	Mechanism	Advantages	Disadvantages	Ref
Lithium bromide or lithium thiocyanate	240	60	Destroying hydrogen bond	Well-established efficient method	High cost; environmental concerns	[30]
Ajisawa's reagent	120	80	Destroying hydrogen bond	Cheap and easy to obtain	Easy to form aggregation when doing the purification; SF degradation	[30]
Tetrabutylammonium hydroxide (TBAOH)	150	25	Breaking hydrogen-bonding; Weakening hydrophobic interactions	Mild temperature; easily recycled; not lead to the degradation	Less reported methods	[35]
$CaCl_2$–formic acid	/	25	Breaking hydrogen-bonding	Facile method; Preserving nanofibril structure	Less reported methods	[36]

to the premature formation of β-sheets. Ajisawa's reagent is the earliest method reported for dissolving silk fibers, but the fibroin significantly aggregates during subsequent dialysis step. However, Ajisawa's reagent cost less than LiBr. In this way, the novel dialysis methods must be established to prevent aggregation of SF under Ajisawa's reagent dissolution. The dissolved solutions were dialyzed against either water or urea solution with a stepwise decrease in concentration. The study found that when the stepwise decrease in concentration urea was adopted, the purified SF had smaller aggregates and lower content of β-sheet compared with dialyzed against water [30]. Above all, the methodology employed during the extraction of SF should be taken into account depending on the use the materials are going to be made for.

8.3 The physicochemical properties of SF

The SF amino acid consists of heavy chain (325–350 KDa), light chain (25 KDa) peptides, and a glycoprotein P25 which are noncovalently connected by a disulfide bond [37]. The molar ratios of the heavy chain, light chain, and P25 are 6:6:1 [38]. The heavy and light chains differ in amino acid sequences. The heavy chain of SF composite of a hydrophilic part Gly-X (X is Ala in 65%, Ser in 23%, and Tyr in 9% of the repeats) and hydrophobic has repeated amino acid blocks, which allows forming stable silk II (β-sheet nanocrystals) from silk I (mixture of random coil, Q-helix, β-sheet) in a different induction method. The light chain consists of ahydrophilic part, due to the nonrepeated sequence (Fig. 8.3A) [39,40]. P25 plays an important role in maintaining the structural integrity of the silk complex.

The mechanical characteristics of SF predominantly originate from the interactions between and within secondary structure. About 55% of raw silk fibers have the antiparallel β-sheet structure of silk II. The most noticeable feature of SF is the secondary structure changing from random coil to β-sheet under chemical or physical triggering methods (Fig. 8.3B). Briefly, the parts of random coil are hydrophilic and hold moisture, while the hydrophobic domain is embedded into the hydrophilic domain. Ethanol, dimethyl sulfoxide (DMSO), acetone and potassium phosphate are the most commonly used chemical reagent for induction β-sheet in SF [41]. Moreover, surfactant such as SDS can accelerate the secondary structure changing [42]. However, taking toxicity into account of organic reagents, some physical methods are also developed, such as infrared, heating, ultrasound, water vapor annealing, and vortex [27, 43–45]. Different methods can be used to differentiate between the silk I and silk II structures (Fig. 8.3C). For example, the X-ray diffraction, Fourier transform infrared spectroscopy (FT-IR), and atomic force microscopy can be used for studying changes in the secondary structure [39,46].

8.3.1 Biocompatibility

After the complete removal of sericin, the SF is proved to be biocompatible. Many cells can grow on SF scaffolds without causing significant immune responses, such as mesenchymal stem cells, smooth muscle cells, cardiomyocytes, human limbal epithelial, and stromal cell [47–50]. For example, the adipose-derived mesenchymal stem cells and bone marrow-derived mesenchymal stem cells (BMSCs) were cultured and grown on SF scaffolds, respectively, for liver therapy. Liver functions of the mice were significantly improved than the control group. Moreover, angiogenesis

A

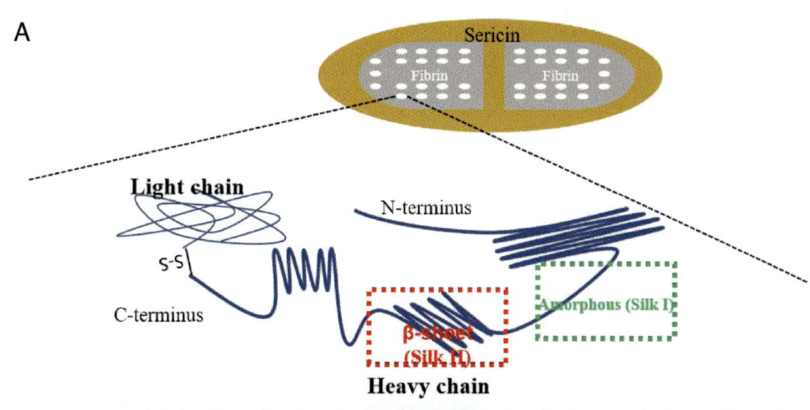

B

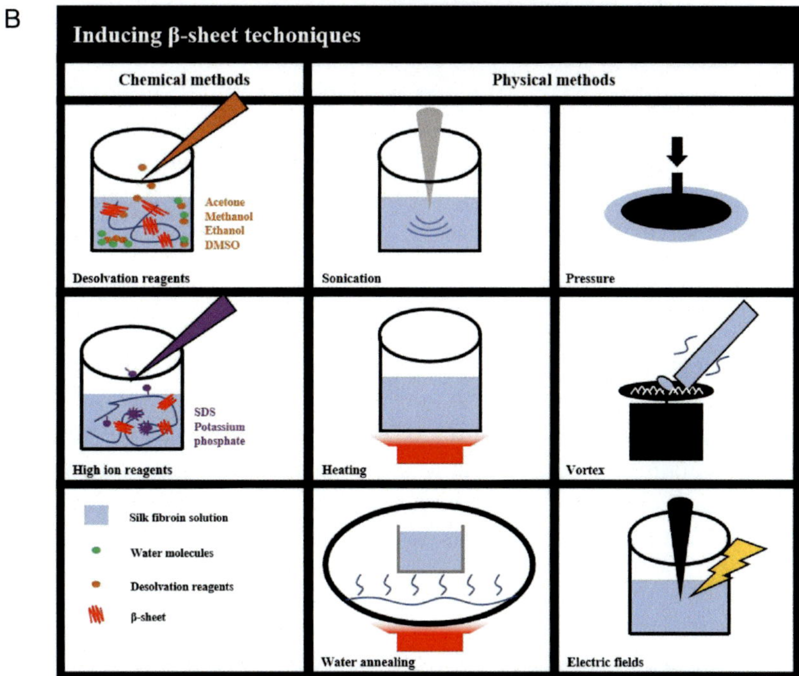

C

Detection methods	From α-helix or random coil to β-sheet
Fourier transform infrared spectroscopy (FT-IR)	Amide I: 1655 cm^{-1} shifted to 1645 cm^{-1}; Amide II: 1540 cm^{-1} shifted to 1520 cm^{-1}.
X-ray diffraction (XRD)	Crystal transformation.
Differential scanning calorimetry (DSC)	Increasing melting point.

FIGURE 8.3 The physicochemical properties of silk fibroin. (A) Structure of raw silk and silk fibroin. Fibroin is covered by sericin. A fibroin fiber consists of heavy chain and light chain, which are connected by a disulfide bond. In the heavy chain, Silk I is characterized by an amorphous, while silk II has a β-sheet nanocrystal conformation. (B) Chemical and physical methods for inducing β-sheet. (C) Detection methods for inducing secondary structure changing from α-helix or random coil to β-sheet.

and hepatocyte-like cells were discovered in the scaffolds. These results proved that SF has good biocompatibility and can be used as a support for cell therapy [49] (Fig. 8.2B). In addition, SF also enhances the biocompatibility of NPs by coating the surface [51,52], which will be discussed in SF nanoshells.

8.3.2 Controllable degradation rate

Nonbiodegradable materials are often permanently implanted into the body as some inert supports while biodegradable materials need to meet the following conditions: (1) Material can be degraded by enzymes. (2) The degradation rate should match the tissue growth rate. (3) The degradation byproducts should be nontoxic [53]. SF sheet can be degraded into amino acid by protease XIV, while Q-chymotrypsin can degrade the dissolved SF [54]. SF was incubated in protease XIV at 37 °C to create an *in-vitro* model system of proteolytic degradation. Gel electrophoresis indicated a decreasing amount of the silk 25 kDa light chain and a shift in the molecular weight of the heavy chain with increasing incubation time in protease XIV [54]. The average molecular weight of SF after degradation follows the order protease XIV < collagenase IA < Q-chymotrypsin. However, the degradation rate of SF still needs to be controlled to suit different application requirements. The degradation rate of SF is directly related to the number of β-sheets. As the number of β-sheet increases, the degradation rate of SF decreases.

8.3.3 Ability to encapsulate drugs

SF as a biological material has a good drug-loading capacity [55–57]. On the one hand, as SF has both hydrophilic and hydrophobic domains, drugs can be encapsulated by hydrophobic interactions [55] (Fig. 8.2B). On the other hand, as SF has the basic unit of amino acids, it can easily react with compounds containing amino groups or carboxyl groups by amide reaction to encapsulate drugs. Studies found that the β-sheet structure of SF has pH sensitivity, which decreases at low pH. Therefore, the hydrophobic drug can be released in a low pH environment [58,59].

8.4 Classification and preparation protocols of SF nanomaterials

8.4.1 SF nanoparticles

The preparation methods of SFNPs include desolvation, salting out, and self-assembling. Desolvation means that the solvation layer of colloidal particles is weakened under the condition of heating or adding other solvents, which has a strong binding force with the original solvent, resulting in colloidal aggregation and particle formation. The solution used in the desolvation method includes methanol, ethanol, dimethyl sulfoxide, and acetone. The acetone possesses stronger desolvation capability than other reagents and is frequently used to induce SF conformational changing. The SFNPs show transparent appearance in methanol, ethanol, and dimethyl sulfoxide while show milk-like appearance under acetone desolvation. Salting out refers to the phenomenon of adding salt to an aqueous protein solution and precipitating the protein as the salt concentration increases. Doxorubicin (DOX) and magnetic NPs (MNPs) were dispersed in potassium phosphate solution. SFNPs were prepared by adding SF solution into potassium phosphate solution. The formation of SFNPs varies considerably and higher DOX concentration, larger amount of MNPs is required to form SFNPs. Otherwise the SF particles aggregated into nondispersible clusters [60]. Self-assembly refers to the process of forming regular NPs spontaneously between molecules under the action of noncovalent bonds. For example, paclitaxel (PTX) solution was added into SF aqueous solution for preparation of SFNPs [61]. The Silk I or random coil structure was the main conformation of SF in PTX-SFNPs. Nevertheless, the use of organic solvents in the preparation process is still a problem, which will eventually lead to failure in clinical approval. Hence, the SFNPs were prepared with the help of supercritical CO_2. A special coaxial nozzle can produce small droplets and improve the mixing effect between compound solution and supercritical CO_2 flow [62]. SFNPs with a controllable particle size less than 100 nm were prepared using this method [62].

8.4.2 SF nanofibers

Unique features of nanofibers provide large potential in the field of biomedical applications. Nanofibers have been applied mainly in regenerative medicine, including skin, bone, and cartilage regeneration. The nanofibers fabrication techniques categories are varied across the study. The mechanical, chemical, and electrospinning methods are utilized for fabrication nanofibers. Among them, the electrospinning is the most commonly used method for the preparation SF nanofibers [63,64]. Briefly, the SF solution was first placed into a syringe and then pushing it to the tip of the syringe by external pumping. When the SF solution droplet is formed at the metallic needle, an electric voltage bias is applied between the needle and collector. The SF solution forms a jet between the electrodes. The SF solution is stretched by high-voltage static electricity, and finally nanofibers are obtained in the collector. The diameter, morphology, and alignment of nanofibers are affected by either SF parameters or spinning parameters. The SF parameters included the concentration and viscosity of SF solution. The spinning parameters included the needle diameter, applied voltage, flow rate, collector, and distance between syringe and collector [63]. In addition, the SF wet spinning process was performed by using $CaCl_2$-formic acid (FA) solution as a solvent. Silk nanofibrils assembled into fibers due to physical shear and water-induced coagulation, which was accompanied by structural transitions from an amorphous state to silk II during these processes [65].

8.4.3 SF nanoshells

The SF is used for NPs coating to improve the intrinsic properties of NPs, such as biocompatibility and drug encapsulating ability. Methods such as electrostatic interaction, coaxial electrospraying, and modified methods are used for SF nanoshells preparation. Cationic core NPs such as chitosan can interact with negatively charged SF directly via electrostatic interaction [66]. Briefly, chitosan NPs were added to the SF solution slowly and stirred for 30 min. The

SF-coated chitosan NPs were self-assembled. PVA and SF were used for the inner and outer polymers, respectively, via coaxial electrospraying. By this method, the model drug, DOX, was successfully encapsulated in the core with greater than 90% drug encapsulation efficiency [67]. SF can also be applied for coating NPs via the modified method [68]. SF was dissolved in phosphate buffered saline (PBS) (pH 7.4) at a certain concentration and then the solutions were placed for several hours to let the SF protein chains stretch thoroughly. Subsequently, the SF solutions were added into the aqueous liposomes to form SF-liposomes mixtures. After gentle agitation, methanol was added in the mixtures to insolubilize the SF. The SF-coated liposomes can be obtained by centrifugation.

8.5 Biological applications and the potential challenges of SF nanomaterials

8.5.1 SF as a targeted drug or imaging agent delivery vehicle

NPs are the most common nanocarriers and have many advantages in drug delivery, such as adjustable particle size, modification of particle surface, enhancing drugs solubility, and coordinated delivery of drugs. In addition, NPs also have obvious advantages in specific diseases diagnosis, and therapy. In particular, NPs can enhance the penetration through the tumor blood vessels and drugs retention in the lesions (EPR effects) so that can enhance the antitumor efficacy. Moreover, in inflammatory disease, such as inflammatory bowel diseases, it is proved that epithelial cells EPR effect can also enhance NPs accumulation in inflammatory intestinal site. To date, there have been many reports of SFNPs used as a delivery vehicle in diseases therapy and diagnosis. In this part, we will discuss the intrinsic properties, colloidal stability, as well as applications of SFNPs.

The intrinsic multiple responsive properties of SFNPs such as pH, H_2O_2, and reactive oxygen species (ROS)-dependent drug release endow its application in disease treatment [58]. As the secondary structures governed the major properties of SFNPs, the change in the secondary structures to different stimulation were examined *in vitro* [58]. The ratio of β-sheet in SFNPs was found to decrease with decreasing pH and increasing H_2O_2 concentration. With the decreasing pH values, the drug releasing rates greatly increased, is which attributed to destroying β-sheet. These SFNPs were able to act as pH-responsive anticancer carriers and serve as a lysosomotropic delivery platform and overcame drug resistance *in vitro* [59]. Conversely, the enhanced drug release induced by glutathione (GSH) was attributed to cleavage of disulfide bonds between heavy chains and light chains of SFNPs, leading to the exposure of β-sheet regions to the releasing buffer. Moreover, due to these properties, SFNPs are good candidates for colonic disease treatment. SFNPs-loaded resveratrol showed immunomodulatory properties and intestinal antiinflammatory effects in mice inflammatory bowel diseases [69].

However, SFNPs lack colloidal stability due to high negative charge of SF surface. When being introduced into the biological media, including PBS and Dulbecco's modified Eagle's medium (DMEM), the SFNPs start aggregation in few minutes. The aggregation will change the *in vitro* cells uptake manner. Moreover, the aggregation of SFNPs *in vivo* will lead to the formation of

Table 8.4 Summary of SFNPs colloidal stability enhancement methods.

Methods	Reagents	Features	Ref
Cross-linking	Genipin	SFNPs had ideal active tumor targeting property and the tumor burden can be reduced noticeably.	[71,77]
	Proanthocyanidins	SFNPs showed stable particle size in physiological medium and photothermal property after near infrared red (NIR) irradiation.	[70]
Surface modification	Polyethylene glycol	PEGylated silk nanoparticles showed excellent drug loading and release capacity.	[72]
	Chitosan	The use of chitosan could transform the lyophobic character of the SFNPs into lyophilic.	[75]
	Glycol chitosan, N,N,N-trimethyl chitosan, polyethylenimine, and PEGylated polyethylenimine	The polymeric coatings significantly enhanced the colloidal stability of SFNPs in biological media and showed better redispersibility after lyophilization.	[74]

agglomeration in blood vessels, which will cause severe side effect. Until now, methods including cross-linking and surface modification have been used for enhancing stability of SFNPs (Table 8.4). Many natural agents such as genipin, proanthocyanidins, and tannic acid have been used as cross-linking agents in tissue engineering because of biological safety [70,71]. Genipin is a natural cross-linking agent that has capability to covalently cross-link between amino groups of protein. Due to its much lower cytotoxicity than chemical cross-linking agent, it has been used for improving stability of SFNPs. The amino residues in SFNPs were cross-linked for 24 h, which endow the SFNPs good stability in DMEM and change the solution color from white into light blue [71]. Proanthocyanidins, another natural agent, was also used for cross-linking SFNPs. The diameter stability of SFNPs in the PBS and DMEM was improved compared with non-cross-linking group [70]. The unmodified SFNPs are tagged by opsonins, subsequently recognized by the mononuclear phagocytic systems, and cleared from blood stream. Polyethylene glycol (PEG) was widely used for prolong NPs circulation time in blood stream. After modified with PEG, the PEG-SFNPs showed better colloidal stability and significant less proinflammatory effects compared with native SFNPs [72]. Moreover, a novel study found that PEGylated SFNPs have immunomodulation effects in macrophages. PEGylation SFNPs reduced the inflammatory and metabolic response initiated by macrophages [73]. Another method to enhance the stability of the SFNPs is to modify the surfaces, mainly with cationic materials. Four cationic polymers including glycol chitosan, N,N,N-trimethyl chitosan, polyethylenimine, and PEGylated polyethylenimine were decorated into the surface of SFNPs through electrostatic interaction [74,75]. Compared with native SFNPs, the modified SFNPs showed higher colloidal stability in biological media. N-[1-(2,3-Dioleoyloxy)propyl]-N,N,N-trimethylammonium methyl-sulfate (DOTAP) and 1,2-dioleoyl-sn-glycero-3-phosphoethan olamine (DOPE) were also used for SFNPs coating [76]. The results showed that coating with cationic lipids can achieve high intracellular delivery efficiency with enhanced capacity and stability.

SFNPs have been widely used as drug delivery agents in many diseases, such as cancer, inflammatory bowel diseases, bone infection, and acute pancreatitis (Table 8.5). As the SFNPs contain both hydrophilic and hydrophobic parts, the drug can be loaded by self-assembly through noncovalent bonding. Indocyanine green (ICG) is an FDA-approved clinical photosensitizer for checking liver function and liver blood flow. Silk was found to be easily dyed with ICG because of strong affinity to many colorants [78]. As such, the ICG was successfully incorporated into SFNPs for tumor photothermal therapy (Fig. 8.4). Because of the EPR effects of SFNPs, the circulation time of ICG in the blood stream was prolonged and ICG can be accumulated into the tumor. After 808 nm near-infrared irradiation, the tumor was completely repressed [79]. Moreover, a multiplatform was developed with ICG and DOX combined function with both diagnosis and therapy. SFNPs remarkably improved tumor inhibitive efficacy through a combination of photothermal/photodynamic/chemotherapy with minimal systemic toxicity or adverse effect [80]. Chemotherapy drugs such as 5-fluorouracil, curcumin, triptolide, and celastrol were also incorporated into SFNPs for treatment of breast carcinoma and pancreatic cancer [11,81]. Tumors could be noticeably reduced after being injected with the drug-entrapped SFNPs. However, the target efficiency still needed improvement. The stable peptides such as SP5-52, cRGDfk, iRGD are conjugated to SFNPs for better targeting to cancer [71,82,83]. A targeted delivery system was developed based on SFNPs for the systemic delivery of gemcitabine to treat lung tumor in a mice model. SP5-52 peptide was conjugated on the SFNPs to improve the targeting ability and SP5-52 conjugated SFNPs were more up-taken by cancerous cells in comparison to the BEAS-2B cells [82]. Other than tumor therapy, the study indicated that SFNPs carry natural antioxidant, quercetin, which can be delivered through gastrointestinal tract [84]. A novel study found that antibiotic vancomycin loaded in SFNPs showed antibacterial activity toward methicillin-resistant *Staphylococcus aureus* [85]. Acute pancreatitis can cause severe inflammatory reaction and lead to systemic inflammation. Bilirubin was found to exert antioxidative, antiinflammatory, and antiapoptotic effects. However, the poor solubility and potential toxicity hinder its application. A novel study used SFNPs to deliver bilirubin to the inflammatory pancreas and showed enzyme response manner (Fig. 8.5). In mice acute pancreatitis model, the bilirubin-encapsulated SFNPs exert strong therapeutic effects through inhibition NF-κB pathway and activating Nrf2/HO-1 pathway [77]. In addition to target delivery, the small diameter SFNPs were proved to penetrate through the skin and can be used for drugs transdermal administration. SFNPs with the mean diameter of 40 nm can penetrate the stratum corneum and delivered deep into the skin [86].

8.5.2 SF nanomaterials for tissue engineering

Fibers materials, which are usually used in textile industry, have been used in biomaterials research. Until now, fibers with micro- or nano-diameter have been developed in tissue engineering [88]. Tissue engineering is an important approach to counteract organ deficit. For example, the skin defects caused by burn, trauma, or disease lead to severe infections and can even be life threatening. Autotransplantation of skin is still the major choice in clinical treatment because of less immunological rejection. However, the available skins are very limited for

Table 8.5 Reported diseases treated with SFNPs loading drugs.

Diseases		Drugs or reagents	Fabrication methods	Size (nm)	Features	Year	Ref
cancer	Gastric cancer	Paclitaxel	Self-assembling	130	SFNPs were fabricated by self-assembling without adding surfactants and excessive toxic organic solvent.	2013	[61]
	Human cervical carcinoma	Doxorubicin	Self-assembling and nucleation	50–200	DOX-loaded Si/SF nanospheres have higher cytotoxicity against HeLa cells than free DOX.	2017	[87]
	Human breast cancer	Doxorubicin	Desolvation (DMSO)	98	The pH-dependent drug release and lysosomal accumulation of SFNPs demonstrate the ability of drug-loaded silk nanoparticles to serve as a lysosomotropic anticancer nanomedicine.	2013	[59]
	Multidrug-resistant breast cancer	Doxorubicin and Fe_3O_4 nanoparticles	One-step potassium phosphate salting-out	130	DOX-loaded magnetic SFNPs work well as a novel drug delivery system (DDS) in multidrug-resistant breast cancer therapy.	2014	[60]
	Lung cancer	Gemcitabine	Desolvation (DMSO)	105–156	SP5-52-conjugated SFNPs are powerful vehicles for delivering Gemcitabine to the tumorigenic lung tissue.	2017	[82]
	Gastric cancer	5-fluorouracil	Desolvation (acetone)	226–365	SFNPs reduced the tumor burden greatly with excellent biocompatibility and safety in vivo.	2018	[71]
	Glioblastoma	Indocyanine green (ICG)	Desolvation (acetone)	209.4	The tumor growth was suppressed significantly after treatment with ICG-SFNPs compared with other control groups.	2018	[79]
	Breast carcinoma	5-flurouracil and curcumin	Desolvation (ethanol)	217	SFNPs showed significant improvement in the cytotoxic activity and bioavailability.	2016	[11]

(continued on next page)

Table 8.5 *(continued)*

Diseases		Drugs or reagents	Fabrication methods	Size (nm)	Features	Year	Ref
	Human pancreatic cancer cells	Triptolide and celastrol	Desolvation (acetone: ethanol=3:2)	170	Combination therapy of triptolide (TPL) and celastrol (CL) encapsulated in SFNPs could be a promising strategy for the treatment of pancreatic cancer.	2017	[81]
	4T1 breast cancer	Indocyanine green and doxorubicin	Desolvation (acetone)	42	SF@MnO_2-based nanocarrier demonstrated a promising potential for integrating multimodal theranostic agents for cancer therapy.	2019	[80]
Inflammatory bowel disease		Resveratrol	Desolvation (methanol)	100	SFNPs constitute an attractive strategy for the controlled release of resveratrol.	2014	[69]
		Curcumin	Self-assembling	175.4	Curcumin-SFNPs could accumulate in colitis tissues after administration, undergo specific internalization by macrophages and exhibit controlled intracellular release of curcumin.	2019	[58]
Severe bone infections		Vancomycin	Desolvation (acetone)	80–90	The vancomycin-loaded SFNPs entrapped in scaffolds reduced bone infections at the defect site with better outcomes than the other treatment groups.	2017	[85]
Acute pancreatitis		bilirubin	Desolvation (acetone)	200–268	Silk fibroin–based nanoplatform for precisely delivery and controlled release of poor solubility bilirubin.	2020	[77]

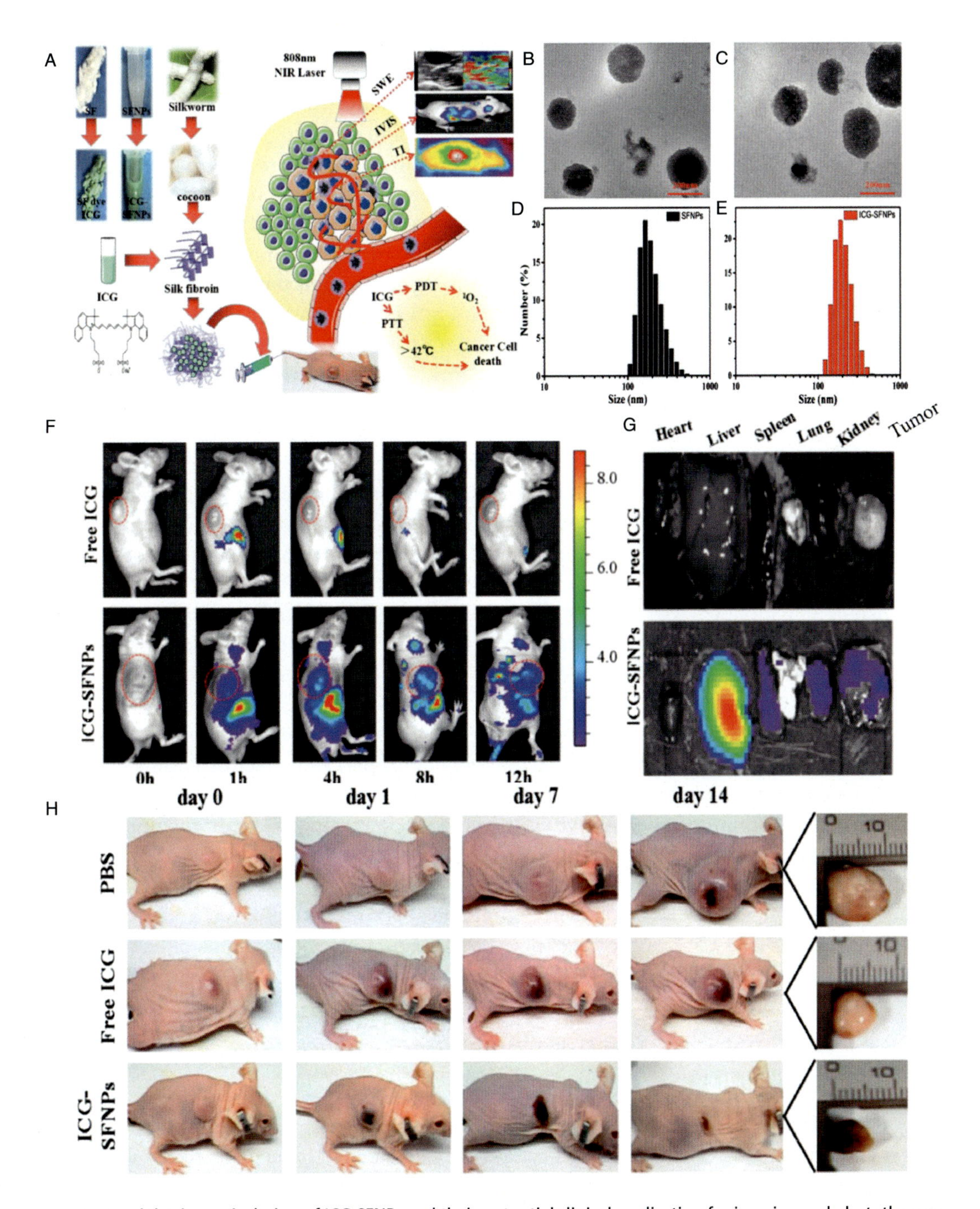

FIGURE 8.4 (A) Schematic design of ICG-SFNPs and their potential clinical application for imaging and phototherapy of glioblastoma. TEM images and particle size distribution of SFNPs (B, D) and ICG-SFNPs (C, E). (F) In-vivo imaging of BalB/C nude mice bearing C6 tumor cells at different time intervals after i.v. injection of free ICG solution or ICG-SFNPs, (G) ex-vivo fluorescent images of major organs and tumors at 12 h after intravenous ICG formulations, (H) photographs of C6 tumor-bearing mice at different intervals after treatment with different formulations [79].

extensive burn patients. Many protein-based fibers have been used for tissue engineering after skin wound such as collagen, keratin, elastin. Of them, collagen is known to be one of the best candidates for skin substitute. However, the collagen meets the problems of rapid degradation and low mechanical strength. Therefore, various natural and synthetic polymers are developed for skin engineering. As describe previously, the SF is a natural polymer that is minimally immunogenic, highly biocompatible, nontoxic, and biodegradable [89]. Two natural biomaterials, de-epithelized human amniotic membrane and SF protein, were used to fabricate artificial skin. The modified amniotic membrane improved the integration with cultured cells *in vitro*, while no sign of toxicity was observed [90]. Fenugreek is an antioxidant natural compound and is incorporated into SF nanofibers by a coelectrospinning method. Wound healing was accelerated in fenugreek-SF nanofibers treated wounds with complete reepithelialization and enhanced collagen deposition [91]. Moreover, study showed that SF nanofibers adjusted SF scaffolds with tunable mechanical strength and supported rat BMSCs adhesion and differentiation [92].

However, the cells are difficult to penetrate through the small pore size of nanofibers thus preventing cells 3D culture on the scaffolds. To solve this problem, the NaCl was added to the SF nanofibers. After leaching out NaCl in H_2O, the pore was formed. The results showed that these SF nanofibers scaffolds supported cells proliferation in the deep layer and more differentiation of keratinocytes in the surface layer, which imitated the human skin structure [93,94]. The diameter of nanofibers was also reported to influence the cells behavior. The diameter of electrospun SF nanofibers was shown to directly influence the proliferation, morphology, and gene expression of primary human dermal fibroblast. Fiber diameters of 250–300 nm were found to support significantly more cell proliferation and cell spreading [95]. Adipose tissue–derived mesenchymal stem cells were effective for cardiac repair after myocardial infraction. However, poor survival of stem cells after grafting limited the therapy efficiency. Cellulose nanofibers modified with chitosan/SF multilayers patches were fabricated with electrospinning via layer-by-layer coating technology. The patches together with stem cells improved the retention of the engrafted stem cells and provide mechanical scaffold for preventing the ventricular remodeling post myocardial infraction [96]. SF nanofibers scaffolds containing bone morphogenetic protein (BMP) 2 and/or NPs of hydroxyapatite were used for *in-vitro* bone formation. The results showed that SF-based scaffolds provided an environment that supported mineralized tissue formation. The incorporation of BMP-2 and/or NPs of hydroxyapatite into the scaffolds enhanced bone formation significantly [97].

8.5.3 SF nanomaterials as a stabilizing agent of bioactive macromolecules

Drugs nanocarriers with unreasonable design will suffer from early drug release, which result in less concentration at target site and affect the drug efficiency. Unexpected drug release also causes severe side effects in other organs. Surface coating is an effective method preventing early drug release. Moreover, due to the intrinsic pH-responsive capability, the SF nanoshells coating NPs can be used as a strategy for lysosomal targeting and escaping. A tumor acidity–responsive nanoplatform was constructed for enhanced chemotherapy by inhibiting premature drug release. DOX-loaded amorphous calcium carbonate NPs were coated with SF. The SF first served as a "gate keeper" to inhibit a drug from prematurely leaking into the circulation. After entering the tumor cells, the SF nanoshells gradually degraded and eventually decomposed inner NPs into Ca^{2+} and CO_2. These result in lysosomal collapse, thus preventing both the efflux of DOX from cancer cells and the protonation of DOX within the lysosome [52]. Moreover, PVA core and SF shell NPs were prepared to encapsulate DOX. The results showed that initial burst release of drugs was minimized by SF coating and large number of drugs remained in the carriers [67]. SF can also enhance the biocompatibility of NPs through surface modification and reduce toxicity. Chitosan NPs have a biological toxicity due to the strong positive charge on the surface. The SF-modified chitosan NPs were developed due to its better stability, low toxicity, mild preparation method [66]. Moreover, SF was developed as a novel mucoadhesive polymer and utilized as the coating of liposomes for ocular drug delivery. Sustained drug release and corneal permeation were observed [68].

Metal NPs, such as quantum dots (QDs) and iron NPs, have poor biocompatibility and can be recognized and eliminated by the immune system, thus preventing reach the target site. QDs are types of NPs with particle size generally less than 20 nanometers, which are mainly composed of II-VI group elements (such as CdS, CdSe, CdTe, ZnSe) and III-V group elements (such as InP, InAs). QDs have attracted tremendous attention as a new optical imaging method for molecular tracing and biomedical diagnostics because of their size-tunable absorption and emission, strong resistance to photobleaching, and longer fluorescent lifetime. However, the biocompatibility and cytotoxicity of nanomaterials might limit their wide application in biology field. SF can be used for coating QDs for enhancing biocompatibility [51]. Co-doped ZnO NPs had a good cellular compatibility due to SF coating [98]. Moreover, excess accumulation of iron NPs was reported to lead to iron-induced oxidative stress and neuronal degeneration in the brain. Magnetic Fe_3O_4 NPs are synthesized via a biomineralization process using SF as a template. The results showed that SF-coated Fe_3O_4 NPs have better neuro-cytocompatibility than raw iron NPs [99].

SF-based materials have been used as carriers for drugs, growth factor, and bioactive agents to wound area [100]. Basic fibroblast growth factor (bFGF) can accelerate wound healing. However, the poor stability and short half-life of bFGF limit its application. A novel liposome with SF hydrogel core was successfully established. Under the sonication, the SF easily formed nano-size hydrogel with the help of liposomal template (Fig. 8.6). The hydrogel core can reduce the rapid leakage in wound fluid and promote the penetrating ability through the skin. *In-vivo* study revealed that bFGF-encapsulated SF hydrogel liposomes can accelerate the wound closure [101]. Further study indicated that after treatment with bFGF-encapsulated SF hydrogel liposomes, the hair growth and hair follicle at wound zone was obviously improved on mice model [102].

8.5.4 SF promote wound healing through NF-κB signaling

SF supported the proliferation, differentiation, and adhesion of various cell types, including epithelial, endothelial, fibroblast, keratinocyte, glial, and osteoblasts. Study found that through

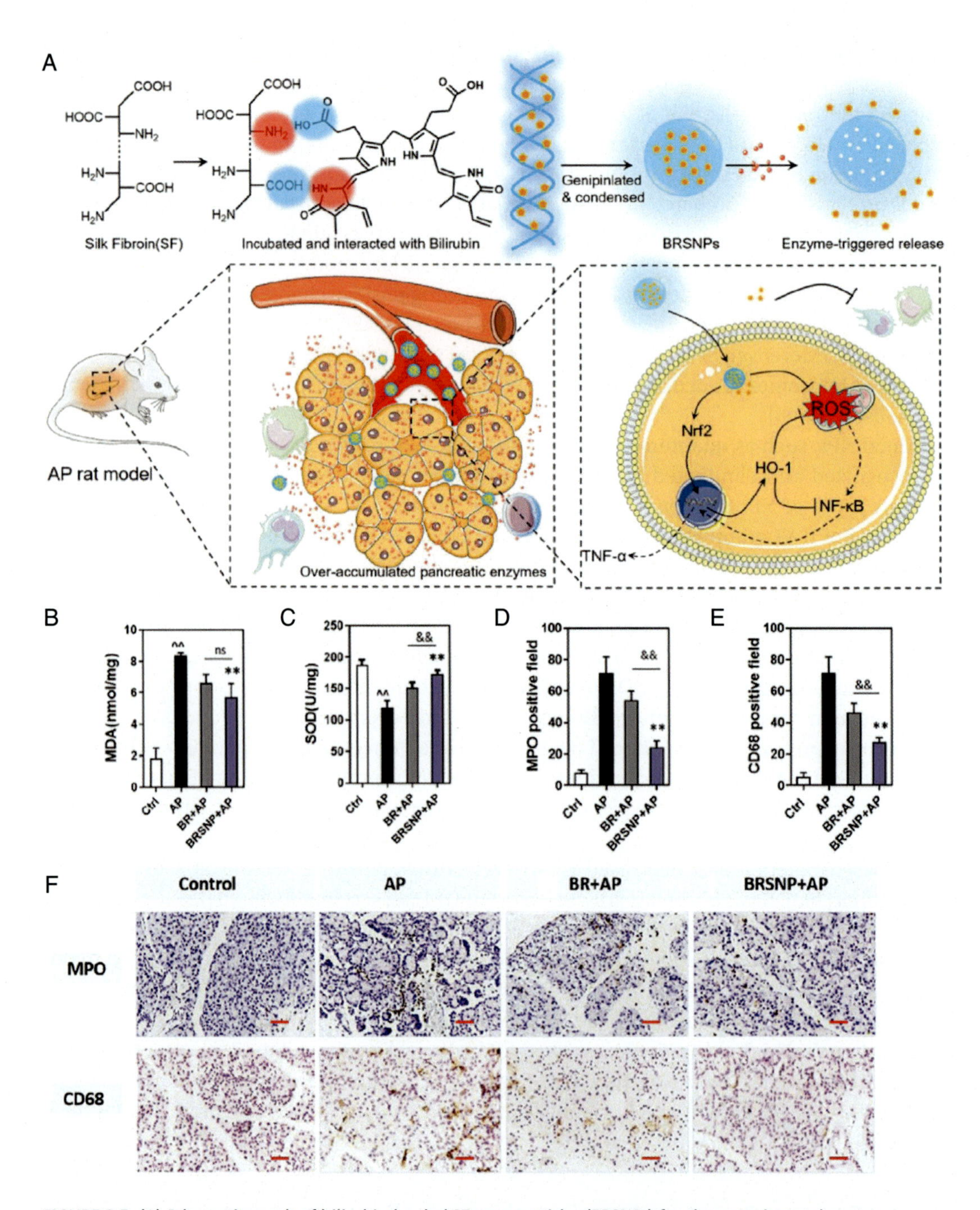

FIGURE 8.5 (A) Schematic graph of bilirubin-loaded SF nanoparticles (BRSNPs) for the experimental acute pancreatitis (AP) application. BRSNP exerted a protective effect in AP by reducing oxidative stress and inflammation via modulation of NF-κB and Nrf2/HO-1 pathway. (B) MDA level of each group rats. (C) SOD level of each group rats. (D) MPO-positive fields' statistical chart of pancreas of each group rats. (E) CD68-positive fields' statistical chart of pancreas of each group rats. (F) The MPO, CD68 immunohistochemical staining of pancreas of each group rats. The scale bar represents 100 μm [77]. MDA, malondialdehyde; SOD, superoxidedismutase; MPO, Myeloperoxidase.

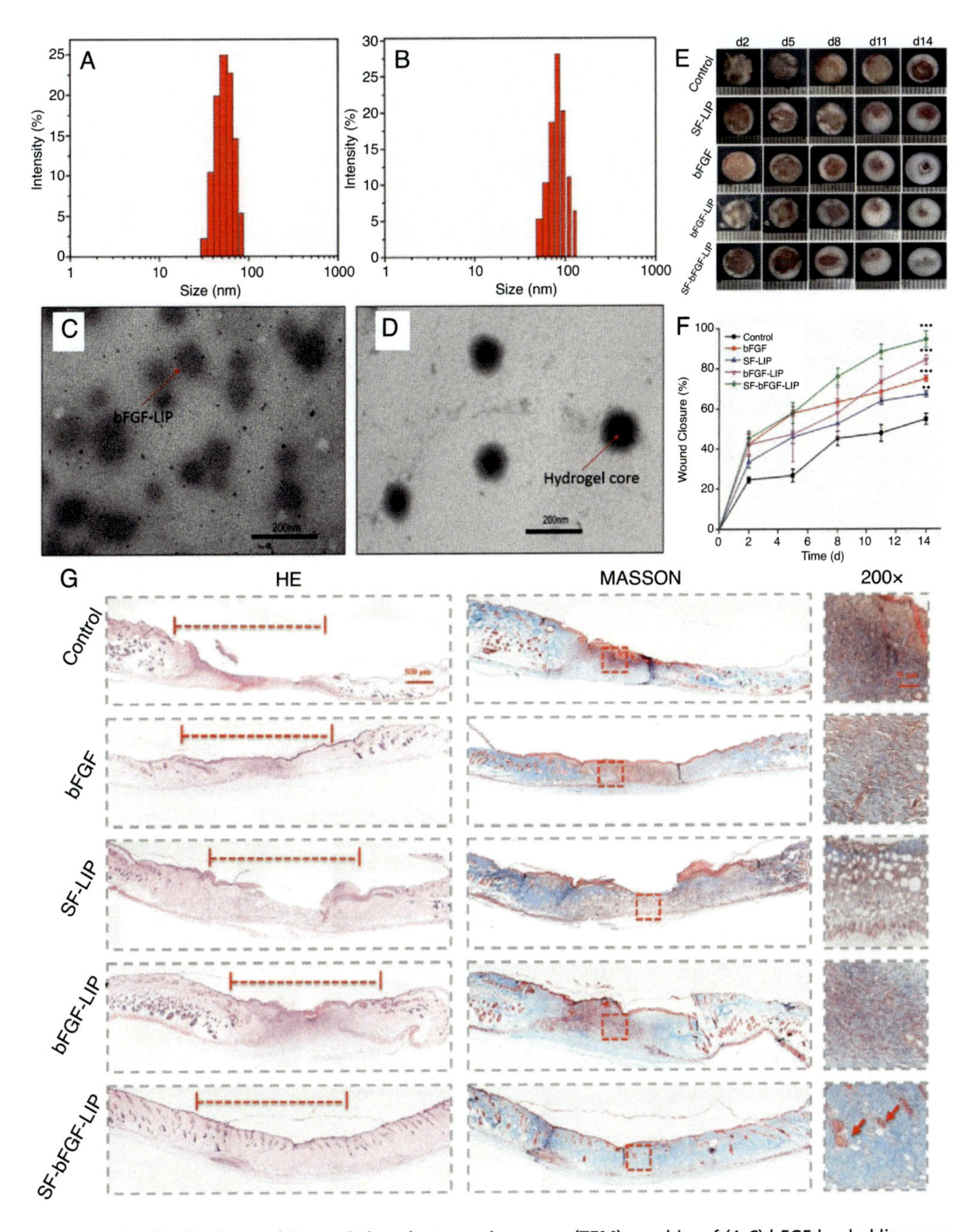

FIGURE 8.6 Size distribution and Transmission electron microscope (TEM) graphics of (A,C) bFGF-loaded liposome (bFGF-LIP) and (B,D) bFGF-loaded liposome with SF hydrogel core (SF-bFGF-LIP). Wound closure of SF-bFGF-LIP-treated wound closure in mouse. (E) Sequential photographs of four types of treated wounds on day 2, 5, 8, 11, and 14. The units are millimeter. (F) Wound closure rates for the five types of the treated wounds. (*** $P < 0.001$, ** $P < 0.01$, * $P < 0.05$, compared to control group, $n = 7$; 7 mice in each group, 14 wounds in total.) (G) Graphics of hematoxylin & eosin (HE) (the wound margin was marked by dashed line) and Masson staining of wound after 14 days of treatments. The red arrow represents skin appendage [101].

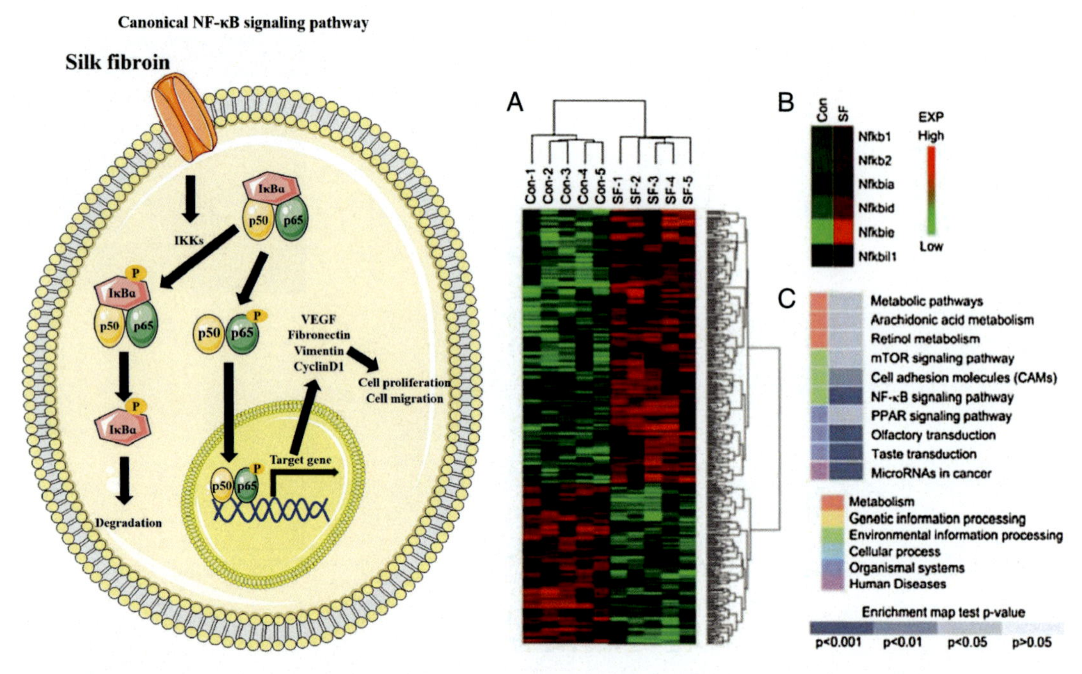

FIGURE 8.7 SF promotes wound healing through NF-κB signaling Screening for the activation of signaling pathways by SF treatment. (A) Heat-map for 368 differently expressed genes ($P<0.01$) between SF-treated cells (SF1-5) and control (Con1-5). The color red or green reflects relative high or low expression levels, respectively, as indicated in the scale bar. (B) The expression levels of NF-κB signaling genes are shown using a heat-map. The expression levels of NF-κB genes were high in SF-treated cells when compared with controls. (C) Pathway enrichment of differentially expressed genes upon treatment with 0.4% SF in the NIH3T3 cells. NF-κB genes expressions were significantly high when compared with controls. All results are representative of five independent experiments [103].

NF-κB signaling pathway, the SF can induce wound healing, which proved that SF itself has certain wound healing ability [103]. It was demonstrated that SF enhanced wound healing effects using NIH3T3 cells by activating the canonical pathway of NF-κB signaling, via modulating its regulated proteins including cyclin D1, vimentin, fibronectin, and vascular endothelial growth factor (VEGF) (Fig. 8.7).

8.5.5 SF hybrid materials

As mentioned earlier, SF has many natural advantages such as biocompatibility, biodegradability, and low immunogenicity as a biological material. However, its properties still have many shortcomings, such as low mechanical strength. For example, in bone tissue regeneration, the material itself needs to provide strong mechanical strength to simulate the bone tissues. Neat SF materials normally are not strong enough for bone repair. In this way, many inorganic materials are added to SF materials to enhance mechanical strength, such as hydroxyapatite, graphene oxide (GO). The chemical structure of hydroxyapatite is mostly similar to inorganic compounds existed in the bone matrix and is mostly used for bone regeneration combining with SF [104].

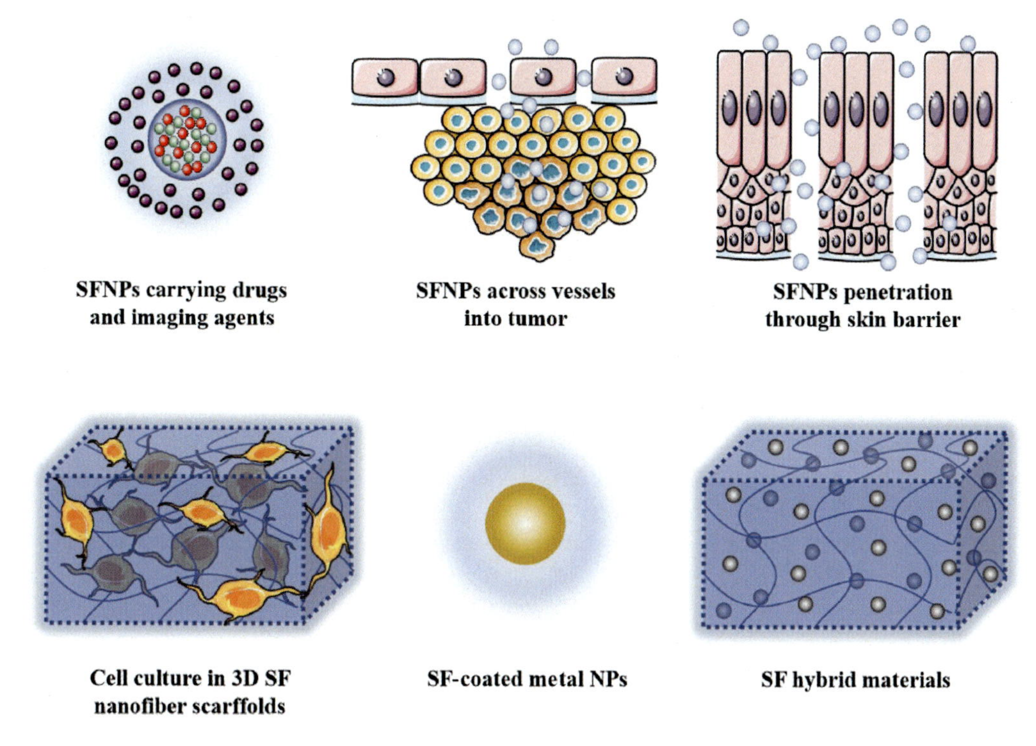

FIGURE 8.8 Summary of SF nanomaterials applications.

However, most of inorganic materials including hydroxyapatite have high weight density and brittle. GO has relative light weight density and can be used as nanofillers for bone regeneration materials. The breaking strength of SF/GO hybrid fibers can be reached at 435.5± 71.6MPa, 72.6% improvement in comparison to pure SF fibers [105] (Fig. 8.8).

8.6 Conclusions

The properties and sizes are both important for designing the available biomaterials for medical application. SF nanomaterials will be particular useful for application as drug delivery and tissue engineering. In this chapter, we first discussed the preparation methods including degumming, desolvation, and purification of regenerated SF. The degumming process is the most important step. The parameters of degumming will affect the quality of the SF peptide chain, which in turn affects the characterization of SF nanomaterials. Moreover, the structure and physicochemical properties SF were also summarized. Based on structure of SF, the SF showed excellent biocompatibility, controllable degradation rate, and ability of encapsulation drugs. Moreover, multiple methods are used for preparation of SF nanomaterials. However, nontoxic methods should be developed in future studies. Finally, the recent studies related to SF-based nanomaterials, including NPs, nanofibers, nanoshells and nanogels, have been applied to drug delivery and tissue engineering. In summary, SF nanomaterials are promising candidates for biomedical applications.

References

[1] T. Adali, M. Uncu, Silk fibroin as a non-thrombogenic biomaterial, Int. J. Biol. Macromol. 90 (2016) 11–19.

[2] J. Melke, S. Midha, S. Ghosh, K. Ito, S. Hofmann, Silk fibroin as biomaterial for bone tissue engineering, Acta Biomater. 31 (2016) 1–16.

[3] B. Kundu, R. Rajkhowa, S.C. Kundu, X. Wang, Silk fibroin biomaterials for tissue regenerations, Adv. Drug Deliv. Rev. 65 (2013) 457–470.

[4] D.G. Harkin, K.A. George, P.W. Madden, I.R. Schwab, D.W. Hutmacher, T.V. Chirila, Silk fibroin in ocular tissue reconstruction, Biomaterials 32 (2011) 2445–2458.

[5] E. Wenk, A.J. Wandrey, H.P. Merkle, L. Meinel, Silk fibroin spheres as a platform for controlled drug delivery, J. Control. Release 132 (2008) 26–34.

[6] M.A. Tomeh, R. Hadianamrei, X. Zhao, Silk fibroin as a functional biomaterial for drug and gene delivery, Pharmaceutics 11 (10) (2019) 494.

[7] A. Golchin, S. Hosseinzadeh, L. Roshangar, The role of nanomaterials in cell delivery systems, Med. Mol. Morphol. 51 (2018) 1–12.

[8] C.T. Ng, G.H. Baeg, L.E. Yu, C.N. Ong, B.H. Bay, Biomedical applications of nanomaterials as therapeutics, Curr. Med. Chem. 25 (2018) 1409–1419.

[9] M. Pirzada, Z. Altintas, Nanomaterials for healthcare biosensing applications, Sensors 19 (23) (2019) 5311.

[10] B.S. Sack, J.R. Mauney, C.R. Estrada Jr., Silk fibroin scaffolds for urologic tissue engineering, Curr. Urol. Rep. 17 (2016) 16.

[11] H. Li, J. Tian, A. Wu, J. Wang, C. Ge, Z. Sun, Self-assembled silk fibroin nanoparticles loaded with binary drugs in the treatment of breast carcinoma, Int. J. Nanomed. 11 (2016) 4373–4380.

[12] F. Danhier, To exploit the tumor microenvironment: since the EPR effect fails in the clinic, what is the future of nanomedicine? J. Control. Release 244 (2016) 108–121.

[13] S.A. MacParland, K.M. Tsoi, B. Ouyang, X.Z. Ma, J. Manuel, A. Fawaz, M.A. Ostrowski, B.A. Alman, A. Zilman, W.C. Chan, I.D. McGilvray, Phenotype determines nanoparticle uptake by human macrophages from liver and blood, ACS Nano 11 (2017) 2428–2443.

[14] D.N. Rockwood, R.C. Preda, T. Yucel, X. Wang, M.L. Lovett, D.L. Kaplan, Materials fabrication from *Bombyx mori* silk fibroin, Nat. Protoc. 6 (2011) 1612–1631.

[15] J. Guan, W. Zhu, B. Liu, K. Yang, F. Vollrath, J. Xu, Comparing the microstructure and mechanical properties of *Bombyx mori* and *Antheraea pernyi* cocoon composites, Acta Biomater. 47 (2017) 60–70.

[16] J. Zhang, J. Kaur, R. Rajkhowa, J.L. Li, X.Y. Liu, X.G. Wang, Mechanical properties and structure of silkworm cocoons: a comparative study of *Bombyx mori, Antheraea assamensis, Antheraea pernyi* and *Antheraea mylitta* silkworm cocoons, Mater. Sci. Eng. C, Mater. Biol. Appl. 33 (2013) 3206–3213.

[17] J. Wang, Y. Chen, G. Zhou, Y. Chen, C. Mao, M. Yang, Polydopamine-coated *Antheraea pernyi* (*A. pernyi*) silk fibroin films promote cell adhesion and wound healing in skin tissue Repair, ACS Appl. Mater. Interfaces 11 (2019) 34736–34743.

[18] M. Yang, Y. Shuai, C. Zhang, Y. Chen, L. Zhu, C. Mao, H. OuYang, Biomimetic nucleation of hydroxyapatite crystals mediated by *Antheraea pernyi* silk sericin promotes osteogenic differentiation of human bone marrow derived mesenchymal stem cells, Biomacromolecules 15 (2014) 1185–1193.

[19] M. Yang, G. Zhou, Y. Shuai, J. Wang, L. Zhu, C. Mao, Ca(2+)-induced self-assembly of *Bombyx mori* silk sericin into a nanofibrous network-like protein matrix for directing controlled nucleation of hydroxylapatite nano-needles, J. Mater. Chem. B 3 (2015) 2455–2462.

[20] S. Du, J. Zhang, W.T. Zhou, Q.X. Li, G.W. Greene, H.J. Zhu, J.L. Li, X.G. Wang, Interactions between fibroin and sericin proteins from *Antheraea pernyi* and *Bombyx mori* silk fibers, J. Coll. Interface Sci. 478 (2016) 316–323.

[21] Y.-Q. Zhang, Applications of natural silk protein sericin in biomaterials, Biotechnol. Adv. 20 (2002) 91–100.

[22] S.D. Aznar-Cervantes, D. Vicente-Cervantes, L. Meseguer-Olmo, J.L. Cenis, A.A. Lozano-Perez, Influence of the protocol used for fibroin extraction on the mechanical properties and fiber sizes of electrospun silk mats, Mater. Sci. Eng. C Mater. Biol. Appl. 33 (2013) 1945–1950.

[23] B. Kundu, N.E. Kurland, V.K. Yadavalli, S.C. Kundu, Isolation and processing of silk proteins for biomedical applications, Int. J. Biol. Macromol. 70 (2014) 70–77.

[24] K. Nultsch, O. Germershaus, Silk fibroin degumming affects scaffold structure and release of macromolecular drugs, Eur. J. Pharma. Sci. 106 (2017) 254–261.

[25] G. Carissimi, A.A. Lozano-Perez, M.G. Montalban, S.D. Aznar-Cervantes, J.L. Cenis, G. Villora, Revealing the influence of the degumming process in the properties of silk fibroin nanoparticles, Polymers 11 (12) (2019) 2045.

[26] H.-Y. Wang, Y.-Q. Zhang, Effect of regeneration of liquid silk fibroin on its structure and characterization, Soft Matter 9 (2013) 138–145.

[27] D. Gupta, A. Agrawal, H. Chaudhary, M. Gulrajani, C. Gupta, Cleaner process for extraction of sericin using infrared, J. Clean. Prod. 52 (2013) 488–494.

[28] K. Haggag, H. El-Sayed, O.G. Allam, Degumming of silk using microwave-assisted treatments, J. Nat. Fibers 4 (2007) 1–22.

[29] H.J. Cho, C.S. Ki, H. Oh, K.H. Lee, I.C. Um, Molecular weight distribution and solution properties of silk fibroins with different dissolution conditions, Int. J. Biol. Macromol. 51 (2012) 336–341.

[30] Z. Zheng, S. Guo, Y. Liu, J. Wu, G. Li, M. Liu, X. Wang, D. Kaplan, Lithium-free processing of silk fibroin, J. Biomater. Appl. 31 (2016) 450–463.

[31] G. Freddi, R. Mossotti, R. Innocenti, Degumming of silk fabric with several proteases, J. Biotechnol. 106 (2003) 101–112.

[32] F. Wang, T.T. Cao, Y.Q. Zhang, Effect of silk protein surfactant on silk degumming and its properties, Mater. Sci. Eng.. C Mater. Biol. Appl. 55 (2015) 131–136.

[33] M.R. Khan, M. Tsukada, Y. Gotoh, H. Morikawa, G. Freddi, H. Shiozaki, Physical properties and dyeability of silk fibers degummed with citric acid, Bioresour. Technol. 101 (2010) 8439–8445.

[34] F. Wang, Y.Q. Zhang, Effects of alkyl polyglycoside (APG) on *Bombyx mori* silk degumming and the mechanical properties of silk fibroin fibre, Mater. Sci. Eng.. C Mater. Biol. Appl. 74 (2017) 152–158.

[35] B. Medronho, A. Filipe, S. Napso, R.L. Khalfin, R.F.P. Pereira, V. de Zea Bermudez, A. Romano, Y. Cohen, Silk fibroin dissolution in tetrabutylammonium hydroxide aqueous solution, Biomacromolecules 20 (2019) 4107–4116.

[36] F. Zhang, X. You, H. Dou, Z. Liu, B. Zuo, X. Zhang, Facile fabrication of robust silk nanofibril films via direct dissolution of silk in CaCl2-formic acid solution, ACS Appl Mater Interfaces 7 (2015) 3352–3361.

[37] Y. Qi, H. Wang, K. Wei, Y. Yang, R.Y. Zheng, I.S. Kim, K.Q. Zhang, A review of structure construction of silk fibroin biomaterials from single structures to multi-level structures, Int. J. Mol. Sci. 18 (2017) 237.

[38] S. Inoue, K. Tanaka, F. Arisaka, S. Kimura, K. Ohtomo, S. Mizuno, Silk fibroin of *Bombyx mori* is secreted, assembling a high molecular mass elementary unit consisting of H-chain, L-chain, and P25, with a 6:6:1 molar ratio, J. Biol. Chem. 275 (2000) 40517–40528.

[39] M.K. DeBari, R.D. Abbott, Microscopic considerations for optimizing silk biomaterials, Wiley interdisciplinary reviews, Nanomed. Nanobiotechnol. 11 (2019) e1534.

[40] Y. Cheng, L.D. Koh, D. Li, B. Ji, M.Y. Han, Y.W. Zhang, On the strength of beta-sheet crystallites of *Bombyx mori* silk fibroin, J. R. Soc. Interface 11 (2014) 20140305.

[41] Y.-Q. Zhang, W.-D. Shen, R.-L. Xiang, L.-J. Zhuge, W.-J. Gao, W.-B. Wang, Formation of silk fibroin nanoparticles in water-miscible organic solvent and their characterization, J. Nanoparticle Res. 9 (2006) 885–900.

[42] X. Wu, J. Hou, M. Li, J. Wang, D.L. Kaplan, S. Lu, Sodium dodecyl sulfate-induced rapid gelation of silk fibroin, Acta Biomater. 8 (2012) 2185–2192.

[43] X. Wang, J.A. Kluge, G.G. Leisk, D.L. Kaplan, Sonication-induced gelation of silk fibroin for cell encapsulation, Biomaterials 29 (2008) 1054–1064.

[44] N. Guziewicz, A. Best, B. Perez-Ramirez, D.L. Kaplan, Lyophilized silk fibroin hydrogels for the sustained local delivery of therapeutic monoclonal antibodies, Biomaterials 32 (2011) 2642–2650.

[45] X. Hu, K. Shmelev, L. Sun, E.S. Gil, S.H. Park, P. Cebe, D.L. Kaplan, Regulation of silk material structure by temperature-controlled water vapor annealing, Biomacromolecules 12 (2011) 1686–1696.

[46] J. Zhong, X. Liu, D. Wei, J. Yan, P. Wang, G. Sun, D. He, Effect of incubation temperature on the self-assembly of regenerated silk fibroin: a study using AFM, Int. J. Biol. Macromol. 76 (2015) 195–202.

[47] L.J. Bray, K.A. George, D.W. Hutmacher, T.V. Chirila, D.G. Harkin, A dual-layer silk fibroin scaffold for reconstructing the human corneal limbus, Biomaterials 33 (2012) 3529–3538.

[48] C. Patra, S. Talukdar, T. Novoyatleva, S.R. Velagala, C. Muhlfeld, B. Kundu, S.C. Kundu, F.B. Engel, Silk protein fibroin from *Antheraea mylitta* for cardiac tissue engineering, Biomaterials 33 (2012) 2673–2680.

[49] L. Xu, S. Wang, X. Sui, Y. Wang, Y. Su, L. Huang, Y. Zhang, Z. Chen, Q. Chen, H. Du, Y. Zhang, L. Yan, Mesenchymal stem cell-seeded regenerated silk fibroin complex matrices for liver regeneration in an animal model of acute liver failure, ACS Appl. Mater. Interfaces 9 (2017) 14716–14723.

[50] M. Zhu, K. Wang, J. Mei, C. Li, J. Zhang, W. Zheng, D. An, N. Xiao, Q. Zhao, D. Kong, L. Wang, Fabrication of highly interconnected porous silk fibroin scaffolds for potential use as vascular grafts, Acta Biomater. 10 (2014) 2014–2023.

[51] S.Q. Chang, Y.D. Dai, B. Kang, W. Han, D. Chen, Gamma-radiation synthesis of silk fibroin coated CdSe quantum dots and their biocompatibility and photostability in living cells, J. Nanosci. Nanotechnol. 9 (2009) 5693–5700.

[52] M. Tan, W. Liu, F. Liu, W. Zhang, H. Gao, J. Cheng, Y. Chen, Z. Wang, Y. Cao, H. Ran, Silk fibroin-coated nanoagents for acidic lysosome targeting by a functional preservation strategy in cancer chemotherapy, Theranostics 9 (2019) 961–973.

[53] Y. Cao, B. Wang, Biodegradation of silk biomaterials, Int. J. Mol. Sci. 10 (2009) 1514–1524.

[54] R.L. Horan, K. Antle, A.L. Collette, Y. Wang, J. Huang, J.E. Moreau, V. Volloch, D.L. Kaplan, G.H. Altman, In vitro degradation of silk fibroin, Biomaterials 26 (2005) 3385–3393.

[55] A.S. Lammel, X. Hu, S.H. Park, D.L. Kaplan, T.R. Scheibel, Controlling silk fibroin particle features for drug delivery, Biomaterials 31 (2010) 4583–4591.

[56] X. Zhao, Z. Chen, Y. Liu, Q. Huang, H. Zhang, W. Ji, J. Ren, J. Li, Y. Zhao, Silk fibroin microparticles with hollow mesoporous silica nanocarriers encapsulation for abdominal wall repair, Adv. Healthc. Mater. 7 (2018) e1801005.

[57] X. Wang, E. Wenk, A. Matsumoto, L. Meinel, C. Li, D.L. Kaplan, Silk microspheres for encapsulation and controlled release, J. Control. Release 117 (2007) 360–370.

[58] S. Gou, Y. Huang, Y. Wan, Y. Ma, X. Zhou, X. Tong, J. Huang, Y. Kang, G. Pan, F. Dai, B. Xiao, Multi-bioresponsive silk fibroin-based nanoparticles with on-demand cytoplasmic drug release capacity for CD44-targeted alleviation of ulcerative colitis, Biomaterials 212 (2019) 39–54.

[59] F.P. Seib, G.T. Jones, J. Rnjak-Kovacina, Y. Lin, D.L. Kaplan, pH-dependent anticancer drug release from silk nanoparticles, Adv. Healthc. Mater. 2 (2013) 1606–1611.

[60] Y. Tian, X. Jiang, X. Chen, Z. Shao, W. Yang, Doxorubicin-loaded magnetic silk fibroin nanoparticles for targeted therapy of multidrug-resistant cancer, Adv. Mater. 26 (2014) 7393–7398.

[61] P. Wu, Q. Liu, R. Li, J. Wang, X. Zhen, G. Yue, H. Wang, F. Cui, F. Wu, M. Yang, X. Qian, L. Yu, X. Jiang, B. Liu, Facile preparation of paclitaxel loaded silk fibroin nanoparticles for enhanced antitumor efficacy by locoregional drug delivery, ACS Appl. Mater. Interfaces 5 (2013) 12638–12645.

[62] M. Xie, D. Fan, Y. Li, X. He, X. Chen, Y. Chen, J. Zhu, G. Xu, X. Wu, P. Lan, Supercritical carbon dioxide-developed silk fibroin nanoplatform for smart colon cancer therapy, Int. J. Nanomed. 12 (2017) 7751–7761.

[63] R. Rasouli, A. Barhoum, M. Bechelany, A. Dufresne, Nanofibers for biomedical and healthcare applications, Macromol. Biosci. 19 (2019) e1800256.

[64] L. Pang, J. Ming, F. Pan, X. Ning, Fabrication of silk fibroin fluorescent nanofibers via electrospinning, Polymers 11 (2019) 986.

[65] F. Zhang, Q. Lu, X. Yue, B. Zuo, M. Qin, F. Li, D.L. Kaplan, X. Zhang, Regeneration of high-quality silk fibroin fiber by wet spinning from $CaCl_2$-formic acid solvent, Acta Biomater. 12 (2015) 139–145.

[66] M.-H. Yang, T.-W. Chung, Y.-S. Lu, Y.-L. Chen, W.-C. Tsai, S.-B. Jong, S.-S. Yuan, P.-C. Liao, P.-C. Lin, Y.-C. Tyan, Activation of the ubiquitin proteasome pathway by silk fibroin modified chitosan nanoparticles in hepatic cancer cells, Int. J. Mol. Sci. 16 (2015) 1657–1676.

[67] Y. Cao, F. Liu, Y. Chen, T. Yu, D. Lou, Y. Guo, P. Li, Z. Wang, H. Ran, Drug release from core-shell PVA/silk fibroin nanoparticles fabricated by one-step electrospraying, Sci. Rep. 7 (2017) 11913.

[68] Y. Dong, P. Dong, D. Huang, L. Mei, Y. Xia, Z. Wang, X. Pan, G. Li, C. Wu, Fabrication and characterization of silk fibroin-coated liposomes for ocular drug delivery,, Eur. J. Pharm. Biopharm. 91 (2015) 82–90.

[69] A.A. Lozano-Perez, A. Rodriguez-Nogales, V. Ortiz-Cullera, F. Algieri, J. Garrido-Mesa, P. Zorrilla, M.E. Rodriguez-Cabezas, N. Garrido-Mesa, M.P. Utrilla, L. De Matteis, J.M. de la Fuente, J.L. Cenis, J. Galvez, Silk fibroin nanoparticles constitute a vector for controlled release of resveratrol in an experimental model of inflammatory bowel disease in rats, Int. J. Nanomed. 9 (2014) 4507–4520.

[70] D.L. ZhuGe, L.F. Wang, R. Chen, X.Z. Li, Z.W. Huang, Q. Yao, B. Chen, Y.Z. Zhao, H.L. Xu, J.D. Yuan, Cross-linked nanoparticles of silk fibroin with proanthocyanidins as a promising vehicle of indocyanine green for photo-thermal therapy of glioma, Artific. cells, Nanomed. Biotechnol. 47 (2019) 4293–4304.

[71] B. Mao, C. Liu, W. Zheng, X. Li, R. Ge, H. Shen, X. Guo, Q. Lian, X. Shen, C. Li, Cyclic cRGDfk peptide and Chlorin e6 functionalized silk fibroin nanoparticles for targeted drug delivery and photodynamic therapy, Biomaterials 161 (2018) 306–320.

[72] T. Wongpinyochit, P. Uhlmann, A.J. Urquhart, F.P. Seib, PEGylated silk nanoparticles for anticancer drug delivery, Biomacromolecules 16 (2015) 3712–3722.

[73] J.D. Totten, T. Wongpinyochit, J. Carrola, I.F. Duarte, F.P. Seib, PEGylation-dependent metabolic rewiring of macrophages with silk fibroin nanoparticles, ACS App.l Mater. Interfaces 11 (2019) 14515–14525.

[74] S. Wang, T. Xu, Y. Yang, Z. Shao, Colloidal stability of silk fibroin nanoparticles coated with cationic polymer for effective drug delivery, ACS Appl. Mater. Interfaces 7 (2015) 21254–21262.

[75] M. Collado-Gonzalez, M.G. Montalban, J. Pena-Garcia, H. Perez-Sanchez, G. Villora, F.G. Diaz Banos, Chitosan as stabilizing agent for negatively charged nanoparticles, Carbohydr. Polym. 161 (2017) 63–70.

[76] W.J. Kim, B.S. Kim, Y.D. Cho, W.J. Yoon, J.H. Baek, K.M. Woo, H.M. Ryoo, Fibroin particle-supported cationic lipid layers for highly efficient intracellular protein delivery, Biomaterials 122 (2017) 154–162.

[77] Q. Yao, X. Jiang, Y.-Y. Zhai, L.-Z. Luo, H.-L. Xu, J. Xiao, L. Kou, Y.-Z. Zhao, Protective effects and mechanisms of bilirubin nanomedicine against acute pancreatitis, J. Control. Release 322 (2020) 312–325.

[78] N.C. Tansil, Y. Li, L.D. Koh, T.C. Peng, K.Y. Win, X.Y. Liu, M.Y. Han, The use of molecular fluorescent markers to monitor absorption and distribution of xenobiotics in a silkworm model, Biomaterials 32 (2011) 9576–9583.

[79] H.L. Xu, D.L. ZhuGe, P.P. Chen, M.Q. Tong, M.T. Lin, X. Jiang, Y.W. Zheng, B. Chen, X.K. Li, Y.Z. Zhao, Silk fibroin nanoparticles dyeing indocyanine green for imaging-guided photo-thermal therapy of glioblastoma, Drug Deliv. 25 (2018) 364–375.

[80] R. Yang, M. Hou, Y. Gao, S. Lu, L. Zhang, Z. Xu, C.M. Li, Y. Kang, P. Xue, Biomineralization-inspired crystallization of manganese oxide on silk fibroin nanoparticles for in vivo MR/fluorescence imaging-assisted tri-modal therapy of cancer, Theranostics 9 (2019) 6314–6333.

[81] B. Ding, M.A. Wahid, Z. Wang, C. Xie, A. Thakkar, S. Prabhu, J. Wang, Triptolide and celastrol loaded silk fibroin nanoparticles show synergistic effect against human pancreatic cancer cells, Nanoscale 9 (2017) 11739–11753.

[82] F. Mottaghitalab, M. Kiani, M. Farokhi, S.C. Kundu, R.L. Reis, M. Gholami, H. Bardania, R. Dinarvand, P. Geramifar, D. Beiki, F. Atyabi, Targeted delivery system based on gemcitabine-loaded silk fibroin nanoparticles for lung cancer therapy, ACS Appl. Mater. Interfaces 9 (2017) 31600–31611.

[83] X. Bian, P. Wu, H. Sha, H. Qian, Q. Wang, L. Cheng, Y. Yang, M. Yang, B. Liu, Anti-EGFR-iRGD recombinant protein conjugated silk fibroin nanoparticles for enhanced tumor targeting and antitumor efficiency, OncoTargets Ther. 9 (2016) 3153–3162.

[84] A.A. Lozano-Perez, H.C. Rivero, M.D.C. Perez Hernandez, A. Pagan, M.G. Montalban, G. Villora, J.L. Cenis, Silk fibroin nanoparticles: efficient vehicles for the natural antioxidant quercetin, Int. J. Pharm. 518 (2017) 11–19.

[85] N. Hassani Besheli, F. Mottaghitalab, M. Eslami, M. Gholami, S.C. Kundu, D.L. Kaplan, M. Farokhi, Sustainable release of vancomycin from silk fibroin nanoparticles for treating severe bone infection in rat tibia osteomyelitis model, ACS Appl. Mater. Interfaces 9 (2017) 5128–5138.

[86] I. Takeuchi, Y. Shimamura, Y. Kakami, T. Kameda, K. Hattori, S. Miura, H. Shirai, M. Okumura, T. Inagi, H. Terada, K. Makino, Transdermal delivery of 40-nm silk fibroin nanoparticles, Coll. Surf. B Biointerfaces 175 (2019) 564–568.

[87] J. Wang, S. Yang, C. Li, Y. Miao, L. Zhu, C. Mao, M. Yang, Nucleation and assembly of silica into protein-based nanocomposites as effective anticancer drug carriers using self-assembled silk protein nanostructures as biotemplates, ACS Appl. Mater. Interfaces 9 (2017) 22259–22267.

[88] K.G. DeFrates, R. Moore, J. Borgesi, G. Lin, T. Mulderig, V. Beachley, X. Hu, Protein-based fiber materials in medicine: a review, Nanomaterials (Basel), 8 (2018).

[89] W. Huang, S. Ling, C. Li, F.G. Omenetto, D.L. Kaplan, Silkworm silk-based materials and devices generated using bio-nanotechnology, Chem. Soc. Rev. 47 (2018) 6486–6504.

[90] S. Arasteh, S. Kazemnejad, S. Khanjani, H. Heidari-Vala, M.M. Akhondi, S. Mobini, Fabrication and characterization of nano-fibrous bilayer composite for skin regeneration application, Methods 99 (2016) 3–12.

[91] S. Selvaraj, N.N. Fathima, Fenugreek incorporated silk fibroin nanofibers-a potential antioxidant scaffold for enhanced wound healing, ACS Appl. Mater. Interfaces 9 (2017) 5916–5926.

[92] S. Bai, H. Han, X. Huang, W. Xu, D.L. Kaplan, H. Zhu, Q. Lu, Silk scaffolds with tunable mechanical capability for cell differentiation, Acta Biomater. 20 (2015) 22–31.

[93] Y.R. Park, H.W. Ju, J.M. Lee, D.K. Kim, O.J. Lee, B.M. Moon, H.J. Park, J.Y. Jeong, Y.K. Yeon, C.H. Park, Three-dimensional electrospun silk-fibroin nanofiber for skin tissue engineering, Int. J. Biol. Macromol. 93 (2016) 1567–1574.

[94] O.J. Lee, H.W. Ju, J.H. Kim, J.M. Lee, C.S. Ki, J.H. Kim, B.M. Moon, H.J. Park, F.A. Sheikh, C.H. Park, Development of artificial dermis using 3D electrospun silk fibroin nanofiber matrix, J. Biomed. Nanotechnol. 10 (2014) 1294–1303.

[95] T. Hodgkinson, X.F. Yuan, A. Bayat, Electrospun silk fibroin fiber diameter influences in vitro dermal fibroblast behavior and promotes healing of ex vivo wound models, J. Tissue Eng. 5 (2014) 1–13.

[96] J. Chen, Y. Zhan, Y. Wang, D. Han, B. Tao, Z. Luo, S. Ma, Q. Wang, X. Li, L. Fan, C. Li, H. Deng, F. Cao, Chitosan/silk fibroin modified nanofibrous patches with mesenchymal stem cells prevent heart remodeling post-myocardial infarction in rats, Acta Biomater. 80 (2018) 154–168.

[97] C. Li, C. Vepari, H.J. Jin, H.J. Kim, D.L. Kaplan, Electrospun silk-BMP-2 scaffolds for bone tissue engineering, Biomaterials 27 (2006) 3115–3124.

[98] Y. Zou, Z. Huang, Y. Wang, X. Liao, G. Yin, J. Gu, Synthesis and cellular compatibility of Co-doped ZnO particles in silk-fibroin peptides, Coll. Surf. B Biointerfaces 102 (2013) 29–36.

[99] M. Deng, Z. Huang, Y. Zou, G. Yin, J. Liu, J. Gu, Fabrication and neuron cytocompatibility of iron oxide nanoparticles coated with silk-fibroin peptides, Coll. Surf. B Biointerfaces 116 (2014) 465–471.

[100] M. Farokhi, F. Mottaghitalab, Y. Fatahi, A. Khademhosseini, D.L. Kaplan, Overview of silk fibroin use in wound dressings, Trends Biotechnol. 36 (2018) 907–922.

[101] H.L. Xu, P.P. Chen, D.L. ZhuGe, Q.Y. Zhu, B.H. Jin, B.X. Shen, J. Xiao, Y.Z. Zhao, Liposomes with silk fibroin hydrogel core to stabilize bFGF and promote the wound healing of mice with deep second-degree scald, Adv. Healthc. Mater. 6 (19) (2017) 1700344.

[102] H.L. Xu, P.P. Chen, L.F. Wang, M.Q. Tong, Z.H. Ou, Y.Z. Zhao, J. Xiao, T.L. Fu, X. Wei, Skin-permeable liposome improved stability and permeability of bFGF against skin of mice with deep second degree scald to promote hair follicle neogenesis through inhibition of scar formation, Coll. Surf. B Biointerfaces 172 (2018) 573–585.

[103] Y.R. Park, M.T. Sultan, H.J. Park, J.M. Lee, H.W. Ju, O.J. Lee, D.J. Lee, D.L. Kaplan, C.H. Park, NF-kappaB signaling is key in the wound healing processes of silk fibroin, Acta Biomater. 67 (2018) 183–195.

[104] M. Farokhi, F. Mottaghitalab, S. Samani, M.A. Shokrgozar, S.C. Kundu, R.L. Reis, Y. Fatahi, D.L. Kaplan, Silk fibroin/hydroxyapatite composites for bone tissue engineering, Biotechnol. Adv. 36 (2018) 68–91.

[105] C. Zhang, Y. Zhang, H. Shao, X. Hu, Hybrid silk fibers dry-spun from regenerated silk fibroin/graphene oxide aqueous solutions, ACS Appl. Mater. Interfaces 8 (2016) 3349–3358.

Lignin-based nanoparticles

Manik Chandra Biswas[a], Debjyoti Banerjee[a], Kowshik Saha[b] and Samin Anjum[c]

[a]FIBER AND POLYMER SCIENCE, WILSON COLLEGE OF TEXTILES, NORTH CAROLINA STATE UNIVERSITY, RALEIGH, UNITED STATES. [b]TEXTILE TECHNOLOGY MANAGEMENT, WILSON COLLEGE OF TEXTILES, NORTH CAROLINA STATE UNIVERSITY, RALEIGH, UNITED STATES. [c]APPLIED CHEMISTRY AND CHEMICAL ENGINEERING, UNIVERSITY OF DHAKA, DHAKA, BANGLADESH.

Chapter outline

9.1 Introduction

Lignin belongs to the class of hydrocarbons where the structure primarily comprises aromatic rings connected together and they are randomly substituted or connected to polar functional groups such as hydroxyl, ether, carboxylic, etc. Lignocellulosic biomass has been known to be

Biopolymeric Nanomaterials: Fundamentals and Applications. DOI: https://doi.org/10.1016/B978-0-12-824364-0.00007-1

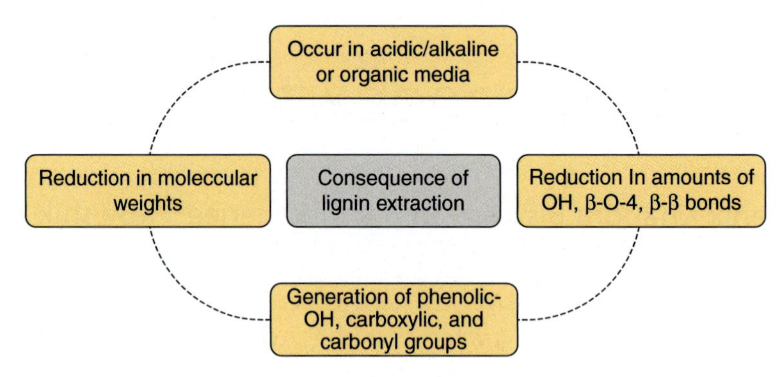

FIGURE 9.1 Some of the important consequences owing to variations in the separation process followed.

one of the most abundant materials that have demonstrated the capability to produce chemicals and biomaterials. Low cost and abundance are the two key factors for the biomass that have made this one of the important materials for research toward making its applicability in our conventional lives. The biomass is composed mainly of cellulose, hemicellulose, and lignin out of which lignin forms the second most abundant component succeeding cellulose. Typically, lignin forms almost about 10%–25% of the lignocellulosic biomass. [1] Although lignin applications are quite understood in present times, nanoparticles derived from lignin form a niche area creating almost a new area of research. One of the key factors that go along post lignin extraction includes the synthesis of lignin nanoparticles that has been adequately demonstrated in the present chapter. Some of the most important applications of lignin nanoparticles lead straight toward environmental and biomedical areas that include tissue engineering/regeneration, antifouling materials, synthetic muscles, etc. Biodegradability, absorption capacity, and nontoxicity of the lignin nanoparticles are some of the key properties that bolster its use in the above-mentioned applications. [2] A thorough discussion about the applications has also been put forward in this chapter to make a complete summary of lignin nanoparticles.

9.2 Sources of technical lignin

Lignocellulosic biomasses act as the source of lignin, extracted using various separation processes and pre-treatment methods. These processes and methods influence the chemical properties of lignin substantially. [3,4] Structure, molecular mass, and chemical composition are some of the factors that are vastly affected owing to various separation processes that are followed. Consequently, the applications of lignin vary according to the chemical nature of the extracted lignin. The pulp mill is the main commercial source of lignin that involves a pulping process to extract lignin. It is the dependence on a different technology that yields lignin with either acidic, alkaline, or some other functionalities. Additionally, lignin undergoes decomposition into smaller molecular mass units [5] also involving reduction and generation of new functional units. Some of the most important variations in lignin extracted using various sources have been demonstrated in Fig. 9.1, and shown lignin monolignols in Fig. 9.2 and Fig. 9.3.

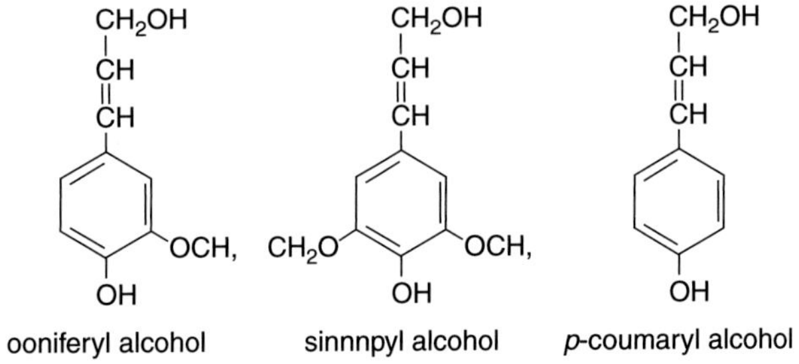

CH_2OH CH CH $OCH,$ OH ooniferyl alcohol

CH_2OH CH CH CH_2O $OCH,$ OH sinnnpyl alcohol

CH_2OH CH CH OH *p*-coumaryl alcohol

FIGURE 9.2 Chemical structures of the monomers used to construct the lignin polymer. These are also called monolignols.

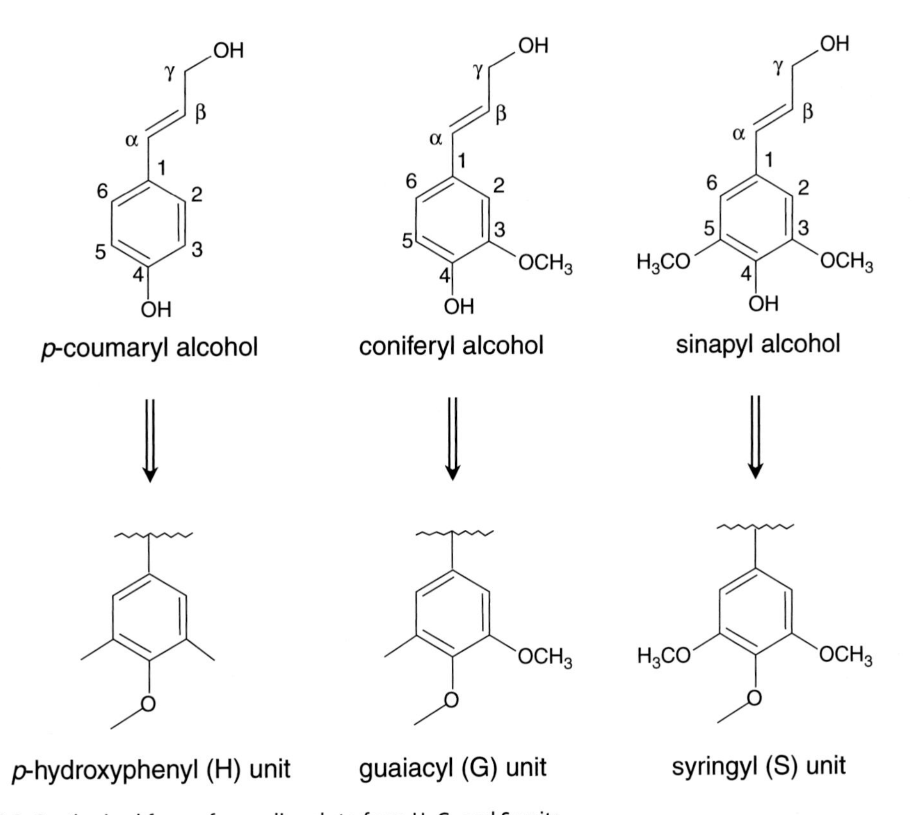

FIGURE 9.3 Synthesized form of monolignols to form H, G, and S units.

9.2.1 Kraft lignin

Kraft lignin is one type of industrial lignin that accounts for almost 85% of the total produced lignin. [6] The Kraft process involves the usage of sodium hydroxide (NaOH) and sodium sulfide (Na_2S) as cooking chemicals called white liquors reagent at a high temperature (140–170 °C). In this Kraft process, almost 90%–95% lignin gets dissolved into the white liquor. The pulping process involved causes deterioration in molecular weight and solubilization in alkaline solution causing it to turn dark brown. The usage of chemicals in this process may cause some generation of impurities in its structure, for instance, sulfur whose presence impedes the effect of catalysis. Kraft lignin contains phenolic -OH groups generated as a result of aryl bonds cleavage during cooking. Quinone and catechol structures may also be formed as a result of an oxidative environment [7]. The fact, that the demand for cellulose is ever-increasing, causes the production of large amounts of Kraft lignin providing us with the opportunity to make its utilization as high-value products.

9.2.2 Lignosulfonate lignin

This specific kind of lignin is obtained from the sulfite pulping process where the main reagent is alkaline earth metal sulfite. The delignification of wood results in the formation of a water-soluble sulfonated system as a result of lignin reacting with the alkaline earth metal sulfide. The reaction causes cleavage of lignin and generation of sulfonated compounds [4]. The lignin thus formed as a result is water-soluble, has higher ash content, and has higher molecular weight compared to kraft lignin [7].

9.2.3 Soda lignin

While the soda lignin process is a sulfur-free process, it is almost identical to the kraft process. Additionally, this process uses vinyl ethers, which marks it different as compared to the kraft process and lignosulfonate process. However, the main reagent still remains to be sodium hydroxide (NaOH). The soda lignin process has been adapted very less compared to other processes because of its low pulp yield [7]. Some of its applications include animal nutrition (due to low sulfur content) [8], production of phenolic resin, dispersants, and polymer synthesis. [7]

9.2.4 Organosolv lignin

Organosolv lignin involves the extraction of lignin from carbohydrates in the lignocellulosic biomasses using organic solvents. The organic solvents dissolve the lignin and the process can be regarded as sulfur-free, has high purity, and chemical reactivity. The structure of lignin is similar to native lignin where they are almost homogenous with low molecular weight and polydispersity. Sometimes, if the process is too severe, the molecular mass reduction of lignin can be almost between 36% and 56%. The lignin thus attained has higher syringeal phenolic-OH content compared to aliphatic-OH content. [9] The nature of this kind of lignin is hydrophobic and to recover this lignin precipitation method is followed [10].

9.2.5 Hydrolysis lignin

Hydrolysis lignin can be mainly classified under two subcategories. This involves enzymatic hydrolysis lignin and acid hydrolysis lignin. Enzymatic hydrolysis lignin shows a higher activity than lignosulfonates and kraft lignin. They can also be used sometimes in the preparation of polymeric materials [11]. Acid hydrolysis lignin involves direct hydrolysis of wood using a dilute acid for the production of ethanol. Immense water consumption and poor recovery rates have somewhat made this pathway out-of-date. Hydrolysis lignin has a high water sorption capacity and it is tedious to make them dewater.

9.2.6 Ionic liquids lignin

Ionic liquids are often referred to here as the organic salts that remain as liquids at a relatively lower temperature. Alkylbenzene sulfonate is one such example that could extract 93% lignin from bagasse [13]. The properties of lignin obtained from this method resemble similar to that of organosolv lignin. This process is not yet available on an industrial scale; however, it shows immense promise.

9.3 Composition, structure, and properties

Lignin is one of the available biopolymers found in nature at higher quantities containing a large portion of carbon [12]. Lignin is a macromolecule having an active site of a distinctive phenyl propane molecule in its structure, which makes it more suitable to be used in the synthesis of biocompatible and biodegradable biopolymer containing nonfossil carbon [13,16,14]. Naturally, lignin shows glue properties remaining in the cellular wall and in intercellular space in plant bodies and provide them with strength [15,16]. Due to the amorphous structure, lignin gives a polymeric arrangement cross-connected with cellulose and hemicellulose by means of phenyl esters, ethers, and hydrogen bond, forming a lignin-carbohydrate composite in the plant cell wall [15–17]. In spite of the high abundance of lignin in nature, a very limited amount of it is being used [18] due to its unstable lignin structure, high polydispersity, and immiscibility with host matrix [18–20].

9.3.1 Composition

Lignin is an unusual biopolymer because of it aromaticity and irregular structure. Due to lack of consistency, defining exact composition of lignin is difficult. Lignin composition contains mostly around 63.4% of carbon, 5.9% of hydrogen, 0.7% of mineral components, and 30% of oxygen [21] corresponding to the formula $(C_{31}H_{34}O_{11})_n$ approximately.

The generally accepted constructing composition of lignin is shaped by means of three fundamental phenylpropanoid monomers which are p-hydroxyphenyl (H), guaiacyl (G), and syringyl (S). These monomers are determined from p-coumaryl, coniferyl, and sinaply alcoholic precursors individually [25]. Ferulates and coumarates are also noticeable in lignin structure [22]. Diverse biomass has a distinctive composition that is dicotyledonous angiosperms could be a blend of G with a really small H and monocotyledonous lignin could be a blend of all three [24], while grasses and palm have G and S respectively [25]. Basically, lignin contains little sums of fragmented or altered monolignols [26]. The respective precursors through biosynthesis undergo coupling reactions to form covalently and noncovalently cross-linked phenolic polymers causing a significant difference in their composition and structure in different biomass [23].

9.3.2 Structure

The structure of the three monomers of lignin is as follows.

The complex polymer structure of lignin includes methoxyl, phenolic hydroxyl, and aldehyde functional groups. Lignin is synthesized by polymerization of the units using an enzyme as a media resulting in the individual unit as p-hydroxyphenyl (H), guaiacyl (G), and syringyl (S) from the respective monomers [24]. Amid polymerization, the precursors produce some alpha and beta positions. These interunit linkages play a vital role in forming the irregular aromatic structure of lignin in the cell wall.

9.3.3 Properties

The properties of lignin are the same as other biopolymers such as microbial and fungal activity. The accessibility of lignin as industrial byproducts, biodegradability, and CO_2 neutral nature is higher than the cellulose and hemicellulose. Lignin is also chemical and bioprocess resistant, which provides the other components protection from being enzymatic degradation and strengthens the cell structure [13]. Lignin also has UV absorbing and fire-retarding properties and some other physicochemical properties, which makes it a more promising material for renewable products [25,26].

9.4 Nanomaterials synthesis from lignin

9.4.1 Conventional methods

9.4.1.1 Lignosulfonate process

The lignosulfonate process is also known as the sulfite process that is a well-known method in pulp and paper processing. It is an acid-pulping approach that is used to extract cellulose from the wood pulp. Sulfite and bisulfite are associated with some well-known cations in the respective process. Through sulfur combustion under a controlled atmosphere and reacting with water, sulfuric acid is produced which in a combination of counterion digest pulp as sulfurous acid–producing lignin polymer. pH can be changed from 1.5 to 5 and the temperature is maintained from 120 to 180 °C. The temperature was kept in the range for about 5 h [27]. Lignin polymers contain sulfur and carbohydrate fraction as impurities via this process.

9.4.1.2 Kraft (sulfate) process

The Kraft process is principally used to produce lignocellulose and produced around 90% of the total production of the world. This method is carried out at a high temperature and high pH. Sodium hydroxide and sulfide are used in this process of treating lignin polymer. The temperature used is 175 °C for 2 h for breaking the lignin-carbohydrate bond and to produce less-sulfur-containing lignin polymer. The Kraft process is more suitable than the sulfite process because of the low sulfur presence and less carbohydrate impurity content [28].

9.4.1.3 Alkaline process

The following process is the most suitable one. It has high efficiency and production capacity of lignin and cellulose. Less woody raw materials are used in this method. The raw materials are cooked at about 200 °C in a sodium hydroxide medium. Saponification takes place which separates lignin from cellulose gradually. This process does not include any use of additional additives, which makes it more preferable than other existing methods [29].

9.4.1.4 Organosolv process

The organic solvent process follows the basic treatment of biomaterials using a solution containing water and organic solvent. Sometimes additives are added. Then heat is applied to the mixture and a fine, high enrich lignin polymer is obtained without the presence of any impurities. Lignins obtained are used in low molecular products. Although pure lignin can be obtained from this method, the production cost is very high. The major drawback of this processing method is that due to the condensation reaction complex lignin structure formed [30,31].

9.4.2 Green methods

Green methods apply renewable sources to produce lignin nanoparticles that show the little corrosive impact on nature and are cost-effective, safer, and simpler methods. Renewable sources produced nanoparticles that are ecofriendly with many sustainable potentials in various application. Lignin nanoparticles can be prepared by the saturation process of carbon-dioxide, flash precipitation, sonication, precipitation with lowering pH with acid, and many other processes [32–34]. Nanoparticles obtained from applying green methods provide us sustainable polymers that can be used in developing new technology while reducing less contamination of our existing ecosystem. These green methods can also be used in breaking down the micromolecules into nanoparticles just by using organic solvents [32].

9.5 Lignin valorization

9.5.1 Lignin-based hydrogels

Lignin-based hydrogels have several applications in delivering drugs for agricultural and medicinal purposes, remediation of water, and sensor because the lignin is biocompatible and biodegradable [35,36]. Lignin-based hydrogels have applications in absorbing the inorganic as well as organic ions. In addition, lignin-based hydrogels have applications for controlled release [36]. As the lignin is less toxic, environmentally friendly, and biocompatible, it can be used to have both a steady discharge of water and functional materials [36].

Technical lignin and Kraft lignin and lignosulfonate, byproducts of pulp mills, have usage as nature absorbents [37–41]. Lignin-based hydrogels exhibited water retention potential in soils [36]. Hydrogels from lignin exhibited lower shrinkage compared to the lignin-free hydrogels reaching the equilibrium swelling ratio [36]. For instance, the hydrogels exhibited better water retention with more cross-linking [42]. Lignin-based hydrogels can overcome the bio-hydrogels' demerits and can be used in the bioelectronic, biomedical, and biocatalytic areas.

Additionally, microcrystalline cellulose has applications in the controlled release of metronidazole and lysozyme. The presence of lignin in the cellulose hydrogel affected the controlled drug release [43]. Lignin hydrogel and silver nanoparticles provided decent antibacterial properties to *Staphylococcus aureus* and *Escherichia coli* [44].

Stimuli-responsive capability response hydrogels are referred to as smart hydrogels [10], such as thermo-response, pH response, and mechanical response hydrogels. As the lignin is soluble in alkaline situations, lignin was innately sensitive to pH. Copolymerization of several lignins, such as kraft, organosolv, has applications in biosensors [36]. In similar ways, lignin-based hydrogel electrodes were made through cross-linking of lignin, silicate, and cellulose [45].

Further, graphene and lignosulfonate hydrogels were combined to make metal-free supercapacitors [46]. The performance of the lignin hydrogels was similar compared to pseudocapacitive supercapacitors. Thus, the prepared hydrogel had more than twice higher conductivity compared to hydrogel from pure graphene [47]. The flexible supercapacitors (FSC) were formed with the assemblage of lignosulfonate/single-walled carbon nanotube hydrogels [47]. And the hydrogels were utilized as electrodes for the FSC formation [47]. More importantly, the formed FSC showed a higher specific capacitance and a good energy density. Lignin also had an application to develop hydrophilic polyether-based polyurethane (HPU) hydrogels by using lignin to fabricate flexible films to make wearable electronics. These materials have potential applications such as sinning fiber, casting, and 3D printing [48].

9.5.2 Lignin nanoparticle and composites

Researchers from various fields such as biomedical engineering, skin-care products, pharmaceuticals, and food industries studied lignin nanoparticles' application (LNPs) [49]. Lignin nanoparticles have applications in the preparation of derma products. For instance, hardwood dioxane and alkali lignin were synthesized by solvent exchange and nanoprecipitation for applications as an active UV protection material and antioxidant [50]. Other applications include a UV absorber, cream lotions, sunscreens due to outstanding UV-vis absorption, antioxidant activity, and UV-blocking property [51–53]. Further, softwood kraft lignin was studied to apply drug delivery especially emphasized on the cancer treatment. Alkali lignin, enzymatic hydrolysis lignin, and organosolv lignin were examined for the applications in hydrophobic drug delivery, hydrophilic drug Doxorubicin Hydrochloride (DOX), and oral drug delivery system, respectively [54–56]. Kraft lignin-based lignin nanoparticles (LNP) can be used as an effective drug carrier specifically for chemotherapeutic applications [57]. Several lignin nanoparticles have been developed and synthesized for supercapacitor applications [58], automotive, acrylic glasses, and lenses [59]. LNP is also synthesized to improve PVA's thermal stability, rubber, and the enhancement of the rubber's mechanical properties [60,61].

Chitosan/nanolignin was used to fabricate a green composite, which provided interaction of nanoscale dimension and hydrophilic functionalities in the composite because of the nanolignin. The green composite had high performance, and the composite was environmentally friendly, which provided an excellent adsorptive platform for lowering pollution [62]. Magnetic lignin–based carbon nanoparticles were synthesized for expanding the application of lignin-based nanoparticles [63]. The advance of lignin nanomaterials made them a potential candidate for various industrial and biomedical applications such as biofuels, chemicals and polymers, and drug delivery.

Lignin-based nanomaterials are antioxidants and UV-absorbents; therefore, they can be used to reinforce nanocomposites, drugs, and gene delivery vehicles in biomedical applications [64]. Multifunctional lignin-based hybrid nanoparticles exhibited superfast adsorption capabilities for Pb^{2+} and Cu^{2+} [65]. LNPs have the potential application for drug delivery applications. More importantly, the superparamagnetic behavior of Fe_3O_4-LNPs made them a possible candidate for cancer therapy and diagnoses, such as magnetic targeting and magnetic resonance imaging [66].

9.5.3 Lignin-based nanoparticles in biomedical applications

Nanomedicine is one of the subclasses of nanotechnology governed by the application of nanomaterials for therapeutics, disease diagnosis, and imaging in biomedical fields. Bioactive agents (e.g., nanoparticles) can be incorporated or encapsulated into the ligand-based nanocarriers for targeted tissue engineering, cell proliferation, and/or diagnosis.

9.5.3.1 Lignin in drug delivery

Tortora et al. developed oil-filled lignin microcapsules from softwood kraft lignin for biomedical applications [67]. These capsules have a diameter of 0.3–1.1 μm and showed prominent results in targeted drug delivery application. Nanocapsules can also be fabricated through a metal-phenolic network where ferric ions (Fe^{3+}) act as the stabilizer for lignin particles. Another promising application of lignin-silver-based nanocapsules is in antimicrobial applications (Fig. 9.4).

9.5.3.2 Lignin in wound healing

The promising fabrication of new nanoparticles with the help of the development of nanotechnology resulted in much attention in their application in wound healing fields. The human skin acts as an excellent protective barrier from an external harsh environment and unexpected injury. As a consequence, fast and irritation-free wound healing is essential and great in demand after any injury. Curcumin (diferuloylmethane) showed prominent performances among various nanomaterials because of having excellent wound-healing capability, nontoxicity, antioxidant and anti-inflammatory properties [2]. But poor solubility and stability of curcumin result in limited skin permeability and bioavailability, and restrict their wide-scale wound healing application. LNPs-loaded curcumin showing prominent wound healing properties suggested the hybrid nanocomposites did not interfere with cell proliferation during the healing process. The nanocomposites exhibited promising *in-vitro* antibacterial performance against common wound bacterium, *S. aureus*, and showed excellent cell proliferation, which is the key parameter

Smart Hydrogels	
Active Hydrogels	Thermoresponsive pH-responsive Dual (Thermo and pH) Mechanically responsive
Biomedical applications	Antibacterial Drug release Aroma Pesticide release
Water remediation	Metals Dyes
Supercapacitators and weareble	Supercapacitators Wearable electronic and 3D printing

FIGURE 9.4

for wound healing. Alqahtani et al. observed in *in vivo* wound healing experiment on rats that curcumin/LNPs composites showed prominent wound healing activities in dermal wound closure compared to the untreated control system (Fig. 9.5) [2].

9.6 Challenges and prospects

Lignin has been demonstrated as one of the important inexpensive materials for functional applications and its abundance has outweighed the advantages of many other macromolecular systems. However, the production of nanoparticles of lignin involves various parameters including varying pH, production at room temperature without modifications, high yields, etc. However, to date, not much research progress has been made to produce regularly shaped and higher yield nanoparticles using a greener and simplistic approach. Also, the properties of the nanoparticles are vastly affected by the way in which lignin extraction has been achieved as well as the processing methods adopted to produce nanoparticles. These could lead to the production of nanoparticles with varying surface properties directed toward various applications [68]. A safe and green approach in synthesizing lignin nanoparticles can broaden the target applications of the nanoparticles. For example, a facile synthesis approach of lignin nanoparticles can allow easy drug loading into the nanoparticles. Additionally, the ability of the surface functional groups to be tuned can allow the nanoparticles to be targeted in specific medical applications [69–71]. One of the important challenges also involves poor solubility in water and its 3D macromolecular structure makes it a challenging biopolymer to be exploited. One of the solutions put forward is to use the lignin in the form of colloidal particles where they are stable in a wide range of pH

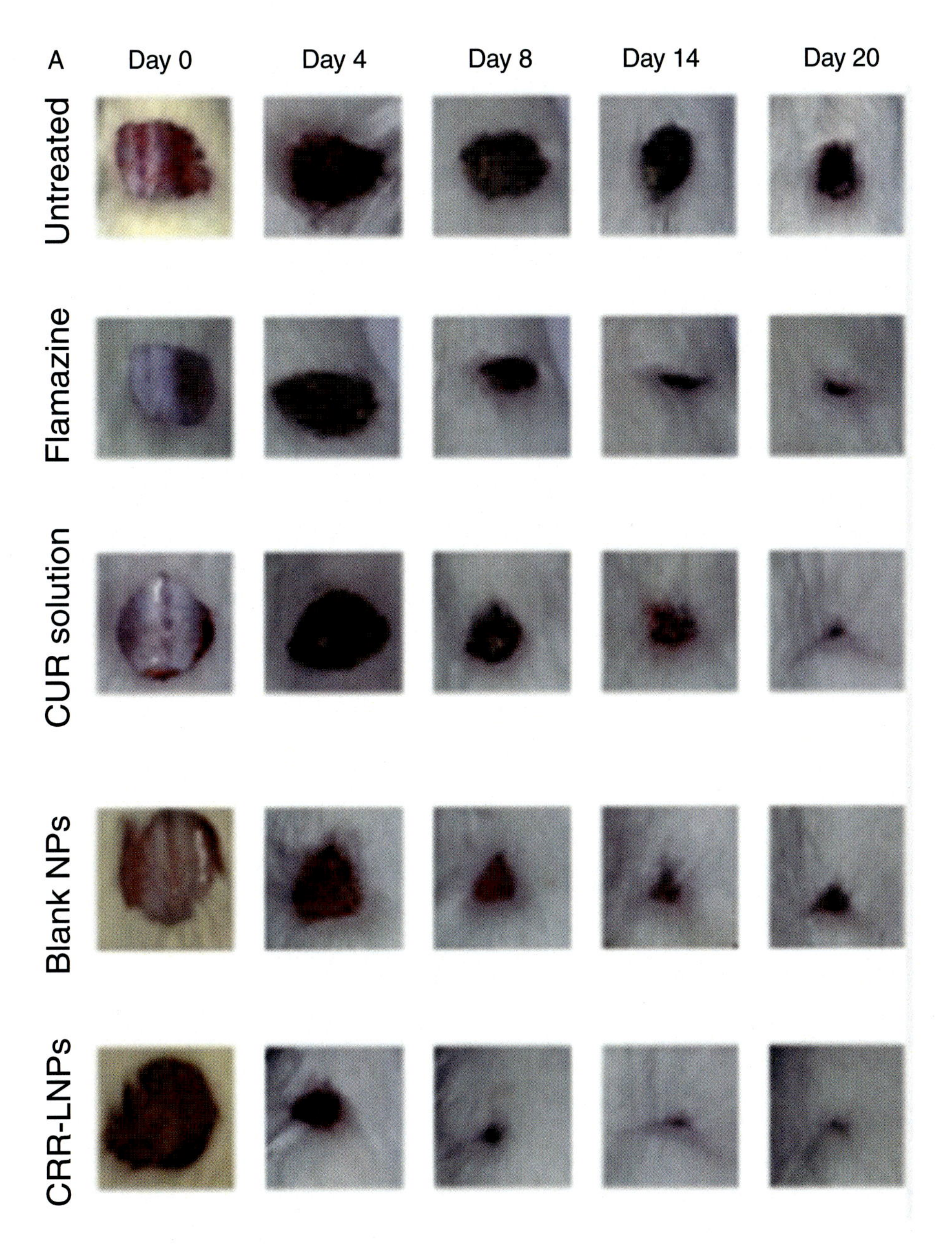

FIGURE 9.5 Curcumin-loaded LNPs accelerate the healing of *in-vivo* excisional wounds. (A) Representative photographs of wounds of the tested groups: Untreated (control), treated with Flamazine, curcumin solution, blank LNPs, and curcumin loaded LNPs. (B) Wound area over time ($n = 6$), (C) Wound contraction measured as a percent of the original area ($n = 6$) data as mean ± SD. *Statistical significance compared with the untreated group. * Significant at $P < 0.05$. [reprinted with permission] [81] .

and can be easily dispersed in organic media. The dispersed colloidal particles with negative surface functionalities can also enable enzymatic and chemical modifications [72]. In addition to all these aspects, one of the important aspects of lignin is the high degree of substitution of lignin in the para positions by an aliphatic chain and ortho positions by methoxy groups [73]. Such substitutions can cause significant steric hindrance which should always be kept in mind. In conclusion, some of the key aspects in fabricating nanoparticles based on lignin, such as adequate handling of solvents, appropriate incorporation of lignin into polymer matrices, and undesirable color of lignin, may pose a hindrance in certain typical applications such as cosmetics, although aesthetically they are one of the most sought after material considering its widespread abundance [74].

9.7 Conclusion

Nature-inspired protective shielding effect of lignin triggered intensive study on lignin-based nanomaterials synthesis and their potential applications. The most viable, sustainable, and eco-friendly processes of delignification is in great demand to synthesize lignin nanomaterials. Xylogenesis is the key parameter to develop sustainable techniques upon understanding the interaction of lignin with other carbohydrates and proteins. The effective synthesis processes of lignin nanomaterials allow biomass utilization, wastes valorization, and remediation.

References

[1] T. Koshijima, T. Watanabe Association Between Lignin and Carbohydrates in Wood and Other Plant Tissues. Springer, Cham (2003).

[2] M.S. Alqahtani, A. Alqahtani, M. Kazi, M.Z. Ahmad, A. Alahmari, M.A. Alsenaidy, R. Syed, Wound-healing potential of curcumin loaded lignin nanoparticles, J. Drug Deliv. Sci. Technol. 60 (2020) 102020. https://doi.org/10.1016/j.jddst.2020.102020.

[3] M.S. Alqahtani, A. Alqahtani, A. Al-Thabit, M. Roni, R. Syed, Novel lignin nanoparticles for oral drug delivery, J. Mater. Chem. B. 7 (2019) 4461–4473. https://doi.org/10.1039/c9tb00594c.

[4] E.D. Bartzoka, H. Lange, K. Thiel, C. Crestini, Coordination complexes and one-step assembly of lignin for versatile nanocapsule engineering, ACS Sustain. Chem. Eng. 4 (2016) 5194–5203. https://doi.org/10.1021/acssuschemeng.6b00904.

[5] M. Baucher, B. Monties, M. Van Montagu, W. Boerjan, Biosynthesis and genetic engineering of lignin, Crit. Rev. Plant Sci. 17 (1998) 125–197. https://doi.org/10.1080/07352689891304203.

[6] B. Baurhoo, C.A. Ruiz-Feria, X. Zhao, Purified lignin: nutritional and health impacts on farm animals-A review, Anim. Feed Sci. Technol. 144 (2008) 175–184. https://doi.org/10.1016/j.anifeedsci.2007.10.016.

[7] W. Boerjan, J. Ralph, M. Baucher, Lignin Biosynthe, sis, Annu. Rev. Plant Biol. 54 (2003) 519–546. https://doi.org/10.1146/annurev.arplant.54.031902.134938.

[8] H.M. Caicedo, L.A. Dempere, W. Vermerris, Template-mediated synthesis and bio-functionalization of flexible lignin-based nanotubes and nanowires, Nanotechnology (2012) 23. https://doi.org/10.1088/0957-4484/23/10/105605.

[9] C.A. Cateto, M.F. Barreiro, A.E. Rodrigues, M.N. Belgacem, Kinetic study of the formation of lignin-based polyurethanes in bulk, React. Funct. Polym. 71 (2011) 863–869. https://doi.org/10.1016/j.reactfunctpolym.2011.05.007.

[10] H. Chen, Lignocellulose Biorefinery Engineering: Principles and Applications, Elsevier Ltd, China, 2015. https://doi.org/10.1016/C2014-0-02702-5.

[11] L. Chen, X. Zhou, Y. Shi, B. Gao, J. Wu, T.B. Kirk, J. Xu, W. Xue, Green synthesis of lignin nanoparticle in aqueous hydrotropic solution toward broadening the window for its processing and application, Chem. Eng. J. 346 (2018) 217–225. https://doi.org/10.1016/j.cej.2018.04.020.

[12] L. Dai, R. Liu, L.Q. Hu, Z.F. Zou, C.L. Si, Lignin nanoparticle as a novel green carrier for the efficient delivery of resveratrol, ACS Sustain. Chem. Eng. 5 (2017) 8241–8249. https://doi.org/10.1021/acssuschemeng.7b01903.

[13] D.P. De Almeida, J.L. Gomide, Anthraquinone and surfactant effect on soda pulping, O Papel. 74 (2013) 53–56. http://www.revistaopapel.org.br/edicoes_impressas/73.pdf.

[14] W.O.S. Doherty, P. Mousavioun, C.M. Fellows, Value-adding to cellulosic ethanol: lignin polymers, Ind. Crops. Prod. 33 (2011) 259–276. https://doi.org/10.1016/j.indcrop.2010.10.022.

[15] E. Dorrestijn, L.J.J. Laarhoven, I. Arends, P. Mulder, J. Anal. Appl. Pyrolysis, (n.d.)

[16] M.Lawoko Duval, A review on lignin-based polymeric, micro- and nano-structured materials, React. Funct. Polym. 85 (2014) 78–96. https://doi.org/10.1016/j.reactfunctpolym.2014.09.017.

[17] H. Erdtman, Lignins: Occurrence, Formation, Structure, and Reactions, Wiley Interscience, New York, USA, 1971.

[18] R.E. Souza, F.J.B. Gomes, E.O. Brito, et al., A review on lignin sources and uses, J. Appl. Biotechnol. Bioeng. 7 (3) (2020) 100–105. https://doi.org/10.15406/jabb.2020.07.00222.

[19] Y. Ge, Z. Li, Application of lignin and its derivatives in adsorption of heavy metal ions in water: a review, ACS Sustain. Chem. Eng. 6 (2018) 7181–7192. https://doi.org/10.1021/acssuschemeng.8b01345.

[20] J.M. Gutiérrez-Hernández, A. Escalante, R.N. Murillo-Vázquez, E. Delgado, F.J. González, G. Toríz, Use of agave tequilana-lignin and zinc oxide nanoparticles for skin photoprotection, J. Photochem. Photobiol. B Biol. 163 (2016) 156–161. https://doi.org/10.1016/j.jphotobiol.2016.08.027.

[21] M.L.Mattinen Henn, Chemo-enzymatically prepared lignin nanoparticles for value-added applications, World J. Microbiol. Biotechnol. 35 (2019). https://doi.org/10.1007/s11274-019-2697-7.

[22] S. Iravani, R.S. Varma, Greener synthesis of lignin nanoparticles and their applications, Green Chem. 22 (2020) 612–636. https://doi.org/10.1039/c9gc02835h.

[23] S. Laurichesse, L. Averous, X. Pan, J. Kadla, K. Ehara, JN.Saddler Gilkes, Organosolv ethanol lignin from hybrid poplar as a radical scavenger: relationship between lignin structure, extraction condition and antioxidant activity, J. Argic. Food Chem. 54 (2000) 5806–5813.

[24] F. Li, X. Wang, R. Sun, A metal-free and flexible supercapacitor based on redox-active lignosulfonate functionalized graphene hydrogels, J. Mater. Chem. A. 5 (2017) 20643–20650. https://doi.org/10.1039/c7ta03789a.

[25] X. Liu, H. Yin, Z. Zhang, B. Diao, J. Li, Functionalization of lignin through ATRP grafting of poly(2-dimethylaminoethyl methacrylate) for gene delivery, Coll. Surf. B. Biointerfaces 125 (2015) 230–237. https://doi.org/10.1016/j.colsurfb.2014.11.018.

[26] Y. Meng, J. Lu, Y. Cheng, Q. Li, H. Wang, Lignin-based hydrogels: a review of preparation, properties, and application, Int. J. Biol. Macromol. 135 (2019) 1006–1019. https://doi.org/10.1016/j.ijbiomac.2019.05.198.

[27] G. Milczarek, M. Nowicki, Carbon nanotubes/kraft lignin composite: characterization and charge storage properties, Mater. Res. Bull. 48 (2013) 4032–4038. https://doi.org/10.1016/j.materresbull.2013.06.022.

[28] H.B.C. Molinari, T.K. Pellny, J. Freeman, P.R. Shewry, R.A.C. Mitchell, Grass cell wall feruloylation: distribution of bound ferulate and candidate gene expression in *Brachypodium distachyon*, Front. Plant. Sci. 4 (2013). https://doi.org/10.3389/fpls.2013.00050.

[29] S.S. Nair, S. Sharma, Y. Pu, Q. Sun, S. Pan, J.Y. Zhu, Y. Deng, A.J. Ragauskas, High shear homogenization of lignin to nanolignin and thermal stability of nanolignin-polyvinyl alcohol blends, ChemSusChem. 7 (2014) 3513–3520. https://doi.org/10.1002/cssc.201402314.

[30] S. Nakagame, R.P. Chandra, J.F. Kadla, J.N. Saddler, The isolation, characterization and effect of lignin isolated from steam pretreated Douglas-fir on the enzymatic hydrolysis of cellulose, Bioresour. Technol. 102 (2011) 4507–4517. https://doi.org/10.1016/j.biortech.2010.12.082.

[31] S. Nethaji, A. Sivasamy, Adsorptive removal of an acid dye by lignocellulosic waste biomass activated carbon: equilibrium and kinetic studies, Chemosphere. 82 (2011) 1367–1372. https://doi.org/10.1016/j.chemosphere.2010.11.080.

[32] B.O. Ogunsile, M.O. Bamgboye, Biosorption of lead (II) onto soda lignin gels extracted from *Nypa fruiticans*, J. Environ. Chem. Eng. 5 (2017) 2708–2717. https://doi.org/10.1016/j.jece.2017.05.016.

[33] F. Oveissi, T. Yen, L. Le, D.F. Fletcher, F. Dehghani, Haldar Duarah, MK. Purkait, Technological advancement in the synthesis and applications of lignin-based nanoparticles derived from agro-industrial waste residues: a review, Int. J. Biol. Macromol. 163 (2018) 1828–1843.

[34] C.P Vance, T.K. Kirk, R.T Sherwood, Lignification as a mechanism of disease resistance, Annu. Rev. Phytopathol. (1980) 259–288. https://doi.org/10.1146/annurev.py.18.090180.001355.

[35] Z. Peng, Y. Zou, S. Xu, W. Zhong, W. Yang, High-performance biomass-based flexible solid-state supercapacitor constructed of pressure-sensitive lignin-based and cellulose hydrogels, ACS Appl. Mater. Interfaces. 10 (2018) 22190–22200. https://doi.org/10.1021/acsami.8b05171.

[36] C.R. Poovaiah, M. Nageswara-Rao, J.R. Soneji, H.L. Baxter, C.N. Stewart, Altered lignin biosynthesis using biotechnology to improve lignocellulosic biofuel feedstocks, Plant. Biotechnol. J. 12 (2014) 1163–1173. https://doi.org/10.1111/pbi.12225.

[37] Y. Qian, X. Zhong, Y. Li, X. Qiu, Fabrication of uniform lignin colloidal spheres for developing natural broad-spectrum sunscreens with high sun protection factor, Ind. Crops. Prod. 101 (2017) 54–60. https://doi.org/10.1016/j.indcrop.2017.03.001.

[38] K. Rajan, J.K. Mann, E. English, D.P. Harper, D.J. Carrier, T.G. Rials, N. Labbé, S.C. Chmely, Sustainable hydrogels based on lignin-methacrylate copolymers with enhanced water retention and tunable material properties, Biomacromolecules. 19 (2018) 2665–2672. https://doi.org/10.1021/acs.biomac.8b00282.

[39] J. Ralph, C. Lapierre, J.M. Marita, H. Kim, F. Lu, R.D. Hatfield, S. Ralph, C. Chapple, R. Franke, M.R. Hemm, J. Van Doorsselaere, R.R. Sederoff, D.M. O'Malley, J.T. Scott, J.J. MacKay, N. Yahiaoui, A.M. Boudet, M. Pean, G. Pilate, L. Jouanin, W. Boerjan, Elucidation of new structures in lignins of CAD- and COMT-deficient plants by NMR, Phytochemistry. 57 (2001) 993–1003. https://doi.org/10.1016/S0031-9422(01)00109-1.

[40] A.P. Richter, J.S. Brown, B. Bharti, A. Wang, S. Gangwal, K. Houck, E.A. Cohen Hubal, V.N. Paunov, S.D. Stoyanov, O.D. Velev, An environmentally benign antimicrobial nanoparticle based on a silver-infused lignin core, Nat. Nanotechnol. 10 (2015) 817–823. https://doi.org/10.1038/nnano.2015.141.

[41] A.P. Richter, B. Bharti, H.B. Armstrong, J.S. Brown, D. Plemmons, V.N. Paunov, S.D. Stoyanov, O.D. Velev, Synthesis and characterization of biodegradable lignin nanoparticles with tunable surface properties, Langmuir. 32 (2016) 6468–6477. https://doi.org/10.1021/acs.langmuir.6b01088.

[42] D. Rico-García, L. Ruiz-Rubio, L. Pérez-Alvarez, S.L. Hernández-Olmos, G.L. Guerrero-Ramírez, J.L. Vilas-Vilela, Lignin-based hydrogels: synthesis and applications, Polymers. 12 (2020). https://doi.org/10.3390/polym12010081.

[43] J. Shigeto, Y. Ueda, S. Sasaki, K. Fujita, Y. Tsutsumi, Enzymatic activities for lignin monomer intermediates highlight the biosynthetic pathway of syringyl monomers in *Robinia pseudoacacia*, J. Plant. Res. 130 (2017) 203–210. https://doi.org/10.1007/s10265-016-0882-4.

[44] L. Siddiqui, J. Bag, D.Mittal Seetha, A. Leekha, H. Mishra, M. Mishra, A.K. Verma, P.K. Mishra, A. Ekielski, Z. Iqbal, S. Talegaonkar, Assessing the potential of lignin nanoparticles as drug carrier: synthesis, cytotoxicity and genotoxicity studies, Int. J. Biol. Macromol. 152 (2020) 786–802. https://doi.org/10.1016/j.ijbiomac.2020.02.311.

[45] S.S.Y. Tan, D.R. MacFarlane, J. Upfal, L.A. Edye, W.O.S. Doherty, A.F. Patti, J.M. Pringle, J.L. Scott, Extraction of lignin from lignocellulose at atmospheric pressure using alkylbenzenesulfonate ionic liquid, Green. Chem. 11 (2009) 339–334. https://doi.org/10.1039/b815310h.

[46] E. Ten, C. Ling, Y. Wang, A. Srivastava, L.A. Dempere, W. Vermerris, Lignin nanotubes as vehicles for gene delivery into human cells, Biomacromolecules. 15 (2014) 327–338. https://doi.org/10.1021/bm401555p.

[47] V.K. Thakur, M.K. Thakur, Recent advances in green hydrogels from lignin: a review, Int. J. Biol. Macromol. 72 (2015) 834–847. https://doi.org/10.1016/j.ijbiomac.2014.09.044.

[48] V.K. Thakur, M.K. Thakur, R.K. Gupta, Review: raw natural fiber-based polymer composites, Int. J. Polym. Anal. Ch. 19 (2014) 256–271. https://doi.org/10.1080/1023666X.2014.880016.

[49] H. Trevisan, C.A. Rezende, Pure, stable and highly antioxidant lignin nanoparticles from elephant grass, Ind. Crop. Prod. (2020) 145. https://doi.org/10.1016/j.indcrop.2020.112105.

[50] M.J.L. Tschan, E. Brulé, P. Haquette, C.M. Thomas, Synthesis of biodegradable polymers from renewable resources, Polym. Chem. 3 (2012) 836–851. https://doi.org/10.1039/c2py00452f.

[50] A.Kraslawski Vishtal, Challenges in industrial applications of technical lignins, BioResources. 6 (2011) 3547–3568. http://www.ncsu.edu/bioresources/BioRes_06/BioRes_06_3_3547_Vishtal_K_Challeng_Indust_Appl_Techn_Lignins_1744.pdf.

[52] G. Vázquez, G. Antorrena, J. González, S. Freire, The influence of pulping conditions on the structure of acetosolv eucalyptus lignins, J. Wood. Chem. Technol. 17 (1997) 147–162. https://doi.org/10.1080/02773819708003124.

[53] D. Watkins, M. Nuruddin, M. Hosur, A. Tcherbi-Narteh, S. Jeelani, Extraction and characterization of lignin from different biomass resources, J. Mater. Res. Technol. 4 (2015) 26–32. https://doi.org/10.1016/j.jmrt.2014.10.009.

[54] W. Yang, M. Rallini, D.Y. Wang, D. Gao, F. Dominici, L. Torre, J.M. Kenny, D. Puglia, Role of lignin nanoparticles in UV resistance, thermal and mechanical performance of PMMA nanocomposites prepared by a combined free-radical graft polymerization/masterbatch procedure, Compos. Pt. A Appl. Sci. Manufact. 107 (2018) 61–69. https://doi.org/10.1016/j.compositesa.2017.12.030.

[55] S.R. Yearla, K. Padmasree, Preparation and characterisation of lignin nanoparticles: evaluation of their potential as antioxidants and UV protectants, J. Exp. Nanosci. 11 (2016) 289–302. https://doi.org/10.1080/17458080.2015.1055842.

[56] G. Yu, B. Li, H. Wang, C. Liu, X. Mu, Preparation of concrete superplasticizer by oxidation- sulfomethylation of sodium lignosulfonate, BioResources. 8 (2013) 1055–1063. https://doi.org/10.15376/biores.8.1.1055-1063.

[57] M. Yáñez-S, B. Matsuhiro, C. Nuñez, S. Pan, C.A. Hubbell, P. Sannigrahi, A.J. Ragauskas, Physicochemical characterization of ethanol organosolv lignin (EOL) from Eucalyptus globulus: effect of extraction conditions on the molecular structure, Polym. Degrad. Stabil. 110 (2014) 184–194. https://doi.org/10.1016/j.polymdegradstab.2014.08.026.

[58] W. Zhao, B. Simmons, S. Singh, A. Ragauskas, G. Cheng, From lignin association to nano-/micro-particle preparation: extracting higher value of lignin, Green. Chem. 18 (2016) 5693–5700. https://doi.org/10.1039/c6gc01813k.

[59] Y. Zhou, Y. Han, G. Li, S. Yang, F. Xiong, F. Chu, Preparation of targeted lignin-based hollow nanoparticles for the delivery of doxorubicin, Nanomaterials. 9 (2019) 188. https://doi.org/10.3390/nano9020188.

[60] S.S. Nair, S. Sharma, Y. Pu, Q. Sun, S. Pan, J.Y. Zhu, Y. Deng, A.J. Ragauskas, High shear homogenization of lignin to nanolignin and thermal stability of nanoligninpolyvinyl alcohol blends, ChemSusChem 7 (12) (2014) 3513–3520.

[61] C. Jiang, H. He, H. Jiang, L. Ma, D. Jia, Nano-lignin filled natural rubber composites: preparation and characterization, Exp. Polym. Lett. 7 (5) (2013).

[62] S. Sohni, R. Hashim, H. Nidaullah, J. Lamaming, O. Sulaiman, Chitosan/nano-lignin based composite as a new sorbent for enhanced removal of dye pollution from aqueous solutions, Int. J. Biol. Macromol. 132 (2019) 1304–1317.

[63] Y.Z. Ma, D.F. Zheng, Z.Y. Mo, R.J. Dong, X.Q. Qiu, Magnetic lignin-based carbon nanoparticles and the adsorption for removal of methyl orange, Coll. Surf A: Physicochem. Eng. Aspects 559 (2018) 226–234 Dec 20.

[64] P. Figueiredo, K. Lintinen, J.T. Hirvonen, M.A. Kostiainen, H.A. Santos, Properties and chemical modifications of lignin: Towards lignin-based nanomaterials for biomedical applications, Progress Mater. Sci. 93 (2018 Apr 1) 233–269.

[65] Y. Zhang, S. Ni, X. Wang, W. Zhang, L. Lagerquist, M. Qin, S. Willför, C. Xu, P. Fatehi, Ultrafast adsorption of heavy metal ions onto functionalized lignin-based hybrid magnetic nanoparticles, Chem. Eng. J. 372 (2019 Sep 15) 82–91.

[66] P. Figueiredo, K. Lintinen, A. Kiriazis, V. Hynninen, Z. Liu, T. Bauleth-Ramos, A. Rahikkala, A. Correia, T. Kohout, B. Sarmento, J. Yli-Kauhaluoma, In vitro evaluation of biodegradable lignin-based nanoparticles for drug delivery and enhanced antiproliferation effect in cancer cells, Biomaterials 121 (2017 Mar 1) 97–108.

[67] M. Tortora, F. Cavalieri, P. Mosesso, F. Ciaffardini, F. Melone, C. Crestini, Ultrasound driven assembly of lignin into microcapsules for storage and delivery of hydrophobic molecules, Biomacromolecules 15 (5) (2014) 1634–1643.

[68] A.P. Richter, B. Bharti, H.B. Armstrong, J.S. Brown, D. Plemmons, V.N. Paunov, S.D. Stoyanov, O.D. Velev, Synthesis and Characterization of Biodegradable Lignin Nanoparticles with Tunable Surface Properties, Langmuir 32 (2016) 6468–6477. doi:10.1021/acs.langmuir.6b01088.

[69] L. Chen, X. Zhou, Y. Shi, B. Gao, J. Wu, T.B. Kirk, J. Xu, W. Xue, Green synthesis of lignin nanoparticle in aqueous hydrotropic solution toward broadening the window for its processing and application, Chem. Eng. J. 346 (2018) 217–225. doi:10.1016/j.cej.2018.04.020.

[70] G. Milczarek, M. Nowicki, Carbon nanotubes/kraft lignin composite: Characterization and charge storage properties, Mater Res. Bull. 48 (2013) 4032–4038. doi:10.1016/j.materresbull.2013.06.022.

[71] L. Dai, R. Liu, L.-Q. Hu, Z.-F. Zou, C.-L. Si, Lignin Nanoparticle as a Novel Green Carrier for the Efficient Delivery of Resveratrol, ACS Sustain. Chem. Eng. 5 (2017) 8241–8249. doi:10.1021/acssuschemeng.7b01903.

[72] A. Henn, M.-L. Mattinen, Chemo-enzymatically prepared lignin nanoparticles for value-added applications, World J. Microbiol. Biotechnol. 35 (2019) 125. doi:10.1007/s11274-019-2697-7.

[73] C.A. Cateto, M.F. Barreiro, A.E. Rodrigues, M.N. Belgacem, Kinetic study of the formation of lignin-based polyurethanes in bulk, Reactive and Functional Polymers. 71 (2011) 863–869. doi:10.1016/j.reactfunctpolym.2011.05.007.

[74] S. Iravani, R.S. Varma, Greener synthesis of lignin nanoparticles and their applications, Green Chem. 22 (2020) 612–636. doi:10.1039/C9GC02835H.

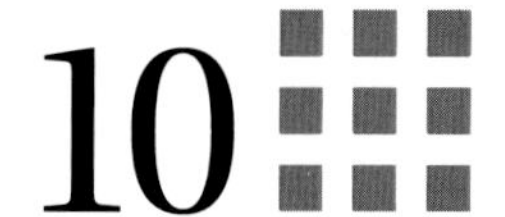

Nano-alginate

Saba Goharshenas Moghadam[a], Hamidreza Parsimehr[a,b] and Ali Ehsani[b]

[a]COLOR AND SURFACE COATINGS GROUP, POLYMER PROCESSING DEPARTMENT, IRAN POLYMER AND PETROCHEMICAL INSTITUTE (IPPI), TEHRAN, IRAN. [b]DEPARTMENT OF CHEMISTRY, FACULTY OF SCIENCE, UNIVERSITY OF QOM, QOM, IRAN.

Chapter outline

Biopolymeric Nanomaterials: Fundamentals and Applications. DOI: https://doi.org/10.1016/B978-0-12-824364-0.00017-4

10.1 Introduction

Renewable biomaterials products [1–4] are swiftly developed because biomaterials have several applications such as biomedical and pharmaceutical applications. The biomaterials were utilized in biomedical applications, such as wound healing or prosthetics, but eventually, there have been replaced by synthetic polymers and ceramics in the previous century. The new trend within the design of biomaterials focuses on the biomimetic approach. The algae species have existed throughout the world. Several aquatic organisms possess antibacterial, antifungal, antiinflammatory, anticancer neuroprotective, analgesic, and immunomodulatory properties. The marine algae especially brown algae species are used for the alginate (ALG) extraction. The marine algae are considered as a major source of natural polysaccharides and various nutrients such as vitamins, sterols, and salts. Alginic acid, also referred as ALG, is a natural anionic carbohydrate and a linear polysaccharide that is illustrated in Fig. 10.1. It is derived from marine algae. The properties of ALG, including biocompatibility, low toxicity, relatively low cost, and simple gelation, make it a perfect biomaterial for various applications, associated with the food, chemical, medical, and agricultural industries. Particularly, the structural similarity of ALG to extracellular matrices of living tissues allows wide applications in wound healing, drug delivery systems, tissue engineering, and cell transplantation [5].

10.2 Chemical structure

ALG contains the irregular blocks of β-D-mannuronic acid (M) and 1-4 linked α-L-guluronic residues (G) which are the water-soluble linear polysaccharide. Its block-like structure is organized in the pattern of homogenous (Poly-G, Poly-M) or heterogeneous (MG) pattern illustrated in Fig. 10.2. The geometries of the G-block regions, M-block regions, and alternating regions are considerably diverse because of the specific profiles of the monomers and their modes of linkage in the polymer. Increasing G blocks in ALG and the molecular weight of the polymer can form stronger or fragile ALG gels. Decreasing pH below the pKa 3.38–3.65 leads to the precipitation of the ALG biopolymer. Ionic strength and gelling ions are the other factors that affect the solubility of the ALG salts.

10.2.1 Conformation

Knowledge of the monomer ring conformations is necessary to understand the polymer properties of ALGs. X-ray diffraction studies of mannuronate- and guluronate-rich ALGs determined

FIGURE 10.1 Alginic acid, derived from marine brown algae, known as *Phaeophyceae*, as well as some microorganisms such as bacterial genera *Pseudomonas* and *Azotobacter*.

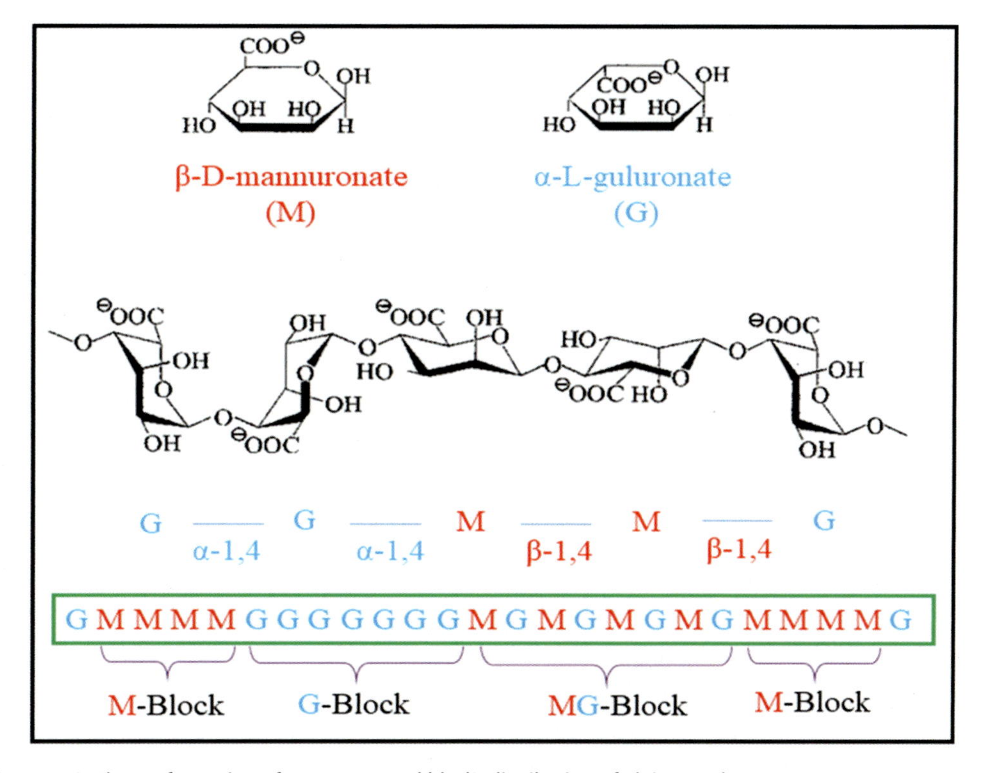

FIGURE 10.2 The conformation of monomers and blocks distribution of alginate salt.

that the guluronate residues in homopolymeric blocks were in the $^1C_4$ conformation while the mannuronate residues had the $^4C_1$ conformation. The viscosity of ALG solutions indicated that the stiffness of the chain blocks increased in the order equatorial-axial (MG) < diequatorial (MM) < diaxial (GG). Hence, ALG contains all four possible glycosidic linkages: MM, GG, MG, and axial-equatorial (GM). The diaxial linkage in G-blocks results in a large, hindered rotation around the glycosidic linkage, which may account for the stiff and extended nature of the ALG chain. Additionally, considering the polyelectrolyte nature of ALG, the electrostatic repulsion between the charged groups on the polymer chain also will increase the chain extension and hence the intrinsic viscosity.

10.3 Sources

ALGs are extracted mainly from three species of brown algae-cell walls including *Ascophyllum nodosum, Laminaria hyperborean,* and *Macrocystis pyrifera,* and also are extracted from several bacteria such as *Azotobacter vinelandii* and *Pseudomonas* spp. The dry weight of ALG in brown algae is more than 40%. Alginic acid is extracted from the algae by using dilute HCl. Then either NaCl or $CaCl_2$ is added to the filtrate extract to perform the fibrous precipitation of sodium or calcium ALG. Finally, sodium ALG powder is obtained after acidic treatment of the precipitate followed by further purification and lyophilization. One of the main properties of sodium ALG is the ability to form hydrogel because of the sodium ions substitution in the guluronic acid residues by different divalent cations (Ca^{2+}, Sr^{2+}, Ba^{2+}, etc.). The resulting alginic acid is converted to hydrophilic sodium salt in the presence of sodium carbonate, which can be easily converted to acid or salt that is illustrated in Fig. 10.3.

10.3.1 Conventional ALGs extraction from brown seaweeds

Generally, the extraction of ALGs is composed of five steps: (1) acidification of the seaweeds, (2) alkaline extraction with sodium carbonate, (3) solid/liquid separation, (4) precipitation, and (5) drying [7]. Algal biomass before extraction of ALGs can be pretreated chemically (e.g., acidification) and physically. First, the biomass used for the ALG extraction must contain more than 80% of a dried matter. Another important parameter is the biomass size. Ground biomass with the size <1 mm and 1–5 mm from *Laminaria digitata* is used for the extraction of ALGs. It was shown that a higher yield was obtained for the biomass with lower particle size which can be explained by its higher surface area. Also, using a low-frequency ultrasound induced seaweed (*Sargassum*) cell wall disruption, causing the improvement of the extraction yield of ALGs [8].

10.3.2 Novel ALGs extraction techniques from brown seaweeds

Nowadays, there is a demand for new, eco-friendly methods that will improve the extraction process (e.g., yield, experimental conditions). Such a technique can be ultrasound-assisted extraction which limits energy consumption by the reduction of the extraction time, as well as the volume of solvents used, thus making it a "greener" process. The extraction yield of ALGs

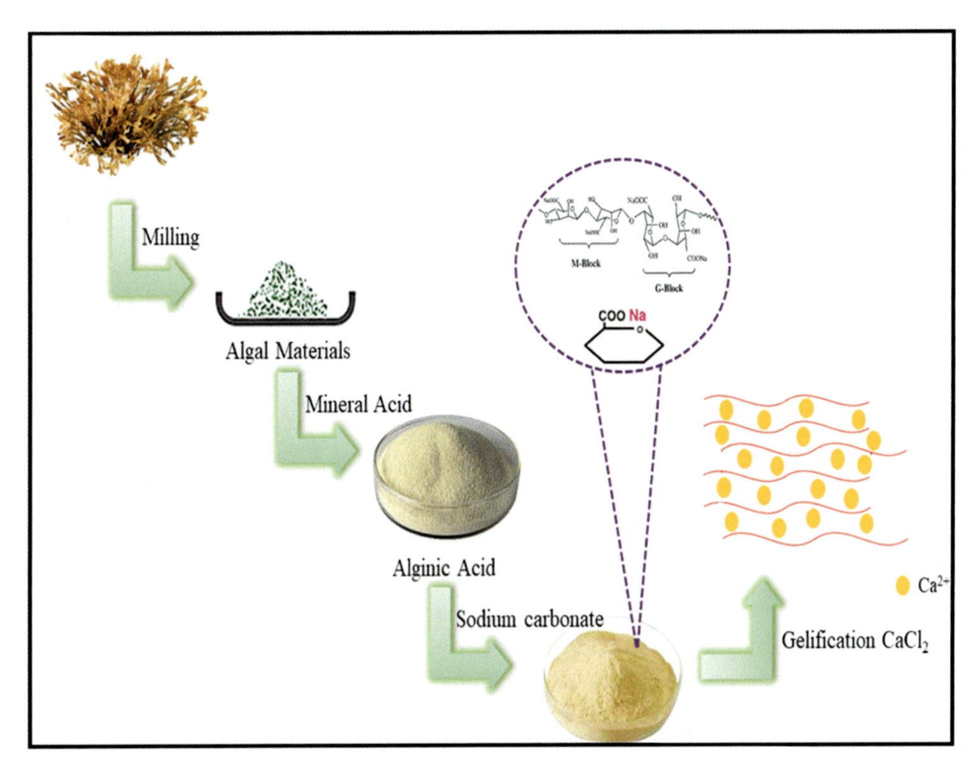

FIGURE 10.3 A typical process for the extrasection of sodium alginate from brown algae is followed by gelification in the presence of $CaCl_2$ [6].

from *Sargassum binderi* and *Turbinaria ornata* depended on algae/water ratio, pH, and the time of exposure to ultrasounds. The highest extraction yield of ALGs equal to 54% was obtained for the following experimental conditions: algae/water ratio of 10 g/L, pH 12, 40-min ultrasound, of 150 W power. Conventional extraction methods permit to obtain 25% ALG yield in 2 h. The time of the extraction of ALGs was significantly reduced from about 2 h for a conventional method to 15–30 min with ultrasound assistance.

10.4 General properties of ALGs

A high degree of physicochemical heterogeneity of ALG influences their quality and determines potential applicability. Industrial grades of ALGs with different molecular weights are available, and also the composition and distribution patterns of M-block and G-block in the chemical structure are responsible for their physicochemical characters such as viscosity, sol-gel transition, and water-uptake ability. The molecular weight of commercial ALG expressed as an average of all the molecules varies between 33,000 and 400,000 g/mol. Alginic acid is insoluble in water and organic solvents, whereas ALG monovalent salts and ALG esters are water-soluble forming stable, viscous solutions. ALG solubility is limited by the solvent pH, ionic strength, gelling ions

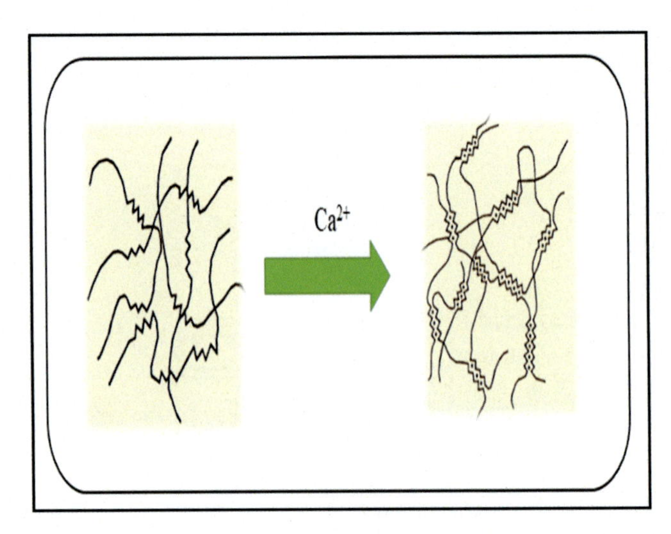

FIGURE 10.4 Schematic illustration of alginates gelation induced by calcium ions in the egg-box structural model.

content [9]. ALG with more heterogeneous components (MG-blocks) is soluble at low pH compared to poly-M or poly-G ALG molecules, at the same conditions. The ALGs are simply formed into diverse semisolid or solid structures under mild conditions due to the ALG unique capability of sol-gel transition. Therefore, ALG possesses high potential applicability as viscosity-increasing agents, thickeners, and suspension and emulsion stabilizers in the food and pharmaceutical industry. ALG gelation can be induced in the presence of divalent ions, which cross-link the polymer chains through the egg-box model [9] or by lowering the pH value below the pKa of ALG monomers using lactones such as d-glucono-δ-lactone. Rapid and uncontrollable ALG gelation occurs with calcium chloride, the main source of Ca^{2+} ions. The gelation rate is a critical parameter in controlling the gelation process [10].

10.5 ALG hydrogels

The hydrogel is the most common form of this natural polymer used in the medical industry. The hydrogels are three-dimensionally cross-linked networks constructed of hydrophilic polymers with a high water content [5]. The hydrogel preparation from ALG depends on the GG block zones. ALG creates gels in the presence of divalent cations, mainly calcium ions that bind GG blocks of aligned ALG chains. This process gives rise to gel-network and commonly called egg-box as shown in Fig. 10.4. ALGs are also found in bioengineering as thermoresponsive and thermo-reversible hydrogels. Thermoreversible ALG hydrogel can form a gel in response to variation of at least two physical parameters at the same time (e.g., pH, temperature, or ionic strength) with the ability to convert to the previous consistency. They can be used as a smart delivery of bioactive agents. Thermoresponsive ALG hydrogel has a potential application in tissue engineering as an injectable cell scaffold [11]. The chemical and physical properties of ALGs have a considerable impact on their molecular characteristic, especially uronic acid ratio (M/G), gelation agent concentration (e.g., calcium ions concentration), molecular weight, degree of polymerization, and the block structure of ALG backbone. Different applications of the alginate nanocomposites have been illustrated in Fig. 10.5.

10.6 ALGs applications

ALGs are one of the most versatile polysaccharides. Their applications span from traditional technical utilization to foods, to biomedicine and biomedical engineering (controlled drug release, cells encapsulation, immunostimulatory agents, scaffolds, etc.) [12], in the pharmaceutical industry (e.g., cosmetic, toothpaste) [13], in agriculture as elicitation of plant growth [14], in textiles (e.g., additives for textile prints), and in medical textiles ALG fibers are used as a wound management material (wound dressing), bandages, pill disintegrators and dental impression material [12], tissue engineering, and ALG scaffolds (e.g., bone, regeneration of cells in soft tissues). ALGs are also known as biosorbents of metal ions such as Cd(II), Pb(II), Cu(II), Cd(II), and Zn(II).

10.7 ALGs biomedical applications

ALGs are increasingly used in the health care industry due to their appropriate properties such as antimicrobial, strong free radical scavenging, antioxidant activity, renoprotective effect, anticancer, and immunostimulatory. Beneficial ALG properties give the possibility to build skin graft and deliver medications in a controlled manner. They can be used in the treatment of patients suffering from diabetes mellitus, liver, and parathyroid disease, as well as in repair and regeneration of tissues, certain cartilages, and organs (e.g., liver) in the case of loss or failure of tissues or organs. It is possible to construct ALG-based scaffolds made for cell growth matrix [5]. Low-cost, wetting and viscoelastic properties make a possibility to use ALG in the dental area, especially as impression materials. Nontoxicity and nonirritating characteristics make them a popular choice to using them in preliminary impressions, for full-arch impressions, partial removable dental prosthesis frameworks, and provisional crown-and-bridge impressions. ALG impressions have good qualities of reproduction surface detail and they are easier to remove compared with elastomeric materials.

10.8 ALG nanoparticle (NP) and nanocomposites

The ALG-based microencapsulation particles have been extensively studied for developing nanomaterials and other functional materials. The hydrogelling ability of ALGs has expanded its uses and research perspectives in biomedicine and multiple disciplines, including the food industry, sewage treatment, and as an adsorptive material for heavy metal removal from contaminated water [5]. ALG polymers are biodegradable, biocompatible mucoadhesive, and hemocompatible. Different types of ALG-based nanomaterials with different sizes, shapes, and compositions have now been synthesized. Several methods such as controlled gelification by Ca^{2+}

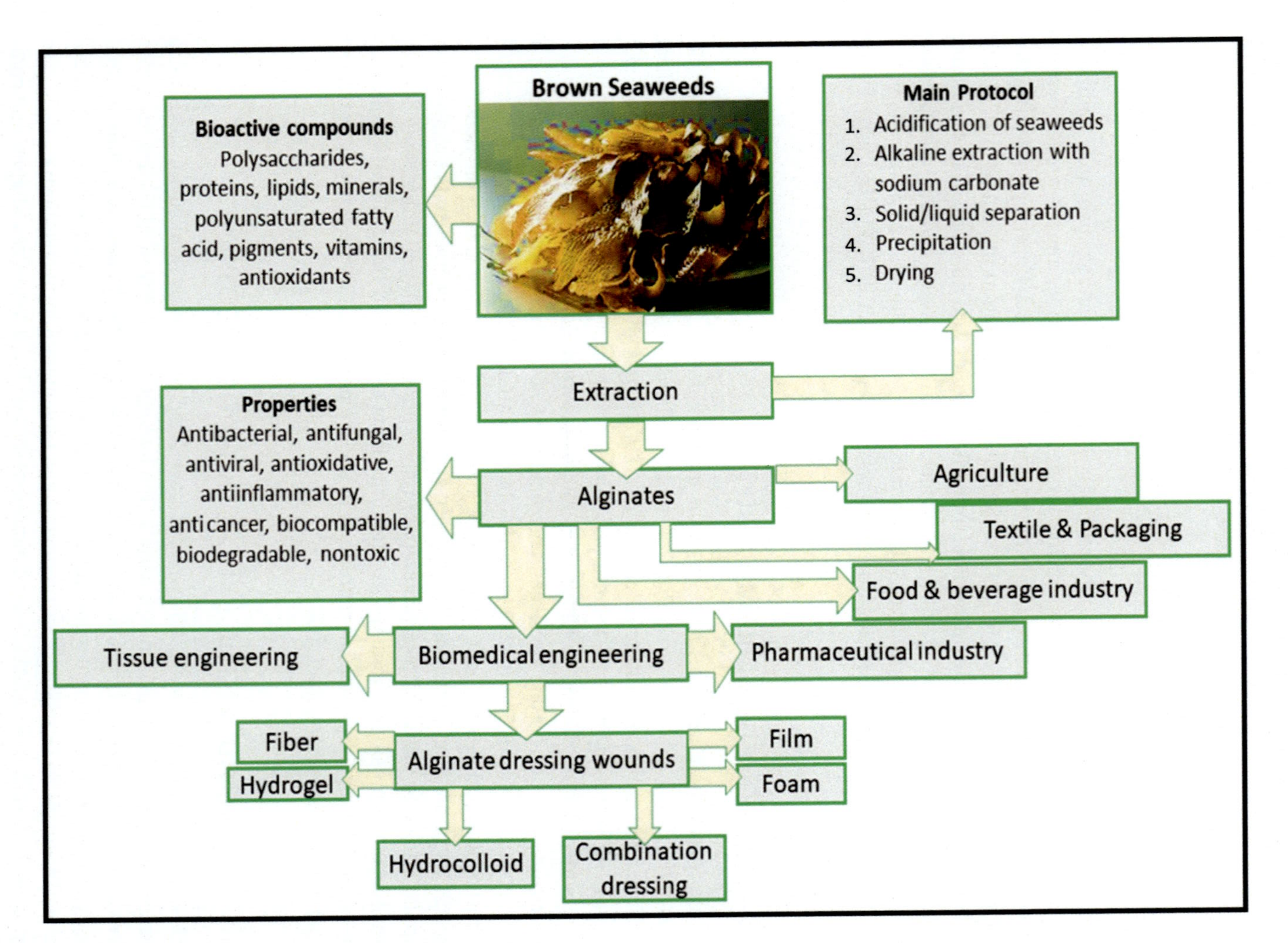

FIGURE 10.5 Alginate processing scheme with potential application.

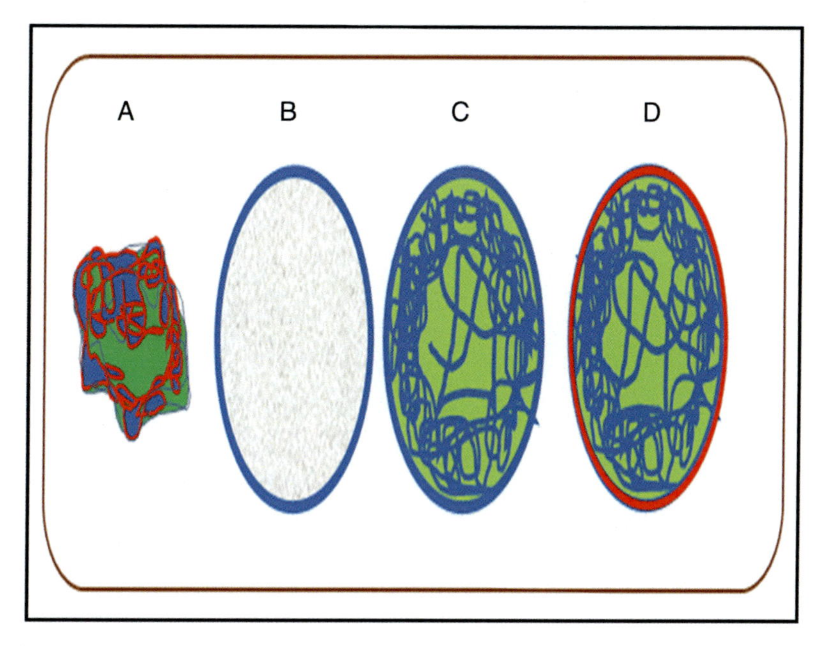

FIGURE 10.6 Schematic representation of (A) nanoaggregates, (B) nanocapsule, (C) nanosphere with the structured interior, and (D) nanocapsule with the structured interior.

ions, ionotropic gelation with positively charged polymers, electrospinning, self-assembly of structurally modified ALG into nanoaggregates, thermally induced phase separation (TIPS), and microfluidics are extensively investigated to prepare ALG-based nanomaterials. The various applications of the ALG-based nanomaterials have extensively been examined.

10.8.1 ALG NPs

Numerous studies have been performed on the development and application of ALG particles. The ALG is one of the most common materials as the hydrogel microparticles. ALG can easily be gelled with multivalent cations under gentle conditions, making it applicable for the entrapment of sensitive materials. Most of the gelled ALG particles have a diameter larger than 100 μm. Small particles of ALG have a higher mechanical strength and a larger specific surface area. The NPs have attracted considerable attention and have several advantages over micron-sized particles especially in drug delivery. Nanoaggregates, nanocapsules, and nanospheres are nano-sized systems with diameters generally ranging from 10 to 1000 nm in size. These systems can hold enzymes, drugs, and other compounds by dissolving or entrapping them in, or attaching them to the particle's matrix that is illustrated in Fig. 10.6. Primarily ALG NP formation is based on the following two methods:

1. *Complexation.* The complex formation occurs in an aqueous solution forming ALG nanoaggregates or on the interface of an oil droplet to prepare ALG nanocapsules. A cross-linker such

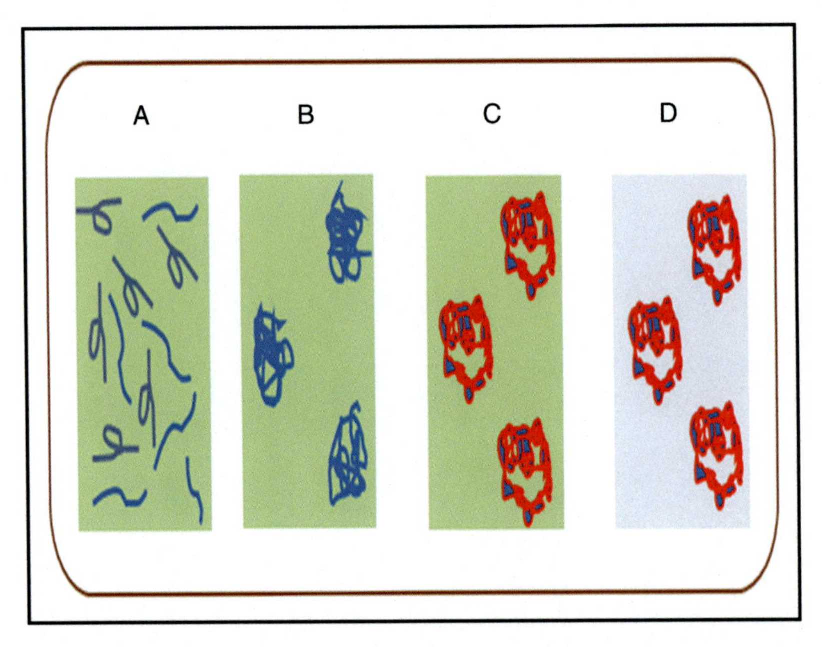

FIGURE 10.7 Schematic representation of the formation of alginate nanoaggregates. (A) Alginate solution, (B) addition of $CaCl_2$ the pregel state is formed, (C) addition of cationic polymer, (D) isolated, washed, and dispersed nanoaggregates.

as calcium chloride is used for the ALG complexation. Complexation can also occur through mixing ALG with an oppositely charged polyelectrolyte such as chitosan [15].

2. *ALG-in-oil emulsification.* It is coupled with external or internal gelation of the ALG emulsion droplet to prepare ALG nanospheres.

10.8.2 ALG nanoaggregate formation by self-assembly and complexation

The ALG nanoaggregates are used as drug carriers with sizes ranging from 250 to 850 nm [16]. The particles were formed in a sodium ALG solution by first adding calcium chloride and then poly-L-lysine. The concentrations of sodium ALG and calcium chloride were lower than those required for typical ALG gel formation. By mixing low concentrations of ALG and calcium chloride a pregel state was formed consisting of nano-sized aggregates dispersed in a water continuous phase (Fig. 10.7B). An aqueous polycationic solution such as poly-L-lysine is added to prepare a polyelectrolyte complex coating of ALG NPs (Fig. 10.7C). As poly-L-lysine is toxic and immunogenic when injected into the human body, chitosan and Eudragit E100 have been used as an alternative cationic polymer. ALG aggregates have also been prepared by solely combining chitosan [15] or calcium ions [17] with a sodium ALG solution. The ALG and cationic polymer concentration, their molecular weight, the calcium chloride concentration, and the order of addition of calcium chloride and cationic polymer to the sodium ALG solution were found to be

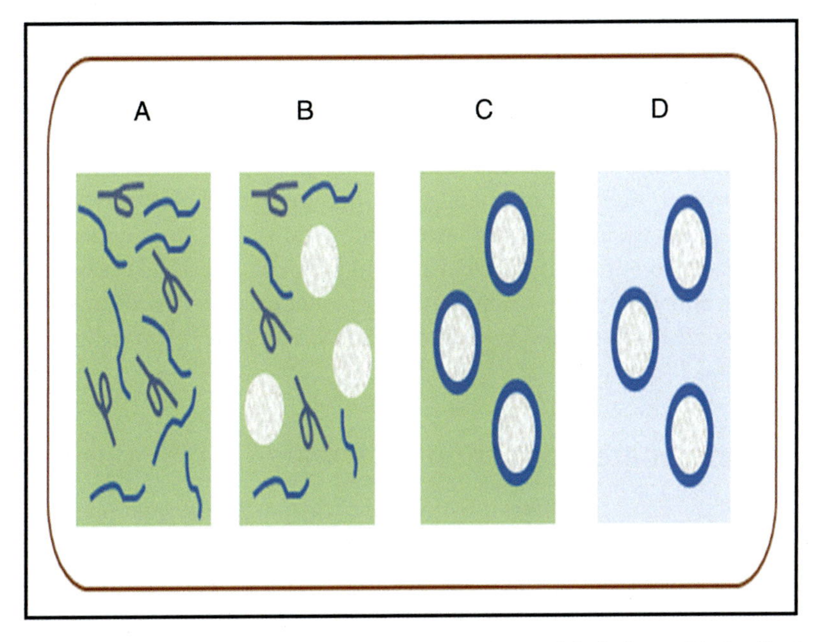

FIGURE 10.8 The alginate nanocapsules formation. (A) Alginate solution, (B) oil phase with the dissolved drug is emulsified in gelation solution, (C) alginate deposits on the interface of the oil droplets after $CaCl_2$ addition, (D) The nanocapsules are isolated, washed, and redispersed after solvent removal.

of great influence on the size and properties of the obtained NPs [16]. Also, ALG nanoaggregates were fabricated through self-assembly that did not require calcium-induced aggregation of ALG or cationic polymers to form a polyelectrolyte complex with ALG. ALG nanoaggregates have successfully been loaded with insulin [18], antisense oligonucleotide [19], and anticancer drugs, such as methotrexate [20] and 5-fluorouracil [21]. The method has been adapted further resulting in ALG nanocomposites with an intracapsular formation of magnetic cobalt silicates that could be obtained by a spray-drying technique [22]. Recently supermagnetic ALG nanoaggregates were prepared for the immobilization of lipase and were subsequently functionalized with oxidic poly(ethylene glycol) [23].

10.8.3 ALG nanocapsules formation on the emulsion droplets interface

One of the most commonly used methods for nanocapsule formation is the polymer deposition on the template droplet interface with subsequent solvent removal. The polymer shell is stabilized by physical or covalent intermolecular cross-linking after the polymer deposition on the droplet interface. ALG has been used to encapsulate various biomaterials such as testosterone [24], acyl derivatives [25], essential oil [26], and *Phyllanthus amarus* [27] in various sizes. First, the drug or components are mixed to be encapsulated with an organic solvent that will act as the capsule interior oil phase. The mixture is slowly added to an aqueous ALG solution containing surfactant. Subsequently, sonication is used to form an oil-in-water (o/w) emulsion (Fig. 10.8B).

Often calcium from a calcium chloride solution is slowly added to the emulsion, and ALG is deposited to form the NP membrane (Fig. 10.8C). Finally, the nanocapsule aqueous suspension is allowed to equilibrate for a certain time before solvent removal (Fig. 10.8D). Chitosan has been used in addition to calcium chloride to form a polyelectrolyte complex between ALG and chitosan that deposits on the interface. It improves capsule stability and reduces the membrane porosity. However, the combined addition of chitosan and calcium chloride results in significantly larger nanocapsules (500–700 nm) compared to NPs prepared without chitosan (150 nm) [28]. The addition order of chitosan and calcium chloride influences the NP size. The addition of chitosan after calcium chloride results in smaller particles. Furthermore, the addition of surfactants and sonication resulted in smaller NPs [28]. Also, the oil phase influences NP size and drug loading [26].

10.8.4 ALG nanospheres formation from water-in-oil emulsions

Emulsion droplets are good templates for the spherical particle formation. The mechanical power used to prepare the emulsion determines the size of the particles. Sonication, membrane emulsification, and self-assembled microemulsions can also be used to form emulsion droplets. Emulsions are relatively easy to produce and emulsion-based NP formation can more easily be scaled up to industrial sizes than nozzle-based methods. ALG-in-oil (w/o) emulsions coupled with internal or external gelation have been used to prepare ALG microspheres and nanospheres. The ALG gels prepared through internal gelation and external gelation have a different homogeneity. The type of gelation used determines the properties and applications of the ALG particles. ALG particles prepared with external gelation have a denser structure with smaller pores at the particle surface due to the gradient concentration of the cations (high concentrations at the surface and low at the core). ALG particles prepared through internal gelation have a higher cation distribution throughout the particle resulting in a higher homogeneity. ALG nanospheres prepared through the emulsification coupled with external gelation are prepared through a two-step procedure. First, an ALG solution containing a certain compound such as drug or enzyme is emulsified in an oil phase to form w/o emulsion. Second a cross-linker often calcium from a calcium chloride solution is added to the emulsion for the ALG emulsion droplet gelation followed by the emulsion demixing. This method has been successfully used to prepare ALG nanospheres containing Green Fluorescent Protein (GFP)-encoding plasmids [29], cytochrome-c as a marker [30], capsaicin as a hydrophobic model drug [31], and doxorubicin with carbon-coated iron NPs to obtain magnetic nanospheres [32]. Oligochitosan has been used as an alternative to calcium chloride to form a complex with low-molecular-weight ALG in ALG NP formation [33].

ALG composite NPs have been prepared though water-in-water emulsification by dispersing an ALG solution under a high shear directly in calcium chloride solution [34]. The resulting particles appear more as random aggregates that are not entirely spherical and have significant surface roughness. ALG nanospheres prepared from a w/o emulsion and demixing with a calcium chloride solution can often result in a clustering of the ALG nanospheres, and multiple emulsion droplets (w/o/w) are formed as a byproduct. Recently a novel method was developed to prepare ALG nanospheres from a w/o emulsion coupled with external gelation. The method does not include a demixing step with a calcium chloride solution but calcium chloride NPs are dispersed in the oil phase for the ALG nanospheres gelation for minimizing the cluster formation and multiple emulsion droplets [35]. The gelled ALG nanospheres are obtained in the oil phase but can be transferred to a water phase if desired. Also, a two-step method for the ALG nanospheres formation through the emulsification coupled with internal gelation has been developed. First, the ALG solution containing an insoluble calcium source such as calcium carbonate is emulsified in an oil phase to form a w/o emulsion. Then an oil-soluble acid is added to the oil phase which lowers the pH inside the ALG particles. It leads to the releasing of Ca^{2+} from the insoluble salt and thereby triggering *in-situ* ALG particle gelation [36].

10.8.5 Interior structured nanocapsules

The porous gels have better transportation of nutrients and oxygen especially in small particles that have a larger specific surface area. The pore size as an important parameter has a direct influence on the performance of the immobilized enzymes and Langerhans islets. The entrapped compounds leach out the ALG particles or the entrapped compounds. These compounds need protection against other compounds such as acids or salts and the ALG gel permeability must be minimized. Then additional shells must be applied to the ALG particles to reduce their permeability. The deposition of additional layers on an ALG nanosphere template results in the nanocapsule formation with a structured interior. It is relatively easy to deposit a cationic polymer on the surface of an ALG particle to form a shell because of the anionic properties of ALG [37]. Poly-L-lysine and chitosan are often used to form a shell on ALG particles.

10.8.6 Controlled gelification using Ca^{2+} ions

Calcium-ALG nanocomposites formation should be done under controlled conditions via the encapsulating different materials such as metal and metal oxide NPs, drugs, and bioanalytical markers. The synthesis of bentonite-supported ZnO NP (20–90 nm) encapsulated ALG nanocomposites via Ca^{2+}-mediated controlled gelification process has been developed. The encapsulation technique follows the incorporation of an ALG solution containing a homogeneous dispersion of bentonite-supported ZnO NPs to a solution of Ca^{2+}. The continuous stirring of the gelation medium (Ca^{2+} solution) ensures uninterrupted bead formation while preventing bead aggregation. Another approach of fabricating nanoshell carrier (250 nm) is the encapsulation of drug-loaded NPs in calcium ALG nanohydrogel shell by w/o microemulsion templates [38].

10.8.7 Polyionic complex formation via ionotropic gelation of intermolecular interactions

ALGs as an anionic polymer have been widely investigated for the nanocomposites production by ionotropic gelation with positively charged polymers. Chitosan ALG polyion nanocomplexes are considered the best choice among the various types of ALG-polycation complexes. Chitosan-ALG complexes are synthesized through the ionic cross-linking between positively charged amino groups of chitosan (pKa = 6.2–7) with negatively charged carboxylate groups of alginic acid. Several studies investigate the efficacy of chitosan-ALG NPs to encapsulate curcumin derivatives such as curcumin diethyl diglutarate [39] and curcumin diglutaric acid by w/o emulsification (˜215 nm) and ionotropic gelification (226 ± 23). Herein the drug is fist dispersed in an ALG solution containing PluronicF-127 with continuous stirring and sonication to achieve the emulsification. Then a $CaCl_2$ solution is drop-wise added to obtain an emulsion with the pregel property (ionic and coordination cross-linking). The resulting emulsion adjusted to pH 4.9 is then combined with chitosan solutions adjusted to pH= 4.6 to optimize the NP synthesis (ionotropic gelation). The curcumin-loaded layer-by-layer assembly of chitosan and sodium ALG platforms (˜200 nm) around citrate coated $MnFe_2O_4$ magnetic NPs (˜12 nm) [40]. The $MnFe_2O_4$ NPs have been synthesized by a thermal decomposition method followed by initial capping with citrate. Layering has been done under monitored conditions at room temperature to prevent the overwhelming complexation of chitosan and ALG in bulk. The NPs precipitate had decanted magnetically suspended in ultrapure water and centrifuged at 10,000 rpm to remove excess biopolymers in the supernatant.

10.8.8 Fabrication of nanocomposite fibrous scaffolds and NPs by electrospinning and electrospraying

Electrohydrodynamic methods including electrospinning and electrospraying lead to the fabrication of fibrous membranes and particles at micron, submicron, and nanoscale levels. Fibrous membranes obtained by electrospinning have a high surface area with an interconnected pore structure. These membranes are desirable for cellular attachment, proliferation, and migration. The electrospinning technique uses an electric field to generate nanofibers from a charged polymeric solution. A jet of polymeric material is sprayed on a surface driven by the application of an electrical potential between a capillary needle and the surface. The electrospun jet undergoes extensive stretching and rapid drying with the evaporation of the solvent. This process leads to the deposition of a porous nonintertwined 3D network of nanofibrils. It could mimic the natural extracellular matrix of animal tissues [41]. Electrospun ALG fiber membranes cannot be fabricated in aqueous solutions because of the low breaking strength of the fibers and brittleness, which limits its applications. An approach to fabricate hydroxyapatite/ALG-based nanocomposite fibrous scaffolds (average diameter of ˜141 nm) for bone tissue regeneration via electrospinning has been developed. A mixture of sodium ALG, polyethylene oxide, and Triton X-100 dissolved in dimethyl sulfoxide has been used to cast the nanocomposite fibers. Sodium phosphate has been used as the precursor ions for the *in-situ* synthesis of hydroxyapatite. The electrospun nanofibers collected at the rotating grounded drum is stirred in a calcium nitrate solution to hydroxyapatite formation while it assists to cross-link ALG [42]. The nanofibrous scaffolds fabrication (diameter range 100–1000 nm) through a nozzle (needle) free electrospinning of a blend of polyvinyl alcohol and ALG has been developed. Nozzle-free electrospinning enables the rapid production of nanofibrous mats. The hydrogen bond formation between polyvinyl alcohol and ALG supposedly improves the thermal stability and mechanical properties of electrospun fiber membranes. The inorganic NPs incorporation such as hydroxyapatite during the fabrication improved mechanical stability, cell adhesion, and proliferation. It stabilizes the electrospun solution preventing agglomeration and precipitation of suspended components in the aqueous spinning solution. Polycaprolactone and ALG coelectrospinning using a dual-jet system for fabricating composite nanofibers (˜691 nm) for *in-situ* transfection applications have been utilized. Polycaprolactone is a biocompatible polymer that can improve cell adhesion. Here DNA is first incorporated with polyethyleneimine which condenses DNA into positively charged particles, facilitating ionic cross-linking with the ALG matrix. Electrospun nanofibers are immersed in absolute ethanol containing 2% $CaCl_2$ to promote the cross-linking of ALG fibers [43].

Electrospraying is a liquid atomization process induced by electrical forces. The difference between the two techniques relies on polymer solution concentration. The fine droplets are formed at a low solution concentration due to jet instability. Consequently, self-dispersing of these highly charged droplets causes particle formation. It prevents droplet agglomeration and coagulation [44].

10.8.9 Nanocomposite fibrous scaffolds fabrication by TIPS

TIPS is a promising technique for fabricating 3D nanofibrous scaffolds. TIPS could be defined as the transformation of a homogenous polymer solution prepared at elevated temperatures. This technique utilizes a low melting point, miscible, and hygroscopic solvent that could dissolve the polymer. The phase separation is achieved with the addition of water followed by rapid mixing to obtain an emulsion. The emulsion is then cast into a mold and rapidly frozen using liquid nitrogen. Then it is freeze-dried to obtain the porous scaffolds [45]. Silk fibroin/sodium ALG composite nanofibrous scaffolds (diameter 50–500 nm) prepared through the TIPS method could be used as a promising extracellular matrix for tissue engineering. The mixture containing degummed and regenerated silk fibroin solution and sodium ALG solution is blended at 60°C and homogeneously mixed with a dioxane/water mixture to obtain the emulsion and then frozen at −80 °C and freeze-dried in 24 h.

10.8.10 ALG NP formation by microfluidics-aided polyelectrolyte complexation

Microfluidic-assisted fabrication can control the size and polydispersity of NPs that is a significant advantage over the conventional methods. A process of devising relatively uniform Na-ALG microdroplets (200 μm) in oil using a microfluidic chip which in turn gelatinized by colliding with a Ca^{2+}-rich phase has been described The particle size could explicitly be controlled by controlling the physical parameters and chemical composition. The generation of ALG-based NPs (380–520 nm) by mixing Ca-ALG pre-gel solution with poly-L-lysine in a microfluidic mixing device has been studied. These NPs enhanced stability against aggregation with control over the particle size by controlling the flow rates. The synthesis of poly lactic-co-glycolic acid/MgO-ALG core-shell microspheres has been described using an o/w emulsion approach. A specially designed microfluidic capillary device with coaxially aligned injection and collection capillaries within a square capillary has been used to produce poly lactic-co-glycolic acid/MgO-ALG core-shell droplets. The polydispersed NPs formed when the preparation process has not controlled. In contrast, produce monodisperse water droplets in oil modernized the fabrication of monodisperse polymer-based NPs through the developments in microfabrication and microreaction technologies [46].

10.9 ALG-based magnetic NPs

Magnetic NPs demonstrate properties such as super-paramagnetism, high saturation field, high field irreversibility, extra anisotropic effects, and exhibit loop shifting after field cooling. It depends on the finite size and the surface properties that control the magnetic behavior of individual NPs [47]. Magnetic ALG NPs have additional benefits over the nonmagnetic ALG nanomaterial. The magnetic property is simply achieved by incorporating ferrimagnetic and ferromagnetic material such as Fe_3O_4, Fe_2O_3, $MnFe_2O_4$, $SrFe_{12}O_{19}$, Fe-C, and alloys such as $SmCo_5$ during the fabrication process [47]. The synthesis of sodium ALG–assisted iron oxide NPs (˜24.5 nm) with ferromagnetic properties via a redox-based hydrothermal process using urea and $FeCl_3.6H_2O$ has been evaluated. Sodium ALG has a dual role in stabilizing and reducing the material. The two major functional groups, carboxylate and hydroxyl, are found to provide coordination sites to Fe^{3+}-forming complexes. The fabrication of a pH-sensitive ALG /maghemite magnetic NPs (˜7.7 nm) by a green synthesis method has adsorption properties on the cationic dye methylene blue. Maghemite obtained by oxidation of magnetite is dispersed in water to obtain a stable magnetic fluid (ferrofluid). Next, ALG solution is gradually introduced to it obtaining ALG-bound maghemite NPs. The use of maghemite instead of magnetite as the magnetic core is much stable. ALG nanomaterials with magnetic properties have revolutionized drug delivery methods due to the magnet-induced drug release capabilities. It localized heat effect upon exposure to an altering magnetic field that could be used in hyperthermia treatment. They can easily remove from treated wastewater in water remediation [48].

10.10 Polymer-based nanomaterials characterization

Several novel techniques are utilized in the polymer-based nanomaterials characterization. The molecular structures mainly characterized by Fourier-Transform Infrared Spectroscopy (FTIR), Nuclear Magnetic Resonance (NMR), and X-ray techniques are also used. The nanomaterial size is analyzed mainly by scanning electron microscopy and photon correlation spectroscopy (dynamic light scattering). Zeta potential is an important parameter that indicates the NP stability in a colloidal dispersion. Dynamic light scattering and zeta potential are the indispensable methods for proper characterization of NPs. Table 10.1 summarizes some of the major NP characterization techniques.

10.11 ALG-based nanomaterials applications

Alginate is one of the biocompatible and biodegradable polymers widely studied for fabricating numerous nanomaterials. This section summarizes a few major applications of ALG-based nanomaterials.

Table 10.1 Commonly used characterization techniques for polymer-based nanomaterials [49].

Technique	Importance
Scanning electron microscopy (SEM)	Shape, size, texture, and aggregation of nanoparticles.
Transmission electron microscopy	Shapes, size, and morphology.
Photon correlation spectroscopy	Average size distribution (granulometry) of the nanoparticles.
Atomic force microscopy	Surface morphology and the size distribution of the nanoparticles.
Scanning electron microscope with energy-dispersive X-ray spectroscopy	Elemental analysis of a selected part of the material while observing through the SEM.
X-ray photoelectron spectroscopy	Analyze the chemical and elemental composition of a surface through measuring the binding energy of electrons associated with atoms.
Differential scanning calorimetry	Thermal behavior characterization (heat capacity as a function of temperature) of polymers and nanoparticles.
Liquid displacement test	Measuring the porosity of the scaffolds.
Superconducting quantum interference device	Determine the magnetic behavior of magnetic nanoparticles.

10.11.1 Drug encapsulation and targeted delivery

Nanoencapsulation is a developing field in drug delivery. Nanoencapsulation increased the bioavailability of drugs and deliver them to the right place with the right dosage. The use of ALG-based nanoencapsulation for controlled delivery of drugs is increased. ALG nanomaterials have shown better ability to withstand compared other particulate carrier systems such as liposomes, microparticles, and some NPs that are rapidly taken up and degraded by mononuclear phagocytes such as the Kupffer cells. ALGs are desirable for encapsulating proteins, enzymes, and some other drugs while enhancing their stability and oral bioavailability. A matrix with an aqueous environment, a higher gel porosity, and biocompatibility are some of the major properties in the majority of ALG NPs [50]. Oral administration of drugs, such as proteins, remains challenging due to their gastric digestibility. Encapsulation in ALG-based NPs can solve this problem due to the indigestible nature and sustained drug release rate (nanoobstruction effect). Delivering a higher drug concentration to a targeted site with minimal toxicity on normal cells is a desirable strategy in cancer chemotherapy. The significance of targeted drug delivery relies on the use of targeting ligands conjugated with NPs and increasing the NP accumulation at the site of tumors.

Drug-entrapped magnetic NPs allow the rapid release of drugs following the application of a magnetic force. Magnetic NPs integrated with targeting ligands are presently used to deliver radionuclide atoms and anticancer drugs to targeted tumors. The heat generated due to oscillations induced by a high-frequency magnetic field could be used to destroy malignant tumors (hyperthermia treatment). It is a bimodal approach against cancer achieving combined chemotherapy and hyperthermia effects [48]. Interparticle interaction in NPs resulting from magnetic field limits magneto-assisted drug delivery and controlled release. Tetra-layered biopolymer (chitosan/ALG) NPs fabricated around $MnFe_2O_4$ NPs via ionotropic gelation have indicated greater thermal stability, without shielding the magnetization while having a decreased interparticle interactions ideal for magneto-assisted drug delivery [40]. Near-infrared laser photothermal therapy is another minimally invasive approach for treating localized cancer. Calcium ALG nanohydrogel encapsulated drug-loaded porous silicon NPs conjugated with gold nanorods (cytocompatible and approved by Food and Drug Administration) found to be sensitive to localized photothermal degradation. These particles too could be utilized in combination therapy [38].

10.11.2 Gene therapy

Gene therapy is a promising approach for cancer treatment and genetic disorders. ALG/$CaCO_3$ hybrid NPs (˜145 nm) have shown higher encapsulation efficiency and delivery capabilities for tumor suppressor gene expression plasmid and doxorubicin hydrochloride (an anticancer drug). Apart from effective delivery into targeted cells, they have shown advantages concerning the biocompatibility and cost over the conventional methods of using viral vectors. Surface-modified ALG NPs have been used as a method of gene transfer to targeted macrophages for antiinflammatory therapy. Tuftsin (a peptide sequence) has been used as targeting ligands that bind with specific receptors on polymorphonuclear leukocytes and macrophages [51].

10.11.3 Regenerative engineering

Temporary scaffolds fabrication capable to provide a biocompatible extracellular matrix is one of the primary necessities in regenerative tissue engineering. Natural polymers possess advantageous characteristics such as hydrophilicity, less-immune resistance, nontoxicity, enhanced cell adhesion, and proliferation which are preferable for tissue engineering. ALG has received much attention as nanocomposite fibrous scaffolds of bone, cartilage, and skin [42] due to its biocompatible properties. ALGs have a structural resemblance to glycosaminoglycans that is a major component of the extracellular matrix in human tissues. The controlled release of Ca^{2+} over a long period is desirable for enhancing osteoblastic activity to stimulate *in-situ* bone regeneration for the bone scaffolds [46]. Hydroxyapatite/ALG nanocomposite fibrous scaffolds are ideal for the adhesion, growth, proliferation, and differentiation of osteoblasts followed by *in-situ* mineralization [52] indicating promising ability to be used in bone tissue engineering. Also, 3D printing is increasingly studied for fabricating tissue scaffolds especially bone tissues with greater control over the size of filaments and pores, complex shapes, and porosity enhancing desirable mechanical properties. ALGs have not participated in the NP formation rather they stabilize hydroxyapatite NPs and provide the hydrogel matrix for cells and initial mechanical strength to the mineralizing tissue. Ca-ALG hydrogel wound dressings are renowned for regulating moisture, antiinflammatory properties, and promoting collagen synthesis. It reduced coagulation time, rapid granulation, reepithelization, and promoted the differentiation of keratinocyte and increased proliferation of fibroblasts mainly due to the controlled release of Ca^{2+}. A cohesive, biodegradable, colloidal gel prepared by using a blend of Poly (Lactic-co-Glycolic Acid) PLGA-chitosan/PLGAALG NPs is desirable for seeding human umbilical cord mesenchymal stem cells. This gel has applications in hematopoietic stem cell transplantation in reconstituted bone tissues [53].

10.11.4 Nanoremediation

Water contamination with industrial effluents is a recent concern requiring an immediate response. The hydrogels composites and NPs have outcompeted conventional adsorbents particularly their adsorption capacity and larger effective surface area. ALGs are extensively investigated for removing metal ions from aqueous solutions. The negatively charged carboxylate groups have a greater affinity/chelating ability toward cations. Encapsulation reduces the agglomeration of NPs by embedding them in the ALG matrix, while increasing the effective surface area, thus increasing their adsorption and catalytic performance. The construction of porous ALG-based beads has shown enhanced adsorption [54].

Properties and potential applications of ALG-based composites depend on the physical and chemical cross-linking methods. Four common methods including ionic cross-linking, emulsification, electrostatic complexation, and self-assembly have been used for the synthesis of the ALG-based composite. Physically cross-linked hydrogels are synthesized by ionic interaction, crystallization, stereocomplex formation, hydrophobized polysaccharides, protein interaction, and hydrogen bond. The chemically cross-linked hydrogels are synthesized by chain-growth polymerization, addition/condensation polymerization, and gamma/electron beam polymerization [55]. The fundamental process involving gel formation is the interaction between sodium ALG and divalent cations (such as calcium ions) or cationic polymers. Sodium ALG has a -COO- group in the molecule. Sodium ALG undergoes a cross-linking reaction when a divalent cation is added to the sodium ALG solution. Na^+ ions from the guluronic acid (G) blocks are exchanged with these divalent cations to form a water-insoluble gel with a characteristic "egg-box" structure. Different cations show a different affinity to ALG. The ability of sodium ALG to bind to multivalent cations follows the following sequence:

$Pb^{2+} > Cu^{2+} > Cd^{2+} > Ba^{2+} > Sr^{2+} > Ca^{2+} > Co^{2+} > Ni^{2+} > Zn^{2+} > Mn^{2+}$ [56].

The solution concentration, pH, and metal ion intensity influenced stability, mechanical strength, shape, and structure of the gel beads in the ionic cross-linking process of sol-gel reaction [57]. Various materials have been incorporated into ALG hydrogel (microspheres) to improve the environmental applications of ALGs. The synthesis of these composites typically starts with mixing the material with sodium ALG solution before the gelation of calcium ALG. The materials encapsulated in ALG for environmental applications include activated carbon (AC), biochar, carbon nanotube (CNT), graphene oxide (GO), NP, magnetic materials, and microorganism. The materials selection to be encapsulated depends on the material functionality and the intended application so that synergetic benefits can be attained by the composite. However, ALG-based composites typically have enhanced physical and mechanical properties for bioengineering applications. First, ALG beards serve as a stable matrix for other types of absorbents that are too fine in particle size and too difficult to separate from aqueous solution. These absorbents are typically carbon based such as AC, biochar, CNTs, and GO [58]. AC has been widely used for wastewater treatment. AC is mostly used as a fine powder that is difficult for separation and regeneration from the effluent, which results in significant loss of the adsorbent. Biochar has been recently used as a low-cost alternative of AC in water/wastewater treatment.

Biochar can be ball milled to increase its surface areas. Like AC powders, ball-milled biochar is difficult to separate from water due to its small particle size [59]. CNTs and GO both have been intensively studied for the removal of organic and inorganic pollutants because of their unique structural features and large specific areas. GO disperses extremely well in water and CNTs are very small and form aggregates. It is difficult to separate them from aqueous solution. Encapsulation of these carbonaceous materials into ALG hydrogels or beads offers ease of separation and regeneration for water/wastewater treatment [59]. Second, fabricating magnetic materials and NPs into ALG brings in nanoeffects and magnetic technology into the composites while attaining excellent absorption performance and reducing the potential environmental risk of NPs. ALG nanomaterial composites enhanced adsorption capacity. A magnetic adsorbent (magsorbent) can be developed by encapsulating magnetic functionalized NPs in ALG beads along with different cross-linking agents [56]. Incorporating maghemite with the ALG in the bead form is very useful in the isolation or recovery process. Magnetic technology has the advantage of simple operation and easy separation. Third, ALG can serve as a microorganism carrier to optimize the microbial processes for environmental and agricultural applications. ALG microorganism composites offer a multitude of advantages such as high biomass, high metabolic activity, and strong resistance to toxic chemicals compared with the conventional suspension system. Moreover, immobilized microorganisms can be used several times without significant loss of activity. Therefore, ALG immobilized microorganism technology has received substantial attention for wastewater treatment [60].

Organic dyes from industrial effluents cause several complications, including cancer, allergy, skin irritation, dermatitis, and mutations. Nanocomposite adsorbents are designed by impregnating nanoscale metallic or bimetallic particles into ALG (or another polymer) based NPs. These absorbents are applied as redox reagents or catalysts for the degradation of environmental contaminants, such as azo dyes, polychlorinated biphenyls, organochlorine pesticides, halogenated aliphatics, nitroaromatics, and halogenated herbicides. These NPs could be integrated with a magnetic core for recovery via a magnetic field. NP immobilization with enzymes such as glucose oxidase and laccase that could degrade organic dyes is proven to be an effective strategy in wastewater treatment [61]. Other than absorption, enzyme immobilization could offer simultaneous adsorption and degradation of the organic pollutants allowing continuous or multiple usage times per application. Table 10.2 confers some specific applications of ALG in nanoremediation. Cost-effective and attractive environmental remediation approaches, as above, may gain increased attention for their easy implementation and effective clean-up procedures as these could be developed at an industrial scale.

10.11.4.1 Ca-ALG beads

ALG with a high M/G ratio extracted from *L. digitata* was evaluated for Cu(II), Cd(II), and Pb(II) sorption in acidic solutions in the form of calcium cross-linked beads. The high M/G ratio of ALG extracted from this algal species is most likely the determining factor for the increased adsorption capacity of the investigated metals, and the mannuronic acid in particular is responsible for the ion exchange mechanism. The presence of carboxyl groups in the ALG structure enhances the

Table 10.2 Applications of alginate in nanoremediation.

Encapsulation system	Encapsulate	Potential applications	Properties
Alginate composite nanoscale adsorbents	Iron sulfide nanoparticles	Selenium (IV) remediation	Improved reactivity due to decreased FeS oxidation
Alginate/Ca^{2+} beads	Fe(0) and magnetite nanoparticles	Nitrate remediation from groundwater	Lower nitrate reduction but higher capacity
Magnetic alginate nanoparticles	MnO-Fe_2O_3 for acid red B removal; Charcoal Fe_2O_3 for triphenylmethane, heteropolycyclic, and azo dyes removal; Copper phthalocyanine-Fe_2O_3 for malachite green and crystal violet removal.	Malachite green, polycyclic dyes, crystal white, and other organic dyes	Higher absorptivity due to increased surface area
Biofilm-coated alginate/Ca^{2+} beads	Nano zero-valent iron	Cr^{6+} removal	Effective reactivity due to reduced agglomeration of Nano zero-valent iron
Alginate/acrylic acid nanocomposites	Calcite or xonotlite	Methomyl (pesticide)	Increased swelling capacity and thermal stability

adsorption of many metal ions compared with other adsorbents. There are differences between the sorption capacities of the ALG beads for different metals with a general order of Pb(II) > Cu(II) > Cd(II) [62]. ALG gel beads showed a high affinity for heavy metal ions of Cu(II) and Mn(II), especially in a low concentration region. The ALG gel beads improved the mechanical strength and resistance to chemical and microbial degradation of the beads without the change in adsorption property after covalently cross-linking with 1,6-diaminohexane bridges. Ca-ALG beads also were applied to remove U(VI) ions from the solution and the results indicated that the interaction between uranium ions and Ca-ALG beads is endothermic. The covalently cross-linked ALG gel beads are expected to be a good candidate for adsorbents to remove heavy metal ions from low heavy metal concentration wastewater [63].

10.11.4.2 AC/Ca-ALG beads

While AC has been used widely to remove organic substances, AC immobilized in ALG beads has been studied for the removal of heavy metals in water and wastewater. Three different adsorbent materials namely KOH-AC-based apricot stone (C), calcium ALG beads (G), and calcium ALG/AC composite beads (GC) for the As removal have been investigated. The results indicated that GC exhibited the maximum As(V) adsorption (66.7 mg/g at 30 °C) [64]. Also, adsorption equilibrium characteristics of Cu and phenol onto powdered AC (PAC), ALG bead, and ALG-AC (AAC) bead have examined. The adsorption capacity of Cu(II) onto different adsorbents was in the following order: ALG bead > AAC bead > AC. The absorption of phenol: AC > AAC bead >

ALG bead. A novel ALG complex by impregnating synthetic zeolite and PAC into ALG gel bead has been fabricated and also the composite could simultaneously remove zinc and toluene from the aqueous solution. The maximum adsorption capacity of ALG complex for zinc and toluene obtained from Langmuir adsorption isotherm is 4.3 g/kg and 13.0 g/kg, respectively [65].

10.11.4.3 Biochar/Ca-ALG composites

The engineered biochar serves as a low-cost AC alternative for heavy metals adsorption [66]. Adsorption of Cd(II) by biochar-ALG bead was studied using batch systems and continuous fixed bed columns and the results indicated that biochar-ALG beads, *Ambrosia trifida* L. var. Trifida biochar-ALG beads (ATLB-AB) can be applied as an eco-friendly and potential adsorbent for the removal of Cd(II) from groundwater [59]. Also, a biochar ALG capsule to remove lead ions Pb(II) from an aqueous solution was synthesized. The maximum adsorption capacity for Pb(II) was found to be 263.158 mg/g at a pH of 5.0 [67].

10.11.4.4 GO/Ca-ALG beads

GO has been studied for the removal of heavy metals, synthetic dyes, and other organic compounds [68]. The regeneration and separation of GO from aqueous media are difficult because it disperses so well in water. Several attempts were made to a couple of magnetic NPs with the fabrication of GO composites to solve this problem. Magnetite GO encapsulated in calcium ALG beads (mGO/beads) were fabricated to absorb Cr(VI) and As(V) from wastewater [69]. The mGO/bead maintained activity in wastewater and exhibited greater adsorption efficiency for both Cr(VI) and As(V) than AC and CNT. FeO NPs embedded GO ALG beads (Fe@GOA beads) are fabricated with the insertion of GO into ALG gel before mixing with zero-valent iron NPs (FeO NPs). It reduced to FeO NPs embedded reduced GO-ALG beads (Fe@GA beads). The Fe@GA beads were used for Cr(VI) removal. The result showed that 1% of ALG and 1.5–2.0% of FeO by weight performed the best with a maximum adsorption capacity of about 34 mg/g [69].

10.11.4.5 CNTs/Ca-ALG beads

CNTs-ALG beads have synergistic effects on the removal of heavy metals [70]. The CNTs/ALG composition can improve the physicochemical properties of ALG-based composites thereby enhancing its ability to adsorb heavy metals [70]. The mixture of CNTs and sodium ALG and also $CaCl_2$ solution lead to prepare CNTs-calcium ALG composites. This composite has a specific surface area and pore size of calcium ALG gel of 28 m^2/g and 0.06 cm^3/g, respectively. The specific surface area and pore size of CNTs-Calcium Alginate (CA) composites were 76 m^2/g and 0.37 $cm^{3/}g$, respectively. The adsorption capacity of Cu (II) on calcium ALG gel was better than that of CNTs. When the equilibrium concentration was 5 mg/L the adsorption capacity was 52.1 mg/g for calcium ALG gel and increased to 67.9 mg/g for CNTs-calcium ALG. The adsorption performance of CNTs-calcium ALG composites to Cu(II) was significantly higher than that of CNTs.

10.11.4.6 Other ALG nanocomposites

ALG nanocomposites have excellent functional properties, biocompatibility, and heavy metal remediation. The natural hydroxyapatite NPs compounded with sodium ALG to prepare the granular and SA film (sodium ALG)/nano-hydroxyapatite composites for Pb(II)adsorption. The SA/nanohydroxyapatite composite membrane exhibited strong Pb(II) adsorption. Pb(II) adsorption with the maximum adsorption capacity of 83.33 mg/g via silica nanopowders/ALG nanocomposite was at pH of 5 [71]. The potential of Hal/ALG nanocomposite beads for the removal of Pb(II) in aqueous solutions through ion exchange with Ca(II) was followed by coordination with carboxylate groups of ALG to physisorption on Hal nanotubes. Adsorption behavior of Cu(II) onto the halloysite nanotube-ALG hybrid bead by continuous fixed-bed column adsorption reached 74.13 mg/g. ALG-goethite sorbent materials have been developed for the removal of trivalent and hexavalent chromium ions from binary aqueous solutions. The sorption capacities for Cr(VI) and Cr(III) increased from 20.5–29.5 mg/g and 20.7–25.3 mg/g, respectively when temperature increased from 20 to 60 °C [72].

10.11.5 ALG applications in nanobiotechnology

A developing trend in biotechnology is the miniaturized biosensing devices development with improved stability, selectivity, and sensitivity. The development of fluorescence-mediated glucose detecting biosensors using glucose oxidase–loaded ALG nanomicrospheres has been developed. The operation basics are based on the use of fluorescence quenching near-infrared radiation of the oxygen-sensitive dye. Aggregation-induced emission (AIE) is a unique fluorescence phenomenon that has gained great interest in biological imaging applications. Oxidized sodium ALG–conjugated PhE-NH_2 polymeric NPs are described to possess such characteristics with desirable properties such as high water dispersibility, good photostability, intense fluorescence, low critical micelle concentrations, desirable cell uptake, and ultrahigh biocompatibility making them promising candidates for biological imaging applications [73]. ALG from marine seaweeds is better than other synthetic polymers used in the synthesis of AIE-active polymeric NPs. Gene therapy is a hopeful approach for treating acquired and inherited diseases based on modifying the gene expression. Genetic material is introduced into the cells to compensate abnormal genes or to synthesize beneficial proteins. Presently there is a high demand for research centered on developing biocompatible gene carriers using nontoxic biopolymeric NPs. As therapeutic genes are introduced as naked DNA, the majority undergo degradation by nucleases, which lower the transfection efficiency. Additionally, the absence of a specific cellular targeting mechanism lowers their transfection efficiency. DNA-loaded chitosan-ALG-dextran sulfate NPs are described to have a higher DNA encapsulation efficiency that makes them ideal as gene transfection vectors [74].

10.11.5.1 Nanoantimicrobials

Nanoantimicrobials is a broad topic receiving increased attention in both biomedical and environmental sciences. Various antituberculosis drugs–loaded NP-based drug delivery systems are studied to reduce the frequency, quantity, and duration of drug administration and to avoid first-pass effects while reducing the side effects. The nanocarrier system preparation for inhalation delivery of antimycobacterial compound rifampicin is evaluated [75]. The nanocapsules are synthesized by coating Tween 80 dissolved rifampicin-loaded sodium ALG emulsion (pH 5.5) with chitosan in an aqueous solution containing acetic acid (pH 4.5) under rapid mixing following the ionic gelation method. The NPs are then separated by centrifugation. Rifampicin is encapsulated in ALG-chitosan NPs with an entrapment efficiency by ionotropic gelation by loading the drug to a pre-gel state of ALG and chitosan. Aqueous honey has been used as a surfactant and a stabilizer. NPs are formed following the drop-wise addition of 1% $CaCl_2$ solution with continuous stirring while maintaining the pH at 5.5. Water remediation by polymer nanocomposites is emerging research for disinfecting bacteria-contaminated water. Zinc oxide NPs–encapsulated sodium ALG nanocomposites have shown efficiency in inactivating bacteria in synthetic and surface water contaminated with *Staphylococcus aureus, Escherichia coli,* and *Pseudomonas aeruginosa* [76]. The inactivation could be improved with dose-dependent nanocomposite supplementation and increasing contact time. Moreover, the leached Zn^{2+} in water was within the recommended limits that are desirable for its application. These NPs possess potential applications for water treatment. The use of electrospun mats of polymer nanofibers incorporated with metal NPs is widely investigated for wound dressing applications due to their antibacterial activities. The fabrication of sodium ALG, polyvinyl alcohol, and nano-ZnO composite nanofibers by electrospinning has adequate antibacterial activity, reduced cytotoxicity, and cell adhesion properties providing the potential to be used in wound dressing applications [77]. Propolis is known to contain numerous polyphenols with a wide range of bioactivities including strong antimicrobial activity, antioxidant, antiinflammatory, and antitumor activities. Encapsulation of Propolis within zein/caseinate/ALG NPs improves its bioavailability in intestinal fluid surpassing the corrosive gastric juice that reduces its bioavailability [78]. These zein/caseinate/ALG-based vehicles broaden the application of propolis and other similar acid-sensitive bioactive compounds in the cosmetic, food, and pharmaceutical industries. Cinnamaldehyde is a natural preservative with good antibacterial and antifungal properties. However, it has poor water solubility, volatility, and instability, which limit its industrial applications. Cinnamaldehyde-loaded sodium ALG-chitosan NPs, which indicate a slow release of the compound, would allow overcoming above limitations expanding its wide applications as a food preservative. Nisin-loaded ALG-poly-l-lysine-chitosan NPs are reported as effective against bacteria and fungus [79].

10.12 Conclusion

Investigation of ALG for nanomaterial development has several advantages such as biocompatibility, mucoadhesive properties, nontoxicity, hydrophilicity, bioavailability, and low-cost. The ALG nanostructures ranged from NPs to nanocomposite scaffolds, nanocolloids, nanofibers, and nanoaggregates. ALG nanomaterials ranging from 25 nm have been fabricated using different methodologies such as controlled gelification using Ca^{2+} ions, polyionic complex formation through ionic gelation, self-assembly into nanoaggregates, nanocomposite fibrous scaffolds by electrospinning or TIPS, and NPs by using microfluidics-aided polyelectrolyte complexation methods. ALG-based nanomaterials have various applications including drug delivery, regenerative engineering, environmental remediation, wound dressing, developing probes and biosensors, genetic material transfection, and various biofunctional materials. The ALG-based nanoencapsulates advancements have been made to overcome limited control of drug release, increasing drug-loading capacity, and reduce pseudo-allergy responses upon *in-vivo* applications. However, more investigations are required to overcome issues such as body accumulation. Developing analytical techniques to study the heterogeneous structure of ALG, which depend on seaweed source and environmental conditions is a major prerequisite. Further studies are required for increasing current understanding of how the monomer composition of ALG, monomer attachment sequence, and ALG molecular weight would affect the physicochemical properties of nanomaterials and biocompatibility.

References

[1] A. Ehsani, H. Parsimehr, H. Nourmohammadi, R. Safari, S. Doostikhah, Environment-friendly electrodes using biopolymer chitosan/poly ortho aminophenol with enhanced electrochemical behavior for use in energy storage devices, Polym. Compos. 40 (2019) 4629–4637. doi:10.1002/pc.25330.

[2] H. Parsimehr, A. Ehsani, Corn-based electrochemical energy storage devices, Chem. Rec. 20 (10) (2020) 1163–1180. doi:10.1002/tcr.202000058.

[3] A. Ehsani, H. Parsimehr, Electrochemical energy storage electrodes via citrus fruits derived carbon: a Minireview, Chem. Rec. 20 (2020) 820–830. doi:10.1002/tcr.202000003.

[4] A. Ehsani, H. Parsimehr, Electrochemical energy storage electrodes from fruit biochar, Adv. Coll. Interface Sci. (2020) 102263. doi:10.1016/j.cis.2020.102263.

[5] K.Y. Lee, D.J. Mooney, Alginate: properties and biomedical applications, Prog. Polym. Sci. 37 (2012) 106–126. doi:10.1016/j.progpolymsci.2011.06.003.

[6] M.C. Catoira, L. Fusaro, D. Di Francesco, M. Ramella, F. Boccafoschi, Overview of natural hydrogels for regenerative medicine applications, J. Mater. Sci. Mater. Med. 30 (2019) 115. doi:10.1007/s10856-019-6318-7.

[7] C. PeteiroB. Rehm, M. Mradali (Eds.), Alginate production from marine macroalgae, with emphasis on kelp farming, Alginates and Their Biomedical Applications. (2018) 27–66. doi:10.1007/978-981-10-6910-9_2.

[8] A.F. Ferreira, A.P. Dias, C.M. Silva, M. Costa, Effect of low frequency ultrasound on microalgae solvent extraction: Analysis of products, energy consumption and emissions, in: A.F. Ferreira (Ed.), Algal Res., 14, Elsevier, 2016, pp. 9–16. 10.1016/j.algal.2015.12.015.

[9] K.I. Draget, G. Skjåk Bræk, O. Smidsrød, Alginic acid gels: the effect of alginate chemical composition and molecular weight, Carbohydr. Polym. 25 (1994) 31–38. doi:10.1016/0144-8617(94)90159-7.

[10] C.K. Kuo, P.X. Ma, Ionically crosslinked alginate hydrogels as scaffolds for tissue engineering: Part 1. Structure, gelation rate and mechanical properties, Biomaterials 22 (2001) 511–521. doi:10.1016/S0142-9612(00)00201-5.

[11] M.B. Łabowska, I. Michalak, J. Detyna, Methods of extraction, physicochemical properties of alginates and their applications in biomedical field – a review, Open. Chem. 17 (2019) 738–762. doi:10.1515/chem-2019-0077.

[12] G. Hernández-Carmona, Y. Freile-Pelegrín, E. Hernández-Garibay, Conventional and alternative technologies for the extraction of algal polysaccharides, 2013. doi:10.1533/9780857098689.3.475.

[13] A. Mazumder, S.L. Holdt, D. De Francisci, M. Alvarado-Morales, H.N. Mishra, I. Angelidaki, Extraction of alginate from *Sargassum muticum*: process optimization and study of its functional activities, J. Appl. Phycol. 28 (2016) 3625–3634. doi:10.1007/s10811-016-0872-x.

[14] T.A. Fenoradosoa, G. Ali, C. Delattre, C. Laroche, E. Petit, A. Wadouachi, P. Michaud, Extraction and characterization of an alginate from the brown seaweed *Sargassum turbinarioides* Grunow, J. Appl. Phycol. 22 (2010) 131–137. doi:10.1007/s10811-009-9432-y.

[15] H.V. Sæther, H.K. Holme, G. Maurstad, O. Smidsrød, B.T. Stokke, Polyelectrolyte complex formation using alginate and chitosan, Carbohydr. Polym. 74 (2008) 813–821. doi:10.1016/j.carbpol.2008.04.048.

[16] M. Rajaonarivony, C. Vauthier, G. Couarraze, F. Puisieux, P. Couvreur, Development of a new drug carrier made from alginate, J. Pharm. Sci. 82 (1993) 912–917. doi:10.1002/jps.2600820909.

[17] C.Y. Yu, H. Wei, Q. Zhang, X.Z. Zhang, S.X. Cheng, R.X. Zhuo, Effect of ions on the aggregation behavior of natural polymer alginate, J. Phys. Chem. B. 113 (2009) 14839–14843. doi:10.1021/jp906899j.

[18] B. Sarmento, A. Ribeiro, F. Veiga, P. Sampaio, R. Neufeld, D. Ferreira, Alginate/chitosan nanoparticles are effective for oral insulin delivery, Pharm. Res. 24 (2007) 2198–2206. doi:10.1007/s11095-007-9367-4.

[19] I. Aynié, C. Vauthier, H. Chacun, E. Fattal, P. Couvreur, Spongelike alginate nanoparticles as a new potential system for the delivery of antisense oligonucleotides, Antisense Nucleic Acid Drug Dev. 9 (1999) 301–312. doi:10.1089/oli.1.1999.9.301.

[20] K. Santhi, S.A. Dhanraj, D. Nagasamyvenkatesh, S. Sangeetha, B. Suresh, Preparation and optimization of sodium alginate nanospheres of methotrexate, Indian J. Pharm. Sci. 67 (2005) 691–696.

[21] C.Y. Yu, L.H. Jia, B.C. Yin, X.Z. Zhang, S.X. Cheng, R.X. Zhuo, Fabrication of nanospheres and vesicles as drug carriers by self-assembly of alginate, J. Phys. Chem. C. 112 (2008) 16774–16778. doi:10.1021/jp806540z.

[22] M. Boissière, P.J. Meadows, R. Brayner, C. Hélary, J. Livage, T. Coradin, Turning biopolymer particles into hybrid capsules: the example of silica/alginate nanocomposites, J. Mater. Chem. 16 (2006) 1178–1182. doi:10.1039/b515797h.

[23] X. Liu, X. Chen, Y. Li, X. Wang, X. Peng, W. Zhu, Preparation of superparamagnetic Fe_3O_4@alginate/chitosan nanospheres for candida rugosa lipase immobilization and utilization of layer-by-layer assembly to enhance the stability of immobilized lipase, ACS Appl. Mater. Interfaces. 4 (2012) 5169–5178. doi:10.1021/am301104c.

[24] B.B. Bhowmik, B. Sa, A. Mukherjee, Preparation and in vitro characterization of slow release testosterone nanocapsules in alginates, Acta Pharm. 56 (2006) 417–429.

[25] D. Grebinişan, M. Holban, V. Şunel, M. Popa, J. Desbrieres, C. Lionte, Novel acyl derivatives of n-(p-aminobenzoyl)-l-glutamine encapsulated in polymeric nanocapsules with potential antitumoral activity, Cellul. Chem. Technol. 45 (2011) 571–577.

[26] P. Lertsutthiwong, K. Noomun, N. Jongaroonngamsang, P. Rojsitthisak, U. Nimmannit, Preparation of alginate nanocapsules containing turmeric oil, Carbohydr. Polym. 74 (2008) 209–214. doi:10.1016/j.carbpol.2008.02.009.

[27] V. Deepa, R. Sridhar, A. Goparaju, P. Neelakanta Reddy, P. Balakrishna Murthy, Nanoemulsified ethanolic extract of *Pyllanthus amarus* Schum & Thonn ameliorates CCl_4 induced hepatotoxicity in Wistar rats, Indian J. Exp. Biol. 50 (2012) 785–794.

[28] P. Lertsutthiwong, P. Rojsitthisak, Chitosan-alginate nanocapsules for encapsulation of turmeric oil, Pharmazie. 66 (2011) 911–915. doi:10.1691/ph.2011.1068.

[29] J.O. You, C.A. Peng, Calcium-alginate nanoparticles formed by reverse microemulsion as gene carriers, Macromol. Symp. 219 (2004) 147–153. doi:10.1002/masy.200550113.

[30] M. Monshipouri, A.S. Rudolph, Liposome-encapsulated alginate: controlled hydrogel particle formation and release, J. Microencapsul. 12 (1995) 117–127. doi:10.3109/02652049509015282.

[31] A. Tachaprutinun, P. Pan-In, S. Wanichwecharungruang, Mucosa-plate for direct evaluation of mucoadhesion of drug carriers, Int. J. Pharm. 441 (2013) 801–808. doi:10.1016/j.ijpharm.2012.12.028.

[32] T. Chen, Q. Yan, F. Li, S. Tang, Nanospheres conjugated with Hab18 as targeting carriers for antitumor drug, Adv. Mater. Res. 535-537 (2012) 2381–2384. doi:10.4028/www.scientific.net/AMR.535-537.2381.

[33] T. Wang, Z. Feng, N. He, Z. Wang, S. Li, Y. Guo, L. Xu, A novel preparation of nanocapsules from alginate-oligochitosan, J. Nanosci. Nanotechnol. 7 (2007) 4571–4574. doi:10.1166/jnn.2007.882.

[34] Y. Yamada, K.I. Kurumada, K. Susa, N. Umeda, G. Pan, Method of fabrication of submicron composite microparticles of hydroxyapatite and ferromagnetic nanoparticles for a protein drug carrier, Adv. Powder Technol. 18 (2007) 251–260. doi:10.1163/156855207780860237.

[35] J.P. Paques, E. Van Der Linden, L.M.C. Sagis, C.J.M. Van Rijn, Food-grade submicrometer particles from salts prepared using ethanol-in-oil mixtures, J. Agric. Food Chem. 60 (2012) 8501–8509. doi:10.1021/jf3023029.

[36] C.P. Reis, A.J. Ribeiro, S. Houng, F. Veiga, R.J. Neufeld, Nanoparticulate delivery system for insulin: design, characterization and in vitro/in vivo bioactivity, Eur. J. Pharm. Sci. 30 (2007) 392–397. doi:10.1016/j.ejps.2006.12.007.

[37] M. George, T.E. Abraham, Polyionic hydrocolloids for the intestinal delivery of protein drugs: alginate and chitosan—a review, J. Control. Release. 114 (2006) 1–14. doi:10.1016/j.jconrel.2006.04.017.

[38] H. Zhang, Y. Zhu, L. Qu, H. Wu, H. Kong, Z. Yang, D. Chen, E. Mäkilä, J. Salonen, H.A. Santos, M. Hai, D.A. Weitz, Gold nanorods conjugated porous silicon nanoparticles encapsulated in calcium alginate nano hydrogels using microemulsion templates, Nano Lett. 18 (2018) 1448–1453. doi:10.1021/acs.nanolett.7b05210.

[39] F.N. Sorasitthiyanukarn, P. Ratnatilaka Na Bhuket, C. Muangnoi, P. Rojsitthisak, P. Rojsitthisak, Chitosan/alginate nanoparticles as a promising carrier of novel curcumin diethyl diglutarate, Int. J. Biol. Macromol. 131 (2019) 1125–1136. doi:10.1016/j.ijbiomac.2019.03.120.

[40] K.V. Jardim, A.F. Palomec-Garfias, B.Y.G. Andrade, J.A. Chaker, S.N. Báo, C. Márquez-Beltrán, S.E. Moya, A.L. Parize, M.H. Sousa, Novel magneto-responsive nanoplatforms based on $MnFe_2O_4$ nanoparticles layer-by-layer functionalized with chitosan and sodium alginate for magnetic controlled release of curcumin, Mater. Sci. Eng. C. 92 (2018) 184–195. doi:10.1016/j.msec.2018.06.039.

[41] W.-J. Li, R.M. Shanti, R.S. Tuan, F.K. Ko, Electrospun nanofibrous structure: A novel scaffold for tissue engineering, J. Biomed. Mater. Res., 60, Willey, 2002, pp. 613–621. 10.1002/jbm.10167.

[42] T. Chae, H. Yang, V. Leung, F. Ko, T. Troczynski, Novel biomimetic hydroxyapatite/alginate nanocomposite fibrous scaffolds for bone tissue regeneration, J. Mater. Sci. Mater. Med. 24 (2013) 1885–1894. doi:10.1007/s10856-013-4957-7.

[43] W.W. Hu, Y.C. Wu, Z.C. Hu, The development of an alginate/polycaprolactone composite scaffold for in situ transfection application, Carbohydr. Polym. 183 (2018) 29–36. doi:10.1016/j.carbpol.2017.11.030.

[44] J. Zhang, J. Huang, K. Huang, J. Zhang, Z. Li, T. Zhao, J. Wu, Egg white coated alginate nanoparticles with electron sprayer for potential anticancer application, Int. J. Pharm. 564 (2019) 188–196. doi:10.1016/j.ijpharm.2019.04.045.

[45] Y.S. Nam, T.G. Park, Biodegradable polymeric microcellular foams by modified thermally induced phase separation method, Biomaterials 20 (1999) 1783–1790. doi:10.1016/S0142-9612(99)00073-3.

[46] Z. Lin, J. Wu, W. Qiao, Y. Zhao, K.H.M. Wong, P.K. Chu, L. Bian, S. Wu, Y. Zheng, K.M.C. Cheung, F. Leung, K.W.K. Yeung, Precisely controlled delivery of magnesium ions thru sponge-like monodisperse PLGA/nano-MgO-alginate core-shell microsphere device to enable in-situ bone regeneration, Biomaterials 174 (2018) 1–16. doi:10.1016/j.biomaterials.2018.05.011.

[47] K.C. Souza, J.D. Ardisson, W.A.A. Macedo, E.M.B. Sousa, Preparation of magnetic nanoparticles in SBA-15 for applications in biomedicine, in: Proc. 8th World Biomater. Congr. 2008, 4, 2008, p. 2325.

[48] G. Ciofani, C. Riggio, V. Raffa, A. Menciassi, A. Cuschieri, A bi-modal approach against cancer: magnetic alginate nanoparticles for combined chemotherapy and hyperthermia, Med. Hypotheses. 73 (2009) 80–82. doi:10.1016/j.mehy.2009.01.031.

[49] J. Grobelny, F.W. DelRio, N. Pradeep, D.-I. Kim, V.A. Hackley, R.F. Cook, NIST-NCL joint assay protocol, PCC-6 size measurement of nanoparticles using atomic force microscopy, Natl. Inst. Stand. Technol. 21702 (2009) 71–82. doi:10.1007/978-1-60327-198-1.

[50] A. Zahoor, S. Sharma, G.K. Khuller, Inhalable alginate nanoparticles as antitubercular drug carriers against experimental tuberculosis, Int. J. Antimicrob. Agents. 26 (2005) 298–303. doi:10.1016/j.ijantimicag.2005.07.012.

[51] S. Jain, M. Amiji, Tuftsin-modified alginate nanoparticles as a noncondensing macrophage-targeted DNA delivery system, Biomacromolecules 13 (2012) 1074–1085. doi:10.1021/bm2017993.

[52] H. Zhang, X. Liu, M. Yang, L. Zhu, Silk fibroin/sodium alginate composite nano-fibrous scaffold prepared through thermally induced phase-separation (TIPS) method for biomedical applications, Mater. Sci. Eng. C. 55 (2015) 8–13. doi:10.1016/j.msec.2015.05.052.

[53] Q. Wang, S. Jamal, M.S. Detamore, C. Berkland, PLGA-chitosan/PLGA-alginate nanoparticle blends as biodegradable colloidal gels for seeding human umbilical cord mesenchymal stem cells, J. Biomed. Mater. Res. 96 (2011) 520–527. doi:10.1002/jbm.a.33000.

[54] Z. Qiusheng, L. Xiaoyan, Q. Jin, W. Jing, L. Xuegang, Porous zirconium alginate beads adsorbent for fluoride adsorption from aqueous solutions, RSC Adv., 5, 2014, pp. 2100–2112.

[55] J. Maitra, V.K. Shukla, Cross-linking in hydrogels—a review, Am. J. Polym. Sci. 4 (2014) 25–31. doi:10.5923/j.ajps.
20140402.01.

[56] G. Russo, R. Malinconico, M. Santagata, Effect of the cross-linking with calcium ions on the structural and thermo-mechanical properties of alginate films, Biomacromolecules 8 (2007) 3193–3197. https://doi.org/10.1021/bm700565h.

[57] L.W. Chan, Y. Jin, P.W.S. Heng, Cross-linking mechanisms of calcium and zinc in production of alginate microspheres, Int. J. Pharm. 242 (2002) 255–258. doi:10.1016/S0378-5173(02)00169-2.

[58] N. Mohammadi, H. Khani, V.K. Gupta, E. Amereh, S. Agarwal, Adsorption process of methyl orange dye onto mesoporous carbon material-kinetic and thermodynamic studies, J. Coll. Interface Sci. 362 (2011) 457–462. doi:10.1016/j.jcis.2011.06.067.

[59] B. Wang, B. Gao, Y. Wan, Entrapment of ball-milled biochar in Ca-alginate beads for the removal of aqueous Cd(II), J. Ind. Eng. Chem. 61 (2018) 161–168. doi:10.1016/j.jiec.2017.12.013.

[60] T. An, L. Zhou, G. Li, J. Fu, G. Sheng, Recent patents on immobilized microorganism technology and its engineering application in wastewater treatment, Recent Patents Eng. 2 (2008) 28–35. doi:10.2174/187221208783478543.

[61] R. Shojaat, N. Saadatjoo, A. Karimi, S. Aber, Simultaneous adsorption-degradation of organic dyes using $MnFe_2O_4$/calcium alginate nano-composites coupled with GOx and laccase, J. Environ. Chem. Eng. 4 (2016) 1722–1730. doi:10.1016/j.jece.2016.02.029.

[62] S.K. Papageorgiou, F.K. Katsaros, E.P. Kouvelos, J.W. Nolan, H.Le Deit, N.K. Kanellopoulos, Heavy metal sorption by calcium alginate beads from *Laminaria digitata*, J. Hazard. Mater. 137 (2006) 1765–1772. doi:10.1016/j.jhazmat.2006.05.017.

[63] T. Gotoh, K. Matsushima, K.I. Kikuchi, Adsorption of Cu and Mn on covalently cross-linked alginate gel beads, Chemosphere 55 (2004) 57–64. doi:10.1016/j.chemosphere.2003.10.034.

[64] A.F. Hassan, A.M. Abdel-Mohsen, H. Elhadidy, Adsorption of arsenic by activated carbon, calcium alginate and their composite beads, Int. J. Biol. Macromol. 68 (2014) 125–130. doi:10.1016/j.ijbiomac.2014.04.006.

[65] S.H. Ching, N. Bansal, B. Bhandari, Alginate gel particles–a review of production techniques and physical properties, Crit. Rev. Food Sci. Nutr. 57 (2017) 1133–1152. doi:10.1080/10408398.2014.965773.

[66] H. Lyu, B. Gao, F. He, A.R. Zimmerman, C. Ding, H. Huang, J. Tang, Effects of ball milling on the physicochemical and sorptive properties of biochar: Experimental observations and governing mechanisms, Environ. Pollut. 233 (2018) 54–63. doi:10.1016/j.envpol.2017.10.037.

[67] X.H. Do, B.K. Lee, Removal of Pb2+ using a biochar-alginate capsule in aqueous solution and capsule regeneration, J. Environ. Manage. 131 (2013) 375–382. doi:10.1016/j.jenvman.2013.09.045.

[68] H. Chen, B. Gao, H. Li, Removal of sulfamethoxazole and ciprofloxacin from aqueous solutions by graphene oxide, J. Hazard. Mater. 282 (2015) 201–207. doi:10.1016/j.jhazmat.2014.03.063.

[69] H.C. Vu, A.D. Dwivedi, T.T. Le, S.H. Seo, E.J. Kim, Y.S. Chang, Magnetite graphene oxide encapsulated in alginate beads for enhanced adsorption of Cr(VI) and As(V) from aqueous solutions: role of crosslinking metal cations in pH control, Chem. Eng. J. 307 (2017) 220–229. doi:10.1016/j.cej.2016.08.058.

[70] B. Wang, B. Gao, A.R. Zimmerman, X. Lee, Impregnation of multiwall carbon nanotubes in alginate beads dramatically enhances their adsorptive ability to aqueous methylene blue, Chem. Eng. Res. Des. 133 (2018) 235–242. doi:10.1016/j.cherd.2018.03.026.

[71] R.D.C. Soltani, G.S. Khorramabadi, A.R. Khataee, S. Jorfi, Silica nanopowders/alginate composite for adsorption of lead (II) ions in aqueous solutions, J. Taiwan Inst. Chem. Eng. 45 (2014) 973–980. doi:10.1016/j.jtice.2013.09.014.

[72] N.K. Lazaridis, C. Charalambous, Sorptive removal of trivalent and hexavalent chromium from binary aqueous solutions by composite alginate-goethite beads, Water Res. 39 (2005) 4385–4396. doi:10.1016/j.watres.2005.09.013.

[73] R. Jiang, M. Liu, H. Huang, L. Mao, Q. Huang, Y. Wen, Q. Yong Cao, J. Tian, X. Zhang, Y. Wei, Ultrafast construction and biological imaging applications of AIE-active sodium alginate-based fluorescent polymeric nanoparticles through a one-pot microwave-assisted Döbner reaction, Dyes Pigment. 153 (2018) 99–105. doi:10.1016/j.dyepig.2018.02.008.

[74] M. Samimi, S. Validov, Characteristics of pDNA-loaded chitosan/alginate-dextran sulfate nanoparticles with high transfection efficiency, Rom. Biotechnol. Lett. 23 (2018) 13996–14006. doi:10.26327/RBL2018.178.

[75] I.R. Scolari, P.L. Páez, M.E. Sánchez-Borzone, G.E. Granero, Promising chitosan-coated alginate-tween 80 nanoparticles as rifampicin coadministered ascorbic acid delivery carrier against mycobacterium tuberculosis, AAPS PharmSciTech. 20 (2019) 67. doi:10.1208/s12249-018-1278-7.

[76] S. Baek, S.H. Joo, M. Toborek, Treatment of antibiotic-resistant bacteria by encapsulation of ZnO nanoparticles in an alginate biopolymer: Insights into treatment mechanisms, J. Hazard. Mater. 373 (2019) 122–130. doi:10.1016/j.jhazmat.2019.03.072.

[77] K.T. Shalumon, K.H. Anulekha, S.V. Nair, S.V. Nair, K.P. Chennazhi, R. Jayakumar, Sodium alginate/poly(vinyl alcohol)/nano ZnO composite nanofibers for antibacterial wound dressings, Int. J. Biol. Macromol. 49 (2011) 247–254. doi:10.1016/j.ijbiomac.2011.04.005.

[78] H. Zhang, Y. Fu, Y. Xu, F. Niu, Z. Li, C. Ba, B. Jin, G. Chen, X. Li, One-step assembly of zein/caseinate/alginate nanoparticles for encapsulation and improved bioaccessibility of propolis, Food Funct 10 (2019) 635–645. doi:10.1039/c8fo01614c.

[79] J. Liu, J. Xiao, F. Li, Y. Shi, D. Li, Q. Huang, Chitosan-Sodium Alginate Nanoparticle as a Delivery System for ε-Polylysine: Preparation, Characterization and Antimicrobial Activity, Elsevier B.V., Amsterdam, 2018. doi:10.1016/j.foodcont.2018.04.020.

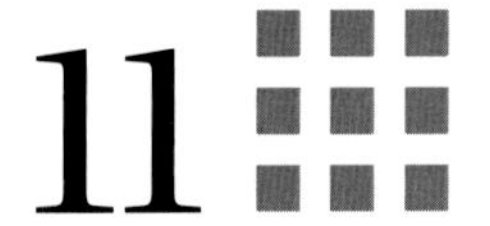

Chitin-based nanomaterials

Rida Badar[a,b], Alap Ali Zahid[a,c,d] and Muhammad Yar[a]

[a]INTERDISCIPLINARY RESEARCH CENTER IN BIOMEDICAL MATERIALS, COMSATS UNIVERSITY ISLAMABAD, LAHORE CAMPUS, PAKISTAN. [b]DEPARTMENT OF CHEMISTRY, COMSATS UNIVERSITY ISLAMABAD, LAHORE CAMPUS, PAKISTAN. [c]DEPARTMENT OF MECHANICAL AND INDUSTRIAL ENGINEERING, COLLEGE OF ENGINEERING, QATAR UNIVERSITY, QATAR. [d]BIOMEDICAL RESEARCH CENTER, QATAR UNIVERSITY, QATAR.

Chapter outline

11.1 Introduction

In nanobiotechnology field, polymeric nanoparticles (NPs) and nanomaterials are getting more attention and have gained great interest for advancing the capabilities and performance of materials in the various fields of science [1,2]. Nowadays, scientists are thoroughly investigating

Biopolymeric Nanomaterials: Fundamentals and Applications. DOI: https://doi.org/10.1016/B978-0-12-824364-0.00013-7

the synthesis, characteristics, and promising applications of nanomaterials. In the field of nanomaterials, chitosan possesses wide interest because of its ease in availability, biocompatibility, biodegradability, nontoxicity, high permeability, and porosity and can easily be molded into various shapes [3,4]. Besides, chitosan has been investigated worldwide for the production of polymeric NPs due to its potential to improve the penetration of large molecules on the mucosal surface and their identification as chitosan mucoadhesive [5]. The term nanotechnology was first coined by the popular scientist Richard Feynman in 1959 [6]. Nanotechnology is the engineering and study of nanoscale material properties in which dimensions of nanomaterials are 100 nm or less. Nanomaterials demonstrate innovative features compared to bulk materials and have a high surface area. Nanomaterials include nanorods, nanobeads, nanosheets, nanofibers, nanobrush, nanopin, and NPs. For nanotechnology, building blocks are nanomaterials. For smaller molecules, the number of particles exposed to the surface is high, while the number of atoms on the surface for the larger molecule is very small. Therefore, nanomaterials are gaining attention in emerging and rapidly growing applications due to high mechanical strength, extremely small size, and high surface to volume ratio.

Chitin ($C_8H_{13}O_5N$), a long-chain polymer of N-acetylglucosamine, is a natural polysaccharide of great importance, first identified in 1884. After cellulose, chitin is the most abundant biopolymer as it is synthesized by an overwhelming number of living organisms [7]. For biomedical applications, chitin is usually converted to its deacetylated derivative, chitosan [8]. Indeed, a remarkable finding is that chitin and chitosan can easily be treated and processed into hydrogels [9-11], beads [12,13], membranes [14-16], nanofibers [17,18], NPs [19-21], sponges [22,23], and scaffolds [24,25] for a range of applications such as drug delivery and gene delivery [26,27], wound healing [7,11,14,28,29], and tissue engineering [30] as shown in Fig. 11.1. Chitin is not only an ideal biomaterial that has been naturally developed, but also an excellent candidate for the production of advanced functional materials for biomedical applications.

11.2 Brief history and description of chitin

11.2.1. Description

Nomenclature: Chitin is derived from the Greek word meaning "tunic" referring to the protective shell. The timeline of early discovery and breakthrough of chitin is displays in Fig. 11.2.

Systematic chemical names
Poly-(acetylamino glucose)
Poly-N-acetyl-D-glucosamine[poly(D-GlcNAc)]
b-(1,4)-Poly-N-acetyl-D-glucosamine
Poly-b-(1,4)-N-acetyl-glucosamine
b-(1,4)-2-Acetamido-2-deoxy-D-glucose
2-Acetamido-2-deoxy-D-glucose
b-(1,4)-2-Amino-2-deoxy-D-glucose
Poly-(N-acetyl-1,4-b-D-glucopyranosamine)
b-(1,4)-2-Acetamido-2-deoxy-D-glucopyranose
Fully acetylated chitosan

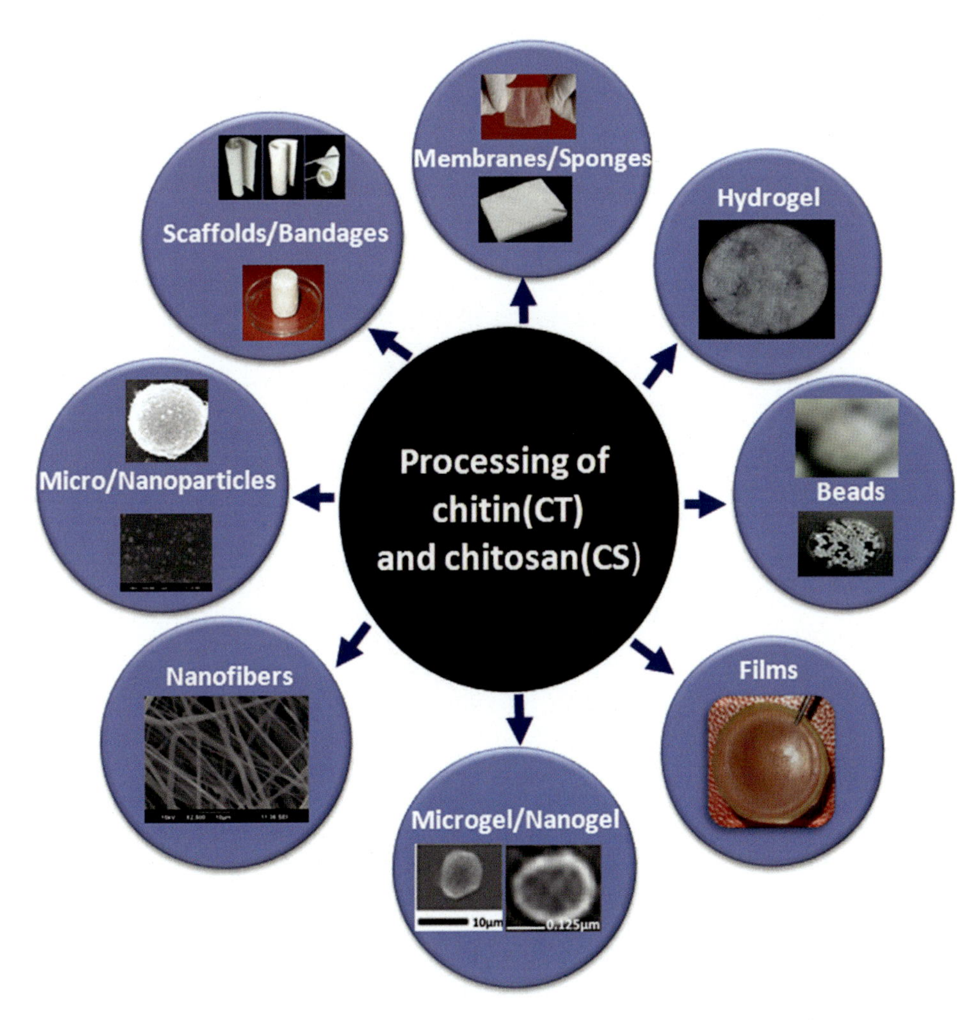

FIGURE 11.1 Graphical view on the ways of processing chitin and chitosan in various forms. Reprinted with permission from [31].

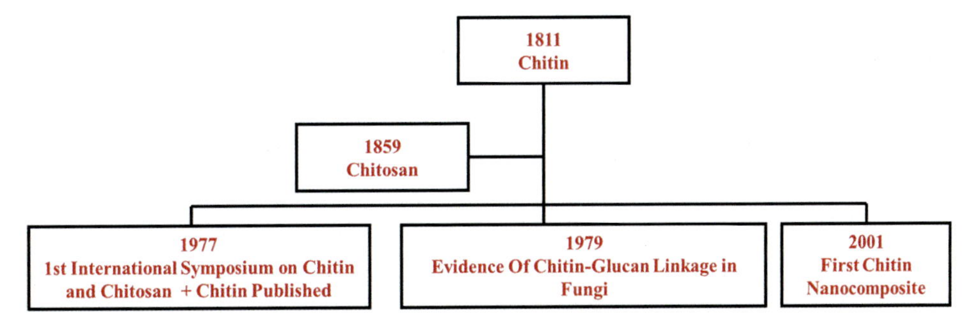

FIGURE 11.2 Timeline of early discovery and breakthrough of chitin.

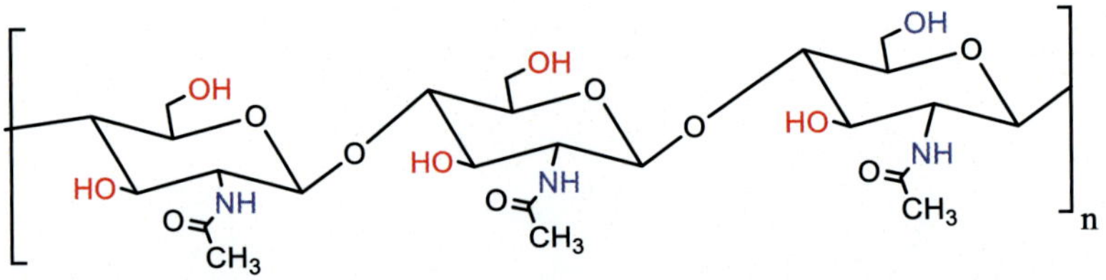

FIGURE 11.3 Chemical structures of chitin and chitosan.

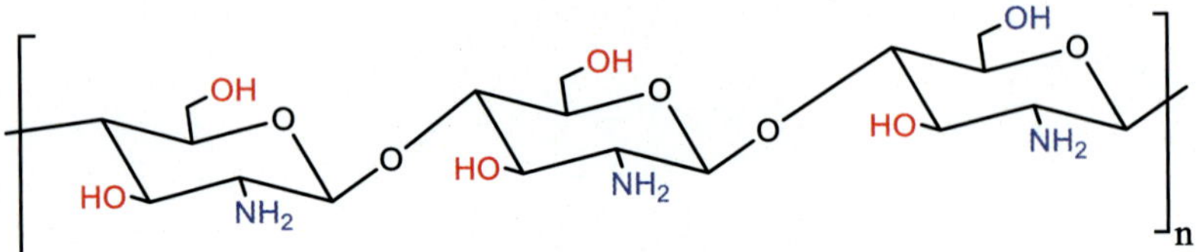

Nonproprietary names

Chitin, poly-acetyl glucosamine; shell protein; poly-acetylglucosaminyltransferase; shell; ming keratinocytes; chitosan; shell protein, b-farming; chitin

Formula

Empirical formula

$[C_8H_{13}NO_5]_n$

Molecular weight

$[203.19]_n$

Chitin has an average molecular weight ranging from 1 million to 2.5 million Da. The difference in the molecular weight depends on the degree of N-acetylation.

Chemical Abstracts Service (CAS) number

1398-61-4

Structural formula

The chemical structure of the monomeric unit chitin is shown in the Fig. 11.3 given.

Elemental investigation of chitin

Elements	Composition (%)
Carbon	47.29
Hydrogen	6.45
Nitrogen	6.89
Oxygen	39.37

Appearance

The appearance of chitin is white, hard, inelastic, nitrogenous polysaccharide found in the outer skeletons of crabs and lobsters and the internal structure of other invertebrates.

11.3 Synthesis of chitin nanofibers

Several mechanical nanofibrillation processes such as high-pressure homogenization, microfluidization, wet shear grinding, or high-speed blending are used for the conversion of extracted chitin into chitin nanofibers [32,33]. All these processes depend on high shearing and impact force exerting on the chitin fiber bundle and causing weak interfaces among nanofibers to be broken. Chitin nanofiber can be produced by the chemical process of oxidation in the presence of catalyst TEMPO (2,2,6,6-tetramethylpiperidine 1-oxyl), sodium bromide NaBr, and an oxidizer sodium hypochlorite NaClO [34,35]. NaClO oxidizes C_6-primary polysaccharide groups in carboxylic acid via an intermediate aldehyde into carboxylic acid moieties using TEMPO. Carboxylate charge promotes electrostatic anionic repulsion that separates individual fibers. This nanofibrillation method was first shown by Isogai and colleagues for cellulose in 2006 [36]. With squid pen β-chitin, TEMPO produces neither nanofibers nor nanowhiskers [37], TEMPO with β-chitin produces nanofibers of width 20–50 nm and with crab α-chitin produces whiskers of 8 mm width only [38].

A simpler method to produce chitin nanofiber from squid pen β-chitin by treating the sample with ultrasonication under acidic conditions was stated by Fan et al. in 2008 [39]. The aim is to cationize free amine groups on the surface of chitin crystallite at pH 3–4. Cationizing contributes to electrostatic repellence similar to repellence occurring during the oxidation of TEMPO but instead of anionic it is cationic. Crab α-chitin nanofibers with a diameter of 10–20 nm are produced with the addition of a sample that has never been dried [32]. Drying tends to cause the fibers to weaken and to lose their ability to expand, making it more difficult to defibrillate. Later it was found that nanofibers of similar width could be produced using dried chitin by grinding in an acidic environment [40]. α-chitin is partially deacetylated into chitosan when treated with 33% NaOH. Due to acid protonation of chitosan, partly deacetylated chitin nanofibers can easily be individualized at pH 3–4 as a result of the positive amino group's cationic repulsion [41]. It was stated that it is possible to rebuild α-chitin (3 nm) nanofibers when the squid pen β-chitin has dissolved in HFIP7 during solvent evaporation. During the precipitation process, the β-chitin dissolved in LiCl/DMAc8 can also be assembled automatically but produces nanofibers of higher diameter (10 nm) [42].

11.4 Synthesis of chitin whiskers (chitin nanocrystals)

Boiling of chitin in HCl followed by ultrasonication produces chitin whiskers also termed as chitin nanocrystals or chitin crystallites. Due to cationic charging on its crystallite surface, these whiskers form steady colloidal suspensions and can be helicoidally reorganized at certain concentrations [44]. Also, the use of HCl does not disturb the thermal stability of chitin. Acid concentration and duration of hydrolysis are major factors that affect the yield and dimension of chitin whiskers. The length-to-width ratio of whiskers is affected by the higher concentration of acid and prolonged exposure [45]. Another process for generating whisker is controlled oxidation by TEMPO. In comparison to traditional hydrolysis, this approach offers several advantages such as the process can be controlled by the amount of oxidizer with no deacetylation of chitin occurring and restoration of whiskers approach to 90% [46]. In a study conducted by Kadokawa et al. where they regenerated chitin from ionic liquid that can resemble a whisker form by soaking the chitin ionic liquid gel into methanol [47]. Fig. 11.4 shows the molecular formula and method of preparation of different chitin nanomaterials that is, ChNF, ChW, CsW, T-ChW, T-ChW and Fig. 11.5 illustrates the isolation of chitin and preparation of chitin nanofibers and nanowhiskers.

11.5 Preparation of chitosan NPs

For the preparation of chitosan-based NPs, the most common methods are coprecipitation, microemulsion, ionotropic gelation, complex coacervation method, and spray drying as shown in Fig. 11.6.

11.5.1 Ionotropic gelation

In this process, when the chitosan solution is dissolved with or without a stabilizing agent such as poloxamer in acetic acid or any polyanionic solution, NPs are readily formed due to complexity. Chitosan spherical particles of various sizes and shapes are produced at room temperature during mechanical stirring. Usually, the reported particle size for NPs is 20–200 and 550–900 nm. NPs of the chitosan-TPP(tripolyphosphate)/vitamin C were produced at room temperature by this gelation technique with continuous stirring for just 1 h among the positively charged amino groups of chitosan-TPP and vitamin C [48,49].

11.5.2 Microemulsion method

In this process, chitosan NPs are formed when the cross-linking process is completed. The mixture of chitosan, acetic acid solution, glutaraldehyde, and hexane was placed on constant stirring at room temperature. Organic solvent that is, hexane was then removed by low pressure evaporation. At this phase, the substance contains an additional surfactant that could be removed by precipitating with calcium chloride, tracked by centrifugation. Finally the suspension of NPs was dialyzed and then lyophilized [50]. With this method, the distribution of very small sizes was seen and the concentration of glutaraldehyde in the NP preparation might control the size [51].

11.5.3 Complex coacervation method

DNA-chitosan NPs were formed by holding the amine groups of chitosan positively charged and the DNA phosphate groups negatively charged [52,53]. In the polyelectrolyte complex (PEC) method, cationic polymer that is, chitosan was added to anionic polymer that is, dextran sulfate DNA solution with mechanical stirring at room temperature followed by charge neutralization [49,54]. The PEC method for insulin produced low-molecular-weight water-soluble chitosan

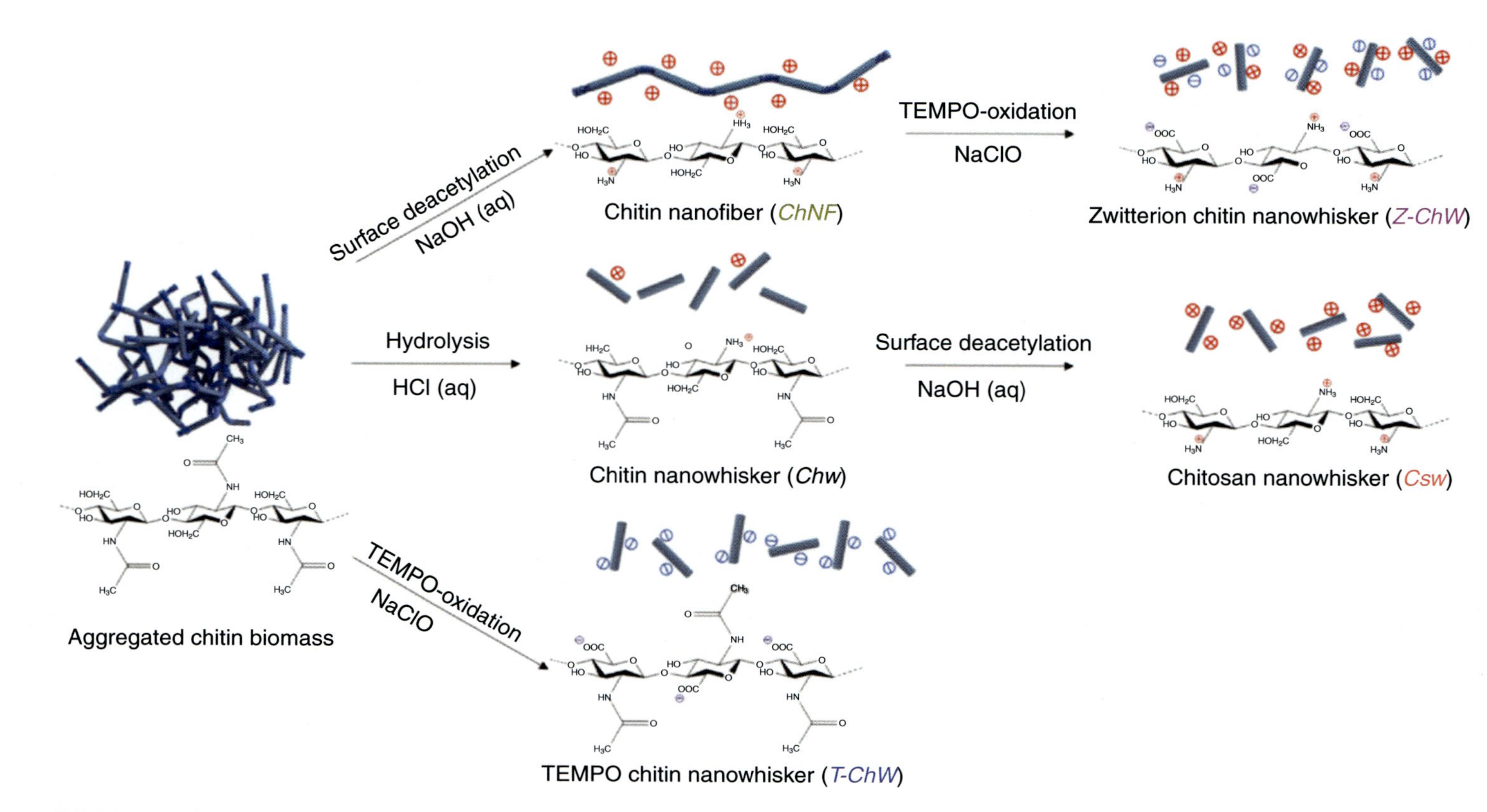

FIGURE 11.4 Schematic illustration of molecular formula and method of preparation of different chitin nanomaterials that is, ChNF, ChW, CsW, T-ChW, T-ChW. Reprinted with permission from [43].

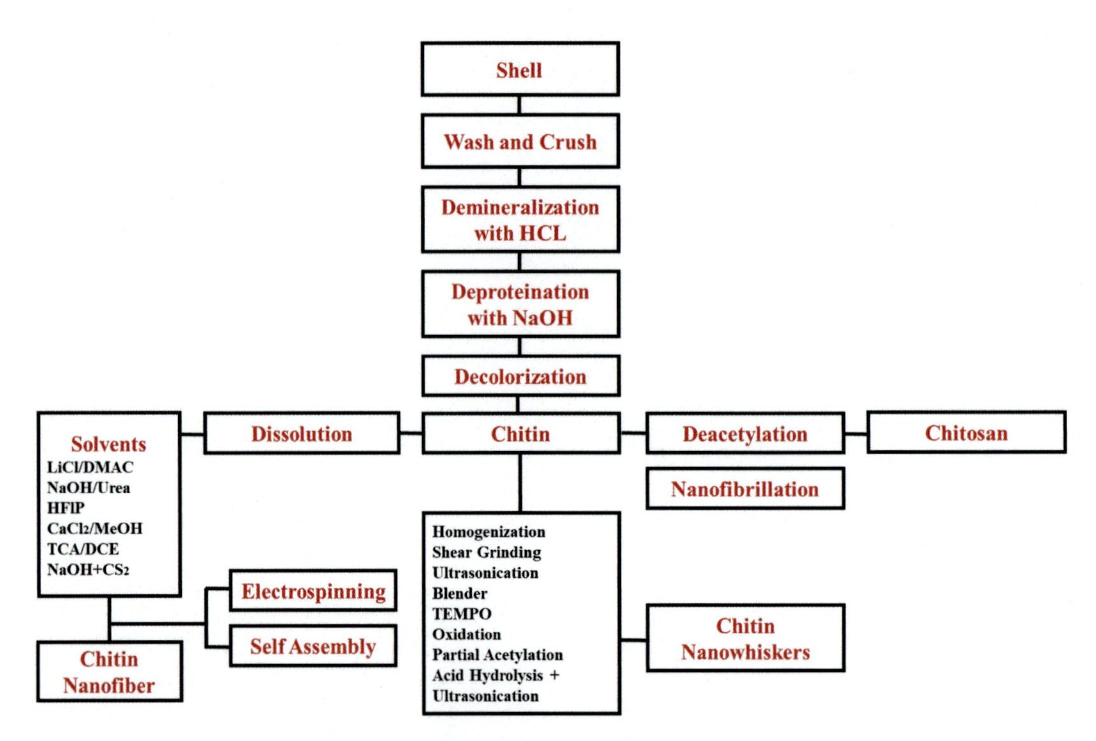

FIGURE 11.5 Isolation of chitin and preparation of chitin nanofibers and nanowhiskers.

nanocarriers, subsequent resulting in insulin-loaded chitosan NPs with a recorded mean diameter of 200 nm and kept constant in an *in-vitro* release profile [49] Figs. 11.4 and 5.

11.5.4 Coprecipitation method

The addition of chitosan solution prepared in a low pH solution of acetic acid to a high pH solution such as ammonium hydroxide results in coprecipitation and the formation of a highly monodisperse population of NPs. NPs of diameter 10 nm can be prepared with high effectiveness encapsulation [55]. Coprecipitation was used to prepare the grafted lactic acid NPs of chitosan in which ammonium hydroxide was used to produce droplets of coacervate. This process yielded spherical-shaped NPs that were distributed evenly [56] Figs. 11.6 and 7.

11.5.5 Spray drying

Spray drying is a very common method to develop chitosan NPs by homogenously mixing drugs and the chitosan in a relevant solvent. The jet nozzle of the spray cone helps to produce small droplets when passing it through a drying chamber that contains hot air to evaporate water and volatile solvents from the droplets to generate NPs. The obtained particles are spherical in shape. Such NPs have been found to have a prolonged and sustained release,

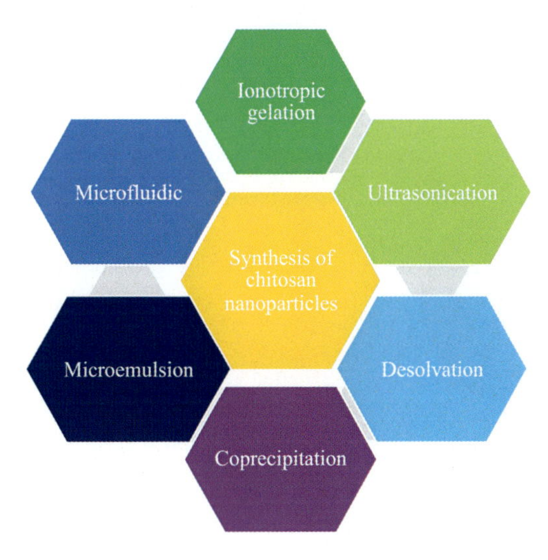

FIGURE 11.6 Schematic representation of methods to prepare chitosan nanoparticles.

also, these NPs may play an important role in pulmonary tuberculosis and neurodegenerative diseases.

11.6 Applications of chitin-based nanomaterials

Chitin nanocrystal/nanofiber is commonly used in polymer nanocomposites manufacturing. In addition, it can be used to interact with inorganic molecules by mineralization chelation, carbonization, and a variety of hybrid composite materials have been developed [57-59].

Additionally, chitosan derivatives and chitosan nanomaterials depicted excellent performance in ophthalmology, dentistry, bioimaging, biosensing, and diagnosis [60]. Historically chitosan derivatives and chitosan NPs are among the most commonly considered class of natural biopolymer material for biomedical applications. Furthermore, chitosan and its derivatives are among the very few biomaterials that can be synthesized in large quantities while still being economically viable. Various applications of chitin and chitosan nanomaterials are shown in Fig. 11.7.

11.6.1 Chitin and chitosan nanomaterials in tissue engineering

Tissue engineering is an interdisciplinary study concerns a broad variety of areas such as materials and genetics, with the goal of creating synthetic replacements for damaged tissues and organs with a high degree of cell proliferation and differentiation. Some biodegradable materials have been explored extensively for biomedical applications and chitin is one of them. They have been widely used to form nanofibers for tissue engineering applications because of their promising properties to adapt native extracellular matrix [61].

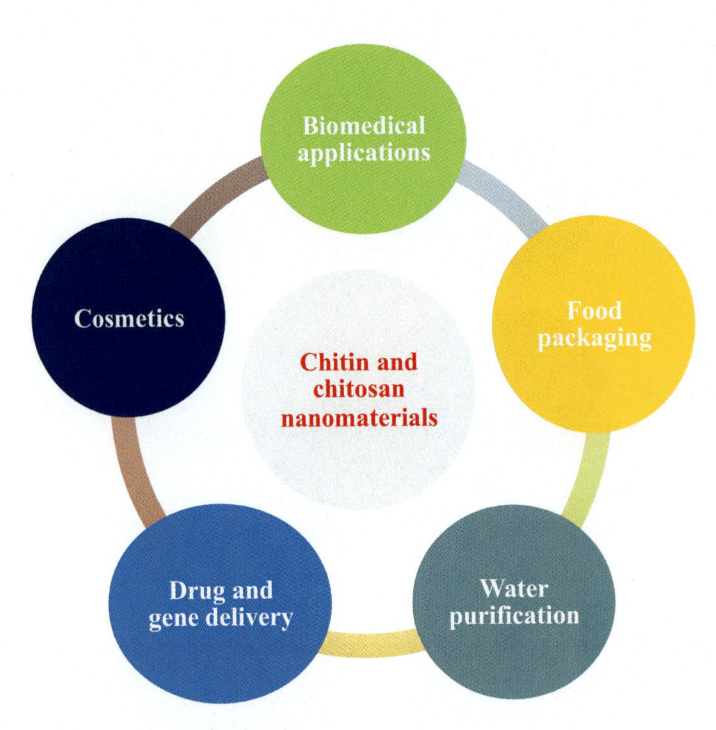

FIGURE 11.7 Applications of chitin and chitosan nanomaterials.

11.6.1.1 Soft tissue engineering

Chitin efficiently stimulates the healing process of wounds. The main biochemical activities of chitin-based nanomaterials found in wound healing are massive cell proliferation and migration, cytokine production, fibroblast activation, and replication of type IV collagen synthesis [62]. Moreover, chitin-based nanomaterials are cytocompatible with the wound tissue shown by studies [63]. Chitin nanofibrous scaffolds have a high surface area as well as 3D featured structures that favored cell growth, proliferation, and cell attachment. Hydrogel scaffolds such as chitin have currently been used extensively to enhance the flexibility of tissue-engineered constructs. Chitin has also proved physiologically compatible with living tissues.

Chitin/chitosan or its derivatives have been widely designed for making hydrogels, membranes, fibers, and sponges [64]. Chitin/chitosan-based nanofibers are commonly used by various types of wound healing materials, because of their outstanding attributes such as high porosity levels, and high surface/volume ratios. These properties facilitated cellular respiration, skin reepithilization, preservation of moisture, removal of exudate, and hemostatic treatment [65]. Chitosan increases cellular granulation and tissue thickness covering the wounds [66,67]. HemCons bandages are made from chitosan fibers in the form of a patch that is used as a commercial wound dressing. Their mechanism of action is due to their adhesive properties when in contact with blood promote wound healing and control bleeding. This mechanism is due to chitosan positive charge as it attracts red blood cells. The RBCs create cover over the wound as they are drawn into the bandage. The major advantage of the chitosan bandage over the

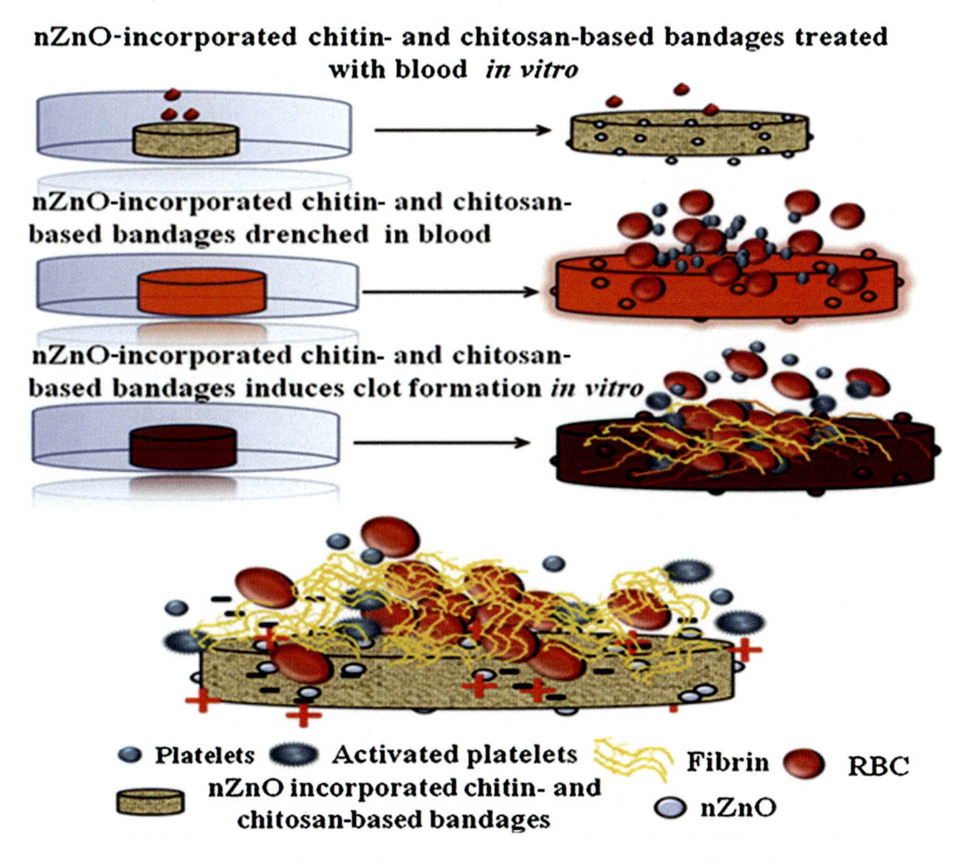

FIGURE 11.8 Schematic illustration of blood clotting caused by the composite bandage of chitin and chitosan integrated into *n*ZnO. Reprinted with permission from [31].

others is that it has no side effects while others may cause tissue damage. Other advantages are bacteriostatic properties [68].

In a study by Sudheesh Kumar et al. they fabricated and characterized β-chitin nanosilver scaffolds for enhancing wound healing. The incorporation of nanosilver in chitin showed good antimicrobial properties by testing with *Staphylococcus aureus* and *Escherichia coli*. Further, they used epithelial cells, which resulted in good cell attachment to the scaffolds [69]. Besides, α-chitin nanosilver scaffolds have also been developed and could be a promising candidate for rapid wound healing applications [70]. Despite numerous efforts to produce usable nanofibers in the biomedical applications, the clinical use of nanofibers based on chitin and chitosan remains a challenge and requires more insight. In the clinical trials, chitosan nanofibers were shown to promote wound healing. Chitosan/PEO (polyethylene oxides) nanofibrous structure was prepared by electrospinning method and applied for III A and III B degree burns as wound dressings [71].

Fig. 11.8 represents the clotting induced by ZnO incorporation on chitin- and chitosan-based bandages. Also, chitosan nanofiber dressings were shown to provide wound ventilation, infection control, successful exudate absorption, and skin tissue regeneration stimulation in patients with III A and III B degree burns.

11.6.1.2 Hard tissue engineering

Bone tissue engineering seems to have become a rapidly expanding research field, offering a new way to repair bone [72-74]. Allografting and autografting are the traditional therapies for the treatment of bone defects; however, these methods may cause problems. On the other hand, bone tissue engineering involves the usage of the scaffold in combination with cells, drugs, and growth factors [75,76]. In the recent years [77], due to their fascinating nature, chitin- and chitosan-based nanofibrous materials showed promising applications in bone tissue engineering applications. They show high biocompatibility, biodegradability, and antibacterial properties. As well as showed good cell attachment and proliferation of osteoblasts, development of mineralized bone matrix, and fast remodeling [78-80]. Hydroxyapatite (HA) is usually used as a composite material with chitin- and chitosan-based nanofibers to be used in bone tissue engineering for increasing the cell viability and activity, also to provide the mechanical strength to the scaffolds [81-85]. HA-containing chitosan nanofiber has been electrospun to test its bone regeneration capability [86], the developed nanofibers possessed the same Young's modulus to the periosteum. The polymer with HA combination has been shown to optimize HA *in vivo*, which can induce osteopathic growth in the implants as the matrix progressively resorbs [87,88]. Therefore, HA or other calcium-containing materials introduced into chitin have been the main research zone where orthopedic or bone substitution was the focus.

11.6.2 Chitin and chitosan nanomaterials in water purification

Clean water is essential, but the available supplies of freshwater are decreasing, the world is facing immense challenges in meeting the growing demand for clean water. There is an increasing interest in emerging, new, and more effective materials and technologies for the removal of various contaminants from wastewater. For this reason, the important removal potential for certain harmful contaminants such as water-soluble organics (phenols, dyes, herbicides, etc.) and heavy metal ions has drawn many scientists' attention for chitin and chitosan [89]. Natural polymers have been attracted by researchers in contrast to synthetic polymers to produce cheaper and more effective adsorbents. The eco-friendly properties include biodegradability, nontoxicity, and biocompatibility [90] of polysaccharides such as chitin [91], chitosan [92], starch [93] cellulose [94], and derivatives. Similarly, their ability to bind physical and chemical interactions with a varied variety of molecules is of significant concern.

The ability of chitin and chitosan to remove pollutants such as organic water-soluble organics and metal ions has been extensively investigated and demonstrated. Chitin and chitosan have been applied as adsorbents to treat wastewaters burdened with organic and heavy metals. Chitin is a natural polymer with a significant number of acetamido and hydroxyl groups. Chitosan (hydrophilic polymer) mainly consists of hydroxyl and amino groups. The mechanical adsorption is based on electrostatic attraction (for anionic organic contaminants) and dative coordination for cations of heavy metals. Acetamide and amino groups in chitin and chitosan are the participants in the adsorption mechanism [95,96]. The occurrence of reactive groups in the chitin and chitosan structure provides multiple adsorption interfaces with various organic contaminants including phenols [97], pesticides [98], herbicides [99], colorants [100], etc.

Peng et al. reported the use of chitosan- and halloysite-based nanotubes-composite hydrogel beads based as absorbing agents for extracting methylene blue and malachite green from aqueous solutions [101]. Chitosan deacetylation plays a key function in eliminating heavy metals from wastewater [102]. In colloidal systems containing algae [103], bacteria [104], and emulsions [105], Chitosan particles of different sizes and types were also used as flocculants. The use of chitin/chitosan nanomaterials has been well reported in the literature for the adsorption of organic contaminants. Consequently, it is reported that chitosan and its derivatives are prospective products for water purification and environmental applications. Nanomaterials are revolutionary drivers in nanotechnology. Nanomaterials offer new possibilities to produce nanostructured and reactive membranes that are more functional and cost effective to purify and desalinate water.

11.6.3 Chitin and chitosan nanomaterials in drug delivery

Recent studies illustrated that drug transportation barriers, along with chitosan nanotechnology and its derivatives, can be overcome by improving the performance of the drug. Chitosan-based nanomaterials are extremely effective in targeted drug therapy [106]. Chitosan NPs are carriers with a slow or controlled drug release that enhances drug delivery efficiently and increases its efficacy and reduce toxicity. *In-vitro* and *in-vivo* experiments have also shown that chitosan has antitumor effects, leading to strong prospects for its use as an antitumor drug and drug carrier supplement [107]. Chitin nanofibers have strong potential and promising contribution for applications in tissue engineering scaffolds, drug delivery, and wound dressing [108]. Microparticles of chitosan have been documented to induce immune responses in mice [109,110]. Chitosan-based nanomaterials have advanced applications in oral drug delivery, gene drug delivery, topical drug delivery, colon-targeted drug delivery, tumors therapy, also in gene and vaccine delivery as shown in Fig. 11.9.

It is stated in previous studies that chitosan NPs have been synthesized as drug carriers [5,111]. Chitosan NPs loaded with insulin could improve the intestinal absorption of insulin and its pharmacological bioavailability will be increased [112]. Chitosan NPs were also used as a gene carrier to improve gene transmission efficiency in cells [113]. Regulated release of active antimicrobials including amoxicillin and metronidazole in the gastric cavity has been prepared with reacetylated chitosan microspheres [114]. Chitosan plays a significant role in drug delivery, which may enhance drug absorption and stabilize drug components to enhance drug target delivery. Chitosan NPs can produce dose-dependent antitumor effects *in vivo*. The performance of particle-based drug delivery system seems to increase by smaller particles [115]. Chitosan microspheres and nanospheres were also formulated for drug delivery

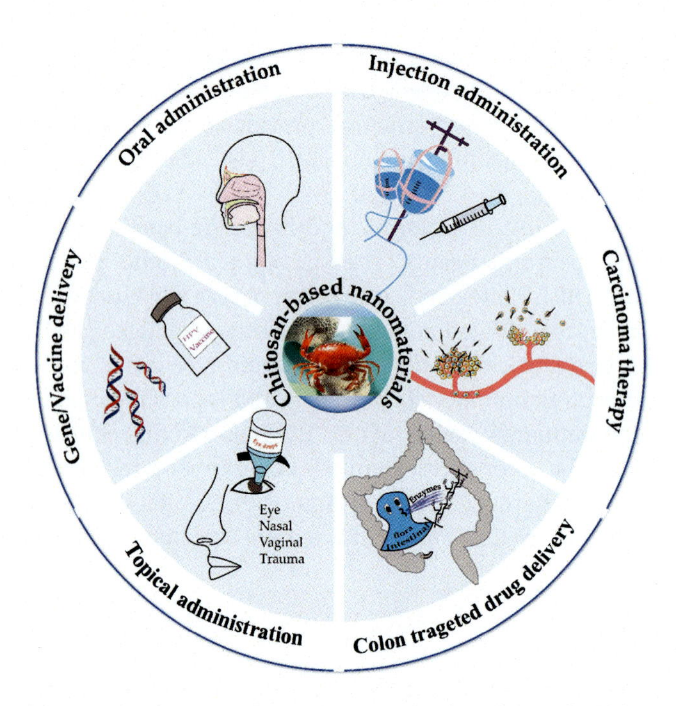

FIGURE 11.9 Chitosan-based nanomaterials in advanced strategy for drug delivery [106] (From an open-access source).

applications. Also, chitosan beads and microgranules are used as a drug carrier for diclofenac disodium [116].

The preparation of nanospheres based on chitosan has also been carried out [117]. Sato et al. studied the transfection efficiency of chitosan and reported that chitosan can be used as a gene delivery vehicle [118]. Nakao et al. stated [119] on the use of chitosan in gene therapy to improve the adenovirus infection rate in mammalian cells. Adenovirus activity was enhanced when lower molecular weight chitosan and its lower concentration was used. Chitosan nanofibers produced by the freeze-thawing method are used as drug carriers and showed the fastest release and it was due to greater permeability, shorter diffusion period, and fast swelling performance [120]. Also in the literature, PLGA (Poly Lactic-co-Glycolic Acid) /chitosan nanofibers were prepared and their potential applications were tested as a drug release system for fenbufen [121]. NPs of chitosan TPP can bind to drugs such as doxorubicin on mucosal surfaces and help to maintain their release profile. Also, water-soluble carboxymethyl chitin prepared through cross-linking method has been used for drug delivery applications. It showed a controlled and sustained drug release profile when hydrophobic anticancer drug 5-fluorouracil was loaded on these NPs [122].

11.6.4 Chitin and chitosan nanomaterials in food packaging

The severe problems regarding human health and environmental influence caused by petroleum-based nonbiodegradable packing resources witnessed great advances in bio-based

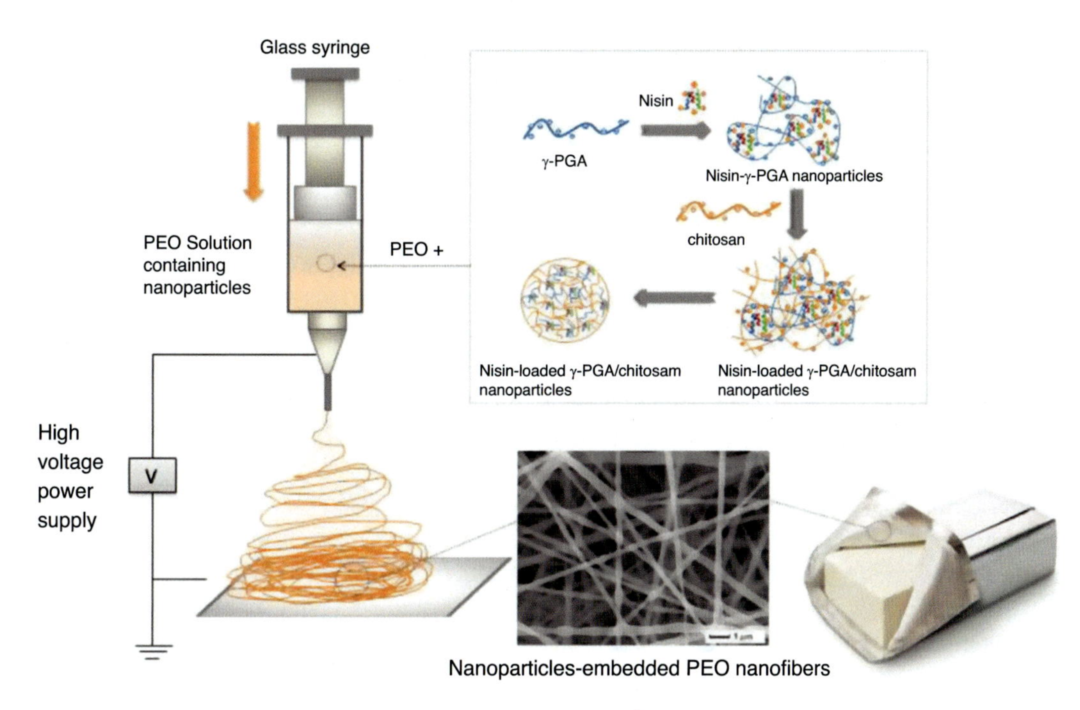

FIGURE 11.10 PEO electrospun nanofibers encapsulated with poly-Ò-glutamic acid chitosan nanoparticles for food packaging. Reprinted from [131,132].

polymer packaging films over recent years [123]. Also, food safety has attracted customers' attention increasingly so in the area of food packaging, specific chitosan films have been developed and applied. Chitosan has potential in a wide range of applications because of its strong film- and gel-forming capabilities. The use of chitosan can therefore also contribute new materials for active food packaging in the formulation of edible films (foodstuffs) or coatings [124]. This approach is a new one aimed at encouraging a more active role in maintaining product quality. As a result, chitosan may be used for other active chemicals, which had been previously introduced into edible chitosan-like coatings, and as an intrinsic active polymer.

There have been several advances in modern food packaging applications from chitosan films including antibacterial films, barrier films, and sensor films. Some researchers discovered that added montmorillonite to chitosan films could also improve mechanical properties and enhance flammability, thereby showing advantages for new food packaging [125-127]. Chitosan shows naturally strong antimicrobial activity with polycationic characterization [128-130]. Chitosan films, therefore, have great potential for food packaging applications, protecting food from microorganisms, as well as food safety monitoring and ensuring health security [123]. Cui et al. showed a successful result in another report with nisin-loaded poly-Ò-glutamic acid chitosan NPs electrospun with PEO nanofibers showing antibacterial activity in food packaging as shown in Fig. 11.10. This composite was used to inhibit the multiplication of bacteria named *Listeria monocytogenes* and showed satisfactory results. Advances in technology are making it possible to plan for chitosan-based multifunctional films with multiple features to be promising for a packaging material.

11.6.5 Chitin and chitosan nanomaterials in cosmetics

Significant advances in the medical sector have led to a large rise in life expectancy, with further efforts aimed at healthy living and rising perceptions about general face and body appearance. In this regard, Morganti et al. established block copolymer NPs consisting of phosphatidylcholine and linoleic acid complexed with chitin nanofibrils [133]. This nanoconstruct was used for skin rejuvenation and acts as an antiaging agent. When the skin was treated with active chitin nanofibrils loaded with BPN (block copolymers NPs), it became softer and more hydrated. Also, the wrinkles and crease lines were reduced.

As nontoxic and nonallergenic chitin and chitosan can be used on the human body, they were used in the manufacturing of emulsifiers, antistatic agents, and emollients to prolong the shelf-life of cosmetic products for example, shampoo and products for hairstyling. As the findings obtained, both *in-vitro* and *in-vivo* studies indicate the efficacy of the injected block-polymer NPs in reducing skin wrinkles and enhancing the signs of aging.

11.7 Antibacterial activity of nanomaterials and chitosan nanomaterials

Naturally, chitosan derivatives have antimicrobial properties that contribute to industrial and topical disinfectants [134,135]. Chitin and chitosan studies have already shown growth inhibition of several bacteria and fungi in particular phytopathogens [122]. Chitosan is also used to cover fresh fruits and vegetables as an antimicrobial film [136]. There is an urgent need for new methods to regulate bacterial behavior while nanoformulations-based materials represent a very promising approach. Silver NPs are renowned antibacterial agents against both the Gram-positive and Gram-negative strains. They have gained considerable interest because of their chemical stability, catalytic activity, and wound healing capability. The combined microbicidal effect of chitosan and metals has been explored to prepare novel nanocomposite materials with enhanced antibacterial properties as shown in Figs. 11.11 and 11.12. The typical antibacterial results shown by chitosan- and silver-based NPs toward *E. coli* a positive control: is the LB (Luria Broth) with incubated bacteria, a 10 μg was the concentration in which there was no bacterial activity. In short silver NPs impregnated on chitosan are promising candidates in biomedical and in general applications.

In research, a wide range of activities against the Gram-negative and Gram-positive bacteria was found for gold-, silver-, and copper-incorporated chitosan NPs. These NPs were developed by mixing chitosan nanosuspension in metal ion solutions. Metal NPs remain embedded in chitosan polymeric matrix and bactericidal activity was checked for *E. coli* and *S. aureus* having

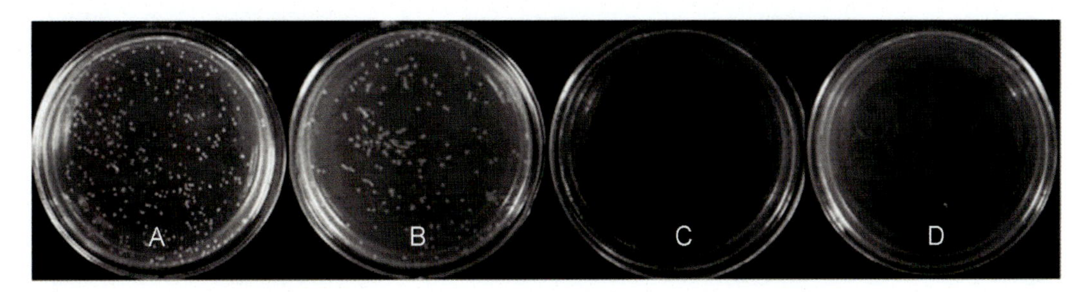

FIGURE 11.11 Representative images of bacterial activity of chitosan-based silver nanoparticles toward *E. coli*. Concentrations of the samples are (A) positive control, (B) 5 μg, (C) 10 μg (D) 01 μg/mL. Reprinted with permission from [137].

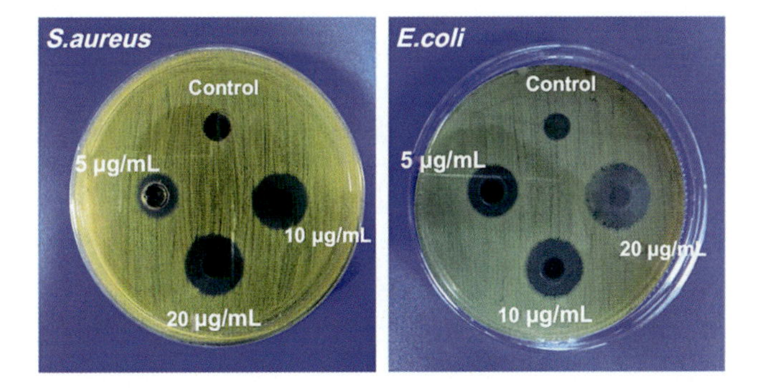

FIGURE 11.12 Disk diffusion test showing inhibition of bacterial growth zone provided by various concentrations of silver-incorporated chitosan nanoparticles against *S. aureus* and *E. coli* [138]. Reprinted with permission from [132].

inhibition diameter from 8 to 10 mm. Bacterial inhibition by disk diffusion method is shown in Fig. 11.12.

Nanomaterials, due to their high surface to volume ratio, have greater bactericidal activity in contrast to their bulk counterparts, offering excellent contact with microorganisms [139]. TiO_2 NPs have recently been found to have a wide range of action against microorganisms, such as bacteria and fungi [140]. In some reports, sulfonate groups or quaternary ammonium groups have modified chitosan and incorporated antibacterial herbs or enzymes into chitosan beads or NPs to improve their antimicrobial capacity.

Chitosan with zinc oxide and with graphene oxide hybrid nanocomposites were prepared by Shanmugam et al. using one-pot chemical strategy; it showed stronger antibacterial activity against *S. aureus* and *E. coli*. Tamara et al. developed protamine-based chitosan NPs, which is a natural peptide composed of arginine residues. Their study shows the protamine-based chitosan NPs increases the antimicrobial activity against *E. coli* [141]. Chitosan is presented as the ideal biomaterial for antimicrobial wound dressings, which can be produced alone in their native form or upgraded and incorporated to increase the antimicrobial effect using antibiotics, metallic

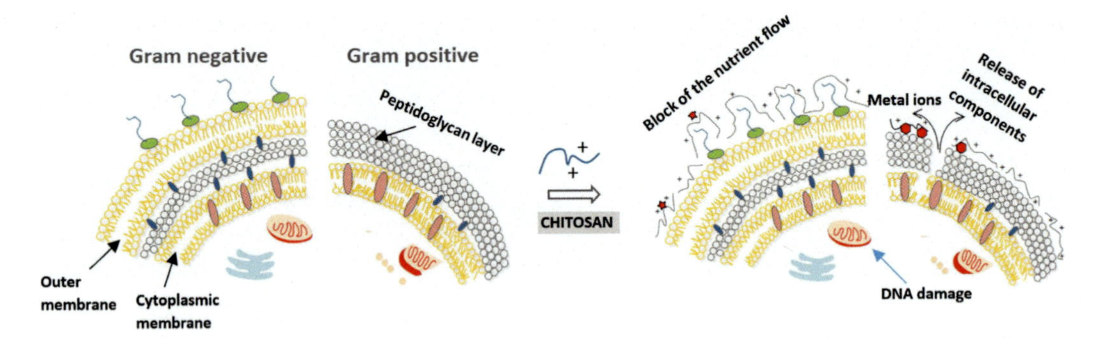

FIGURE 11.13 Four suggested models for chitosan's behavior tested against the Gram-positive and Gram-negative bacteria. Chitosan causes intercellular components to be released, binds to bacterial DNA, blocks nutrient supply, and chelates with essential metals. [142] (Reused from an open access source).

NPs, natural compounds, and extracts [142]. Chitosan NPs showed better antimicrobial activity compared to chitosan due to their small size [143].

The mechanism of antibacterial activity involves cell lysis, breakdown of cytoplasmic membrane barrier, and chelation of trace metal cations by the chitosan. Cationic chitosan must disrupt and react with both inner and outer bacterial cell membranes in the killing of the Gram-negative bacteria [144,145]. Finally, it should be noted that chitosan offers the best potential for the design of sustainable antimicrobial nanomaterials. Even for use in food processing and preservation, antimicrobial activities of chitosan and N-sulfobenzoyl chitosan against various pathogens and food spoilage microorganisms were investigated [146].

While the precise antimicrobial function of chitosan and its derivatives is not yet completely known, many theories have recently been suggested and accepted regarding the mode of action as shown in Fig. 11.13. The most widely accepted theory is that glucosamine's positive-charged amine groups NH^{3+} interact with the bacteria's negatively charged surface, triggering intracellular constituent leakage that results in cell death [147]. Other possibilities include chitosan-DNA binding which prevents mRNA after it enters nucleus of the microorganism. A confocal laser microscope analysis of *E. coli* indicates that chitooligomers have been present within the cell and may be prevent DNA transcription [148]. The chelating effect from chitosan, which binds essential metals to prevent microbial growth, is another alternative. Chitosan is well known for its remarkable activity of metal-binding abilities in which amino charged groups coupled with metals [149]. The interaction of amino groups with divalent ions such as Ca^{+2} and Mg^{+2} in the cell wall of microorganisms inhibits toxin development and restricts bacterial growth. These four proposed models introduce chitosan as a blocking agent that prevents the entry of nutrients and oxygen into the cell.

11.8 Future recommendations and way forward

Chitosan possess widespread applications in various fields of science and technology. Specifically, in biomedical applications chitosan nanomaterials will be further explored and advanced toward the more precise development of wound dressings, and cancer diagnostics as well as drug delivery. Researchers have enormous scope to establish an effective and eco-friendly production technology focused on innovative methods for nanomaterials and nanocomposites for use in biomedical applications. The toxicity of NPs must also be overcome before clinical applications *in vivo*. In future studies, there are many possibilities to use chitosan in combination with natural compounds and extracts for many applications. Development can be done by introducing sensors and therapeutic molecules in smart dressings that will be released at the same time. It is worthy to mention that chitin- and chitosan-based nanomaterials and also their derivatives as nanobiodegradable carriers show a significant part in brain drug delivery. The toxicity of some chitosan derivatives, which may serve as carriers for drug delivery applications, remains a challenge. The near future is expected to be an exciting time in study, production, and applications of chitin-based nanomaterials, pursuing new processing techniques and functional materials.

11.9 Conclusions

In conclusions, chitin- and chitosan-based nanomaterials are being used in clinical applications today and will likely influence emerging biomedical applications even more. Chitin and chitosan, with their special biological properties, are used in many biomedicine research applications as they are nearly ideal biopolymers. In this chapter, chitin and chitosan have been presented as the ideal agents for manufacturing of antibiotic wound dressings, either unaided or upgraded to include metallic NPs as antibiotics. Metallic NPs have been described as a new concept to increase and maintain antimicrobial effects in chitosan matrices and smart chitosan-based nanomaterials are of great interest to researchers. Studies also showed that chitosan-based NPs exhibit remarkable antibacterial activity compared to chitosan. Chitosan-based materials also play a significant role in food packaging applications. Recent *in-vivo* studies have shown that chitin is highly biocompatible and biodegradable. Chitin is more biocompatible compared to synthetic polymers.

References

[1] M.M. Al-Remawi, Properties of chitosan nanoparticles formed using sulfate anions as crosslinking bridges, Am. J. Appl. Sci. 9 (7) (2012) 1091.

[2] S.K. Shukla, et al., Chitosan-based nanomaterials: a state-of-the-art review, Int. J. Biol. Macromol. 59 (2013) 46–58.

[3] M. Prabaharan, J. Mano, Chitosan derivatives bearing cyclodextrin cavitiesas novel adsorbent matrices, Carbohydr. Polym. 63 (2) (2006) 153–166.

[4] M. Prabaharan, R. Reis, J. Mano, Carboxymethyl chitosan-graft-phosphatidylethanolamine: amphiphilic matrices for controlled drug delivery, React. Funct. Polym. 67 (1) (2007) 43–52.

[5] Y. Xu, Y. Du, Effect of molecular structure of chitosan on protein delivery properties of chitosan nanoparticles, Int. J. Pharma. 250 (1) (2003) 215–226.

[6] M.X. Faraday, The Bakerian Lecture—Experimental relations of gold (and other metals) to light, Philos. Trans. R. Soc. Lond. (147) (1857) 145–181.

[7] R. Jayakumar, et al., Biomedical applications of chitin and chitosan based nanomaterials—A short review, Carbohydr. Polym. 82 (2) (2010) 227–232.

[8] K. Azuma, et al., Chitin, chitosan, and its derivatives for wound healing: old and new materials, J. Funct. Biomater. 6 (1) (2015) 104–142.

[9] H. Nagahama, et al., Preparation of biodegradable chitin/gelatin membranes with GlcNAc for tissue engineering applications, Carbohydr. Polym. 73 (3) (2008) 456–463.

[10] H. Nagahama, et al., Novel biodegradable chitin membranes for tissue engineering applications, Carbohydr. Polym. 73 (2) (2008) 295–302.

[11] H. Tamura, et al., Biomedical applications of chitin hydrogel membranes and scaffolds, Carbohydr. Polym. 84 (2) (2011) 820–824.

[12] N.L.B.M. Yusof, L.Y. Lim, E. Khor, Preparation and characterization of chitin beads as a wound dressing precursor, J. Biomed. Mater. Res. 54 (1) (2001) 59–68.

[13] R. Jayakumar, R. Reis, J. Mano, Phosphorous containing chitosan beads for controlled oral drug delivery, J. Bioactive Compat. Polym. 21 (4) (2006) 327–340.

[14] R. Jayakumar, et al., Sulfated chitin and chitosan as novel biomaterials, Int. J. Biol. Macromol. 40 (3) (2007) 175–181.

[15] R. Jayakumar, et al., Synthesis, characterization and biospecific degradation behavior of sulfated chitin, Macromolecular Symposia, 264, 1st ed., Wiley Online Library, Weinheim: WILEY-VCH Verlag., 2008, pp. 163–167.

[16] K. Madhumathi, et al., Wet chemical synthesis of chitosan hydrogel–hydroxyapatite composite membranes for tissue engineering applications, Int. J. Biol. Macromol. 45 (1) (2009) 12–15.

[17] K. Shalumon, et al., Electrospinning of carboxymethyl chitin/poly (vinyl alcohol) nanofibrous scaffolds for tissue engineering applications, Carbohydr. Polym. 77 (4) (2009) 863–869.

[18] K. Shalumon, et al., Single step electrospinning of chitosan/poly (caprolactone) nanofibers using formic acid/acetone solvent mixture, Carbohydr. Polym. 80 (2) (2010) 413–419.

[19] M. Prabaharan, Chitosan derivatives as promising materials for controlled drug delivery, J. Biomater. Appl. 23 (1) (2008) 5–36.

[20] A. Anitha, et al., Synthesis, characterization, cytotoxicity and antibacterial studies of chitosan, O-carboxymethyl and N, O-carboxymethyl chitosan nanoparticles, Carbohydr. Polym. 78 (4) (2009) 672–677.

[21] A. Dev, et al., Novel carboxymethyl chitin nanoparticles for cancer drug delivery applications, Carbohydr. Polym. 79 (4) (2010) 1073–1079.

[22] A. Portero, et al., Development of chitosan sponges for buccal administration of insulin, Carbohydr. Polym. 68 (4) (2007) 617–625.

[23] K. Muramatsu, et al., In vitro degradation behavior of freeze-dried carboxymethyl-chitin sponges processed by vacuum-heating and gamma irradiation, Polym. Degrad. Stabil. 81 (2) (2003) 327–332.

[24] M. Peter, et al., Nanocomposite scaffolds of bioactive glass ceramic nanoparticles disseminated chitosan matrix for tissue engineering applications, Carbohydr. Polym. 79 (2) (2010) 284–289.

[25] M. Prabaharan, R. Jayakumar, Chitosan-graft-β-cyclodextrin scaffolds with controlled drug release capability for tissue engineering applications, Int. J. Biol. Macromol. 44 (4) (2009) 320–325.

[26] M. Prabaharan, J. Mano, Chitosan-based particles as controlled drug delivery systems, Drug Deliv. 12 (1) (2004) 41–57.

[27] R. Jayakumar, et al., Chitosan conjugated DNA nanoparticles in gene therapy, Carbohydr. Polym. 79 (1) (2010) 1–8.

[28] R. Jayakumar, et al., Graft copolymerized chitosan—present status and applications, Carbohydr. Polym. 62 (2) (2005) 142–158.

[29] R. Jayakumar, et al., Novel chitin and chitosan nanofibers in biomedical applications, Biotechnol. Adv. 28 (1) (2010) 142–150.

[30] R. Jayakumar, et al., Novel carboxymethyl derivatives of chitin and chitosan materials and their biomedical applications, Progr. Mater. Sci. 55 (7) (2010) 675–709.

[31] A. Anitha, et al., Chitin and chitosan in selected biomedical applications, Progr. Mater. Sci. 39 (9) (2014) 1644–1667.

[32] S. Ifuku, et al., Preparation of chitin nanofibers with a uniform width as α-chitin from crab shells, Biomacromolecules 10 (6) (2009) 1584–1588.

[33] M.I. Shams, H. Yano, Simplified fabrication of optically transparent composites reinforced with nanostructured chitin, J. Polym. Environ. 21 (4) (2013) 937–943.

[34] P. Bragd, H. Van Bekkum, A. Besemer, TEMPO-mediated oxidation of polysaccharides: survey of methods and applications, Top. Catal. 27 (1-4) (2004) 49–66.

[35] A. Isogai, T. Saito, H. Fukuzumi, TEMPO-oxidized cellulose nanofibers, Nanoscale 3 (1) (2011) 71–85.

[36] T. Saito, et al., Homogeneous suspensions of individualized microfibrils from TEMPO-catalyzed oxidation of native cellulose, Biomacromolecules 7 (6) (2006) 1687–1691.

[37] Y. Fan, T. Saito, A. Isogai, TEMPO-mediated oxidation of β-chitin to prepare individual nanofibrils, Progr. Mater. Sci. 77 (4) (2009) 832–838.

[38] Y. Fan, T. Saito, A. Isogai, Chitin nanocrystals prepared by TEMPO-mediated oxidation of α-chitin, Biomacromolecules 9 (1) (2008) 192–198.

[39] Y. Fan, T. Saito, A. Isogai, Preparation of chitin nanofibers from squid pen β-chitin by simple mechanical treatment under acid conditions, Biomacromolecules 9 (7) (2008) 1919–1923.

[40] S. Ifuku, et al., Fibrillation of dried chitin into 10–20 nm nanofibers by a simple grinding method under acidic conditions, Progr. Mater. Sci. 81 (1) (2010) 134–139.

[41] Y. Fan, T. Saito, A. Isogai, Individual chitin nano-whiskers prepared from partially deacetylated α-chitin by fibril surface cationization, Progr. Mater. Sci. 79 (4) (2010) 1046–1051.

[42] C. Zhong, et al., A facile bottom-up route to self-assembled biogenic chitin nanofibers, Soft Matter 6 (21) (2010) 5298–5301.

[43] T.H. Tran, et al., Five different chitin nanomaterials from identical source with different advantageous functions and performances, Progr. Mater. Sci. 205 (2019) 392–400.

[44] J.-F. Revol, R. Marchessault, In vitro chiral nematic ordering of chitin crystallites, Int. J. Biol. Macromol. 15 (6) (1993) 329–335.

[45] X.M. Dong, J.-F. Revol, D.G. Gray, Effect of microcrystallite preparation conditions on the formation of colloid crystals of cellulose, Cellulose 5 (1) (1998) 19–32.

[46] J.-B. Zeng, et al., Chitin whiskers: an overview, Biomacromolecules 13 (1) (2012) 1–11.

[47] J.-i. Kadokawa, et al., Preparation of chitin nanowhiskers using an ionic liquid and their composite materials with poly (vinyl alcohol), Carbohydr. Polym. 84 (4) (2011) 1408–1412.

[48] A. Alishahi, et al., Chitosan nanoparticle to carry vitamin C through the gastrointestinal tract and induce the non-specific immunity system of rainbow trout (*Oncorhynchus mykiss*), Carbohydr. Polym. 86 (1) (2011) 142–146.

[49] M.A. Mohammed, et al., An overview of chitosan nanoparticles and its application in non-parenteral drug delivery, Pharmaceutics 9 (4) (2017) 53.

[50] A. Maitra, et al., Process for the preparation of highly monodispersed polymeric hydrophilic nanoparticles, Google Patents, 1999.

[51] Y. Wang, et al., Adsorption of bovin serum albumin (BSA) onto the magnetic chitosan nanoparticles prepared by a microemulsion system, Bioresour. Technol. 99 (9) (2008) 3881–3884.

[52] K. Zhao, et al., Preparation and immunological effectiveness of a swine influenza DNA vaccine encapsulated in chitosan nanoparticles, Vaccine 29 (47) (2011) 8549–8556.

[53] Y. Zhuo, et al., Quenching of the emission of peroxydisulfate system by ferrocene functionalized chitosan nanoparticles: a sensitive “signal off” electrochemiluminescence immunosensor, Sens. Actuators B Chem. 192 (2014) 791–795.

[54] K. Nagpal, S.K. Singh, D.N. Mishra, Chitosan nanoparticles: a promising system in novel drug delivery, Chem. Pharma. Bull. 58 (11) (2010) 1423–1430.

[55] W. Tiyaboonchai, Chitosan nanoparticles: a promising system for drug delivery, Naresuan University Journal: Sci. Technol. 11 (3) (2013) 51–66.

[56] N. Bhattarai, et al., Chitosan and lactic acid-grafted chitosan nanoparticles as carriers for prolonged drug delivery, Int. J. Nanomed. 1 (2) (2006) 181.

[57] W. Suginta, P. Khunkaewla, A. Schulte, Electrochemical biosensor applications of polysaccharides chitin and chitosan, Chem. Rev. 113 (7) (2013) 5458–5479.

[58] J.L. Arias, M.a.S. Fernández, Polysaccharides and proteoglycans in calcium carbonate-based biomineralization, Chem. Rev. 108 (11) (2008) 4475–4482.

[59] B.J. Mcafee, et al., Biosorption of metal ions using chitosan, chitin, and biomass of *Rhizopus oryzae*, Sep. Sci. Technol. 36 (14) (2001) 3207–3222.

[60] R. Ramya, et al., Biomedical applications of chitosan: an overview, J. Biomater. Tissue Eng. 2 (2) (2012) 100–111.

[61] J.-K.F. Suh, H.W. Matthew, Application of chitosan-based polysaccharide biomaterials in cartilage tissue engineering: a review, Biomaterials 21 (24) (2000) 2589–2598.

[62] S. Thomas, Alginate dressings in surgery and wound management—Part 1, J. wound care 9 (2) (2000) 56–60.

[63] H.K. Noh, et al., Electrospinning of chitin nanofibers: degradation behavior and cellular response to normal human keratinocytes and fibroblasts, Biomaterials 27 (21) (2006) 3934–3944.

[64] R. Jayakumar, et al., Biomaterials based on chitin and chitosan in wound dressing applications, Biotechnol. Adv. 29 (3) (2011) 322–337.

[65] K.A. Rieger, N.P. Birch, J.D. Schiffman, Designing electrospun nanofiber mats to promote wound healing–a review, J. Mater. Chem. B 1 (36) (2013) 4531–4541.

[66] H. Ueno, et al., Accelerating effects of chitosan for healing at early phase of experimental open wound in dogs, Biomaterials 20 (15) (1999) 1407–1414.

[67] H. Ueno, T. Mori, T. Fujinaga, Topical formulations and wound healing applications of chitosan, Adv. Drug Deliv. Rev. 52 (2) (2001) 105–115.

[68] T. Kean, M. ThanouP.A. Williams (Ed.), Chitin and chitosan: sources, production and medical applications, Renewable Resources for Functional Polymers and Biomaterials, The Royal Society of Chemistry (2011) 292–318. 10.1039/9781849733519-00292.

[69] P.S. Kumar, et al., Preparation and characterization of novel β-chitin/nanosilver composite scaffolds for wound dressing applications, Carbohydr. Polym. 80 (3) (2010) 761–767.

[70] K. Madhumathi, et al., Development of novel chitin/nanosilver composite scaffolds for wound dressing applications, J. Mater. Sci. Mater. Med. 21 (2) (2010) 807–813.

[71] L. Kossovich, Y. Salkovskiy, I. Kirillova, Electrospun chitosan nanofiber materials as burn dressing, In: Proc. 6th World Congress of Biomechanics (WCB 2010), Singapore, Springer, 2010 August 1–6, 2010.

[72] J.-H. Jang, O. Castano, H.-W. Kim, Electrospun materials as potential platforms for bone tissue engineering, Adv. Drug Deliv. Rev. 61 (12) (2009) 1065–1083.

[73] Q. Wang, et al., Injectable PLGA based colloidal gels for zero-order dexamethasone release in cranial defects, Biomaterials 31 (18) (2010) 4980–4986.

[74] Q. Wang, et al., Biodegradable colloidal gels as moldable tissue engineering scaffolds, Adv. Mater. 20 (2) (2008) 236–239.

[75] X. Li, et al., Nanostructured scaffolds for bone tissue engineering, J. Biomed. Mater. Res. A 101 (8) (2013) 2424–2435.

[76] X. Liu, P.X. Ma, Polymeric scaffolds for bone tissue engineering, Ann. Biomed. Eng. 32 (3) (2004) 477–486.

[77] M.P. Prabhakaran, J. Venugopal, S. Ramakrishna, Electrospun nanostructured scaffolds for bone tissue engineering, Acta Biomater. 5 (8) (2009) 2884–2893.

[78] J. Venkatesan, S.-K. Kim, Chitosan composites for bone tissue engineering—an overview, Marine Drugs 8 (8) (2010) 2252–2266.

[79] A. Di Martino, M. Sittinger, M.V. Risbud, Chitosan: a versatile biopolymer for orthopaedic tissue-engineering, Biomaterials 26 (30) (2005) 5983–5990.

[80] S.K.L. Levengood, M. Zhang, Chitosan-based scaffolds for bone tissue engineering, J. Mater. Chem. B 2 (21) (2014) 3161–3184.

[81] W. Thein-Han, R. Misra, Biomimetic chitosan–nanohydroxyapatite composite scaffolds for bone tissue engineering, Acta Biomater. 5 (4) (2009) 1182–1197.

[82] L. Kong, et al., A study on the bioactivity of chitosan/nano-hydroxyapatite composite scaffolds for bone tissue engineering, Eur. Polym. J. 42 (12) (2006) 3171–3179.

[83] H. Liu, et al., The promotion of bone regeneration by nanofibrous hydroxyapatite/chitosan scaffolds by effects on integrin-BMP/Smad signaling pathway in BMSCs, Biomaterials 34 (18) (2013) 4404–4417.

[84] Y. Zhang, et al., Electrospun biomimetic nanocomposite nanofibers of hydroxyapatite/chitosan for bone tissue engineering, Biomaterials 29 (32) (2008) 4314–4322.

[85] G. Toskas, et al., Chitosan (PEO)/silica hybrid nanofibers as a potential biomaterial for bone regeneration, Carbohydr. Polym. 94 (2) (2013) 713–722.

[86] D.V.H. Thien, et al., Electrospun chitosan/hydroxyapatite nanofibers for bone tissue engineering, J. Mater. Sci. 48 (4) (2013) 1640–1645.

[87] S. Higashi, et al., Polymer-hydroxyapatite composites for biodegradable bone fillers, Biomaterials 7 (3) (1986) 183–187.

[88] M. Ito, In vitro properties of a chitosan-bonded hydroxyapatite bone-filling paste, Biomaterials 12 (1) (1991) 41–45.

[89] P. Samoila, et al.L.A.M. Van Den Broek, C.G. Boeriu (Eds.), Chitin and Chitosan for Water Purification, Chitin and Chitosan: Properties and Applications (2019) 429–460. https://doi.org/10.1002/9781119450467.ch17.

[90] V.K. Thakur, M.K. Thakur, R.K. Gupta, Graft copolymers of natural fibers for green composites, Carbohydr. Polym. 104 (2014) 87–93.

[91] S.A. Figueiredo, J. Loureiro, R. Boaventura, Natural waste materials containing chitin as adsorbents for textile dyestuffs: batch and continuous studies, Water Res. 39 (17) (2005) 4142–4152.

[92] H.-Y. Zhu, et al., A novel magnetically separable γ-Fe_2O_3/crosslinked chitosan adsorbent: preparation, characterization and adsorption application for removal of hazardous azo dye, J. Hazard. Mater. 179 (1-3) (2010) 251–257.

[93] F. Renault, et al., Cationized starch-based material as a new ion-exchanger adsorbent for the removal of CI Acid Blue 25 from aqueous solutions, Bioresour. Technol. 99 (16) (2008) 7573–7586.

[94] M. Liu, et al., Adsorption and desorption of copper (II) from solutions on new spherical cellulose adsorbent, J. Appl. Polym. Sci. 84 (3) (2002) 478–485.

[95] W.W. Ngah, L. Teong, M.M. Hanafiah, Adsorption of dyes and heavy metal ions by chitosan composites: A review, Carbohydr. Polym. 83 (4) (2011) 1446–1456.

[96] I. Anastopoulos, et al., Chitin adsorbents for toxic metals: a review, Int. J. Mol. Sci. 18 (1) (2017) 114.

[97] G. Pigatto, et al., Chitin as biosorbent for phenol removal from aqueous solution: equilibrium, kinetic and thermodynamic studies, Chem. Eng. Process. 70 (2013) 131–139.

[98] K. Yoshizuka, Z. Lou, K. Inoue, Silver-complexed chitosan microparticles for pesticide removal, Reactive Funct. Polym. 44 (1) (2000) 47–54.

[99] R. Celis, et al., Montmorillonite–chitosan bionanocomposites as adsorbents of the herbicide clopyralid in aqueous solution and soil/water suspensions, J. Hazard. Mater. 209 (2012) 67–76.

[100] T. Rêgo, et al., Statistical optimization, interaction analysis and desorption studies for the azo dyes adsorption onto chitosan films, J. Coll. Interface Sci. (411) (2013) 27–33.

[101] Q. Peng, et al., Adsorption of dyes in aqueous solutions by chitosan–halloysite nanotubes composite hydrogel beads, Micropor. Mesopor. Mater. 201 (2015) 190–201.

[102] K. Bhavani, et al., Chitosan a low cost adsorbent for electroplating waste water treatment, J. Bioremediat. Biodegrad. 7 (2016) 346.

[103] X. Lu, et al., UV-initiated synthesis of a novel chitosan-based flocculant with high flocculation efficiency for algal removal, Sci. Total Environ. 609 (2017) 410–418.

[104] S.P. Strand, T. Nordengen, K. Østgaard, Efficiency of chitosans applied for flocculation of different bacteria, Water Res. 36 (19) (2002) 4745–4752.

[105] S. Bratskaya, et al., Enhanced flocculation of oil-in-water emulsions by hydrophobically modified chitosan derivatives, Coll. Surf. A 275 (1-3) (2006) 168–176.

[106] J. Li, et al., Chitosan-based nanomaterials for drug delivery, Molecules 23 (10) (2018) 2661.

[107] J.J. Wang, et al., Recent advances of chitosan nanoparticles as drug carriers, Int. J. Nanomed. 6 (2011) 765.

[108] R.A. Muzzarelli, et al., Chitin nanofibrils/chitosan glycolate composites as wound medicaments, Carbohydr. Polym. 70 (3) (2007) 274–284.

[109] K. Roy, et al., Oral gene delivery with chitosan–DNA nanoparticles generates immunologic protection in a murine model of peanut allergy, Nat. Med. 5 (4) (1999) 387–391.

[110] M. Kumar, A.K. Behera, R.F. Lockey, J. Zhang, G. Bhullar, C.P. De La Cruz, L.C. Chen, K.W. Leong, S.K. Huang, S.S. Mohapatra, Intranasal gene transfer by chitosan-DNA nanospheres protects BALB/c mice against acute respiratory syncytial virus infection, Hum. Gene Ther. 13 (2002) 1415–1425.

[111] K.A. Janes, et al., Chitosan nanoparticles as delivery systems for doxorubicin, J. Control. Release 73 (2-3) (2001) 255–267.

[112] Y. Pan, et al., Bioadhesive polysaccharide in protein delivery system: chitosan nanoparticles improve the intestinal absorption of insulin in vivo, Int. J. Pharma. 249 (1-2) (2002) 139–147.

[113] T.H. Kim, et al., Galactosylated chitosan/DNA nanoparticles prepared using water-soluble chitosan as a gene carrier, Biomaterials 25 (17) (2004) 3783–3792.

[114] D.F. Kendra, L.A. Hadwiger, Characterization of the smallest chitosan oligomer that is maximally antifungal to *Fusarium solani* and elicits pisatin formation in *Pisum sativum*, Exp. Mycol. 8 (3) (1984) 276–281.

[115] L. Qi, et al., Cytotoxic activities of chitosan nanoparticles and copper-loaded nanoparticles, Bioorg. Med. Chem. Lett. 15 (5) (2005) 1397–1399.

[116] K. Gupta, M.R. Kumar, Drug release behavior of beads and microgranules of chitosan, Biomaterials 21 (11) (2000) 1115–1119.

[117] Y. Hu, et al., Synthesis and characterization of chitosan–poly (acrylic acid) nanoparticles, Biomaterials 23 (15) (2002) 3193–3201.

[118] T. Sato, T. Ishii, Y. Okahata, In vitro gene delivery mediated by chitosan. Effect of pH, serum, and molecular mass of chitosan on the transfection efficiency, Biomaterials 22 (15) (2001) 2075–2080.

[119] Y. Kawamata, et al., Receptor-independent augmentation of adenovirus-mediated gene transfer with chitosan in vitro, Biomaterials 23 (23) (2002) 4573–4579.

[120] A. Ahmed, et al., Dual-tuned drug release by nanofibrous scaffolds of chitosan and mesoporous silica microspheres, J. Mater. Chem. 22 (48) (2012) 25027–25035.

[121] Z. Meng, et al., Fabrication, characterization and in vitro drug release behavior of electrospun PLGA/chitosan nanofibrous scaffold, Mater. Chem. Phys. 125 (3) (2011) 606–611.

[122] D. Elieh-Ali-Komi, M.R. Hamblin, Chitin and chitosan: production and application of versatile biomedical nanomaterials, Int. J. Adv. Res. 4 (3) (2016) 411.

[123] H. Wang, J. Qian, F. Ding, Emerging chitosan-based films for food packaging applications, J. Agric. Food Chem. 66 (2) (2018) 395–413.

[124] V. Coma, A. BartkowiakL.A.M. Van Den Broek, C.G. Boeriu (Eds.), Potential of chitosans in the development of edible food packaging, Chitin and Chitosan: Properties and Applications (2019) 349–369. https://doi.org/10.1002/9781119450467.ch14.

[125] J.F. Rubilar, et al., Effect of nanoclay and ethyl-Nα-dodecanoyl-l-arginate hydrochloride (LAE) on physico-mechanical properties of chitosan films, LWT Food Sci. Technol. 72 (2016) 206–214.

[126] M. Zhou, et al., Starch/chitosan films reinforced with polydopamine modified MMT: effects of dopamine concentration, Food Hydrocoll. 61 (2016) 678–684.

[127] M. Vlacha, et al., On the efficiency of oleic acid as plasticizer of chitosan/clay nanocomposites and its role on thermo-mechanical, barrier and antimicrobial properties–comparison with glycerol, Food Hydrocoll. 57 (2016) 10–19.

[128] W. Ma, et al., Fabrication and characterization of kidney bean (*Phaseolus vulgaris* L.) protein isolate-chitosan composite films at acidic pH, Food Hydrocoll. 31 (2) (2013) 237–247.

[129] Z. Song, et al., Combination of nisin and ε-polylysine with chitosan coating inhibits the white blush of fresh-cut carrots, Food Control 74 (2017) 34–44.

[130] H. Wang, et al., Rheological properties, antimicrobial activity and screen-printing performance of chitosan-pigment ($FeO(OH)\cdot xH_2O$) composite edible ink, Prog. Org. Coat. 111 (2017) 75–82.

[131] H. Cui, et al., Improving anti-listeria activity of cheese packaging via nanofiber containing nisin-loaded nanoparticles, LWT Food Sci. Technol. 81 (2017) 233–242.

[132] D.R. Perinelli, et al., Chitosan-based nanosystems and their exploited antimicrobial activity, Eur. J. Pharma. Sci. 117 (2018) 8–20.

[133] P. Morganti, et al., A phosphatidylcholine hyaluronic acid chitin–nanofibrils complex for a fast skin remodeling and a rejuvenating look, Clinic. Cosm. Investigat. Dermatol. 5 (2012) 213.

[134] Y.-C. Chung, et al., Effect of abiotic factors on the antibacterial activity of chitosan against waterborne pathogens, Bioresour. Technol. 88 (3) (2003) 179–184.

[135] C.-Y. Chen, Y.-C. Chung, Antibacterial effect of water-soluble chitosan on representative dental pathogens *Streptococcus mutans* and *Lactobacilli brevis*, J. Appl. Oral Sci. 20 (6) (2012) 620–627.

[136] F. Devlieghere, A. Vermeulen, J. Debevere, Chitosan: antimicrobial activity, interactions with food components and applicability as a coating on fruit and vegetables, Food Microbiol. 21 (6) (2004) 703–714.

[137] D. Wei, et al., The synthesis of chitosan-based silver nanoparticles and their antibacterial activity, Carbohydr. Polym. 344 (17) (2009) 2375–2382.

[138] X. Huang, et al., Green synthesis of silver nanoparticles with high antimicrobial activity and low cytotoxicity using catechol-conjugated chitosan, RSC Adv. 6 (69) (2016) 64357–64363.

[139] M. Sweet, I. Singleton, Silver nanoparticles: a microbial perspective, Advances in Applied Microbiology, Editors: Laskin, Allen I. Sariaslani, Sima Gadd, Geoffrey M., Elsevier, 2011, pp. 115–133. https://doi.org/10.1016/B978-0-12-387044-5.00005-4.

[140] A. Kubacka, M. Ferrer, M. Fernández-García, Kinetics of photocatalytic disinfection in TiO_2-containing polymer thin films: UV and visible light performances, Appl. Catal. B Environ. 121 (2012) 230–238.

[141] F.R. Tamara, et al., Antibacterial effects of chitosan/cationic peptide nanoparticles, Nanomaterials 8 (2) (2018) 88.

[142] G. Kravanja, et al., Chitosan-based (nano) materials for novel biomedical applications, Molecules 24 (10) (2019) 1960.

[143] L. Qi, et al., Preparation and antibacterial activity of chitosan nanoparticles, Carbohydr. Polym. 339 (16) (2004) 2693–2700.

[144] Y.-C. Chung, et al., Relationship between antibacterial activity of chitosan and surface characteristics of cell wall, Acta Pharmacol. Sin. 25 (7) (2004) 932–936.

[145] I. Helander, et al., Chitosan disrupts the barrier properties of the outer membrane of Gram-negative bacteria, Int. J. Food Microbiol. 71 (2-3) (2001) 235–244.

[146] H. Kauss, W. Jeblick, A. Domard, The degrees of polymerization and N-acetylation of chitosan determine its ability to elicit callose formation in suspension cells and protoplasts of *Catharanthus roseus*, Planta 178 (3) (1989) 385–392.

[147] N. Sudarshan, D. Hoover, D. Knorr, Antibacterial action of chitosan, Food Biotechnol. 6 (3) (1992) 257–272.

[148] X. Fei Liu, et al., Antibacterial action of chitosan and carboxymethylated chitosan, J. Appl. Polym. Sci. 79 (7) (2001) 1324–1335.

[149] A. Varma, S. Deshpande, J. Kennedy, Metal complexation by chitosan and its derivatives: a review, Carbohydr. Polym. 55 (1) (2004) 77–93.

12

Polymer-coated magnetic nanoparticles

Adnan Khan[a], Sumeet Malik[a], Muhammad Bilal[b], Nisar Ali[a], Lingli Ni[c], Xiaoyan Gao[c] and Kun Hong[c]

[a]INSTITUTE OF CHEMICAL SCIENCES, UNIVERSITY OF PESHAWAR, KHYBER PAKHTUNKHWA, PAKISTAN. [b]SCHOOL OF LIFE SCIENCE AND FOOD ENGINEERING, HUAIYIN INSTITUTE OF TECHNOLOGY, HUAI'AN, CHINA. [c]KEY LABORATORY FOR PALYGORSKITE SCIENCE AND APPLIED TECHNOLOGY OF JIANGSU PROVINCE, NATIONAL & LOCAL JOINT ENGINEERING RESEARCH CENTRE FOR DEEP UTILIZATION TECHNOLOGY OF ROCK-SALT RESOURCE, FACULTY OF CHEMICAL ENGINEERING, HUAIYIN INSTITUTE OF TECHNOLOGY, HUAI'AN, CHINA.

Chapter outline

12.1 Introduction

Nanotechnology has emerged as an advanced field of science, and currently, it builds the basis of knowledge in many areas [1]. The nanomaterials tend to be made at the atomic or molecular level with a size range of 1–100 nm, prompting to have the unique property of small size and large surface-to-volume ratio. Such nanomaterials have proved to be potent agents taking over various fields such as optics, catalysis, biomedical fields, biotech and microbiology, ecological restoration (wastewater treatment being the prominent one), and engineering [2,3]. Nanomaterials constitute a whole class with a variety of compounds in different morphologies being exploited by researchers as well as the industrial community as nanocatalysts [4,5],

Biopolymeric Nanomaterials: Fundamentals and Applications. DOI: https://doi.org/10.1016/B978-0-12-824364-0.00019-8

nanosorbents [6–9], nanobiosensors [10], oil-water separation [11–20], etc. However, magnetic nanoparticles (MNPs) are prominent due to their magnetic properties as well as nontoxicity, and simple synthesis methods make it liable toward utilization in many fields. The magnetic property associated with iron-based NPs allows the more convenient separation through magnetic decantation [21]. The iron-based NPs are categorized into various classes depending upon their final forms, including alloys [22], metal oxides [7], and metal ferrites [23]. The most common forms of iron-based MNPs that have a greater scope in separation and wastewater remediation include nanoscale zero-valent iron (nZVI) [24], magnetite (Fe_3O_4) [25], and maghemite (γ-Fe_2O_3) [26]. These forms of iron NPs differ based on the oxidation state difference of iron as well as their activity toward environmental remediation. Both magnetite and maghemite possess ferromagnetic properties based on their spine structure. One common practice is the MNPs tend to agglomerate, hence, increasing their size to a larger extent reducing the availability of the active sites. Another issue associated with bare MNPs is their vulnerability toward oxidation [27]. Taking into consideration these oddities of the MNPs, the fabrication of these MNPs has offered a promising way of minimizing such concerns. The development of super paramagnetic iron oxide nanoparticles (SPIONs) turns to one's advantage as such MNPs possess the least remnant magnetization thusly reducing the risk of agglomeration and maintaining the hydrodynamic diameter of the MNPs. On that account, SPIONs could be thought of as the best choice in separation and environment remediation strategies based on their stability (electrostatic and satirical), nontoxicology, smaller size being the most outstanding features [28]. The surface coating/fabrication of such MNPs is necessary, due to their pyrophoric nature, which further leads to their stability and shield from oxidation, preserving their useful properties. Many organic and inorganic compounds have been used for the coating of the SPIONs[29]. One example is using long chains of alkyl surfactants including oleic acid, which helps to inhibit the agglomeration through steric shielding [30]. Derivatives of materials such as phosphoric acid and dopamine provide extra functionalities rendering their stability [31,32]. Yet another common functionalization has been the silica coating of the MNPs providing stabilization and omitting any extra interactions of the MNPs with the attached materials. The silica coating is mostly obtained using tetraethoxysilane as the precursor agent. The surface of the MNPs becomes functionalized with the silanol groups, enhancing the activity of the bare MNPs [33]. The above-discussed materials pose to be suitable fabricating materials for the safety of the MNPs, yet there are some ambiguities related to such functionalities. Like in the case of silica coating, the absence of hydroxyl groups on metal nanoparticle (NP) surfaces may require an additional need of a primer [34]. An appropriate alternative to the above-mentioned fabrications could be polymeric materials that provide the MNPs with plenty of new functional groups, which also stabilize them.

Polymeric materials have proved to be the best choice for the fabrication of MNPs based upon their easy availability, economic stability, extensive usage, etc. The most commonly used polymers for the fabrication of MNPs include synthetic polymers such as polyethylene, polyester, Teflon, nylon as well as natural polymers such as chitosan, dextrin, proteins, cellulose, etc. [35]. The coating of the suitable polymer onto the surface of the MNPs is done either by absorbing the coating materials onto the surface of MNPs, resulting in a core-shell type of structure with the nano-range sizes (20–200 nm) or by dispersing the MNPs in the matrix of the polymer resulting

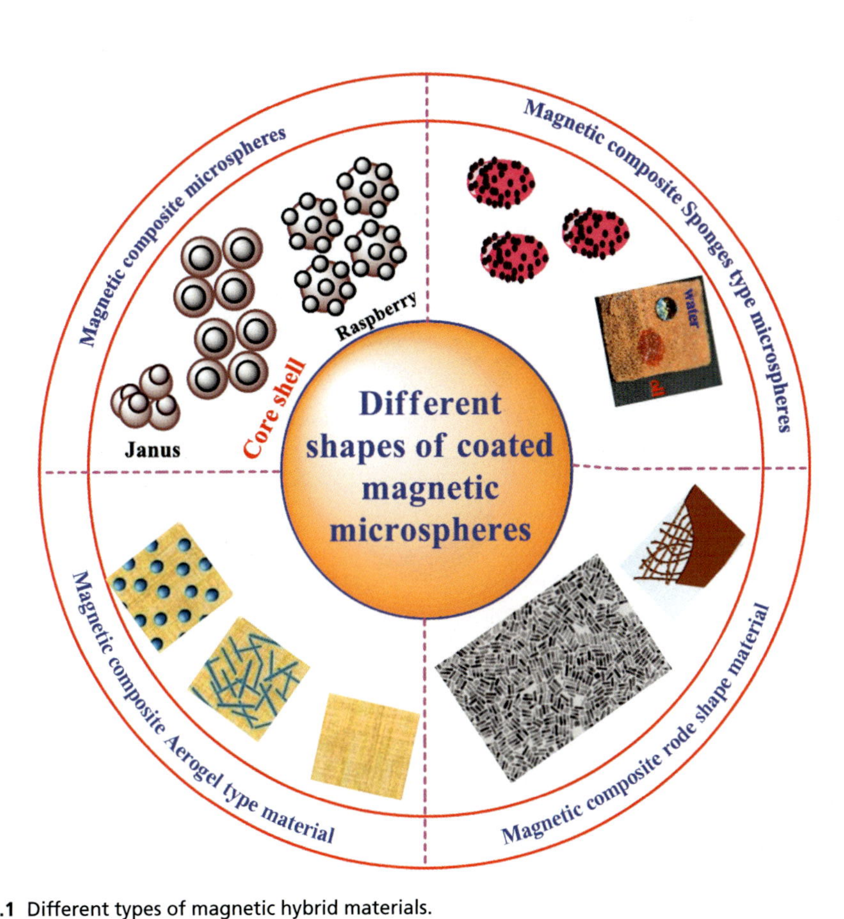

FIGURE 12.1 Different types of magnetic hybrid materials.

in the formation of beads with a micrometer-size range. The size of the beads could be reduced by homogenizing them with appropriate pressures leading to the conversion into nano-range particles (Fig 12.1).

12.2 Components of polymer-coated MNPs

12.2.1 Magnetic NPs

The MNPs are being extensively employed in every field presently, biomedical [36], optics [37], magnetic fluids [38], and wastewater remediation techniques [12,39–41]. The most commonly utilized forms of MNPs include metallic, bimetallic, and SPIONs. These forms of MNPs have non-toxic nature and reactive surfaces, which makes them the right choice for applications in different fields [42]. Variety of reports are available for the synthesis of MNPs by different methodologies that include thermal decomposition [43], hydrothermal synthesis [44], sonochemical synthesis [45], coprecipitation [46], etc. These methods have widely been utilized and have their points of interest, but lately, coprecipitation has emerged as a massively followed strategy due to its

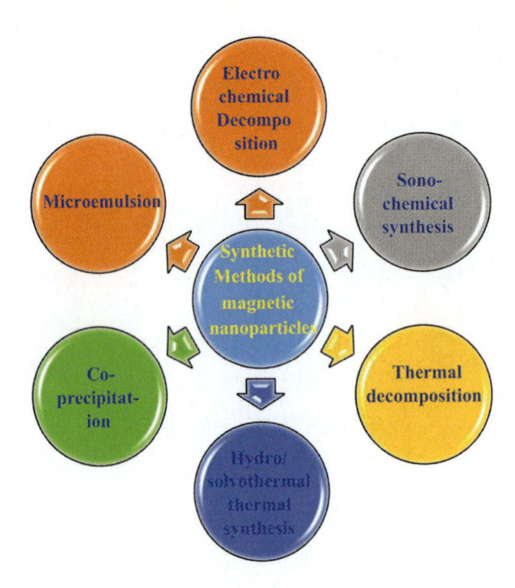

FIGURE 12.2 Different methods for the synthesis of magnetic nanoparticles.

feasibility, cost-effectiveness, and high yield [47]. Although MNPs have proved to be useful in various fields, yet a problem is still to be faced, which is due to the internal instability of the MNPs, which causes their agglomeration, thereby reducing the availability of the active sites. Also, such aggregation leads to decreased surface energy by lowering the high surface to volume ratios. Besides, the surface of the MNPs is highly reactive, which causes its oxidation if exposed to air. Taking into consideration all these factors, measures are necessary to be taken to overcome these issues, which divert the attention of the researchers to the surface coating or functionalization with proper materials. The surface fabrication of MNPs protects the surface of the particles, inhibits leaching as well as provide additional functionalities enhancing their properties in many applications (Fig. 12.2).

12.2.2 Synthetic polymers

Owing to their useful characteristics, a variety of synthetic polymers are being used for the coating/fabrication of the MNPs. The polymers stabilize and enhance the activity of the MNPs up to the degree of their capacity and characteristics [48]. Some commonly used polymers include chitosan [11], dextran [49], polyvinyl alcohol (PVA) [50], polyacrylic acid [51], polyethyleneimine [52], etc. Chitosan is a linear copolymer with β-(1$\rightarrow$4) linked 2-amino-2-deoxy-D-glucopyranose residues, which is hydrophilic and cationic [53]. Based upon its biocompatibility, and ease in preparation, it has dramatically been used in various fields. The most common usage of chitosan is for the fabrication of the MNPs producing chitosan-coated MNPs, which could be utilized for multiple purposes explicitly speaking, the wastewater treatment purposes [54]. Dextran is also a biocompatible polymer that has dramatically been utilized for the fabrication of polymers for its stability inducing properties [55]. PVA exhibits emulsifying and adhesive properties, which render strength to the MNPs inhibiting their agglomeration [56]. Saad et al. [57] studied the preparation of hematite@chitosan core/organically shell nanocomposite (NC). They checked the efficiency of the prepared polymer-coated NC for the removal of toxic heavy metals like Cu^{2+}, Pb^{2+}, and Cd^{2+} with an excellent sorption capacity of 76%, 94%, and 83%, respectively. The morphological studies revealed that the hematite NPs are well-dispersed inside the encapsulation of the chitosan having a rhombus shape and a size range of 40–80 nm. The NC exhibits a crystalline structure. Rezazadeh et al. [58] synthesized a ferromagnetic NC of PVA-coated magnetite Fe_3O_4@PVA for the successive removal of mercury (Hg^{2+}). The morphological features suggested the almost spherical shape of the PVA-coated MNPs with the diameter range of 4–19 nm. The surface area of the Fe_3O_4@PVA was found to be 79.73 m^2/g. The polymer-coated MNPs exhibited an excellent sorption capacity of 73.27 mg/g. Zheng et al. [59] prepared polyacrylamide modified chitosan (CS-PAA) coated on to the surface of silica combined MNPs (FS). The MNPs were synthesized by the solvothermal method, while the polymer coating on to the surface was done via mechanical stirring. The morphological studies confirmed a core-shell structure of FS@CS-PAA with a diameter thickness of 65 nm. The prepared polymer-coated MNPs were used as a sorbent for the removal of cationic dyes and metal ions. The FS@CS-PAA NC presented a high sorption capacity of 2330.17 and 1044.06 mg/g, respectively, for crystal violet and methylene blue dyes while the removal percentage for Cu^{2+} and Ni^{2+} was found to be >86% and >75%.

12.3 Construction of polymer-coated MNPs

The polymer coating of the MNPs presents some impressive points of interest such as stability, minimal agglomeration, additional functionalities, mechanical and thermal properties, electroactivity, etc. To obtain these plus points of polymer coating onto the surface of the MNPs, various pathways have been followed to date. But particular challenges are yet to be faced while synthesizing polymer-coated magnetic NCs due to high costs and the aggregation of the MNPs, thereby reducing the actives sites and diminishing their useful properties. Hence, there is an utmost need to develop methods for the fabrication of polymers onto the MNPs with high proportions of homogeneity and well dispersion. The casting of the fabrication of polymers onto the surface of MNPs falls into two main categories, including the *ex-situ* pathway and the *in-situ* strategy. A generalized view of both approaches is given as follows. Fig. 12.3 explains different fabrication techniques of polymer-coated magnetic materials.

12.3.1 Fabrication of polymers through *ex-situ* approaches

The *ex-situ* approach is the classical procedure for the synthesis of the polymer-coated MNPs composites. In simpler words, the technique could be called as the blending of direct compounding strategy. The main idea behind the approach is the direct mixing of both the components together, leading to product formation. The benefits of following such methods are their cost-effectiveness as well as good efficiency irrespective of the polymer or NPs being utilized [60]. The most common forms of this technique include hot-melt extrusion (HME), solution, emulsion, and by applying mechanical strength. One of the most common methods is the HME process that

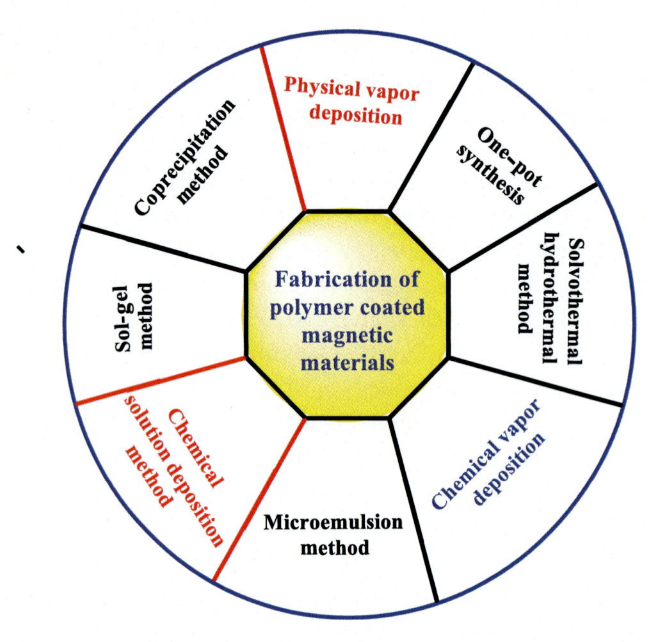

FIGURE 12.3 Different fabrication techniques of polymer-coated magnetic materials.

tends to be a cost-effective way toward the production of high quantity products [61]. Koo et al. [62] utilized the HME strategy to reduce the size of $FeSO_4$ by fabricating its surface with nonionic surfactants such as Span 80 and Tween 80 as well as a hydrophilic polymer, poly(ethylene glycol) 6000. The whole process intended to reduce the size of the $FeSO_4$ powder by the coating of a specific ratio of the surfactant and the polymeric material based on their useful properties such as surface tension reduction. The HME strategy provided some valuable properties such as continuity in the process, no need for solvent, convenient scale-up, and versatile applications, including 3D printing. The HME proceeds with blending the NPs and the fabricating components in proper proportions through various steps such as melting, conveying, dispersing, and homogenizing leading to the formation of nano-sized particles dispersed in distilled water. The inductively coupled plasma-optical emission spectroscopy (ICP-OES) speculations confirmed 100% encapsulation of the NPs without any loss through the HME technique. The finally obtained $FeSO_4$-based (ISNPs) possess a hydrodynamic size of 350–400 nm and a spherical morphology. The ISNPs were then analyzed for their antiproliferation properties and apoptosis potentiality. Ebrahim et al. [63] followed the mechanical mixing of the prepared magnetic NPs (β-FeOOH) and polyaniline conducting polymer (PANI). The first step involved the preparation of Akaganéite NPs by the coprecipitation method, while polyaniline was prepared by oxidative chemical polymerization of pristine aniline. The prepared NPs and PANI were mixed in a ratio of PANI (70%) and NPs (30%) and were ground together by using a mortar leading to the formation of a composite of PANI/akaganéite NC. The morphological features of the NC showed that PANI was fabricated as a white layer onto the surface of blackish NPs, giving a granular particle structure with the particle size range of 8.95 to 16.21 nm. The prepared NC was then utilized for the successful removal of

FIGURE 12.4 Different shapes of polymer-coated magnetic materials.

hazardous Cr(VI) ions with a removal efficiency of 17.36 mg/g. Zu at al. [64] prepared an NC of propylene and silica through a hot emulsion sol-gel technique incorporating a network of polymer interpenetrated inside silica. The surface morphology analysis confirmed the irregular shapes of the prepared NCs with a size range of 150–172 nm and a degree of crystallinity of 10.5%. The thermal stability was also observed to be the highest with a limit of 490 °C.

The morphology analysis showed that there was no presence of the polymer onto the surface of silica, thus confirming its entrapment inside the silica particles. Although these methods come off as handy due to their economic features and feasibility, one of the problems associated with these methods is the agglomeration of the NPs, thus resulting in a nonuniform structure of the composite. This led to the development of various *in-situ* approaches. Fig. 12.4 explains different shapes of polymer-coated MNPs.

12.3.2 Fabrication of polymers through *in-situ* approaches

The primary issue associated with the synthesis of polymer-coated MNPs is the agglomeration of the MNPs, which decreases the active sites minimizing their applicability in wastewater remediation purposes. This problem could be overcome by following the *in-situ* methods for the preparation of polymer-coated magnetic NCs. The *in-situ* methods follow the strategy of mixing the components *in-situ*, developing a chemical reaction of the interaction between them, resulting in a homogenous and uniform NC. The *in-situ* techniques are generally based on the principles of combining the already synthesized MNPs with the polymer counterparts or mixing of previously formed polymeric materials with the MNPs precursors, or both the MNPs and the polymeric materials could be synthesized simultaneously. Based on these principles, a variety of techniques have been utilized for the preparation of polymer-coated nanomaterials, including sol-gel, dispersion, mini-emulsion, suspension cross-linking, surface-initiated atom transfer radical polymerization (ATRP), reversible addition-fragmentation chain transfer polymerization (RAFT), etc. Li et al. [65] studied the preparation of silver NPs–loaded-magnetic biochar/polydopamine NC through the *in-situ* reduction method and checked the efficiency of the prepared NC for the catalytic degradation of hazardous dyes such as methylene blue, rhodamine B, and methyl orange with a removal efficiency of up to 90%. The magnetic property of the catalyst led to its good recyclability for up to five cycles ensuring its cost-effectiveness and efficiency. The morphological features of the prepared Ag nanoparticles loaded on polydopamine coated magnetic biochar (MC-PDA-Ag NC) showed an irregular spherical shape with a uniform and crystalline structure. Rani and Shankar [66] also studied the *in-situ* preparation of poly(methyl-methacrylate)/metal oxide NC and utilized it for the photocatalytic degradation of an organic dye methylene blue. The spherical and semi-crystalline Fe_3O_4-PMMA (poly(methyl methacrylate)), ZnO-PMMA, CuO-PMMA, and Ni_2O_3-PMMA NCs were self-assembled following a green route utilizing plant extract of *Sapindus mukorossi*. The prepared NC showed a high degradation efficiency in the range of 90–99%. The incorporation of PMMA enhances the degradation efficiency by providing stability, enhanced active sites, and ease for $OH^{\bullet}$ radical access leading to more degradation. Wang et al. [67] exploited the preparation of boronic acid functionalized Fe_3O_4 MNPs *through* surface-initiated ATRP. The whole process was completed in two steps, first the MNPs were synthesized using the solvothermal method and second, the poly(acrylamide phenylboronic acid) (AAPBA) functionalized Fe_3O_4 MNPs (Fe_3O_4@PAAPBA) were synthesized through the surface-initiated ATRP technique. The prepared sorbent was used for the removal of proteins and showed excellent sorption efficiency. The morphological analysis showed that the prepared NC was spherical with a diameter of 315 ± 35 nm that was increased due to the fabrication of the polymer.

12.3.3 Covalent approach for fabrication of polymers on to MNPs

The covalent fabrication of polymers onto the surface of the MNPs requires preceding functionalization with a proper material to introduce additional functionalities to capture the polymer coating better. Mostly, SiO_2 or thiols are incorporated with the MNPs, which are then further functionalized using polymer coating. The common pathway is the grafting of the polymer, which is then further subdivided as grafting-to, grafting-from, and grafting-through approach. Table 12.1 summarizes the fabrication of MNPs with polymers through different techniques.

12.3.3.1 Grafting method

While following the grafting-to method, the end-point of the polymer contains functionalities that can be attached to the modified NP surface, like the MNPs modified with silica. The grafting-from approach proceeds through the surface functionalization of the MNPs using a

Table 12.1 Fabrication of MNPs with polymers through different techniques.

Polymer	MNPs	Fabrication technique	Application	Reference
Chitosan	Fe_3O_4	Reverse micro-emulsion precipitation	Pb^{2+} removal	[73]
Chitosan	Fe_3O_4	*In-situ* coprecipitation	Cu^{2+} removal	[74]
Poly (acrylamido-amineamidoxime-co-2-acrylamido-2-methylpropane sulfonic acid (AO/AMPS)	Fe_3O_4	Radical solution polymerization technique	U(VI) removal	[75]
Acrylonitrile and AMPS) monomers	Fe_3O_4	*In-situ* solution polymerization	Co^{2+} and Ni^{2+}	[76]
Poly(N-isopropylacrylamide) (PNIPAM)	Fe_3O_4	Grafting-through	Treatment of emulsified oily water	[77]
Poly(acrylic acid)	Fe_3O_4	*In-situ* chemical precipitation	Pb^{2+} AND crystal violet	[78]
Poly[2-(dimethylamino)ethyl methacrylate]	Fe_3O_4	*In-situ* coprecipitation	Pickering emulsion system preparation	[79]
Poly(sodium p-styrenesulfonate hydrate) and carboxymethyl chitosan	Fe_3O_4	Radical polymerization	Ciprofloxacin	[80]
Polystyrene	Fe_3O_4	Grafting	Oil removal	[81]
Bronsted acidic I.L., 1-vinyl-3-(3-sulfopropyl)imidazolium hydrogen sulfate	Fe_3O_4/SiO_2	Radical grafting copolymerization	Esterification	[82]
Melamine ring covalent organic framework	Fe_3O_4	*In-situ* grafting	Auramine O and rhodamine B removal	[83]
Poly(methyl methacrylate) (PMMA)	Super-paramagnetic iron oxide nanoparticles (SPIONs)	Grafting	Pb(II), Hg(II), Cu(II) and Co(II)	[84]
Poly(γ-glutamic acid)	Fe_3O_4	Co-precipitation	Cr^{3+}, Cu^{2+}, Pb^{2+}, and Ni^{2+} removal	[85]

(*continued on next page*)

Table 12.1 *(continued)*

Polymer	MNPs	Fabrication technique	Application	Reference
3-aminopropyltriethoxysilane, acrylic acid and crotonic acid	Fe_3O_4	Free radical polymerization	Crystal violet, methylene blue and alkali blue 6B	[86]
[2-methacryloyloxy)ethyl trimethyl-ammonium chloride	Fe_3O_4	ATRP	Cr(VI)	[87]
Polystyrene	Fe_3O_4	Emulsion polymerization	Oil absorption	[88]
Poly(2-aminoethyl methacrylate hydrochloride)	Fe_3O_4	Surface-initiated conventional radical polymerization	Hg(II)	[89]
Catechol and branched polyethylenimine	Fe_3O_4	Copolymerization	Methyl blue, orange G and amaranth	[90]
Polyvinylpyrrolidone (PVP)	Fe_3O_4	One step solvothermal method	Emulsified oil removal	[91]
β-CD, succinic anhydride (S.A.), 2, 3-glycidyltrimethylammonium chloride	Fe_3O_4	Ring-opening polymerization	Congo red and Cr(VI)	[92]
Chitosan	Fe_3O_4	Mechanical method	Heavy metals	[93]
Poly(ionic liquid)	Fe_3O_4	Reversible-deactivation radical polymerization method	Alizarin red (A.R.), Thionin acetate (T.A.), Malachite green (M.G.), and Acid orange II (A.O.)	[94]
3-aminopropyl trimethoxysilane and dendrimer-like polyamidoamine	Fe_3O_4	Controlled/living polymerization	Cd(II), Co(II), Zn(II), Pb(II), and Cu(II)	[95]
PVP	Fe_3O_4	Hydrothermal synthesis	Oil remediation	[96]
Extracellular polymeric substance of *Klebsiella species*	Fe_3O_4	Coprecipitation	Ag^+ remediation	[97]
Alginate	γ-Fe_2O_3	Coprecipitation	Methylene blue sorption	[98]
Carboxyl functionalized	Magnetic porous organic framework	*In-situ* surface polymerization	Malachite green	[99]
Poly(vinylbenzyl chloride)	Fe_3O_4	*In-situ* method	Pb^{2+} removal	[100]

suitable initiator, which then initiates the fabrication of the polymer on to the surface of the MNPs through polymerization approach, let us say, ATRP. The typical initiators include *N*-(2-aminoethyl)-2-bromo-2-methylpropanamide [68]. The grafting-through approach follows the condensation of the polymerizable groups such as γ-methacryloxy-propyl-trimethoxysilane [69]. Afzal et al. [70] studied the preparation of epoxy-functionalized polymer grafted MNPs. The epoxy functionalized polymers were attached to the surface of the MNPs by surface-initiated ATRP utilizing the initiators obtained through electron transfer of glycidyl methacrylate (GMA). The prepared polymer-coated MNPs were used to check their efficiency in the immobilization of an enzyme Candida Antarctica lipase B (Cal-B). The morphological features of the prepared composite showed a core-shell structure of the having superparamagnetic properties and a saturation magnetization of 20.02 emu/g. Biehl and Schacher [71] studied the fabrication of polymers on the thiol-functionalized MNPs using the free radical polymerization technique (RAFT). In the first step, the MNPs were silanized using (3-mercaptopropyl) trimethoxy silane to introduce the thiols groups in the MNPs. The thiol-based MNPs were then polymerized by introducing a core shell of the polymers using different monomers such as (*tert*-butyl acrylate (*t*BA), methyl methacrylate, styrene, 2-vinyl pyridine (2VP), and *N*-isopropylacrylamide (NIPAAm). The morphological features showed the presence of 10–30% of the polymer in the polymer-fabricated thiol-based magnetic nanoparticles (MNPS). Pourjavadi et al. [72] studied the fabrication of Poly (N-isopropylacrylamide) (PNIPAM)/GMA on to the silica-modified MNPs (Fe_3O_4@SiO_2). The fabrication of the polymer brushes onto the surface of the MNPs followed the surface-initiated RAFT (SI-RAFT) polymerization. The prepared Fe_3O_4@SiO_2@PNG-Hy was used for drug delivery purposes using doxorubicin (DOX) as a model drug. For better binding of DOX, the polymer brushes were functionalized using hydrazine groups. The hydrazine groups cause the opening of oxirane groups of PNG (poly(N-isopropyl acrylamide-co-glycidyl methacrylate)) on the surface of MNPs, which allows better capture of the DOX.

12.4 Magnetic properties and related functions of polymer-MNP hybrids

The MNPs are so defined based on the magnetic properties they exhibit. These magnetic properties are attributed to the magnetic susceptibility (X), which could be described as the ratio of the degree of magnetization(M) to the applied magnetic field(H). Based on their magnetization and the alignment of the magnetic moments, the MNPs could be categorized as ferromagnetic, ferrimagnetic, paramagnetic, or superparamagnetic NPs [101]. In the case of ferromagnetic particles, the magnetic moments are correctly aligned, having equal magnitude, and ordered states. The small-sized (nano-sized) particles often exhibit a single domain. But the high temperature applied to such NPs may lead to the rotation of the particles and may cause the relaxation of the magnetic moments (Brown relaxation and Neel relaxation), causing a loss in the net magnetization to result in the paramagnetic behavior of the particles. In the case of paramagnetic particles, the magnetization is lost after the removal of the external magnetic field, but they still exhibit colloidal stability and can inhibit agglomeration or aggregation [102]. In the case of ferromagnetic NPs, the atoms or ions tend to have anti-parallel alignment in the absence of the magnetic field below Neel temperature in the magnetic domain [103]. Superparamagnetic behavior could be observed when no remnant magnetization is observed in quasistatic measurements (e.g., vibrating sample magnetometry) but a well-defined hysteresis when a strong external magnetic field is applied. Another kind of magnetic behavior is observed when the superparamagnetic particles form clusters. These clusters do not show any remnant magnetization in the absence of the external magnetic field, but the application of the strong magnetic field shows a ferromagnetic behavior of the NPs depending upon the interaction level among them and the phenomenon is termed as super ferrimagnetism [104,105].

12.5 Conclusion and future perspectives

Nanomaterials, specifically MNPs, gave an outburst of utilization in various fields, including wastewater remediation techniques. Although the MNPs possess key features such as small sizes and high surface to volume ratios, making them worthy of utilizing in the removal of contaminants, the inadequacy of the usage of MNPs may arise due to some reasons. One such reason is their tendency to agglomerate due to high internal energies making them bulky and reducing surface exposure inhibiting their applicability. Also, the high reactivity and unstable nature causes the oxidation of MNPs deploying their activity. To overcome such issues, protective fabrication with suitable polymeric materials has been entertained enormously, providing excellent results. Different fabrication techniques are being followed depending on the means and demands as the fabrication procedure also affects the morphology and, ultimately, the properties of the MNPs. Yet, pretty much expertise is still lacking in the field of the production of the polymer-coated MNPs. Some steps are required to be taken to carry this operation at the industrial levels to enhance the industrial yield, in turn boosting up the economy. The innovative polymerization techniques with a complete hold over the size, shape, and quantity of the product are still needed to be practiced. The given chapter highlights the importance of the polymer-fabricated MNPs, explicitly covering the environmental remediation through wastewater treatment and also infers upon the need to improve this area further so that more exploitation of the beneficial properties of polymer-coated MNPs could be made sure.

Acknowledgment

Authors are grateful to their representative universities/institutes for providing literature facilities.

Conflict of interests

Authors declare no conflict of interest in any capacity, including financial and competing.

References

[1] S.K. Mehta, Impact of functionalized nanomaterials towards the environmental remediation: challenges and future needs, in: M.H. Chaudhery (Ed.), Handbook of Functionalized Nanomaterials for Industrial Applications, Elsevier, Amsterdam, 2020, pp. 505–524.

[2] P. Biehl, M. Von der Lühe, S. Dutz, F.H. Schacher, Synthesis, characterization, and applications of magnetic nanoparticles featuring polyzwitterionic coatings, Polymers 10 (1) (2018) 91.

[3] E. Yilmaz, M. Soylak, Functionalized nanomaterials for sample preparation methods, in: M.H. Chaudhery (Ed.), Handbook of Nanomaterials in Analytical Chemistry, Elsevier, Amsterdam, 2020, pp. 375–413.

[4] N. Ali, A. Khan, M. Bilal, S. Malik, S. Badshah, H. Iqbal, Chitosan-based bio-composite modified with thiocarbamate moiety for decontamination of cations from the aqueous media, Molecules 25 (1) (2020) 226.

[5] O.V. Vodyankina, G.V. Mamontov, V.V. Dutov, T.S. Kharlamova, M.A. Salaev, Ag-containing nanomaterials in heterogeneous catalysis: advances and recent trends, in: V. Sadykov (Ed.), Advanced Nanomaterials for Catalysis and Energy, Elsevier, Cham, 2019, pp. 143–175.

[6] N. Ali, T. Kamal, M. Ul-Islam, A. Khan, S.J. Shah, A. Zada, Chitosan-coated cotton cloth supported copper nanoparticles for toxic dye reduction, Int. J. Biol. Macromol. 111 (2018) 832–838.

[7] A. Khan, S. Badshah, C. Airoldi, Biosorption of some toxic metal ions by chitosan modified with glycidylmethacrylate and diethylenetriamine, Chem. Eng. J. 171 (1) (2011) 159–166.

[8] A. Khan, S. Badshah, C. Airoldi, Dithiocarbamated chitosan as a potent biopolymer for toxic cation remediation, Coll. Surf. B 87 (1) (2011) 88–95.

[9] M. Wen, G. Li, H. Liu, J. Chen, T. An, H. Yamashita, Metal–organic framework-based nanomaterials for adsorption and photocatalytic degradation of gaseous pollutants: recent progress and challenges, Environ. Sci. Nano 6 (4) (2019) 1006–1025.

[10] P. Bondavalli, Nanomaterials for biosensors: fundamentals and applications, MRS Bull. 44 (2019) 317.

[11] N. Ali, S. Ahmad, A. Khan, S. Khan, M. Bilal, S.U. Din, H. Khan, Selenide-chitosan as high-performance nanophotocatalyst for accelerated degradation of pollutants, Chem. Asian J. 15 (17) (2020) 2660–2673.

[12] N. Ali, M. Bilal, A. Khan, F. Ali, Y. Yang, M. Khan, H.M. Iqbal, Dynamics of oil-water interface demulsification using multifunctional magnetic hybrid and assembly materials, J. Mol. Liq. 312 (2020) 113434.

[13] N. Ali, S. Uddin, A. Khan, S. Khan, S. Khan, N. Ali, ..., M. Bilal, Regenerable chitosan-bismuth cobalt selenide hybrid microspheres for mitigation of organic pollutants in an aqueous environment, Int. J. Biol. Macromol. 161 (15) (2020) 1305–1317..

[14] N. Ali., B. Zhang., H. Zhang., et al., Key synthesis of magnetic Janus nanoparticles using a modified facile method, Particuology 17 (2014) 59–65.

[15] N. Ali., B. Zhang., H. Zhang., Novel Janus magnetic microparticle synthesis and its applications as a demulsifier for breaking heavy crude oil and water emulsion, Fuel 141 (2015) 258–267.

[16] N. Ali., B. Zhang., H. Zhang., Interfacially active and magnetically responsive composite nanoparticles with raspberry-like structure; synthesis and its applications for heavy crude oil/water separation, Coll. Surf. A Physicochem. Eng. Asp. 472 (2015) 38–49.

[17] N. Ali., B. Zhang., H. Zhang., et al., Iron oxide-based polymeric magnetic microspheres with a core shell structure: from controlled synthesis to demulsification applications, J. Polym. Res. 22 (11) (2015) 219.

[18] N. Ali., B. Zhang., H. Zhang., et al., Monodispers and multifunctional magnetic composite core shell microspheres for demulsification applications, J. Chin. Chem. Soc. 62 (8) (2015) 695–702.

[19] N. Ali., H Zaman., M. Bilal., A.A Shah., M.N. Nazir., H.M.N. Iqbal., Environmental perspectives of interfacially active and magnetically recoverable composite materials-a review, Sci. Total Environ. 670 (2019) 523–538.

[20] H Zaman., N Ali., A.A Shah., X Gao., S Zhang., K Hong., M Bilal., Effect of pH and salinity on stability and dynamic properties of magnetic composite amphiphilic demulsifier molecules at the oil-water interface, J. Mol. Liq. 290 (2019) 111186.

[21] M.A. Maksoud, A.M. Elgarahy, C. Farrell, H. Ala'a, D.W. Rooney, A.I. Osman, Insight on water remediation application using magnetic nanomaterials and biosorbents, Coord. Chem. Rev. 403 (2020) 213096.

[22] C. Zhang, Q. Chi, J. Zhang, Y. Dong, A. He, X. Zhang, ..., B. Shen, Correlation among the amorphous forming ability, viscosity, free-energy difference and interfacial tension in Fe–Si–B–P soft magnetic alloys, J. Alloys Compd. 831 (2020) 154784.

[23] M. Zahid, N. Nadeem, M.A. Hanif, I.A. Bhatti, H.N. Bhatti, G. Mustafa, Metal ferrites and their graphene-based nanocomposites: synthesis, characterization, and applications in wastewater treatment, in: H.S. Nalwa (Ed.), Magnetic Nanostructures, Springer, Cham, 2019, pp. 181–212.

[24] D. Vollprecht, L.M. Krois, K.P. Sedlazeck, P. Müller, R. Mischitz, T. Olbrich, R. Pomberger, Removal of critical metals from waste water by zero-valent iron, J. Clean. Prod. 208 (2019) 1409–1420.

[25] A. Masudi, G.E. Harimisa, N.A. Ghafar, N.W.C Jusoh, Magnetite-based catalysts for wastewater treatment, Environ. Sci. Pollut. Res. 27 (5) (2020) 4664–4682.

[26] I. Badran, R. Khalaf, Adsorptive removal of alizarin dye from wastewater using maghemite nanoadsorbents, Sep. Sci. Technol. 55 (5) (2019) 1-16.

[27] K.R. Reddy, P.A. Reddy, C.V. Reddy, N.P. Shetti, B. Babu, K. Ravindranadh, S. Naveen, Functionalized magnetic nanoparticles/biopolymer hybrids: synthesis methods, properties and biomedical applications, in: Methods in Microbiology, Elsevier, Academic Press, 2019, vol. 46, pp. 227–254.

[28] A.V. Samrot, C.S. Sahithya, J. Selvarani, S. Pachiyappan, Surface-engineered super-paramagnetic iron oxide nanoparticles for chromium removal, Int. J. Nanomed. 14 (2019) 8105.

[29] A.V. Samrot, P. Senthilkumar, S. Rashmitha, P. Veera, C.S. Sahithya, Azadirachta indica influenced biosynthesis of super-paramagnetic iron-oxide nanoparticles and their applications in tannery water treatment and X-ray imaging, J. Nanostruct. Chem. 8 (3) (2018) 343–351.

[30] A.A. Baharuddin, B.C. Ang, N.A.A. Hussein, A. Andriyana, Y.H Wong, Mechanisms of highly stabilized ex-situ oleic acid-modified iron oxide nanoparticles functionalized with 4-pentynoic acid, Mater. Chem. Phys. 203 (2018) 212–222.

[31] S. Mumtaz, L.S. Wang, M. Abdullah, S.Z. Hussain, Z. Iqbal, V.M. Rotello, I. Hussain, Facile method to synthesize dopamine-capped mixed ferrite nanoparticles and their peroxidase-like activity, J. Phys. D Appl. Phys. 50 (11) (2017) 11LT02.

[32] P. Piotrowski, A. Krogul-Sobczak, A. Kaim, Magnetic iron oxide nanoparticles functionalized with C60 phosphonic acid derivative for catalytic reduction of 4-nitrophenol, J. Environ. Chem. Eng. 7 (3) (2019) 103147.

[33] C. Meng, W. Zhikun, L. Qiang, L. Chunling, S. Shuangqing, H. Songqing, Preparation of amino-functionalized Fe_3O_4@ $mSiO_2$ core-shell magnetic nanoparticles and their application for aqueous Fe^{3+} removal, J. Hazard. Mater. 341 (2018) 198–206.

[34] Q.M. Kainz, O. Reiser, Polymer-and dendrimer-coated magnetic nanoparticles as versatile supports for catalysts, scavengers, and reagents, Acc. Chem. Res. 47 (2) (2014) 667–677.

[35] L. Mohammed, D. Ragab, H. Gomaa, Bioactivity of hybrid polymeric magnetic nanoparticles and their applications in drug delivery, Curr. Pharma. Des. 22 (22) (2016) 3332–3352.

[36] Y. Chen, X. Ding, Y. Zhang, A. Natalia, X. Sun, Z. Wang, H. Shao, Design and synthesis of magnetic nanoparticles for biomedical diagnostics, Quant. Imaging Med. Surg. 8 (9) (2018) 957.

[37] Y.L. Lin, J.H. Chu, H.J. Lu, N. Liu, Z.Q. Wu, Facile synthesis of optically active and magnetic nanoparticles carrying helical poly (phenyl isocyanide) arms and their application in enantioselective crystallization, Macromol. Rapid Commun. 39 (5) (2018) 1700685.

[38] C. Vasilescu, M. Latikka, K.D. Knudsen, V.M. Garamus, V. Socoliuc, R. Turcu, L. Vékás, High concentration aqueous magnetic fluids: structure, colloidal stability, magnetic and flow properties, Soft Matter 14 (32) (2018) 6648–6666.

[39] A. Khan, N. Ali, M. Bilal, S. Malik, S. Badshah, H. Iqbal, Engineering functionalized chitosan-based sorbent material: characterization and sorption of toxic elements, Appl. Sci. 9 (23) (2019) 5138.

[40] S. Sohni, K. Gul, F. Ahmad, I. Ahmad, A. Khan, N. Khan, S. Bahadar Khan, Highly efficient removal of acid red-17 and bromophenol blue dyes from industrial wastewater using graphene oxide functionalized magnetic chitosan composite, Polym. Compos. 39 (9) (2018) 3317–3328.

[41] Y.Q. Zhang, X.B. Yang, Z.X. Wang, J. Long, L. Shao, Designing multifunctional 3D magnetic foam for effective insoluble oil separation and rapid selective dye removal for use in wastewater remediation, J. Mater. Chem. A 5 (16) (2017) 7316–7325.

[42] J. Kudr, Y. Haddad, L. Richtera, Z. Heger, M. Cernak, V. Adam, O. Zitka, Magnetic nanoparticles: from design and synthesis to real world applications, Nanomaterials 7 (9) (2017) 243.

[43] K. Sartori, F. Choueikani, A. Gloter, S. Begin-Colin, D. Taverna, B.P. Pichon, Room temperature blocked magnetic nanoparticles based on ferrite promoted by a three-step thermal decomposition process, J. Am. Chem. Soc. 141 (25) (2019) 9783–9787.

[44] M.T.H. Siddiqui, S. Nizamuddin, H.A. Baloch, N.M. Mubarak, D.K. Dumbre, A.M. Asiri, G.J Griffin, Synthesis of magnetic carbon nanocomposites by hydrothermal carbonization and pyrolysis, Environ. Chem. Lett. 16 (3) (2018) 821–844.

[45] M.A. Almessiere, Y. Slimani, A.D. Korkmaz, N. Taskhandi, M. Sertkol, A. Baykal, B.E.K.İ.R. Ozcelik, Sonochemical synthesis of Eu^{3+} substituted $CoFe_2O_4$ nanoparticles and their structural, optical and magnetic properties, Ultrason. Sonochem. 58 (2019) 104621.

[46] H. Liu, A. Li, X. Ding, F. Yang, K. Sun, Magnetic induction heating properties of Mg1-xZnxFe_2O_4 ferrites synthesized by co-precipitation method, Solid State Sci. 93 (2019) 101–108.

[47] H. Khan, A.K. Khalil, A. Khan, K. Saeed, N. Ali, Photocatalytic degradation of bromophenol blue in aqueous medium using chitosan conjugated magnetic nanoparticles, Kor. J. Chem. Eng. 33 (10) (2016) 2802–2807.

[48] A.S. Jawad, A.F. Al-Alawy, Synthesis and characterization of coated magnetic nanoparticles and its application as coagulant for removal of oil droplets from oilfield produced water, In: Proc. AIP *Conference Proceedings*, (Vol. 2213, No. 1, p. 020174). AIP Publishing LLC. (2020, March).

[49] A.M. Predescu, E. Matei, A.C. Berbecaru, C. Pantilimon, C. Drăgan, R. Vidu, V. Kuncser, Synthesis and characterization of dextran-coated iron oxide nanoparticles, R.Soc. Open Sci. 5 (3) (2018) 171525.

[50] A. Maleki, M. Niksefat, J. Rahimi, R. Taheri-Ledari, Multicomponent synthesis of pyrano [2, 3-d] pyrimidine derivatives via a direct one-pot strategy executed by novel designed copperated Fe_3O_4@ polyvinyl alcohol magnetic nanoparticles, Mater. Today Chem. 13 (2019) 110–120.

[51] Sanchez, L. M., Ochoa Rodríguez, P. A., Actis, D. G., Elías, V. R., Eimer, G. A., Lassalle, V. L. & Alvarez, V. A. (2020). Ecofriendly-developed polyacrylic acid-coated magnetic nanoparticles as catalysts in photo-Fenton processes. Adv. Mater. Lett., 11(3), 20031486

[52] H. Zhao, C. Zhang, D. Qi, T. Lü, D. Zhang, One-Step synthesis of polyethylenimine-coated magnetic nanoparticles and its demulsification performance in surfactant-stabilized oil-in-water emulsion, J. Disp. Sci. Technol. 40 (2) (2019) 231–238.

[53] A. Khan, S. Badshah, C. Airoldi, Environmentally benign modified biodegradable chitosan for cation removal, Polym. Bull. 72 (2) (2015) 353–370.

[54] Radwa, M.A. n, M.A. Rashad, M.A. Sadek, H.A. Elazab, Synthesis, characterization and selected application of chitosan-coated magnetic iron oxide nanoparticles, J. Chem. Technol. Metall. 54 (2) (2019) 303–310.

[55] A.L. Lungoci, M. Pinteala, A.R. Petrovici, I. Rosca, I.A. Turin-Moleavin, A. Fifere, Biosynthesized dextran coated magnetic nanoparticles with antifungal activity, Rev. Roum. Chim. 63 (5-6) (2018) 497–503.

[56] A. Altinisik, K. Yurdakoc, Chitosan-/PVA-coated magnetic nanoparticles for Cu (II) ions adsorption, Desalin. Water Treat. 57 (39) (2016) 18463–18474.

[57] A.A.H. Saad, A. M Azzam, S. T El-Wakeel, B. B Mostafa, M. B Abd El-latif, Industrial wastewater remediation using hematite@ chitosan nanocomposite, Egypt. J. Aquat. Biol. Fish. 24 (1) (2020) 13–29.

[58] L. Rezazadeh, S. Sharafi, M. Schaffie, M. Ranjbar, Application of oxidation-reduction potential (ORP) as a controlling parameter during the synthesis of Fe_3O_4@ PVA nanocomposites from industrial waste (raffinate), Environmental Science and Pollution Research 27 (25) (2020) 32088–32099.

[59] X. Zheng, H. Zheng, Z. Xiong, R. Zhao, Y. Liu, C. Zhao, C. Zheng, Novel anionic polyacrylamide-modify-chitosan magnetic composite nanoparticles with excellent adsorption capacity for cationic dyes and pH-independent adsorption capability for metal ions, Chem. Eng. J. 392 (2020) 123706.

[60] J.S. Koo, S.Y. Lee, S. Nam, M.O.K. Azad, M. Kim, K. Kim, H.J Cho, Preparation of cupric sulfate-based self-emulsifiable nanocomposites and their application to the photothermal therapy of colon adenocarcinoma, Biochem. Biophys. Res. Commun. 503 (4) (2018) 2471–2477.

[61] S.Y. Lee, S. Nam, Y. Choi, M. Kim, J.S. Koo, B.J. Chae, H.J. Cho, Fabrication and characterizations of hot-melt extruded nanocomposites based on zinc sulfate monohydrate and soluplus, Appl. Sci. 7 (9) (2017) 902.

[62] J.S. Koo, S.Y. Lee, M.O.K. Azad, M. Kim, S.J. Hwang, S. Nam, H.J Cho, Development of iron (II) sulfate nanoparticles produced by hot-melt extrusion and their therapeutic potentials for colon cancer, Int. J. Pharma. 558 (2019) 388–395.

[63] S. Ebrahim, A. Shokry, H. Ibrahim, M. Soliman, Polyaniline/akaganéite nanocomposite for detoxification of noxious Cr (VI) from aquatic environment, J. Polym. Res. 23 (4) (2016) 79.

[64] L. Zu, R. Li, L. Jin, H. Lian, Y. Liu, X. Cui, Preparation and characterization of polypropylene/silica composite particle with interpenetrating network via hot emulsion sol–gel approach, Prog. Nat. Sci. 24 (1) (2014) 42–49.

[65] H. Li, D. Jiang, Z. Huang, K. He, G. Zeng, A. Chen, G. Chen, Preparation of silver-nanoparticle-loaded magnetic biochar/poly (dopamine) composite as catalyst for reduction of organic dyes, J. Coll. Interface Sci. 555 (2019) 460–469.

[66] M. Rani, U. Shanker, Sun-light driven rapid photocatalytic degradation of methylene blue by poly (methyl methacrylate)/metal oxide nanocomposites, Coll. Surf. A 559 (2018) 136–147.

[67] J. Wang, X. He, L. Chen, Y. Zhang, Boronic acid functionalized magnetic nanoparticles synthesized by atom transfer radical polymerization and their application for selective enrichment of glycoproteins, RSC Adv. 6 (52) (2016) 47055–47061.

[68] L.G. Bach, M.R. Islam, J.T. Kim, S. Seo, K.T. Lim, Encapsulation of Fe_3O_4 magnetic nanoparticles with poly (methyl methacrylate) via surface functionalized thiol-lactam initiated radical polymerization, Appl. Surf. Sci. 258 (7) (2012) 2959–2966.

[69] Y. Chen, Z. Xiong, L. Zhang, J. Zhao, Q. Zhang, L. Peng, H. Zou, Facile synthesis of zwitterionic polymer-coated core–shell magnetic nanoparticles for highly specific capture of N-linked glycopeptides, Nanoscale 7 (7) (2015) 3100–3108.att type="article"?>

[70] H.A. Afzal, R.V. Ghorpade, A.K. Thorve, S. Nagaraja, B.E. Al-Dhubiab, G. Meravanige, T.S. Roopashree.

[71] P. Biehl, F.H. Schacher, Surface Functionalization of magnetic nanoparticles using a thiol-based grafting-through approach, Surfaces 3 (1) (2020) 116–131.

[72] A. Pourjavadi, M. Kohestanian, C. Streb, pH and thermal dual-responsive poly (NIPAM-co-GMA)-coated magnetic nanoparticles via surface-initiated RAFT polymerization for controlled drug delivery, Mater. Sci. Eng. C 108 (2020) 110418.

[73] R.G. López, M.G. Pineda, G. Hurtado, R.D.D. León, S. Fernández, H. Saade, D Bueno, Chitosan-coated magnetic nanoparticles prepared in one step by reverse microemulsion precipitation, Int. J. Mol. Sci. 14 (10) (2013) 19636–19650.

[74] C. Yuwei, W. Jianlong, Preparation and characterization of magnetic chitosan nanoparticles and its application for Cu (II) removal, Chem. Eng. J. 168 (1) (2011) 286–292.

[75] Z.F. Akl, S.M. El-Saeed, A.M. Atta, In-situ synthesis of magnetite acrylamide amino-amidoxime nanocomposite adsorbent for highly efficient sorption of U (VI) ions, J. Ind. Eng. Chem. 34 (2016) 105–116.

[76] A.M. Atta, H.A. Al-Lohedan, A.O. Ezzat, Z.A. Issa, A.B. Oumi, Synthesis and application of magnetite polyacrylamide amino-amidoxime nano-composites as adsorbents for water pollutants, J.Polym. Res. 23 (4) (2016) 69.

[77] T. Lü, S. Zhang, D. Qi, D. Zhang, H. Zhao, Thermosensitive poly (N-isopropylacrylamide)-grafted magnetic nanoparticles for efficient treatment of emulsified oily wastewater, J. Alloys Compd. 688 (2016) 513–520.

[78] M.A. Mudassir, S.Z. Hussain, A. Jilani, H. Zhang, T.M. Ansari, I. Hussain, Magnetic hierarchically macroporous emulsion-templated poly (acrylic acid)–iron oxide nanocomposite beads for water remediation, Langmuir 35 (27) (2019) 8996–9003.

[79] L.E. Low, C.W. Ooi, E.S. Chan, B.H. Ong, B.T. Tey, Dual (magnetic and pH) stimuli-reversible Pickering emulsions based on poly (2-(dimethylamino) ethyl methacrylate)-bonded Fe_3O_4 nanocomposites for oil recovery application, J. Environ. Chem. Eng. 8 (2) (2020) 103715.

[80] C. Zheng, H. Zheng, C. Hu, Y. Wang, Y. Wang, C. Zhao, Q. Sun, Structural design of magnetic biosorbents for the removal of ciprofloxacin from water, Bioresour. Technol. 296 (2020) 122288.

[81] P.M. Reddy, C.J. Chang, J.K. Chen, M.T. Wu, C.F. Wang, Robust polymer grafted Fe_3O_4 nanospheres for benign removal of oil from water, Appl. Surf. Sci. 368 (2016) 27–35.

[82] W. Xie, H. Wang, Immobilized polymeric sulfonated ionic liquid on core-shell structured Fe_3O_4/SiO_2 composites: a magnetically recyclable catalyst for simultaneous transesterification and esterifications of low-cost oils to biodiesel, Renew. Energy 145 (2020) 1709–1719.

[83] S. Shakeri, Z. Rafiee, K. Dashtian, Fe_3O_4-based melamine-rich covalent organic polymer for simultaneous removal of Auramine O and Rhodamine B, J. Chem. Eng. Data 65 (2) (2020) 696–705.

[84] Y. Wanna, A. Chindaduang, G. Tumcharern, D. Phromyothin, S. Porntheerapat, J. Nukeaw, S. Pratontep, Efficiency of SPIONsfunctionalized with polyethylene glycol bis (amine) for heavy metal removal, J. Magn. Magn. Mater. 414 (2016) 32–37.

[85] J. Chang, Z. Zhong, X.U. Hong, Y.A.O. Zhong, C.H.E.N Rizhi, Fabrication of poly (γ-glutamic acid)-coated Fe_3O_4 magnetic nanoparticles and their application in heavy metal removal, Chinese J. Chem. Eng. 21 (11) (2013) 1244–1250.

[86] F. Ge, H. Ye, M.M. Li, B.X. Zhao, Efficient removal of cationic dyes from aqueous solution by polymer-modified magnetic nanoparticles, Chem. Eng. J. 198 (2012) 11–17.

[87] S. Hanif, A. Shahzad, Removal of chromium (VI) and dye Alizarin Red S (ARS) using polymer-coated iron oxide (Fe_3O_4) magnetic nanoparticles by co-precipitation method, J. Nanopart. Res. 16 (6) (2014) 2429.

[88] L. Yu, G. Hao, J. Gu, S. Zhou, N. Zhang, W. Jiang, Fe_3O_4/PS magnetic nanoparticles: synthesis, characterization and their application as sorbents of oil from wastewater, J. Magn. Magn. Mater. 394 (2015) 14–21.

[89] A. Farrukh, A. Akram, A. Ghaffar, S. Hanif, A. Hamid, H. Duran, B. Yameen, Design of polymer-brush-grafted magnetic nanoparticles for highly efficient water remediation, ACS Appl. Mater. Interfaces 5 (9) (2013) 3784–3793.

[90] Y. Long, L. Xiao, Q. Cao, Co-polymerization of catechol and polyethylenimine on magnetic nanoparticles for efficient selective removal of anionic dyes from water, Powder Technol. 310 (2017) 24–34.

[91] S. Shao, Y. Li, T. Lü, D. Qi, D. Zhang, H. Zhao, Removal of emulsified oil from aqueous environment by using polyvinylpyrrolidone-coated magnetic nanoparticles, Water 11 (10) (2019) 1993.

[92] D. Cai, T. Zhang, F. Zhang, X. Luo, Quaternary ammonium β-cyclodextrin-conjugated magnetic nanoparticles as nano-adsorbents for the treatment of dyeing wastewater: synthesis and adsorption studies, J. Environ. Chem. Eng. 5 (3) (2017) 2869–2878.

[93] A. Esmaeili, N.T. Farrahi, The efficiency of a novel bioreactor employing bacteria and chitosan-coated magnetic nanoparticles, J. Taiwan Inst.Chem. Eng. 59 (2016) 113–119.

[94] H. Yang, J. Zhang, Y. Liu, L. Wang, L. Bai, L. Yang, ..., H. Chen, Rapid removal of anionic dye from water by poly (ionic liquid)-modified magnetic nanoparticles, J. Mol. Liq. 284 (2019) 383–392.

[95] Y. Harinath, D.H.K. Reddy, L.S. Sharma, K Seshaiah, Development of hyperbranched polymer encapsulated magnetic adsorbent (Fe_3O_4@ SiO_2-NH_2-PAA) and its application for decontamination of heavy metal ions, J. Environ. Chem. Eng. 5 (5) (2017) 4994–5001.

[96] S. Mirshahghassemi, A.D. Ebner, B. Cai, J.R. Lead, Application of high gradient magnetic separation for oil remediation using polymer-coated magnetic nanoparticles, Sep. Purific. Technol. 179 (2017) 328–334.

[97] W. Wei, A. Li, S. Pi, Q. Wang, L. Zhou, J. Yang, ..., B.J. Ni, Synthesis of core-shell magnetic nanocomposite Fe_3O_4@ microbial extracellular polymeric substances for simultaneous redox sorption and recovery of silver ions as silver nanoparticles, ACS Sustain. Chem. Eng. 6 (1) (2018) 749–756.

[98] D. Talbot, S. Abramson, N. Griffete, A. Bee, pH-sensitive magnetic alginate/γ-Fe_2O_3 nanoparticles for adsorption/desorption of a cationic dye from water, J. Water Process Eng. 25 (2018) 301–308.

[99] L. Huang, J. Miao, Q. Shuai, Carboxyl-functionalized magnetic porous organic polymers as efficient adsorbent for wastewater remediation, J. Taiwan Inst. Chem. Eng. 109 (2020) 97–102.

[100] Ş. Yılmaz, A. Zengin, Y. Akbulut, T. Şahan, Magnetic nanoparticles coated with aminated polymer brush as a novel material for effective removal of Pb (II) ions from aqueous environments, Environ. Sci. Pollut. Res. 26 (20) (2019) 20454–20468.

[101] C. Blanco-Andujar, D. Ortega, P. Southern, Q.A. Pankhurst, N.T.K Thanh, High performance multi-core iron oxide nanoparticles for magnetic hyperthermia: microwave synthesis, and the role of core-to-core interactions, Nanoscale 7 (5) (2015) 1768–1775.

[102] U.M. Engelmann, C. Shasha, E. Teeman, I. Slabu, K.M. Krishnan, Predicting size-dependent heating efficiency of magnetic nanoparticles from experiment and stochastic Néel-Brown Langevin simulation, J. Magn. Magn. Mater. 471 (2019) 450–456.

[103] A. Akbarzadeh, M. Samiei, S. Davaran, Magnetic nanoparticles: preparation, physical properties, and applications in biomedicine, Nanoscale Res. Lett. 7 (1) (2012) 144.

[104] S. Kalia, S. Kango, A. Kumar, Y. Haldorai, B. Kumari, R. Kumar, Magnetic polymer nanocomposites for environmental and biomedical applications, Coll. Polym. Sci. 292 (9) (2014) 2025–2052.

[105] M. Lazaratos, K. Karathanou, L. Mainas, A. Chatzigoulas, N. Pippa, C. Demetzos, Z. Cournia, Coating of magnetic nanoparticles affects their interactions with model cell membranes, Biochim. Biophys. Acta 1864 (11) (2020) 129671.

Biopolymer-nanoparticles hybrids

Federica Catania[a], Mattia Bartoli[a,b] and Alberto Tagliaferro[a,b,c]

[a]DEPARTMENT OF APPLIED SCIENCE AND TECHNOLOGY, POLITECNICO DI TORINO, TORINO, ITALY. [b]CONSORZIO INTERUNIVERSITARIO NAZIONALE PER LA SCIENZA E TECNOLOGIA DEI MATERIALI (INSTM), FLORENCE, ITALY. [c]FACULTY OF SCIENCE, ONTARIOTECH UNIVERSITY, OSHAWA, ONTARIO, CANADA.

Chapter outline

13.1 Polymers and nanoparticles: long history

In the early years of the twentieth century Ostwald proposed a very catchy definition for colloidal phases namely them after "neglected dimensions." Since then, nanoscience has risen as the new frontier of material science [1]. Nowadays, nanoparticles applications cover a wide range of field from biomedicine [2] to catalysis [3]. Since the pioneering study of Derjaguin, Landau and Verwey and Overbeek [4,5], the stability of nanoparticles colloidal phase was correlated with polymers through the DLVO theory [6]. DLVO theory described the stabilization of a colloidal phase accordingly with the potential profile reported in Fig. 13.1.

DLVO theory combines the effects of the van der Waals attraction and the electrostatic repulsion together with the effect of polymers acting as surfactants. This is the main accountable effect for the nanoparticles suspension stability. The indissoluble bond between nanoparticles and polymers lays the foundation for modern advancement in nanoscience.

Nowadays, the world transition toward a sustainable development paradigm has led to the exploration of environmentally friendly solution instead of the traditional oil-based ones. In this new approach, bioderivates and biopolymers act as main players. Biopolymers are not only offering advantages as cheap and widely available raw materials but also in end-of-life disposal. Furthermore, the complexity of biopolymers represents an unreachable aim for current synthetic

Biopolymeric Nanomaterials: Fundamentals and Applications. DOI: https://doi.org/10.1016/B978-0-12-824364-0.00008-3

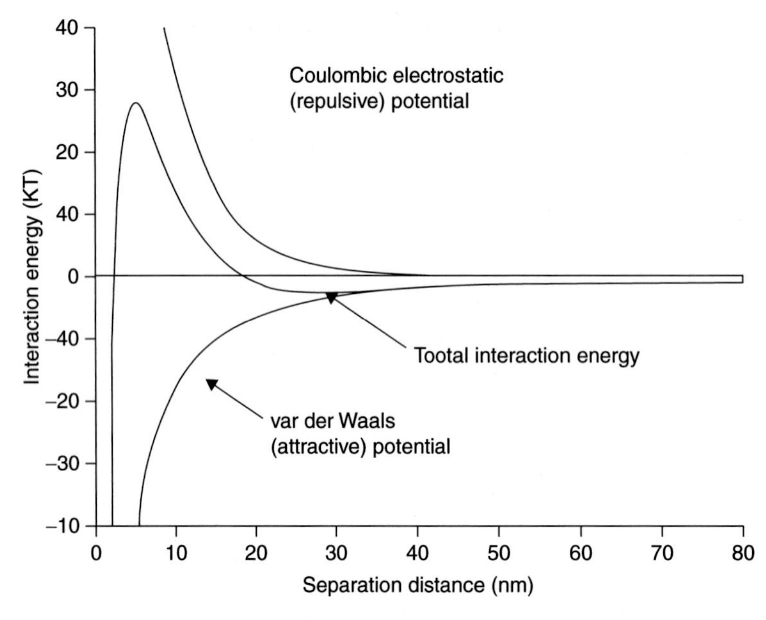

FIGURE 13.1 DLVO potential of colloidal suspension. Potential profile of a colloidal suspension accordingly with DLVO theory.

chemistry. This great complexity and the low environmental impact are counterbalanced by complex purification processes and high cost-price. For these two reasons, biopolymers are struggling to compete with the far cheaper oil derived polymers.

Since 1970s, it was experimentally proved how polymers can behave as a drug delivery system. Theranostic research takes improvements from the development of hybrid systems based on surface-modified nanoparticles. Indeed, biopolymers act as vehicles by coating nanoparticles and improving the specificity on target site. This leads to more efficient drug release and side-effects reduction. Further advantages of biopolymers employment come from the longer circulation time compared with synthetic ones. Finally, they also favored the accumulation on targeted tissues thanks to the enhanced permeation and retention effect.

Nonetheless, the production of hybrid materials represents a sound solution for the use of biopolymers. By the combination between biopolymers and inorganic structures, new highly biocompatibility materials could be produced. Furthermore, the interest increment in metallopolymers has let the development of new materials that find a wide range of applications due to tunability of both their chemical and physical properties [7,8]. The highly complex biopolymers architectures have not been totally exploited its full potential as support for the production of versatile metal hybrid catalysts, adsorptive materials, and biological scaffolds.

In this chapter, we present the most outstanding achievements in the production of hybrid biopolymers materials trying to overview this immense field. We focused our discussion on the most recent advances in the use of several biopolymers such as cellulose, lignin, proteins, and chitosan for a wide range of applications including drugs production, catalysis, environmentally remediation, and composites materials.

13.2 Nanoparticles supported onto biopolymers

13.2.1 Cellulose-based hybrid materials

Cellulose is a fibrous, highly crystalline, water-insoluble polymer that represents the major constituent of lignocellulosic biomasses [9] and one of the most common utilized feedstock for biorefinery and chemical processes [10]. Cellulose has also found plenty of applications in materials science as a starting material for the production of nanoparticles supported species [11]. The hydroxylic functionalities and the high molecular weight of cellulose are two of the main strong points of its use as support for the nanostructures as proved by Musino et al.[12]. In this study, authors proved that the hydroxylic functionalities are able to promote the nucleation of silver nanoparticles. By using unmodified and surface-tailored cellulose nanocrystals, authors also established that size and shape of nanoparticles could be tuned by tuning the amount of hydroxylic residues. A partial conjugation with hydrophobic molecules reduced the size down to 90 nm while the nanostructure sizes increased above 90 nm after a reformation of -OH residues.

The other great advantage in the use of cellulose for hybrid materials production is represented by its high biocompatibility. Accordingly, several authors have used cellulose as a basal structure for the assembling of nano-sized and nanostructured materials for biological applications. Remarkably, Mohammed et al. [13] used density functional theory calculation to evaluate the interactions occurring between cellulose and several nanoscale materials such as single-walled boron nitride nanotube, fullerene, and boron nitride nanosheet. Based on interaction energy, authors speculated that cellulose could be used as a chelating agent for the development of nanovaccines. Nonetheless, they did not provide any evidence of such possibility. A more consistent proof of cellulose hybrid materials for biological applications was reported by Chaabane et al. [14] based on ferrite nanoparticles as sketched in Fig. 13.2.

In this research, authors modified a bacterial cellulose by forming a tetraaza macrocyclic Schiff using ethylenediamine, benzyl chloride, and Fe(II) ions. The resulting material contained Fe_3O_4 particles able to exploit cytotoxic activity toward peripheral blood mononucleocyte with remarkable antitumoral activity *in vivo* by using mice as tester cells. Authors ascribed the effect of the synthesized materials to a Fenton and Fenton-like reactivity induced by the presence of iron on the cellulose surface. Hu et al. [15] evaluated the effect of bimetallic cellulose materials as an antimicrobial agent by decorating unmodified cellulose with silver and gold. Through deeply structural investigation, authors showed that hybrid materials were decorated with bimetallic nanoparticles with an average size of up to around 10 nm with an ultrathin silver-rich outermost shell around a gold core. This multilayered structure magnified the bactericidal effect without the improved eukaryote cytotoxicity. Similarly, Abdelgawad et al. [16] combined silver nanoparticles with carrageenan and cellulose nanocrystal via a solid-state synthetic technique. The cryogel prepared by the authors was characterized by a porous structure with a native antimicrobial effect mainly exploited by the silver cations on the nanoparticles surface. Authors achieved a 100% reduction for *Staphylococcus aureus* and *Escherichia coli* proliferation by using the prepared hybrid material suggesting that the silver-carrageenan/cellulose could be a sound wound dressing material. Bundjaja et al. [17] prepared a cellulose carbamate hydrogel

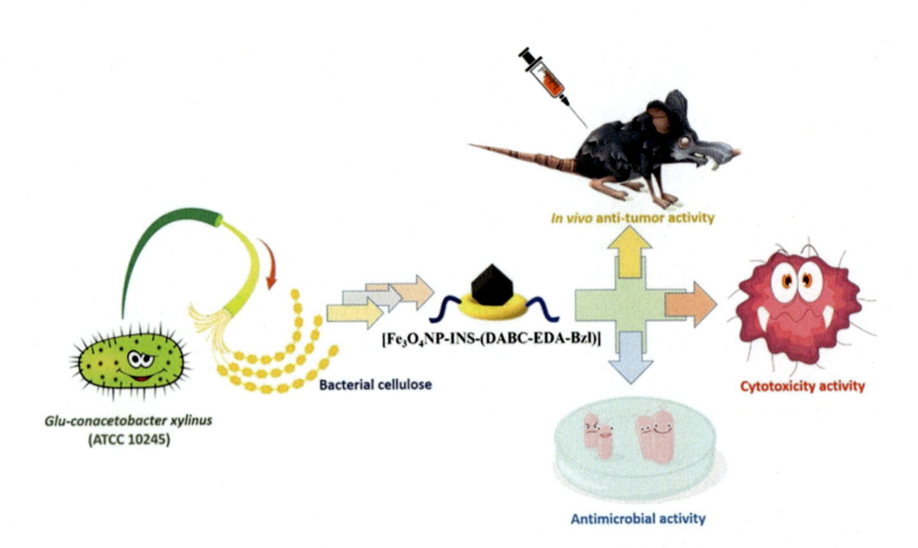

FIGURE 13.2 Effect of magnetite nanoparticles. Schematic effect of ferrite particles cytotoxicity supported onto cellulose From L. Chaabane, H. Chahdoura, R. Mehdaoui, M. Snoussi, E. Beyou, M. Lahcini, M.H. V Baouab, Functionalization of developed bacterial cellulose with magnetite nanoparticles for nanobiotechnology and nanomedicine applications, Carbohydr. Polym. 247 (2020) 116707. https://doi.org/10.1016/j.carbpol.2020.116707.

wound dressing tailored with silver nanoparticles and stabilized by using rarasaponin as a surfactant for enhancing the silver loading of up to 10 mg/g. In the very same field, Madla-Cruz et al.[18] described the antimicrobial activity and inhibition of bacterial biofilm formation *in vitro* and *in vivo* on human dentine of a silver nanoparticles-carboxymethyl-cellulose hybrid material. Authors produced spherical silver/cellulose nanocomposite with an average size of up to 22 nm with a cytotoxicity of up to 89%, by using concentrations $> 15\ \mu g/mL$. An *in-vitro* study showed a good antimicrobial activity of up to 67%, 66%, and 96% using, respectively, *Candida albicans, Escherichia faecalis,* and *Fusobacterium nucleatum*. An *in-vivo* study proved also a reduction of up to 58% of biofilm formation on human dentine proving the suitability for endodontic treatment. El-Naggar et al. [19] proved that silver/titania nanoparticles supported onto cellulose could be used as fungicide by adsorbing and releasing tebuconazole fungicide.

Cellulose hybrid materials have also been used as catalysts. Recently, Xie et al. [20] developed a gold nanoparticles/cellulose monolith hybrid material operating in continuous-flow conditions. Authors used these new hybrid materials for reducing 4-nitrophenol in a watery phase with good conversion values. Furthermore, they applied the same materials in the hydroxylation of dimethylphenylsilane at 200 °C achieving also in this case very promising outputs. Moving on different cutting-edge applications, Suo et al. [21] modified a cellulose-based substrate by using ionic liquids and magnetic nanoparticles. Authors further tailored the surface of these hybrid materials through the immobilization of lipase PPL-IL-MCMC. Immobilized lipase showed an activity 2.8 higher than free enzymes. According to the authors' hypothesis, the activity enhancement was due to the positive effect of cellulose microenvironment leading to a change of the secondary structure of enzyme causing a greater exposure of lipase active sites.

Cellulose hybrid materials have also used in plenty of environmental remediation processes. Elrhman et al. [22] modified cellulose nanocrystals through several graftings by using citric, carbamic, and acroleic acids. The neat and grafted materials acted as support for superparamagnetic magnetite particles used for the removal of thallium 232 from large volume of watery solutions. Ghazitabar et al. [23] produced a graphene aerogel containing both cellulose fibers and magnetite nanoparticles for the recovery of gold from cyanide solutions. Authors claimed a high gold adsorption capacity of up to 130 mg/g. Magnetic cellulose hybrid materials have also been used to dewatering of crude oil emulsions [24], organic dyes [25], or for the removal of emerging pollutants such ibuprofen [26]. Guo et al. [27] tailored cellulose by using CuS nanoparticle for the selective sanification of mercury ions from a watery phase. Authors achieved an astonishing mercury adsorption capability of up to 1040 mg/g after only 2 min of contact time.

Furthermore, cellulose hybrid materials have exited from research labs reaching real market with very promising products. Abdollahi et al. [28] produced an invisible high-security anti-counterfeiting inks based on tailored latex nanoparticles with spiropyran derivatives supported onto cellulose. Authors reported both photochromism and fluorescence of the material upon UV irradiation at the same time. Xu and coworkers [29] exploited the cellulose hybrid materials properties by using them for cultural heritage applications. Authors grafted alkaline nanoparticles on the surface of cellulose creating a hybrid material very useful for both strengthening and deacidification of cellulosic artworks.

13.2.2 Lignin-based hybrid materials

Lignin is a cross-linked biopolymer containing several aromatic units linked together by at least 10 different C—C and C—O bonds [30] forming up to 30% of wood weight. Lignin is a unique biological material because it is the only aromatic biopolymer naturally synthesized. This reason laid on the great interest that lignin has harvested during the years. Consequently, Koch et al.[31] evaluated the life-cycle assessment of lignin nanoparticle production by a biorefinery platform. Authors considered a wide range of parameters (i.e., eutrophication potential, global warming potential, human toxicity potential, water use and scarcity index) to build up a solid and trustworthy instrument for a decision-making approach useful for real case applications. This was particularly interesting because lignin hybrid materials are quite similar for biocompatibility with cellulose one but they are not used for fermentations and generally burned as waste-streams. So, the lignin valorization through a high added cost transformation such as those involved in the hybrid materials production is a very attractive way.

In this field, Matsakas et al. [32] described a very interesting procedure for the production of spherical lignin nanoparticles from different sources as shown in Fig. 13.3.

Authors used biomass biorefinery waste-stream lignin from fermentation of birch and spruce for isolation of spherical lignin particles with an average size of up to 100 nm. Furthermore, authors developed a very simple two-stage isolation procedure based on the collecting of a solvent lignin-rich fraction and a freeze-drying step. Neat lignin nanoparticles could be also produced by enzymatic oxidation [33] or through a sonochemical route [34]. This last approach lead to the production of nanoparticles ranging from 10 nm of up to 50 nm that could be

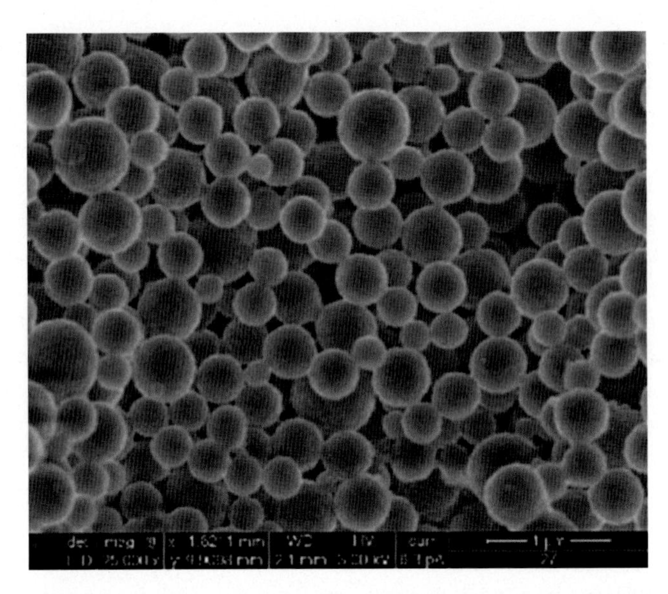

FIGURE 13.3 Lignin nanoparticles. Scanning electron microscope of lignin particles isolated by dialysis coupled with centrifugation from birch From L. Matsakas, M. Gerber, L. Yu, U. Rova, P. Christakopoulos, Preparation of low carbon impact lignin nanoparticles with controllable size by using different strategies for particles recovery, Ind. Crop. Prod. 147 (2020) 112243. https://doi.org/10.1016/j.indcrop.2020.112243.

used for the production of highly cross-linked hydrogel structures useful for drug delivery, food packaging, and wound dressing.

Consequently, Chen et al. [35] produced high mechanical properties of a hydrogel based on poly(acrylamide)/lignin nanoparticle hybrid nanocomposite hydrogel. Authors claimed a magnification of compressive and tensile strengths by using a ratio of poly(acrylamide)/lignin nanoparticle close to 0.1. Furthermore, lignin hybrid material showed an impressive recoverability after multiple solicitations. Authors ascribed the mechanical properties magnification to the synergistic effect of nanocomposite network structure together with hydrogen bonding between poly(acrylamide) chains and lignin nanoparticles. These effects combination promoted an excellent mechanism of redistribution of the applied load.

Hydrogels based on lignin hybrid materials have been also used for the production of antimicrobial materials. Li et al. [36] tailored the surface of a poly(vinyl alcohol)/lignin hydrogel through an *in-situ* reduction of silver nitrate. The nanoparticles hybrid material showed good antibacterial properties in *in-vitro* test on both *S. aureus* and *E. coli* together with a good biocompatibility estimated by using L929 cells. Similarly, Aadil et al. [37] used poly(vinyl alcohol) and acacia-derived lignin nanofibers to create a silver nanoparticles–based hybrid hydrogel. Authors tested the tailored hydrogel against *Bacillus circulans* and *E. coli* showing a good antimicrobial activity in both cases. Wang et al. [38] modified the lignin as quaternized materials and decorated it with silver nanoparticle through microwave-assisted technique. Positively charged quaternized lignin endowed the adhesion through electrostatic effect of negatively charged *E. coli* and *S. aureus* cells magnifying the bacterial activity of up to 99.999% after 30 min without silver leaching. The same approach was used by Yin et al. [39] developing a flocculant formulation based on lignin nanoparticles/gelatin produced from switchgrass for the capture of *S. aureus* and *E. coli.* Authors achieved in both cases a remarkable flocculation efficiency over up to 95% in 30 min at pH 4.5, while the flocculation efficiency of 90% was achieved in 60 min at pH 5.

Antimicrobial effect is not the only one biological feature of lignin hybrid materials that have been found application in drug delivery as biocompatible agents. Bertolo et al. [40] extracted organosolv lignin from sugarcane bagasse for using as pickering emulsions stabilizer and encapsulated antioxidant compounds such as curcumin and polyphenols. This formulation showed a high retaining of curcumin of up to 73% after 96 h. Curcumin-loaded lignin hybrid materials have also been used to healing skin wounds [41]. After 12 days, authors observed wound size reduction of up to 43% with an advanced granulation tissue formation characterized by collagen deposition. Additionally, authors reported a reduction of myeloperoxidase activity associated to a decrement of inflammatory infiltration. Lower expression of matrix metalloproteinases (MMPs), especially MMP9, was characteristic of wounds treated with curcumin-loaded lignin nanopartcles (LNPs).

Siddiqui et al. [42] comprehensively evaluated irinotecan-loaded lignin nanoparticles as drug carriers considering plenty of parameters through an *in-vivo* study based on *Drosophila melanogaster.* This spherical shaped hybrid material has an average size of up to 150 nm with a negative surface charge close to -23 mV. A preliminary *in-vitro* cytotoxicity evaluation showed significant toxicity for human breast adenocarcinoma up to 74.38 ± 4.74%, a lower cytotoxicity in human alveolar epithelial adenocarcinoma 38.8 ± 4.70%, and an insignificant toxicity for human embryonic kidney. An *in-vivo* test showed that the cytotoxicity strongly depended on concentration without inducing nuclei fragmentation or gut cell damage but preventing irinotecan effects. Figueiredo et al. [43] realized a peptide-guided resiquimod-loaded lignin nanoparticles been proved as a favorable therapeutic route for the modulation of the tumor microenvironment. Lignin loaded with resiquimod targeted the CD206-positive M2-like TAMs by using the "mUNO" peptide. This approach allowed to revert M_2 pro-tumor phenotype into antitumor M_1-like macrophages in the tumor microenvironment as proved by *in-vitro* test on an aggressive triple-negative breast cancer. Figueiredo et al. [44] produced a ferrite-tailored lignin hybrid material for *in-vitro* evaluation of antiproliferation effect of cancer cells. Apart from the possibility of use as magnetic resonance imaging agents, ferrite/lignin hybrid materials outperformed poor water-soluble anticancer such as sorafenib and benzazulene.

Furthermore, lignin hybrid materials could be used as a vaccine adjuvant as described by Alqahtani et al. [45]. Author encapsulated a model antigen ovalbumin into lignin nanoparticles with an efficiency of up to 81.6%. *In-vitro* studies on mice proved solid time-dependent systemic immune responses after the administration. Another cutting-edge application was reported by Amini and coworkers [46]. Authors created a hybrid electrospun material based on poly(caprolactone) fibers embedding into lignin nanoparticles for peripheral nerve regeneration. Lignin-based materials showed a good cell viability and adhesion of PC12 and human adipose-derived stem cells showing. Furthermore, cell viability and differentiation along with neurite length extension increased by increasing the lignin content with neural markers expression for differentiated cells.

Lignin hybrid materials have also found an appreciable use for environmentally remediation purposes as proved by Chen et al. [47]. Authors synthesized lignin nanospheres tailored with palladium nanoparticles from kraft lignin isolated from bamboo. This high surface area material was used as a redox catalyst for the removal of chromium and organic dyes. Adsorption was also a very profitable application for lignin hybrid materials tailored with magnetic ferrite-based nanoparticles. This class of hybrid materials has found plenty of applications in adsorption of chromium [48], lead, copper [49], and organic dyes [50,51]. Another interesting application was proposed by Li et al. [52] based on lignin/ferrite hybrid material for the recovery of phosphates. Removal of phosphate from watery waste-streams is a capital issue and its reuse as a fertilizer is a very sound and sustainable approach. The proposed material showed both a high capability to adsorb phosphates and to release them in the soil. Furthermore, on-field real applications are possible due to its easy recoverability from watery phase and the absence of iron leaching in the soil.

Another interesting use of lignin hybrid materials was proposed by Padilha and coworkers [53]. Authors produced hollow poly(caprolactone) microcapsules together with soda lignin nanoparticles for removal of emulsified oil. Soda lignin can form nanoparticles with a high capability of pickering emulsification. Furthermore, soda lignin is unstable enough to precipitate in the opportune conditions inducing demulsification problems. By combining these properties, authors achieved a de-oil of watery phase of up to 95 %.

13.2.3 Protein-based hybrid materials

Proteins and peptides are the main focus of the postgenomic era [54] and represent a precious tool for the realization of bio-derived hybrid materials. Recently, Thalhauser et al. [55] comprehensively studied the production of protein hybrid material produced by the conjugation between aminated silica nanoparticles and the single accessible thiol in human serum albumin as reported in Fig. 13.4.

Authors evaluated the effect of amino groups' density on silica surface on protein conjugation. An amount of upt to 20,000 amino residues has been found as the optimum for an efficient conjugation through chemical bonds while the hybrid materials is mainly assembled by weak interactions at lower function group density. Magro et al. [56] investigated the role of carboxylic groups on protein surface in the recognition of iron oxide nanoparticles and in the production of related hybrid composites. This study deep investigates the interactions between proteins and nanoparticles because protein corona on nanomaterial is responsible for all the physiological response of the organism, influencing cell processes, from transport to accumulation. Authors hypothesized that specific recognition sites are responsible of the noncovalent interaction promoting the nanoparticle harboring and allowing the formation of functional protein coronas. The combination of protein and nanoparticles are particularly interesting also for protein purification processes as described by Chen et al. [57]. Authors used silica nanoparticles to promote lysozyme crystallization effectively against high concentrations of protein impurity bovine serum albumin.

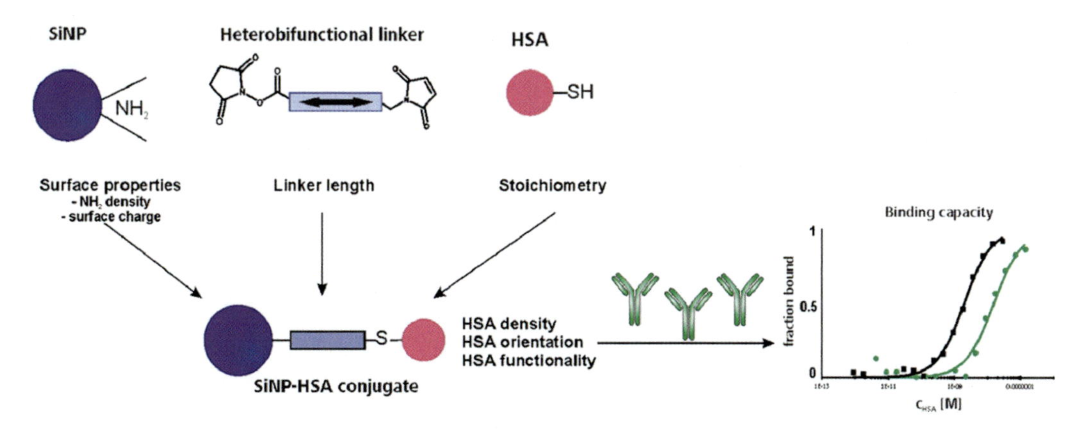

FIGURE 13.4 Synthesis of human serum albumin hybrid materials. Synthesis of human serum albumin hybrid materials through thiol capping with aminated silica nanoparticles. From S. Thalhauser, M. Breunig, Considerations for efficient surface functionalization of nanoparticles with a high molecular weight protein as targeting ligand, Eur. J. Pharma. Sci. 155 (2020) 105520. https://doi.org/10.1016/j.ejps.2020.105520.

Each well-defined protein represents a very powerful tool for the production of biosensors with high sensibility and specificity. Agrawal et al. [58] tailored microtubule-associated proteins (MAP) recombinant protein with gold nanoparticles for the rapid diagnosis of *Mycobacterium avium* subspecies paratuberculosis infection. Authors reached a detection improvement of up to 84% to indirect enzyme-linked immunosorbent assay. Furthermore, protein hybrids were used as an agent for biological drug delivery as overviewed by Martínez-López et al.[59]. In this field several examples have been reported by using protein hybrid materials for delivering antioxidants such as resveratrol [60] or curcumin [61].

At last protein hybrid materials have been widely used as antibacterial materials. Simon et al. [62] described the interaction of gold nanoparticles with bovine serum albumin, human serum albumin, and hen egg white lysozyme. Authors tested the antibacterial activity of protein hybrid gold nanoparticles against *Bacillus pumilus, E. coli, Pseudomonas aeruginosa,* and *S. aureus.* Hen egg white lysozyme based hybrid material showed a tunable antimicrobial activity against *P. aeruginosa* due to the maximum binding of hen egg-white lysozyme (HEWL) with gold nanoparticles while the other materials showed similar performances.

13.2.4 Chitin-derived materials: chitosan-based materials

Chitin is a structural polymer and the main component of cell walls of fungi, crustacean, and insects [63]. Neat chitin is poorly used in material science application and it is generally converted into chitosan. Chitosan is a linear polysaccharide composed of randomly distributed β-(1→4)-linked D-glucosamine and N-acetyl-D-glucosamine (acetylated unit) produced by hydrolysis of chitin shells of shrimp and other crustaceans by using alkali media [64]. Chitosan is widely used as a polymeric matrix for the production of plenty of hybrid materials due to the presence of aminic functionalities in its structure. Due to its chemical structure peculiarity, chitosan is used for the production of hybrid membrane as reported by Jiamjirangkul et al. [65]. Authors produced a metal organic framework tailored chitosan/poly(vinyl alcohol) nanofibrous membrane through electrospinning. This hybrid materials was used for CO_2 capture and filtration. Similarly, Ali et al. [66] produced a chitosan-CuO hybrid nanocomposite as thin film able to selectively detect H_2S at a low temperature. This hybrid membrane showed a fast response together with good sensitivity at 40 °C.

Chitosan hybrid materials have led the recent advancement in bio-based materials for environment remediation. Aadnan and coworkers [67] developed a very active photocatalysts by tailoring the surface of chitosan with zinc oxide nanoparticles promoting the efficient degradation of both methyl orange and paranitrophenol under UV-A and visible light irradiation. Ali et al. [68] synthesized a trimetallic bismuth/cobalt/selenide hybrid microspheres supported onto chitosan for Congo red removal from watery phases by adsorption. Similarly, Banu et al. [69] encapsulated lanthanum into a chitosan/kaolin clay hybrid composite for the recovery of nitrate and phosphate from water. The combination of chitosan with clays was also described by da Silva and coworkers [70]. Authors mixed chitosan with bentonite under microwave irradiation for the removal of Reactive Violet 5R dye from aqueous solutions reaching an adsorption capacity of up to 282 mg/g. Pereira et al. [71] mixed hydroxyapatite with chitosan reticulated with glutaraldehyde for diclofenac removal. Authors claimed a pollutant removal of up to 125 mg/g at pH 3 after only 15 min. The recoverability of chitosan hybrids was improved by adding magnetic particles as reported by Wang et al. [72] for organic dyes and Zeng et al. for arsenic anions [73]. In both cases, the decoration of chitosan with iron nanoparticles improved the recoverability and adsorption properties by both supplying additional surface area and providing specific interactions as in the case of arsenic anions adsorption.

Antimicrobial effects of chitosan-based hybrid materials were widely explored. Bueloni, et al. [74] produced a nalidixic acidic vanadium complex loaded onto chitosan hybrid nanoparticles from 170 to 330 nm assembled by using myristyl myristate. Authors achieved an encapsulation of up to 97% and tested the effects on *E. coli, Bacillus cereus, S. aureus*, and *P. aeruginosa* showing a particularly selectivity for inhibition of *P. aeruginosa*. Similarly, Diosa et al. [75] produced a chitosan/silica hybrid material loaded with KR-12 peptide exhibiting proteolytic stability and good antimicrobial effects. Authors suggested that it could be represented as a viable solution for oral drug administration. Molybdenum sulfide was used by Kasinathan et al. [76] to tailor biogenic chitosan and using it for *in-vitro* antibacterial and anticancer tests. Authors reported antibacterial zones of inhibition against *S. aureus* of up to 32 mm and against *E. coli* of up to 35 mm. Anticancer activity was also promising with a 65% growth inhibition of MCF-7 cell line.

Chitosan hydrogels are very promising as medium for drug delivery and experimental medicine as nerve repair. Furthermore, the inorganic chitosan-based hybrid materials are also very attractive as reported by Wang et al. Authors tailored a carboxymethyl chitosan matrix with a layered double hydroxide hybrid nanocomposite used as drug delivery agent to the posterior ocular segment. Prior to supporting the inorganic species, a peptide transporter-1 and glutathione were used to modify chitosan. The *in-vitro* experiment proved the scarce cytotoxicity together with an enhanced permeability of an ocular model membrane. The *in-vivo* experiments were run using loaded drops involved clathrin-mediated endocytosis and peptide transporter-1 mediated active targeting transport. Results showed diffusion into choroid-retina reaching at very high concentrations.

Baktash et al. [77] developed a hybrid chitosan-based graphene oxide–grafted material tailored with magnetic nanoparticle for theranostic applications. Authors loaded doxorubicin showing that it was adsorbed by the simultaneous contribution of π-π and hydrogen bond formation with chitosan and graphene oxide. Furthermore, authors clearly show that the doxorubicin release could be controlled by pH changing. Additionally, authors reported a T_2 contrast efficacy increased by grafting the iron nanoparticles on the surface of chitosan-based material. The effects on healthy L929 cell lines also showed a very low cytotoxicity proving their possible use as theranostic agents.

Another field of application of chitosan-based hybrid materials is represented by coatings as antimicrobial layer bioimplant such as jugular vein–valved conduit [78] or as an anticorrosive protective layer of mild steel [79]. In both cases, the superiors filming properties of chitosan were combined with entrapped oxide nanoparticles (i.e., zinc oxide, silica) that represent the real protective agent uniformly distributed and immobilized in the chitosan matrix.

13.3 Conclusions

Driven by new environmental regulations, biopolymers have risen in attention as sustainable nonoil-based platforms for materials sciences. Their combination with inorganic species for the realization of hybrid materials represents a groundbreaking event in the mind set of material scientists. This new vision has let to astonishing discoveries in biomedicine, environmental remediation, and productive processes. Even if biopolymers are not easy to manage as traditional polymeric matrix, the amazing complexity derived by their biosynthesis counterbalances the great part of the drawback due to the isolation and purification procedure. Complexity is also the key for the biopolymers hybrid material exploitation representing a powerful tool to tune the biological effects.

References

[1] D.B. Kittelson, Engines and nanoparticles: a review, J. Aerosol Sci. 29 (1998) 575–588. https://doi.org/10.1016/S0021-8502(97)10037-4.

[2] N. Tran, T.J. Webster, Magnetic nanoparticles: biomedical applications and challenges, J. Mater. Chem. 20 (2010) 8760–8767. https://doi.org/10.1039/C0JM00994F.

[3] D. Astruc, Nanoparticles and Catalysis, Wiley-VCH, France, 2008. https://doi.org/10.1002/9783527621323.

[4] V.E.J. W., Theory of the Stability of Lyophobic Colloids, J. Phys. Coll. Chem. (1947) 631–636. https://doi.org/10.1021/j150453a001.

[5] B. Deraguin, L. Landau, Theory of the stability of strongly charged lyophobic sols and of the adhesion of strongly charged particles in solution of electrolytes, Acta Physicochim. 14 (1941) 633–662.

[6] M.Borkovec Trefalt, Overview of DLVO theory, Laboratory of Colloid and Surface Chemistry, University of Geneva, Switzerland, 2014.

[7] G.R. Whittell, M.D. Hager, U.S. Schubert, I. Manners, Functional soft materials from metallopolymers and metallosupramolecular polymers, Nat. Mater. 10 (2011) 176–188. https://doi.org/10.1038/nmat2966.

[8] C. Haensch, M. Chiper, C. Ulbricht, A. Winter, S. Hoeppener, U.S. Schubert, Reversible supramolecular functionalization of surfaces: Terpyridine ligands as versatile building blocks for noncovalent architectures, Langmuir. 24 (2008) 12981–12985. https://doi.org/10.1021/la8026682.

[9] E. Sjostrom, Wood Chemistry: Fundamentals and Applications, Gulf professional publishing, Houston, TX, 1993.

[10] Y.H.P. Zhang, Reviving the carbohydrate economy via multi-product lignocellulose biorefineries, J. Ind. Microbiol. Biotechnol. 35 (2008) 367–375. https://doi.org/10.1007/s10295-007-0293-6.

[11] B. Gabriela, C. Sergiu, Cellulose: a ubiquitous platform for ecofriendly metal nanoparticles preparation, Coord. Chem. Rev. (2019) 155–173. https://doi.org/10.1016/j.ccr.2019.01.007.

[12] D. Musino, C. Rivard, G. Landrot, B. Novales, T. Rabilloud, I. Capron, Hydroxyl groups on cellulose nanocrystal surfaces form nucleation points for silver nanoparticles of varying shapes and sizes, J. Coll. Interface Sci. 584 (2021) 360–371. https://doi.org/10.1016/j.jcis.2020.09.082.

[13] M.H. Mohammed, B.A. Jarullah, F.H. Hanoon, Using of cellulose with various nanoparticles as chelating factors in nanovaccines: density functional theory investigations, Solid State Commun. 316–317 (2020) 113945. https://doi.org/10.1016/j.ssc.2020.113945.

[14] L. Chaabane, H. Chahdoura, R. Mehdaoui, M. Snoussi, E. Beyou, M. Lahcini, M.H.V Baouab, Functionalization of developed bacterial cellulose with magnetite nanoparticles for nanobiotechnology and nanomedicine applications, Carbohydr. Polym. 247 (2020) 116707. https://doi.org/10.1016/j.carbpol.2020.116707.

[15] X. Hu, X. Xu, F. Fu, B. Yang, J. Zhang, Y. Zhang, S.S. Binte Touhid, L. Liu, Y. Dong, X. Liu, J. Yao, Synthesis of bimetallic silver-gold nanoparticle composites using a cellulose dope: tunable nanostructure and its biological activity, Carbohydr. Polym. 248 (2020) 116777. https://doi.org/10.1016/j.carbpol.2020.116777.

[16] A.M. Abdelgawad, M.E. El-Naggar, D.A. Elsherbiny, S. Ali, M.S. Abdel-Aziz, Y.K. Abdel-Monem, Antibacterial carrageenan/cellulose nanocrystal system loaded with silver nanoparticles, prepared via solid-state technique, J. Environ. Chem. Eng. 8 (2020) 104276. https://doi.org/10.1016/j.jece.2020.104276.

[17] V. Bundjaja, S.P. Santoso, A.E. Angkawijaya, M. Yuliana, F.E. Soetaredjo, S. Ismadji, A. Ayucitra, C. Gunarto, Y.-H. Ju, M.-H. Ho, Fabrication of cellulose carbamate hydrogel-dressing with rarasaponin surfactant for enhancing adsorption of silver nanoparticles and antibacterial activity, Mater. Sci. Eng. C. 118 (2021) 111542. https://doi.org/10.1016/j.msec.2020.111542.

[18] E. Madla-Cruz, M. De la Garza-Ramos, C.I. Romo-Saenz, P. Tamez-Guerra, M.A. Garza-Navarro, V. Urrutia-Baca, M.A. Martínez-Rodríguez, R. Gomez-Flores, Antimicrobial activity and inhibition of biofilm formation in vitro and on human dentine by silver nanoparticles/carboxymethyl-cellulose composites, Arch. Oral Biol. (2020) 104943. https://doi.org/10.1016/j.archoralbio.2020.104943.

[19] M.E. El-Naggar, M. Hasanin, A.M. Youssef, A. Aldalbahi, M.H. El-Newehy, R.M. Abdelhameed, Hydroxyethyl cellulose/bacterial cellulose cryogel dopped silver@titanium oxide nanoparticles: antimicrobial activity and controlled release of Tebuconazole fungicide, Int. J. Biol. Macromol. 165 (2020) 1010–1021. https://doi.org/10.1016/j.ijbiomac.2020.09.226.

[20] Z.-T. Xie, T.-A. Asoh, Y. Uetake, H. Sakurai, H. Uyama, Dual roles of cellulose monolith in the continuous-flow generation and support of gold nanoparticles for green catalyst, Carbohydr. Polym. 247 (2020) 116723. https://doi.org/10.1016/j.carbpol.2020.116723.

[21] H. Suo, L. Xu, Y. Xue, X. Qiu, H. Huang, Y. Hu, Ionic liquids-modified cellulose coated magnetic nanoparticles for enzyme immobilization: Improvement of catalytic performance, Carbohydr. Polym. 234 (2020) 115914. https://doi.org/10.1016/j.carbpol.2020.115914.

[22] H.M.A. Elrhman, Synthesis and characterization of core-shell magnetite nanoparticles with modified nano-cellulose for removal of radioactive ions from aqueous solutions, Results Mater. 8 (2020) 100138. https://doi.org/10.1016/j.rinma.2020.100138.

[23] A. Ghazitabar, M. Naderi, D. Fatmehsari Haghshenas, D. Alijani Ashna, Graphene aerogel/cellulose fibers/magnetite nanoparticles (GCM) composite as an effective Au adsorbent from cyanide solution with favorable electrochemical property, J. Mol. Liquids 314 (2020) 113792. https://doi.org/10.1016/j.molliq.2020.113792.

[24] X. He, Q. Liu, Z. Xu, Cellulose-coated magnetic janus nanoparticles for dewatering of crude oil emulsions, Chem. Eng. Sci. (2020) 116215. https://doi.org/10.1016/j.ces.2020.116215.

[25] A.H. Nordin, S. Wong, N. Ngadi, M. Mohammad Zainol, N.A.F. Abd Latif, W. Nabgan, Surface functionalization of cellulose with polyethyleneimine and magnetic nanoparticles for efficient removal of anionic dye in wastewater, J. Environ. Chem. Eng. 9 (1) (2020) 104639. https://doi.org/10.1016/j.jece.2020.104639.

[26] A. Khadir, M. Motamedi, M. Negarestani, M. Sillanpää, M. Sasani, Preparation of a nano bio-composite based on cellulosic biomass and conducting polymeric nanoparticles for ibuprofen removal: kinetics, isotherms, and energy site distribution, Int. J. Biol. Macromol. 162 (2020) 663–677. https://doi.org/10.1016/j.ijbiomac.2020.06.095.

[27] J. Guo, H. Tian, J. He, Integration of CuS nanoparticles and cellulose fibers towards fast, selective and efficient capture and separation of mercury ions, Chem. Eng. J. 408 (2020) 127336. https://doi.org/10.1016/j.cej.2020.127336.

[28] A. Abdollahi, A. Herizchi, H. Roghani-Mamaqani, H. Alidaei-Sharif, Interaction of photoswitchable nanoparticles with cellulosic materials for anticounterfeiting and authentication security documents, Carbohydr. Polym. 230 (2020) 115603. https://doi.org/10.1016/j.carbpol.2019.115603.

[29] Q. Xu, G. Poggi, C. Resta, M. Baglioni, P. Baglioni, Grafted nanocellulose and alkaline nanoparticles for the strengthening and deacidification of cellulosic artworks, J. Coll. Interface Sci. 576 (2020) 147–157. https://doi.org/10.1016/j.jcis.2020.05.018.

[30] H.H. H. T., Lignin structure, properties, and applications, Biopolymers (2009) 1–63.

[31] D. Koch, M. Paul, S. Beisl, A. Friedl, B. Mihalyi, Life cycle assessment of a lignin nanoparticle biorefinery: decision support for its process development, J. Clean. Prod. 245 (2020) 118760. https://doi.org/10.1016/j.jclepro.2019.118760.

[32] L. Matsakas, M. Gerber, L. Yu, U. Rova, P. Christakopoulos, Preparation of low carbon impact lignin nanoparticles with controllable size by using different strategies for particles recovery, Ind. Crops. Prod. 147 (2020) 112243. https://doi.org/10.1016/j.indcrop.2020.112243.

[33] M.-L. Mattinen, J.J. Valle-Delgado, T. Leskinen, T. Anttila, G. Riviere, M. Sipponen, A. Paananen, K. Lintinen, M. Kostiainen, M. Österberg, Enzymatically and chemically oxidized lignin nanoparticles for biomaterial applications, Enzyme Microb. Technol. 111 (2018) 48–56. https://doi.org/10.1016/j.enzmictec.2018.01.005.

[34] K. Ingtipi, V.S. Moholkar, Sonochemically synthesized lignin nanoparticles and its application in the development of nanocomposite hydrogel, in: In: Proc. International Conference on Advanced Materials, Energy & Environmental Sustainability (ICAMEES-2018), at UPES - Dehradun on 14th -15th December 2018, 17, 2019, pp. 362–370. https://doi.org/10.1016/j.matpr.2019.06.443.

[35] Y. Chen, K. Zheng, L. Niu, Y. Zhang, Y. Liu, C. Wang, F. Chu, Highly mechanical properties nanocomposite hydrogels with biorenewable lignin nanoparticles, Int. J. Biol. Macromol. 128 (2019) 414–420. https://doi.org/10.1016/j.ijbiomac.2019.01.099.

[36] M. Li, X. Jiang, D. Wang, Z. Xu, M. Yang, In situ reduction of silver nanoparticles in the lignin based hydrogel for enhanced antibacterial application, Coll. Surf. B. 177 (2019) 370–376. https://doi.org/10.1016/j.colsurfb.2019.02.029.

[37] K.R. Aadil, S.I. Mussatto, H. Jha, Synthesis and characterization of silver nanoparticles loaded poly(vinyl alcohol)-lignin electrospun nanofibers and their antimicrobial activity, Int. J. Biol. Macromol. 120 (2018) 763–767. https://doi.org/10.1016/j.ijbiomac.2018.08.109.

[38] Y. Wang, Z. Li, D. Yang, X. Qiu, Y. Xie, X. Zhang, Microwave-mediated fabrication of silver nanoparticles incorporated lignin-based composites with enhanced antibacterial activity via electrostatic capture effect, J. Coll. Interface Sci. 583 (2021) 80–88. https://doi.org/10.1016/j.jcis.2020.09.027.

[39] H. Yin, L. Liu, X. Wang, T. Wang, Y. Zhou, B. Liu, Y. Shan, L. Wang, X. Lü, A novel flocculant prepared by lignin nanoparticles-gelatin complex from switchgrass for the capture of *Staphylococcus aureus* and *Escherichia coli*, Coll. Surf. A. 545 (2018) 51–59. https://doi.org/10.1016/j.colsurfa.2018.02.033.

[40] M.R.V. Bertolo, L.B. Brenelli de Paiva, V.M. Nascimento, C.A. Gandin, M.O. Neto, C.E. Driemeier, S.C. Rabelo, Lignins from sugarcane bagasse: renewable source of nanoparticles as Pickering emulsions stabilizers for bioactive compounds encapsulation, Ind. Crops. Prod. 140 (2019) 111591. https://doi.org/10.1016/j.indcrop.2019.111591.

[41] M.S. Alqahtani, A. Alqahtani, M. Kazi, M.Z. Ahmad, A. Alahmari, M.A. Alsenaidy, R. Syed, Wound-healing potential of curcumin loaded lignin nanoparticles, J. Drug Deliv. Sci. Technol. 60 (2020) 102020. https://doi.org/10.1016/j.jddst.2020.102020.

[42] L. Siddiqui, J. Bag, D.Mittal Seetha, A. Leekha, H. Mishra, M. Mishra, A.K. Verma, P.K. Mishra, A. Ekielski, Z. Iqbal, S. Talegaonkar, Assessing the potential of lignin nanoparticles as drug carrier: synthesis, cytotoxicity and genotoxicity studies, Int. J. Biol. Macromol. 152 (2020) 786–802. https://doi.org/10.1016/j.ijbiomac.2020.02.311.

[43] P. Figueiredo, A. Lepland, P. Scodeller, F. Fontana, G. Torrieri, M. Tiboni, M. Shahbazi, L. Casettari, M.A. Kostiainen, J. Hirvonen, T. Teesalu, H.A. Santos, Peptide-guided resiquimod-loaded lignin nanoparticles convert tumor-associated macrophages from M2 to M1 phenotype for enhanced chemotherapy, Acta Biomater. (2020). https://doi.org/10.1016/j.actbio.2020.09.038.

[44] P. Figueiredo, K. Lintinen, A. Kiriazis, V. Hynninen, Z. Liu, T. Bauleth-Ramos, A. Rahikkala, A. Correia, T. Kohout, B. Sarmento, J. Yli-Kauhaluoma, J. Hirvonen, O. Ikkala, M.A. Kostiainen, H.A. Santos, In vitro evaluation of biodegradable lignin-based nanoparticles for drug delivery and enhanced antiproliferation effect in cancer cells, Biomaterials 121 (2017) 97–108. https://doi.org/10.1016/j.biomaterials.2016.12.034.

[45] M.S. Alqahtani, M. Kazi, M.Z. Ahmad, R. Syed, M.A. Alsenaidy, S.A. Albraiki, Lignin nanoparticles as a promising vaccine adjuvant and delivery system for ovalbumin, Int. J. Biol. Macromol. 163 (2020) 1314–1322. https://doi.org/10.1016/j.ijbiomac.2020.07.026.

[46] S. Amini, A. Saudi, N. Amirpour, M. Jahromi, S.S. Najafabadi, M. Kazemi, M. Rafienia, H. Salehi, Application of electrospun polycaprolactone fibers embedding lignin nanoparticle for peripheral nerve regeneration: in vitro and in vivo study, Int. J. Biol. Macromol. 159 (2020) 154–173. https://doi.org/10.1016/j.ijbiomac.2020.05.073.

[47] S. Chen, G. Wang, W. Sui, A.M. Parvez, L. Dai, C. Si, Novel lignin-based phenolic nanosphere supported palladium nanoparticles with highly efficient catalytic performance and good reusability, Ind. Crops. Prod. 145 (2020) 112164. https://doi.org/10.1016/j.indcrop.2020.112164.

[48] L. Dai, Y. Li, R. Liu, C. Si, Y. Ni, Green mussel-inspired lignin magnetic nanoparticles with high adsorptive capacity and environmental friendliness for chromium(III) removal, Int. J. Biol. Macromol. 132 (2019) 478–486. https://doi.org/10.1016/j.ijbiomac.2019.03.222.

[49] Y. Zhang, S. Ni, X. Wang, W. Zhang, L. Lagerquist, M. Qin, S. Willför, C. Xu, P. Fatehi, Ultrafast adsorption of heavy metal ions onto functionalized lignin-based hybrid magnetic nanoparticles, Chem. Eng. J. 372 (2019) 82–91. https://doi.org/10.1016/j.cej.2019.04.111.

[50] C.E. de Araújo Padilha, C. da Costa Nogueira, D.F. de Santana Souza, J.A. de Oliveira, E.S. dos Santos, Organosolv lignin/Fe_3O_4 nanoparticles applied as a β-glucosidase immobilization support and adsorbent for textile dye removal, Ind. Crops. Prod. 146 (2020) 112167. https://doi.org/10.1016/j.indcrop.2020.112167.

[51] Y. Ma, D. Zheng, Z. Mo, R. Dong, X. Qiu, Magnetic lignin-based carbon nanoparticles and the adsorption for removal of methyl orange, Coll. Surf. A. 559 (2018) 226–234. https://doi.org/10.1016/j.colsurfa.2018.09.054.

[52] T. Li, S. Lü, Z. Wang, M. Huang, J. Yan, M. Liu, Lignin-based nanoparticles for recovery and separation of phosphate and reused as renewable magnetic fertilizers, Sci. Total Environ. (2020) 142745. https://doi.org/10.1016/j.scitotenv.2020.142745.

[53] C.E. de A. Padilha, C. da C. Nogueira, S.C.B. Matias, J.D.B. da Costa Filho, D.F. de S. Souza, J.A. de Oliveira, E.S. dos Santos, Fabrication of hollow polymer microcapsules and removal of emulsified oil from

aqueous environment using soda lignin nanoparticles, Coll. Surf. A. 603 (2020) 125260. https://doi.org/10.1016/j.colsurfa.2020.125260.

[54] D. Eisenberg, E.M. Marcotte, I. Xenarios, T.O. Yeates, Protein function the post-genomic era, Nature 405 (2000) 823–826. https://doi.org/10.1038/35015694.

[55] S. Thalhauser, M. Breunig, Considerations for efficient surface functionalization of nanoparticles with a high molecular weight protein as targeting ligand, Eur. J. Pharma. Sci. 155 (2020) 105520. https://doi.org/10.1016/j.ejps.2020.105520.

[56] M. Magro, G. Cozza, S. Molinari, A. Venerando, D. Baratella, G. Miotto, L. Zennaro, M. Rossetto, J. Frömmel, M. Kopečná, M. Šebela, G. Salviulo, F. Vianello, Role of carboxylic group pattern on protein surface in the recognition of iron oxide nanoparticles: A key for protein corona formation, Int. J. Biol. Macromol. 164 (2020) 1715–1728. https://doi.org/10.1016/j.ijbiomac.2020.07.295.

[57] W. Chen, T.N.H. Cheng, L.F. Khaw, X. Li, H. Yang, J. Ouyang, J.Y.Y. Heng, Protein purification with nanoparticle-enhanced crystallisation, Sep. Purif. Technol. 255 (2021) 117384. https://doi.org/10.1016/j.seppur.2020.117384.

[58] A. Agrawal, R. Varshney, A. Gattani, P. Kirthika, M.H. Khan, R. Singh, S. Kodape, S.K. Patel, P. Singh, Gold nanoparticle based immunochromatographic biosensor for rapid diagnosis of Mycobacterium avium subspecies paratuberculosis infection using recombinant protein, J. Microbiol. Methods 177 (2020) 106024. https://doi.org/10.1016/j.mimet.2020.106024.

[59] A.L. Martínez-López, C. Pangua, C. Reboredo, R. Campión, J. Morales-Gracia, J.M. Irache, Protein-based nanoparticles for drug delivery purposes, Int. J. Pharma. 581 (2020) 119289. https://doi.org/10.1016/j.ijpharm.2020.119289.

[60] Y. Fan, X. Zeng, J. Yi, Y. Zhang, Fabrication of pea protein nanoparticles with calcium-induced cross-linking for the stabilization and delivery of antioxidative resveratrol, Int. J. Biol. Macromol. 152 (2020) 189–198. https://doi.org/10.1016/j.ijbiomac.2020.02.248.

[61] S. Wang, Y. Lu, X. Ouyang, J. Ling, Fabrication of soy protein isolate/cellulose nanocrystal composite nanoparticles for curcumin delivery, Int. J. Biol. Macromol. 165 (2020) 1468–1474. https://doi.org/10.1016/j.ijbiomac.2020.10.046.

[62] J. Simon, S. Udayan, E.S. Bindiya, S.G. Bhat, V.P.N. Nampoori, M. Kailasnath, Optical characterization and tunable antibacterial properties of gold nanoparticles with common proteins, Anal. Biochem. 612 (2021) 113975. https://doi.org/10.1016/j.ab.2020.113975.

[63] R.A. Muzzarelli, Chitin, Oxford U.K, Pergamon Press, 1970.

[64] I. Ahmed, Chitosan: Derivatives, Composites and Applications, Hoboken U.S., John Wiley & Sons, 2017.

[65] P. Jiamjirangkul, T. Inprasit, V. Intasanta, A. Pangon, Metal organic framework-integrated chitosan/poly(vinyl alcohol) (PVA) nanofibrous membrane hybrids from green process for selective CO_2 capture and filtration, Chem. Eng. Sci. 221 (2020) 115650. https://doi.org/10.1016/j.ces.2020.115650.

[66] F.I.M. Ali, S.T. Mahmoud, F. Awwad, Y.E. Greish, A.F.S. Abu-Hani, Low power consumption and fast response H_2S gas sensor based on a chitosan-CuO hybrid nanocomposite thin film, Carbohydr. Polym. 236 (2020) 116064. https://doi.org/10.1016/j.carbpol.2020.116064.

[67] I. Aadnan, O. Zegaoui, I. Daou, J.C.G. Esteves da Silva, Synthesis and physicochemical characterization of a ZnO-Chitosan hybrid-biocomposite used as an environmentally friendly photocatalyst under UV-A and visible light irradiations, J. Environ. Chem. Eng. 8 (2020) 104260. https://doi.org/10.1016/j.jece.2020.104260.

[68] N. Ali, S. Uddin, A. Khan, S. Khan, S. Khan, N. Ali, H. Khan, H. Khan, M. Bilal, Regenerable chitosan-bismuth cobalt selenide hybrid microspheres for mitigation of organic pollutants in an aqueous environment, Int. J. Biol. Macromol. 161 (2020) 1305–1317. https://doi.org/10.1016/j.ijbiomac.2020.07.132.

[69] H.A.T. Banu, P. Karthikeyan, S. Vigneshwaran, S. Meenakshi, Adsorptive performance of lanthanum encapsulated biopolymer chitosan-kaolin clay hybrid composite for the recovery of nitrate and phosphate from water, Int. J. Biol. Macromol. 154 (2020) 188–197. https://doi.org/10.1016/j.ijbiomac.2020.03.074.

[70] J.C.S. da Silva, D.B. França, F. Rodrigues, D.M. Oliveira, P. Trigueiro, E.C. Silva Filho, M.G. Fonseca, What happens when chitosan meets bentonite under microwave-assisted conditions? Clay-based hybrid nanocomposites for dye adsorption, Coll. Surf. A. 609 (2021) 125584. https://doi.org/10.1016/j.colsurfa.2020.125584.

[71] M.B.B. Pereira, D.B. França, R.C. Araújo, E.C. Silva Filho, B. Rigaud, M.G. Fonseca, M. Jaber, Amino hydroxyapatite/chitosan hybrids reticulated with glutaraldehyde at different pH values and their use for diclofenac removal, Carbohydr. Polym. 236 (2020) 116036. https://doi.org/10.1016/j.carbpol.2020.116036.

[72] Y. Wang, K. Wang, J. Lin, L. Xiao, X. Wang, The preparation of nano-MIL-101(Fe)@chitosan hybrid sponge and its rapid and efficient adsorption to anionic dyes, Int. J. Biol. Macromol. 165 (2020) 2684–2692. https://doi.org/10.1016/j.ijbiomac.2020.10.073.

[73] H. Zeng, F. Wang, K. Xu, J. Zhang, D. Li, Optimization and regeneration of chitosan-alginate hybrid adsorbent embedding iron-manganese sludge for arsenic removal, Coll. Surf. A. 607 (2020) 125500. https://doi.org/10.1016/j.colsurfa.2020.125500.

[74] B. Bueloni, D. Sanna, E. Garribba, G.R. Castro, I.E. León, G.A. Islan, Design of nalidixic acidvanadium complex loaded into chitosan hybrid nanoparticles as smart strategy to inhibit bacterial growth and quorum sensing, Int. J. Biol. Macromol. 161 (2020) 1568–1580. https://doi.org/10.1016/j.ijbiomac.2020.07.304.

[75] J. Diosa, F. Guzman, C. Bernal, M. Mesa, Formation mechanisms of chitosan-silica hybrid materials and its performance as solid support for KR-12 peptide adsorption: impact on KR-12 antimicrobial activity and proteolytic stability, J. Mater. Res. Technol. 9 (2020) 890–901. https://doi.org/10.1016/j.jmrt.2019.11.029.

[76] K. Kasinathan, B. Murugesan, N. Pandian, S. Mahalingam, B. Selvaraj, K. Marimuthu, Synthesis of biogenic chitosan-functionalized 2D layered MoS2 hybrid nanocomposite and its performance in pharmaceutical applications: In-vitro antibacterial and anticancer activity, Int. J. Biol. Macromol. 149 (2020) 1019–1033. https://doi.org/10.1016/j.ijbiomac.2020.02.003.

[77] M.S. Baktash, A. Zarrabi, E. Avazverdi, N.M. Reis, Development and optimization of a new hybrid chitosan-grafted graphene oxide/magnetic nanoparticle system for theranostic applications, J. Mol. Liquids (2020) 114515. https://doi.org/10.1016/j.molliq.2020.114515.

[78] I.S. Chaschin, G.A. Khugaev, S.V. Krasheninnikov, A.A. Petlenko, G.A. Badun, M.G. Chernysheva, K.M. Dzhidzhikhiya, N.P. Bakuleva, Bovine jugular vein valved conduit: a new hybrid method of devitalization and protection by chitosan-based coatings using super- and subrcritical CO_2, J. Supercritc. Fluids 164 (2020) 104893. https://doi.org/10.1016/j.supflu.2020.104893.

[79] H. Joz Majidi, A. Mirzaee, S.M. Jafari, M. Amiri, M. Shahrousvand, A. Babaei, Fabrication and characterization of graphene oxide-chitosan-zinc oxide ternary nano-hybrids for the corrosion inhibition of mild steel, Int. J. Biol. Macromol. 148 (2020) 1190–1200. https://doi.org/10.1016/j.ijbiomac.2019.11.060.

14

Silk protein and its nanocomposites

K.M. Faridul Hasan, Péter György Horváth and Tibor Alpár

SIMONYI KÁROLY FACULTY OF ENGINEERING, UNIVERSITY OF SOPRON, SOPRON, HUNGARY

Chapter outline

14.1 Introduction

Nanotechnology has brought revolutionary changes to the technological and scientific world for last few decades. There are novel inventions continuously getting reported by the researchers throughout the globe. However, initially some of the nano-based studies were not highly concerned about the harmful/toxic effects of the NPs [1-3]. But, with span of time peoples are getting more conscious about the environment and health hazardous material issues [4,5]. So, manufactures and researchers are trying to focus more on environment-friendly NPs. In this regard, green synthesis or natural resource–mediated NPs are getting more popularity [6,7]. Silk is one of the most prominent naturally originated materials that is getting attentions by the researchers for multifaceted potentialities.

Silk is a biologically originated protein fiber having superior mechanical strength. Generally, silk comprises two different types of protein: (1) 70%–80% fibroin and (2) 20%–30% sericin [8]. Silk has extensive applications as a prominent biomaterial for artificial tendon, suture, ligament, and cell structure substrate. Fibroin is comparatively better biomaterial than sericin in terms of biodegradability, water and oxygen vapor permeability, biocompatibility, lowest inflammatory reactivity, wound healing, safety perspective [9,10], and so on. Sericin creates some critical

Biopolymeric Nanomaterials: Fundamentals and Applications. DOI: https://doi.org/10.1016/B978-0-12-824364-0.00027-7

problems, such as with hypersensitivity and biocompatibility when used as a biomaterial [11,12]. Silk fibroin has been used for hundreds of years for its outstanding features; which is used nowadays as feasible biomaterials such as scaffolds of tissue engineering and biomedical tools. In some recent studies, NPs are also incorporated with the silk to develop innovative nanocomposites [13-16].

The polymeric NP composites are generally considered a double-faced system (one is reinforcement and another is matrix). There are several methods that assist for combining/compositing these two systems together. This work has been reported the nanocomposites from silk sericin and fibroin and their structural properties and associated manufacturing technologies as well. The biological and physicochemical properties of these two different types of materials play an important role for the integration of them during composites formations. Although there are several biopolymers available in nature such as collagen [17], chitosan [18], silk [19,20], and alginate [21] but silk fibroin is getting significant attentions for the superior performances. The mild processing technologies and optical transparency of silk had also pushed the scientist and made them eager to develop several fabrication methods with NPs through novel innovative routes.

14.2 Structural analysis of polymeric protein

Biopolymers (polysaccharides, polyphenols, polynucleic acid, polyamino acid, and so on) are used since long times especially for catalysis and energy storage or information [22]. Most of the silks from spider contains vast amount of amino acids that are termed as "motif." Amino acids of polymeric proteins are bonded with an amide group [23,24]. The silk protein/polyamino acid is produced by the silkworms to protect them at metamorphosis stage into the moths [22]. The polyamino acid comprises 19 monomers of amino acids (-NH-CHR-CO-) and an imino acid (-NR-CHR-CO-) through linked with amide bonding between them [22]. The silk proteins have versatility for aqueous solutions, controlled biodegradability, processing methodology, biocompatibility, and after all the mechanical characteristics that made it suitable for biomedical applications [25]. The motifs (short amino acid sequence) form the secondary structure that determines the shape of overall protein, thus ensures the interactions of silk protein into the core fiber structure [26]. The hierarchical and molecular structure and functioning influence the natural proteins to develop new polymeric composites. Some types of silks are even much more stronger than even synthetic substrates such as nylon and kevlar [27]. The mechanical features of various protein-based materials, nylon 66, and Kevlar 49, are provided in Table 14.1. It happened because the outstanding characteristics of silk fibers (extensibility and strength) enable it to absorb more energy before the breakage when the load is applied on it [28].

14.2.1 Silk protein

Silk is a protein-based biopolymer originated from natural spider or silkworms through different technical or biological methods with various types of morphologies such as films, capsules, fibers, gels, and foams. Humans are harvesting silk from cocoons for centuries to produce clothing materials with bright luster, good strength, and moisture absorbance [29,30]. The

Table 14.1 Mechanical properties of different types of silk, proteins, and other synthetic fibers.

Fiber materials	Density (kg/m³)	Strength (MPa)	Toughness (MJ/m³)	Modulus of elasticity (GPa)	References
Silk (*B. Mori*)	1300	500–600	70	7	[27,33]
Silk (*B. Mori*) fibroin	–	610–690	–	15-17	[34]
Silk (*B. Mori*) sericin	–	500	–	5–12	[34]
Silk (*A. yamamai*)	1300	650	113	9	[27]
Silk (*A. viscid*)	1300	500	150	0.5	[35,36]
Silk (*A. dragline*)	1300	1100	160	10	[35]
Kevlar 49 fiber	1440	3600	130	50	[33]
Nylon 66	–	750–950	–	80	[33]
Tendon collagen	–	120	6	1.2	[33]
Bone	1800–2080	160	4	20	[36]
Collagen	–	120	–	6	[33]
Wool	–	200	60	–	[33]

cultivation of silk (*Bombyx mori*) started in China nearly 4000 BCE or earlier. However, the technology did not transfer outside from China for long times. But, later on technology started to be headed up in Korea and Japan. Sericulture of silk (operation associated to mulberry plants cultivations, silkworm rearing, silk filaments reeling from cocoons) has started to gradually develop in other parts of the world. The silk has become popular and accepted by a big community for its gorgeous and luxury appearances and features. However, with the expansion of technology and science the usage of silk has also been enlarged, which is not only limited to textiles but also in biomedical and so many other sectors [24,31,32]. Silk protein-based NP is one of the latest inventions and technologies.

Animal fiber originated silk also exhibited superior strength, toughness, and elongation which could be suitably applied for engineered structures as well. The reason behind this successful advancement is the nanoarchitecture achieved through hierarchical arrangement of self-assembled silk proteins. Recently, some innovations are reported which were developed through controlling the nanostructured silks or functionalizing nanosilk materials [37-39] using various fabrication techniques. The unique characteristics of protein silk (2–4 nm β-crystal nanosheet, around 20–80 nm nanofibrils (diameter), and few hundreds nanometer bundle width of nanofibrils) has made it capable of generating higher strength and stiffness [27]. The structural protections and resistance against fracture are also developed for these outstanding features of silk nanomaterials through integrating the strength and toughness [33]. In this regard, nanostructured silks could be directly utilized in the forms of membranes, fibers, and hydrogels for functionalizing the materials. According to different models and theories (Maxwell model, cross-linking model, and mean field theory model) silk is considered as the homogeneous biopolymer, whereas the mechanical performances are correlated with the sequence of amino acids in their primary/secondary structures [27,40]. Besides, the noncrystalline area and β-crystal nanosheets determine the elasticity, stiffness/strength, and other mechanical features. Silk fibroin could be self-assembled both in 1D (one-dimensional) and 3D (three-dimensional) nanointerfaces in oil in water, cellulose nanofibrils, and inorganic/water nanosheets, which influences the improved biocompatible and mechanical features in nanocomposites/constructed nanoequipments/devices [27,41,42]. Feng et al. [14] has reported TiO_2 (80 nm diameter) NP-loaded silk fibroin nanocomposites with improved mechanical and thermal performances. The same study [14] also revealed that, elastic properties of the reported nanocomposites were increased with the increase of TiO_2 nanoloading. In general, the elasticity has a close correlation with the silk fibroins crystallinity and rigidity reported by Huang et al. [43].

14.2.2 Silk treatment/degumming

Silk degumming is performed by using several methods: (1) alkaline boiling by using Na_2CO_3, (2) soap degumming, (3) autoclaving, (4) enzymatic degumming, (5) extraction degumming using water, (5) urea degumming, (6) ultrasonic treatment (high power), and so on [44,45]. The degumming of silk (Fig. 14.1) is highly important to remove silk sericin from silk cocoons for getting purified silk fibroin, which is the prerequisite to be used in the field of biomedical [44]. But, alkaline and urea degumming techniques are also popularly used. Alkaline method of degumming is the efficient and most simple way for dissolution of the silk sericin. Besides, alkaline method could damage the silk fiber if the alkali ($NaOH/Na_2CO_3$) concentrations are not carefully controlled, although it provides whiteness appearances to the silk through removing natural color [46]. In contrast, degumming is more feasible as it does not degrade the silk strength and property [47]. A 3D silk fibroin/TiO_2 scaffold was reported by Johari et al. [48] for bone tissue manufacturing purposes from degummed silk and TiO_2 NP. This study has revealed the incorporation of TiO_2 on degummed silk fibroin has facilitated with enhanced mechanical strength [48].

14.3 Protein NPs and nanocomposites

Protein is an important biopolymer having an excellent biodegradability property. The main features of biodegradable polymer is the release of harmless water and carbon dioxide when degraded with the natural microorganisms to the environment at convenient and feasible atmospheric conditions (temperature, humidity, pH, oxygen (O_2), and pressure) [50]. Depending on intended applications, different proteins have been synthesized for producing protein-based NPs. The examples of some successfully synthesized proteins are: silk protein, keratin, fish protein, collagen (hydrolyzed), milk protein (hydrolyzed), albumins, protamines, lysozymes, wheat gluten, cruciferin, polypeptides, and plant-based legumin, soy protein, and peanut protein [51,52]. The nanocomposites fabricated from protein could be categorized in two ways: one is protein NP and another one is protein-conjugated NP. In the case of the first category, the NPs are totally made from proteins, whereas proteins are conjugated with NPs made from different sources for the second category [52].

When there is at least one chemical ingredient/polymer present into the matrix exhibiting nanoscale range dimensions and shape is termed as nanocomposites [53]. There are different processes (Fig. 14.2) developed for nanocomposites production such as solutions casting, desolvations, melt blending, electrospraying, and *in-situ* method. The difference in polymerization

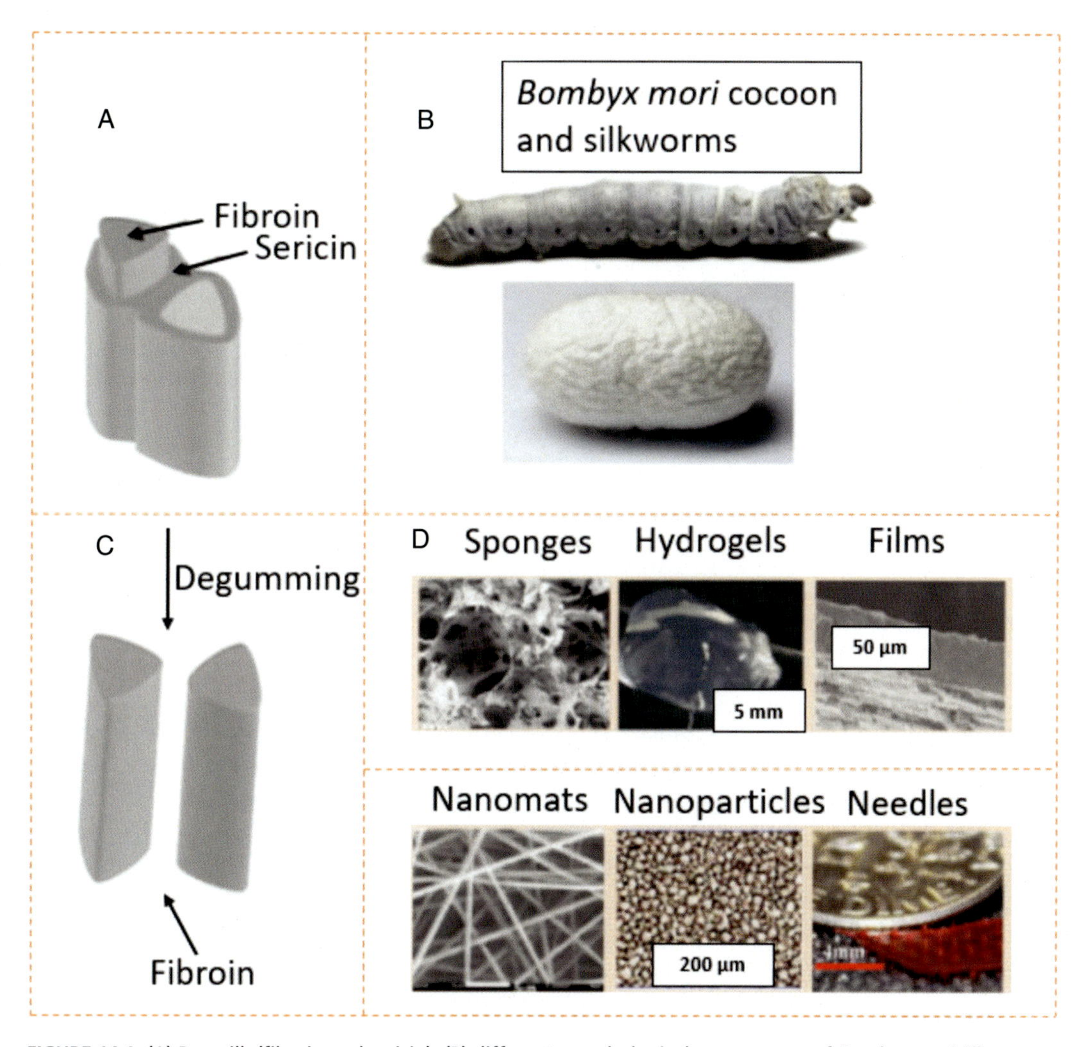

FIGURE 14.1 (A) Raw silk (fibroin and sericin), (B) different morphological appearances of *Bombyx mori* silkworms, and (C) sponges, hydrogels, films, nanomats, nanoparticles, and needles produced from the solution of regenerated silk fibroin. Adapted with permission from Elsevier [49]. Copyright Elsevier, 2015.

methods performed maybe because of every individual polymer has unique chemical and physical characteristics needed different processing routes [54].

14.4 Silk fibroin NPs

Silk fibroin from *B. mori* silkworm has applications since long times. Silk fibroin was used for textiles, biomedical, and cosmetic sectors widely [49,55]. However, silk fibroin is also showing potentiality for pharmaceutical, optical, food, biosensor, and electrical sectors with the expansion

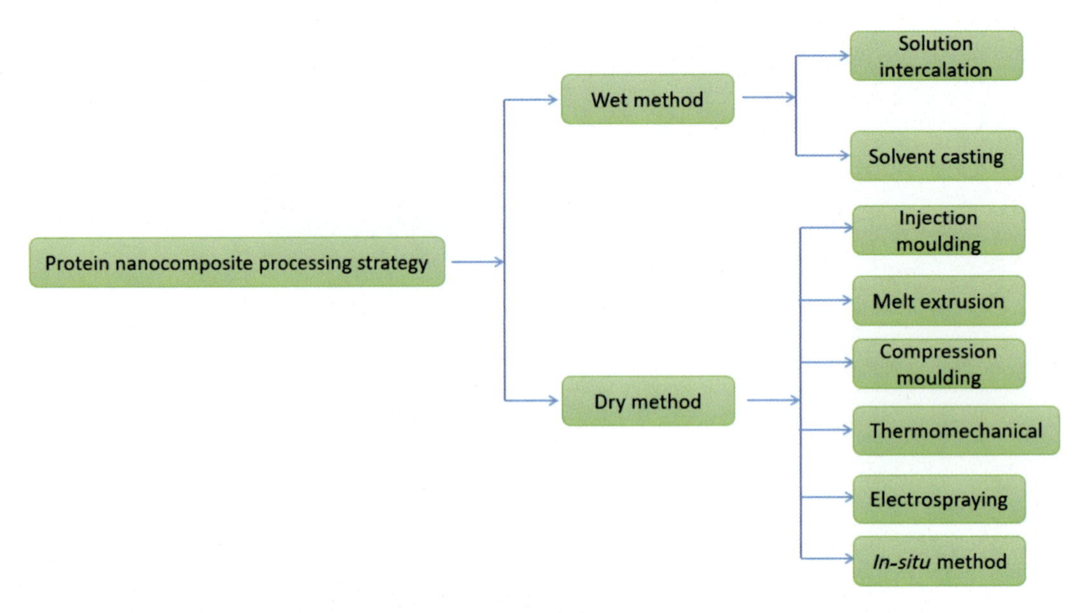

FIGURE 14.2 Processing routes of protein-based nanocomposites.

of technology [55-57]. Silk fibroin could be prepared in several forms such as hydrogels, scaffolds, films, nanofibers, microspheres, and so on [58]. Silk fibroin possesses higher tensile strength ranging from 340 to 740 MPa [49]. Besides, because of having chemical versatility, silk fibroin NPs exhibit better interfacial adhesions when composited along with enhanced performances in nano- or microstructures. However, several studies also revealed that silk degumming or sericin removal from cocoons of silk strongly influences the integrity of silk fibroin as well as their biochemical or mechanical features from processed fiber or scaffold biomaterials [44]. A multifunctional silk fibroin originated carrier was developed for antioxidant agents [59]. A one-step desolvation technique was used to produce sulforaphane-incorporated silk fibroin nanocomposites using cerium oxide NP and carbon dots (Fig. 14.3).

Nanocomposites could be produced from silk fibroin through applying various methods by utilizing self-assembly governed by hydrophobic and hydrophilic chain interactions [60]. However, the NPs reported from silk (mulberry and nonmulberry) did not exhibit any notable differences for particle geometry, cellular uptake, and size [61]. Besides, it is really challenging to achieve the NPs within 100 nm dimensions by using protein-based medium through controlling standard production protocol [60]. The size of different silk fibroin–based NPs are provided in Table 14.2. The electrospray, milling method, and supercritical fluid–mediated technology are some of the niche technology used for developing silk fibroin–based nanosystem. However, there are also capillary-microdot and salting methods used for silk fibroin NPs preparations. Pillai et al. [63] have functionalized silk fibroin proteins coating with CNF (carbon nanofiber) dispersed on poly-ł-caprolactone to develop braided conduits that could be utilized for the recovery and regeneration of peripheral nerves. In this regard, the silk threads were degummed

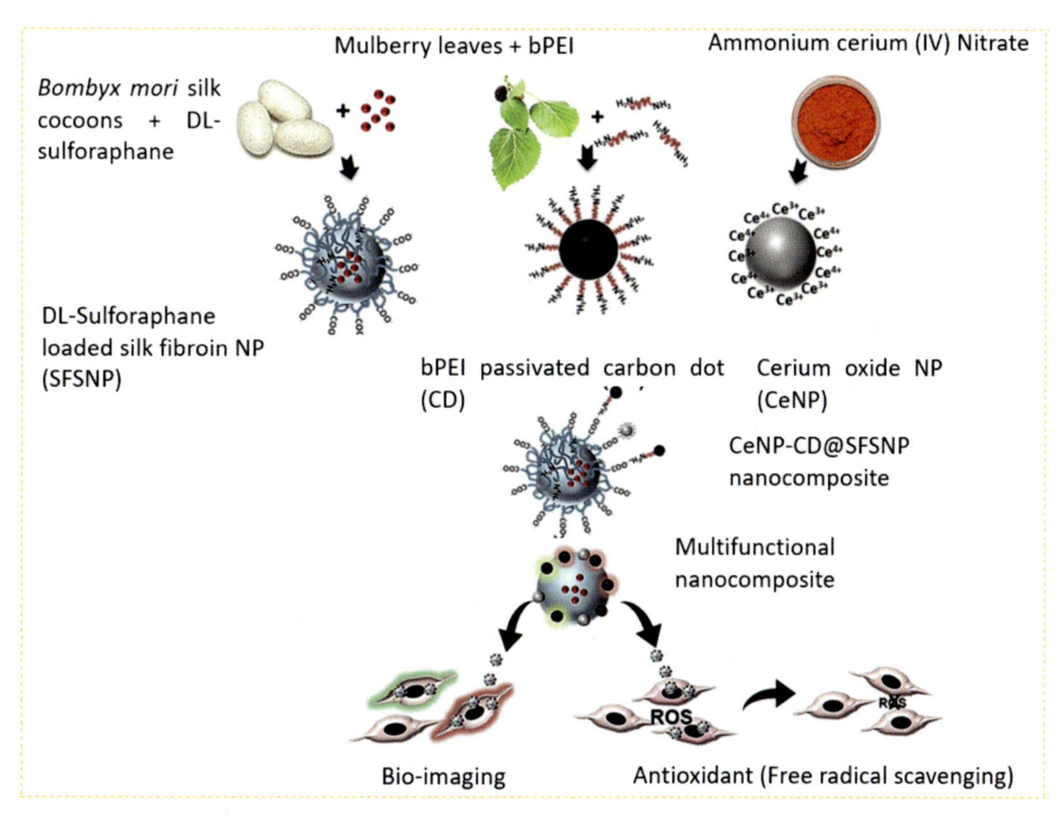

FIGURE 14.3 Cerium oxide NP and carbon dots coupled with sulforaphane incorporated silk fibroin NPs. Adapted with permission from Elsevier [59]. Copyright Elsevier, 2020.

initially. The developed conduits also exhibited better mechanical performances for CNF-coated samples [63]. A laminated nanocomposite was reported for graphene-, reduced graphene oxide (GO)-, and silicon dioxide-based NPs with silk fibroin through using spin-assisted technology at various concentrations, speeds, and spinning dynamics to observe the effects for developing effective fabrication and design of nanocomposites through manipulating reconstituted silk fibroin morphology and structures on different surfaces [64]. These silk fibroin synthesis methods have some benefits such as low cost, high yield, controlled NP size, used chemical ingredients are nontoxic, no requirement of using organic solvents, simple and safe operations, and so on [65].

14.5 Silk sericin NPs

Silk sericin is hydrophilic in nature which is collected from silkworms of the cocoons. Silk sericin is highly rich with serine which is nearly 32% along with higher number of hydroxyl groups. So, sericin retains water into the cutane surface. However, it is produced during the processing of

Table 14.2 Properties of silk-based nanoparticles.

Type of silk	Type of silk sericin/fibroin	Compositing material	Size of NPs (nm)	Application fields	References
	Silk sericin-Atorvastatin	Genipin	166	Biomedical	[66]
Silk sericin	Silk sericin-Fenofibrate	Poly(ethyl cyanoacrylate)	175	Enhancing fenofibrate bioefficacy	[67]
	Silk sericin-paclitaxel	Pluronic F-87 and F-127	100–110	Drug delivery/tissue engineering	[68]
	Silk sericin-doxorubicin	Folate	≈50	Hydrophobic chemotherapy agent for cancer treatments	[69]
	Silk fibroin–paclitaxel	Coumarin-6	130	Cancer treatment, carriers for drug delivery	[70]
	Silk fibroin–doxorubicin	–	130	Chemotherapy as multidrug-resistant cancer	[71]
Silk fibroin	Silk fibroin–cisplatin	–	59–75	Antitumor functioning	[72]
	Silk fibroin–Curcumin	Chitosan	100	Cancer therapy	[73]

silk, but it is considered as the discarded or wastage byproduct. However, silk sericin is studied by the researchers significantly because of antioxidant property, moisturizing characteristics, and having very good capability to resist tumor growth if applied in organic matrix [74,75]. Veiga et al. [76] have reported a comparative study to investigate the performance of produced nanocomposite from hydroxyapatite (bone-like) and silk sericin through using a stirred tank and meso-oscillatory flow reactor. They have concluded that nanocomposite obtained from a meso-oscillatory flow reactor is four times faster and efficient in contrast to a stirred tank reactor. The same study also revealed that, the concentration of silk sericin determines the morphology and shape of NPs [76]. The usage of silk sericin to produce nanocomposites reduces the wastage significantly. The formation of self-assembled spherical NP was reported by Cho et al. [75] where silk sericin was reacted with the poly (ethylene glycol). As they used cosmetics, so sericin was exposed to a challenge with instability to water or insoluble characteristics in organic solvent. So, they have introduced PEG into silk sericin to overcome this difficulty [75]. The reported NP sizes were within 200–400 nm (Fig. 14.4A) and spherical (Fig. 14.4B).

14.6 Applications and future development

The bioinspired materials are becoming more popular nowadays for their environmental sustainability in terms of controlled biodegradable and biocompatible characteristics beside the chemical versatility [77]. The hierarchical structures and assembly of silk have emerged with some competitive benefits of fabricated or tailored silk protein structures to nanoscale range [78]. Besides, the traditional biomedical applications; silk protein-based NPs are also exhibiting

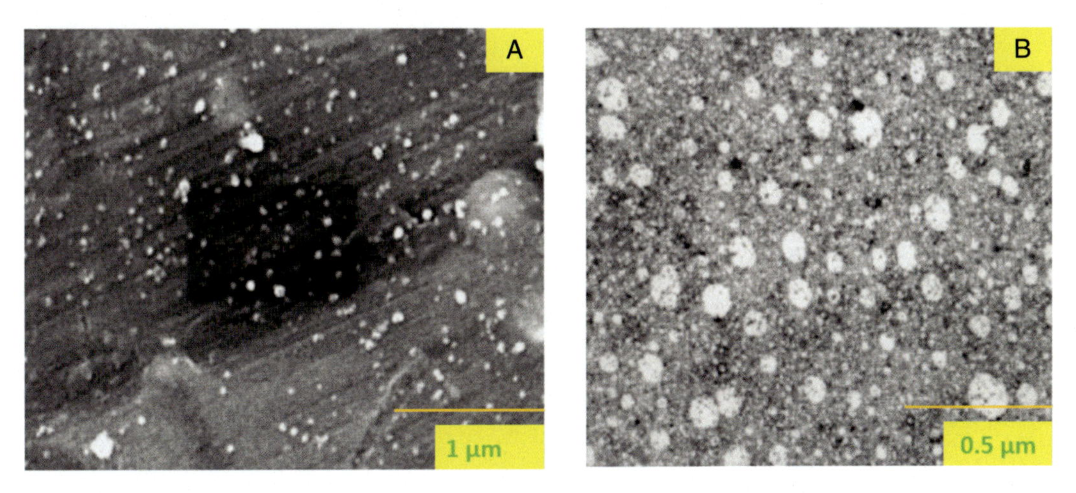

FIGURE 14.4 Photographs of sericin-poly(ethylene glycol) nanoparticles: (A) scanning electron microscope and (B) transmission electron microscope. Adapted with permission from Elsevier [75]. Copyright Elsevier, 2003.

superior potentiality for tissue engineering, drug delivery, implantable biomonitoring devices (Fig. 14.5) [79] and wearable electronic sensors for biosensing [80,81] purposes. Silk-based nanocomposites also showed potentiality with optical, electrical, and mechanical advantages for producing biocompatible devices through using silk fibroin-based NPs in flexible electronics items [81]. Besides, the combination of silk with the NPs such as GO provides superior strengths to the nanocomposites [82]. Silk-based nanocomposites also created new avenues for drug delivery system and advanced nanomedicine field [83]. Although silk protein is widely available in nature, their usages were limited for textile-based products since long times. The functionalization of silk protein NP could improve the chemical and physical properties and above all performances of the nanocomposites. However, many scientists are reporting attractive and feasible features of structural materials such as silk through incorporating with different NPs which could be applicable for biomedical sector in a larger extent as a prominent biomaterial. One of the major challenges for protein-based silk is their multifaceted structure along with processing, synthesis, design, and purifications [84]. However, more attention and studies by the scientists are needed in the near future for establishing and innovating new routes of silk-based NPs and their potential applications.

14.7 Conclusion

Silk is an important protein material. Nanocomposites based on silk are mainly prepared from the sericin and fibroin parts. However, fibroin parts are mostly used compared to the sericin. Various manufacturing routes contribute to the specific shape and size of silk-based NPs. The incorporation and introduction of silk with NPs bring superior performances and functionalities (biological, thermal, electrical, mechanical, magnetic, and fluorescent). Silk fibroin contains

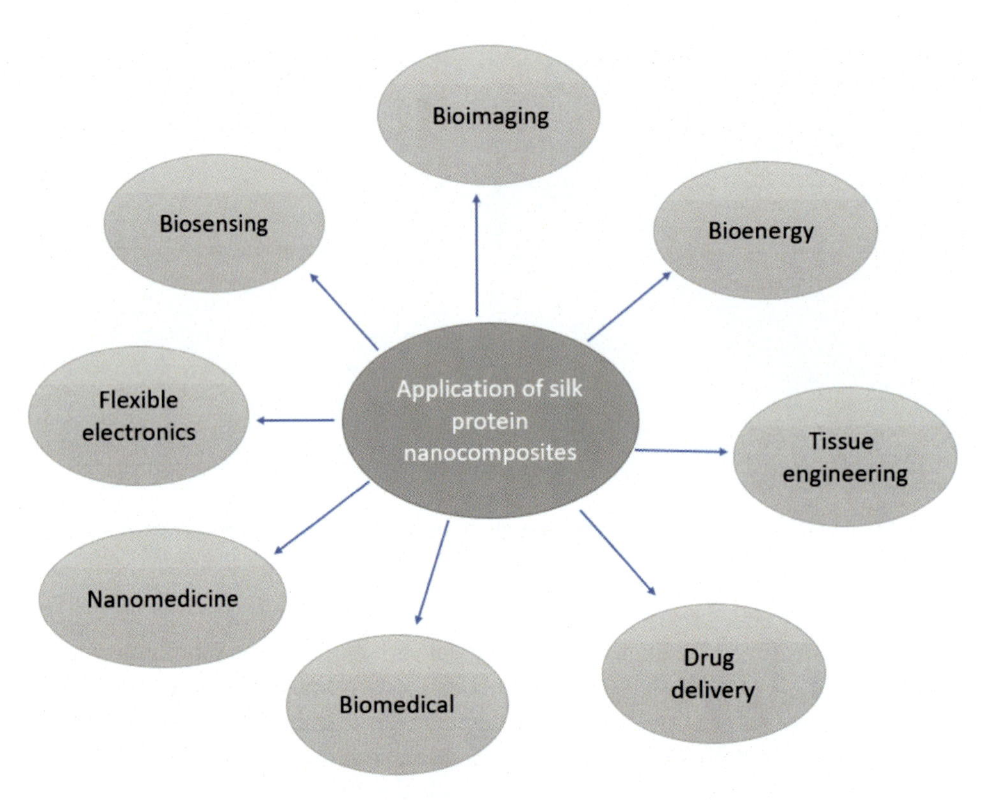

FIGURE 14.5 Silk protein–based nanocomposites applications.

physicochemical and biological features, which has made it more feasible biopolymer for developing and designing nanosystems. Silk sericin exhibits less structural characteristics but it has higher biological features as well and potentialities in contrast to the fibroin. Both the silk sericin and fibroin systems possess some benefits and limitations, but this could be overcome through exploring and innovating new technologies. Overall, this review work entailed a detailed study about silk and its structural overview along with associated chemistry, NPs preparation, methods, and potential applications with future scopes of research and improvements.

References

[1] W.S. Vedakumari, et al., Fibrin nanoparticles as Possible vehicles for drug delivery, Biochim. Biophys. Acta 1830 (8) (2013) 4244–4253.

[2] G. Shams, M. Ranjbar, A. Amiri, Effect of silver nanoparticles on concentration of silver heavy element and growth indexes in cucumber (*Cucumis sativus* L. negeen), J. Nanopart. Res. 15 (5) (2013) 1630.

[3] M. Sivakumaran, M. Platt, Tunable resistive pulse sensing: potential applications in nanomedicine, Nanomedicine 11 (16) (2016) 2197–2214.

[4] S. Talebian, et al., Nanotechnology-based disinfectants and sensors for SARS-CoV-2, Nat. Nanotechnol. 15 (8) (2020) 618–621.

[5] K. Hasan, P.G. Horváth, T. Alpár, Potential natural fiber polymeric nanobiocomposites: a review, Polymers 12 (5) (2020) 1072.

[6] K.F. Hasan, et al., Coloration of aramid fabric via in-situ biosynthesis of silver nanoparticles with enhanced antibacterial effect, Inorg. Chem. Commun. (2020) 108115.

[7] K. Hasan, et al., A novel coloration of polyester fabric through green silver nanoparticles (G-AgNPs@ PET), Nanomaterials 9 (4) (2019) 569.

[8] X. Kong, et al., Preparation of hydroxyapatite-fibroin nanocomposites, in: J.H. Kim (Ed.), Key Engineering Materials, Trans Tech Publ, Zurich, 2005.

[9] W. Zhang, et al., Silk fibroin biomaterial shows safe and effective wound healing in animal models and a randomized controlled clinical trial, Adv. Healthc. Mater. 6 (10) (2017) 1700121.

[10] M.A. de Moraes, M.M. Beppu, Biocomposite membranes of sodium alginate and silk fibroin fibers for biomedical applications, J. Appl. Polym. Sci. 130 (5) (2013) 3451–3457.

[11] G.H. Altman, et al., Silk-based biomaterials, Biomaterials 24 (3) (2003) 401–416.

[12] P. Aramwit, Bio-response to silk sericin, in: S.C. Kundu (Ed.), Silk Biomaterials for Tissue Engineering and Regenerative Medicine, Elsevier, Amsterdam, 2014, pp. 299–329.

[13] K. Hu, et al., Ultra-robust graphene oxide-silk fibroin nanocomposite membranes, Adv. Mater 25 (16) (2013) 2301–2307.

[14] X.-X. Feng, et al., Preparation and characterization of novel nanocomposite films formed from silk fibroin and nano-TiO_2, Int. J. Biol. Macromol. 40 (2) (2007) 105–111.

[15] P.C. Mane, et al., Highly sensitive label-free bio-interfacial colorimetric sensor based on silk fibroin-gold nanocomposite for facile detection of chlorpyrifos pesticide, Sci. Rep. 10 (1) (2020) 1–14.

[16] J. Mobika, et al., Fabrication of bioactive hydroxyapatite/silk fibroin/gelatin cross-linked nanocomposite for biomedical application, Mater. Chem. Phys. 250 (2020) 123187.

[17] P.S. Kaparekar, S. Pathmanapan, S.K. Anandasadagopan, Polymeric scaffold of Gallic acid loaded chitosan nanoparticles infused with collagen-fibrin for wound dressing application, Int. J. Biol. Macromol. 15 (A) (2020) 930–947.

[18] K.F. Hasan, et al., Wool functionalization through AgNPs: coloration, antibacterial, and wastewater treatment, Surf. Innov. 9 (1) (2020) 25–36.

[19] P. Asapur, et al., Spectroscopic analysis of Muga silk nanoparticles synthesized by microwave method, Biotechnol. Appl. Biochem. 68 (2) (2020) 345–355.

[20] V. Pandey, et al., Surface modified silk fibroin nanoparticles for improved delivery of doxorubicin: development, characterization, in-vitro studies, Int. J. Biol. Macromol. 164 (2020) 2018–2027.

[21] S. Mahmud, et al., In situ synthesis of green AgNPs on ramie fabric with functional and catalytic properties, Emerg. Mater. Res (2019) 1–11.

[22] J.G. Hardy, L.M. Römer, T.R. Scheibel, Polymeric materials based on silk proteins, Polymer 49 (20) (2008) 4309–4327.

[23] N.J.-A. Chan, et al., Spider-silk inspired polymeric networks by harnessing the mechanical potential of β-sheets through network guided assembly, Nat. Commun. 11 (1) (2020) 1–14.

[24] G. Carissimi, et al., On the secondary structure of silk fibroin nanoparticles obtained using ionic liquids: an infrared spectroscopy study, Polymers 12 (6) (2020) 1294.

[25] P. Ducheyne, Comprehensive biomaterials, Vol. 1, Elsevier, Amsterdam, 2015.

[26] T.A. Blackledge, M. Kuntner, I. Agnarsson, The form and function of spider orb webs: evolution from silk to ecosystems, in: R. Jurenka (Ed.), Advances in Insect Physiology, Elsevier, Amsterdam, 2011, pp. 175–262.

[27] Y. Wang, et al., Design, fabrication, and function of silk-based nanomaterials, Adv. Funct. Mater. 28 (52) (2018) 1805305.

[28] S. Ling, D.L. Kaplan, M.J. Buehler, Nanofibrils in nature and materials engineering, Nat. Rev. Mater. 3 (4) (2018) 1–15.

[29] R. Marsh, R. Corey, L. Pauling, The structure of tussah silk fibroin (with a note on the structure of β-poly-L-alanine), Acta Crystallogr. 8 (11) (1955) 710–715.

[30] C. Zanier, The Silk Cycle in China and its Migration. In book, Seri-Technics: Historical Silk Technologies, Max Planck Institute for the History of Science (2020) 13–31.

[31] S. Nagarajan, et al., Overview of protein-based biopolymers for biomedical application, Macromol. Chem. Phys. 220 (14) (2019) 1900126.

[32] A. Jain, et al., Protein nanoparticles: promising platforms for drug delivery applications, ACS Biomater. Sci. Eng. 4 (12) (2018) 3939–3961.

[33] J.L. Yarger, B.R. Cherry, A. Van Der Vaart, Uncovering the structure–function relationship in spider silk, Nat. Rev. Mater. 3 (3) (2018) 1–11.

[34] B. Galateanu, et al., Silk-based hydrogels for biomedical applications, Polym. Polym. Compos. A (2019) 1791–1817.

[35] J. Gosline, et al., The mechanical design of spider silks: from fibroin sequence to mechanical function, J. Exp. Biol. 202 (23) (1999) 3295–3303.

[36] M. Ashby, The CES EduPack Database of Natural and Man-made Materials, Cambridge University and Granta Design, Cambridge, UK, 2008.

[37] M. Humenik, T. Scheibel, Nanomaterial building blocks based on spider silk–oligonucleotide conjugates, ACS Nano 8 (2) (2014) 1342–1349.

[38] T. Giesa, et al., Nanoconfinement of spider silk fibrils begets superior strength, extensibility, and toughness, Nano Lett. 11 (11) (2011) 5038–5046.

[39] N. Srinivasan, S. Kumar, Ordered and disordered proteins as nanomaterial building blocks, Wiley Interdiscip. Rev. Nanomed. Nanobiotechnol. 4 (2) (2012) 204–218.

[40] L. Niu, et al., A novel nanocomposite particle of hydroxyapatite and silk fibroin: biomimetic synthesis and its biocompatibility, J. Nanomater. 2010 (2010).

[41] J. Von Fraunhofer, W. Sichina, Characterization of surgical suture materials using dynamic mechanical analysis, Biomaterials 13 (10) (1992) 715–720.

[42] U. Shimanovich, et al., Silk micrococoons for protein stabilisation and molecular encapsulation, Nat. Commun. 8 (1) (2017) 1–9.

[43] Y. Huang, et al., Swelling behaviours and mechanical properties of silk fibroin–polyurethane composite hydrogels, Compos. Sci. Technol. 84 (2013) 15–22.

[44] G. Carissimi, et al., Revealing the influence of the degumming process in the properties of silk fibroin nanoparticles, Polymers 11 (12) (2019) 2045.

[45] Z. Wang, et al., Effect of silk degumming on the structure and properties of silk fibroin, J. Textile Inst. 110 (1) (2019) 134–140.

[46] J. Kim, M. Kwon, S. Kim, Biological degumming of silk fabrics with proteolytic enzymes, J. Nat. Fibers 13 (6) (2016) 629–639.

[47] H. Yang, et al., Structure and properties of silk fibroin aerogels prepared by non-alkali degumming process, Polymer 192 (2020) 122298.

[48] N. Johari, H.M. Hosseini, A. Samadikuchaksaraei, Optimized composition of nanocomposite scaffolds formed from silk fibroin and nano-TiO_2 for bone tissue engineering, Mater. Sci. Eng. C, 79 (2017) 783–792.

[49] L.-D. Koh, et al., Structures, mechanical properties and applications of silk fibroin materials, Prog. Polym. Sci. 46 (2015) 86–110.

[50] A. Sorrentino, G. Gorrasi, V. Vittoria, Potential perspectives of bio-nanocomposites for food packaging applications, Trends Food Sci. Technol. 18 (2) (2007) 84–95.

[51] X. Li, et al., The effect of peanut protein nanoparticles on characteristics of protein-and starch-based nanocomposite films: a comparative study, Ind. Crops Prod. 77 (2015) 565–574.

[52] S. Sarkar, et al., Protein nanocomposites: special inferences to lysozyme based nanomaterials, Int. J. Biol. Macromol. (2020) 467–482.

[53] C. Xia, et al., Property enhancement of soy protein isolate-based films by introducing POSS, Int. J. Biol. Macromol. 82 (2016) 168–173.

[54] S. Gautam, B. Sharma, P. Jain, Green Natural Protein Isolate based composites and nanocomposites: A review, Polym. Test. 99 (2020) 106626.

[55] Z. Zhu, et al., High-strength, durable all-silk fibroin hydrogels with versatile processability toward multifunctional applications, Adv. Funct. Mater. 28 (10) (2018) 1704757.

[56] J.A. Delezuk, et al., Silk fibroin organization induced by chitosan in layer-by-layer films: application as a matrix in a biosensor, Carbohydr. Polym. 155 (2017) 146–151.

[57] C. Shivananda, B.L. Rao, Structural, thermal and electrical properties of silk fibroin–silver nanoparticles composite films, J. Mater. Sci. 31 (1) (2020) 41–51.

[58] Z. Xu, et al., Preparation and biomedical applications of silk fibroin-nanoparticles composites with enhanced properties-a review, Mater. Sci. Eng. C 95 (2019) 302–311.

[59] M. Passi, V. Kumar, G. Packirisamy, Theranostic nanozyme: Silk fibroin based multifunctional nanocomposites to combat oxidative stress, Mater. Sci. Eng. C 107 (2020) 110255.

[60] H.-J. Jin, D.L. Kaplan, Mechanism of silk processing in insects and spiders, Nature 424 (6952) (2003) 1057–1061.

[61] J. Kundu, et al., Silk fibroin nanoparticles for cellular uptake and control release, Int. J. Pharma. 388 (1-2) (2010) 242–250.

[62] Z. Zhao, Y. Li, M.-B. Xie, Silk fibroin-based nanoparticles for drug delivery, Int. J. Mol. Sci. 16 (3) (2015) 4880–4903.

[63] M.M. Pillai, et al., Effect of nanocomposite coating and biomolecule functionalization on silk fibroin based conducting 3D braided scaffolds for peripheral nerve tissue engineering, Nanomed. Nanotechnol. Biol. Med. 24 (2020) 102131.

[64] A.M. Grant, et al., Silk fibroin–substrate interactions at heterogeneous nanocomposite interfaces, Adv. Funct. Mater. 26 (35) (2016) 6380–6392.

[65] O. Gianak, et al., A review for the synthesis of silk fibroin nanoparticles with different techniques and their ability to be used for drug delivery, Curr. Anal. Chem. 15 (4) (2019) 339–348.

[66] J. Kanoujia, et al., Novel genipin crosslinked atorvastatin loaded sericin nanoparticles for their enhanced antihyperlipidemic activity, Mater. Sci. Eng. C 69 (2016) 967–976.

[67] K. Luo, Y. Yang, Z. Shao, Physically crosslinked biocompatible silk-fibroin-based hydrogels with high mechanical performance, Adv. Funct. Mater. 26 (6) (2016) 872–880.

[68] B.B. Mandal, S. Kundu, Self-assembled silk sericin/poloxamer nanoparticles as nanocarriers of hydrophobic and hydrophilic drugs for targeted delivery, Nanotechnology 20 (35) (2009) 355101.

[69] L. Huang, et al., Design and fabrication of multifunctional sericin nanoparticles for tumor targeting and pH-responsive subcellular delivery of cancer chemotherapy drugs, ACS Appl. Mater. Interfaces 8 (10) (2016) 6577–6585.

[70] P. Wu, et al., Facile preparation of paclitaxel loaded silk fibroin nanoparticles for enhanced antitumor efficacy by locoregional drug delivery, ACS Appl. Mater. Interfaces 5 (23) (2013) 12638–12645.

[71] Y. Tian, et al., Doxorubicin-loaded magnetic silk fibroin nanoparticles for targeted therapy of multidrug-resistant cancer, Adv. Mater. 26 (43) (2014) 7393–7398.

[72] J. Qu, et al., Silk fibroin nanoparticles prepared by electrospray as controlled release carriers of cisplatin, Mater. Sci. Eng. C 44 (2014) 166–174.

[73] V. Gupta, et al., Fabrication and characterization of silk fibroin-derived curcumin nanoparticles for cancer therapy, Int. J. Nanomed. 4 (2009) 115.

[74] R.I. Kunz, et al., Silkworm sericin: properties and biomedical applications, BioMed Res. Int. 2016 (2016) 8175701.

[75] K.Y. Cho, et al., Preparation of self-assembled silk sericin nanoparticles, Int. J. Biol. Macromol. 32 (1-2) (2003) 36–42.

[76] A. Veiga, et al., High efficient strategy for the production of hydroxyapatite/silk sericin nanocomposites, J. Chem. Technol. Biotechnol. 96 (2021) 241–248.

[77] H. Tao, D.L. Kaplan, F.G. Omenetto, Silk materials–a road to sustainable high technology, Adv. Mater. 24 (21) (2012) 2824–2837.

[78] S. Mehrotra, et al., Comprehensive review on silk at nanoscale for regenerative medicine and allied applications, ACS Biomater. Sci. Eng. 5 (5) (2019) 2054–2078.

[79] D.-H. Kim, et al., Dissolvable films of silk fibroin for ultrathin conformal bio-integrated electronics, Nat. Mater. 9 (6) (2010) 511–517.

[80] B. Zhu, et al., Silk fibroin for flexible electronic devices, Adv. Mater. 28 (22) (2016) 4250–4265.

[81] L.-D. Koh, et al., Advancing the frontiers of silk fibroin protein-based materials for futuristic electronics and clinical wound-healing (invited review), Mater. Sci. Eng. C 86 (2018) 151–172.

[82] W. Xie, et al., Extreme mechanical behavior of nacre-mimetic graphene-oxide and silk nanocomposites, Nano Lett. 18 (2) (2018) 987–993.

[83] F.P. Seib, Silk nanoparticles-an emerging anticancer nanomedicine, AIMS Bioeng. 42 (2) (2017) 239–258.

[84] K. Numata, How to define and study structural proteins as biopolymer materials, Polym. J. 52 (2020) 1043–1056.

15

Tissue engineering applications

Daniel López-Ângulo[a], Jeannine Bonilla[a] and Paulo J.A. Sobral[a,b]

[a]LABORATORY OF FOOD TECHNOLOGY, FACULTY OF ANIMAL SCIENCE AND FOOD ENGINEERING—FZEA/USP, UNIVERSITY OF SÃO PAULO, PIRASSUNUNGA, SÃO PAULO, BRAZIL. [b]FOOD RESEARCH CENTER (FORC), UNIVERSITY OF SÃO PAULO, RUA DO LAGO, SÃO PAULO, BRAZIL.

Chapter outline

15.1 Introduction

The human body has a limited ability to correctly autoregenerate most, if not all, of its major tissues and organs in the event that the original tissue integrity has been seriously damaged as a result of medical disorders involving tissue dysfunction or devastating deficits [1–3]. According to statistics reported by the U.S Department of Health and Human Services, as of January 2019, more than 113,000 people were on the waiting list for a life-saving organ, meanwhile, only 36,528 organ transplant procedures were conducted in the year 2018 [4]. The report also mentions that an alarming up to 20 people die every day as they wait for an organ. The deficit of organs, as demonstrated by these statistics, is unlikely to be resolved solely by organ donation and it is becoming increasingly clear that an alternative solution is of a dire need.

Current clinical therapies for the treatment of organ failure, tissue loss, and traumatic injuries are based on the use of auto and allografts [5]. Such therapies have shown to be insufficient, and researchers have been on the search for developing new strategies to solve this major medical challenge [6].

Biopolymeric Nanomaterials: Fundamentals and Applications. DOI: https://doi.org/10.1016/B978-0-12-824364-0.00028-9

In the past few years, many biomedical researchers have focused on fabricating multidisciplinary platforms to mimic the structural and physicochemical features of natural tissues [7–9]. Researchers in many fields of study such as medicine, nanotechnology, biology, and engineering collaborate closely to develop some techniques for mimicking natural tissues [10,11]. Additionally, in most cases, this type of research seeks to assist and accelerate the regenerative process by stimulating the patient's own inherent healing potential or, alternatively, to create replacement biological tissues (or, more challengingly, whole organs) to replace damaged, deteriorated, or lost body parts [3].

Faced with an ever-increasing burden of trauma, congenital abnormalities and degenerative diseases, tissue engineering (TE) and regenerative medicine promise to develop new biological therapeutics to treat a diverse range of diseases that are currently intractable [1]. TE is an emerging field in the human health care sector. The term TE has been coined for the first time in the year 1988 during the meeting of the U.S. National Science Foundation, and it was defined as "the application of the principles and methods of engineering and the life sciences toward the fundamental understanding of structure/function relationships in normal and pathological mammalian tissues and the development of biological substitutes to restore, maintain or improve functions." However, practicing of tissue replacement was not new. In 1968, the first bone marrow replacement was reported [12]. In 1970s, attempts were made to prepare artificial skin and biohybrid pancreas by using cells combined with biomaterials [13]. However, after 1980, a much organized scientific consciousness was observed globally with the setup of new TE-based organizations, regulatory bodies, scientific societies and, forums and journals. Over the last two decades, TE research has picked up the speed and become a matter of intense interest in both academia and industry [7]. TE's general strategies can be classified into three groups: (1) implantation into the organism of isolated cells or cell replacements (Fig. 15.1), (2) delivering tissue that induces substances such as growth factor, where traditionally, growth factor refers to proteins or polypeptides that can promote tissue growth, and (3) placing cells in or on various matrices [14]. TE is majorly classified into two types: (1) soft TE that deals with skin, blood vessel, tendon/ligament, cardiac patch, nerve, and skeletal muscle, (2) hard TE that deals with bone [15].

15.2 Biopolymeric matrix for obtain scaffolds

This field currently offers innovative strategies to implement biological substitutes in different medical fields. Major scientific advances in biomaterials, stem cells, growth and differentiation factors, and tissue microenvironment have fallowed "making" tissue in the laboratory by combining extracellular matrices, cells, and biologically active molecules [17]. Cells are seeded in artificial structures called scaffolds that are able to behave as templates for tissue formation (Fig. 15.2).

These scaffolds are critical and the success of tissue regeneration depends on both, their macro and microstructure, as well as on the constituent biomaterials. Scaffolds must be able to reproduce the existing *in-vivo* environment and to allow cells to exert influence. Recently, López et al. [18] reported the fabrication, characterization, and *in-vitro* cell study of

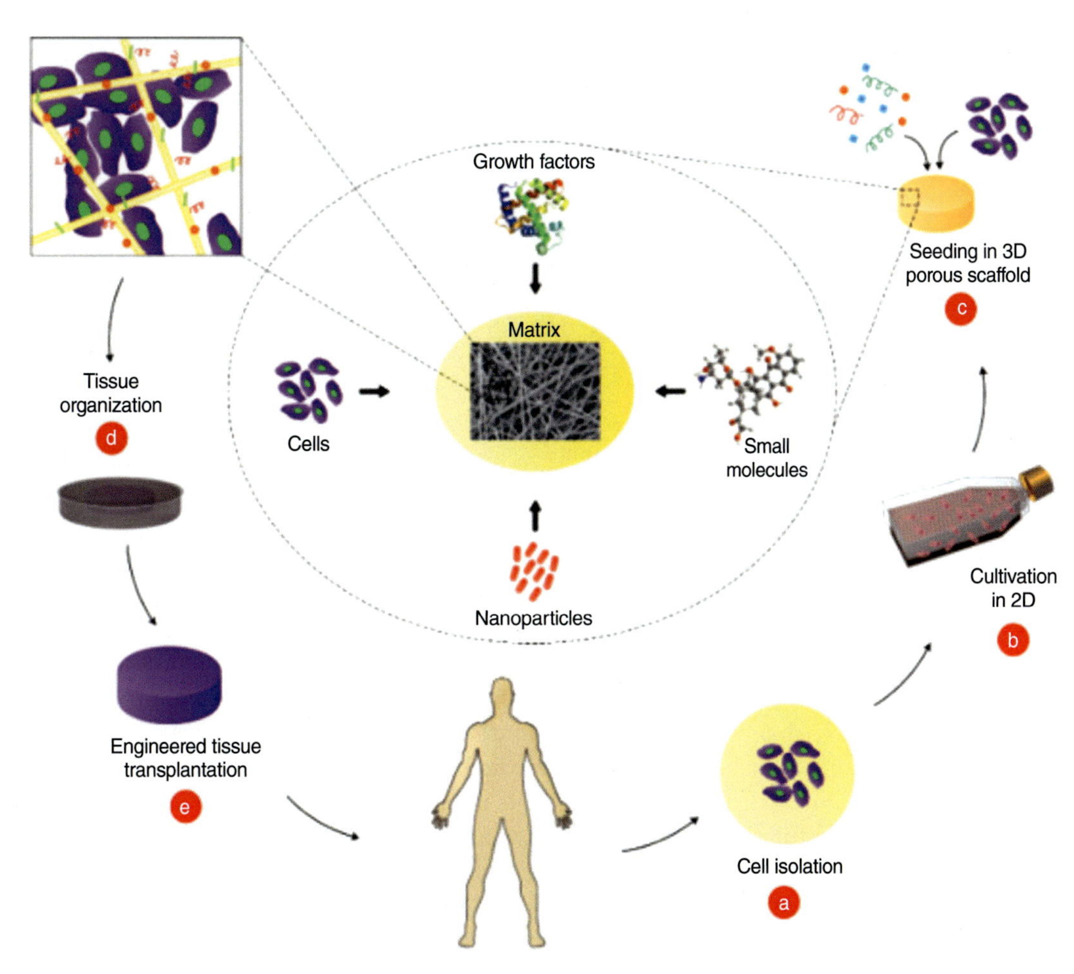

FIGURE 15.1 An example of a tissue engineering concept that involves seeding cells within porous biomaterial scaffolds. Adapted from Dvir et al. [16].

freeze-dried-based gelatin-chitosan scaffolds. These material presented compatible and adequate intrinsic stiffness, and high swelling capacity. Regarding bioactivity, the scaffolds were successful in both fibroblast and mesenchymal stem cells (MSC) growth, with extensive cell proliferation and expansion throughout the material during the 4-week incubation period, confirming in this way the degradation process of the material [18].

At first, it was thought that scaffolds are fundamental for cells' physical support, the biomaterial or scaffold can now be loaded with biological factors to facilitate tissue recovery [19–21]. Because of the diverse recovery limits of various tissues, some tissues do not demand cells but rather simply the biomaterial and biological molecules. On the other hand, other tissues have restricted recovery limits and demand the biomaterial, biomolecules, and cells for recovery to happen. There are tissues and organs with constrained or no possibility for recovery such as

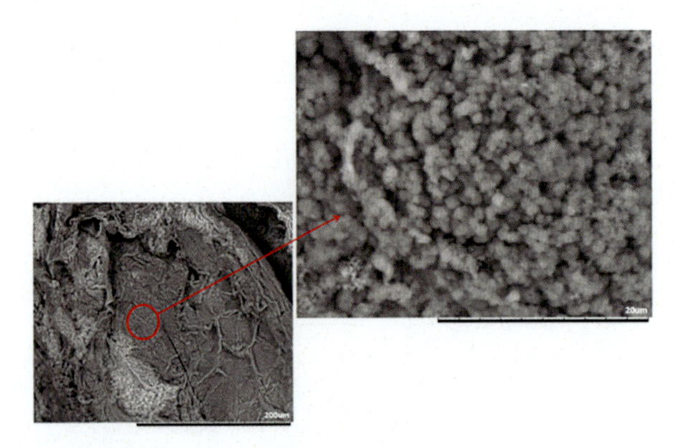

FIGURE 15.2 Scanning electron microscope micrographs (at ×50 and ×5000 magnification) of rabbit fibroblast interaction with matrix gelatin-chitosan and snail mucus scaffolds after 28 days of cell growing. Original image of our own work group at FZEA-USP [18].

ligament and cornea while others have great recovery capabilities such as the liver and the lungs tissues [22,23].

Other important parameters are the biodegradation and biocompatible properties of the scaffold materials. Any cytotoxic or immunogenic effect must be avoided. Therefore, the use of biomaterials for scaffold production is mandatory. Moreover, as scaffolds are conceived as temporary substitutes, their degradation into nontoxic reaction products must be pursued. Depending on the target tissue, scaffold must be completely reabsorbed once the regeneration process has been completed. This aspect highlights the importance of using biomaterials with specific mechanical and chemical properties for scaffold production. Their mechanical strength should be high enough to maintain the original scaffold porous architecture within the time frame of tissue regeneration, as well as to mechanically sustain stresses and loadings generated during *in-vitro* or *in-vivo* regeneration of tissues. However, the biomaterial should be also able to degrade in a predictable way, as far as the tissue regeneration process is going to the end [24].

Both naturally derived and synthetic biodegradable polymers represent a very interesting solution for the development of biocompatible scaffolds, showing several advantages [25]. The main natural and synthetic bioresorbable biopolymers and their biomedical applications are summarized in Tables 15.1 and 15.2. They can be prepared by a huge number of fabrication processes, into various shapes, and with the desired porous fraction and pore morphologies. The comparison of various methods for scaffold fabrication is summarized in Table 15.3. The corresponding mechanical and degradation properties may be properly customized with respect to the target tissue that needs regeneration. Moreover, biodegradable polymers generally degrade as the new tissue is formed, and often into nontoxic degradation products that may be easily metabolized and eliminated via the human secretion system. Furthermore, several biological moieties useful for promoting tissue regeneration, such as growth factors and pharmaceutical agents, may be properly anchored to the biopolymer scaffold architecture, thanks to the presence of reactive functional groups on their outer surface in many cases [26]. Ideally, the biomaterials selected to obtain scaffolds should allow one or more of the following functions: cell adhesion

Table 15.1 The main natural bioresorbable biopolymers and their biomedical applications.

Biopolymers	Characteristic	Applications	References
Collagen	Insoluble fibrous protein of animal connective tissue, containing three polypeptide chains, α-chains, which are coiled around each other into a triple helix, collagen monomers is (Glycine-X Y), X and Y are often proline and hydroxyproline.	Scaffolds for cardiovascular, musculoskeletal and nervous tissue engineering, Bioengineered skin equivalents.	[28,29]
Chitosan	β-(1-4)e linked D-glucosamine and N-acetyl-D-glucosamine, biologically renewable, biodegradable, biocompatible, nontoxic and biofunctional. (fully or partially deacetylated form of chitin)	Tissue engineering ranging from skin, bone, cartilage, improve blood coagulation, accelerating wound healing.	[30,31]
Hyaluronic acid	Isolated from brown seaweed (Phaeophyceae) units of α-1,4-D-glucuronic acid and β-1,3-N-acetyl-D-glucosamine, linked by β (1→3) bonds.	For tissue engineering on cartilage, bone, and osteochondral applications, regeneration, wound healing, tumor invasion, culture media for *in-vitro* fertilization.	[32,33]
Dextran	α1→6linked glucose polymers, synthesized from sucrose by lactic acid bacteria.	Antithrombotic (antiplatelet), antifibrin agent, as scaffolds for soft tissue, coatings for neural implants.	[34–36]
Gelatine	Obtained by controlled hydrolysis of collagen from various sources (hides, skins, ossein from bones, or sinews).	Pharmaceutical industry, for tissue engineering on cartilage, bone, and osteochondral applications, regeneration, wound healing.	[37,38]
Alginate	Copolymers of L-glucuronic and D-mannuronic acid residues connected by 1:4 glycosidic linkages, excellent biocompatibility and biodegradability.	Applicable 3D scaffolding materials such as hydrogels, microspheres, microcapsules, sponges, foams, and fibers, drug delivery systems, promotion of angiogenesis, wound dressing.	[39,40]
Silk fibroin	Derived from *Bombyx mori* cocoon, noninflammatory responses, low antigenicity, good biocompatibility, controllable biodegradability.	Surgical suture material fabricate 3D biodegradable scaffolds, skeletal tissue such as cartilage, bone, connective skin, and ligament tissue.	[41,42]
Shellac (aliphatic polyesters)	Resin, mixture of monoesters and polyesters (secreted by female lac bug), backbone consisting of aleuritic acid, terpenic acids, and minor fatty acids.	Pharmaceutical glaze. Coating material for tablets and capsules, time-released or delayed-action pills,.	[36]

Table 15.2 The main synthetic bioresorbable biopolymers and their biomedical applications.

Biopolymers	Characteristic	Applications	References
Poly(lactide-co-glycolide. PLGA	Random copolymerization of polyglycolic acid (PLA) (both L- and D,L-lactide forms) and PGA, known as poly (lactide-co-glycolide) PLGA, great cell adhesion and proliferation properties.	Sutures (Vicryl Rapide), drug delivery devices and tissue engineering scaffolds,	[43]
Poly(3-hydroxybutyrate)	Semi-crystalline isotactic polymer, good biocompatibility, processability and degradability, T_g: 5 °C and T_m: from 160 to 180 °C.	Tissue engineering applications.	[44]
Poly(vinyl alcohol). Polyvinyl acetate	Produced via hydrolysis of poly(vinyl acetate), biocompatibility, high water solubility, resistance to most organic solvents.	Soft contact lenses, eye drops, embolization particles, tissue adhesion barriers and in cartilage replacement.	[45]
PLA	Polyester, poly(α-hydroxy acids), hard and semicrystalline, high-molecular-weight polymer with thermoplastic properties.	Biomedical material in tissue engineering and drug delivery. Medical devices, sutures and bone fillings.	[46]
PGA	Polyester, poly(α-hydroxy acids) simplest linear aliphatic polyester, produced by polycondensation of glycolic acid, high melting point and high crystalline ratio, insoluble in water.	First biodegradable suture (Dexon).	[36]
PCL	Aliphatic polyester, most flexible synthetic biodegradable polymers T_g: -60°C), low melting point (60 °C)	Surgical material, controlled drug delivery systems.	[47]
Poly(ester-ether) Polydioxanone	Multiple repeating ether-ester units, ether oxygen group in the backbone is responsible for its flexibility, colorless, crystallinity 55%, T_g: between -10 °C and 0 °C.	Surgical sutures. Orthopedics (e.g., fixation screws and absorbable pins), plastic surgery, drug delivery, cardiovascular devices, and ophthalmology.	[48]
Poly(propylene fumarate)	Unsaturated linear polyester, *in-situ* cross-linking characteristics, biodegradability, and injectability.	Bone repair; in particular, it has been used for fabricating preformed scaffolds.	[49]
Polyanhydrides	Anhydride bonds that connect repeat units of the polymer backbone chain, very reactive and hydrolytically unstable.	Vehicles for the short-term release of drugs or bioactive agents, bone replacement and polyanhydride copolymers as vehicles for vaccine delivery.	[50]

Table 15.3 Comparison of various methods for scaffold fabrication [51–53].

Scaffold fabrication method	Advantage	Disadvantage
Electrospinning	Uniform, aligned fiber, strong interconnectivity of porosity, 80%–95% porosity, 100–1100 nm fiber diameter, <80% cell viability, extracellular matrix (ECM)-like structure, superior mechanical properties, large surface area, and simple method.	Need high voltage apparatus, solvents can be toxic, packaging, shipping, and handling.
Self-assembly	80%–90% porosity, 70%–90% cell viability, 5–300 nm.	Using peptides, complex process, not scalable, poor control over fiber dimension.
Phase separation	Simple method, 60%–95% porosity.	Potentially toxic solvents, poor control over architecture, and restricted range of pore sizes.
Gas foaming	Solvent free, do not loss bioactive molecules in the scaffold matrix.	Need high pressure, poor interconnectivity of porosity, the existence of skimming film layers on the scaffold surface.
Solvent casting	Simple method, high mechanical stability	Lacks reproducibility, uncontrolled structure.
Freeze drying	30%–80% porosity, 50–450 nm, cell viability <90%, it needs neither high temperature nor a separate leaching step.	Need freeze-dryer, limited to small pore size, irregular porosity, long processing time.
3D printing	Fabricate desired structure.	Need 3D printer, using toxic organic solvents, lack of mechanical strength

and migration, diffusion of vital cell nutrients and secreted products, vascularization, support of mechanical and biological functions in particular situations. New biomaterials also interact with the tissue in specific ways at the molecular and cellular levels, combining the properties of bioabsorbability and bioactivity [27].

15.3 Strategies for enhanced biopolymers materials for TE

Despite showing numerous advantages, many of the currently available biodegradable polymers do not satisfy all these requirements at the same time. Synthetic biodegradable polymers may show good mechanical strength, controlled degradation rates, and low production costs. However, their hydrophobic behavior together with lack of biological sites for cell recognition somehow limits their successful application for tissue regeneration [54,55]. On the other hand, natural biodegradable polymers are more hydrophilic and may be more easily recognized by the ingrowing cells, hence favoring cells adhesion. Nevertheless, they are more expensive due to the limited amount of material source in some cases, and generally suffer from poor mechanical properties [55]. To overcome these limitations, different strategies have been developed. One of the most intriguing is represented by the use of nanocomposite systems [55–57].

Table 15.4 Some examples of nanocomposites investigated in bone tissue engineering applications.

Nanocomposites in bone tissue engineering	References
HAp/Ti bionanocomposites	[67,68]
HAp-TiO_2-based nanocomposites	[69]
Zirconia-alumina nanocomposite	[70]
β-tricalcium phosphate-PCL nanocomposites	[71]
Graphene-HAp nanocomposite hydrogels	[72]
Bacterial cellulose-collagen nanocomposite	[73]
Bacterial cellulose-HAp nanocomposites	[74]
Chitosan/HAp nanocomposites	[75]
Nanocomposites of HAp with aspartic acid and glutamic acid	[76]
HAp-collagen-chondroitin sulfate nanocomposite	[77]
Nanocomposite scaffolds of bioactive glass ceramic nanoparticles disseminated chitosan matrix	[78]
Scaffold containing first otoliths/collagen/bacterial cellulose nanocomposites	[79]
Nanocomposite scaffold containing porous K-carrageenan/$CaPO_4$	[80]
Bone-like composite of self-assembled collagen fibers/HAp nanocrystals	[81]
Bone-like zirconium oxide nanoceramic modified chitosan-based porous nanocomposites	[82]
Bioactive glass nanofiber-collagen nanocomposite	[83]
Nanocomposites of bioactive glass nanofiber-filled PLA	[84]
Nanocomposites of HAp with aspartic acid and glutamic acid	[76]
Ferrimagnetic glass-ceramic nanocomposites	[85]

Nanocomposites are materials loaded with reinforcing materials such as nanoparticles, nanofibers, or nanocrystals [58]. These nanocharges belong to a broad class of materials with, at least, one dimension < 100 nm [59]. Several approaches are available to fabricate or synthesize different kinds of nanocomposites. Important and widely employed approaches of fabrication or synthesis of nanocomposites comprise physically mixing, *in-situ* preparation, film casting, dip coating, ionotropic gelation, coprecipitation, covalent coupling, electrospinning, colloidal assembly, and layer-by-layer assembly [60]. Almost all these approaches facilitate benefits of hydrogen bonding, hydrophobic effects, ionic interactions, electrostatic interactions, and coulombic interactions. Challenges in fabrication or synthesis of nanocomposites comprise controlling of fabrication/synthesis procedures; guaranteeing compatibility of diverse material constituents; and attaining desirable and unique material characteristics [59,61]. Nanocomposites differ from the conventional composite materials because of the remarkably higher surface-to-volume ratio of the reinforcing-phase and/or their extremely higher characteristic ratio, increased ductility with no decrease of strength, and scratching resistance [62]. Reinforcing materials generally improve the physical as well as mechanical characteristics of the matrix. Reinforcing materials can be composed of inorganic materials (e.g., metals, metal oxides, other inorganic components), fibers (e.g., electrospun fibers, carbon nanotubes (CNTs), etc.), or sheets (e.g., exfoliated clay stacks) [63–65]. Thus, numerous nanocomposites-containing biomolecules, multiple polysaccharides, functional polymers, metals, metal oxides, other inorganic components, structured carbons, etc., have been investigated and developed [59,60,66] (Table 15.4).

15.4 Applications of nanocomposites in TE

Reinforcement of inorganic nanomaterials inside the organic polymeric matrices is always proposed to enhance the cell adhesion as well as proliferation [86]. These are also essential for the bone TE purposes. During past few decades, numerous researchers have designed nanocomposites through incorporating nano-hydroxyapatite (nano-HAp) along with other biocompatible polymers for the use in bone TE applications [87–92]. HAp is a bioceramic extensively utilized in a variety of orthopedics as well as dentistry applications by reason of possessing the chemical composition very close to the inorganic mineral phase of the natural bone and teeth [93]. It possesses excellent biocompatibility and osteoconductivity properties [94]. Inorganic-inorganic nanocomposites have been designed and evaluated for their bone tissue regeneration applications as inorganic materials are able to add sufficient mechanical strength to the scaffold for load-bearing capability [94]. Several inorganic–inorganic bone scaffolds and implants have already reported in the literature, which have shown their prospects in the bone tissue repairing along with excellent load-bearing characteristics [95,96].

Nath et al. [96] have also designed nano-HAp-mullite composites for orthopedic application and evaluated their *in-vivo* responses up to 12 weeks of implantation using the rabbit model [95]. Bone growth was noticed over the 12 weeks of implantation period. Thus, the *in-vivo* study of nano-HAp-mullite composites clearly suggested their use in bone repair application.

Very recently, Swain et al. [94] have reported dense bioresorbable β-tricalcium phosphate-(iron-silver) nanocomposites, which were fabricated by means of attrition milling technique subsequently high-pressure processing consolidation (cold sintering at 2.50 GPa) [94]. They have incorporated β-tricalcium phosphate within the nanocomposite structure as it is chemically comparable to the composition of bone minerals. Therefore, β-tricalcium phosphate has been regarded as a suitable bioinorganic material for designing of various bioresorbable bone repairing scaffolds. Yet, intrinsic brittleness as well as low bending strength makes β-tricalcium phosphate as inappropriate material for the use in load-bearing bone sites. The mechanical properties, before and after immersion, were tested in compression and bending. All the compositions of these bioresorbable nanocomposites exhibited high mechanical strength (the strength in bending). Partial substitution of iron with silver led to an increase in both strength and ductility. The degradation behavior of the developed β-tricalcium phosphate-(iron-silver) nanocomposites was studied by immersion in Ringer's solutions and saline solutions for up to 1 month. After 1-month immersion, the composites retained about 50% of their initial bending strength. The *in-vitro* cytotoxicity evaluation of these developed bioresorbable β-tricalcium phosphate-(iron-silver) nanocomposites on the primary human osteoblast cell lines was carried out. In *in-vitro* cell culture experiments, these nanocomposites exhibited no signs of cytotoxicity toward human osteoblast cell lines indicating that these can be used as an implant material [97].

Recently, Bhowmick et al. [98] have developed bone-like organic–inorganic hybrid porous nanocomposites of chitosan, poly(ethylene glycol) (PEG), and nano-HAp, where zirconium oxide nanoparticles were incorporated. The ratio of chitosan, PEG, and nano-HAp-zirconium oxide in three formulations of zirconium oxide nanoparticles-doped chitosan-PEG-nano-HAp was 55:40:5, 55:35:10, and 55:30:15, respectively. Scanning electron microscope pictures of the surface morphologies of these nanocomposites exhibited porous morphology. Numerous pores of different sizes were observed on the surface of nanocomposites. In addition, numerous macrovoids were seen on these bone-like nanocomposites. These nanocomposites have porosities of 65.71%–73.02%. Therefore, zirconium oxide nanoparticles-doped chitosan-PEG-nano-HAp was highly porous in nature. The porosity of these nanocomposites was almost similar to that of the human cancellous bone, and Thus, helpful for permitting cell migration, exchanges of nutrients, etc. Various mechanical characteristics, such as tensile strength and stiffness of these bone-like organic-inorganic hybrid porous nanocomposites, were obtained in tensile mode. Tensile strength and stiffness of these nanocomposites were found to be improved gradually from 21.69 to 24.34 and from 17,829.93 to 55,304.40 MPa, respectively, with the increment of inorganic materials in the nanocomposite formula (nano-HAp-zirconium oxide) [98].

Recently, Sajjad et al. [99] reported works on modified montmorillonite-bacterial cellulose (BC) nanocomposites as a novel substitute for burn skin and tissue regeneration. BC is a promising biopolymer with wound healing and tissue regenerative properties but lack of antimicrobial property limits its biomedical applications. Therefore, this study was proposed to combine wound healing property of BC with antimicrobial activity of montmorillonite (MMT) and modified MMTs (Cu-MMT, Na-MMT, and Ca-MMT) to design novel artificial substitute for burns. The results concluded that loading of BC with modified MMTs imparts significant antimicrobial activity against burn-associated pathogens. In addition, modified MMTs-BC nanocomposites improved wound healing and tissue regeneration in comparison to BC and MMT-BC nanocomposites in animal models (Fig. 15.3). The flexibility of BC will allow to apply the modified MMTs-BC nanocomposites on moving parts of the body such as knee and elbow. It will also ensure better patient compliance as it requires less frequent changeovers or one-time application, reduces pain after wounds as it has soothing property and painless removal. Thus, modified MMTs-BC can be used as an artificial skin substitute for burn wounds after further preclinical and clinical studies that will ensure proper tissue regeneration at cellular level and enhance the healing of wounds [99].

15.5 Strategies for using nanocomposites in 3D print and hydrogels for TE

Scaffolds designed for biomedical use are supporting structures employed for recreating the extracellular environment and allowing cells survival and growth inside a well-determined volume [100]. To combat the limitations of conventional techniques for scaffolding, 3D printing provides a more controlled and versatile technology with the capability necessary to create biomimetic scaffolds to promote the formation of functional tissue [101]. 3D printing was soon discovered to be a valid strategy to build scaffolds with customized geometries, depending on the case object of the study. Moreover, 3D printing allows for processing of different materials, which not only should have the required physical properties to be processed and to hold a shape after processing, but should also induce specific responses or, in some case, do not induce adverse reactions in the hosted cells. Indeed, pore size and shape, total porosity, chemical and physical

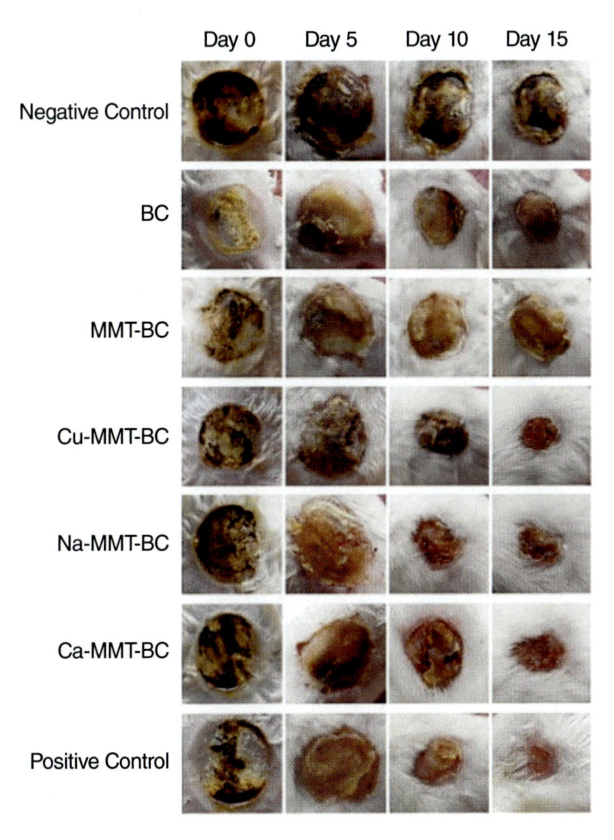

FIGURE 15.3 Representative wound photographs of Cu-MMT-BC, Na-MMT-BC, and Ca-MMT-BC treated group, negative control, BC, MMT-BC group, and positive control group during course of treatment. With permission from Sajjad et al. [99].

properties are crucial for TE scaffolds [100]. 3D printing, also known as additive manufacturing, is the placement and layering of the same or various materials in succession through an automated layer-by-layer process to fabricate 3D structure [101]. 3D-printed polymer nanocomposites with metal nanoparticle, metal nanowire, CNT, and ceramic nanoparticle fillers were reported [102,103].

Metal nanoparticles such as silver (Ag) were found to improve the electrical properties of the polymer matrix, whereas CNT provided high tensile strength with good dispersibility in polymer matrix resulting in mechanically improved nanocomposites. With the addition of titanium dioxide (TiO_2) nanoparticles, tensile strength, flexural strength, and hardness of the nanocomposites were improved [104]. Thermal stability of this nanocomposite was also found to be enhanced through the addition of TiO nanoparticles. For centuries, Ag and Ag compounds have been known as high performance antimicrobial materials [105]. They have been used both in the treatment of human diseases and in the disinfection of medical devices [106].

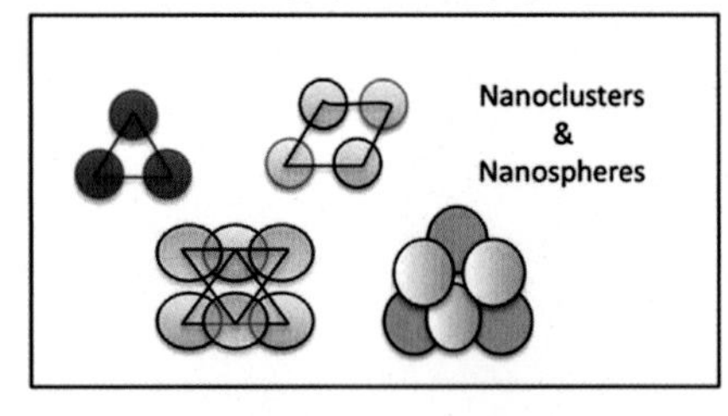

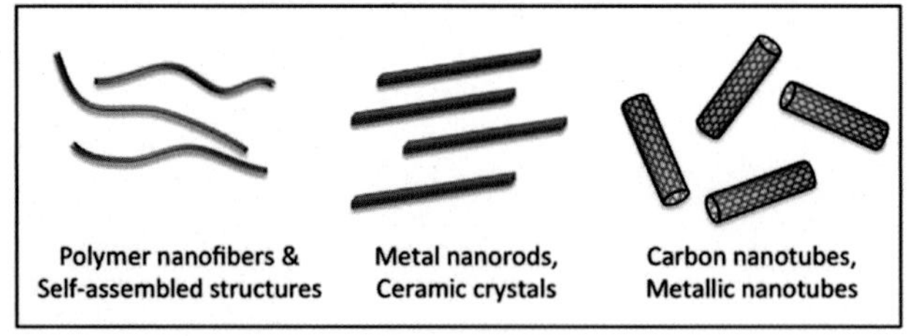

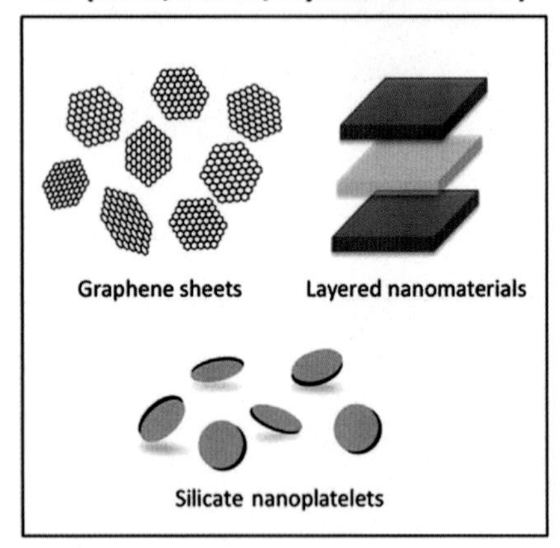

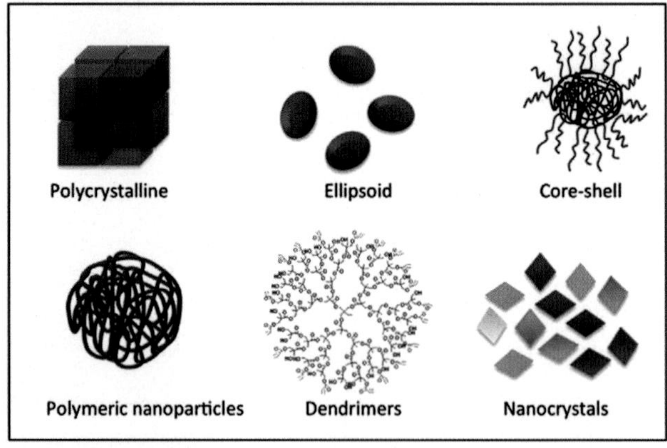

FIGURE 15.4 Illustrated are the various dimensions (0D, 1D, 2D, 3D) and corresponding materials available for implementation within polymeric nanocomposite designs. J.K. Carrow and A.K. Gaharwar [109].

Very recently, Bayraktar et al. [107] have reported work with silver nanoware (Ag NW) used for the first time as nanofillers in 3D printing. Functional Ag NW/Poly(lactic acid) (PLA) nanocomposites were fabricated and their antimicrobial properties were investigated. It was found that Ag NWs were homogenously distributed within the PLA matrix by solution mixing and highly aligned in the direction of shear force. After twin screw extrusion and 3D printing processes, it was noted that NW did get shorter and their alignment was lost. During time kill assay and adhesion tests against *Staphylococcus aureus* and *Escherichia coli*, Ag NW dramatically improved the antibacterial performances even at low loading levels. Time kill test results indicated a killing capacity of 100% against both *S. aureus* and *E. coli* in 2 h. Moreover, 100% bactericidal effect was found to continue even after 24 h against *E. coli* for all Ag NW loadings. On the other hand, 100% killing capacity against *S. aureus* was observed up to 8 h at a NW loading of 4 wt%. Adhesion test and time kill test results were found to be in good agreement. In 3D-printed nanocomposites, Ag NW successfully prevented the bacterial reproduction and killed bacterial cells. Nanocomposites showed higher performance against *E. coli* as Gram-negative bacilli rather than *S. aureus* due to the cell wall structure of the bacteria groups [107].

Nanofibrillated cellulose (NFC) is a natural nanoparticle in the form of a fibril, as the name suggests, thus classified as a 1D (Fig. 15.4) object with one dimension in the nanometric range.

NFC is highly biocompatible, as well as biodegradable, and characterized by high tensile characteristics [108]. They consist in fibrils very similar to collagen, thus particularly appropriate for applications in the field of cartilage tissue regeneration. NFC is commonly integrated in an alginate-based bioink; alginate is a natural polymer that falls into the category of polysaccharides and its gelation might be induced by the addiction of gelling agents, like for instance divalent cations. After gelation, alginate can be considered a hydrogel, thus a 3D polymer network is able to hold a large quantity of water [109].

The most employed method to print NFC-filled alginate bioinks is the μ-extrusion. Markstedt [110] and Martínez [111] have exploited the 3DDiscovery, a syringe μ-extrusion printer from regenHU Ltd., trying to replicate the cartilaginous tissue by using an alginate matrix enriched with NFC. The considered process consists in depositing the bioink creating the desired shape and then immersing it in the aqueous solution containing the cross-linking agent for 10 min.

In the first case [110], an NFC/alginate ink with 80:20 w/w ratio (named Ink8020) was printed; the cross-linking agent in this case was calcium chloride ($CaCl_2$), which, when added to water, dissociates into Ca^{2+} (the divalent cation) and $2Cl^-$. As pure alginate has low zero-shear viscosity, the shape fidelity after printing is also low; NFC helped to improve the printing results. Indeed, when NFC was added to the alginate bioink, a typical shear-thinning behavior could be observed. Shear-thinning behavior is strictly related to the viscosity of the ink and highly recommended when bioprinting via extrusion methods. Materials showing such behavior are characterized by lower viscosity when extruded (higher shear rates) and become tougher with higher viscosity when at rest (lower shear rate). Therefore, the ink can be easily extruded and, after deposition, maintain the conferred shape resulting in high-fidelity printing and, if cells are present, cell viability [110]. After bioink optimization, Markstedt et al. [110] were able to print not only simple structures such as small grids (Fig. 15.5A–C), but also shapes mimicking cartilage tissues, such a human ear (Fig. 15.5D) and a sheep meniscus (Fig. 15.5E, F).

Martínez et al. [111] printed a cell-laden auricular construct similar to Markstedt [110], but with a wider inner structure as reported in Fig. 15.6, trying to improve the supply of nutrients to embedded cells. The computer-aided designs model of the auricular structure was extrapolated directly from the patient MRI scan, and then was translated in an *.STL file (Stereolithography) to be managed by the printer program (Blueprint). The ink in this study was prepared by adding 2 wt% NFC and 0.5 wt% sodium alginate to Cellink bioink, a commercial hydrogel suitable for cartilage and skin tissue-engineering applications. Moreover, human nasoseptal chondrocytes were added to the printable blend to obtain a final concentration of 20×10^6 cells/mL.

15.5.1 Nanosilica/nanoclay (SiO_2)

Nanosilica particles, also referred to as silicon dioxide nanoparticles (SiO_2) and nanoclays, which consist of layered silicates stacked together [112], are commonly used fillers in the field of biomedicine due to their stability and low-toxicity properties. Moreover, nanosilicates have demonstrated to induce osteogenic differentiation if added to human mesenchymal stem cells (hMSCs) and adipose stem cells [113]. As regards 3D printing bioinks, nanosilica/nanoclays can be employed in combination with different materials, and designed to serve both as fillers and

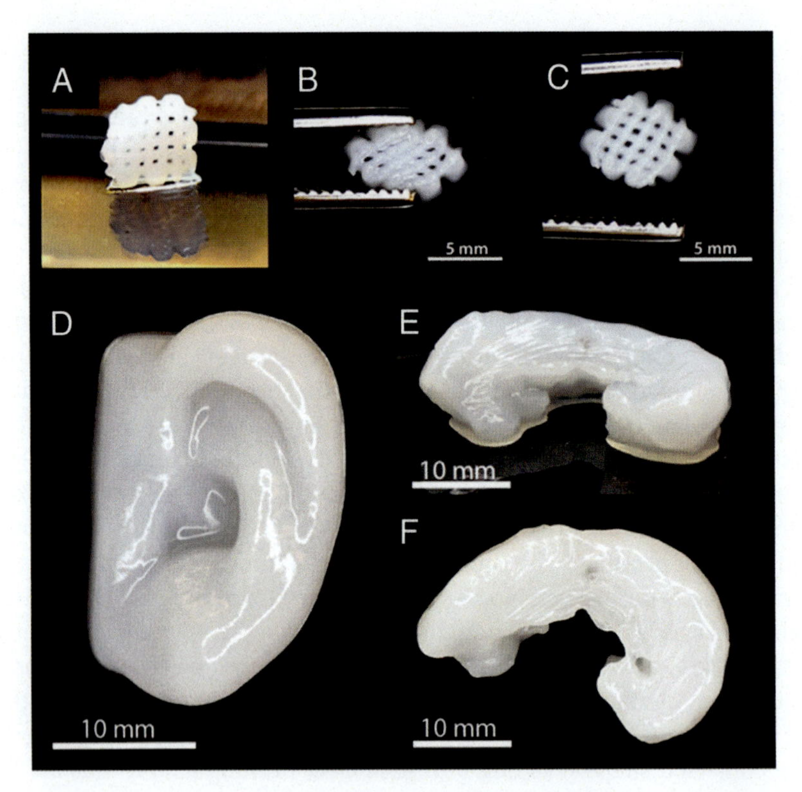

FIGURE 15.5 (A) 3D-printed small grids (7.2 × 7.2 mm^2) with Ink8020 after cross-linking. (B) The shape of the grid deformed upon squeezing, and (C) restored after squeezing. (D) 3D-printed human ear and (E and F) sheep meniscus (side view and top view, respectively). Adapted from K. Markstedt et al. [110].

as coatings to improve physical, chemical, and biological functionalities thanks to their uniform shape and charged surface [113]. An example of nanosilica/nanoclay coating is in the study carried out by Lee et al. [114]; in this case, the aim was to enhance the mechanical properties of scaffolds printed by natural polymer deposition. A biomimetic scaffold printed with a mixture

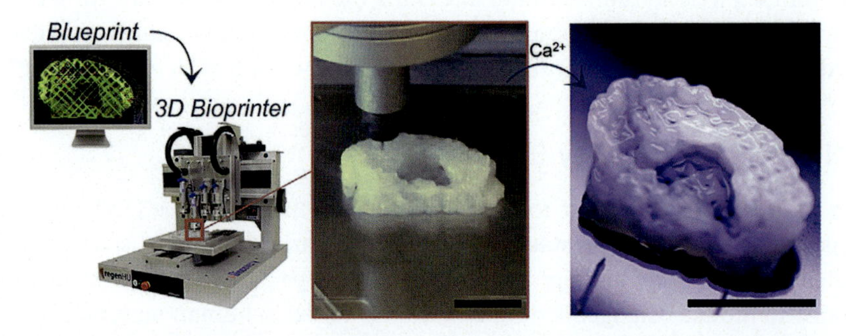

FIGURE 15.6 3D bioprinting process of chondrocyte-laden bioink of auricular constructs with open porosity. "Blueprint" software shows the toolpaths of the object material. With permission from H. Martińez et al. [111].

of collagen and alginate by syringe extrusion (AD-3000C—Ugin-tech) at 40 °C was then cross-linked in two separate steps (one for collagen and one for alginate reticulation). After that, the scaffold was soaked in a silica coating solution prepared by mixing tetraethyl-orthosilicate, nano-sized silica (SiO_2), ethanol, isopropanol, and deionized water. The printed and coated scaffolds were efficiently *in-vitro* mineralized (by incubating them in simulated body fluid—SBF—for 7 days) showing an increased mineralized layer deposition with increasing nanosilica concentrations. It was found that a 6.6 wt% of nanosilica in the scaffold could enhance the mechanical properties not only directly, but also indirectly inducing mineralization and generation of hydroxyapatite—HA—if soaked in SBF. Tests with cells showed that mouse preosteoblasts covered the scaffold in a thick layer, showing higher cell proliferation than a simple collagen-alginate scaffold.

15.5.2 Carbon-based nanoparticles

When reporting nanofillers for polymers mechanical properties enhancement, it is worth mentioning about carbon-based particles specifically. Graphene nanoflakes and CNTs are probably the most frequently employed, especially when electrical conductivity of polymers is required. But, as we will see, carbon-based nanoparticles represent an effective aid in enriching typical TE materials, which often suffer from poor mechanical resistance and self-sustainment, aiming to improve structural integrity of the printed parts. For example, graphene, made up of carbon atoms arranged in monolayers, can be employed as nanoflakes in the form of graphene oxide (GO) to improve GelMA (UV cross-linked gelatin methacryloyl) viscosity [115], allowing for better stability after deposition through syringe extrusion and without affecting its shear-thinning behavior. Moreover, GO nanoflakes have shown to induce osteogenic differentiation, even if no osteogenic supplements are added to the culture media in which the GelMa-GO scaffold is immersed.

If graphene sheets are wrapped into a cylindrical tube, they turn into CNT; and when a single layer of graphene composes the nanotube, single-walled CNTs (SWCNTs) are obtained. Otherwise, multiwalled CNTs occur when more than one graphene sheet is cylindrically wrapped in a concentric manner. SWCNTs were employed to enhance the mechanical properties of alginate [116] to be extruded by a customized multinozzle biopolymer deposition system. In this system, one syringe was employed to deposit alginate with 1 wt% SWCNTs, while the other one contained a calcium chloride solution acting as cross-linking agent. Such printed material has a tensile strength of 24% higher than pure alginate and is also biocompatible. Indeed, the presence of SWCNTs in the printed scaffolds (Fig. 15.7A), different from pure alginate ones (Fig. 15.7B), favors the generation of material defects and, as a consequence, alters the energy surface improving cellular adhesion and proliferation.

15.5.3 Hydrogels-based nanocomposite

Hydrogels are cross-linked 3D polymeric networks that contain a large quantity of water and have been extensively studied for a range of applications such as, TE [117], drug delivery [118],

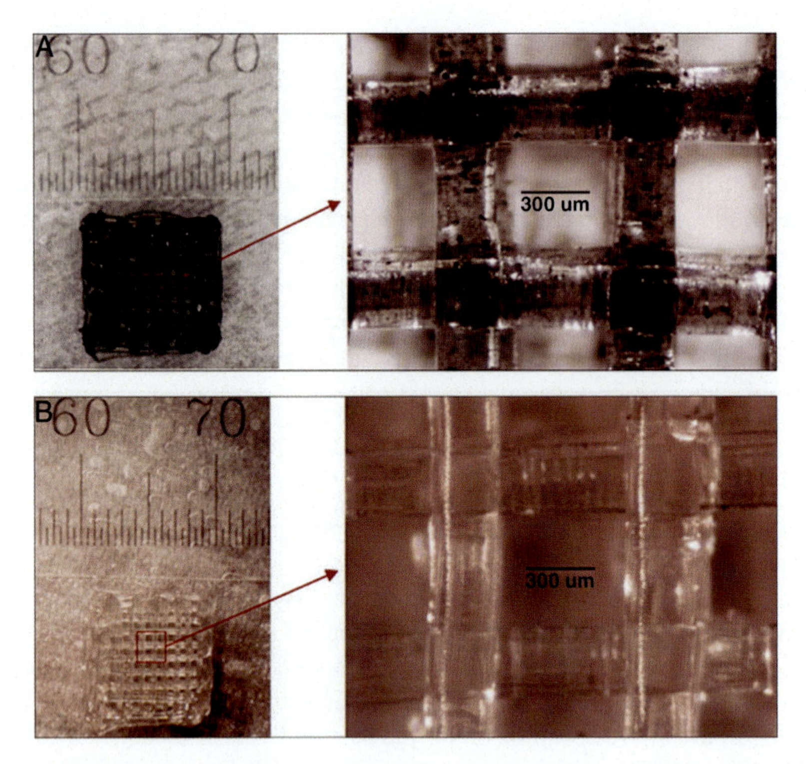

FIGURE 15.7 Scaffold morphology of 3D (A) alginate/SWCNT composite scaffold and (B) pure alginate scaffold. With permission from E.D. Yildirim et al. [116].

contact lenses [119], sensors [120], supercapacitors[120], and for wastewater treatment [121]. Additive manufacturing (AM) allows fabrication of complex 3D structures and different AM techniques have been used throughout the literature. Extrusion-based 3D printing is the most commonly used technique for hydrogels, but other methods have also been used, including ink-jet techniques [122] laser-based techniques (2-photon polymerization [123] and stereolithography [122]). Further information about different printing techniques used for hydrogels can be found in the literature [124].

Several studies have been published on extrusion printing of nanoclay-based hydrogels previously for different materials. Zhai et al. [125] developed a nanoclay hydrogel using poly(N-acryloyl glycinamide) and used a syringe-based dispensing technique to print scaffolds for bone regeneration [125]. Gao et al. [126] extruded a nanoclay gelatine methacrylate (GelMa) for complex scaffold structures such as a branched vessel and bionic ear. Jin et al. [127] extruded PEG diacrylate-laponite hydrogels with good biocompatibility and low degradation rates [127]. They reported that laponite effectively helps provide structural support and retain the shape of structures (Fig. 15.8) as deposited during printing, cross-linking of the entire structure can be delayed until printing is complete. This direct-write approach greatly broadens the range of materials suitable for extrusion-based 3D printing by blending laponite nanoclays into applicable

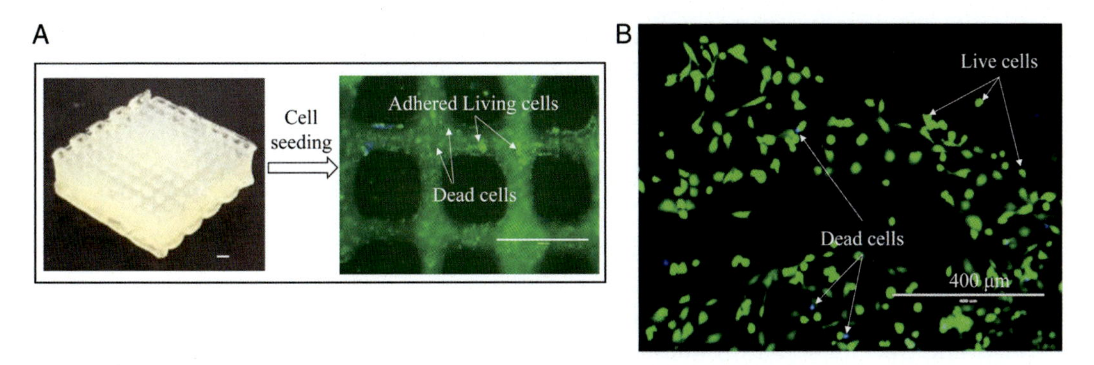

FIGURE 15.8 Effects of nanoclay on the biological properties. Cell adhesion and proliferation on the surface of (A) 3D-printed hydrogel composite scaffold (scale bars: 1.0 mm) and (B) hydrogel composite sheet with living cells in green and dead cells in blue. With permission from Jin et al. 2017 [127].

polymer precursors, which are hydrophilic and have no detrimental chemical interaction with laponite.

The addition of nanoclays can not only facilitate the printing of various hydrogel precursors but also improve the mechanical properties of printed structures. The cytocompatibility and degradation resistance of laponite-based hydrogel composites enables them to be widely used *in vivo* and *in vitro*. The proposed approach can certainly be extended to the fabrication of nonbiological structures, such as electronic devices, soft robots, microfluidic devices, etc. As laponite nanoclay colloids are hydrophilic materials, it should be tuned to facilitate the printing of hydrophobic liquid build materials.

15.6 Final remarks

In recent years, the use of nanocomposites has been increasingly highlighted in the most prominent areas of medicine, such as drug delivery, TE, wound dressings, stem cell therapy among others. Nanocomposites vary from the conventional composite materials because of the remarkably higher surface to volume ratio of the reinforcing phase and/or extremely higher characteristic ratio, increased ductility with no decrease of strength and scratching resistance, the structure of nanocomposites have positive and negative charges enables it to be mixed with many chemically and physically cross-linked network. Reinforcing materials generally improve the physical as well as mechanical characteristics of the matrix. The search for nanoparticles to be incorporated into biopolymeric matrices such as scaffolds represent a potential solution for giving to the scaffold materials new functionalities, such as promising antibacterial properties, achieve fast degradation rates due to the increase of thermal conductivity, optimize the surface roughness and wetting properties of the composite material with the aim of further improving cell adhesion phenomena and control bacterial adhesion. However, a strict control over the amount of nanoparticles incorporated must be carried out due to their cytotoxic effects.

As the human body is delicate, how to match the degradation rate of scaffolds with the formation rate of tissues is one of the most difficult problems restricting their use at present. With the advent of 3D printing technology and artificial intelligence, the manufacture of scaffolds charged with nanoparticles will be more convenient and intelligent, and could overcome the existing problems of materials and benefit more patients with organ damage.

Acknowledgments

The authors acknowledge the financial support from the São Paulo Research Foundation (FAPESP) for the Grant (2013/07914-8 and 2016/24209-4) and the Brazilian National Council for Scientific and Technological Development (CNPq) for the Research fellowship of Paulo J.A. Sobral (30.0799/2013-6).

References

[1] F. Chen, X. Liu, Advancing biomaterials of human origin for tissue engineering, Prog. Polym. Sci. 53 (2016) 86–168.

[2] G.C. Gurtner, M.J. Callaghan, M.T. Longaker, Progress and potential for regenerative medicine, Annu. Rev. Med. 58 (2007) 299–312.

[3] G.C. Gurtner, S. Werner, Y. Barrandon, M.T. Longaker, Wound repair and regeneration, Nature 453 (2008) 314–321.

[4] U.S. Department of Health & Human Services, National Data - OPTN (2019)(Accessed July 11, 2019).

[5] Z.J. Balogh, et al., Advances and future directions for management of trauma patients with musculoskeletal injuries, Lancet. 380 (9847) (2012) 1109–1119.

[6] N. Annabi, et al., 25th anniversary article: rational design and applications of hydrogels in regenerative medicine, Adv. Mater. 26 (1) (2014) 85–124.

[7] R. Langer, J.P. Vacanti, Tissue engineering, Science. 260 (1993) 920–926.

[8] M. Rahmati, E.A. Silva, J.E. Reseland, C. AH, H.J. Haugen, Biological responses to physicochemical properties of biomaterial surface, Chem. Soc. Rev. 49 (2020) 5178–5224.

[9] S. Pina, V.P. Ribeiro, C.F. Marques, F.R. Maia, T.H. Silva, R.L. Reis, Scaffolding strategies for tissue engineering and regenerative medicine applications., Materials (Basel) 12 (2019) 1824.

[10] G. Khang, Handbook of Intelligent Scaffold for Tissue Engineering and Regenerative Medicine, CRC Press,, Boca Raton, FL, 2012.

[11] M. Fathi-Achachelouei, H. Knopf-Marques, C.E.R. da Silva, J. Barthes, E. Bat, A. Tezcaner, Use of nanoparticles in tissue engineering and regenerative medicine, Fronti. Bioeng. Biotechnol. 7 (2019) 113.

[12] N. Karak, Dynamics of Advanced Sustainable Nanomaterials and Their Related Nanocomposites at the Bio-Nano Interface, Elsevier, Amsterdam, 2019.

[13] E.S. Place, N.D. Evans, M.M. Stevens, Complexity in biomaterials for tissue engineering, Nat. Mater. 8 (6) (2009) 457.

[14] Y.Yoshida Dhandayuthapani, T. Maekawa, D.S. Kumar, Polymeric scaffolds in tissue engineering application: a review, Int. J. Polym. Sci. 2011 (2011) 290602.

[15] S.G. Kumbar, R. James, S.P. Nukavarapu, C.T. Laurencin, Electrospun nanofiber scaffolds: engineering soft tissues, Biomed. Mater. 3 (2008) 15.

[16] T. Dvir, B.P. Timko, D.S. Kohane, R. Langer, Nanotechnological strategies for engineering complex tissues, Nat. Nanotechnol. 6 (2011) 13–22.

[17] A.P. Moreno, M.S. Mariel, V. Sanchez, A.P. Rodriguez, Advances in additive manufacturing for bone tissue engineering scaffolds, Mater. Sci. Eng. C. 100 (2019) 631–644.

[18] D.E. López-Angulo, C.E. Ambrosio, R. Lourenço, N.J. Nardelli Gonçalves, F.S. Cury, P.J. Sobral do Amaral, Fabrication, characterization and in vitro cell study of gelatin-chitosan scaffolds: new perspectives of use of aloe vera and snail mucus for soft tissue engineering, Mater. Chem. Phys. v. 234 (2019) 268–280.

[19] K. Dzobo, T. Turnley, A. Wishart, Fibroblast-derived extracellular matrix induces chondrogenic differentiation in human adipose-derived mesenchymal stromal/stem cells in vitro, Int. J. Mol. Sci. 17 (8) (2016) 1–20.

[20] N.D. Evans, E. Gentleman, X. Chen, C.J. Roberts, J.M. Polak, M.M. Stevens, Extracellular matrix-mediated osteogenic differentiation of murine embryonic stem cells, Biomaterials 31 (12) (2010) 3244–3252.

[21] K. Sadtler, A. Singh, M.T. Wolf, X. Wang, D.M. Pardoll, J.H. Elisseeff, Design, clinical translation and immunological response of biomaterials in regenerative medicine, Nat. Rev. Mater. 1 (7) (2016) 16040.

[22] A. Atala, Regenerative medicine strategies Journal of Pediatric Surgery 47 (1) (2012) 17–28.

[23] D.N. Kotton, E.E. Morrisey, Lung regeneration: mechanisms, applications and emerging stem cell populations, Nat. Med. 20 (8) (2014) 822–832.

[24] M. Laurenti, V. CaudaV. Grumezescu, A.M. Grumezescu (Eds.), Biodegradable polymer nanocomposites for tissue engineering: synthetic strategies and related applications, Materials for Biomedical Engineering: Absorbable Polymers 6 (2019) 157–198. https://doi.org/10.1016/B978-0-12-818415-8.00006-1.

[25] B. Dhandayuthapani, Y. Yoshida, T. Maekawa, D. Sakthi Kumar, Polymeric scaffolds in tissue engineering application: a review, Int. J. Polym. Sci. 2011 (2011) 1–20.

[26] X. Liu, J.M. Holzwarth, P.X. Ma, Functionalized synthetic biodegradable polymer scaffolds for tissue engineering, Macromol. Biosci. 12 (2012) 911–919.

[27] S.G. Kumbhar, S.H. Pawar, Self-functionalized, oppositely charged chitosan-alginate scaffolds for biomedical applications, Biotechnol. Ind. J. 13 (2) (2017) 130.

[28] P. Sai, M. Babu, Collagen based dressings—a review, Burns 26 (1) (2000) 54–62.

[29] M. Yousefi, F. Ariffin, N. Huda, An alternative source of type I collagen based on by-product with higher thermal stability, Food Hydrocoll. 63 (2017) 372–382.

[30] T. Minagawa, Y. Okamura, Y. Shigemasa, S. Minami, Y. Okamoto, Effects of molecular weight and deacetylation degree of chitin/chitosan on wound healing, Carbohydr. Polym. 67 (2007) 640–644.

[31] D. López, P. Sobral, Characterization of gelatin/chitosan scaffold blended with aloe vera and snail mucus for biomedical purpose, International Journal of Biological Macromolecules 92 (2016) 645–653.

[32] S. Liao, L. He, Y. Zhang, C. Zeng, M. Ngiam, D. Quan, Y. Zeng, J. Lu, S. Ramakrishna, Manufacture of PLGA multiple-channel conduits with precise hierarchical pore architectures and in vitro/vivo evaluation for spinal cord injury. Tissue Eng. C: Methods 15 (2009) 243–255.

[33] J.F. Mano, G.A. Silva, H.S. Azevedo, P.B. Malafaya, R.A. Sousa, S.S. Silva, L.F. Boesel, J.M. Oliveira, T.C. Santos, A.P. Marques, N.M. Neves, R.L. Reis, Natural origin biodegradable systems in tissue engineering and regenerative medicine: present status and some moving trends, J. R. Soc. Interface 4 (2007) 999–1030.

[34] A. Steinbuechel, S.K. Rhee, Polysaccharides and Polyamides in the Food Industry: Properties, Production, and Patents, Wiley-VCH, Weinheim, 2005.

[35] S.P. Massia, M.M. Holecko, G.R. Ehteshami, In vitro assessment of bioactive coatings for neural implant applications, J. Biomed. Mater. Res. 68A (2004) 177–186.

[36] G. Pertici, Introduction to bioresorbable polymers for biomedical applications, in: G. Perale, J. Hilborn (Eds.), Bioresorbable Polymers for Biomedical Applications, University of Applied Sciences and Arts of Southern Switzerland, SUPSI, Manno, Switzerland, 2017 Industrie Biomediche Insubri SA, Mezzovico-Vira, Switzerland.

[37] T. Ahmad, A. Ismail, S. Ahmad, Khalilah A. Khalil, Y. Kumar, Kazeem D. Adeyemi, Awis Q. Sazili, Recent advances on the role of process variables affecting gelatin yield and characteristics with special reference to enzymatic extraction: a review, Food Hydrocoll. 63 (2017) 85–96.

[38] D.E. López-Angulo, C.E. Ambrosio, R. Lourenço, N.J. Nardelli Gonçalves, F.S. Cury, P.J. Sobral do Amaral, Fabrication, characterization and in vitro cell study of gelatin-chitosan scaffolds: new perspectives of use of aloe vera and snail mucus for soft tissue engineering, Mater. Chem. Phys. 234 (2019) 268–280.

[39] I. Donati, S. Paoletti, Material properties of alginates, Alginates: Biol. Appl. 13 (2009) 1–53.

[40] E. Ruvinov, J. Leor, S. Cohen, The promotion of myocardial repair by the sequential delivery of IGF-1 and HGF from an injectable alginate biomaterial in a model of acute myocardial infarction, Biomaterials 32 (2011) 565–578.

[41] Ming-Te Cheng, Yu-Ru V. Shih, K. Lee OscarA. Vishwakarma, P. Sharpe, S. Songtao, M. Ramalingam (Eds.), Tendon and ligament tissue engineering, Stem Cell Biology and Tissue Engineering in Dental Sciences (2015).

[42] D.H. Lee, N. Tripathy, J.H. Shin, J.E. Song, J.G. Cha, K.D. Min, Ch.H. Park, Enhanced osteogenesis of β-tricalcium phosphate reinforced silk fibroin scaffold for bone tissue biofabrication, Int. J. Biol. Macromol. 95 (2017) 14–23.

[43] D. Ulery Bret, S. Nair Lakshmi, T Laurencin Cato, Biomedical applications of biodegradable polymers, J. Polym. Sci. B. Polym. Phys. 49 (12) (2011) 832–864.

[44] T. Ahmed, H. Marcal, M. Lawless, N.S. Wanandy, A. Chiu, L.J.R. Foster, Polyhydroxybutyrate and its copolymer with polyhydroxyvalerate as biomaterials: Influence on progression of stem cell cycle, Biomacromolecules 11 (2010) 2707–2715.

[45] M.I. Baker, S.P. Walsh, Z. Schwartz, B.D. Boyan, A review of polyvinyl alcohol and its uses in cartilage and orthopedic applications, J. Biomed. Mater. Res. Part B. 100B (2012) 1451–1457.

[46] B. Lembeck, N. W¬ulker, Severe cartilage damage by broken polyeLelactic acid (PLLA) interference screw after ACL reconstruction, Knee Surg. Sports Traumatol. Arthrosc. 13 (4) (2005) 283–286.

[47] M.H. Perez, C. Zinutti, A. Lamprecht, N. Ubrich, A. Aster, M. Hoffman, R. Bodmeier, P. Maincent, The preparation and evaluation of poly(ε-caprolactone) microparticles containing both a lipophilic and a hydrophilic drug, J. Control. Release 65 (2000) 429–438.

[48] S. Kumbar, C. Laurencin, M. Deng, Natural and Synthetic Biomedical Polymers, Elsevier Science,, Amsterdam, 2014.

[49] K.-W. Lee, S. Wang, L. Lu, E. Jabbari, B.L. Currier, M.J. Yaszemski, Fabrication and characterization of poly(propylene fumarate) scaffolds with controlled pore structures using 3-dimensional printing and injection molding, Tissue Eng. 12 (10) (2006) 2801–2811.

[50] M. Westphal, Z. Ram, V. Riddle, D. Hilt, E. Bortey, On behalf of the Executive Committee of the Gliadel® Study Group, Gliadel® wafer in initial surgery for malignant glioma: long-term follow-up of a multicentre controlled trial, Acta Neurochirurgica (Wien) 148 (2006) 269–275.

[51] T. Lu, Y. Li, T. Chen, Techniques for fabrication and construction of three-dimensional scaffolds for tissue engineering, Int. J. Nanomed 8 (2013) 337–350.

[52] F. Gervaso, A. Sannino, G.M. Peretti, The biomaterialist's task: scaffold biomaterials and fabrication technologies, Joints 1 (2013) 130–137.

[53] T. Garg, O. Singh, S. Arora, R. Murthy, Scaffold: a novel carrier for cell and drug delivery, Crit. Rev. Ther. Drug Carrier Syst. 29 (2012) 1–63.

[54] B. Guo, P.X. Ma, Synthetic biodegradable functional polymers for tissue engineering: a brief review, Sci. China. Chem. 57 (2014) 490–500.

[55] I. Armentano, M. Dottori, E. Fortunati, S. Mattioli, J.M. Kenny, Biodegradable polymer matrix nanocomposites for tissue engineering: a review, Polym. Degrad. Stab. 95 (2010) 2126–2146.

[56] M. Okamoto, B. John, Synthetic biopolymer nanocomposites for tissue engineering scaffolds, Prog. Polym. Sci. 38 (2013) 1487–1503.

[57] S. Pina, J.M. Oliveira, R.L. Reis, Natural-based nanocomposites for bone tissue engineering and regenerative medicine: a review, Adv. Mater. 27 (2015) 1143–1169.

[58] E.T. Thostenson, C. Li, T.W. Chou, . Nanocomposites in context, . Compos. Sci. Technol. 65 (2005) 491–516.

[59] P.M. Ajayan, Nanocomposite Science and Technology, Wiley-VCH Verlag, Weinheim, 2000.

[60] Y. Zheng, J. Monty, R.J. Linhardt, Polysaccharide-based nanocomposites and their applications, Carbohydr. Res. 405 (2015) 23–32.

[61] P.H.C. Camargo, K.G. Satyanarayana, F. Wypych, Nanocomposites: synthesis, structure, properties and new application opportunities, Mater. Res. 12 (2009) 1–39.

[62] S.K. Kumar, R. Krishnamoorti, Nanocomposites: structure, phase behavior, and properties, Annu. Rev. Chem. Biomol. Eng. 1 (2010) 37–58.

[63] E. Manias, Nanocomposites: stiffer by design, Nat. Mater. 6 (2007) 9–11.

[64] Y. Kasirga, A. Oral, C. Caner, Preparation and characterization of chitosan/montmorillonite-K10 nanocomposites films for food packaging applications, Polym. Compos. 33 (2012) 1874–1882.

[65] W. Hu, S. Chen, J. Yang, Z. Li, H. Wang, Functionalized bacterial cellulose derivatives and nanocomposites, Carbohydr. Polym. 101 (2014) 1043–1060.

[66] A. Ashori, S. Sheykhnazari, T. Tabarsa, A. Shakeri, M. Golalipour, Bacterial cellulose/silica nanocomposites: preparation and characterization, Carbohydr. Polym. 90 (2012) 413–418.

[67] A.J. Nathanael, J.H. Lee, D. Mangalaraj, S.I. Hong, Y.H. Rhee, Multifunctional properties of hydroxyapatite/titania bio-nano-composites: bioactivity and antimicrobial studies, Powder Technol. 228 (2012) 410–415.

[68] C.Q. Ning, Y. Zhou, In vitro bioactivity of a biocomposite fabricated from HA and Ti powder metallurgy method, Biomaterials 23 (2002) 2909–2915.

[69] M. Salarian, W.Z. Xu, Z. Wang, T.K. Sham, P.A. Charpentier, Hydroxyapatite-TiO(2)-based nanocomposites synthesized in supercritical CO(2) for bone tissue engineering: physical and mechanical properties, ACS Appl. Mater. Interfaces 6 (2014) 16918–16931.

[70] Y.-M. Kong, C.-J. Bae, S.-H. Lee, H.-W. Kim, H.-E. Kim, Improvement in biocompatibilty of zirconia-alumina nanocomposite by addition of HA, Biomaterials 26 (2005) 509–517.

[71] M. Bernstein, I. Gotman, C. Makarov, A. Phadke, S. Radin, P. Ducheyne, E.Y. Gutmanas, Low temperature fabrication of β-TCP-PCL nanocomposites for bone implants, Adv. Eng. Mater. 12 (2010) B341–B347.

[72] X. Xie, K. Hu, D. Fang, L. Shang, S.D. Tran, M. Cerruti, Graphene and hydroxyapatite self-assemble into homogeneous, free standing nanocomposite hydrogels for bone tissue engineering, Nanoscale 7 (2015) 7992–8002.

[73] S. Saska, L.N. Teixeira, P.T. de Oliveira, A.M.M. Gaspar, S.J.L. Ribeiro, Y. Messaddeq, Bacterial cellulose-collagen nanocomposite for bone tissue engineering, J. Mater. Chem. 22 (2012) 22102–22112.

[74] S. Saska, H. Barud, A. Gaspar, R. Marchetto, S. Ribeiro, Y. Messaddeq, Bacterial cellulose-hydroxyapatite nanocomposites for bone regeneration, Int. J. Biomater. 2011 (2011) 175362 article ID.

[75] I. Yamaguchi, K. Tokuchi, H. Fukuzaki, Preparation and mechanical properties of chitosan/hydroxyapatite nanocomposites, Key Eng. Mater. 192-195 (2011) 673–679.

[76] E. Boanini, P. Torricelli, M. Gazzano, R. Giardino, A. Bigi, Nanocomposites of hydroxyapatite with aspartic acid and glutamic acid and their interaction with osteoblast-like cells, Biomaterials 27 (2006) 4428–4433.

[77] S.H. Rhee, J. Tanaka, Synthesis of a hydroxyapatite/collagen/chondroitin sulphate nanocomposite by a novel precipitation method, J. Am. Ceram. Soc. 84 (2001) 459–461.

[78] M. Peter, N.S. Binulal, S. Soumya, S.V. Nair, T. Furuike, H. Tamura, R. Jayakumar, Nanocomposite scaffolds of bioactive glass ceramic nanoparticles disseminated chitosan matrix for tissue engineering applications, Carbohydr. Polym. 79 (2010) 284–289.

[79] G.M. Olyveira, D.P. Valido, L.M.M. Costa, P.B.P. Gois, L. Xavier Filho, P. Basmaji, First otoliths/collagen/bacterial cellulose nanocomposites as apotential scaffold for bone tissue regeneration, J. Biomater. Nanobiotechnol. 2 (2011) 239–243.

[80] A.L. Daniel-Da-Silva, A.B. Lopes, A.M. Gil, R.N. Correia, Synthesis and characterization of porous K-carrageenan/calcium phosphate nanocomposite scaffolds, J. Mater. Sci. 42 (2007) 8581–8591.

[81] A. Tampieri, G. Celotti, E. Landi, M. Sandri, G. Falini, N. Roveri, Biologically inspired synthesis of bone-like composite: self-assembled collagen fibers/hydroxyapatite nano- crystals, J. Biomed. Mater. Res. 67A (2003) 618–625.

[82] A. Bhowmick, N. Pramanik, P. Jana, T. Mitra, A. Gnanamani, M. Das, P.P. Kundu, Development of bone-like zirconium oxide nanoceramic modified chitosan based porous nanocomposites for biomedical application, Int. J. Biol. Macromol. 95 (2017) 348–356.

[83] H.W. Kim, J.H. Song, H.E. Kim, Bioactive glass nanofiber–collagen nanocomposite as a novel bone regeneration matrix, J. Biomed. Mater. Res. A 79 (2006) 698–705.

[84] H.W. Kim, H.H. Lee, G.S. Chun, Bioactivity and osteoblast responses of novel bio-medical nanocomposites of bioactive glass nanofiber filled poly(lactic acid), J. Biomed. Mater. Res. A 85 (2008) 651–663.

[85] A.M. Gamal-Eldeen, S.A. Abdel-Hameed, S.M. El-Daly, M.A. Abo-Zeid, M.M. Swellam, Cytotoxic effect of ferrimagnetic glass-ceramic nanocomposites on bone osteosarcoma cells, Biomed. Pharmacother. 88 (2017) 689–697.

[86] L.M. Famá, V. Pettarin, S.N. Goyanes, C.R. Bernal, Starch/multi-walled carbon nanotubes composites with improved mechanical properties, Carbohydr. Polym. 83 (2011) 1226–1231.

[87] C. Du, F.Z. Cui, Q.L. Feng, X.D. Zhu, K. de Groot, Tissue response to nano-hydroxyapatite/collagen composite implants in marrow cavity, J. Biomed. Mater. Res. 42 (1998) 540–548.

[88] C. Du, F.Z. Cui, Q.L. Feng, X.D. Zhu, K. de Groot, Three-dimensional nano-HAp/collagen matrix loading with osteogenic cells in organ culture, J. Biomed. Mater. Res. 44 (1999) 407–415.

[89] M. Kikuchi, S. Itoh, S. Ichinose, K. Shinomiya, J. Tanaka, Self-organization mechanism in a bone like hydroxyapatite/collagen nanocomposite synthesized in vitro and its biological reaction in vivo, Biomaterials 22 (2001) 1705–1711.

[90] H.W. Kim, H.E. Kim, V. Salih, Stimulation of osteoblast responses to biomimetic nanocomposites of gelatin–hydroxyapatite for tissue engineering scaffolds, Biomaterials 26 (2005) 5221–5230.

[91] H.W. Kim, J.C. Knowles, H.-E. Kim, Gelatin/hydroxyapatite nanocomposite scaffolds for bone repair, J. Biomed. Mater. Res. 72A (2005) 136–145.

[92] D.A. Wahl, J.T. Czernuszka, Collagen-hydroxyapatite composites for hard tissue repair, Eur. Cell. Mater. 11 (2006) 43–56.

[93] Y.M. Kong, C.-J. Bae, S.-H. Lee, H.-W. Kim, H.-E. Kim, Improvement in biocompatibilty of zirconia-alumina nanocomposite by addition of HA, Biomaterials 26 (2005) 509–517.

[94] S.K. Swain, I. Gotman, R. Unger, E.Y. Gutmanas, Bioresorbable β-TCP-FeAg nanocomposites for load bearing bone implants: high pressure processing, properties and cell compatibility, Mater. Sci. Eng. C 78 (2017) 88–95.

[95] A.J. Nathanael, J.H. Lee, D. Mangalaraj, S.I. Hong, Y.H. Rhee, Multifunctional properties of hydroxyapatite/titania bio-nano-composites: bioactivity and antimicrobial studies, Powder Technol 228 (2012) 410–415.

[96] S. Nath, B. Basu, M. Mohanty, P.V. Mohanan, In vivo response of novel hydroxyapatite-mullite composites: results up to 12 weeks of implantation, J. Biomed. Mater. Res. B 90 (2009) 547–557.

[97] M.S. HasnainA.M.Asiri Inamuddin, A. Mohammad (Eds.), Amit Kumar Nayak. Nanocomposites for improved orthopedic and bone tissue engineering applications, Applications of Nanocomposite Materials in Orthopedics (2019) 145–177.

[98] A. Bhowmick, N. Pramanik, P. Jana, T. Mitra, A. Gnanamani, M. Das, P.P. Kundu, Development of bone-like zirconium oxide nanoceramic modified chitosan based porous nanocomposites for biomedical application, Int. J. Biol. Macromol. 95 (2017) 348–356.

[99] W. Sajjad, T. Khan, M. Ul-Islam, R. Khand, Z. Hussain, F A.Khalida, Development of modified montmorillonite-bacterial cellulose nanocomposites as a novel substitute for burn skin and tissue regeneration, Carbohydr. Polym 206 (2019) 548–556.

[100] S.J. Hollister, R.D. Maddox, J.M. Taboas, Optimal design and fabrication of scaffolds to mimic tissue properties and satisfy biological constraints, Biomaterials 23 (20) (2002) 4095–4103.

[101] D.J.K. Ying (Ed.), Woodhead Publishing Series in Biomaterials, Functional 3D Tissue Engineering Scaffolds Materials, Technologies, and Applications, Elsevier Ltd, Amsterdam, 2018.

[102] X. Wang, M. Jiang, Z. Zhou, J. Gou, D. Hui, 3D printing of polymer matrix composites: a review and prospective, Compos. B Eng. 110 (2017) 442–458.

[103] Z. Weng, J. Wang, T. Senthil, L. Wu, Mechanical and thermal properties of ABS/montmorillonite nanocomposites for fused deposition modeling 3D printing, Mater. Des. 102 (2016) 276–283.

[104] M.R. Skorski, J.M. Esenther, Z. Ahmed, A.E. Miller, M.R. Hartings, The chemical, mechanical, and physical properties of 3D printed materials composed of TiO_2-ABS nanocomposites, Sci. Technol. Adv. Mater. 17 (2016) 89–97.

[105] C. Elliott, The effects of silver dressings on chronic and burns wound healing, Br. J. Nurs 19 (2013) S32–S36.

[106] N.P. Aditya, P.G. Vathsala, V. Vieira, R.S.R. Murthy, E.B. Souto, Advances in nanomedicines for malaria treatment, Adv. Coll. Interface Sci. 201 (2013) 1–17.

[107] B. Ipek, D. Doganay, S. Coskun, C. Kaynak, G. Akca, 3D printed antibacterial silver nanowire/polylactide nanocomposites, Compos. Part B 172 (2019) 671–678.

[108] S. Jeong, et al., High efficiency, transparent, reusable, and active PM2.5 filters by hierarchical Ag nanowire percolation network, Nano Lett. 17 (2017) 4339–4346.

[109] J.K. Carrow, A.K. Gaharwar, Bioinspired polymeric nanocomposites for regenerative medicine, Macromol. Chem. Phys. 216 (3) (2015) 248–264.

[110] K. Markstedt, A. Mantas, I. Tournier, H. Martińez Ávila, D. Ha¬gg, P. Gatenholm, 3D bioprinting human chondrocytes with nanocellulose–alginate bioink for cartilage tissue engineering applications, Biomacromolecules 16 (5) (2015) 1489 –96.

[111] M. Ávila, S. Schwarz, N. Rotter, P. Gatenholm, 3D bioprinting of human chondrocyte-laden nanocellulose hydrogels for patient-specific auricular cartilage regeneration, Bioprinting 1–2 (2016) 22–35.

[112] S. Jeong, et al., High efficiency, transparent, reusable, and active PM2.5 filters by hierarchical Ag nanowire percolation network, Nano Lett. 17 (2017) 4339–4346.

[113] J.R. Xavier, T. Thakur, P. Desai, M.K. Jaiswal, N. Sears, E. Cosgriff-Hernandez, et al., Bioactive nanoengineered hydrogels for bone tissue engineering: a growth-factor-free approach, ACS Nano 9 (3) (2015) 3109–3118.

[114] H. Lee, Y. Kim, S. Kim, G. Kim, Mineralized biomimetic collagen/alginate/silica composite scaffolds fabricated by a low-temperature bio-plotting process for hard tissue regeneration: fabrication, characterisation and in vitro cellular activities, J. Mater. Chem. B 2 (35) (2014) 5785.

[115] M. Nair, D. Nancy, A.G. Krishnan, G.S. Anjusree, S. Vadukumpully, S.V. Nair, Graphene oxide nanoflakes incorporated gelatin–hydroxyapatite scaffolds enhance osteogenic differentiation of human mesenchymal stem cells, Nanotechnology 26 (16) (2015) 161001.

[116] E.D. Yildirim, X. Yin, K. Nair, W. Sun, Fabrication, characterization, and biocompatibility of single-walled carbon nanotube-reinforced alginate composite scaffolds manufactured using freeform fabrication technique, J. Biomed. Mater. Res. Part B: Appl. Biomater. 87B (2) (2008) 406–414.

[117] K. Yue, G. Trujillo-de Santiago, M.M. Alvarez, A. Tamayol, N. Annabi, A. Khademhosseini, Synthesis, properties, and biomedical applications of gelatinmethacryloyl (GelMA) hydrogels, Biomaterials 73 (2015) 254–271.

[118] T.R. Hoare, D.S. Kohane, Hydrogels in drug delivery: progress and challenges, Polymers. 49 (8) (2008) 1993–2007.

[119] A. Mühlebach, B. Müller, C. Pharisa, M. Hofmann, B. Seiferling, D. Guerry, New water-soluble photo crosslinkable polymers based on modified poly(vinyl alcohol), J. Polym. Sci. Part A Polym. Chem. 35 (16) (1997) 3603–3611.

[120] A. Baldi, M. Lei, Y. Gu, R.A. Siegel, B. Ziaie, A microstructured silicon membrane with entrapped hydrogels for environmentally sensitivefluid gating, Sens. Actuators B. Chem. 114 (1) (2006) 9–18.

[121] R.M. Shamsuddin, C.J.R. Verbeek, M.C. Lay, Settling of bentonite particles in gelatin solutions for stickwater treatment, Procedia Eng. [Internet] 148 (2016) 194–200.

[122] V.A. Liu, S.N. Bhatia, Three-dimensional photopatterning of hydrogels containing living cells, Biomed. Microdev. 4 (4) (2002) 257–266.

[123] A.A. Pawar, G. Saada, I. Cooperstein, L. Larush, J.A. Jackman, S.R. Tabaei, High-performance 3D printing of hydrogels by water-dispersible photo initiator nanoparticles, Sci. Adv. 2 (4) (2016) 1–8.

[124] T. Billiet, M. Vandenhaute, J. Schelfhout, S. Van Vlierberghe, P. Dubruel, A review of trends and limitations in hydrogel-rapid prototyping for tissue engineering, Biomaterials 33 (26) (2012) 6020–6041.

[125] X. Zhai, Y. Ma, C. Hou, F. Gao, Y. Zhang, C. Ruan, 3D-printed high strength bioactive supramolecular polymer/clay nanocomposite hydrogel scaffold for boneregeneration, ACS Biomater. Sci. Eng. 3 (6) (2017) 1109–1118.

[126] Q. Gao, X. Niu, L. Shao, L. Zhou, Z. Lin, A. Sun, 3D printing of complex GelMA-based scaffolds with nanoclay, Biofabrication 11 (3) (2019) 035006.

[127] Y. Jin, C. Liu, W. Chai, A. Compaan, Y. Huang, Self-supporting nanoclay as internal scaffold material for direct printing of soft hydrogel composite structures in air, ACS Appl. Mater. Interfaces (2017) 17456–17465.

16

Drug delivery

Elena Marras[a], Mattia Bartoli[a,b] and Alberto Tagliaferro[a,b]

[a]DEPARTMENT OF APPLIED SCIENCE AND TECHNOLOGY, POLITECNICO DI TORINO, TORINO, ITALY. [b]CONSORZIO INTERUNIVERSITARIO NAZIONALE PER LA SCIENZA E TECNOLOGIA DEI MATERIALI (INSTM), FLORENCE, ITALY.

Chapter outline

16.1 An introduction to drug delivery

Nowadays, frontiers of medicine are moving forward day by day [1]. Regenerative medicine, theranostic, and genetic drugs are herein to move from fiction to reality conquering a place in the medical practice [2,3]. Nonetheless, traditional drugs have found a way to survive beyond the past state of the art by combining their effect with new smart drug delivery systems (DDS) [4].

During the years, several systems with a variety of release mechanisms (i.e., diffusion, hydrolysis) have been developed trying to achieve a controlled release of drugs as summarized in Fig. 16.1.

The choice of the mechanism and the design depends on the properties of the drug, the pharmacokinetic, and the administration way. One of the major issues is represented by dose-dumping that is a consequence of matrix design [5]. DDS based on polymers have been driven by advances in materials science, medical chemistry, and conjugate chemistry leading to the production of controlled release systems (CRS).

CRS represent the last frontier of DDS that are designed to deliver drugs for days to years with a well-defined release profile. Ideally, these systems maintain drug concentration within the therapeutic window, and they are accumulated in the required site of action to both increasing the efficacy and increasing potency [6]. The other key CRS requirement is the rapidly cleared or degraded when administered on their own. Engineering all of these properties is very challenging requiring a material that can store a sufficient quantity of drug, protect it from chemically cleavage during administration, and have a very controlled time release. Even if just one of these properties is missed, the DDS could induce severe drawbacks such as cytotoxicity, drug degradation, and leaching and reduce effectiveness.

Biopolymeric Nanomaterials: Fundamentals and Applications. DOI: https://doi.org/10.1016/B978-0-12-824364-0.00001-0

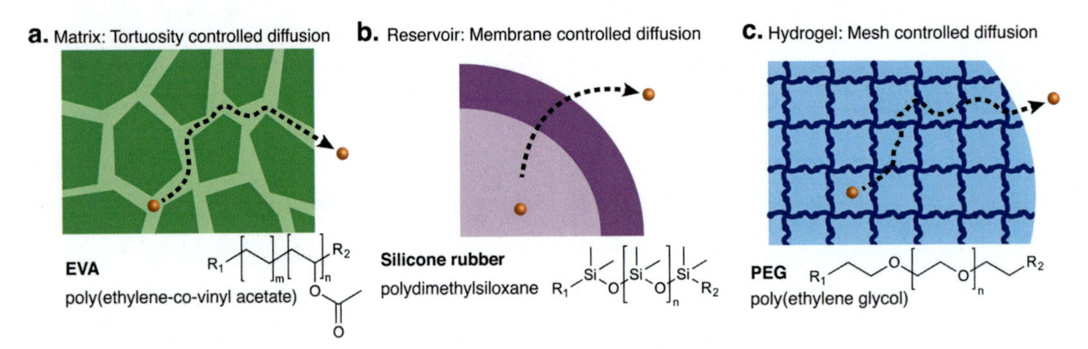

FIGURE 16.1 Mechanisms of drugs controlled release. Different release mechanisms relate to most common drug delivery approaches. Copyright 2016 American Chemical Society. From M. Tibbitt, J. Dahlman, R. Langer, Emerging Frontiers in Drug Delivery, J. Am. Chem. Soc. 138 (2016) 704-717. https://doi.org/10.1021/jacs.5b09974.

The nanotechnology, first introduced by the Nobel prize winner Richard Feynman in the late 1960s, could be a way to improve DDS applications from mere DDS to theranostic field. Theragnostic is a nanotechnology strategy that combines a diagnostic approach with a therapeutic one; it allows us to monitor the real-time response to the therapy. Thanks to the functional groups on the surface of the nanoparticles we can conjugate biomolecules for multifunctionalization which leads to targeting ligand cells, drugs for cancer therapy, and genes. Theragnostic nanoparticles are made of some fundamental components: a signal emitter that performs the imaging, a therapeutic load that can be a chemotherapeutic medication or nucleic acids, and a carrier, usually a polymeric material.

Trying to match all the requirements, innovative DDS are designed using a wide array of materials and chemical strategies [7,8]. Biopolymers represent a game-changing event in DDS development [9] due the possibility of use biological scaffold such as cellulose, proteins, or chitosan as matrix for DDS.

Accordingly in this chapter, we are presenting a brief overview on the use of biopolymers as DDS and as CRS. We also reported a brief overview on carbon dots (CDs) as a bridge between biopolymers and highly engineered solutions.

16.2 Drug delivery and biopolymers: an overview on recent advancements

Polysaccharides are biopolymers with a high degree of biocompatibility, biodegradability, and ability to mimic the natural extracellular matrix environment [10,11] and are mainly represented by cellulose, starch, and glycogen.

Cellulose is a polysaccharide made of a linear chain of D-glucose units bound to each other by $\beta(1\rightarrow4)$ bonds [12]. Drug loading of neat cellulose is quite low and requires further modifications to enhance its properties [13–15].

Muchová et al. [16] produced a 2,3-dialdehyde cellulose derivative grafted with poly(vinyl alcohol) (PVA). Authors proposed the hydrogel film based on the network formed by grafted cellulose as drug-loaded patches or wound dressings applications. An optimization of hydrogel properties was achieved by the variation of two factors—the amount of cross-linker and the weight-average molecular weight (Mw) of the source PVA. Hydrogels prepared were biocompatible, with a high porosity, and without cytotoxicity. Authors loaded the materials with caffeine showing the very efficient release in time.

Nazari et al. [17] developed a cellulose-based electrospun mats for buccal delivery of indomethacin, a nonsteroidal antiinflammatory drug. Authors used hydroxypropyl methylcellulose derivative to load and release by using a simple single-step production procedure. Moving on hydrogels, Chen et al. [18] produced a highly biocompatible, tunable, and injectable hydrogels embedded with pH-responsive diblock copolymer micelles. Authors used three cellulose derivatives (carboxymethyl cellulose, hydrazide modified carboxymethyl cellulose, and oxidized carboxymethyl cellulose) conjugate with micelles formed by using poly (ethylene oxide)-poly (2-(diisopropylamino) ethyl methacrylate) copolymers through atom transfer radical polymerization. Authors achieved a highly controllable pH-triggering release together with a prolonged and slow-release profile. Anirudhan et al. [19] proposed a hybrid conjugate based on cellulose tailored with gold nanoparticles for transdermal drug delivery of diltiazem hydrochloride. This drug is quite challenging to administer through the skin due to slower penetration rate. Authors highlighted that the skin transportation was influenced by storage time and temperature and was relatively good compared to similar materials proposed in the literature additionally showing a good stability but only if it is stored at low temperature. Skin adhesion test showed a good peeling force together with a cell viability of up to 80.0%, sufficient to be employed as non-without skin irritation. Furthermore, the materials inhibit microbial growth even after long time. Similarly, Mohd. Amin et al. [20] produced a thermo and pH-responsive bacterial cellulose derived poly(acrylic acid) hydrogels for drug delivery through a direct grafting process induced by light irradiation. Treesuppharat et al. [21] synthesized a bacterial cellulose/gelatin-based hydrogel cross-linked by using glutaraldehyde as a cross-linking agent with a swelling ratio of up to 600%.

Yang et al. [22] developed different methods based on the supramolecular assembling of cellulose nanoparticles driven by host-guest interactions of adamantane-grafted carboxyethyl hydroxyethyl cellulose and β-cyclodextrin-grafted glycerol ethoxylate that showed a quasi-spherical shape with an average diameter of around 25 nm. Authors loaded the supramolecular complex with doxorubicin linked to a β-cyclodextrin and tested it by using a Hela cell line. The materials showed a very promising pH-responsive drug release behavior together with a high *in-vitro* cytocompatibility. Furthermore, authors proved the cellular uptake mechanisms through under endocytosis.

Bhatt and Kumar [23,24] thoroughly studied the production of cellulose nanocapsule as DDS highlighting the main mechanisms of drug release. Author showed that osmotic and diffusive mechanisms are involved in the release process due intrinsically permeability of the capsules together with their porous network.

Cellulose derivatives could be a sound alternative to neat cellulose for DDS development. Wang et al. [25] used cellulose triacetate for the production of a 3D aerogel through supercritical antisolvent process. Authors loaded paracetamol up to 62 wt% (w/w), showing a higher drug concentration close to the surface of aerogel. Cellulose derivative could be also used together with advance techniques. Infanger et al. [26] used hydroxypropyl cellulose as a solid binder for the production of 3D-printed DDS of highly loaded drug delivery devices. Hydroxy ethyl cellulose could be also used as a matrix for the production of colloidal systems for the realization of efficient DDS as reported by Bekaroğlu et al. [27]. Authors loaded the cellulose derivatives with ferrite showing promising features as a theranostic platform. Maver et al. [28] used carboxymethyl cellulose loaded with diclofenac for the production of DDS coatings with improved osteogenic potential. Authors reported a high biocompatibility on human osteoblast cells together with improved cell viability up to 17% compared to blank. This material was used to protect AISI 316LVM stainless steel used in prosthetic production with an appreciable decrement of inflammatory processes. Among all cellulose derivatives, carboxymethyl cellulose is one of the most used due to the tuneable hydrophobic properties and the possibility to produce spherical size controlled nanoparticles ranging from spheres [29] to nanofibers [30,31]. Furthermore, Barkhordari et al. [32] used carboxymethyl cellulose to encapsulate cephalexin. Authors also add a gastroprotector for the oral administration and gastrointestinal release. Pooresmaeil et al. [33] combined carboxymethyl cellulose with mesoporous ferrite tailored graphene oxide for the realization of an efficient ibuprofen DDS through simple diffusion processes, while Rao et al. [34] produced a similar system but with a pH-responsive release system.

Chemical modifications are not the only tailoring process for neat cellulose improvement. Morphological changes with the production of cellulose nanocrystals could also lead to a further improvement of the DDS properties as reported by several authors [35,36]. In this field, Kim et al. [37] modified cellulose nanocrystals (CNCs) for delivered siRNA as reported in Fig. 16.2.

Authors introduced sulfonic residue on CNC surface and anchored polymeric siRNA produced through electrostatic interaction. Authors reported a significantly enhanced enzymatic stability, gene knockdown efficacy, and selective cell-induced apoptosis. Li et al. [38] produced rod-like cellulose CNCs for load and release cis-aconityl-doxorubicin upon pH changes.

Starch is a complex polysaccharide made of D-glucose units bound each other by $\alpha(1\rightarrow4)$ and $\alpha(1\rightarrow4)$ bonds assembled in two subunits: amylose and amylopectin [39]. Several procedures have been developed to tailoring the cellulose surface for enhancing the DDS ability. Grafting with polymers is one of the most explored due to the high tunability of functions inserted and the wide range of chemical approaches available [40].

Queiroz et al. [41] used a corn-derived starch for producing a DDS film loaded with chlorhexidine gluconate. Authors conducted an *in-vitro* drug release test showing a release of up to 0.02% in the cultivation medium. Chin et al. [42] realized an antimicrobial hydrogel by using starch loaded with penicillin G. Authors combined starch with PVA and poly(ethylene glycol) (PEG) via the freeze-thaw technique for creating a cross-linked network. The final starch-based material show good results against several bacterial strains such as *Escherichia coli, Staphylococcus pyogenes, Salmonella typhimurium,* and *Streptococcus aureus.* Furthermore, authors reported an excellent activity against bacterium such as *Klebsiella pneumoniae* that is resistant to multiple antibodies. Authors observed that a release profile was stable for sustained release for over 7 days. Similarly, Nallasamy et al. [43] extracted nanoparticles from a mixture of herbs, which have good DDS properties. Polyherbal drug-loaded starch nanoparticles are a promising DDS. Authors claimed relevant antibacterial activity against *Salmonella typhi* and *Shigella dysenteriae,* and

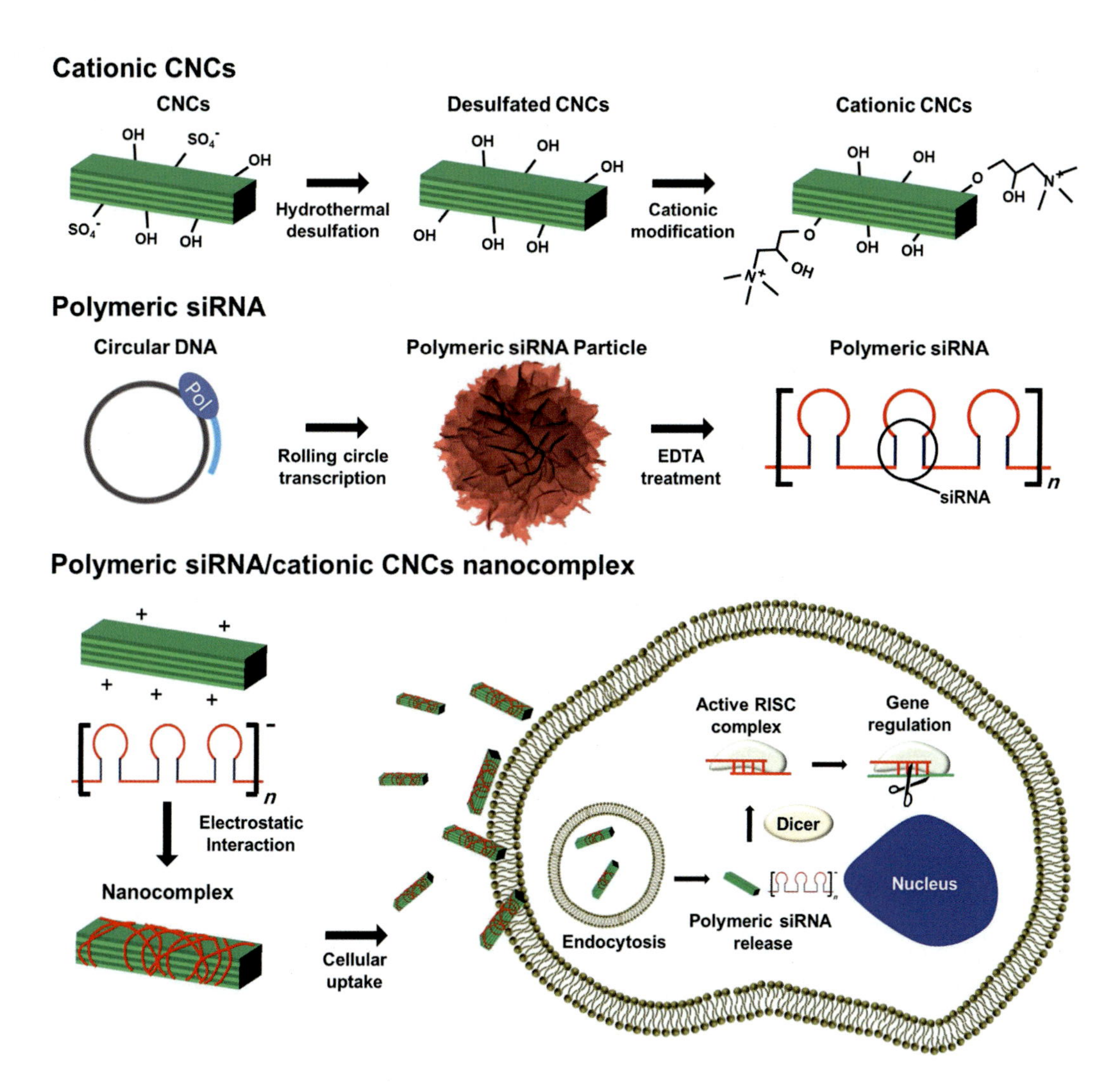

FIGURE 16.2 Modified CNCs for siRNA. Scheme of preparation and action of modified CNCs for siRNA delivery to cancer cells. From Y.M. Kim, Y.S. Lee, T. Kim, K. Yang, K. Nam, D. Choe, Y.H. Roh, Cationic cellulose nanocrystals complexed with polymeric siRNA for efficient anticancer drug delivery, Carbohydr. Polym. 247 (2020) 116684. https://doi.org/10.1016/j.carbpol.2020.116684.

antibiofilm activity against ATCC MRSA 33591 and clinical strain N7. Those properties combine with high porosity strongly point to the use of these materials as DDS. Mariadoss et al. [44], encapsulated copper oxide nanoparticles for targeted drug delivery in breast cancer therapy into folic acid–functionalized starch. Starch-based scaffold showed a hexagonal, oval-shape with an average size of 109 nm and up to 241 nmol/mg of folic acid on the surface. After copper nanoparticles loading, this DDS showed a high cytotoxicity to human breast cancer (MDA-MB-231 cell line) through the reactive oxygen species generation that induced nuclear damage and the reduction of mitochondrial membrane potential, activating the apoptosis-related protein expression. Overall, the results proved that folic acid and starch decoration increased the nanoparticles penetration in cell through folate receptor–based endocytosis for enhanced breast cancer therapy. The combination with folic acid was also used by Sun et al. [45], for the production via sonochemistry of starch microcapslules.

Starch was also used for the production of hybrid DDS as reported by Massoumi et al. [46]. Authors tailored starch with Fe_3O_4/poly(N-isopropylacrylamide-co-maleic anhydride) and loaded it with doxorubicin. The resulting material showed a good drug loading and encapsulation efficiencies as well as pH- and temperature-responsive drug release. Similarly Nezami et al. [47] produced a ferrite-tailored starch hydrogel cross-linked using itaconic acid. Authors used their system to load guaifenesin for wound healing achieving a fast release of up to 90% after 24 h at pH 7. Xu et al. [48] used oxidized cholesterol/imidazole modified starch to load and release curcumin upon pH changes showing a significantly inhibition of cancer cells.

Chitosan is a linear polysaccharide composed of randomly distributed β-(1→4)-linked D-glucosamine linked to N-acetyl-D-glucosamine through a β-(1→4) bond [49]. Chitosan has been used a lot for plenty of biological application due to its high biocompatibility and chemical tunability.

Accordingly, El-Alfy et al. [50] prepared biocompatible chitosan nanoparticles loaded by antibiotics (i.e., tetracycline, gentamycin, and ciprofloxacin) as novel DDS and added to cellulose fabrics for improving the antimicrobial activity.

Iglesias et al. [51] produced a chitosan hydrogel by using citric acid and a diiodo-trehalose derivative as cross-linker agents loaded with diclofenac. Authors observed total dissolution of the hydrogel after 96 h. A more stable cross-linked hydrogel was proposed by Karpkird et al. [52] by using chitosan cross-linked with citric acid and β-cyclodextrin. Authors used this DDS to encapsulate curcumin and were able to release it by changing pH. Karthik et al. [53] proved that chitosan-based nanospheres DDS outperformed cellulose-based ones for the release of ibuprofen after mandibular third molar surgery.

Chitosan could be combined with liposomes as reported by Grozdova et al. [54]. Authors produced a "positively charged" linear chitosan molecules cross-linked with sulfate-anions to form chitosan nanoparticles used as a scaffold of negatively charged cardiolipin/egg lecithin liposomes loaded with doxorubicin. This approach led to a fivefold increment in doxorubicin delivery, demonstrating its effectiveness against 3T3 line of mouse fibroblasts, drug-sensitive tumor MCF-7, and drug-resistant human ovarian carcinoma OVCAR-8 cell lines. Ailincai et al. [55] prepared citrylamine-PEGylated chitosan hydrogels with an increased resistance to hydrolysis induced by lysozyme but with a mass loss of 47 wt% after 21 days.

Unsoy et al. [56] proved that also chitosan-based materials could be combined with metal nanoparticles for the production of hybrid DDS loaded with doxorubicin as sketched in Fig. 16.3.

Authors used this material for *in-vitro* targeting of the MCF-7 cell lines proving doxorubicin release relation with pH and finding a best value close to 4.2 while the chitosan-based material was stable up to pH 7.4. Furthermore, the presence of iron nanoparticles allowed to develop a real theragnostic platform combining chemotherapy with fluorescence analysis. Similarly Oh et al. [57] produced a pluronic acid–modified chitosan tailored with gold nanoparticles for hydrophobic drug delivery. Shih et al. [58] showed that chitosan could be combined with gold

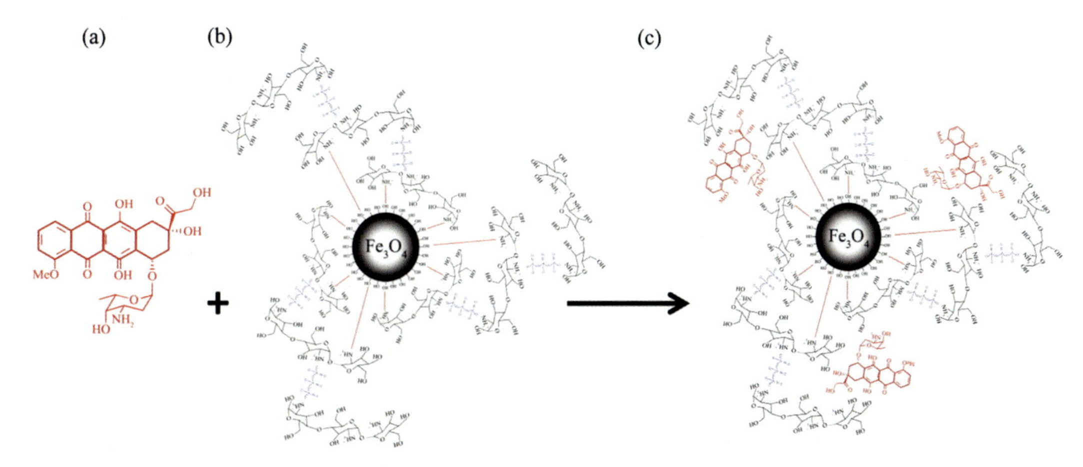

FIGURE 16.3 Chitosan hybrids loaded with Doxorubicin. Schematic representation of Doxorubicin-loaded chitosan hybrid materials tailored with ferrite. From G. Unsoy, R. Khodadust, S. Yalcin, P. Mutlu, U. Gunduz, Synthesis of Doxorubicin loaded magnetic chitosan nanoparticles for pH responsive targeted drug delivery, Eur. J. Pharma. Sci. 62 (2014) 243-250. https://doi.org/10.1016/j.ejps.2014.05.021.

suspensions for the production of a plethora of different morphologies selecting the best fit for the DDS used. Wang et al. [59] combined chitosan nanocarriers with bismuth sulfides for the production of a theragnostic platform able to perform both fluorescent imaging and photothermal therapy.

Proteinaceous materials are the last great family of biopolymers used for drug delivery. Their use has been slowed down due to their elevated Mw and difficulty in storage. Nonetheless, several attempts have been reported about the protein use as DDS [60] but it is remain an open field full with unsolved challenges.

In the next section, we described DDS based on materials able to partially mimic all the properties of biopolymers with very complex but simply to produce structures: CDs.

16.3 CDs: a new DDS approach based on highly biocompatible polymers

CDs are carbon-based nanoparticles whose dimensions range in the order of few nanometers, commonly defined as zero-dimension. Properties and applications of these nanoparticles brought a growing interest in the field in the last years. CDs are mainly constituted of carbon's sp^2 and other heteroatoms as oxygen and nitrogen, and the superficial structure determines the essential property of these materials: the photoluminescence (PL). They are the "green" alternative of the predecessors, quantum dots; they have been discovered in the early 2000s by the Carlson group during the purification of carbon nanotubes as the preparation for electrophoresis. Since their discovery, a lot of synthesis methods have been tested, as well as a lot of precursors bringing a diversification of the products were obtained.

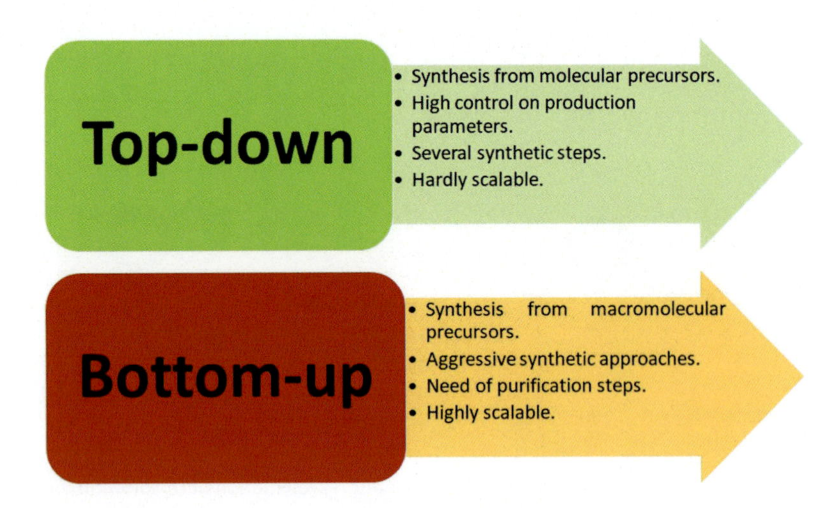

FIGURE 16.4 Classification of Carbon dot. Schematic classification of c-dots.

A first classification of CDs has been done by Valcárcel research group [61] based on CDs' nature, crystalline structure, and quantum confinement that is a parameter in the luminescent class of materials as shown in Fig. 16.4. According to this classification, there are: carbon nanodots are amorphous, semi-spherical, and do not present quantum confinement; carbon quantum dots (CQDs) have a full spherical structure with quantum confinement; graphene quantum dots are made of few graphene layers π-conjugated and the also have quantum confinement.

The synthesis methods are of two types: bottom-up and top-down. Top-down methods need bulk carbon precursors that have to be cut down in the nanoscale, an example of bulk carbon materials is graphene. The bottom-up methods involve, as precursor, organic molecules that have to be polymerized and carbonized during the process. The energetic request for the approaches is very different as the top-down one has a much higher request because of the physical process required to reduce the dimension of the raw material. The principal advantage of the top-down technique is the abundance of the precursors and the products obtained are strongly functionalized on the surface. The bottom-up technique uses chemical reaction such as a hydrothermal approach, microwave, or ultrasonication, the products present fewer flaws and especially the dimensions can be controlled better. Recently, "green" methods for the synthesis of CDs are investigated, both for the precursor and the approach itself: a source of carbon could be used as biomass such as organisms or waste materials decreasing the energy demand to follow an increasingly sustainable path.

An objective differentiation could be done based on the precursor for the synthesis of CDs and their chemical properties [62,63]. Carbon-based nanoparticles consist of two major groups: Fullerenes and carbon nanotubes. They have interesting characteristics such as electrical conductivity, strength, and electron affinity that find space in the commercial market as, for example, fillers or as a support medium. The structure resembles a graphitic sheet, this consciousness brings to another differentiation: the rolled sheet can be single-walled, double-walled, and multi-walled. Metal-based nanoparticles have as precursors noble metals like copper, silver, or gold; thanks to their optical properties the applications are in the technologies for spectroscopy. Ceramic nanoparticles are inorganic and nonmetallic. Some nanoparticles are semiconductors, they have properties of both metallic and nonmetallic materials; therefore they have several interesting applications in photocatalysis and electronics. The lipid-based nanoparticles are the one involved more in biomedical applications, as a drug carrier, because the core of the nanoparticle is made of lipid and on the surface, there are lipophilic molecules.

In 2004 Xu et al. [64] prepared CDs starting from single-walled nanotubes. A couple of years later, Zhongke et al. [65] used laser ablation and a boron/carbon target producing c-dots with enhanced luminescence emissions due to surface passivation. A further class of CDs, polymer CDs (PCDs), leads to new structures of nonconjugated groups and metal-free products, the properties of PCDs are unique, presenting high luminescence and great interaction between molecules [66].

Ionic liquids are solvent already used in chemical synthesis and extractions, but studies have shown a new application for ionic liquids in bioapplications and as precursors for nanomaterials [67]. For example, Jiang et al. [68] reported how to produce CDs starting from amino acids in acid or alkali medium. What makes ionic liquids good precursors for the synthesis of CDs with luminescent activity is the nitrogen-functionalized bulk.

Based on the synthetic route chosen, we will have different CDs, different defects, and properties. The characterization that follows the synthesis is carried out with the X-ray diffraction, the transmission electronic microscopy, UV-vis, and the Fourier transform infrared spectroscopy.

Among all properties, we find excellent biocompatibility, ease of preparation, and a relatively low production cost [69]. CDs are both soluble and stable in water because of the carboxylic, hydroxylic, aminic group that characterize their surface, but more importantly the stability permits their application on the cellular imaging. The principal and most fascinating property of the CDs is the PL. PL is a process in which a molecule absorbs a photon in the visible region, exciting one of its electrons to a higher electronic excited state, and then radiates a photon as the electron returns to a lower energy state, in other words, the incident light is going to be absorbed and successively reemitted with a little delay, in the order of 10^{-8} s, in other wavelengths [70].

What started as a research on an alternative of the older zinc or cadmium-doped quantum dots brought to the acknowledgement of the property of CDs surface-passivated with organic biomolecules. They become fluorescent in the visible regions and once functionalized they are both physicochemically and photochemically stable, enabling the possibility of cellular imaging [71].

Fluorescent nanoparticles can play a role in cellular imaging due to their cytotoxicity and *in-vivo* toxicity as an optical agent. The diagnostic function of the optical agents consists in providing information about the position of certain cells, such as tumoral ones, giving information also on the dimensions and type; this could be game-changing on the diagnosis of tumors [72].

To better understand how the diagnostic with bioimaging works, let us analyze the cell. The cell membrane is defined as "a double layer of lipids and proteins that surround a cell" protecting the cytoplasm, which is the interior part; analyzing the proteins that constitute the membrane, the cells can be distinguished by type. Biomarkers become central in this kind of investigation using receptor linked to CDs.

Carbonaceous materials are considered as nontoxic and various evaluations have been made about the functionalized CDs, with PEG for instance. *In-vivo* investigation brought about the awareness of the biocompatibility of these materials, raising interest in the prospect of using them as a contrast agent [73].

Eskalen et al. [74] synthesized CQDs with cotton linter as a precursor with a hydrothermal approach paired with a microwave process; this ultra-fast method provided not only a cheaper and easier approach to the synthesis of this product but also more efficient CDs for cancer-imaging applications. After the characterization with the technologies listed above, the cell viability and the proliferation were analyzed and it was suggested that the CDs were cytotoxic against the cell, inhibiting the cellular growth depending on the dose. Ding et al. [75] synthesized nitrogen and sulfur CDs (N/S-CDs), once characterized the team analyzed the wavelength and emissions, thus produced dots were applied for the detection of MnO_4 and $Cr_2O_7^{2-}$. Furthermore, N/S-CDs, for their biocompatibility, have proved to be a good choice for the application in cellular imaging.

Drug delivery faces hard challenges in the blood-brain barrier (BBB) due to the impediment to entering drugs into the central nervous system, reducing treatment options for diseases such as brain cancer and Alzheimer [76]. CDs, however, find an application in this sense thanks to their outstanding properties and dimensions. Hettiarachchi et al. [77] used EDA (1,2-ethylenediamine) and urea during the synthesis of CDs and shown how the CD-EDA could penetrate the BBB. In the case of an intercranial hemorrhage due to the damages in the membrane, CDs may be used for the evaluation of the damage and even to administer a therapy; for example, could be used to carry a thrombolytic agent.

CDs can simultaneously carry more than one therapeutic agent on their surface thanks to covalent bonds of the functional groups, such as carboxylic and hydroxylic groups. To combine optical properties and therapeutic performance, Ge et al. [78] passivated CDs with oxaliplatin, which is a chemotherapeutic agent that can interfere with the DNA life cycle. The Oxa-C-dots produced shown in *in-vitro* and *in-vivo* studies how the fluorescent tracking leads to the drug dosage distribution in the cells.

16.4 Conclusions

Drug delivery is a very vast realm and one of the most interesting fields in medicine. The use of biopolymers represents the real frontier in drug delivery advance due the use of bio-derived matrix able to mimic the biological structures. This deception could represent the event that will drive the drug delivery from its present to its bright future. Cellulose, starch, chitosan are largely available and cheap to isolated in a pure form providing a ground for the production of plenty of different shapes and tailored materials based on DDS requirement. The combination with nonbiopolymers such as PVA or PEG could be a further step forward in the realization of highly performing materials. While the great properties of CDs, including PL and cytotoxicity, allow these nanomaterials to have a central role in various bioapplications, there are still obstacles to further purposes. The principal impediments are the difficulties on the production on a large scale, because of the nonuniformity of the product features, and the no clarity about the origin of the PL. To overcome these limitations, efforts still have to be made in developing high-yield production routes and for a better understanding on the physical mechanisms that are involved, and by doing that this could open the way to a world of new astonishing applications.

References

[1] P. Jayaraman, A.Morshed Forkan, P. Haghighi, YB. Kang, Healthcare 4.0: A review of frontiers in digital health, Wiley Interdiscipl. Rev 10 (2020) e1350.

[2] J. Delhove, I. Osenk, I. Prichard, M. Donnelley, Public acceptability of gene therapy and gene editing for human use: a systematic review, Hum. Gene Ther 31 (2020) 20–46. https://doi.org/10.1089/hum.2019.197.

[3] G.C. Gurtner, S. Werner, Y. Barrandon, M.T. Longaker, Wound repair and regeneration, Nature 453 (2008) 314–321. https://doi.org/10.1038/nature07039.

[4] M. Shahriari, V.P. Torchilin, S.M. Taghdisi, K. Abnous, M. Ramezani, M. Alibolandi, Smart" self-assembled structures: toward intelligent dual responsive drug delivery systems, Biomaterials Science 8 (2020) 5787–5803.

[5] O. Anand, L.X. Yu, D.P. Conner, B.M. Davit, Dissolution testing for generic drugs: an FDA perspective, AAPS J 13 (2011) 328–335. https://doi.org/10.1208/s12248-011-9272-y.

[6] K.E. Uhrich, S.M. Cannizzaro, R.S. Langer, K.M. Shakesheff, Polymeric systems for controlled drug release, Chem. Rev 99 (1999) 3181–3198. https://doi.org/10.1021/cr940351u.

[7] T. Gaurav, T. Ruchi, B. SaurabhK, B. L, P. S, P. P, S. Birendra, Drug delivery systems: an updated review, Int. J. Pharma. Investig 2 (2012). https://doi.org/10.4103/2230-973x.96920.

[8] A. Samad, Y. Sultana, M. Aqil, Liposomal drug delivery systems: an update review, Curr. Drug Deliv 4 (2007) 297–305. https://doi.org/10.2174/156720107782151269.

[9] J. Jacob, J.T. Haponiuk, S. Thomas, S. Gopi, Biopolymer based nanomaterials in drug delivery systems: a review, Mater. Today Chem 9 (2018) 43–55. https://doi.org/10.1016/j.mtchem.2018.05.002.

[10] S.S. Ferreira, C.P. Passos, P. Madureira, M. Vilanova, M.A. Coimbra, Structure-function relationships of immunostimulatory polysaccharides: a review, Carbohydr. Polym 132 (2015) 378–396. https://doi.org/10.1016/j.carbpol.2015.05.079.

[11] J. Liu, S. Willför, C. Xu, A review of bioactive plant polysaccharides: biological activities, functionalization, and biomedical applications, Bioact. Carbohydr. Diet. Fibre 5 (2015) 31–61. https://doi.org/10.1016/j.bcdf.2014.12.001.

[12] L.Godbout Ven, Cellulose: Fundamental Aspects, BoD–Books on Demand GmbH, Norderstedt, Germany, 2013.

[13] A.N. Zelikin, Drug releasing polymer thin films: New era of surface-mediated drug delivery, ACS Nano 4 (2010) 2494–2509. https://doi.org/10.1021/nn100634r.

[14] M. Badshah, H. Ullah, A.R. Khan, S. Khan, J.K. Park, T. Khan, Surface modification and evaluation of bacterial cellulose for drug delivery, Int. J. Biol. Macromol 113 (2018) 526–533. https://doi.org/10.1016/j.ijbiomac.2018.02.135.

[15] U. Beekmann, L. Schmölz, S. Lorkowski, O. Werz, J. Thamm, D. Fischer, D. Kralisch, Process control and scale-up of modified bacterial cellulose production for tailor-made anti-inflammatory drug delivery systems, Carbohydr. Polym 236 (2020) 116062. https://doi.org/10.1016/j.carbpol.2020.116062.

[16] M. Muchová, L. Münster, Z. Capáková, V. Mikulcová, I. Kuřitka, J. Vícha, Design of dialdehyde cellulose crosslinked poly(vinyl alcohol) hydrogels for transdermal drug delivery and wound dressings, Mater. Sci. Eng. C 116 (2020) 111242. https://doi.org/10.1016/j.msec.2020.111242.

[17] K. Nazari, E. Kontogiannidou, R.H. Ahmad, A. Gratsani, M. Rasekh, M.S. Arshad, B.S. Sunar, D. Armitage, N. Bouropoulos, M.-W. Chang, X. Li, D.G. Fatouros, Z. Ahmad, Development and characterisation of cellulose based electrospun mats for buccal delivery of non-steroidal anti-inflammatory drug (NSAID), Eur. J. Pharma. Sci 102 (2017) 147–155. https://doi.org/10.1016/j.ejps.2017.02.033.

[18] N. Chen, H. Wang, C. Ling, W. Vermerris, B. Wang, Z. Tong, Cellulose-based injectable hydrogel composite for pH-responsive and controllable drug delivery, Carbohydr. Polym. 225 (2019) 115207. https://doi.org/10.1016/j.carbpol.2019.115207.

[19] T.S. Anirudhan, S.S. Nair, C. Sekhar. V, Deposition of gold-cellulose hybrid nanofiller on a polyelectrolyte membrane constructed using guar gum and poly(vinyl alcohol) for transdermal drug delivery, J. Membr. Sci 539 (2017) 344–357. https://doi.org/10.1016/j.memsci.2017.05.054.

[20] M.C.I. Mohd Amin, N. Ahmad, N. Halib, I. Ahmad, Synthesis and characterization of thermo- and pH-responsive bacterial cellulose/acrylic acid hydrogels for drug delivery, Carbohydr. Polym 88 (2012) 465–473. https://doi.org/10.1016/j.carbpol.2011.12.022.

[21] W. Treesuppharat, P. Rojanapanthu, C. Siangsanoh, H. Manuspiya, S. Ummartyotin, Synthesis and characterization of bacterial cellulose and gelatin-based hydrogel composites for drug-delivery systems, Biotechnol. Rep 15 (2017) 84–91. https://doi.org/10.1016/j.btre.2017.07.002.

[22] X. Yang, X. Jiang, H. Yang, L. Bian, C. Chang, L. Zhang, Biocompatible cellulose-based supramolecular nanoparticles driven by host–guest interactions for drug delivery, Carbohydr. Polym 237 (2020) 116114. https://doi.org/10.1016/j.carbpol.2020.116114.

[23] B. Bhatt, V. Kumar, Regenerated cellulose capsules for controlled drug delivery: Part III. Developing a fabrication method and evaluating extemporaneous utility for controlled-release, Eur. J. Pharma. Sci 91 (2016) 40–49. https://doi.org/10.1016/j.ejps.2016.05.021.

[24] B. Bhatt, V. Kumar, Regenerated cellulose capsules for controlled drug delivery: Part IV. In-vitro evaluation of novel self-pore forming regenerated cellulose capsules, Eur. J. Pharma. Sci 97 (2017) 227–236. https://doi.org/10.1016/j.ejps.2016.11.027.

[25] C. Wang, S. Okubayashi, 3D aerogel of cellulose triacetate with supercritical antisolvent process for drug delivery, J. Supercrit. Fluids 148 (2019) 33–41. https://doi.org/10.1016/j.supflu.2019.02.026.

[26] S. Infanger, A. Haemmerli, S. Iliev, A. Baier, E. Stoyanov, J. Quodbach, Powder bed 3D-printing of highly loaded drug delivery devices with hydroxypropyl cellulose as solid binder, Int. J. Pharma 555 (2019) 198–206. https://doi.org/10.1016/j.ijpharm.2018.11.048.

[27] M.G. Bekaroğlu, Y. İşçi, S. İşçi, Colloidal properties and in vitro evaluation of Hydroxy ethyl cellulose coated iron oxide particles for targeted drug delivery, Mater. Sci. Eng. C 78 (2017) 847–853. https://doi.org/10.1016/j.msec.2017.04.030.

[28] U. Maver, K. Xhanari, M. Žižek, L. Gradišnik, K. Repnik, U. Potočnik, M. Finšgar, Carboxymethyl cellulose/diclofenac bioactive coatings on AISI 316LVM for controlled drug delivery, and improved osteogenic potential, Carbohydr. Polym 230 (2020) 115612. https://doi.org/10.1016/j.carbpol.2019.115612.

[29] S. Butun, F.G. Ince, H. Erdugan, N. Sahiner, One-step fabrication of biocompatible carboxymethyl cellulose polymeric particles for drug delivery systems, Carbohydr. Polym 86 (2011) 636–643. https://doi.org/10.1016/j.carbpol.2011.05.001.

[30] A. Allafchian, H. Hosseini, S.M. Ghoreishi, Electrospinning of PVA-carboxymethyl cellulose nanofibers for flufenamic acid drug delivery, Int. J. Biol. Macromol 163 (2020) 1780–1786. https://doi.org/10.1016/j.ijbiomac.2020.09.129.

[31] A. Esmaeili, M. Haseli, Optimization, synthesis, and characterization of coaxial electrospun sodium carboxymethyl cellulose-graft-methyl acrylate/poly(ethylene oxide) nanofibers for potential drug-delivery applications, Carbohydr. Polym 173 (2017) 645–653. https://doi.org/10.1016/j.carbpol.2017.06.037.

[32] S. Barkhordari, M. Yadollahi, Carboxymethyl cellulose capsulated layered double hydroxides/drug nanohybrids for Cephalexin oral delivery, Appl. Clay Sci 121–122 (2016) 77–85. https://doi.org/10.1016/j.clay.2015.12.026.

[33] M. Pooresmaeil, S. Javanbakht, S. Behzadi Nia, H. Namazi, Carboxymethyl cellulose/mesoporous magnetic graphene oxide as a safe and sustained ibuprofen delivery bio-system: synthesis, characterization, and study of drug release kinetic, Coll. Surf. A 594 (2020) 124662. https://doi.org/10.1016/j.colsurfa.2020.124662.

[34] Z. Rao, H. Ge, L. Liu, C. Zhu, L. Min, M. Liu, L. Fan, D. Li, Carboxymethyl cellulose modified graphene oxide as pH-sensitive drug delivery system, Int. J. Biol. Macromol 107 (2018) 1184–1192. https://doi.org/10.1016/j.ijbiomac.2017.09.096.

[35] K. Löbmann, A.J. Svagan, Cellulose nanofibers as excipient for the delivery of poorly soluble drugs, Int. J. Pharma 533 (2017) 285–297. https://doi.org/10.1016/j.ijpharm.2017.09.064.

[36] R.D. Gupta, N. Raghav, Nano-crystalline cellulose: Preparation, modification and usage as sustained release drug delivery excipient for some non-steroidal anti-inflammatory drugs, Int. J. Biol. Macromol 147 (2020) 921–930. https://doi.org/10.1016/j.ijbiomac.2019.10.057.

[37] Y.M. Kim, Y.S. Lee, T. Kim, K. Yang, K. Nam, D. Choe, Y.H. Roh, Cationic cellulose nanocrystals complexed with polymeric siRNA for efficient anticancer drug delivery, Carbohydr. Polym 247 (2020) 116684. https://doi.org/10.1016/j.carbpol.2020.116684.

[38] N. Li, W. Lu, J. Yu, Y. Xiao, S. Liu, L. Gan, J. Huang, Rod-like cellulose nanocrystal/cis-aconityl-doxorubicin prodrug: a fluorescence-visible drug delivery system with enhanced cellular uptake and intracellular drug controlled release, Mater. Sci. Eng. C 91 (2018) 179–189. https://doi.org/10.1016/j.msec.2018.04.099.

[39] Starch in Food: Structure, Function and Applications, CRC Press, Cambridge, 2004.

[40] D. Roy, M. Semsarilar, J.T. Guthrie, S. Perrier, Cellulose modification by polymer grafting: a review, Chem. Soc. Rev 38 (2009) 2046–2064. https://doi.org/10.1039/b808639g.

[41] V.M. Queiroz, I.C.S. Kling, A.E. Eltom, B.S. Archanjo, M. Prado, R.A. Simão, Corn starch films as a long-term drug delivery system for chlorhexidine gluconate, Mater. Sci. Eng. C 112 (2020) 110852. https://doi.org/10.1016/j.msec.2020.110852.

[42] S.F. Chin, A.N.B. Romainor, S.C. Pang, S. Lihan, Antimicrobial starch-citrate hydrogel for potential applications as drug delivery carriers, J. Drug Deliv. Sci. Technol 54 (2019) 101239. https://doi.org/10.1016/j.jddst.2019.101239.

[43] P. Nallasamy, T. Ramalingam, T. Nooruddin, R. Shanmuganathan, P. Arivalagan, S. Natarajan, Polyherbal drug loaded starch nanoparticles as promising drug delivery system: antimicrobial, antibiofilm and neuroprotective studies, Process Biochem 92 (2020) 355–364. https://doi.org/10.1016/j.procbio.2020.01.026.

[44] A.V.A. Mariadoss, K. Saravanakumar, A. Sathiyaseelan, K. Venkatachalam, M.-H. Wang, Folic acid functionalized starch encapsulated green synthesized copper oxide nanoparticles for targeted drug delivery in breast cancer therapy, Int. J. Biol. Macromol 164 (2020) 2073–2084. https://doi.org/10.1016/j.ijbiomac.2020.08.036.

[45] Y. Sun, C. Shi, J. Yang, S. Zhong, Z. Li, L. Xu, S. Zhao, Y. Gao, X. Cui, Fabrication of folic acid decorated reductive-responsive starch-based microcapsules for targeted drug delivery via sonochemical method, Carbohydr. Polym 200 (2018) 508–515. https://doi.org/10.1016/j.carbpol.2018.08.036.

[46] B. Massoumi, Z. Mozaffari, M. Jaymand, A starch-based stimuli-responsive magnetite nanohydrogel as de novo drug delivery system, Int. J. Biol. Macromol 117 (2018) 418–426. https://doi.org/10.1016/j.ijbiomac.2018.05.211.

[47] S. Nezami, M. Sadeghi, H. Mohajerani, A novel pH-sensitive and magnetic starch-based nanocomposite hydrogel as a controlled drug delivery system for wound healing, Polym. Degrad. Stabil 179 (2020) 109255. https://doi.org/10.1016/j.polymdegradstab.2020.109255.

[48] Y. Xu, Y. Zi, J. Lei, X. Mo, Z. Shao, Y. Wu, Y. Tian, D. Li, C. Mu, pH-Responsive nanoparticles based on cholesterol/imidazole modified oxidized-starch for targeted anticancer drug delivery, Carbohydr. Polym 233 (2020) 115858. https://doi.org/10.1016/j.carbpol.2020.115858.

[49] M.N.V.R. Kumar, R.A.A. Muzzarelli, C. Muzzarelli, H. Sashiwa, A.J. Domb, Chitosan chemistry and pharmaceutical perspectives, Chem. Rev 104 (2004) 6017–6084. https://doi.org/10.1021/cr030441b.

[50] E.A. El-Alfy, M.K. El-Bisi, G.M. Taha, H.M. Ibrahim, Preparation of biocompatible chitosan nanoparticles loaded by tetracycline, gentamycin and ciprofloxacin as novel drug delivery system for improvement the antibacterial properties of cellulose based fabrics, Int. J. Biol. Macromol 161 (2020) 1247–1260. https://doi.org/10.1016/j.ijbiomac.2020.06.118.

[51] N. Iglesias, E. Galbis, C. Valencia, M.J. Díaz-Blanco, B. Lacroix, M.-V. de-Paz, Biodegradable double cross-linked chitosan hydrogels for drug delivery: Impact of chemistry on rheological and pharmacological performance, Int. J. Biol. Macromol 165 (2020) 2205–2218. https://doi.org/10.1016/j.ijbiomac.2020.10.006.

[52] T. Karpkird, A. Manaprasertsak, A. Penkitti, C. Sinthuvanich, T. Singchuwong, T. Leepasert, A novel chitosan-citric acid crosslinked beta-cyclodextrin nanocarriers for insoluble drug delivery, Carbohydr. Res 498 (2020) 108184. https://doi.org/10.1016/j.carres.2020.108184.

[53] K.P. Karthik, R. Balamurugan, Evaluation and comparison of anti-inflammatory properties of Ibuprofen using two drug delivery system after third molar surgery— using chitosan microspheres as a carrier for local drug delivery in to the third molar socket and through oral route, Br. J. Oral Maxillofac. Surg (2020). https://doi.org/10.1016/j.bjoms.2020.08.025.

[54] I. Grozdova, N. Melik-Nubarov, A. Efimova, A. Ezhov, G. Krivtsov, E. Litmanovich, A. Yaroslavov, Intracellular delivery of drugs by chitosan-based multi-liposomal complexes, Coll. Surf. B 193 (2020) 111062. https://doi.org/10.1016/j.colsurfb.2020.111062.

[55] D. Ailincai, L. Mititelu-Tartau, L. Marin, Citryl-imine-PEG-ylated chitosan hydrogels—promising materials for drug delivery applications, Int. J. Biol. Macromol 162 (2020) 1323–1337. https://doi.org/10.1016/j.ijbiomac.2020.06.218.

[56] G. Unsoy, R. Khodadust, S. Yalcin, P. Mutlu, U. Gunduz, Synthesis of Doxorubicin loaded magnetic chitosan nanoparticles for pH responsive targeted drug delivery, Eur. J. Pharma. Sci 62 (2014) 243–250. https://doi.org/10.1016/j.ejps.2014.05.021.

[57] K.S. Oh, R.S. Kim, J. Lee, D. Kim, S.H. Cho, S.H. Yuk, Gold/chitosan/pluronic composite nanoparticles for drug delivery, J. Appl. Polym. Sci 108 (2008) 3239–3244. https://doi.org/10.1002/app.27767.

[58] C.M. Shih, Y.T. Shieh, Y.K. Twu, Preparation of gold nanopowders and nanoparticles using chitosan suspensions, Carbohydr. Polym 78 (2009) 309–315. https://doi.org/10.1016/j.carbpol.2009.04.008.

[59] K. Wang, J. Zhuang, Y. Liu, M. Xu, J. Zhuang, Z. Chen, Y. Wei, Y. Zhang, PEGylated chitosan nanoparticles with embedded bismuth sulfide for dual-wavelength fluorescent imaging and photothermal therapy, Carbohydr. Polym 184 (2018) 445–452. https://doi.org/10.1016/j.carbpol.2018.01.005.

[60] B.J. Bruno, G.D. Miller, C.S. Lim, Basics and recent advances in peptide and protein drug delivery, Therap. Deliv 4 (2013) 1443–1467. https://doi.org/10.4155/tde.13.104.

[61] A. Cayuela, M.L. Soriano, C. Carrillo-Carrión, M. Valcárcel, Semiconductor and carbon-based fluorescent nanodots: the need for consistency, Chem. Commun 52 (2016) 1311–1326. https://doi.org/10.1039/c5cc07754k.

[62] C.L. Shen, Q. Lou, K.K. Liu, L. Dong, C.X. Shan, Chemiluminescent carbon dots: Synthesis, properties, and applications, Today 35 (2020) 100954. https://doi.org/10.1016/j.nantod.2020.100954.

[63] I. Khan, K. Saeed, I. Khan, Nanoparticles: Properties, applications and toxicities, Arab. J. Chem 12 (2019) 908–931. https://doi.org/10.1016/j.arabjc.2017.05.011.

[64] Q. Xu, W. Li, L. Ding, W. Yang, H. Xiao, W.J. Ong, Function-driven engineering of 1D carbon nanotubes and 0D carbon dots: mechanism, properties and applications, Nanoscale 11 (2019) 1475–1504. https://doi.org/10.1039/c8nr08738e.

[65] W. Zhongke, S. Yoshiki, S. Takeshi, K. Kazuhiro, K. Kenji, K. Kaoru, K. Naoto, Fabrication of crystallized boron films by laser ablation, J. Solid State Chem (2004) 1639–1645. https://doi.org/10.1016/j.jssc.2003.12.018.

[66] S. Tao, S. Lu, Y. Geng, S. Zhu, S.A.T. Redfern, Y. Song, T. Feng, W. Xu, B. Yang, Design of metal-free polymer carbon dots: a new class of room-temperature phosphorescent materials, Angew. Chem. Int. Ed 57 (2018) 2393–2398. https://doi.org/10.1002/anie.201712662.

[67] A. Zhao, C. Zhao, M. Li, J. Ren, X. Qu, Ionic liquids as precursors for highly luminescent, surface-different nitrogen-doped carbon dots used for label-free detection of Cu^{2+}/Fe^{3+} and cell imaging, Anal. Chim. Acta 809 (2014) 128–133. https://doi.org/10.1016/j.aca.2013.10.046.

[68] J. Jiang, Y. He, S. Li, H. Cui, Amino acids as the source for producing carbon nanodots: microwave assisted one-step synthesis, intrinsic photoluminescence property and intense chemiluminescence enhancement, Chem. Commun 48 (2012) 9634–9636. https://doi.org/10.1039/c2cc34612e.

[69] X.T. Zheng, A. Ananthanarayanan, K.Q. Luo, P. Chen, Glowing graphene quantum dots and carbon dots: properties, syntheses, and biological applications, Small 11 (2015) 1620–1636. https://doi.org/10.1002/smll.201402648.

[70] Basic mechanisms of photoluminescence, in: K.N. Shinde, S.J. Dhoble, H.C. Swart, K. Park (Eds.)Phosphate Phosphors for Solid-State Lighting, Springer Nature, Cham, 2012.

[71] S.T. Yang, X. Wang, H. Wang, F. Lu, P.G. Luo, L. Cao, M.J. Meziani, J.H. Liu, Y. Liu, M. Chen, Y. Huang, Y.P. Sun, Carbon dots as nontoxic and high-performance fluorescence imaging agents, Journal of Physical Chemistry C 113 (2009) 18110–18114. https://doi.org/10.1021/jp9085969.

[72] J. Du, N. Xu, J. Fan, W. Sun, X. Peng, Carbon dots for in vivo bioimaging and theranostics, Small 15 (32) 2019, 1805087.

[73] S.T. Yang, L. Cao, P.G. Luo, F. Lu, X. Wang, H. Wang, M.J. Meziani, Y. Liu, G. Qi, Y.P. Sun, Carbon dots for optical imaging in vivo, J. Am. Chem. Soc 131 (2009) 11308–11309. https://doi.org/10.1021/ja904843x.

[74] H. Eskalen, S. Uruş, S. Cömertpay, A.H. Kurt, Ş. Özgan, Microwave-assisted ultra-fast synthesis of carbon quantum dots from linter: Fluorescence cancer imaging and human cell growth inhibition properties, Ind. Crop. Prod (2020) 147. https://doi.org/10.1016/j.indcrop.2020.112209.

[75] C. Ding, Z. Deng, J. Chen, Y. Jin, One-step microwave synthesis of N,S co-doped carbon dots from 1,6-hexanediamine dihydrochloride for cell imaging and ion detection, Coll. Surf. B 189 (2020) 10838. https://doi.org/10.1016/j.colsurfb.2020.110838.

[76] K.J. Mintz, G. Mercado, Y. Zhou, Y. Ji, S.D. Hettiarachchi, P.Y. Liyanage, R.R. Pandey, C.C. Chusuei, J. Dallman, R.M. Leblanc, Tryptophan carbon dots and their ability to cross the blood-brain barrier, Coll. Surf. B 176 (2019) 488–493. https://doi.org/10.1016/j.colsurfb.2019.01.031.

[77] S.D. Hettiarachchi, R.M. Graham, K.J. Mintz, Y. Zhou, S. Vanni, Z. Peng, R.M. Leblanc, Triple conjugated carbon dots as a nano-drug delivery model for glioblastoma brain tumors, Nanoscale 11 (2019) 6192–6205. https://doi.org/10.1039/C8NR08970A.

[78] J. Ge, Q. Jia, W. Liu, L. Guo, Q. Liu, M. Lan, H. Zhang, X. Meng, P. Wang, Red-emissive carbon dots for fluorescent, photoacoustic, and thermal theranostics in living mice, Advanced Mater 27 (2015) 4169–4177. https://doi.org/10.1002/adma.201500323.

17

Applications in food products

Maria Jaízia dos Santos Alves[a], Wilson Daniel Caicedo Chacon[a], Kennya Thayres dos Santos Lima[a], Talita Ribeiro Gagliardi[b], Alcilene Rodrigues Monteiro[a] and Germán Ayala Valencia[a]

[a]DEPARTMENT OF CHEMICAL AND FOOD ENGINEERING, FEDERAL UNIVERSITY OF SANTA CATARINA, FLORIANÓPOLIS, SC, BRAZIL. [b]DEPARTMENT OF CELL BIOLOGY, EMBRYOLOGY AND GENETICS, FEDERAL UNIVERSITY OF SANTA CATARINA, FLORIANÓPOLIS, SC, BRAZIL.

Chapter outline

17.1 Introduction

Biopolymers can be classified in three categories, such as biopolymers isolated from biomass, biopolymers chemically synthesized using monomers from agro-resources, and biopolymers synthetized by means of microbial routes [1,2]. Particularly, most studies have been addressed to explore the applications of biopolymers isolated from biomass in the food sector because biomass is an abundant, renewable, and inexpensive raw material [2]. The most common biopolymers isolated from biomass are carbohydrates (e.g., cellulose, starch, and chitin/chitosan) and proteins (e.g., collagen/gelatin, gluten, soybean, zein, and whey) [2]. These biopolymers are "generally recognized as safe" and commonly used for food applications due to their natural sources, biocompatibility, nontoxicity, biodegradability, and acceptable mechanical properties [3]. More details about biopolymer classification and applications can be found in the specialized literature [2,4].

In the last years, several research studies have been focused on the development of biopolymer nanoparticles using carbohydrates and proteins as these materials have unique physical and chemical characteristics due to their quantum size, surface, and microquantum tunnel effects [5–7]. However, in the reviewed literature, no review paper has classified systematically the potential applications of nanoparticles based on biopolymers in the food industry. Therefore,

Biopolymeric Nanomaterials: Fundamentals and Applications. DOI: https://doi.org/10.1016/B978-0-12-824364-0.00011-3

this chapter aims to review the state of the art with respect to the application of biopolymeric nanoparticles in food products.

17.2 Main applications of biopolymer nanomaterials in food products

17.2.1 Antimicrobials

Some biopolymer nanoparticles have natural antimicrobial properties and they can be used to control microbial growth in foods. In general, the particle size reduction of biopolymers can turn these materials more reactive due to the high exposure of functional groups. Also, biopolymer nanoparticles can encapsulate compounds with antimicrobial properties in its structure (Fig. 17.1). In both alternatives, biopolymer nanoparticles with or without antimicrobial compounds can cause physical damage to microorganisms associated with food spoilage [6].

Chitosan has been the main biopolymer used to produce nanoparticles with antimicrobial properties [7]. The particle size reduction of chitosan increases the exposure of amino groups that can interact with the electronegative charges on the surface of the microbial cell, resulting in the leakage of components [8]. Furthermore, chitosan nanoparticles have been used to encapsulate antimicrobial compounds by means of several approaches such as ionic gelation, ultrasonication, high pressure, and stirring, among others (Table 17.1).

Essential oils are generally encapsulated with chitosan nanoparticles to manufacture nanomaterials with high antimicrobial properties. Recently, Das et al. [7] produced chitosan nanoparticles encapsulating essential oil from *Coriandrum sativum* by ionic gelation and observed that these materials can be used to control the fungal and aflatoxin contamination of stored rice. In other research, chitosan nanoparticles encapsulating nutmeg seed oil [8], as well as limonene, linalool, menthol, and thymol [5] have been used to manufacture edible coatings for strawberries and minced meat. In these studies, it was observed that food packaging containing chitosan nanoparticles encapsulating essential oils can be used to retard the spoilage caused by *Escherichia coli*. Chitosan nanoparticles have been used to stabilize metabolites from lactic acid bacteria for the production of cheese coatings [9] and nano-silicon oxides films [10]. The developed materials displayed antimicrobial activity against *Staphylococcus aureus, Listeria. monocytogenes, E. coli, Staphylococcus sciuri, Bacillus cereus, Salmonella enterica*, and *Pseudomonas aeruginosa* (Table 17.1).

Nanoparticles based on cellulose are other materials showing future food applications. Cellulose nanoparticles do not have antimicrobial properties; hence, these nanomaterials are used to encapsulate antimicrobial compounds [11]. The applications of cellulose nanoparticles in the food area have been observed in its derivatives, such as nanocomposites with carboxymethylcellulose, cellulose nanofibers (CNFs), hydroxypropyl methylcellulose (HPMC), and TEMPO (2,2,6,6-tetramethylpiperidine-1-oxyl radical)-oxidized cellulose. These materials have been used as nanocarriers of several antimicrobial compounds (Table 17.1) [12–14]. In this way, Li et al. [15] concluded that HPMC emulsions with cinnamon essential oil can reduce the microbial growth of *E. coli, S. aureus*, and *L. monocytogenes*. Similar behavior was observed in

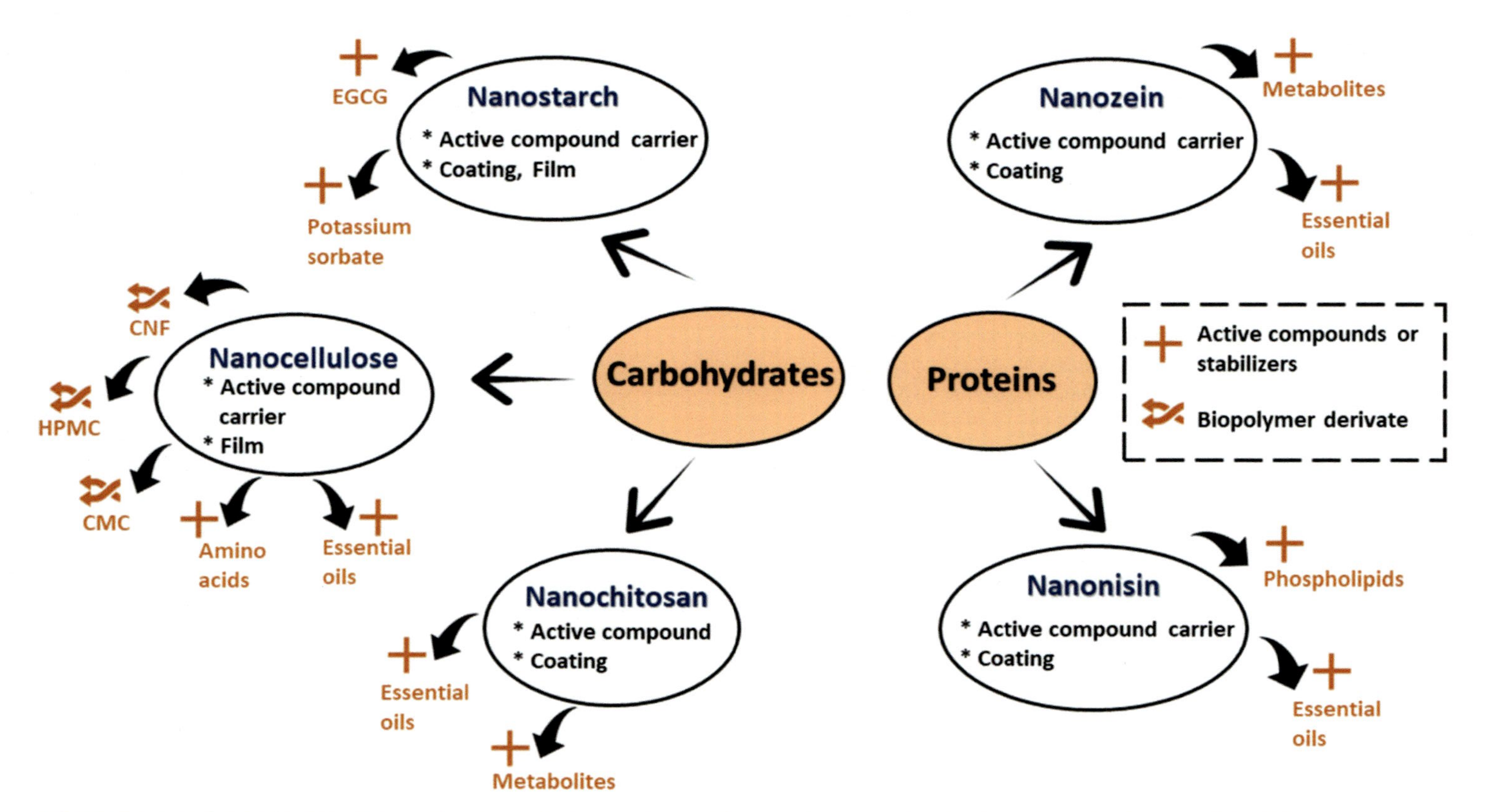

FIGURE 17.1 Biopolymeric nanoparticles with antimicrobial properties and their applications in food systems. Epigallocatechin gallate, CNF, carboxymethylcellulose, hydroxypropyl methylcellulose.

Table 17.1 Different types of biopolymeric nanoparticles with antimicrobial properties.

Materials	Production approach	Presentation type	Main result and application	References
Chitosan/monoterpenes	Ultrasonication	Coating	Coatings with antimicrobial activity against *E. coli* and applied in meat products	[5]
Chitosan/*Coriandrum sativum* essential oil	Ionic gelation	Coating	Coatings reduced fungal growth and they were used to extend the rice shelf-life	[7]
Chitosan/nutmeg seed oil	High pressure and stirring	Coating	Coatings reduced microbial growth in strawberries	[8]
Chitosan/metabolites from lactic acid bacteria	Mechanical stirring	Additive	Chitosan nanoparticles loaded with metabolites showed antimicrobial activity against *Salmonella enterica* and *P. chrysogenum* in Egyptian cheese	[9]
Chitosan/nano-silicon oxides	Ultrasonication	Coating	Coatings with antimicrobial activity against *E. coli* and *S. aureus* applied in green tomato	[10]
Chitosan/clove essential oil	Ionic gelation	Film	Films with antibacterial activity against *L. monocytogenes* and *S. aureus*	[17]
Chitosan/nisin	Ionic gelation	Film	Films with antimicrobial properties against *L. monocytogenes* in refrigerated and vacuum-sealed lean meat for 10 days	[18]
Chitosan/*Satureja hortensis* L. essential oil	Ionic gelation	Additive	Nanoparticles with microbial activity against *E. coli, S. aureus*, and *L. monocytogenes*	[19]
Chitosan/bitter orange oil	Ionic gelation	Additive	Nanoparticles with microbial activity against mesophilic and psychrophilic bacteria, yeasts, and molds	[20]
Chitosan	Ionic cross-linking	Additive	Nanoparticles with microbial activity against *P. fluorescens, E. carotovora*, and *E. coli*	[21]
Chitosan/phoenix dactylifera tree extract	Ionic gelation	Additive	Nanoparticles with microbial activity against *B. cereus, B. subtilis*, and *C. albicans*	[22]
Chitosan/zinc oxides	Sol-gel	Film	Films with microbial activity against *S. aureus* and *E. coli*	[23]

(*continued on next page*)

Table 17.1 *(continued)*

Materials	Production approach	Presentation type	Main result and application	References
Carboxymethyl cellulose/glycine/copper	Thermal treatment with stirring	Additive	Nanoparticles with microbial activity against *B. subtilis, C. albicans, E. coli, P. aeruginosa*, and *S. aureus*	[11]
Cellulose nanocrystal/polyvinyl alcohol	Mechanical treatment and esterification	Film	Films with antimicrobial activity against *C. Albicans, E. Carotovora, P. Vulgaris, B. subtilis*, and *S. aureus*	[12]
Cellulose/starch/eucalyptus globulus leaf extract	Acid hydrolysis	Film	Films reduced the growth of *E. coli, L. monocytogenes, S. typhimurium* and *Penicillium spp.*, in grapes stored for 28 days at 25 °C	[13]
TEMPO-oxidized cellulose/L-phenyl alanine or L-tryptophan	Mechanical stirring	Additive	Nanoparticles with microbial activity against *E. coli, P. aeruginosa, S. aurous, B. subtilis*, and *C. albicans*	[14]
Hydroxypropyl methyl cellulose/cinnamon essential oil	Emulsification	Additive	Nanoparticles with antimicrobial activity against *E. coli, S. enterica* subsp. Enterica serovar *Typhi, S. aureus*, and *L. monocytogenes*	[15]
Cellulose nanofibers/ginger essential oil and citric acid	Thermal treatment with stirring	Coating	Coatings increased the shelf life of meat products up to 6 days	[16]
Zein/carboxymethyl chitosan/natamycin	Antisolvent precipitation	Coating	Coatings with antimicrobial activity against *B. cinerea* spores	[24]
Zein/Arabic gum/tymol	Antisolvent precipitation	Additive	Nanoparticles with antimicrobial activity against *E. coli*	[25]
Zein/*Thymbra capitata* [L.] Cav. essential oil	Self-assembly	Additive	Nanoparticles with antimicrobial activity against *E. coli* and *L. monocytogenes*	[26]
Starch/epigallocatechin gallate	Antisolvent precipitation	Additive	Nanoparticles with antimicrobial activity against *E. coli* and *S. aureus*	[27]
Starch/silver	Acid hydrolysis	Film	Films with antimicrobial activity against *S. aureus, S. typhi*, and *E. coli*	[28]

cellulose nanoparticles containing *Eucalyptus globulus* leaf [13]. In this same line, Khaledia et al. [16] developed a bioactive coating based on CNF and ginger essential oil and concluded that this material significantly reduced the microbial growth of ready-to-cook barbecue chicken meat.

Cellulose nanoparticles associated with amino acids also have been studied. A large number of antimicrobial amino acids show broad-spectrum activities against different types of microorganisms such as Gram-positive and Gram-negative bacteria, fungi, and viruses (Table 17.1) [14]. In this context, Hasasnin et al. [14] associated TEMPO-oxidized cellulose with two amino acids (L-phenylalanine and L-tryptophan) and concluded that those nanomaterials had antimicrobial activity against *Bacillus subtilis, E. coli, P. aeruginosa,* and *S. aureus,* and antifungal activity against *Candida albicans.*

Starch is another polysaccharide used as a carrier of antimicrobial compounds, this biopolymer has particular properties such as higher solubility, faster diffusion rates, higher absorption capacity, and higher penetration rates through biological barriers [29]. Normally, starch nanoparticles are produced by means of nanoprecipitation or antisolvent precipitation method [27,30]. Some researchers have encapsulated potassium sorbate [30] and epigallocatechin gallate [27], so that the resulting nanomaterials showed antimicrobial properties against *S. aureus* and *E. coli.*

Recently, the use of protein nanomaterials has been getting considerable attention, such as zein which is being applied in the food and pharmaceutical areas [31]. In the literature, some researchers have encapsulated metabolites from bacteria and plant extracts with zein by the antisolvent precipitation method. Based on antimicrobial properties, these colloidal nanomaterials have future applications in the food packaging area [24,26]. Recently, Li et al. [25] developed pickering nanoemulsions based on zein and thymol and observed that these systems can reduce the growth of *E. coli.* Furthermore, natamycin has been used as an antimicrobial compound to manufacture zein nanoparticles with antimicrobial properties against yeasts and molds [24,32].

17.2.2 Encapsulation of food ingredients

Polysaccharides and proteins can be used to preserve and stabilize bioactive compounds by means of different techniques (Fig. 17.2). Among all polysaccharides, chitosan is the only naturally occurring cationic polysaccharide where the amino groups can form stable covalent bonds with other chemical groups from bioactive compounds [1]. In the reviewed literature, several studies have produced chitosan nanoparticles by means of electrostatic deposition, ultrasonication, ionic gelation, antisolvent precipitation, and liquid–liquid dispersion to stabilize bioactive components such as vitamins and anthocyanins [20,33–38]. In this way, Sun et al. [39] fabricated nanoparticles to stabilize cyanidin-3-oglucoside by means of ionic cross-linking using chitosan, chitosan oligosaccharides, and carboxymethyl chitosan united with ionic cross-linking agent γ-polyglutamic acid or calcium chloride ($CaCl_2$). The resulting nanomaterials had encapsulation efficiency and loading capacity of approximately 54% and 5%, respectively, because the encapsulated anthocyanins retained their biological activity. In another research, Karimirad et al. [20] used the ionic gelation method to produce chitosan nanoparticles encapsulating cyminum

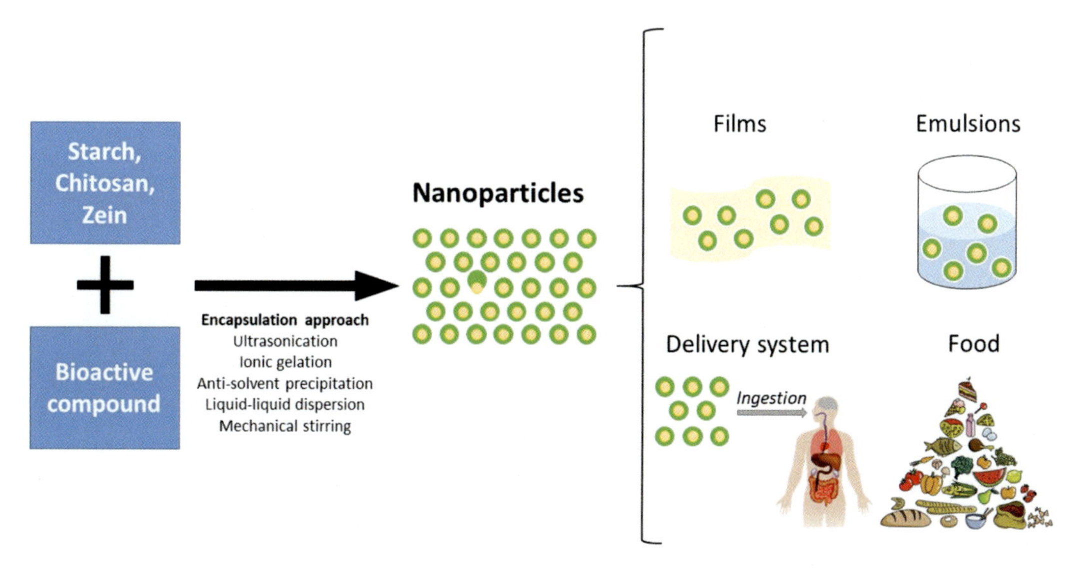

FIGURE 17.2 Schematic illustration for encapsulation of bioactive compounds using biopolymeric nanoparticles and their applications in food systems.

oil with average particle size between 30 and 80 nm. The authors concluded these nanoparticles can reduce the mushrooms oxidation during storage.

Zein is another biopolymer used as shell material to encapsulate hydrophobic bioactive compounds. This fact is due to the presence of hydrophilic amino acids in the zein chains which are soluble in ethanol solutions (60–90%) [38]. In this way, nanoparticles based on zein have been used to encapsulate hydrophobic nutrients such as resveratrol [40], curcumin [41,42], β-caroteno [43], lutein [44], and quercetin [45] by means of antisolvent precipitation or liquid-liquid dispersion (Table 17.2). These materials have shown a broad range of encapsulation efficiencies (21–51%), because of that the stability of the encapsulated biological compound against light or UV radiation has been improved. Furthermore, the digestibility of carotenoids such as lutein was improved almost 28% after its encapsulation using zein nanoparticles by antisolvent precipitation method [38]. In this same line, Ma et al. [46] optimized the encapsulation efficiency of lutein using zein and tea saponin, because the increase of mass ratio of zein to lutein from 25:1 (% w/w) to 50:1 (% w/w) increased the efficiency of encapsulation from 89.3% to 92,9%. Also, the same nanoparticles showed high thermal and chemical stability in a temperature and pH range from 37 °C to 80 °C, and from 4.0 to 9.0, respectively.

Some studies have explored the use of starch nanoparticles as shell wall to encapsulate bioactive compounds (Table 17.2). Yang et al. [62] encapsulated approximately 97% of linoleic acid with octenyl succinic anhydride starch and xanthan gum. *In vivo* animal experiments using rats showed that starch-based nanoparticles were able to protect conjugated linoleic acid during digestion, because a high amount of this bioactive compound was released in the small intestine. In this same line. Tangeretin was encapsulated using debranched starch and β-cyclodextrin. The nanomaterials obtained had an encapsulation efficiency of 83% and high physical stability when

Table 17.2 Different types of biopolymeric nanoparticles encapsulating bioactive compounds.

Materials	Encapsulation approach	Main result and application	References
Chitosan/cuminum cyminum oil	Ionic gelation	Nanoparticles with particle size around 50 nm and antioxidant properties used to extent the shelf-life of mushrooms	[20]
Chitosan/curcumin	Mechanical stirring	The nanoparticles showed high *in vitro* release and cytotoxicity in A549 cell line, as well as antimicrobial activity against *S. aureus* and *E. coli*	[33]
Chitosan/Arabic gum/curcumin	Antisolvent precipitation	Nanoparticles with particle size around 100 nm encapsulating approximately 94% of curcumin were obtained	[34]
Chitosan/hesperidin	Pickering emulsion	Encapsulated hesperidin using chitosan nanoparticles showed high surface stability and long shelf-life	[35]
Chitosan/catechin	Ultrasonication	Chitosan nanoparticles can transport tea catechins and increase polyphenol stability and bioavailability	[36]
Chitosan/ cinnamaldehyde	Ionic gelation	Nanoparticles with particle size around 400 nm and encapsulation efficiency of 73%	[37]
Chitosan/ anthocyanins	Ionic crosslinking	Nanoparticles with particle size around 180 nm and encapsulation efficiency of 54%	[39]
Chitosan/zein/ resveratrol	Liquid–liquid dispersion	Nanoparticles with particle size around 300 nm and encapsulation efficiency of 51%	[40]
Zein/curcumin	Antisolvent precipitation	Nanoparticles with encapsulation efficiency between 18% and 98% and controlled curcumin released in simulated gastrointestinal fluids	[41]
Zein/curcumin	Stirring	Nanoparticles with particle size around 100 nm and physical stability in pH 5–9	[42]
Chitosan/ zein/alginate/ resveratrol	Electrostatic deposition	The nanoparticles were stable in gastric solutions, because the polyphenol bioaccessibility increased to 81% in encapsulated resveratrol	[47]
Chitosan/β-galactosidase	Ionic gelation	Nanoparticles with particle size around 143 nm encapsulating approximately 68% of β-galactosidase were obtained	[48]
Chitosan/ protocatechuic acid	Ionic gelation	Nanoparticles with particle size around 400 nm and antioxidant properties	[49]
Zein/lutein	Antisolvent precipitation	Nanoparticles with particle size around 75 nm and high stability when exposed with gastric enzymes	[38]
Zein/carboxymethyl chitosan/Tea polyphenol/β-caroteno	Antisolvent precipitation	Nanoparticles with encapsulation efficiency of 93% and physical stability in simulated gastrointestinal conditions	[43]

(*continued on next page*)

Table 17.2 *(continued)*

Materials	Encapsulation approach	Main result and application	References
Zein/lutein	Antisolvent precipitation	Nanoparticles with particle size around 200 nm and encapsulation efficiency of 80%	[44]
Zein/quercetin	Antisolvent precipitation	Nanoparticles with particle size around 200 nm, encapsulation efficiency of 82% and physical stability at pH 2–8	[45]
Zein/tea saponin/lutein	Antisolvent precipitation	Nanoparticles with particle size around 200 nm, encapsulation efficiency of 93%, and physical stability at pH 4–9	[46]
Zein/curcumin	Antisolvent precipitation	Nanoparticles with particle size around 230 nm and encapsulation efficiency of 90%	[50]
Zein/pectin/alginate/curcumin	Electrostatic deposition	Nanoparticles with particle size around 200 nm and antioxidant properties	[51]
Zein/caseinate/sodium alginate/curcumin	Antisolvent precipitation	Nanoparticles with particle size around 90 nm, encapsulation efficiency of 76%, antioxidant properties and controlled curcumin released in simulated gastrointestinal fluids	[52]
Zein/casein/ferulic acid	Liquid–liquid dispersion	Nanoparticles with particle size around 200 nm and encapsulation efficiency of 23%	[53]
Zein/ι-carrageenan/curcumin	Antisolvent precipitation	Nanoparticles with particle size around 100 nm, antioxidant properties and controlled curcumin released in simulated gastrointestinal fluids	[54]
Zein/eugenol	Complexation	Nanoparticles with particle size around 600 nm and antimicrobial properties	[55]
Zein/curcumin	Antisolvent precipitation	Nanoparticles with particle size around 100 nm, encapsulation efficiency of 85%, and physical stability at pH 2–9	[56]
Zein/whey protein isolate/curcumin	Stirring	Nanoparticles with particle size around 90 nm and thermal stability until 80 °C	[57]
Gelatin/fish oil	Pickering emulsions	Nanoparticles with particle size around 25 nm and physical stability at pH 3–11	[58]
Starch/lutein	Antisolvent precipitation	Nanoparticles with particle size around 350 nm and physical stability during storage	[59]
Modified starch/tangeretin	Ultrasonication	Nanoparticles with particle size around 400 nm and encapsulation efficiency of 99%	[60]
Starch/gallic acid	Antisolvent precipitation	Nanoparticles with particle size around 900 nm and encapsulation efficiency of 90%	[61]
Modified starch/linoleic acid	Stirring	Nanoparticles with encapsulation efficiency of 97% and good resistance against gastric acid	[62]
Gliadin/curcumin	Antisolvent precipitation	Nanoparticles with encapsulation efficiency of 91% and physical stability at pH 7–9	[63]
Zein/alginate/gelatin/curcumin	Antisolvent precipitation	Nanoparticles with particle size around 250 nm with physical and thermal stability at pH 3–7 and until 80 °C, respectively	[64]

exposed to different simulated human microenvironments, including ionic strength (physiological saline), temperature (37 °C), and acidic stomach conditions (pH = 2.1) [60].

In recent years, some studies have been addressed to encapsulate curcumin, a natural compound with antioxidant, antiinflammatory, anticarcinogenic, antitumor, and antimicrobial activity (Table 17.2). The encapsulation of curcumin with hydroxyethyl starch can improve the thermal stability and antioxidant properties of this bioactive compound [54]. Other macromolecules such as gliadins [63], pectins [65], carboxymethylated corn fiber gum [66], Konjac glucomannan octenyl succinate [67], and κ-carrageenan [68] have been used as shell wall materials of biopolymeric nanoparticles to preserve curcumin. For all the studies mentioned above, nanoparticles have been shown to be potent and suitable delivery systems for different food bioactive. The progress of nano-designed particles has systematically promoted food delivery systems and contributed to expanding the application of these compounds.

17.2.3 Enhancement of physical properties in foods

There are few studies about the effect of biopolymeric nanoparticles on the physical properties of foods. Some researchers have demonstrated that chitosan nanoparticles can retain some physicochemical properties in foods. Chouljenko et al. [69] concluded that chitosan nanoparticles containing sodium tripolyphosphate can be used to retain the color, texture, and moisture content, as well as to reduce the lipid oxidation of shrimps stored at −20 °C during 120 days. In this same line, Karimirad et al. [20] concluded that chitosan nanoparticles containing *Cuminum cyminum* oil improved the firmness and inhibited the formation of brown patches through reduction of microbial count of packed mushrooms during 20 days. In another study, Hu et al. [70] concluded that chitosan nanoparticles loaded with cinnamon essential oil reduced the lipid oxidation of the chilled pork during the storage (15 days). Finally, Ma et al. [71] produced chitosan/gelatin nanoparticles loaded with anthocyanins and observed that these nanomaterials exhibited a reddish color in spoiled milk, whereas the color of fresh milk did not change. The authors observed that these nanomaterials were able to detect the milk freshness during storage as a consequence of the color change of anthocyanins in different pH values.

Regarding the starch nanoparticles, Chacon et al. [72] produced potato starch nanoparticles by antisolvent precipitation and observed that these nanomaterials are soluble in water at room temperature. Also, the solutions based on starch nanoparticles there was not phase separation in water. The authors speculated that these biopolymeric nanoparticles can be used as a raw material in food products such as sauces and instant soups which the starch must be solubilized at ambient temperature. Future research will systematically study the effect of these biopolymeric nanoparticles on the physical properties of foods.

17.2.4 Food packaging

Food packaging has several functions, including those related to containment, information, and marketing. Their primary function, however, is to separate foods from the surrounding environment, reducing exposure to spoilage and avoiding losses, retaining food quality during the

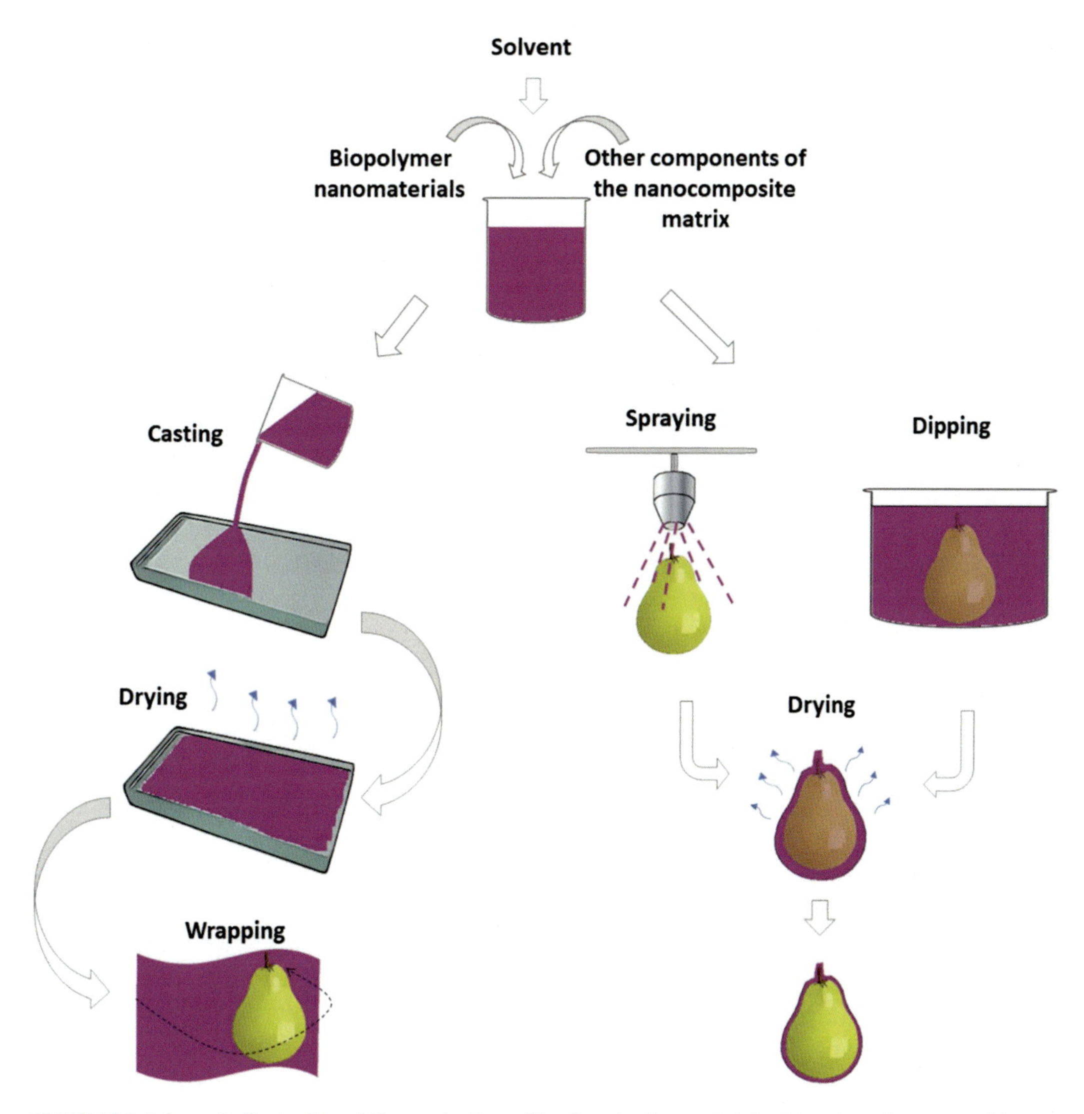

FIGURE 17.3 Schematic illustration of the production of food packaging containing biopolymeric nanoparticles.

storage period and distribution to the final consumers [73,74]. The commonly used packaging is based on nonbiodegradable synthetic polymers causing serious environmental pollution [75]. In this way, biodegradable films and coatings represent an alternative to reduce the use of synthetic packaging. Films and coatings are primary packaging, generally they can be obtained from the same formulation, but they are different in the method of application (Fig. 17.3). Coatings are applied in a liquid form before forming the coating (e.g., dipping and/or spraying) meanwhile films are obtained as solid sheets and then applied to food products (e.g., casting and extrusion) [76–78]. More detailed characteristics about packaging formation and applications can be found in the specialized literature [2,4,79].

Biopolymeric materials based on polysaccharides and proteins have been applied to produce biodegradable and edible films and coatings for food packaging [75,80]. Also, biopolymeric nanoparticles for packaging applications have been studied in recent years as those nanomaterials usually have better properties than their unmodified biopolymers (Table 17.3) [81]. Biopolymeric nanomaterials can be used to provide active and/or intelligent properties for food packaging; however, the greatest emphasis to use these nanomaterials is to manufacture packaging with improved mechanical and barrier properties [82]. So, this section addresses the review of the effect of biopolymeric nanoparticles on the mechanical and barrier (water vapor, O_2/CO_2, and light) properties, as well as thermal stability and biodegradability of food packaging [81].

Nanocellulose is the main biopolymer nanoparticles applied in food packaging due to their many interesting characteristics such as nontoxicity, ability to produce fully biodegradable packaging, high intrinsic mechanical strength along with good flexibility, low thermal expansion coefficient, and low density. Such characteristics allow the nanocellulose to be used as reinforcement in various polymer matrices [83–85]. Thus, Shahbazi et al. [86] prepared the okra mucilage-quince seed mucilage (OM-QSM) coatings containing bacterial CNFs (0, 0.5, and 1%) and eryngium planum extract (0, 0.25, and 0.5%) and applied the coatings to extend the shelf-life of fresh strawberries during cold storage. The addition of 1% CNFs increased the thickness from 147.2 to 189.9 μm, whereas the resistance to water vapor increased from 73.46 to 91.12 s/cm and the vapor permeability decreased from 12.2 to 5.1 $[cm^3/m^2\ s\ Pa] \times 10^{-12}$ when compared with control films. In another study, Pacaphol et al. [87] also prepared coatings containing nanocellulose; however, in this case, CNFs were applied in fresh-cut spinach. The authors observed that the respiration rate decreased after 12 h of storage in all samples coated with nanocellulose (between 0.1 and 0.5% based on the weight of the matrix) as consequence of the reduction in the CO_2 permeability through the food packaging.

Films based on soy protein isolate containing cellulose nanocrystals and zinc oxide nanoparticles were applied to extend the shelf-life of fresh pork. The presence of cellulose nanocrystals increased the thermal decomposition temperature from 312 to 319°C, being correlated with a higher thermal stability. Furthermore, cellulose nanocrystals enhanced the water resistance and reduced the migration of zinc oxide nanoparticles from the film to the food system [88]. Cellulose nanocrystals associated with silver nanoparticles have been used to manufacture films. In general, cellulose nanocrystals can reduce the vapor water and oxygen permeabilities of polylactic acid films in approximately 59% and 60%, respectively [89].

Cellulose nanoparticles can be isolated from agri-food wastes. In this context, Coelho et al. [90] manufactured starch films containing cellulose nanocrystals from grape pomace. The authors noted that the incorporation of nanocellulose was able to decrease water vapor permeability values in starch films. Films containing 5–15% of nanocellulose (based on the weight of starch mass) showed greater tensile strength and Young's modulus and they appeared opaque, being potential candidates to be applied as food packaging. Improvements in the physical polymer of films after the addition of cellulose nanocrystal have been credited to the interactions of these nanoparticles with the macromolecules by means of hydrogen bonds interactions.

Table 17.3 Different types of biopolymeric nanoparticles used to manufacture food packaging.

Nanoparticle	Nanoparticle synthesis	Film production approach	Main result and application	References
Cellulose nanocrystals	Acid hydrolysis	Casting	Cellulose nanocrystals improved the thermal stability of polyvinyl alcohol films	[85]
Bacterial cellulose nanofibers	Bacterial cellulose was purchased from Nano Novin [Gorgan, Iran]	Dipped in coating solutions	Cellulose nanofibers increased water vapor resistance and decreased oxygen permeability value of coated samples. The coating increased the shelf-life of strawberries	[86]
Cellulose nanocrystals/silver nanoparticles	Acid hydrolysis	Casting	The combination of surfactant-modified cellulose nanocrystals and silver nanoparticles increased the barrier properties of the produced films. However, only cellulose nanocrystals had a positive effect on water vapor permeability	[89]
Cellulose nanocrystals/zinc oxide nanoparticles	Acid hydrolysis	Casting	Cellulose nanocrystals improved the tensile strength, barrier properties, and thermal stability of films. Films were used to packed fresh pork	[88]
Cellulose nanocrystals	Acid hydrolysis	Casting	The incorporation of cellulose nanocrystals decreased the water vapor permeability and increased the tensile strength and Young's modulus in the films	[90]
Cellulose nanocrystals from microcrystalline cellulose	Acid hydrolysis	Casting	The addition of cellulose nanocrystals did not alter the water vapor barrier of the films, but they became slightly stiffer and more stretchable	[95]
Cellulose nanofibers from banana peel	Acid hydrolysis followed by mechanical treatment with high-pressure homogenizer	Casting	Cellulose nanofibers formed films with improved mechanical, optical, and water barrier properties	[96]
Cellulose nanofibers from softwood	Esterification	Casting	Addition of modified cellulose nanofibers improved the barrier, thermal, and mechanical properties of nanocomposites	[97]

(*continued on next page*)

Table 17.3 *(continued)*

Nanoparticle	Nanoparticle synthesis	Film production approach	Main result and application	References
Cellulose nanoparticles	Nanocellulose was obtained from Blue Goose Biorefineries Inc. [BGB ULTRA, Canada].	Casting	Films-containing cellulose nanoparticles showed better water vapor barrier and mechanical properties. The release of bergamot oil from the films was slower when stabilized with nanocellulose	[98]
Nanofibrillated cellulose	Defibrillation	Coating	Films-containing nanofibrillated cellulose decreased the respiration rate in terms of CO_2 content and they were used to extend the shelf-life of fresh-cut spinach	[87]
Cellulose nanocrystals/lignin nanoparticles	Acid hydrolysis	Casting	Cellulose nanocrystals improved the thermal stability and mechanical behavior of nanocomposite films	[99]
Corn starch nanoparticles	Enzymatic hydrolysis followed by recrystallization	Casting	Films with starch nanoparticles showed less permeability to water vapor, higher thermal stability, and better tensile strength when compared with control films	[94]
Debranched starch nanoparticles from waxy corn starch	Reverse emulsification	Casting	The presence of starch nanoparticles increased the tensile strength of nanocomposite films, and decreased water vapor permeability and oxygen transmission rate when compared with control films	[93]
Green sago starch nanoparticles	High pressure homogenization	Casting	Films-containing starch nanoparticles had better mechanical and barrier properties when compared with control films	[92]
Starch nanocrystals	Acid hydrolysis and ultrasound irradiation	Casting	Starch nanoparticles reduced the elongation at break and increased the Young's modulus in the films	[100]
Potato starch nanoparticles	Enzymatic hydrolysis followed by recrystallization	Casting	Films containing starch nanoparticles had low water vapor permeability values and improved thermal stability and mechanical properties	[101]

(continued on next page)

Table 17.3 *(continued)*

Nanoparticle	Nanoparticle synthesis	Film production approach	Main result and application	References
Chitosan nanofiber	Chitosan nanofiber was purchased from Nano Novin Polymer Co. [Iran]	Casting	The addition of chitosan nanofiber increased the water vapor permeability and significantly increased the values of ultimate tensile strength and Young's modulus in the films	[102]
Chitosan nanoparticle encapsulating pomegranate peel extract	Ionic gelation	Casting	Chitosan nanoparticles improved the thermal stability of the film, however, reduced elongation at break from 2.6 to 1.2%. The films inhibited the growth of *L. monocytogenes* in pork during the storage	[103]
Chitosan nanoparticles/thymol	Ionic gelation	Coating	The incorporation of the nanoparticles decreased the water vapor permeability. The presence of the nanoparticles (1% w/w) enhanced the film matrix without impacting its elongation at break. Films were used to packed blueberries and cherry tomatoes	[104]
Chitosan nanoparticles/tripolyphosphate	Ionic gelation	Casting	The addition of chitosan-tripolyphosphate nanoparticles in the films improved tensile properties, thermal stability, and water vapor permeability	[105]
Chitosan nanoparticles/gelatin nanofibers/moringa oil	Ionic crosslinking	Electrospinning	Chitosan nanoparticles reduced the water vapor permeability of films. Films were applied in cheese and reduced the growth of *L. monocytogenes* and *S. aureus* during the storage	[106]
Zein nanoparticles	Antisolvent precipitation	Casting	Zein nanoparticles did not impacted the mechanical properties (tensile strength, elongation at break and Young's modulus] and water vapor permeability of films	[107]

(*continued on next page*)

Table 17.3 *(continued)*

Nanoparticle	Nanoparticle synthesis	Film production approach	Main result and application	References
Zein nanoparticles	Antisolvent precipitation	Casting	Methylcellulose films and zein nanoparticles at pH 6.5 exhibited decreased tensile strength and elongation at break, also increasing water vapor permeability, polysaccharide and surfactant were used to restore tensile strength and vapor permeability	[108]
Zein nanoparticles	Antisolvent	Emulsification	The presence of zein nanoparticles increased the mechanical resistance and reduced approximately 50% of the water vapor permeability in the film	[109]
Zein nanoparticles/Rutin	Acid hydrolysis	Casting	Zein/rutin nanoparticles decreased the water vapor permeability and increased the tensile strength and elongation at break of films	[110]

Starch is an alternative material for food packaging applications due to its environmental compatibility, wide availability, and low cost. Recently it was demonstrated that starch nanoparticles have better biodegradability than starch granules in aqueous environments as a consequence of the higher surface area of starch nanoparticles [81,91]. In this context, Ahmad et al. [92] manufactured green composited containing sago starch nanoparticles as reinforcing material. The authors observed that an increment in the concentration of starch nanoparticles from 0 to 8% (w/w) decreased significantly the film transparency from 76 to 63% in starch films. At the same time, films containing sago starch nanoparticles had the lowest water vapor permeability.

In another research, debranched starch nanoparticles via reverse emulsification were applied in films based on corn starch. The results showed that 5% of starch nanoparticles increased the tensile strength by 86% and decreased the water vapor permeability and oxygen transmission rate at 30.9% and 79.3%, respectively, in the films [93]. Similarly, Liu et al. [94] studied the effect of self-assembled starch nanoparticles in edible corn starch nanocomposite films. The authors concluded that the incorporation of starch nanoparticles (15% w/w) in the films led to a decreasing in the water vapor permeability from 5.9×10^{-12} to 3.1×10^{-12} g/m s Pa, as well as an increasing in the tensile strength from 1.4 to 2.3 MPa of nanocomposite films when compared with control films.

Chitosan nanomaterials have been used as fillers to improve the mechanical, gas barrier, and thermal properties of films. Also, the incorporation of chitosan nanoparticles in food-packaging systems can prevent microbial growth, being used to extend the food shelf-life [77,78].

Recently, Cui et al. [103] produced a zein-active film incorporated with pomegranate peel extract encapsulated in chitosan nanoparticles and observed that the addition of chitosan nanoparticles significantly increased the tensile strength from 12.2 to 21 MPa. Furthermore, the chitosan nanoparticles reduced the *L. monocytogenes* growth in stored pork. In another research, Medina et al. [104] manufactured chitosan-quinoa protein films containing chitosan nanoparticles encapsulating thymol. The authors applied these films on blueberries and cherry tomatoes and observed that the nanocomposite films inhibited the mycelia growth of *Botrytis cinerea* and improved the water vapor permeability.

In the reviewed literature, some studies have been using zein nanoparticles to manufacture biopolymeric films. Zein is the major storage protein of corn and is only soluble in organic solvents. This biopolymer has been widely investigated for its film forming and barrier properties as it has low water vapor permeability when compared with other edible protein films [107,109]. In this way, zein nanoparticles have been used as reinforcements in starch films [109].

17.3 Conclusions and future aspects

Biopolymers can be used to manufacture nanoarchitectures by different approaches such as antisolvent precipitation, high pressure, ionic gelation, liquid-liquid dispersion, ultrasonication, among others. These nanomaterials have antimicrobial properties, and they could be used to reduce the spoilage in food products. Furthermore, biopolymeric nanoparticles can be used as material wall for encapsulating hydrophilic and hydrophobic compounds or as additives in foods and food packaging materials. However, future studies can be addressed to study the toxicity and the effect of biopolymeric nanoparticles on the sensory properties in foods, as well as the effect of these nanomaterials on the biodegradability of films and coatings.

Conflicts of interest

The author declares no conflict of interest.

Acknowledgments

The authors gratefully acknowledge the Federal University of Santa Catarina (UFSC) and to CAPES (Coordination for the Improvement of Higher Education Personnel) for their support. G.A. Valencia gratefully acknowledged to CNPq (National Council for Scientific and Technological Development) for the research grant (405432/2018-6).

References

[1] M.G.A. Vieira, Silva M.A. da, Santos.L.O. dos, M.M. Beppu, Natural-based plasticizers and biopolymer films : a review, Eur. Polym. J. 47 (3) (2011) 254–263. http://dx.doi.org/10.1016/j.eurpolymj.2010.12.011.

[2] G.A. Valencia, A. Sobral PJ do, Recent trends on nano-biocomposite polymers for food packaging, in: Tomy J. Gutiérrez (Ed.), Polymers for Food Applications, 1st ed., Springer, Gewerbestrasse, 2018, pp. 101–130.

[3] G.A. Valencia, E.N. Zare, P. Makvandi, T.J. Gutiérrez, Self-assembled carbohydrate polymer for food applications: a review, Compr. Rev. Food Sci. Food Saf. 18 (6) (2019) 2009–2024.

[4] G.A. Valencia, C.G. Luciano, A.R. Monteiro, Smart and active edible coatings based on biopolymers, in: Tomy J. Gutiérrez (Ed.), Polymers For Agri-Food Applications, Springer, Gewerbestrasse, 2019, pp. 391–416.

[5] M.E.I. Badawy, T.M.R. Lotfy, S.M.S. Shawir, Facile synthesis and characterizations of antibacterial and antioxidant of chitosan monoterpene nanoparticles and their applications in preserving minced meat, Int. J. Biol. Macromol. 156 (2020) 127–136.

[6] T. Niaz, S. Shabbir, T. Noor, A. Rahman, H. Bokhari, M. Imran, Potential of polymer stabilized nanoliposomes to enhance antimicrobial activity of nisin Z against foodborne pathogens, LWT 96 (2018) 98–110.

[7] S. Das, V.K. Singh, A.K. Dwivedy, A.K. Chaudhari, N. Upadhyay, P. Singh, et al., Encapsulation in chitosan-based nanomatrix as an efficient green technology to boost the antimicrobial, antioxidant and in situ efficacy of *Coriandrum sativum* essential oil, Int. J. Biol. Macromol. 133 (2019) 294–305.

[8] H. Horison, F.O. Sulaiman, D. Alfredo, A.A. Wardana, Physical characteristics of nanoemulsion from chitosan/nutmeg seed oil and evaluation of its coating against microbial growth on strawberry, Food. Res. 3 (2019) 821–827.

[9] O.M. Sharaf, M.S. Al-Gamal, G.A. Ibrahim, N.M. Dabiza, S.S. Salem, M.F. El-ssayad, et al., Evaluation and characterization of some protective culture metabolites in free and nano-chitosan-loaded forms against common contaminants of Egyptian cheese, Carbohydr. Polym. 223 (2019) 115094.

[10] Y. Zhu, D. Li, T. Belwal, L. Li, H. Chen, T. Xu, et al., Effect of nano-SiOx/chitosan complex coating on the physicochemical characteristics and preservation performance of green tomato, Molecules. 24 (24) (2019) 1–16.

[11] M. Hasanin, A. El-Henawy, W.H. Eisa, H. El-Saied, M. Sameeh, Nano-amino acid cellulose derivatives: eco-synthesis, characterization, and antimicrobial properties, Int. J. Biol. Macromol. 132 (2019) 963–969.

[12] E.M.A. Bary, A. Fekri, Y.A. Soliman, A.N. Harmal, E.M.A. Bary, A. Fekri, et al., Characterisation and swelling-deswelling properties of superabsorbent membranes made of PVA and cellulose nanocrystals, Int. J. Environ. Stud. 76 (1) (2018) 118–135.

[13] G. Ghoshal, D. Singh, Synthesis and characterization of starch nanocellulosic films incorporated with *Eucalyptus globulus* leaf extract, Int. J. Food. Microbiol. 332 (2020) 108765.

[14] M.S. Hasanin, G.O. Moustafa, New potential green, bioactive and antimicrobial nanocomposites based on cellulose and amino acid, Int. J. Biol. Macromol. 144 (2020) 441–448.

[15] S. Li, J. Zhou, Y. Wang, A. Teng, K. Zhang, Z. Wu, et al., Physicochemical and antimicrobial properties of hydroxypropyl methylcellulose-cinnamon essential oil emulsion: effects of micro-and nanodroplets, Int. J. Food. Eng. 15 (9) (2019) 1–11.

[16] Y. Khaledian, M. Pajohi-Alamoti, B. Bazargani-Gilani, Development of cellulose nanofibers coating incorporated with ginger essential oil and citric acid to extend the shelf life of ready-to-cook barbecue chicken, J. Food. Process. Preserv. 43 (10) (2019) 1–13.

[17] M. Hadidi, S. Pouramin, F. Adinepour, S. Haghani, S.M. Jafari, Chitosan nanoparticles loaded with clove essential oil: characterization, antioxidant and antibacterial activities, Carbohydr. Polym. 236 (2020) 116075.

[18] P. Zimet, Á.W. Mombrú, R. Faccio, G. Brugnini, I. Miraballes, C. Rufo, et al., Optimization and characterization of nisin-loaded alginate-chitosan nanoparticles with antimicrobial activity in lean beef, LWT Food Sci. Technol. 91 (2018) 107–116.

[19] G.C. Feyzioglu, F. Tornuk, Development of chitosan nanoparticles loaded with summer savory (*Satureja hortensis* L.) essential oil for antimicrobial and antioxidant delivery applications, LWT Food Sci. Technol. 70 (2016) 104–110.

[20] R. Karimirad, M. Behnamian, S. Dezhsetan, Application of chitosan nanoparticles containing Cuminum cyminum oil as a delivery system for shelf life extension of *Agaricus bisporus*, LWT 106 (2019) 218–228.

[21] A. Mohammadi, M. Hashemi, S. Masoud Hosseini, Effect of chitosan molecular weight as micro and nanoparticles on antibacterial activity against some soft rot pathogenic bacteria, LWT Food Sci. Technol. 71 (2016) 347–355.

[22] H.A. Sahyon, S.A. Al-Harbi, Antimicrobial, anticancer and antioxidant activities of nano-heart of *Phoenix dactylifera* tree extract loaded chitosan nanoparticles: in vitro and in vivo study, Int. J. Biol. Macromol. 160 (2020) 1230–1241.

[23] X. Hu, X. Jia, C. Zhi, Z. Jin, M. Miao, Improving the properties of starch-based antimicrobial composite films using ZnO-chitosan nanoparticles, Carbohydr. Polym. 2019 (210) (2018) 204–209.

[24] M. Lin, S. Fang, X. Zhao, X. Liang, D. Wu, Natamycin-loaded zein nanoparticles stabilized by carboxymethyl chitosan: evaluation of colloidal/chemical performance and application in postharvest treatments, Food. Hydrocoll. 106 (2020) 105871.

[25] J. Li, X. Xu, Z. Chen, T. Wang, Z. Lu, W. Hu, et al., Zein/gum Arabic nanoparticle-stabilized Pickering emulsion with thymol as an antibacterial delivery system, Carbohydr. Polym. 200 (2018) 416–426.

[26] N. Merino, D. Berdejo, R. Bento, H. Salman, M. Lanz, F. Maggi, et al., Antimicrobial efficacy of *Thymbra capitata* (L.) Cav. essential oil loaded in self-assembled zein nanoparticles in combination with heat, Ind. Crops. Prod. 133 (2019) 98–104.

[27] Y. Qin, L. Xue, Y. Hu, C. Qiu, Z. Jin, X. Xu, et al., Green fabrication and characterization of debranched starch nanoparticles via ultrasonication combined with recrystallization, Ultrason. Sonochem. 66 (2020) 105074.

[28] M. Amirsoleimani, M.A. Khalilzadeh, F. Sadeghifar, H. Sadeghifar, Surface modification of nanosatrch using nano silver: a potential antibacterial for food package coating, J. Food Sci. Technol. 55 (2018) 899–904.

[29] Y. Qin, C. Liu, S. Jiang, L. Xiong, Q. Sun, Characterization of starch nanoparticles prepared by nanoprecipitation: influence of amylose content and starch type, Ind. Crop. Prod. 87 (2016) 182–190. http://dx.doi.org/10.1016/j.indcrop.2016.04.038.

[30] P. Alzate, M.M. Zalduendo, L. Gerschenson, S.K. Flores, Micro and nanoparticles of native and modified cassava starches as carriers of the antimicrobial potassium sorbate, Starch/Staerke 68 (11–12) (2016) 1038–1047.

[31] M.R. Kasaai, Trends in Food Science & Technology Zein and zein-based nano-materials for food and nutrition applications: a review, Trends. Food Sci. Technol. 79 (2018) 184–197.

[32] J.F. Aparicio, E.G. Barreales, T.D. Payero, C.M. Vicente, A. de Pedro, J. Santos-Aberturas, Biotechnological production and application of the antibiotic pimaricin: biosynthesis and its regulation, Appl. Microbiol. Biotechnol. 100 (2016) 61–78.

[33] N. Alizadeh, Malakzadeh S. Antioxidant, antibacterial and anti-cancer activities of β-and γ-CDs/curcumin loaded in chitosan nanoparticles, Int. J. Biol. Macromol. 147 (2020) 778–791.

[34] J. Han, F. Chen, C. Gao, Y. Zhang, X. Tang, Environmental stability and curcumin release properties of pickering emulsion stabilized by chitosan/gum arabic nanoparticles, Int. J. Biol. Macromol. 157 (2020) 202–211.

[35] I. Dammak, José do Amaral Sobral P. Formulation optimization of lecithin-enhanced pickering emulsions stabilized by chitosan nanoparticles for hesperidin encapsulation, J. Food. Eng. 229 (2018) 2–11.

[36] P. Chanphai, H.A. Tajmir-Riahi, Conjugation of tea catechins with chitosan nanoparticles, Food Hydrocoll. 84 (2018) 561–570 Nov.

[37] M. Ji, X. Sun, X. Guo, W. Zhu, J. Wu, L. Chen, et al., Green synthesis, characterization and in vitro release of cinnamaldehyde/sodium alginate/chitosan nanoparticles, Food Hydrocoll. 90 (2019) 515–522.

[38] C.J. Cheng, M. Ferruzzi, O.G. Jones, Fate of lutein-containing zein nanoparticles following simulated gastric and intestinal digestion, Food Hydrocoll. 87 (2019) 229–236 Feb.

[39] J. Sun, J. Chen, Z. Mei, Z. Luo, L. Ding, X. Jiang, et al., Synthesis, structural characterization, and evaluation of cyanidin-3-O-glucoside-loaded chitosan nanoparticles, Food Chem. 330 (2020) 127239.

[40] D. Pauluk, A.K. Padilha, N.M. Khalil, R.M. Mainardes, Chitosan-coated zein nanoparticles for oral delivery of resveratrol: formation, characterization, stability, mucoadhesive properties and antioxidant activity, Food Hydrocoll. 94 (2019) 411–417.

[41] L. Dai, R. Li, Y. Wei, C. Sun, L. Mao, Y. Gao, Fabrication of zein and rhamnolipid complex nanoparticles to enhance the stability and in vitro release of curcumin, Food Hydrocoll. 77 (2018) 617–628.

[42] L. Dai, H. Zhou, Y. Wei, Y. Gao, D.J. McClements, Curcumin encapsulation in zein-rhamnolipid composite nanoparticles using a pH-driven method, Food Hydrocoll. 93 (2019) 342–350.

[43] M. Wang, Y. Fu, G. Chen, Y. Shi, X. Li, H. Zhang, et al., Fabrication and characterization of carboxymethyl chitosan and tea polyphenols coating on zein nanoparticles to encapsulate β-carotene by anti-solvent precipitation method, Food Hydrocoll. 77 (2018) 577–587.

[44] H. Li, Y. Yuan, J. Zhu, T. Wang, D. Wang, Y. Xu, Zein/soluble soybean polysaccharide composite nanoparticles for encapsulation and oral delivery of lutein, Food Hydrocoll. 103 (2020) 105715.

[45] H. Li, D. Wang, C. Liu, J. Zhu, M. Fan, X. Sun, et al., Fabrication of stable zein nanoparticles coated with soluble soybean polysaccharide for encapsulation of quercetin, Food Hydrocoll. 87 (2019) 342–351.

[46] M. Ma, Y. Yuan, S. Yang, Y. Wang, Z. Lv, Fabrication and characterization of zein/tea saponin composite nanoparticles as delivery vehicles of lutein, LWT 125 (2020) 109270.

[47] M.A. Khan, C. Yue, Z. Fang, S. Hu, H. Cheng, A.M. Bakry, et al., Alginate/chitosan-coated zein nanoparticles for the delivery of resveratrol, J. Food Eng. 258 (2019) 45–53.

[48] Z. Deng, K. Zhu, R. Li, L. Zhou, H. Zhang, Cellulose nanocrystals incorporated β-chitosan nanoparticles to enhance the stability and in vitro release of β-galactosidase, Food Res. Int. 137 (2020) 109380.

[49] A.R. Madureira, A. Pereira, M. Pintado, Chitosan nanoparticles loaded with 2,5-dihydroxybenzoic acid and protocatechuic acid: properties and digestion, J. Food. Eng. 174 (2016) 8–14.

[50] T. Cai, P. Xiao, N. Yu, Y. Zhou, J. Mao, H. Peng, et al., A novel pectin from *Akebia trifoliata* var. australis fruit peel and its use as a wall-material to coat curcumin-loaded zein nanoparticle, Int. J. Biol. Macromol. 152 (2020) 40–49.

[51] X. Huang, X. Huang, Y. Gong, H. Xiao, D.J. McClements, K. Hu, Enhancement of curcumin water dispersibility and antioxidant activity using core-shell protein-polysaccharide nanoparticles, Food Res. Int. 87 (2016) 1–9.

[52] Q. Liu, Y. Jing, C. Han, H. Zhang, Y. Tian, Encapsulation of curcumin in zein/caseinate/sodium alginate nanoparticles with improved physicochemical and controlled release properties, Food Hydrocoll. 93 (2019) 432–442.

[53] G. Heep, A. Almeida, R. Marcano, D. Vieira, R.M. Mainardes, N.M. Khalil, et al., Zein-casein-lysine multicomposite nanoparticles are effective in modulate the intestinal permeability of ferulic acid, Int. J. Biol. Macromol. 138 (2019) 244–251.

[54] S. Chen, J. Wu, Q. Tang, C. Xu, Y. Huang, D. Huang, et al., Nano-micelles based on hydroxyethyl starch-curcumin conjugates for improved stability, antioxidant and anticancer activity of curcumin, Carbohydr. Polym. 228 (2020) 115398.

[55] M. Veneranda, Q. Hu, T. Wang, Y. Luo, K. Castro, J.M. Madariaga, formation and characterization of zein-caseinate-pectin complex nanoparticles for encapsulation of eugenol, LWT Food Sci. Technol. 89 (2018) 596–603.

[56] Y. Yuan, H. Li, J. Zhu, C. Liu, X. Sun, D. Wang, et al., Fabrication and characterization of zein nanoparticles by dextran sulfate coating as vehicles for delivery of curcumin, Int. J. Biol. Macromol. 151 (2020) 1074–1083.

[57] X. Zhan, L. Dai, L. Zhang, Y. Gao, Entrapment of curcumin in whey protein isolate and zein composite nanoparticles using pH-driven method, Food Hydrocoll. 106 (2020) 105839.

[58] M. Ding, T. Zhang, H. Zhang, N. Tao, X. Wang, J. Zhong, Effect of preparation factors and storage temperature on fish oil-loaded crosslinked gelatin nanoparticle pickering emulsions in liquid forms, Food Hydrocoll. 95 (2019) 326–335.

[59] Y. Fu, J. Yang, L. Jiang, L. Ren, J. Zhou, Encapsulation of lutein into starch nanoparticles to improve its dispersity in water and enhance stability of chemical oxidation, Starch/Stärke 71 (5-6) (2018) 1800248.

[60] Y. Hu, Y. Qin, C. Qiu, X. Xu, Z. Jin, J. Wang, Ultrasound-assisted self-assembly of β-cyclodextrin/debranched starch nanoparticles as promising carriers of tangeretin, Food Hydrocoll. 108 (2020) 106021.

[61] N.R. de Oliveira, B. Fornaciari, S. Mali, G.M. Carvalho, Acetylated starch-based nanoparticles: synthesis, characterization, and studies of interaction with antioxidants, Starch/Stärke 70 (3-4) (2018) 1700170.

[62] J. Yang, H. He, Z. Gu, L. Cheng, C. Li, Z. Li, et al., Conjugated linoleic acid loaded starch-based emulsion nanoparticles: in vivo gastrointestinal controlled release, Food Hydrocoll. 101 (2020) 105477.

[63] Y. Wang, W. Yan, R. Li, X. Jia, Y. Cheng, Impact of deamidation on gliadin-based nanoparticle formation and curcumin encapsulation, J. Food Eng. 260 (2019) 30-39.

[64] K. Yao, W. Chen, F. Song, D.J. McClements, K. Hu, Tailoring zein nanoparticle functionality using biopolymer coatings: impact on curcumin bioaccessibility and antioxidant capacity under simulated gastrointestinal conditions, Food Hydrocoll. 79 (2018) 262-272.

[65] P. Ezati, J.W. Rhim, pH-responsive pectin-based multifunctional films incorporated with curcumin and sulfur nanoparticles, Carbohydr. Polym. 230 (2020) 115638.

[66] Y. Wei, Z. Cai, M. Wu, Y. Guo, P. Wang, R. Li, et al., Core-shell pea protein-carboxymethylated corn fiber gum composite nanoparticles as delivery vehicles for curcumin, Carbohydr. Polym. 240 (2020) 116273.

[67] F.B. Meng, Q. Zhang, Y.C. Li, J.J. Li, D.Y. Liu, L.X. Peng, Konjac glucomannan octenyl succinate as a novel encapsulation wall material to improve curcumin stability and bioavailability, Carbohydr. Polym. 238 (2020) 116193.

[68] L. Youssouf, A. Bhaw-Luximon, N. Diotel, A. Catan, P. Giraud, F. Gimié, et al., Enhanced effects of curcumin encapsulated in polycaprolactone-grafted oligocarrageenan nanomicelles, a novel nanoparticle drug delivery system, Carbohydr. Polym. 217 (2019) 35-45.

[69] A. Chouljenko, A. Chotiko, F. Bonilla, M. Moncada, V. Reyes, S. Sathivel, Effects of vacuum tumbling with chitosan nanoparticles on the quality characteristics of cryogenically frozen shrimp, LWT Food Sci. Technol. 75 (2017) 114-123.

[70] J. Hu, X. Wang, Z. Xiao, W. Bi, Effect of chitosan nanoparticles loaded with cinnamon essential oil on the quality of chilled pork, LWT Food Sci. Technol. 63 (1) (2015) 519-526.

[71] Y. Ma, S. Li, T. Ji, W. Wu, D.E. Sameen, S. Ahmed, et al., Development and optimization of dynamic gelatin/chitosan nanoparticles incorporated with blueberry anthocyanins for milk freshness monitoring, Carbohydr. Polym. 247 (1) (2020) 116738. https://doi.org/10.1016/j.snb.2019.127065.

[72] W.D.C. Chacon, S. Lima KT dos, G.A. Valencia, A.C.A. Henao, Physicochemical properties of potato starch nanoparticles produced by anti-solvent precipitation, Starch/Stärke (2020) 2000086.

[73] C.G. Otoni, R.J. Avena-Bustillos, H.M.C. Azeredo, M.V. Lorevice, M.R. Moura, L.H.C. Mattoso, et al., Recent advances on edible films based on fruits and vegetables—a review, Compr. Rev. Food Sci. Food Saf. 16 (5) (2017) 1151-1169.

[74] F. Topuz, T. Uyar, Antioxidant, antibacterial and antifungal electrospun nanofibers for food packaging applications, Food Res. Int. 130 (2020) 108927.

[75] K. Kraśniewska, S. Galus, M. Gniewosz, Biopolymers-based materials containing silver nanoparticles as active packaging for food applications-a review, Int. J. Mol. Sci. 21 (3) (2020) 698.

[76] S.A.A. Mohamed, M. El-Sakhawy, M.A.M. El-Sakhawy, Polysaccharides, protein and lipid-based natural edible films in food packaging: a review, Carbohydr. Polym. 238 (2020) 116178.

[77] P. Cazón, G. Velazquez, J.A. Ramírez, M. Vázquez, Polysaccharide-based films and coatings for food packaging: a review, Food. Hydrocoll. 68 (2017) 136-148.

[78] S. Kumar, A. Mukherjee, J. Dutta, Chitosan based nanocomposite films and coatings: emerging antimicrobial food packaging alternatives, Trends Food Sci. Technol. 97 (2020) 196-209.

[79] A.R.M. Fritz, M. Fonseca J de, T.C. Trevisol, C. Fagundes, G.A. Valencia, Active, eco-friendly and edible coatings in the post-harvest - a critical discussion, Polymers for Agri-Food Applications (2019) 433–463.

[80] J.A. Aguirre-Joya, M.A. De Leon-Zapata, O.B. Alvarez-Perez, C. Torres-León, D.E. Nieto-Oropeza, J.M. Ventura-Sobrevilla, et al., Basic and applied concepts of edible packaging for foods, in: A.M. Grumezescu, A.M. Holban (Eds.), Food Packaging and Preservation, Elsevier, Amsterdam, 2018, pp. 1–61.

[81] N. Kumar, P. Kaur, S. Bhatia, Advances in bio-nanocomposite materials for food packaging: a review, Nutr. Food Sci. 47 (4) (2017) 591–606.

[82] Rhim J.W., Park H.M., Ha C.S. Bio-nanocomposites for food packaging applications. Prog. Polym. Sci. 38 (10-11) 1629-1652.

[83] S. Boufi, I. González, M. Delgado-Aguilar, Q. Tarrès, M.À. Pèlach, P. Mutjé, Nanofibrillated cellulose as an additive in papermaking process: a review, Carbohydr. Polym. 154 (2016) 151–166.

[84] U. Qasim, A.I. Osman, A.H. Al-Muhtaseb, C. Farrell, M. Al-Abri, M. Ali, et al., Renewable cellulosic nanocomposites for food packaging to avoid fossil fuel plastic pollution: a review, Environ. Chem. Lett. 19 (2020) 613–641.

[85] W. Yang, X. He, F. Luzi, W. Dong, T. Zheng, J.M. Kenny, et al., Thermomechanical, antioxidant and moisture behaviour of PVA films in presence of citric acid esterified cellulose nanocrystals, Int. J. Biol. Macromol. 161 (2020) 617–626.

[86] Y. Shahbazi, N. Shavisi, N. Karami, Development of edible bioactive coating based on mucilages for increasing the shelf life of strawberries, J. Food. Meas. Charact. 15 (2020) 394–405.

[87] K. Pacaphol, K. Seraypheap, D. Aht-Ong, Development and application of nanofibrillated cellulose coating for shelf life extension of fresh-cut vegetable during postharvest storage, Carbohydr. Polym. 224 (2019) 115167.

[88] Y. Xiao, Y. Liu, S. Kang, K. Wang, H. Xu, Development and evaluation of soy protein isolate-based antibacterial nanocomposite films containing cellulose nanocrystals and zinc oxide nanoparticles, Food Hydrocoll. 106 (2020) 105898.

[89] E. Fortunati, M. Peltzer, I. Armentano, A. Jiménez, J.M. Kenny, Combined effects of cellulose nanocrystals and silver nanoparticles on the barrier and migration properties of PLA nano-biocomposites, J. Food Eng. 118 (1) (2013) 117–124.

[90] S. Coelho CC de, R.B.S. Silva, C.W.P. Carvalho, A.L. Rossi, J.A. Teixeira, O. Freitas-Silva, et al., Cellulose nanocrystals from grape pomace and their use for the development of starch-based nanocomposite films, Int. J. Biol. Macromol. 159 (2020) 1048–1061.

[91] H. Li, Y. Qi, Y. Zhao, J. Chi, S. Cheng, Starch and its derivatives for paper coatings: a review, Progress in Organic Coatings, Elsevier B.V., Amsterdam, Vol. 135, 2019, pp. 213–227.

[92] A.N. Ahmad, S.A. Lim, N. Navaranjan, Y.I. Hsu, H. Uyama, Green sago starch nanoparticles as reinforcing material for green composites, Polymer. (Guildf) 202 (2020) 122646.

[93] Q. Lin, N. Ji, M. Li, L. Dai, X. Xu, L. Xiong, et al., Fabrication of debranched starch nanoparticles via reverse emulsification for improvement of functional properties of corn starch films, Food. Hydrocoll. 104 (2020) 105760.

[94] C. Liu, S. Jiang, S. Zhang, T. Xi, Q. Sun, L. Xiong, Characterization of edible corn starch nanocomposite films : the effect of self-assembled starch nanoparticles, Starch 68 (2016) 239–248.

[95] A. Cano, E. Fortunati, M. Cháfer, C. González-Martínez, A. Chiralt, J.M. Kenny, Effect of cellulose nanocrystals on the properties of pea starch–poly(vinyl alcohol) blend films, J. Mater. Sci. 50 (21) (2015) 6979–6992.

[96] H. Tibolla, A. Czaikoski, F.M. Pelissari, F.C. Menegalli, R.L. Cunha, Starch-based nanocomposites with cellulose nanofibers obtained from chemical and mechanical treatments, Int. J. Biol. Macromol. 161 (2020) 132–146.

[97] H. Almasi, B. Ghanbarzadeh, J. Dehghannya, A.A. Entezami, A.K. Asl, Novel nanocomposites based on fatty acid modified cellulose nanofibers/poly(lactic acid): morphological and physical properties, Food Packag Shelf Life 5 (2015) 21–31.

[98] E. Sogut, Active whey protein isolate films including bergamot oil emulsion stabilized by nanocellulose, Food Packag. Shelf Life 23 (2020) 100430.

[99] Yang W., Qi G., Puglia D. Effect of cellulose nanocrystals and lignin nanoparticles on mechanical, antioxidant and water vapour barrier properties of glutaraldehyde crosslinked PVA films. Polymers 12(6) (2020) 1364.

[100] F. Luzi, E. Fortunati, A. Di Michele, E. Pannucci, E. Botticella, L. Santi, et al., Nanostructured starch combined with hydroxytyrosol in poly(vinyl alcohol) based ternary films as active packaging system, Carbohydr. Polym. [Internet] 193 (2018) 239–248. https://doi.org/10.1016/j.carbpol.2018.03.079.

[101] S. Jiang, C. Liu, X. Wang, L. Xiong, Q. Sun, Physicochemical properties of starch nanocomposite films enhanced by self-assembled potato starch nanoparticles, LWT Food Sci. Technol. 69 (2016) 251–257. http://dx.doi.org/10.1016/j.lwt.2016.01.053.

[102] S. Amjadi, S. Emaminia, S. Heyat Davudian, S. Pourmohammad, H. Hamishehkar, L. Roufegarinejad, Preparation and characterization of gelatin-based nanocomposite containing chitosan nanofiber and ZnO nanoparticles, Carbohydr. Polym. 216 (2019) 376–384.

[103] H. Cui, D. Surendhiran, C. Li, L. Lin, Biodegradable zein active film containing chitosan nanoparticle encapsulated with pomegranate peel extract for food packaging, Food Packag. Shelf Life 24 (2020) 100511.

[104] E. Medina, N. Caro, L. Abugoch, A. Gamboa, M. Díaz-Dosque, C. Tapia, Chitosan thymol nanoparticles improve the antimicrobial effect and the water vapour barrier of chitosan-quinoa protein films, J. Food. Eng. 240 (2019) 191–198.

[105] M.R. de Moura, F.A. Aouada, R.J. Avena-Bustillos, T.H. McHugh, J.M. Krochta, L.H.C. Mattoso, Improved barrier and mechanical properties of novel hydroxypropyl methylcellulose edible films with chitosan/tripolyphosphate nanoparticles, J. Food. Eng. 92 (4) (2009) 448–453.

[106] L. Lin, Y. Gu, H. Cui, Moringa oil/chitosan nanoparticles embedded gelatin nanofibers for food packaging against *Listeria monocytogenes* and *Staphylococcus aureus* on cheese, Food Packag. Shelf Life 19 (2019) 86–93.

[107] L. Spasojević, J. Katona, S. Bučko, S.M. Savić, L. Petrović, J. Milinković Budinčić, et al., Edible water barrier films prepared from aqueous dispersions of zein nanoparticles, LWT 109 (2019) 350–358.

[108] C.J. Cheng, O.G. Jones, Effect of drying temperature and extent of particle dispersion on composite films of methylcellulose and zein nanoparticles, J. Food. Eng. 250 (2019) 26–32.

[109] R. Farajpour, Z. Emam Djomeh, S. Moeini, H. Tavahkolipour, S. Safayan, Structural and physico-mechanical properties of potato starch-olive oil edible films reinforced with zein nanoparticles, Int. J. Biol. Macromol. 149 (2020) 941–950.

[110] S. Zhang, H. Zhao, Preparation and properties of zein-rutin composite nanoparticle/corn starch films, Carbohydr. Polym. 169 (2017) 385–392.

18 Applications in cosmetics

Cătălina Bogdan and Mirela Liliana Moldovan

FACULTY OF PHARMACY, DEPARTMENT OF DERMOPHARMACY AND COSMETICS, IULIU HAŢIEGANU, UNIVERSITY OF MEDICINE AND PHARMACY, CLUJ-NAPOCA, ROMANIA.

Chapter outline

18.1 Skin as the target site for actives

Skin is one of the largest organs of the body, accounting for about 16% of the body weight. It is the external layer that covers the entire body, acting as an interface with the environment. It has a surface of about 2 m^2 and a variable thickness, between 1 and 4 mm, depending on the body site [1,2]. The skin structure is complex, being composed of three layers: epidermis, dermis, and the subcutaneous tissue—hypodermis, together with the skin appendages: glandular (sebaceous

Biopolymeric Nanomaterials: Fundamentals and Applications. DOI: https://doi.org/10.1016/B978-0-12-824364-0.00022-8

and sweat glands) and corneous (hair follicle and nails). Each layer has specific functions and characteristics, but the main role of the skin is to protect the whole body against environmental factors and external aggressions. Other important functions of the skin are: sensorial, thermoregulation, metabolic and also an important role in social interactions, the face being able to reflect diverse emotions [2,3].

Epidermis is located at the outermost level and is formed of the following five layers: basal, spinous, granular, lucid and corneous, or keratinous layer. The keratinocytes, also referred to as corneocytes in the stratum corneum (SC), are the main cells in the epidermis and their major components are the keratin filaments that ensure the structural support. Keratinocytes are produced by stem cells in the basal layer, where they mainly are responsible for the renewal of epidermis. In the spinous layer, keratinocytes have a greater mechanical strength due to keratin formation and are strongly linked by desmosomes. Lamellar bodies containing enzymes (proteases, lipases, and glycosidases) and lipids (ceramides, fatty acids, and cholesterol) are also formed at this level. In the granular layer the "granules" are represented by keratohyalin granules filled with profilaggrin; the granular cells have anabolic properties, here being synthetized several elements such as filaggrin, high-molecular-weight keratins, and proteins of the cornified cell envelope. The keratinous layer is composed of corneocytes, the completely cornified mature keratinocytes, cells without organelles, and rich in proteins, which are embedded in a lipid matrix with a bilayer structural arrangement. This organization is known as "brick and mortar" model. The process during which the maturation and differentiation of keratinocytes takes place is called keratogenesis, cell cycle, or cell turnover and lasts from 28 to 42 days. The last stage of keratinization is cell corneocytes desquamation, when individual cells or small groups of cells are removed. This process ensures the maintenance of normal skin thickness, the disturbances of this process leading to cell accumulation or hyperkeratosis. Besides keratinocytes, the epidermis also contains other cells such as melanocytes, specialized cells synthetizing the melanin pigments which ensure the protection against ultraviolet radiation effects, Merkel cells with a sensorial role, and Langerhans cells involved in skin immunity [3,4].

Dermis, a dense conjunctive tissue, is responsible for skin thickness, being composed of several cell types and the extracellular matrix which surround the cells. Fibroblasts are the primary cells in the dermis. They synthetize collagen, elastin, and other proteins as well as enzymes such as matrix metallopeptidases. Immune cells are also found at this level, consisting of mast cells, lymphocytes, polymorphonuclear monocytes, and macrophages. The extracellular matrix is composed of fibrous proteins such as collagen and elastin, glycoproteins, glycosaminoglycans, and proteoglycans. Vessels, nerve fibers, sweat glands, sebaceous glands, and hair follicles are also present in extracellular matrix of the dermis. The dermis ensures the nutrition of both dermis and epidermis, is involved in the thermoregulation through the presence of blood vessels in the percutaneous absorption due to blood and lymphatic vessels, ensures a protection against mechanical aggressions due to the collagen and elastin fibers, contributes to the sensory perception by the nerve endings, and participates in skin defense in the immune inflammatory process as well as the phagocytic process through its specialized cells [2–4].

Hypodermis is the innermost layer of the skin, a loose conjunctive tissue that mainly contains adipocytes, mast cells, and histiocytes, and its vascularization and innervation is dependent on the anatomical site. Its main role is protection and fat storage [3].

The protection role of the skin is mainly realized by the epidermis through its outermost layer—the SC (the keratinous layer)—which is the principal barrier against penetration of exogenous substances [3]. Even if the protection function of the skin is very efficient, skin is not an impermeable layer, being hence used as an application site for different active delivery from drug or cosmetics [5].

In most cases, epidermis is the target of surface active cosmetic products, while in case of drugs the action can be targeted at any of the three layers. In all cases, actives are not applied alone on the skin, they are incorporated in vehicles. A proper formulation of topically applied products will ensure the delivery of actives to the desired site of action: local, regional or systemic. The skin penetration of an active may be controlled, minimized, or maximized through an optimization of formulation, by studying the active-vehicle and active-vehicle-skin interactions. The actives may penetrate across the skin by intracellular route, intercellular route, or via hair follicles or eccrine sweat glands, depending on its physicochemical properties. It is generally accepted that, usually, molecules with a molecular weight below 500 Da have the ability to penetrate the skin, while molecules with higher molecular weight do not penetrate it. The partition coefficient between water and octanol of an active is also a modality to predict its skin penetration through keratinous layer, a very hydrophobic structure [6].

The skin delivery of an active may be improved by increasing its concentration in the keratinous layer and the rate of transport in this layer from the vehicle. Thus, increasing the thermodynamic activity by supersaturation or by balancing the active solubility (higher enough to achieve the desired dose, but lower enough to promote its delivery from the vehicle), the addition of chemical penetration enhancers that may alter the keratinous layer lipids or proteins, may alter corneodesmosomes or the corneocytes, all leading to an increased diffusivity in a granular layer. Also, pH adjustment, the formation of eutectic mixtures, and the presence of solvents and cosolvents may improve active solubilization and thus its penetration [5].

In case of nanomaterials used in cosmetics, skin penetration is not desired, as their absorption may trigger pathological effects at the skin surface—such as irritation, deeper skin effects—such as inflammation and allergic reactions, or even systemic effects including systemic inflammation, toxicity for different organs or even cancer. For these reasons, the nanomaterials safety has to be investigated taking into account both their systemic absorption and the level of exposure, before placing them on the market [7].

18.2 Nanotechnology in the cosmetic field

Nanotechnology has had a significant impact on the advancement of cosmetic field and a growing interest has been shown in the development of formulations with enhanced physicochemical and organoleptic characteristics that are likely to impact the overall performance of the product. The innovation targets both the cosmetic delivery of actives and the improvement of cosmetic properties such as solubility, color, transparency, texture, or durability [8].

18.2.1 Nanocosmetics and cosmetic regulations

According to the European Union definition, a cosmetic product consists of "a substance or mixture intended to be placed in contact with the external parts of the human body or with the teeth and the mucous membranes of the oral cavity for the purpose of cleaning them, perfuming them, changing their appearance, protecting them, keeping them in good condition or correcting body odours" [9].

The Regulation (EC) No 1223/2009 of the European Parliament and of the Council of 30 November 2009 on cosmetic products defines the nanomaterials as "insoluble or biopersistent and intentionally manufactured materials with one or more external dimensions, or an internal structure, on the scale from 1 to 100 nm" [9].

Upon placement of the first formulation on the market, nanomaterials have greatly influenced the development of the cosmetic sector. Their superior characteristics have ensured a significantly higher efficacy of the formulations compared to conventional cosmetics. A plethora of nanocosmetics is available today and provides a wide range of benefits [10].

The rapidly expanding use of nanomaterials in cosmetics has led to increased concerns as regards the potential risk to health and environment and the need for a safety revaluation strategy to efficiently assess the risk related to nanomaterials. The European Scientific Committee on Consumer Safety (SCCS) has recently identified several characteristics of nanomaterials in relation to the risk for consumers' health, including: (1) physicochemical aspects, which are related to the small dimensions of the particles, to the chemical nature and toxicity of the nanomaterial, and to the surface chemistry and characteristics; (2) exposure aspects, which are related to the risk for the consumers following the systemic exposure to nanoparticles; (3) other aspects are related to novel properties of nanomaterials, which can reach other parts of the body than those initially targeted as a result of their altered biokinetics, or related to the type of application, such as the inhalation risk following application of nanomaterials conditioned as loose powders or as sprays [11].

In particular, the risk of cutaneous toxicity correlated with a higher absorption of small particle size nanoparticles mainly from cosmetic formulations such as powders, foundations, and concealers raises more and more concerns. Important efforts are made by both the academic environment and the industry to characterize the nanomaterials and to evaluate their safety [12].

18.2.2 Safety of biopolymeric nanomaterials

Regarding the toxicity of nanocarriers it is important to discriminate between the intrinsic toxicity of the nanocarrier itself and that of the payload (that might be subject to reformulation) [6]. Nano-specific risk assessment may not be needed in particular situations when: (1) the nanomaterial totally disintegrates, (2) there is no systemic exposure of particulate form of the nanomaterial, or (3) the nanoform of the material has not been shown to have toxic effects. In this case, the nanomaterials are evaluated through the conventional risk assessment methods [11].

An important parameter to be considered when speaking about nanomaterials safety is their solubility. Safety assessments must be performed only on the insoluble materials, when all the nanomaterials or only a part of it is not completely solubilized in the cosmetic product. As the term "insoluble" is not actually defined, the following categories agreed by the Pharmacopoeias, both European Pharmacopoeia 10th edition and U.S. Pharmacopoeia 38—the in-force editions:

"sparingly soluble," "very slightly soluble," and "practically insoluble" are considered in the nanomaterial category. In Europe, a safety dossier must be provided 6 months before placing on the market of a nanomaterial-containing product, which includes precise information about the nanomaterial such as: the physicochemical characterization (solubility, surface interactions, etc.), the evaluation of the exposure level, and the assessment of the toxicity and of the risk of the nanomaterial. In case of nanocarriers and of nano-encapsulated actives, the safety assessment must be performed on the final encapsulated product for which the safety data of the individual components being insufficient. They have to be thoroughly characterized in terms of their physicochemical properties and stability, their chemical composition and purity, the concentration and dermal penetration for components and nanomaterial as well, together with their toxicological effects (such as cytotoxicity, skin or eye irritation, skin sensitization, mutagenicity, genotoxicity) and the exposure estimation in normal conditions of use. Currently, several nanomaterials are submitted to evaluation on scientific bases by SCCS to formulate an official opinion regarding their safety such as hydroxyapatite [nano], copper [nano], colloidal copper [nano] [11].

The use of "generally recognized as safe" (GRAS) materials is an important goal for obtaining safe cosmetics. The safety of each new product obtained through nanotechnology should be assessed individually [13]. Thus, nanocellulose has shown low toxicity on several *in-vitro* and *in-vivo* models [14]. Chitosan is considered a nontoxic polymer or exhibits a low toxicity according to a high number of studies reported elsewhere [13,15]. Several studies revealed the correlation between the degree of deacetylation, the molecular weight, and the toxicity of chitosan [16]. The safety of use in cosmetic products has been evaluated for 34 polysaccharide gums by the Expert Panel of The Cosmetic Ingredient Review [17], but the safety of their nanoparticulate form for both chitosan and polysaccharides has to be also proven. The analysis of available animal and clinical data underlined the safety of those ingredients used in cosmetic formulations, according to current practices and concentration [17].

In this respect, the extensive research for each ingredient along with the constant update of cosmetic regulations allow the consumers to benefit from the advantages of the new delivery systems that are deemed to be safe to use.

18.3 Characteristics of skin delivery of cosmetic ingredients from nanoparticulate systems

The uppermost layer of the epidermis, SC is the main protective barrier of the skin against xenobiotics. A topically applied active substance penetrates the skin by gradually crossing the lipophilic SC to more hydrophilic layers of epidermis, dermis, and microcirculation [3]. They penetrate the skin through more than one of the three known penetration pathways via hair follicles, through intercellular pathway, or transcellular pathway.

The topically applied nanoparticles are widely used for cosmetic purposes [18]. Different nanoparticles have been described elsewhere to increase the rate of penetration of actives in deeper layers of the skin even if the polymeric nanoparticles do not pass through intact SC [19,20]. In case of disrupted skin barrier, nanoparticles penetration may increase [21].

In most cases, polymeric nanoparticles remain at the skin surface from where the active ingredient is gradually released. Laser scanning confocal microscopy images revealed that 30-nm-diameter nanoparticles do not penetrate through intercellular route even though the width of intercellular channels is 100 nm, because of the presence of intercellular lipid bilayers [22]. Due to their small dimensions, nanoparticles allow a close contact with skin surface and tend to accumulate in superficial junctions and furrows between corneocytes. Additionally, the water evaporation from the delivery system leads to skin occlusion and increases permeation of actives in superficial layers of the skin [18].

The follicular pathway of absorption had been recognized as an important pathway of polymeric nanoparticles penetration. After topical application, nanoparticles accumulate at the follicular opening and in the hair follicle [18]. The mechanism of penetration process is mainly driven by the hair movement. In addition, due to the particular architecture, the hair follicle may act as a reservoir for delivery of active ingredients. Evidence suggests that the optimal size for follicular penetration is 600 nm, with the correlation between the particle size and follicular penetration being also demonstrated. The follicular penetration is a complex process divided in intrafollicular and follicular penetration. Depending on the desired effect, the delivery system may be tailored to reach the target site: nanocapsules may be designed for an intrafollicular accumulation to target sebaceous gland or the bulge region. This effect is beneficial in different hair conditions, such as alopecia. These observations are supported by several studies that describe the targeted delivery of minoxidil to hair follicles [23].

The delivery pathway of nanospheres (500 nm in diameter) in hairless rat abdominal skin, porcine ear skin, and in a three-dimensional cultured human skin model was thoroughly evaluated by Todo et al. Skin surface distribution of nanospheres was mainly observed on the surface of SC and around the openings of hair follicle. Cross-section images of the hair follicle area have shown the presence of nanospheres in hair follicles, namely at the surface of hair shaft and connective tissue follicles in a decreasing amount up to a maximum 200-mm depth from the skin [24].

In the case of transfollicular delivery, the active ingredients are released from the carrier at a specific time and location. Transfollicullar penetration of nanocarriers themselves is low, polymeric particles remain in the infundibulum where the active molecule is released followed by transfollicular absorption [21]. Therefore, the nanoparticulate systems can deliver actives in deeper layers of the skin and into the systemic circulation [18]. Based on these observations, follicular delivery is an innovative approach that can be explored by the cosmetic industry to design nanoparticles that penetrate through follicular pathway into viable tissues. Several cosmetics applications are of interest, including the delivery of macromolecules, antiacne ingredient that targets the pilosebaceous unit, or antiperspirant ingredients that target eccrine glands [18].

18.4 Polymeric nanomaterials in cosmetics

18.4.1 Natural biodegradable polymers used in cosmetic applications

As consumers have become more concerned about the quality of life and the potential undesirable side effects of some synthetic ingredients, the use of natural compound–based

Table 18.1 Examples of natural biodegradable polymers used in cosmetic applications(nonexhaustive list) [25,26,60].

Polymer class	Examples
Polysaccharides	Chitosan, chitin, hyaluronic acid, xanthan gum, guar gum, pullulan, starch, alginate, chondroitin sulfate, agar, carrageenan, pectin, dextran, cyclodextrins, cellulose derivatives: carboxymethylcellulose, ethylcellulose, cellulose acetate, cellulose acetate propionate, hydroxypropyl methyl cellulose
Protein-based polymers	Gelatin, collagen, albumin, keratin, elastin, wheat proteins, soy protein, placental proteins

cosmetics has widely increased. As natural polymers are easily available and inexpensive, they are broadly used in the cosmetic field [25]. Natural biopolymers are biodegradable and biocompatible compounds that may show a certain degree of variability in composition and could be moderately immunogenic [26]. The variation in purity and in batch-to-batch consistency may lead to variation in the encapsulation efficiency and release of actives [27]. Unlike the natural polymers, the synthetic polymers may present superior advantages in terms of controlled chemical composition. However, the residues from synthesis process may contaminate the final product [26].

Several biopolymers such as polysaccharides and proteins are broadly used in cosmetic formulation. Protein-based polymers are obtained from animals, fungi, or algae and are used in various cosmetic applications for a long time [25,26]. Many polysaccharides such as cellulose, chitosan, hyaluronic acid, xanthan gum are also used in cosmetic applications (Table 18.1). The polysaccharides play an important role in cosmetic formulations due to their multifunctional properties in the improvement of stability of the cosmetic product (emulsifiers, thickeners, etc.) but also in product performance by acting as active ingredients (moisturizers, antiwrinkle agents, etc.) [25,26]. The cellulose and its derivatives are among the most commonly used polysaccharides in cosmetics. The cellulose gum or carboxymethyl cellulose, the most important cellulose derivative is obtained by introducing carboxymethyl groups on the cellulose backbone. It is used in a wide range of applications as a thickening agent and film-forming agent. The specific ability to react with charged molecules is pH-dependent [25]. The alginates, another important class of biopolymers, are natural hydrophilic heteropolysaccharides obtained from brown seaweed. They are composed of D-mannuronic acid and L-guluronoic acid linkages being available as sodium, ammonium, and potassium derivatives. Sodium alginate is well known for its safety and it has been used for a longtime in the cosmetic industry as a gelling agent [28–30]. Hyaluronic acid is a natural polysaccharide with high water binding capacity. Numerous cosmetic products contain hyaluronic acid due to its excellent properties to increase water amount in SC and to accelerate epidermal regeneration [25]. Also, hyaluronic acid stimulates collagen synthesis by dermal fibroblasts leading to an improved elasticity and an antiwrinkle effect [31].

18.4.2 Innovative biopolymers used in cosmetic design

In addition to traditional biopolymers, various novel natural biopolymers have been constantly investigated in the cosmetic field. Phytoglycogen is a nano-sized natural biopolymer described in the patent US10172946B2, assigned to Mirexus Biotechnologies Inc. This multifunctional compound exhibits high water retention and favorable rheological properties with many applications, especially in moisturizers formulation [28,32].

Bioligin is a phenolic polymeric-compound with multiple biological activities such as antioxidant and antiinflammatory properties [33]. An innovative product that consists of lignin nanoparticles was recently described by Lee et al. Spherical nanoparticles of light-colored lignin were prepared under mild conditions and further tested as a natural sunscreen agent. When added in a cream base, lignin nanoparticles exhibited broad UV barrier properties [34]. Morganti et al. described another innovative formulation, a cosmetic mask as nonwoven tissues containing chitin nanofibril-lignin block copolymeric nanoparticles and several active ingredients, such as sodium ascorbyl phosphate, melatonin, beta-glucan, and nicotinamide entrapped in chitin nanofibril-lignin network. The *in-vivo* evaluation on human volunteers has shown the protective and rejuvenating activity of the biomasks [35].

18.5 Delivery systems for cosmetic ingredients

Emergent nanotechnologies facilitate the formulation of delivery systems for cosmetic ingredients with better characteristics and performances. The cosmetic delivery systems based on biopolymers from renewable and green resources are rapidly developing and make it possible to customize the performance of cosmetic care products. Due to versatile properties of biopolymers, a wide range of applications are possible. The biopolymers exhibit film-forming properties, while few biopolymers such as cellulose, chitosan, and zein fiber exhibit additional fiber-forming properties [13]. Moreover, positively charged chitosan has the ability to form polyelectrolyte complexes that can be applied in nanocarriers with multilayer architecture [36].

18.5.1 Nanoparticles

Polymeric nanoparticles are submicron-sized carriers obtained from biodegradable or non-biodegradable polymers. Among different types of nanocarriers, nanoparticles seem to have the highest resistance toward degradation in the skin tissue [7].

The advantages of using polymeric nanoparticles for skin delivery of cosmetic actives arise from the high surface-to-volume ratio, the ease of preparation, and the ability to efficiently entrap active ingredients. Over the past years, nanoparticles have been extensively studied for cosmetic applications because they provide: (1) an increased stability, polymeric shell protecting the actives against denaturation/degradation, (2) an enhanced penetration and distribution, or (3) a controlled/sustained release of actives [37–39].

Apart from the vehicle type and the concentration of both active ingredients and nanoparticles, several physical and chemical properties, such as size, surface charge, and surface chemistry, have been identified to influence skin diffusion [37].

Based on their structure, the main types of polymeric nanoparticles are the nanospheres (matrix type) and the nanocapsules (reservoir type) [40].

18.5.1.1 Nanocapsules

In the cosmetic industry where innovation is a constant process, nanocapsules have been considered as a valuable tool back since 1995 when L'Oréal first marketed nanocapsules in the innovative formulations, Primordiale Intense and Hydra Zen Serum. Nanocapsules are core-shell structures consisting of a polymeric membrane and a liquid core. The active ingredients encapsulated can be a liquid, solid, or molecular dispersion. Commonly, nanocapsules size varies from 5–10 nm to 100–500 nm. In addition to the encapsulation of the actives in the core, depending on solubility, nanocapsules can transport the active substances at the surface or embedded in the polymer coating [41]. Therefore, lipophilic substances are dissolved in the core and amphiphilic ingredients are attached to the membrane [42].

Various nanocapsules have been proposed for protecting cosmetic ingredients, such as antioxidants, unsaturated fatty acids, or vitamins but also to obtain a sustained release of actives, avoiding incompatibility among ingredients, or improving the lasting performances of the fragrances [10,20]. Moreover, positively charged nanocapsules present properties of bioadhesion to the SC [41]. Nanocapsules suspension can be applied topically *per se*, or incorporated into a semi-solid vehicle [43]

Natural polymers used for shell synthesis include chitosan, gelatin, sodium alginate, and albumin. The polymeric shell significantly contributes to the physicochemical and biological properties of nanocapsules. The resemblance to biological membranes and their biodegradability is beneficial for cosmetic applications.

The core of nanocapsules is usually made of lipophilic compounds, such as saturated or unsaturated vegetal or synthetic oils such as triglycerides, Miglyol 812, ethyl laureate, ethyl oleate, sesame oil, corn oil, or sunflower oil and also surfactants compatible with active ingredient. The actives are either solubilized within the lipid core or loaded via conjugation to the polymer chains and further released at a controlled rate [7,44].

18.5.1.2 Nanospheres

Nanospheres are crystalline or amorphous core-shell nanoparticles with an average particle size of 10–200 nm in which active compound is dispersed or covalently bound to the polymer matrix [44,45]. Applications of nanospheres include antiwrinkle creams, moisturizing creams, and antiacne creams. Nanospheres are used to deliver the actives into deeper levels of the skin, but also to protect the active ingredient from enzymatic or chemical degradation or to obtain a controlled release [46]. In the case of fragrances, this delivery system had been shown to prolong the release of actives. The polymers employed for the preparation of nanospheres are similar to those used for nanocapsules. As the active ingredient is dissolved or dispersed in the polymer organic solution, a network organization of the polymeric chains is obtained and hence a slow release of actives [42,47].

18.5.2 Liquid crystals

At sufficiently high concentrations, and if the molecules form fibrils of sufficient anisotropy, biopolymers can self-assemble in a liquid crystals phase [48]. Depending on the differences in the positional and orientation order, under the influence of external factors, thermotropic and lyotropic liquid crystals may be obtained. Based upon the structural organization, lyotropic liquid crystals are classified into three types, that is lamellar, cubic, or hexagonal.[49]

Liquid crystals are considered as multifunctional cosmetic vehicles as they ensure the protection of active ingredients against photo- and thermodegradation, as well as an enhanced physicochemical stability. They have moisturizing properties by increasing the retention of water in the SC and by releasing of actives due to the structural similarities with epidermis [49,50].

18.5.3 Hydrogels

Hydrogels are 3D, porous, and chemically or physically cross-linked networks of water-soluble polymers. They swell in biological fluids because of a cross-link formation. Numerous biopolymers such as chitosan, hyaluronic acid, sodium alginate, collagen, gelatin, xanthan gum, starch, pectin, or cellulose derivatives may in included in hydrogels marketed as "beauty masks." Recently Cho et al. described the preparation of thanaka cellulose hydrogel films and facial masks as environmental friendly and biocompatible cosmetic product with increased water retention properties [51]. Likewise, carrageenan-based hydrogels containing sorbitol and glycerin with high ability to retain water and high resistance to degradation were reported as promising cosmetic preparations [52].

Different hydrogels were designed for the skin delivery of varied compounds such as arbutin, niacinamide, or adenosine in skin-whitening and antiwrinkle applications. The active ingredients are incorporated directly into the hydrogel structure or, after a preliminary step that includes the nanoencapsulation of actives in biopolymers [53]. The delivery of actives from hydrogels is pH dependent. At a pH of 6 or higher the penetrability of active ingredients increases and the active ingredient is delivered into skin layers [10]. Bioadhesive hydrogels have shown advantages over conventional hydrogels due to the increased persistence to the application site [53]. In this sense, the preparation, characterization, release, and antioxidant activity of caffeic acid–loaded collagen and chitosan hydrogel composites were reported by Thongchai et al. [54].

Hydrogels have several drawbacks including the low entrapment of the hydrophobic compounds and the fast release of active ingredients. Until this date, several experimental approaches have tried to overcome this inconvenience and to enhance the consistency, adhesiveness, and release time at the application site by adding bioadhesive polymers such chitosan, hyaluronic acid, or sodium alginate. The development of bigels represents an advantageous approach for delivery of both hydrophobic and hydrophilic ingredients. The future perspectives in skin care and hair care target the preparations of "smart hydrogels", stimuli-sensitive hydrogels that respond to the changes of external environment [53].

18.5.4 Films

The use of nanostructured films based on biopolymers represents an innovative approach aimed at ensuring superior delivery profile of actives while providing an easy application. Different film compositions containing chitosan neutralized in citrate or acetate films, with/without adding glycerol as plasticizer have been prepared and analyzed in terms of physical and chemical properties. Chitosan films neutralized in citrate buffer have shown superior properties in terms of films thickness, lower moisture absorbance, and better swelling capacity. The results have shown that the neutralized chitosan citrate films prepared without glycerol promoted SC desquamation by decreasing the cell cohesion with consequent detachment. The study emphasized the potential application of chitosan films as exfoliating agents within an antiaging skin routine [55]. Consistent with these observations, Afonso et al. described the development of an antiaging skin mask by entrapping antioxidant ingredients such as vitamin C and annatto (*Bixa Orellana L.*) in a chitosan matrix. Reacetylated chitosan films exhibited high flexibility, selective permeability, high water affinity, and favorable releasing properties of the actives [56].

18.5.5 Nanofibers

Nanofibers represent the widest class of nanomaterials with important applications in the cosmetic field. Different properties of nanofibers such as low diameter, increase surface-area-to-volume ratio, increase strength value, high porosity, and small pore size represent important features for cosmetic applications [57]. Electrospinning technique is used to obtain nanofibers from various polymers including chitosan, collagen, gelatin, cellulose, and carboxymethylcellulose [10,57]. The small pores and high increase surface area may lead to increased penetration of active ingredients into skin. The application domain targets the manufacturing of nonwoven masks for skin care [10].

18.6 Applications of biopolymers in cosmetics

18.6.1 Chitosan

Chitosan is a cationic polysaccharide derived from the *N* deacetylation of chitin, the second most abundant natural polymer. The main advantages of using chitosan in cosmetic formulations consist of its biocompatibility and biodegradability [10]. The rigid chemical structure represents a favorable property for the encapsulation of the active ingredients [58]. In addition, positive charges of chitosan favor the interaction with negatively charged polymers to obtain multilayer structures or electrostatic complexes [59]. Positively charged chitosan and zein particles may increase the dermal accumulation of active ingredients [13].

According to CosIng database, the European Commission database for information on cosmetic substances and ingredients, the functions assigned for chitosan are film-forming and hair-fixing properties [60]. However, many functions and cosmetic applications of chitosan are described elsewhere. Chitosan's multiple properties make it a valuable compound in skin care, hair care, or oral care products as thickener, film-forming, bioadhesive ingredient, or delivery system [10].

18.6.1.1 Applications of chitosan in oral care products

In oral care products, the mucoadhesivity represents an advantageous approach to increase the contact time of the products with the site of application, a premise for an increased concentration gradient of active ingredients at oral mucosa level. The mucoadhesivity properties of chitosan is related to the interaction of positively charged amino groups with the negatively charged sialic acid from the mucosal surfaces [61].

Dental caries and periodontal diseases as microbial-mediated oral pathologies are among the most prevalent infectious diseases affecting humans [62]. Chitosan has been described to have a broad antimicrobial spectrum against several oral microorganisms. This effect was thoroughly demonstrated by *in-vitro* and *in-vivo* studies. The effect against oral *streptococci* was explored on several bacterial strains, the studies showing that chitosan is highly effective to inhibit *Streptococcus mutans* adhesion [63]. As regards the periodontal anaerobic pathogens including *Prevotella buccae, Prevotella intermedia, Porphyromonas gingivalis, Tannerella forsythensis*, and *Aggregatibacter actinomycetemcomitans*, Costa et al. investigated the effect of chitosan in biofilm formation by both single and dual species. The results showed that the inhibition was strain dependent and molecular weight–dependent, with better results in one-species biofilm compared to dual-species biofilm. Chitosan showed an increased activity against periodontal pathogens through a mechanism that may interfere with bacterial coaggregation [64].

The antimicrobial activity of nanochitosan against dental caries–associated microorganisms, *S. mutans* and *Candida albicans* has been recently reported [65]. The superior anticariogenic effect of nanochitosan compared to chitosan has been previously found in oral *streptococci* for four bacterial species: *S. mutans, Streptococcus sobrinus, Streptococcus sanguis*, and *Streptococcus salivarius*. The enhanced antibacterial activity compared to chitosan and the increased affinity for bacterial cells is probably due to the small size of the particles [66]. Apart from the antimicrobial activity, Silva et al. demonstrated the proliferative response induced by chitosan in primary cultures of human gingival fibroblasts and hence the favorable effect in periodontal healing and regeneration [67].

Another application of chitosan in oral care consists of the protective effect against enamel erosion when used in an F/Sn toothpaste. The impact of chitosan on the inhibition of tissue loss depends on the molecular weight and specific properties of the molecule such as viscosity and reactivity [68]. The antimicrobial activity of chitosan was also assessed after the incorporation in oral care formulations, specifically in toothpastes and mouthwashes. The assessment of a toothpaste formulation containing biosurfactants and chitosan demonstrated an inhibitory effect on biofilm formed by *S. mutans*. The inclusion of chitosan in the formulation has been proven to play an important role in the inhibition of dental biofilm [69]. Taken together, these studies suggest the potential use of chitosan in oral care products.

18.6.1.2 Applications of chitosan in skin care

The growing interest of the researchers to exploit the properties of chitosan is reflected by the high number of studies, patents, or patent applications filed for skin care products containing chitosan. In recent years, there has been an increasing interest in the encapsulation of UV filters in biopolymeric delivery systems. Cefali et al. described the development of a sunscreen containing chitosan/tripolyphosphate (TPP) nanoparticles loaded with flavonoids-enriched vegetable extracts (*Ginkgo biloba* L., *Dimorphandra mollis* Benth, *Ruta graveolens*, and *Vitis vinifera* L.).The formulation exhibited photostability and a favorable release of flavonoids from nanocarrier [70].

Table 18.2 Applications of chitosan nanoparticles.

No.	Delivery system	Polymeric materials	Active ingredients	Activity	Reference
1	Nanocapsules	Chitosan, alginate	Turmeric oil	Antimicrobial, antioxidant, insect repellent	[74]
2	Nanoparticles	Chitosan	*Centella asiatica* extract	Antiaging activity by inducing skin cell proliferation and AQP3 expression	[75]
3	Nanoparticles	Chitosan	Annatto, ultrafiltrated annatto, saffron, ultrafiltrated saffron	Protection against UV radiation—preparation of sunscreen emulsions	[76]
4	Nanoparticles	Chitosan hydrochloride and sodium TPP	*Ilex paraguariensis* extract	Antioxidant, antiinflamatory antiaging	[77]
5	Nanoparticles	Chitosan (high molecular weight) Zein (low molecular weight chitosan)	Retinol	Antioxidant, antiwrinkle	[78]

The application of nanocapsules in sunscreen formulations aims for the encapsulation of highly lipophilic ingredients and retention at the skin surface without penetration into viable tissues [41]. The development of sunscreen formulation containing benzophenone-3-loaded-chitosan-coated polymeric nanocapsules was described by Siquera et al. The hydrogels containing nanoparticles were characterized and the results demonstrated that the positive coating of nanocapsules using chitosan was able to control the permeation of the sunscreen through the skin and to maintain the sunscreen at skin surface for a longer period [71]. Consistent with these observations, the presence of chitosan in a photoprotective and antioxidant nanoemulsion increased the retention of the topical formulation in SC, thus increasing the substantivity and the safety of the cosmetic product [72]

The encapsulation of fragrances in chitosan nanoparticles represents a valuable strategy for maintaining the highly volatile ingredients in cosmetic formulations. In this respect, Xiao et al. investigated chitosan nanoparticles as a promising controlled release carrier for tuberose fragrance [73]. Other applications of chitosan nanoparticles in cosmetics are shown in Table 18.2.

18.6.1.3 Applications of chitosan in hair care

The ability of chitosan to bind to hair has been exploited in hair care cosmetics. Chitosan blends with other polymers were used to obtain thin films for conditioning of the hair [79]. Triple component of chitosan, collagen, and hyaluronic acid was obtained in the thin film form through solvent evaporation and further evaluated in terms of mechanical parameters film and forming properties on the surface of hair. The addition of chitosan improved several characteristics such as the stability in aqueous solutions, surface free energy but also the mechanical parameters of the hair. Overall, the thin films containing chitosan, collagen, and hyaluronic acid increased hair thickness, enhanced hair mechanical properties, and led to an improvement of hair appearance [79].

The research conducted by Matos et al. described the preparation of minoxidil sulfate nanoparticles using low-molecular-weight chitosan and TPP as a cross-link agent. The results revealed the intrafollicular accumulation of minoxidil sulfate nanoparticles and the sustained release of active ingredient over 12 h [23].

Considering the above-mentioned studies, it can be concluded that chitosan exhibits many valuable properties such as active ingredient in oral, hair, or skin care. Furthermore, the inclusion in nano-delivery systems promotes the achievement of innovative cosmetics with better performances.

18.6.2 Cellulose

The use of microbial nanocellulose in the cosmetic field is due to multiple benefits in terms of elasticity, mechanical strength, skin adhesion, but also to the easiness of use and disposal. Bacterial nanocellulose has a high chemical purity, without other impurities from lignocellulosic biomass such as lignin or pectin. It forms a dense network of twisting ribbons with the average diameters of 20–100 nm connected through hydrogen bonds. This structure with hydrogel-like properties and a highly water content of the membrane (above 95%) provides highly moisturizing properties. These useful characteristics are exploited in cosmetic procedures such as peeling or microdermabrasion, for atopic skin care or in moisturizing facial masks [80,81]. The importance of this biocompatible compound is emphasized by the numerous patents and patent applications filed for nanocellulose-based products. In this regard, the patent U.S. 9018189 describes an innovative composition containing plant-derived microfibrillated cellulose for topical care and/or treatment of skin inflammation conditions and disorders for example atopic dermatitis, psoriasis, skin burn, etc. According to the patent description, the compound may be used for dermatological or cosmetic indications as liquid, semi-solid, or solid topical products [82].

18.6.3 Alginates

The nanoforms of alginates are obtained through the pregelation process with calcium. Further addition of polycations such as chitosan leads to the formation of a polyelectrolyte complex [28]. Low-molecular-weight alginate has better biodegradation properties while high-molecular-weight alginate has an increased mechanical strength. The encapsulation of hydrophobic compounds in alginate-based nanocapsules represents a useful strategy to protect the active molecules, followed by the inclusion of nanocarriers' suspensions in a suitable vehicle with sensory attributes. This approach may represent an important factor for success of the cosmetic product. The active ingredients are entrapped in oily core protected by a hydrophilic matrix. In this regard, core-shell nanocarriers wherein the organic core consists of triglycerides and the shell is made of alginates have been prepared as nanosystems intended for dermal preparations. Alginates-based nanocarriers exhibited good stability in carbomed-based hydrogel [30]. Other applications of alginates aimed the preparation of hydrogel, nanofibers, or nanocomposites [28].

18.6.4 Hyaluronic acid

Antiwrinkle and skin-rejuvenating effects of hyaluronic acid have been demonstrated to be molecular weight dependent. High-molecular-weight hyaluronic acid is unable to penetrate beyond the outermost layer of epidermis to ensure antiwrinkle effects. For this reason, nano-sized hyaluronic acid has been shown to have an increased percutaneous absorption and enhanced antiwrinkle efficacy. These effects were documented during *in-vivo* studies on human volunteers [31].

Recent findings reported deeper penetration of hyaluronic acid nanoparticles into the skin as nanoparticulate polyion complexes. The nanoparticulate polyion complexes were obtained as a result of electrostatic interactions between hyaluronic acid as anionic compound and protamine as cationic compound. Skin penetration of hyaluronic acid nanoparticles was compared through *in-vitro* and *in-vivo* studies with hyaluronic acid in a nonnanoparticulate form. The results highlighted the beneficial effect of hyaluronic nanoparticles to restore skin-barrier function and to reduce UV-induced skin damage [83]. Similarly, Shigefuji et al. reported the preparation of nanoparticulate polyion complexes based on hyaluronic acid and poly-L-lysine hydrochloride as a cationic component. The mechanism of penetration investigated through an *in-vivo* skin penetration study has revealed that polyion complexes enhanced the skin permeability because of stronger interaction with biological membranes [84]. Nanoparticles of quaternized cyclodextrin-grafted chitosan associated with hyaluronic acid have been also prepared for the effective skin delivery of hyaluronic acid. The safety for dermal applications was assessed in a range of 0.01–0.1 mg/mL on dermal fibroblasts [85].

Besides the use on hyaluronic acid–based nanoparticles, other cosmetic applications have been proposed. Cross-linking of the hyaluronic acid aims to avoid the degradation and enhanced penetration. Thus, stable colloid systems may be obtained for potential applications as fillers used for facial wrinkles or depressed scars [86,87]. Thin films obtained from collagen, hyaluronic acid, and chitosan by using dialdehyde starch as cross-linking agent exhibited suitable bioadhesive and mechanical properties for cosmetic applications [88].

18.7 Future perspectives

The field of biopolymers-based nanocosmetics is currently subject to change and innovation. However, despite the extensive research on biopolymeric nanomaterials, only relatively few results have been used into cosmetic applications. The reason may be related to the difficulty to convert the *in-vitro* results in *in-vivo* applications. Moreover, the reevaluation of safety, stability, and effectiveness of these nanocosmetics is required for the already used actives in the context of the emergent properties or activities. While the science of nanotechnology-related cosmetic products continuously evolves, there is a continuous need to update the regulatory framework relative to the distinctive properties of nanomaterials.

18.8 Conclusions

Given the continuous expanding of nanotechnology-related cosmetic products, the use of biodegradable, biocompatible, easily available with no toxicity or low toxicity biopolymers may represent the future of the cosmetic sector to obtain multifunctional cosmetics with enhanced efficacy and optimal sensorial properties [13].

References

[1] R. Penzer, R. Ersser, Principles of Skin Care: A Guide for Nurses and Health Care Practitioners, Willey-Blackwell, Oxford, 2010, pp. 1–28.

[2] Avi Shai, Howard Maibach, R. Baran, Handbook of Cosmetic Skin Care, 2nd Edition, CRC press, Boca Raton, FL, 2009, pp. 4–12.

[3] A.O. Barel, M. Paye, H.I. Maibach, Handbook of Cosmetic Science and Technology, CRC press, Boca Raton, FL, 2014.

[4] L. Baumann, Cosmetic dermatology. Principle and Practice, 2nd edn, McGraw-Hill Professional Publishing, New York, 2009, pp. 3–12.

[5] R. Kuswahyuning, J.E. Grice, H.R. Moghimi, M.S. Roberts, Formulation effects in percutaneous absorption, in: N. Dragicevic, H.I. Maibach (Eds.), Percutaneous Penetration Enhancers Chemical Methods in Penetration Enhancement: Drug Manipulation Strategies and Vehicle Effects, Springer, Berlin Heidelberg, 2015, pp. 109–134.

[6] C.A. Poland, R.e.a.d.S.A.K. Larsen, S.M. Hankin, H.r. Lam, Assessment of Nano-enabled Technologies in Cosmetics [Internet], 2016. Available from. www.mst.dk/english.

[7] C.A. Poland, S.A.K. Read, Varet -Iom, Dermal Absorption of Nanomaterials, 2016. Available from. www.mst.dk/english.

[8] T.A. Nguyen, S. Rajendran, Current commercial nanocosmetic products, in: A. Nanda, S. Nanda, T.A. Nguyen, S. Rajendran, Y. Slimani (Eds.), Nanocosmetics, Elsevier, Amsterdam, 2020, pp. 445–453.

[9] Regulation (EC) No 1223/2009 of the European Parliament and of the Council of 30 November 2009 on cosmetic products [Internet]. Available from https://eur-lex.europa.eu/eli/reg/2009/1223/oj.

[10] S. Nafisi, H.I. Maibach, Nanotechnology in cosmetics, in: K. Sakamoto, R.Y. Lochhead, H.I. Maibach, Y. Yamashita (Eds.), Cosmetic Science and Technology: Theoretical Principles and Applications, Elsevier Inc., Amsterdam, 2017, pp. 337–361.

[11] Guidance on the Safety Assessment of Nanomaaterials in Cosmetics [Internet]. Available from https://ec.europa.eu/health/sites/health/files/scientific_committees/consumer_safety/docs/sccs_o_239.pdf.

[12] A.O. Barel, M. Paye, H.I. Maibach, M. Paye, H.I. Maibach, Impact of formula structure to skin delivery, in: A.O. Barel, M. Paye, H.I. Maibach (Eds.), Handbook of Cosmetic Science and Technology, CRC Press, Boca Raton, FL, 2014, pp. 623–632.

[13] M.R. Kasaai, Biopolymer-based nanomaterials for food, nutrition, and healthcare sectors: an overview on their properties, functions, and applications, in: M.H. Chaudhery (Ed.), Handbook of Functionalized Nanomaterials for Industrial Applications, Elsevier, Amsterdam, 2020, pp. 167–184.

[14] A.B. Seabra, J.S. Bernardes, W.J. Fávaro, A.J. Paula, N. Durán, Cellulose nanocrystals as carriers in medicine and their toxicities: a review, Carbohydr. Polym. 181 (2018) 514–527.

[15] S.K. Shukla, A.K. Mishra, O.A. Arotiba, B.B. Mamba, Chitosan-based nanomaterials: a state-of-the-art review, Int. J. Biol. Macromol. 59 (2013) 46–58.

[16] A. Matica, G. Menghiu, V. Ostafe, Toxicity of chitosan based products, Former. Ann. West. Univ. Timisoara-Series. Chem. 26 (1) (2017) 65–74.

[17] W.F. Bergfeld, F.A.C.P. Donald, V. Belsito, R.A. Hill, C.D. Klaassen, D.C. Liebler et al., Safety Assessment of Microbial Polysaccharide Gums as Used in Cosmetics The 2012 Cosmetic Ingredient Review Expert [Internet]. Available from https://www.cir-safety.org/sites/default/files/microb092012rep.pdf.

[18] S.K. Jain, A. Verma, A. Jain, P. Hurkat, Transfollicular drug delivery: current perspectives, Res. Rep. Transdermal. Drug. Deliv. 5 (2016) 1–17.

[19] K.L. Mao, Z.L. Fan, J.D. Yuan, P.P. Chen, J.J. Yang, J. Xu, et al., Skin-penetrating polymeric nanoparticles incorporated in silk fibroin hydrogel for topical delivery of curcumin to improve its therapeutic effect on psoriasis mouse model, Coll. Surf. B. Biointerfaces 160 (2017) 704–714.

[20] L. Rigano, N. Lionetti, Polymeric nanoparticles: nanospeheres and nanocapsules, in: A. Grumezescu (Ed.), Nanobiomaterials in Galenic Formulations and Cosmetics: Applications of Nanobiomaterials, Elsevier Inc., Amsterdam, 2016, pp. 121–148.

[21] A. Patzelt, J. Lademann, Recent advances in follicular drug delivery of nanoparticles, Exp. Opin. Drug Deliv. 17 (2020) 49–60.

[22] X. Wu, G.J. Price, R.H. Guy, Disposition of nanoparticles and an associated lipophilic permeant following topical application to the skin, Mol. Pharm. 6 (5) (2009) 1441–1448.

[23] B.N. Matos, T.A. Reis, T. Gratieri, G.M. Gelfuso, Chitosan nanoparticles for targeting and sustaining minoxidil sulphate delivery to hair follicles, Int. J. Biol. Macromol. 75 (2015) 225–229.

[24] H. Todo, E. Kimura, H. Yasuno, Y. Tokudome, F. Hashimoto, Y. Ikarashi, et al., Permeation pathway of macromolecules and nanospheres through skin, Biol. Pharm. Bull. 33 (8) (2010) 1394–1399.

[25] R. Augustine, R. Rajendran, U. Cvelbar, M. Mozetič, A. George, Biopolymers for health, food, and cosmetic applications, in: S. Thomas, D. Durand, C. Chassenieux, P. Jyotishkumar (Eds.), Handbook of Biopolymer-Based Materials: From Blends and Composites to Gels and Complex Networks, Wiley-VCH,, Weinheim, Germany, 2013, pp. 801–849.

[26] P. Severino, J.F. Fangueiro, M.V. Chaud, J. Cordeiro, A.M. Silva, E.B. Souto, Advances in nanobiomaterials for topical administrations: new galenic and cosmetic formulations, in: A. Grumezescu (Ed.), Nanobiomaterials in Galenic Formulations and Cosmetics: Applications of Nanobiomaterials, Elsevier Inc., Amsterdam, 2016, pp. 1–23.

[27] R. Jijie, A. Barras, R. Boukherroub, S. Szunerits, Nanomaterials for transdermal drug delivery: beyond the state of the art of liposomal structures, J. Mater. Chem. B. 5 (44) (2017) 8653–8675.

[28] M.L. Verma, B.S. Dhanya, Rani V. Sukriti, M. Thakur, J. Jeslin, et al., Carbohydrate and protein based biopolymeric nanoparticles: current status and biotechnological applications, Int. J. Biol. Macromol. 154 (2020) 390–412.

[29] H.A.E. Benson, M.S. Roberts, V.R. Leite-Silva, K.A. Walters, S. Trehan, R. Soskind, et al., Natural products and stem cells and their commercial aspects in cosmetics, in: H.A.E. Benson, M.S. Roberts, V.R. LeiteSilva, K. Walters (Eds.), Cosmetic Formulation, CRC Press, Boca Raton, FL, 2019, pp. 221–250.

[30] H.T.P. Nguyen, E. Munnier, M. Souce, X. Perse, S. David, F. Bonnier, et al., Novel alginate-based nanocarriers as a strategy to include high concentrations of hydrophobic compounds in hydrogels for topical application, Nanotechnology 26 (25) (2015) 255101.

[31] S.N.A. Bukhari, N.L. Roswandi, M. Waqas, H. Habib, F. Hussain, S. Khan, et al., Hyaluronic acid, a promising skin rejuvenating biomedicine: a review of recent updates and pre-clinical and clinical investigations on cosmetic and nutricosmetic effects, Int. J. Biol. Macromol. 120 (2018) 1682–1695.

[32] United States Patent: 10172946 [Internet], 2019. Available from http://patft.uspto.gov/netacgi/nph-Parser?Sect1=PTO2&Sect2=HITOFF&p=1&u=%2Fnetahtml%2FPTO%2Fsearch-bool.html&r=1&f=G&l=50&co1=AND&d=PTXT&s1=Korenevski&OS=Korenevski&RS=Korenevski.

[33] J. Luis Espinoza-Acosta, P.I. Torres-Chávez, B. Ramírez-Wong, C. María López-Saiz, Montaño-Leyva B. Antioxidant, Antimicrobial, and antimutagenic properties of technical lignins and their applications, BioRes 11 (2) (2016) 5452–5481.

[34] S.C. Lee, E. Yoo, S.H. Lee, K. Won, Preparation and application of light-colored lignin nanoparticles for broad-spectrum sunscreens, Polymers (Basel) 12 (3) (2020) 699.

[35] P. Morganti, M. Palombo, F. Carezzi, M. Nunziata, G. Morganti, M. Cardillo, et al., Green nanotechnology serving the bioeconomy: natural beauty masks to save the environment, Cosmetics. 3 (4) (2016) 41.

[36] V. Paşcalău, C. Bogdan, E. Pall, L. Matroş, S.L. Pandrea, M. Suciu, et al., Development of BSA gel/Pectin/Chitosan polyelectrolyte complex microcapsules for Berberine delivery and evaluation of their inhibitory effect on Cutibacterium acnes, React. Funct. Polym. (2020) 147.

[37] A.H. Mota, A. Sousa, M. Figueira, M. Amaral, B. Sousa, J. Rocha, et al., Natural-based consumer health nanoproducts: medicines, cosmetics, and food supplements, in: M.H. Chaudhery (Ed.), Handbook of Functionalized Nanomaterials for Industrial Applications, Elsevier, Amsterdam, 2020, pp. 527–578.

[38] R. Alvarez-Román, A. Naik, Y.N. Kalia, R.H. Guy, H. Fessi, Skin penetration and distribution of polymeric nanoparticles, J. Control. Release 99 (1) (2004) 53–62.

[39] Z. Zhang, P.C. Tsai, T. Ramezanli, B.B. Michniak-Kohn, Polymeric nanoparticles-based topical delivery systems for the treatment of dermatological diseases, Wiley Interdiscip. Rev. Nanomed. Nanobiotechnol, 5, 2013, pp. 205–218.

[40] F.U. Din, W. Aman, I. Ullah, O.S. Qureshi, O. Mustapha, S. Shafique, et al., Effective use of nanocarriers as drug delivery systems for the treatment of selected tumors, Int. J. Nanomed. 12 (2017) 7291–7309.

[41] A. Kapuscinska, A. Olejnik, I. Nowak, Nanocapsules as carriers of active substances, in: A. Grumezescu (Ed.), Nanobiomaterials in Galenic Formulations and Cosmetics: Applications of Nanobiomaterials, Elsevier Inc., Amsterdam, 2016, pp. 175–199.

[42] B.G. Chiari-Andréo, M.G.J. De Almeida-Cincotto, J.A. Oshiro, C.Y.Y. Taniguchi, L.A. Chiavacci, V.L.B. Isaac, Nanoparticles for cosmetic use and its application, in: A. Grumezescu (Ed.), Nanoparticles in Pharmacotherapy, Elsevier, Amsterdam, 2019, pp. 113–146.

[43] G. Fytianos, A. Rahdar, G.Z. Kyzas, Nanomaterials in cosmetics: recent updates, Nanomaterials 10 (5) (2020) 979.

[44] I. Armentano, N. Rescignano, E. Fortunati, S. Mattioli, F. Morena, S. Martino, et al., Multifunctional nanostructured biopolymeric materials for therapeutic applications, in: D. Ficai, A. Grumezescu (Eds.), Nanostructures for Novel Therapy: Synthesis, Characterization and Applications, Elsevier Inc., Amsterdam, 2017, pp. 107–135.

[45] S. Dhawan, P. Sharma, S. NandaA. Nanda, S. Nanda, T.A. Nguyen, S. Rajendran, Y. Slimani (Eds.), Cosmetic nanoformulations and their intended use, Elsevier, Nanocosmetics (2020) 141–169.

[46] S. Kaul, N. Gulati, D. Verma, S. Mukherjee, U. Nagaich, Role of nanotechnology in cosmeceuticals: a review of recent advances, J. Pharm. 2018 (2018) 1–19.

[47] S.S. Guterres, M.P. Alves, A.R. Pohlmann, Polymeric nanoparticles, nanospheres and nanocapsules, for cutaneous applications, Drug Target Insights 2 (2007) 147–157.

[48] I.W. Hamley, Liquid crystal phase formation by biopolymers, Soft. Matter 6 (2010) 1863–1871.

[49] V.K. Rapalli, T. Waghule, N. Hans, A. Mahmood, S. Gorantla, S.K. Dubey, et al., Insights of lyotropic liquid crystals in topical drug delivery for targeting various skin disorders, J. Mol. Liq. 315 (2020) 113771.

[50] L. Bonato Alves Oliveira, R. Oliveira, C. Oliveira, N. Raposo, M. Brandão, A. Ferreira, et al., Cosmetic potential of a liotropic liquid crystal emulsion containing resveratrol, Cosmetics 4 (4) (2017) 54.

[51] C. Cho, T. Kobayashi, Advanced cellulose cosmetic facial masks prepared from Myanmar Thanaka heartwood, Curr. Opin. Green. Sustain. Chem. 27 (2020) 100413.

[52] J. Kozlowska, K. Pauter, A. Sionkowska, Carrageenan-based hydrogels: effect of sorbitol and glycerin on the stability, swelling and mechanical properties, Polym. Test 67 (2018) 7–11.

[53] S. Mitura, A. Sionkowska, A. Jaiswal, Biopolymers for hydrogels in cosmetics: review, J. Mater. Sci. Mater. Med. 31 (6) (2020) 50.

[54] K. Thongchai, P. Chuysinuan, T. Thanyacharoen, S. Techasakul, S. Ummartyotin, Characterization, release, and antioxidant activity of caffeic acid-loaded collagen and chitosan hydrogel composites, J. Mater. Res. Technol. 9 (3) (2020) 6512–6520.

[55] I.C. Libio, R. Demori, M.F. Ferrão, M.I.Z. Lionzo, N.P. da Silveira, Films based on neutralized chitosan citrate as innovative composition for cosmetic application, Mater. Sci. Eng. C. 67 (2016) 115–124.

[56] C.R. Afonso, R.S. Hirano, A.L. Gaspar, E.G.L. Chagas, R.A. Carvalho, F.V. Silva, et al., Biodegradable antioxidant chitosan films useful as an anti-aging skin mask, Int. J. Biol. Macromol. 132 (2019) 1262–1273.

[57] F. Yilmaz, G. Celep, G. Tetik, Nanofibers in cosmetics, in: .M. Rehman, A.M. Asiri (Eds.), Nanofiber Research - Reaching New Heights, InTechopen, London, 2016, pp. 127–140.

[58] Z.A. Raza, S. Khalil, A. Ayub, I.M. Banat, Recent developments in chitosan encapsulation of various active ingredients for multifunctional applications, Carbohydr. Res. 492 (2020) 108004.

[59] M.E. Abd El-Hack, M.T. El-Saadony, M.E. Shafi, N.M. Zabermawi, M. Arif, G.E. Batiha, et al., Antimicrobial and antioxidant properties of chitosan and its derivatives and their applications: A review, Int. J. Biol. Macromol. 164 (2020) 2726–2744.

[60] CosIng - Cosmetics - GROWTH - European Commission [Internet]. Available from https://ec.europa.eu/growth/tools-databases/cosing/index.cfm?fuseaction=search.results.

[61] A. Kumar, A. Vimal, A. Kumar, Why Chitosan? From properties to perspective of mucosal drug delivery, Int. J. Biol. Macromol. 91 (2016) 615–622.

[62] C. Bogdan, A. Pop, S.M. Iurian, D. Benedec, M.L. Moldovan, Research advances in the use of bioactive compounds from *Vitis vinifera* by-products in oral care, Antioxidants 9 (6) (2020) 502.

[63] E.M. Costa, S. Silva, M. Veiga, F.K. Tavaria, M.M. Pintado, A review of chitosan's effect on oral biofilms: perspectives from the tube to the mouth, J. Oral Biosci. 59 (4) (2017) 205–210.

[64] M. Pintado, Antimicrobial effect of chitosan against periodontal pathogens biofilms, SOJ Microbiol. Infect. Dis. 2 (1) (2014) 1–6.

[65] R. Ikono, A. Vibriani, I. Wibowo, K.E. Saputro, W. Muliawan, B.M. Bachtiar, et al., Nanochitosan antimicrobial activity against *Streptococcus mutans* and *Candida albicans* dual-species biofilms, BMC Res. Notes 12 (1) (2019) 383.

[66] A. Aliasghari, M.R. Khorasgani, S. Vaezifar, F. Rahimi, H. Younesi, M. Khoroushi, Evaluation of antibacterial efficiency of chitosan and chitosan nanoparticles on cariogenic streptococci: an in vitro study, Iran. J. Microbiol. 8 (2) (2016) 93–100.

[67] D. Silva, R. Arancibia, C. Tapia, C. Acuña-Rougier, M. Diaz-Dosque, M. Cáceres, et al., Chitosan and platelet-derived growth factor synergistically stimulate cell proliferation in gingival fibroblasts, J. Periodontal Res. 48 (6) (2013) 677–686.

[68] N.I.P. Pini, D.A.N.L. Lima, B. Luka, C. Ganss, N. Schlueter, Viscosity of chitosan impacts the efficacy of F/Sn containing toothpastes against erosive/abrasive wear in enamel, J. Dent. 92 (2020) 103247.

[69] A.H.M. Resende, J.M. Farias, D.D.B. Silva, R.D. Rufino, J.M. Luna, T.C.M. Stamford, et al., Application of biosurfactants and chitosan in toothpaste formulation, Coll. Surf. B Biointerfaces 181 (2019) 77–84.

[70] L.C. Cefali, J.A. Ataide, S. Eberlin, F.C. da Silva Gonçalves, A.R. Fernandes, J. Marto, et al., In vitro SPF and photostability assays of emulsion containing nanoparticles with vegetable extracts rich in flavonoids, AAPS. Pharm Sci. Tech. 20 (1) (2019) 9.

[71] N.M. Siqueira, R.V. Contri, K. Paese, R.C.R. Beck, A.R. Pohlmann, S.S. Guterres, Innovative sunscreen formulation based on benzophenone-3-loaded chitosan-coated polymeric nanocapsules, Skin Pharmacol. Physiol. 24 (3) (2011) 166–174.

[72] C. Cerqueira-Coutinho, R. Santos-Oliveira, E. dos Santos, C.R. Mansur, Development of a photoprotective and antioxidant nanoemulsion containing chitosan as an agent for improving skin retention, Eng. Life. Sci. 15 (6) (2015) 593–604.

[73] Z. Xiao, T. Tian, J. Hu, M. Wang, R. Zhou, Preparation and characterization of chitosan nanoparticles as the delivery system for tuberose fragrance, Flavour. Fragr. J. 29 (1) (2014) 22–34.

[74] P. Lertsutthiwong, P. Rojsitthisak, Chitosan-alginate nanocapsules for encapsulation of turmeric oil, Pharmazie 66 (12) (2011) 911–915.

[75] L. Yulianti, K. Bramono, E. Mardliyati, H.-J. Freisleben, Effects of *Centella asiatica* ethanolic extract encapsulated in chitosan nanoparticles on proliferation activity of skin fibroblasts and keratinocytes, type I and III collagen synthesis and aquaporin 3 expression in vitro, J. Pharm. Biomed. Sci. 6 (5) (2016).

[76] S. Ntohogian, V. Gavriliadou, E. Christodoulou, S. Nanaki, S. Lykidou, P. Naidis, et al., Chitosan nanoparticles with encapsulated natural and UF-purified annatto and saffron for the preparation of UV protective cosmetic emulsions, Molecules 23 (9) (2018) 2107.

[77] R. Harris, E. Lecumberri, I. Mateos-Aparicio, M. Mengíbar, A. Heras, Chitosan nanoparticles and microspheres for the encapsulation of natural antioxidants extracted from *Ilex paraguariensis*, Carbohydr. Polym. 84 (2) (2011) 803–806.

[78] C.E. Park, D.J. Park, B.K. Kim, Effects of a chitosan coating on properties of retinol-encapsulated zein nanoparticles, Food Sci. Biotechnol. 24 (5) (2015) 1725–1733.

[79] A. Sionkowska, B. Kaczmarek, M. Michalska, K. Lewandowska, S. Grabska, Preparation and characterization of collagen/chitosan/hyaluronic acid thin films for application in hair care cosmetics, Pure Appl. Chem. 89 (12) (2017) 1829–1839.

[80] P. Phanthong, P. Reubroycharoen, X. Hao, G. Xu, A. Abudula, G. Guan, Nanocellulose: extraction and application, Carbon Resourc. Convers 1 (1) (2018) 32–43.

[81] K. Ludwicka, M. Jedrzejczak-Krzepkowska, K. Kubiak, M. Kolodziejczyk, T. Pankiewicz, S. Bielecki, Medical and cosmetic applications of bacterial nanocellulose, in: M. Gama, F. Dourado, S. Bielecki (Eds.), Bacterial Nanocellulose: From Biotechnology to Bio-Economy, Elsevier Inc., Amsterdam, 2016, pp. 145–165.

[82] United States Patent: 9018189 [Internet]. Available from http://patft.uspto.gov/netacgi/nph-Parser?Sect1=PTO2&Sect2=HITOFF&p=1&u=%2Fnetahtml%2FPTO%2Fsearch-bool.html&r=41&f=G&l=50&co1=AND&d=PTXT&s1=nanocellulose&s2=cosmetic&OS=nanocellulose+AND+cosmetic&RS=nanocel lulose+AND+cosmetic.

[83] Y. Tokudome, T. Komi, A. Omata, M. Sekita, A new strategy for the passive skin delivery of nanoparticulate, high molecular weight hyaluronic acid prepared by a polyion complex method, Sci. Rep. 8 (1) (2018) 2336.

[84] M. Shigefuji, Y. Tokudome, Nanoparticulation of hyaluronic acid: a new skin penetration enhancing polyion complex formulation: mechanism and future potential, Materialia 14 (2020) 100879.

[85] S. Sakulwech, N. Lourith, U. Ruktanonchai, M. Kanlayavattanakul, Preparation and characterization of nanoparticles from quaternized cyclodextrin-grafted chitosan associated with hyaluronic acid for cosmetics, Asian J. Pharm. Sci. 13 (5) (2018) 498–504.

[86] M.M.A. Abdel-Mottaleb, H. Abd-Allah, R.I. El-Gogary, M. Nasr, Versatile hyaluronic acid nanoparticles for improved drug delivery, in: R. Shegokar (Ed.), Drug Delivery Aspects, Elsevier, Amsterdam, 2020, pp. 1–18.

[87] S. Berkó, M. Maroda, M. Bodnár, G. Eros, P. Hartmann, K. Szentner, et al., Advantages of cross-linked versus linear hyaluronic acid for semisolid skin delivery systems, Eur. Polym. J. 49 (9) (2013) 2511–2517.

[88] A. Sionkowska, M. Michalska-Sionkowska, M. Walczak, Preparation and characterization of collagen/hyaluronic acid/chitosan film crosslinked with dialdehyde starch, Int. J. Biol. Macromol. 149 (2020) 290–295.

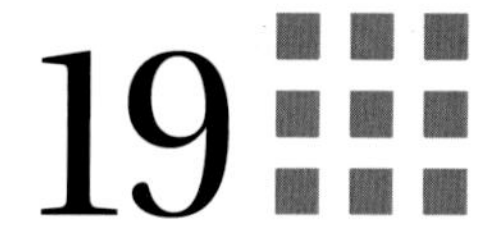

Biopolymeric nanomaterials: water purification

Ankita Dhillon[a] and Dinesh Kumar[b]

[a]DEPARTMENT OF CHEMISTRY, BANASTHALI UNIVERSITY, RAJASTHAN, INDIA [b]SCHOOL OF CHEMICAL SCIENCES, CENTRAL UNIVERSITY OF GUJARAT, GANDHINAGAR, GUJARAT, INDIA

Chapter outline

19.1 Introduction

Water is an essential part of daily life, yet the quality of major water resources is rapidly declining due to rapid population growth, industrial development, agricultural activities, geographical and ecological changes, etc. Therefore, it requires the remediation of various inorganic and organic pollutants passed into our water bodies. Membrane-based filtration, adsorption, catalytic pollutants degradation, flocculation, and decontamination are comprehensively well-performed water purification practices in toxicant remediation [1,2]. However, the employment of

Biopolymeric Nanomaterials: Fundamentals and Applications. DOI: https://doi.org/10.1016/B978-0-12-824364-0.00018-6

cost-effective, environmentally friendly alternatives, including high performance and low carbon footprint is essential for a sustainable future.

In this context, the use of profuse biopolymer nanomaterials can efficiently conquer the above problems. Further, the surface alterations or functionalization have resulted in enhanced adsorption of different contaminations [3,4]. Regarding environmental disquiets, the chief biopolymeric nanomaterials rely on biodegradable materials such as chitosan, cellulose, cellulosic derivatives, gelatin (GE), clay, and alginate polymers. The chapter includes valued information on water purification using several biopolymeric nanomaterials and their adsorption properties and performance. It gives a beneficial role of biopolymeric nanomaterials, including chitosan-based biopolymeric nanomaterials, cellulose-based biopolymeric nanomaterials, GE hydrogel–based biopolymeric nanomaterials, clay polymer–based biopolymeric nanomaterials, alginate hydrogels–based biopolymeric nanomaterials applications in water purification. The chapter also focuses on the specific properties of these biopolymer nanomaterials needed to speed up large-scale water treatment applications for the effective remediation of various pollutants.

19.2 Classification of biopolymeric nanomaterials

19.2.1 Chitosan-based biopolymeric nanomaterials

In comparison to naturally occurring popular polysaccharides such as pectin, agarose, cellulose, etc., chitosan polysaccharide is very basic polysaccharide. Also, it displays distinct properties such as viscosity, versatility in solubility, formation of polyoxy salt, film formation abilities, chelation of metal, microbial cell binding, etc. The glucosamine and acetyl-glucosamine units form the hetero-polymer type of chitosan. Based on acetylation, these units are offered in various grades that determine many polymer properties such as solubility and acid–base activities [5,6]. The cellulose's chitosan second carbon atom alcoholic group is substituted by the acetyl amino group. Consequently, chitosan is a copolymeric form comprising two repeating units of N-acetyl-2-amino-2-d-glucopyranose and 2-amino-2-deoxy-d-glucopyranose linked by β-(1$\rightarrow$4)-glycosidic bond. This organization imparts a firm crystalline construction of chitosan via inter- and intramolecular hydrogen bonding [7,8]. Because of the incidence of one -NH_2 group and two -OH groups on every glycosidic residue, chitosan has wonderful chemical and biological characteristics [9]. Because of the functionality of repeating units, chitosan has a chelating ability to bind several metal ions.

19.2.2 Cellulose-based biopolymeric nanomaterials

Cellulose holds many appropriate properties, for instance, eco-friendliness, biocompatibility, and availability of high proportion adjustable–OH groups. It is classified as a carbohydrate polymer owing to repeating β-D glucopyranose units. Fibers of cellulose include both lightly packed amorphous domains and firmly packed crystalline chains [10].

Cellulose-based nanomaterials comprise two main sets, that is, cellulose nanocrystals (CNCs) and cellulose nanofibrils (CNFs) [11]. Cellulose-based nanomaterials have many properties, for instance, enhanced surface area, great aspect ratio, possibility of excellent surface functionalization, and abundant availability which present them an exceptional applicant for water purification processes [12].

19.2.3 GE hydrogel–based biopolymeric nanomaterials

Animal tissue is a source of GE and was first used as a glue in 6000 BC. Later, because of several applications, GE found its industrial-scale fabrication. GE is produced from pig skin, bones of cattle, and bovine hides [13] and has various applications in pharmacological industries, nutrition, etc. It has a similar composition to that of collagen, and few composition alterations are because of its fabrication process. The accurate measurements of quantity and form of amino acids of GE are not known.

19.2.4 Clay polymer–based biopolymeric nanomaterials

Clays and their various adapted forms' free availability and cost-effective nature have openly assisted in their widespread applications as metal ions adsorbents. An extensive review of clay minerals and organo-clay minerals for the heavy metal adsorption and organic micro pollutants remediation, respectively, has been given by Bhattacharyya and Gupta [14] and de Paiva et al. [15]. It has been noticed that neat clay minerals have low efficiency in removing micropollutants because of their reduced surface area difficulty in recovering clay particles during regeneration. Therefore, their intercalation with polymer forming clay polymer–based nanomaterials has gained increasing attention in the near past [16-19] Compared to the discrete components, clay polymer–based nanomaterials have improved mechanical strength and high regenerability, resulting in high effectiveness in removing contaminants during wastewater treatment.

19.2.5 Alginate hydrogels–based biopolymeric nanomaterials

Alginate is a biopolymer having brown algae (30%–60%) content and has been accounted comprehensive because of its gel's formation ability and microparticles properties [20-23] The irregular blocks of mannuronic and guluronic acid units present it as an anionic biopolymer [24]. The guluronic acid and mannuronic acid forms α (1$\rightarrow$4) linkages and β (1$\rightarrow$4) linkages, respectively and are joined by glycosidic linkages. Similarly, alginate hydrogels can be planned by polymer chains cross-linking [25]. Intermolecular cross-linking is one of the appropriate processes for the alginate hydrogels formation involving only alginate guluronic groups that participate with cations such as calcium [26]. Further improvement in alginate hydrogels properties, for instance, solvent miscibility, hydrophobic nature, and pharmacological activity can occur by modifying via existing hydroxyl and carboxyl groups [27].

19.3 Biopolymeric nanomaterials' applications in water purification

19.3.1 Chitosan-based biopolymeric nanomaterials in water purification

By a large quantity of $-NH_2$ and -OH groups compared to other biopolymers; chitosan has gained superiority in water decontamination. Both modified and unmodified forms of chitosan can grab poisonous contaminants such as heavy metals and colorants impurities [28,29]. Recently, a casting method was employed by Karthikeyan et al. for preparing lanthanum-incorporated chitosan membrane (La@CS) membrane to remediate nitrate and phosphate ions wastewater [30]. The developed membrane showed good removal performance toward both the ions from wastewater. The adsorption of anionic species occurred by ion-exchange mechanism after electrostatic attraction. Therefore, the prepared membrane system showed an ecofriendly and cost-effective phosphate and nitrate ions remediation from aqueous systems [30]. Amine-based materials have attraction toward anionic dyes and therefore, chitosan was cross-linked with bifunctional amines such as diammonium tartrate and urea/diammonium tartrate for the enhanced congo red dye uptake from textile wastewater. The functionalized chitosan showed an excellent adsorption capacity toward both the targeted contaminants. The chemical bond establishment between dye and adsorbent was realized as supported by thermodynamic studies [31].

Vakili et al. changed chitosan beads using diepoxyoctane and spermine for the Cr(VI) remediation from wastewater [32]. There was an increased amount of adsorption sites and surface properties after the modification of CS beads. The changed beads exhibited high removal performance of 352.0 mg/g toward Cr(VI) at acidic pH. Repeated regeneration and reusability even after five cycles presented a good adsorption performance of 192.31 mg/g [32]. Zia et al. used electrospinning method for the development of a green nanofibrous adsorbent membrane namely, porous poly(L-lactic acid) for Cu^{2+} ions remediation from wastewater [33]. Straight immersion and coating of chitosan was achieved onto a porous membrane. The copper ions chelation was achieved using hydroxyl and amine groups of chitosan adsorbent. The adsorbent presented highest remediation performance of 128.53 mg/g for copper ions at pH 7, within 10 min at 25 °C. The developed adsorbent presented larger adsorption performance than other reported chitosan adsorbents due to enhanced surface properties and abundant surface groups to bind copper ions [33].

Similarly, Trikkaliotis et al. also carried out Cu^{2+} ions remediation from wastewater by developing a hydrogel bead made of Cs, polyvinyl alcohol (PVA), and polyethylene glycol (PEG) [34]. The developed beads carried out high adsorption (99.99 %) of Cu^{2+} at 25 mg/L copper concentration, pH 5, at 45 °C, and composite dose 1 g/L. Langmuir experimental data supported Langmuir adsorption performance of ˜45 mg/g. Further improvement in adsorption performance can be achieved by chitosan deacetylation reaction, hydrogel's freeze-drying, and pH adjustment to about 5.85 value [34]. Karpuraranjith and Thambidurai synthesized a novel chitosan/zinc oxide-polyvinylpyrrolidone nanocomposite (chitosan/ZnO-PVP) in one-step method [35]. The developed hybrid biopolymer had layered rod-like arrangements between nanometers. The hybrid polymer showed higher thermal stability compared to precursor materials. Also, the chitosan/ZnO-PVP nanocomposite performed better antimicrobial activity by membrane damage than PVP-ZnO composite [35]. Although chitosan has large-scale availability, biodegradability, and economical nature, yet it requires its further modification to increase surface area and chemical stability. Salih et al. physically modified chitosan involving diatomaceous earth (DE) as dispersion medium having high porosity, high surface area, etc. [36]. The modified chitosan/DE composite (CSD) presented high adsorption performance of 106 and 88 mg/g for As(III) and As(V), respectively, from wastewater. The presence of phosphate ions negatively affected the adsorption process [36]. In a study by Min et al. iron-doped chitosan electrospun nanofiber mat (Fe@CTS ENM) was developed for the remediation of arsenite at neutral pH water [37]. The adsorption experiments presented high adsorption performance of 36.1 mg/g at an adsorbent dose of over 0.3 g/L in a wide pH range within 2 h. The presence of coexisting ions such as carbonate, sulfate, chloride, fluoride, etc. had negligible effect on adsorption performance. Also, regeneration studies showed no considerable adsorption performance degradation [37].

19.3.2 Cellulose-based biopolymeric nanomaterials in water purification

Adsorbents including activated carbons have been extensively used in wastewater treatment technologies. But, emissions of greenhouse gas, expensive production, and regeneration of activated carbons require development of alternative inexpensive adsorbents [38,39]. In this situation, utilization of cellulose-based adsorbents has attracted researchers globally [40,41]. The application of adsorption coupled membrane filtration avoids direct contact of solute with the membrane, thereby reducing its fouling. Therefore, Gago et al. prepared dicarboxymethyl cellulose (DCMC) based cross-linked polymer for methylene blue (MB) dye uptake [42]. The number of the equivalent of sodium 2-bromomalonate during the fabrication was adjusted to increase functionalization. The Langmuir isotherm was best fitted in acidic medium signifying the formation of adsorption monolayer on a homogenous adsorbent's surface. However, further work is needed using DCMC in real wastewater large-scale samples to evaluate marketable feasibility of the offered method [42]. Cellulose fibrils have been proved as excellent bioadsorbents due to robustness, easy chemical alterability, eco-friendly, nonhazardous nature, etc. In this manner, easily available and inexpensive *Phragmites australis* was used for the development of cellulose microfibrils using chemical extraction procedure and then used for the biosorption of MB. The crystallinity value increased to 69% because of considerable loss of hemicellulose and lignin. Adsorption experiments presented 54.9 mg/g biosorption capacity over a wide pH range [43].

In another study, Chen et al. fabricated environmentally friendly polymer composite aerogel by adding TiO_2 in the cellulose/graphene oxide (GO) hydrogels [44]. Though the adsorbent matrix has GO and TiO_2 nanoparticles, it upholds the hydrogel structure throughout the catalytic interaction minimizing the secondary pollution during MB remediation. Because of exclusive porous structure, the developed hydrogel presented greater performance and reproducibility.

This effort increased the possibility of bio-template fabrication of future nanomaterials for many uses [44].

Valencia et al. used carboxylated CNFs-based films for the remediation of heavy metal ions from wastewater [45]. The work involved the formation of nanoparticles of copper oxide on developed films besides the enhanced performance toward dye remediation and antimicrobial activity. Therefore, the process allowed the use of materials based on nanocellulose for numerous uses, therefore enhancing its ecological viability and competence during the metal ions uptake [45].

Quaternized ammonium functionalized poly(diallyldimethylammonium chloride) (PDADMAC) presented an inexpensive choice for the cationic functionalization of cellulose membrane. They developed cellulose-g-PDADMAC membranes for enhanced uptake of anionic methyl orange dye from wastewater. During the synthesis, grafting of PDADMAC onto the filter membrane surface was achieved by well-regulated polymerization. The mechanism involved the electrostatic forces among membrane's quaternary ammonium groups and dye's sulfonic groups. Also, the developed material performed efficient antimicrobial performance against *Staphylococcus aureus* and *Escherichia coli.* [46]. Even though nanomaterials have been proven to overtake many conventional adsorbents, they have limited real-world applicability at industrial scale because of their difficult separation from medium after treatment process. The issue was addressed by their incorporation in various matrices, for instance, hydrogel beads that have wide-scale applicability in batch and column studies.

Tam et al. developed alginate hydrogel beads of bovine serum albumin-protected gold nanoclusters (Au@BSA NCs)-loaded CNC nanocomposite for the instantaneous detection and remediation of toxic Hg^{2+} ions from wastewater [47]. The nanocomposite having alginate in double amounts to that of CNCs acts as a supreme sensor forager system. Because of high Hg^{2+} and Au^{+} metallophilic interaction on the developed nanocomposite surface, the nanocomposite's fluorescence was totally quenched by Hg^{2+} ions in the solution. Also, there was a visible change in color by the nanocomposite after the adsorption of Hg^{2+} ions. Therefore, the study supported the development of such a sensor scavenger system for similar applications [47].

Wang et al. prepared $KMnO_4$-deoxidized nano-MnO_2 by bamboo cellulose fibers as a nonagglomerated MnO_2/CNFs hybrid material for the MB removal [48]. CNFs provided both support and reduction, resulting in high adsorption performance toward MB in solution. pH played an important role as acidic pH stimulated the oxidative decolorization and alkaline pH enabled physical adsorption [48].

As a result of special structure, biocompatibility, and cost-effective nature, β-cyclodextrin (β-CD) is widely used in water treatment. Therefore, Chen et al. grafted β-CD onto CNCs@Fe_3O_4@SiO_2 hybrids for procaine hydrochloride and imipramine hydrochloride remediation [49]. The developed β-CD hybrids presented high adsorption performance toward selected pharmacological remains. The coating of silica improved the thermal constancy of CNCs as supported by thermal gravimetric analysis (TGA) studies [49].

Yang et al. used an *in-situ* coprecipitation method for the synthesis of Mg-Al layered double hydroxide (LDH)/cellulose nanocomposite beads (LDH@CB) adsorbents for toxic amoxicillin from municipal wastewater [50]. The adsorbent presented enhanced surface properties, and high hydration for largest adsorption performance (138.3 mg/g) in wastewater. The mechanism behind the process involved the electrostatic force of attraction between anionic amoxicillin and cationic adsorbent [50]. Mwafy et al. successfully prepared 2,2,6,6-tetramethylpiperidine 1-oxyl (TEMPO)-oxidized cellulose nanofibers (TOCN) by a green method to develop CdO-TOCN nanocomplexes in a single step [51]. The nanocomposite showed wide-ranging antibacterial performance toward both Gram-negative and Gram-positive bacteria bacterial strains [51].

19.3.3 GE hydrogel–based biopolymeric nanomaterials in water purification

The unique performance of hydrogels has been used as an adsorbent for the remediation of various pollutants from the environment. GE is a polyampholite that shows both positive and negatively charged domains and also contains uncharged hydrophilic and hydrophobic moieties. The formation of a steric barrier by GE can stabilize surfaces as a result its chief purpose is to act as a stabilizing agent.

Mohseni et al. synthesized gelatin-carbon nanotube (GE-CNT) embedded magnetic nanoparticles for the dual remediation of anionic direct red 80 (DR) dye and cationic MB dye from wastewater [52]. The magnetic properties of the adsorbent beads facilitated easy recovery from the reaction medium. The developed adsorbent presented 96.1% adsorption performance toward DR and 76.3% adsorption performance toward MB dye [52]. Peter et al. developed highly porous chitosan–GE /nanophase hydroxyapatite composite scaffolds for applications in tissue engineering [53]. Where, nanohydroxyapaties (nHA) supported reduced degradation rate of composite scaffolds and increased mineralization. Also, the presence of nHA in nanocomposite enhanced protein uptake and cell adherence on the scaffold surfaces. Compared to chitosan–GE scaffold, the developed nanocomposite has greater biological reaction toward MG-63 cells in terms of advanced propagation, and dispersal [53]. Another study synthesized composite hydrogels, namely cellulose-*graft*-polyacrylamide/hydroxyapatite for copper (II) ions remediation. The authors utilized hydroxyapatite because of its excellent biocompatibility and good adsorption performance in wastewater. The adsorbent showed good performance against copper (II) ions remediation with a maximum adsorption performance of 175 mg/g of composite hydrogel [54].

GE has been an excellent immobilization support because of its easy gelation without encapsulated enzyme inactivation. Therefore, Bilal et al. encapsulated manganese peroxidase (MnP) on a GE matrix using glutaraldehyde (GA) as a cross-linking agent [55]. The highest immobilization of 82.5% was achieved using 20% GE, 0.25% of GA with activation time of 2 h, and concentration of protein as 0 6 mg/mL. The encapsulated enzyme showed highest performance and good thermal stability at pH 6.0 and 60 °C. The encapsulated enzyme activity was tested against $MnSO_4$ as a substrate, and it was found that successful reactive red dye decolorization (90%) was achieved with good activity of more than 50% after six cycles [55]. Acharya et al. successfully fabricated silver nanoparticles (AgNPs) by using sodium alginate as a reductant for Ag^+ ions and the colloidal mixture was further stabilized using GE [56]. The authors performed an effectual and inexpensive fabrication of biocompatible silver-based nanocomposite using green solvents, an environmental benign reducing agent, and a bio-based stabilizing agent for antibacterial performance. Using sodium alginate for reduction and GE for stabilization during the fabrication of nanocomposite presented a smart approach. Antibacterial analyses showed the efficacy of developed composite toward Gram-negative bacteria [56].

To advance the property of material and biological activity of the PEG-based bioadhesive, Li et al. incorporated near about 7.5 wt% of cross-linked GE microgel into dopamine-modified PEG [57]. GE microgel incorporation in adhesive network decreased treatment time, though it increased the elastic modulus and cross-linking concentration. Also, incorporating GE microgel drastically enhanced the viscous degeneracy aptitude of the adhesive because of the development of revocable physical links into the adhesive system. Further, *in-vitro* cell culture practices noncytotoxic behavior of composite adhesive [57].

Similarly, Kankeu et al. used GE hydrogel and its hybrid clinoptilolite nanocomposite for Cd^{2+} ions remediation from multimetal ions mine wastewater [58]. The copolymerization grafting of acrylamide (AAm) onto GE was carried out to synthesize hydrogel, and incorporation of clinoptilolite within the hydrogel matrix was carried out for hybrid hydrogel nanocomposite fabrication. Kinetics studies of adsorption presented pseudo-second-order kinetic, while the isotherm studies supported the Freundlich and Langmuir isotherm. The GE hydrogel and its hybrid clinoptilolite nanocomposite showed the highest remediation performance of 54.95 and 78.13 mg/g toward Cd^{2+} ions [58].

Another study by Thakur and Arotiba fabricated hydrogel nanocomposites by acrylic acid (AA) polymerization utilizing biopolymeric properties of sodium alginate and TiO_2 nanoparticle functioned as an inorganic cross-linker and N, N-methylene-bisacrylamide as an organic cross-linker [59]. The developed nanocomposites carried out methyl violet dye uptake from wastewater. Kinetics studies of adsorption presented pseudo-second-order rate, while the isotherm experiments supported the Langmuir isotherm with the maximum performance 1156.61 mg/g, presenting the efficacy of composite toward cationic dyes selective remediation from wastewater [59].

19.3.4 Clay polymer–based biopolymeric nanomaterials in water purification

The last few decades involved growing interest in the clays used as adsorbents to remediate various environmental contaminations of both organic and inorganic origins. A representative clay mineral, namely, bentonite, has a sheet-like structure involving mineral montmorillonite (MMT) as the chief part. This clay mineral has noble swelling ability, enhanced specific surface area, and cation exchange performance, presenting it as an excellent adsorbent for contaminants. Therefore, a study involved intercalation polymerization technique for the development of a composite of polyacrylonitrile/organo bentonite having amidoxime functionality. The mechanism of interaction involved both adsorption-complexation and ion-exchange behavior. The adsorbent showed a maximum performance of 99.8%, 98.9%, 97.4% for Cu(II), Zn(II), and Cd (II), respectively. Adsorption kinetics presented pseudo-second-order reaction rate, while the isotherm studies supported the Langmuir isotherm [60]. Anirudhan and Suchithra explored humic acid-immobilized-amine modified polyacrylamide/bentonite composite (HA-Am-PAA-B) for heavy metal ions uptake from wastewater [61]. The maximum remediation of Cu(II) occurred at pH 5.0, Zn(II) pH 9.0 and Co(II) at pH 8.0. The mechanism of interaction involved both adsorption-complexation and ion exchange behavior. The kinetics studies of adsorption presented pseudo-second-order rate, while the isotherm experiments supported the Langmuir isotherm [61]. Unuabonah et al. prepared clay polymer–based composite using kaolinite clay and PVA for the remediation of Pb^{2+} ions [62]. Increase in adsorption performance toward Pb^{2+} ions was observed on increasing bed height and metal ion initial concentration but the coincidence of Ca^{2+}/Pb^{2+} and Na^{+}/Pb^{2+} reduced the adsorption capacity [62]. In a study, a two-layer artificial neural networks model was proposed for Pb(II) uptake on a Cloisite C20A polycaprolactone (C20A-PCL) polymer-clay composite [63]. The model presented a cost-effective study of the effect of analyte concentration, solution pH, and temperature parameters for Pb(II) uptake. The developed composite showed remarkable performance such as 3% of filler content remediated 87% of Pb(II) ions from wastewater [63].

Bleiman and Mishael carried out successful adsorption of selenate from wastewater on chitosan–MMT polymer–clay composites [64]. The composite showed higher Langmuir adsorption performance (18.4 mg/g) compared to Al-oxide (17.2 mg/g) and Fe-oxide (8.2 mg/g). Compared to oxides adsorbent showing decreased performance at high pH, the developed composite presented pH-independent selenium adsorption. Also, the developed composite brought selenium levels of contaminated well water below than WHO (World Health Organization) limit and showed selenium selectively with coexisting sulfur [64]. Zadaka et al. carried out successful remediation of atrazine from wastewater by poly (4-vinylpyridine-co-styrene)-MMT (PVP-co-S90%-mont) polymer–clay composite [65]. Adsorption studies revealed 90%–99% of atrazine adsorption within 20–40 min. Also, the column treatment method remediated 93%–96% of atrazine, while the identical quantity of granular activated carbon (GAC) presented lower adsorption performance (83%–75%). A comparative study involving the presence of dissolved organic matter showed a considerable reduction in adsorption performance of GAC filter than PVP-co-S90%-mont filter. Therefore, the developed filter successfully eliminated atrazine beneath the present US EPA regulation [65]. Phenolic compounds present serious health and ecological risks. Therefore, Ganigar et al. carried out trinitrophenol (picric acid—PA) and trichlorophenol (TCP) uptake from water reservoirs [66]. Poly-4-vinylpyridine-co-styrene (PVPcoS) adsorption on MMT was considerably quicker compared to sepiolite. Therefore, polycation–MMT composites were further tested for phenolic pollutant remediation. A complete PA adsorption was achieved by the PVPcoS–MMT composite while only 40%– 60% of TCP remediation occurred. The mechanism of interaction involved both hydrophobic interactions and electrostatic interactions. The binding of PA to wet composites was found to be because of hydration properties of PA while the binding of TCP to dry composites was attributed to its hydrophobic properties [66]. In a study by Ravikumar and Udayakumar, a composite clay polymer forming a coagulant in the form of *Moringa oleifera* seed and an adsorbent in the form of bentonite clay was prepared for cadmium, chromium, and lead remediation [67]. The composite exhibited almost 99.99% removal efficiency for heavy metals with maximum Cd adsorption (pH range 6–8), Cr (pH range 2–4), and Pb (pH range 5–7) [67].

19.3.5 Alginate hydrogels–based biopolymeric nanomaterials in water purification

The pH-sensitive biopolymer, that is, alginate has carboxylic group moieties and by the variation of the solution pH, these moieties can act as both proton donor and acceptor. Because of these pH-dependent properties, alginate centered materials are examined in enzyme immobilization and other adsorption applications. Therefore, Mahdavinia et al. prepared cross-linked magnetic hydrogel beads by mixing sodium alginate and PVA containing magnetic laponite RD (Rapid Dispersion) for the rapid adsorption of BSA [68]. To generate magnetic properties and strengthening of hydrogels, magnetic laponite RD nanoparticles were integrated into the structure. The developed hydrogel beads presented good stability in a wide pH range without degeneration. At pH 4.5, the nanocomposite beads proved good Langmuir adsorption performance of 127.3 mg/g [68]. In a study, granular alginate–based hydrogels were developed by insertion and cross-linked reactions between sodium alginate (SA), AA, PVP, and GE [69]. A bulk gel was produced by grafting and cross-linking of SA and AA, followed by the PVP and GE addition into the reaction mixture that resulted in granule formation. The key driving forces in granule formation were both electrostatic forces and hydrogen-bonding between all the constituents. The developed composite presented reasonable adsorption capacities toward target metal ions. Further, competitive adsorption studies proposed strong affinity of the hydrogel toward Cu^{2+} ion ions [69]. Hydrogel microspheres of SA/PVA/GO were developed for the remediation of $Cu^{2+}$and U^{6+} ions from wastewater. Physical cross-linking of sodium alginate was achieved by Ca^{2+} ions and the encapsulation of GO into the composite was done to reinforce the hydrogels. The role of PVA was the dispersion of GO in SA. The isotherm experiments supported the Langmuir isotherm with the maximum performance of 247.16 and 403.78 mg/g for Cu^{2+} and ${UO_2}^{2+}$ ions, correspondingly. Further, there was no substantial loss in adsorption performance after repeated adsorption-desorption cycles [70].

Zhuang et al. synthesized alginate/reduced GO (RGO) double-network (GAD) hydrogel through a simplistic process [71]. The developed GAD's were then compared for their mechanical properties, constancy, and adsorption performance with an alginate/RGO hydrogel (GAS). The hydrogel was prepared by forming an alginate network with arbitrarily dispersed GO that resulted in development of GAS. After that, prepared GAS was hydrothermally reduced involving GO reduction and self-assembling into a second RGO network, which resulted in the formation of GAD. The developed GAD had greater Young's modulus, and minor swelling ratio compared to GAS, thereby resulting in enhanced gel constancy in pH solutions. The developed beads showed excellent adsorption performance of 169.5 and 72.5 mg/g for Cu^{2+}, and ${Cr_2O_7}^{2-}$ ions. Further, there was no substantial loss in adsorption performance after 10 repeated adsorption-desorption cycles [71].

Thakur et al. performed the remediation of MB cationic dye from wastewater using hydrogel nanocomposite of inorganic titania incorporated organic SA cross-linked AA (SA-cl-poly(AA)-TiO_2) [72]. Using a cross-linking agent such as TiO_2 nanoparticles, and a free radical initiator, the nanocomposite was developed by copolymerization of AA onto SA biopolymer. The developed hydrogel nanocomposite showed high swelling aptitude (412.98 g/g). Kinetics studies presented pseudo-second-order kinetic, while the isotherm experiments supported the Langmuir isotherm with the maximum performance of 2257.36 mg/g [72].

Ren et al. prepared alginate–carboxymethyl cellulose (CMC) gel beads [73]. The developed SA–CMC beads demonstrated 99% Pb(II) uptake much advanced than traditional adsorbents. Kinetics studies of adsorption presented pseudo-second-order rate, while the isotherm studies supported the Langmuir isotherm. The physical, chemical, and electrostatic interactions were found to be involved in Pb(II) removal where chemical adsorption was the chief adsorption mechanism [73]. CarAlg/MMt nanocomposite hydrogels made of *kappa*-carrageenan (Car) and SA biopolymers were manufactured by incorporating sodium MMT (Na-MMT) nanoclay [74]. The developed hydrogels carried out efficient remediation of cationic crystal violet dye from wastewater. On increasing clay concentration, there was an increased adsorption performance of nanocomposites. The isotherm studies supported the Langmuir isotherm with the maximum performance of 88.8 mg/g at instances of acidic pH [74].

Lu et al. carried out the sizable fabrication of SA beads of controlled swelling activities, pH responsiveness, and good remediation performance for MB and heavy metal ions at industrial scale using a post-cross-linking method [75]. The beads were prepared using GA, acetic acid, and hydrochloric acid, the solution as the coagulating agent. Kinetics studies of adsorption presented pseudo-second-order rate, while the isotherm studies supported the Langmuir isotherm with the maximum performance of 572 mg/g for MB and heavy metal ions [75].

Liu et al. used an ecological, cheap, and plentiful plant protein, soy protein isolate (SPI) as a matrix, polyethyleneimine (PEI) for functionalization for preparing SPI/PEI composite hydrogels via a cross-linking technique [76]. The developed SPI/PEI composite hydrogels with 50% PEI concentration efficiently remediated Cu^{2+} ions from wastewater. This material was then proven to function as a catalyst for a model reaction, like in 4-nitrophenol reduction [76].

19.4 Conclusion

Because of their good mechanical properties, biodegradability, and ecofriendly nature, biopolymeric nanomaterials have gained attention in a wide variety of water purification applications. The characteristic properties of these biopolymeric nanomaterials essential to speed up the large-scale water treatment applications were discussed for the effective remediation of different pollutants. In this chapter, a beneficial role of chitosan, cellulosic, GE hydrogel, clay-polymer, and alginate hydrogels based biopolymeric nanomaterials in water remediation applications is discussed. Using these biopolymers pioneers an alternative possibility to generate innovative biopolymeric nanomaterials with enhanced characteristics for the utilization in many applications.

Acknowledgment

Dinesh Kumar is thankful to DST, New Delhi, for the financial support offered to this work (sanctioned vide project Sanction Order F. No. DST/TM/WTI/WIC/2K17/124(C).

References

[1] S. Ahuja, Handbook of Water Purity and Quality, Academic Press, MA, 2009.

[2] DW. Hendricks, Water Treatment Unit Processes: Physical and Chemical, CRC Press, Boca Raton, FL, 2018.

[3] B. Pan, Q. Zhang, F. Meng, X. Li, X. Zhang, J. Zheng, W. Zhang, B. Pan, J. Chen, Sorption enhancement of aromatic sulfonates onto an aminated hyper-cross-linked polymer, Environ. Sci. Technol. 39 (9) (2005) 3308–3313.

[4] K. Zheng, B. Pan, Q. Zhang, W. Zhang, B. Pan, Y. Han, Q. Zhang, D. Wei, Z. Xu, Q. Zhang, Enhanced adsorption of p-nitroaniline from water by a carboxylated polymeric adsorbent, Sep. Purif. Technol. 57 (2) (2007) 250–256.

[5] F. Hoppe-Seiler, Über Chitin und Zellulose, Berichte der Deutschen Chemischen Gesellschaft. 27 (1894) 3329-3331.

[6] P. Sorlier, A. Denuzière, C. Viton, A. Domard, Relation between the degree of acetylation and the electrostatic properties of chitin and chitosan, Biomacromolecules. 2 (3) (2001) 765–772.

[7] H. Kargarzadeh, M. Mariano, D. Gopakumar, I. Ahmad, S. Thomas, A. Dufresne, J. Huang, N. Lin, Advances in cellulose nanomaterials, Cellulose. 25 (4) (2018) 2151–2189.

[8] M. Dash, F. Chiellini, R.M. Ottenbrite, E. Chiellini, Chitosan—A versatile semi-synthetic polymer in biomedical applications, Prog. Polym. Sci. 36 (8) (2011) 981–1014.

[9] P. Agrawal, G.J. Strijkers, K. Nicolay, Chitosan-based systems for molecular imaging, Adv. Drug Deliv. Rev. 62 (1) (2010) 42–58.

[10] B.L. Peng, N. Dhar, H.L. Liu, KC. Tam, Chemistry and applications of nanocrystalline cellulose and its derivatives: a nanotechnology perspective, Can. J. Chem. Eng. 89 (5) (2011) 1191–1206.

[11] S. Bi, J. Pang, L. Huang, M. Sun, X. Cheng, X. Chen, The toughness chitosan-PVA double network hydrogel based on alkali solution system and hydrogen bonding for tissue engineering applications, Int. J. Biol. Macromol. 146 (2020) 99–109.

[12] N. Mahfoudhi, S. Boufi, Nanocellulose as a novel nanostructured adsorbent for environmental remediation: a review, Cellulose. 24 (3) (2017) 1171–1197.

[13] M.C. Gómez-Guillén, M. Pérez-Mateos, J. Gómez-Estaca, E. López-Caballero, B. Giménez, P. Montero, Fish gelatin: a renewable material for developing active biodegradable films, Trends Food Sci. Technol. 20 (1) (2009) 3–16.

[14] K.G. Bhattacharyya, SS. Gupta, Kaolinite and montmorillonite as adsorbents for Fe (III), Co (II) and Ni (II) in aqueous medium, Appl. Clay Sci. 41 (1-2) (2008) 1–9.

[15] L.B. De Paiva, A.R. Morales, FR. Díaz, Organoclays: properties, preparation and applications, Appl. Clay Sci. 42 (1-2) (2008) 8–24.

[16] F. Bergaya, C. Detellier, J.F. Lambert, G. Lagaly, Introduction to Claypolymer Nanocomposites (CPN), Developments in Clay Science, Editor(s): Faïza Bergaya and Gerhard Lagaly, 5, Elsevier, Amsterdam, 2013, pp. 655–677.

[17] P. Kiliaris, CD. Papaspyrides, Polymer/layered silicate (clay) nanocomposites: an overview of flame retardancy, Progr. Polym. Sci. 35 (7) (2010) 902–958.

[18] P. Liu, Polymer modified clay minerals: a review, Appl. Clay Sci. 38 (1-2) (2007) 64–76.

[19] S. Pavlidou, CD. Papaspyrides, A review on polymer–layered silicate nanocomposites, Prog. Polym. Sci. 33 (12) (2008) 1119–1198.

[20] C.H. Goh, P.W. Heng, LW. Chan, Alginates as a useful natural polymer for microencapsulation and therapeutic applications, Carbohydr. Polym. 88 (1) (2012) 1–2.

[21] S.N. Pawar, KJ. Edgar, Alginate derivatization: a review of chemistry, properties and applications, Biomaterials. 33 (11) (2012) 3279–3305.

[22] R. Pereira, A. Carvalho, D.C. Vaz, M.H. Gil, A. Mendes, P. Bártolo, Development of novel alginate based hydrogel films for wound healing applications, Int. J. Biol. Macromol. 52 (2013) 221–230.

[23] S. Thakur Sodium alginate, xanthan gum biopolymer composites: synthesis, characterisation and application in organic dye removal from water. (Doctoral dissertation), University of Johannesburg.

[24] J.O. You, S.B. Park, H.Y. Park, S. Haam, C.H. Chung, WS. Kim, Preparation of regular sized Ca-alginate microspheres using membrane emulsification method, J. Microencapsul. 18 (4) (2001) 521–532.

[25] L.W. Chan, H.Y. Lee, PW. Heng, Mechanisms of external and internal gelation and their impact on the functions of alginate as a coat and delivery system, Carbohydr. Polym. 63 (2) (2006) 176–187.

[26] IM. El-Sherbiny, Enhanced pH-responsive carrier system based on alginate and chemically modified carboxymethyl chitosan for oral delivery of protein drugs: preparation and in-vitro assessment, Carbohydr. Polym. 80 (4) (2010) 1125–1136.

[27] U. Zimmermann, G. Klöck, K. Federlin, K. Hannig, M. Kowalski, R.G. Bretzel, A. Horcher, H. Entenmann, U. Sieber, T. Zekorn, Production of mitogen-contamination free alginates with variable ratios of mannuronic acid to guluronic acid by free flow electrophoresis, Electrophoresis. 13 (1) (1992) 269–274.

[28] M.R. Gandhi, N. Viswanathan, S. Meenakshi, Preparation and application of alumina/chitosan biocomposite, Int. J. Biol. Macromol. 47 (2) (2010) 146–154.

[29] G.Z. Kyzas, M. Kostoglou, NK. Lazaridis, Copper and chromium (VI) removal by chitosan derivatives—equilibrium and kinetic studies, Chem. Eng. J. 152 (2-3) (2009) 440–448.

[30] P. Karthikeyan, H.A. Banu, S. Meenakshi, Removal of phosphate and nitrate ions from aqueous solution using La3+ incorporated chitosan biopolymeric matrix membrane, Int. J. Biol. Macromol. 124 (2019) 492–504.

[31] A. Zahir, Z. Aslam, M.S. Kamal, W. Ahmad, A. Abbas, RA. Shawabkeh, Development of novel cross-linked chitosan for the removal of anionic Congo red dye, J. Mol. Liq. 244 (2017) 211–218.

[32] M. Vakili, S. Deng, D. Liu, T. Li, G. Yu, Preparation of aminated cross-linked chitosan beads for efficient adsorption of hexavalent chromium, Int. J. Biol. Macromol. 139 (2019) 352–360.

[33] Q. Zia, M. Tabassum, Z. Lu, M.T. Khawar, J. Song, H. Gong, J. Meng, Z. Li, J. Li, Porous poly (L–lactic acid)/chitosan nanofibres for copper ion adsorption, Carbohydr. Polym. 227 (2020) 115343.

[34] D.G. Trikkaliotis, A.K. Christoforidis, A.C. Mitropoulos, GZ. Kyzas, Adsorption of copper ions onto chitosan/poly (vinyl alcohol) beads functionalized with poly (ethylene glycol), Carbohydr. Polym. 234 (2020) 115890.

[35] M. Karpuraranjith, S. Thambidurai, Chitosan/zinc oxide-polyvinylpyrrolidone (CS/ZnO-PVP) nanocomposite for better thermal and antibacterial activity, Int. J. Biol. Macromol. 104 (2017) 1753–1761.

[36] S.S. Salih, A. Mahdi, M. Kadhom, TK. Ghosh, Competitive adsorption of As (III) and As (V) onto chitosan/diatomaceous earth adsorbent, J. Environ. Chem. Eng. 7 (5) (2019) 103407.

[37] L.L. Min, L.M. Yang, R.X. Wu, L.B. Zhong, Z.H. Yuan, YM. Zheng, Enhanced adsorption of arsenite from aqueous solution by an iron-doped electrospun chitosan nanofiber mat: preparation, characterization and performance, J. Coll. Interface Sci. 535 (2019) 255–264.

[38] P. Sharma, H. Kaur, M. Sharma, V. Sahore, A review on applicability of naturally available adsorbents for the removal of hazardous dyes from aqueous waste, Environ. Monit. Assess. 183 (1-4) (2011) 151–195.

[39] I. Ali, New generation adsorbents for water treatment, Chem. Rev. 112 (10) (2012) 5073–5091.

[40] T. Kamal, I. Ahmad, S.B. Khan, AM. Asiri, Bacterial cellulose as support for biopolymer stabilized catalytic cobalt nanoparticles, Int. J. Biol. Macromol. 135 (2019) 1162–1170.

[41] S. Hokkanen, A. Bhatnagar, M. Sillanpää, A review on modification methods to cellulose-based adsorbents to improve adsorption capacity, Water Res. 91 (2016) 156–173.

[42] D. Gago, R. Chagas, L.M. Ferreira, S. Velizarov, I. Coelhoso, A novel cellulose-based polymer for efficient removal of methylene blue, Membranes. 10 (1) (2020) 13.

[43] G.B. Kankılıç, AÜ. Metin, Phragmites australis as a new cellulose source: extraction, characterization and adsorption of methylene blue, J. Mol. Liq. 312 (2020) 113313.

[44] Y. Chen, Z. Xiang, D. Wang, J. Kang, H. Qi, Effective photocatalytic degradation and physical adsorption of methylene blue using cellulose/GO/TiO_2 hydrogels, RSC Adv. 10 (40) (2020) 23936–23943.

[45] L. Valencia, S. Kumar, E.M. Nomena, G. Salazar-Alvarez, AP. Mathew, In-situ growth of metal oxide nanoparticles on cellulose nanofibrils for dye removal and antimicrobial applications, ACS Appl. Nano Mater. 3 (7) (2020) 7172–7181.

[46] S. Lu, Z. Tang, W. Li, X. Ouyang, S. Cao, L. Chen, L. Huang, H. Wu, Y. Ni, Diallyl dimethyl ammonium chloride-grafted cellulose filter membrane via ATRP for selective removal of anionic dye, Cellulose. 25 (12) (2018) 7261–7275.

[47] N. Mohammed, A. Baidya, V. Murugesan, A.A. Kumar, M.A. Ganayee, J.S. Mohanty, K.C. Tam, T. Pradeep, Diffusion-controlled simultaneous sensing and scavenging of heavy metal ions in water using atomically precise cluster–cellulose nanocrystal composites, ACS Sustain. Chem. Eng. 4 (11) (2016) 6167–6176.

[48] Y. Wang, X. Zhang, X. He, W. Zhang, X. Zhang, C. Lu, In situ synthesis of MnO_2 coated cellulose nanofibers hybrid for effective removal of methylene blue, Carbohydr. Polym. 110 (2014) 302–308.

[49] L. Chen, R.M. Berry, KC. Tam, Synthesis of β-cyclodextrin-modified cellulose nanocrystals (CNCs)@ Fe_3O_4@ SiO_2 superparamagnetic nanorods, ACS Sustain. Chem. Eng. 2 (4) (2014) 951–958.

[50] C. Yang, L. Wang, Y. Yu, P. Wu, F. Wang, S. Liu, X. Luo, Highly efficient removal of amoxicillin from water by Mg-Al layered double hydroxide/cellulose nanocomposite beads synthesized through in-situ coprecipitation method, Int. J. Biol. Macromol. 149 (2020) 93–100.

[51] E.A. Mwafy, M.S. Hasanin, AM. Mostafa, Cadmium oxide/TEMPO-oxidized cellulose nanocomposites produced by pulsed laser ablation in liquid environment: synthesis, characterization, and antimicrobial activity, Opt. Laser Technol. 120 (2019) 105744.

[52] S. Saber-Samandari, S. Saber-Samandari, H. Joneidi-Yekta, M. Mohseni, Adsorption of anionic and cationic dyes from aqueous solution using gelatin-based magnetic nanocomposite beads comprising carboxylic acid functionalized carbon nanotube, Chem. Eng. J. 308 (2017) 1133–1144.

[53] M. Peter, N. Ganesh, N. Selvamurugan, S.V. Nair, T. Furuike, H. Tamura, R. Jayakumar, Preparation and characterization of chitosan–gelatin/nanohydroxyapatite composite scaffolds for tissue engineering applications, Carbohydr. Polym. 80 (3) (2010) 687–694.

[54] S. Saber-Samandari, S. Saber-Samandari, M. Gazi, Cellulose-graft-polyacrylamide/hydroxyapatite composite hydrogel with possible application in removal of Cu (II) ions, React. Funct. Polym. 73 (11) (2013) 1523–1530.

[55] M. Bilal, M. Asgher, H. Hu, X. Zhang, Kinetic characterization, thermo-stability and Reactive Red 195A dye detoxifying properties of manganese peroxidase-coupled gelatin hydrogel, Water Sci. Technol. 74 (8) (2016) 1809–1820.

[56] C. Acharya, C.R. Panda, P.K. Bhaskara, A. Sasmal, S. Shekhar, AK. Sen, Physicochemical and antimicrobial properties of sodium alginate/gelatin-based silver nanoformulations, Polym. Bull. 74 (3) (2017) 689–706.

[57] Y. Li, H. Meng, Y. Liu, A. Narkar, BP. Lee, Gelatin microgel incorporated poly (ethylene glycol)-based bioadhesive with enhanced adhesive property and bioactivity, ACS Appl. Mater. Interfaces. 8 (19) (2016) 11980–11989.

[58] E. Fosso-Kankeu, H. Mittal, F. Waanders, SS. Ray, Thermodynamic properties and adsorption behaviour of hydrogel nanocomposites for cadmium removal from mine effluents, J.Indus. Eng. Chem. 48 (2017) 151–161.

[59] S. Thakur, O. Arotiba, Synthesis, characterization and adsorption studies of an acrylic acid-grafted sodium alginate-based TiO_2 hydrogel nanocomposite, Adsorp. Sci. Technol. 36 (1-2) (2018) 458–477.

[60] T.S. Anirudhan, G.S. Lekshmi, F. Shainy, Synthesis and characterization of amidoxime modified chitosan/bentonite composite for the adsorptive removal and recovery of uranium from seawater, J. Coll. Interface Sci. 534 (2019) 248–261.

[61] T.S. Anirudhan, PS. Suchithra, Heavy metals uptake from aqueous solutions and industrial wastewaters by humic acid-immobilized polymer/bentonite composite: kinetics and equilibrium modeling, Chem. Eng. J. 156 (1) (2010) 146–156.

[62] E.I. Unuabonah, M.I. El-Khaiary, B.I. Olu-Owolabi, A. KO, Predicting the dynamics and performance of a polymer-clay based composite in a fixed bed system for the removal of lead (II) ion, Chem. Eng. Res. Des. 90 (8) (2012) 1105–1115.

[63] D.S. Dlamini, A.K. Mishra, BB. Mamba, ANN modeling in Pb (II) removal from water by clay-polymer composites fabricated via the melt-blending, J. Appl. Polym. Sci. 130 (6) (2013) 3894–3901.

[64] N. Bleiman, YG. Mishael, Selenium removal from drinking water by adsorption to chitosan–clay composites and oxides: batch and columns tests, J. Hazard. Mater. 183 (1-3) (2010) 590–595.

[65] D. Zadaka, S. Nir, A. Radian, YG. Mishael, Atrazine removal from water by polycation–clay composites: effect of dissolved organic matter and comparison to activated carbon, Water Res. 43 (3) (2009) 677–683.

[66] R. Ganigar, G. Rytwo, Y. Gonen, A. Radian, YG. Mishael, Polymer–clay nanocomposites for the removal of trichlorophenol and trinitrophenol from water, Appl. Clay Sci. 49 (3) (2010) 311–316.

[67] K. Ravikumar, J. Udayakumar, Preparation and characterisation of green clay-polymer nanocomposite for heavy metals removal, Chem. Ecol. 36 (3) (2020) 270–291.

[68] G.R. Mahdavinia, S. Mousanezhad, H. Hosseinzadeh, F. Darvishi, M. Sabzi, Magnetic hydrogel beads based on PVA/sodium alginate/laponite RD and studying their BSA adsorption, Carbohydr. Polym. 147 (2016) 379–391.

[69] W. Wang, Y. Kang, A. Wang, One-step fabrication in aqueous solution of a granular alginate-based hydrogel for fast and efficient removal of heavy metal ions, J. Polym. Res. 20 (3) (2013) 101.

[70] X. Yi, F. Sun, Z. Han, F. Han, J. He, M. Ou, J. Gu, X. Xu, Graphene oxide encapsulated polyvinyl alcohol/sodium alginate hydrogel microspheres for Cu (II) and U (VI) removal, Ecotoxicol. Environ. Saf. 158 (2018) 309–318.

[71] Y. Zhuang, F. Yu, H. Chen, J. Zheng, J. Ma, J. Chen, Alginate/graphene double-network nanocomposite hydrogel beads with low-swelling, enhanced mechanical properties, and enhanced adsorption capacity, J. Mater. Chem. A. 4 (28) (2016) 10885–10892.

[72] S. Thakur, S. Pandey, OA. Arotiba, Development of a sodium alginate-based organic/inorganic superabsorbent composite hydrogel for adsorption of methylene blue, Carbohydr. Polym. 153 (2016) 34–46.

[73] H. Ren, Z. Gao, D. Wu, J. Jiang, Y. Sun, C. Luo, Efficient Pb (II) removal using sodium alginate–carboxymethyl cellulose gel beads: preparation, characterization, and adsorption mechanism, Carbohydr. Polym. 137 (2016) 402–409.

[74] G.R. Mahdavinia, H. Aghaie, H. Sheykhloie, M.T. Vardini, H. Etemadi, Synthesis of CarAlg/MMt nanocomposite hydrogels and adsorption of cationic crystal violet, Carbohydr. Polym. 98 (1) (2013) 358–365.

[75] T. Lu, T. Xiang, X.L. Huang, C. Li, W.F. Zhao, Q. Zhang, CS. Zhao, Post-crosslinking towards stimuli-responsive sodium alginate beads for the removal of dye and heavy metals, Carbohydr. Polym. 133 (2015) 587–595.

[76] J. Liu, D. Su, J. Yao, Y. Huang, Z. Shao, X. Chen, Soy protein-based polyethylenimine hydrogel and its high selectivity for copper ion removal in wastewater treatment, J. Mater. Chem. A. 5 (8) (2017) 4163–4171.

Nanomaterials for packaging application

Ewelina Jamróz

DEPARTMENT OF CHEMISTRY, FACULTY OF FOOD TECHNOLOGY, UNIVERSITY OF AGRICULTURE, CRACOW, POLAND

Chapter outline

20.1 Introduction

The main task of packaging is to ensure appropriate physical and chemical conditions for food products. Recent work on packaging is aimed at achieving better quality and food safety. Nanotechnology in packaging could contribute to extending shelf-life as well as monitoring the safety and quality of food products. Moreover, the packaging material can implement the basic functions of containment and convenience: protection and preservation, marketing, and communication [1] (Fig. 20.1).

The use of packaging materials that are made of plastics, glass, and metals has negative impact on the environment. Huge amounts of packaging materials are landfilled and incinerated, which is a huge threat to our health. Moreover, dwindling oil reserves have forced the

Biopolymeric Nanomaterials: Fundamentals and Applications. DOI: https://doi.org/10.1016/B978-0-12-824364-0.00010-1

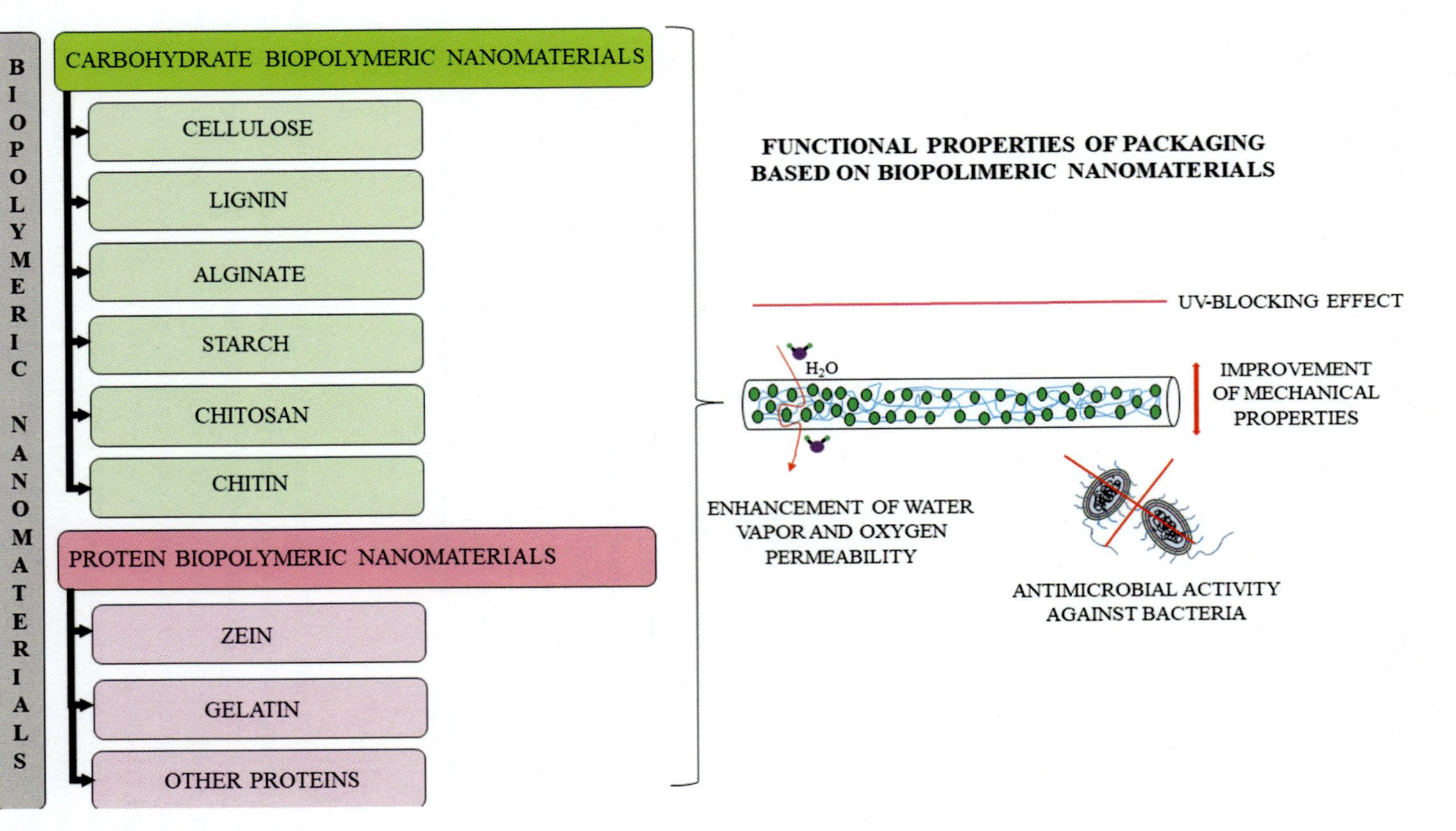

FIGURE 20.1 Basic functions of packaging materials.

economy to look for alternatives in resources of forest and agricultural origin. Particularly, in the area of packaging materials for food products, materials are sought that are safe for human health, environmentally friendly, and have an added value in the form of active and intelligent properties. Active packaging is a material that is enriched with additives with functional properties (e.g., essential oils, plant extracts, nanofillers). This type of packaging can release active ingredients into the package, Thus, maintaining quality and extending the shelf-life of the packed food. Active packaging includes oxygen absorption, carbon absorption/emission, moisture absorption, ethylene capture, and antioxidant and antimicrobial release systems [2, 3]. Intelligent packaging is a system that can monitor the condition of stored food and provide food quality information to consumers throughout the food supply chain. To date, several intelligent packaging techniques have been developed, including O_2 and CO_2, detectors, pH indicators, humidity and time-temperature sensors, and biosensors for pathogenic bacteria. Overall, these intelligent packaging systems are attached as labels, affixed to or printed on food packaging materials to monitor the quality of food products [4]. Currently, the greatest challenge for scientists is to design a packaging material that would extend shelf-life (active packaging) and provide information on the quality of stored food (intelligent packaging).

The group of raw materials that can be used in the production of packaging materials as an alternative to plastic ones includes biopolymers. Specifically, proteins (casein, whey, and gluten) and polysaccharides (e.g. starch and cellulose derivatives, chitosan and alginates) are of great interest. Among the biopolymers, starch and chitosan are the most frequently used natural ingredients for the production of packaging materials. Due to a number of functional properties (thermal and biological stability—antibacterial and antioxidant activity), chitosan seems very promising. However, it has a number of disadvantages, that is, low mechanical strength, high sensitivity to moisture, and limited barrier properties, which causes numerous limitations in industrial application, particularly including food packaging [5]. Starch-based packaging is characterized by very good organoleptic-, optical-, and gas-barrier properties; however, its mechanical strength parameters are low. The functional properties of starch-based packaging are influenced by many factors, such as starch type, storage conditions, preparation time and temperature, type and concentration of the plasticizer [6]. Protein-based packaging is characterized by better mechanical strength than that polysaccharide-based packaging. The problem with the commercialization of protein films is their too high sensitivity to moisture [7].

The solution to these limitations concerning the disadvantages of biopolymers is their mixing with other biopolymers, because of which the obtained material is not only improved, but also gains new properties. Moreover, nanotechnology is certainly a promising tool for improving the mechanical and barrier properties of biopolymer-based materials [8]. In addition, this may contribute to providing the packaging with additional functions, such as antibacterial and antioxidant activity. Nanomaterials have been incorporated into protein- and/or polysaccharide-based packaging to improve flexibility, strength, low volatility, gas barrier, moisture stability, and temperature. In addition to application related to the packaging of food materials, packaging materials enriched with nanomaterials are also intended for monitoring stored products in terms of safety and quality, as well as improving packaging biodegradability [7].

Packaging materials enriched with nanofillers are characterized by improved efficiency, increased strength, and barrier properties, but also many other advantages (e.g., antibacterial

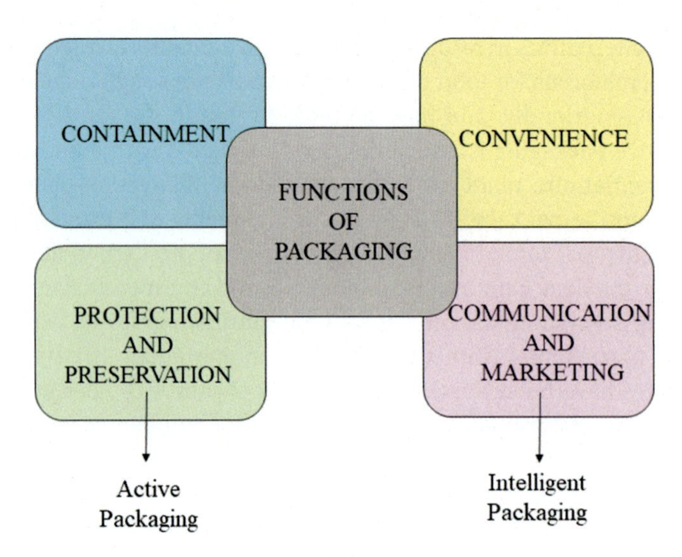

FIGURE 20.2 Types of biopolymeric materials and their effects on functional properties of packaging materials.

activity). Moreover, this type of packaging has "smart" properties that can prevent or react to spoilage of food products (e.g., indicators that change colors at different stages of product deterioration) [9]. In this chapter, all biopolymeric nanomaterials are discussed that enrich packaging materials, in particular biopolymer films, thus affecting their functional properties.

20.2 Immobilization of biopolymeric nanomaterials for packaging materials

Biopolymer nanomaterials can be divided into nanomaterials based on polysaccharides and proteins. The advantages of this type of biomaterials include low immunogenicity, biodegradability, biocompatibility, and antibacterial activity. The group of carbohydrate or polysaccharide nanomaterials includes alginate, starch, chitosan, and cellulose; while the group of protein nanomaterials includes collagen, gelatin, albumin, and silk fibroin [10]. Biopolymer nanomaterials are often added to packaging materials to give them specific properties (Fig. 20.2).

In this section, the influence of biopolymeric nanomaterials on the functional properties of packaging materials is discussed, in particular, those based on biopolymers. It can be seen that there have been more attempts to use biopolymeric carbohydrate nanomaterials as nanofillers in packaging materials. This is probably due to the fact that both proteins and their biopolymeric protein nanomaterials have worse functional properties, that is, higher solubility and poor parameters of mechanical properties. In Table 20.1, the influence of biopolymeric nanomaterials on the functional properties of packaging materials based on biopolymers is presented.

20.2.1 Cellulose-based nanomaterials

The discovery of cellulose-based nanomaterials is seen as a concrete achievement in the scientific field of material design. Nanocellulose is one of the groups of nanomaterials that is

Table 20.1 Effects of biopolymeric nanomaterials on the functional properties of packaging materials based on biopolymers.

Type of biopolymer matrix	Type of nanomaterials	Functional properties of packaging	Reference
Cellulose-based nanomaterials			
Gelatin	CNC (10, 20, and 30 wt%)	-Improvement in TS by up to ~70%; in YM by up to ~43%	[34]
Gelatin	BCNC (1, 2, 3, 4, and 5 wt%)	-Improvement in TS by up to ~30%; in YM by up to ~7%; in stiffness by up to ~13%; in WVP by up to 15%	[35]
Starch	CNC (1, 2, 5, 10, and 15 wt%)	-Improvement in WVP by up to ~43%; in TS by up to ~41%; in YM by up to ~83%	[36]
Cassava starch	Carboxymethyl cellulose nanocrystals (N-CMCs)	-Improvement in TS by up to ~554%; in EAB by up to ~41%; in solubility by up to ~123% -Reduction in WVP by up to ~42.7% and in moisture absorption by up to ~15.9%	[37]
Sago starch	Carboxymethyl cellulose nanoparticles with (1, 2, 3, 4 and 5 wt%)	-Improvement in TS by up to ~31%	[38]
Lignin-based nanomaterials			
Wheat gluten	Lignin nanoparticles (0,1, and 3 wt%)	-Improvement in YM by up to ~207%; in TS by up to ~142%; in moisture content (MC) by up to ~20% -Improvement in thermal stability	[30]
Starch-based nanomaterials			
Corn starch	Starch nanoparticles (2 and 5 wt%)	-Improvement in TS by up to ~86%; in YM by up to ~145% -Reduction in WVP by up to ~31% and in OTR by up to 79% -Decrease in EAB by up to ~16%	[39]
Chia mucilage	Starch nanocrystals (3 and 6%)	-Improvement in thermal stability -Reduction in TS by up to ~8% and in EAB by up to ~13% -Improvement in YM by up to ~38% -Antimicrobial activity against *E. coli, S. aureus, S. typhimurium, S. mutans, B. thuringiensis, P. aeruginosa*	[40]
Corn starch	Spray-dried starch nanoparticles (SDSN) (0.5 wt.%) vacuum freeze–dried starch nanoparticles (VFDSN) (0.5 wt%)	-Reduction in WVP by up to ~42% (SDSN) and ~44% (VFDSN)	[41]
Zein-based nanomaterials			
Whey protein isolate (WPI)	Zein nanoparticles (0.2, 0.4, 0.8, and 1.2 (w/w of WPI)	-Reduction in WVP by up to ~84% -Improvement in TS by up to ~300%	[42]

(*continued on next page*)

Table 20.1 *(continued)*

Type of biopolymer matrix	Type of nanomaterials	Functional properties of packaging	Reference
Chitin-based nanomaterials			
Gelatin	Chitin nanoparticles (3,5 and 10 wt%)	-Improvement in thermal stability -Reduction in WVP by up to ~15%; in solubility by up to ~10% -Improvement in TS by up to ~83 %; in EAB by up to ~41%; in YM by up to ~92% -Antifungal activity against *Aspergillus niger*	[43]
Carboxymethyl cellulose	Chitin nanocrystals (0, 1, 5, and 10 wt%)	-Improvement in TS by up to ~88%; in EM by up to ~243% -Reduction in WVP by up to ~7%	[44]
Starch	Chitin nanoparticles (1, 2, 3, 4 or 5 wt%)	-Improvement in TS by up to ~174% -Decrease in EAB by up to ~67%	[45]
Maize starch	Chitin nanowhiskers (0.5, 1.0, 2.0, and 5.0 wt%)	-Improvement in TS by up to ~125% -Reduction in WVP by up to ~58% -Improvement in thermal stability -Antimicrobial activity against *L. monocytogenes* and *E. coli*	[46]
Chitosan-based nanomaterials			
Gelatin	Chitosan nanoparticles (2, 4, 6 and 8 wt%)	-Improvement in TS by up to ~52%; in EM by up to ~63% -Reduction in water solubility (WS) by up to ~11% and in WVP by up to ~50%	[47]
Carboxymethyl cellulose	Chitosan nanoparticles (0.2, 0.5, and 0.8 wt%)	-Reduction in WS by up to ~3%; in EM by up to ~22%; in EAB by up to ~5% -Improvement in TS by up to ~540% -Reduction in WVP by up to ~42%	[48]
Low-methyl pectins (LDM) high-methyl pectins (HDM)	Chitosan nanoparticles (3 wt%)	-Improvement in TS by up to ~52% (HDM films) and ~125% (LDM films); in EAB by up to ~24% (HDM films) and ~210% -Increase in WVP by up to ~35% (HDM films) and ~25% (LDM films)	[49]

Abbreviations: TS, tensile strength; EAB, elongation at break; EM, elastic modulus; YM, Young's modulus; WVP, water vapor permeability; OTR, oxygen transition rate.

obtained from the most significant source of biological material, given its predominant physical, mechanical, nontoxic, inexhaustible, biodegradable, recyclable, and low-cost properties. These novel properties of nanocellulose increase its range of applications in many areas, including biomedicine, films and packaging, water purification, composites, energy, and hydrogels [11].

Recently, an increase of interest in nanocellulose materials can be observed, due to their very good influence on mechanical and barrier properties for packaging materials. Nanocrystalline cellulose or cellulose nanowhiskers are pure nano-sized crystalline cellulose, and they are obtained by enzymatic hydrolysis or under moderate acid hydrolysis conditions by eliminating amorphous areas. Nanocrystalline cellulose is one of the strongest and nonelastic materials of natural origin. Important properties of nanocrystalline cellulose include high tensile strength (TS) and aspect ratio, very good optical and electrical properties, high stiffness, and a large surface area [1]. It is worth noting that the use of cellulose nanofibers may reduce the production costs of packaging materials because they are widely available. Moreover, packaging materials using this raw material are safe for the environment because they are reusable and recyclable [12].

Nanofibrillated cellulose, which has cellulose fibers embedded in a polymer matrix, can improve the mechanical properties of packaging materials by improving stiffness, breaking strength and flexibility. Another feature is its very good oxygen barrier, as the dense network of cellulose fibers makes it difficult for gas molecules to penetrate [13]. The use of nanocellulose in the production of packaging materials is a very interesting solution, as it ensures the production of packaging with improved properties that are environmentally friendly. Chi and Catchmark [14] used nanocellulose to strengthen chitosan/carboxymethyl cellulose film. The developed films showed improved mechanical and barrier properties, which is attributed to the homogeneous distribution and good compatibility of nanocellulose in the polysaccharide matrix. In the research, it was shown that the chitosan/carboxymethyl cellulose coatings with the addition of nanocellulose (<5 wt%), were a very good barrier against fats, oils, and water for a cardboard substrate. A designed barrier material that is environmentally friendly has high potential for packaging applications. Cellulose nanocrystals (CNC) were extracted from flax fibers by acid hydrolysis and enriched with this type of nanofiller in chitosan films. As the concentration of CNC increases (5 to 30%), the values of TS (by up to ~ 24%) and Young's modulus (YM) (by up to ~ 141%) also increase; however, no significant improvement was noted in for the water vapor permeability (WVP) parameter, which could be related to the overall lower crystallinity of the film [15]. Hänninen et al. [16] prepared films based on chitosan, nanocellulose, and their mixtures. Interestingly, chitosan/nanocellulose films had the lowest values of TS. The authors concluded that the cause of such behavior is the process of producing the films. Drying the chitosan-nanocellulose film likely led to internal stress in the film-forming matrix that may have contributed to their brittle behavior. The mechanical and water vapor barrier properties of the chitosan films have been improved by adding cellulose nanofibers. The nanocomposite film with 15% cellulose nanofibers and plasticized with 18% glycerin was comparable to synthetic polymers in terms of strength and stiffness, but their flexibility parameters were worse, which indicates that the film can only be used in applications where high flexibility and/or a water vapor barrier are not necessary [17]. CNC containing aldehyde groups were prepared by periodate oxidation and used to strengthen gelatin-based biocomposites. The aldehyde group of the nanocellulose obtained in this way formed a covalent bond with the amino group of the gelatin, which showed a synergistic effect and led to significant improvement in the mechanical properties and water stability of the obtained films [18]. The addition of CNC to the bacterial cellulose (BC) film produced from cashew juice caused increased TS but decreased WVP [19].

To develop a biodegradable and renewable active packaging material, Maliha et al. [20] created a method of producing nanocellulose sheets enriched with the nontoxic complex, phenyl bismuth bis (diphenylphosphinato) on an industrial scale using a spray system. The composites had good water vapor barrier properties, and the bismuth complex did not have any significant effect on the performance of this barrier. The composite sheets were able to inhibit the growth of bacteria (*Staphylococcus aureus*, Vancomycin-resistant *enterococcus,* and Methicillin-resistant *S. aureus, Escherichia coli* and *Pseudomonas aeruginosa*) and fungi (*Candida albicans* and *Candida glabrata*). An environmentally friendly composite paper with an antimicrobial effect was obtained which consisted of a poorly soluble bismuth complex dispersed in a nanocellulose matrix, indicating a high potential for use as an active packaging material. Currently, a solution is being sought to improve the properties of cardboard packaging. For this purpose, Bideau et al. [21] applied (2,2,6,6-Tetramethylpiperidin-1-yl) oxyl (TEMPO) oxidized cellulose nanofibers (TOCN) and polypyrrole coating on such cardboard. The mechanical properties and the reduced gas permeability of the coated cardboard were greatly improved due to the dense network formed by TOCN and polypyrrole particles. Moreover, the biodegradable nature could also reduce the amount of packaging waste generated by plastics, as it is possible to separate the cardboard from the coating for recycling. Zulfiana et al. [22] designed an antimicrobial paper from *Imperata cylindrica*-coated anionic nanocellulose cross-linked cationic to create a system with the ability to actively control microbe growth in the packaging materials. The coated paper was antimicrobial against *E. coli,* Salmonella *typhi, S. aureus,* and *Bacillus subtilis*. On the other hand, the treatment with Al^{3+} cations improved the tear strength parameter, which was 38.53 gf. The preparation of nano- and micropaper as a packaging material was undertaken. For this purpose, nanofibers were prepared using milled cellulose nanofibers (GC) from rapeseed straw and BC nanofibers, as well as micropaper from as-bleached cellulose microfibers of rapeseed straw, which served as a control sample to evaluate the impact of grinding on the properties of nanopaper. The GC nanopaper had lower thermal stability than both the micro- and BC nanopaper. GC and BC nanopapers showed higher parameters of barrier properties than micropaper, which was permeable to air. Grinding had a strong influence on the mechanical properties. GC and BC nanopapers had very good TS parameters (114 MPa and 185 MPa, respectively) and YM (13.6 GPa and 17.3 GPa, respectively), which indicates an average 11-fold increase in relation to micropaper. The authors believe that GC and BC nanopapers are very durable, fully biological. and biodegradable and can act as multifunctional materials [23].

On a laboratory scale, nanocellulose films are produced by casting or vacuum filtration [24]. Nanocellulose films are considered a prospective alternative to synthetic packaging materials; however their limited moisture barrier properties are considered a huge limitation in their commercialization. To improve barrier-related properties, carboxymethyl cellulose was added to the nanocellulose films. Noteworthy is the fact that the obtained films have improved moisture barrier properties (by up to $\sim$92%) compared to nanocellulose films, which makes them packaging materials comparable to polyethylene terephthalate, polycarbonates (PC), etc. Considering the environmental aspect, nanocellulose/carboxymethyl cellulose (NC/CMC) films generally have low environmental impact due to their ease of recycling [25]. Chen et al. [26] prepared films based on nanocellulose from corn husks, then enriched them with silica using two methods: surface coating and internal grafting. The silica nanocellulose films obtained by the surface coating method contained less nanosilica and had better properties, including higher transparency and surface roughness, as well as excellent hydrophobic and antifouling properties than silica-enriched nanocomposite films via the internal grafting method.

20.2.2 Lignin-based nanomaterials

From lignin, the second most abundant biopolymer after cellulose, lignin nanoparticles (LNP) can be obtained, which, thanks to their unique properties, can influence the functional properties of packaging materials. The use of lignin in the form of nanoparticles (NPs) eliminates most of the drawbacks of using lignin (heterogeneity and low solubility) [27]. LNPs can be easily surface modified due to the availability of a large number of functional groups (e.g., thiols, aliphatic hydroxyl, phenolic groups), because of which their application range can be increased [28]. The use of LNP as a reinforcing agent in packaging materials improves the mechanical, thermal, and biocompatibility properties of the formed nanocomposites. Yang et al. [29] developed two- and three-component films based on chitosan, poly(vinyl alcohol) (PVA), and LNPs (1 and 3 wt%). The enrichment of the film with LNP increased the TS and YM parameters, causing a toughness effect in the chitosan matrix. The very good dispersion of LNP in the matrix not only increased the crystallinity value, but also significantly improved the thermal stability of the obtained nanocomposites. Additionally, while testing the antioxidant activity (measured by the (2,2-diphenyl-1-picryl-hydrazyl-hydrate) DPPH free radical method), a synergistic effect of LNP and chitosan was noticed in the antioxidant properties of the nanocomposite. Moreover, in wheat gluten films enriched with LNP, even distribution of NPs in the biopolymer matrix was noted, which had a positive effect on the parameters of mechanical strength and thermal stability [30]. LNPs show barrier properties against UV radiation. Compared to raw lignin, the addition of LNPs in PVA film can increase UV protection efficiency by 13.3% in the ultraviolet spectrum (250 nm) [31]. Similar conclusions were reached by Xiong et al. (2018), who developed nanocomposite films based on PVA and lignin nanospheres. The obtained nanocomposites were characterized by very good absorption of UV radiation and transparency, which makes them an interesting candidate as a material for medicine bottles and food packaging [32].

The type of lignin used is also important in the preparation of nanocomposite packaging materials. Nevarez et al. [33] used three types of LNPs: organosolv, hydrolytic, and kraft, with or without acetylation, as a strengthening agent for a polymer matrix based on cellulose triacetate. The acetylation process influenced the increase in molecular weight, greater dispersion of particles in the polymer, as well as the replacement of hydroxyl groups with a nonpolar structure which, in turn, influenced the morphology and mechanical properties of nanocomposites. Sulfate lignin, due to its hydrophilic nature, did not mix well with the polymer. The acetylation process contributed to the good dispersion of sulfate lignin particles. However, the acetylation process had no effect on the mechanical properties of the matrix based on cellulose triacetate and the sulfate LNPs. The organosolv lignin acetylation improved the mechanical properties of the composite only at the highest relative humidity (70%) and at a temperature of 45 °C. The highest acetylation was achieved for the hydrolytic lignin, with significant changes in the lignin particle size, molecular weight, and particle dispersion during membrane formation. Mechanical properties improved when only hydrolytic lignin was used in the nanocomposite.

20.2.3 Alginate-based nanomaterials

Alginate is nontoxic, biodegradable, inexpensive, and readily available. Additionally, it is a mucoadhesive, biocompatible, and nonimmunogenic substance [50]. Given the area of packaging research, some of the materials currently in use are undesirable due to their nondegradable nature, causing environmental problems. Alginate is an environmentally friendly ingredient that can be recycled and degraded. Alginate nanomaterials are a widely and rapidly developing field [51].

Mokhena and Luyt [52] obtained alginate nanofibers that were prepared by electrospinning of alginate with synthetic electrospun polyethylene oxide and ionically cross-linked with calcium chloride, and then, chemically cross-linked with glutaraldehyde. Next, a high-flow three-layer composite membrane was developed, consisting of a coating layer based on chitosan and chitosan with silver NPs, electrospun alginate nanofibers as an intermediate layer, and a nonwoven fabric as a mechanical support substrate. The obtained membranes with microbiological activity are a promising material for the removal of dyes and the separation of oils.

Amjadi et al. [53] developed nanofibers based on zein/sodium alginates with the addition of titanium dioxide (TiO_2 NPs) and betanin NPs using the electrospinning technique for food packaging applications. Compatibility between zein, alginate, and active substances was observed, which translated into very good mechanical properties and hydrophobicity of the surface. The obtained nanocomposite with antimicrobial and antioxidant activity is a good step in the direction of durability and quality of food products.

20.2.4 Chitin-based nanomaterials

Materials based on nanofibers are gaining more and more significance in packaging. Dresvyanina et al. [54] assessed the interaction between water and a chitosan-based nanocomposite and chitin nanofibrils. It has been shown that the introduction of chitin nanofibrils into the chitosan matrix leads to the formation of ordered structures consisting of chitosan macromolecules on the surface of chitin nanofibrils, which reduces the sorption capacity of composite films. Moreover, the highest affinity between chitosan and chitin nanofibrils occurs when the concentration of chitin nanofibrils ranges from 1 to 5 wt%. Similar conclusions were reached by Sahraee et al. [43] who enriched gelatin films with chitin nanofibers. Although both components were hydrophilic, the addition of nanochitin to the gelatin films reduced the film's tendency to absorb water. Chitin nanocrystals were isolated from shrimp shell powder using acid hydrolysis and ammonium persulfate methods. The addition of chitin nanocrystals to the carboxymethylcellulose film improved the mechanical properties (TS by up to ~20% and elastic modulus (EM) by up to ~ 59%) and reduced the WVP of the CMC film by 27% [55].

The type, content, shape, and size of chitin nanoobjects play an important role in the end properties of packaging materials. Salaberria et al. [56] compared the effects of different nanochitin morphologies (nanocrystals (CHNC) and nanofibers (CHNF) at concentrations of 5%–20%) on the structural and functional properties of cast-based, starch-based thermoplastic films. Both types of films had better functional properties than thermoplastic starch films.

However, films with the addition of CHNF were characterized by the best parameters of mechanical and barrier properties, as well antifungal activity, which can be attributed to the web-like morphology of the CHNF. Chitin nanowhiskers were also used to coat the paper sheets via the dip-coating method. It was noticed that the coated material was characterized by an increase in the parameters of elongation at break and TS in wet and dry conditions [57].

20.2.5 Chitosan-based nanomaterials

Chitosan NPs, obtained from natural material, are bioactive and environmentally friendly. Their use as a stabilizing agent in packaging materials has a positive effect on the functional properties of nanocomposites.

Ju et al. [58] evaluated how chitosan and chitosan NPs affect the functional properties of BC/poly(vinyl alcohol) film. The obtained results indicated that chitosan NPs improve water barrier properties. This behavior can be attributed to the dense structure of the NPs, which hinders the diffusion of water, thereby extending its path in the film, and consequently, causing a decrease in water and WVP solubility. Higher antibacterial activity of the film with the addition of NPs against *E. coli* and *S. aureus* was observed. Compared to chitosan, their NPs have a higher surface charge density, which can lead to more interactions with bacterial cells. Additionally, films with the addition of chitosan NPs showed better barrier properties against UV radiation.

Due to the incorporation of chitosan NPs into the gelatin matrix, an increase in the thermal stability of nanocomposites was observed. The use of chitosan nanoparticles (CNP) synthesized by the ion cross-linking method resulted in a slight change regarding the crystal structure of the nanocomposite, and also influenced the segmental mobility of gelatin chains [59].

Temperature and relative humidity have great impact on water sorption and mechanical strength of packaging materials, and Thus, contribute to increasing the stability and shelf-life of packaging. Othman et al. [60] observed that films based on starch and chitosan NPs should be stored at low relative humidity and high temperature, such conditions reduce water sorption and improve the mechanical strength of nanocomposites. The authors also drew attention to the fact that the storage temperature should not be extremely high, as at 40 °C, the chitosan NPs become unstable which, in turn, translates into deterioration of the functional properties of starch-based packaging materials.

20.2.6 Starch-based nanomaterials

Compared to native starch, starch NPs (SNPs) have a higher surface area to weight ratio. Currently, great development in the field of the use of nanotechnology in the preparation of packaging materials has contributed to the use of SNPs, due to their impact on improving mechanical properties and biodegradability.

SNPs were obtained by acid hydrolysis of starch extracted from water chestnut (*Trapa bispinosa*). The inclusion of SNPs in the native starch-based film improved TS values while decreasing the water vapor transmission rate and solubility values, which is the basic feature of good packaging [61,62] adopted a new strategy to obtain covalently cross-linked SNPs using boron ester bonds formed between debranched starch and borax. Compared to the pure starch film, the TS of the starch film with 10% SNPs increased by about 45%, and elongation at break of the starch films with 5% SNPs increased by about 20%. Kristo and Biliaderis [63] reached similar conclusions when they enriched pullulan films with waxy maize starch nanocrystal (SNC). The relative enhancing effect of SNCs was related to the key role of the strong interfacial bonding between the starch nanofiller and the pullulan matrix.

Banana flour films have been enriched with NPs of banana starch and montmorillonite. The best strengthening (increase by 11% in TS) and water barrier properties (increase by 41%) were found in films with 5% montmorillonite content and NPs of banana starch [64].

Starch was extracted from mango seeds (with a relatively high yield of approx. 38.5%), which was used to obtain a bionanocomposite and was further enriched with SNCs. Compared to corn flour films, nanocomposite films with a 5% content of SNCs were characterized by higher TS parameters (by 90%), EM (by 120%), and reduced WVP parameters (by 15%) [65].

Nanocomposite films were obtained based on phosphated or methylated pumpkin flour (*Cucurbita maxima*) with or without the addition of "huesito" plum flour (*Spondias purpurea*). The films prepared from methylated flour had lower sensitivity to water, greater thermal stability, and a higher percentage of crystallinity than the phosphated flour films. However, these materials were very brittle and also ecotoxic, which limited their compostability. Thus, while both phosphated and methylated pumpkin flour based films were biodegradable, only phosphated films could be considered environmentally friendly. In addition, the intelligent properties of the films with the addition of plum flour were evaluated. Unfortunately, the tested films were not sensitive to changes in pH value [66].

20.2.7 Zein-based nanomaterials

Zein is a hydrophobic protein that has high propensity to form nanofibers. The obtained zein nanofibers showed much better thermal stability than pure zein, therefore, they were used as nanofillers for gelatin films. Karim et al. [67] developed films based on gelatin and zein nanofibers, which were characterized by reduced moisture content, water solubility, swelling percentage, vapor permeability, elongation at the break, and transparency. Zein is insoluble in water, Thus, zein films and coatings are usually prepared by spraying or pouring its aqueous ethanol solutions onto the contact surface. Spasojević et al. [68] prepared films from NP zein through antisolvent precipitation from 90% v/v aqueous ethanol zein solutions. The films based on zein NPs were characterized by granular structure and surface roughness. There were no differences in the parameters of mechanical properties between films based on zein NPs and those from zein. Under the influence of a neutral pH environment and too high concentration of zein NPs, the films based on methylcellulose are weakened. For this purpose, the influence of stabilizing additives and the drying temperature on the physical properties of methylcellulose-based films with the addition of high concentrations of zein NPs were tested. Among all the stabilizers used (poly (ethylene glycol), oleic acid, carrageenan, and Tween 20), the addition of carrageenan and the Tween 20 surfactant improved TS parameters, elongation at break, and WVP of the films prepared at a low temperature. Additionally, films with these types of stabilizers had the best dispersion of zein nanoparticles (ZNP) in films. The addition of carrageenan and Tween 20 significantly improved the functional parameters of the films based on methylcellulose and ZNP, while making this type of active packaging material potentially applicable in many areas [69].

The action of zein NPs in packaging materials can be enhanced by other types of nanofillers. Huang et al. [70] enriched chitosan films with the addition of hydrophilic palygorskite nanoclay (PAL) and hydrophobic zein particles (ZP). The homogeneous dispersion of the combined nanofillers in the chitosan matrix resulted in a significant increase in TS, water resistance, and surface hydrophobicity. The obtained results confirmed the synergistic effect of hydrophilic PAL and hydrophobic ZP. Farajpour et al. [71] developed films based on potato starch, glycerol, olive oil, and zein-dispersed NPs. The addition of ZNP significantly improved water vapor barrier properties and mechanical endurance. Moreover, a dynamic mechanical test showed that the parameters of the storage module increase along with the concentration of ZNP.

20.2.8 Gelatin-based nanomaterials

Zhang et al. [72] developed the bovine serum albumin (BSA) encapsulation method in gelatin NPs, then the obtained NPs were placed in a biopolymer film with a thickness around 150 μm composed of poly (2-hydroxyethyl methacrylate) (pHEMA) through photopolymerization. The obtained nanocomposite was characterized by a very good BSA release profile from the matrix. Research results indicate that gelatin NPs can encapsulate water-soluble proteins, while the pHEMA film contributes to the extension of the protein release profile. Thus, the obtained nanocomposite can be an alternative material consisting of a medical implant for protein therapy, and is also suitable for evaluation in the field of active packaging materials.

Gelatin/chitosan nanofibers were produced using the nozzle-less electrospinning technique. Then, the obtained nanomaterials were used for the production of edible gluten films. The resulting packaging material had lower WVP and very good mechanical properties, while the maximum force required for the tensile test of the film containing nanofibers was five times greater than the control sample [73].

20.2.9 Other protein-based nanomaterials

Li et al. [74] enriched soy protein and corn starch films into peanut protein NPs (PNPs). The addition of PNP at the level of ~4% significantly improved TS, water vapor barrier, and thermal stability of both the protein and starch films. The PNP-enriched protein films demonstrated a more homogeneous structure than the starch films, which may indicate better compatibility between the protein matrix and PNPs.

20.2.10 Packaging materials with a few types of biopolymeric nanomaterials

In the literature on the subject, scientific works can be found that focused on the influence of two types of biopolymeric nanomaterials regarding the functional properties of packaging materials. Jannatyha et al. [75] found how chitosan or nanocellulose NPs affect the functional properties of films based on carboxymethyl cellulose. The best effect was found in nanocomposites containing chitosan NPs. This type of nanomaterial caused a significant reduction in the parameters of water solubility, moisture content, and absorption. Moreover, chitosan NPs improved thermal stability and mechanical strength. Nanocomposites containing chitosan NPs showed antibacterial activity against *P. aeruginosa, Salmonella enteritidis, E. coli, Bacillus cereus,* and *S. aureus,* while films based on carboxymethyl cellulose enriched with nanocellulose did not have an inhibitory effect. CNC and SNPs were obtained, which were then wet-coated on polylactic acid (PLA) as well as paper substrates, and incorporated into starch films. Reduced oxygen permeability was achieved thanks to cellulose nanocrystal coatings and SNPs on PLA. The paper barrier improvement was not successful due to holes in the substrate. There was no improvement in WVP parameters or mechanical properties of cast starch layers containing CNC or SNP. Based on the results, the authors found that the agglomeration of NPs during the preparation of the cast film and the separation of the nanofiller from the matrix during drying negatively affects technical and functional properties. It is suggested to further purify the hydrolyzed biopolymeric nanomaterials for ionic residues and agglomerates to improve the technical and functional properties of nanocomposite coatings and cast films [76]. Currently, an interesting solution is to combine various biopolymeric and inorganic nanofillers in a biopolymer matrix. BC pellicles were synthesized through *Gluconacetobacter xylinus* in molasses, which were then subjected to acid hydrolysis to obtain CNC. Chitosan films were enriched with isolated bacterial CNC (BCNC) and silver NPs. Evenly distributed nanofillers (BCNC 20–30 nm and AgNPs 35–50 nm) improved solubility (by up to ~26%), WVP (by up to ~45%), and mechanical properties (by up to ~26%) is ~ 104% [77]. Hybrid NPs based on chitin nanowhiskers and ZnO-AgNPs were obtained and then added to the CNC film. Nanocomposite film based on CMC containing 5 wt% ChNW/ZnO-Ag NPs was homogeneous and showed a high barrier to UV radiation. The TS and EM parameters of the packaging material increased by 18%–32% and 55%–100%, respectively, while the elongation at break (EAB) decreased by 23%–33% [78]. Also, the combination of nanoadditives in the biopolymer matrix can cause their synergistic effect. Amjadi et al. [79] developed films based on gelatin, chitosan nanofiber, and ZnO NPs. The obtained results showed that the synergistic effect between ZnO NPs and chitosan nanofibers contributed to the improvement of the nanocomposite antimicrobial activity (against *E. coli, P. aeruginosa,* and *S. aureus*). Packaging based on biopolymers can also be enriched with several types of nanomaterials simultaneously. Silva et al. [80] extracted starch and SNC from seeds and nanocrystals of cellulose (CNC) from mango seed shells. The obtained starch-based films had different contents of SNCs and CNC. The optimized content of nanofillers (1.5 wt% CNC and 8.5 wt% SNC on a starch basis) contributed to improvement of mechanical strength and water barrier properties.

20.3 Bionanocomposites and their packaging applications in the food industry

Although the addition of nanofillers to packaging improves the mechanical properties of these materials, there is still a well-founded concern among consumers about the effects of consuming nanomaterials. It is very important to control the migration of nanofillers to food products, as well as to determine toxicity by understanding the dynamics of action in the human body. Moreover, it is very important to pay attention to the biodegradation process of nanomaterials, as well as their impact on the environment [1].

Nanotechnology is now expected to have a positive impact on many areas of food science. Great attention is paid to extending the shelf-life of food as well as protection against foodborne diseases. That is why it is so important to develop packaging materials with antioxidant and/or antimicrobial properties based on biopolymeric nanomaterials. There is also an on-going effort to design materials that will provide information about the conditions of stored food [10]. In this section, packaging materials based on biopolymeric nanomaterials that have active and intelligent properties will be discussed.

20.3.1 Active packaging

Extending the shelf-life of food products is of great importance for the food industry. The mission of food packaging materials is to maintain the quality and properties of food products as long as possible, and thus, extend their shelf-life. BC, which consists of nanofibers with very good film-forming properties, has no antibacterial or antioxidant effects, which limits its use in the field of active packaging. Tsai et al. [81] developed NPs based on silymarin and zein, in which they enriched the films with BC. Due to the formation of the NP/nanofiber complex, improved wettability and swelling properties of nanocomposite films were observed. The zein–silymarin NPs increased the release of sparingly soluble silymarin, which positively influenced antibacterial and antioxidant activity. The obtained active packaging materials assumed a protective function against lipid degradation and oxidation during salmon storage in refrigerated conditions.

Shapi'i et al. [82] obtained nanochromic chitosan, which they then placed in a starch film (0%–20% w/w CNP). The obtained nanocomposites served as active packaging materials for cherry tomatoes. The obtained results indicated that the nanocomposite film with a 20% CNP content was the most effective in protecting these vegetables by inhibiting the growth of microorganisms (7 × 102 CFU/g) compared to the film without the addition of CNP (2.15 × 103 CFU/g) [82]. Correa-Pacheco et al. [83] developed packaging materials based on chitosan, chitosan NPs, and propolis ethanol extract. The addition of active ingredients to the chitosan matrix decreased the solubility value and the degree of swelling of the bionanocomposites. The obtained bionanocomposites were characterized by very good antimicrobial activity against *E. coli, Listeria monocytogenes*, and *S. enteritidis*, while also acting as an active packaging material for strawberries. In Table 20.2, the effect of active materials based biopolymeric nanomaterials on the quality of food storage is presented.

20.3.2 Intelligent packaging

Intelligent indicators based on biopolymeric nanomaterials are gaining more and more popularity as they become an interesting material that has added value in the form of the ability to monitor, for example, the freshness of food products. Lu et al. [93] designed a hydrogel based

Table 20.2 Examples of packaging based on biopolymeric nanomaterials.

Type of biopolymeric nanomaterials	Tested model food	Influence on food	Reference
Chitosan films with nanocellulose	Ground meat	After 6 days of meat storage, the lactic acid bacteria population in the meat packed in nanocomposite films was reduced by 1.3 (at 3 °C) and 3.1 (at 25 °C) log cycles, respectively, compared to samples packed in nylon	[84]
Carboxymethyl cellulose films with cellulose nanocrystals	Red chilies	During 7 days of storage, the nanocomposite film effectively reduced loss in mass and kept the vitamin C content at a constant level	[85]
CMC/cellulose nanocrystals/AgNP coated paper	Strawberries	-Extending the shelf-life of strawberries to 7 days -Reduction of total aerobic bacteria counts by 4.11 logCFU/g -Reducing the rate of mass loss over the storage period -The vitamin C content was still high after 7 days of storage -A slow downward trend in titratable acids (TA) content (index related to strawberry flavor) was observed. The coated paper had an air barrier creating a high CO_2 environment which slowed down the respiration of the strawberries during storage and Thus, reduced the consumption of TA as a respiratory substrate -Delaying the degradation of a soluble solid and maintaining a higher total soluble solid (TSS) level (quality rating index and affects the sweet taste of strawberries), thus extending the shelf-life of strawberries -On the fifth day of storage, unpackaged strawberries in the control group showed mound spots, signs of shrinkage, and dark color -The strawberries in the coated paper still were bright red in color, and were free from mound spots until the seventh day of storage	[86]
Zein films with pomegranate peel extract (PE) encapsulated in chitosan nanoparticles (CSNPs)	Pork	-The nanocomposite film showed a slower PE release rate and effectively limited the growth of *L. monocytogenes*, compared to the control film	[87]
PVA/cellulose nanocrystals/chitosan nanoparticles	Mango fruits	-After 20 days of storage, physiological mass losses of fruit in packaging from cellulose nanocrystals and chitosan nanoparticles were observed, which amounted to 31% and 42%, respectively -Black spots were observed in the control samples after 16 days of storage	[88]

(*continued on next page*)

Table 20.2 (*continued*)

Type of biopolymeric nanomaterials	Tested model food	Influence on food	Reference
Sugarcane bagasse nanocellulose films with nisin	Ready-to-eat ham	-Hybrid film was used as a liner of low-density polyethylene plastic packaging -During 7 days of storage at 4 °C, complete inhibition of *Listeria monocytogenes* was noted	[89]
Nanochitosan coating with chitosan (0.2 and 0.5 wt. %)	Apple cv. Golab Kohanz	-0.5% chitosan nanochitosan coating significantly eliminated peak climacteric respiration, and then lowered the maximal ethylene production to approx. 33% -controlling enzyme activity -during 9 weeks of storage, the color quality of the fruit was improved, the softening of the fruit slowed down and the mass loss reduced	[90]
Coating of chitosan (2 wt. %) or chitosan nanoparticles (2 wt. %)	Silver carp (*Hypoph-thalmicthys molitrix*) fillets	-Comparing both coatings, those with nanochitosan showed higher antimicrobial activity -Nanochitosan demonstrated a stronger ability to inhibit TVB-N content (from 11.4 to 24.6 mg N/100 g of meat—on day 9; and up to 30.8 mg N/100 g of meat—on day 12 of storage -No significant differences in TBA values were observed between chitosan and nanochitosan coatings during storage, while clear differences were noted between control samples and those coated with chitosan and nanochitosan	[91]
Chitosan nanoparticles loaded with moringa oil embedded on gelatin nanofiber	Cheese	-During 10 days of storage, the nanocomposites showed strong antibacterial activity against *L. monocytogenes* on cheese at 4 °C (population reduction by 78.63%) and 25 °C (98.67%), as well as a slight effect on surface color and sensory quality of the cheese	[92]

Abbreviations: TVB-N, Total volatile basic nitrogen; TBA, Thiobarbituric acid index.

on nanocellulose from sugarcane pomace as a colorimetric freshness indicator for monitoring chicken breast spoilage. Nanocellulose was prepared from sugarcane cellulose fibers by TEMPO-mediated oxidation and then used to form a strong, self-contained hydrogel matrix by Zn^{2+} cross-linking. The hydrogel acted as a carrier for the pH reactive dyes (bromothymol blue/methyl red), which turned from green to red during the third day of storage of the chicken breast when the log CFU/g exceeded the acceptable limit for human consumption. The obtained results confirmed that the developed indicator based on nanocellulose provides a quick response to food spoilage. Taherkhani et al. [94] introduced grape anthocyanins (GA) that were embedded in bacterial nanocellulose (BNC) *ex situ* to produce an easy-to-use colorimetric label. On the first day of storage of ground beef at 4 °C, the indicator was bright red in color. On the third and fifth day of storage, the indicator turned purple-red, while at 7 days, it turned blue, representing spoiled condition. All these color changes could be seen with the naked eye, which proves very good compatibility of BNC and GA.

Yang et al. [95] developed a multilayer composite film made of nanofibers based on hybridization of BC (BCNF) and chitin nanofibers (CNF). The nanofibers were responsible for the formation of well-dispersed micro- and NPs of curcumin (Cur) in nanocomposite films. The release of Cur from nanocomposite films was influenced by CNF and the size of the Cur particles generated *in situ*. CNF improved the mechanical strength and barrier properties of the Cur/BCNF/CNF nanocomposite film. Moreover, the multilayer nanofiber composite film demonstrated excellent dynamic antioxidant capacity and antibacterial activity, and was also able to monitor pH changes as well as trace amounts of boric acid. The results of this study suggest that Cur/BCNF/CNF composite films may be used as an intelligent and active food packaging materials. Similar conclusions were reached by Wu et al. [96, 97] who obtained novel intelligent films based on chitosan/chitin nanocrystals enriched with curcumin to monitor the freshness of seafood, and films based on oxidized-chitin nanocrystals/konjac glucomannan enriched with red cabbage-based anthocyanins. The resulting intelligent nanocomposites have great potential in monitoring the freshness of food products.

20.4 Future perspectives

The use of biopolymeric carbohydrate and protein nanomaterials is an excellent and cost-effective approach to improving packaging quality. Many activities were undertaken to enhance the functional properties and give active and intelligent features to packaging materials based on biopolymeric nanomaterials. It has been noted that in many scientific works, biopolymeric carbohydrate nanomaterials are more often added to reinforce packaging materials than biopolymeric protein nanomaterials. This is due to the fact that protein-based nanomaterials do not cause any specific changes in the performance properties of the packaging materials. On the other hand, a very strong impact on the functional properties of packaging was noted in the case of incorporating biopolymeric carbohydrate nanomaterials, which results from their unique features. However, there are still many obstacles in fundamental and applied research that need to be overcome in order for the developed packaging materials to be commercially applicable:

- It is necessary to overcome the sensitivity of packaging materials based on biopolymeric nanomaterials to moisture, which is the main obstacle in their application, for example, in the contact of products with high humidity.
- Good weather resistance for packaging materials under harsh indoor/outdoor conditions and thermal stability at high temperatures are highly desirable but still a challenge.
- The biodegradation process of nanocomposites based on biopolymeric nanomaterials should be further specified.
- Further research is needed to accurately determine the migration of nanofillers and determine their potential toxicity.

- Currently, multidisciplinary research on packaging materials based on biopolymeric nanomaterials led to improved internal properties, but also demonstrate a response to environmental stimuli, which is likely to open a new era of sustainable, inexpensive, high-performance, active and intelligent packaging for advanced applications.

References

[1] A.M. Youssef, SM. El-Sayed, Bionanocomposites materials for food packaging applications: concepts and future outlook, Carbohydr. Polym. 193 (2018) 19–27.

[2] H. Yong, J. Liu, Recent advances in the preparation, physical and functional properties, and applications of anthocyanins-based active and intelligent packaging films, Food Packag. Shelf Life. 26 (2020) 100550.

[3] L. Marangoni, R.P. Vieira, E. Jamróz, CAR. Anjos, Furcellaran: an innovative biopolymer in the production of films and coatings, Carbohydr. Polym. (2020) 117221.

[4] N. Bhargava, V.S. Sharanagat, R.S. Mor, K. Kumar, Active and intelligent biodegradable packaging films using food and food waste-derived bioactive compounds: a review, Trends Food Sci. Technol. 105 (2020) 385–401.

[5] H. Haghighi, F. Licciardello, P. Fava, H.W. Siesler, A. Pulvirenti, Recent advances on chitosan-based films for sustainable food packaging applications, Food Packag. Shelf Life. 26 (2020) 100551.

[6] R. Thakur, P. Pristijono, C.J. Scarlett, M. Bowyer, S.P. Singh, QV. Vuong, Starch-based films: major factors affecting their properties, Int. J. Biol. Macromol. 132 (2019) 1079–1089.

[7] S.A.A. Mohamed, M. El-Sakhawy, El-SMA-M. Polysaccharides, Protein and lipid-based natural edible films in food packaging: a review, Carbohydr. Polym. 238 (2020) 116178.

[8] E. Jamróz, P. Kulawik, P. Kopel, The effect of nanofillers on the functional properties of biopolymer-based films: a review, Polymers (Basel). 11 (4) (2019).

[9] M.-J. Khalaj, H. Ahmadi, R. Lesankhosh, G. Khalaj, Study of physical and mechanical properties of polypropylene nanocomposites for food packaging application: Nano-clay modified with iron nanoparticles, Trends Food Sci. Technol. 51 (2016) 41–48.

[10] M.L. Verma, B.S. Dhanya, R.V. Sukriti, M. Thakur, J. Jeslin, et al., Carbohydrate and protein based biopolymeric nanoparticles: current status and biotechnological applications, Int. J. Biol. Macromol. 154 (2020) 390–412.

[11] C. Zinge, B. Kandasubramanian, Nanocellulose based biodegradable polymers, Eur. Polym. J. 133 (2020) 109758.

[12] P. Thomas, T. Duolikun, N.P. Rumjit, S. Moosavi, C.W. Lai, M.R. Bin Johan, et al., Comprehensive review on nanocellulose: Recent developments, challenges and future prospects, J. Mech. Behav. Biomed. Mater. 110 (2020) 103884.

[13] S.S. Nair, J. Zhu, Y. Deng, AJ. Ragauskas, High performance green barriers based on nanocellulose, Sustain. Chem. Proces. 2 (1) (2014) 23.

[14] K. Chi, JM. Catchmark, Improved eco-friendly barrier materials based on crystalline nanocellulose/chitosan/carboxymethyl cellulose polyelectrolyte complexes, Food Hydrocoll. 80 (2018) 195–205.

[15] M. Mujtaba, A.M. Salaberria, M.A. Andres, M. Kaya, A. Gunyakti, J. Labidi, Utilization of flax (*Linum usitatissimum*) cellulose nanocrystals as reinforcing material for chitosan films, Int. J. Biol. Macromol. 104 (2017) 944–952.

[16] A. Hänninen, E. Sarlin, I. Lyyra, T. Salpavaara, M. Kellomäki, S. Tuukkanen, Nanocellulose and chitosan based films as low cost, green piezoelectric materials, Carbohydr. Polym. 202 (2018) 418–424.

[17] H.M. Azeredo, L.H.C. Mattoso, R.J. Avena-Bustillos, G.C. Filho, M.L. Munford, D. Wood, et al., Nanocellulose reinforced chitosan composite films as affected by nanofiller loading and plasticizer content, J. Food Sc. 75 (1) (2010) N1–N7.

[18] H.W. Kwak, H. Lee, S. Park, M.E. Lee, H-J. Jin, Chemical and physical reinforcement of hydrophilic gelatin film with di-aldehyde nanocellulose, Int. J. Biol. Macromol. 146 (2020) 332–342.

[19] N. Sá, A.L.A. Mattos, L.M.A. Silva, E.S. Brito, M.F. Rosa, HMC. Azeredo, From cashew byproducts to biodegradable active materials: Bacterial cellulose-lignin-cellulose nanocrystal nanocomposite films, Int. J. Biol. Macromol. 161 (2020) 1337–1345.

[20] M. Maliha, M. Herdman, R. Brammananth, M. McDonald, R. Coppel, M. Werrett, et al., Bismuth phosphinate incorporated nanocellulose sheets with antimicrobial and barrier properties for packaging applications, J. Clean. Prod. 246 (2020) 119016.

[21] B. Bideau, E. Loranger, C. Daneault, Nanocellulose-polypyrrole-coated paperboard for food packaging application, Prog. Org. Coat. 123 (2018) 128–133.

[22] D. Zulfiana, A. Karimah, S.H. Anita, N. Masruchin, K. Wijaya, L. Suryanegara, et al., Antimicrobial Imperata cylindrica paper coated with anionic nanocellulose crosslinked with cationic ions, Int. J. Biol. Macromol. 164 (2020) 892–901.

[23] H. Yousefi, M. Faezipour, S. Hedjazi, M.M. Mousavi, Y. Azusa, AH. Heidari, Comparative study of paper and nanopaper properties prepared from bacterial cellulose nanofibers and fibers/ground cellulose nanofibers of canola straw, Ind. Crop. Prod. 43 (2013) 732–737.

[24] K. Shanmugam, S. Varanasi, G. Garnier, W. Batchelor, Rapid preparation of smooth nanocellulose films using spray coating, Cellulose. 24 (7) (2017) 2669–2676.

[25] H. Nadeem, M. Naseri, K. Shanmugam, M. Dehghani, C. Browne, S. Miri, et al., An energy efficient production of high moisture barrier nanocellulose/carboxymethyl cellulose films via spray-deposition technique, Carbohydr. Polym. 250 (2020) 116911.

[26] Q. Chen, J. Xiong, G. Chen, T. Tan, Preparation and characterization of highly transparent hydrophobic nanocellulose film using corn husks as main material, Int. J. Biol. Macromol. 158 (2020) 781–789.

[27] P. Duarah, D. Haldar, MK. Purkait, Technological advancement in the synthesis and applications of lignin-based nanoparticles derived from agro-industrial waste residues: a review, Int. J. Biol. Macromol. 163 (2020) 1828–1843.

[28] PS. Chauhan, Lignin nanoparticles: eco-friendly and versatile tool for new era, Bioresour. Technol. Rep. 9 (2020) 100374.

[29] W. Yang, J.S. Owczarek, E. Fortunati, M. Kozanecki, A. Mazzaglia, G.M. Balestra, et al., Antioxidant and antibacterial lignin nanoparticles in polyvinyl alcohol/chitosan films for active packaging, Ind. Crop. Prod. 94 (2016) 800–811.

[30] W. Yang, J.M. Kenny, D. Puglia, Structure and properties of biodegradable wheat gluten bionanocomposites containing lignin nanoparticles, Ind. Crops Prod. 74 (2015) 348–356.

[31] T. Ju, Z. Zhang, Y. Li, X. Miao, J. Ji, Continuous production of lignin nanoparticles using a microchannel reactor and its application in UV-shielding films, RSC Adv. 9 (43) (2019) 24915–24921.

[32] F. Xiong, Y. Wu, G. Li, Y. Han, F. Chu, Transparent nanocomposite films of lignin nanospheres and poly (vinyl alcohol) for UV-absorbing, Ind. Eng. Chem. Res. 57 (4) (2018) 1207–1212.

[33] L.A.M. Nevárez, L.B. Casarrubias, A. Celzard, V. Fierro, V.T. Muñoz, A.C. Davila, et al., Biopolymer-based nanocomposites: effect of lignin acetylation in cellulose triacetate films, Sci. Technol. Adv. Mater. 12 (4) (2011) 045006.

[34] L.S.F. Leite, C.M. Ferreira, A.C. Corrêa, F.K.V. Moreira, LHC. Mattoso, Scaled-up production of gelatin-cellulose nanocrystal bionanocomposite films by continuous casting, Carbohydr. Polym. 238 (2020) 116198.

[35] J. George, Siddaramaiah, High performance edible nanocomposite films containing bacterial cellulose nanocrystals, Carbohydr. Polym. 87 (3) (2012) 2031–2037.

[36] C. CCdS, R.B.S. Silva, C.W.P. Carvalho, A.L. Rossi, J.A. Teixeira, O. Freitas-Silva, et al., Cellulose nanocrystals from grape pomace and their use for the development of starch-based nanocomposite films, Int. J. Biol. Macromol. 159 (2020) 1048–1061.

[37] X. Ma, Y. Cheng, X. Qin, T. Guo, J. Deng, X. Liu, Hydrophilic modification of cellulose nanocrystals improves the physicochemical properties of cassava starch-based nanocomposite films, LWT. 86 (2017) 318–326.

[38] M. Tabari, Characterization of a new biodegradable edible film based on sago starch loaded with carboxymethyl cellulose nanoparticles, Nanomed. Res. J. 3 (1) (2018) 25–30.

[39] Q. Lin, N. Ji, M. Li, L. Dai, X. Xu, L. Xiong, et al., Fabrication of debranched starch nanoparticles via reverse emulsification for improvement of functional properties of corn starch films, Food Hydrocoll. 104 (2020) 105760.

[40] M. Mujtaba, B. Koc, A.M. Salaberria, S. Ilk, D. Cansaran-Duman, L. Akyuz, et al., Production of novel chia-mucilage nanocomposite films with starch nanocrystals; an inclusive biological and physicochemical perspective, Int. J. Biol. Macromol. 133 (2019) 663–673.

[41] S. A-m, W. L-j, D. Li, B. Adhikari, Characterization of starch films containing starch nanoparticles: Part 1: Physical and mechanical properties, Carbohydr. Polym. 96 (2) (2013) 593–601.

[42] P. Oymaci, SA. Altinkaya, Improvement of barrier and mechanical properties of whey protein isolate based food packaging films by incorporation of zein nanoparticles as a novel bionanocomposite, Food Hydrocoll. 54 (2016) 1–9.

[43] S. Sahraee, J.M. Milani, B. Ghanbarzadeh, H. Hamishehkar, Physicochemical and antifungal properties of bio-nanocomposite film based on gelatin-chitin nanoparticles, Int. J. Biol. Macromol. 97 (2017) 373–381.

[44] A.A. Oun, J-W. Rhim, Effect of oxidized chitin nanocrystals isolated by ammonium persulfate method on the properties of carboxymethyl cellulose-based films, Carbohydr. Polym. 175 (2017) 712–720.

[45] P.R. Chang, R. Jian, J. Yu, X. Ma, Starch-based composites reinforced with novel chitin nanoparticles, Carbohydr. Polym. 80 (2) (2010) 420–425.

[46] Y. Qin, S. Zhang, J. Yu, J. Yang, L. Xiong, Q. Sun, Effects of chitin nano-whiskers on the antibacterial and physicochemical properties of maize starch films, Carbohydr. Polym. 147 (2016) 372–378.

[47] S.F. Hosseini, M. Rezaei, M. Zandi, F. Farahmandghavi, Fabrication of bio-nanocomposite films based on fish gelatin reinforced with chitosan nanoparticles, Food Hydrocoll. 44 (2015) 172–182.

[48] M.R. de Moura, M.V. Lorevice, L.H.C. Mattoso, V. Zucolotto, Highly stable, edible cellulose films incorporating chitosan nanoparticles, J. Food Sci. 76 (2) (2011) N25–N29.

[49] M.V. Lorevice, C.G. Otoni, M. MRd, LHC. Mattoso, Chitosan nanoparticles on the improvement of thermal, barrier, and mechanical properties of high- and low-methyl pectin films, Food Hydrocoll. 52 (2016) 732–740.

[50] J.P. Paques, E. van der Linden, C.J.M. van Rijn, LMC. Sagis, Preparation methods of alginate nanoparticles, Adv. Coll. Interface Sci. 209 (2014) 163–171.

[51] I.P.S. Fernando, W. Lee, E.J. Han, G. Ahn, Alginate-based nanomaterials: fabrication techniques, properties, and applications, Chem. Eng. J. 391 (2020) 123823.

[52] T.C. Mokhena, AS. Luyt, Development of multifunctional nano/ultrafiltration membrane based on a chitosan thin film on alginate electrospun nanofibers, J. Clean. Prod. 156 (2017) 470–479.

[53] S. Amjadi, H. Almasi, M. Ghorbani, S. Ramazani, Preparation and characterization of TiO_2NPs and betanin loaded zein/sodium alginate nanofibers, Food Packag. Shelf Life. 24 (2020) 100504.

[54] E.N. Dresvyanina, S.F. Grebennikov, V.Y. Elokhovskii, I.P. Dobrovolskaya, E.M. Ivan'kova, V.˘. Yudin, et al., Thermodynamics of interaction between water and the composite films based on chitosan and chitin nanofibrils, Carbohydr. Polym. 245 (2020) 116552.

[55] A.A. Oun, J-W. Rhim, Preparation of multifunctional carboxymethyl cellulose-based films incorporated with chitin nanocrystal and grapefruit seed extract, Int. J. Biol. Macromol. 152 (2020) 1038–1046.

[56] A.M. Salaberria, R.H. Diaz, J. Labidi, SCM. Fernandes, Role of chitin nanocrystals and nanofibers on physical, mechanical and functional properties in thermoplastic starch films, Food Hydrocoll. 46 (2015) 93–102.

[57] Z. Li, R. Yang, F. Yang, M. Zhang, B. Wang, Structure and properties of chitin whisker reinforced papers for food packaging application, BioResources. 10 (2) (2015) 2995–3004.

[58] S. Ju, F. Zhang, J. Duan, J. Jiang, Characterization of bacterial cellulose composite films incorporated with bulk chitosan and chitosan nanoparticles: a comparative study, Carbohydr. Polym. 237 (2020) 116167.

[59] S.F. Hosseini, M. Rezaei, M. Zandi, F. Farahmandghavi, Preparation and characterization of chitosan nanoparticles-loaded fish gelatin-based edible films, J. Food Process Eng. 39 (5) (2016) 521–530.

[60] S.H. Othman, N.R. Kechik, R.A. Shapi'i, R.A. Talib, I.S. Tawakkal, Water sorption and mechanical properties of starch/chitosan nanoparticle films, J. Nanomater. 2019 (2019) Article ID 3843949.

[61] C. Dularia, A. Sinhmar, R. Thory, A.K. Pathera, V. Nain, Development of starch nanoparticles based composite films from non-conventional source—water chestnut (*Trapa bispinosa*), Int. J. Biol. Macromol. 136 (2019) 1161–1168.

[62] H. Lu, N. Ji, M. Li, Y. Wang, L. Xiong, L. Zhou, et al., Preparation of borax cross-linked starch nanoparticles for improvement of mechanical properties of maize starch films, J. Agric. Food Chem. 67 (10) (2019) 2916–2925.

[63] E. Kristo, CG. Biliaderis, Physical properties of starch nanocrystal-reinforced pullulan films, Carbohydr. Polym. 68 (1) (2007) 146–158.

[64] A. Orsuwan, R. Sothornvit, Development and characterization of banana flour film incorporated with montmorillonite and banana starch nanoparticles, Carbohydr. Polym. 174 (2017) 235–242.

[65] A.V. Oliveira, A.P.M. da Silva, M.O. Barros, M. de sá, M. Souza Filho, M.F. Rosa, HMC. Azeredo, Nanocomposite films from mango kernel or corn starch with starch nanocrystals, Starch/Stärke. 70 (11-12) (2018) 1800028.

[66] TJ. Gutiérrez, Are modified pumpkin flour/plum flour nanocomposite films biodegradable and compostable? Food Hydrocoll. 83 (2018) 397–410.

[67] M. Karim, M. Fathi, S. Soleimanian-Zad, Incorporation of zein nanofibers produced by needle-less electrospinning within the casted gelatin film for improvement of its physical properties, Food Bioprod. Process. 122 (2020) 193–204.

[68] L. Spasojević, J. Katona, S. Bučko, S.M. Savić, L. Petrović, J. Milinković Budinčić, et al., Edible water barrier films prepared from aqueous dispersions of zein nanoparticles, LWT. 109 (2019) 350–358.

[69] C.J. Cheng, OG. Jones, Effect of drying temperature and extent of particle dispersion on composite films of methylcellulose and zein nanoparticles, J. Food Eng. 250 (2019) 26–32.

[70] D. Huang, Y. Zheng, Z. Zhang, Q. Quan, X. Qiang, Synergistic effect of hydrophilic palygorskite and hydrophobic zein particles on the properties of chitosan films, Mater. Des. 185 (2020) 108229.

[71] R. Farajpour, Z. Emam Djomeh, S. Moeini, H. Tavakolipour, S. Safayan, Structural and physico-mechanical properties of potato starch-olive oil edible films reinforced with zein nanoparticles, Int. J. Biol. Macromol. 149 (2020) 941–950.

[72] J. Zhang, J. Zhu, K.F. Akhter, AA. Thomas, Encapsulation of BSA within gelatin nanoparticles-laden biopolymer film, MRS Proc. 1237 (2009) 1102.

[73] S. Ebrahimi, M. Fathi, M. Kadivar, Production and characterization of chitosan-gelatin nanofibers by nozzle-less electrospinning and their application to enhance edible film's properties, Food Packag. Shelf Life. 22 (2019) 100387.

[74] X. Li, N. Ji, C. Qiu, M. Xia, L. Xiong, Q. Sun, The effect of peanut protein nanoparticles on characteristics of protein- and starch-based nanocomposite films: a comparative study, Ind. Crop Prod. 77 (2015) 565–574.

[75] N. Jannatyha, S. Shojaee-Aliabadi, M. Moslehishad, E. Moradi, Comparing mechanical, barrier and antimicrobial properties of nanocellulose/CMC and nanochitosan/CMC composite films, Int. J. Biol. Macromol. 164 (2020) 2323–2328.

[76] C. Metzger, S. Sanahuja, L. Behrends, S. Sängerlaub, M. Lindner, H. Briesen, Efficiently extracted cellulose nanocrystals and starch nanoparticles and techno-functional properties of films made thereof, Coatings. 8 (4) (2018) 142.

[77] M. Salari, M. Sowti Khiabani, R. Rezaei Mokarram, B. Ghanbarzadeh, H. Samadi Kafil, Development and evaluation of chitosan based active nanocomposite films containing bacterial cellulose nanocrystals and silver nanoparticles, Food Hydrocoll. 84 (2018) 414–423.

[78] A.A. Oun, J-W. Rhim, Preparation of multifunctional chitin nanowhiskers/ZnO-Ag NPs and their effect on the properties of carboxymethyl cellulose-based nanocomposite film, Carbohydr. Polym. 169 (2017) 467–479.

[79] S. Amjadi, S. Emaminia, S. Heyat Davudian, S. Pourmohammad, H. Hamishehkar, L. Roufegarinejad, Preparation and characterization of gelatin-based nanocomposite containing chitosan nanofiber and ZnO nanoparticles, Carbohydr. Polym. 216 (2019) 376–384.

[80] A.P.M. Silva, A.V. Oliveira, S.M.A. Pontes, A.L.S. Pereira, S.F. MdsM, M.F. Rosa, et al., Mango kernel starch films as affected by starch nanocrystals and cellulose nanocrystals, Carbohydr. Polym. 211 (2019) 209–216.

[81] Y.-.H. Tsai, Y.-.N. Yang, Y.-.C. Ho, M.-.L. Tsai, F.-.L. Mi, Drug release and antioxidant/antibacterial activities of silymarin-zein nanoparticle/bacterial cellulose nanofiber composite films, Carbohydr. Polym. 180 (2018) 286–296.

[82] R.A. Shapi'i, S.H. Othman, N. Nordin, R. Kadir Basha, M. Nazli Naim, Antimicrobial properties of starch films incorporated with chitosan nanoparticles: in vitro and in vivo evaluation, Carbohydr. Polym. 230 (2020) 115602.

[83] Z.N. Correa-Pacheco, S. Bautista-Baños, R.-.G. MdL, M.-.G. MdC, J. Hernández-Romano, Physicochemical characterization and antimicrobial activity of edible propolis-chitosan nanoparticle films, Prog. Org. Coat. 137 (2019) 105326.

[84] D. Dehnad, H. Mirzaei, Z. Emam-Djomeh, S.M. Jafari, S. Dadashi, Thermal and antimicrobial properties of chitosan-nanocellulose films for extending shelf life of ground meat, Carbohydr. Polym. 109 (2014) 148–154.

[85] H. Li, H. Shi, Y. He, X. Fei, L. Peng, Preparation and characterization of carboxymethyl cellulose-based composite films reinforced by cellulose nanocrystals derived from pea hull waste for food packaging applications, Int. J. Biol. Macromol. 164 (2020) 4104–4112.

[86] Y. He, H. Li, X. Fei, L. Peng, Carboxymethyl cellulose/cellulose nanocrystals immobilized silver nanoparticles as an effective coating to improve barrier and antibacterial properties of paper for food packaging applications, Carbohydr. Polym. 252 (2021) 117156.

[87] H. Cui, D. Surendhiran, C. Li, L. Lin, Biodegradable zein active film containing chitosan nanoparticle encapsulated with pomegranate peel extract for food packaging, Food Packag. Shelf Life. 24 (2020) 100511.

[88] D. Dey, V. Dharini, S. Periyar Selvam, E. Rotimi Sadiku, M. Mahesh Kumar, J. Jayaramudu, U.N. Gupta, et al., Physical, antifungal, and biodegradable properties of cellulose nanocrystals and chitosan nanoparticles for food packaging application, in: Materials Today: Proceedings, 38, Elsevier, 2021, pp. 860–869.

[89] Y. Yang, H. Liu, M. Wu, J. Ma, P. Lu, Bio-based antimicrobial packaging from sugarcane bagasse nanocellulose/nisin hybrid films, Int. J. Biol. Macromol. 161 (2020) 627–635.

[90] A. Sahraei Khosh Gardesh, F. Badii, M. Hashemi, A.Y. Ardakani, N. Maftoonazad, AM. Gorji, Effect of nanochitosan based coating on climacteric behavior and postharvest shelf-life extension of apple cv. Golab Kohanz., LWT. 70 (2016) 33–40.

[91] Z. Ramezani, M. Zarei, N. Raminnejad, Comparing the effectiveness of chitosan and nanochitosan coatings on the quality of refrigerated silver carp fillets, Food Control. 51 (2015) 43–48.

[92] L. Lin, Y. Gu, H. Cui, Moringa oil/chitosan nanoparticles embedded gelatin nanofibers for food packaging against *Listeria monocytogenes* and *Staphylococcus aureus* on cheese, Food Packag. Shelf Life. 19 (2019) 86–93.

[93] P. Lu, Y. Yang, R. Liu, X. Liu, J. Ma, M. Wu, et al., Preparation of sugarcane bagasse nanocellulose hydrogel as a colourimetric freshness indicator for intelligent food packaging, Carbohydr. Polym. 249 (2020) 116831.

[94] E. Taherkhani, M. Moradi, H. Tajik, R. Molaei, P. Ezati, Preparation of on-package halochromic freshness/spoilage nanocellulose label for the visual shelf life estimation of meat, Int. J. Biol. Macromol. 164 (2020) 2632–2640.

[95] Y.-.N. Yang, K.-.Y. Lu, P. Wang, Y.-.C. Ho, M.-.L. Tsai, F.-.L. Mi, Development of bacterial cellulose/chitin multi-nanofibers based smart films containing natural active microspheres and nanoparticles formed in situ, Carbohydr. Polym. 228 (2020) 115370.

[96] C. Wu, J. Sun, M. Chen, Y. Ge, J. Ma, Y. Hu, et al., Effect of oxidized chitin nanocrystals and curcumin into chitosan films for seafood freshness monitoring, Food Hydrocoll. 95 (2019) 308–317.

[97] C. Wu, Y. Li, J. Sun, Y. Lu, C. Tong, L. Wang, et al., Novel konjac glucomannan films with oxidized chitin nanocrystals immobilized red cabbage anthocyanins for intelligent food packaging, Food Hydrocoll. 98 (2020) 105245.

21

Polysaccharide-derived biopolymeric nanomaterials for wastewater treatment

Muhammad Bilal Asif[a], Zhenghua Zhang[a], Sidra Iftekhar[b] and Vesa-Pekka Lehto[b]

[a]INSTITUTE OF ENVIRONMENTAL ENGINEERING AND NANO-TECHNOLOGY, TSINGHUA-SHENZHEN INTERNATIONAL GRADUATE SCHOOL, TSINGHUA UNIVERSITY, SHENZHEN, GUANGDONG, CHINA. [b]DEPARTMENT OF APPLIED PHYSICS, UNIVERSITY OF EASTERN FINLAND, KUOPIO, FINLAND

Chapter outline

21.1 Introduction

The water-energy-food nexus approach was propelled to the international limelight by the Bonn conference of 2011 [1,2]. The nexus approach has now become an integral component of the discussions on sustainable development. The main aim of introducing the nexus approach was to advocate and promote the inseparable link between the use of natural resources and the provision of the basic right to water, energy, and food security [1,3]. Although the nexus approach was initially presented from a securities perspective in the Bonn conference, its subsequent versions have considered various facets with alternative components such as land-energy-water, and food-land-water. Depending on the type of linkages under consideration, the nexus approach can be divided into core and central components. For instance, water reserves are the central component, while core components of the nexus include food, energy, and land [4–6]. Water is the central component of most nexus linkages mainly because of limited freshwater reserves. Due to the rapid population increase and urbanization, water demand for domestic,

Biopolymeric Nanomaterials: Fundamentals and Applications. DOI: https://doi.org/10.1016/B978-0-12-824364-0.00012-5

industrial, and agricultural activities is increasing at an alarming rate [7,8]. Less than 3% of all the water on the earth is categorized as freshwater; however, most of the freshwater are frozen in glaciers, ice, and snow with less than 1% presenting as fresh groundwater and soil moisture and less than 0.01% presenting as surface water in lakes, swamps, and rivers. The situation is exacerbated by erratic rainfall patterns due to climate change, which imposes new challenges to the already water-stressed areas [8–10]. In addition to effective water resource management, wastewater treatment and reuse are important strategies because wastewater can serve as an alternative nonconventional source of water for diverse end-user applications (e.g., irrigation and nonpotable reuse), particularly in water-scarce regions.

Various physicochemical and biological processes such as coagulation, flocculation, membrane-based separation, advanced oxidation, and adsorption as well as their combinations have been used for the treatment of water and wastewater [11–20]. However, the current treatment has several limitations: ineffective removal of certain pollutants (e.g., micropollutants and heavy metals); generation of toxic sludge; and/or energy-intensive/expensive [21]. In this context, the development of "greener," ecofriendly, and cost-effective treatment has gained significant attention in recent years [21–24]. Among green technologies, the fabrication of naturally occurring polymeric nanomaterials, which are renewable, biodegradable, and cost-effective, has become an attractive and viable option for application in water/wastewater [21–23].

Biopolymers are present in a variety of sources such as plants and animals and could be extracted using physicochemical or biological methods. Based on their monomeric units, they can be divided into three categories: (1) polynucleotides, (2) polysaccharides, and (3) polypeptides. Although biopolymers have been investigated for a range of applications, polysaccharides have been predominantly investigated as an adsorbent or photocatalyst for water and wastewater treatment [21–23]. Importantly, among the polysaccharide-derived materials, cellulose [25–35] and chitosan [36–46] have been mostly studied for water treatment applications. It is noteworthy that the adsorption capacity of naturally occurring polysaccharides may not be commercially viable because water molecules limit the interaction of biopolymers with pollutants, for example, dyes. In addition, the recovery of biopolymer from reaction media may not be possible [21–23]. Therefore, polysaccharide-derived biopolymers are commonly combined with nanomaterials. This combination improves the physicochemical and mechanical properties of the biopolymers, and at the same time facilitates the stability of the nanomaterials [47–55]. In this chapter, recent advances in polysaccharide-derived biopolymeric nanomaterials for sustainable water and treatment are presented. The performance of cellulose- and chitosan-based nanomaterials for the removal of different pollutants (e.g., dyes, heavy metal, and micropollutants) is mainly focused because they have been most frequently investigated. In addition to features of the polysaccharide-derived biopolymers, a few future research directions are presented.

21.2 Application of cellulose-based nanomaterials in water treatment

Cellulose is the most abundant biopolymer in nature as compared to other polysaccharide-derived biopolymers. A range of cellulose-based nanomaterials has been synthesized and

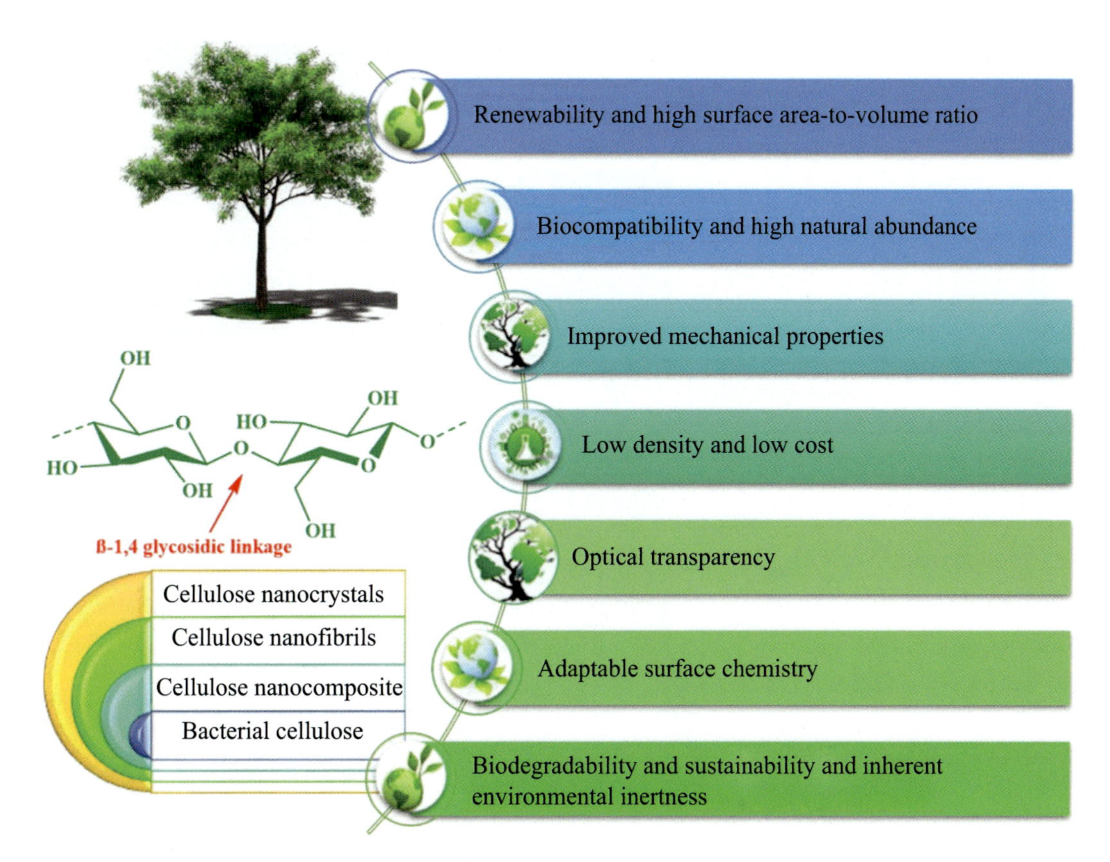

FIGURE 21.1 Key features and properties of cellulose-based nanomaterials. Reproduced with permission from Ref. [21] Copyright 2020 Elsevier Ltd.

investigated in recent years for applications in bioplastics, medicine, electronics, supercapacitors, and cosmetic products [21,56]. Owing to their peculiar physicochemical properties, cellulose-based nanomaterials have also attracted significant attention, particularly for developing novel nanosorbents, membranes, and catalysts, which could be used for application in water treatment [21,57–62]. Key features of cellulose-based nanomaterials are presented in Fig. 21.1.

Numerous cellulose-based nanomaterials have been isolated using different synthesis methods [58], and, owing to their applications in different fields (e.g., medicine and electronics), there exists a lack of consistent nomenclature [58,63]. For instance, different names such as nanocellulose, polysaccharide nanocrystals, cellulose microfibrils, and bacterial cellulose (BC) have been used in the existing literature. In this context, a consortium of interested American and Canadian organizations has proposed that cellulose-based nanomaterials could be classified into: (1) cellulose nanofibrils; and (2) cellulose nanocrystals [64]. Production of cellulose nanofibrils can be achieved either from the cellulose-based feedstock or from certain species of bacteria such as *Gluconacetobacter xylinus*. Compared to feedstock (plants and wood fibers), cellulose nanofibrils produced by bacteria have different surface properties; higher crystallinity

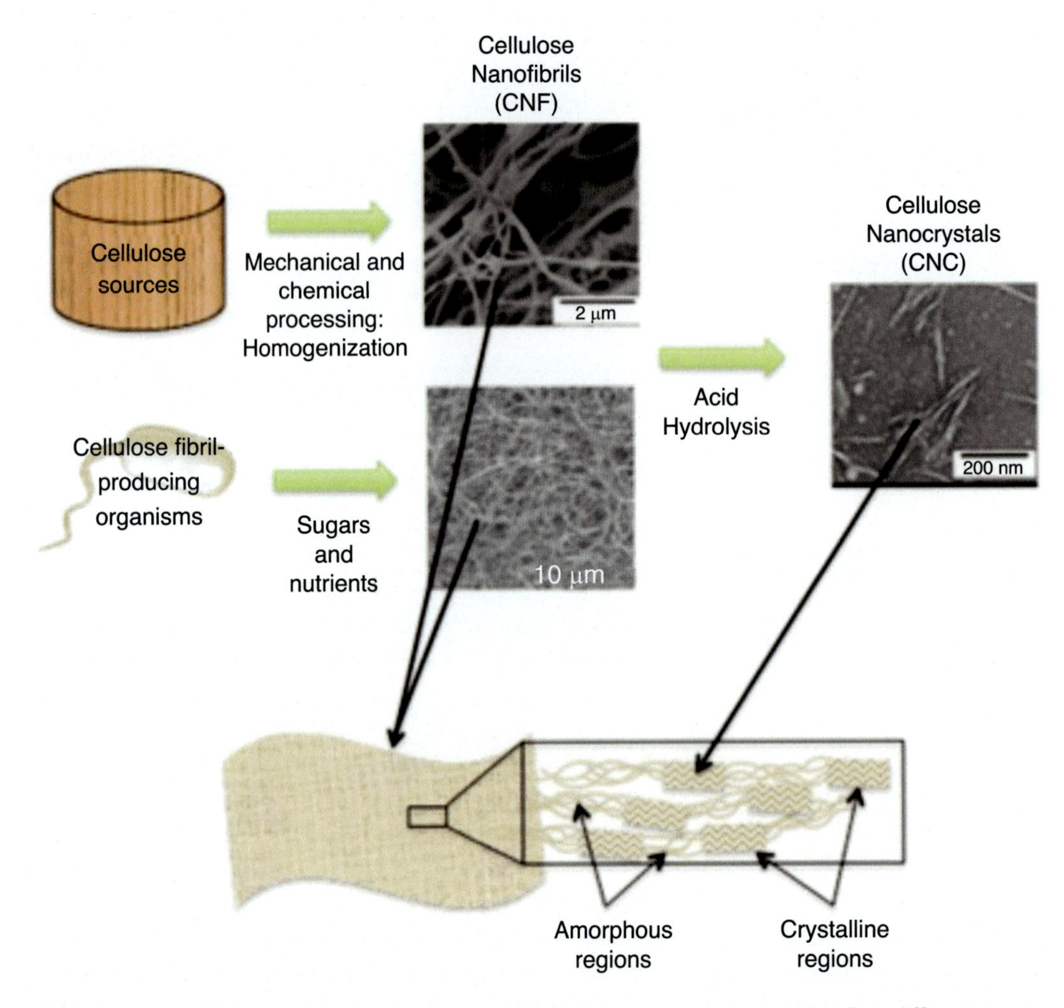

FIGURE 21.2 A simplified illustration of methods for cellulose-based nanomaterial isolation from different source material. Reproduced with permission in part from Ref. [67] (Copyright 2013 Elsevier Ltd.), Ref. [66] (Copyright 2011 Cambridge University Press) and Ref. [58] (Copyright 2015, American Chemical Society).

and molecular weight [65,66]. The diameter of cellulose nanofibrils can range between 5 and 60 nm, and both the amorphous and crystalline regions of cellulose are present in cellulose nanofibrils [58]. Notably, to produce cellulose nanocrystals, acid hydrolysis of cellulose nanofibrils is required (Fig. 21.2). Depending on the source and synthesis method, cellulose nanocrystals can have a diameter ranging from 5 to 70 nm (100–250 nm in length) and have fewer amorphous regions of cellulose as compared to cellulose nanofibrils [58,67].

Isolation and production of cellulose-based nanomaterials have been widely studied in the last decade [58,65]. However, the extensive energy consumption required for the isolation of cellulose nanofibrils as well as acidic waste generated during cellulose nanocrystals could limit their widespread application [68–70]. Therefore, efforts are required to develop innovative and cost-effective methods for the production of cellulose-based nanomaterials [58,69]. In this context, enzymatic methods with low-energy requirement could be considered, and research on the optimization of enzyme dose for the effective polymerization and without compromising the structure of cellulose-based nanomaterials should be carried out [71]. In addition, to reduce the energy consumption during the production of cellulose-based nanomaterials, a charge on the surface of fibers can be provided, which would not only reduce friction among fibers but would also limit fiber aggregation [71]. Despite the recent advances, the minimum energy required to produce one ton of cellulose nanofibrils is approximately 500 kWh, while the energy requirement to produce cellulose nanocrystals is still orders of magnitude higher [72]. Although the quest to develop environmentally friendly and cost-effective methods for their production is still on-going, cellulose-based nanomaterials have been assessed for the removal (Table 21.1) of a range of pollutants such as heavy metals, dyes, and pesticides during water treatment [25–35].

21.2.1 Removal of heavy metals

The presence of heavy metals in freshwater water is a growing concern, and their effective removal is essential for safe water use applications. Heavy metals, which have a high molecular weight with a density higher (up to five times) than water, are naturally occurring elements and can be toxic to the aquatic ecosystem and human life [75–77]. Cellulose-based nanomaterials can remove a range of heavy metals via sorption, but their modification is often carried out for efficient performance. For instance, in a study by Yu et al. [31], modification of cellulose nanocrystals using succinic anhydride transformed the carboxylic acid groups to sodiated carboxylates, which led to improved removal (60–90%) of Pb^{2+} and Cd^{2+}. In another study, the amendment of the cellulose nanofibrils using COO^- functional group improved the sorption of four heavy metals (Ni^{2+}, Pb^{2+}, Cd^{2+}, and Cr^{3+}) by up to 10%, and also the regeneration cycle from three to five [78]. Similarly, a modification of cellulose nanofibrous membrane using cysteine resulted in the formation of thiol functionalities, leading to improved sorption of Pb^{2+} and Cr^{6+} from an aqueous solution [79]. Following the functionalization of cellulose nanocrystals with amine groups, sorption of Cr^{3+} and Cr^{6+} from an aqueous solution was above 98%, and importantly, the modified cellulose nanocrystals achieved stable removal for at least five cycles [27]. In a study by Ma et al. [80], metal-organic framework (MOF) aerogels were synthesized via *in-situ* growth and were assembled on BC. Compared to original MOF aerogels that is, zeolitic imidazolate framework-8 (ZIF-8), the performance of BC@ZIF-8 was observed to be 1.2 times better for the removal of Pb^{2+} (overall removal 81% within 24 h). Notably, magnetite nanoparticles can be assembled into cellulose-based nanomaterials and could be easily recovered using magnetic separation [81,82]. Indeed, Zhu et al. [83] incorporated magnetite nanoparticles onto BC, and the resulting nanocomposite was observed to achieve an effective removal of heavy metals. The sorption of the selected heavy metals were reported to be in the following order: $Pb^{2+}>Mn^{2+}>Cr^{3+}$, while their elution for the reuse of nanocomposite was as follows: $Mn^{2+}> Pb^{2+}> Cr^{3+}$ [83].

Table 21.1 A selection of studies on cellulose-based nanomaterials for water treatment.

Cellulose-based nanomaterials	Pollutant/experimental conditions	Maximum sorption capacity (mg/g)	Reference
Virgin cellulose nanofibers	Ag^{+}/pH = 5.45, Temperature = 298 K	34.4	[25]
Virgin cellulose nanocrystal	Ag^{+}/pH = 6.39, Temperature = 298 K	15.5	[25]
Phosphorylated cellulose nanofibers	Cu^{+2}/pH = 3.4–4.5, Temperature = 298 K	72.7	[26]
Phosphorylated cellulose nanocrystal	Cu^{+2}/pH = 3.4–4.5, Temperature = 298 K	72.8	[26]
Succinic anhydride modified cellulose nanocrystal	Cr^{3+}/pH = 6.5, Temperature = 298 K	2.8	[27]
Sulfonated cellulose nanocrystal	Au^{3+}/pH = 3.26, Temperature = 295 K	57–60	[28]
Sodium substituted succinic anhydride- amended cellulose nanocrystal	Pb^{2+} and Cd^{2+}/pH = 5.5 and 6.5, Temperature = 295 K	465 and 345	[31]
Carboxylated cellulose nanofibers/polyvinyl alcohol hybrid aerogels	Pb^{2+}, Hg^{2+}, Cu^{2+}, and Ag^{+}	110.6, 158.1, 151.3, 114.3	[30]
Amino functionalized cellulose nanocrystal	Acid red GR/pH = 4.7, Temperature = 298 K	555.6	[29]
MnO_2-coated cellulose nanocrystal	Methylene blue/pH = 9.6, Temperature = 298 K	Above 99.8% removal	[32]
Cellulose nanocrystal@polydopamine	Methylene blue/pH = 10, Temperature = 298 K	2066.7	[34]
Cellulose nanocrystal	Victoria Blue 2B, methyl Violet 2B, and rhodamine 6 G/pH = 10, Temperature = 298 K	98, 90, and 78% removal	[33]
Carboxylated cellulose fabric filter	Methylene blue and Pb^{2+}/pH = 5, Temperature = 298 K	76.9, 81.3	[35]
Cu-modified microcrystalline cellulose	Prometryn	97.8	[73]
Cu-BTC@cotton nanocomposite	Ethion	182	[74]

21.2.2 Removal of micropollutants

Micropollutants such as dyes, pharmaceuticals, and pesticides are recalcitrant, and their removal by water/wastewater treatment process has been reported to be ineffective or unstable [8,11,15,84–86], which could lead to their ubiquitous occurrence in freshwater sources [87]. Different physicochemical advanced treatment technologies such as membrane distillation, nanofiltration, advanced oxidation, and hybrid membrane-based processes are commonly used to ensure the effective micropollutant removal [12,15,88–91]. The performance of cellulose-based nanomaterials has predominantly been assessed for the removal of pesticides and dyes as discussed below.

According to a study by Garba et al. [73], the removal of the herbicide prometryn was investigated using a copper-modified microcrystalline cellulose. The synthesized composite was efficient for prometryn removal and had a sorption capacity of 97.8 mg/g. Importantly, the synthesized composite remained for six sorption and desorption cycles [73]. Similarly, triolein embedded into the cellulose sphere achieved significant sorption of organochlorinated pesticides such as dieldrin [92], while the nanocomposite prepared by embedding cadmium sulfide nanoparticles onto cellulose nanofibers maintained effective and stable removal of one insecticide (chlorpyrifos) and two dyes (safranin O and methylene blue) for up to six runs [93]. In the existing literature, a combination of cellulose-based nanomaterials and MOFs has been investigated to gain full benefits from their properties such as the high sorption potential of MOFs and eco-friendly/sustainability of cellulose. Abdelhameed et al. [74] prepared Cu-BTC@cotton nanocomposites by allowing the interaction of Cu in the BTC (copper-benzene-1,3,5-tricarboxylic acid) with the functional groups of cellulose in cotton. The maximum sorption capacity of the Cu-BTC@cotton nanocomposite for the insecticide ethion was reported to be 182 mg/g with an overall removal of above 97%, which remained above 85% even after five runs. In a study by Gan et al. [94], $CoFe_2O_4$-coated cellulose-based carbon nanofibers were fabricated for the activation of peroxymonosulfate during the degradation of the pesticide dimethyl phthalate. The results showed that activation by $CoFe_2O_4$ or carbon nanofibers achieved up to 50% removal of dimethyl phthalate, while the activation by the $CoFe_2O_4$-coated carbon nanofibers achieved 20%–50% better removal of dimethyl phthalate. However, the efficacy of $CoFe_2O_4$-coated carbon nanofibers for peroxymonosulfate activation remained stable after five cycles/runs, resulting in consistent removal of the selected pesticide. Cellulose-based nanomaterials have rarely been studied for the removal of a range of micropollutants. In a study by Herrera-Morales et al. [95], sorption of three pharmaceuticals, namely N,N-diethyl-meta-toluamide, sulfamethoxazole, and acetaminophen using polyethylene glycol (PEG) functionalized cellulose nanocrystals, was investigated. They reported that the extent of sorption is influenced by the hydrophobicity of the selected pharmaceutical. The modification of cellulose nanocrystals using PEG could facilitate the interaction between the pharmaceuticals and the sorbent, which would lead to improved removal efficiency [95].

According to an estimate, approximately 1.6 million tons of textile dyes are produced every year, and around 10%–15% of these dyes would find their way into wastewater [21]. Removal of dyes using various cellulose-based nanomaterials can be achieved following their modifications with anionic or cationic moieties. In this context, carboxylation of cellulose nanofibrils or nanocrystals is commonly carried out for improving the sorption potential [21]. Indeed, He et al. [96] used an ammonium sulfate hydrolysis process for the fabrication of carboxylated-cellulose nanocrystals. They demonstrated effective removal (90%, up to seven desorption cycles) of a cationic dye (methylene blue) via sorption onto negatively charged carboxylate groups. In another study, carboxylated-cellulose nanocrystals were prepared using acid hydrolysis (HCl/critic acid, sulfuric acid, or formic acid) method for photocatalytic degradation of methylene blue [97]. Compared to the nanocatalysts produced using sulfuric acid or formic acid, carboxylated-cellulose nanocrystals obtained after HCl/critic acid hydrolysis performed better, and UV-assisted degradation of methylene blue was above 99.9% removal after 4 h [97]. In other studies, the efficient removal (80%–99%) of methylene blue, Rhodamine B, and methyl orange via sorption and/or degradation has been reported using different cellulose-based nanomaterials such as CoPc@BC and amphoteric poly(vinyl amine)-cellulose nanocrystals [98–100]. Notably, Derami et al. [101] developed a robust hybrid polydopamine/bacterial nanocellulose membrane and assessed the sorption of two heavy metals (Pb^{2+} and Cd^{2+}) and three dyes (methylene blue, Rhodamine B, and methyl orange). The hybrid membrane exhibited effective removal of the selected pollutants and maintain its performance even after 10 regeneration cycles [101].

21.3 Application of chitin- and chitosan-based nanomaterials in water treatment

Chitin and/or chitosan is the second abundant biopolymer in nature after cellulose. It is essentially an amino carbohydrate, which can be extracted from a range of sources such as shellfish, microorganisms (e.g., yeast and green/brown algae), and insects (e.g., spiders) [102]. On the other hand, deacetylation of chitin is required for the preparation of chitosan, which features both the acetylated and deacetylated units linked via glycosidic bonds (Fig. 21.3A). Chitin extraction from shrimp shells and other sources can be achieved by using alkaline solvents of different strength such as 30%–60% NaOH (w/w) at 80–160 °C [103–105], while chitosan of different molecular weights could be produced via hydrolysis using chitin deacetylase enzyme [105]. Chitin and chitosan are cheap and renewable biopolymers, and the main source is shellfish waste that is produced in massive quantities by the shellfish industry [102].

It is important to note that the availability of nitrogen contents (approximately 7% by weight) in chitin/chitosan makes them ideal for the preparation of N-doped carbon (nano)materials [102]. A few examples of N-doped carbon (nano)materials include porous carbon, carbon dots, and graphene (Fig. 21.3B). Physicochemical properties (Fig. 21.3C) of chitosan are governed by the functional groups (e.g., hydroxy and amino) present in its structure as well as its polymeric properties [105]. On the other hand, the chitosan source and extraction method could influence its molecular weight and deacetylation, as well as the crystallinity [106]. Due to the presence of different functional groups, chitosan-based materials exhibit strong chelating properties and gelation ability [102,107]. In addition, despite the presence of hydrophilic hydroxyl and amino functional groups, chitosan remains significantly hydrophobic, which allows its use as a sorbent and/or modification [108]. To date, chitin- and chitosan-based nanomaterials have been assessed for a range of applications such as water/wastewater treatment, drug delivery, biocatalysis, and cell immobilization [109–111].

As mentioned above, chitosan is a cheap and renewable biopolymer. Various chitosan-based nanomaterials (Table 21.2) have been developed and have demonstrated promising results for the removal of organic pollutants [36–46]. During conventional water treatment processes, removal of pollutants such as heavy metals can be achieved in the coagulation/flocculation process that requires the addition of coagulants such as inorganic salts (e.g., alum and $FeCl_3$) [112]. In the last decade, naturally occurring polymers have replaced inorganic salts and/or synthetic polymers as coagulants and have become a preferred option [111,113]. Indeed, the performance

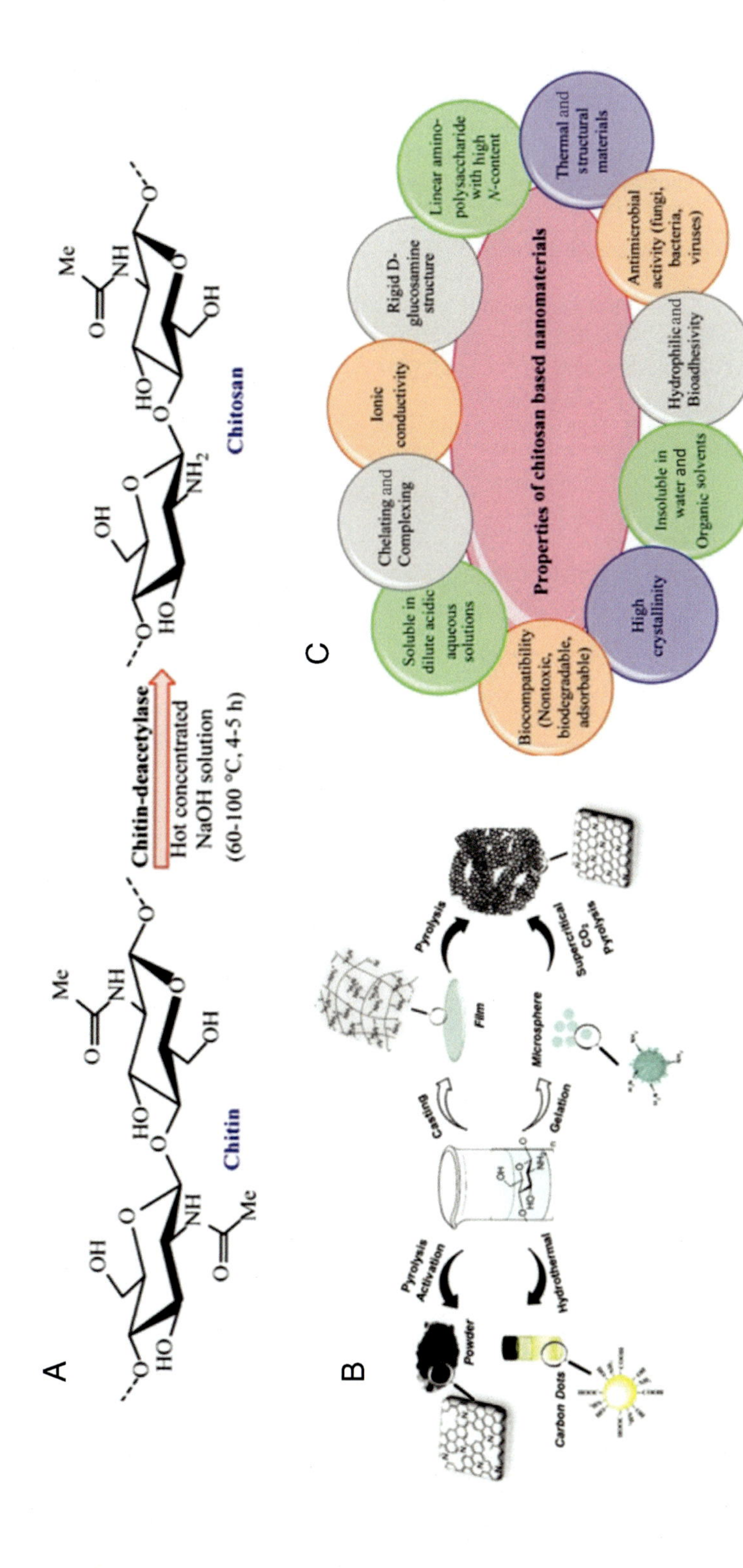

FIGURE 21.3 Chitosan production via chitin deacetylation (A), synthesis of N-doped carbon (nano)materials using chitosan (B), and physicochemical properties of chitosan-based nanomaterials (C). Reproduced with permission in part from Ref. [21] and Ref. [102] (Copyright 2020 Elsevier Ltd.).

Table 21.2 A selection of studies on chitosan-based nanomaterials for organic pollutant removal during water/wastewater treatment.

Chitosan-based nanomaterials	Pollutant and experimental conditions	Maximum sorption capacity (mg/g)	Isotherm model/reaction kinetics	Reference
Chitosan/graphene nanoplates	Methyl orange and acid red 1 pH = 3–4, Temperature = 298 K, Time = 60 min	230 and 133	Freundlich/pseudo second-order	[36]
β-cyclodextrin/chitosan/GO hydrogel	Methylene blue pH = 12, Temperature = 298 K, Time = 2 h	1134	Freundlich/pseudo second-order	[37]
Graphene-chitosan hydrogel	Congo red pH = 7, Temperature = r.t., Time = 10 min	384	Langmuir/pseudo second-order	[38]
GO/chitosan	Picric acid pH = 7, Temperature = 298 K, Time = 5 h	263	Freundlich/pseudo second-order	[39]
Magnetic chitosan/GO	Fluoxetine pH = 4.5, Temperature = 298 K, Time = 10 min	66	Langmuir/pseudo second-order	[40]
Ionic liquid–modified magnetic chitosan/GO	Methylene blue pH = 12, Temperature = 303 K, Time = 1 h	243	Langmuir/pseudo second-order	[41]
Graphene/Fe_3O_4/chitosan	Methylene blue pH = 9, Time = 5 min	249	Langmuir/pseudo second-order	[42]
GO/chitosan nanofibers	Fuchsin acid Time = 750 min	175	Langmuir/pseudo second-order	[43]
Magnetic β-cyclodextrin–Chitosan/GO	Methylene blue pH = 2–11, Temperature = 298 K, Time = 80 min	84	Langmuir/pseudo second-order	[44]
GO/chitosan	Reactive Black 5 pH = 2, Temperature = 298 K, Time = 24 h	277	Langmuir and Freundlich/Brouers–Sotolongo	[45]
GO/magnetic cyclodextrin–Chitosan	Hydroquinone pH = 6, Temperature = 303 K, Time = 40 min	458	Freundlich/pseudo second-order	[46]

GO, graphene oxide; r.t., room temperature.

of chitosan (an ecofriendly coagulant) has been assessed for the removal of dyes, heavy metals, and other pollutants [114,115]. Chitosan is positively charged in acidic media and interacts with pollutants via ion exchange and electrostatic interaction, while the nonprotonated amino groups in chitosan facilitate the formation of complexes with metal and organic pollutants [21]. In a study by Rajeswari et al. [116], two chitosan-based composites were prepared for the removal of nitrate ions. They reported a sorption capacity of approximately 35 mg/g for polyvinyl alcohol-chitosan and 51 mg/g for PEG-chitosan [116]. Similarly, selective removal of Pb^{2+} and Cr^{6+}/Cd^{2+} using goethite/chitosan nanocomposites [117] and carboxymethyl chitosan [118], respectively, was reported. Chitosan-based polymers may not be highly water-soluble. Nevertheless, a highly water-soluble (2-methacryloyloxyethyl) trimethyl ammonium chloride grafted-chitosan flocculant was prepared by Wang et al. [119], which outperformed polyacrylamide for the removal of bulk organics, lignin, and turbidity during the treatment of textile wastewater.

Notably, owing to its ion exchange properties, chitosan can be used to improve the stability of conductive polymers during capacitive deionization (CDI), which could be due to the interaction between $-COOH/-NH_2$ functional groups of chitosan and conductive polymers such as polypyrrole [120–122]. In a study by Zhang et al. [122], the nanoelectrodes for the CDI process prepared using polypyrrole, chitosan, and carbon nanotubes were reported to have an adsorption capacity of 16.83 mg/g for Cu^{2+}. Importantly, the fabricated nanocomposites showed stability and effective performance even after 50 cycles. In addition to improving the performance and stability of conductive materials, impregnation of nanoparticles on biopolymers such as chitosan can also improve their stability and could also prevent the aggregation of nanoparticles [21,123]. Notably, the sorption capacity of chitosan can be enhanced by a fusion of nanoparticles on their surface [124]. In previous studies, a fusion of Fe_3O_4 and TiO_2 in chitosan-based materials has been reported [125,126]. For instance, TiO_2/chitosan/Cu beads were studied for the removal of arsenic from an aqueous solution [127]. They reported that the photolysis of the media resulted in the transformation of As^{3+} to As^{5+} as well as effective sorption of both As^{3+} and As^{5+} in the presence of phosphate [127].

The performance of chitosan-based materials has been extensively studied for the removal of dyes [36–46]. According to the available literature, functional groups of chitosan, particularly -OH functional groups are effective for the sorption of dyes [21]. In a study by Gibbs et al. [128], an increase in the relative protonation of amine groups by diminishing the extent of acetylation of chitosan was observed to favor the adsorption of acid green 25 with a maximum adsorption capacity of 525 mg/g [128]. Nevertheless, other factors such as the properties of dyes and repartitioning of acetyl groups can influence the performance of the chitosan-based nanomaterials [21,129,130]. In a study by Marrakchi et al. [131], chitosan/sepiolite composites were prepared via a cross-linking method for the removal of methylene blue and reactive orange 16 dyes. The developed chitosan/sepiolite composite had a maximum adsorption capacity of approximately 191 mg/g for methylene blue and 41 mg/g for reactive orange 16. Although the comparison based on the adsorption capacity for different chitosan-based materials may not be appropriate due to the difference in operating conditions, the performance of chitosan/sepiolite composite [131] in terms of maximum adsorption capacity (191 mg/g) for methylene blue was better than chitosan/zeolite [132], chitosan/organic rectorite [133] and clay/biochar composites [134] with a reported maximum adsorption capacity of 37, 25, and 12 mg/g, respectively.

21.4 Application of other biopolymer-derived nanomaterials in water treatment

Various other biopolymer-based materials such as starch-, alginate-, and gum-based have been developed for the removal of pollutants during water treatment [47–55]. In this section, the

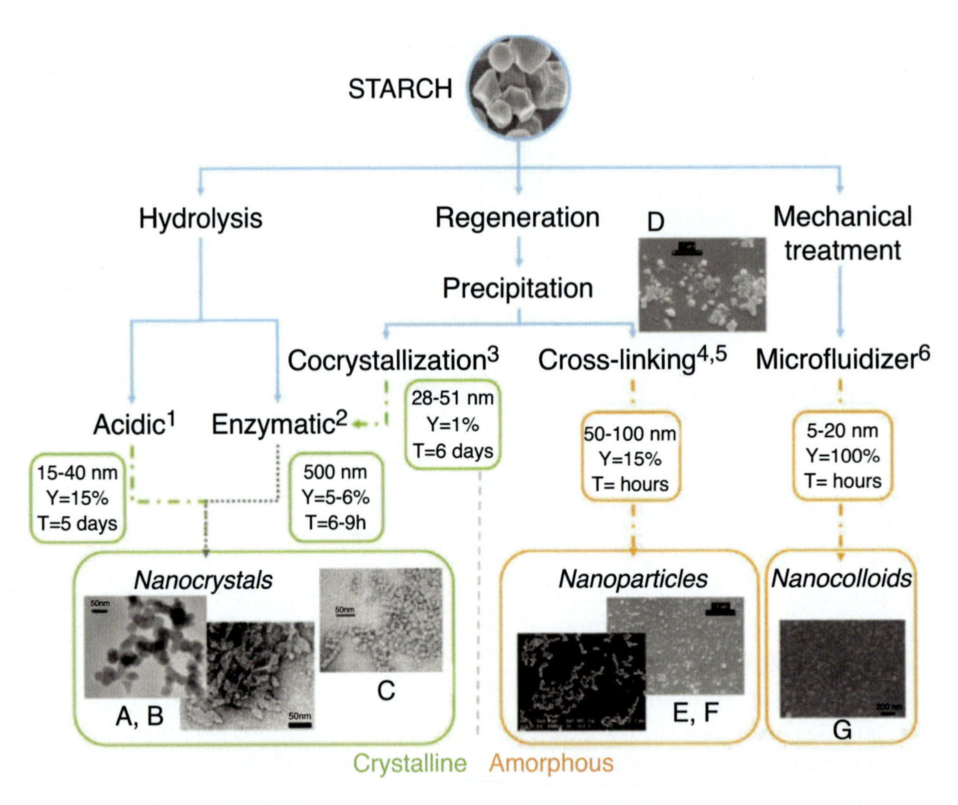

FIGURE 21.4 Methods for the preparation of starch-based amorphous and crystalline materials. Transmission electron microscopy images of starch nanocrystals produced via acidic hydrolysis (A), enzymatic hydrolysis, (B), and co-crystallization/enzymatic hydrolysis (C) are presented. In addition, starch nanoparticles before cross-linking (D), the finished product of starch nanoparticles (E), starch nanoparticles cross-linked using citric acid (F), and starch-based nanocolloids (G) are shown. References for 1, 2, 3, 4, 5, and 6 are [137,142,143,144,145], and [146], respectively. Reproduced with permission from Ref. [137] (Copyright 2010 American Chemical Society).

current status and properties of starch- and gum-based materials are discussed. Starch is one of the naturally occurring biopolymers, which is abundantly available and has attracted attention due to its renewability, biocompatibility, and biodegradability [21]. Starch is a reserve carbohydrate that can be extracted from different parts (e.g., roots and stalks) of plants/crops such as wheat, maize, rice, or corn. According to the available literature, starch has a 3D structure along with a range of crystallinity (15%–45%) [21,135,136]. All starch-based nanomaterials such as starch nanocrystal and hydrolyzed starch contain crystallinity and could be produced via hydrolysis [137]. Different methods for the preparation of starch-based amorphous and crystalline materials are illustrated in Fig. 21.4. In general, a modification of starch by mixing fibrous clays to improve its physicochemical properties has been preferred, and the resultant starch-based composites have been recognized as environmentally friendly and biodegradable sorbents for application in water treatment (Table 21.3) [138,139]. It is noteworthy that starch-based

Table 21.3 A selection of studies on other biopolymer-based nanomaterials for organic pollutant removal during water/wastewater treatment.

Material name	Pollutant and experimental conditions	Maximum sorption capacity (mg/g)	Isotherm model/reaction kinetics	Reference
Xanthan gum-cl-poly (acrylic acid)/rGO Hydrogel	Methyl blue pH = 5, Temperature = 298 K, Time = 90 min	793	Langmuir/pseudo second-order	[48]
Starch/poly(acrylamide)/ GO/n-HAP hydrogel	Malachite green pH = 10, Temperature = 298 K, Time = 60 min	297	Langmuir/pseudo second-order	[49]
Biochar-supported rGO	Atrazine pH = 6, Temperature = 298 K, Time = 24 h	67	Langmuir/pseudo second-order	[50]
Agricultural waste/GO	Methyl blue pH = 12, Temperature = 298 K, Time = 3 h	414	Temkin/pseudo second-order	[51]
PVA/CMC/GO/bentonite hydrogels	Methyl blue pH = 8, Temperature = 298 K, Time = 2.5 h	172	Langmuir/pseudo second-order	[52]
Magnetic β-cyclodextrin-GO	p-Phenylenediamine pH = 8, Temperature = 318 K, Time = 2 h	1120	Langmuir/pseudo second-order	[53]
β-cyclodextrin/poly (l-glutamic acid) supported magnetic GO	17b-estradiol pH = 7, Temperature = 298 K, Time = 12 h	86	Langmuir/pseudo second-order	[54]
GO/calcium alginate	Ciprofloxacin pH = 6, Temperature = 293 K, Time = 24 h	39	Langmuir/pseudo second-order	[55]
$GO_x/MnFe_2O_4$/Calcium alginate	Methyl blue pH = 10, Temperature = 303 K, Time = 2.5 h	74% removal	-	[47]

GO, graphene oxide; n-HAP, nanohydroxyapatite; PVA, polyvinyl alcohol; CMC, carboxymethyl cellulose.

nanomaterials, particularly starch nanocrystals can act as an ideal support in catalysis reactions due to their cost-effectiveness and eco-friendliness [140,141].

Gums are another type of polysaccharide-derived biopolymers characterized as environmentally friendly and have been used for different applications in pharmaceuticals and food industries [21]. Guar gum, which consists of linear chains of mannopyranosyl units, can be extracted from guar beans, while gum arabic (also known as acacia gum) is one of the oldest known gums and could be extracted from trees such as *Acacia seyal* [147,148]. Notably, owing to their ability to modify rheological properties, both guar gum and gum Arabic can act as a thickening agent, binder, and ion-exchange resin [149]. In addition, stabilization of nanomaterials and their reduction using gums have been reported in the literature, while gums could be used for nanocatalysts preparation and biosorbents during water/wastewater treatment [24,149–153].

21.5 Conclusion and perspectives

Naturally occurring polymers, particularly polysaccharide-derived resources extracted from a range of derivatives are futuristic and have been used for different applications such as food, cosmetics, drug delivery, and water treatment. This is because biopolymers have peculiar properties and are nontoxic, biodegradable, and renewable as well as easily amenable during the preparation of polymeric nanomaterials. This chapter mainly presented the main features of polysaccharide-based nanomaterials (such as cellulose and chitosan), and their applications in water treatment as biosorbents and nanocatalysts are summarized. The review of the existing literature suggests that polysaccharide-based biopolymers can be considered attractive support for the fabrication of nanocomposites and nanocatalysts. These nanocomposites and nanocatalysts can achieve an effective removal of different pollutants such as heavy metals, dyes, and micropollutants. Despite the advances in the utilization of biopolymers for application in water treatment, future research should consider the following aspects:

- Extraction of biopolymers (e.g., cellulose) and preparation of biopolymeric nanomaterials could be energy intensive. Process optimization and scalable fabrication of biopolymeric nanomaterials should be focused on wide-scale applications.
- Synthesis of nanocomposites and nanocatalysts using natural supports such as clays and montmorillonite may be considered for cost effectiveness.
- Waste feedstock (biowaste) and animal waste could be used for the synthesis of nanocomposites and nanocatalysts, which would not only reduce the solid waste but also would be a cost-effective approach.
- Biopolymer-based magnetic polymeric nanomaterials/nanocatalysts require attention in future studies because of the ease of separation during water treatment.

Acknowledgment

This research is supported by the project funded by China Postdoctoral Science Foundation (2021M691773). Funding by the National Natural Science Foundation of China (51708325), and the Committee of Science and Technology Innovation of Shenzhen (KQJSCX20180320171226768; JCYJ20190813163401660) is also appreciated.

References

[1] H. Hoff, in: Understanding the Nexus, Background paper for the Bonn Conference 2011: the Water-Energy-Food Security Nexus, Stockholm, Stockholm Environment Institute, 2011.

[2] C.A. Scott, M. Kurian, J.L. Wescoat, The water-energy-food nexus: enhancing adaptive capacity to complex global challenges, in: M. Kurian, R. Ardakanian (Eds.), Governing the nexus: Water, Soil and Waste Resources Considering Global Change, Springer, Basel, Switzerland, 2015, pp. 15–38.

[3] M. Kurian, The water-energy-food nexus: trade-offs, thresholds and transdisciplinary approaches to sustainable development, Environ. Sci. Policy 68 (2017) 97–106.

[4] E.M. Biggs, E. Bruce, B. Boruff, J.M. Duncan, J. Horsley, N. Pauli, K. McNeill, A. Neef, F. Van Ogtrop, J. Curnow, Sustainable development and the water–energy–food nexus: a perspective on livelihoods, Environ. Sci. Policy 54 (2015) 389–397.

[5] C. Ringler, A. Bhaduri, R. Lawford, The nexus across water, energy, land and food (WELF): potential for improved resource use efficiency? Curr. Opin. Environ. Sustain. 5 (2013) 617–624.

[6] M. Howells, S. Hermann, M. Welsch, M. Bazilian, R. Segerström, T. Alfstad, D. Gielen, H. Rogner, G. Fischer, H. Van Velthuizen, Integrated analysis of climate change, land-use, energy and water strategies, Nat. Clim. Change 3 (2013) 621.

[7] C.Y. Tang, Z. Yang, H. Guo, J.J. Wen, L.D. Nghiem, E. Cornelissen, Potable water reuse through advanced membrane technology, Environ. Sci. Technol. 52 (2018) 10215–10223.

[8] F.I. Hai, L.D. Nghiem, S.J. Khan, M.B. Asif, W.E. Price, K. Yamamoto, Removal of emerging trace organic contaminants (TrOC) by MBR, in: F.I. Hai, K. Yamamoto, C. Lee (Eds.), Membrane Biological Reactors: Theory, Modeling, Design, Management and Applications to Wastewater Reuse, DOI, IWA publishing, London, United Kingdom, 2019, pp. 413–468. (ISBN: 9781780409177; DOI:https://doi.org/9781780409110.9781780402166/9781780409177).

[9] J.B. Zimmerman, J.R. Mihelcic, J. Smith, Global stressors on water quality and quantity, Environ. Sci. Technol. 42 (2008) 4247–4254.

[10] N. Mancosu, R. Snyder, G. Kyriakakis, D. Spano, Water scarcity and future challenges for food production, Water 7 (2015) 975–992.

[11] M.B. Asif, C. Li, B. Ren, T. Maqbool, X. Zhang, Z. Zhang, Elucidating the impacts of intermittent in-situ ozonation in a ceramic membrane bioreactor: micropollutant removal, microbial community evolution and fouling mechanisms, J. Hazard. Mater. 402 (2021) 123730.

[12] M.B. Asif, Z. Fida, A. Tufail, J.P. van de Merwe, F.D.L. Leusch, B.K. Pramanik, W.E. Price, F.I. Hai, Persulfate oxidation-assisted membrane distillation process for micropollutant degradation and membrane fouling control, Sep. Purif. Technol. 222 (2019) 321–331.

[13] M.B. Asif, R. Habib, S. Iftekharb, Z. Khan, N. Majeedd, Optimization of the operational parameters in a submerged membrane bioreactor using box behnken response surface methodology: membrane fouling control and effluent quality, Desalin. Water Treat. 82 (2017) 26–38.

[14] M.B. Asif, Z. Khan, Characterization and treatment of flour mills wastewater for reuse – a case study of Al-kausar flour mills, Pakistan, Desalin. Water Treat. 57 (2016) 3881–3890.

[15] M.B. Asif, B. Ren, C. Li, T. Maqbool, X. Zhang, Z. Zhang, Powdered activated carbon – membrane bioreactor (PAC-MBR): Impacts of high PAC concentration on micropollutant removal and microbial communities, Sci. Total Environ. 745 (2020) 141090.

[16] M.B. Asif, B. Ren, C. Li, T. Maqbool, X. Zhang, Z. Zhang, Evaluating the impacts of a high concentration of powdered activated carbon in a ceramic membrane bioreactor: mixed liquor properties, hydraulic performance and fouling mechanism, J. Membr. Sci. 616 (2020) 118561.

[17] M.B. Asif, J.P. van de Merwe, F.D.L. Leusch, B.K. Pramanik, W.E. Price, F.I. Hai, Elucidating the performance of an integrated laccase- and persulfate-assisted process for degradation of trace organic contaminants (TrOCs), Environ. Sci. Water Res. Technol. 6 (2020) 1069–1082.

[18] R. Habib, M.B. Asif, S. Iftekhar, Z. Khan, K. Gurung, V. Srivastava, M. Sillanpää, Influence of relaxation modes on membrane fouling in submerged membrane bioreactor for domestic wastewater treatment, Chemosphere 181 (2017) 19–25.

[19] S. Iftekhar, M.U. Farooq, M. Sillanpää, M.B. Asif, R. Habib, Removal of Ni(II) using multi-walled carbon nanotubes electrodes: relation between operating parameters and capacitive deionization performance, Arab. J. Sci. Eng. 42 (2017) 235–240.

[20] S. Iftekhar, D.L. Ramasamy, V. Srivastava, M.B. Asif, M. Sillanpää, Understanding the factors affecting the adsorption of Lanthanum using different adsorbents: a critical review, Chemosphere 204 (2018) 413–430.

[21] M. Nasrollahzadeh, M. Sajjadi, S. Iravani, R.S. Varma, Starch, cellulose, pectin, gum, alginate, chitin and chitosan derived (nano)materials for sustainable water treatment: a review, Carbohydr. Polym. 251 (2021) 116986.

[22] R. Gusain, N. Kumar, S.S. Ray, Recent advances in carbon nanomaterial-based adsorbents for water purification, Coord. Chem. Rev. 405 (2020) 213111.

[23] R.S. Varma, Journey on greener pathways: from the use of alternate energy inputs and benign reaction media to sustainable applications of nano-catalysts in synthesis and environmental remediation, Green Chem. 16 (2014) 2027–2041.

[24] S. Iftekhar, V. Srivastava, A. Casas, M. Sillanpää, Synthesis of novel GA-g-PAM/SiO_2 nanocomposite for the recovery of rare earth elements (REE) ions from aqueous solution, J. Clean. Prod. 170 (2018) 251–259.

[25] P. Liu, H. Sehaqui, P. Tingaut, A. Wichser, K. Oksman, A.P. Mathew, Cellulose and chitin nanomaterials for capturing silver ions (Ag^+) from water via surface adsorption, Cellulose 21 (2014) 449–461.

[26] P. Liu, P.F. Borrell, M. Božič, V. Kokol, K. Oksman, A.P. Mathew, Nanocelluloses and their phosphorylated derivatives for selective adsorption of Ag^+, Cu^{2+} and Fe^{3+} from industrial effluents, J. Hazard. Mater. 294 (2015) 177–185.

[27] K. Singh, J.K. Arora, T.J.M. Sinha, S. Srivastava, Functionalization of nanocrystalline cellulose for decontamination of Cr(III) and Cr(VI) from aqueous system: computational modeling approach, Clean Technol. Environ. Policy 16 (2014) 1179–1191.

[28] A.D. Dwivedi, S.P. Dubey, S. Hokkanen, M. Sillanpää, Mechanistic investigation on the green recovery of ionic, nanocrystalline, and metallic gold by two anionic nanocelluloses, Chem. Eng. J. 253 (2014) 316–324.

[29] L. Jin, W. Li, Q. Xu, Q. Sun, Amino-functionalized nanocrystalline cellulose as an adsorbent for anionic dyes, Cellulose 22 (2015) 2443–2456.

[30] Q. Zheng, Z. Cai, S. Gong, Green synthesis of polyvinyl alcohol (PVA)–cellulose nanofibril (CNF) hybrid aerogels and their use as superabsorbents, J. Mater. Chem. A 2 (2014) 3110–3118.

[31] X. Yu, S. Tong, M. Ge, L. Wu, J. Zuo, C. Cao, W. Song, Adsorption of heavy metal ions from aqueous solution by carboxylated cellulose nanocrystals, J. Environ. Sci. 25 (2013) 933–943.

[32] Y. Wang, S. Yadav, T. Heinlein, V. Konjik, H. Breitzke, G. Buntkowsky, J.J. Schneider, K. Zhang, Ultra-light nanocomposite aerogels of bacterial cellulose and reduced graphene oxide for specific absorption and separation of organic liquids, RSC Adv. 4 (2014) 21553–21558.

[33] Z. Karim, A.P. Mathew, M. Grahn, J. Mouzon, K. Oksman, Nanoporous membranes with cellulose nanocrystals as functional entity in chitosan: removal of dyes from water, Carbohydr. Polym. 112 (2014) 668–676.

[34] G. Wang, J. Zhang, S. Lin, H. Xiao, Q. Yang, S. Chen, B. Yan, Y. Gu, Environmentally friendly nanocomposites based on cellulose nanocrystals and polydopamine for rapid removal of organic dyes in aqueous solution, Cellulose 27 (2020) 2085–2097.

[35] C. Li, H. Ma, S. Venkateswaran, B.S. Hsiao, Highly efficient and sustainable carboxylated cellulose filters for removal of cationic dyes/heavy metals ions, Chem. Eng. J. 389 (2020) 123458.

[36] C. Zhang, Z. Chen, W. Guo, C. Zhu, Y. Zou, Simple fabrication of chitosan/graphene nanoplates composite spheres for efficient adsorption of acid dyes from aqueous solution, Int. J. Biol. Macromol. 112 (2018) 1048–1054.

[37] Y. Liu, S. Huang, X. Zhao, Y. Zhang, Fabrication of three-dimensional porous β-cyclodextrin/chitosan functionalized graphene oxide hydrogel for methylene blue removal from aqueous solution, Coll. Surf. A Physicochem. Eng. Asp. 539 (2018) 1–10.

[38] S. Omidi, A. Kakanejadifard, Eco-friendly synthesis of graphene–chitosan composite hydrogel as efficient adsorbent for Congo red, RSC Adv. 8 (2018) 12179–12189.

[39] M. Mohseni Kafshgari, H. Tahermansouri, Development of a graphene oxide/chitosan nanocomposite for the removal of picric acid from aqueous solutions: study of sorption parameters, Coll. Surf. B Biointerfaces 160 (2017) 671–681.

[40] A. Barati, E. Kazemi, S. Dadfarnia, A.M. Haji Shabani, Synthesis/characterization of molecular imprinted polymer based on magnetic chitosan/graphene oxide for selective separation/preconcentration of fluoxetine from environmental and biological samples, J. Ind. Eng. Chem. 46 (2017) 212–221.

[41] L. Li, F. Liu, H. Duan, X. Wang, J. Li, Y. Wang, C. Luo, The preparation of novel adsorbent materials with efficient adsorption performance for both chromium and methylene blue, Coll. Surf. B Biointerfaces 141 (2016) 253–259.

[42] N. Van-Hoa, T.T. Khong, T.Thi Hoang Quyen, T.Si Trung, One-step facile synthesis of mesoporous graphene/Fe_3O_4/chitosan nanocomposite and its adsorption capacity for a textile dye, J. Water Process Eng. 9 (2016) 170–178.

[43] Y. Li, J. Sun, Q. Du, L. Zhang, X. Yang, S. Wu, Y. Xia, Z. Wang, L. Xia, A. Cao, Mechanical and dye adsorption properties of graphene oxide/chitosan composite fibers prepared by wet spinning, Carbohydr. Polym. 102 (2014) 755–761.

[44] L. Fan, C. Luo, M. Sun, H. Qiu, X. Li, Synthesis of magnetic β-cyclodextrin–chitosan/graphene oxide as nanoadsorbent and its application in dye adsorption and removal, Coll. Surf. B Biointerfaces 103 (2013) 601–607.

[45] N.A. Travlou, G.Z. Kyzas, N.K. Lazaridis, E.A. Deliyanni, Graphite oxide/chitosan composite for reactive dye removal, Chem. Eng. J. 217 (2013) 256–265.

[46] L. Li, L. Fan, M. Sun, H. Qiu, X. Li, H. Duan, C. Luo, Adsorbent for hydroquinone removal based on graphene oxide functionalized with magnetic cyclodextrin–chitosan, Int. J. Biol. Macromol. 58 (2013) 169–175.

[47] P. Zolfaghari, R. Shojaat, A. Karimi, N. Saadatjoo, Application of fluidized bed reactor containing GOx/$MnFe_2O_4$/calcium alginate nano-composite in degradation of a model pollutant, J. Environ. Chem. Eng. 6 (2018) 6414–6420.

[48] E. Makhado, S. Pandey, J. Ramontja, Microwave assisted synthesis of xanthan gum-cl-poly (acrylic acid) based-reduced graphene oxide hydrogel composite for adsorption of methylene blue and methyl violet from aqueous solution, Int. J. Biol. Macromol. 119 (2018) 255–269.

[49] H. Hosseinzadeh, S. Ramin, Fabrication of starch-graft-poly(acrylamide)/graphene oxide/hydroxyapatite nanocomposite hydrogel adsorbent for removal of malachite green dye from aqueous solution, Int. J. Biol. Macromol. 106 (2018) 101–115.

[50] Y. Zhang, B. Cao, L. Zhao, L. Sun, Y. Gao, J. Li, F. Yang, Biochar-supported reduced graphene oxide composite for adsorption and coadsorption of atrazine and lead ions, Appl. Surf. Sci. 427 (2018) 147–155.

[51] S. Liu, H. Ge, C. Wang, Y. Zou, J. Liu, Agricultural waste/graphene oxide 3D bio-adsorbent for highly efficient removal of methylene blue from water pollution, Sci. Total Environ. 628-629 (2018) 959–968.

[52] H. Dai, Y. Huang, H. Huang, Eco-friendly polyvinyl alcohol/carboxymethyl cellulose hydrogels reinforced with graphene oxide and bentonite for enhanced adsorption of methylene blue, Carbohydr. Polym. 185 (2018) 1–11.

[53] D. Wang, L. Liu, X. Jiang, J. Yu, X. Chen, X. Chen, Adsorbent for p-phenylenediamine adsorption and removal based on graphene oxide functionalized with magnetic cyclodextrin, Appl. Surf. Sci. 329 (2015) 197–205.

[54] L. Jiang, Y. Liu, S. Liu, X. Hu, G. Zeng, X. Hu, S. Liu, S. Liu, B. Huang, M. Li, Fabrication of β-cyclodextrin/poly (l-glutamic acid) supported magnetic graphene oxide and its adsorption behavior for 17β-estradiol, Chem. Eng.J. 308 (2017) 597–605.

[55] S. Wu, X. Zhao, Y. Li, C. Zhao, Q. Du, J. Sun, Y. Wang, X. Peng, Y. Xia, Z. Wang, L. Xia, Adsorption of ciprofloxacin onto biocomposite fibers of graphene oxide/calcium alginate, Chem. Eng. J. 230 (2013) 389–395.

[56] C.M. Ewulonu, X. Liu, M. Wu, H. Yong, Lignin-containing cellulose nanomaterials: a promising new nanomaterial for numerous applications, J. Bioresour. Bioprod. 4 (2019) 3–10.

[57] M. Muqeet, R.B. Mahar, T.A. Gadhi, N.Ben Halima, Insight into cellulose-based-nanomaterials—a pursuit of environmental remedies, Int. J. Biol. Macromol. 163 (2020) 1480–1486.

[58] A.W. Carpenter, C.-F. de Lannoy, M.R. Wiesner, Cellulose nanomaterials in water treatment technologies, Environ. Sci. Technol. 49 (2015) 5277–5287.

[59] P.Z. Ray, H.J. Shipley, Inorganic nano-adsorbents for the removal of heavy metals and arsenic: a review, RSC Adv. 5 (2015) 29885–29907.

[60] H. Bessaies, S. Iftekhar, M.B. Asif, J. Kheriji, C. Necibi, M. Sillanpää, B. Hamrouni, Characterization and physicochemical aspects of novel cellulose-based layered double hydroxide nanocomposite for removal of antimony and fluoride from aqueous solution, J. Environ. Sci. 102 (2021) 301–315.

[61] H. Bessaies, S. Iftekhar, B. Doshi, J. Kheriji, M.C. Ncibi, V. Srivastava, M. Sillanpää, B. Hamrouni, Synthesis of novel adsorbent by Intercalation of biopolymer in LDH for the removal of arsenic from synthetic and natural water, J. Environ. Sci. 91 (2020) 246–261.

[62] S. Iftekhar, V. Srivastava, M. Abdul Wasayh, M. Hezarjaribi, M. Sillanpää, Incorporation of inorganic matrices through different routes to enhance the adsorptive properties of xanthan via adsorption and membrane separation for selective REEs recovery, Chem. Eng. J. 388 (2020) 124281.

[63] F. Hansen, V. Brun, E. Keller, W. Nieh, T. Wegner, M. Meador, L. Friedersdorf, Cellulose Nanomaterials—A Path Towards Commercialization Workshop Report, 2014, USDA Forest Service, Washington D.C, 2014, p. 21.

[64] W.L.-S. Nieh, Current international standards development activities for cellulose nanomaterials, in: M.T. Postek, R.J. Moon, A.W. Rudie, M.A. Bilodeau (Eds.), Production and Applications of Cellulose Nanomaterials, TAPPI Press, Peachtree Corners, GA, 2013, pp. 2013–2014.

[65] R.J. Moon, A. Martini, J. Nairn, J. Simonsen, J. Youngblood, Cellulose nanomaterials review: structure, properties and nanocomposites, Chem. Soc. Rev. 40 (2011) 3941–3994.

[66] P. Gatenholm, D. Klemm, Bacterial nanocellulose as a renewable material for biomedical applications, MRS Bull. 35 (2011) 208–213.

[67] F. Jiang, Y.-L. Hsieh, Chemically and mechanically isolated nanocellulose and their self-assembled structures, Carbohydr. Polym. 95 (2013) 32–40.

[68] C.J. Chirayil, L. Mathew, S. Thomas, Review of recent research in nano cellulose preparation from different lignocellulosic fibers, Rev. Adv. Mater. Sci. 37 (1) (2014) 20–28.

[69] K.H.P.S. Khalil, Y. Davoudpour, M.N. Islam, A. Mustapha, K. Sudesh, R. Dungani, M. Jawaid, Production and modification of nanofibrillated cellulose using various mechanical processes: a review, Carbohydr. Polym. 99 (2014) 649–665.

[70] L. Brinchi, F. Cotana, E. Fortunati, J.M. Kenny, Production of nanocrystalline cellulose from lignocellulosic biomass: technology and applications, Carbohydr. Polym. 94 (2013) 154–169.

[71] D. Klemm, F. Kramer, S. Moritz, T. Lindström, M. Ankerfors, D. Gray, A. Dorris, Nanocelluloses: a new family of nature-based materials, Angew. Chem. Int. Ed. 50 (2011) 5438–5466.

[72] Q. Li, S. McGinnis, C. Sydnor, A. Wong, S. Renneckar, Nanocellulose life cycle assessment, ACS Sustain. Chem. Eng. 1 (2013) 919–928.

[73] Z.N. Garba, W. Zhou, I. Lawan, M. Zhang, Z. Yuan, Enhanced removal of prometryn using copper modified microcrystalline cellulose (Cu-MCC): optimization, isotherm, kinetics and regeneration studies, Cellulose 26 (2019) 6241–6258.

[74] R.M. Abdelhameed, H. Abdel-Gawad, M. Elshahat, H.E. Emam, Cu–BTC@cotton composite: design, removal of ethion insecticide from water, RSC Adv. 6 (2016) 42324–42333.

[75] W. Huang, N. Liu, X. Zhang, M. Wu, L. Tang, Metal organic framework g-C3N4/MIL-53(Fe) heterojunctions with enhanced photocatalytic activity for Cr(VI) reduction under visible light, Appl. Surf. Sci. 425 (2017) 107–116.

[76] S.Z.N. Ahmad, W.N. Wan Salleh, A.F. Ismail, N. Yusof, M.Z. Mohd Yusop, F. Aziz, Adsorptive removal of heavy metal ions using graphene-based nanomaterials: Toxicity, roles of functional groups and mechanisms, Chemosphere 248 (2020) 126008.

[77] D.R. Wallace, A. Buha Djordjevic, Heavy metal and pesticide exposure: a mixture of potential toxicity and carcinogenicity, Curr. Opin. Toxicol. 19 (2020) 72–79.

[78] S. Srivastava, A. Kardam, K.R. Raj, Nanotech reinforcement onto cellulosic fibers: green remediation of toxic metals, Int. J. Green Nanotechnol. 4 (2012) 46–53.

[79] R. Yang, K.B. Aubrecht, H. Ma, R. Wang, R.B. Grubbs, B.S. Hsiao, B. Chu, Thiol-modified cellulose nanofibrous composite membranes for chromium (VI) and lead (II) adsorption, Polymer 55 (2014) 1167–1176.

[80] X. Ma, Y. Lou, X.-B. Chen, Z. Shi, Y. Xu, Multifunctional flexible composite aerogels constructed through in-situ growth of metal-organic framework nanoparticles on bacterial cellulose, Chem. Eng. J. 356 (2019) 227–235.

[81] C. Zhou, Q. Wu, T. Lei, I.I. Negulescu, Adsorption kinetic and equilibrium studies for methylene blue dye by partially hydrolyzed polyacrylamide/cellulose nanocrystal nanocomposite hydrogels, Chem. Eng. J. 251 (2014) 17–24.

[82] Y. Zhou, S. Fu, L. Zhang, H. Zhan, M.V. Levit, Use of carboxylated cellulose nanofibrils-filled magnetic chitosan hydrogel beads as adsorbents for Pb(II), Carbohydr. Polym. 101 (2014) 75–82.

[83] H. Zhu, S. Jia, T. Wan, Y. Jia, H. Yang, J. Li, L. Yan, C. Zhong, Biosynthesis of spherical Fe_3O_4/bacterial cellulose nanocomposites as adsorbents for heavy metal ions, Carbohydr. Polym. 86 (2011) 1558–1564.

[84] M.B. Asif, T. Maqbool, Z. Zhang, Electrochemical membrane bioreactors: state-of-the-art and future prospects, Sci. Total Environ. 741 (2020) 140233.

[85] M.B. Asif, A.J. Ansari, S.-S. Chen, L.D. Nghiem, W.E. Price, F.I. Hai, Understanding the mechanisms of trace organic contaminant removal by high retention membrane bioreactors: a critical review, Environ. Sci. Pollut. Res. 26 (2019) 34085–34100.

[86] F. Hai, S. Yang, M. Asif, V. Sencadas, S. Shawkat, M. Sanderson-Smith, J. Gorman, Z.-Q. Xu, K. Yamamoto, Carbamazepine as a possible anthropogenic marker in water: occurrences, toxicological effects, regulations and removal by wastewater treatment technologies, Water 10 (2018) 107.

[87] Y. Luo, W. Guo, H.H. Ngo, L.D. Nghiem, F.I. Hai, J. Zhang, S. Liang, X.C. Wang, A review on the occurrence of micropollutants in the aquatic environment and their fate and removal during wastewater treatment, Sci. Total Environ. 473 (2014) 619–641.

[88] M.B. Asif, F.I. Hai, J. Kang, J.P. Van De Merwe, F.D. Leusch, K. Yamamoto, W.E. Price, L.D. Nghiem, Degradation of trace organic contaminants by a membrane distillation—enzymatic bioreactor, Appl. Sci. 7 (2017) 879.

[89] M.B. Asif, L.N. Nguyen, F.I. Hai, W.E. Price, L.D. Nghiem, Integration of an enzymatic bioreactor with membrane distillation for enhanced biodegradation of trace organic contaminants, Int. Biodeteriorat. Biodegrad. 124 (2017) 73–81.

[90] M.B. Asif, F.I. Hai, J. Kang, J.P. Van De Merwe, F.D. Leusch, W.E. Price, L.D. Nghiem, Biocatalytic degradation of pharmaceuticals, personal care products, industrial chemicals, steroid hormones and pesticides in a membrane distillation-enzymatic bioreactor, Bioresour. Technol. 247 (2018) 528–536.

[91] M.B. Asif, J. Hou, W.E. Price, V. Chen, F.I. Hai, Removal of trace organic contaminants by enzymatic membrane bioreactors: role of membrane retention and biodegradation, J. Membr. Sci. 611 (2020) 118345.

[92] H. Liu, R. Dai, J. Qu, J. Ru, Preparation and characterization of a novel adsorbent for removing lipophilic organic from water, Sci. China Ser. B Chem. 48 (2005) 600–604.

[93] K.Gupta Komal, V. Kumar, K.B. Tikoo, A. Kaushik, S. Singhal, Encrustation of cadmium sulfide nanoparticles into the matrix of biomass derived silanized cellulose nanofibers for adsorptive detoxification of pesticide and textile waste, Chem. Eng. J. 385 (2020) 123700.

[94] L. Gan, Q. Zhong, A. Geng, L. Wang, C. Song, S. Han, J. Cui, L. Xu, Cellulose derived carbon nanofiber: a promising biochar support to enhance the catalytic performance of $CoFe_2O_4$ in activating peroxymonosulfate for recycled dimethyl phthalate degradation, Sci. Total Environ. 694 (2019) 133705.

[95] J. Herrera-Morales, K. Morales, D. Ramos, E.O. Ortiz-Quiles, J.M. Lopez-Encarnacion, E. Nicolau, Examining the use of nanocellulose composites for the sorption of contaminants of emerging concern: An experimental and computational study, ACS omega 2 (2017) 7714–7722.

[96] X. He, K.B. Male, P.N. Nesterenko, D. Brabazon, B. Paull, J.H. Luong, Adsorption and desorption of methylene blue on porous carbon monoliths and nanocrystalline cellulose, ACS Appl. Mater. Interfaces 5 (2013) 8796–8804.

[97] H.-Y. Yu, D.-Z. Zhang, F.-F. Lu, J. Yao, New approach for single-step extraction of carboxylated cellulose nanocrystals for their use as adsorbents and flocculants, ACS Sustain. Chem. Eng. 4 (2016) 2632–2643.

[98] J. Yang, J. Yu, J. Fan, D. Sun, W. Tang, X. Yang, Biotemplated preparation of CdS nanoparticles/bacterial cellulose hybrid nanofibers for photocatalysis application, J. Hazard. Mater. 189 (2011) 377–383.

[99] S. Chen, Y. Huang, Bacterial cellulose nanofibers decorated with phthalocyanine: preparation, characterization and dye removal performance, Mater. Lett. 142 (2015) 235–237.

[100] L. Jin, Q. Sun, Q. Xu, Y. Xu, Adsorptive removal of anionic dyes from aqueous solutions using microgel based on nanocellulose and polyvinylamine, Bioresour. Technol. 197 (2015) 348–355.

[101] H.G. Derami, Q. Jiang, D. Ghim, S. Cao, Y.J. Chandar, J.J. Morrissey, Y.S. Jun, S. Singamaneni, A robust and scalable polydopamine/bacterial nanocellulose hybrid membrane for efficient wastewater treatment, ACS Appl. Nano Mater. 2 (2019) 1092–1101.

[102] N. Hammi, S. Chen, F. Dumeignil, S. Royer, A. El Kadib, Chitosan as a sustainable precursor for nitrogen-containing carbon nanomaterials: synthesis and uses, Mater. Today Sustain. 10 (2020) 100053.

[103] M.N.V.R. Kumar, R.A.A. Muzzarelli, C. Muzzarelli, H. Sashiwa, A.J. Domb, Chitosan chemistry and pharmaceutical perspectives, Chem. Rev. 104 (2004) 6017–6084.

[104] W. Suginta, P. Khunkaewla, A. Schulte, Electrochemical biosensor applications of polysaccharides chitin and chitosan, Chem. Rev. 113 (2013) 5458–5479.

[105] R. Jayakumar, D. Menon, K. Manzoor, S.V. Nair, H. Tamura, Biomedical applications of chitin and chitosan based nanomaterials—a short review, Carbohydr. Polym. 82 (2010) 227–232.

[106] A. El Kadib, M. Bousmina, D. Brunel, Recent progress in chitosan bio-based soft nanomaterials, J. Nanosci. Nanotechnol. 14 (2014) 308–331.

[107] A. El Kadib, M. Bousmina, Chitosan bio-based organic-inorganic hybrid aerogel microspheres, Chemistry (Weinheim an der Bergstrasse, Germany), 18 (2012) 8264-8277.

[108] J. Wang, S. Zhuang, Removal of various pollutants from water and wastewater by modified chitosan adsorbents, Crit. Rev. Environ. Sci. Technol. 47 (2017) 2331–2386.

[109] H. Bagheri, A. Roostaie, M.Y. Baktash, A chitosan–polypyrrole magnetic nanocomposite as μ-sorbent for isolation of naproxen, Anal. Chim. Acta 816 (2014) 1–7.

[110] C. Xu, M. Nasrollahzadeh, M. Sajjadi, M. Maham, R. Luque, A.R. Puente-Santiago, Benign-by-design nature-inspired nanosystems in biofuels production and catalytic applications, Renew. Sustain. Energy Rev. 112 (2019) 195–252.

[111] M.B. Asif, N. Majeed, S. Iftekhar, R. Habib, S. Fida, S. Tabraiz, Chemically enhanced primary treatment of textile effluent using alum sludge and chitosan, Desalin. Water Treat. 57 (2016) 7280–7286.

[112] A.G. El Samrani, B.S. Lartiges, F. Villiéras, Chemical coagulation of combined sewer overflow: heavy metal removal and treatment optimization, Water Res. 42 (2008) 951–960.

[113] H. Zemmouri, M. Drouiche, A. Sayeh, H. Lounici, N. Mameri, Chitosan application for treatment of Beni-Amrane's water dam, Energy Procedia 36 (2013) 558–564.

[114] A.J. Sami, M. Khalid, S. Iqbal, M. Afzal, A.R. Shakoori, Synthesis and application of chitosan-starch based nanocomposite in wastewater treatment for the removal of anionic commercial dyes, Pak. J. Zool. 49 (2017) 21–26.

[115] P. Kanmani, J. Aravind, M. Kamaraj, P. Sureshbabu, S. Karthikeyan, Environmental applications of chitosan and cellulosic biopolymers: a comprehensive outlook, Bioresour. Technol. 242 (2017) 295–303.

[116] A. Rajeswari, A. Amalraj, A. Pius, Adsorption studies for the removal of nitrate using chitosan/PEG and chitosan/PVA polymer composites, J. Water Process Eng. 9 (2016) 123–134.

[117] S. Rahimi, R.M. Moattari, L. Rajabi, A.A. Derakhshan, Optimization of lead removal from aqueous solution using goethite/chitosan nanocomposite by response surface methodology, Coll. Surf. A Physicochem. Eng. Asp. 484 (2015) 216–225.

[118] F.G.L. Medeiros Borsagli, A.A.P. Mansur, P. Chagas, L.C.A. Oliveira, H.S. Mansur, O-carboxymethyl functionalization of chitosan: complexation and adsorption of Cd (II) and Cr (VI) as heavy metal pollutant ions, React. Funct. Polym. 97 (2015) 37–47.

[119] J.-P. Wang, Y.-Z. Chen, S.-J. Yuan, G.-P. Sheng, H.-Q. Yu, Synthesis and characterization of a novel cationic chitosan-based flocculant with a high water-solubility for pulp mill wastewater treatment, Water Res. 43 (2009) 5267–5275.

[120] S. Abdi, M. Nasiri, A. Mesbahi, M.H. Khani, Investigation of uranium (VI) adsorption by polypyrrole, J. Hazard. Mater. 332 (2017) 132–139.

[121] H. Huang, J. Wu, X. Lin, L. Li, S. Shang, M.C.-w. Yuen, G. Yan, Self-assembly of polypyrrole/chitosan composite hydrogels, Carbohydr. Polym. 95 (2013) 72–76.

[122] Y.-J. Zhang, J.-Q. Xue, F. Li, J.I.Z. Dai, X.-Z.-Y. Zhang, Preparation of polypyrrole/chitosan/carbon nanotube composite nano-electrode and application to capacitive deionization process for removing Cu^{2+}, Chem. Eng. Process. 139 (2019) 121–129.

[123] L. Chen, W. Cao, P.J. Quinlan, R.M. Berry, K.C. Tam, Sustainable catalysts from gold-loaded polyamidoamine dendrimer-cellulose nanocrystals, ACS Sustain. Chem. Eng. 3 (2015) 978–985.

[124] Y. Qiu, Z. Ma, P. Hu, Environmentally benign magnetic chitosan/Fe_3O_4 composites as reductant and stabilizer for anchoring Au NPs and their catalytic reduction of 4-nitrophenol, J. Mater. Chem. A 2 (2014) 13471–13478.

[125] L.M. Anaya-Esparza, J.M. Ruvalcaba-Gómez, C.I. Maytorena-Verdugo, N. González-Silva, R. Romero-Toledo, S. Aguilera-Aguirre, A. Pérez-Larios, E. Montalvo-González, Chitosan-TiO_2: a versatile hybrid composite, Materials 13 (2020) 811.

[126] W. Li, L. Xiao, C. Qin, The characterization and thermal investigation of chitosan-Fe_3O_4 nanoparticles synthesized via a novel one-step modifying process, J. Macromol. Sci. A 48 (2010) 57–64.

[127] L.N. Pincus, F. Melnikov, J.S. Yamani, J.B. Zimmerman, Multifunctional photoactive and selective adsorbent for arsenite and arsenate: evaluation of nano titanium dioxide-enabled chitosan cross-linked with copper, J. Hazard. Mater. 358 (2018) 145–154.

[128] G. Gibbs, J.M. Tobin, E. Guibal, Sorption of Acid Green 25 on chitosan: influence of experimental parameters on uptake kinetics and sorption isotherms, J. Appl. Polym. Sci. 90 (2003) 1073–1080.

[129] T.K. Saha, H. Ichikawa, Y. Fukumori, Gadolinium diethylenetriaminopentaacetic acid-loaded chitosan microspheres for gadolinium neutron-capture therapy, Carbohydr. Res. 341 (2006) 2835–2841.

[130] M. Rinaudo, Chitin and chitosan: Properties and applications, Prog. Polym. Sci. 31 (2006) 603–632.

[131] F. Marrakchi, W.A. Khanday, M. Asif, B.H. Hameed, Cross-linked chitosan/sepiolite composite for the adsorption of methylene blue and reactive orange 16, Int. J. Biol. Macromol. 93 (2016) 1231–1239.

[132] J. Xie, C. Li, L. Chi, D. Wu, Chitosan modified zeolite as a versatile adsorbent for the removal of different pollutants from water, Fuel 103 (2013) 480–485.

[133] L. Zeng, M. Xie, Q. Zhang, Y. Kang, X. Guo, H. Xiao, Y. Peng, J. Luo, Chitosan/organic rectorite composite for the magnetic uptake of methylene blue and methyl orange, Carbohydr. Polym. 123 (2015) 89–98.

[134] Y. Yao, B. Gao, J. Fang, M. Zhang, H. Chen, Y. Zhou, A.E. Creamer, Y. Sun, L. Yang, Characterization and environmental applications of clay-biochar composites, Chem. Eng. J. 242 (2014) 136–143.

[135] Y. Fang, J. Fu, P. Liu, B. Cu, Morphology and characteristics of 3D nanonetwork porous starch-based nanomaterial via a simple sacrifice template approach for clove essential oil encapsulation, Ind. Crop. Prod. 143 (2020) 111939.

[136] Y. Fang, X. Lv, X. Xu, J. Zhu, P. Liu, L. Guo, C. Yuan, B. Cui, Three-dimensional nanoporous starch-based material for fast and highly efficient removal of heavy metal ions from wastewater, Int. J. Biol. Macromol. 164 (2020) 415–426.

[137] D. Le-Corre, J. Bras, A. Dufresne, Starch nanoparticles: a review, Biomacromolecules 11 (2010) 1139–1153.

[138] E. Ruiz-Hitzky, M. Darder, F.M. Fernandes, B. Wicklein, A.C.S. Alcântara, P. Aranda, Fibrous clays based bionanocomposites, Prog. Polym. Sci. 38 (2013) 1392–1414.

[139] F. Chivrac, E. Pollet, M. Schmutz, L. Avérous, Starch nano-biocomposites based on needle-like sepiolite clays, Carbohydr. Polym. 80 (2010) 145–153.

[140] G. Arash, G. Mohammad, F. Habib, Palladium deposited on naturally occurring supports as a powerful catalyst for carbon-carbon bond formation reactions, Curr. Org. Chem. 20 (2016) 327–348.

[141] M. Gholinejad, F. Saadati, S. Shaybanizadeh, B. Pullithadathil, Copper nanoparticles supported on starch micro particles as a degradable heterogeneous catalyst for three-component coupling synthesis of propargylamines, RSC Adv. 6 (2016) 4983–4991.

[142] J.-Y. Kim, D.-J. Park, S.-T. Lim, Fragmentation of waxy rice starch granules by enzymatic hydrolysis, Cereal Chem. 85 (2008) 182–187.

[143] J.-Y. Kim, S.-T. Lim, Preparation of nano-sized starch particles by complex formation with n-butanol, Carbohydr. Polym. 76 (2009) 110–116.

[144] Y. Tan, K. Xu, L. Li, C. Liu, C. Song, P. Wang, Fabrication of size-controlled starch-based nanospheres by nanoprecipitation, ACS Appl. Mater. Interfaces 1 (2009) 956–959.

[145] X. Ma, R. Jian, P.R. Chang, J. Yu, Fabrication and Characterization Of Citric Acid-Modified Starch Nanoparticles/Plasticized-Starch Composites, Biomacromolecules 9 (2008) 3314–3320.

[146] D. Liu, Q. Wu, H. Chen, P.R. Chang, Transitional properties of starch colloid with particle size reduction from micro- to nanometer, J. Coll. Interface Sci. 339 (2009) 117–124.

[147] M.P. Yadav, J. Manuel Igartuburu, Y. Yan, E.A. Nothnagel, Chemical investigation of the structural basis of the emulsifying activity of gum arabic, Food Hydrocoll. 21 (2007) 297–308.

[148] V.V.T. Padil, C. Senan, M. Černík, "Green" polymeric electrospun fibers based on tree-gum hydrocolloids: fabrication, characterization and applications, in: V. Grumezescu, A.M. Grumezescu (Eds.), Materials for Biomedical Engineering, Elsevier, Amsterdam, 2019, pp. 127–172.

[149] Q. Miao, H. Jiang, L. Gao, Y. Cheng, J. Xu, X. Fu, X. Gao, Rheological properties of five plant gums, Am. J. Anal. Chem. 9 (2018) 210–223.

[150] R.S. Soumya, S. Ghosh, E.T. Abraham, Preparation and characterization of guar gum nanoparticles, Int. J. Biol. Macromol. 46 (2010) 267–269.

[151] Z. Zamani, S.M.A. Razavi, Physicochemical, rheological and functional properties of nettle seed (*Urtica pilulifera*) gum, Food Hydrocoll. 112 (2021) 106304.

[152] G. Sharma, S. Sharma, A. Kumar, A.a.H. Al-Muhtaseb, M. Naushad, A.A. Ghfar, G.T. Mola, F.J. Stadler, Guar gum and its composites as potential materials for diverse applications: a review, Carbohydr. Polym. 199 (2018) 534–545.

[153] S. Iftekhar, V. Srivastava, D.L. Ramasamy, W.A. Naseer, M. Sillanpää, A novel approach for synthesis of exfoliated biopolymeric-LDH hybrid nanocomposites via in-stiu coprecipitation with gum Arabic: application towards REEs recovery, Chem. Eng. J. 347 (2018) 398–406.

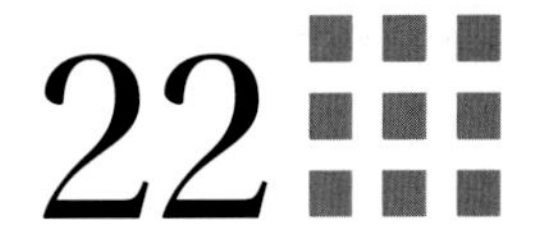

Nitrogen-based green flame retardants for bio-polyurethanes

Felipe M. de Souza and Ram K. Gupta

DEPARTMENT OF CHEMISTRY, KANSAS POLYMER RESEARCH CENTER, PITTSBURG STATE UNIVERSITY, PITTSBURG, KS, UNITED STATES

Chapter outline

22.1 Introduction

Polyurethanes (PUs) are one of the most prominent polymeric materials which is synthesized through the exothermic reaction between isocyanate and hydroxyl components. PUs can be incorporated in many applications because of the wide-ranging modular-like polymer structure. The versatility as well as its compatibility lead to replace other scare materials and consume in many applications such as foams, paints, elastomers, insulators, liquid coatings, etc. [1–3]. The foundation of PUs was invented in the 1930s by Professor Otto Bayer and his coworkers [1]. PU was first developed as an alternative for rubber and was commonly found by the middle of the 1940s as elastomer, coating, and adhesives. Toward the latter part of the 1950s, its development continued for flexible foams for applications in comfortable cushions, upholstery, automotive along the development of rigid foams. Since then, they have been continuously improving in many applications, and now all around us, PUs appear in an astonishing variety of forms, making them the most versatile of any family of plastic materials [4–6].

It is necessary to understand the main components along with the involved reactions for the synthesis of PUs. The chemistry of PU is a reaction between at least two functional groups typically hydroxyl (-OH) and isocyanate (-NCO) groups that can produce PU in either a single-step or prepolymer method. Other chemicals such as blowing agents, surfactants, and catalysts may be required depending on the applications. The process is called a single-step method when

Biopolymeric Nanomaterials: Fundamentals and Applications. DOI: https://doi.org/10.1016/B978-0-12-824364-0.00020-4

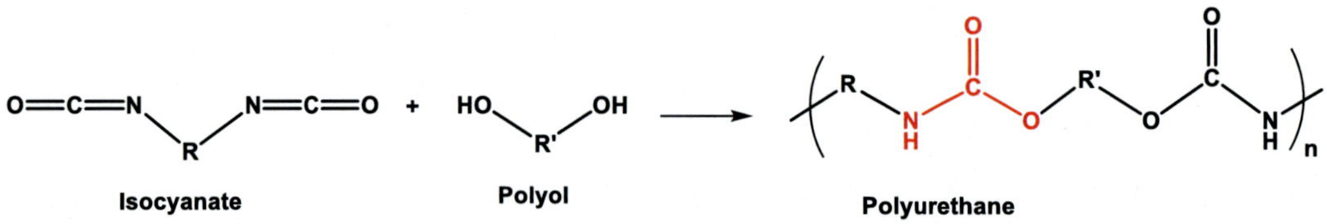

FIGURE 22.1 A general reaction between a polyol and isocyanate to form a polyurethane.

all the chemicals are mixed together, while the prepolymer method involves the mixing of all the chemicals but isocyanate. The urethane group is the major repeating unit in the molecular backbone (Fig. 22.1). However, a typical PU may contain groups such as urethane linkage, aliphatic, aromatic rings, esters, ethers, amides, urea in its structure. The molecular structure of PU depends on the specific polyol and isocyanate used for the synthesis of PU. Therefore, a broad range of PUs for different applications can be prepared by using polyols with different molecular weights, functionality, functional groups, and different isocyanates. The properties of PU strongly depend on chemical materials and their relative ratio, synthesis, and processing conditions [7–9].

Nowadays, polymers became part of modern life, as it is in almost every object, device, or equipment that people deal with regularly. Some direct examples of PUs in daily use are the elastomers in shoe dampers, soft foams in mattress, furniture, car seats, rigid foams for thermal insulation such as in buildings and refrigerators just to name a few [2]. Therefore, PUs landed a huge impact on the current way of life. However, that convenience came with the cost of large exploitation of nonrenewable sources to harvest raw materials that would be later used for PU synthesis. In the beginning, the use of petrochemical for PUs was not of much concern but with growing awareness of environment protection, alternative materials for petrochemicals are much needed as petrochemicals cause emission of toxic and greenhouse gases during the purification and synthetic steps to obtain starting materials. The traditional use of these petroleum-based materials was mostly derived from ethylene and propylene. These reagents were then epoxidized in the presence of oxygen or hydrogen peroxide to yield ethylene oxide, propylene oxide, and glycerol. Furthermore, these materials go through alkoxylation in the presence of potassium hydroxide to form a polyether polyol. The latter is then suitable to react with isocyanate to make PU foams. The reaction to the process is described in Fig. 22.2 [3].

Accompanied to that factor another environmental concern that arose for the production of PUs was the use of blowing agents, which are the components responsible for controlling the bubble formation in the microstructure of the foam with a simultaneous decrease of density and cost [4]. Among them, the chlorofluorocarbons (CFCs) and carbon dioxide were largely used, but mostly the first one. The main reason for its large employment was due to its low cost, molecular weight, and toxicity along with nonflammability and boiling point around room temperature [5]. However, Molina and Rowland in 1974 provided scientific evidence that states that chlorine radical atoms originated from the CFCs were ascending to the stratosphere and catalyzing the decomposition of ozone into oxygen through photodissociation mechanism [6]. This process became commonly known as the ozone layer depletion, which prompts the United Nations to

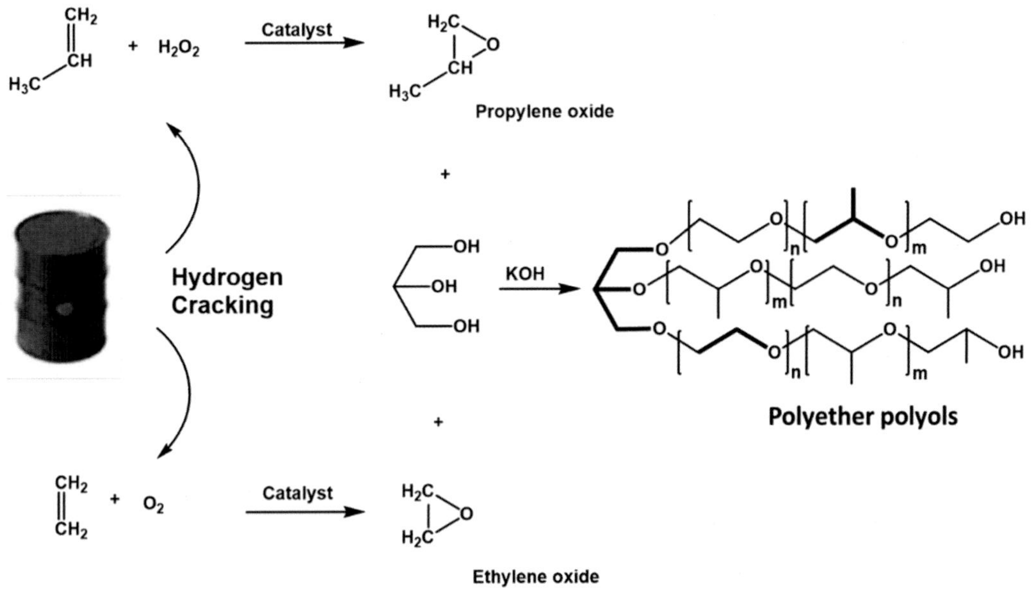

FIGURE 22.2 Industrial synthetic route for the production of petroleum-based polyether polyols through the process of hydrocarbon cracking followed by epoxidation and alkoxylation to form the polyether polyol. *"Adapted with permission from [3]. Copyright (2011) Springer Nature."*

develop international regulations such as the Montreal Protocol [7]. Part of the demands for the protocol included the drastic reduction of carbon dioxide emission and practically the banishment of CFC-based compounds. These factors were important marks in the PU industry that drastically reshaped its structure to attend the requirements. After that, the scientific community focused efforts on developing new approaches that were bio-friendly. It has proven to be an extremely challenging task as these new green materials need to surpass or at least even out the properties of petrochemical-based PUs such as mechanical, thermal stability, processing, and cost to be applicable for large-scale production. Despite the difficulties, some biopolymers achieved those features becoming more competitive in the market. The main examples are polylactide, bio polyethylene terephthalate, and bio polyethylene [8]. Nevertheless, there is still a long road for biopolymers to conquer a fair share in the global market, as they represent only about 0.56% of the currently commercialized polymers [8]. Even though it may look like a far-reaching possibility, the use of biorenewable sources is a largely rewarding achievement as the source of starting materials usually comes from biomass or biowastes, which provide an inherently lower cost and economic stability. These starting materials are not subjected to international frictions and can be replenished within a year or so, which is a promising advantage for their industrial applications.

Many types of starting materials can be used for the synthesis of biopolymers. Starting materials can be divided into three main categories. The first category is carbohydrates that include starchy monomers, lignin, cellulose, and sugars such as glucose and fructose. The second group

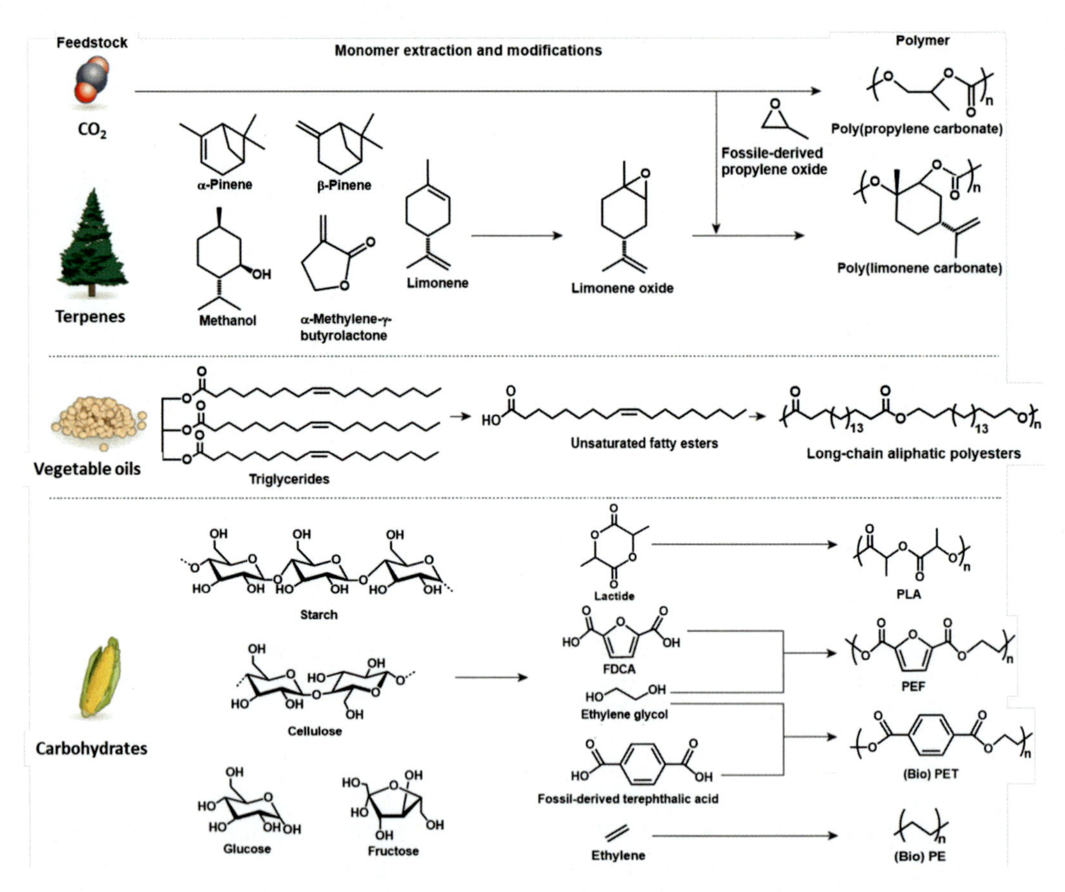

FIGURE 22.3 Schematics of biomaterials that can be employed for the synthesis of biopolymers. Carbon dioxide can be used as a monomer along with propylene yielding propylene carbonate polyols. In the same principle, terpenes and terpenoids can be carbonated yielding various polymers with different structures. Finally, starchy and cellulosic type of monomer can be broken down into smaller molecules and be polymerized to yield biopolymers such as bio-polylactide, bio-polyethylene terephthalate, and even bio-polyethylene. *"Adapted with permission from [12]. Copyright (2016) Springer Nature."*

is related to vegetable oils such as soybean, corn, palm, castor, Jatropha, other triglycerides, and unsaturated fats. The third group of biosources is the terpenes and terpenoids, which are olefins, usually found as essential oils extracted from plants, mostly conifers, and citrus trees [9–11]. A general scheme of these compounds and their main polymeric forms are described in Fig. 22.3 [12]. The suitability of these materials for polyol, a starting material for the synthesis of bio-polyurethane, depends on the various factors such as the amount of unsaturation, size of the carbon chain, the presence of substituent groups, etc. The presence of unsaturation in these materials offers a variety of methods to convert them into polyols for PUs. One of the largely exploited possibilities is the epoxidation of the double bonds of a glyceride, followed by ring-opening of the oxirane by alcohol (alcoholysis) to yield a polyol (Fig. 22.4). Another example of

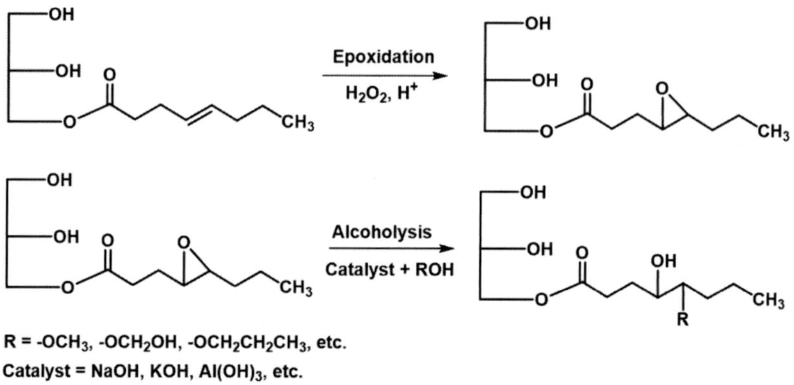

R = $-OCH_3$, $-OCH_2OH$, $-OCH_2CH_2CH_3$, etc.
Catalyst = NaOH, KOH, $Al(OH)_3$, etc.

FIGURE 22.4 Conversion of a general unsaturated monoglyceride through epoxidation followed by alcoholysis.

a synthesis of the bio-based polyol is through thiol-ene reaction, which is a facile method that can be performed at room temperature in the presence of a photocatalyst and UV light [13–18]. It consists of a radical addition reaction to the double bond through a photo-generated sulfonyl radical that monolithically breaks the unsaturation, usually follows an anti-Markovnikov type of mechanism [19]. Many reports have demonstrated its use as a viable tool to synthesize polyols by using starting materials such as unsaturated vegetable oils [11], lignin [20], and terpenes [15]. A synthetic route of a corn oil-based polyol through thiol-ene reactions is described in Fig. 22.5 [11]. Limonene, an extract from the orange peel, was used as a starting material for the synthesis of polyol using thiol-ene chemistry [18,21,22]. Other derivatives of limonene such as limonene dimercaptan can be also used to synthesize biopolyols using thiol-ene click chemistry as shown in Fig. 22.6. As mentioned, there are many possible ways to use bio-based materials as viable sources for the synthesis of polyols for further foaming process with isocyanate. The following sections discuss the main aspects of PUs and the importance of flame retardancy property that can be introduced into these polymers.

As discussed earlier, PUs are extremely versatile polymers that can be used in many applications, which is one of the main reasons why PUs increase their market share every year providing more capital, employment, and technological development [23]. The global value of the PU industry was about $69 billion in 2019 and is expected to grow over 5% per year until 2025 [24]. Seeing the importance of biopolymers, there is constant growth in global bio-polyurethanes. The global bio-polyurethane market is expected to reach about $37 million by 2020, according to a new study by Grand View Research, Inc. [25]. Environmental concerns and increased demand for sustainable products are among some of the reasons for the increasing demand for bio-polyurethanes. More than half of the total use of PUs is dedicated to mattresses, furniture, packing, buildings, households, car seats, etc. Other segments that complement the overall use of these polymeric materials are related to footwear, elastomers, coatings, and adhesives. Fig. 22.7 demonstrates the distribution of the global market for PUs. Besides the general purpose of thermal insulation and comfort, PUs carry an inherently sustainable and economic factor in

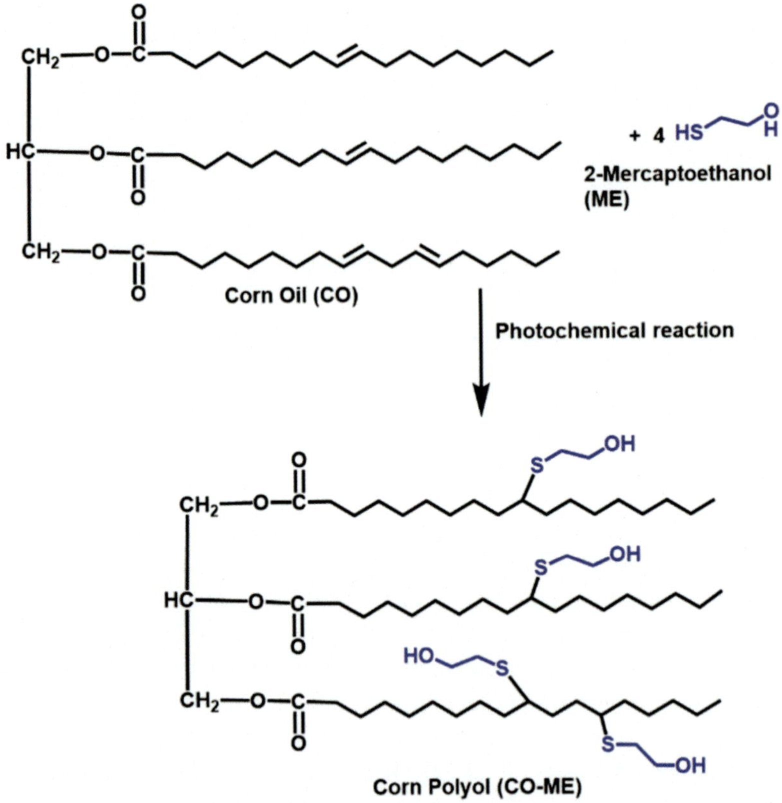

FIGURE 22.5 Thiol-ene reaction for the conversion of corn oil into polyol suitable for the synthesis of polyurethane [11].

their applications. It prevents waste of electrical energy from heaters and air conditioning, in the same that it improves efficiency of fridges and coolers. In addition, the automobile industry utilizes PU foams in many parts of the vehicle due to their light weight and relatively high mechanical properties, improving both efficiencies of fuel consumption as well as safety measures [26]. The compilation of all these properties in one type of polymer is one of the factors that make PUs so important to the economy.

PUs are valuable materials for the industry due to the large scope of synthetic approaches, which are facile to perform. This condition is accompanied by a large set of starting materials providing a variety of polymeric structures that allow them to be used in many applications. Some properties of PU such as light-weight, spray applications, low thermal conductivity, and relatively high mechanical strength are among some factors that make them very suitable for applications in buildings and constructions. However, poor resistance against fire is one of the main drawbacks of PUs, which restricts their wide applications. The high flammability of PU foams is due to their organic nature, the presence of high surface area, and porosity which ease the diffusion of oxygen within the microstructure making them susceptible to fire. This

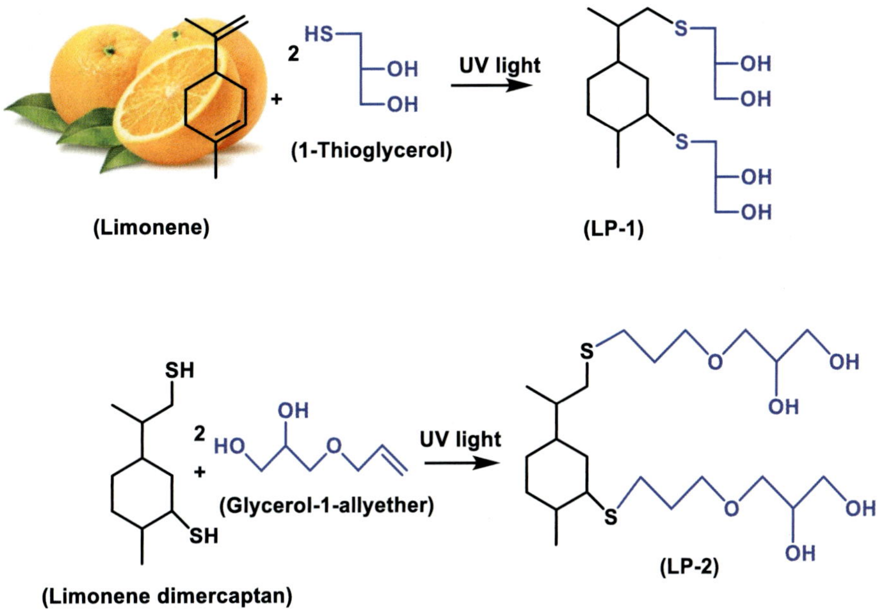

FIGURE 22.6 Synthesis of biopolyol using thiol-ene reaction for bio polyurethanes. *"Adapted with permission from [15]. Copyright (2014) Springer Nature."*

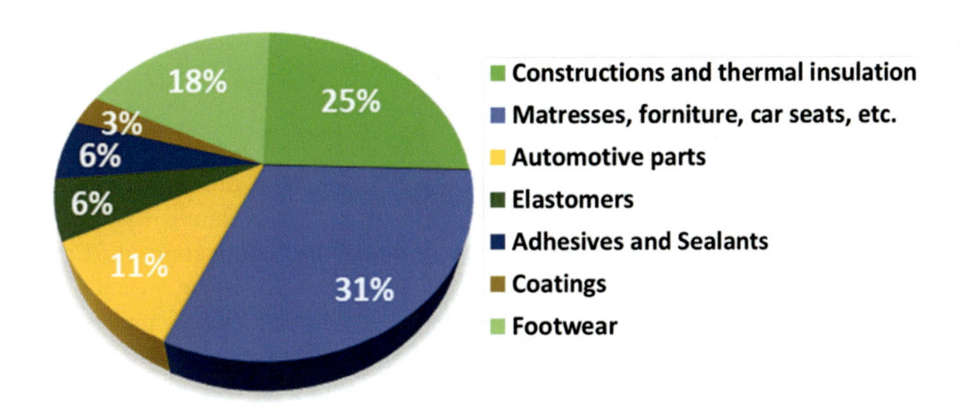

FIGURE 22.7 Global market distribution of polyurethanes.

becomes a concern in the case of a fire incident due to the release of toxic smokes, dripping, and high heat that is released during the combustion of PUs. This drawback limits the potential applications of PUs, thus demanding a response to revert this scenario that causes a major damage to buildings and jeopardize people's lives [27]. Another concern regarding some PUs is the presence of dandling groups that prevent the polymeric chains to be tightly packed, hence providing a plasticizing effect that diminishes the mechanical properties. Depending on the applications of PUs, these features may be undesirable. One of the ways to avoid these issues is the addition of a suitable filler that can provide a proper interaction within the polymeric matrix as well as fits within the cellular structure of the PU. By attending these conditions the filler can be effectively blended with the PU and therefore enhance the mechanical properties [28,29]. Further details will be discussed in the following sections regarding the role of flame retardants in bio-polyurethanes along with their pros and cons with emphasis on the mechanism of nitrogen-based flame retardants.

22.2 Need for flame retardants

The use of flame retardants is a requirement to guarantee safety in close environments in case of a fire incident. Flame retardants are capable to suppress the fire and smoke as a quick and automatic response preventing both heat and toxic smokes to spread [30]. In 2018, the National Fire Protection Association estimates 1.32 million fires resulting in deaths of 3655 civilians, injuries to 15,200 civilians, and direct property loss of about $13.2 billion [31]. Among them, fatal deaths of 2720 civilians, injuries to 12,700 civilians, and loss worth $11.1 billion to the property was due to a structural fire. Rigid PU foam is one of the important construction materials used as thermal insulation in the house, furniture, vibration insulator, electrical insulation, and industrial and household packing; hence, it is essential to develop efficient fire-retardant PUs that can prevent fire hazards at the source [32–34]. Efficient and cost-effective flame retardants can play an important role in the PU industry to prevent harsh situations and property damage [30].

22.3 Types of flame retardants

Two types of flame retardants can be used during the formulation process of PUs: additive and reactive. The additive or nonreactive flame retardants are compounds that do not participate in chemical bonding with starting materials during the preparation of PUs and are just physically blended with the polymeric matrix. Such flame retardants are usually preferred for large-scale applications due to economic viability as they mostly need a proper mixing to work efficiently [29]. However, to optimize the efficacy of blended flame retardants, they must be evenly dispersed throughout the polymeric matrix as well as present a proper interaction with the matrix. The latter is referred to as an optimum size of the flame retardants' particle that can fit into the cellular microstructures of the PU foams. Otherwise, it may create defects and voids into the arrangement of the cells that deteriorate the mechanical properties [29]. Another issue with additive flame retardants is their migration with time which can affect their effectiveness as flame retardants. Also, a high load of additive flame retardants may be required to achieve a satisfactory result [35]. However, as the nonreactive flame retardants do not require an active site to be chemically bonded into the polymers, many materials can be used for this purpose. Some widely used additive flame retardants are melamine and its derivatives (such as melamine phosphate (MP), melamine polyphosphate (MPP), melamine cyanurate (MC)), expandable graphite, organophosphates (such as dimethyl phosphate, diethyl phosphate, ammonium polyphosphate), aluminum trihydroxide, aluminum hypophosphite, and many more [11,36–42].

The reactive or inherent flame retardants are components that are chemically bonded with the structure of the polymeric matrix. An extra synthesis step may be required to synthesize reactive flame retardants that could increase the cost; however, they usually present a higher flame retardancy compared with blended flame retardants. Their main advantages are related to the even distribution of flame retardants in the polymeric matrix, which improves the fire-quenching capabilities and provides stability over time while maintaining other properties [13]. As there is no physical entrapment of particles within the matrix, the cellular structure of the reactive flame retardant-based polymers is more regular than blended ones. This feature prevents a decrease of mechanical properties and in some cases even enhancing it [13]. However, if the reactive flame retardants act as a bulky dandling group in the main chain it can lead to a plasticizing effect, which may then decrease the mechanical properties. Hence, one of the current challenges of the scientific community is to find compounds that can be physically blended as well as have reactive components that can optimize both flame retardancy and mechanical properties. The mechanism that takes place to quench the fire occurs through solid-phase or gas-phase which may happen simultaneously but in different degrees depending on the flame retardant. The solid-phase mechanism occurs by the decomposition of flame retardant when exposed to a source of energy (heat) that forms a thermostable char layer over the polymer's surface. The effectiveness of this layer is defined by how well packed it is to prevent oxygen from diffusing inside the polymer's matrix and continue to oxidize the polymer. Besides, it should also decrease the thermal conductivity to diminish heat transfer [43,44]. The gas-phase mechanism occurs when the flame retardant releases radical scavenger species that can capture the radical fragments originated from the thermal decomposition of the polymer as expressed in Eqs. (22.1) and (22.2), converting into thermostable substances [45]. This process then interrupts the transfer of heat back to the polymer, ceasing the flame and cooling down the system. Hence, preventing further combustion with oxygen by decreasing its concentration in the media, thus suffocating the fire [40,46].

$$H^{\cdot} + O_2 \rightarrow OH^{\cdot} + O^{\cdot} \tag{22.1}$$

$$O^{\cdot} + H_2 \rightarrow OH^{\cdot} + H^{\cdot} \tag{22.2}$$

Several materials that contain carbon, halogens, phosphorus, nitrogen in their structure can be used as flame retardants. The carbon-based flame retardants have been widely reported in the literature for decades but are still researched due to the variety of allotropic structures of carbon that can fit for this purpose. Materials such as carbon black, graphite, and expandable graphite were widely used; however, they require higher loading to be effective flame retardants. Nowadays, new materials such as graphene, carbon nanotubes (CNT), and fullerene are gaining more attention as a lower quantity of these materials can be used along with possibilities of attaching various functional groups to react with the monomers during polymerization process that was not considered earlier [47].

Among various allotropes of carbon, fullerene holds a great opportunity as a flame-retardant material. It is believed that its structure accompanied by its aromaticity gives the properties of heat absorption, which aids the thermal shielding avoiding the heat transfer for the underneath polymeric matrix [47]. Another feature of fullerene which is capable of imparting to a polymeric composite is its high reactivity toward radical species, acting as a sponge that quenches the fire during the gas-phase mechanism [47]. The other route of action for fullerene is observed when it reacts with polypropylene. During the pyrolysis process, the fullerene was capable of bonding with polymeric matrix fragments creating a cross-linked structure, referred to as gelled-ball [48]. In a way, this structure created a shielding layer preventing heat flux and mass transfer [48]. The current studies have demonstrated that the addition of fullerene decreases the heat absorption to almost 50% compared with the neat polymeric matrix, thus making fullerene as an efficient flame retardant [47]. Similarly, CNTs have also been reported as promising materials to deduce the flammability. The functionalization of CNTs through a variety of approaches offers enhanced mechanical strength and improved flame retardancy of the polymeric materials [49–51]. CNTs-based flame retardants mostly act in the condensed phase by forming a stable and tridimensional char layer network that effectively prevents the flame from penetrating and scaping of volatiles from the underlying polymer [52]. Graphene, which is an excessively researched material, also presents interesting flame retardancy. When graphene is exposed to flames for a short time, it becomes red, like an incandescent effect, without actually catching the fire [52]. However, graphene must be pure to act as an efficient flame retardant. Impurities such as potassium nitrate that is usually present during the synthesis of reduced graphene oxide reduce its effectiveness as a flame retardant [52]. Other approaches utilize a direct blend of graphene with polymeric matrices through a hot-melt extrusion process [52]. The incorporation of graphene with red phosphorus has been reported and presented effective results to reduce the flammability [52]. Graphene as a flame retardant differs from the other carbon-based flame retardants as graphene disperse more evenly due to the presence of functional groups through the polymeric chains, which provides more stability and better flame-retardancy [52]. Expandable graphite is a material with intercalated sulfuric acid and works as a flame retardant. When expandable graphite is heated, it releases gases such as carbon dioxide and sulfur dioxide that burst out from the edges of the graphitic structure leading to permanent expansion. The main reaction that takes place during this process can be expressed as

$$C + 2H_2SO_4 \rightarrow CO_2 + 2H_2O + 2SO_2 \tag{22.3}$$

This reaction describes the flame-retardant mechanism of expandable graphite. During heating, the gases that are released dilute the concentration of oxygen in the media. Besides, the expansion of the carbon flakes creates a physical barrier preventing oxygen and free radical species to get into contact with the polymer that quickly quenches the fires over the material's surface [43]. Even though requiring a high load of material to be functional, carbon-based flame retardants are low-cost and environmental-friendly as most of the smokes that are released from their combustion are carbon dioxide and water vapor. This explains in part why these materials have been discovered for a long time and are still being used as flame retardants.

The halogen-based flame retardants are very efficient in preventing and spread of the flame. The action of halogen is emphasized on quickly scavenging $H^{\bullet}$ and $OH^{\bullet}$ [53–56]. The effective smoke-suppressing process yields thermostable volatile products that present the mass transfer from these radicals back to the polymer as well as dilutes oxygen in the media. Most of the halogens-based flame retardants are bromine-carbon-based compounds that release the radical bromine to scavenge the radical pyrolysis products due to the weak bonding between bromine and carbon. Some examples of this type of flame retardants are tetrabromo bisphenol A (TBBPA), hexabromocyclododecane (HBCD), penta, octa, and deca-brominated diphenyl ether [45]. The main concern about halogen-based flame retardants is their inherent toxicity along with the generation of harmful smokes such as hydrogen halides that can worsen the situation in the case of fire due to the corrosive nature of these compounds, causing health issues and deterioration of the environment. Therefore, the use of halogen-based flame retardants has been limited [45].

The flame retardant market has its largest share devoted to phosphorus-based compounds that include many varieties such as red phosphorus, phosphate salts, organophosphates, phosphine oxides, and others. The action of phosphorus-based compounds as a flame retardant is also divided in both condensed and gas phase. However, is believed that it is more pronounced at the condensed-phase by dehydrating and forming a well-packed char layer due to the formation of polyphosphoric acid [57]. The water vapor that is released in the gas phase can dilute the flammable species in the environment and thus preventing the spread of the fire. Along with that, there is a release of phosphorus-based scavenger species analogous to the halogen-based flame retardants which can capture $H^{\bullet}$ and $OH^{\bullet}$. These species are mostly free radicals such as $HPO^{\bullet}$, $PO^{\bullet}$, and $PO_2^{\bullet}$ [58]. The phosphorus-based flame retardants are costlier than other flame retardants. Despite that, phosphorus-based flame retardants are extremely valuable components to decrease the flammability of polymers along with providing a range of approaches for physical blending as well as reactive flame retardants for PU foams [13,59].

One group of flame retardants that are emerging among the industry as well as the scientific community is nitrogen-based compounds. Despite the slightly lower efficiency compared to the halogen-based flame retardants, their lower cost, low emission of toxic smokes, eco-friendly features, and possibilities of functionalization with other elements such as phosphorus to introduce enhanced synergetic effects are strong factors that prompt their wide applications. The variety of compounds that can be used to broaden the formulations and synthetic approaches include mostly melamine and its derivatives (cyanurate, phosphate, polyphosphate, and pyrophosphate), urea, ammonium polyphosphates, and triazines [59]. The following section discusses the approaches that can be adopted for additive and reactive nitrogen-based flame retardants (NFRs) and their variations. The presentation of synthetic routes and mechanisms are also demonstrated.

22.4 Nitrogen-based green flame retardants

The application of NFRs is rapidly growing due to their eco-friendly characteristics, low cost, low smoke emission, versatile use in terms of functionalization, and can be used in many polymeric materials including PUs. Their range of efficiency falls in between halogen-based and metal oxides/hydroxides-based flame retardants [60]. Compared to halogen-based flame retardants, the NFRs present an inherent advantage of not releasing halogenic gas to the environments as well as suppressing the fire almost as quickly as halogen-based flame retardants. The contrast with metal oxides/hydroxides such as aluminum trihydroxide or magnesium hydroxide is that NFRs can perform similarly while requiring a lesser load during the formulation. This is an important factor as it aids to maintain the mechanical properties of the polymeric materials while also maintaining a green aspect. By the end of their shelf-life, these materials can be properly incinerated to be converted into fertilizers for the soil due to the high concentration of N and, in some cases, P, which are essential nutrients for plants in agricultural activities [60]. All these factors accompanied by a reasonable cost turn NFRs into valuable components to prevent fire hazards. The versatility of NFRs comes by the fact that most of them are solid compounds used in the form of a fine powder, which can be dispersed in a prepolymer mixture to make PUs. However, due to their chemical functionality, many synthetic approaches can be developed to chemically attach the NFRs into the polymeric chain of PUs. Usually, this process is performed in the polyol's chain, as there is a larger variety of starting materials for this class rather than isocyanates. Largely reported NFRs are melamine and its derivatives. This material is obtained on the industrial scale by utilizing two main synthetic routes. The first consists of heating urea catalyzed by ammonia, which yields melamine and carbon dioxide. The mechanism is expressed in Fig. 22.8 [61]. The second synthetic route can be performed by the heating of dicyandiamide in the presence of ammonium to obtain melamine (Fig. 22.9) [62].

Melamine is used as an effective flame retardant by physically mixing in a prepolymer mixture with the polyol that is later used to make PUs. It can quench fire by acting as a heat sink, which undergoes an endothermic decomposition when exposed to heat usually around 320–350 °C. It then releases ammonia that diminishes the concentration of oxygen in the media, which works as a gas-phase mechanism. Although NFRs are less effective compared with phosphorus based retardant that uses radical scavengers, emission of low toxic smoke is an advantage. Simultaneously, the released ammonia from melamine absorbs heat to form melam around 350 °C, followed by conversion into melem around 450 °C, and melon around 600 °C [63,64]. Upon further heating, it fully converts into graphitic carbon nitride. The flame retardancy of melamine works in a solid-phase mechanism where a compact char layer is formed after each decomposition process. When it is converted into a different compound, more ammonia is released as well as energy is absorbed (endothermic process). These phenomena describe both gas and condensed-phase mechanisms [63]. The chemical reactions involved in the process are described in Fig. 22.10 [63].

A recurrent strategy adopted in most studies relies on combining compounds to create a synergy effect and improve the overall fire resistance behavior of the PU. The use of N with P-based compounds is a viable approach, vastly reported in the literature [65–67]. For example, a facile

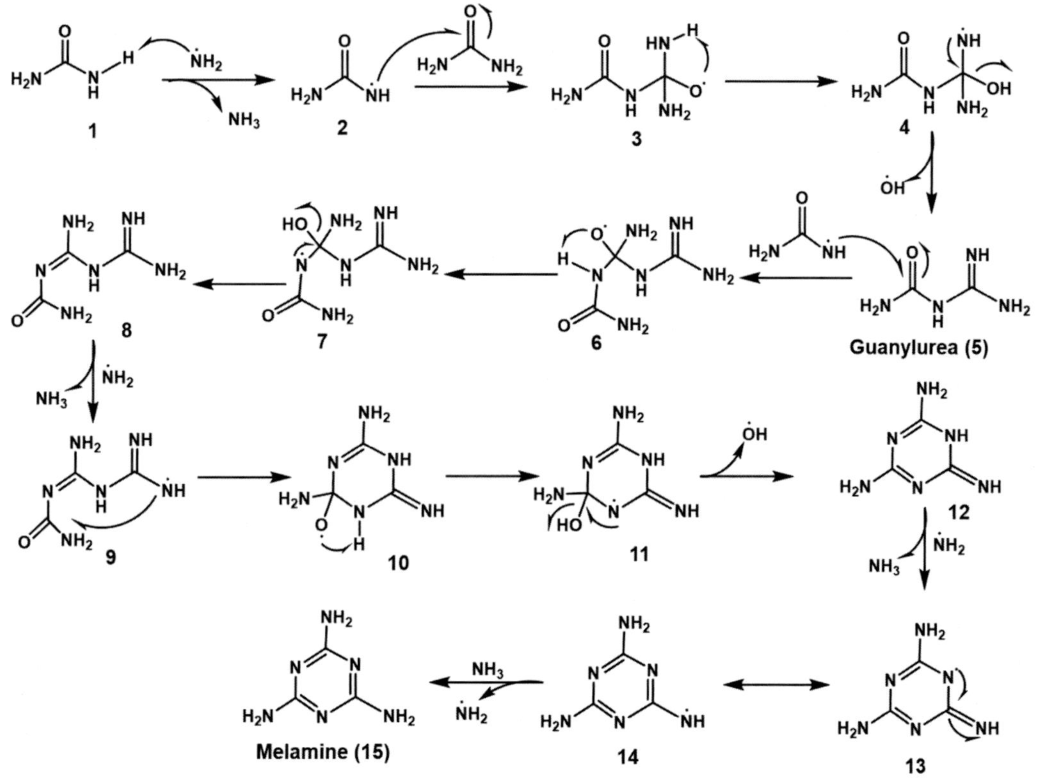

FIGURE 22.8 A mechanism for the synthesis of melamine through the urea trimerization mechanism catalyzed by ammonia. *"Adapted with permission from [61]. Copyright (2014) The Royal Society of Chemistry."*

method for the synthesis of MP was used where phosphoric acid was added to the dispersion of melamine in water [68,69]. The chemical reaction is demonstrated in Eq. (22.4). MPP is usually synthesized via calcination of MP to about 300–350°C as shown in Eq. (22.5) [68,69].

$$C_3N_6H_6 + H_3PO_4 \rightarrow C_3N_6H_6.H_3PO_4 \tag{22.4}$$

$$nC_3N_6H_6 \cdot H_3PO_4 \rightarrow (C_3N_6H_6 \cdot H_3PO_4)_n + nH_2O \tag{22.5}$$

This process causes the MP to polymerize into MPP. The latter is usually reported as a better flame retardant because the polymeric chain of phosphates creates a more compact char layer when exposed to flame, hence providing a faster effect against flame. Another representative of the melamine derivative is MC, which is a 1:1 complex of melamine and cyanuric acid. It undergoes an endothermic decomposition around 350 °C, which breaks the hydrogen bonding reversing the chemical structure back to melamine and cyanuric acid [70,71]. The hydrogen bonding that is formed between melamine and cyanuric acid is the main difference regarding MC in comparison with the other derivatives. This attractive force between these molecules leads to the formation of complex decomposition reactions that can lead to cross-linking with

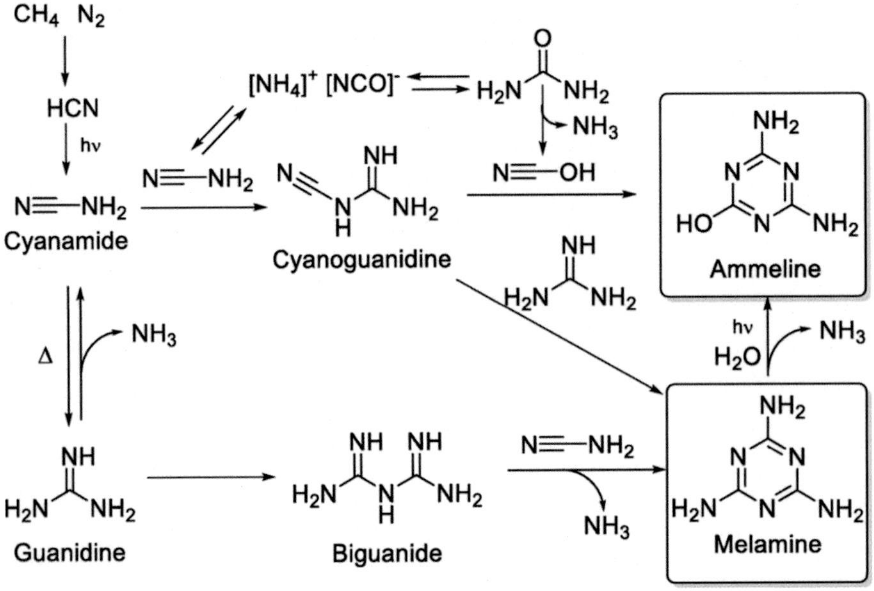

FIGURE 22.9 Synthesis of melamine through heating of cyanamide showing its different routes from starting substances up to final products. *"Adapted with permission from [62]. Copyright (2018) Springer International Publishing AG, part of Springer Nature."*

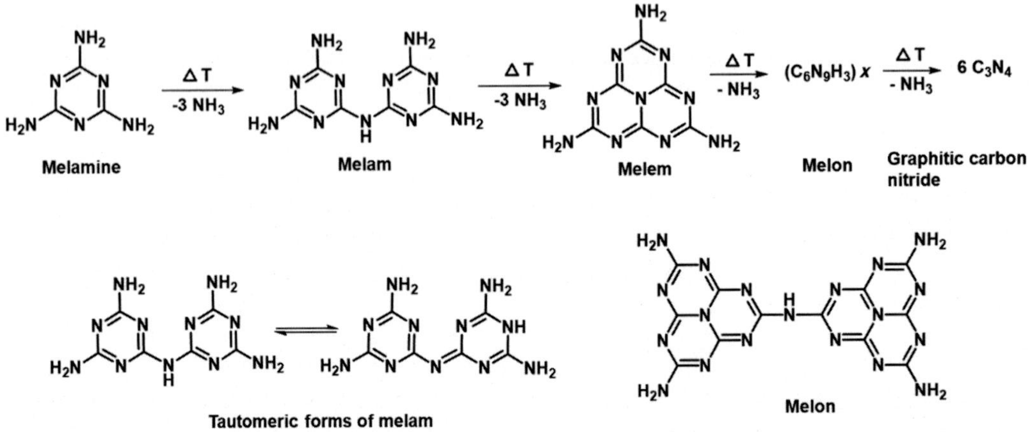

FIGURE 22.10 Thermal decomposition of melamine into melam, melem, melon, and graphitic carbon nitride accompanied by the evolution reaction of ammonia [63].

the polymeric matrix when exposed to fire, creating a shielding layer that decreases the temperature and prevents oxygen from reacting with the polymer [71]. The chemical structure of MC and the hydrogen bonding formed between the two compounds is demonstrated in Fig. 22.11 [72].

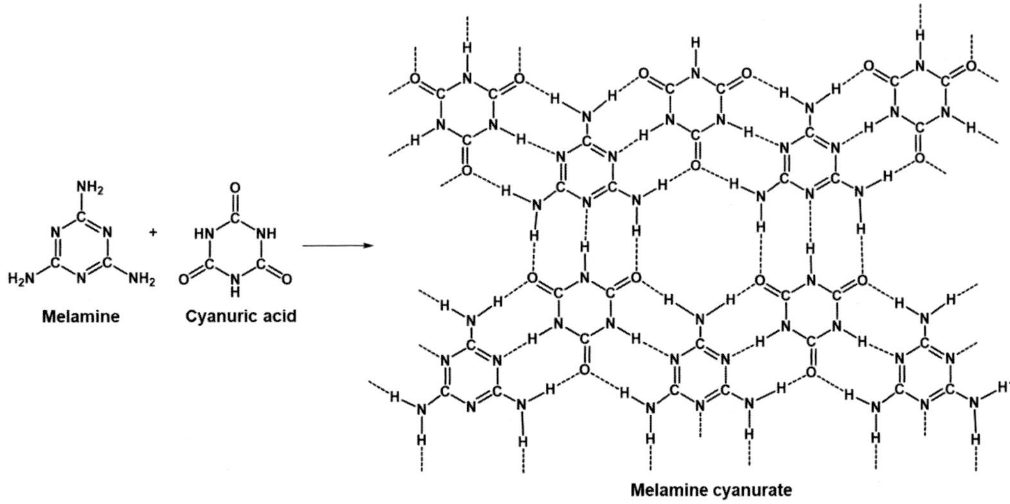

FIGURE 22.11 Melamine and cyanurate acid forming the hydrogen bonding that defines the self-assembly structure of melamine cyanurate.

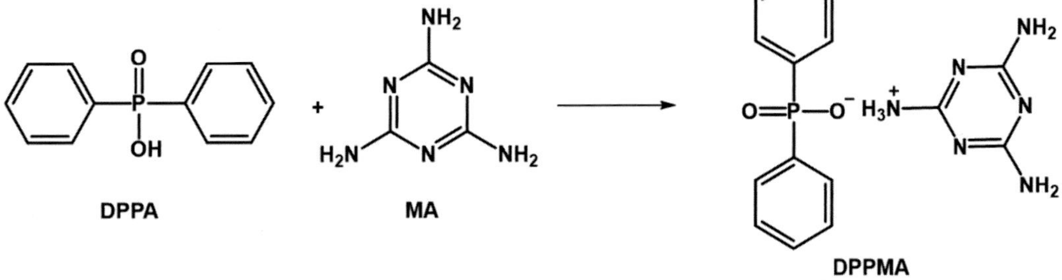

FIGURE 22.12 The chemical reaction for the synthesis of diphenylphosphine melamine by using diphenyl phosphinic acid and melamine. *"Adapted with permission from [73]. Copyright (2017) American Chemical Society."*

An interesting approach was developed by reacting melamine with diphenyl phosphinic acid (DPPA) to obtain a melamine and phosphorus-based salt, named as diphenyl phosphinic melamine (DPPMA) to be used as a flame retardant that could provide better flame-retardancy due to synergism effect between nitrogen and phosphorous as described in Fig. 22.12 [73]. The flame retardant was blended into a PU matrix to impart flame-retardancy. The characterizations conducted into the composite PU demonstrated an enhanced fire-retardancy. One of the standard tests such as limiting oxygen index (LOI), which describes how easily a material can burn, was performed. It improved the LOI from 21% to 25% which means that under an oxygen concentration of 21% the material without the flame retardant will burn with ease. However, after adding around 30 parts per hundred (pph) of the synthesized flame retardant the amount of oxygen concentration in the media required to burn the composite polymer increased to 25%, demonstrating that the material requires more oxygen to combust. Another fire behavior analysis performed was the fire growth rate, which is a ratio between the peak heat release rate and time heat release rate. The information from these data describes that obtaining a low value relates to a slower release of heat or a longer time for the material to reach the flashpoint. A value of 6.67 was reported for the neat material which dropped to 3.6 after the addition of 30 pph of flame retardant. The vertical burning test was also performed for the set of samples and demonstrated that without the flame retardant the PU burned almost completely and with the addition of flame retardant the time of burning was reduced to less than 15 s, preserving most of the material occasioning low weight loss.

The use of an N and P-based compounds created a synergy effect, which improves the overall efficiency of the material if they were used separately. This effect was observed after the extraction of data from thermal gravimetric analysis and Fourier Transformed infra-red, along with many other techniques that can be employed to understand how the chemical mechanism quenches the fire. It was proposed that with the increase in temperature, the DPPMA decomposed into its starting materials and release signature fragments during thermal degradation. In the case of DPPA, benzene and phosphorus gases were formed. The latter can capture radical species and act on the condensed phase to form polyphosphoric acid, which aids in the condensed-phase mechanism [74]. In simultaneous process melamine also acts on that process by undergoing endothermic decomposition releasing ammonia and forming its polymeric products known as melam, melem, and melon [75–78]. The scheme for this fire behavior mechanism is shown in Fig. 22.13 [73]. The mechanical properties also improved presenting a tensile strength from 110 to 115 kPa and an elongation at break from 320% to 357% with an increasing quantity of flame retardant. It shows that despite some blended flame retardants cause deterioration of mechanical properties, that is not always the case.

Another experiment was conducted by utilizing a mix between several NFRs such as MC, MPP along with an aluminum organophosphorus-based flame retardant named as aluminum diethyl phosphinate in a thermoplastic PU matrix [79]. The ratio of these flame retardants was alternated in a set of samples to determine which ratio was more effective to improve the flame-retardancy while maintaining constant the amount of 30 wt% load of flame retardant. By finding the amount of 12, 10, and 8 wt% for MC, MPP, and aluminum diethyl phosphinate, respectively, a great improvement in fire inhibition was observed. After performing the cone calorimeter test, the peak heat release rate dropped from 2660 to 452 kW/m^2. Along with that the burning time which was above 30 s for the neat material dropped to less than 10 s after blending with the flame retardants. This study also observed an improvement in the mechanical properties for the thermoplastic PU after the addition of flame retardants which also improved the tensile strength from 20 to 25.7 kPa and elongation at break from 312.6% to 421.5 % [79].

It is notable that despite the complexity of the chemical reaction that takes place during the fire decomposition, facile approaches that require a direct physical mixture of NFRs with the PUs can be performed to impart enhanced flame-retardancy into these materials, usually requiring lower quantities of NFRs, while presenting a greener alternative in comparison with other flame retardants. However, more approaches are possible by elaborating synthetic procedures to chemically bond the NFRs into the structure of the final PUs, allowing even more possibilities to obtain new materials.

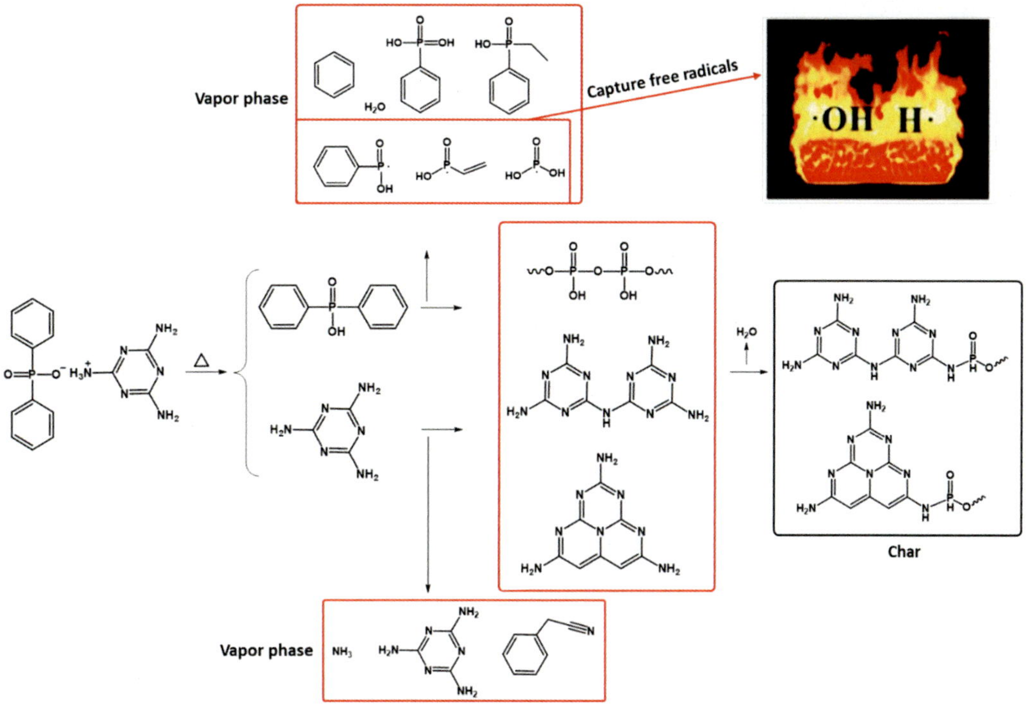

FIGURE 22.13 Fire decomposition mechanism for diphenyl phosphinic melamine describing the thermal decomposition products and synergistic effect between N and P. *"Adapted with permission from [73]. Copyright (2017) American Chemical Society."*

Melamine is a widely used material as a flame retardant due to its effective and eco-friendly properties regarding fire inhibiting, but also, it is a compound that allows many possibilities regarding chemical functionality. One of the largely used approaches is the Mannich reaction [80]. It consists of a methylene bridge that connects an amino group to an acidic α-carbon, that functions as a nucleophile. A study reported the synthesis of a cardanol functionalized with melamine through the Mannich reaction. The obtained polyol was then used to make inherent flame-retardant PUs. The reaction was carried out in a two-step process. In the first step, cardanol reacted with the melamine-formaldehyde intermediate along with diethanolamine to increase the N% as well as increase the functionality by introducing more hydroxyl groups. The second step consists of the propoxylation reaction (Fig. 22.14).

This strategy to synthesize a reactive polyol used mostly environmental-friendly starting materials that yielded effective results. Despite that, the flame-retardancy of the PU was further enhanced by blending other flame retardants such as expandable graphite, ammonium polyphosphate, and diethyl ethyl phosphate. The results demonstrated a general improvement in both flame-retardancy as well as mechanical properties, which for this case were highlighted mostly from expandable graphite. The composited flame retardant bio-polyurethane demonstrated a high value of LOI reaching 29.8%, slight improvement of mechanical properties, an increase of residue after thermostability test from around 18% to 29%, and general improvement

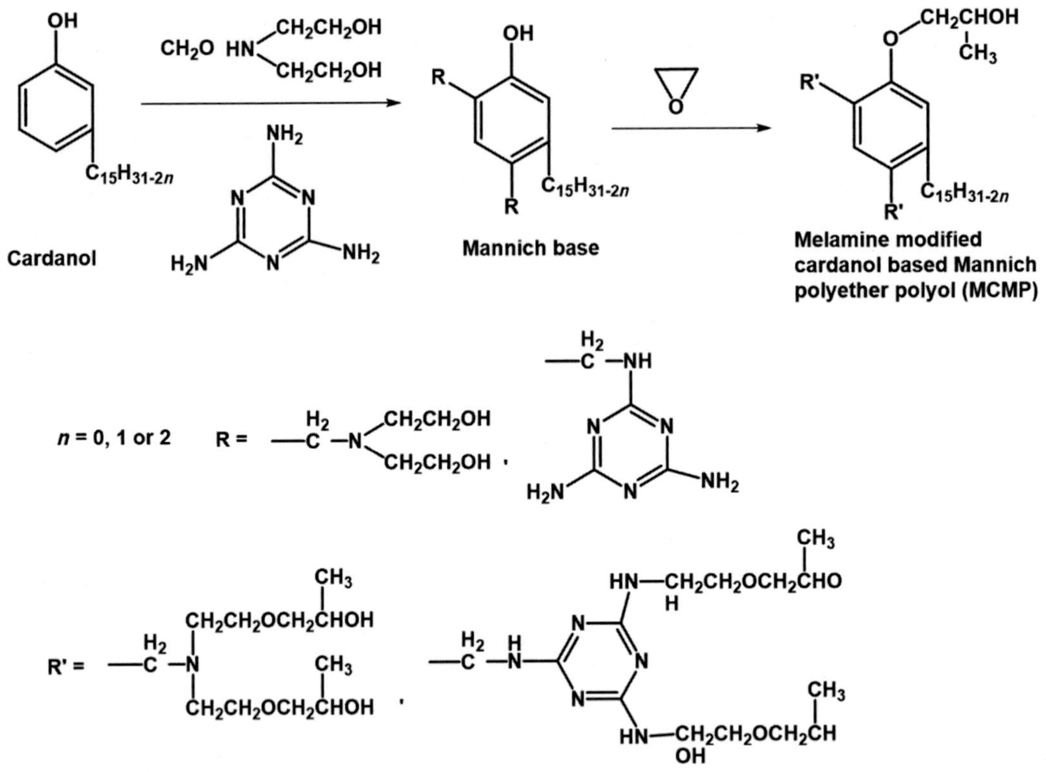

FIGURE 22.14 Reaction scheme for the synthesis of cardanol-melamine-based polyol. *"Adapted with permission from [80]. Copyright (2014) Elsevier."*

in peak heat release rate, total heat release, and char residue. It demonstrates that there are viable green approaches to obtain flame retardant bio-polyurethanes from low-cost sources and yet satisfactory properties. Another study demonstrated other possible paths by preparing a melamine-formaldehyde Mannich base prepared by reacting with ethylene glycol and obtaining a resin to make PU. The synthesis for this resin is expressed in Fig. 22.15 [81].

A common approach usually performed by most researchers consists of synthesizing a reactive polyol or resin to make PU and use a blended flame retardant to further improve the properties of the final polymer. For example, a P-based flame retardant was added as a blend. This approach avoids the overloading of the PU with an extra amount of flame retardant, hence preventing it from deteriorating other properties. Along with that it also decreases the total cost of the production. Thus, enabling several different approaches and synergetic studies that can be tested. Finally, the combination of flame retardant in this way imparts a more effective synergy effect among them, optimizing the performance for the PU [81]. Industrial applications have also been filed for the use of polyol-based melamine, showing promising uses for PU foams [82,83].

A widely used approach to enhance the fire behavior into the PU is by taking advantage of available reactive sites into the polymeric structure that allows the chemical attachment of a

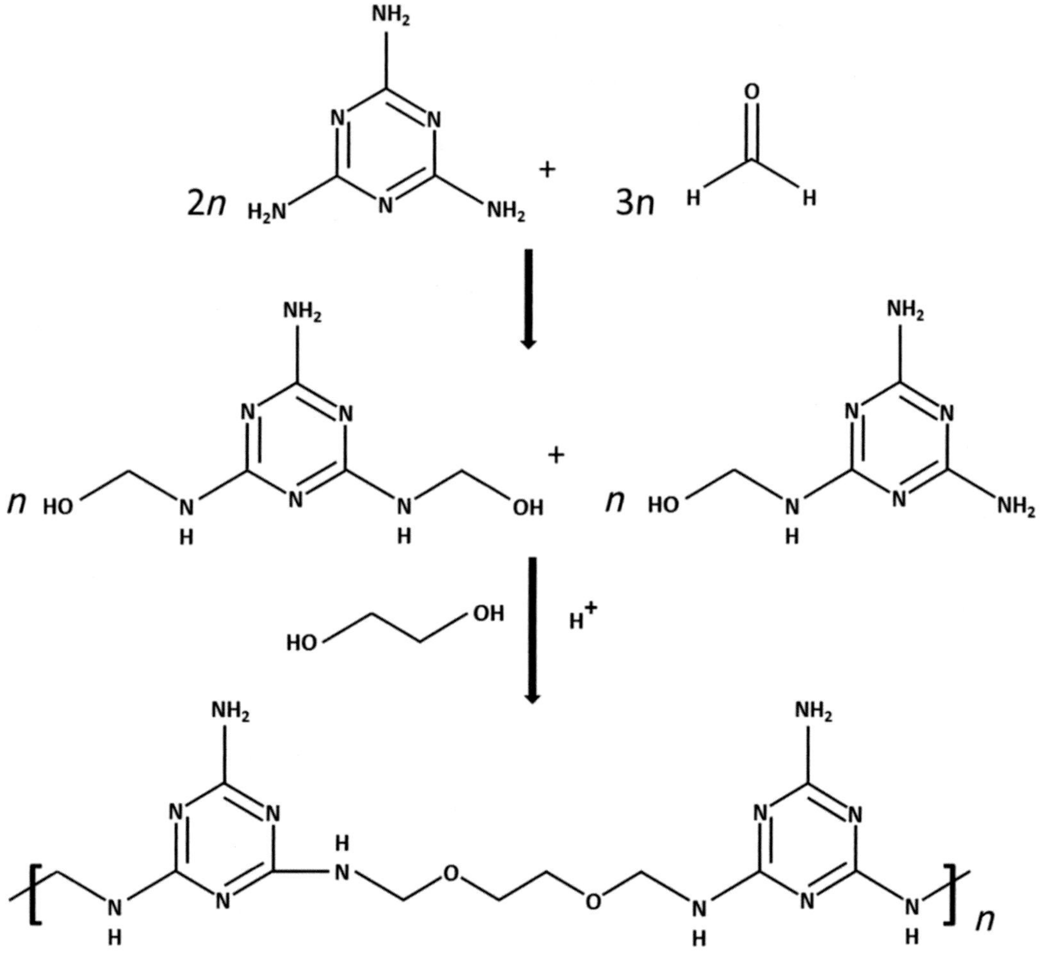

FIGURE 22.15 The chemical reaction for the synthesis of melamine, formaldehyde, and ethylene glycol for an inherent flame-retardant resin [81].

flame-retardant compound. This strategy allows the material to become inherently flame retardant and usually demands a lower amount of flame retardant compared with procedures that use the physical blending of flame retardants. Despite the extra step of the synthesis, the reactive flame retardant tends to be more efficient and usually improves the mechanical properties of the PU, broadening the scope of applications as well as increasing the number of materials available for that use. A report described the synthesis procedure of a novel inherent flame retardant polyol based on the "click" chemistry between phosphate and triazole groups yielding a phosphorus triazole flame-retardant monomer [84]. The novel flame-retardant polyol was used with a castor oil–based polyol as a starting material to compose the backbone structure of the PU. Hence, yielding an inherent flame retardant bio-polyurethane. The general scheme for the procedure is demonstrated in Fig. 22.16 [84].

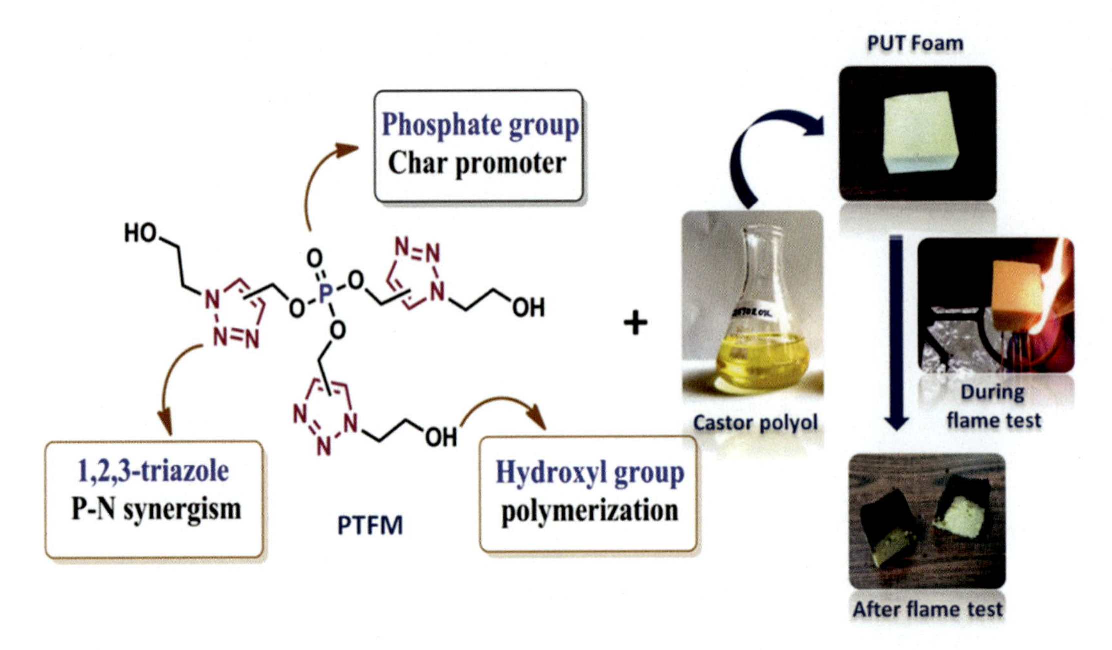

FIGURE 22.16 Schematics for the production of a bio-polyurethane with inherent flame-retardant properties based on a novel phosphorus triazole flame retardant monomer and castor oil–based polyol [84].

The N-P synergy effect for this inherent flame-retardant monomer works in two simultaneous steps. First, the triazole group, when exposed to fire, converts into nitrogen gas, which is an inert gas that can dilute the concentration of oxygen in the media and swell the char layer that is formed during combustion, which effectively decreases the heat transfer [84]. Along with that, the phosphate group forms a char layer of polyphosphoric acid creating a physical barrier preventing oxygen and radical species to get into contact with the PU as well as decreasing the heat transfer. In addition to that, the polyphosphoric acid char layer can undergo a cross-link reaction with the carbonaceous layer from the PU, which may create a more compact physical layer to protect against the fire [85,86]. The flame-retardancy of the bio-polyurethane was analyzed through many techniques such as Fourier-transform infrared (FTIR), thermal analysis, X-ray photoelectron spectroscopy, cone calorimeter, LOI, and many others. In this work, an increase in LOI was obtained reaching up to 27%. Also, the thermal decomposition observed in thermal analysis presents a great improvement leading to almost 30% residue after exposure to 700 °C compared with nearly 5% for the neat sample. The analysis obtained through the cone calorimeter showed that in use of flame-retardant monomer into the backbone of the PU increase almost all the properties of the PU in comparison with the neat sample. It increases the time of ignition from 3 to 12 s, decreased the peak heat release rate from 611.15 to 235 kW/m^2, along with CO, CO_2 emissions, and dripping. Hence, it demonstrated that the implementation of a flame retardant directly into the backbone of the PU is a viable and effective approach to improve flame retardancy.

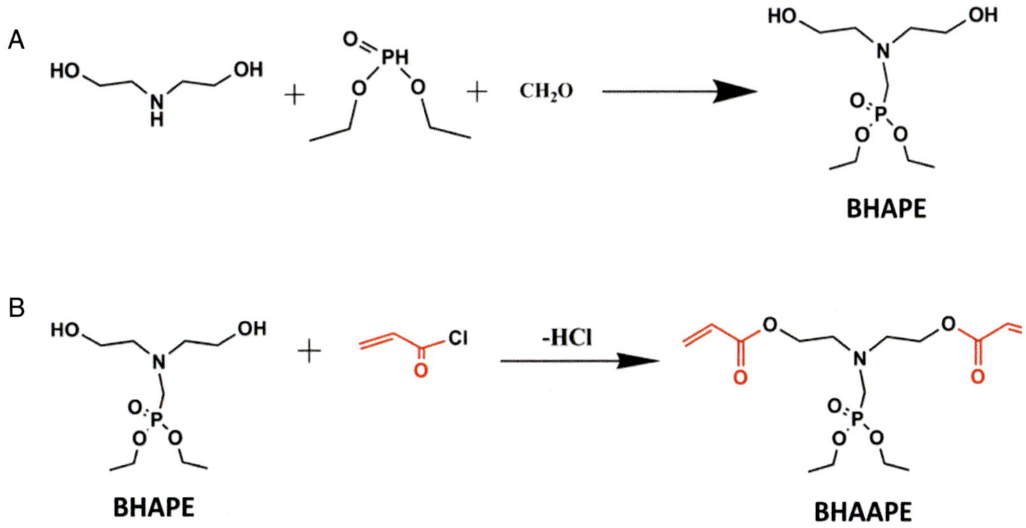

A

HO OH N H + O PH O O + CH_2O → HO OH N O P O O BHAPE

B

HO OH N O P O O BHAPE + Cl O -HCl → O O N O O O P O O BHAAPE

FIGURE 22.17 Chemical reactions for the structure of BHAPE (N,N-bis(2-hydroxyethyl)-aminomethylphosphonic acid diethyl ester)) and BHAAPE N,N-bis(2-hydroxyethyl)-aminomethylphosphonic acid diethyl ester). *"Adapted with permission from [89]. Copyright (2013) American Chemical Society."*

Another interesting strategy used to develop inherent flame-retardant compounds was performed through the Kabachnik–Fields reaction, which is known to provide a compound with nitrogen and phosphorus connected by a methylene segment. It takes place by reacting three reactive groups: amine, formaldehyde, and hydrophosphoryl [87,88]. A report described the synthesis of two compounds used through this reaction. The first one was N,N-bis(2-hydroxyethyl)-aminomethylphosphonic acid diethyl ester) named BHAPE and the second one was (N,N-bis(2-hydroxyethyl acrylate) aminomethyl phosphonic acid diethyl ester named as BHAAPE. The chemical reactions for these compounds are described in Fig. 22.17 [89].

This synthetic approach allows many possibilities for further reaction to other monomers to yield different categories of polymers. The authors of this work opted to implement an acrylate group and promote polymerization through UV curing, which is a viable and green procedure. It proved to be successful as the polymer presented general satisfactory fire behavior properties such as increment in LOI from 21% to 31% and decreasing the total heat release by 30.1% to 27.4% after implementation of 30 wt% of BHAAPE. Following the same line, BHAPE is a compound that presents a primary hydroxyl group, which can easily react with diisocyanates to make PU foams. As demonstrated, many strategies can be adopted to develop a variety of different compounds. Among the published studies, one performed a synthesis of a flame-retardant polyol containing both N and P using castor oil as a starting material [90]. Along with that, BHAPE was also employed as another source of flame retardancy properties. The presence of hydroxyl groups in both compounds allows their reaction with diisocyanates to make PUs. The synthetic route for these series of reactions is described in Fig. 22.18 [90]. The implementation of N and P imparts a synergy effect. The analysis of heat release rate and total heat released showed synergism as

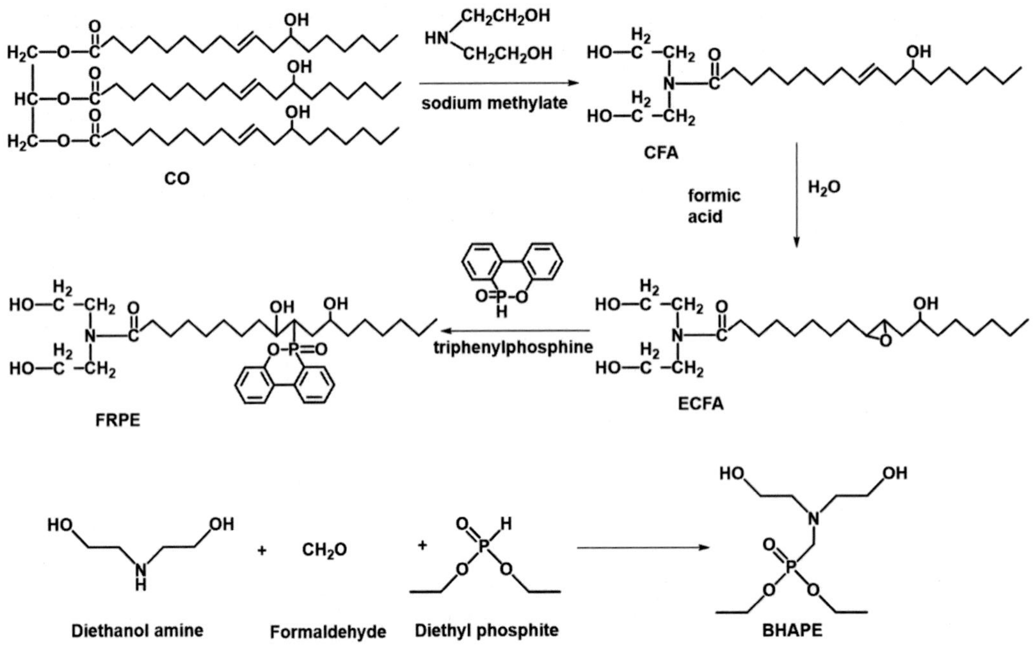

FIGURE 22.18 Synthetic route for the functionalization of castor oil (CO) into an inherent flame-retardant polyol. First, obtaining the fatty acid amide (CFA). Second, epoxidation of CFA into epoxidized castor oil fatty acid amide (ECFA). Third, ring-opening of ECFA with triphenylphosphine to obtain the flame-retardant polyol. Along with that BHAPE was also added into the formulation of the polyurethane. *"Adapted with permission from [90]. Copyright (2017) Elsevier."*

there was a decrease of more than 50% of the initial values compared with the neat PU. It was observed after the implementation of the N and P flame-retardant compounds into the fatty acid chain of castor oil and the use of BHAPE to form a PU with copolymeric structure. The property relation of the PU in the function of the concentration of BHAPE was also performed. The results demonstrated that as the concentration of BHAPE increased so did the LOI reaching a maximum of 24%. However, a decrease in mechanical properties was observed by drastic decrease from 0.26 MPa to 0.14 MPa for the concentrations of 20 to 50% of BHAPE, respectively [90]. These results demonstrate the challenge that accompanies the synthesis of flame-retardant PUs, which is finding effective materials to improve the properties, but also that can prevent other properties from deteriorating.

22.5 Summary and outlook

Through the analysis of the scientific trend, it is notable that a great amount of published work presents an extremely high quality of knowledge regarding the focal points for the understanding of how safety is an important concern in direct relation with the greener approaches for the synthesis of PUs utilizing low cost and renewable sources containing NFRs. Now the foresight directs on finding the optimal condition to improve the flame-retardant properties of these materials by finding the middle ground between the quantity of additives and maintenance of properties. Along with that, many unique approaches have emerged in the scientific community that prompt new innovative ideas demonstrating how vast this research area is. In all regards, it brings an extremely important concern by showing how science can improve the quality of life as well as insert a sustainable process. In conclusion, the use of NFRs for bio-PUs demonstrated quick progress and still can adopt many novel approaches to improve the quality of works for further large-scale applications.

References

[1] M. Ionescu, Chemistry and technology of polyols for polyurethanes, Polimeri 26 (2006) 218 218.

[2] J.O. Akindoyo, M. Beg, S. Ghazali, M.R. Islam, N. Jeyaratnam, A.R. Yuvaraj, Polyurethane types, synthesis and applications–a review, RSC Adv. 6 (2016) 114453–114482.

[3] D.A. Babb, Polyurethanes from renewable resources, in: DA Babb (Ed.), Synthetic Biodegradable Polymers, Springer, Verlag Berlin, Verlag, 2011, pp. 315–360.

[4] Y. Savelyev, V. Veselov, L. Markovskaya, O. Savelyeva, E. Akhranovich, N. Galatenko, L. Robota, T. Travinskaya, Preparation and characterization of new biologically active polyurethane foams, Mater. Sci. Eng. C 45 (2014) 127–135. https://doi.org/10.1016/j.msec.2014.08.068.

[5] S.N. Singh, Blowing Agents for Polyurethane Foams, Smithers Rapra Publishing, Shrewsbury, Shropshire, United Kingdom, 2001.

[6] M.J. Molina, F.S. Rowland, Stratospheric sink for chlorofluoromethanes: chlorine atom-catalysed destruction of ozone, Nature 249 (1974) 810–812. https://doi.org/10.1038/249810a0.

[7] M. Protocol, Montreal Protocol on Substances that Deplete the Ozone Layer, 26, US Gov Print Off, Washington, DC, 1987, pp. 128–136.

[8] L. Shen, E. Worrell, M. Patel, Present and future development in plastics from biomass, Biofuels. Bioprod. Biorefining 4 (2010) 25–40. https://doi.org/10.1002/bbb.189.

[9] I. Javni, W. Zhang, Z.S. Petrović, Effect of different isocyanates on the properties of soy-based polyurethanes, J. Appl. Polym. Sci. 88 (2003) 2912–2916. https://doi.org/10.1002/app.11966.

[10] S.F.M. de, C.J, S. Bhoyate, P.K. Kahol, R.K. Gupta, Expendable graphite as an efficient flame-retardant for novel partial bio-based rigid polyurethane foams. J. Carbon Res. C 20 (2020) 6–27. https://doi.org/10.3390/c6020027.

[11] S. Ramanujam, C. Zequine, S. Bhoyate, B. Neria, P. Kahol, R. Gupta, Novel biobased polyol using corn oil for highly flame-retardant polyurethane foams. J. Carbon Res. C 5 (2019) 13. https://doi.org/10.3390/c5010013.

[12] Y. Zhu, C. Romain, C.K. Williams, Sustainable polymers from renewable resources, Nature 540 (2016) 354–362. https://doi.org/10.1038/nature21001.

[13] S. Bhoyate, M. Ionescu, P.K. Kahol, J. Chen, S.R. Mishra, R.K. Gupta, Highly flame-retardant polyurethane foam based on reactive phosphorus polyol and limonene-based polyol, J. Appl. Polym. Sci. 135 (2018) 16–19. https://doi.org/10.1002/app.46224.

[14] C. Zhang, S. Bhoyate, M. Ionescu, P.K. Kahol, R.K. Gupta, Highly flame retardant and bio-based rigid polyurethane foams derived from orange peel oil, Polym. Eng. Sci. 58 (2018) 2078–2087. https://doi.org/10.1002/pen.24819.

[15] R.K. Gupta, M. Ionescu, D. Radojcic, X. Wan, Z.S. Petrovic, Novel renewable polyols based on limonene for rigid polyurethane foams, J. Polym. Environ. 22 (2014) 304–309. https://doi.org/10.1007/s10924-014-0641-3.

[16] R.K. Gupta, M. Ionescu, X. Wan, D. Radojcic, Z.S. Petroviˆ, Synthesis of a novel limonene based Mannich polyol for rigid polyurethane foams, J. Polym. Environ. 23 (2015) 261–268. https://doi.org/10.1007/s10924-015-0717-8.

[17] C.K. Ranaweera, M. Ionescu, N. Bilic, X. Wan, P.K. Kahol, R.K. Gupta, Biobased polyols using thiol-ene chemistry for rigid polyurethane foams with enhanced flame-retardant properties, J. Renew. Mater 5 (2017) 1–12. https://doi.org/10.7569/jrm.2017.634105.

[18] N. Elbers, C.K. Ranaweera, M. Ionescu, X. Wan, P.K. Kahol, R.K. Gupta, Synthesis of novel biobased polyol via thiol-ene chemistry for rigid polyurethane foams, J. Renew. Mater 5 (2017) 74–83. https://doi.org/10.7569/jrm.2017.634137.

[19] C.E. Hoyle, T.Y. Lee, T. Roper, Thiol-enes: Chemistry of the Past with Promise for the Future, Department of Polymer Science, Hattiesburg, Mississippi, 2004.

[20] H. Liu, H. Chung, Visible-light induced thiol-ene reaction on natural lignin, ACS Sustain. Chem. Eng. 5 (2017) 9160–9168. https://doi.org/10.1021/acssuschemeng.7b02065.

[21] R.K. Gupta, M. Ionescu, X. Wan, D. Radojcic, Z.S. Petroviˆ, Synthesis of a novel limonene based mannich polyol for rigid polyurethane foams, J. Polym. Environ. 23 (2015) 261–268. https://doi.org/10.1007/s10924-015-0717-8.

[22] C.K. Ranaweera, M. Ionescu, N. Bilic, X. Wan, P.K. Kahol, R.K. Gupta, Biobased polyols using thiol-ene chemistry for rigid polyurethane foams with enhanced flame-retardant properties, J. Renew. Mater. 5 (2017) 1–12. https://doi.org/10.7569/jrm.2017.634105.

[23] M.E. Collier, Pressure-reducing mattresses, J. Wound Care 5 (1996) 207–211. https://doi.org/10.12968/jowc.1996.5.5.207.

[24] A. Bhatnagar, A. Sharma, Polyurethane Foam Stabilizer Market Size, Industry Analysis Report (2019) 2018–2025. https://dotblogs.com.tw/businessnews24/2019/02/04/182347.

[25] Grand View ResearchGlobal Bio Polyurethane (PU) Market By Product (Rigid Foams, Flexible Foams, CASE), By End-Use (Furniture & Interiors, Construction, Automotive, Footwear), Grand View Research, San Francisco, CA, 2015.

[26] J.O. Akindoyo, M.D.H. Beg, S. Ghazali, M.R. Islam, N. Jeyaratnam, A.R. Yuvaraj, Polyurethane types, synthesis and applications-a review, RSC Adv. 6 (2016) 114453–114482. https://doi.org/10.1039/c6ra14525f.

[27] V. Babrauskas, D. Lucas, D. Eisenberg, V. Singla, M. Dedeo, A. Blum, Flame retardants in building insulation: a case for re-evaluating building codes, Build. Res. Inf. 40 (2012) 738–755. https://doi.org/10.1080/09613218.2012.744533.

[28] B. Czupryński, J. Paciorek-Sadowska, J. Liszkowska, Properties of rigid polyurethane-polyisocyanurate foams modified with the selected fillers, J. Appl. Polym. Sci. 115 (2010) 2460–2469. https://doi.org/10.1002/app.30937.

[29] A. Agrawal, R. Kaur, R.S. Walia, Investigation on flammability of rigid polyurethane foam-mineral fillers composite, Fire. Mater. 43 (2019) 917–927. https://doi.org/10.1002/fam.2751.

[30] J. Troitzsch, Flame retardants, Kunststoffe. Ger. Plast. 77 (1987) 90–91.

[31] M.J. Kenter, Fire loss in the United States during 1989, Fire. J. Boston. Mass 84 (1990) 1–268.

[32] J.J. Shea, Polymeric foams: mechanisms and materials, IEEE. Electr. Insul. Mag. 21 (2005) 56. https://doi.org/10.1109/MEI.2005.1412232.

[33] S. Bhoyate, M. Ionescu, D. Radojcic, P.K. Kahol, J. Chen, S.R. Mishra, R.K. Gupta, Highly flame-retardant bio-based polyurethanes using novel reactive polyols, J. Appl. Polym. Sci. 135 (12) (2018) 46027. https://doi.org/10.1002/app.46027.

[34] M. Ionescu, Chemistry and Technology of Polyols for Polyurethanes, Rapra Technology, Shrewsbury UK, 2007.

[35] M. Modesti, A. Lorenzetti, Halogen-free flame retardants for polymeric foams, Polym. Degrad. Stab. 78 (2002) 167–173. https://doi.org/10.1016/S0141-3910(02)00130-1.

[36] E.D. Weil, SV. Levchik, Flame retardants in commercial use or development for polyolefins. J. Fire Sci. 26 (2008) 5–43.

[37] L. Wang, X. He, C.A. Wilkie, The utility of nanocomposites in fire retardancy, Materials (Basel) 3 (2010) 4580–4606. https://doi.org/10.3390/ma3094580.

[38] W. Yang, G. Tang, L. Song, Y. Hu, R.K.K. Yuen, Effect of rare earth hypophosphite and melamine cyanurate on fire performance of glass-fiber reinforced poly(1,4-butylene terephthalate) composites, Thermochim. Acta 526 (2011) 185.

[39] L. Zhang, M. Zhang, Y. Zhou, L. Hu, The study of mechanical behavior and flame retardancy of castor oil phosphate-based rigid polyurethane foam composites containing expanded graphite and triethyl phosphate, Polym. Degrad. Stab. 98 (2013) 2784–2794. https://doi.org/10.1016/j.polymdegradstab.2013.10.015.

[40] P. Khalili, K.Y. Tshai, D. Hui, I. Kong, Synergistic of ammonium polyphosphate and alumina trihydrate as fire retardants for natural fiber reinforced epoxy composite, Compos. Part B. Eng. 114 (2017) 101–110. https://doi.org/10.1016/j.compositesb.2017.01.049.

[41] C. Hoffendahl, G. Fontaine, S. Duquesne, F. Taschner, M. Mezger, S. Bourbigot, The combination of aluminum trihydroxide (ATH) and melamine borate (MB) as fire retardant additives for elastomeric ethylene vinyl acetate (EVA), Polym. Degrad. Stab. 115 (2015) 77–88. https://doi.org/10.1016/j.polymdegradstab.2015.03.001.

[42] S.S. Xiao, M.J. Chen, L.P. Dong, C. Deng, L. Chen, Y.Z. Wanga, Thermal degradation, flame retardance and mechanical properties of thermoplastic polyurethane composites based on aluminum hypophosphite, Chin. J. Polym. Sci. 32 (2014) 98–107 (English Ed.). https://doi.org/10.1007/s10118-014-1378-0 .

[43] G. Camino, S. Duquesne, R. Delobel, B. Eling, C. Lindsay, T. Roels, Mechanism of expandable graphite fire retardant action in polyurethanes, ACS Symp. Ser 797 (2001) 90–109. https://doi.org/10.1021/bk-2001-0797.ch008.

[44] Y. Wang, F. Wang, Q. Dong, M. Xie, P. Liu, Y. Ding, S. Zhang, M. Yang, G. Zheng, Core-shell expandable graphite@aluminum hydroxide as a flame-retardant for rigid polyurethane foams, Polym. Degrad. Stab. 146 (2017) 267–276. https://doi.org/10.1016/j.polymdegradstab.2017.10.017.

[45] A. Dasari, Z.-.Z. Yu, G.-.P. Cai, Y.-.W. Mai, Recent developments in the fire retardancy of polymeric materials, Prog. Polym. Sci. 38 (2013) 1357–1387. https://doi.org/10.1016/j.progpolymsci.2013.06.006.

[46] B. Yuan, C. Bao, Y. Guo, L. Song, K.M. Liew, Y. Hu, Preparation and characterization of flame-retardant aluminum hypophosphite/poly(vinyl alcohol) composite, Ind. Eng. Chem. Res. 51 (2012) 14065–14075. https://doi.org/10.1021/ie301650f.

[47] X. Wang, E.N. Kalali, J.-.T. Wan, D.-.Y. Wang, Carbon-family materials for flame retardant polymeric materials, Prog. Polym. Sci. 69 (2017) 22–46. https://doi.org/10.1016/j.progpolymsci.2017.02.001.

[48] B.B. Troitskii, L.S. Troitskaya, A.A. Dmitriev, A.S. Yakhnov, Inhibition of thermo-oxidative degradation of poly(methyl methacrylate) and polystyrene by C60, Eur. Polym. J. 36 (2000) 1073–1084. https://doi.org/10.1016/S0014-3057(99)00156-1.

[49] S. Bourbigot, G. Fontaine, A. Gallos, S. Bellayer, Reactive extrusion of PLA and of PLA/carbon nanotubes nanocomposite: processing, characterization and flame retardancy, Polym. Adv. Technol. 22 (2011) 30–37. https://doi.org/10.1002/pat.1715.

[50] B. Schartel, P. Pötschke, U. Knoll, M. Abdel-Goad, Fire behaviour of polyamide 6/multiwall carbon nanotube nanocomposites, Eur. Polym. J. 41 (2005) 1061–1070. https://doi.org/10.1016/j.eurpolymj.2004.11.023.

[51] T. Kashiwagi, F. Du, K.I. Winey, K.M. Groth, J.R. Shields, S.P. Bellayer, H. Kim, J.F. Douglas, Flammability properties of polymer nanocomposites with single-walled carbon nanotubes: effects of nanotube dispersion and concentration, Polymer (Guildf) 46 (2005) 471–481. https://doi.org/10.1016/j.polymer.2004.10.087.

[52] F. Gao, G. Beyer, Q. Yuan, A mechanistic study of fire retardancy of carbon nanotube/ethylene vinyl acetate copolymers and their clay composites, Polym. Degrad. Stab. 89 (2005) 559–564. https://doi.org/10.1016/j.polymdegradstab.2005.02.008.

[53] P. Georlette, J. Simons, L. Costa, Halogen-containing fire-retardant compounds, in: A. Grand and C. Wilkie (Eds.), Fire Retardancy, Marcel Dekker Inc., 2000, pp. 246–281.

[54] J. Green, Mechanisms for flame retardancy and smoke suppression—a review, J. Fire. Sci. 14 (1996) 426–442. https://doi.org/10.1177/073490419601400602.

[55] G. Camino, L. Costa, M.P. Luda di Cortemiglia, Overview of fire retardant mechanisms, Polym. Degrad. Stab. 33 (1991) 131–154. https://doi.org/10.1016/0141-3910(91)90014-I.

[56] M. Lewin, E.D. Weil, Chapter 2: Mechanisms and modes of action in flame retardancy of polymers, in: A.R. Horrocks and D. Price (Eds.), Fire Retardant Materials, Woodhead Publishing, Cambridge, England, 2001, pp. 31–68.

[57] S.-.Y. Lu, I. Hamerton, Recent developments in the chemistry of halogen-free flame retardant polymers, Prog. Polym. Sci. 27 (2002) 1661–1712. https://doi.org/10.1016/S0079-6700(02)00018-7.

[58] S.V. Levchik, E.D. Weil, A review of recent progress in phosphorus-based flame retardants, J. Fire Sci. 24 (2006) 345–364. https://doi.org/10.1177/0734904106068426.

[59] S. BK, New thinking on flame retardants, Environ. Health Perspect. 116 (2008) A210–A213. https://doi.org/10.1289/ehp.116-a210.

[60] H. Horacek, R. Grabner, Advantages of flame retardants based on nitrogen compounds, Polym. Degrad. Stab. 54 (1996) 205–215. https://doi.org/10.1016/S0141-3910(96)00045-6.

[61] Y.A. Jeilani, T.M. Orlando, A. Pope, C. Pirim, M.T. Nguyen, Prebiotic synthesis of triazines from urea: a theoretical study of free radical routes to melamine, ammeline, ammelide and cyanuric acid, RSC. Adv. 4 (2014) 32375–32382. https://doi.org/10.1039/C4RA03717K.

[62] C. Menor-Salván, From the dawn of organic chemistry to astrobiology: urea as a foundational component in the origin of nucleobases and nucleotides, in: C. Menor-Salvan (Ed.), Prebiotic Chemistry and Chemical Evolution of Nucleic Acids, Springer, Cham, 2018, pp. 85–142.

[63] X. Liu, J.-.W. Hao, S. Gaan, Recent studies of decomposition and strategies of smoke and toxicity suppression for polyurethane based materials, RSC. Adv. 6 (2016) 74742–74756. https://doi.org/10.1039/C6RA14345H.

[64] A. König, U. Fehrenbacher, E. Kroke, T. Hirth, Thermal decomposition behavior of the flame retardant melamine in slabstock flexible polyurethane foams, J. Fire. Sci. 27 (2009) 187–211. https://doi.org/10.1177/0734904108099329.

[65] M.-.J. Chen, Y.-.J. Xu, W.-.H. Rao, J.-.Q. Huang, X.-.L. Wang, L. Chen, Y.-.Z. Wang, Influence of valence and structure of phosphorus-containing melamine salts on the decomposition and fire behaviors of flexible polyurethane foams, Ind. Eng. Chem. Res. 53 (2014) 8773–8783. https://doi.org/10.1021/ie500691p.

[66] W. Xing, W. Yang, W. Yang, Q. Hu, J. Si, H. Lu, B. Yang, L. Song, Y. Hu, R.K.K. Yuen, Functionalized carbon nanotubes with phosphorus- and nitrogen-containing agents: effective reinforcer for thermal, mechanical, and flame-retardant properties of polystyrene nanocomposites, ACS. Appl. Mater. Interfaces 8 (2016) 26266–26274. https://doi.org/10.1021/acsami.6b06864.

[67] Y. Chen, Z. Jia, Y. Luo, D. Jia, B. Li, Environmentally friendly flame-retardant and its application in rigid polyurethane foam, Int. J. Polym. Sci. (2014) 2014: Article ID 263716. https://doi.org/10.1155/2014/263716 .

[68] B. Cichy, E. Kuzdzał, Kinetic model of melamine phosphate precipitation, Ind. Eng. Chem. Res. 51 (2012) 16531–16536. https://doi.org/10.1021/ie3020928.

[69] B. Cichy, D. Łuczkowska, M. Nowak, M. Władyka-Przybylak, Polyphosphate flame retardants with increased heat resistance, Ind. Eng. Chem. Res. 42 (2003) 2897–2905. https://doi.org/10.1021/ie0208570.

[70] Y. Liu, Q. Wang, The investigation on the flame retardancy mechanism of nitrogen flame retardant melamine cyanurate in polyamide 6, J. Polym. Res. 16 (2009) 583–589. https://doi.org/10.1007/s10965-008-9263-6.

[71] P. Gijsman, R. Steenbakkers, C. Fürst, J. Kersjes, Differences in the flame retardant mechanism of melamine cyanurate in polyamide 6 and polyamide 66, Polym. Degrad. Stab. 78 (2002) 219–224. https://doi.org/10.1016/S0141-3910(02)00136-2.

[72] Y. Liu, Q. Wang, Melamine cyanurate-microencapsulated red phosphorus flame retardant unreinforced and glass fiber reinforced polyamide 66, Polym. Degrad. Stab. 91 (2006) 3103–3109. https://doi.org/10.1016/j.polymdegradstab.2006.07.026.

[73] W.-.H. Rao, Z.-.Y. Hu, H.-.X. Xu, Y.-.J. Xu, M. Qi, W. Liao, S. Xu, Y.-.Z. Wang, Flame-retardant flexible polyurethane foams with highly efficient melamine salt, Ind. Eng. Chem. Res. 56 (2017) 7112–7119. https://doi.org/10.1021/acs.iecr.7b01335.

[74] A.R. Horrocks, D. Price, D. Price, Fire Retardant Materials, Woodhead Publishing, Cambridge, England, 2001.

[75] Z.-.M. Zhu, Y.-.J. Xu, W. Liao, S. Xu, Y.-.Z. Wang, Highly flame retardant expanded polystyrene foams from phosphorus–nitrogen–silicon synergistic adhesives, Ind. Eng. Chem. Res. 56 (2017) 4649–4658. https://doi.org/10.1021/acs.iecr.6b05065.

[76] D. Allan, J.H. Daly, J.J. Liggat, Thermal volatilisation analysis of a TDI-based flexible polyurethane foam containing ammonium polyphosphate, Polym. Degrad. Stab. 102 (2014) 170–179. https://doi.org/10.1016/j.polymdegradstab.2014.01.016.

[77] S. Duquesne, M. Le Bras, S. Bourbigot, R. Delobel, G. Camino, B. Eling, C. Lindsay, T. Roels, H. Vezin, Mechanism of fire retardancy of polyurethanes using ammonium polyphosphate, J. Appl. Polym. Sci. 82 (2001) 3262–3274. https://doi.org/10.1002/app.2185.

[78] B.V. Lotsch, W. Schnick, New light on an old story: formation of melam during thermal condensation of melamine, Chem. Eur. J. 13 (2007) 4956–4968. https://doi.org/10.1002/chem.200601291.

[79] A. Sut, E. Metzsch-Zilligen, M. Großhauser, R. Pfaendner, B. Schartel, Synergy between melamine cyanurate, melamine polyphosphate and aluminum diethylphosphinate in flame retarded thermoplastic polyurethane, Polym. Test 74 (2019) 196–204. https://doi.org/10.1016/j.polymertesting.2019.01.001.

[80] M. Zhang, J. Zhang, S. Chen, Y. Zhou, Synthesis and fire properties of rigid polyurethane foams made from a polyol derived from melamine and cardanol, Polym. Degrad. Stab. 110 (2014) 27–34. https://doi.org/10.1016/j.polymdegradstab.2014.08.009.

[81] H. Zhu, S. Xu, Preparation of flame-retardant rigid polyurethane foams by combining modified melamine–formaldehyde resin and phosphorus flame retardants, ACS Omega 5 (2020) 9658–9667. https://doi.org/10.1021/acsomega.9b03659.

[82] C. Scaccia, D.H. Fisher, P.E. Throckmorton, Stable polyol-melamine blend for use in the manufacture of fire retardant flexible urethane foam. Patent no. US4644015A, (1987).

[83] D. Nissen, M. Marx, W. Jarre, E. Schoen, W. Decker, Stable melamine polyol dispersions, a process for their manufacture and for the preparation of foamed polyurethane plastics. Patent no. US-4293657-A (1981).

[84] K. Sykam, K.K.R. Meka, S. Donempudi, Intumescent phosphorus and triazole-based flame-retardant polyurethane foams from castor oil, ACS Omega 4 (2019) 1086–1094. https://doi.org/10.1021/acsomega.8b02968.

[85] F. Gao, L. Tong, Z. Fang, Effect of a novel phosphorous–nitrogen containing intumescent flame retardant on the fire retardancy and the thermal behaviour of poly(butylene terephthalate), Polym. Degrad. Stab. 91 (2006) 1295.

[86] S. Gaan, G. Sun, K. Hutches, M.H. Engelhard, Effect of nitrogen additives on flame retardant action of tributyl phosphate: phosphorus–nitrogen synergism, Polym. Degrad. Stab. 93 (2008) 99–108. https://doi.org/10.1016/j.polymdegradstab.2007.10.013.

[87] P.R. Varga, G. Keglevich, Synthesis of α-Aminophosphonates and Related Derivatives; The Last Decade of the Kabachnik–Fields Reaction, Molecules, 26 (2021), 2511.

[88] E.K. Fields, The synthesis of esters of substituted amino phosphonic acids, J. Am. Chem. Soc. 74 (1952) 1528–1531. https://doi.org/10.1021/ja01126a054.

[89] S. Jiang, Y. Shi, X. Qian, K. Zhou, H. Xu, S. Lo, Z. Gui, Y. Hu, Synthesis of a novel phosphorus- and nitrogen-containing acrylate and its performance as an intumescent flame retardant for epoxy acrylate, Ind. Eng. Chem. Res. 52 (2013) 17442–17450. https://doi.org/10.1021/ie4028439.

[90] H. Ding, K. Huang, S. Li, L. Xu, J. Xia, M. Li, Synthesis of a novel phosphorus and nitrogen-containing bio-based polyol and its application in flame retardant polyurethane foam, J. Anal. Appl. Pyrolysis 128 (2017) 102–113. https://doi.org/10.1016/j.jaap.2017.10.020.

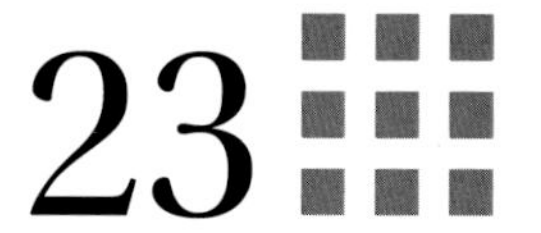

Highly flame-retardant and efficient bio-based polyurethane foams via addition of melamine-based intumescent flame-retardants

Niloofar Arastehnejad, Felipe De Souza and Ram K. Gupta

DEPARTMENT OF CHEMISTRY, KANSAS POLYMER RESEARCH CENTER, PITTSBURG STATE UNIVERSITY, PITTSBURG, UNITED STATES

Chapter outline

23.1 Introduction

Polyurethanes (PUs) are prominent sorts of polymeric materials that can be found within a plethora of commercial and industrial applications ranging from furniture, sound insulation, automotive, and coatings. The ubiquity of PUs can be attributed to the diverse range of synthetic feedstocks, and preparatory methods used to manufacture both rigid and flexible foam derivatives, elastomers, and adhesives [1]. Depending on the progenitor polyol species, the physical and mechanical properties of PUs, for example density, flexibility, rigidity, etc., are remarkably variable [2]. Traditionally PUs are synthesized from a multifunctional, hydroxy rich $(\text{-ROH})_n$, oligomeric polyol, and a corresponding di- or polyisocyanate species $(\text{-RCNO})_n$, resulting in a highly cross-linked polymer network [3]. In light of recent policy initiatives by both the European

Biopolymeric Nanomaterials: Fundamentals and Applications. DOI: https://doi.org/10.1016/B978-0-12-824364-0.00024-1

Chemicals Agency (ECHA) and the United States Environmental Protection Agency (EPA) a renewed focus has been applied toward finding environmentally friendly, bio-based, renewable alternatives to petrochemically sourced polyols and isocyanates [4]. Plant-based bio-polyols, including a sundry of common vegetable oils, such as palm oil, soybean oil, rapeseed oil, sunflower oil, and linseed oil have garnered significant attention as viable alternatives to conventional petrochemical-based polyols [5]. Bio-polyols derived from plant sources are plentiful in polyunsaturated fatty acids, which can be efficiently, economically, and chemically modified to hydroxyl moieties used for PU synthesis [6].

An alternative approach is to produce a novel bio-polyol by derivatizing pre-existing commercially available natural resources [7]. A conceptualization of this method is the mercaptenization of the olefin-rich hydrocarbon β-myrcene, or monoterpene, into a polyol. β-myrcene is an oil derivative found within several plants including but not limited to the bay, cannabis, thyme, lemongrass, and hops [8]. The alkenes present within β-myrcene exhibit high reactivity with thiols (-SH) and can be converted in one step via a UV catalyzed thiol-ene reaction. Utilization of a one-step thiol-ene reaction allows for an immediate, high purity, high conversion synthetic method that produces a derivatized bio-based polyol with a narrow molecular weight distribution Furthermore, mercaptanized cross-linkers can be used in conventional PU synthesis to enhance thermomechanical properties [9]. The primary drawback of rigid PU foams (RPUF) is that they are overtly flammable. Small ignition sources can instigate fires with accelerative rates of heat release, which further liberates voluminous amounts of smoke containing toxic vapors [10]. The noteworthy flammability of RPUFs originates from the architectural configuration of the foam matrix, in particular low composite density and open cell structure. After the initial ignition event, thermal decomposition is further propagated by elevated carbon, hydrogen, and oxygen content within the aliphatic chains in the polymeric backbone. More often, the susceptibility of ignition is exacerbated by the porous architecture of the foam which facilitates oxygen diffusion throughout the polymer matrix. Increased internal oxygen concentrations subsequently promote the occurrence of an ignition. Mitigation of the flammability of RPUF's, without sacrificing mechanical integrity, is a preeminent concern [11].

To mitigate flammability chemical additives known as flame retardants (FRs) have been incorporated into commercial RPUFs to decrease their flammability. FR encompasses a diverse range of materials including halogens, inorganics, organophosphates, arenes, etc. [12]. Mechanistically these materials either prevent or impede successive combustion after the initial ignition event. Despite their pervasive industrial use, halogenated FRs are notorious for their deleterious environmental effects and bioaccumulation. Hence, a cheap, effective, environmentally benign, halogen-free FRs have been a sought after additive by the PU industry [13].

Recently, phosphorous-nitrogen containing additives have been investigated as a possible replacement for conventional halogenated FRs [14]. Intumescent FR (IFR) additives have rapidly become the de-facto halogen-free FR due to their lower toxicity, minimal smoke generation, and superior fire protection [15]. In principle, FRs swell and expand when heated beyond a critical threshold. This expansion produces a foamed cellular carbonaceous char layer that protects the preceding layer from additional heat flux and direct contact with open flames, effectively quarantining the combustible material [16]. Melamine (ME) and phosphoryl derivatives are known FRs. When heated ME acts as a spumific agent, producing nonflammable gases upon decomposition, along with a carbonaceous insulation layer [17]. Phosphoryl compounds act similarly by producing an isolating carbon layer, but also behave as a dehydrating agent by scavenging radical hydrogen (H*) and hydroxy (OH*) groups, resulting in the esterification of hydroxyl groups. Concomitant use of ME and phosphoryl materials results in the synergism of both mechanisms. Employment of IFRs improves thermal resistance, increases the ignition temperature, reduces the combustion rate, and decreases the heat released comparable to that of conventional FRs [13,18]. The successful implementation of additive FRs, especially, IFRs, is reliant upon the compatibility with the host RPUF matrix and the impact on physicomechanical properties. In our previous studies, it was conclusively shown that minimal loading of dimethyl methyl phosphonate dramatically reduced the self-extinguishing time (SET), while simultaneously generating a carbonaceous char layer in the host foam [19].

Herein, unmodified β-myrcene was chosen to synthesis a novel natural oil-based polyol via a single-step method, solvent-less, and UV catalyzed thiol-ene reaction. FR capabilities of the resulting RPUFs were enhanced via the application of all-in-one IFRs ME cyanurate (MC) and ME phosphonate (MP), using pure ME as a comparative control. The ramifications of varying concentrations of both IFRs were investigated about the thermomechanical and physicomechanical properties of the RPUFs. It was shown that increasing the minimal concentration of MP into the bio-based RPUFs significantly reduced the time of self-extinguishment, duration of heat release, and the totality of heat evolved.

23.2 Materials and experimental details

23.2.1 Materials

β-Myrcene (M), 2-hydroxy-2-methyl propiophenone, and ME (99%) were purchased from Sigma-Aldrich, (St. Louis, MO, USA). 2-Mercaptoethanol (2-ME) (99%) was purchased from Acros Organic (Pittsburgh, PA, USA). DABCO T-12 ($>$ 95%) and NIAX A-1 as a role of catalyst were bought from Air Products (Allentown, PA, USA). Jeffol 522 and Rubinate M isocyanate (methylene diphenyl diisocyanate (MDI), isocyanate (NCO) content of 31%) were received from Huntsman (The Woodlands, TX, USA). Tegostab B-8404 as a silicone-based surfactant was obtained from Evonik, USA. Distilled water was consumed as a blowing agent and obtained from Walmart (Pittsburg, KS, USA). MC was ordered from (JLS Chemical Inc., Pomona, California, USA), o-phosphoric acid, 85% was bought from Fisher Chemical in the United States.

23.2.2 Synthesis of myrcene-based polyol

The synthesis of myrcene-based polyol (M-2ME-Polyol) was performed through a single-step reaction between myrcene and 2-mercaptoethanol via thiol-ene click chemistry in the presence of UV light. Myrcene (0.5 mol) and 2-mercaptoethanol (1.5 mol) were mixed with 1:3 molar ratio along with 2.0 wt% of 2-hydroxy-2-methyl propiophenone as a photoinitiator in a 500 mL reaction vessel. Stirring was applied to the reaction using a magnetic stir bar and the mixture was

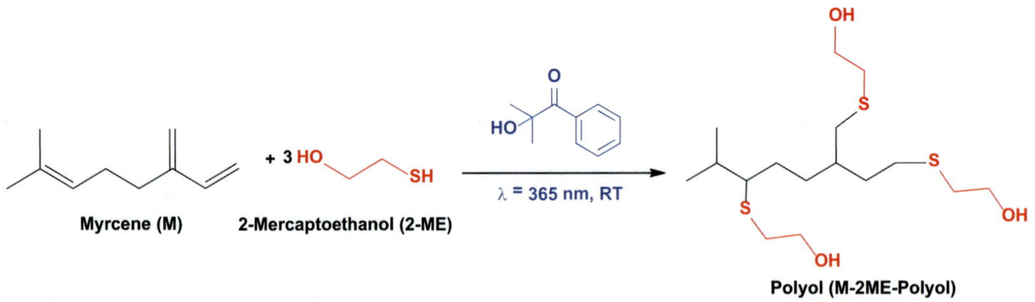

FIGURE 23.1 The chemical reaction for the synthesis of myrcene-based polyol.

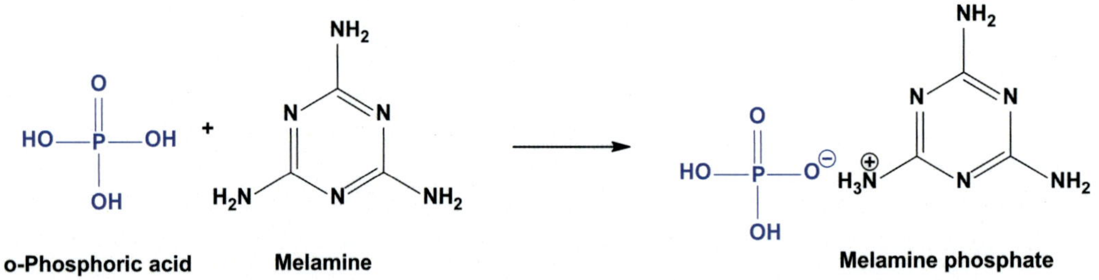

FIGURE 23.2 The chemical reaction for the synthesis of melamine phosphate.

placed under a UVP Blak-Ray UV Benchtop Lamp for 8 h under ultraviolet radiation ($\lambda = 365$ nm) and ambient conditions. The chemical reaction for the synthesis of the polyol is shown in Fig. 23.1.

23.2.3 Preparation of FRs

MP was prepared via reacting ME and orthophosphoric acid in a 1:1 molar ratio. This compound contains both nitrogen (37.5%) and phosphorus (13.8%) which can provide flame retardancy in the PUs. ME and orthophosphoric acid were mixed with the help of ultrasonication for 30 min followed by drying of water at 110 °C. The synthesized complex was grounded and used as a blended FR in the myrcene-based PU foams. The reaction for the synthesis of MP is demonstrated in Fig. 23.2.

23.2.4 Preparation of rigid FR PU foams

The composition of polyol, isocyanate, catalyst, FR, surfactant, and blowing agent are given in Table 23.1. PUs are made by the exothermic reaction between polyols and isocyanates. The polyurethane foams (PUFs) were prepared by adding the polyols and FR into a 500 mL plastic cup. The mixture was stirred vigorously (3000 rpm) until a homogenous system was formed. Then, all the components except MDI were added and stirred. Then, the equivalent weight of

Table 23.1 The formula for rigid polyurethane foams along with FR. (All the weights are listed in grams).

Compound								
Jeffol-522	10.00	10.00	10.00	10.00	10.00	10.00	10.00	10.00
M-2ME-Polyol	10.00	10.00	10.00	10.00	10.00	10.00	10.00	10.00
Tegostab B-8404	0.40	0.40	0.40	0.40	0.40	0.40	0.40	0.40
Niax-A1	0.14	0.14	0.14	0.14	0.14	0.14	0.14	0.14
T-12	0.04	0.04	0.04	0.04	0.04	0.04	0.04	0.04
Water	0.80	0.80	0.80	0.80	0.80	0.80	0.80	0.80
MDI	35.81	35.81	35.81	35.81	35.81	35.81	35.81	35.81
Flame retardant	0.00	0.50	1.50	3.00	5.00	8.00	10.00	12.00
Wt% of FR	0.00	0.86	2.55	4.98	8.03	12.27	14.88	17.34

MDI was charged into the cup and mixed around 20 s. The foaming process took place and the obtained samples were named as 0, 0.5, 1.5, 3, 5, 8, 10, and 12 based on the amount of FR added.

23.2.5 Characterization of the polyol and foams

The synthesized M-2ME-Polyol was analyzed with standard characterization methods. The hydroxyl value was measured using a phthalic anhydride/pyridine (PAP) technique based on ASTM-D 4274. Olefin concentration of synthesized polyol was determined via the Hanus titration method. Incorporated iodine concentration was determined via subsequent titration using aqueous sodium thiosulfate. PerkinElmer Spectrum Two spectrophotometer was utilized for collecting Fourier Transform Infrared (FT-IR) spectra. Dynamic viscosity (η) of the polyol was determined using a AR 2000 dynamic stress rheometer (TA Instruments, USA) at room temperature with shear stress linearly ramping from 1 to 2000 Pa. Dynamic viscosity measurements were carried out using a 2° cone plate with a diameter of 25 mm. Polyol molecular weight was determined using gel permeation chromatography (GPC) using a Waters (Milford, MA) instrument equipped with four 300 × 7.8 mm Phenogel 5 μm columns with pore sizes of 50, 102, 103, and 104 ú. Tetrahydrofuran was chosen as a standard eluent solvent, with experiments carried out at a flow rate of 1 mL per min at 30 °C.

The foam samples were cut in a cylinder of 45 × 30 mm (diameter × height). Both apparent density and closed-cell content were measured by using these cylindrical foams in Ultrapycnometer (Ultrafoam 1000) based on ASTM D1622 and ASTM D2856. Compressive strength was analyzed following ASTM D 1621 by cutting the rectangle prims specimen with the dimension of 50 × 50 × 25 mm^3 (width × length × height). The samples were analyzed by a universal electronic tensile tester named "Q test 2-tensile machine." The detailed images of the cell surfaces and morphology of the foams were analyzed by scanning electron microscopy (SEM) via reliable Thermo Scientific Phenom Pure desktop SEM from Sioux manufacture in the Netherlands. The samples were sliced in cubical shapes and size of 0.5 cm^3. Samples were coated with a thin gold layer on the surfaces by a magnetron sputtering instrument from Kurt J. Lesker company. The gold sputtering technique was required before taking the images to facilitate the electron

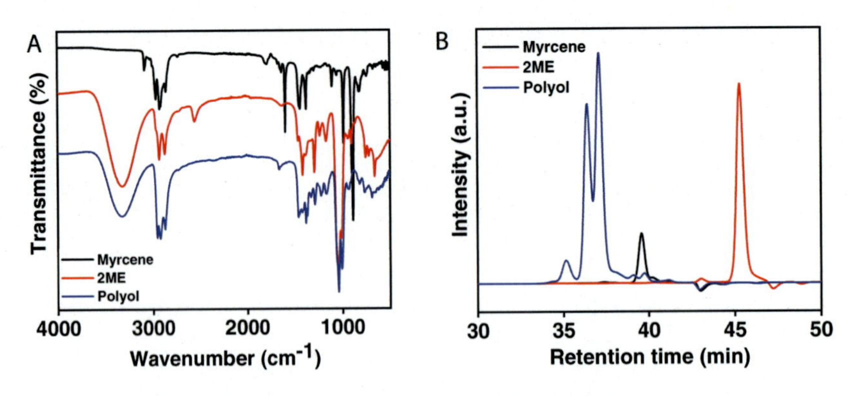

FIGURE 23.3 (A) FT-IR and (B) GPC for myrcene, 2-mercaptoethanol, and polyol.

conduction at the surface. Thermogravimetric analysis (TGA) was used to study the thermal behavior of the foams that were determined by the TA instrument (TGA Q-500). Typical thermogravimetric studies consumed 10 mg of the target PUF, placed onto a Pt pan. Foams were heated at a ramp rate of 10 °C/min in N_2. The flammability behavior of FR foams was studied using the horizontal burning test (ASTM D4986-98). Rectangle prism-shaped specimens with a dimension of 150 × 50 × 12.5 mm^3 (length × width × thickness) were submitted to a direct flame for 10 s and then SET and weight loss were measured after the burn test.

23.3 Results and discussion

The hydroxyl value and dynamic viscosity of the prepared M-2ME-Polyol were 466 mg KOH/g and 1.52 Pa.s, respectively, at 25 °C. The synthesis of PU is contingent upon the equimolar reaction between a hydroxyl and an isocyanate group. As a result, the hydroxyl number represents the reactivity of the entirety of the polyol. The dynamic viscosity value of 1.52 Pa.s certifies that the synthesized polyol has minimal resistance to flow, allowing for facile handling of material while also ensuring homogenous mixing. On top of that, the initial viscosity of myrcene as a starting material was 0.00436 Pa.s and increased up to 1.52 Pa.s providing empirical evidence for the occurrence of the reaction. The amount of unreacted double bond could be determined based on the iodine number of synthesized polyol, which was zero suggesting that all double bonds reacted with the mercaptan groups. Structural composition and architecture were verified via GPC and FT-IR. The GPC further corroborates the successful synthesis of the M-2ME-Polyol. Fig. 23.3A comparatively details the FT-IR spectra of the initial myrcene and 2-ME starting materials concerning the final polyol. Representative peaks for myrcene are as follows: 3089 cm^{-1} alkenyl C-H stretching (w), 2968–2858 cm^{-1} alkyl CH_3 and CH_2 C-H stretching (m), 1632 cm^{-1} C=C stretching (w), 1595 cm^{-1} C=C vibration (m), 1442 cm^{-1} alkyl C-H bending (m), 1377 CH_3 and C-H bending (m), 989 cm^{-1} monosubstituted alkenyl C-H bending (m), and 890 cm^{-1} symmetric disubstituted alkenyl C-H bending (s). The diagnosed peaks for 2-ME are as follows: 3333 cm^{-1} H-bonded O-H stretching (s, broad), 2929, and 2872 CH_2 C-H stretching (m), 2551 cm^{-1} S-H stretching (w), and 1049 cm^{-1} C-O (primary alcohol) stretching (s) [20,21].

The absence of peaks corresponding to vinylic protons and carbons (3089 cm^{-1}, 1632 cm^{-1}, 1595 cm^{-1}, 989 cm^{-1}, and 890 cm^{-1}) present within myrcene coincides with the disappearance of the alkyl thiol group (2551 cm^{-1}) in 2-ME during the synthesis of the polyol. Consumption of both functional groups indicates a successful mercaptonization of myrcene. All FT-IR peaks are in congruence with prior results. The compiled chromatograms in Fig. 23.3B succinctly illustrates that both starting materials, myrcene (40 min) and 2-ME (46 min), are consumed over the course of the reaction. The resulting higher molecular species showed a decreased retention time of approximately 36 min. The trimodal distribution presented in the M-2ME-polyol chromatogram may likely be attributed to the formation of oligomers such as dimers or trimers, which is a common thiol-ene side reaction where $C^•$ species may react with another $C^•$ instead of an $H^•$, which is backed by several authors. Also, the hydroxyl number, which represents the amount of -OH groups available, was 466 mg KOH/g suggesting a functionality around 3 as it was theoretically expected. Other factors such as the disappearance of double bonds stretch around 1632 cm^{-1} and iodine values close to zero for the polyol strongly suggest complete consumption of these reactive groups. Finally, the considerable viscosity's increase from myrcene to M-2ME-polyol also supports the satisfactory completion of polyol synthesis [22,23].

One of the significant parameters that may impact physicomechanical properties of the foam is the density. All mechanical and thermal tests were compared for the foams having different ME derivatives as FR. As observed in Fig. 23.4, it is noticeable the foam density increased by adding a similar amount of each of different FRs into the foam. The density range of ME and MC foams are nearly similar from 39 to 58 kg/m^3 and 38 to 60 kg/m^3, respectively. In comparison, MP foam became denser than ME and MC, reaching around 37–67 kg/m^3. The increase in density for the MP-based foams was most likely due to the slow rising time during the foam making. It occurred because a reminiscent of phosphoric acid from MP reacted with the amine-based catalysts, which decreased the rising rate. This effect resulted in a higher density for the MP foam. Despite that, all the foams are in an acceptable range for use as RPUF for buildings and constructions according to building standards [23]. Similarly, the high closed-cell content as given in Fig. 23.5, shows the value of 93%–95% for all the foams suggesting less diffusion of air, which contributes to thermal insulation by restricting airflow within the cellular structure [24].

The mechanical properties of PU can vary based on many parameters such as density, stability of cell structure, and its size along with proper interaction between additives and the PU matrix. Fig. 23.6 illustrates the compressive strength of all the foams with various amounts of FR at yield a break. The neat foam showed a compressive strength of ˜ 211 kPa. The obtained foams with various amounts of FR did not show any trend, which could be due to the incompatibility of FR in the PU. The addition of ME particles resulted in a stiffening effect and a relative increase of compression strength [25]. Similar patterns were observed for ME and MC, reaching 184–257 and 170–267 kPa, respectively. But, for the case of MP samples, it was 140–221 kPa. The compressive behavior for ME and MC demonstrated an increase when it went up to 10 g of each FR [26]. It suggests that FR was able to arrange properly within the PU matrix leading to foam's strengthening [14,27]. On the other hand, MP-blended foams had its highest compressive strength when 3 g of FR was added. Over larger amounts, there was a diminishment of values, which could be related to higher interaction between dispersed solid particles that do not provide

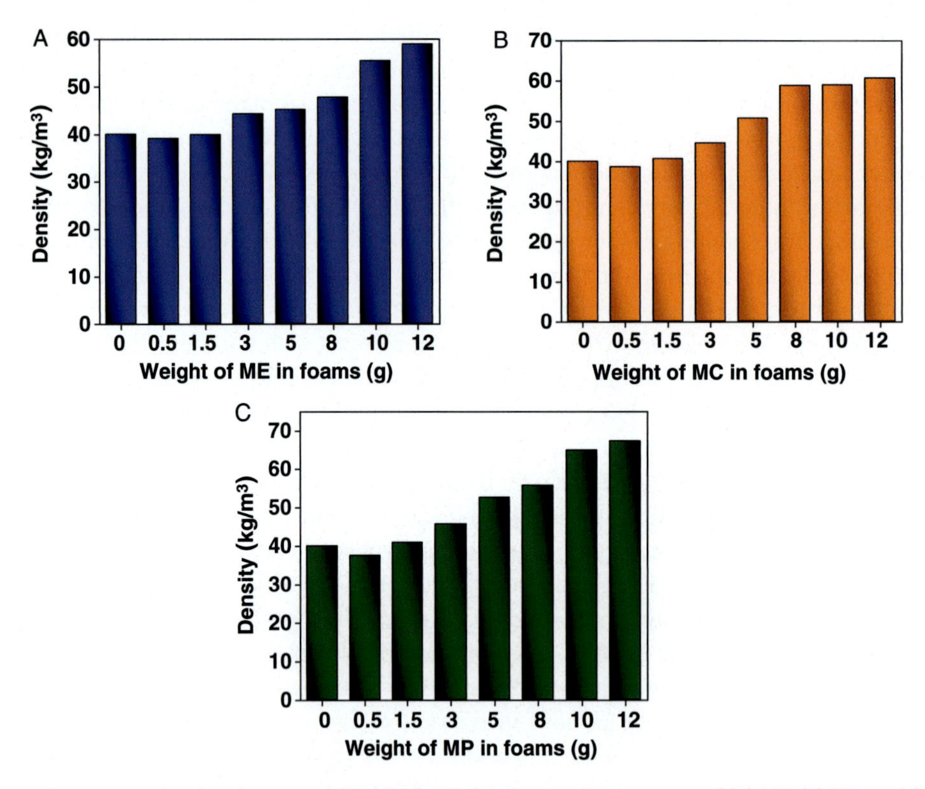

FIGURE 23.4 Apparent density of prepared rigid PU foams having a various amount of (A) ME, (B) MC, and (C) MP as flame retardants.

effective support against an applied load. From an overall perspective, the highest values for compressive strength among all the samples were 10 MC, 10 ME, and 3 MP with 267, 257, and 221 kPa, respectively. Despite the oscillation of values, there was an increase in compressive strength for optimum amounts of FR for all cases, showing results comparable to other reports.

The morphology of PU is an important source of information to understand cellular structure and regularity of pores, which can be correlated with mechanical properties and density. For that, the micrographs of SEM images for all the samples are given in Fig. 23.7. The neat PU presented an average cell size of 145 μm. For the case of ME-blended PU there was a slight increase in cell size for the samples 0.5 and 1.5 ME. This effect led to a decrease in cell number. It means that an applied load over the foam's surface may be in contact with fewer cell walls in a unit of area; hence providing less support, which correlates to the observed decrease in compressive strength. On the other hand, for the case of 3–10 ME samples, there was a general decrease in average cell size roughly from 141 μm to 115 μm, which provided a higher number of cells exposed to a load. This effect may explain the increase in the compressive strength for that range of ME concentration. A different pattern was observed regarding 12 ME because there was a decrease in average cell size to 110 μm as well as compressive strength. This may be attributed to the drastic increase of viscosity of the premixture of polyol and a high quantity of ME. It could have formed large

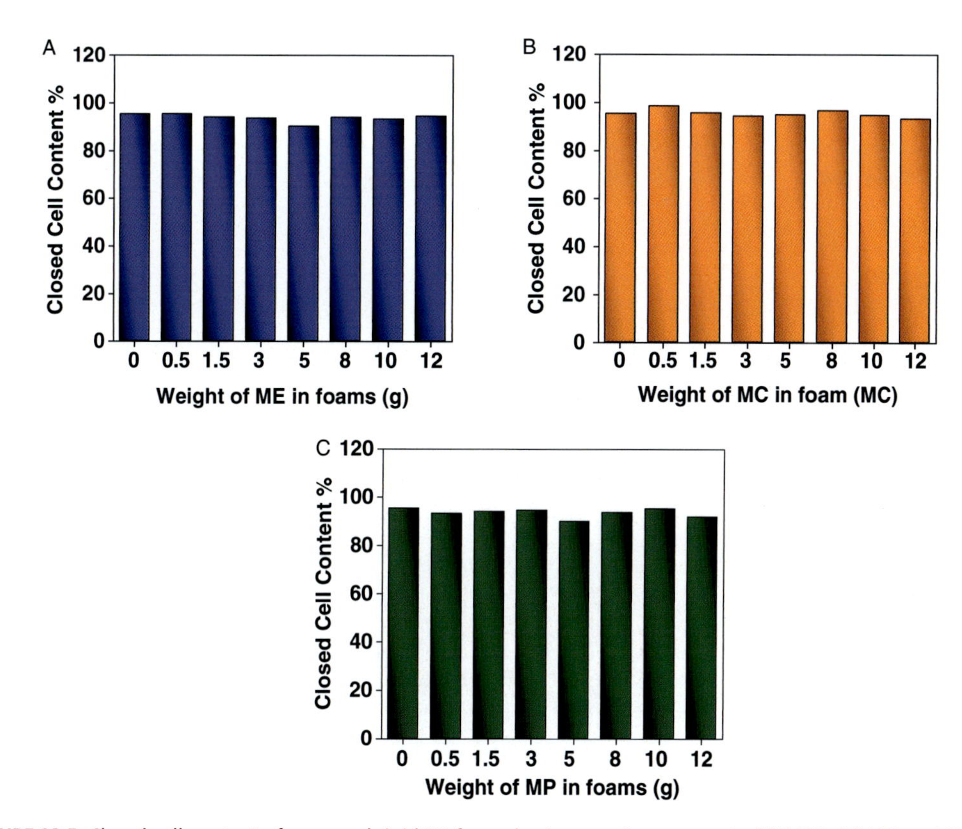

FIGURE 23.5 Closed-cell content of prepared rigid PU foams having a various amount of (A) ME, (B) MC, and (C) MP as flame retardants.

bubbles that did not coalesce during the foaming process, which caused a decrease in cell size and failures into the cell structure, occasioning the deterioration of mechanical properties [28].

The cell morphology of MC was more regular as it presented a correlation between the increase of compressive strength and density while a relative decrease of cell size from 132 to 107μm, which may imply a larger number of cells to withstand a load per unit of area. Hence, MC had a better interaction with the PU matrix. For the samples blended with MP, it could be observed that 3 MP presented the smallest cell size, which can be correlated with the highest compressive strength for that set. On the other hand, the MP samples presented an irregular cellular structure pattern as well as large cell size from 195 to 240 μm, which can be correlated to the observed decrease in compressive strength [29,30].

The thermal stability of the synthesized rigid foams with different FRs was analyzed. The TGA plots of all the samples are shown in Fig. 23.8. The results are summarized in Table 23.2 by $T_{d5\%}$ and $T_{d50\%}$ which stand for 5 wt% and 50 wt% loss in TGA analysis. It is noticeable that the degradation process under nitrogen occurs in a two-stage process and led to changes in the PU's thermal behavior [31]. For the case of ME-based FR, the general effect is that it

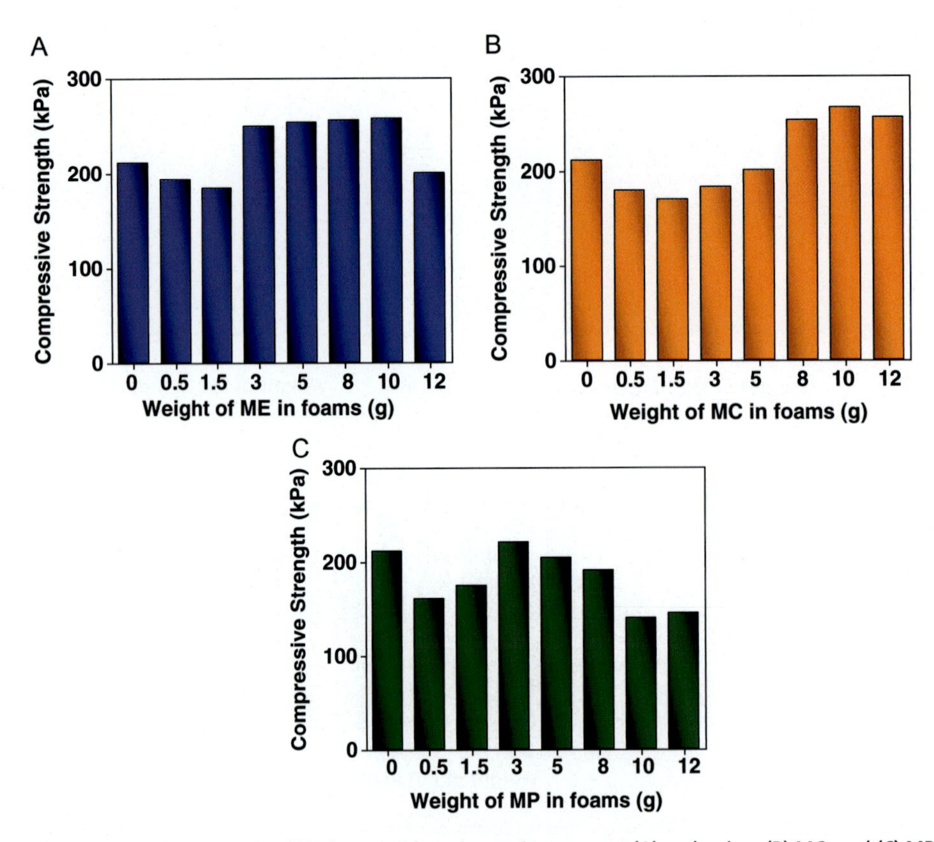

FIGURE 23.6 Compressive strength of PU foams with an increasing amount (A) melamine, (B) MC, and (C) MP-based flame retardants.

absorbs a significant amount of heat to decrease the surface temperature of polymer degradation, which according to several authors, is an endothermic process that yields oligomers of ME through the simultaneous release of ammonia. Through this is observable that all samples had its first thermal transition around 240 °C which was also similar to the $T_{d5\%}$ values. This effect was referred to in the literature as the thermal decomposition of ME-based compounds as well as partial degradation of the PU matrix itself as it has been observed in previous reports [11,32].

The second transition that occurred nearly around 500 °C demonstrated a considerable improvement in the thermal stability of the foams as the concentration of FR increased. This effect was more pronounced for the case of MP that presented a higher amount of residue, even surpassing the thermal stability of other N-P-based FR. This large improvement of thermal stability may be due to the synergy effect of the oligomerization of ME to create char along with the formation of polyphosphoric acid. According to previous studies, the polyphosphoric acid can dehydrate the PU matrix inducing a cross-linking process, that despite causing its degradation it also yields a more compact char layer [27]. The amount of residue left for MP

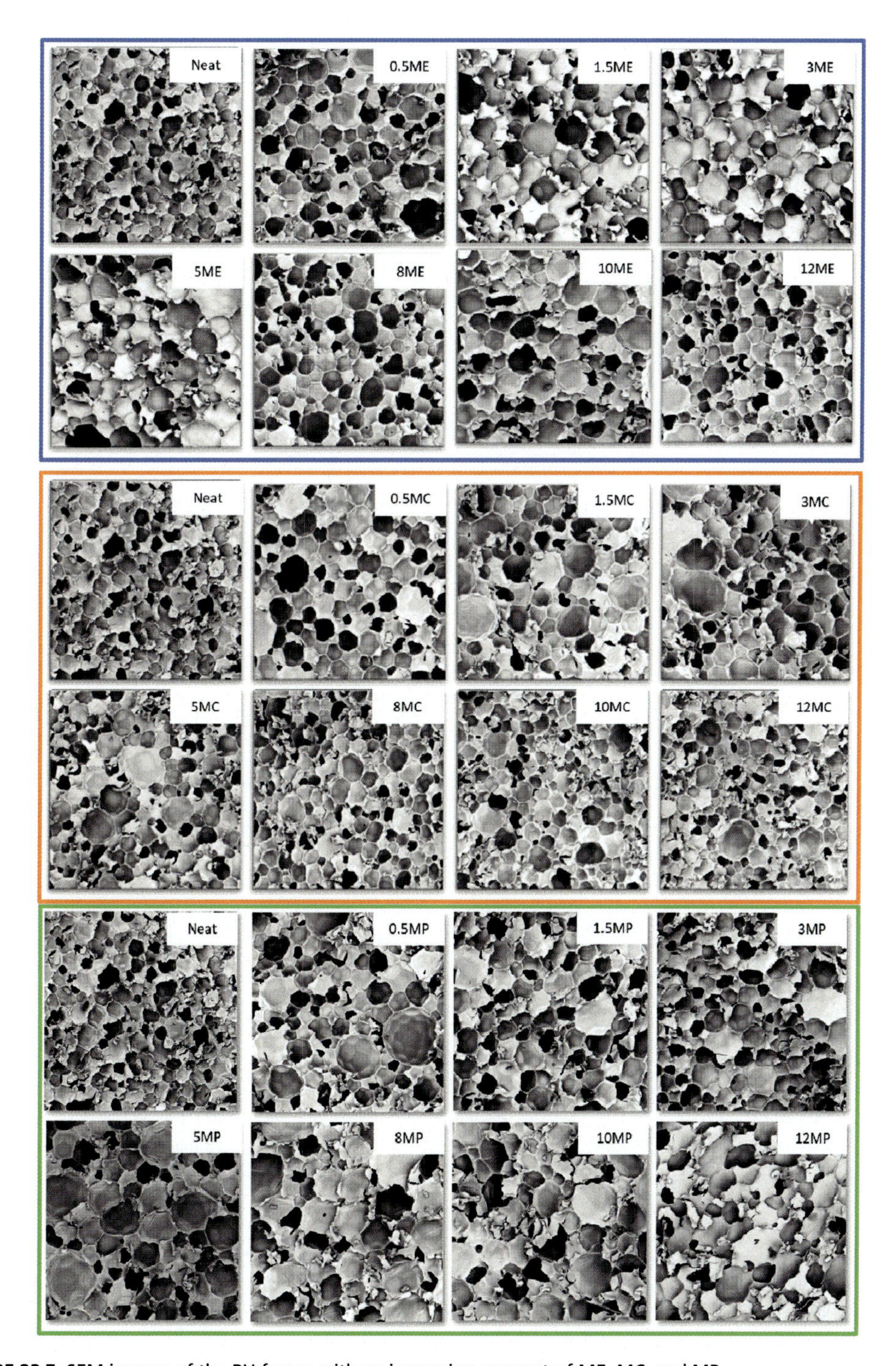

FIGURE 23.7 SEM images of the PU foams with an increasing amount of ME, MC, and MP.

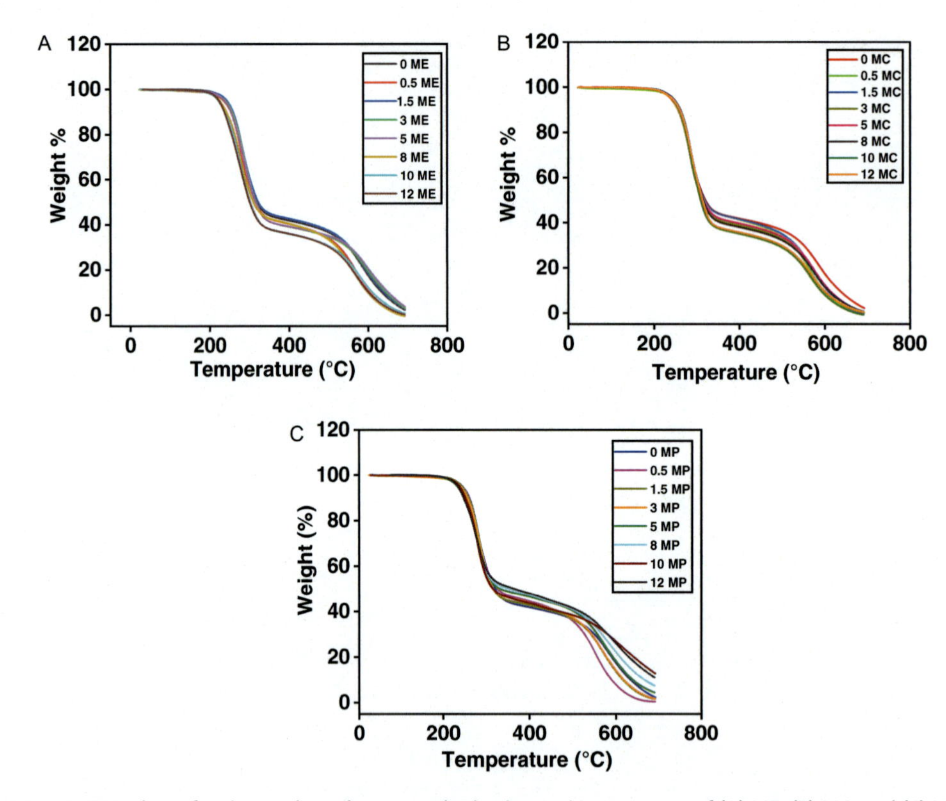

FIGURE 23.8 TGA plots of various polyurethane samples having various amounts of (A) ME, (B) MC, and (C) MP.

Table 23.2 $T_{d5\%}$ and $T_{d50\%}$ and residue for ME-, MC-, and MP-based polyurethane foams.

	ME			MC			MP		
Amount of FR (g)	$T_{d5\%}$	$T_{d50\%}$	Residue (%)	$T_{d5\%}$	$T_{d50\%}$	Residue (%)	$T_{d5\%}$	$T_{d50\%}$	Residue (%)
0	240.22	320.02	0	240.22	320.02	0	240.22	320.02	0
0.5	237.41	315.40	0	240.77	319.68	0.181	245.88	327.91	0.476
1.5	245.49	323.38	2.981	245.16	320.40	0	244.18	316.03	1.686
3	240.76	318.27	3.196	241.82	313.83	0.562	240	315.51	1.640
5	239.31	316.76	3.839	245.15	316.03	0.499	233.84	332.89	4.470
8	226.88	313.29	0	245.16	315.52	0.678	235.66	351.83	7.429
10	223.23	304.63	0.792	240.76	309.13	0	238.47	358.30	12.790
12	224.69	302.30	0.286	241.35	313	0.436	234.99	364.98	11.054

also supports this thermal stability improvement due to the implementation of P, as it presented 10.54% while ME and MC had 0.286 and 0.436%, respectively.

Fig. 23.9 demonstrates the thermal behavior of the pristine powder of each FR used in this work. The behavior of ME is described as its sublimation that occurs around 280 °C as provided in Table 23.3, which has been commonly observed in previous studies. However, when in contact

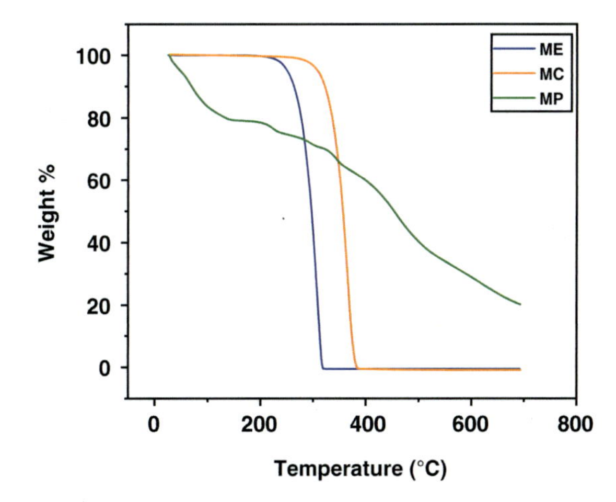

FIGURE 23.9 TGA plots for the pure ME-based FRs.

Table 23.3 $T_{d5\%}$ and $T_{d50\%}$ and residue for ME, MC, and MP powders.

ME powder			MC powder			MP powder		
$T_{d5\%}$	$T_{d50\%}$	Residue (%)	$T_{d5\%}$	$T_{d50\%}$	Residue (%)	$T_{d5\%}$	$T_{d50\%}$	Residue (%)
247.94	279.64	0	306.29	337.30	0	44.67	452.43	20.16

with a polymeric matrix it could absorb energy to perform its endothermic decomposition to form other derivatives such as melam at 360 °C, melem at 450 °C, and melon at 600 °C [33]. For this case, however, there was a progressive release of ammonia throughout the range of temperature. The thermal degradation of MC followed a similar trend that showed a shift in the degradation curve that presented the onset temperature around 306 °C [34]. This increase compared to ME occurred due to the required energy to break the intermolecular interactions of hydrogen-bonding to convert the MC into ME and cyanuric acid. After that, the decomposition occurred similarly to the thermal decomposition observed for pristine ME. The thermal degradation of MP showed an early decomposition step that occurred around 45 °C. This effect may be due to an excess of phosphoric acid that could be released in the form of volatile species. However, as the range of temperature increased the high content of phosphate could be converted to its polymeric form, which accounted for the 20% residue. This is an important phenomenon that described the FR mechanism of MP, which forms a protective and compact layer that shields and prevents the degradation of the PU matrix. According to the report of Sandia National Lab.[35], this effect was also observed in the foams mixed with MP, which presented a better thermal stability behavior if compared with ME- or MC-containing foams.

The horizontal flame tests for the rectangle prism-shaped foams specimen are shown in Fig. 23.10 along with the samples before and after the burning test that are demonstrated in photocopies given in Fig. 23.11. The neat foam had a burning time of 75 s and lost around 40% of its weight. After progressive addition of FR, an effective diminishment in burning time

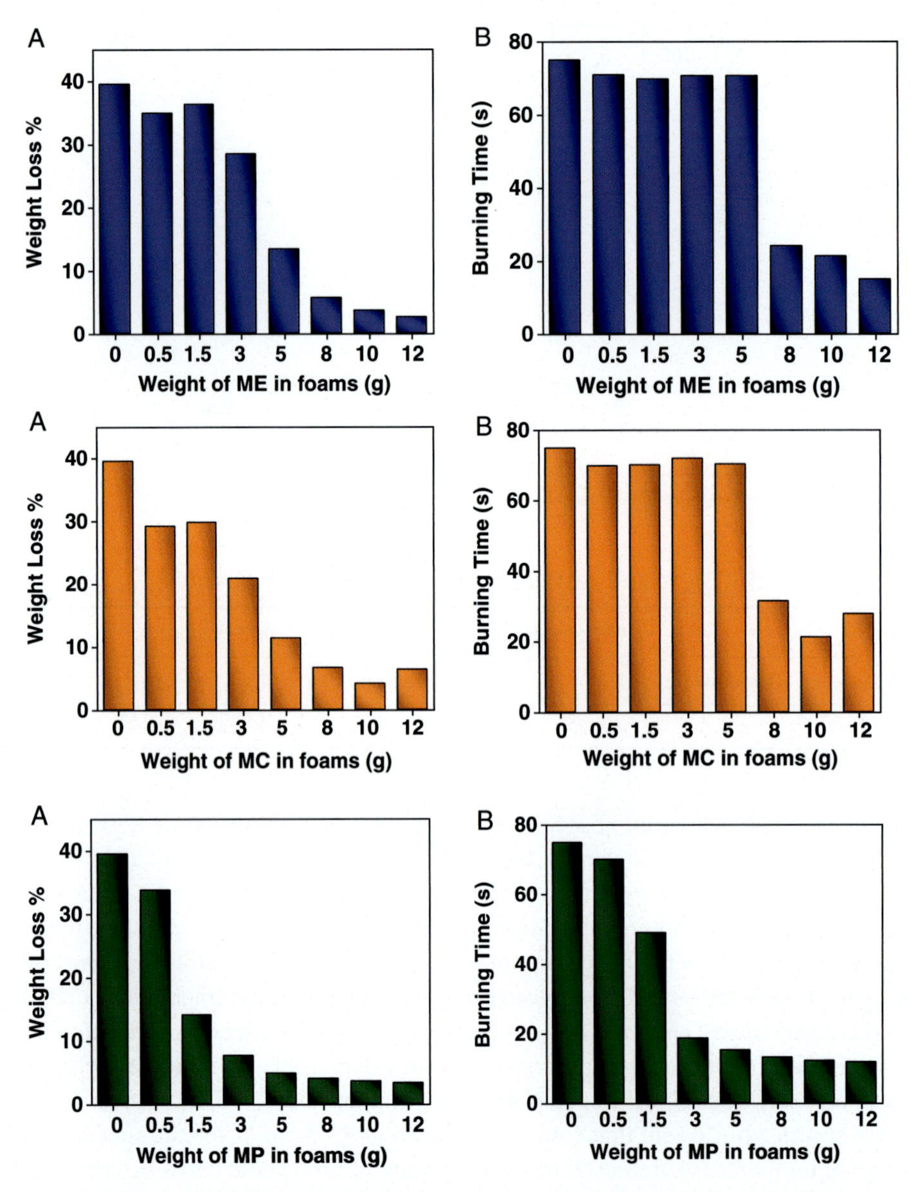

FIGURE 23.10 Comparison of weight loss percentage and burning time with different weight of ME, MC, and MP.

and consequently weight loss was observed. For the ME set of foams, the sample containing 17.3 wt% (12 ME) achieved 14 s of burning time and weight loss of 2.5%. The MC-containing foams also presented satisfactory results for the 12 MC sample reaching 21 s and 4.1% of burning time and weight loss, respectively. The best performance was achieved using MP for the sample that contained 12.27 wt% (8 MP) by reaching the lowest burning time of 11 s and slightly higher weight loss of around 3.3%. This difference in fire resistance may be attributed to the fire-quenching

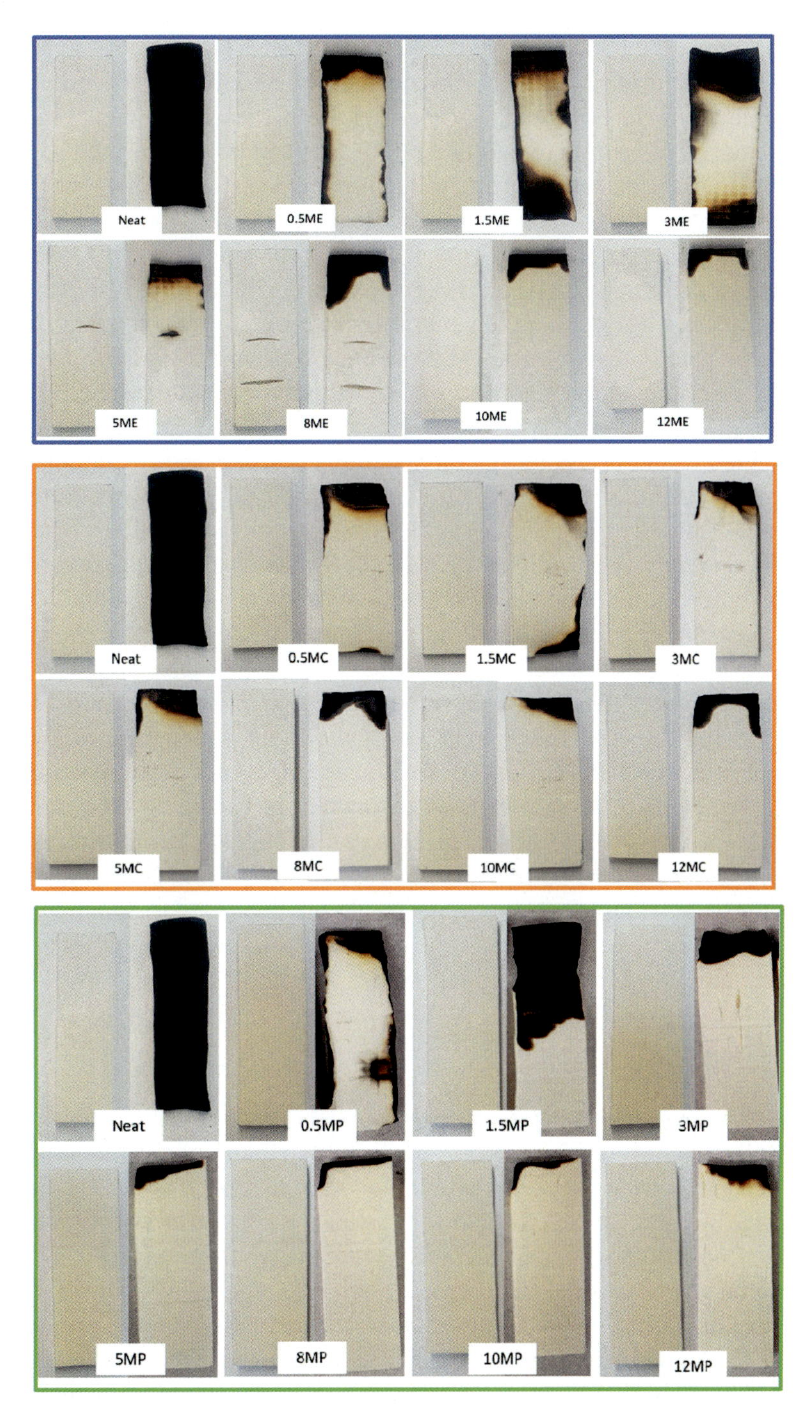

FIGURE 23.11 Polyurethane foam pictures after the horizontal burning test with different loadings of ME, MC, and MP.

mechanism for each of the FR used in the work. Pristine ME is known to act in both solid and gaseous phases. The solid-phase action is described as the endothermic decomposition of ME that is responsible for the formation of oligomeric and compact derivatives, known as melam, melem, and melon, as previously mentioned [33]. These compounds formed a compact char layer over the PU's surface that blocked the passage of oxygen or any reactive species such as $H^{Ï}$ and $OH^{Ï}$, which could further catalyze the combustion of the PU matrix. Along with that, as the decomposition process of ME is endothermic it also absorbs energy lowering the temperature and placing it below the ignition of the PU. The gas-phase mechanism consists of the progressive release of ammonia as a byproduct of the endothermic process that converts ME into melam, melem, and melon [36,37]. According to the literature, MC also acts similarly if compared with ME. However, it offers one more energy barrier to prevent the ignition of the PU, which is the endothermic reversion of its structure to ME and cyanuric acid. These compounds can then cross-link with the PU matrix and form a more compact carbonaceous char layer, preventing the fire and comburent from further deteriorating the foam [38]. This work, however, did not present the most effective flame retardancy as the occurrence of flickering perpetuated the flame over the foam longer than expected. For the case of MP, there was a satisfactory synergy effect between the N source from ME and P source from phosphate [14]. The addition of phosphate promoted the formation of polyphosphoric acid, which created an effective and more compact char layer, that sealed the underneath PU from the flames. In addition to that other studies in the literature demonstrated that FRs that contains P in its composition are also effective in the gaseous phase [19,29]. The decomposition of phosphate could lead to the formation of radical-scavenge species such as $HPO^{Ï}$, $PO^{Ï}$, or $PO_2^{Ï}$ which are capable of quickly reacting with $H^{Ï}$ and $OH^{Ï}$. Hence, by removing these species from the media the fire could be quickly quenched. Therefore, the synergy effect arose from the combination of both ME and phosphate compounds that presented an enhanced fire behavior if compared to the other FRs. Besides the fact that none of the samples presented dripping, the optimum results obtained for the sample containing 12.27 wt% of MP (8 MP) presented a flame retardancy that managed to surpass other nitrogen-based FRs incorporated into PU which had been reported.

23.4 Conclusions

Our group was able to successfully synthesize a bio-based RPUF by utilizing myrcene, an unsaturated essential oil, and converted it into a bio polyol through thiol-ene click chemistry procedure that provided a facile, solvent-free, and room temperature synthetic route. The obtained RPUF was mixed with ME-based FR (ME, MC, and MP) which presented an overall improvement in their properties. Density had an expected increase reaching values around 60–70 kg/m^3, which were within the range of industrial standards. Closed-cell content had an average of 90%–95% and improvement in thermal degradation was observed for all the samples after increasing the addition of FR. The highest compressive strength of 257 kPa was achieved for the sample that contained 14.8 wt% of ME (10 ME) most likely due to proper compatibility between ME and the PU matrix. The most effective flame resistance result was obtained for the sample that contained 12.27 wt% of MP (8 MP) which was able to self-quench the fire within 11 s along with relatively low weight loss of around 3%. The pronounced flame retardancy for MP in relation to ME and MC was due to the synergy effect between nitrogen and phosphorus that improved both solid phase by forming a more compact char layer serving to block oxygen entering in contact with the underneath PU matrix as well as gaseous phase due to the release of ammonia to dilute oxygen and radical-scavenge species originated from phosphate. The results were able to surpass other materials that come from petrochemical sources, which demonstrated that using eco-friendly materials such as myrcene and nitrogen-based FR offered a viable option to improve the overall properties of RPUF.

References

[1] J.O. Akindoyo, M.D.H. Beg, S. Ghazali, M.R. Islam, N. Jeyaratnam, A.R. Yuvaraj, Polyurethane types, synthesis and applications-a review, RSC. Adv. 6 (2016) 114453–114482.

[2] M. Ionescu, Chemistry and Technology of Polyols for Polyurethane, Volume 2, iSmithers Rapra Publishing, Shrewsbury, Shropshire, United Kingdom, 2016.

[3] B. Foams, M. Alinejad, S. Nikafshar, A. Gondaliya, S. Bagheri, N. Chen, S.K. Singh, D.B. Hodge, M. Nejad Lignin-based polyurethanes: opportunities for and adhesives. Polymers 11(7) 1202.

[4] C. Zhang, S. Bhoyate, M. Ionescu, P.K. Kahol, R.K. Gupta, Highly flame retardant and bio-based rigid polyurethane foams derived from orange peel oil, Polym. Eng. Sci. 58 (2018) 2078–2087.

[5] G. Lligadas, J.C. Ronda, M. Galiá, V. Cádiz, Plant oils as platform chemicals for polyurethane synthesis: current state-of-the-art, Biomacromolecules 11 (2010) 2825–2835.

[6] H.P. Benecke, B.R. Vijayendran, D.B. Garbark, K.P. Mitchell, Low cost and highly reactive biobased polyols: a co-product of the emerging biorefinery economy, Clean Soil Air Water 36 (2008) 694–699.

[7] MA Sawpan, Polyurethanes from vegetable oils and applications: a review, J. Polym. Res. 25 (2018) 184.

[8] A. Behr, L. Johnen, Myrcene as a natural base chemical in sustainable chemistry: a critical review, ChemSusChem 2 (2009) 1072–1095.

[9] M. Desroches, S. Caillol, V. Lapinte, R. Auvergne, B. Boutevin, Synthesis of biobased polyols by thiol-ene coupling from vegetable oils, Macromolecules 44 (2011) 2489–2500.

[10] R. Kaur, M. Kumar, Addition of anti-flaming agents in castor oil based rigid polyurethane foams: Studies on mechanical and flammable behaviour, Mater. Res. Exp. 7 (1) (2020) 015333.

[11] D.K. Chattopadhyay, D.C. Webster, Thermal stability and flame retardancy of polyurethanes, Prog. Polym. Sci. 34 (2009) 1068–1133.

[12] P. Kiliaris, C.D. Papaspyrides, Polymers on fire, in: C.D. Papaspyrides, P. Kiliaris (Eds.), Polymer Green Flame Retardants, Elsevier B.V., Amsterdam, 2014, pp. 1–43.

[13] A.B. Morgan, C.A. Wilkie, Non-Halogenated Flame Retardant Handbook, John Wiley & Sons,, NJ, 2014.

[14] T.T. Li, M. Xing, H. Wang, S.Y. Huang, C. Fu, C.W. Lou, J.H. Lin, Nitrogen/phosphorus synergistic flame retardant-filled flexible polyurethane foams: Microstructure, compressive stress, sound absorption, and combustion resistance, RSC Adv. 9 (2019) 21192–21201.

[15] G. Camino, S. Lomakin, Intumescent Materials, Woodhead Publishing Ltd, Cambridge, England, 2001.

[16] G. Camino, L. Costa, G. Martinasso, Intumescent fire-retardant systems, Polymer (Guildf) 23 (1989) 359–376.

[17] A. Konig, U. Fehrenbacher, T. Hirth, E. Kroke, Flexible polyurethane foam with the flame-retardant melamine, J. Cell. Plast. 44 (2008) 469–480.

[18] M. Thirumal, D. Khastgir, G.B. Nando, Y.P. Naik, N.K. Singha, Halogen-free flame retardant PUF: Effect of melamine compounds on mechanical, thermal and flame retardant properties, Polym. Degrad. Stab. 95 (2010) 1138–1145.

[19] C.K. Ranaweera, M. Ionescu, N. Bilic, X. Wan, P.K. Kahol, R.K. Gupta, Biobased polyols using thiol-ene chemistry for rigid polyurethane foams with enhanced flame-retardant properties, J. Renew. Mater. 5 (2017) 1–12.

[20] RM. Silverstein, W. FX, K. DJ, Spectrometric Identification of Organic Compounds, 7th ed, John Wiley & Sons,, NJ, 2014.

[21] D. Lin-Vien, N.B. Colthup, W.G. Fateley, J.G. Grasselli, Alkanes. In: The Handbook of Infrared and Raman Characteristic Frequencies of Organic Molecules, Elsevier, Amsterdam, 1991, pp. 9–28.

[22] F.M. de Souza, J. Choi, S. Bhoyate, P.K. Kahol, R.K. Gupta, Expendable graphite as an efficient flame-retardant for novel partial bio-based rigid polyurethane foams, J. Carbon Res. C. 6 (2020) 27.

[23] S. Ramanujam, C. Zequine, S. Bhoyate, B. Neria, P. Kahol, R. Gupta, Novel biobased polyol using corn oil for highly flame-retardant polyurethane foams. J. Carbon Res. C 5 (2019) 13.

[24] M.V. Gravit, O. Ogidan, E. Znamenskaya, Methods for determining the number of closed cells in rigid sprayed polyurethane foam. 03027 (2018) 1–8.

[25] H. Zhu, S. Xu, Preparation of flame-retardant rigid polyurethane foams by combining modified melamine-formaldehyde resin and phosphorus flame retardants, ACS Omega 5 (2020) 9658–9667.

[26] S. Członka, A. Strąkowska, K. Strzelec, A. Kairytė, A. Kremensas, Melamine, silica, and ionic liquid as a novel flame retardant for rigid polyurethane foams with enhanced flame retardancy and mechanical properties, Polym. Test. 87 (2020) 106511.

[27] A. Sut, E. Metzsch-Zilligen, M. Großhauser, R. Pfaendner, B. Schartel, Synergy between melamine cyanurate, melamine polyphosphate and aluminum diethylphosphinate in flame retarded thermoplastic polyurethane, Polym. Test. 74 (2019) 196–204.

[28] M. Kageoka, Y. Tairaka, K. Kodama, Effects of melamine particle size on flexible polyurethane foam properties, J. Cell. Plast. 33 (1997) 219–236.

[29] S. Bhoyate, M. Ionescu, P.K. Kahol, R.K. Gupta, Castor-oil derived nonhalogenated reactive flame-retardant-based polyurethane foams with significant reduced heat release rate, J. Appl. Polym. Sci. 136 (2019) 1–7.

[30] C. Wang, Y. Wu, Y. Li, Q. Shao, X. Yan, C. Han, Z. Wang, Z. Liu, Z. Guo, Flame-retardant rigid polyurethane foam with a phosphorus-nitrogen single intumescent flame retardant, Polym. Adv. Technol. 29 (2018) 668–676.

[31] H. Horacek, R. Grabner, Advantages of flame retardants based on nitrogen compounds, Polym. Degrad. Stab. 54 (1996) 205–215.

[32] X. Liu, J.-.W. Hao, S. Gaan, Recent studies of decomposition and strategies of smoke and toxicity suppression for polyurethane based materials, RSC. Adv. 6 (2016) 74742–74756.

[33] B. Bann, S.A. Miller, Melamine and derivatives of melamine, Chem. Rev. 58 (1958) 131–172.

[34] V. Sangeetha, N. Kanagathara, R. Sumathi, N. Sivakumar, G. Anbalagan, Spectral and thermal degradation of melamine cyanurate, J. Mater. 2013 (2013) 1–7.

[35] A.E. Abelow, A. Nissen, L. Massey, L. Whinnery Effectiveness of flame retardants in tuffoam. Technical Report SAND-2017-13391 659436 Albuquerque, NM (United States) (2017).

[36] W.H. Rao, Z.Y. Hu, H.X. Xu, Y.J. Xu, M. Qi, W. Liao, S. Xu, Y.Z. Wang, Flame-retardant flexible polyurethane foams with highly efficient melamine salt, Ind. Eng. Chem. Res. 56 (2017) 7112–7119.

[37] E.D. Weil, W. Zhu, Some practical and theoretical aspects of melamine as a flame retardant, ACS Sym. Ser. 599 (1995) 199–216.

[38] H.T. Nhung, P.D. Linh, N.T. Hanh, N.T. Nhan, H.T. Oanh, Effect of the incorporation of organoclay and melamine cyanurate on the flame retardancy and mechanical property of polyurethane foam. Vietnam J. Chem. 57 (2019) 368–374.

24

Nanolignin in materials science and technology—does flame retardancy matter?

H. Vahabi[a], N. Brosse[b], N.H. Abd Latif[c], W. Fatriasari[d], N.N. Solihat[d], R. Hashim[e], M. Hazwan Hussin[c], F. Laoutid[f] and M.R. Saeb[a]

[a]UNIVERSITÉ DE LORRAINE, CENTRALESUPÉLEC, LMOPS, METZ, FRANCE. [b]UNIVERSITÉ DE LORRAINE, INRAE, LERMAB, NANCY, FRANCE. [c]MATERIALS TECHNOLOGY RESEARCH GROUP (MATREC), SCHOOL OF CHEMICAL SCIENCES, UNIVERSITI SAINS MALAYSIA, MINDEN, PENANG, MALAYSIA. [d]RESEARCH CENTER FOR BIOMATERIALS, INDONESIAN INSTITUTE OF SCIENCES (LIPI), CIBINONG, INDONESIA. [e]SCHOOL OF INDUSTRIAL TECHNOLOGY, UNIVERSITI SAINS MALAYSIA, MINDEN, PENANG, MALAYSIA. [f]LABORATORY OF POLYMERIC & COMPOSITE MATERIALS, MATERIA NOVA RESEARCH CENTER, MONS, BELGIUM.

Chapter outline

24.1 Introduction

Polymeric materials are building blocks of a wide variety of common and complex materials and systems for diverse applications. Although they bring about a tremendous advantages to

Biopolymeric Nanomaterials: Fundamentals and Applications. DOI: https://doi.org/10.1016/B978-0-12-824364-0.00003-4

a wide range of uses from general-purpose to advanced products, the statistics on fire accidents which arise from flammability of polymer materials indicate human and economy losses due to the massive use of polymeric materials from the eighties [1]. Polymers burn easily due to their significant calorific capacity. It is therefore necessary to improve their fire behavior by adding additives, called flame retardants (FRs). The aim of the aforementioned operation is to neutralize (at best) or reduce (at least) the amount of the heat and also smoke normally release during the combustion of polymeric materials [2].

Looking at the fire retardancy in its historical context can help demonstrate the evolution in this field of research. Steps taken are: making wood and cotton fire retardant by Egyptian (about 500 B.C.); using alum solution to make wooden battleships FR by Greek (about 100 B.C.); chemical synthesis of an FR by British Chemist, O. Wyld (in 1735) [3]; and scientifically developed FR by French chemist, J.L Gay-Lussac (in 1821). In the present time, a wide variety of FR additives and FR materials are commercially available for diverse fields and uses, among which are halogenated, phosphorus, nitrogen, intumescent, and inorganic based FRs. Moreover, nanoparticles (NPs) have also showed their efficiency in flame retardancy in some cases alone or in combination with conventional FRs [4]. In line with developments in the knowledge and industrial production of FRs, two dimensions are given to flame retardancy: (1) new instructions denoting the ban on the use of toxic FRs and (2) sever fire safety legislations. In this regard, using bio-based and renewable polymers and additives for flame retardancy improvement has progressed in recent years because of sustainability and environmental requirements [5]. In view of their origin, two families of bio-based FRs are available (Fig. 24.1), including those obtained from the biomass (lignin, starch, phytic acid, cellulose, tannins, proteins, and oils), and those obtained from animals (DNA and chitosan (CH)) [6].

Among bio-based macromolecules, lignin has received significant attention to improve flame retardancy of polymers during the last decades [7]. Lignocellulosic materials represent a vast majority of biomass and are commonly made up of 40%–60% cellulose, 20%–40% hemicellulose, and 10%–25% lignin. Lignin is the second most abundant bio-based natural polymer after cellulose. It is predicted that around 50 million tons per year of lignin is produced worldwide. Under an inert atmosphere, lignin decomposes through a two-stage thermal degradation reaction into compounds of lower molar mass (first stage between 280 and 390 °C and the main stage in the vicinity of 420 °C). At higher temperatures (> 500 °C), the rearrangement and condensation of the aromatic structures leads to the formation of 30–50 wt% of char, the rate of which still remains high (ca. 35%–38%) at 900 °C [8]. Therefore, lignin has high capacity to produce char, an efficient candidate for improving the flame retardancy of polymers. To understand the necessity and benefits of char production during the combustion of polymers in flame retardancy, the mechanism of FR actions in the condensed phase is briefly described hereafter.

There are several scenarios accepted by the experts to explain the flame retardancy of polymers. Polymers decompose in the course of combustion, while fire retardancy action goes to the condensed and gas (and sometimes both) phases. The chemical structure (polymer backbone, functional groups, side chains, functionality, and molecular weight) of polymers considerably governs their FR action. When an FR acts in the condensed phase, it (1) retards thermal

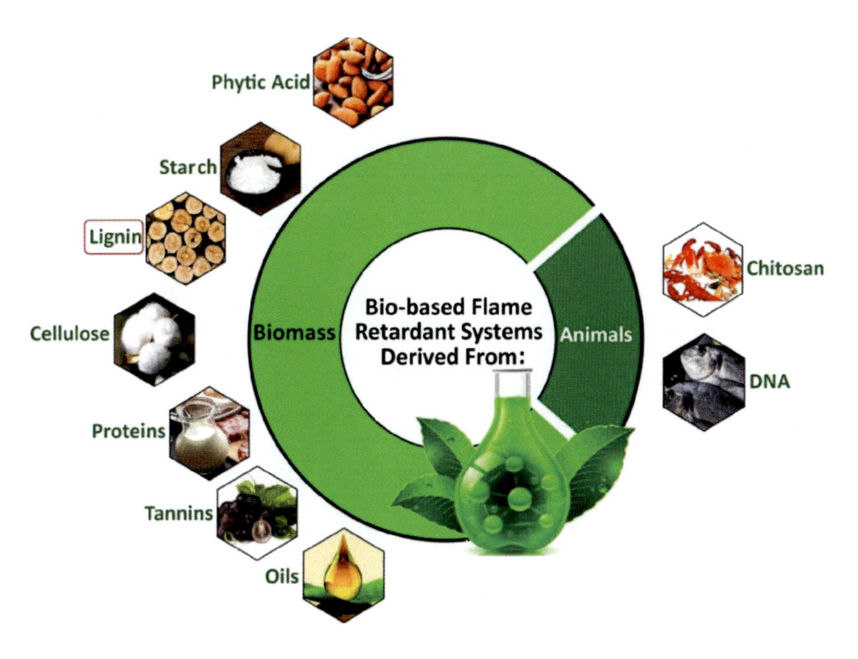

FIGURE 24.1 Schematic of various bio-based flame retardants from two main sources; biomass (left-hand side), and animals (right-hand side).

decomposition of chains by producing less-combustible volatiles and additional char that protects polymer from fire; and/or (2) plays the role of coolant and lowers the temperature of fire; and/or (3) causes melt dripping, which may be a benefit and a detriment during fire. The FR properties of polymers may be reflected in the barrier properties of layers formed on the substrate surface, physically or chemically, during combustion, and are commonly carbonaceous (char), mineral, vitreous, ceramic, or intumescent. Protective layers are barriers against heat, mass, or oxygen, which keep two phases separated. The use of some FR increases the amount of char by assisting the intermolecular combustion. Thus, polymers known for being highly charring are good candidates to insulate the substrate from combustion through reduction in the amount of fuel available for combustion.

Lignin is a bio-based polymer and can play a key role as FR in view of its abundance, reasonable costs, and charring character with respect to the synthetic conventional FRs. In particular, in view of natural character of lignin, it can be applied as FR in developing sustainable polymer materials when a green polymer is used. It is worth mentioning that the use of lignin and nanolignin is usually associated with intumescent system because of high amount of char they form during combustion [6]. An intumescent system acts in the condensed phase by formation of high quantity of an expandable char. The microstructure of the char layer formed on the surface of the material is generally porous with low thermal conductivity, which limits heat transfer from the flame to the material. Such a system should generally combine the following components: a source of acid such as inorganic acids which causes dehydration of the carbonaceous agent;

a carbon source agent, in general a polyhydric compound or the polymer itself, and a swelling agent which releases noncombustible gases (NH_3 or CO_2) at elevated temperatures and allows the char layer for expansion. In this sense, lignin or nanolignin can play the role of carbon sources in intumescent systems. Finally, it has been shown that the incorporation of NPs into the polymers can additionally improve their flame behavior [6]. In the present chapter, first, the structure, preparation methods, and applications of nanolignin are comprehensively described. Then, we focus on the use of nanolignin as a part of FR system in polymers. We believe that such a survey is among first reports on the early stage of progress in the field of materials science and technology devoted to the potential application of nanolignin as an FR material for polymer systems.

24.2 Structure and availability of lignin

After cellulose, lignin is the most abundant biopolymer on the earth but also the most recently emerged in the evolution 380 million years ago. Lignin is a result of an adaptation of algal ancestors to terrestrial environment and its primarily roles are to protect the plant tissues against UV-B radiation and to provide structural supports [9]. Lignin is an aromatic polymer resulting from the radical polymerization of three monolignols (hydroxycinnamoyl, p-coumaryl and coniferyl alcohols, Fig. 24.2A) leading to three aromatic units (respectively H, G, and S) in the resulting random 3D structures (Fig. 24.2B). A lignin polymer in plant retains different linkages between the units (β-O-4, $\beta-\beta$, β-5, etc.) [10].

Lignin, which comprises between 15% and 30% of lignocellulosic biomass, potentially represents an abundant resource of aromatics. However, in the cell wall, lignin is embedded in a complex and recalcitrant network including cellulose and hemicelluloses. The extraction of lignin from lignocellulosic resources is a challenging task and generally results in tremendous chemical modifications with the formation of C-C interunit bonds and condensation reactions [10,11]. There is a broad range of different lignins but the term "lignin" is often misleading, as it can designate polymers of very different natures and be very variable according to the raw materials used and especially the extraction process implemented. In contrast to other biopolymers such as DNA, proteins or cellulose, lignin has a complex structure that changes significantly in function of the biomass source and the isolation technique. As a result, despite the potential benefits of this renewable aromatic resource, its integration in chemical industry is a challenging task. An overview of the most important types of lignin classified according to their extraction processes is given in Table 24.1 and discussed next.

Kraft lignin is a coproduct of the Kraft pulping, the dominant chemical pulping process. As a result, Kraft pulping is currently the largest source of industrial lignin. The Kraft pulping consists in cooking wood chips in an aqueous solution of NaOH and Na_2S at a temperature range of 150–180 °C. After cellulosic pulp recovery, the lignin-containing black liquor is concentrated and essentially incinerated for in-site energy production [12,13]. However, Kraft lignin can be isolated from black liquors on industrial scale by acidic precipitation, filtration, water washing, and drying (Lignoboost, Lignoforce processes) [14]. Chemically, Kraft lignin has a very different

A

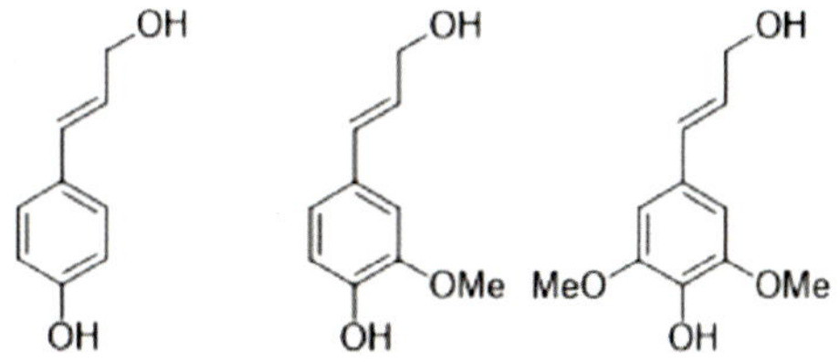

	lignin %	coumaryl alcohol	coniferyl alcohol	sinapyl alcohol
Herbaceous	9–20%	5%	75%	25%
Softwood	27–33%	-	90–95%	-
Hardwood	18–25%	-	50%	50%

B

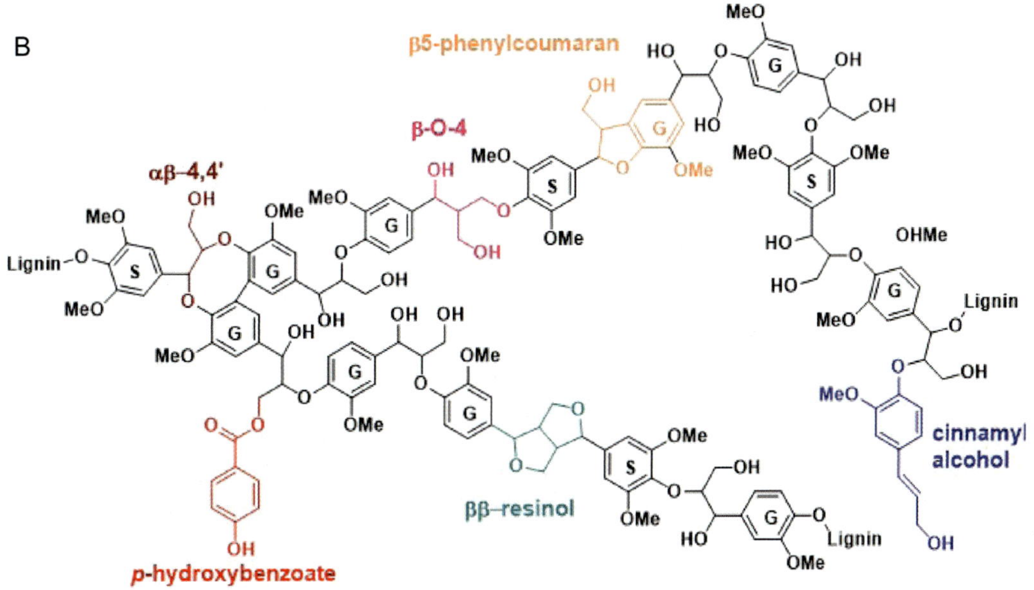

FIGURE 24.2 (A) Structure of monolignols and percent lignin and monolignols in various biomass sources. (B) Representative structure of native lignin from hardwood.

structure from native lignin. It is composed of fragments of small molecular masses, free of aryl ether interunit moieties, recondensed and soluble in alkali.

The lignosulfonates are produced by the sulfite pulping process, which represents a small portion of the market pulp (<10%). Lignosulfonates display a higher molecular weight than Kraft lignin and are grafted with sulfonate groups on the lateral chain (α position) making them water soluble [15]. Lignosulfonates like Kraft lignins are generally impure materials and contaminated by sulfur, sugars, degraded sugars, and inorganics. However, thanks to their particular properties and contrary to Kraft lignins, lignosulfonates find many industrial applications (cement, animal

Table 24.1 Available lignin overview.

Lignin	Raw materials	scale	Technology	Structure	T_g (°C)	Solubility	Trade names	Suppliers	Ref
Kraft	Hardwood, softwood	Commercial	Not specify Lignoboost Lignoforce Lignoboost	1–3 KDa recondensed	140–150	Alkali	Indulin AT, Biochoice Amalin Lineo	Ingevity Domtar West fraser Stora Enso	[12-14]
Lignosulfonate	Hardwood, softwood	Commercial		15–50 KDa α–sulfonated	130	Water	Borresperse, Reax	Borregaard	[15]
Sulfonated Kraft	Kraft lignin	Commercial	Sulfomethylation	Ar–sulfonated		Water		MeadWestvaco	[16]
Soda	Annual plants	Commercial		1–3 KDa	140	alkali	Protobind	Greenvalue	[17]
Organosolv	Hardwood, softwood, annual plants	Pilot	Formic/acetic Ethanol acetone	1–5 KDa high β-O-4 content	90–110	Alkali, organic solvents	Biolignin Lignol Fabiola	CIMV Fraunhofer CBP TNO	[18, 19]
Hydrolysis	Hardwood,	Pilot	Extrusion		75–90		Sunburst	Sweetwater	
	annual plants	pilot	Steam explosion	β-O-4 and recondensed		Alkali, organic solvents	sunliquid	Clariant	[19]
	Hardwood	lab	Steam explosion						
Solubilization	all	Lab	Ionic liquid, deep eutectic	1 KDa		Alkali, organic solvents	-	-	[20]
Analytical	all	lab	MWL CEL EMAL	8–15 KDa "native"	120–130	Alkali, organic solvents	-	-	[17, 21]

feed, etc.). To increase the industrial attractiveness of Kraft lignin, its sulfonation to produce sulfonated Kraft lignins (Table 24.1) with different degrees of sulfonation can be performed [22].

Besides pulping processes, an increasing attention is given to biorefinery processes that offer new opportunities for supplying lignins [23,24]. Most of these technologies have not yet reached commercial scale, but there are multiple examples of demonstrations at pilot scale. However, the biorefinery lignins described below generally present characteristics that are more suitable for subsequent utilizations. Biorefinery lignins are less recondensed and sulfur-free, have lower levels of impurities, and generally retain much of their original structure [25].

Protobind lignin is a sulfur-free soda lignin extracted from wheat straw [17]. This lignin is commercially available and very broadly experimented in the literature for new end-use applications.

Organosolv lignins are produced from organosolv pulpings which receive much attention in the context of biofuels and biorefineries [18]. Organosolv pulpings usually involve a mixture of organic solvent (e.g., ethanol, acetic acid glycerol, etc.), water, and eventually a catalyst allowing simultaneous hydrolytic breakdown and lignin solubilization in the medium. Organosolv lignins are described to retain high amount of aryl-ether linkages and to be little recondensed and relatively pure [19].

Hydrolysis processes (including steam explosion, dilute acid hydrolysis, supercritical water) are under development for cellulosic bioethanol production. Among them, steam explosion is one of the more promising with current industrial developments. After the hydrolytic pretreatment step, the sugar fraction is hydrolyzed with enzymes and fermented using microorganisms. The lignin fraction is usually recovered at the end of the whole transformation as an insoluble and impure residue primarily incinerated to produce energy. However, lignin first approaches have been proposed at a lab-scale for a high-quality lignin recovery just after the pretreatment step [20].

Another approach, based on a total biomass solubilization, has been proposed but only at lab-scale. This requires a solvent with a very high ionic strength such as ionic liquids or deep eutectic liquids. The lignin is further precipitated by the addition of a nonsolvent [21].

Analytical procedures have been also reported to extract lignin from lignocellulose minimizing structural changes, the goal being to produce fractions at low yields but being consider to be representative of lignin in wood. These methods require extensive ball milling to degrade the cell wall matrix and to extract lignin-carbohydrate complex fragments. A solid–liquid extraction of milled wood with 96% dioxane followed by purification steps produces milled wood lignin (MWL) [17]. An enzymatic hydrolysis of milled wood prior to the dioxane extraction can also be done to produce cellulolytic enzyme lignin (CEL) or enzymatic mild acidolysis lignin (EMAL) using slightly acidified dioxane-water solvent [26].

The thermal degradation of isolated lignins has been studied for the elucidation of lignin structure but also for other purposes including the production of lignin-based carbon materials (carbon fibers) and lignin-based composites and polymer blends. The thermal behavior of lignin is highly complex and irreproducible. It is function of (1) the structure of lignin and recondensation reactions occurred at elevated temperatures, (2) the composition and impurities content including sugars and inorganics, (3) the processing conditions (heating rate, gaz flow, etc.). The thermal degradation of lignin takes place in a broad range of temperature from

200 °C to 700 °C but the presence of large amount of impurities in technical lignin strongly affects its decomposition behavior, such as residual carbohydrates in hydrolysis lignin (dilute acid, steam explosion) or inorganics in Kraft lignin [27,28]. The chemical structure of the lignin such as molecular mass and condensation degree also influences its thermal stability. Lignins with low S/G ratio or highly recondensed industrial lignins (higher carbon-carbon linkages content) have shown to be more resistant to pyrolytic bond breakage than lignins rich in aryl-ether bonds [29-31].

24.3 Methods of production of nanolignin

Recently, the conversion of lignin into nanolignin or NPs has become interesting studies among the researchers because the specific surface area and surface active sites will be increased and lead to better improvement properties such as solubility, antioxidant activity, and UV protection activity. Although the research of nanolignin field is still in infancy, lignin can be successful utilized through the various advanced applications in the nanotechnology field. Therefore, many kinds of methods using a combination of chemical and physical methods have been published for the preparation of nanolignin.

24.3.1 Chemical methods

24.3.1.1 Acid precipitation method

Acid precipitation method was developed by Frangville et al. [32] to produce novel nontoxic and biodegradable NPs of lignin for microalgae and yeast. In this study, the lignin NPs (LNPs) were produced using two different methods to compare its stability upon change in pH solution before precipitation as shown in Fig. 24.3. For the first method (Fig. 24.3A), the precipitation of LNPs is obtained from a solution of Indulin AT (IAT) in ethylene glycol (EG) by slowly addition of aqueous hydrochloric acid (HCl) solution. The precipitated LNPs obtained are pH-stable up to pH 10 after dialysis even without the presence of cross-linking step because it has very dense amount of lignin domains. As for the second method (Fig. 24.3B), the lignin is dissolved in high pH aqueous sodium hydroxide (NaOH) solution and LNPs are precipitated when the nitric acid (HNO_3) solution is added. Thus, the resulted LNPs have stability only at acidic range (above pH 5) and it was highly porous that consist of smaller lignin domains.

24.3.1.2 Alkaline precipitation

According to Gutierrez-Hernandez et al. [33] LNPs can be generated using alkaline precipitation method by using soda and organosolv lignin from *Agave tequilana* Weber bagasse. Briefly, 17 wt% of lignin slurry was stirred at 1 h followed by the addition of NaOH (amount calculated on basis of lignin content). Then, ammonium hydroxide was added after 2 h and proceeded with high intensity mixing at 24,000 rpm using Ultra-turrax (IKA, T10) for 5 min. After that, active

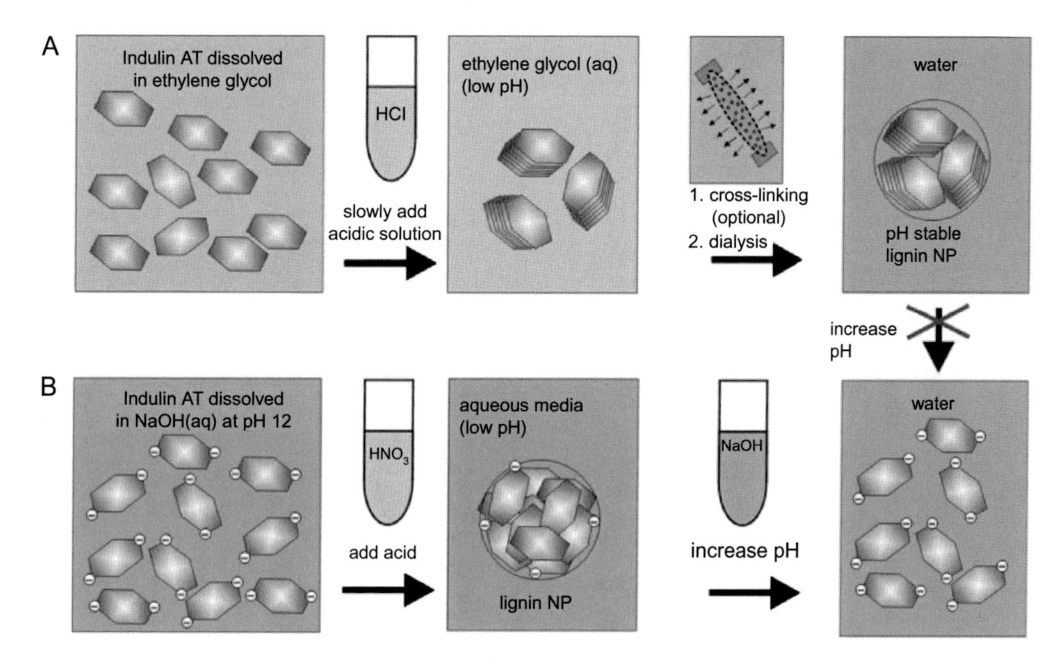

FIGURE 24.3 The proposed mechanism of lignin nanoparticles (A) precipitation from Indulin AT (IAT) dissolved in ethylene glycol with HCl (aq.) and possible subsequent cross-linking and dialysis, and by (B) precipitation from IAT dissolved in basic to acidic aqueous medium [32].

formaldehyde was added followed by increasing of temperature up to 85 °C for 2 h. Lastly, the suspension undergo cross-linking under magnetic stirring at 600 rpm and NPs will be formed.

24.3.1.3 Chemical modification method

LNPs can be obtained based on hydroxymethylation of lignins under special condition (50% NaOH solution, 25% NH_4OH solution as a catalyst, 37% formaldehyde, precipitation with HCl solution). The hydroxymethylated lignin-based NPs were studied by dimensional distribution. In addition, the NPs also can be obtained after performing the epoxidation reaction of the Protobind-commercial products in the separated supernatant [34]. By the presence of copper ions, the modified LNPs have been obtained with increase in biological wood stability [35]. The veneer samples showed low mass loss values and high contact angle values after buried in soil for 6 months. Due to this chemical treatment, the hydrophobicity properties increased thus resulting in long-term stability.

24.3.2 Physical methods

24.3.2.1 Ultrasonication

Ultrasonication method appears to be a simple and low requirement for equipment as well as an ease operation that does not use any hazardous chemical (no solvent fractionation required) to produce LNPs. Gonzalez et al. [36] used an ultrasonication treatment at different periods

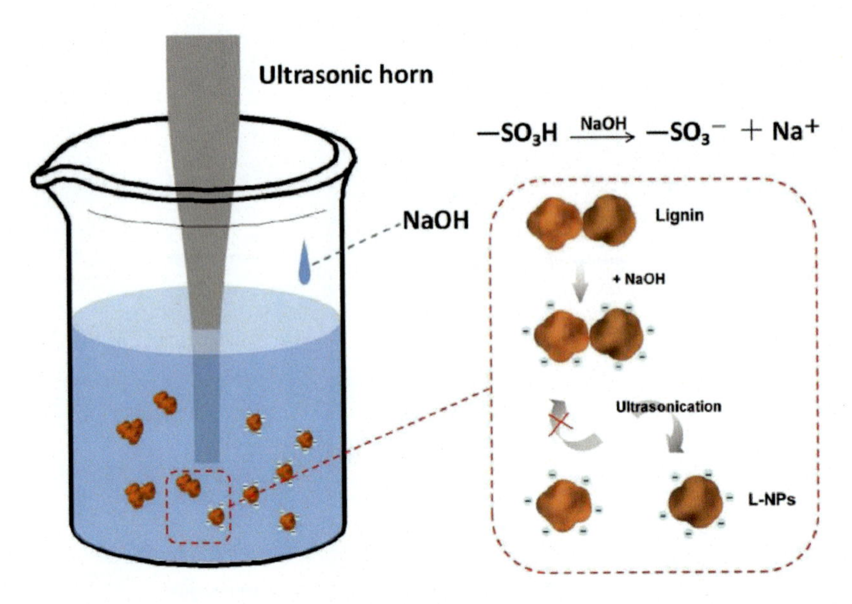

FIGURE 24.4 Schematic diagram on the preparation of lignin-nanoparticles by using ultrasonic-assisted alkali method [37].

of time to obtain a stable colloidal suspension of nanolignin particles at smaller size such as 10–50 nm. The nanolignin was prepared using a Sonic & Materials VCX130 sonicator tip at frequency of 20 kHz, 130 W power, and 95 % oscillation amplitude to get 0.1 wt% lignin-water dispersion. In addition, Yin et al. [37] have produced lignin-NPs by using ultrasonic-assisted alkali method. The lignin isolated from switchgrass was dissolved in deionized water containing sodium hydroxide (NaOH) and then sonicated for 1 h at 400 W as shown in Fig. 24.4 to obtain LNPs with the average size around 220.2 nm.

24.3.2.2 Supercritical antisolvent (SAS) method

The supercritical antisolvent (SAS) method is a patent "compressed/supercritical fluid" method based technology that gain attention in the production of polymeric NPs. This method usually involved steps of dissolution and precipitation by using carbon dioxide (CO_2) as SAS. According to Lu et al. [38], the precipitation chamber first must filled with supercritical CO_2 (SC-CO_2) at 30 MPa pressure under 35 °C by pressing pure acetone (solvent) into the chamber. After obtained stable conditions, the dissolved lignin/acetone solution was totally injected into the precipitation chamber and SC-CO_2 continuously flow to remove residual organic solvent. Finally, the precipitated nano-scale lignin was removed from stainless steel frit vessel in precipitation chamber. From the result, it showed that the solubility of nano-scale lignin enhanced in the water. This method does not produce any pollution as the solvent can be recycled and no solvent residue is obtained as well as a high performance of nanolignin is obtained.

24.3.2.3 Freeze-drying and thermal stabilization

The synthesis of carbon NPs was conducted by Gonugunta et al. [39] through combination of freeze-drying process, thermal stabilization, and carbonization by using lignin as a renewable source. At first, the lignin was dissolved with different amounts of KOH under sonication before solidified using liquid nitrogen. The solidified lignin samples were freeze-dried to form porous microstructure lignin as confirmed by the scanning electron microscope (SEM) analysis. By heating up to 250 °C, it led to thermal stabilization of freeze-dried lignins. Thus, it helped to retain the porous microstructure during the carbonization process. Then, thermo-stabilized lignin samples were carbonized in a tube furnace at 700 °C under nitrogen atmosphere until carbon NPs obtained. From the transmission electron microscopy (TEM) analysis, it showed that the thermal stabilization of freeze-dried lignin inhibited agglomeration of carbon NPs at the time of carbonization. The formation of carbon NPs had a size range between 25 and 150 nm when using lignin precursor and influence of KOH.

24.3.2.4 Self-assembly

Novel green NPs have been developed based on lignin recovered from corn cob in the absence of chemical modification. Alkali lignin (AL) was used in the preparation of NPs with perfect spherical-shaped and well dispersibility characteristics. The NPs were prepared using a simple self-assembly method by adding water with methanol solution of AL [40]. Liu et al. [41] produced lignin-NPs by using ethanol on the sequential organosolv fragmentation approach (SOFA) and explored various stages of catalyst for selective-dissolution of lignins from corn stover. The fabrication of LNPs can be done through self-assembly by altering its chemical characteristics and generated multiple uniform lignin streams as well as its reactivity. As results, the smallest effective diameter was approximately around 130 nm from SOFA that utilized ethanol as well as sulfuric acid.

24.4 Characterization of nanolignin

Surface morphology studies are very important for characterization of nanolignin in order to assess information on their particle size and shape. The common microscopy methods used for morphological characterization are TEM, atomic force microscopy (AFM), and SEM. Because the size of particles is in nanometer scale, the TEM analysis is needed to study the morphology and dimension of the particles. From the TEM images, the information on topographical, morphological, compositional, and crystalline can be analyzed. From Fig. 24.5, the TEM images of LNPs are obtained by the acid precipitation from EG solutions. Fig. 24.5A shows the high-resolution image of lignin molecules that might be clustered in stack-like structures after being negatively stained with uranyl acetate. Meanwhile, the other image in Fig. 24.5B showed low magnification of TEM image that indicated that the domains of small lignin molecules clusters form highly porous NPs [32]. Other than TEM analysis, AFM analysis is also widely used as a tool to measure, imaging, and manipulate matter in a nanoscale range. AFM technique has been used to characterize the

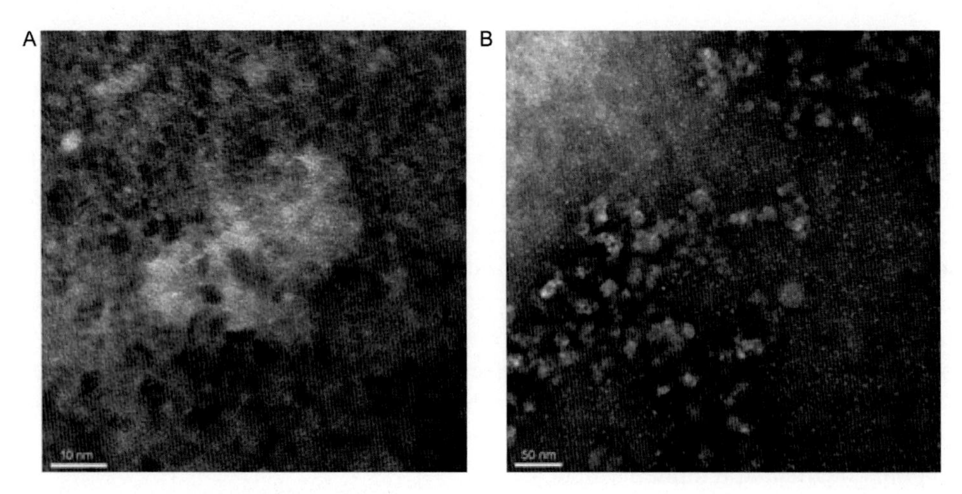

FIGURE 24.5 The lignin nanoparticles obtained by acid precipitation at (A) high resolution of TEM image and (B) low magnification of TEM image [32].

surface roughness of LNPs before and after the fungal laccases treatment and dispersion of the particles in water and tetrahydrofuran (THF). By using these techniques, researchers can study the effect of laccase treatments on the surface-modified LNPs. In Fig. 24.6, the AFM images of unmodified lignin-NPs had a smooth surface and after treated with high redox of laccase at low concentration of lignin-NPs, the surface roughness occurred as well as some aggregates due to intercross-linking of the particles [42]. As for the SEM analysis, it seems a better choice to study the images in a micrometer range as it gives a clear perception details on the fine structure. SEM images gave an evaluation of filler particle distribution in the bio-based nanocomposite materials. Fig. 24.7 shows the SEM images of formation of large agglomerates visible on the top surface of polyurethane-lignin composite (PU/IND) but PU-nanolignin (PU/NL) composite showed a smooth surface morphology, which means an excellent level of dispersion and distribution of nanolignin particles into polymer matrix [36].

In addition, thermal analysis such as differential scanning calorimetry (DSC), thermogravimetric analysis (TGA), dynamic mechanical thermal analysis (DMTA), and thermal mechanical analysis (TMA) are the common instrument used to study the thermal behavior of nanopolymers and nanocomposite materials. DSC analysis has been widely used to identify the melting, crystallization, and glass transition temperatures (T_g) of the nanofillers and polymer nanocomposites. Gonugunta et al. [39] had used DSC analysis to confirm the increasing of T_g and glassy state of freeze-dried lignin before and after thermal stabilization. For solid-state samples, usually TGA analysis is used to analyze the thermal stability of the nanocomposite materials when nanofillers added in polymer matrix [36]. The coefficients of thermal expansion of nanocomposite materials were mainly measured by TMA analysis. On the other hand, DMTA measured the stiffness and energy loss as a function of temperature in nanocomposite materials. DMTA analysis of nanocomposites displayed an α-peak in the tan δ curve, which originated from the movement of longer molecular chains in the amorphous region that correspond to the T_g [43].

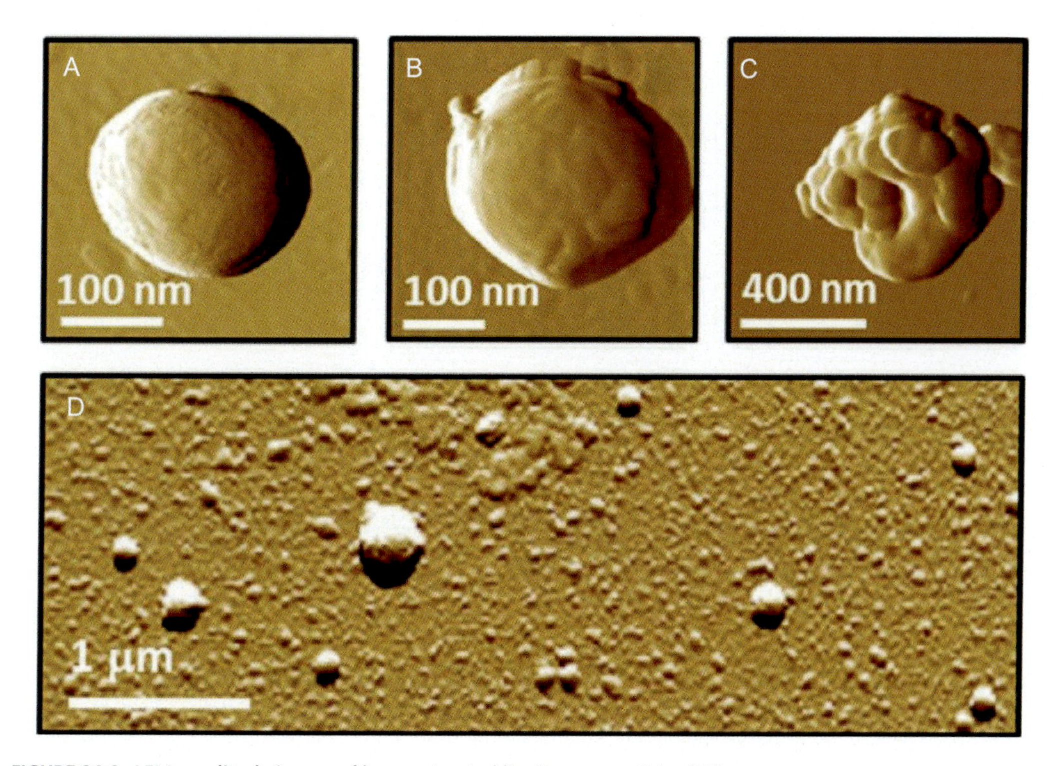

FIGURE 24.6 AFM amplitude image of laccase-treated lignin nanoparticles [42].

Besides, the structure of nanolignin has been characterized using X-ray diffraction (XRD). Lu et al. [38] used XRD to determine the amorphous structure of nanoscale lignin prepared by an SAS process. Furthermore, size is also an important physical parameter to characterize the behavior of the NPs. The particle size distributions, zeta potential values, molecular weight, and second viral coefficient can be characterized using electrical mobility particle size analyzers. The most common analyzer device is differential mobility analyzer that used to classify and measure the nanometer-sized aerosol particles in the diameter range 1 nm to 1 μm based on their electrical mobility [42,44].

24.4.1 Recent application of nanolignin

Nanolignin has been explored in newly many industrial processes that are figured out in Fig. 24.8.

24.4.1.1 Hybrid composite

LNPs are often used as reinforcing agents in composites. The addition of LNPs into polymer as a reinforcing agent enhanced the mechanical, thermal, and biocompatible properties compared to original polymers [45–48]. Phenolic foams reinforced by LNPs result low density and smaller cell sizes of foams and improvement in compressive modulus and high compressive strength

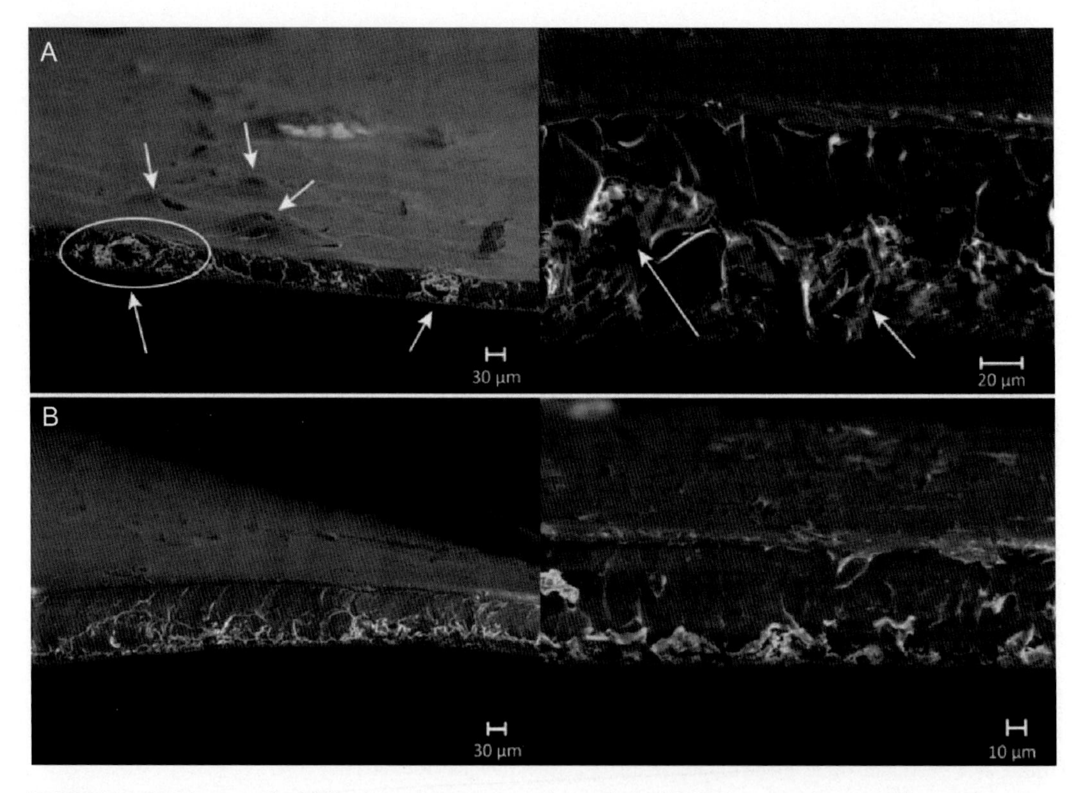

FIGURE 24.7 SEM images of a (A) lignin agglomerates on the surface of PU/IND composites and (B) smooth surface of PU/NL nanocomposites [36].

[49]. There is an increase contact angle of lignin reverse micelles (LRMs) films [35] that leads to improve the miscibility. The mechanical properties of LRM also increase. The LRM was prepared from cyclohexane, AL/dioxane solution, and high-density polyethylene (HDPE). Incorporation of LNPs into wheat gluten bio-nanocomposites was performed by Yang et al. [50]. Even though there is the reduction of transparency on the composites, the thermal stability, glass transition, and mechanical properties tend to increase. Melt extrusion and solvent casting of polylactic acid (PLA)/LNPs bio-nanocomposites were reported by Yang et al. [50]. Melt extrusion of bio-nanocomposites enhanced remarkably the nucleation effect in homogenous dispersion of 1 wt% LNPs in PLA matrix. Incorporation of 1 wt% LNPs hinders disintegration of PLA matrix. Aggregation of LNPs/PLA film occurred in increase of 3% LNPs that does not support the crystallization behavior. With this method, the LNPs' presence increases the elongation at break of film. The casting method gave inhomogeneous dispersion which can cause weak interaction in LNPs/PLA matrix, thus the mechanical properties did not improve. The lignin-rubber-g-poly (D-lactide) copolymer particles and commercial poly(L-lactide) (PLLA) in chloroform were prepared by Sun et al. [51]. The elongation at break of PLLA/lignin-rubber-D increase toughness about sevenfold and there is an improvement in tensile strength and Young's modulus. Lignin can disperse well

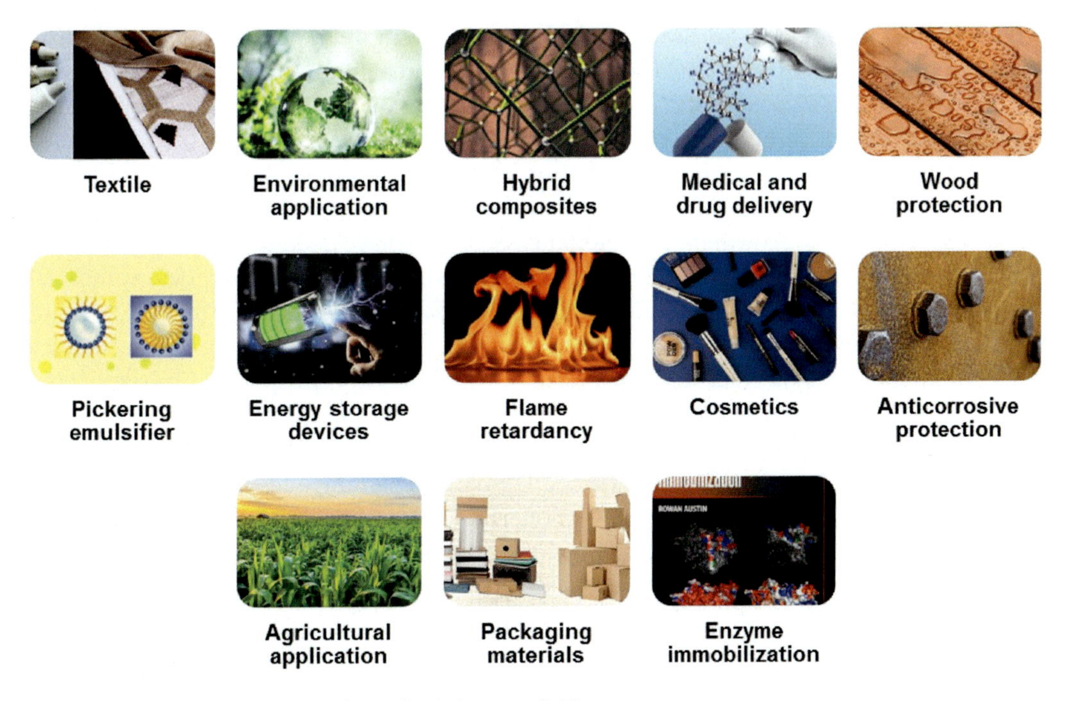

FIGURE 24.8 Potential application of nanolignin in many fields.

and rubber-initiated crazing is a toughening mechanism. This nanocomposite has also desirable UV light barrier properties with excellent mechanical properties that have high potency as a packaging material.

Chung et al. [52] performed graft polymerization of lactide lignin by copolymerization reaction with triazabicyclodecene as a catalyst to produce lignin-g-PLA copolymer. Preacetylation or lignin and lactide ratio was used to adjust the chain length of PLA. The range of T_g of lignin-g-PLA copolymers was 45–85°C with multiphase melting behavior. The addition of 10% lignin-g-PLA improved the UV absorbance and mechanical properties of PLA bioplastics. To enhance the UV absorption and reduce brittleness without a loss in tensile modulus of PLA-based materials, the lignin-g-PLA-copolymers used dispersion modifiers. Lignin plays as a cross-linker to cure epoxidized natural rubber and reinforcing filler to improve the mechanical characteristics of lignin-epoxidized natural fiber composites [53]. Excellent compatibility to rubber was caused by reaction hydroxyl and carbonyl groups of lignin with epoxy sites in epoxidized natural rubber by ring-opening reaction under temperature of 180 °C. In this condition lignin can disperse well in epoxidized natural rubber matrix. Preparation of nanolignin-poly (diallyldimethylammonium chloride) (PDADMAC) and complexes (LPCs) was reported by Jiang et al. [54]. Nanolignin size of less than 100 nm could stably disperse in aqueous solution. The homogeneous dispersion presented in the blending of nanolignin with natural rubber (NR) latex. As result, the thermal stability and mechanical properties of LPCs/NR composites increased.

Liu et al. [55] reported the fabrication by gel-spinning technology of lignin-polyacrylonitrile (PAN)-carbon nanotubes (CNT) that was carried out in preparation of composite fibers. The PAN, PAN/lignin, and PAN/lignin/CNT were carbonized at 1100 °C under identical stabilization condition to produce carbon fibers. And the mechanical properties of lignin carbon fibers were comparable to PAN carbon fibers. LNPs-reinforced poly(methyl methacrylate) (PMMA) nanocomposites prepared via combining solvent-free radical polymerization, micro extrusion, and hot press methods following the masterbatch approach have been successfully fabricated. LNPs were dispersed well in PMMA that contributed in improvement of hardness values, UV resistance, thermal and scratch resistance of LNPs-PMMA nanocomposites. Nair et al. [56] prepared nanokraft lignin by a simple high shear homogenizer without causing major chemical composition changes, molecular weight distribution as well as its polydispersity. The size of LNPs was less than 100 nm after 4 h of mechanical shearing. The blending LNPs with polyvinyl alcohol (PVA) increased the thermal stability of LNPs/PVA film compared to lignin/PVA composites.

The LRMs were introduced into HDPE as a UV-blocking additive for thermoplastics. LRM film was formed by the addition of 7% vol cyclohexane into AL/dioxane solution. The addition of 5% LRM increased significantly the elongation at break and Young's modulus. The increase of cyclohexane caused aggregation in LRM and then it will get separated from solution. The miscibility of LRM with HDPE improved significantly compared to lignin-HDPE. An excellent UV-absorbing performance of LRM-HDPE was observed [35].

Gupta et al. [57] produced bio-poly (trimethylene terephthalate) (bio-PTT) hybrid nanocomposite consisted of 1.5% LNPs and vapor-grown carbon fibers (VGCF) into hybrid nanocomposite through melt extrusion followed by a microinjection molding technique. In preparation of bio-PTT hybrid nanocomposites in which LNPs were used as a reinforcing agent enhanced the mechanical properties and biodegradation characteristic of nanocomposite. The hybrid nanocomposite (VGCF/bio-PTT/LNPs) has a positive effect on the modulus, flexural strength, modulus, and impact strength. The thermal stability of bio-PTT and its hybrid nanocomposite improved and heat deflection temperature increases threefold by reinforcing with VGCF. Introduction of LNPs also increased deflection temperature almost 120%. Thus, the bio-PTT/LNPs/VGCF hybrid nanocomposites also exhibited good biodegradation characteristics that are prospective for various end-use applications.

Kai et al. [58] reported the grafting lignin (lignin-poly(methyl methacrylate [PMMA]) copolymer through atom-transfer radical polymerization (ATRP) improved the miscibility and compatibility of lignin with other plastics. These copolymers were then blended with poly(ε-caprolactone) (PCL) and engineered into nanofibrous composites using electrospinning. Incorporation of lignin–PMMA copolymers obviously enhanced the tensile strength, Young's modulus, and storage modulus of the resulting nanofibrous composites. The PCL/lignin-MMA nanofibers showed good biocompatibility. These results bring to apply the copolymer in the biomedical application. Similarly, the synthesis of PMMA-grafted LNPs (PMMA-g-LNPs) nanocomposites was prepared by solvent-free radical polymerization and further blended into the commercial PMMA. The presence of LNPs as cross-linking agents during polymerization process into PMMA-g-LNPs affected to result in higher thermal stability, hardness, UV resistance, scratch resistance and molecular weight compared to the commercial PMMA, due to the presence of LNPs that could serve as cross-linking agents during the polymerization process. By this result, it is possible to use as coating, automotive, flooring, acrylic glasses, and lenses [59]. LNP nanocomposites have been used as natural UV barriers and antibacterial agents in the finishing process of protective textiles. LNPs incorporation significantly improved the compressive modulus and compressive strength of the foams [60].

Qi et al. [61] developed PU nanocomposites that were prepared by prepolymerization of poly EG and diisocyanates and curing process in the presence of LNPs of 1%–7%. High reactive LNPs were prepared by an acidolysis process from pristine lignin acted as bio-based polyol and cross-linker. LNPs enhanced thermal stability and mechanical behavior of PU nanocomposites, increased cross-linking densities, thus the hydrophobicity of PU nanocomposites increased. The ultraviolet blocking capability of the nanocomposites enhanced by the addition of well-dispersed LNPs in the nanocomposites, at the same time PU visible light transparency altered. LNPs-based PU nanocomposites presented good thermal reprocess ability because the transcarbamoylation reaction leads to use it for novel multifunctional bio-based materials.

24.4.1.2 Wood protection

LNPs prepared by a nonsolvent method were used to perform wood surface treatment by using a dip-coating technique. LNPs-mixed sawdust presented better UV resistance to wood affected by the presence of aromatic extractive compounds in Iroko lignin macromolecules. The dip-coated wood samples with LNPs showed prospective surface modification technique appear as a fused LNPs film [62]. Colloidal lignin particle (CLP) was prospective to be applied in industrial wood coatings. CLPs were changing the wood beneath. The wood protection against weathering has been reported by the use of pure LNPs (PLNPs) dispersions [62]. LNPs that entrapped essential oil from cinnamon bark (*Cinnamomum zeylanicum* Blume), common thyme (*Thymus vulgaris* L.), and wild thyme (*Thymus serpyllum* L.) using a fast antisolvent method were developed by Zikeli et al. [63]. The utilization of essential oil (EO) entrapped inside LNPs contributes to extensive π-stacking between aromatic compounds in EOs on one side and aromatic lignin units on the other side. This result opens up new bio-based biocide delivery systems for wood preservation [63].

24.4.1.3 Medical and drug delivery system

Nanotechnology application in modification of drug delivery becomes prospective method to show better results. One of the best candidates for drug delivery systems due to its dispersibility, high porosity, and specific surface area was polymer NPs. These properties have positive effect on the permeability and retention for drug in body and therefore the therapeutic effect [64]. Drug dissolution improved by reducing the particle size thus its bioavailability increased. In cancer chemotherapy, the utilization of NP drug delivery enhanced therapeutic effectiveness and reduced side effects of the drug payloads by improving their pharmacokinetics. LNPs are a prospective biopolymer as drug delivery regarding their antioxidant properties and versatile functional moieties on the surface [65].

Dai et al. [40] reported AL LNPs (AL LNPs) with bioactive molecule resveratrol (RSV) and Fe_3O_4 magnetic NPs for anticancer drug delivery. The cytological and animal tests of AL/RSV/Fe_3O_4 NPs significantly improved the RSV stability, accumulation, and anticancer efficacy in comparison to free drugs. Compared to free drugs, it has also good anticancer effects, drug accumulation, better tumor reduction, and lower adverse effects. AL LNPs application is a new and highly efficient nanodelivery. The utilization of LNPs becomes very effective method for the delivery of insoluble drugs.

Qian et al. [66] assembled colloidal spheres from the acetylated lignin in the mixed solvent of THF and water, which showed potential applications in drug delivery and controlled release, and pesticide microencapsulation areas. Figueiredo et al. [47] developed LNPs that is pLNPs, Fe-LNPs, and Fe_3O_4-LNPs with low cytotoxicity. pLNPs were able to effectively load some drugs or other cytotoxic agents with poor water solubility and improved their release profiles at different pH values. However, benzazulene (BZL)-pLNPs showed an enhanced antiproliferation effect than the pure BZL.

Li et al. [67] studied pH-responsive complex micelles synthesized from quaternized AL and sodium dodecyl benzenesulfonate in green solvents and used to encapsulate the oral drug Ibuprofen through hydrophobic interactions. This study was successful as a novel approach to fabricate oral drug delivery carriers to the intestine with reduced drug leakage in the stomach. It is also prospective for controlled delivery of hydrophobic oral drugs, such as indomethacin, nifedipine, aspirin, etc. Mishra and Wimmer [68] prepared lignin colloidal particles using ultrasonic spray-freezing route without any material functionalization and stabilized it by electrostatic route. By application of this method, the hollow/solid lignin colloids have good particle size control to be a good candidate in drug delivery.

Chen et al. [69] prepared LNPs in non-toxic aqueous sodium p-toluenesulfonate (pTsONa) solutions at room temperature. LNPs fabricated in water with a wide pH range at room temperature. By this system, various drugs with high efficiency can be produced. Various water-soluble or water-insoluble drugs can be dissolved and encapsulated in the LNPs with an encapsulation efficiency of up to 90% because of the hydrotropic system. The drug-encapsulated LNPs showed excellent properties, with sustained drug-releasing capability and biocompatibility. LNPs could be used for versatile drug/bioactive in the biomedical application.

Adhesive hydrogels have been popular in biomedical applications. Gan et al. [70] developed adhesive hydrogel in the dynamic catechol redox system by using Ag-LNPs. It creates long-term adhesiveness, high toughness, and antibacterial ability for long time. This redox system–generated catechol groups continuously make the hydrogel with long-term and repeatable adhesiveness. The hydrogel had high toughness, durable adhesiveness, good cell affinity, antiinfection, high antibacterial activity, biocompatible and is not harmful to skin tissue. This result opens the prospective challenge in producing tough adhesive hydrogel suitable for surgical operation or other biomedical applications [70].

Figueiredo et al. [47] prepared PLNPs, iron (III)-complexed LNPs (Fe-LNPs), and Fe_3O_4-infused LNPs (Fe_3O_4-LNPs) that have round shape, narrow size distribution, reduced polydispersity, very good stability at pH 7.4, and negative surface charge. The LNPs had low cytotoxicity at concentrations of up to 100 mg/mL in different cell lines, low *in-vitro* hemolysis, and no significant intracellular hydrogen peroxide production within 6-h incubation. pLNPs presented to load poorly water-soluble cytotoxic agents and drugs and improved the release profiles of drugs at different pH values. The addition of pLNPs enhanced the antiproliferative effect of BZL in different cell lines. Based on the results, LNPs became a promising candidate for drug delivery and biomedical applications. LNPs can next be modified with targeting moieties to increase the cellular interaction with specific cells, for example, cancer therapy, diagnosis. In the LNPs pH-sensitive polymers can be added to form pH-responsive release of drugs and load hydrophilic drugs.

Lignin hollow NPs were developed as vehicles for the antineoplastic, antibiotic drug doxorubicin hydrochloride. The NPs played an important role in the cellular uptake and the accumulation of the drug within HeLa cells. LNPs with a spherical hollow structure became potential vehicles for compounds with benzene rings because of their exceptional absorption capacity, biodegradability, and nontoxicity [71]. LNPs combine with NP and lignin as a superior potential candidate for drug delivery [47]. Blank LNPs (BLNPs) with controllable particle size and maximized yield were evaluated as drug carrier. *In-vitro* cell line studies showed that BLNPs were compatible with normal and lung cancer cells even at very high concentrations and thus are safe for use as a drug carrier. Irinotecan-loaded LNPs decreased the IC_{50} of irinotecan by threefold, thus become an efficient drug carrier such as for chemotherapeutic applications [72].

24.4.1.4 Agricultural application

Deng et al. [73] reported encapsulation pesticide avermectin (AVM) from hollow lignin azo colloid sphere. Hollow lignin azo colloidal spheres have a high AVM encapsulation efficiency indicated an excellent controlled release performance for AVM. Li et al. [74] developed synthesis of novel lignin-based microsphere that is, uniform colloidal spheres (SL-CTAB) from the mixture of sodium lignosulfonate (SL) and cetyltrimethylammonium bromide (CTAB) for the encapsulation of photosensitive AVM, a photosensitive pesticide. SL-CTAB has exerted reversible aggregation behavior with a solid spherical morphology and can shell material to prepare microspheres with uniform spherical structure. The performance of a controlled release and antiphotolysis properties could be adjusted by the proportion of SL-CTAB. By this finding, SL-CTAB has a great application for photosensitive pesticides against photodegradation and to obtain a controlled release. Yiamsawas et al. [75] reported hollow nanocapsule with diameters of 150–200 nm which was stable in organic or aqueous dispersion for several weeks to months. They were prepared by interfacial polyaddition in inverse miniemulsions. These cross-linked lignin nanocontainers could be loaded with hydrophilic substances that can be released by an enzymatic trigger from natural plant extracts. They have potential for agricultural applications.

24.4.1.5 Environmental application

During past decade, the water contamination by heavy metal ions caused by civilization, population, and industrial development has become a crucial environmental issue. These ions such as copper, zinc, lead, silver, cadmium, etc. can remain for long time in the environment that is dangerous to human health and aquatic life [76]. Lignin could also act as an adsorbent to remove heavy metal ions and dyes by the binding method [45]. The adsorption capability can be improved by using nano-sized particle because of enhancement diffusion effect and contacting frequency on heavy metal ions. Functionalized lignin-based nano-trap (LBNT) was used to capture heavy metal ions by a simple inverse-emulsion copolymerization method. The versatile adsorbent of some heavy metal ions can be carried out by LBNT. LBNT also becomes as new metal ion host to silver ions thus can be used as an efficient antimicrobial material. In the LBNT, surface-contained ionic silver could be released into the culture media directly and interact with the bacterial organisms through binding to the cell walls or even penetrating the cells. As a result, the bactericide rates of the silver-loaded nanocomposite toward *Escherichia coli* and *Staphylococcus aureus* were almost 100%. The removal degree of heavy metal ions after LBNT use was up to 99% with effective adsorption performance for soft acid (silver, mercury and cadmium ions) and borderline acid (lead, copper and zinc ions). As a result, the residual concentration that allowed content regulated in drinking water by the World Health Organization (WHO) can be fulfilled [77].

Azimvand et al. [78] used copolymerization reactions between LNPs and polyacrylic acid in the presence of radical initiator namely potassium persulfate for fabricating LNPs-g-polyacrylic acid adsorbent. The adsorption capacity of LNPs-g-polyacrylic (138.88 mg/g) was higher than LNPs (99 mg/g). Compared to adsorption capacity of AL for the removal of Basic Red 2 (BR2) from aqueous solutions, LNPs had excellent performance [78]. Besides, heavy metal spilled oil can cause environmental problem in aquatic ecosystem. Yang et al. [79] tried to develop xerogel to absorb high amount of oil from the oil/water (O/W) mixture. Xerogel from modified diissocyanate (MMDI)-PU with modified lignin as a precursor was prepared by a sol-gel process and ambient pressure drying method. This xerogel has high characteristic in self-cleaning and super hydrophobicity. It has high potency as absorbents, coating and spilled oil removal [79].

Porous lignosulfonate spheres (PLS) with high porosity to adsorb prospective ions were prepared through a feasible gelation-solidification method [80]. The excellent adsorption capacity of PLS even at initial concentration of 25.0 mg/L leads to its application in industrial wastewater treatment containing heavy metals. The utilization of LNPs-g-polyacrylic acid and LNPs as adsorbents was to remove Safranin-O from aqueous solutions was reported previously by Azimvand et al. [78]. LNPs-g-polyacrylic acid was prepared by copolymerization reactions between polyacrylic acid and LNPs using potassium persulfate as a radical initiator. There was increase 1.4-fold of adsorption capacity in removing Safranin-O dye from wastewater by using a lignin nanoparticles (LNPs)-g-polyacrylic acid (PAA) copolymer compared to LNPs. Luo et al. [81] reported modification of lignin by triethylenetetramine with the Mannich reaction, and then Fe(III) was chelated onto the aminated lignin. The resulted adsorbent was uniformly ball-shaped with particle size of 450 nm. This study indicated that biomass-based lignin can become a potential adsorbent to remove low-concentration phosphate from waste or wastewater.

24.4.1.6 Anticorrosive of metal protection

Rahman et al. [82] have successfully prepared LNPs through a polyol route in three different polyol media: castor oil (CO), EG, and water (W). LNPs were then used as anticorrosive nanofillers for epoxy coatings to protect carbon steel (CS) substrate. Spherical LNPs were

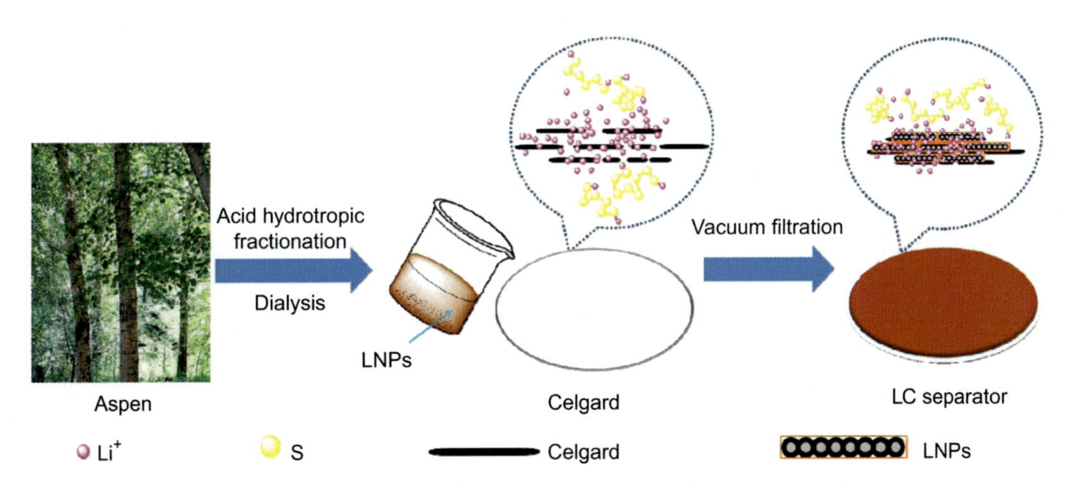

FIGURE 24.9 The production process of lignin nanoparticles (LNPs)-coated Celgard (LC) separator [83].

resulted in use of CO and EG. To observe the anticorrosive behavior, LNPs were dispersed in an epoxy matrix to formulate nanocomposite coatings. Based on impedance spectroscopy and its salt mist test, LNPs can act as anticorrosive nanofillers to protect the underlying metal substrate and had better protection than epoxy coating. As anticorrosive nanofiller, LNPs can reduce the infrastructure loss due to corrosion.

24.4.1.7 Energy storage devices

Zhang et al. [83] reported LNPs-coated Celgard (LC) membrane for a novel separator for Li-S batteries presented in Fig. 24.9. LNPs played a role of temporary electrolyte depositor to restrain the soluble polysulfides from directly diffusing into the bulk electrolyte. LNPs also have abundant electron-donating groups to form chemical binding of polysulfides and improve the cycling stability. LNPs can absorb electrons and release protons to facilitate the transport of lithium ions therefore a long lifecycle of the Li–S battery can be achieved.

The preparation commercial lignin-based microsphere via reverse phase polymerization was carried out by Yu et al. [84]. Then application of lignin-based microsphere on activated carbon was investigated. Size distribution and surface morphology of lignin-based microsphere were affected by lignin emulsion. Without preoxidiation process, the carbonization of lignin-based microsphere occurred directly. The reverse-phase polymerization process affected to modify the morphology, surface appearance, and size distribution of lignin-based microsphere by coordinating the ratio of solid content and dispersed phase content. Lignin-based microsphere had very high thermal stability. Based on the gravimetric capacitance, lignin-based activated carbon microsphere is good candidate for high performance of supercapacitor.

24.4.1.8 Sunscreen active substances

Because of its excellent oxidation resistance capability, lignin has great potential as a candidate as UV absorbent or sun block. By using the sunscreen cream, the skin and negative effect due to exposure the UV radiation for long time can be minimized (Chauhan 2020, [45]. Development of lignin-based sunscreens by the amalgamation of LNPs and pure cream was reported. The poor dispersibility of lignin can be overcome by utilizing LNPs because of increase of LNPs higher surface area. LNPs also presented light color of product that it is more desirable. The incorporation of LNPs in sunscreen creams increased the sun protection factor (SPF) value compared to SPF of pure cream. The preparation of LNPs was first carried out by microwave acetylation process without a catalyst and solvent. Acetic anhydride played a role as a reaction reagent and dispersion solvent. The subsequent process, that is, a solvent shifting combined ultrasonic-assisted technique was performed to prepare high-yield LNPs with a regular shape. The smaller LNPs size, the higher the sunscreen performance that caused by the conjugated system in lignin. The functional groups of lignin that is, methoxyl groups, S and G-type lignin perform easy UV-absorbing function. Besides, $\pi - \pi$ stacking between the sunscreen cream and aromatic rings in LNPs affected the performance of the sunscreen creams [85].

Tian et al. [86] produced two types of LNPs (deep eutectic solvent (DES) and ethanol-organosolv extracted technical lignin from two fractionation pretreatment). They had spherical morphology with unique core–shell nanostructure with more uniform particle size distribution found in DES-LNPs. The incorporation of LNPs in poly (vinyl alcohol) (PVA) resulted in a transparent nanocomposite film with high UV-shielding, antioxidant activity, and biocompatibility properties. The improvement in mechanical and thermal properties of LNPs/PVA nanocomposite film was attributed to good interfacial adhesion LNPs with PVA matrix through the hydrogen bonding network.

Ju et al. [87] developed sphere-like LNPs prepared by using a T-shaped microchannel reactor with polyvinylpyrrolidone)/sodium dodecyl sulfate as stabilizers to prevent the LNPs agglomeration [88]. LNPs had good stability for 60-day storage and used as an additive to produce UV-shielding PVA-composite films. The utilization of LNPs improved the UV-shielding efficacy of composites compared to raw lignin. These LNPs nanocomposite film was prospective as packaging and medicine [87]. The photoprotection characteristics of *A. tequilana* LNPs in association with Zinc oxide (ZnO) NPs as main UVA radiation blocking was reported by Gutiérrez-Hernández et al. [33]. Compared to sunscreens without LNPs (only ZnO NPs) the LNPs showed absorption in UV-B and UV-C regions, which can enhance the SPF value of sunscreens.

Xing et al. [89] developed melanin-like polydopamine thin layer in incorporation with LNPs to form UV-blocking-core-shell lignin-melanin NP ((LMNP) with higher compatibility and durability. To increase the UV-barrier capability and photostability of poly (butylene adipate-coterephthalate) PBAT films, the PBAT was mixed with LNPs, LMNPs, melanin NPs (MNPs), and a mix of LNPs and MNPs (MixNP). The films had excellent UV-blocking capacity up to 80% of UV-B light with LNPs of 0.5–5 wt%. After UV exposure for 40 h, the PBAT-LMNP films had high UV-shielding stability and the best retention in mechanical properties. Based on the properties, the biodegradable PBAT films had potency to apply in agricultural and packaging materials that required high UV resistance. Yang et al. [90] prepared PLA/LNPs films through premixed 1% wt. LNPs into PLA or glycidyl methacrylate (GMA) grafted PLA (g-PLA) by using a novel master batch method. The films were observed after accelerated UV weathering (up to 480 h). GMA grafting of matrix leads to disperse LNPs in a PLA matrix. LNPs behave better as a UV light barrier in g-PLA.

LNPs and GMA accelerate the oxidation effect thus deformability of grafted PLA nanocomposites has improved.

Qian et al. [35] reported the application of LRM as a UV-blocking additive for thermoplastics. The water contact of LRM was higher than an AL film as a control sample. The utilization of LRM improved the miscibility significantly with HDPE because of hydrophobicity. The HDPE/LRM composite had excellent UV-absorbing performance and improved the mechanical properties of HDPE. By addition of 5% LRM, Young's modulus increased almost twofold while the elongation at break improved 65%. Fabrication of different sizes and structure lignin colloidal spheres (LCS) by self-assembly method and blending them with pure skin creams to develop lignin-based sunscreens were reported in many previous studies. Compared with original lignin, the addition LCS in creams enhanced the sunscreen performance. The higher size of colloidal spheres, the lower sunscreen performance. Lignin nanospheres had the best sunscreen performance. Phenolic hydroxyl groups affected sun-blocking process in which the acetylation of lignin for shielding hydroxyl groups decreased SPF value significantly. The increase of hydroxyl group content linearly affected increase of SPF value. The results indicated that it is an easy method to prepare high UV-blocking properties of lignin-based sunscreens [91].

Currently, the effectiveness of Kraft lignin, LNPs, and its modification for application of sunscreen has been compared. A demethylation reaction to increase unit of catechol and auxochromes was carried out for both lignin and LNPs by thermal treatment. UV transmittance showed the efficacy in the following order: original lignin < demethylated lignin < LNPs < demethylated LNPs. Particle size of lignin was a substantial factor to improve sunscreen performance. Excellent UV transmittance, 0.5–3.8% in UVA-UVB region with SPF >21, was obtained from demethylated LNPs. This performance was significantly better compared to commercial sunscreen, UV transmittance 2.7–51.1% [92]. Another study found that SPF value was twice higher in LNPs-based moisturizing cream than lignin-based. This may be caused by interaction of noncovalent bond in NP [93].

The synthesis of lignin colloidal particles by ultrasonic spray-freezing route without any prechemical modification of material was reported by Mishra and Wimmer [68]. The result was stable aqueous suspension in alkaline pH (10.5). By this route, they produced hollow/solid lignin colloid that used six-layered coating by layer-by-layer deposition on quartz slide with the aid of negligible UV-absorbing polyelectrolyte aqueous solution of PDADMAC [Poly (diallyldimethylammonium chloride)]. The increase in UV-absorbing ability of lignin resulted in the increase in number of layers [68].

Trevisan et al. [94] prepared lignin and lignin acetate NPs dispersed in water-basis by antisolvent addition. Pure lignin was isolated from elephant grass (*Pennisetum purpureum*) using a dilute acid and alkali solutions and then the obtained LNPs from elephant grass were incorporated into a neutral cream. The yield of conversion LNPs of elephant grass was 37%. LNPs were stable in a wide pH range (5–11) and the ionic strength lower than 0.01. The LNPs were then incorporated in a neutral cream that results in a tinted sunscreen formulation caused by lignin biocompatibility and biodegradability, great UV-vis and visible absorption. The UV-vis light shielding performance of blended creams/LNPs was increased by increasing percentage of LNPs in the creams.

To improve safety and efficiency of UV-blocking for humans, Zhou et al. [71] developed polydopamine-grafted AL (AL-PDA) through the free radical addition of AL and dopamine and then encapsulated chemical sunscreen activities. The formed AL-PDA capsules were prepared through ultrasonic cavitation; thus the nanoscale with spherical morphology was obtained. The utilization of AL-PDA nanocapsules with strong bio-adhesion as the sole active ingredient (dosage of 10 wt.%) to formulate sunscreen, the SPF reached 195.33 lasting for over 8 h under UV radiation. The AL-PDA nanocapsules could be used securely in the sunscreen with good photostability.

24.4.1.9 Pickering emulsifier

Micro- and NPs of lignin became an effective solid emulsifier for Pickering emulsions [80,95-97]. O/W Pickering emulsion stabilized by LNPs as a precursor to PLA/lignin composites enhanced mechanical properties and UV-shielding [80]. LNPs could decrease surface and interfacial tensions in mixtures of water and hydrophobic liquids and create stable Pickering emulsions [98]. Qian et al. [99] reported the use of lignin-g–(diethylamino)ethyl methacrylate (DEAEMA) as a surfactant for Pickering emulsion. By N_2 bubbling, the dispersed LNPs could also be precipitated out quickly. An emulsion of decane was prepared with the LNPs and it could be demulsified and reemulsified by bubbling of the two gases. The emulsification/demulsification processes are highly reversible and easily repeatable.

Wang et al. [85] developed Pickering emulsion stabilized by LNPs to microencapsulate 1-tetradecanol (TDA) via polymerization of acrylates for thermal management. LNPs with an average size of 160 nm were formed by using a nanoprecipitation method. TDA was successfully microencapsulated using PMMA (methyl methacrylate) as shell based on the O/W Pickering emulsions stabilized by LNPs. The leakage and accelerated thermal cycling tests presented that the microcapsules had good thermal and chemical stability. The combination of microcapsules with gypsum resulted in good thermal storage composite as good candidate for thermal management in the construction field.

Qian et al. [99] fabricated N,N-diethylaminoethyl methacrylate (DEAEMA)-grafted LNPs with sizes ranging from 237 to 404 nm via ATRP. The DEAEMA-g-LNPs were used as a surfactant for CO_2/N_2-switchability Pickering emulsions that was correlated with the graft density and chain length of the DEAEMA. The preparation of spherical LNPs by using an aerosol flow reactor, a high-yield and high-throughput manufacturing approach that rendered particles with sizes of ca. 30 nm to 2 mm was reported by Ago et al. [100]. The LNPs can stabilize O/W Pickering emulsions with a tunable droplet size, the particle size, concentration, and water contact angle (WCA) affected on the properties [100].

Nypelö et al. [101] prepared self-assembly method for supracolloidal structure. Pickering emulsions were demonstrated and their properties were compared against those obtained by using traditional inorganic particles. Spherical particles (90 nm to 1 mm) were resulted from W/O microemulsions as a mixture of nonionic surfactant and colloidal dispersion of low molecular weight of AL. The lignin-based particles were assembled at O/W Pickering emulsions. A larger particle size presented an increased emulsion drop size and stability.

Silmore et al. [102] studied grafting polyacrylamide onto kraft LNPs using reversible addition–fragmentation chain transfer (RAFT) chemistry to form polymer-grafted LNPs that tune aggregation strength while retaining interfacial activities in forming Pickering emulsions. Radical polymerization controlled well is a key tool in polymer grafting that helps in the intrinsic interfacial functions of lignin to form Pickering emulsions.

24.4.1.10 Textile

Increasing global demand of natural textile increased alongside growth of population [103]. Lignin as the second plentiful bio-polymer next to cellulose has many potencies in textile industry such as antimicrobial, antioxidant, and antistatic properties. Recently, efficient antimicrobial and antioxidant of LNPs on linen and cotton fabrics had been discussed by Juikar and Nadanathangam [104]. Bulk lignin was extracted from cotton stalk by kraft process. LNPs were produced by two different processes: mechanical and biological. Mechanical process to prepare lignin was conducted by subjecting bulk lignin to the high shear homogenization for 60 min at 10,000 rpm. LNPs as supernatant were obtained after 1 h settle. For another mechanical process, bulk lignin was sonicated at 37 kHz and 30 W for 60 min. Biological process was carried out for isolate *Aspergillus oryzae* in basal medium under shaker with the following condition: 100 rpm, 31 °C, 15 days. Subsequently, filtrate LNPs were separated from supernatant containing nonlignin substance through filtration 1 μm. Prior to coating process, linen and cotton fiber were boiled in autoclave with 1% NaOH at 121 °C, 4 h, and pressure 15 psi. Thereafter, the fabrics were bleached with peroxide containing metasilicate nonahydrate and dried for 1 h, 80 °C to reduce moisture content. Finally, the dry fabrics were immersed in 0.4% LNPs and 4% binder. The excess solution was removed by passing fabric through padding mangle. Size analysis of LNPs showed that the smallest particle size (nm) was obtained from biological process, followed by ultrasonication and homogenization. However, T_g of all LNPs were typically similar around 156 °C. The antibacterial activity at 0.5% concentration LNPs showed inhibition zone against *Klebsiella pneumoniae* and *S. aureus*. The antibacterial properties were typically stable after fabrics were washed 10 times. UV protection factor (UPF) steadily elevated in LNPs-treated fabrics, yet significantly decreased after washed 10 times. Antioxidant activity was measured by changing color from pink to yellow at 518 nm. Reducing absorbance value indicated the efficiency LNPs as antioxidant in the fabrics. However, washing process reduced antioxidant activity due to removal LNPs on surface fabrics. Apart from common function in apparel, this functionalized fabric can be used both as technical and medical textile.

Another report utilized LNPs from coconut fiber through microbial hydrolysis process [105] on linen and cotton fabrics as multifunctional textile. Size of LNPs from coconut fiber (27.5 nm) was slightly lower compared to cotton stalk (27.9 nm). Furthermore, LNPs were coated onto fabrics surface by using pad-dry-cure method. The minimal inhibitory concentration showed that concentration of LNPs above 0.4% can inhibit colony of *S. aureus* and *K. pneumoniae*. LNPs concentration below 0.2%–0.3% showed antimicrobial activity against *S. aureus* only. Both UPF and antioxidant test demonstrated excellent result after LNPs coated on fabrics where UPF value increased about five times after LNPs coating. However, washing machine of fabrics decreased UV protecting and radical scavenging properties. This result showed potency of LNPs as a multifunctional agent in textile industries.

Nair and colleagues reported blending of LNPs and PVA to increase its thermal stability that can be used as coating and sizing in textile industries [56]. Lignin was isolated through acid precipitation from black liquor as kraft pulp mill wastewater. Subsequently, lignin was diluted in deionized water and the suspension was treated in a high shear homogenizer at 15,000 rpm for 1, 2, and 4 h. The chemical structure of LNPs was typically comparable after mechanical treatment where no cleavage bond of aryl-O-ether, carboxylic content, phenolic, and hydroxyl group. PLA-LNPs composite was prepared by mixing PVA in water and LNPs through manual stirring. Thermal analysis resulted lignin after mechanical treatment had higher T_{max} than original lignin, indicating higher thermal stability of lignin NPs. Increasing duration of homogenizer treatment decreased the thermal degradation rate. The temperature of T_g for 1- and 2-h treatment duration was lower than 4 h due to retained larger particle in short period of mechanical treatment assuming that most of lignin particle was transformed into NPs. Appropriate concentration of LNPs was an essential factor to increase thermal stability of nanocomposites. In this work, the highest T_{max} (276 °C) was obtained from 10% LPNs/PVA. Excellent dispersion of LNPs PVA matrix with low agglomeration was observed compared to original lignin with PVA matrix that had poor dispersion. High thermal property of textile was widely used as firefighters clothing and sleeping bags [106,107].

Besides used as antimicrobial and anti-UV radiation, LNPs in textile industries also played a role as antistatic. Kozlowski et al. [108] applied LNPs in silicon emulsion with level concentration 5, 25, and 50 g/L onto cellulose textile using a padding technique at temperature 18–20 °C for 2–5 min. Evaluation of efficiency covering textile with LNPs was determined by properties of UV blocking factor, antimicrobial, and antielectrostatic. The UPF result showed very good protection according to European Standard EN 13758-1:2001, while antimicrobial activity showed protection against *Corynebacterium xerosis, Bacillus licheniformis, Mariniluteicoccus flavus, Staphylococcus haemolyticus, S. aureus, K. pneumoniae, E. coli,* and *Pseudomonas aeruginosa.* Surface resistance was a measurement for antistatic properties where textile covered by LNPs was under 2×10^{10} &. The washing-resistance analysis showed nanocoating did not increase stiffness of fabric and did not change their both bio-physical properties.

In utilization of garment in an extremely high UV index area, the fabric from natural fibers could be protected by UV absorbers in which LNPs included as the most ecological UV absorber. Application of LNPs as a UV absorber in the textile finishing process is a very good solution for the problems of producing human and environmentally friendly UV protective clothing. To contribute in the reduction of medical and human cost of skin cancers, the textiles with high UPF rating was one solution against skin cancer [60]. Zimniewska et al. [109] used LNPs as a UV blocker for linen fabrics, in which LNPs were prepared by ultrasonic treatment and then was padded on linen fabrics. The results showed that LNPs improved fabric UV barrier properties. The higher LNPs on linen fabric, the higher level of UPF. The use of LNPs did not influence negatively on its physical and bio-physical properties. Mixed LNPs with silicone emulsion for lignocellulosic fabrics had product with excellent UV protection, bactericidal activity, and maintaining antistatic properties. Zimniewska et al. [109] reported that textiles coated by LNPs had excellent UV protection and good washing resistance with same air permeability. Coating with LNPs did not increase the stiffness and affect the color. Also, the biophysical properties did not worsen of fabrics' comfort and gave positive effect on the human body. The coating textiles by LNPs contributed in reduction of the use of chemical absorbents of UV radiation [60].

24.4.1.11 Cosmetics

Due to environmental concern, sustainable material is intriguing researcher attention into all industries including pharmaceutical and cosmetics. Nowadays, natural-based cosmetics become popular due to the fact that synthetic-based cosmetics could have unfavorable effect both on human and environment [93]. Lignin consists of an aromatic polymer with functional groups of hydroxyls and methoxyl that can act as radical scavenging or antioxidant in cosmetics. Besides, the phenolic and methoxyl group in lignin as auxochromes which conjugated with aromatic rings as a chromophore makes lignin potential as a sun-blocker agent for sunscreen [92,110,111]. Converting technical lignin to nanoscale increased performance of cosmetics product because an active ingredient in nano-size can efficiently be delivered into the skin cell [112]. In cosmetics sector, LNPs also played roles as antioxidant, antibacterial, and antiwrinkle agent [110,113].

Besides as UV protection, antioxidant properties of LNPs that can be used in cosmetics industry had been reviewed by Yearla and colleagues. They used two different origins of LNPs: dioxane LNPs (DN) from subabul stems and commercial alkali LNPs (AN) from softwood with size of 80 ± 27 nm by a transmission electron microscope. Antioxidant capacity determined by monitoring 2,2-diphenyl-1-picrylhydrazyl (DPPH) absorbance relied on the phenolic group in lignin. The antioxidant ability of DN was slightly higher than AN, yet both of them were higher compared to their parent types, original alkali (AL) and dioxane lignin. The antioxidant results were in agreement with UV protection result where nano-size of lignin was more efficient than its parent polymer by observing mortality of *E. coli* toward inducing UV-irradiation [114]. Efficacy of antimicrobial activity of LNPs had been examined against Gram-negative: *Xanthomonas arboricola, Xanthomonas axonopodis*, and *Pseudomonas syringae* [115].

Nanocomposite of LNPs (NL) and chitin (CN) was successfully used as green functional agents in cosmetic product for skin regeneration [113]. Electronegativity in LNPs can capture active ingredient (lipophilic/hydrophilic molecules) that is usually used as an anti-wrinkle agent, anti-inflammatory, and antioxidant. However, LNPs should be combined with electropositive substance such as CN to increase the ability of entrapping microcapsule to both lipophilic and hydrophilic molecules [111]. The effectiveness of nanocomposite of NL-CN was compared to load of glycyrrehitinic acid (GA) as a bioactive molecule into a nanocomposite. The result showed NL-CN-GA was more thermally stable than NL-CN due to radical scavenger activity of GA. Besides, the mass loss of NL-CN-GA was slightly lower because of the presence of GA increased stability of nanocomposite. Analyses of *in-vitro* cultures of hMSCs (human mesenchymal stromal cells) and HaCaT cells (human keratinocytes) were studied for checking potential application of the nanocomposites in skin contact. HaCaT cell aims to study reaction of each layer in epidermis by examining cytokines array against immune response and inflammation. *In-vitro* study of HaCaT cell showed that NL-CN and NL-CN-GA did not decrease cell viability at concentration 0.2 μg/mL and 0.5 μg/mL, respectively, which means these green nanocomposites are safe to be used as skincare. The goal of hMSCs analysis was to assess safety and cytocompatibility of nanocomposites. Used concentration in hMSCs analysis was the recommendation from HaCaT cell cultures. Treatment with these NPs did not change the morphology of hMSCs at all the concentrations. The viable cell of hMSCs indicated that these nanocomposites did not change capability of the osteo-differentiation. NL-CN was able to regulate expression of antimicrobial peptide while expression of proinflammatory cytokines was effectively modulated by NL-CN-GA. The hMSCs presented excellent cell affinity and antiinflammatory activity of these nanocomposites, indicating these green nanocomposites are safe for skin care application and skin regeneration [113].

24.4.1.12 Packaging materials

Eco-friendly food packaging demand predictably can reach USD 249.5 billion by 2025 due to boosting consumer awareness about hazard of fossil fuel–based packaging [116]. Lignin, plentiful biopolymer with hydrophobic properties, is possible sustainable material to substitute fossil-based material in packaging production. PLA is widely used as natural-based material in food packaging industry, yet it has major drawback such as low stability in thermal, low crystallinity, and poor protection against microbial attack. Therefore high-thermal substance is needed to enhance performance of PVA matrix such as lignin [117,118]. Nano-size of lignin had been demonstrated more enhanced thermal, mechanical, and radical scavenging properties because its properties depend on particle size [119,120].

Recently, properties of matrix bionanocomposite of PLA-LNPs (PLA-LNPs) have been evaluated. Besides, evaluation of the effect of metal oxide in the matrix was conducted. LNPs with size of 50 ± 20 nm were synthesized from AL by mechanical process, stirring. Furthermore, the characterized LNPs or metal oxide was mixed with PLA in chloroform by using sonication, called binary nanocomposites. The ternary nanocomposites consist of PLA, LNPs, and metal oxide (Ag_2O, TiO_2, WO_3, Fe_2O_3, $ZnFe_2O_4$). According to thermal analysis, the presence of metal oxide reduced thermal stability of nanocomposite because of metal depolymerized surface of nanocomposite, thus enhancing thermo-degradation process. The maximum rate (T_{peak}) was lower than pure PLA, yet the presence of LNPs in ternary nanocomposite can counteract up the thermal temperature to pure PLA. On the other hand, the presence of LNPs enhanced thermal stability of binary nanocomposite (PLA-LNPs). WCA analysis was an important toll to control characteristic film. PLA/LNPs increased WCA value from 68–70°. The highest WCA value was obtained from PLA-LNPs-Ag_2O that can relate to the morphology of film, smooth and uniform. The antioxidant analysis presented that ternary-based system gave antioxidant activity better than binary based (PLA-LNPs) because metal oxide scavenged free radical. On the other hand, antibacterial activity of PLA-LNPs was more effective toward *S. aureus* (as Gram positive) than *E. coli* (Gram negative), yet the presence of metal oxide inhibited growth both Gram-positive and Gram-negative bacteria. This phenomenon can be explained by significant bacterial activity of metal oxide itself which has positive charge to disrupt negative charge of bacterial surface. Of this, the ternary nanocomposites had better antimicrobial activity against *S. aureus* and *E. coli* at 6 h. Moreover, the presence of metal oxide corroborated UV-protection properties of film [118].

Antibacterial, thermal, and UV-blocker properties of binary nanocomposites: PLA-LNPs, PLA-cellulose nanocrystal (PLA-CNC) and ternary nanocomposite PLA-LNPs-CN in food packaging were examined by Yang and colleague. A synergic effect of two lignocellulosic nanofillers (LNPs and CNC) reinforced PLA has been confirmed by capability of UV light blocking. Furthermore, crystallinity of ternary system was more higher compared to the binary which proved effectiveness in enhanced tendency to crystallize [121]. Low crystallinity of PLA-based food packaging had main disadvantage on gas barrier, high elongation break, and brittle of film [122]. Compared to binary, ternary nanocomposite also showed higher values of modulus, strength, and antibacterial activity against tomato bacterial pathogen (*P. syringae pv. tomato*). Particularly, the nanocomposite containing LNPs indicated how this green-novel nanocomposite can be safely applied in the food packaging industry [121].

Capacity of LNPs to inhibit bacterial plant/fruit pathogen which is able to damage fruit/vegetable after harvest has been successfully investigated by blending LNPs with PVA and CH at 25 °C for 1 h. In addition, the presence of LNPs in this nanocomposite increased tensile strength, antioxidant activities, UV-protection, migration, dispersion, and interfacial adhesion. In conclusion, nanocomposite of PVA-CH-LNPs has potential to be used in fresh food packaging sectors [123].

To increase flexibility and UV barrier in the packaging, another report demonstrated that LNPs were reacted with PLA and PCL, denoted as NL-PLA/PCL. First, NL and ε-caprolactone (PCL) were diluted in toluene and the solution was purged by N_2 gas to obtain homogenous solution. After L-lactide addition, reaction was continued for 2 days at 120 °C. Finally, NL-poly (lactic acid - ε-caprolactone), NL-p(LA-CL), as copolymer was precipitated by pouring methanol into solution. Nanocomposite of NL-PLA/PCL was prepared by mixing PLA/PCL (settled at 20%) with variation amounts of NL-p(LA-CL). The mechanical properties were conducted to evaluate effect of inserting copolymer NL-p(LA-CL) into PLA/PCL. PLA had high tensile strength of 40 MPa and elongation at break of 33%. However, the elongation break gained to 185% and the tensile strength degraded after introducing PCL. The elongation at break number gradually increased to 280% after inserting copolymer NL-p(LA-CL), yet the toughness was increased 1.5 folds compared to PLA/PCL. It might attribute to interfacial compatibility improvement of PLA/PCL. Furthermore, the presence of NL-p(LA-CL) also enhanced crystallization and UV protection of nanocomposites. Through this result, NL-PLA/PCL can be used as UV-protector and impact resistance material in food packaging sector [117].

24.4.1.13 Enzyme immobilization

In general, enzyme immobilization technique is used in industry as a catalyst due to some benefits such as increased enzyme stability and activity. However, most of functional platforms for immobilization supports are fossil-based synthetic polymers that can leak and be harmful in food industry. Consequently, development natural-based support material is needed. The functional platform should be insoluble in water, mechanically stable, inexpensive, and safe for human [124]. The feasibility of lignin as a sustainable polymer had been evaluated as a functional platform in enzyme immobilization [125-130].

Organosolv LNPs (OLN) was successfully used as a platform for immobilization of tyrosinase by deposition of layer-by-layer, encapsulation, and direct adsorption procedure. Thanks to presence of OLN which enhanced stability of tyrosinase. OLN was also stable in molecular weight after tyrosinase treatment. Deposition layer-by-layer procedure was the most active system for phenol oxidation, the highest electrocatalytic responsiveness, activity, and kinetic parameters. The order of affectivity procedure was layer-by-layer>direct adsorption>encapsulation [129].

Piccinino et al. [126] also used compared technology of layer-by-layer, physical adsorption, and encapsulation to immobilized laccase from *Trametes versicolor* with OLN as an immobilization support agent. Again, layer-by-layer technique showed highest reactivity value, kinetic parameters, and activity that might cause electrochemical active lignin exertion specific effect. Besides, the presence of OLN gave higher selectivity in alcohol oxidation and activity of catalyst compared to silica as immobilization support, indicating this system is attractive for biotechnological industries.

Layer-by-layer procedure was also effective to immobilize glucose oxidase and horseradish system where OLN acted as a green functional platform in the immobilization system. Reusability, stability, electrochemical properties, and application of the system as glucose detection have been examined. The activity of the immobilized enzyme was typically reduced about 50% after reused for five times. The presence of OLN was boosting electrocatalytic activity of the system, indicating retained activity of enzyme after immobilization. Moreover, the system was stable due to electron transfer occurred in aromatic structure of OLN. Efficacy of system showed successful oxidation of β-glucose with a lower limit of detection compared to other system without OLN [130].

Another research reported cationic lignin nanosphere (CLP) in calcium alginate as an inexpensive enzyme carrier through adsorption procedure owing to large surface area of nanolignin for protein adsorption. Immobilization of amano lipase and humicola insolens (HiC) was started by adsorption the hydrolases on CLP. Higher activity of Lipase and HiC after immobilization in CLP-CaAlginate was achieved compared to immobilization in acrylic resin which indicated CLP increased activity of enzymes. In particular, fabrication of a sustainable biocatalyst worked well for ester synthesis in aqueous condition [128].

24.4.1.14 Antimicrobial

In this part authors pointed out influencing factor of LNPs as an antimicrobial agent such as type of microbial and concentration [115]. Methoxy and hydroxy phenolic functional groups in lignin play a role as an antimicrobial agent that can damage and lysis cell of bacteria [131]. However, some authors reported that lignin and its nano-size were stronger to inhibit Gram-positive bacteria than Gram-negative bacteria [132,133]. However, LNPs had a synergistic effect on CH and resulted in effectiveness against Gram-negative bacteria [134] because amino and hydroxyl chains in CH act as an antimicrobial agent [133]. LNPs concentration in nanocomposite also has critical effect on microbial activity. Higher concentration of LNPs resulted in a higher inhibition zone of *P. syringae pv tomato, X. axonopodis vesicatoria,* and *X. arboricola pv pruni* [115].

24.5 Lignin and nanolignin in flame retardancy

Lignin has been widely used in various polymers as a part of FR systems. Several review papers have been already published on this subject [7,135,136]. However, the use of nanolignin is limited and; therefore, there are several unanswered questions about the potential and the efficiency of using nanolignin as FR of polymer systems. Hereafter, some of the research papers dealing with lignin and flame retardancy will be reviewed. Then, the focus is placed on the use of nanolignin-incorporated polymers as a recently developed field of research on FR systems.

24.5.1 Lignin

Lignin has been used in bio-based polymers, such as PLA [137], poly(3-hydroxybutyrate) (PHB) [138], poly(butylene succinate) [139,140], polyamide 11 [141,142] as well as in non-bio-based polymers such as Acrylonitrile Butadiene Styrene (ABS) [143,144], Polyurethanes (PUs) [145-147], poly(ethylene-co-vinyl acetate) (EVA)[148], polypropylene [149,150], polyethylene terephthalate (PET) [151], Polyvinyl alcohol (PVA) [152], unsaturated polyester [153], and epoxy [154-156]. As previously mentioned, the main mechanism of action of lignin is in the condensed phase through increasing the char residue making a barrier against flame. The combination of lignin and phosphorus FRs increases further the rate of char residue and keeps the fire and polymers in reaction. In some cases, the chemical modification of lignin with phosphorus or nitrogen moieties can also help the increase in the amount of char residue. Moreover, the chemical modification of lignin is necessary to avoid the degradation of polymer in the presence of lignin due to the unavoidable hydrolysis of some polymers [157]. The chemical modification can also improve the dispersion state and the interfacial adhesion between the lignin and the polymer matrix leading to a higher resistance against flame. A combination of lignin and inorganic fillers such as montmorillonite [158] or metallic hydroxides [148] can also increase the amount of char residue and; therefore, improve the flame retardancy of polymers. For example, Laoutid et al. [148] investigated the combination of a kraft lignin and magnesium hydroxide (MDH) to improve the flame retardancy of EVA. They reported boosted flame behavior of EVA by combining the charring effect of lignin and endothermic effect of MDH. As a result, the peak of heat release rate (pHRR) decreased from 400 kW/m^2 for pure EVA to the 230 kW/m^2 for composite in which 20 wt% lignin and 40 wt% MDH were used.

Due to the importance of using lignin in bio-based polymers, some examples of these works are briefly described. It is worth mentioning that the origin of lignin has an important effect on its efficiency as a part of FR system. For example, two different types of behavior in flame retardancy are observed for organosolv and kraft lignins. Costes et al. [159] compared the effect of two lignins obtained by different extraction methods on thermal decomposition and fire behavior of poly(lactic acid) (PLA). It is revealed that Kraft lignin is thermally more stable than the organosolv lignin. The virgin PLA totally degraded starting from 400 °C in a very short temperature interval, but the incorporation of 20 wt% of kraft or organosolv lignin led to the formation of char residue, 17 wt% and 10 wt%, respectively. However, the presence of lignin led to decrease in thermal stability of PLA due to the presence of carboxylic acid and phenolic functions. The results obtained in a cone calorimeter test showed decrease in time to ignition (TTI) of 47 s for kraft lignin and 60 s for organosolv lignin, with respect to neat PLA. Overall, however, an improvement in fire behavior has been observed, in view of decrease in the pHRR and the total heat release, 21% and 23%, respectively, for the kraft lignin and 33% and 30% for the organosolv lignin. These results were achieved because of the formation of a thin char layer and its barrier effect protecting samples. Then, these lignins have been chemically modified by grafting of a phosphorus chemical, 10 wt% of phosphorus. This strategy appeared very effective to decrease the flammability of PLA. By incorporation of the modified lignin at 20 wt%, V0 classification has been obtained in UL 94 test, while PLA-containing unmodified lignin was unrated. The TTI also increased by the chemical modification of lignin due to the presence of ammonium functions, which allowed the dilution of the gas phase by the release of ammonia and phosphorus groups before ignition. Vahabi et al. [160] incorporated lignin in (PHB). They observed that though lignin decreases the pHRR in a microcalorimeter of a combustion test, it degrades PHB when it is used solely, as demonstrated by a fall in the temperature of pHRR. However, its combination with ammonium polyphosphate (APP) overcame this problem. Réti et al. [161] studied the effect of lignin in an intumescent FR system for PLA. The selected intumescent FR system was a mixture of APP at 30 wt% and pentaerythritol (PER) at 10 wt%. They substituted PER, which is a non-bio-based source of carbon, by lignin. Their results revealed that although lignin is not as efficient as PER, an acceptable level of flame retardancy can be obtained. The proof was the high value of limiting oxygen index around 32%, V0 rating in UL94 test, and 47% decrease in pHRR with respect to the neat PLA.

The combination of lignin and FR elements can also be performed via chemical methods by chemical modification of lignin. Due to the presence of reactive phenolic hydroxyl and aliphatic hydroxyl groups on lignin, its chemical modification is possible. Fig. 24.10 displays some types of lignin modification using phosphorus and/or nitrogen, as well as metal ions to bring FR character to lignin [136]. Metal ions can improve the FR effect of phosphorus/nitrogen structures via catalytic charring effect. All in all, it can be understood that lignin is a good candidate for flame-retardancy improvement of polymers.

24.5.2 Nanolignin

As previously mentioned, lignin can be used as a carbon source in FR systems based on polymers and acts in the condensed phase by creating char residue. The main mechanism of action of nanolignin is similar to that of lignin, as their chemical structure is similar. However, nanolignin brings a new dimension to the flame retardancy—reason why we underline its merit for fire retardancy in this chapter. Actually, due to its nanometric size which can be dispersed at nanometric level in polymer, nanolignin is of higher potential compared to lignin. Cholet et al. [162] prepared LNPs from Kraft lignin microparticles (LMPs) using dissolution-precipitation. They modified LNPs using diethyl chlorophosphate and diethyl (2-(triethoxysilyl)ethyl) phosphonate (SiP) and then incorporated them at 5 wt% and 10 wt% into PLA (Fig. 24.11). They investigated flame retardancy of these samples and also compared the results with similar samples containing modified-lignin microparticles. The results obtained from cone calorimetry tests showed that

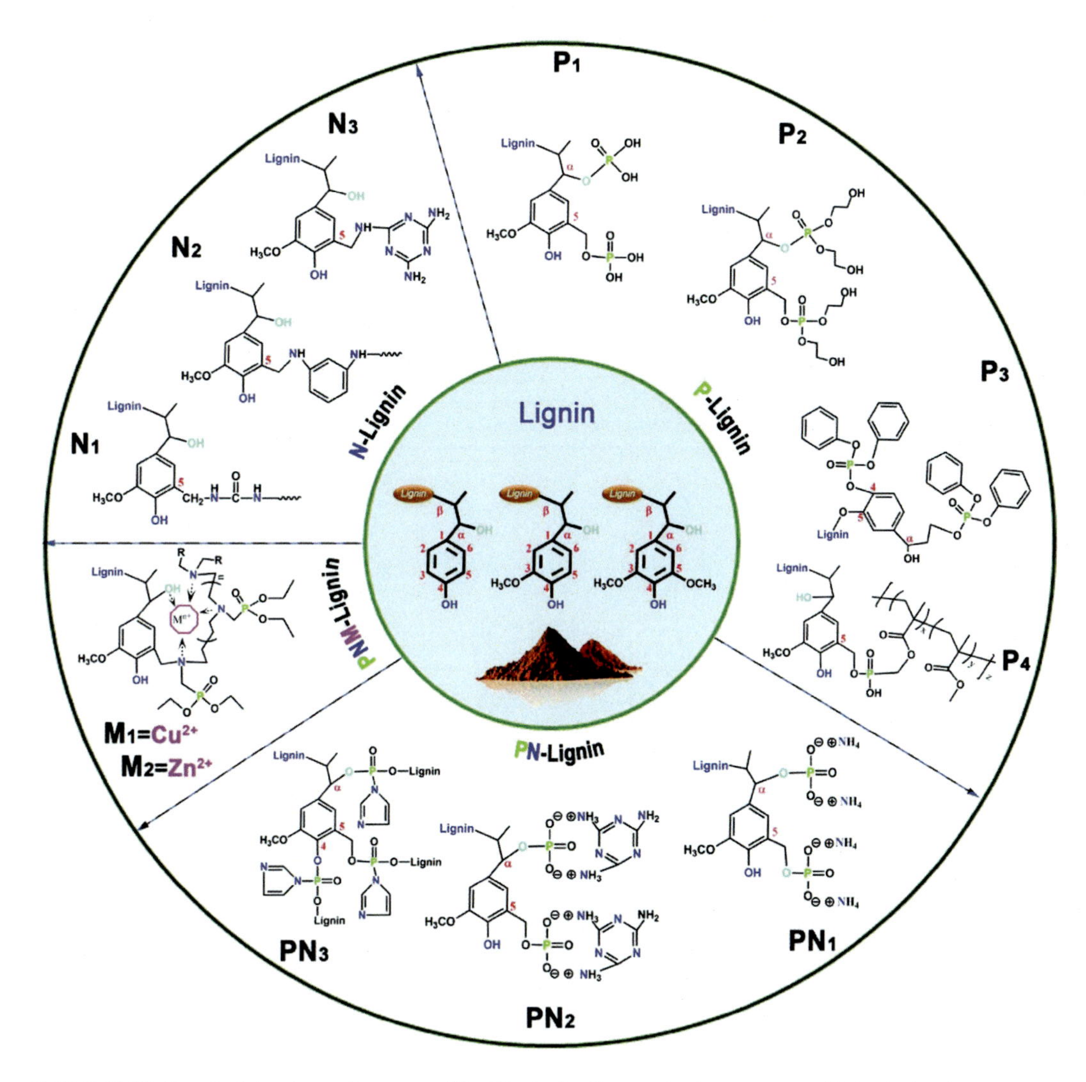

FIGURE 24.10 Chemical structure of some modified lignin for flame retardancy purpose (P_1 to P_4 structures correspond to modified lignin with phosphorus. N_1 to N_3 structures are related to modified lignin with nitrogen. PN_1 to PN_3 involve the chemical modification of lignin with both phosphorus and nitrogen. PNM structure corresponds to a modified lignin with phosphorus, nitrogen, and metal ions.) [136].

both pure LNPs and LMPs, respectively at 5 wt% and 10 wt%, had no positive effect on flame retardancy of PLA (Fig. 24.12). One can see that there was an increase in the pHRR and decrease in the TTI, as evidence.

Interestingly, however, the modification of LNPs with SiP led to decrease in the pHRR and increase in the TTI, for both samples containing 5 wt% and 10 wt% of modified nanolignin (Fig. 24.12). The pHRR has been decreased around 11% and 18% for the PLA containing 5 wt% and 10 wt% of modified nanolignin, respectively, with respect to the neat PLA. The increase in

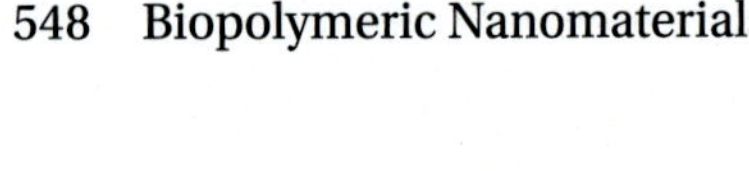

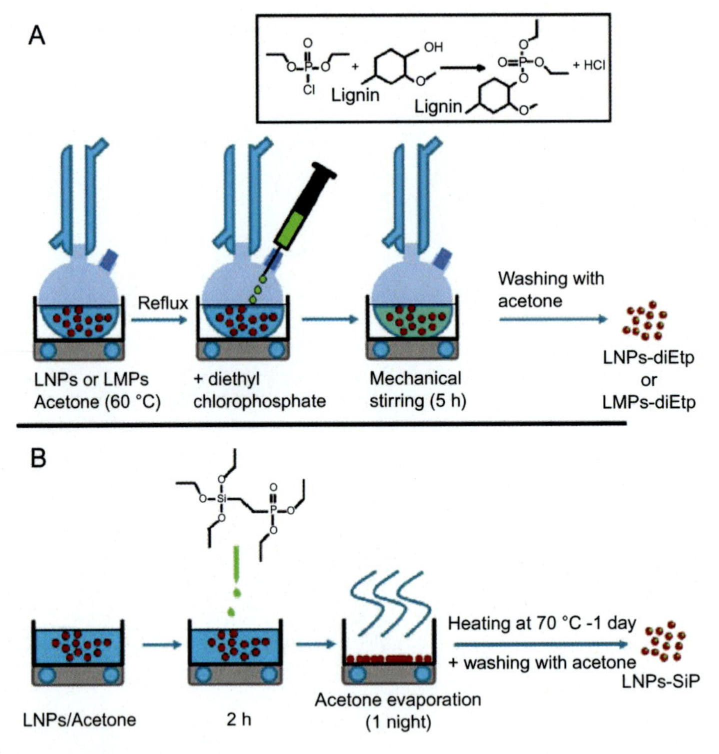

FIGURE 24.11 Chemical modification of micro and nanolignin by (A) diethyl chlorophosphate, and (B) diethyl (2-(triethoxysilyl)ethyl) phosphonate [162].

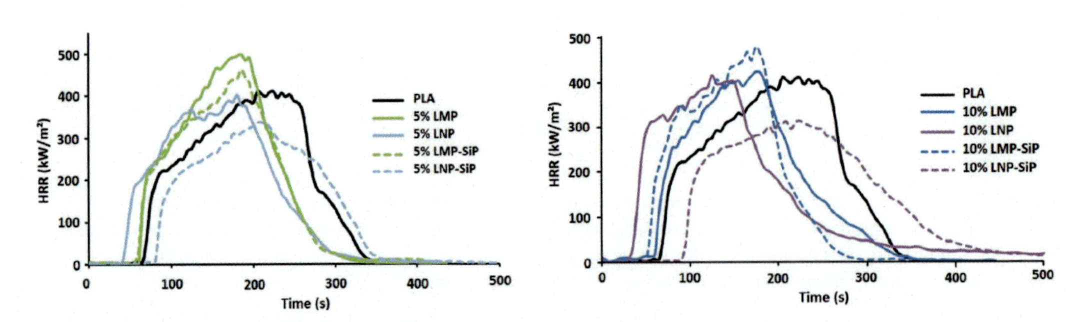

FIGURE 24.12 Curves of heat release rate (HRR) as a function of time for PLA and PLA composites containing micro (LMPs) and nano-lignin (LNPs) and modified (with diethyl (2-(triethoxysilyl)ethyl) phosphonate (SiP)) micro and nano-lignin particles, obtained in cone calorimeter test [162].

the TTI was around 20 s for both samples with respect to pure PLA (68 s). They also showed that the modification with diethyl chlorophosphate of LNPs and LMPs was not beneficial to flame retardancy of PLA.

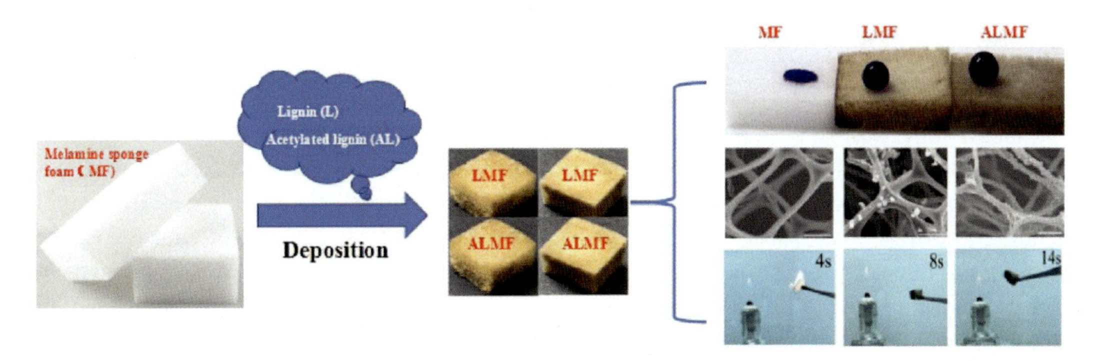

FIGURE 24.13 Deposition of nanolignin (L) and acylated nanolignin (AL) particles on melamine- formaldehyde (MF) sponge foams for improving flame retardancy (LMF: lignin modified MF, ALMF: acylated nanolignin modified MF) [163].

Yu et al. [163] investigated the FR effect of nanolignin and acylated nanolignin deposited on commercially available melamine-formaldehyde (MF) sponge foams (Fig. 24.13). The acylation of nanolignin performed in the mixture solution of acetic acid and acetyl bromide at 55 °C for 3 h. Then, they prepared a solution of nanolignin and/or acylated nanolignin containing 100 mL THF/H_2O (80:20, vol%). Then, 1 cm^3 MF sponges have been immersed in the aforementioned solution for 20 min. After removal of THF and drying the sample, the flame characterization has been performed using a simple ignition of samples by alcohol lamp and measuring the mass remained after combustion. It is revealed that in the case of MF modified by two types of nanolignin, a weak flame appeared for 3 s and 90% of the initial mass was remained. They also showed that the treated MF had hydrophobic properties.

24.6 Conclusions

Natural polymers are the most preferred macromolecules due to their environment-friendly character and abundance in the nature. Although they appeared excellent for a wide range of applications due to their versatile character, their poor mechanical properties are considered as a weak point to be improved. Furthermore, reports on their flame retardancy have not been yet properly integrated into systematic instructions. Thus, still new methodologies/technologies should be examined in the way commercialization of the FR biodegradable polymers. The multiplicity of types of behavior as well as lack of enough experimental data should particularly be noticed in this regard. Lignin, as the second most abundant biopolymer, has a high potential to be used in various applications, ranging from cosmetics and textile to medical and drug systems. In contrast to the general behavior of natural polymers, lignin is a good candidate for FR applications because of being a charring bio-based polymer. Herein, we provided a global view over synthesis, characterization, and applications of nanolignin in diverse fields such as development of hybrid composites, applications in medical and drug systems, enzyme immobilization, surface active substrates, energy storage devices, agricultural uses, environmental uses, anti-corrosive substrate developments, textile industry, and anti-bacterial applications. Besides, some other uses such as cosmetics, Pickering emulsifiers, and packaging industry based on nanolignin were reviewed. Lignin-based FRs have been widely considered in materials science and technology. Moreover, modification of the lignin structure as well as the combination of lignin and FR additive have been widely addressed by the researchers. Nevertheless, FR features of nanolignin have been rarely studied, despite huge potential of this biopolymer. The use of nanolignin brings about some new benefits to flame retardancy improvement with respect to the lignin in micron size. Manipulating the size of nanolignin, modifying its chemical structure and its usage together with FR additives are possible ways for improving flame retardancy of nanolignin such that the amount and compactness of the char residue are changed. Despite such superior properties of nanolignin over lignin in flame retardancy applications, reports on this field are a few. This chapter highlights the importance of working in this field.

References

[1] J. Zhuang, V.M. Payyappalli, A. Behrendt, K. Lukasiewicz, Total Cost of Fire in the United States, Fire Protection Research Foundation, Buffalo, NY, USA, 2017.

[2] F. Seidi, E. Movahedifar, G. Naderi, V. Akbari, F. Ducos, R. Shamsi, et al., Flame retardant polypropylenes: A review, Polymers 12 (2020) 1701.

[3] RR. Hindersinn, Historical Aspects of Polymer Fire Retardance, ACS Publications, Washington DC, 1990.

[4] J. Chen, J. Han, Comparative performance of carbon nanotubes and nanoclays as flame retardants for epoxy composites, Results Phys. 14 (2019) 102481.

[5] N. Mattar, A.R. de Anda, H. Vahabi, E. Renard, V. Langlois, Resorcinol-based epoxy resins hardened with limonene and eugenol derivatives: from the synthesis of renewable diamines to the mechanical properties of biobased thermosets, ACS Sustain. Chem. Eng. 8 (2020) 13064–13075.

[6] L. Costes, F. Laoutid, S. Brohez, P. Dubois, Bio-based flame retardants: when nature meets fire protection, Mater. Sci. Eng. R. Rep. 117 (2017) 1–25.

[7] N. Mandlekar, A. Cayla, F. Rault, S. Giraud, F. Salaün, G. Malucelli, et al. in: An overviewon the use of lignin and its derivatives in fire retardant polymer systems, M. Poletto (Ed.), Lignin-Trends and Applications Intechopen, London, UK, 2018.

[8] S.M. Liu, J.Y. Huang, Z.J. Jiang, C. Zhang, J.Q. Zhao, J. Chen, Flame retardance and mechanical properties of a polyamide 6/polyethylene/surface-modified metal hydroxide ternary composite via a master-batch method, J. Appl. Polym. Sci. 117 (2010) 3370–3378.

[9] J.K. Weng, C. Chapple, The origin and evolution of lignin biosynthesis, New Phytol. 187 (2010) 273–285.

[10] J.J. Liao, N.H. Abd Latif, D. Trache, N. Brosse, M.H. Hussin, Current advancement on the isolation, characterization and application of lignin, Int. J. Biol. Macromol. 162 (2020) 985–1024.

[11] A.G. Vishtal, A. Kraslawski, Challenges in industrial applications of technical lignins, Bio. Resources. 6 (2011) 3547–3568.

[12] H. Tran, EK. Vakkilainnen, The kraft chemical recovery process, Tappi Kraft Pulping Short Course (2008) 1–8.

[13] I.F. Demuner, J.L. Colodette, A.J. Demuner, CM. Jardim, Biorefinery review: wide-reaching products through kraft lignin, Bio. Resources. 14 (2019) 7543–7581.

[14] P. Tomani, The lignoboost process, Cellulose Chem. Technol. 44 (2010) 53.

[15] T. Aro, P. Fatehi, Production and application of lignosulfonates and sulfonated lignin, Chem. Sus. Chem. 10 (2017) 1861–1877.

[16] V.K. Garlapati, A.K. Chandel, S.J. Kumar, S. Sharma, S. Sevda, A.P. Ingle, et al., Circular economy aspects of lignin: towards a lignocellulose biorefinery, Renew. Sustain. Energy Rev. 130 (2020) 109977.

[17] W. Pd, W. Huijgen, R. Linden, U. Hd, J. Snelders, B. Benjelloun-Mlayah, Organosolv fractionation of ligno-cellulosic biomass for an integrated biorefinery, NPT Procestechnol 1 (2015) 10–11.

[18] R. El Hage, N. Brosse, L. Chrusciel, C. Sanchez, P. Sannigrahi, A. Ragauskas, Characterization of milled wood lignin and ethanol organosolv lignin from miscanthus, Polym. Degrad. Stabil. 94 (2009) 1632–1638.

[19] S.N. Obame, I. Ziegler-Devin, R. Safou-Tchima, N. Brosse, Homolytic and heterolytic cleavage of β-ether linkages in hardwood lignin by steam explosion, J. Agric. Food Chem. 67 (2019) 5989–5996.

[20] Q. He, I. Ziegler-Devin, L. Chrusciel, S.N. Obame, L. Hong, X. Lu, et al., Lignin-first integrated steam explosion process for green wood adhesive application, ACS Sustain. Chem. Eng. 8 (2020) 5380–5392.

[21] H. Ji, P. Lv, Mechanistic insights into the lignin dissolution behaviors of a recyclable acid hydrotrope, deep eutectic solvent (DES), and ionic liquid (IL), Green Chem. 22 (2020) 1378–1387.

[22] L. Kazmerski, Renewable and sustainable energy reviews, Renew. Sustain. Energy Rev. 38 (2016) 834–847.

[23] F. Cotana, G. Cavalaglio, A. Nicolini, M. Gelosia, V. Coccia, A. Petrozzi, et al., Lignin as co-product of second generation bioethanol production from ligno-cellulosic biomass, Energy Procedia 45 (2014) 52–60.

[24] J.H. Lora, WG. Glasser, Recent industrial applications of lignin: a sustainable alternative to nonrenewable materials, J. Polym. Environ. 10 (2002) 39–48.

[25] Y. Cao, S.S. Chen, S. Zhang, Y.S. Ok, B.M. Matsagar, K.C.-.W. Wu, et al., Advances in lignin valorization towards bio-based chemicals and fuels: lignin biorefinery, Bioresour. Technol. 291 (2019) 121878.

[26] A. Tolbert, H. Akinosho, R. Khunsupat, A.K. Naskar, AJ. Ragauskas, Characterization and analysis of the molecular weight of lignin for biorefining studies, Biofuel. Bioprod. Biorefin. 8 (2014) 836–856.

[27] J. Sameni, S. Krigstin, D. dos Santos Rosa, A. Leao, M. Sain, Thermal characteristics of lignin residue from industrial processes, Bio. Resources 9 (2014) 725–737.

[28] H. Hatakeyama, Y. Tsujimoto, M. Zarubin, S. Krutov, T. Hatakeyama, Thermal decomposition and glass transition of industrial hydrolysis lignin, J. Therm. Anal. Calorim. 101 (2010) 289–295.

[29] T.C.F. Silva, R.B. Santos, H. Jameel, J.L. Colodette, LA. Lucia, Quantitative molecular structure–pyrolytic energy correlation for hardwood lignins, Energy Fuels 26 (2012) 1315–1322.

[30] R. Sun, J. Tomkinson, GL. Jones, Fractional characterization of ash-AQ lignin by successive extraction with organic solvents from oil palm EFB fibre, Polym. Degrad. Stabil. 68 (2000) 111–119.

[31] S. Wang, H. Lin, B. Ru, W. Sun, Y. Wang, Z. Luo, Comparison of the pyrolysis behavior of pyrolytic lignin and milled wood lignin by using TG–FTIR analysis, J. Anal. Appl. Pyrolysis. 108 (2014) 78–85.

[32] C. Frangville, M. Rutkevičius, A.P. Richter, O.D. Velev, S.D. Stoyanov, VN. Paunov, Fabrication of environ-mentally biodegradable lignin nanoparticles, Chem. Phys. Chem. 13 (2012) 4235.

[33] J.M. Gutiérrez-Hernández, A. Escalante, R.N. Murillo-Vázquez, E. Delgado, F.J. González, G. Toríz, Use of *Agave tequilana*-lignin and zinc oxide nanoparticles for skin photoprotection, J. Photochem. Photobiol. B. Biol. 163 (2016) 156–161.

[34] V.I. Popa, A.-.M. Capraru, S. Grama, T. Malutan, Nanoparticles based on modified lignins with biocide properties, Cellul. Chem. Technol. 45 (2011) 221.

[35] Y. Qian, X. Qiu, X. Zhong, D. Zhang, Y. Deng, D. Yang, et al., Lignin reverse micelles for UV-absorbing and high mechanical performance thermoplastics, Ind. Eng. Chem. Res. 54 (2015) 12025–12030.

[36] M.N. Garcia Gonzalez, M. Levi, S. Turri, G. Griffini, Lignin nanoparticles by ultrasonication and their incorporation in waterborne polymer nanocomposites, J. Appl. Polym. Sci. 134 (2017) 45318.

[37] H. Yin, L. Liu, X. Wang, T. Wang, Y. Zhou, B. Liu, et al., A novel flocculant prepared by lignin nanoparticles-gelatin complex from switchgrass for the capture of *Staphylococcus aureus* and *Escherichia coli*, Coll. Surf. A Physicochem. Eng. Asp. 545 (2018) 51–59.

[38] Q. Lu, M. Zhu, Y. Zu, W. Liu, L. Yang, Y. Zhang, et al., Comparative antioxidant activity of nanoscale lignin prepared by a supercritical antisolvent (SAS) process with non-nanoscale lignin, Food Chem. 135 (2012) 63–67.

[39] P. Gonugunta, S. Vivekanandhan, A.K. Mohanty, M. Misra, A study on synthesis and characterization of biobased carbon nanoparticles from lignin, World J. Nano Sci. Eng. 2 (2012) 148–153.

[40] L. Dai, R. Liu, L.-.Q. Hu, Z.-.F. Zou, C-L. Si, Lignin nanoparticle as a novel green carrier for the efficient delivery of resveratrol, ACS Sustain. Chem. Eng. 5 (2017) 8241–8249.

[41] Z.-.H. Liu, N. Hao, S. Shinde, Y. Pu, X. Kang, A.J. Ragauskas, et al., Defining lignin nanoparticle properties through tailored lignin reactivity by sequential organosolv fragmentation approach (SOFA), Green Chem. 21 (2019) 245–260.

[42] M.-.L. Mattinen, J.J. Valle-Delgado, T. Leskinen, T. Anttila, G. Riviere, M. Sipponen, et al., Enzymatically and chemically oxidized lignin nanoparticles for biomaterial applications, Enzyme. Microb. Technol. 111 (2018) 48–56.

[43] L. Shen, Y. Lin, Q. Du, W. Zhong, Studies on structure–property relationship of polyamide-6/attapulgite nanocomposites, Compos. Sci. Technol. 66 (2006) 2242–2248.

[44] P. Intra, N. Tippayawong, An overview of differential mobility analyzers for size classification of nanometer-sized aerosol particles, Songklanakarin J. Sci. Technol. (2008) 30.

[45] P.S. Chauhan, Lignin nanoparticles: Eco-friendly and versatile tool for new era, Bio. Tech. Rep. 9 (2020) 100374.

[46] D. Feldman, Lignin nanocomposites, J. Macromol. Sci. A. 53 (2016) 382–387.

[47] P. Figueiredo, K. Lintinen, A. Kiriazis, V. Hynninen, Z. Liu, T. Bauleth-Ramos, et al., In vitro evaluation of biodegradable lignin-based nanoparticles for drug delivery and enhanced antiproliferation effect in cancer cells, Biomaterials 121 (2017) 97–108.

[48] M. Yang, W. Zhao, S. Singh, B. Simmons, G. Cheng, On the solution structure of kraft lignin in ethylene glycol and its implication for nanoparticle preparation, Nanoscale Adv. 1 (2019) 299–304.

[49] B. Del Saz-Orozco, M. Oliet, M. Alonso, E. Rojo, F. Rodríguez, Formulation optimization of unreinforced and lignin nanoparticle-reinforced phenolic foams using an analysis of variance approach, Compos. Sci. Technol. 72 (2012) 667–674.

[50] W. Yang, E. Fortunati, F. Dominici, J. Kenny, D. Puglia, Effect of processing conditions and lignin content on thermal, mechanical and degradative behavior of lignin nanoparticles/polylactic (acid) bionanocomposites prepared by melt extrusion and solvent casting, Eur. Polym. J. 71 (2015) 126–139.

[51] Y. Sun, L. Yang, X. Lu, C. He, Biodegradable and renewable poly (lactide)–lignin composites: synthesis, interface and toughening mechanism, J. Mater. Chem. A. 3 (2015) 3699–3709.

[52] Y.-.L. Chung, J.V. Olsson, R.J. Li, C.W. Frank, R.M. Waymouth, S.L. Billington, et al., A renewable lignin-lactide copolymer and application in biobased composites, ACS Sustain. Chem. Eng. 1 (2013) 1231–1238.

[53] C. Jiang, H. He, X. Yao, P. Yu, L. Zhou, D. Jia, Self-crosslinkable lignin/epoxidized natural rubber composites, J. Appl. Polym. Sci. 131 (23) (2014) 41166.

[54] C. Jiang, H. He, H. Jiang, L. Ma, D. Jia, Nano-lignin filled natural rubber composites: preparation and characterization, Exp. Polym. Lett. 7 (5) (2013) 480-493.

[55] H.C. Liu, A.-.T. Chien, B.A. Newcomb, Y. Liu, S. Kumar, Processing, structure, and properties of lignin-and CNT-incorporated polyacrylonitrile-based carbon fibers, ACS Sustain. Chem. Eng. 3 (2015) 1943–1954.

[56] S.S. Nair, S. Sharma, Y. Pu, Q. Sun, S. Pan, J. Zhu, et al., High shear homogenization of lignin to nanolignin and thermal stability of nanolignin-polyvinyl alcohol blends, Chem. Sus. Chem. 7 (2014) 3513–3520.

[57] A.K. Gupta, S. Mohanty, S. Nayak, Influence of addition of vapor grown carbon fibers on mechanical, thermal and biodegradation properties of lignin nanoparticle filled bio-poly (trimethylene terephthalate) hybrid nanocomposites, RSC Adv. 5 (2015) 56028–56036.

[58] D. Kai, S. Jiang, Z.W. Low, XJ. Loh, Engineering highly stretchable lignin-based electrospun nanofibers for potential biomedical applications, J. Mater. Chem. B. 3 (2015) 6194–6204.

[59] W. Yang, E. Fortunati, F. Bertoglio, J. Owczarek, G. Bruni, M. Kozanecki, et al., Polyvinyl alcohol/chitosan hydrogels with enhanced antioxidant and antibacterial properties induced by lignin nanoparticles, Carbohydr. Polym. 181 (2018) 275–284.

[60] M. Zimniewska, J. Batog, Ultraviolet-blocking properties of natural fibres., in: R.M. Kozłowski (Ed.), (Ed.) Handbook of Natural Fibres: Processing and Applications, Elsevier,, Amsterdam, 2012, pp. 141–167.

[61] G. Qi, W. Yang, D. Puglia, H. Wang, P. Xu, W. Dong, et al., Hydrophobic, uv resistant and dielectric polyurethane-nanolignin composites with good reprocessability, Mater. Des. 196 (2020) 109150.

[62] F. Zikeli, V. Vinciguerra, A. D'Annibale, D. Capitani, M. Romagnoli, G. Scarascia Mugnozza, Preparation of lignin nanoparticles from wood waste for wood surface treatment, Nanomaterials 9 (2019) 281.

[63] F. Zikeli, V. Vinciguerra, S. Sennato, G. Scarascia Mugnozza, M. Romagnoli, Preparation of lignin nanoparticles with entrapped essential oil as a bio-based biocide delivery system, ACS Omega 5 (2019) 358–368.

[64] A. Wicki, D. Witzigmann, V. Balasubramanian, J. Huwyler, Nanomedicine in cancer therapy: challenges, opportunities, and clinical applications, J. Control. Release. 200 (2015) 138–157.

[65] A. Rangan, M. Manjula, K. Satyanarayana, R. Menon, Lignin/nanolignin and their biodegradable composites, Biodegrad. Green Compos. (2016) 167–198.

[66] Y. Qian, Y. Deng, X. Qiu, H. Li, D. Yang, Formation of uniform colloidal spheres from lignin, a renewable resource recovered from pulping spent liquor, Green Chem. 16 (2014) 2156–2163.

[67] Y. Li, X. Qiu, Y. Qian, W. Xiong, D. Yang, pH-responsive lignin-based complex micelles: preparation, characterization and application in oral drug delivery, Chem. Eng. J. 327 (2017) 1176–1183.

[68] P.K. Mishra, R. Wimmer, Aerosol assisted self-assembly as a route to synthesize solid and hollow spherical lignin colloids and its utilization in layer by layer deposition, Ultrason. Sonochem. 35 (2017) 45–50.

[69] L. Chen, X. Zhou, Y. Shi, B. Gao, J. Wu, T.B. Kirk, et al., Green synthesis of lignin nanoparticle in aqueous hydrotropic solution toward broadening the window for its processing and application, Chem. Eng. J. 346 (2018) 217–225.

[70] D. Gan, W. Xing, L. Jiang, J. Fang, C. Zhao, F. Ren, et al., Plant-inspired adhesive and tough hydrogel based on Ag-Lignin nanoparticles-triggered dynamic redox catechol chemistry, Nat. Commun. 10 (2019) 1–10.

[71] Y.J. Zhou, Y. Qian, Y.J. Wang, X.Q. Qiu, H.B. Zeng, Bioinspired lignin-polydopamine nanocapsules with strong bioadhesion for long-acting and high-performance natural sunscreens, Biomacromolecules 21 (2020) 3231–3241.

[72] L. Siddiqui, J. Bag, M.D. Seetha, A. Leekha, H. Mishra, et al., Assessing the potential of lignin nanoparticles as drug carrier: synthesis, cytotoxicity and genotoxicity studies, Int. J. Biol. Macromol. 152 (2020) 786–802.

[73] Y. Deng, H. Zhao, Y. Qian, L. Lü, B. Wang, X. Qiu, Hollow lignin azo colloids encapsulated avermectin with high anti-photolysis and controlled release performance, Ind. Crop. Prod. 87 (2016) 191–197.

[74] Y. Li, M. Zhou, Y. Pang, X. Qiu, Lignin-based microsphere: preparation and performance on encapsulating the pesticide avermectin, ACS Sustain. Chem. Eng. 5 (2017) 3321–3328.

[75] D. Yiamsawas, G. Baier, E. Thines, K. Landfester, F.R. Wurm, et al., Biodegradable lignin nanocontainers, RSC Adv. 4 (2014) 11661.

[76] B.O. Okesola, DK. Smith, Applying low-molecular weight supramolecular gelators in an environmental setting – self-assembled gels as smart materials for pollutant removal, Chem. Soc. Rev. 45 (2016) 4226–4251.

[77] D. Xiao, W. Ding, J. Zhang, Y. Ge, Z. Wu, Z. Li, Fabrication of a versatile lignin-based nano-trap for heavy metal ion capture and bacterial inhibition, Chem. Eng. J. 358 (2019) 310–320.

[78] J. Azimvand, K. Didehban, S. Mirshokraie, Safranin-O removal from aqueous solutions using lignin nanoparticle-g-polyacrylic acid adsorbent: synthesis, properties, and application, Adsorp. Sci. Technol. 36 (2018) 1422–1440.

[79] Y. Yang, Y. Deng, Z. Tong, C. Wang, Renewable lignin-based xerogels with self-cleaning properties and superhydrophobicity, ACS Sustain. Chem. Eng. 2 (2014) 1729–1733.

[80] Z. Li, Y. Ge, L. Wan, Fabrication of a green porous lignin-based sphere for the removal of lead ions from aqueous media, J. Hazard. Mater 285 (2015) 77–83.

[81] X. Luo, C. Liu, J. Yuan, X. Zhu, S. Liu, Interfacial solid-phase chemical modification with Mannich reaction and Fe(III) chelation for designing lignin-based spherical nanoparticle adsorbents for highly efficient removal of low concentration phosphate from water, ACS Sustain. Chem. Eng. 5 (2017) 6539–6547.

[82] R. Ou, S. Shi, J. Ding, D. Wang, S. Ahmad, H. Yu, Lignin nanoparticles: synthesis, characterization and corrosion protection performance, New J. Chem. 42 (2018) 3415–3425.

[83] Z. Zhang, S. Yi, Y. Wei, H. Bian, R. Wang, Y. Min, Lignin nanoparticle-coated celgard separator for high-performance lithium-sulfur batteries, Polymers 11 (2019) 1946.

[84] B. Yu, Z. Chang, Y. Zhang, C. Wang, Preparation and formation mechanism of size-controlled lignin based microsphere by reverse phase polymerization, Mater. Chem. Phys. 203 (2018) 97–105.

[85] B. Wang, D. Sun, H.-.M. Wang, T.-.Q. Yuan, R-C. Sun, Green and facile preparation of regular lignin nanoparticles with high yield and their natural broad-spectrum sunscreens, ACS Sustain. Chem. Eng. 7 (2019) 2658–2666.

[86] D. Tian, J. Hu, J. Bao, R.P. Chandra, J.N. Saddler, C. Lu, Lignin valorization: lignin nanoparticles as high-value bio-additive for multifunctional nanocomposites, Biotechnol. Biofuel. 10 (2017) 192.

[87] T. Ju, Z. Zhang, Y. Li, X. Miao, J. Ji, Continuous production of lignin nanoparticles using a microchannel reactor and its application in UV-shielding films, RSC Adv. 9 (2019) 24915–24921.

[88] A. Pongpeerapat, K. Itoh, Y. Tozuka, K. Moribe, T. Oguchi, K. Yamamoto, Formation and stability of drug nanoparticles obtained from drug/PVP/SDS ternary ground mixture, J. Drug Deliv. Sci. Technol. 14 (2004) 441–447.

[89] Q. Xing, P. Buono, D. Ruch, P. Dubois, L. Wu, W-J. Wang, Biodegradable UV-blocking films through core-shell lignin–melanin nanoparticles in poly(butylene adipate-co-terephthalate), ACS Sustain. Chem. Eng. 7 (2019) 4147–4157.

[90] W. Yang, F. Dominici, E. Fortunati, J.M. Kenny, D. Puglia, Effect of lignin nanoparticles and masterbatch procedures on the final properties of glycidyl methacrylate-G-poly (lactic acid) films before and after accelerated UV weathering, Ind. Crop. Prod. 77 (2015) 833–844.

[91] Y. Qian, X. Zhong, Y. Li, X. Qiu, Fabrication of uniform lignin colloidal spheres for developing natural broad-spectrum sunscreens with high sun protection factor, Ind. Crop. Prod. 101 (2017) 54–60.

[92] P. Widsten, T. Tamminen, T. Liitia, Natural sunscreens based on nanoparticles of modified kraft lignin (catlignin), ACS Omega 5 (2020) 13438–13446.

[93] S.C. Lee, E. Yoo, S.H. Lee, K. Won, Preparation and application of light-colored lignin nanoparticles for broad-spectrum sunscreens, Polymers (Basel) (2020) 12.

[94] H. Trevisan, CA. Rezende, Pure, stable and highly antioxidant lignin nanoparticles from elephant grass, Ind. Crop. Prod. 145 (2020) 112105.

[95] W. Boerjan, J. Ralph, M. Baucher, Lignin biosynthesis, Annu. Rev. Plant Biol. 54 (2003) 519–546.

[96] I. Spiridon, CE. Tanase, Design, characterization and preliminary biological evaluation of new lignin-PLA biocomposites, Int. J. Biol. Macromol. 114 (2018) 855–863.

[97] X. He, F. Luzi, X. Hao, W. Yang, L. Torre, Z. Xiao, et al., Thermal, antioxidant and swelling behaviour of transparent polyvinyl (alcohol) films in presence of hydrophobic citric acid-modified lignin nanoparticles, Int. J. Biol. Macromol. 127 (2019) 665–676.

[98] W. Gao, P. Fatehi, Lignin for polymer and nanoparticle production: current status and challenges, Can. J. Chem. Eng. 97 (2019) 2827–2842.

[99] Y. Qian, Q. Zhang, X. Qiu, S. Zhu, CO_2-responsive diethylaminoethyl-modified lignin nanoparticles and their application as surfactants for CO_2/N_2-switchable Pickering emulsions, Green Chem. 16 (2014) 4963–4968.

[100] M. Ago, S. Huan, M. Borghei, J. Raula, E. Kauppinen, O. Rojas, High-throughput synthesis of lignin particles (approximately 30 nm to approximately 2 mum) via aerosol flow reactor: size fractionation and utilization in Pickering emulsions, ACS Appl. Mater. Interfaces 8 (2016) 23302–23310.

[101] T.E. Nypelö, C.A. Carrillo, OJ. Rojas, Lignin supracolloids synthesized from (W/O) microemulsions: use in the interfacial stabilization of Pickering systems and organic carriers for silver metal, Soft Matter 11 (2015) 2046–2054.

[102] K.S. Silmore, C. Gupta, NR. Washburn, Tunable Pickering emulsions with polymer-grafted lignin nanoparticles (PGLNs), J. Coll. Interface Sci. 466 (2016) 91–100.

[103] C. Stone, F.M. Windsor, M. Munday, I. Durance, Natural or synthetic—how global trends in textile usage threaten freshwater environments, Sci. Total Environ. 718 (2020) 134689.

[104] S.J. Juikar, V. Nadanathangam, Microbial production of nanolignin from cotton stalks and its application onto cotton and linen fabrics for multifunctional properties, Waste Biomass Valori. 11 (2019) 6073–6083.

[105] S.J. Juikar, N. Vigneshwaran, Microbial production of coconut fiber nanolignin for application onto cotton and linen fabrics to impart multifunctional properties, Surf. Interfaces 9 (2017) 147–153.

[106] V. Glombikova, P. Komarkova, E. Hercikova, A. Havelka, How high-loft textile thermal insulation properties depend on compressibility, Autex Res. J. 20 (2020) 338–343.

[107] G. Song, S. Mandal, RM. Rossi, Effects of various factors on performance of thermal protective clothing, in: G. Song, S. Mandal, R.M. Rossi (Eds.), Thermal Protective Clothing for Firefighters, Elsevier Ltd., Amsterdam, 2017, pp. 163–182.

[108] R. Kozlowski, M.J. Zimniewska, Batog Cellulose fibre textiles containing nanolignins, a method of applying nanolignins onto textiles and the use of nanolignins in textile production, French Patent No. WO2008140337A1. (2008).

[109] M. Zimniewska, R. Kozłowski, J. Batog, Nanolignin modified linen fabric as a multifunctional product, Mol. Cryst. Liq. Cryst. 484 (2008) 43 /[409]-50/[16].

[110] P. Morganti, M-B. Coltelli, A new carrier for advanced cosmeceuticals, Cosmetics 6 (1) (2019) 10.

[111] P. Morganti, P. Febo, M. Cardillo, G. Donnarumma, A. Baroni, Chitin nanofibril and nanolignin: natural polymers of biomedical interest, J. Clin. Cosmet. Dermatol. 1 (2017).

[112] S. Kaul, N. Gulati, D. Verma, S. Mukherjee, U. Nagaich, Role of nanotechnology in cosmeceuticals: a review of recent advances, J. Pharm 2018, (2018) 3420204.

[113] S. Danti, L. Trombi, A. Fusco, B. Azimi, A. Lazzeri, P. Morganti, et al., Chitin nanofibrils and nanolignin as functional agents in skin regeneration, Int. J. Mol. Sci. (2019) 20.

[114] S.R. Yearla, K. Padmasree, Preparation and characterisation of lignin nanoparticles: evaluation of their potential as antioxidants and UV protectants, J. Exp. Nanosci. 11 (2015) 289–302.

[115] W. Yang, E. Fortunati, D. Gao, G.M. Balestra, G. Giovanale, X. He, et al., Valorization of acid isolated high yield lignin nanoparticles as innovative antioxidant/antimicrobial organic materials, ACS Sustain. Chem. Eng. 6 (2018) 3502–3514.

[116] S. Gupta, Market Leader - Eco-friendly Food Packaging Market. USA, MarketandMarket, 2020. https://www.marketsandmarkets.com/ResearchInsight/eco-friendly-food-packaging-market.asp.

[117] W. Yang, G. Qi, H. Ding, P. Xu, W. Dong, X. Zhu, et al., Biodegradable poly (lactic acid)-poly (ε-caprolactone)-nanolignin composite films with excellent flexibility and UV barrier performance, Compos. Commun. 22 (23) (2020) 100497.

[118] E. Lizundia, I. Armentano, F. Luzi, F. Bertoglio, E. Restivo, L. Visai, et al., Synergic effect of nanolignin and metal oxide nanoparticles into poly(l-lactide) bionanocomposites: material properties, antioxidant activity, and antibacterial performance, ACS Appl. Bio. Mater 3 (2020) 5263–5274.

[119] B. Del Saz-Orozco, M. Oliet, M.V. Alonso, E. Rojo, F. Rodríguez, Formulation optimization of unreinforced and lignin nanoparticle-reinforced phenolic foams using an analysis of variance approach, Compos. Sci. Technol. 72 (2012) 667–674.

[120] Y. Ge, Q. Wei, Z. Li, Preparation and evaluation of the free radical scavenging activities of nanoscale lignin biomaterials, Bio. Resource 9 (2014) 6699–6706.

[121] W. Yang, E. Fortunati, F. Dominici, G. Giovanale, A. Mazzaglia, G.M. Balestra, et al., Synergic effect of cellulose and lignin nanostructures in PLA based systems for food antibacterial packaging, Eur. Polym. J. 79 (2016) 1–12.

[122] A. Guinault, C. Sollogoub, S. Domenek, A. Grandmontagne, V. Ducruet, Influence of crystallinity on gas barrier and mechanical properties of PLA food packaging films, Int. J. Mater. Form 3 (2010) 603–606.

[123] W. Yang, J.S. Owczarek, E. Fortunati, M. Kozanecki, A. Mazzaglia, G.M. Balestra, et al., Antioxidant and antibacterial lignin nanoparticles in polyvinyl alcohol/chitosan films for active packaging, Ind. Crop. Prod. 94 (2016) 800–811.

[124] M. Kahraman, G. Bayramoglu, N. Kayamanapohan, A. Gungor, α-Amylase immobilization on functionalized glass beads by covalent attachment, Food Chem. 104 (2007) 1385–1392.

[125] T. Tay, E. Kose, R. Kecili, R. Say, Design and preparation of nano-lignin peroxidase (nanoLiP) by protein block copolymerization approach, Polymers (Basel) 8 (6) (2016) 223.

[126] D. Piccinino, E. Capecchi, L. Botta, P. Bollella, R. Antiochia, M. Crucianelli, et al., Layer by layer supported laccase on lignin nanoparticles catalyzes the selective oxidation of alcohols to aldehydes, Catal. Sci. Technol. 9 (2019) 4125–4134.

[127] W. Gong, Z. Ran, F. Ye, G. Zhao, Lignin from bamboo shoot shells as an activator and novel immobilizing support for alpha-amylase, Food Chem. 228 (2017) 455–462.

[128] M.H. Sipponen, M. Farooq, J. Koivisto, A. Pellis, J. Seitsonen, M. Osterberg, Spatially confined lignin nanospheres for biocatalytic ester synthesis in aqueous media, Nat. Commun. 9 (2018) 2300.

[129] E. Capecchi, D. Piccinino, I. Delfino, P. Bollella, R. Antiochia, R. Saladino, Functionalized tyrosinase-lignin nanoparticles as sustainable catalysts for the oxidation of phenols, Nanomaterials (Basel) 8 (2018) 438.

[130] E. Capecchi, D. Piccinino, E. Tomaino, B.M. Bizzarri, F. Polli, R. Antiochia, et al., Lignin nanoparticles are renewable and functional platforms for the concanavalin a oriented immobilization of glucose oxidase–peroxidase in cascade bio-sensing, RSC Adv. 10 (2020) 29031–29042.

[131] G. Cazacu, M. Capraru, VI. Popa, Advances concerning lignin utilization in new materials, in: S Thomas, P.M. Visakh, A.P. Mathew (Eds.), Advances in Natural Polymers, Springer, Cham, 2013, pp. 255–312.

[132] H. Setälä, H.-.L. Alakomi, A. Paananen, G.R. Szilvay, M. Kellock, M. Lievonen, et al., Lignin nanoparticles modified with tall oil fatty acid for cellulose functionalization, Cellulose 27 (2019) 273–284.

[133] A. Alzagameem, S.E. Klein, M. Bergs, X.T. Do, I. Korte, S. Dohlen, et al., Antimicrobial activity of lignin and lignin-derived cellulose and chitosan composites against selected pathogenic and spoilage microorganisms, Polymers (Basel) 11 (4) (2019) 670.

[134] W. Yang, E. Fortunati, F. Bertoglio, J.S. Owczarek, G. Bruni, M. Kozanecki, et al., Polyvinyl alcohol/chitosan hydrogels with enhanced antioxidant and antibacterial properties induced by lignin nanoparticles, Carbohydr. Polym. 181 (2018) 275–284.

[135] CE. Hobbs, Recent advances in bio-based flame retardant additives for synthetic polymeric materials, Polymers 11 (2019) 224.

[136] H. Yang, B. Yu, X. Xu, S. Bourbigot, H. Wang, P. Song, Lignin-derived bio-based flame retardants toward high-performance sustainable polymeric materials, Green Chem. 22 (2020) 2129–2161.

[137] A. Cayla, F. Rault, S. Giraud, F. Salaün, V. Fierro, A. Celzard, PLA with intumescent system containing lignin and ammonium polyphosphate for flame retardant textile, Polymers 8 (2016) 331.

[138] F. Bertini, M. Canetti, A. Cacciamani, G. Elegir, M. Orlandi, L. Zoia, Effect of ligno-derivatives on thermal properties and degradation behavior of poly (3-hydroxybutyrate)-based biocomposites, Polym. Degrad. Stabil. 97 (2012) 1979–1987.

[139] L. Dumazert, D. Rasselet, B. Pang, B. Gallard, S. Kennouche, JM. Lopez-Cuesta, Thermal stability and fire reaction of poly (butylene succinate) nanocomposites using natural clays and FR additives, Polym. Adv. Technol. 29 (2018) 69–83.

[140] L. Ferry, G. Dorez, A. Taguet, B. Otazaghine, J. Lopez-Cuesta, Chemical modification of lignin by phosphorus molecules to improve the fire behavior of polybutylene succinate, Polym. Degrad. Stabil. 113 (2015) 135–143.

[141] A. Cayla, F. Rault, S. Giraud, F. Salaün, R. Sonnier, L. Dumazert, Influence of ammonium polyphosphate/lignin ratio on thermal and fire behavior of biobased thermoplastic: the case of polyamide 11, Materials 12 (2019) 1146.

[142] N. Mandlekar, G. Malucelli, A. Cayla, F. Rault, S. Giraud, F. Salaün, et al., Fire retardant action of zinc phosphinate and polyamide 11 blend containing lignin as a carbon source, Polym. Degrad. Stabil. 153 (2018) 63–74.

[143] B. Prieur, M. Meub, M. Wittemann, R. Klein, S. Bellayer, G. Fontaine, et al., Phosphorylation of lignin to flame retard acrylonitrile butadiene styrene (abs), Polym. Degrad. Stabil. 127 (2016) 32–43.

[144] P. Song, Z. Cao, S. Fu, Z. Fang, Q. Wu, J. Ye, Thermal degradation and flame retardancy properties of ABS/lignin: effects of lignin content and reactive compatibilization, Thermochim. Acta. 518 (2011) 59–65.

[145] Y. Zhang, Q. Zhao, L. Li, R. Yan, J. Zhang, J. Duan, et al., Synthesis of a lignin-based phosphorus-containing flame retardant and its application in polyurethane, RSC Adv. 8 (2018) 32252–32261.

[146] H. Vahabi, H. Rastin, E. Movahedifar, K. Antoun, N. Brosse, MR. Saeb, Flame retardancy of bio-based polyurethanes: opportunities and challenges, Polymers 12 (2020) 1234.

[147] W. Xing, H. Yuan, H. Yang, L. Song, Y. Hu, Functionalized lignin for halogen-free flame retardant rigid polyurethane foam: preparation, thermal stability, fire performance and mechanical properties, J. Polym. Res. 20 (2013) 234.

[148] F. Laoutid, V. Duriez, L. Brison, S. Aouadi, H. Vahabi, P. Dubois, Synergistic flame-retardant effect between lignin and magnesium hydroxide in poly (ethylene-co-vinyl acetate), Flame Retardancy Therm. Stabil. Mater 2 (2019) 9–18.

[149] Y. Yu, S. Pa, C. Jin, S. Fu, L. Zhao, Q. Wu, et al., Catalytic effects of nickel (cobalt or zinc) acetates on thermal and flammability properties of polypropylene-modified lignin composites, Ind, Eng. Chem. Res. 51 (2012) 12367–12374.

[150] A. De Chirico, M. Armanini, P. Chini, G. Cioccolo, F. Provasoli, G. Audisio, Flame retardants for polypropylene based on lignin, Polym. Degrad. Stabil. 79 (2003) 139–145.

[151] M. Canetti, F. Bertini, Influence of the lignin on thermal degradation and melting behaviour of poly (ethylene terephthalate) based composites, e-Polymers 9 (1) (2009) 049.

[152] X. Wang, S.-.L. Ji, X.-.Q. Wang, H.-.Y. Bian, L.-.R. Lin, H.-.Q. Dai, et al., Thermally conductive, super flexible and flame-retardant BN-OH/PVA composite film reinforced by lignin nanoparticles, J. Mater. Chem. C. 7 (2019) 14159–14169.

[153] S. Farishi, A. Rifathin, BF. Ramadhoni, Phosphorus/nitrogen grafted lignin as a biobased flame retardant for unsaturated polyester resin, in: Proc. International Manufacturing Engineering Conference & The Asia Pacific Conference on Manufacturing Systems, Springer, 2019, pp. 429–434.

[154] D. Liang, X. Zhu, P. Dai, X. Lu, H. Guo, H. Que, et al., Preparation of a novel lignin-based flame retardant for epoxy resin, Mater. Chem. Phys. (2020) 124101.

[155] G.P. Mendis, S.G. Weiss, M. Korey, C.R. Boardman, M. Dietenberger, J.P. Youngblood, et al., Phosphorylated lignin as a halogen-free flame retardant additive for epoxy composites, Green Mater 4 (2016) 150–159.

[156] S. Zhou, R. Tao, P. Dai, Z. Luo, M. He, Two-step fabrication of lignin-based flame retardant for enhancing the thermal and fire retardancy properties of epoxy resin composites, Polym. Compos. 41 (2020) 2025–2035.

[157] L. Costes, F. Laoutid, F. Khelifa, G. Rose, S. Brohez, C. Delvosalle, et al., Cellulose/phosphorus combinations for sustainable fire retarded polylactide, Eur. Polym. J. 74 (2016) 218–228.

[158] R. Zhang, X. Xiao, Q. Tai, H. Huang, J. Yang, Y. Hu, The effect of different organic modified montmorillonites (OMMTs) on the thermal properties and flammability of PLA/MCAPP/lignin systems, J. Appl. Polym. Sci. 127 (2013) 4967–4973.

[159] L. Costes, F. Laoutid, M. Aguedo, A. Richel, S. Brohez, C. Delvosalle, et al., Phosphorus and nitrogen derivatization as efficient route for improvement of lignin flame retardant action in PLA, Eur. Polym. J. 84 (2016) 652–667.

[160] H. Vahabi, L. Michely, G. Moradkhani, V. Akbari, M. Cochez, C. Vagner, et al., Thermal stability and flammability behavior of Poly (3-hydroxybutyrate)(PHB) based composites, Materials 12 (2019) 2239.

[161] C. Reti, M. Casetta, S. Duquesne, S. Bourbigot, R. Delobel, Flammability properties of intumescent PLA including starch and lignin, Polym. Adv. Technol. 19 (2008) 628–635.

[162] B. Chollet, J.-.M. Lopez-Cuesta, F. Laoutid, L. Ferry, Lignin nanoparticles as a promising way for enhancing lignin flame retardant effect in polylactide, Materials 12 (2019) 2132.

[163] H. Yu, W. Zhan, Y. Liu, Engineering lignin nanoparticles deposition on melamine sponge skeleton for absorbent and flame retardant materials, Waste Biomass Valoriz 11 (2020) 4561–4569.

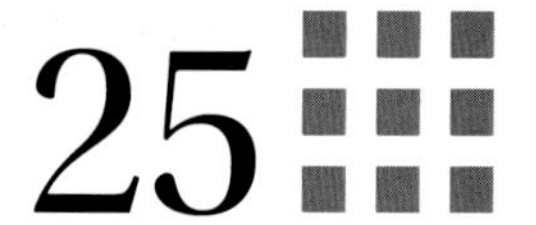

Natural polymer-based magnetic nanohybrids toward biomedical applications

Rachel Auzély-Velty and Anna Szarpak

UNIVERSITY OF GRENOBLE ALPES, CNRS, CERMAV, GRENOBLE, FRANCE

Chapter outline

25.1 Introduction

Magnetic hybrid nanoparticles (NPs) have gained much attention in the field of biomedical applications. The nanohybrids composed of magnetic core coated with polymer chains are highly attractive because of the combined properties that can be offered by both components.

Magnetic core may guide drugs to a site of interest in the body with the aid of a magnetic field, it can be used as a contrast agent for magnetic resonance imaging (MRI) or can be heated upon application of an alternating magnetic field (AMF) to destroy cancer cells in hyperthermia treatment. Most commonly investigated magnetic nanoparticles (MNPs) for biomedical applications are magnetite (Fe_3O_4), maghemite (γ-Fe_2O_3), and ferrites (mixed oxides of iron and other transition metals). The MNPs smaller than 30 nm exhibit superparamagnetic behavior, which means that in the absence of an external magnetic field, these NPs have zero magnetization and

Biopolymeric Nanomaterials: Fundamentals and Applications. DOI: https://doi.org/10.1016/B978-0-12-824364-0.00023-X

less tendency to agglomerate. Such superparamagnetic iron oxide NPs (SPIONs) can form stable colloidal suspensions that can be crucial for biomedical applications, especially *in vivo* [1,2].

The successful application of MNPs in biological media is directly dependent on the nature of the magnetic surface, biocompatibility, and biodegradability. The proper modification of the MNP surfaces should provide stability against aggregation at physiological pH and ionic strength, reduce capture by the body's immune system, and provide reactive sites suitable for further binding of drugs and biological ligands [3]. Although many studies focused on the functionalization of MNPs, with organic materials (synthetic polymers), noble metals (gold), or oxide materials (silica or alumina) to achieve good dispersion, nanohybrids containing natural polymers such as polysaccharides or proteins have attracted considerable attention [4,5].

Polysaccharides are a large family of polymeric carbohydrates comprised of long chains of monosaccharide units bound together by glycosidic linkages. They are derived from renewable resources, such as plants, animals and microorganisms, and are therefore widely distributed in nature. Polysaccharides also have attractive biological and chemical properties for biomaterial design, for example, biocompatibility, biodegradability, low toxicity, and high chemical reactivity. The hydrophilic polysaccharide shell can reduce aggregation of MNPs, increase the stability, and their storage life either by electrostatic or steric repulsions. Due to the presence of different reactive groups on the macromolecular chains, they can be easily modified with specific components, that is, drugs and/or targeting ligands, and release their content during shell degradation. The most common polysaccharides used for modifying the surface of MNPs are chitosan, dextran, alginate, carrageenan, hyaluronic acid (HA), heparin, and pullulan.

Proteins are naturally occurring macromolecules derived from plants and animals, easily accessible renewable resources. Their primary structure is composed of amino acids connected to one another via peptide bonds. They are biocompatible, generally biodegradable, and may also exhibit biological activity. There are a variety of different protein polymers suitable for the development of biohybrid NPs. Proteins are very important candidates to improve the *in-vivo* fate of MNPs, as they can be used for site-specific active targeting, as well as the enhancement of colloidal stability in physiological conditions.

In this chapter, we describe the most common preparation methods of polysaccharide-, and protein-based hybrid MNPs used for biomedical applications. The following sections will describe different polysaccharides and proteins studied for the fabrication of hybrid NPs together with various approaches used for their conjugation to the magnetic surface. The formation of self-assembled multiparticle magnetic core stabilized with amphiphilic polysaccharides and proteins will be also described. Finally, the achievement in MRI diagnosis, drug delivery, and hyperthermia treatment will be demonstrated.

25.2 Natural polymers used for the modification of MNPs

The MNPs are chemically active, oxidize easily in air, resulting in a loss of magnetism. The coating of a magnetic nucleus with polymer offers the protection from oxidation but also may reduce toxicity, aggregation, increase stability, and extend the storage life. The biocompatible natural

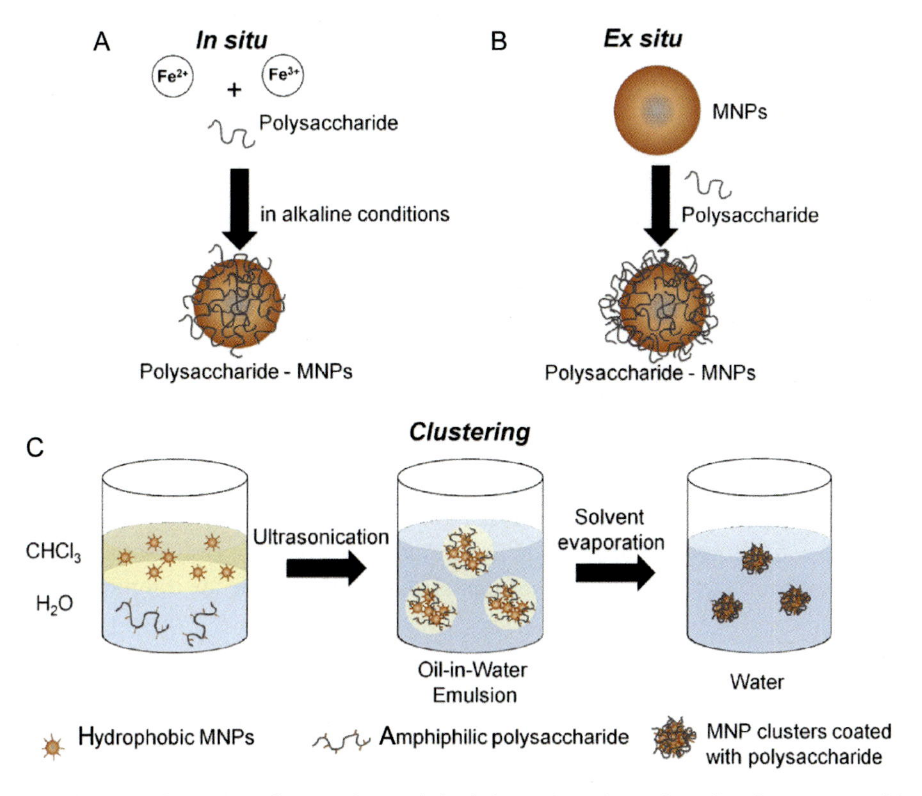

FIGURE 25.1 Schematic illustration of magnetic nanohybrids formation using polysaccharides: *in-situ* modification (A), *ex-situ* coating (B), and emulsion technique for nanoclusters formation from hydrophobic MNPs and amphiphilic polysaccharide (C).

polymers commonly used and preferred for coating/functionalizing the surface of MNPs for biomedical applications include polysaccharides and proteins. Due to the presence of amino, hydroxyl, carboxyl functions on their chains, this natural coat can be used for further functionalization to bind drug molecule, dyes, peptides, antibodies, etc. [6]

25.2.1 Polysaccharides

Generally, the MNPs modification with polysaccharides can be obtained in three different ways: (1) *in situ*, with polymers being added during NPs coprecipitation method; (2) *ex situ*, by direct coating of NP surface, or (3) by emulsion technique allowing stabilization of magnetic nanoassemblies.

1. The polysaccharide stabilized MNP surface can be prepared in one step by the coprecipitation method, by adding a polymer solution into the system. In the typical synthesis, a polymer solution is mixed with ferrous ($FeCl_2$) and ferric ($FeCl_3$) salts followed by alkaline treatment (NaOH or NH_4OH) (Fig. 25.1A)

Table 25.1 Polysaccharides commonly employed for the preparation of hybrid magnetic nanoparticles.

Polysaccharide	Natural source	Reactive groups	Charge	Reference
Hyaluronic acid	Animal tissues, produced by bacteria *Streptococcus zooepidemicus*	–COOH, –OH	Anionic	[7]
Alginate	Cell walls of brown algae	–COOH, –OH		[8]
Carrageenans	Red seaweeds	–OH, $-OSO_3H$		[9]
Heparin	Animal tissues	–OH, $-OSO_3H$		[10]
Chitosan	Exoskeletons of shrimps and other crustaceans	–OH, $-NH_2$	Cationic	[11]
Dextran	Produced by lactic acid bacteria	–OH	Neutral	[12]
Pullulan	Produced by fungus *Aureobasidium pullulans*	–OH		[13]

2. Once MNPs are synthesized, they can be stabilized by simple dispersion of NPs in the solution of polysaccharide under stirring or via ultrasonication (Fig. 25.1B)
3. In the emulsion technique (oil-in-water), hydrophobic MNPs are enclosed in nanodroplets of organic solvent dispersed in an aqueous phase containing an amphiphilic polymer (Fig. 25.1C)

Table 25.1 summarizes the common polysaccharides used for the coating of MNPs. Most of them are either negatively charged or neutral; chitosan is the only natural polymer with positive charges.

25.2.1.1 Hyaluronic acid

HA is a naturally occurring linear polysaccharide composed of disaccharide unit of β-D-glucuronic acid and N-acetyl-β-D-glucosamine, linked together by alternating β-1,3 and β-1,4 glycosidic bonds. HA is known to be a principal constituent of connective tissues and the most widespread glycosaminoglycan (GAG) of the extracellular matrix where it plays an essential role in biological processes, such as cell adhesion, migration, and proliferation [14]. By exploitation of its available carboxylic and hydroxyl groups, it can be easily modified and cross-linked to obtain different physicochemical properties. This biodegradable, biocompatible, nontoxic, and nonimmunogenic natural polymer has been recognized as an important building block for the engineering of new biomaterials for various biomedical applications in tissue engineering and regenerative medicine [15–18]. HA is also identified as a targeting ligand based on its recognition by the CD44 receptor overexpressed on the surface of some tumor cells. Therefore, HA has been widely used as a cancer-targeting moiety for imaging [19,20], gene delivery [21,22], and drug delivery [23–25] applications.

MNPs can be modified with HA either via covalent bond formation or electrostatic interactions.

El-Dakdouki et al. used covalent conjugation of low-molecular-weight HA to amine-containing SPIONs for MRI and doxorubicin (DOX) delivery. HA (31 kg/mol) was coupled using

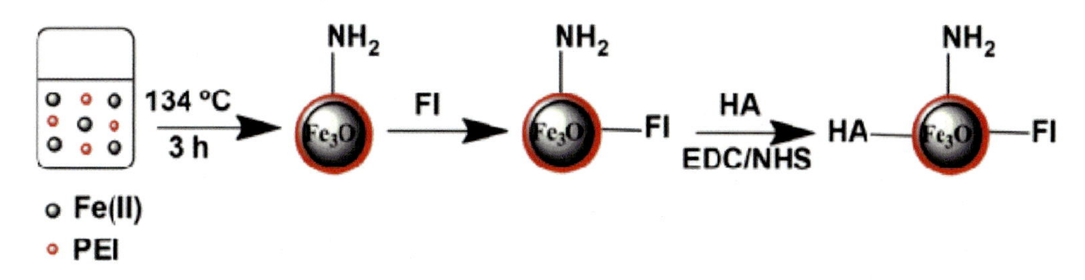

FIGURE 25.2 Schematic representation of the synthesis of Fe_3O_4-PEI-FI-HA nanoparticles. "Reprinted from Biomaterials, 35(11), Li et al., Hyaluronic acid-modified hydrothermally synthesized iron oxide nanoparticles for targeted tumor MR imaging, 3666–77. Copyright (2014), with permission from Elsevier".

2-chloro-4,6-dimethoxy-1,3,5-triazine. Next, the fluorescein isothiocyanate (FITC) was covalently immobilized onto HA-SPIONs through reaction with the residual amines and/or hydroxyl groups, generating the fluorescent FITC-HA-SPIONs [26]. Another applied strategy for anchoring HA chains onto MNPs relies on the conjugation chemistry using polyethyleneimine (PEI)-coated Fe_3O_4 [27]. The amide linkages are formed between amino groups of PEI and carboxyl functions of HA in the presence of 1-ethyl-3-(3-dimethylaminopropyl)carbodiimide (EDC)/*N*-hydroxysuccinimide (Fig. 25.2). The hydrothermally synthesized Fe_3O_4 NPs showed good water-solubility, colloid stability, and biocompatibility after being modified by HA. It has been shown that HA-PEI-Fe_3O_4 can be uptaken by cells overexpressing CD44 receptors but also be monitored by MRI [28].

Negatively charged HA can be electrostatically bound to individual MNPs previously modified with aminosilane. The layer of aminosilane on the surfaces of iron oxide NPs obtained as the result of modification with *N*-[(3-trimethoxysilyl-propyl)ethylene diamine] (TMSPEDA) provides a reactive platform for the further grafting of organic molecules onto NPs. The reactive amino groups on MNP induce positive surface charge to facilitate further modification via the layer-by-layer (LbL) assembly. In the initial step, the negatively charged curcumin-HA conjugate was adsorbed onto TMSPEDA@MNPs with a positive surface charge. The HA layer, providing a negative surface charge, helps the subsequent adsorption of a positively charged macromolecules. The LbL assembly allows controlling the number of layers, thus tailoring the amount of drug conjugated with polymer chains onto the hybrid NP [29].

Recently, several reports described the polydopamine (PDA)-mediated adsorption of HA. The Fe_3O_4 NPs were prepared by the chemical coprecipitation method and, coated with dopamine by mild stirring under alkaline conditions. In these conditions, dopamine self-polymerize resulting in the formation of a PDA layer. The PDA shell allows the association of HA via hydrogen bond interactions and electrostatic adsorption due to the presence of catechol and amino groups, respectively [30]. The iron oxide NPs can also be directly coated after synthesis with dopamine-HA conjugate. While dopamine has a high affinity for the γ-Fe_2O_3 surface (strong affinity to diverse metal oxide surfaces through coordination bonding), HA ensures colloidal stability and boosts the relaxivity of the iron oxide. A low concentration of dopamine-HA-γ-Fe_2O_3 NPs was sufficient to achieve visible contrast in MRI [31].

25.2.1.2 Dextran

Dextran is one of the most used polymers to coat MNPs. It is a polysaccharide produced by lactic acid bacteria (*Streptococcus, Lactobacillus*), composed of α-D-(1→6) linked glucose units and some α-D-(1→3) linked glucose branch units. This neutral polysaccharide offers attractive properties to the magnetic NPs such as water solubility, low toxicity, biocompatibility, and improves the blood circulation by reducing interactions between NPs and plasma proteins [32]. Several NPs with an iron core and dextran coatings have been approved for human use. The United States Food and Drug Administration (FDA) approved Feridex I.V. (ferumoxides) as the iron oxide imaging agent for the detection of liver lesions, Combidex (ferumoxtran-10) for imaging the prostate cancer lymph-node metastases, and Feraheme (ferumoxytol) for the treatment of iron deficiency anemia with chronic kidney disease. Ferumoxytol is also under clinical investigation for the detection of the central nervous system inflammation, brain neoplasms, and cerebral metastases from lung or breast cancer [33]. The commercial Ferumoxtran-10 and ferumoxides are prepared by the coprecipitation method with *in-situ* coating by dextran, while ferucarbotran and ferumoxytol by carboxydextran and carboxymethyl dextran, respectively [34].

Dextran-coated MNPs can be obtained *in situ* during the coprecipitation reaction (under alkali conditions) or by adsorption onto the MNPs surface after crystallization [35]. The physisorption of dextran macromolecule to magnetite NPs occurs via noncovalent interactions with hydroxyl groups present on iron oxide NPs. In 1982, Molday and Mackenzie were the first to report the formation of magnetite NPs in the presence of dextran. Fe^{2+} and Fe^{3+} salts were mixed with dextran polymer under alkaline conditions and purified by gel filtration chromatography to obtain 30–40 nm particles [32]. The physisorption of dextran on the NP surfaces during coprecipitation depends on the reaction conditions such as adsorption time, molecular weight, temperature, and the concentration of polymer [36]. Analysis of dextran nanohybrids showed that the polymer limits particle size compared to uncoated particles [37]. Moreover, it has been shown that the presence of dextran ensures the stability against aggregation in physiological conditions [38], decreases the cytotoxicity of nanohybrids [39], and prolongs blood circulation time, which allows those MNPs to access macrophages located in deep and pathological tissues (such as lymph nodes, kidney, brain, osteoarticular tissues) [34].

To prevent dextran dissociation, the NPs can be treated with epichlorohydrin to cross-link the polysaccharide coating film [40]. The post-synthesis covalent binding of dextran to the MNP surface can be achieved through the incorporation of aldehyde groups onto the dextran backbone. Aldehyde groups can be obtained by oxidation of hydroxyl groups of dextran with sodium periodate [41]. The covalent binding is carried out via Schiff base formation between the aldehyde groups and amine ($-NH_2$) functions introduced onto MNPs by condensation of aminopropylsilane (APS) [42]. MNPs modified with APS have also been used for the covalent binding of carboxymethyldextran having carboxylic acid groups (-COOH) via carbodiimide-mediated amidation [43].

25.2.1.3 Chitin and chitosan

Chitin is a natural polymer present in the exoskeleton of many invertebrates (e.g., shrimps and crabs) and is an essential component of cell walls of fungi and yeast. This polysaccharide consists of β-(1→4) linked *N*-acetyl-glucosamine residues and can be converted into the water-soluble chitosan upon deacetylation in a strongly alkaline environment [11]. The physical properties of chitosan (composed of randomly distributed β-(1→4)-linked D-glucosamine and *N*-acetyl-D-glucosamine) can be controlled by changing its molecular weight and degree of deacetylation, and finally by modifying both hydroxyl and amine groups that are present in the molecular backbone [44]. It can be dissolved in acidic aqueous conditions thanks to the protonation of its amino groups, which confers positive charges. Chitosan has many advantages, such as biocompatibility, biodegradability, bioadhesion, non-toxicity, nonimmunogenicity, antibacterial properties, and antifungal bioactivity [45]. It has been explored as a promising biomaterial for the generation of biomedical systems for cell culture, bioimaging, and therapy [46,47]. Especially, NPs made of chitosan have been undergoing extensive exploitation for delivery of drugs, proteins/peptides, genes, DNA [48–50].

The most straightforward approach for the preparation of chitosan-based MNPs is the *ex-situ* coating. Iron oxide NPs obtained by the coprecipitation method are coated with chitosan shell. The neutralization of an acidified magnetite-chitosan suspension induces precipitation of the polysaccharide on the NP surface. It is well known that amine groups on chitosan may interact with Fe_3O_4 [51,52]. Additionally, chitosan chains can be cross-linked via a glutaraldehyde cross-linker [53]. To minimize the protein adsorption on the NP surface, Fe_3O_4 was coated with polyethylene glycol (PEG)-chitosan. PEG plays here the role in protein resistance, thereby prolonging the circulation time *in vivo* [54]. Carboxymethyl chitosan was also used for the coating of amino-functionalized MNPs using EDC as a coupling agent [55]. Recently, chitosan-coated SPIONs showed antimicrobial activity against Gram-positive (*Bacillus subtilis*) and Gram-negative (*Escherichia coli*) bacteria. Chitosan was demonstrated to enhance reactive oxygen species (ROS) production on the microbial environment, hence the final antimicrobial activity [56].

In *in-situ* approach, the iron oxide NPs are formed from a mixture of $FeCl_3$ and $FeCl_2$ salts by reacting with ammonia in the presence of chitosan. As chitosan is insoluble in alkali conditions, it was first modified with charged quaternary amino groups leading to a chitosan derivative soluble in basic conditions. Next, the LbL technique based on the alternating adsorption of oppositely charged polyelectrolytes was employed to add anionic chitosan modified with sulfonate groups [57].

25.2.1.4 Alginate

Alginate is a naturally occurring anionic polymer derived from brown algae cell walls and several bacteria strains. It is a linear copolymer consisting of 1,4-linked β-D-mannuronic acid (M) and 1,4 α-L-guluronic acid (G) residues. The sugar units are arranged in consecutive G (GGGGGG), consecutive M (MMMMMM), or alternating M and G residues (GMGMGM). This polysaccharide has been extensively investigated for many biomedical applications due to its favorable properties, including biocompatibility, low toxicity, and easy physical gelation with divalent cations such as Ca^{2+}. Only G blocks participate in the physical cross-linking; thus, the composition of polymer (i.e., M/G ratio), G-block length, and molecular weight are critical factors influencing the physical properties of alginate hydrogels [8]. Besides ionic cross-linking, alginate-based gels can be obtained by cross-linking the chains via covalent bonds. Alginate hydrogels are

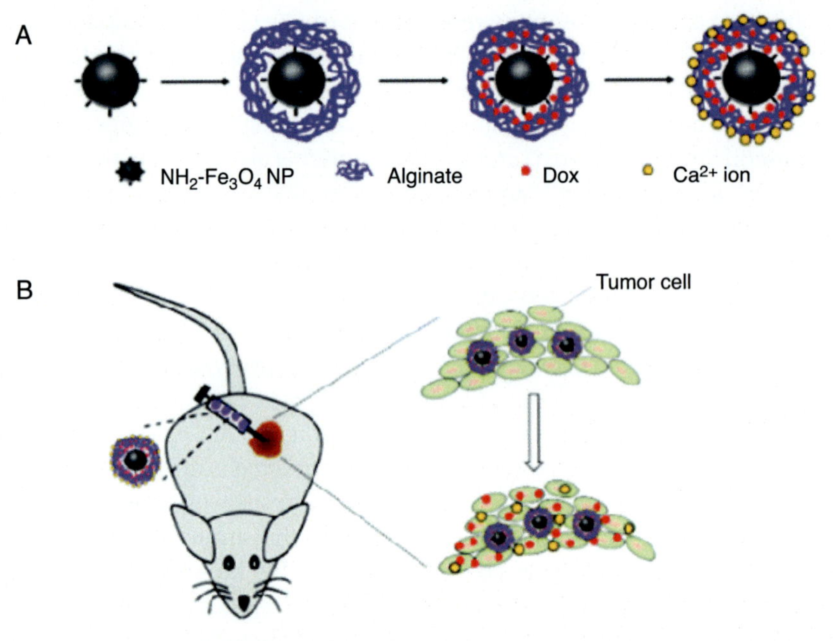

FIGURE 25.3 Schematic mechanism of DOX-encapsulating nanocarriers in alginate-based magnetic nanohybrid (A), and DOX release from nanocarriers to tumor region (B). "Republished with permission of Royal Society of Chemistry, In vitro and in vivo applications of alginate/iron oxide nanocomposites for theranostic molecular imaging in a brain tumor model, Chia-Hao Sua and Fong-Yu Cheng, 5, 2015, 90061".

attractive in a wide range of applications. They include *in-situ* gel formation [58–60], delivery of bioactive agents and controlled release [61–63], and cell transplantation [64]. This polysaccharide is nondegradable in mammals due to the lack of the enzyme alginase, typically present in marine algae, which cleaves the polymer chains. However, the release of complexing divalent ions can destroy the physically cross-linked gels [8]. Some examples of stimuli-responsive alginate nanocontainers have been reported for theranostic applications [65–67]. Kroll at al. [68] and Llanes et al. [69] reported the preparation of alginate-maghemite NPs by the addition of alginate solution to a solution containing Fe^{2+}. The G blocks of alginate chains selectively bind Fe^{2+} to form cross-linked domains. The oxidation of ferrous ions occurs *in situ* after treatment with a solution of NaOH, leading to hybrid NPs with diameters of 4–15 nm. The *ex-situ* coating could be achieved by covalent conjugation of alginate to the surface of NH_2-Fe_3O_4 NPs. The addition of Ca^{2+} solution led to the cross-linking of the alginate shell and entrapment of the active molecule. Both *in-vitro* and *in-vivo* experimental results showed that the DOX/Alg-Fe_3O_4 NPs inhibited C6 tumor cell (glioma cell) growth and killed them without damaging healthy nontumor cells (Fig. 25.3) [70].

25.2.1.5 Carrageenan

Carrageenans are sulfated polysaccharides extracted from *Rhodophyta red* edible seaweeds. Carrageenans are linear sulfated polysaccharides, composed of alternating α-1,3-linked

D-galactopyranose and β-1,4-linked D-galactopyranose units, and 3,6-anhydrogalactose residues. The most known are kappa (κ)-carrageenan with one sulfate group per disaccharide, iota (ι)-carrageenan with two, and lambda (λ)-carrageenan with three sulfate groups per disaccharide. Carrageenan is allowed under FDA regulations as a food additive and is considered safe when used as a gelling, thickening, or stabilizing agent in food. This anionic polysaccharide is also used in cosmetics, pharmaceutical, textile formulations, and printings [9]. κ-carrageenan exhibits gelation properties, whereas λ-carrageenan is known as a nongelling polysaccharide. A double-helical structure of κ-carrageenan chains can be stabilized by various multivalent cations [71,72] which is advantageous for the build-up of "cages" around the iron oxide particles protecting and stabilizing the final metal oxide NPs [73]. In this mechanism, at low pH, the Fe^{2+} and Fe^{3+} cations are first attracted to the functional sulfate group or coordinated by hydroxyl groups of polysaccharide, leading to the physical cross-linking of chains. While pH increases, the iron cations start to detach from polysaccharide and migrate toward the particle nucleation sites, leaving a number of unbound sulfates behind. Some sulfate moieties still remain coordinated to the NP surface, ensuring the high colloidal stability [74] .The magnetic nanobeads based on carrageenan and a positively charged chitosan were also reported. As the pK_a of the sulfate groups is lower than 2, carrageenan can easily interact with cationic polymer by electrostatic interactions in a wide range of pH. The Fe_3O_4 NPs were obtained by the coprecipitation method in the presence of κ-carrageenan solution followed by chitosan coating [75–77].

25.2.1.6 Heparin

Heparin sulfate is a component of the extracellular matrix and is one of the members of the GAG family that can covalently attach to proteins to form proteoglycans. The most common unit making up heparan sulfate is glucuronic acid (GlcA) linked to *N*-acetyl glucosamine (GlcNac). This GAG has the highest negative charge density among the biological polysaccharides. It has been widely used in drug delivery and in tissue engineering to improve the biocompatibility and blood compatibility of biomaterials [78]. Besides, heparin can inhibit cancer cell angiogenesis and thus inhibit tumor growth and metastasis [79].

Heparin-coated MNPs are generally prepared by alkaline coprecipitation [80]; however, they have also been fabricated through the *ex-situ* coating [81]. Heparin coating increases the efficiency of MNP uptake because it increases the hydrophilic properties. The heparin-based nanohybrids were also studied for simultaneous targeted drug delivery and MRI. DOX was conjugated with heparin before coating the SPIONs. Such prepared NPs enhanced anticancer activity and showed potential used as an MRI contrast agent [81]. In another example, Fe_3O_4 NPs were synthesized in a solvothermal reaction in the presence of positively charged PEI, followed by self-assembly with negatively charged heparin. Thanks to the presence of heparin, these nanostructures showing anticoagulant properties and magnetically guided have potential to be applied for hemodialysis [82].

25.2.1.7 Pullulan

Pullulan is a water-soluble neutral linear polysaccharide consisting of α-(1,6)-linked maltotriose residues, produced by the fungus *Aureobasidium pullulans*. Pullulan is nontoxic, edible and biodegradable, nonantigenic, and nonimmunogenic. Due to its properties, it is therefore widely used in food, drug and gene delivery, and tissue engineering [83,84].

In 2004, Gupta and Gupta demonstrated the fabrication of pullulans-coated MNPs using glutaraldehyde for cross-linking with MNPs. Although the authors observed improved stabilization of NPs, the residues of glutaraldehyde in the human body can have an adverse effect [85]. Alternatively, another derivative of pullulan was proposed to stabilize and control the size of MNPs, that is, pullulan acetate. Its amphiphilic properties allow the adsorption onto the oleic acid–coated MNPs and render them stable in aqueous solution. [86] Besides, Jo et al. functionalized MNPs with ethylenediamine (cationic) and succinic (anionic) derivative of pullulan to obtain charged MNPs and study their effect on cellular labeling [87].

25.2.2 Proteins

Proteins are naturally occurring, linear, unbranched polymer chains composed of amino acids connected to one another via peptide bonds. Apart from safety, biodegradability, nonantigenicity, they offer many interesting features. Thanks to the presence of multiple functional groups (NH_2, COOH, OH), they offer high drug binding potential. Some proteins can enhance tumor targeting of the anticancer drug via specific receptors overexpressed on tumor cells and show great cancer cell penetration ability [88,89]. β-Lactoglobulin, gelatin, and elastin showed pH-, enzyme- and thermo-responsive drug release, respectively, so they can be exploited in the development of stimuli-responsive nanosystems. Some proteins, such as zein and gliadin, can be used for encapsulation and controlled delivery of poorly soluble hydrophobic drugs [90]. Additionally, they are attractive candidates to increase MRI contrast. Serum albumin, casein, or gelatin coatings can increase the r2 relaxivity of MNPs due to their high hydrophilicity and ability to retain water molecules. They are widely utilized as carriers for fluorophores, enzymes, and therapeutics. Albumin coatings decrease the nonspecific adsorption of serum proteins on the NPs surface [91].

Various techniques can be successfully employed for the development of protein-based hybrid MNPs [92]. The most often used include the chemical conjugation (via covalent binding between reactive moieties of protein and the surface of MNPs), *in-situ* coating (MNPs coated with a protein layer via electrostatic or hydrophobic interactions), emulsion method, or desolvation-chemical cross-linking (inorganic NPs are added to the aqueous protein solution before desolvation by adding either ethanol or acetone).

Carbodiimide coupling is often used reaction for protein-NP conjugation via amide or ester bond formation. To introduce bovine serum albumin (BSA) onto MNP surface, arginine was employed first to decorate the surface of iron oxide NPs to provide free amine groups giving an opportunity for amide bond formation with the carboxylic groups of BSA. BSA conjugation improves the stealth characteristics of iron oxide NPs and hence prolongs the blood circulation time [93]. In another example, ferromagnetic $FeNi_3$@BSA core@shell structured microspheres have been prepared by the desolvation technique (Fig. 25.4). $FeNi_3$ inorganic NPs were added to the BSA prior to desolvation with ethyl alcohol. As albumin is negatively charged above its pI (isoelectric point) (4.8), BSA molecules were deposited onto $FeNi_3$ NPs with its anionic

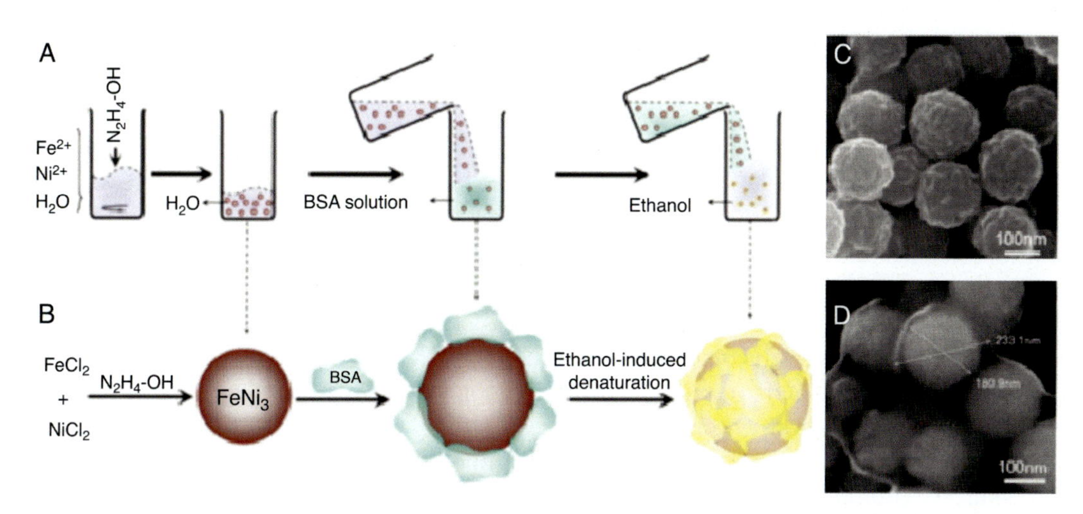

FIGURE 25.4 Schematic illustration of the synthesis of $FeNi_3$@BSA core@shell composite particles (A, B), and scanning electron microscope images of bare $FeNi_3$ (C) and $FeNi_3$@BSA (D) particles prepared in 30 mg/mL aqueous solution of BSA. "Reprinted from Colloids Surf. Physicochem. Eng. 414 (2012) 168, Lu et al., Protein-passivated $FeNi_3$ particles with low toxicity and high inductive heating efficiency for thermal therapy. Copyright (2012), with permission from Elsevier".

carboxylate groups. Upon adding ethanol, the attached BSA molecules were fixed on the $FeNi_3$ NPs exterior, which leads to the formation of stable core-shell $FeNi_3$-BSA nanohybrids. Due to the deposition of the protein layer, the $FeNi_3$@BSA composite particles exhibit low toxicity and superior biocompatibility [94].

In another example, desolvation method with glutaraldehyde as a cross-linking agent was applied for fabrication of BSA-MNPs. The resulting nanohybrids were further conjugated to folic acid. The magnetic core was used for hyperthermia effect which, together with DOX chemotherapy, can increase efficacy against cancer cells. Here, BSA played the role of the protector to avoid uptake by reticuloendothelial system (RES) and removal by macrophages, while folic acid improved the targeting effect [95].

Silk fibroin protein is among the most popular natural polymers used for the creation of biomaterials due to its acceptance by the FDA, low cost, and abundance. Silk is a protein fiber produced by silkworms, spiders, mites, fleas, and scorpion. Proteins extracted from spider silk, *Bombyx mori* has demonstrated extraordinary promise in biomedical fields due to their good biocompatibility and tunable biodegradability. The silk-based NPs were studied for encapsulation of both hydrophobic and hydrophilic drugs, growth factors, or DNA [96]. Tian et al. showed a simple "salting-out" method for the preparation of DOX-loaded magnetic silk fibroin NPs. First, citrate-coated MNPs interact with positively charged DOX, both dispersed in potassium phosphate solution (1.25 M, pH 8). When silk solution is added under low-temperature, its elongated conformation at basic pH changes under potassium phosphate and protein assembly occurs due to the enhancement of hydrophobic interactions. During protein assembly, MNPs together with DOX are encapsulated interior of spherical nanostructures. The generation of hybrid NPs and control of their size can be regulated by the electrostatic and/or hydrophobic interactions among negative MNPs, positive DOX, and negative silk fibroin. At higher DOX concentration, larger amount of MNPs is required, otherwise particles aggregate into nondispersible clusters in the salting-out process [97].

Gelatin is biodegradable, nontoxic, nonimmunogenic, and FDA approved. It is mainly extracted from bovine or porcine sources by a hydrolysis procedure, shows good stability at high temperature, and has a wide range of pH. Besides reactive -NH_2, -COOH, and -OH groups available for modification, it possesses native hydrophilic and hydrophobic segments in polypeptide chains. Hydrophobic interaction between the hydrophobic amino acids within the gelatin sequence and the hydrophobic capping agent of the NPs could also be used as a hybridization mechanism. The MNPs can be transferred from chloroform to water by interactions with gelatin. Owing to the numerous active groups in gelatin, the obtained hybrid NPs can be functionalized with FITC and Pt(IV) prodrug. The multifunctional drug delivery system can be used as T_2-weighted MRI contrast agent and demonstrate the anticancer ability by releasing drug within the cellular environment [98].

It should be highlighted that using amphiphilic proteins for coating of oleic acid–coated MNPs required harsh conditions (heating, sonication) which can easily denaturate proteins.

Mild conditions were applied for coating of MNPs with casein. Casein is a main ingredient of bovine milk that contains several related phosphoproteins. Naturally, casein self-assembles into micelles acting as natural nanovehicle for delivery of different biomolecules. This assembly is driven by hydrophobic/hydrophilic domains of proline-rich phosphoproteins. In formation of casein-based hybrid NPs, the oleic acid capping ligand was first replaced by glucose-based oligosaccharide to be subsequently coated with casein to obtain water-soluble nanostructures stable in aqueous solution. It was explained that the high MRI contrast was related to high water diffusion through permeable casein coat, facilitating the exchange between bulk water and those residing on the casein surface [99].

Different examples of protein-based hybrid NPs with potential biomedical applications are presented in Table 25.2.

25.3 Stabilization of magnetic nanoclusters

25.3.1 Magnetic nanoclusters

The particles with superparamagnetic properties become magnetized on applying the magnetic field up to their saturation magnetization, but display negligible or no remanent magnetization when the external magnetic field is removed. The superparamagnetic properties are essential for homogenous dispersion in liquid media or separation processes. However, under external magnetic field, many MNPs (generally lower than 20 nm) show low magnetization at saturation due to the low iron oxide content, which becomes complicated in separation processes or magnetic targeting. They cannot be used to target tumors located at a depth of more than 2 cm beneath the skin, because the magnetic fields that can currently be generated cannot exert enough force to manipulate NPs at such depth [116–118]. To improve the response upon exposure to an external magnetic field, the self-assembly of magnetic nanocrystals into higher colloids based

Table 25.2 Hybrid protein-MNPs studied for biomedical applications.

Protein	Inorganic NPs	Drug	Application	Reference
Casein	Fe_3O_4 NPs		Improved T_2-MRI contrast	[99]
BSA	Fe_3O_4/BSA/Sia		Detection of β-amyloid/MRI contrast	[100]
BSA	Fe_3O_4/SiO_2		Magnetic guiding, fluorescent tracing	[101]
BSA	Fe_3O_4 NPs	KaempferolPTX	Chemotherapy	[102]
BSA	Fe_3O_4 NPs	Doxorubicin	Combined chemotherapy and hyperthermia	[95]
BSA	SPIONs		MRI, ultrasound imaging	[103]
BSA	Fe_3O_4 NPs	20(s)-ginsenoside Rg3	Combined chemotherapy and hyperthermia	[104]
HSA	Fe_3O_4 NPs	20(s)-ginsenoside Rg3	Combined chemotherapy and hyperthermia	[105]
HSA	Fe_3O_4 NPs	5-flurouracil	Combined chemotherapy and magnetic targeting	[106]
HSA	IO NP-mSiO_2	DOX	Drug release, MRI	[107]
Transferrin	$Fe_3O_4/NaYF_4$ NPs		Magnetic separation/fluorescent imaging	[108]
Transferrin	Fe_3O_4/PTBA/mSiO_2	PTX	Imaging and therapy	[109]
Gelatin	Fe_3O_4/CaP NPs	DOX	T_2-MRI contrast, drug delivery	[110]
Gelatin	Fe_3O_4 NPs	Pt(IV)	MRI contrast, drug delivery	[98]
Lactoferrin	MIONs	PTX	Thermo-chemo-therapy	[111]
Gelatin	Fe_3O_4 NPs	Cis-Pt	Magnetic targeting, drug delivery	[112]
Gelatin	Fe_3O_4 NPs	DOX	T_2-MRI contrast	[113]
Gelatin	MMSNs	PTX	Magnetic targeting, chemotherapy	[114]
Albumin	Fe_3O_4 NPs	5-flurouracil	Magnetic targeting, chemotherapy	[106]
Silk fibroin	Fe_3O_4 NPs	Curcumin	Magnetic targeting, chemotherapy	[115]
Silk fibroin	Fe_3O_4 NPs	DOX	Magnetic targeting, chemotherapy	[97]

on a dense inorganic core of several MNPs prior to the coating is a promising strategy. Recent studies demonstrated the potential of such condensed colloidal crystals, also named clusters, in advanced applications, such as bioseparation, targeted delivery, and imaging. Clusters are aggregates of individual MNPs that still display superparamagnetic properties but profoundly enhance the magnetic moment of the overall assembly, making magnetic manipulation more effective [116]. Several strategies have been demonstrated for the construction of assembled nanoclusters. Most often the synthetic polymers have been employed in stabilization of magnetic nanoclusters [80,119–126]. The stability of MNPs assemblies is mainly mediated by molecular interactions, including hydrophobic interactions, hydrogen bonding, or electrostatic interactions. Here, we present different methods of nanoclusters formation following by examples of coating with polysaccharides and proteins.

25.3.1.1 Emulsification/solvent evaporation

Hydrophobic reaction–directed assembly is a common method to assemble hydrophobic MNPs and stabilize the resulting cluster with amphiphilic polymers. A typical emulsion process for the formation of hybrid nanobeads with a magnetic cluster and polymer coat can be divided

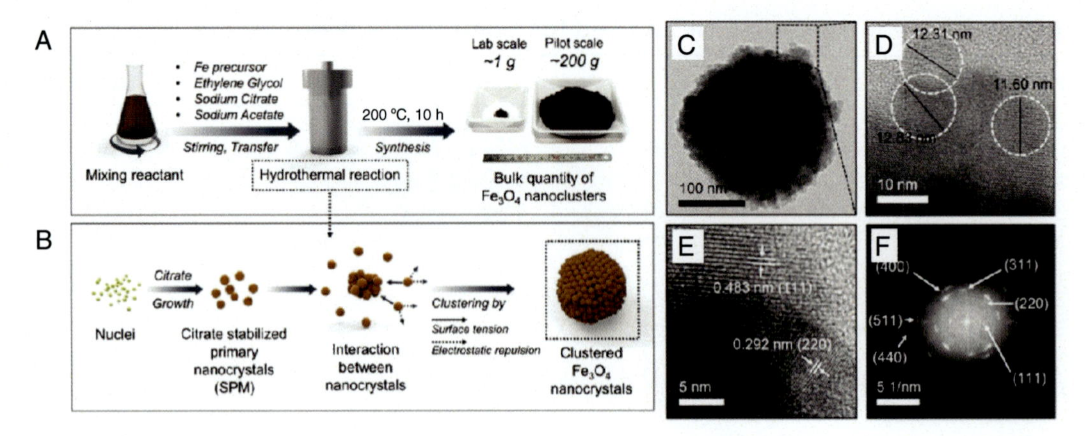

FIGURE 25.5 Schematic illustration of the synthesis of Fe_3O_4 nanoclusters (A, B), transmission electron microscopy images of magnetic nanoclusters at different magnifications (C–E), selected area electron diffraction pattern acquired from a single Fe_3O_4 nanocluster (F). "Reprinted with permission from Xuan et al., Tuning the Grain Size and Particle Size of Superparamagnetic Fe_3O_4 Microparticles. ACS Appl. Mater. Interfaces 2018, 10, 41935–41946. Copyright (2018) American Chemical Society."

into three steps. First, MNPs dispersed in organic solvent are emulsified in an aqueous solution containing a surfactant (i.e., sodium dodecyl sulfate), producing "oil" droplets. Next, the droplets are condensed into clusters of MNPs by evaporating the organic solvent within the emulsion droplet. Furthermore, the clusters are encapsulated with a polymer shell [127]. Very often, the surfactant is replaced directly with the amphiphilic polymer, which is added directly to the emulsion mixture (Fig. 25.1C).

25.3.1.2 Solvothermal synthesis

Solvo-/hydrothermal synthesis methods have been extensively developed as they provide a simple, versatile, and cost-effective route for producing large quantities of highly crystalline magnetic clusters with controlled morphology. Magnetic clusters can be obtained by solvothermal reduction of $FeCl_3$ with ethylene glycol or diethylene glycol in the presence of sodium acrylate or sodium acetate. The growth of Fe_3O_4 particles follows the two-stage growth model in which the primary nanocrystals nucleate first in supersaturated solution and then aggregate into larger secondary particles. [128]

Fig. 25.5 presents the synthesis of hydrophilic magnetic clusters using a facile one-step solvothermal method by reduction of $FeCl_3$ with ethylene glycol at 200 °C in the presence of sodium acetate as an alkali source and biocompatible trisodium citrate dihydrate as a stabilizer. Ethylene glycol acts as both the solvent and reductant at a relatively high boiling point, whereas sodium acetate provides strongly alkaline conditions that promote hydrolysis of iron oxide in solution. Trisodium citrate plays here the role of the surfactant due to its strong coordination affinity to Fe^{3+} ions, which favors the attachment of citrate groups on the magnetite nanocrystals and prevents them from aggregation into large crystals [129].

25.3.1.3 Solvent displacement

Colloidal nanoclusters can also be formed via the solvent displacement method, which exploits a change of solvent polarity. Pellegrino et al. reported that hydrophobic MNPs dissolved in tetrahydrofuran (THF) tend to self-assemble after the addition of less apolar acetonitrile (ACN). The presence of poly(maleic anhydride-*alt*-1-octadecene), which undergoes partial hydrolysis in water, stabilized nanoclusters in aqueous conditions [126,130].

25.3.2 Polyelectrolyte/MNP self-assemblies

Direct mixing of solutions containing anionic superparamagnetic NPs and cationic polyelectrolytes generates aggregates with controlled shape and morphology. The mechanism is based on electrostatic interactions and on the compensation between the opposite charges. Poly(acrylic acid)-coated iron oxide NPs form the clusters under direct mixing with poly(diallyldimethylammonium chloride) (PDADMAC) or PEI solutions at an appropriate ionic strength [131].

25.3.3 Coating with polysaccharide and proteins

In the emulsion technique (oil-in-water), hydrophobic MNPs are entrapped in nanodroplets of organic solvent dispersed in an aqueous phase containing an amphiphilic polymer. Before employing the polysaccharide, the grafting of hydrophobic moieties to the principal polysaccharide chain is essential. While the hydrophilic polysaccharide backbone ensures the colloidal stability in aqueous solutions, the grafted hydrophobic chains enable interactions with hydrophobic MNPs. For example, the introduction of hydrophobic oleyl- or pyrenyl functional groups onto the HA backbone was crucial for the encapsulation of hydrophobic MNPs and nonpolar compounds [132,133]. The amphiphilic HA derivatives are dissolved in water (aqueous phase) and mixed with hydrophobic MNPs dissolved in hexane (organic phase) under sonication. After evaporation of the organic solvent, the polysaccharide-coated magnetic nanoclusters can be dispersed in water. The presence of HA allows to study the selective recognition of CD44 receptor and breast cancer diagnosis using MRI. The clusters sizes of hybrid nanobeads could be readily controlled by varying the degree of substitution of 1-pyrenylbutyric acid in pyrenyl-HA conjugates. This had a critical effect not only on the size of the cluster (number of aggregated SPIONs), but also on the loading efficiency of magnetic nanocrystals and their magnetization [132]. The sonication used for homogenization of emulsions should be used with precaution, especially for natural polymers, which can be sensitive and undergo depolymerization under ultrasound treatment [134]. As an alternative, the coating with the amphiphilic polysaccharide can be performed after cluster formation. In our group, HA grafted with poly(di(ethylene glycol) methacrylate-*co*-butyl methacrylate) (HA-*g*-poly(DEGMA-*co*-BMA)) was used to stabilize the clusters of assembled oleic acid-coated MNPs. The strategy included (1) synthesis of hydrophobic superparamagnetic NPs, (2) formation of nanoclusters, and (3) assembly of HA-*g*-poly(DEGMA-*co*-BMA) on the nanocluster, as shown in Fig. 25.6.

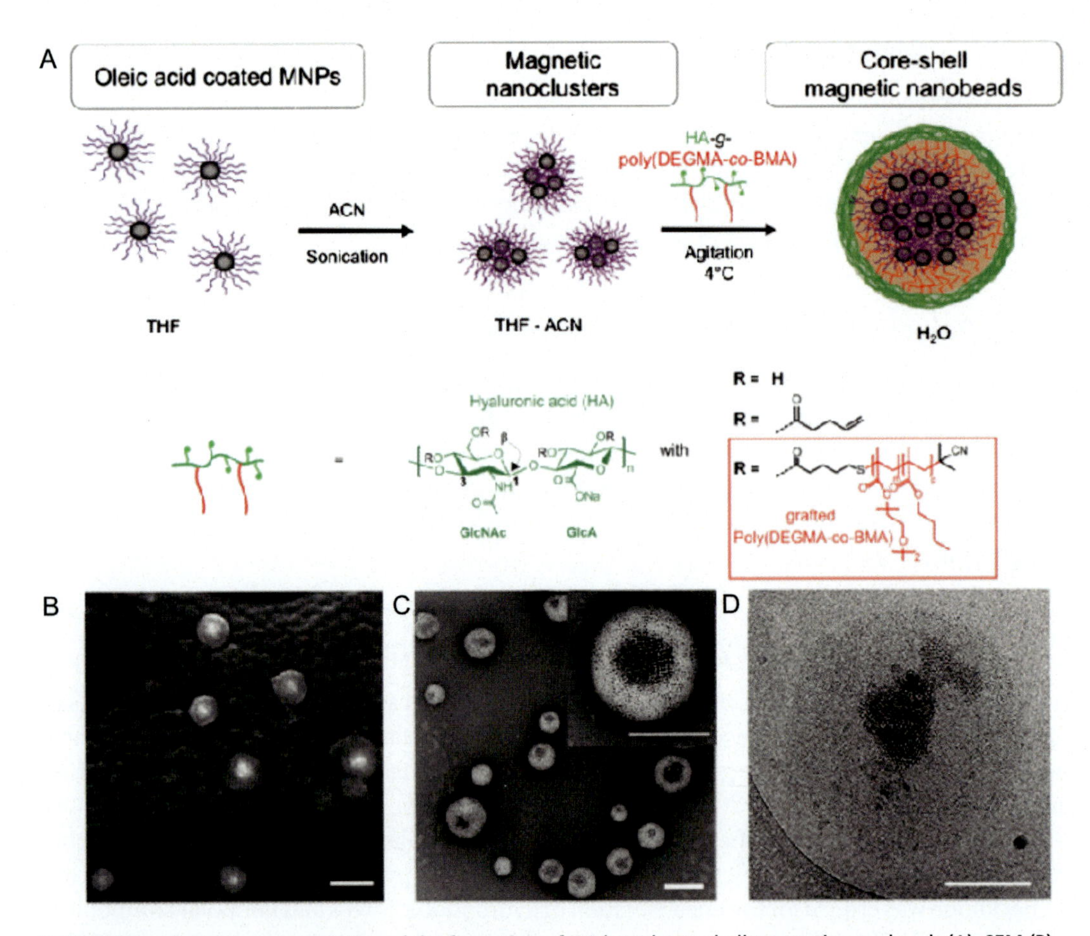

FIGURE 25.6 Schematic representation of the formation of HA-based core-shell magnetic nanobeads (A), SEM (B), transmission electron microscopy (TEM) with negative staining (C), and cryo-TEM (D) images of HA-*g*-poly(DEGMA-*co*-BMA)-coated magnetic nanoclusters. Scale bar: 200 nm. "Reprinted from Appl. Surf. Sci., 510 (2020) 145354, Rippe et al. Synthesis and magnetic manipulation of hybrid nanobeads based on Fe_3O_4 nanoclusters and hyaluronic acid grafted with an ethylene glycol-based copolymer. Copyright (2020), with permission from Elsevier".

The hydrophobic MNPs were dissolved in THF, followed by slow addition of ACN, under sonication. As the MNPs are not stable in polar solvents, thus the presence of ACN induces the formation of hydrophobic nanoclusters (solvent displacement method). After most of the THF has evaporated, the nanocluster suspension was then gently mixed with a solution of HA-*g*-poly(DEGMA-*co*-BMA), to avoid sonication and possible chain cleavage. Due to the high concentration of MNPs within the bead core, a bulk magnet could be used for sample purification and nanobeads separation. The electron microscopy images (Fig. 25.6B-D) showed that core-shell structure of obtained hybrid nanostructures, with cluster, constituted close-packed MNPs encased in an outer polymer shell. The strong hydrophobic interactions between the hydrophobic magnetic core and the copolymer present on HA chains make the structure intact. The HA protects the magnetic nanoclusters from uncontrolled aggregation, ensures stability, and easy dispersion in aqueous conditions. Additionally, the possibility of magnetic manipulation of drug-loaded HA nanobeads was demonstrated. [135]

Zheng et al. synthesized HA-hexadecyl (C_{16}) derivatives capable to encapsulate the anti-cancer drug docetaxel and SPIONs. The specific targeting capability based on CD44 receptor-mediated endocytosis and the enhanced targeting efficacy in the presence of external magnetic field of these hybrid magnetic nanoclusters were investigated [136].

Dextrans modified with stearic acid [137] or benzoporphyrin [138] were also employed for the stabilization of clusters. The multifunctional micelles could be fabricated by means of hydrophobic benzoporphyrin derivative (BPD) covalently conjugated to dextran-*b*-oligo (amidoamine) dendron copolymer. These amphiphilic dextran-BPD conjugates self-assemble onto densely packed magnetic core using a microemulsion method leading to dual-functional nanoplatform for the delivery and photodynamic therapy (Fig. 25.7) [138].

L-3,4-dihydroxyphenylalanine (DOPA)-conjugated chitosan oligosaccharide (chitosan-DOPA) was used to coat ferrimagnetic iron oxide nanocubes showing excellent heating efficiency in hyperthermia treatment. As the catechol side chain of DOPA exhibits strong affinity to diverse metal oxide surfaces through coordination bonding, the chitosan-DOPA can be efficiently immobilized onto the surface of ferrimagnetic iron oxide. First, the magnetic nanocubes were dispersed in chloroform and mixed with the aqueous solution of chitosan-DOPA by ultrasonication to form an oil-in-water emulsion. Upon evaporation of the residual solvent in the droplets, the multivalent binding of chitosan-DOPA to the iron oxide surfaces led to the immediate formation of water-dispersible nanobeads [139]. Or else, chitosan grafted with (*N*-palmitoyl-*N*-monomethyl-*N,N*-dimethyl-*N,N,N*-trimethyl-6-O-glycol) was successfully applied for entrapment of MNPs assemblies increasing T_2-weighted MRI contrast [140].

It has been reported that the hydrophobicity of protein could be increased by employing the linear aliphatic ligand, that is, hexanoyl groups on the amino groups along the fibrous gelatin. As the hydrophobic hexanoyl groups replace part of the amino groups along the gelatin molecule, this hydrophobically modified gelatin possesses two different groups anchored along the chain. Such modified gelatin facilitates interaction with the adsorbed oleic acid on magnetic nanocrystallites. The ultrasound emulsification of iron oxide NPs dispersed in chloroform with amphiphilic gelatin dissolved in water, followed by evaporation of the organic solvent, resulted in aggregated hydrophobic magnetic core (cluster) surrounded by carboxylic groups of gelatin [141]. These amphiphilic gelatin assembled iron oxide NPs that were coated with a thin shell of CaP as an efficient drug reservoir (Fig. 25.8) [110].

Recently, Khramtsov et al. presented the comparison of coating the magnetic nanoclusters with different proteins such as albumin, casein, and gelatin. The MNPs (Fe@C) were first functionalized with amino groups and added to the aqueous protein solution under sonication, followed by adjusting pH up to 7.2–7.6. Aminated Fe@C NP is stable at pH 4 and aggregates at neutral pH. Two simultaneous processes occurred: a pH-dependent aggregation of Fe@C-NH_2 and stabilization of aggregates by protein molecules (Fe@C-NH_2/Protein). The formed surface

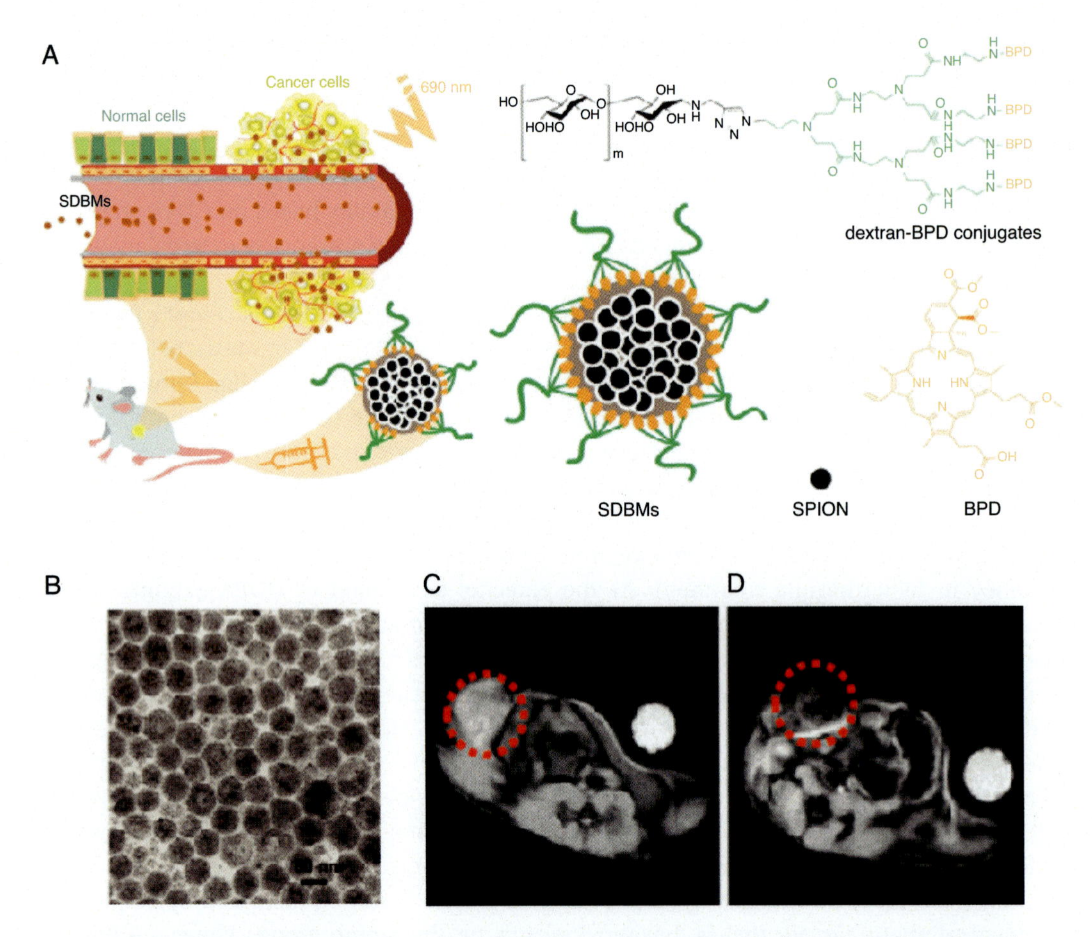

FIGURE 25.7 Schematic diagram of self-assembled SPION-loaded dextran-BPD micelles. Micelles were formed through the coassembly of the small hydrophobic SPIONs and the dextran-BPD conjugates (A), transmission electron microscopy image of the self-assembled SPION-loaded dextran-BPD micelles (B), and T_2-weighted magnetic resonance images in the axial plane prior to injection (precontrast, C) and 24-h after intravenous injection (postcontrast, D) of nanoclusters. "Reprinted (adapted) with permission from Bioconjug. Chem. 30 (2019) 2974, Yan et al., Dextran-Benzoporphyrin Derivative (BPD) Coated Superparamagnetic Iron Oxide Nanoparticle (SPION) Micelles for T_2-Weighted Magnetic Resonance Imaging and Photodynamic Therapy. Copyright (2019) American Chemical Society."

protein layer was cross-linked with glutaraldehyde to stabilize the coat. Their size was dependent only on the coating conditions (pH, ionic strength, protein/NP ratio). All types of protein coatings provide the nanoclusters with excellent long-term storage stability. The gelatin layer prevents nanoclusters from aggregating over a wide range of pH: from 4 to 10, whereas casein- and BSA-coated nanoclusters were stable at pH 6–10. The protein-coated nanoclusters withstand salt concentrations up to 2M without a significant change in size. The developed protein-coated nanoclusters can be applied for not only *in-vitro* but also *in-vivo* diagnostics (e.g., as a T2 contrast in MRI) due to their good stability in physiological media and high relaxivity [91].

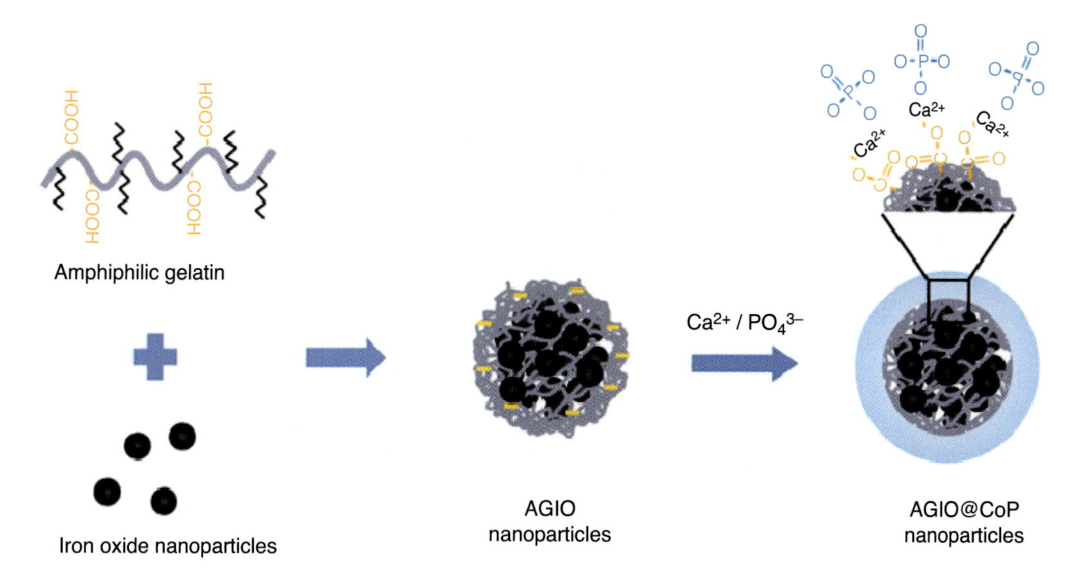

FIGURE 25.8 Schematic representation of self-assembly of iron oxide NPs in the presence of amphiphilic gelatin by the emulsification method and corresponding to amphiphilic gelatin iron oxide (AGIO), and AGIO@CaP nanoparticles transmission electron microscopy images. "Reprinted from Acta Biomater., 9 (2013) 5360, Li et al., In situ doxorubicin-CaP shell formation on amphiphilic gelatin–iron oxide core as a multifunctional drug delivery system with improved cytocompatibility, pH-responsive drug release and MR imaging. Copyright (2013), with permission from Elsevier".

25.4 Application in biomedical field

25.4.1 Drug delivery

Drug delivery methods using MNPs in combination with polysaccharides or proteins have been growing considerably in the past decades. Compared to the uncoated magnetic nanocrystals, MNPs assisted by natural polymers offer novel physicochemical, biological, toxicological properties, and multifunctionalities. The coating of inorganic MNPs with polymers reduces toxicity, offers prolonged circulation time (limits opsonization and their recognition by macrophages (liver/spleen)), protection of the entrapped payload and premature release, and improves tumor uptake associated with the enhanced permeability and retention (EPR) effect as well as targeting capability [115]. Proteins, like polysaccharides, improve cytocompatibility by protecting from the leakage of free metal ions causing tissue inflammation, cell apoptosis, and DNA damage [142]. They can reduce the immunotoxicity, that is, albumin coating of MNPs led to reduced phagocytosis by macrophages [143] and improved colloidal stability when highly charged [99].

Thanks to the presence of MNPs, such drug nanocarriers can be accumulated in the desired site using the very strong external magnetic field gradient, thus maximize the efficiency of treatment and reduce side-effects, which is one of the major challenges in cancer treatment. The magnetic drug targeting/delivery can be realized upon the application of an external magnetic field from electromagnetic coils or various types of permanent magnets. Additionally, drug-loaded magnetic nanohybrids after being heated in a magnetic field can promote drug release, and have advantages of MRI monitoring during drug distribution [54]. Among polysaccharide or protein magnetic hybrid NPs studied as drug carriers, the polymer-coated magnetic clusters have attracted more and more interest as they offer control of magnetic properties by controlling the amount of enclosed MNPs, and possibility to entrap higher amount of drug.

The drug can be incorporated into the nanohydrid system by physical adsorption (hydrophobic or electrostatic interactions), conjugation between chemical functionalities presented on the coat and active molecule, or be entrapped in the core during formation of magnetic cluster. To enable on-demand drug release, various stimuli such as pH, enzymes, or magnetic field can be applied.

Table 25.3 presents various examples of polysaccharide-coated either individual MNPs or clusters for biomedical applications. The most often tested anticancer drug encapsulated in hybrid NPs is DOX. For example, in FITC-HA-MNPs system, a hydrazone linker that can be formed at pH 7.4 and hydrolyzed rapidly in acidic conditions ($pH<5$) of tumor cells, was used for attachment of DOX to HA. These multifunctional hybrid NPs with magnetic and fluorescent properties can be exploited for MRI and fluorescent imaging, and play the role of magnetic vehicle for drug delivery [26].

Further, the cystamine-grafted alginate was applied to engineer stimuli-responsive DOX-loaded SPIONAlgSS (Fig. 25.9A). *In-vitro* DOX release studies in different pH environments and glutathione (GSH) concentrations were demonstrated (Fig. 25.9B,C). Disulfide bonds can be cleaved into free thiols under reductive environments, such as in the presence of glutatione, and release the encapsulated drug on demand. In acidic conditions, the faster release is related to the protonation of the un-cross-linked carboxylic groups of alginate, which weakens the electrostatic interactions with DOX and promotes the drug release from SPIONAlgSS [154].

In another example, DOX/Alg-Fe_3O_4 NPs inhibited glioma cells (C6 tumor cells) growth and killed them without damaging healthy nontumor cells. The growth of C6 tumor cells was not inhibited in phosphate-buffered saline (PBS), Alg-Fe_3O_4 NPs, or in the presence of free DOX in comparison to the drug-loaded alginate-based hybrid magnetic NPs. Apparently, the C6 cells can uptake DOX-loaded magnetic nanohybrids and then, most of the drug is released in the cytoplasm and enters into the nuclei to cause apoptosis. By contrast, C6 cells did not show uptake of free DOX. Blood brain barrier permeating NPs based on DOX/alg-Fe_3O_4 are expected to be developed [70].

The hybrid cluster "bomb" based on PEG-chitosan derivative (PEG-CS) was fabricated for tumor-specific theranostics for targeted DOX delivery and MRI. A mixture of DOX and oleic acid-Fe_3O_4 was added to the acidic solution of PEG-CS under sonication. PEG-CS plays the role of surfactant in the emulsification process. After adjusting pH up to 10, PEG-CS stabilizes the clusters due to deprotonation of amine groups of chitosan. At low pH, the magnetic nanohybrids are destroyed and enable the drug release. [151]

Chu et al. encapsulated hydrophobic MNP and DOX within alkylated N-(2-hydroxy) propyl-3-trimethyl ammonium chitosan chloride (alkyl-HTCC) or alkylated PEG alkyl-PEG-HTCC polymer, respectively, using a typical oil-in-water emulsion method. The emulsifying ability of octyl-HTCC was better than that of octyl-PEG-HTCC due to its stronger hydrophobicity, and thus

Table 25.3 Role of polysaccharides in the build-up of magnetic nanohybrids for biomedical applications.

Polysaccharide	Magnetic core	Role of polysaccharide	Application	Reference
Hyaluronic acid	Fe_3O_4 NPs	Hydrophilicity, stability, CD44 targeting	Cellular targeting, MRI	[28,31,144–146]
	Fe_3O_4 clusters	Stability, biocompatibility	Magnetic guiding, drug encapsulation	[135]
	SPIONs clusters	Biocompatibility, CD44 targeting	MRI	[132,133]
	PEI-MNPs	Reducing cytotoxicity, recognition of CD44	DNA delivery	[147]
	Fe_3O_4 clusters	Biocompatibility, biodegradability, CD44 recognition	MRI, photothermal-chemotherapy	[136]
Chitosan	Fe_3O_4 clusters	Stable coating, long-term interaction with tumor due to the positive charge	Magnetic guiding, hyperthermia	[139]
	Fe_3O_4 clusters	Stabilization of hydrophobic clusters	MRI	[140,148]
	MNPs	Positive charge for complexing	Gene delivery	[149]
	Fe_3O_4 NPs	Enhance ROS production	Antibacterial activity	[56]
	Fe_3O_4 clusters	Biocompatibility	Hyperthermia	[150]
	Fe_3O_4 clusters	Positive charge for electrostatic interactions	Drug delivery, MRI	[151]
Dextran	Fe_3O_4 clusters	Steric stabilization and the reduction of the opsonization process	MRI, drug delivery	[138]
	Fe_3O_4 clusters	Biocompatibility, Stability	Hyperthermia	[35]
	SPIONs	Steric stabilization, prolonged circulation in blood	MRI	[152]
	Fe_3O_4 NPs	Recognition of macrophage surface receptors	MRI	[153]
Alginate	Fe_3O_4 NPs	Direct coordination of G-blocks with Fe^{2+},	Drug release	[70]
	SPIONs	Biocompatibility, cross-linking using Ca^{2+}	Responsive drug release, MRI	[154]
Heparin	SPIONs	Strong hydrophilic properties, low toxicity, passive targeting	Anticancer drug delivery, MRI	[81]
	Fe_3O_4 NPs	Anticoagulant	Blood clotting prevention for hemodialysis	[82]
Pullulan	Fe_3O_4 NPs	Biocompatibility	Hyperthermia treatment	[86]
	IONPs	Biocompatibility	Cellular labeling	[87]

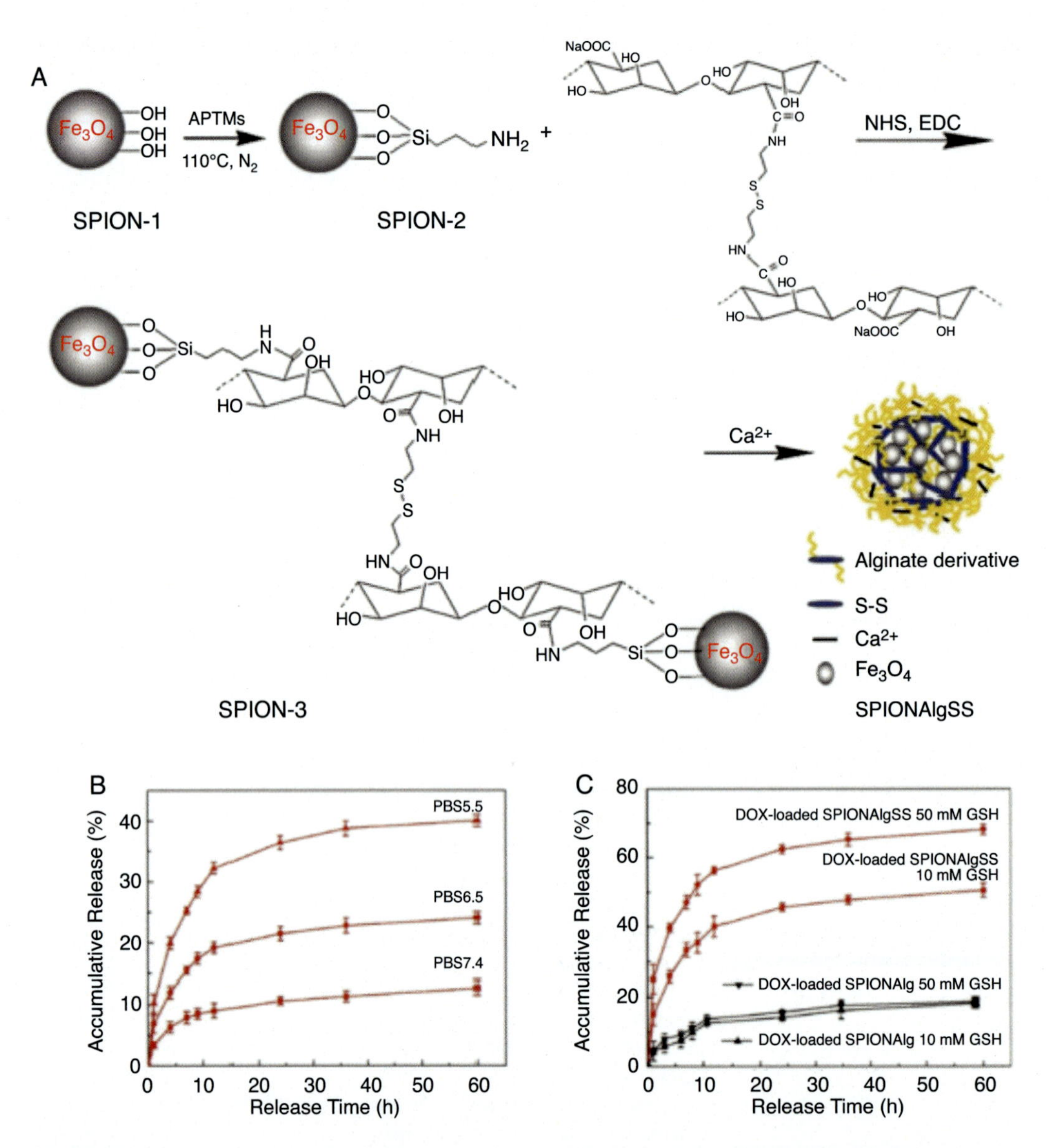

FIGURE 25.9 Preparation of dual-responsive Alg- modified SPIONs (SPIONAlgSS) (A) pH-triggered release of DOX from DOX-loaded SPIONAlgSS in PBS (B), and redox-triggered release of DOX from DOX-loaded SPIONAlg and SPIONAlgSS in PBS (pH 7.4) with GSH (C). "Reprinted from Carbohydr Polym., 204 (2019) 32, Peng et al., Novel dual responsive alginate-based magnetic nanogels for onco-theranostics. Copyright (2019), with permission from Elsevier".

carried more oleic acid-coated MNPs. The amphipathic derivatives of chitosan improved entrapment efficiency of paclitaxel (PTX) which could be released at low pH due to the electrostatic repulsion between positive charges and shell loosening [155].

DOX-conjugated heparin was also used for targeted anticancer drug delivery. The hydrophilic MNPs were mixed with heparin in water under sonication resulting in cluster formation. Their cellular uptake efficiency was higher than of free DOX in A549 cancer cells. The heparin-modified SPIONs were able to target solid tumors through the passive targeting effect and inhibit the tumor growth. [81]

The MNPs can also be used for carrying therapeutic genes injected intravenously (*in-vivo* magnetofection). Once they are captured at the desired site, they are uptaken by the cells, followed by the enzymatic cleavage or degradation of biomacromolecules, resulting in gene release. DNA was mixed with HA solution to interact subsequently with PEI-modified MNPs. HA played the role in reducing cytotoxicity and facilitating recognition of CD44-mediated endocytosis of the particles into dendritic cells [147]. Besides, the MNPs coated with positively charged chitosan can be directly used to promote interaction with different genes such as enhanced green fluorescent protein plasmid DNA, Viral gene (Ad/LacZ), or MDR1 siRNA [149]. The MNPs containing Fe_3O_4-dextran-anti-β-human chorionic gonadotropin were prepared using the chemical coprecipitation method. Anti-β-human chorionic gonadotropin monoclonal antibody was conjugated to the dextran-coated MNPs via the Schiff base reaction after oxidation of dextran hydroxyl groups to aldehydes. The transfection efficiency of these NPs was found to be significantly greater than the efficiency of liposomes [156].

Some polysaccharides can be used for tumor cell targeting through recognition by cancer cell surface glycoproteins. As mentioned above, HA can be recognized by the CD44 receptor overexpressed in tumor cells and many types of cancer such as breast, ovarian, or colon cancer. Based on the selective HA-CD44 interaction, HA-coated MNPs were able to detect leukemia cells and extract them from the cell mixture [157], recognize and enhance the efficacy of drug release in human breast cancer cells [132,158], or were selectively cytotoxic toward colon adenocarcinoma (HT-29) [133]. Modification of MNPs with HA is an up-and-coming technique that can be used in the development of MNPs that target tumor cells and simultaneously can be used for imaging tumors using MRI.

Proteins, similarly to polysaccharides, can offer cellular targeting and anticancer drug delivery with controlled release under different stimuli, hyperthermia treatment, or MRI diagnosis (Table 25.2). The human serum albumin (HSA) was used for engineering the magnetic hybrid NPs with physically enclosed 5-fluorouracil. HSA is known from the enhanced uptake into cancer cells via the binding to the albondin receptor or binding to "secreted protein, acidic and rich in cysteine" (SPARC)— highly expressed in malignant and stromal cells. However, an external magnetic field was used as a global guiding mechanism, the local targeting and cellular uptake by squamous cell carcinoma was ensured by the HSA presence [106]. Drugs which are chemically conjugated to the amine groups of proteins through carbodiimide coupling can be released after bond cleavage within the tumor cells containing enzymes capable of digesting proteins [98]. When a drug is physically encapsulated, that is, during desolvation method of HSA NPs preparation, its release depends on the degradation rate by proteolytic enzymes [92]. The release of Pt prodrug from gelatin-coated MNPs can be triggered by using the pancreatic enzyme causing degradation and detachment of protein segments from the surface [98]. The protease enzyme present in intestine was able to degrade layer of casein which earlier protected iron oxide NPs and DOX in the gastric acidic medium [159]. By using magnetic field as an external stimulus, the expansion of the protein chains and drug release was possible due to the changes in alignment of MNPs [142]. Additionally, it has been observed, that the magnetically induced hyperthermia enhanced the drug release from gelatin-modified MNPs. The increased rate of cisplatine release could be related to the movement of MNPs at higher temperature, thus loosening the polymer matrix [112].

25.4.2 Hyperthermia

Hyperthermia is a type of cancer treatment in which specific areas of the body undergo a local increase of the temperature to damage and kill cancer cells. After injection, MNPs accumulate at the tumor either by using the magnetic guiding, or thanks to the selective binding of ligands on the MNPs surface to receptors on cancer cells, or by the EPR effect. The MNPs can convert the electromagnetic energy into heat by fluctuating the external magnetic field, which results in the local increase of the temperature [160]. This effect provides a promising cancer therapy strategy by raising the cell temperature to 41–45 °C. The damage of healthy cells in this temperature range is reversible, whereas it is irreversible for cancer cells [161].

For hyperthermia treatment, MNPs should have high magnetization to generate high thermal energies, be nontoxic and biocompatible, and protected against NPs oxidation in air.

The nanoclusters of assembled ferrimagnetic iron oxide nanocubes coated with chitosan-DOPA polymer shells were designed for two aims: magnetically guided targeting and localized heating (Fig. 25.10). These nanohybrids displayed higher saturation magnetization which was advantageous for magnetic guiding applications as it allows to respond rapidly to an external magnetic field. To confirm their applicability for magnetic hyperthermia, the tests of heating efficiency were performed in comparison with a commercially available superparamagnetic iron oxide NP (Feridex). When chitosan-coated ferrimagnetic iron oxide nanoclusters (Chito-Fion) solution was exposed to the AMF, the temperature increased rapidly and reached the therapeutic threshold required for cancer hyperthermia ($T > 42$ °C) within 10 min. In contrast, the temperature of the Feridex solution was not raised above 42 °C even after 20 min of exposure to an alternating current magnetic field at a frequency of 1 MHz. The tumor treatment under an AMF using the Fe_3O_4@Chito NPs decreased the tumor volume by ˜70%, while no significant tumoricidal effect of Feridex was observed [139].

Pullulan acetate–coated MNPs showed good heating properties in an AMF and induced a decrease of cell viability in comparison to the magnetic field alone [86].

The protein-MNPs conjugates showed also potential for hyperthermia treatment. The biocompatible BSA-MNPs were capable to induce heating for thermal therapy. It was observed that due to better colloidal stability, the specific absorption rate (SAR) value under a magnetic field of the BSA-iron oxide is higher than the iron oxide NPs without BSA, irrespective of the size of the iron oxide NPs and method of conjugation [94,162]. However, this protein application is problematic due to its heterogeneous structure that can induce adverse immune response of patients. Chang et al. showed rapid heating capacity under AMF exposure for MNPs coated with the human-like collagen protein (HLC-MNPs). In this case, the SAR values were higher when MNPs were conjugated with protein in comparison to NPs alone. Magnetic NPs after surface modification have good dispersibility thanks to the presence of protein shell, which is important to keep high values of energy absorption [163]. Despite research effort, it is not very

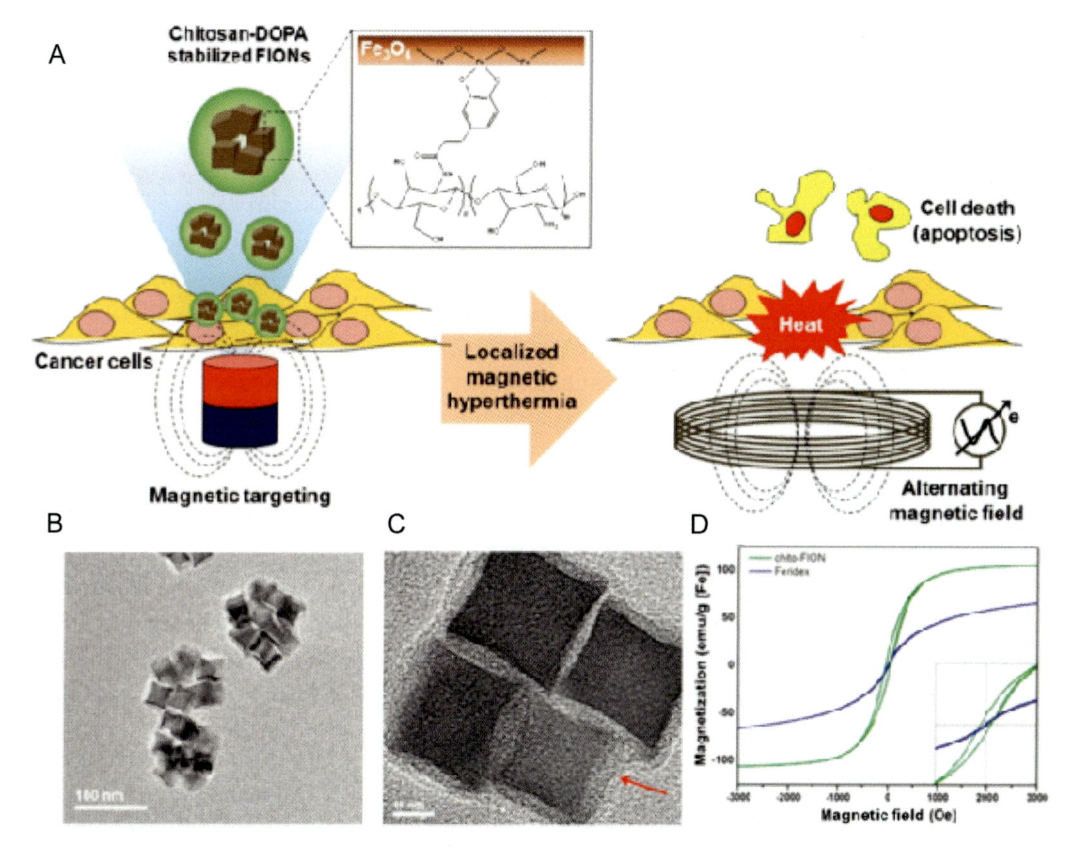

FIGURE 25.10 Schematic illustration of Chito-FIONs and their applications for the localized magnetic hyperthermia of cancer cells (A) with transmission electron microscopy images of Chito-FIONs (the red arrow indicates the presence of the polymer coating layers) (B,C), and saturation magnetization curves of Chito-FIONs and commercial Feridex (C). "Reprinted (adapted) with permission from ACS Nano., 6 (2012) 5266, Bae et al., Chitosan Oligosaccharide-Stabilized Ferrimagnetic Iron Oxide Nanocubes for Magnetically Modulated Cancer Hyperthermia. Copyright (2012) American Chemical Society."

clear whether protein can have significant effect in improving SAR. The effect of MNP size, surface protein, and biocompatibility should be profoundly investigated to provide more information on the design parameters allowing to produce a perfect magnetic hyperthermia agent in clinical applications.

25.4.3 Magnetic resonance imaging

MRI, based on the relaxation of protons in tissues, is one of the most powerful noninvasive imaging methods in the clinic. The unique superparamagnetic properties of SPIONs are very advantageous after introduction into living systems; the particles are only active in the presence of an external magnetic field. When accumulated in tissues, SPIONs enhance the proton relaxation of one tissue compared to others, thus serve as an MRI contrast agent.

SPIONs have long been used as contrast-enhancing agents for MRI. Today, among various SPION-based contrast agents tested clinically, we can find dextran and carboxymethyldextran-coated iron oxide NPs. Ferumoxtran (dextran-coated MNPs) was clinically tested for MR imaging of carotid plaques in patients with atherosclerosis. Despite initial very promising data, it was withdrawn from use because of unsatisfactory statistical evidence of efficacy. Carboxymethyldextran-coated SPIONs (Sienna+), approved in Europe (2011), are applied locally into the interstitial tissue of patients with breast cancer and are detected by SentiMAG (not MRI). The combination of Sienna+ with the SentiMAG magnetometer has also shown promising preliminary results in the detection of lymph nodes in patients with prostate cancer. [164,165] The role of dextran corona is steric stabilization in water and physiological medium and reduction of the opsonization process *in vivo*. Dextran sulfate–coated SPIONs were synthesized as targeted contrast agents for cardiovascular imaging, because the negatively charged dextran sulfate can be recognized and endocytosed by macrophage surface receptors [153].

Similarly, HA-modified SPIONs were prepared as targeted MRI agents. A dopamine derivative of HA was tested for the fabrication of magnetic hybrid NPs, which can play the role of MRI contrast agent. [31,144] T_2-weighted gradient-echo MRI was performed using the CD44-positive cell line, HCT116, and a CD44-negative cell line, NIH3T3. The results of this study confirmed the targeting capability of the HA-coated MNPs via the CD44-HA receptor-ligand mechanism [144]. The higher relaxivity of HA-coated MNPs in comparison to MNPs coated only with dopamine is most probably related with the amount of water surrounding the particles due to the extremely high hydrophilicity of HA. HA carries large amount of water close to MNPs, thus increase MRI contrast [31]. High relaxivity was also observed for multifunctional Fe_3O_4/Au–PEI-HA and pheophorbide—a conjugated acetylated HA onto MNPs (AHP@MNPs) NPs, both revealing good water dispersibility, colloidal stability, cytocompatibility, and targeting properties to CD44 receptor overexpressing cancer cells [145,146].

To assess potential clinical relevance of chitosan-coated SPIONs micelles, a commercially available SPION-negative contrast agent (Ferucarbotran) was tested as a control. Darkening of the liver was observed in both cases; however, many of the fine hepatic liver vessels observed with the chitosan-coated MNPs were not seen with commercially available Ferucarbotran [140]. Oleic acid–coated MNPs can be also encapsulated into the NPs of chitosan–linoleic acid. The encapsulation procedure resulted in the clustering of SPIONs in the cores, which dramatically improved T2 relaxivity. After *in-vivo* intravascular injection, the resulting magnetic hybrid NPs showed relative signal enhancement in the liver [148]. The stimuli-responsive hybrid clusters of PEGylated chitosan encapsulating DOX and MNPs in the core revealed the high saturation magnetization which is promising for MRI and drug delivery [151]. Shi *et al.* tested carboxymethyl-chitosan-grafted superparamagnetic iron oxide nanoparticles (IONPs) (Fe_3O_4@Chito) for MRI of stem cells. The high relaxivity ratio R_2/R_1 of Fe_3O_4@Chito was calculated to be about 40 higher than those of the commercial Feridex and Resovist. The NPs have shown no significant cytotoxicity and were internalized by stem cells *via* nonspecific adsorptions [55].

Among different proteins, casein, gelatin, or BSA show enhancement of MRI contrast, due to the favorable water diffusion and interaction with the inner layer adjacent to the MNP surface [92].

25.5 Conclusion

Employing natural polymers in the buildup of magnetic nanohybrids for biomedical applications can provide many excellent properties including stabilization in biological fluids, protection of the magnetic core, biocompatibility, drug and gene transport and delivery. So far, a number of formulations of polysaccharide or protein magnetic nanohybrids exhibiting multifunctional properties have been investigated.

This chapter presented different methods for modification of MNP using natural polymers, either polysaccharides or proteins. Both *in-situ* modification and *ex-situ* coating of MNPs result in hydrophilic coats playing the role in the stabilization of NPs in physiological medium. Besides, magnetic nanoclusters obtained by the assembly of individual MNPs in the presence of surfactants or polymers are attracting more and more attention due to the better control of magnetic properties. Therefore, a part of this chapter was dedicated to methods for their preparation and coating with natural polymers.

The performance of such nanomaterials relies strongly on the quality of initially used building blocks, MNPs, and polymer chains. Compared to synthetic polymers, proteins and polysaccharides are often heterogeneous in sizes/molecular weights. The final nanohybrids' size and their dispersity can thus vary from batch to batch. A special attention should be devoted to the initial step of natural polymers preparation such as source of extraction, purification, and chain size separation. Nevertheless, while several dextran-coated MNPs have been commercially used as clinical contrast agents for MRI because of the role of dextran as steric stabilizer, the research studies show great promise of other polysaccharide-coated MNPs for targeted cancer imaging and therapy. HA has thus been shown a promising building block for designing such nanohybrids due to its ability to recognize surface receptors overexpressed by tumor cells. Moreover, its commercial availability in a wide range of molecular weights and its chemical structure, allowing selective chemical modifications, open doors for the development of MNPs for advanced therapeutic applications including drug delivery, MRI, and hyperthermia treatment. Though the magnetic guiding of hybrid NPs is still challenging in the large animals and humans, the bio-coated MNPs, especially magnetic nanoclusters should be considered as the one of the most promising platforms for the next stage of cancer treatment.

References

[1] J. Chomoucka, J. Drbohlavova, D. Huska, V. Adam, R. Kizek, J. Hubalek, Magnetic nanoparticles and targeted drug delivering, Pharmacol. Res. 62 (2010) 144–149. https://doi.org/10.1016/j.phrs.2010.01.014.

[2] Y. Xiao, J. Du, Superparamagnetic nanoparticles for biomedical applications, J. Mater. Chem. B. 8 (2020) 354–367. https://doi.org/10.1039/C9TB01955C.

[3] A. Shkilnyy, E. Munnier, K. Hervé, M. Soucé, R. Benoit, S. Cohen-Jonathan, P. Limelette, M.-L. Saboungi, P. Dubois, I. Chourpa, Synthesis and evaluation of novel biocompatible super-paramagnetic iron oxide nanoparticles as magnetic anticancer drug carrier and fluorescence active label, J. Phys. Chem. C. 114 (2010) 5850–5858. https://doi.org/10.1021/jp9112188.

[4] M.M. Lin, D.K. Kim, A.J. El Haj, J. Dobson, Development of superparamagnetic iron oxide nanoparticles (SPIONS) for translation to clinical applications, IEEE Trans. NanoBiosci. 7 (2008) 298–305. https://doi.org/10.1109/TNB.2008.2011864.

[5] G.A. Marcelo, C. Lodeiro, J.L. Capelo, J. Lorenzo, E. Oliveira, Magnetic, fluorescent and hybrid nanoparticles: from synthesis to application in biosystems, Mater. Sci. Eng. C. 106 (2020) 110104. https://doi.org/10.1016/j.msec.2019.110104.

[6] K. Raghava Reddy, P.A. Reddy, C.V. Reddy, N.P. Shetti, B. Babu, K. Ravindranadh, M.V. Shankar, M.C. Reddy, S. Soni, S. NaveenV. Gurtler, A.S. Ball, S. Soni (Eds.), Functionalized magnetic nanoparticles/biopolymer hybrids: synthesis methods, properties and biomedical applications, Methods in Microbiology (2019) 227–254. https://doi.org/10.1016/bs.mim.2019.04.005.

[7] S. Dumitriu, Polysaccharides: Structural Diversity and Functional Versatility, 2nd Edition, CRC Press, FL, 2004.

[8] K.Y. Lee, D.J. Mooney, Alginate: properties and biomedical applications, Prog. Polym. Sci. 37 (2012) 106–126. https://doi.org/10.1016/j.progpolymsci.2011.06.003.

[9] K.M. Zia, S. Tabasum, M. Nasif, N. Sultan, N. Aslam, A. Noreen, M. Zuber, A review on synthesis, properties and applications of natural polymer based carrageenan blends and composites, Int. J. Biol. Macromol. 96 (2017) 282–301. https://doi.org/10.1016/j.ijbiomac.2016.11.095.

[10] D.L. Rabenstein, Heparin and heparan sulfate: structure and function, Nat. Prod. Rep. 19 (2002) 312–331. https://doi.org/10.1039/B100916H.

[11] I. Younes, M. Rinaudo, Chitin and chitosan preparation from marine sources. structure, properties and applications, Mar. Drugs. 13 (2015) 1133–1174. https://doi.org/10.3390/md13031133.

[12] C.E. Ioan, T. Aberle, W. Burchard, Structure properties of dextran. 2. dilute solution, Macromolecules 33 (2000) 5730–5739. https://doi.org/10.1021/ma000282n.

[13] R.S. Singh, G.K. Saini, J.F. Kennedy, Pullulan: Microbial sources, production and applications, Carbohydr. Polym. 73 (2008) 515–531. https://doi.org/10.1016/j.carbpol.2008.01.003.

[14] M.N. Collins, C. Birkinshaw, Hyaluronic acid based scaffolds for tissue engineering—a review, Carbohydr. Polym. 92 (2013) 1262–1279. https://doi.org/10.1016/j.carbpol.2012.10.028.

[15] J.A. Burdick, G.D. Prestwich, Hyaluronic acid hydrogels for biomedical applications, Adv. Mater. 23 (2011) H41–H56. https://doi.org/10.1002/adma.201003963.

[16] S. Bowman, M.E. Awad, M.W. Hamrick, M. Hunter, S. Fulzele, Recent advances in hyaluronic acid based therapy for osteoarthritis, Clin. Transl. Med. 7 (2018). https://doi.org/10.1186/s40169-017-0180-3.

[17] D. Tarus, L. Hamard, F. Caraguel, D. Wion, A. Szarpak-Jankowska, B. van der Sanden, R. Auzély-Velty, Design of hyaluronic acid hydrogels to promote neurite outgrowth in three dimensions, ACS Appl. Mater. Interfaces. 8 (2016) 25051–25059. https://doi.org/10.1021/acsami.6b06446.

[18] P. Zhai, X. Peng, B. Li, Y. Liu, H. Sun, X. Li, The application of hyaluronic acid in bone regeneration, Int. J. Biol. Macromol. 151 (2020) 1224–1239. https://doi.org/10.1016/j.ijbiomac.2019.10.169.

[19] H.-J. Cho, Recent progresses in the development of hyaluronic acid-based nanosystems for tumor-targeted drug delivery and cancer imaging, J. Pharm. Investig. 50 (2020) 115–129. https://doi.org/10.1007/s40005-019-00448-w.

[20] Y. Zhu, X. Wang, J. Chen, J. Zhang, F. Meng, C. Deng, R. Cheng, J. Feijen, Z. Zhong, Bioresponsive and fluorescent hyaluronic acid-iodixanol nanogels for targeted X-ray computed tomography imaging and chemotherapy of breast tumors, J. Control, Release. 244 (2016) 229–239. https://doi.org/10.1016/j.jconrel.2016.08.027.

[21] X.-Y. He, B.-Y. Liu, C. Xu, R.-X. Zhuo, S.-X. Cheng, A multi-functional macrophage and tumor targeting gene delivery system for the regulation of macrophage polarity and reversal of cancer immunoresistance, Nanoscale 10 (2018) 15578–15587. https://doi.org/10.1039/C8NR05294H.

[22] R. Ran, Y. Liu, H. Gao, Q. Kuang, Q. Zhang, J. Tang, K. Huang, X. Chen, Z. Zhang, Q. He, Enhanced gene delivery efficiency of cationic liposomes coated with PEGylated hyaluronic acid for anti P-glycoprotein siRNA: a potential candidate for overcoming multi-drug resistance, Int. J. Pharm. 477 (2014) 590–600. https://doi.org/10.1016/j.ijpharm.2014.11.012.

[23] Y. Zhu, Z. PangM.S. Hasnain, A.K. Nayak (Eds.), Hyaluronic acid in drug delivery applications, Natural Polysaccharides in Drug Delivery and Biomedical Applications (2019) 307–325. https://doi.org/10.1016/B978-0-12-817055-7.00013-3.

[24] S. Trombino, C. Servidio, F. Curcio, R. Cassano, Strategies for hyaluronic acid-based hydrogel design in drug delivery, Pharmaceutics (2019) 11. https://doi.org/10.3390/pharmaceutics11080407.

[25] H. Kim, M. Shin, S. Han, W. Kwon, S.K. Hahn, Hyaluronic acid derivatives for translational medicines, Biomacromolecules 20 (2019) 2889–2903. https://doi.org/10.1021/acs.biomac.9b00564.

[26] M.H. El-Dakdouki, D.C. Zhu, K. El-Boubbou, M. Kamat, J. Chen, W. Li, X. Huang, Development of multifunctional hyaluronan-coated nanoparticles for imaging and drug delivery to cancer cells, Biomacromolecules 13 (2012) 1144–1151. https://doi.org/10.1021/bm300046h.

[27] H. Cai, X. An, J. Cui, J. Li, S. Wen, K. Li, M. Shen, L. Zheng, G. Zhang, X. Shi, Facile hydrothermal synthesis and surface functionalization of polyethyleneimine-coated iron oxide nanoparticles for biomedical applications, ACS Appl. Mater. Interfaces. 5 (2013) 1722–1731. https://doi.org/10.1021/am302883m.

[28] J. Li, Y. He, W. Sun, Y. Luo, H. Cai, Y. Pan, M. Shen, J. Xia, X. Shi, Hyaluronic acid-modified hydrothermally synthesized iron oxide nanoparticles for targeted tumor MR imaging, Biomaterials 35 (2014) 3666–3677. https://doi.org/10.1016/j.biomaterials.2014.01.011.

[29] S. Manju, K. Sreenivasan, Enhanced drug loading on magnetic nanoparticles by layer-by-layer assembly using drug conjugates: blood compatibility evaluation and targeted drug delivery in cancer cells, Langmuir 27 (2011) 14489–14496. https://doi.org/10.1021/la202470k.

[30] Q. Li, Y. Chen, X. Zhou, D. Chen, Y. Li, J. Yang, X. Zhu, Hyaluronic acid-methotrexate conjugates coated magnetic polydopamine nanoparticles for multimodal imaging-guided multistage targeted chemo-photothermal therapy, Mol. Pharm. 15 (2018) 4049–4062. https://doi.org/10.1021/acs.molpharmaceut.8b00473.

[31] M. Babic, D. Horak, P. Jendelova, V. Herynek, V. Proks, V. Vanecek, P. Lesny, E. Sykova, The use of dopamine-hyaluronate associate-coated maghemite nanoparticles to label cells, Int. J. Nanomedicine. 7 (2012) 1461–1474. https://doi.org/10.2147/IJN.S28658.

[32] R.S. Molday, D. Mackenzie, Immunospecific ferromagnetic iron-dextran reagents for the labeling and magnetic separation of cells, J. Immunol. Methods. 52 (1982) 353–367. https://doi.org/10.1016/0022-1759(82)90007-2.

[33] C. Tassa, S.Y. Shaw, R. Weissleder, Dextran-coated iron oxide nanoparticles: a versatile platform for targeted molecular imaging, molecular diagnostics, and therapy, Acc. Chem. Res. 44 (2011) 842–852. https://doi.org/10.1021/ar200084x.

[34] S. Laurent, D. Forge, M. Port, A. Roch, C. Robic, L. Vander Elst, R.N. Muller, Magnetic iron oxide nanoparticles: synthesis, stabilization, vectorization, physicochemical characterizations, and biological applications, Chem. Rev. 108 (2008) 2064–2110. https://doi.org/10.1021/cr068445e.

[35] P.H. Linh, N.X. Phuc, L.V. Hong, L.L. Uyen, N.V. Chien, P.H. Nam, N.T. Quy, H.T.M. Nhung, P.T. Phong, I.-J. Lee, Dextran coated magnetite high susceptibility nanoparticles for hyperthermia applications, J. Magn. Magn. Mater. 460 (2018) 128–136. https://doi.org/10.1016/j.jmmm.2018.03.065.

[36] X.Q. Xu, H. Shen, J.R. Xu, J. Xu, X.J. Li, X.M. Xiong, Core-shell structure and magnetic properties of magnetite magnetic fluids stabilized with dextran, Appl. Surf. Sci. 252 (2005) 494–500. https://doi.org/10.1016/j.apsusc.2005.01.027.

[37] H. Pardoe, W. Chua-anusorn, T.G. St. Pierre, J. Dobson, Structural and magnetic properties of nanoscale iron oxide particles synthesized in the presence of dextran or polyvinyl alcohol, J. Magn. Magn. Mater. 225 (2001) 41–46. https://doi.org/10.1016/S0304-8853(00)01226-9.

[38] T. Kawaguchi, T. Hanaichi, M. Hasegawa, S. Maruno, Dextran-magnetite complex: conformation of dextran chains and stability of solution J. Mater. Sci. Mater. Med. 12, 121–127.

[39] Z. Shaterabadi, G. Nabiyouni, M. Soleymani, High impact of in situ dextran coating on biocompatibility, stability and magnetic properties of iron oxide nanoparticles, Mater. Sci. Eng. C. 75 (2017) 947–956. https://doi.org/10.1016/j.msec.2017.02.143.

[40] E.Y. Sun, L. Josephson, K.A. Kelly, R. Weissleder, Development of nanoparticle libraries for biosensing, Bioconjug. Chem. 17 (2006) 109–113. https://doi.org/10.1021/bc050290e.

[41] X. Hong, W. Guo, H. Yuan, J. Li, Y. Liu, L. Ma, Y. Bai, T. Li, Periodate oxidation of nanoscaled magnetic dextran composites, J. Magn. Magn. Mater. 269 (2004) 95–100. https://doi.org/10.1016/S0304-8853(03)00566-3.

[42] S. Mornet, J. Portier, E. Duguet, A method for synthesis and functionalization of ultrasmall superparamagnetic covalent carriers based on maghemite and dextran, J. Magn. Magn. Mater. 293 (2005) 127–134. https://doi.org/10.1016/j.jmmm.2005.01.053.

[43] A.P. Herrera, C. Barrera, C. Rinaldi, Synthesis and functionalization of magnetite nanoparticles with aminopropylsilane and carboxymethyldextran, J. Mater. Chem. 18 (2008) 3650–3654. https://doi.org/10.1039/B805256E.

[44] S. Islam, M.A.R. Bhuiyan, M.N. Islam, Chitin and chitosan: structure, properties and applications in biomedical engineering, J. Polym. Environ. 25 (2017) 854–866. https://doi.org/10.1007/s10924-016-0865-5.

[45] A. Verlee, S. Mincke, C.V. Stevens, Recent developments in antibacterial and antifungal chitosan and its derivatives, Carbohydr. Polym. 164 (2017) 268–283. https://doi.org/10.1016/j.carbpol.2017.02.001.

[46] H. Wang, J. Qian, F. Ding, Recent advances in engineered chitosan-based nanogels for biomedical applications, J. Mater. Chem. B. 5 (2017) 6986–7007. https://doi.org/10.1039/C7TB01624G.

[47] H. Hamedi, S. Moradi, S.M. Hudson, A.E. Tonelli, Chitosan based hydrogels and their applications for drug delivery in wound dressings: a review, Carbohydr. Polym. 199 (2018) 445–460. https://doi.org/10.1016/j.carbpol.2018.06.114.

[48] Ö. Tezgel, A. Szarpak-Jankowska, A. Arnould, R. Auzély-Velty, I. Texier, Chitosan-lipid nanoparticles (CS-LNPs): application to siRNA delivery, J. Coll. Interface Sci. 510 (2018) 45–56. https://doi.org/10.1016/j.jcis.2017.09.045.

[49] U. Garg, S. Chauhan, U. Nagaich, N. Jain, Current Advances in chitosan nanoparticles based drug delivery and targeting, Adv. Pharm. Bull. 9 (2019) 195–204. https://doi.org/10.15171/apb.2019.023.

[50] Y. Cao, Y.F. Tan, Y.S. Wong, M.W.J. Liew, S. Venkatraman, Recent advances in chitosan-based carriers for gene delivery, Mar. Drugs. (2019) 17. https://doi.org/10.3390/md17060381.

[51] Z. Marková, K. Šišková, J. Filip, K. Šafářová, R. Prucek, A. Panáček, M. Kolář, R. Zbořil, Chitosan -based synthesis of magnetically-driven nanocomposites with biogenic magnetite core, controlled silver size, and high antimicrobial activity, Green Chem 14 (2012) 2550–2558. https://doi.org/10.1039/C2GC35545K.

[52] S. Honary, P. Ebrahimi, H.A. Rad, M. Asgari, Optimization of preparation of chitosan-coated iron oxide nanoparticles for biomedical applications by chemometrics approaches, Int. Nano Lett. 3 (2013) 48. https://doi.org/10.1186/2228-5326-3-48.

[53] A. Kong, P. Wang, H. Zhang, F. Yang, S. Huang, Y. Shan, One-pot fabrication of magnetically recoverable acid nanocatalyst, heteropolyacids/chitosan/Fe_3O_4, and its catalytic performance, Appl. Catal. Gen. 417–418 (2012) 183–189. https://doi.org/10.1016/j.apcata.2011.12.040.

[54] X. Song, X. Luo, Q. Zhang, A. Zhu, L. Ji, C. Yan, Preparation and characterization of biofunctionalized chitosan/Fe_3O_4 magnetic nanoparticles for application in liver magnetic resonance imaging, J. Magn. Magn. Mater. 388 (2015) 116–122. https://doi.org/10.1016/j.jmmm.2015.04.017.

[55] Z. Shi, K.G. Neoh, E.T. Kang, B. Shuter, S.-C. Wang, C. Poh, W. Wang, Carboxymethyl)chitosan-modified superparamagnetic iron oxide nanoparticles for magnetic resonance imaging of stem cells, ACS Appl. Mater. Interfaces. 1 (2009) 328–335. https://doi.org/10.1021/am8000538.

[56] M. Arakha, S. Pal, D. Samantarrai, T.K. Panigrahi, B.C. Mallick, K. Pramanik, B. Mallick, S. Jha, Antimicrobial activity of iron oxide nanoparticle upon modulation of nanoparticle-bacteria interface, Sci. Rep. 5 (2015) 14813. https://doi.org/10.1038/srep14813.

[57] A. Szpak, G. Kania, T. Skórka, W. Tokarz, S. Zapotoczny, M. Nowakowska, Stable aqueous dispersion of superparamagnetic iron oxide nanoparticles protected by charged chitosan derivatives, J. Nanoparticle Res. 15 (2012) 1372. https://doi.org/10.1007/s11051-012-1372-9.

[58] H. Chen, X. Xing, H. Tan, Y. Jia, T. Zhou, Y. Chen, Z. Ling, X. Hu, Covalently antibacterial alginate-chitosan hydrogel dressing integrated gelatin microspheres containing tetracycline hydrochloride for wound healing, Mater. Sci. Eng. C. 70 (2017) 287–295. https://doi.org/10.1016/j.msec.2016.08.086.

[59] S. Deepthi, R. Jayakumar, Alginate nanobeads interspersed fibrin network as in situ forming hydrogel for soft tissue engineering, Bioact. Mater. 3 (2018) 194–200. https://doi.org/10.1016/j.bioactmat.2017.09.005.

[60] S.T. Bendtsen, S.P. Quinnell, M. Wei, Development of a novel alginate-polyvinyl alcohol-hydroxyapatite hydrogel for 3D bioprinting bone tissue engineered scaffolds, J. Biomed. Mater. Res. A. 105 (2017) 1457–1468. https://doi.org/10.1002/jbm.a.36036.

[61] J.M. Unagolla, A.C. Jayasuriya, Drug transport mechanisms and in vitro release kinetics of vancomycin encapsulated chitosan-alginate polyelectrolyte microparticles as a controlled drug delivery system, Eur. J. Pharm. Sci. 114 (2018) 199–209. https://doi.org/10.1016/j.ejps.2017.12.012.

[62] N. Kahya, F.B. Erim, Surfactant modified alginate composite gels for controlled release of protein drug, Carbohydr. Polym. 224 (2019) 115165. https://doi.org/10.1016/j.carbpol.2019.115165.

[63] S. Choudhary, J.M. Reck, A.J. Carr, S.R. Bhatia, Hydrophobically modified alginate for extended release of pharmaceuticals, Polym. Adv. Technol. 29 (2018) 198–204. https://doi.org/10.1002/pat.4103.

[64] S.J. Bidarra, C.C. Barrias, P.L. Granja, Injectable alginate hydrogels for cell delivery in tissue engineering, Acta Biomater 10 (2014) 1646–1662. https://doi.org/10.1016/j.actbio.2013.12.006.

[65] M. Pei, X. Jia, X. Zhao, J. Li, P. Liu, Alginate-based cancer-associated, stimuli-driven and turn-on theranostic prodrug nanogel for cancer detection and treatment, Carbohydr. Polym. 183 (2018) 131–139. https://doi.org/10.1016/j.carbpol.2017.12.013.

[66] K. Podgórna, K. Szczepanowicz, M. Piotrowski, M. Gajdošová, F. Štěpánek, P. Warszyński, Gadolinium alginate nanogels for theranostic applications, Coll. Surf. B Biointerfaces. 153 (2017) 183–189. https://doi.org/10.1016/j.colsurfb.2017.02.026.

[67] E. Lengert, M. Saveleva, A. Abalymov, V. Atkin, P.C. Wuytens, R. Kamyshinsky, A.L. Vasiliev, D.A. Gorin, G.B. Sukhorukov, A.G. Skirtach, B. Parakhonskiy, Silver alginate hydrogel micro- and nanocontainers for theranostics: synthesis, encapsulation, remote release, and detection, ACS Appl. Mater. Interfaces. 9 (2017) 21949–21958. https://doi.org/10.1021/acsami.7b08147.

[68] E. Kroll, F.M. Winnik, R.F. Ziolo, In situ preparation of nanocrystalline γ-Fe_2O_3 in iron(II) cross-linked alginate gels, Chem. Mater. 8 (1996) 1594–1596. https://doi.org/10.1021/cm960095x.

[69] F. Llanes, D.H. Ryan, R.H. Marchessault, Magnetic nanostructured composites using alginates of different M/G ratios as polymeric matrix, Int. J. Biol. Macromol. 27 (2000) 35–40. https://doi.org/10.1016/S0141-8130(99)00115-4.

[70] C.-H. Su, F.-Y. Cheng, In vitro and in vivo applications of alginate/iron oxide nanocomposites for theranostic molecular imaging in a brain tumor model, RSC Adv 5 (2015) 90061–90064. https://doi.org/10.1039/C5RA20723A.

[71] E.R. Morris, D.A. Rees, G. Robinson, Cation-specific aggregation of carrageenan helices: domain model of polymer gel structure, J. Mol. Biol. 138 (1980) 349–362. https://doi.org/10.1016/0022-2836(80)90291-0.

[72] K. Oya, T. Tsuru, Y. Teramoto, Y. Nishio, Nanoincorporation of iron oxides into carrageenan gels and magnetometric and morphological characterizations of the composite products, Polym. J. 45 (2013) 824–833. https://doi.org/10.1038/pj.2012.221.

[73] F. Jones, H. Cölfen, M. Antonietti, Interaction of κ##, Biomacromolecules 1 (2000) 556–563. https://doi.org/10.1021/bm0055089.

[74] A.L. Daniel-da-Silva, T. Trindade, B.J. Goodfellow, B.F.O. Costa, R.N. Correia, A.M. Gil, In situ synthesis of magnetite nanoparticles in carrageenan gels, Biomacromolecules 8 (2007) 2350–2357. https://doi.org/10.1021/bm070096q.

[75] G.R. Mahdavinia, A. Mosallanezhad, M. Soleymani, M. Sabzi, Magnetic- and pH-responsive κ-carrageenan/chitosan complexes for controlled release of methotrexate anticancer drug, Int. J. Biol. Macromol. 97 (2017) 209–217. https://doi.org/10.1016/j.ijbiomac.2017.01.012.

[76] M.H. Karimi, G.R. Mahdavinia, B. Massoumi, A. Baghban, M. Saraei, Ionically crosslinked magnetic chitosan/κ-carrageenan bioadsorbents for removal of anionic eriochrome black-T, Int. J. Biol. Macromol. 113 (2018) 361–375. https://doi.org/10.1016/j.ijbiomac.2018.02.102.

[77] J. Long, X. Yu, E. Xu, Z. Wu, X. Xu, Z. Jin, A. Jiao, In situ synthesis of new magnetite chitosan/carrageenan nanocomposites by electrostatic interactions for protein delivery applications, Carbohydr. Polym. 131 (2015) 98–107. https://doi.org/10.1016/j.carbpol.2015.05.058.

[78] U. Bhaskar, E. Sterner, A.M. Hickey, A. Onishi, F. Zhang, J.S. Dordick, R.J. Linhardt, Engineering of routes to heparin and related polysaccharides, Appl. Microbiol. Biotechnol. 93 (2012) 1–16. https://doi.org/10.1007/s00253-011-3641-4.

[79] S.M. Smorenburg, C.J.F.V. Noorden, The complex effects of heparins on cancer progression and metastasis in experimental studies, Pharmacol. Rev. 53 (2001) 93–106.

[80] J. Lee, M.J. Jung, Y.H. Hwang, Y.J. Lee, S. Lee, D.Y. Lee, H. Shin, Heparin-coated superparamagnetic iron oxide for in vivo MR imaging of human MSCs, Biomaterials 33 (2012) 4861–4871. https://doi.org/10.1016/j.biomaterials.2012.03.035.

[81] Y. Yang, Q. Guo, J. Peng, J. Su, X. Lu, Y. Zhao, Z. Qian, Doxorubicin-conjugated heparin-coated superparamagnetic iron oxide nanoparticles for combined anticancer drug delivery and magnetic resonance imaging, J. Biomed. Nanotechnol. 12 (2016) 1963–1974. https://doi.org/10.1166/jbn.2016.2298.

[82] W. Zhao, Q. Liu, X. Zhang, B. Su, C. Zhao, Rationally designed magnetic nanoparticles as anticoagulants for blood purification, Coll. Surf. B Biointerfaces. 164 (2018) 316–323. https://doi.org/10.1016/j.colsurfb.2018.01.050.

[83] A. Grenha, S. Rodrigues, Pullulan-based nanoparticles: future therapeutic applications in transmucosal protein delivery, Ther. Deliv. 4 (2013) 1339–1341. https://doi.org/10.4155/tde.13.99.

[84] Rekha R., C.P. Sharma, Pullulan as a promising biomaterial for biomedical applications: a perspective, (2007) 20(2)111-116.

[85] A.K. Gupta, M. Gupta, Cytotoxicity suppression and cellular uptake enhancement of surface modified magnetic nanoparticles, Biomaterials 26 (2005) 1565–1573. https://doi.org/10.1016/j.biomaterials.2004.05.022.

[86] F. Gao, Y. Cai, J. Zhou, X. Xie, W. Ouyang, Y. Zhang, X. Wang, X. Zhang, X. Wang, L. Zhao, J. Tang, Pullulan acetate coated magnetite nanoparticles for hyper-thermia: preparation, characterization and in vitro experiments, Nano Res 3 (2010) 23–31. https://doi.org/10.1007/s12274-010-1004-6.

[87] J. Jo, I. Aoki, Y. Tabata, Design of iron oxide nanoparticles with different sizes and surface charges for simple and efficient labeling of mesenchymal stem cells, J. Control. Release. 142 (2010) 465–473. https://doi.org/10.1016/j.jconrel.2009.11.014.

[88] A.O. Elzoghby, M.W. Helmy, W.M. Samy, N.A. Elgindy, Novel ionically crosslinked casein nanoparticles for flutamide delivery: formulation, characterization, and in vivo pharmacokinetics, Int. J. Nanomedicine. 8 (2013) 1721–1732. https://doi.org/10.2147/IJN.S40674.

[89] A.O. Elzoghby, B.Z. Vranic, W.M. Samy, N.A. Elgindy, Swellable floating tablet based on spray-dried casein nanoparticles: near-infrared spectral characterization and floating matrix evaluation, Int. J. Pharm. 491 (2015) 113–122. https://doi.org/10.1016/j.ijpharm.2015.06.015.

[90] A.O. Elzoghby, W.M. Samy, N.A. Elgindy, Protein-based nanocarriers as promising drug and gene delivery systems, J. Control. Release. 161 (2012) 38–49. https://doi.org/10.1016/j.jconrel.2012.04.036.

[91] P. Khramtsov, I. Barkina, M. Kropaneva, M. Bochkova, V. Timganova, A. Nechaev, I. Byzov, S. Zamorina, A. Yermakov, M. Rayev, Magnetic nanoclusters coated with albumin, casein, and gelatin: size tuning, relaxivity, stability, protein corona, and application in nuclear magnetic resonance immunoassay, Nanomaterials 9 (2019) 1345. https://doi.org/10.3390/nano9091345.

[92] A.O. Elzoghby, A.L. Hemasa, M.S. Freag, Hybrid protein-inorganic nanoparticles: from tumor-targeted drug delivery to cancer imaging, J. Control. Release. 243 (2016) 303–322. https://doi.org/10.1016/j.jconrel.2016.10.023.

[93] T.T.-D. Tran, T. Van Vo, P.H.-L. Tran, Design of iron oxide nanoparticles decorated oleic acid and bovine serum albumin for drug delivery, Chem. Eng. Res. Des. 94 (2015) 112–118. https://doi.org/10.1016/j.cherd.2014.12.016.

[94] X. Lu, J. Wu, G. Huo, Q. Sun, Y. Huang, Z. Han, G. Liang, Protein-passivated $FeNi_3$ particles with low toxicity and high inductive heating efficiency for thermal therapy, Coll. Surf. Physicochem. Eng. Asp. 414 (2012) 168–173. https://doi.org/10.1016/j.colsurfa.2012.08.062.

[95] R. Yang, Y. An, F. Miao, M. Li, P. Liu, Q. Tang, Preparation of folic acid-conjugated, doxorubicin-loaded, magnetic bovine serum albumin nanospheres and their antitumor effects in vitro and in vivo, Int. J. Nanomedicine. 9 (2014) 4231–4243. https://doi.org/10.2147/IJN.S67210.

[96] K.G. DeFrates, R. Moore, J. Borgesi, G. Lin, T. Mulderig, V. Beachley, X. Hu, Protein-based fiber materials in medicine: a review, Nanomaterials 8 (2018) 457. https://doi.org/10.3390/nano8070457.

[97] Y. Tian, X. Jiang, X. Chen, Z. Shao, W. Yang, Doxorubicin-loaded magnetic silk fibroin nanoparticles for targeted therapy of multidrug-resistant cancer, Adv. Mater. 26 (2014) 7393–7398. https://doi.org/10.1002/adma.201403562.

[98] Z. Cheng, Y. Dai, X. Kang, C. Li, S. Huang, H. Lian, Z. Hou, P. Ma, J. Lin, Gelatin-encapsulated iron oxide nanoparticles for platinum (IV) prodrug delivery, enzyme-stimulated release and MRI, Biomaterials 35 (2014) 6359–6368. https://doi.org/10.1016/j.biomaterials.2014.04.029.

[99] J. Huang, L. Wang, R. Lin, A.Y. Wang, L. Yang, M. Kuang, W. Qian, H. Mao, Casein-coated iron oxide nanoparticles for high MRI contrast enhancement and efficient cell targeting, ACS Appl. Mater. Interfaces 5 (11) (2013) 4632–4639.

[100] S.H. Nasr, H. Kouyoumdjian, C. Mallett, S. Ramadan, D.C. Zhu, E.M. Shapiro, X. Huang, Detection of β-amyloid by sialic acid coated bovine serum albumin magnetic nanoparticles in a mouse model of Alzheimer's disease, Small 14 (2018) 1701828. https://doi.org/10.1002/smll.201701828.

[101] Z. Li, S. Li, X. Zhou, L. Sun, Q. Zhang, Y. Pan, Q. Zhao, Synthesis of multifunctional nanocomposites and their application in imaging and targeting tumor cells in vitro, Artif. Cells Nanomed. Biotechnol. 44 (2016) 1236–1246. https://doi.org/10.3109/21691401.2015.1019667.

[102] X. Zhang, Q. Pan, L. Hao, Q. Lin, X. Tian, Z. Zhang, S. Wang, H. Wang, Preparation of magnetic fluorescent dual-drug nanocomposites for codelivery of kaempferol and paclitaxel, J. Wuhan Univ. Technol. Mater Sci Ed. 33 (2018) 256–262. https://doi.org/10.1007/s11595-018-1814-z.

[103] T.-Y. Liu, M.-Y. Wu, M.-H. Lin, F.-Y. Yang, A novel ultrasound-triggered drug vehicle with multimodal imaging functionality, Acta Biomater 9 (2013) 5453–5463. https://doi.org/10.1016/j.actbio.2012.11.023.

[104] L. Wang, Y. An, C. Yuan, H. Zhang, C. Liang, F. Ding, Q. Gao, D. Zhang, GEM-loaded magnetic albumin nanospheres modified with cetuximab for simultaneous targeting, magnetic resonance imaging, and double-targeted thermochemotherapy of pancreatic cancer cells, Int. J. Nanomedicine. 10 (2015) 2507–2519. https://doi.org/10.2147/IJN.S77642.

[105] R. Yang, D. Chen, M. Li, F. Miao, P. Liu, Q. Tang, 20(s)-ginsenoside Rg3-loaded magnetic human serum albumin nanospheres applied to HeLa cervical cancer cells in vitro, Biomed. Mater. Eng. 24 (2014) 1991–1998. https://doi.org/10.3233/BME-141008.

[106] H. Misak, N. Zacharias, Z. Song, S. Hwang, K.-P. Man, R. Asmatulu, S.-Y. Yang, Skin cancer treatment by albumin/5-Fu loaded magnetic nanocomposite spheres in a mouse model, J. Biotechnol. 164 (2013) 130–136. https://doi.org/10.1016/j.jbiotec.2013.01.003.

[107] M. Ménard, F. Meyer, C. Affolter-Zbaraszczuk, M. Rabineau, A. Adam, P.D. Ramirez, S. Bégin-Colin, D. Mertz, Design of hybrid protein-coated magnetic core-mesoporous silica shell nanocomposites for MRI and drug release assessed in a 3D tumor cell model, Nanotechnology 30 (2019) 174001. https://doi.org/10.1088/1361-6528/aafe1c.

[108] C. Mi, J. Zhang, H. Gao, X. Wu, M. Wang, Y. Wu, Y. Di, Z. Xu, C. Mao, S. Xu, Multifunctional nanocomposites of superparamagnetic (Fe_3O_4) and NIR-responsive rare earth-doped up-conversion fluorescent (NaYF 4 : Yb,Er) nanoparticles and their applications in biolabeling and fluorescent imaging of cancer cells, Nanoscale 2 (2010) 1141–1148. https://doi.org/10.1039/C0NR00102C.

[109] Y. Jiao, Y. Sun, X. Tang, Q. Ren, W. Yang, Tumor-targeting multifunctional rattle-type theranostic nanoparticles for MRI/NIRF bimodal imaging and delivery of hydrophobic drugs, Small 11 (2015) 1962–1974. https://doi.org/10.1002/smll.201402297.

[110] W.-M. Li, S.-Y. Chen, D.-M. Liu, In situ doxorubicin–CaP shell formation on amphiphilic gelatin–iron oxide core as a multifunctional drug delivery system with improved cytocompatibility, pH-responsive drug release and MR imaging, Acta Biomater 9 (2013) 5360–5368. https://doi.org/10.1016/j.actbio.2012.09.023.

[111] Y.-L. Su, J.-H. Fang, C.-Y. Liao, C.-T. Lin, Y.-T. Li, S.-H. Hu, Targeted mesoporous iron oxide nanoparticles-encapsulated perfluorohexane and a hydrophobic drug for deep tumor penetration and therapy, Theranostics 5 (2015) 1233–1248. https://doi.org/10.7150/thno.12843.

[112] H. Yılmaz, S.H. Sanlıer, Preparation of magnetic gelatin nanoparticles and investigating the possible use as chemotherapeutic agent, Artif. Cells Nanomed. Biotechnol. 41 (2013) 69–77. https://doi.org/10.3109/21691401.2012.745863.

[113] A. Qi, L. Deng, X. Liu, S. Wang, X. Zhang, B. Wang, L. Li, Gelatin-encapsulated magnetic nanoparticles for pH, redox, and enzyme multiple stimuli-responsive drug delivery and magnetic resonance imaging, J. Biomed. Nanotechnol. 13 (2017) 1386–1397. https://doi.org/10.1166/jbn.2017.2433.

[114] E. Che, Y. Gao, L. Wan, Y. Zhang, N. Han, J. Bai, J. Li, Z. Sha, S. Wang, Paclitaxel/gelatin coated magnetic mesoporous silica nanoparticles: preparation and antitumor efficacy in vivo, Microporous Mesoporous Mater 204 (2015) 226–234. https://doi.org/10.1016/j.micromeso.2014.11.013.

[115] W. Song, M. Muthana, J. Mukherjee, R.J. Falconer, C.A. Biggs, X. Zhao, Magnetic-silk core–shell nanoparticles as potential carriers for targeted delivery of curcumin into human breast cancer cells, ACS Biomater. Sci. Eng. 3 (2017) 1027–1038. https://doi.org/10.1021/acsbiomaterials.7b00153.

[116] A. Nacev, I.N. Weinberg, P.Y. Stepanov, S. Kupfer, L.O. Mair, M.G. Urdaneta, M. Shimoji, S.T. Fricke, B. Shapiro, Dynamic inversion enables external magnets to concentrate ferromagnetic rods to a central target, Nano Lett 15 (2015) 359–364. https://doi.org/10.1021/nl503654t.

[117] S. Takeda, F. Mishima, S. Fujimoto, Y. Izumi, S. Nishijima, Development of magnetically targeted drug delivery system using superconducting magnet, J. Magn. Magn. Mater. 311 (2007) 367–371. https://doi.org/10.1016/j.jmmm.2006.10.1195.

[118] M. Chorny, I. Fishbein, S. Forbes, I. Alferiev, Magnetic nanoparticles for targeted vascular delivery, IUBMB Life 63 (2011) 613–620. https://doi.org/10.1002/iub.479.

[119] A.L. Glover, J.B. Bennett, J.S. Pritchett, S.M. Nikles, D.E. Nikles, J.A. Nikles, C.S. Brazel, Magnetic heating of iron oxide nanoparticles and magnetic micelles for cancer therapy, IEEE Trans. Magn. 49 (2013) 231–235. https://doi.org/10.1109/TMAG.2012.2222359.

[120] L. Zhu, J. Ma, N. Jia, Y. Zhao, H. Shen, Chitosan-coated magnetic nanoparticles as carriers of 5-Fluorouracil: Preparation, characterization and cytotoxicity studies, Coll. Surf. B Biointerfaces. 68 (2009) 1–6. https://doi.org/10.1016/j.colsurfb.2008.07.020.

[121] D.G. You, G. Saravanakumar, S. Son, H.S. Han, R. Heo, K. Kim, I.C. Kwon, J.Y. Lee, J.H. Park, Dextran sulfate-coated superparamagnetic iron oxide nanoparticles as a contrast agent for atherosclerosis imaging, Carbohydr. Polym. 101 (2014) 1225–1233. https://doi.org/10.1016/j.carbpol.2013.10.068.

[122] R.Y. Hong, J.H. Li, J.M. Qu, L.L. Chen, H.Z. Li, Preparation and characterization of magnetite/dextran nanocomposite used as a precursor of magnetic fluid, Chem. Eng. J. 150 (2009) 572–580. https://doi.org/10.1016/j.cej.2009.03.034.

[123] H. Ma, X. Qi, Y. Maitani, T. Nagai, Preparation and characterization of superparamagnetic iron oxide nanoparticles stabilized by alginate, Int. J. Pharm. 333 (2007) 177–186. https://doi.org/10.1016/j.ijpharm.2006.10.006.

[124] Z. Cheng, S. Liu, H. Gao, W. Tremel, N. Ding, R. Liu, P.W. Beines, W. Knoll, A facile approach for transferring hydrophobic magnetic nanoparticles into water-soluble particles, Macromol. Chem. Phys. 209 (2008) 1145–1151. https://doi.org/10.1002/macp.200800085.

[125] L. Jiang, Q. Zhou, K. Mu, H. Xie, Y. Zhu, W. Zhu, Y. Zhao, H. Xu, X. Yang, pH/temperature sensitive magnetic nanogels conjugated with Cy5.5-labled lactoferrin for MR and fluorescence imaging of glioma in rats, Biomaterials 34 (2013) 7418–7428. https://doi.org/10.1016/j.biomaterials.2013.05.078.

[126] S.R. Deka, A. Quarta, R.D. Corato, A. Riedinger, R. Cingolani, T. Pellegrino, Magnetic nanobeads decorated by thermo-responsive PNIPAM shell as medical platforms for the efficient delivery of doxorubicin to tumour cells, Nanoscale 3 (2011) 619–629. https://doi.org/10.1039/C0NR00570C.

[127] C. Paquet, L. Pagé, A. Kell, B. Simard, Nanobeads highly loaded with superparamagnetic nanoparticles prepared by emulsification and seeded-emulsion polymerization, Langmuir 26 (2010) 5388–5396. https://doi.org/10.1021/la903815t.

[128] S. Xuan, Y.-X.J. Wang, J.C. Yu, K. Cham-Fai Leung, Tuning the grain size and particle size of superparamagnetic Fe_3O_4 microparticles, Chem. Mater. 21 (2009) 5079–5087. https://doi.org/10.1021/cm901618m.

[129] J. Kim, V.T. Tran, S. Oh, C.-S. Kim, J.C. Hong, S. Kim, Y.-S. Joo, S. Mun, M.-H. Kim, J.-W. Jung, J. Lee, Y.S. Kang, J.-W. Koo, J. Lee, Scalable solvothermal synthesis of superparamagnetic Fe_3O_4 nanoclusters for bioseparation and theragnostic probes, ACS Appl. Mater. Interfaces. 10 (2018) 41935–41946. https://doi.org/10.1021/acsami.8b14156.

[130] R. Di Corato, P. Piacenza, M. Musarò, R. Buonsanti, P.D. Cozzoli, M. Zambianchi, G. Barbarella, R. Cingolani, L. Manna, T. Pellegrino, Magnetic–fluorescent colloidal nanobeads: preparation and exploitation in cell separation experiments, Macromol. Biosci. 9 (2009) 952–958. https://doi.org/10.1002/mabi.200900154.

[131] H. Li, M.J. Henderson, K. Wang, X. Tuo, Y. Leng, K. Xiong, Y. Liu, Y. Ren, J. Courtois, M. Yan, Colloidal assembly of magnetic nanoparticles and polyelectrolytes by arrested electrostatic interaction, Coll. Surf. Physicochem. Eng. Asp. 514 (2017) 107–116. https://doi.org/10.1016/j.colsurfa.2016.11.049.

[132] E.-K. Lim, H.-O. Kim, E. Jang, J. Park, K. Lee, J.-S. Suh, Y.-M. Huh, S. Haam, Hyaluronan-modified magnetic nanoclusters for detection of CD44-overexpressing breast cancer by MR imaging, Biomaterials 32 (2011) 7941–7950. https://doi.org/10.1016/j.biomaterials.2011.06.077.

[133] D. Šmejkalová, K. Nešporová, G. Huerta-Angeles, J. Syrovátka, D. Jirák, A. Gálisová, V. Velebný, Selective in vitro anticancer effect of superparamagnetic iron oxide nanoparticles loaded in hyaluronan polymeric micelles, Biomacromolecules 15 (2014) 4012–4020. https://doi.org/10.1021/bm501065q.

[134] R. Stern, G. Kogan, M.J. Jedrzejas, L. Šoltés, The many ways to cleave hyaluronan, Biotechnol. Adv. 25 (2007) 537–557. https://doi.org/10.1016/j.biotechadv.2007.07.001.

[135] M. Rippe, M. Michelas, J.-L. Putaux, M. Fratzl, G.G. Eslava, N.M. Dempsey, R. Auzély-Velty, A. Szarpak, Synthesis and magnetic manipulation of hybrid nanobeads based on Fe_3O_4 nanoclusters and hyaluronic acid grafted with an ethylene glycol-based copolymer, Appl. Surf. Sci. 510 (2020) 145354. https://doi.org/10.1016/j.apsusc.2020.145354.

[136] S. Zheng, J. Han, Z. Jin, C.-S. Kim, S. Park, K. Kim, J.-O. Park, E. Choi, Dual tumor-targeted multifunctional magnetic hyaluronic acid micelles for enhanced MR imaging and combined photothermal-chemotherapy, Colloids Surf. B Biointerfaces. 164 (2018) 424–435. https://doi.org/10.1016/j.colsurfb.2018.02.005.

[137] H. Su, Y. Liu, D. Wang, C. Wu, C. Xia, Q. Gong, B. Song, H. Ai, Amphiphilic starlike dextran wrapped superparamagnetic iron oxide nanoparticle clsuters as effective magnetic resonance imaging probes, Biomaterials 34 (2013) 1193–1203. https://doi.org/10.1016/j.biomaterials.2012.10.056.

[138] L. Yan, L. Luo, A. Amirshaghaghi, J. Miller, C. Meng, T. You, T.M. Busch, A. Tsourkas, Z. Cheng, Dextran-benzoporphyrin derivative (BPD) coated superparamagnetic iron oxide nanoparticle (SPION) micelles for T_2-weighted magnetic resonance imaging and photodynamic therapy, Bioconjug. Chem. 30 (2019) 2974–2981. https://doi.org/10.1021/acs.bioconjchem.9b00676.

[139] K.H. Bae, M. Park, M.J. Do, N. Lee, J.H. Ryu, G.W. Kim, C. Kim, T.G. Park, T. Hyeon, Chitosan oligosaccharide-stabilized ferrimagnetic iron oxide nanocubes for magnetically modulated cancer hyperthermia, ACS Nano 6 (2012) 5266–5273. https://doi.org/10.1021/nn301046w.

[140] N.J. Hobson, X. Weng, B. Siow, C. Veiga, M. Ashford, N.T. Thanh, A.G. Schätzlein, I.F. Uchegbu, Clustering superparamagnetic iron oxide nanoparticles produces organ-targeted high-contrast magnetic resonance images, Nanomedicine 14 (2019) 1135–1152. https://doi.org/10.2217/nnm-2018-0370.

[141] W.-M. Li, D.-M. Liu, S.-Y. Chen, Amphiphilically-modified gelatin nanoparticles : self-assembly behavior, controlled biodegradability, and rapid cellular uptake for intracellular drug delivery, J. Mater. Chem. 21 (2011) 12381–12388. https://doi.org/10.1039/C1JM10188A.

[142] M. Deng, Z. Huang, Y. Zou, G. Yin, J. Liu, J. Gu, Fabrication and neuron cytocompatibility of iron oxide nanoparticles coated with silk-fibroin peptides, Colloids Surf. B Biointerfaces. 116 (2014) 465–471. https://doi.org/10.1016/j.colsurfb.2014.01.021.

[143] M. He, A. R, G. Js, M. Kp, Z. Nm, W. Ph, Y. Sy, Albumin-based nanocomposite spheres for advanced drug delivery systems, Biotechnol. J. 9 (2014) 163–170. https://doi.org/10.1002/biot.201300150.

[144] Y. Lee, H. Lee, Y.B. Kim, J. Kim, T. Hyeon, H. Park, P.B. Messersmith, T.G. Park, Bioinspired surface immobilization of hyaluronic acid on monodisperse magnetite nanocrystals for targeted cancer imaging, Adv. Mater. 20 (2008) 4154–4157. https://doi.org/10.1002/adma.200800756.

[145] Y. Hu, J. Yang, P. Wei, J. Li, L. Ding, G. Zhang, X. Shi, M. Shen, Facile synthesis of hyaluronic acid-modified Fe_3O_4/Au composite nanoparticles for targeted dual mode MR/CT imaging of tumors, J. Mater. Chem. B. 3 (2015) 9098–9108. https://doi.org/10.1039/C5TB02040A.

[146] K.S. Kim, J. Kim, J.Y. Lee, S. Matsuda, S. Hideshima, Y. Mori, T. Osaka, K. Na, Stimuli-responsive magnetic nanoparticles for tumor-targeted bimodal imaging and photodynamic/hyperthermia combination therapy, Nanoscale 8 (2016) 11625–11634. https://doi.org/10.1039/c6nr02273a.

[147] F.M. Nawwab AL-Deen, C. Selomulya, Y.Y. Kong, S.D. Xiang, C. Ma, R.L. Coppel, M. Plebanski, Design of magnetic polyplexes taken up efficiently by dendritic cell for enhanced DNA vaccine delivery, Gene Ther 21 (2014) 212–218. https://doi.org/10.1038/gt.2013.77.

[148] C.-M. Lee, D. Jang, J. Kim, S.-J. Cheong, E.-M. Kim, M.-H. Jeong, S.-H. Kim, D.W. Kim, S.T. Lim, M.-H. Sohn, Y.Y. Jeong, H.-J. Jeong, Oleyl-chitosan nanoparticles based on a dual probe for optical/MR imaging in vivo, Bioconjug. Chem. 22 (2011) 186–192. https://doi.org/10.1021/bc100241a.

[149] S. Uthaman, S.J. Lee, K. Cherukula, C.-S. Cho, I.-K. Park, Polysaccharide-coated magnetic nanoparticles for imaging and gene therapy, BioMed. Res. Int. 2015 (2015) e959175. https://doi.org/10.1155/2015/959175.

[150] V. Zamora-Mora, M. Fernández-Gutiérrez, J.S. Román, G. Goya, R. Hernández, C. Mijangos, Magnetic core-shell chitosan nanoparticles: rheological characterization and hyperthermia application, Carbohydr. Polym. 102 (2014) 691–698. https://doi.org/10.1016/j.carbpol.2013.10.101.

[151] P. Xie, P. Du, J. Li, P. Liu, Stimuli-responsive hybrid cluster bombs of PEGylated chitosan encapsulated DOX-loaded superparamagnetic nanoparticles enabling tumor-specific disassembly for on-demand drug delivery and enhanced MR imaging, Carbohydr. Polym. 205 (2019) 377–384. https://doi.org/10.1016/j.carbpol.2018.10.076.

[152] H. Ersoy, P. Jacobs, C.K. Kent, M.R. Prince, Blood pool MR angiography of aortic stent-graft endoleak, Am. J. Roentgenol. 182 (2004) 1181–1186. https://doi.org/10.2214/ajr.182.5.1821181.

[153] B.R. Jarrett, M. Frendo, J. Vogan, A.Y. Louie, Size-controlled synthesis of dextran sulfate coated iron oxide nanoparticles for magnetic resonance imaging, Nanotechnology 18 (2007) 035603. https://doi.org/10.1088/0957-4484/18/3/035603.

[154] N. Peng, X. Ding, Z. Wang, Y. Cheng, Z. Gong, X. Xu, X. Gao, Q. Cai, S. Huang, Y. Liu, Novel dual responsive alginate-based magnetic nanogels for onco-theranostics, Carbohydr. Polym. 204 (2019) 32–41. https://doi.org/10.1016/j.carbpol.2018.09.084.

[155] L. Chu, Y. Zhang, Z. Feng, J. Yang, Q. Tian, X. Yao, X. Zhao, H. Tan, Y. Chen, Synthesis and application of a series of amphipathic chitosan derivatives and the corresponding magnetic nanoparticle-embedded polymeric micelles, Carbohydr. Polym. 223 (2019) 114966. https://doi.org/10.1016/j.carbpol.2019.06.005.

[156] C. Jingting, L. Huining, Z. Yi, Preparation and characterization of magnetic nanoparticles containing Fe3O4-dextran-anti-β-human chorionic gonadotropin, a new generation choriocarcinoma-specific gene vector, Int. J. Nanomed. 6 (2011) 285–294. https://doi.org/10.2147/IJN.S13410.

[157] Y. Zhou, Q. Xie, Hyaluronic acid-coated magnetic nanoparticles-based selective collection and detection of leukemia cells with quartz crystal microbalance, Sens. Actuators B Chem. 223 (2016) 9–14. https://doi.org/10.1016/j.snb.2015.09.063.

[158] V.D. Nguyen, S. Zheng, J. Han, V.H. Le, J.-O. Park, S. Park, Nanohybrid magnetic liposome functionalized with hyaluronic acid for enhanced cellular uptake and near-infrared-triggered drug release, Coll. Surf. B Biointerfaces. 154 (2017) 104–114. https://doi.org/10.1016/j.colsurfb.2017.03.008.

[159] J. Huang, Q. Shu, L. Wang, H. Wu, A.Y. Wang, H. Mao, Layer-by-layer assembled milk protein coated magnetic nanoparticle enabled oral drug delivery with high stability in stomach and enzyme-responsive release in small intestine, Biomaterials 39 (2015) 105–113. https://doi.org/10.1016/j.biomaterials.2014.10.059.

[160] S.S. Khiabani, M. Farshbaf, A. Akbarzadeh, S. Davaran, Magnetic nanoparticles: preparation methods, applications in cancer diagnosis and cancer therapy, Artif. Cells Nanomed. Biotechnol. 45 (2017) 6–17. https://doi.org/10.3109/21691401.2016.1167704.

[161] L. Mohammed, H.G. Gomaa, D. Ragab, J. Zhu, Magnetic nanoparticles for environmental and biomedical applications: a review, Particuology 30 (2017) 1–14. https://doi.org/10.1016/j.partic.2016.06.001.

[162] V. Kalidasan, X.L. Liu, T.S. Herng, Y. Yang, J. Ding, Bovine serum albumin-conjugated ferrimagnetic iron oxide nanoparticles to enhance the biocompatibility and magnetic hyperthermia performance, Nano-Micro Lett 8 (2016) 80–93. https://doi.org/10.1007/s40820-015-0065-1.

[163] L. Chang, X.L. Liu, D. Di Fan, Y.Q. Miao, H. Zhang, H.P. Ma, Q.Y. Liu, P. Ma, W.M. Xue, Y.E. Luo, H.M. Fan, The efficiency of magnetic hyperthermia and in vivo histocompatibility for human-like collagen protein-coated magnetic nanoparticles, Int. J. Nanomed. 11 (2016) 1175–1185. https://doi.org/10.2147/IJN.S101741.

[164] Y. Min, J.M. Caster, M.J. Eblan, A.Z. Wang, Clinical translation of nanomedicine, Chem. Rev. 115 (2015) 11147–11190. https://doi.org/10.1021/acs.chemrev.5b00116.

[165] Y.X.J. Wáng, J.-M. Idée, A comprehensive literatures update of clinical researches of superparamagnetic resonance iron oxide nanoparticles for magnetic resonance imaging, Quant. Imaging Med. Surg. 7 (2017) 88–122. https://doi.org/10.21037/qims.2017.02.09.

26

Aluminum hypophosphite–based highly flame-retardant rigid polyurethanes for industrial applications

Felipe M. de Souza[a], Seongwoo Hong[a], Tuan Anh Nguyen[b] and Ram K. Gupta[a]

[a]DEPARTMENT OF CHEMISTRY AND KANSAS POLYMER RESEARCH CENTER, PITTSBURG STATE UNIVERSITY, PITTSBURG, KS, UNITED STATES. [b]INSTITUTE FOR TROPICAL TECHNOLOGY, VIETNAM ACADEMY OF SCIENCE AND TECHNOLOGY, HANOI, VIETNAM.

Chapter outline

26.1 Introduction

Polymers such as polyurethanes (PU) are extremely versatile materials that can be synthesized through a variety of polyols and polyisocyanates, which enables them to cover a broad range of applications. Rigid PU foams are mostly applied in constructions and automobiles because of their high thermal insulation, lightweight, and high mechanical properties. However, this class of polymers also presents a range of other properties such as high flexibility, applications in resin, adhesives, and sealants [1–4]. The range of properties is possible by using various types of polyols and diisocyanates used in the synthesis of PU. Several chemistries can be used to modify the polyols and diisocyanates to cover a range of properties [5]. However, most of the PU foams are flammable when they come in contact with fire. Their porous structure allows oxygen to diffuse within the foam facilitating flames to spread, leading to combustion that releases toxic smoke composed of carcinogenic furans and dioxins, bringing safety and environmental issues [6,7]. To avoid this scenario many approaches have been made such as synthesizing foams that have flame-retardant (FR) compounds covalently bonded in their structure, named as inherent or reactive FR. This type of component delivers many benefits such as homogenous dispersion

Biopolymeric Nanomaterials: Fundamentals and Applications. DOI: https://doi.org/10.1016/B978-0-12-824364-0.00005-8

throughout the polymeric matrix, stable properties over time, and enhanced flame retardancy. Nowadays many phosphorous and nitrogen-based compounds have been receiving more attention to be used as reactive FR [8,9]. However, due to synthesis requirements and energy usage, it may increase production costs [10,11]. Hence, another viable option is to physically mix the FR compounds into the PU matrix during the foaming process, named as blended or nonreactive FR, which is a cheaper method that does not require the synthesis step and yields properties similar or even better than reactive FR [12–14]. Nevertheless, it may demand high quantities of FR to obtain satisfactory flame retardancy, which can lead to deterioration of mechanical properties because of the disruption of foam's cellular structure and irregular dispersion of FR [8,15]. Many materials are used for this purpose, some examples are expandable graphite, zinc borate, melamine, ammonium phosphate, dimethyl phosphate, and aluminum hypophosphite (AHP) [13,16–20]. The combination of these two processes can boost properties due to the advent of synergetic effects, which was studied in many reports [15,19,21].

In this work, two novel reactive polyols were synthesized using 1,3,5-triallyl isocyanurate and triallyl cyanurate by reacting their terminal double bonds with 2-mercaptoethanol in a synthesis that required only UV-light and stirring under room temperature, commonly known as thiol-ene click chemistry [22]. This type of radical addition called the attention of many authors because of the great number of polyfunctional thiols available, which in this case enabled the attachment of primary hydroxyl groups that are known to be more reactive toward isocyanates to make PU [23–26]. Allyl triazine and allyl isocyanurate are the groups chemically bonded into the polyol's structure synthesized in this work, which acts in both gaseous and condensed phases. The formation of free radicals can neutralize active radical species such as $^{\cdot}OH$, $^{\cdot}H$, and hydrocarbon fragments, which restrict the combustion process and prevent the polymeric matrix from decomposing [27]. To further enhance the flame retardancy properties, AHP was used as a nonreactive FR, also acting in both gaseous and condensed phases by forming PH_3 that can promptly react with oxygen to create a char layer of polyphosphoric acid that quenches the fire [20,28,29]. Therefore, with the combination of different mechanisms, our group was able to make unique FR PU foams.

26.2 Synthesis and characterizations

1,3,5-triallyl isocyanurate (T1), triallyl cyanurate (T2), 2-hydroxy-2-methyl propiophenone (photoinitiator), aluminum chloride hexahydrate, and sodium hypophosphite dihydrate were purchased from Sigma-Aldrich (USA). Rubinate M (methylene diisocyanate, MDI) and Jeffol SG-520 were provided from Huntsman (USA). NIAX A-1 (amine-based catalyst) and DABCO T-12 (blowing agent catalyst) were both purchased from Air Products (USA). Tegostab B-8404 (silicon-based surfactant) was purchased from Evonik (USA). DI water was purchased from the local Walmart (Pittsburg, USA). 2-mercaptoethanol (2ME) was purchased from Acros Organic (USA).

T1- and T2-based polyols were synthesized using thiol-ene click chemistry. In a typical synthesis, 1 mol of T1 was added in 3 mol of 2ME in a beaker containing 3 wt% of photoinitiator.

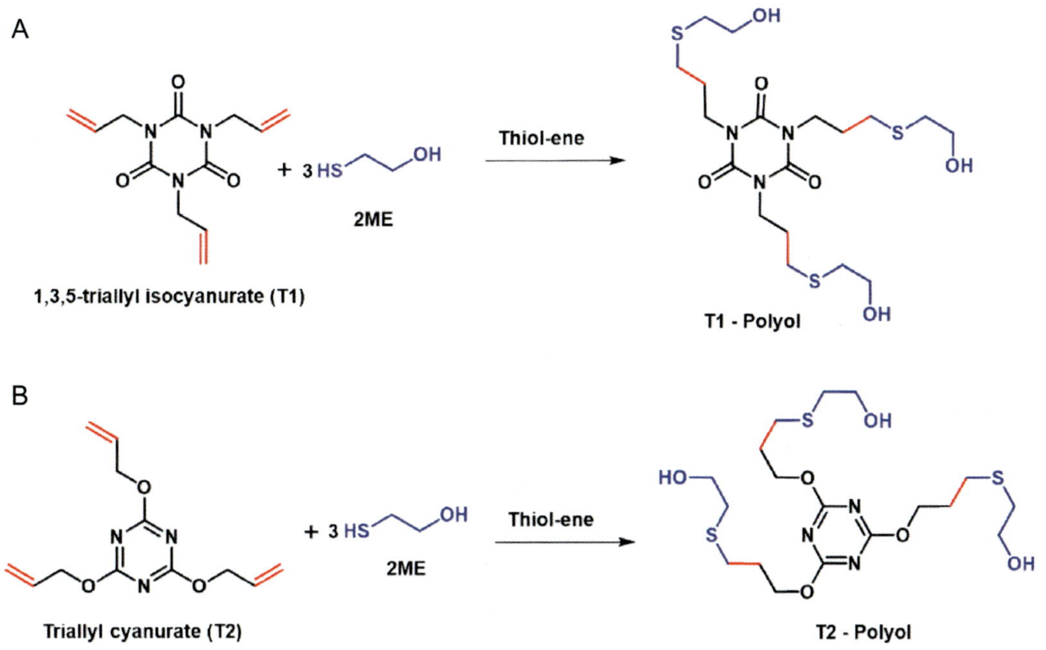

FIGURE 26.1 Thiol-ene reactions for the synthesis of (A) T1- Polyol and (B) T2- Polyol.

The reaction was carried out in the presence of UV-light (365 nm) for 6 h with constant stirring. The same procedure was adopted for the synthesis of polyol using T2. The chemical reactions for the synthesis of T1- and T2-based polyols are shown in Fig. 26.1.

AHP as FR was synthesized using a facile method. For the synthesis of AHP, 50.88 g of sodium hypophosphite dihydrate ($NaPO_2H_2.2H_2O$) was dissolved in 30 mL of distilled water in a three-necked flask at 50 °C and stirred for 15 min. After the solution became transparent, the temperature was raised to 85 °C for 25 min. A fresh solution containing 38.54 g of aluminum chloride hexahydrate ($AlCl_3.6H_2O$) was added dropwise into a solution of sodium hypophosphite dihydrate. A white precipitate was formed after the addition and the entire mixture was left at 85 °C for 1 h to ensure the completion of the reaction. The white precipitate was filtrated at room temperature, washed several times with distilled water, and dried in the oven for 24 h at 100 °C [30].

PU foams using T1- and T2-based polyols and AHP as FR were prepared and tested. A one-pot foaming process was used for the preparation of all the foams. In a typical preparation, 10 g of T1-polyol, 10 g of Jeffol SG-520, 0.14 g of A-1, 0.8 g of water, 0.4 g of B8404, 0.04 g of T-12, and various amounts of AHP (0, 1, 3, 5, 8, and 10 g) were added into a 500-mL plastic cup and mixed using mechanical stirrer at 3000 rpm. After homogenizing, 33.37 g of MDI was added and stirred for around 10 s more to make the foam rise. The foams containing 0, 1, 3, 5, 8, and 10 g of AHP in T1-polyol were named as T1-0, T1-AHP1, T1-AHP2, T1-AHP3, T1-AHP4, and T1-AHP5, respectively. The same procedure was adopted for the preparation of the foams using T2-polyol and AHP. The foams containing 0, 1, 3, 5, 8, and 10 g of AHP in T2-polyol were named as T2-0, T2-AHP1, T2-AHP2, T2-AHP3, T2-AHP4, and T2-AHP5, respectively.

The synthesized polyols and foams were characterized using ASTM (American Society for Testing and Materials), ISO (International Organization for Standardization), and IUPAC (International Union of Pure and Applied Chemistry) methods. First, the starting materials T1 and T2 were characterized to determine unsaturation (double bonds) using the iodine value technique (Hanus method) which consisted in the mix with a known volume of BrI (0.2 N) solution (Hanus reagent) in an Erlenmeyer for 3 h in the dark. Then the leftover of Hanus reagent was titrated with sodium thiosulfate (0.1 N) and a 1% starch solution was used as an indicator. The starting components and polyols were characterized by gel permeation chromatography (GPC) run by Waters System Software, Milford, MA. Fourier Transformed Infrared (FTIR) spectra were recorded using a PerkinElmer Spectrum from Waltham, MD. The viscosity was obtained by using an AR 2000 ex rheometer from New Castle, DE, at room temperature using a cone of 25 mm of ϕ with a 2° angle, starting from shear stress of 0.1–100 Pa. The hydroxyl number was determined by reacting the polyols with p-toluenesulfonyl isocyanate that was titrated by an automatic 888 Titrando machine from Tiamo Software system by Metrohm, AG, Switzerland, according to ASTM E- 1899-97 standard.

Gravimetric density, closed-cell content, and compressive strength of the PU foams were determined using a cylindrical shaped foam having dimensions of 45 × 30 (mm^3). The volumetric density was determined using ASTM 1622-14 method. The closed-cell content was estimated using a Humi Pyc Volumetric and Relative Humidity instrument. The compressive strength was measured using an Instron Compressive instrument (Raleigh, NS, USA). The sheer force was applied from the top of the sample at strain overtime of 5 cm/min. The thermal behavior of the foams was determined using thermal gravimetric analysis (TGA) through a Q500 Discovery run by TRIOS System (New Castle, DE, USA). The tests were performed under a nitrogen atmosphere at a ramp temperature of 10 °C per min, from 25 to 700 °C. The flame test was performed by cutting the foams into a parallelepiped shape of 12.5 × 50 × 150 (mm^3), thickness, width, and length, respectively. Then the samples were exposed to a flame in a horizontal position for 10 s and the self-extinguish and weight loss were measured after the test. The microstructures of the foams were captured using a scanning electron microscope.

26.3 Results and discussion

The completion of thiol-ene reaction was determined using GPC. Fig. 26.2 shows the GPC of starting materials and the synthesized polyols. As seen in Fig. 26.2, the lowest elution time was observed for 2ME (˜44 min) as it is the smallest molecule among the others. Both T1 and T2 presented the same elution time (˜38 min) by the fact they are isomers. The accomplishment of the thiol-ene reaction could be confirmed for both T1 and T2 polyols due to the equal decrease of elution time (˜36 min) related to the single peak in the chromatograms, which demonstrated not only accomplishment but also regioselectivity for this reaction as expected for some thiol-ene synthesis [22]. To further confirm the attachment of hydroxyl groups, the hydroxyl

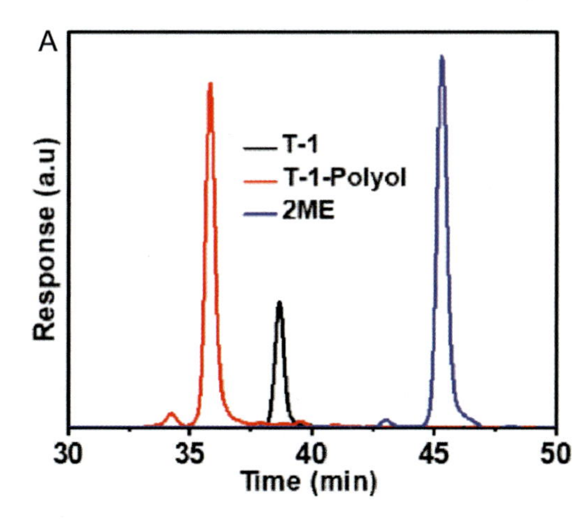

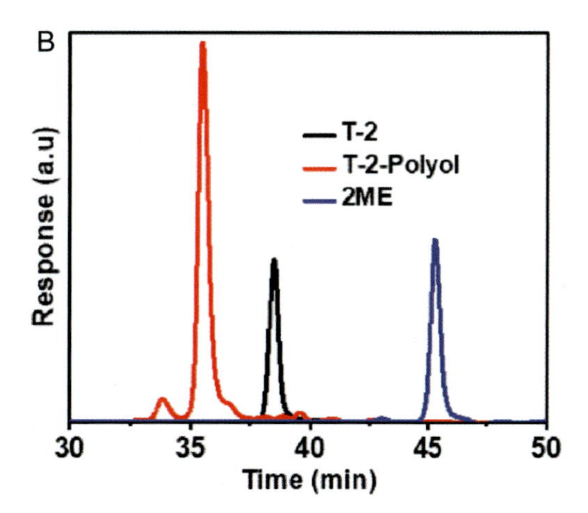

FIGURE 26.2 GPC of starting materials and their polyols.

number for the polyols was measured, which is a common procedure that represents the number of available hydroxyl groups. These are the main reactive function to make PU foams by reacting with isocyanates [3]. For T1 and T2 polyols, the hydroxyl number was 362 and 365 mg KOH/g, respectively, that is close to the theoretical value of 366 mg KOH/g, which implied a high level of conversion of starting materials to polyols. Viscosity is another important measurement because it points out the processing for polyols, in a way that low-to-medium viscosities values are desired [3]. In the case of T1, the starting viscosity was 0.16 Pa.s and after its conversion to polyol, it increased to 20.22 Pa.s. For the case of T2, the initial value was 0.021 Pa.s and after the thiol-ene reaction it went up to 11.32 Pa.s. The values obtained were lower than previously reported studies. It is an important parameter due to the inherent lack of miscibility between polyols and isocyanates. Thus, synthesizing polyols that do not require higher amounts of shear force to quickly disperse in isocyanate is an important aspect for large-scale production [31,32].

To confirm the polyol's chemical bonds, the FTIR spectra for the starting materials and polyols were recorded (Fig. 26.3). The objective in performing this thiol-ene reaction was to chemically attach the hydroxyl groups to T1 and T2 structures to convert them into polyols. This process can be observed in both polyols spectra due to the appearance of a large band of -O-H stretch at 3500 cm^{-1} and the simultaneous disappearance of C=C stretch at 900 cm^{-1} [22,33–35]. Furthermore, the disappearance of a weak peak of S-H stretch at 2580 cm^{-1} and its conversion into C-S bond usually found in between 500 and 700 cm^{-1} also confirm the conversion of T1 and T2 into T1 and T2 polyols, respectively [23,26,36]. These observations strongly suggested the chemical attachment of hydroxyl groups through the thiol-ene reaction. As a complement, the signature peaks for the chemical stretching of T1 are observed in a sharp peak at 1460 cm^{-1} related to C-N stretch and at 1690 cm^{-1} that represents the carbonyl (C=O) vibration [27]. For T2 the vibrational bands of the triazine ring can be observed at 1570 cm^{-1}. Also, the peaks at 1460 and 1330 cm^{-1} correspond to C-N and C-O, respectively [37].

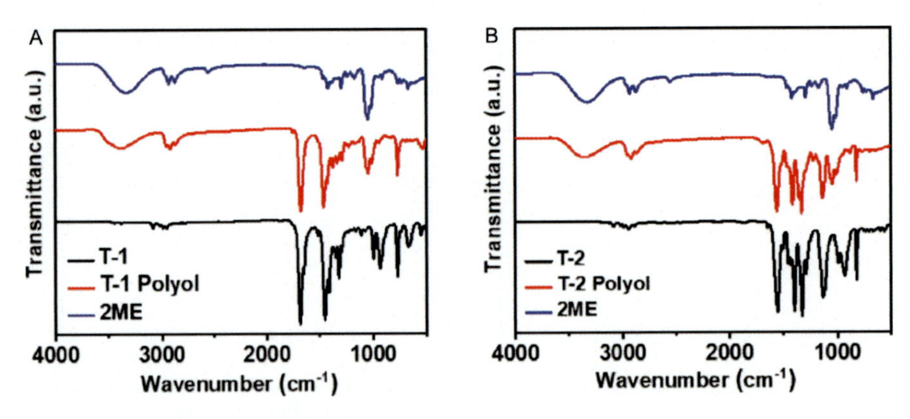

FIGURE 26.3 FTIR spectra of materials and polyols.

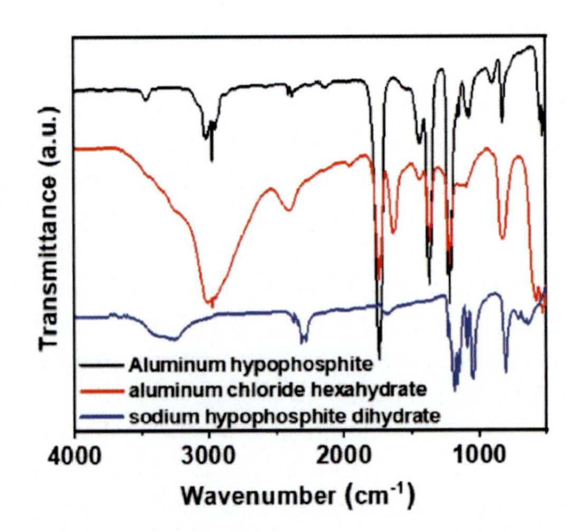

FIGURE 26.4 FTIR spectra of staring materials and AHP.

The synthesis of AHP powder was confirmed through FTIR analysis (Fig. 26.4). The main peaks that can be observed for the AHP structure are related to the weak bands around 3285–2400 cm^{-1} which are related to P-H stretch. Also, the short peaks at 1200 and 1090 cm^{-1} represent the P=O and symmetric vibrational stretching of P-O, respectively [12,38–40]. The Al-O stretch at 500 cm^{-1} is also observed for AHP and $AlCl_3.6H_2O$ [39,41]. The large band observed around 3000 cm^{-1} is related to OH stretch due to the molecules of water used to stabilize the crystal structure of $AlCl_3.6H_2O$ [41]. Hence, the observed peaks confirmed the chemical stretching present in the AHP molecule.

The digital photographs of the foams prepared using T1 and T2 polyols along with AHP as FR is shown in Fig. 26.5. The foams are free of major defects such as crack, suggesting their high quality. The foam's morphology is an important characterization to understand the physical properties and effect of additive FR into the arrangement of cellular structure. To

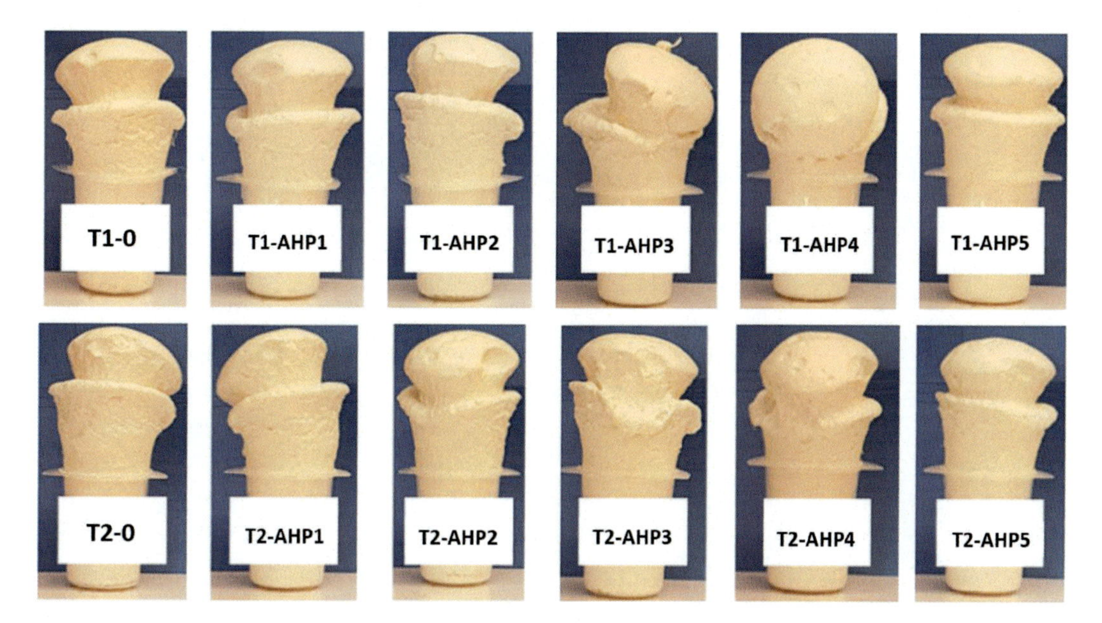

FIGURE 26.5 PhotoELSST045-26-graphs of PU foams with increasing amounts of AHP.

understand that, scanning electron microscope micrographs for all the foams were obtained and shown in Fig. 26.6. All PU foams showed a uniform morphology. The micrographs also demonstrated that despite the variation in the cell's diameter across the set of PU, each sample presented a relatively constant cell size. Hence, it suggested that the foaming process was not disturbed by the addition of AHP. The effect of the addition of AHP demonstrated that T1 foams had notable cell size distribution of around 100 μm. Besides, with the increasing amount of AHP, the cell size increased to in the range of 250–150 μm. The behavior for the T2 foam's series demonstrated a slight decrease, at which T2-0 had a uniform cell size around 150 μm and with the increase of AHP concentration, the cell size also decreased to the range of 100–80 μm.

The measurement of closed-cell content relates to the diffusion of air that can pass through a Polyurethane foam (PUF). For better flame retardancy, a closed-cell structure is desirable as it would be able to block the passage of air. Hence, it would diminish the propagation of fire [42]. Another factor related to high closed-cell content is the increase of thermal insulation because of the low passage of air [43]. Fig. 26.7 shows that the PUF obtained in this work presented a closed-cell content above 90%, indicating that the macrostructure of the foams remained closed for the passage of air after increasing the addition of AHP.

Density is an important physical property of PUF because it directs some of its applications. For example, low density, thermal insulation, high rigidity, and mechanical energy absorption are properties required for construction materials used in buildings [44–46]. Hence, the use of low-density material implies a reduced overall weight of the construction. Also, even though the density of PUF can be controlled by varying the quantity of blowing agents, it has to maintain its

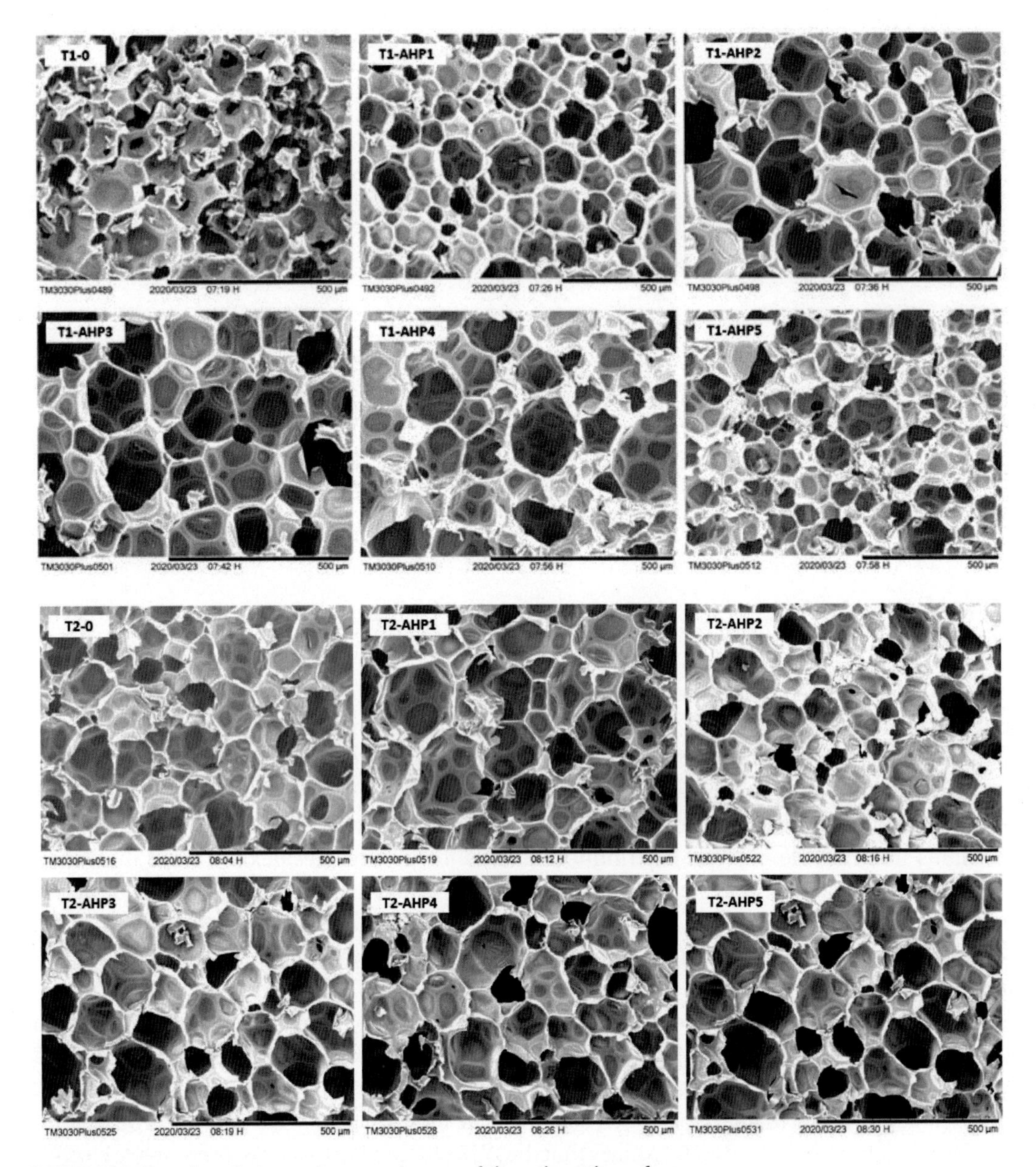

FIGURE 26.6 Scanning electron microscope images of the polyurethane foams.

properties to not become too friable or too rigid [44]. The density values are given in Fig. 26.8. It showed an obvious increase in density as the concentration of AHP increased. For the case of the T1-based PU foams, the values were around 29–49 kg/m^3, whereas, for the T2-based foams, it was around 34–45 kg/m^3. Both of these ranges are comparable to previous reports that synthesized other types of PU foams [47–49].

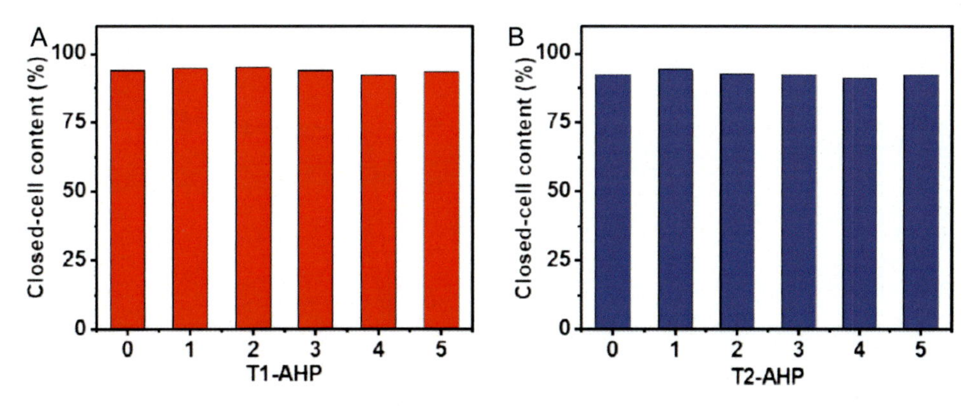

FIGURE 26.7 Closed-cell content of the rigid polyurethane foams having different amounts of AHP for (A) T1-AHP (B) T2-AHP.

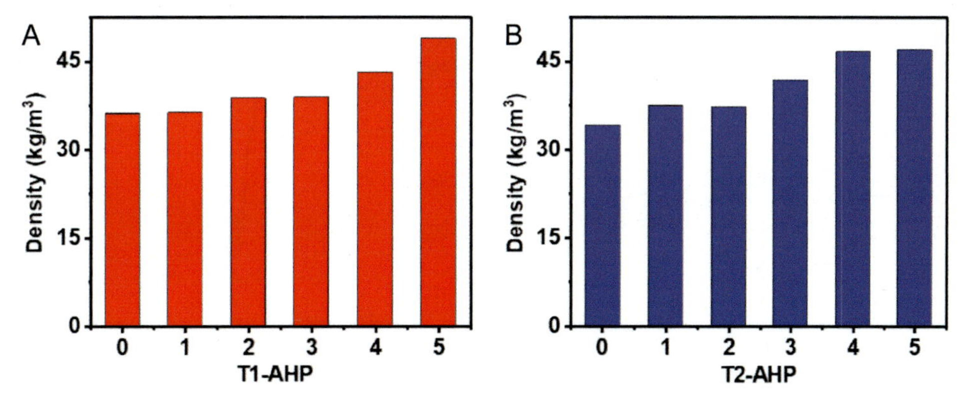

FIGURE 26.8 The density of the rigid PUF with different amounts of AHP for (A) T1-AHP and (B) T2-AHP.

The deterioration of compressive strength with the addition of blended FR has been widely reported in other studies that used materials such as expandable graphite, dimethyl methyl phosphonate, or melamine [50–53]. The value variation in compressive strength was also observed in this work as showed in Fig. 26.9, which is following observed behavior previously reported in studies that implemented AHP into a polymeric matrix [8,20,28]. However, for the case of T1-based foams, there was a fluctuation in values from 214.3 to 150 kPa at a yield at the break. This effect could be related to the increase in cell size after the addition of AHP. The increase in cell size leads to a decrease in cell number, causing each cell to withstand a higher amount of pressure when a load is applied; hence, decreasing the overall compressive strength of the foam. On the other hand, the T2-based foams presented a slight increase in compressive strength with the addition of AHP that went up from around 150 to 197.5 kPa. Oppositely of T1-based foams, the T2-based foams had their cell size decreased, which allowed a higher number of cells to be exposed to a load per the same amount of area [54–56]. Despite the oscillatory behavior of AHP addition into the PU matrix, the compressive strength was higher than other studies [32,57,58].

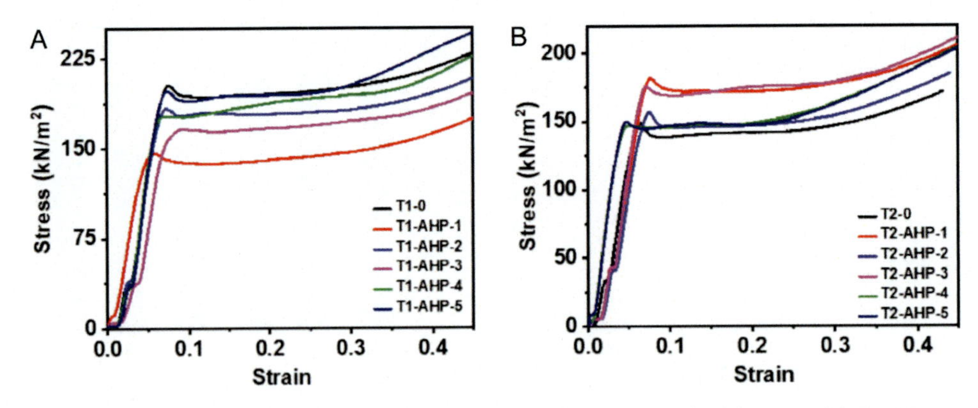

FIGURE 26.9 Effect on compression strength for PUF after the addition of AHP for (A) T1-AHP (B) T2-AHP.

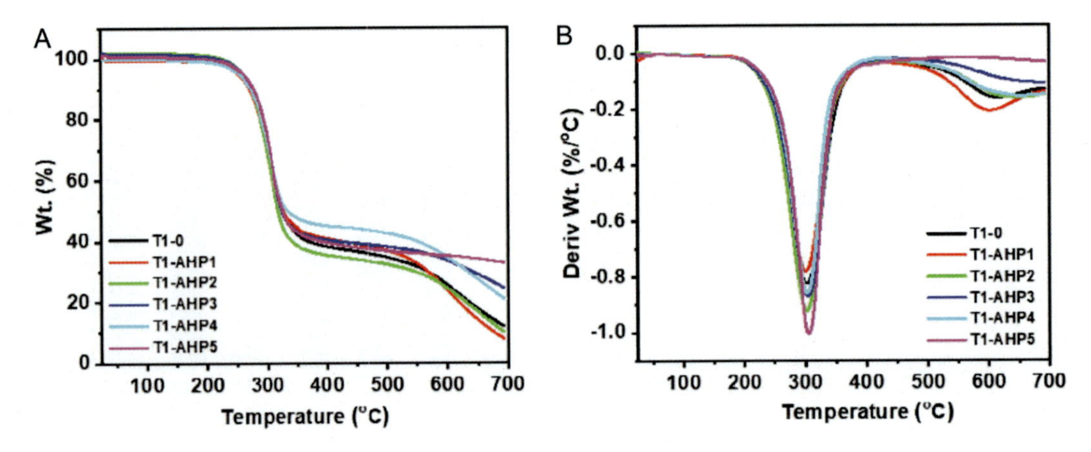

FIGURE 26.10 (A) TGA and (B) DTGA plots for T1-based foams.

This phenomenon variation in properties may have occurred due to the partial dispersion of AHP particles into the PU matrix [54].

The thermal stability of the foams was analyzed through TGA and Differential Thermogravimetric Analysis (DTGA) as demonstrated in Figs. 26.10 and 26.11. Both sets of foams presented two similar thermal decompositions curves. The addition of AHP did not show much influence in the first decomposition process in neither of the PU which presented its max decomposition at 300 °C, as it can be observed in the DTGA of Figs. 26.10 and 26.11, which also correlates with previous studies [8,20,28]. The first decomposition that occurred in the range of 300 °C for both sets of foams is related to the disruption of the urethane linkage reverting it to aromatic isocyanates and alcohols [44,59,60]. The improvement in thermal stability was notable during the second decomposition behavior that occurred within the range of 500–600°C, as the amount of residue increased during that transition with increasing load of AHP. It represented the thermal degradation of AHP that released phosphine that further converted into phosphoric acid [20,61,62]. Phosphoric acid had the role of dehydrate the PU matrix, which formed a carbonaceous char

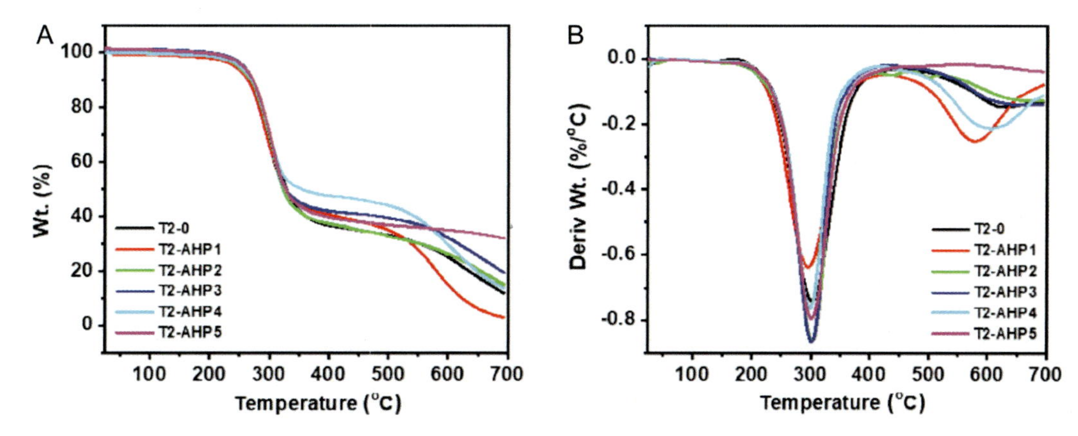

FIGURE 26.11 (A) TGA and (B) DTGA plots for T2-based foams.

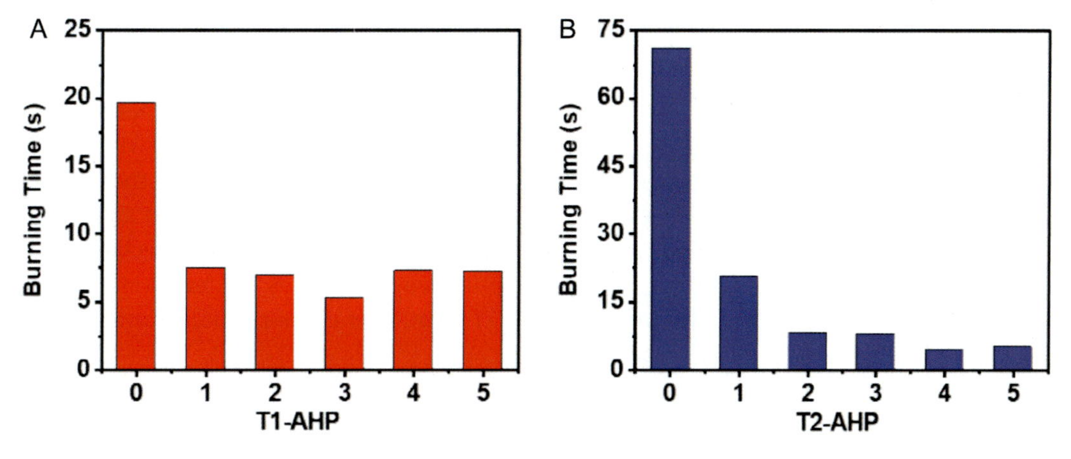

FIGURE 26.12 Burning time for (A) T1-AHP and (B) T2-AHP.

layer and a cross-linked structure that was more compact and stable, which prevented the inner layers of PU from decomposing [8,28]. This effect explained a larger amount of residue when a higher concentration of AHP was added, suggesting the formation of a more compact char layer. Hence, as the load of AHP increased the thermal stability of the foams also increased. Thus, the addition of AHP suggested an improvement in the thermal stability of both foams.

The flame behavior of the samples was tested by performing the horizontal burning test as per ASTM D 4986-98, which consisted of exposing the foams directly to a flame during 10 s. After that, the time taken for the sample to self-quench the fire, and the weight loss (% wt loss) were recorded. Figs. 26.12 and 26.13 describe the burning time and weight loss for both sets of foams, respectively. The FR mechanism is a complex event that evolves many steps, in this case, there was a difference in the decomposition process of T1 and T2 foams that led to widely different behaviors. The chemical structure of T1 is composed of a triallyl ring, which is formed through

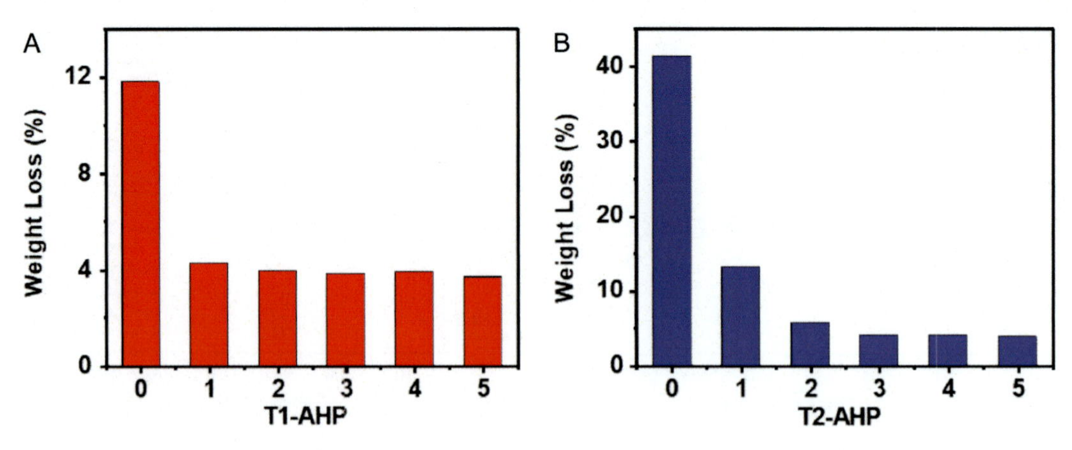

FIGURE 26.13 Weight loss percentage for (A) T1-AHP and (B) T2-AHP.

the simultaneous reaction of three isocyanate groups [44,63,64]. Previous studies have been performed by Qian et. al. [27] which explained that when this ring is exposed to high oxidative energy such as flames it decomposes into radicals fragments with low reactivity. These radical species can scavenge other radicals such as $H^{\bullet}$ and $HO^{\bullet}$, which otherwise would react with the PU matrix leading to highly exothermic reactions that propagate the fire. On top of that, the radicals originated from the decomposition of the triallyl ring can also promote a compact and cross-linked char by reacting with other segments of the polymer [27]. Hence, the FR mechanism of ring decomposition into stable fragments able to capture reactive species and form a stable char layer is in accordance with the behavior of the neat T1 foam. After the horizontal burning test, it reached a burning time of 20 s and a weight loss of 12%.

The fire behavior observed for the T2 neat foam was less effective as if compared with the T1 neat foam. A study performed by Guo et. al. [37] demonstrated the chemical fragments obtained from the thermal decomposition of a 1,3,5-triazine ring that presented an equal structure of T2. It demonstrated that the fragments presented high conjugated double bonds able to stabilize the radical fragments, likely decreasing their reactivity. This effect led to the formation of stable volatile byproducts that acted as diluting agents rather than radical scavengers, which could then allow highly reactive species to enter in contact with the foam and further catalyze the burning process. This fire behavior can be further suggested by the burning time of the T2 neat sample that went up to 70 s and around 40% wt loss.

The addition of AHP in both foams demonstrated a drastic improvement in their flammability. As it can be observed in Fig. 26.14, the burnt area decreased substantially. This improvement may be explained due to the FR mechanism of AHP. It consists of action in both the solid and gas phase. In the solid phase, AHP is known to decompose as a carbonaceous char layer that is formed in a two-stage thermal decomposition process that forms $Al_2(HPO_4)_3$ and later degrades into $Al_4(P_2O_7)_3$ [20]. These compounds accompanied by carbon and phosphate byproducts constitute the char layer that protects the inner PU from oxygen and radicals [28]. Simultaneous to that process occurs the gas phase action, which consists of the release of PH_3 that acts as a

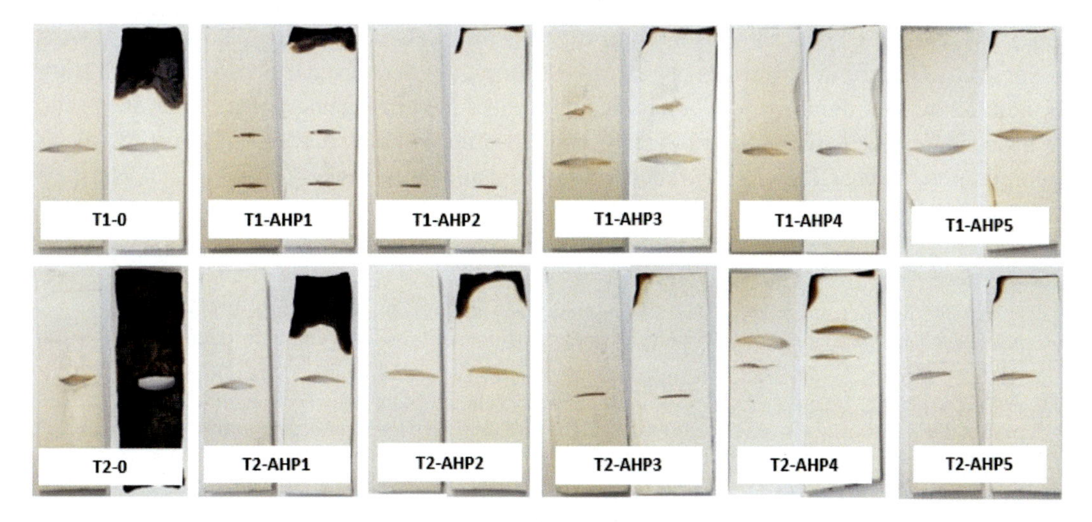

FIGURE 26.14 Photographs of the polyurethane foams prepared before and after the burning test.

radical scavenger as well as a suffocating agent by reacting with oxygen [8,20]. The latter reaction leads to the condensation of polyphosphoric acid that dehydrates the PU matrix and therefore enhancing the char yield as it has also been observed on TGA given in Figs 26.10 and 26.11. Thus, the implementation of AHP provided an effective synergy effect for the foams leading to stable values of burning time and wt loss% of 7 s and 10%, respectively. For T1 foam these values remained fairly constant regardless of further addition of AHP, showing that high flammability could be achieved even at lower additions. This is the desired quality that shows a potential application for large-scale production as a lesser quantity of additives required to achieve certain properties implies cost reduction. On top of that none of the samples had dripping, which adds another safety measure besides the quick extinguishment of the fire.

26.4 Conclusion

The synthesis of two novel polyols based on 1,3,5-triallyl isocyanurate and triallyl cyanurate through thiol-ene reaction and their applications to prepare high-performance PUs was performed. The results demonstrated an overall improvement in the properties of the foams. Density was in an acceptable range of industry standards (30–50 Kg/m^3) for foams prepared using both polyols and FR. The closed-cell content was around 95% for both sets of foams. The addition of AHP caused different effects on the compressive strength for each of the foams. There was a slight decrease in compressive strength with an increase in AHP for the foams prepared using T1-polyols, while foams prepared using T2-based polyol showed an increase in compressive strength. The flame retardancy received a great improvement for both foams after the addition of AHP. For T1-based foams, the burning time and weight loss for the neat sample were 19.6 s and 11.82 wt%, respectively. After the addition of 1.79 wt% of AHP, the values decreased to 7 s and 3.97%. For the T2 neat sample, the burning time and weight loss were 71.2 s and 41.41 wt%, respectively. After the addition of 5.19 wt% of AHP, the values decreased to 8.2 s and 14 wt%. Thus, the addition of AHP demonstrated a significant improvement in the flame retardancy of T1- and T2-based foams due to the synergy effect between the inherent FR structure of T1 and T2 polyols accompanied to the action of AHP in both condensed and gas phases.

Referencses

[1] N. Wu, F. Niu, W. Lang, J. Yu, G. Fu, Synthesis of reactive phenylphosphoryl glycol ether oligomer and improved flame retardancy and mechanical property of modified rigid polyurethane foams, Mater. Des. 181 (2019) 107929.

[2] H. Pan, W. Wang, Y. Pan, L. Song, Y. Hu, K.M. Liew, Formation of layer-by-layer assembled titanate nanotubes filled coating on flexible polyurethane foam with improved flame retardant and smoke suppression properties, ACS. Appl. Mater. Interfaces 7 (2015) 101–111.

[3] M. Ionescu, Chemistry and Technology of Polyols for Polyurethanes., iSmithers Rapra Publishing,, Shrewsbury, Shropshire, United Kingdom, 2006.

[4] Z. Li, R.W. Zhang, K.S. Moon, Y. Liu, K. Hansen, T.R. Le, C.P. Wong, Highly conductive, flexible, polyurethane-based adhesives for flexible and printed electronics, Adv. Funct. Mater. 23 (2013) 1459.

[5] M.J. Chen, Z.B. Shao, X.L. Wang, L. Chen, Y.Z. Wang, Halogen-free flame-retardant flexible polyurethane foam with a novel nitrogen-phosphorus flame retardant, Ind. Eng. Chem. Res. 51 (2012) 9769–9776.

[6] Y. Kiuchi, M. Iji, H. Nagashima, T. Miwa, Increase in flame retardance of glass-epoxy laminates without halogen or phosphorous compounds by simultaneous use of incombustible-gas generator and charring promoter, J. Appl. Polym. Sci. 101 (2006) 3367–3375.

[7] L. Ye, X.-Y. Meng, X.-M. Liu, J.-H. Tang, Z.-M. Li, Flame-retardant and mechanical properties of high-density rigid polyurethane foams filled with decabrominated dipheny ethane and expandable graphite, J. Appl. Polym. Sci. 111 (2009) 2372–2380.

[8] X. Zhou, J. Li, Y. Wu, Synergistic effect of aluminum hypophosphite and intumescent flame retardants in polylactide, Polym. Adv. Technol. 26 (2015) 255–265.

[9] H.Y. Ding, C.L. Xia, J.F. Wang, C.P. Wang, F.X. Chu, Inherently flame-retardant flexible bio-based polyurethane sealant with phosphorus and nitrogen-containing polyurethane prepolymer, J. Mater. Sci. 51 (2016) 5008.

[10] H. Ge, W. Wang, Y. Pan, X. Yu, W. Hu, Y. Hu, An inherently flame-retardant polyamide containing a phosphorus pendent group prepared by interfacial polymerization, RSC. Adv. 6 (2016) 81802–81808.

[11] S. Bhoyate, M. Ionescu, P.K. Kahol, J. Chen, S.R. Mishra, R.K. Gupta, Highly flame-retardant polyurethane foam based on reactive phosphorus polyol and limonene-based polyol, J. Appl. Polym. Sci. 135 (2018) 16–19.

[12] H. Ge, G. Tang, W.Z. Hu, B.B. Wang, Y. Pan, L. Song, Y. Hu, Aluminum hypophosphite microencapsulated to improve its safety and application to flame retardant polyamide, J. Hazard. Mater. 294 (2015) 186–194.

[13] L. Wang, X. He, C.A. Wilkie, The utility of nanocomposites in fire retardancy, Materials. (Basel) 3 (2010) 4580–4606.

[14] W.J. Wang, K. He, Q.X. Dong, Y. Fan, N. Zhu, Y.B. Xia, H.F. Li, J. Wang, Z. Yuan, E.P. Wang, X. Wang, H.W. Ma, Influence of aluminum hydroxide and expandable graphite on the flammability of polyisocyanurate-polyurethane foams, Appl. Mech. Mater. 368 (2013) 741.

[15] Z. Yang, H. Peng, W. Wang, T. Liu, Crystallization behavior of poly(ε-caprolactone)/layered double hydroxide nanocomposites, J. Appl. Polym. Sci. 116 (2010) 2658–2667.

[16] X. Zheng, G. Wang, W. Xu, Roles of organically-modified montmorillonite and phosphorous flame retardant during the combustion of rigid polyurethane foam, Polym. Degrad. Stab. 101 (2014) 32–39.

[17] E.D. Weil, S.V. Levchik, Flame retardants in commercial use or development for polyolefins, J. Fire Sci., 26, SAGE Publications, 2008, pp. 243–281.

[18] A. Castrovinci, G. Camino, C. Drevelle, S. Duquesne, C. Magniez, M. Vouters, Ammonium polyphosphate-aluminum trihydroxide antagonism in fire retarded butadiene-styrene block copolymer, Eur. Polym. J. 41 (2005) 2023–2033.

[19] Y. Liu, J. He, R. Yang, Effects of dimethyl methylphosphonate, aluminum hydroxide, ammonium polyphosphate, and expandable graphite on the flame retardancy and thermal properties of polyisocyanurate-polyurethane foams, Ind. Eng. Chem. Res. 54 (2015) 5876–5884.

[20] F. Luo, K. Wu, M. Lu, L. Yang, J. Shi, Surface modification of aluminum hypophosphite and its application for polyurethane foam composites, J. Therm. Anal. Calorim. 129 (2017) 767–775.

[21] Y. Wang, F. Wang, Q. Dong, M. Xie, P. Liu, Y. Ding, S. Zhang, M. Yang, G. Zheng, Core-shell expandable graphite @ aluminum hydroxide as a flame-retardant for rigid polyurethane foams, Polym. Degrad. Stab. 146 (2017) 267–276.

[22] C.E. Hoyle, T.Y. Lee, T. Roper, Thiol-enes: Chemistry of the Past with Promise for the Future, Department of Polymer Science, Hattiesburg, Mississippi, 2004.

[23] M. Desroches, S. Caillol, V. Lapinte, A.R. m., B. Boutevin, Synthesis of biobased polyols by thiol-ene coupling from vegetable oils, Macromolecules 44 (2011) 2489.

[24] L.M. Campos, K.L. Killops, R. Sakai, J.M.J. Paulusse, D. Damiron, E. Drockenmuller, B.W. Messmore, C.J. Hawker, Development of thermal and photochemical strategies for thiol—ene click polymer functionalization, Macromolecules 41 (2008) 7063–7070.

[25] G. Boutevin, B. Ameduri, B. Boutevin, J.-P. Joubert, Synthesis and use of hydroxyl telechelic polybutadienes grafted by 2-mercaptoethanol for polyurethane resins, J. Appl. Polym. Sci. 75 (2000) 1655–1666.

[26] J. Samuelsson, M. Jonsson, T. Brinck, M. Johansson, Thiol-ene coupling reaction of fatty acid monomers, J. Polym. Sci. A. Polym. Chem. 42 (2004) 6346–6352.

[27] L. Qian, Y. Qiu, N. Sun, M. Xu, G. Xu, F. Xin, Y. Chen, Pyrolysis route of a novel flame retardant constructed by phosphaphenanthrene and triazine-trione groups and its flame-retardant effect on epoxy resin, Polym. Degrad. Stab. 107 (2014) 98–105.

[28] S.S. Xiao, M.J. Chen, L.P. Dong, C. Deng, L. Chen, Y.Z. Wanga, Thermal degradation, flame retardance and mechanical properties of thermoplastic polyurethane composites based on aluminum hypophosphite, Chinese. J. Polym. Sci. 32 (2014) 98–107.

[29] S. Wu, D. Deng, L. Zhou, P. Zhang, G. Tang, Flame retardancy and thermal degradation of rigid polyurethane foams composites based on aluminum hypophosphite, Mater. Res. Express 6 (2019) 105365.

[30] B. Zhao, Z. Hu, L. Chen, Y. Liu, Y. Liu, Y.-Z. Wang, A phosphorus-containing inorganic compound as an effective flame retardant for glass-fiber-reinforced polyamide 6, J. Appl. Polym. Sci. 119 (2011) 2379–2385.

[31] M. Modesti, A. Lorenzetti, Halogen-free flame retardants for polymeric foams, Polym. Degrad. Stab. 78 (2002) 167–173.

[32] M. Ionescu, Z.S. Petrović, High functionality polyether polyols based on polyglycerol, J. Cell. Plast. 46 (2010) 223–237.

[33] C.E. Hoyle, C.N. Bowman, Thiol-ene click chemistry, Angew. Chem,. Int. Ed. 49 (2010) 1540.

[34] C.R. Morgan, F. Magnotta, A.D. Ketley, Thiol/ene photocurable polymers, J. Polym. Sci. Polym. Chem. Ed. 15 (1977) 627–645.

[35] H. Liu, H. Chung, Visible-light induced thiol-ene reaction on natural lignin, ACS. Sustain. Chem. Eng. 5 (2017) 9160–9168.

[36] C. Zhang, S. Bhoyate, M. Ionescu, P.K. Kahol, R.K. Gupta, Highly flame retardant and bio-based rigid polyurethane foams derived from orange peel oil, Polym. Eng. Sci. 58 (2018) 2078–2087.

[37] S Guo, M Bao, X Ni, The synthesis of meltable and highly thermostable triazine-DOPO flame retardant and its application in PA66, Polym. Adv. Technol. 32 (2) (2020) 815–828.

[38] W. Yang, G. Tang, L. Song, Y. Hu, R.K.K. Yuen, Effect of rare earth hypophosphite and melamine cyanurate on fire performance of glass-fiber reinforced poly(1,4-butylene terephthalate) composites, Thermochim. Acta 526 (2011) 185.

[39] G. Tang, X. Wang, R. Zhang, B. Wang, N. Hong, Y. Hu, L. Song, X. Gong, Effect of rare earth hypophosphite salts on the fire performance of biobased polylactide composites, Ind. Eng. Chem. Res. 52 (2013) 7362–7372.

[40] W. Wu, S. Lv, X. Liu, H. Qu, H. Zhang, J. Xu, Using TG–FTIR and TG–MS to study thermal degradation of metal hypophosphites, J. Therm. Anal. Calorim. 118 (2014) 1569–1575.

[41] K. Wilpiszewska, T. Spychaj, W. Paździoch, Carboxymethyl starch/montmorillonite composite microparticles: Properties and controlled release of isoproturon, Carbohydr. Polym. 136 (2016) 101–106.

[42] R.K. Gupta, M. Ionescu, D. Radojcic, X. Wan, Z.S. Petrovic, Novel renewable polyols based on limonene for rigid polyurethane foams, J. Polym. Environ. 22 (2014) 304–309.

[43] C.K. Ranaweera, M. Ionescu, N. Bilic, X. Wan, P.K. Kahol, R.K. Gupta, Biobased polyols using thiol-ene chemistry for rigid polyurethane foams with enhanced flame-retardant properties, J. Renew. Mater. 5 (2017) 1.

[44] D.K. Chattopadhyay, D.C. Webster, Thermal stability and flame retardancy of polyurethanes, Prog. Polym. Sci. 34 (2009) 1068–1133.

[45] K. Kulesza, K. Pielichowski, Thermal decomposition of bisphenol A-based polyetherurethanes blown with pentane: part II—Influence of the novel $NaH_2PO_4/NaHSO_4$ flame retardant system, J. Anal. Appl. Pyrolysis 76 (2006) 249–253.

[46] W. Zatorski, Z.K. Brzozowski, A. Kolbrecki, New developments in chemical modification of fire-safe rigid polyurethane foams, Polym. Degrad. Stab 93 (2008) 2071–2076.

[47] Y. Li, A.J. Ragauskas, Kraft lignin-based rigid polyurethane foam, J. Wood. Chem. Technol. 32 (2012) 210–224.

[48] J. Peyrton, C. Chambaretaud, A. Sarbu, L. Avérous, Biobased polyurethane foams based on new polyol architectures from microalgae oil, ACS Sustain. Chem. Eng. 8 (2020) 12187–12196.

[49] A. Wolska, M. Goździkiewicz, J. Ryszkowska, Thermal and mechanical behaviour of flexible polyurethane foams modified with graphite and phosphorous fillers, J. Mater. Sci. 47 (2012) 5627–5634.

[50] W. Xi, L. Qian, Y. Chen, J. Wang, X. Liu, Addition flame-retardant behaviors of expandable graphite and [bis(2-hydroxyethyl)amino]-methyl-phosphonic acid dimethyl ester in rigid polyurethane foams, Polym. Degrad. Stab. 122 (2015) 36–43.

[51] A. Gharehbaghi, R. Bashirzadeh, Z. Ahmadi, Polyurethane flexible foam fire resisting by melamine and expandable graphite: industrial approach, J. Cell. Plast. 47 (2011) 549–565.

[52] A. Zhang, Y. Zhang, F. Lv, P.K. Chu, Synergistic effects of hydroxides and dimethyl methylphosphonate on rigid halogen-free and flame-retarding polyurethane foams, J. Appl. Polym. Sci. 128 (2013) 347–353.

[53] A. König, A. Malek, U. Fehrenbacher, G. Brunklaus, M. Wilhelm, T. Hirth, Silane-functionalized flame-retardant aluminum trihydroxide in flexible polyurethane foam, J. Cell. Plast. 46 (2010) 395–413.

[54] A. Agrawal, R. Kaur, R.S. Walia, Investigation on flammability of rigid polyurethane foam-mineral fillers composite, Fire Mater. 43 (2019) 917–927.

[55] B. Czupryński, J. Paciorek-Sadowska, J. Liszkowska, Modifications of the rigid polyurethane - polyisocyanurate foams, J. Appl. Polym. Sci. 100 (2006) 2020–2029.

[56] A. Nik Pauzi NNP, R. Majid, M.H. Dzulkifli, M.Y. Yahya, Development of rigid bio-based polyurethane foam reinforced with nanoclay, Compos. B. Eng 67 (2014) 521–526.

[57] Y. Yuan, C. Ma, Y. Shi, L. Song, Y. Hu, W. Hu, Highly-efficient reinforcement and flame retardancy of rigid polyurethane foam with phosphorus-containing additive and nitrogen-containing compound, Mater. Chem. Phys. 211 (2018) 42–53.

[58] K.H. Choe, D.S. Lee, W.J. Seo, W.N. Kim, Properties of rigid polyurethane foams with blowing agents and catalysts, Polym. J. 36 (2004) 368–373.

[59] J. Ferguson, Z. Petrovic, Thermal stability of segmented polyurethanes, Eur. Polym. J. 12 (1976) 177–181.

[60] Y. Zhang, S. Shang, X. Zhang, D. Wang, D.J. Hourston, Influence of structure of hydroxyl-terminated maleopimaric acid ester on thermal stability of rigid polyurethane foams, J. Appl. Polym. Sci. 58 (1995) 1803–1809.

[61] G. Tang, X. Wang, R. Zhang, W. Yang, Y. Hu, L. Song, X. Gong, Facile synthesis of lanthanum hypophosphite and its application in glass-fiber reinforced polyamide 6 as a novel flame retardant, Compos. A. Appl. Sci. Manuf. 54 (2013) 1–9.

[62] G. Tang, X. Wang, W. Xing, P. Zhang, B. Wang, N. Hong, W. Yang, Y. Hu, L. Song, Thermal degradation and flame retardance of biobased polylactide composites based on aluminum hypophosphite, Ind. Eng. Chem. Res. 51 (2012) 12009–12016.

[63] Y. Taguchi, I. Shibuya, M. Yasumoto, T. Tsuchiya, K. Yonemoto, The synthesis of isocyanurates on the trimerization of isocyanates under high pressure, Bull. Chem. Soc. Jpn. 63 (1990) 3486–3489.

[64] S.M. Raders, J.G. Verkade, An electron-rich proazaphosphatrane for isocyanate trimerization to isocyanurates, J. Org. Chem. 75 (2010) 5308–5311.

27

Waterborne polyurethane-based electrode nanomaterials

Saadat Majeed[a], Tahir Rasheed[b], Sameera Shafi[b], Ahmad Reza Bagheri[c], Tuan Anh Nguyen[d], Najam ul Haq[a] and Muhammad Bilal[e]

[a]DIVISION OF ANALYTICAL CHEMISTRY, INSTITUTE OF CHEMICAL SCIENCES, BAHAUDDIN ZAKARIYA UNIVERSITY, MULTAN, PAKISTAN. [b]SCHOOL OF CHEMISTRY AND CHEMICAL ENGINEERING, STATE KEY LABORATORY OF METAL MATRIX COMPOSITES, SHANGHAI JIAO TONG UNIVERSITY, SHANGHAI, CHINA. [c]DEPARTMENT OF CHEMISTRY, YASOUJ UNIVERSITY, YASOUJ, IRAN. [d]MICROANALYSIS DEPARTMENT, INSTITUTE FOR TROPICAL TECHNOLOGY, VIETNAM ACADEMY OF SCIENCE AND TECHNOLOGY, HANOI, VIETNAM. [e]SCHOOL OF LIFE SCIENCE AND FOOD ENGINEERING, HUAIYIN INSTITUTE OF TECHNOLOGY, HUAIAN, CHINA.

Chapter outline

Biopolymeric Nanomaterials: Fundamentals and Applications. DOI: https://doi.org/10.1016/B978-0-12-824364-0.00025-3

27.1 Introduction

Polyurethanes (PUs) consider as a type of thermosetting polymer consists of materials attached via carbamate groups. These types of polymers have the ability to maintain their shapes and do not melt by heating [1]. PUs were first introduced in 1937 by Otto Bayer in in collaboration with Leverkusen at IG Farben. The PU polymer is conventionally or more commonly made via reacting a di- and triisocyanate from polyol [2]. Hence, PUs include two kinds of monomers that polymerize to form alternating copolymers. Alike the polyols and isocyanates utilized to form PUs incorporating an average of two and many functional groups on molecules [1].

27.2 Synthesis of PUs

PUs can be fabricated using various methods [3]. One of the versatile methods for the synthesis of PUs is the polyol method [4,5]. Fig. 27.1 represents the fabrication of anionic waterborne PUs (WPUs). Other materials can also be useful for the PU synthesis. These materials are include surfactants, fillers, and so on. Polyol is categorized into two hydroxyl units; it is assorted with any methylene diphenyl isocyanate (MDI) and toluene diisocyanate (TDI). As a result, a linear-shaped polymer is created [8]. For example, linear PUs are created within a reaction through the simplest form of diol-(ethane-1,2-diol) and diisocyanate. Under polymerization, condensation occurs [9]. Some other PUs are created with polyol and TDI, and also originate from epoxypropane [10].

If polyol possesses greater than two activated hydroxyl units, connected via a wide range of molecules, these units are joined at middle ends. These cross-links make an immobile polymer structure through developed automatic features used during "stiff" PUs [11]. Thereby, diisocyanate, such as TDI and MDI, that interact from a polyol over three -OH units, for example, originated from epoxyethane and propane-1,2,3-trio, experience a cross-linking and also formulate a hard, thermosetting polymer [12].

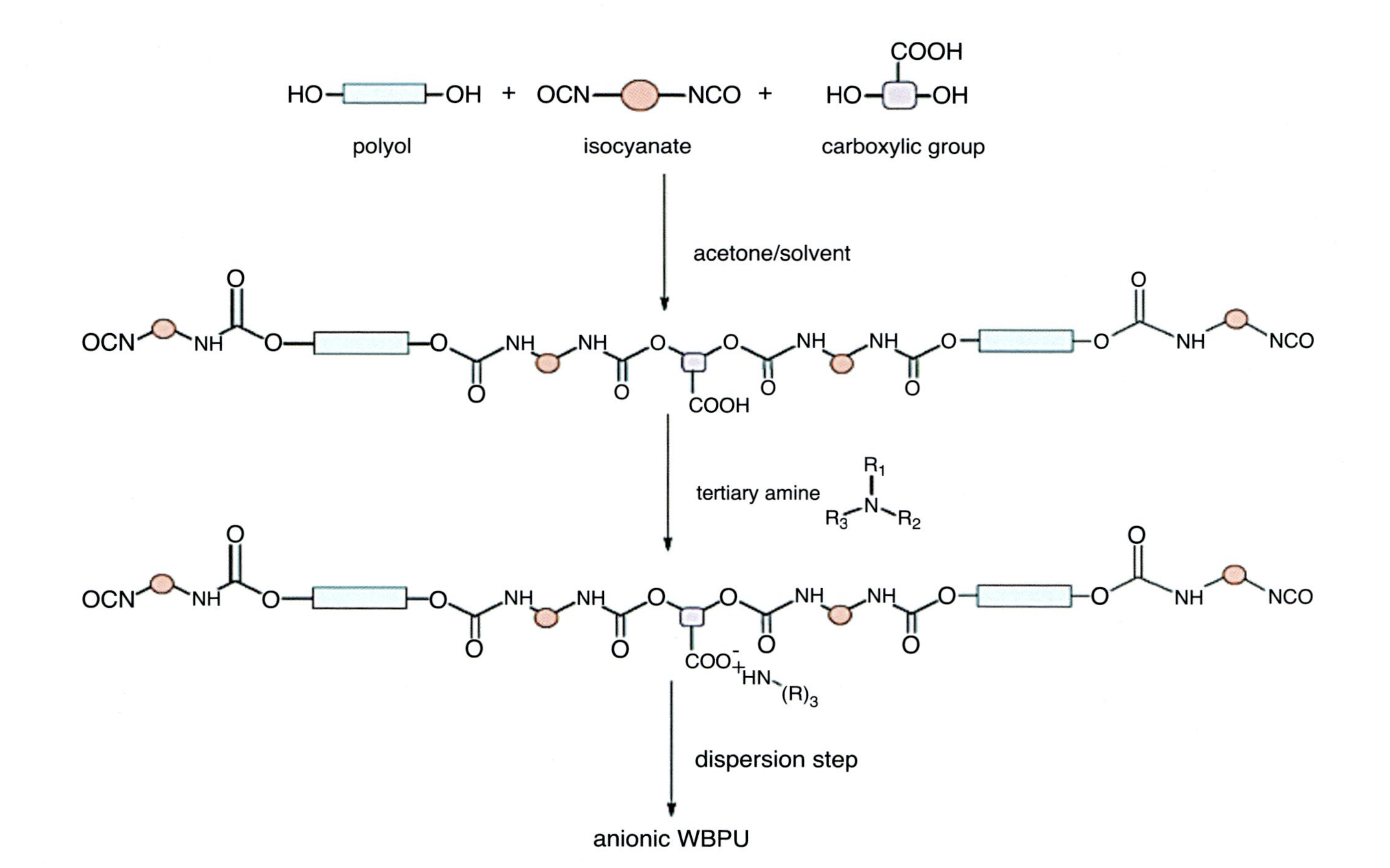

FIGURE 27.1 Synthetic route of anionic waterborne polyurethanes (Mucci et al. [5]; an open-access paper licensed under a Creative Commons Attribution 4.0 International License).

27.3 Uses of PUs

PUs are considered synthetic substances, such as plastic and foam, which occurs in several different shapes [3]. It may be modified whether firm, dense, and elastic and is the material for a wide scope of end-client applications, for instance: insulation of refrigerators and freezers [4]. PUs are used in the manufacturing of different materials such as gaskets, and microporous foam seals; sprinkle foam, long-lasting elastomeric tires and wheels, such as an escalator, skateboard wheels and roller coaster; car suspension bushings; electrically powered cooking utensil; highly efficient glutinous materials, sealing and fullerene-coated surface; polysynthetic materials similar to fibers, textiles and rubbers; floor covering such as carpets; a rigid plastic piece in particular to electronic tools, prophylactics, and hosepipes [5]. PUs are a man-made resin, and sort of paint-like material applied in mop-up of floors, compartments, and several woodworks. It has a desirable quality of paint because it is clear, water-, fungus- and decaying-resistive, and escape erosions [6].

27.4 Polyurethane nanocomposites (PUNCs) as electrode materials

WPUs have excellent properties in terms of eco-friendliness and nonflammability, which are green class materials applied in different fields, such as adhesives, coatings, drug delivery, and tissue engineering [7]. PUs, especially thermoplastic PUs (TPUs) represent amazing properties in terms of unique physical and chemical features. One of the properties of TPUs is their high resistance toward chemical processes and corrosion. Other main properties of TPUs are their flexibility at low-temperature, high mechanical strength, and toughness [8]. However, WPUs faced some drawbacks such as low thermal stability, low water resistance, and low mechanical stability. To address these main limitations, many efforts have been directed [9]. The synthesis and application of different PUNCs is one of the main directions for addressing these limitations [10]. WPUs composites (WPUCs) have excellent physical, chemical, and hydrolytic features. Fabrication of WPUCs is based on the incorporation of different types of materials with WPUs [10]. One of the interesting materials for incorporation with WPUs is nanomaterials (NMs). NMs have exclusive properties in high chemical and physical stability, high surface area, high adsorption capacity, and high conductivity, making them excellent materials for application in various fields [11]. Till date, many research imputes have been made for the construction of nano-based WPUCs [12]. One of the applications of WPUCs is their use as electrode materials [13]. Recently, El-Raheem and co-workers synthesized PU-doped platinum nanoparticles modified carbon paste electrode to determine copper ions [14]. The fabricated electrode showed high performance. The applied method was the first report for the synthesis of PU-modified electrodes to determine copper-free ions in biological samples. As PU has low electrical conductivity, PU alone showed low electrochemical signals. To remove this main limitation, PUNC based on Au, Ag, and Pt was synthesized and applied as electrode. This PUNC represent a low detection limit. In another work, a cheap and eco-friendly graphene oxide/PU foam reduced electrode was

Table 27.1 Some other reports for the application of polyurethane nanocomposites as electrode materials.

Material	Function	References
Surface-modified $BaTiO_3$ nanoparticles/PUNCs	Dielectric elastomer generators	[18]
Encapsulated core-sheath carbon nanotube–graphene/PU composite fiber	Strain sensor	[19]
Graphite-PU composite electrode modified with a molecularly imprinted polymer	Determination of tetracycline	[20]
Three-dimensional flexible PU sponge decorated with nickel hydroxide	Nonenzymatic glucose detection	[21]
NC of PU/reduced graphene oxide/silver nanoparticles	Electrode	[22]
PU foam doubly coated with conformal silicone rubber and CNT/TPU NC	Sensor	[23]
Graphite–PU composite electrode	Determination of dopamine	[24]
Disposable PU nanospiked gold electrode	Clostridium difficile detection	[25]
Graphite-polyurethane composite electrode modified with gold	Determination of tryptophan	[26]
Graphite oxide–Polyurethane composite electrode modified with cyclodextrin	Determination of environmental contaminants	[27]
PU modified with Ag	Electrode	[28]
PU/carbon fiber composite	Electrode	[29]
Nanostructured Ag:TiN thin films produced by glancing angle deposition on PU substrates	Electrode	[30]

constructed and used for the electrochemical applications [15]. The synthesized electrode had high surface area and high electrochemical properties that make it an excellent candidate for the electrochemical uses. The applied method for synthesizing proposed electrode was simple, facile, and green. The proposed electrode showed high supercapacitors (0.33 W h/kg) and long-term stability. Es'haghi and Moeipour developed an electrode-based carbon nanotube/PU-modified hollow fiber-pencil graphite and used this for the determination of anticancer drugs [16]. The applied electrode had high sensitivity and low detection limit, making it a suitable method for application in sensing of different materials in complex matrixes. In another study, carbon nanotube/TPUNCs were synthesized using noncovalent interactions and applied as a strain sensor [17]. The proposed sensor was fabricated using fused deposition modeling. Simmons' tunneling theory was used for investigating the mechanism of reaction. Table 27.1 shows some reports that have applied PUNCs as electrode materials.

27.5 Different reagents for the production of PUs

The primary elements used to produce PUs are polyols and di- and triisocyanates. Different substances are inserted to assist the manufacturing of polymers that alter polymers' characteristics [31].

27.5.1 Isocyanates

PUs are formed by isocyanates, and more than two isocyanate groups are attached to particular molecules. Generally, applied isocyanates are diisocyanates, TDI, and MDI. MDI or TDI is considered as the most reactive and less costly compared to other isocyanates. The commercial-scale MDI and TDI are a combination of isomers, and methylene diphenyl diisocyanate frequently includes polymeric substances [32]. Consequently, MDI is employed to produce a foam elastic and resilience (such as wrought foam for car seating and slabstock foam for bed-bottoms), elastomers (soles of shoes), hard foam (such as foam insulation inside freezers), etc. [33]. The shape of isocyanates could be altered by introducing them with polyols and inserting certain materials to decrease the toxicity, the volatility of isocyanates, and their freezing points, and to improve the features of ultimate polymers [34]. Cycloaliphatic and aliphatic isocyanates are utilized into litter quantities, most commonly in coatings or diverse petitions wherever their transparency and color are much significant because PUs create through aromatic isocyanates incline to brighten upon the influence of light. Most frequently used cycloaliphatic and aliphatic isocyanates, are 1,6-hexamethylene diisocyanate, 4,4′-diisocyanato di-cyclohexyl methane, and hydrogenated MDI (H_{12}-MDI) [35].

27.5.2 Polyols

Polyols are categorized into two main classes used in the PUs industry: polyesters and polyethers [36]. Polyols are a class of polyether polyols produced through a mixture of epoxides from reactive hydrogen, including mixtures. Polyols-based polyester is formed from polycondensation of multivalent carboxylic acids (-RCOOH), polyhydroxyl compounds such as sugars, cyclitols, and acyclic polyols [37]. Furthermore, it is categorized by end-user. Polyols have molecular weights (MWs) in the range of 2,000–10,000. These higher MW polyols make better PUs, whereas lower MW polyols make the most rigid and stable products. For more elastic applications, polyols utilize fewer operability initiators similar to dipropylene glycol, glycerin, and sorbitol ($f = 2, 3, 2.75$) in water resolution. Consequently, the flexible implementation uses advanced functionality initiators such as toluene diamine, Mannich bases, sorbitol, and sucrose ($f = 4, 4, 6, 8$). Several other oxides, such as ethylene or propylene oxides, are inserted into initiators until desirable MW is attained [38]. The state of insertion and quantity of any oxides impact different polyols properties such as water solubility, reactivity, and comparability. Fig. 27.2 illustrates the chemical structures for the synthesis of polyols. They are concluded over secondary -OH groups or considered less activated than polyols capped over ethylene oxide to incorporate to particular -OH groups. Polymer-based on polyols further occupy graft polyols that holds well and disperses acrylonitrile, styrene-acrylonitrile, and polyurea polymer, contains solids structures and are chemically grafted to a higher MW polyether skeleton [39]. They are made to enhance materials' supporting features, including toughness to microporous foams or formed elastomers. Initiators, that is triethanolamine and ethylenediamine, have been applied to produce low MW hard foams materials. Consequently, poly(tetramethylene ether) glycols are one of the particular class of polyether polyols that are made by polymerizing tetrahydrofuran, which is applied in higher processing of coating phenomena, elastomers, and micturition use [40].

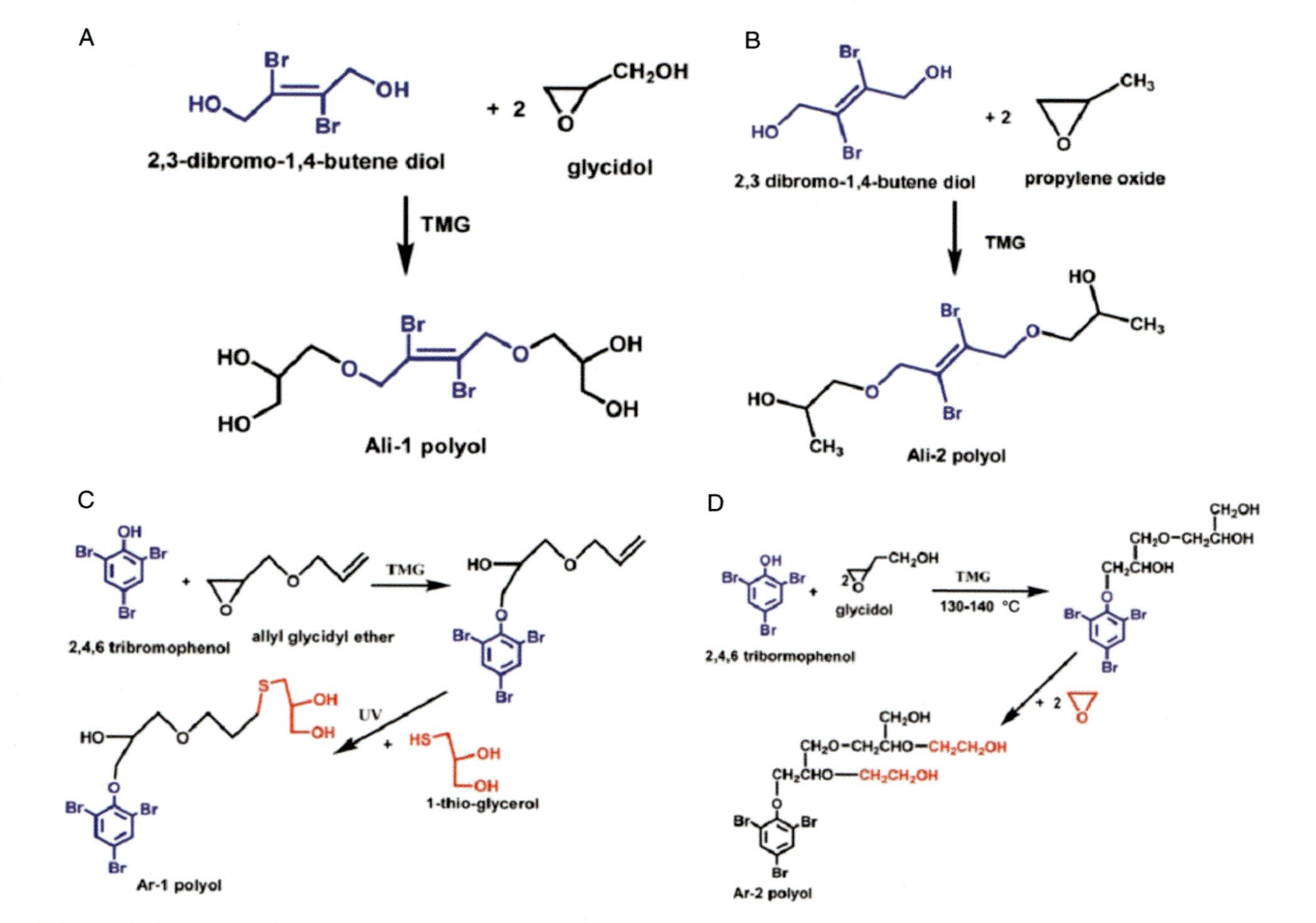

FIGURE 27.2 Chemical structure of the synthesized polyols (A) Ali-1, (B) Ali-2, (C) Ar-1, and (D) Ar-2[41]. Reproduced with permission from Bhoyate et al. [37]. Copyright © 2017 Wiley Periodicals, Inc.

Traditional polyester polyols are composed of pure raw matter. They are fabricated via direct polyesterification of significant purity diacids along with glycols. Polyesters are much expensive and are highly thick than polyether, which make PUs the best solvent for cut and abrasion resistance, [41]. Some other polyols of polyester rely on rescued raw materials. They are produced through the transesterification (glycolysis) of recycled poly (ethylene terephthalate) and dimethyl terephthalate from glycols by diethylene glycol. Similarly, the lower MW aromatic polyester is exploited under the rigid and firm foam and carries less expensive or distinctive superior combustibility [42]. The products are applied in the sealant, adhesive uses, and elastomers that demand resistance to ecological and chemical attack and have superior weather applicability [43]. The natural oils of polyols originated from castor oil and different vegetable oils are applicable to elastic, elastomers, or flexible molded foams [44].

27.6 Bio-derived materials

Different oils, such as castor and soybean seeds, are utilized to prepare polyols for PUs [45]. Vegetable oils may be characterized via different methods or varied to alkyds, polyether, etc. Similarly, renewable resources are applied to produce polyols [46]. Both bio-based and isocyanate-free PUs undergo the chemical reaction among cyclic carbonates or polyamines to create polyhydroxyurethanes [47]. PUs mixed with reused materials has been processed in a few pieces of the literature. Polyols based on fish oil and castor oil (CO) are utilized by bio-derived raw products [48]. Recycled polyols are formulated through the chemical-based decomposition of PUs elastic foam. Depending on their characteristics with recycled and natural polyols, end products of PUs are designed as hard foams and assemblage systems. Therefore, the fish-based oil of recycled polyols is appropriate for producing hard PU foam [49]. Owing to many secondary hydroxyl groups, the castor oil-based recycled polyols are less responsive to isocyanates than recycled polyols of fish-based oils. Accordingly, recycled polyols of castor oil are exploited to prepare thick cast PUs products [50]. Large aliphatic chains of recycled polyols can be well employed to produce elastic PUs flooring. Then, castor oil applications are applied to get the least reactive recycled polyol that could be used to prepare thick and solid cast PUs [51]. Among various organic materials, the chemical constitution of *Syzygium aromaticum*, usually called cloves, has a high possibility than renewable reinforcement to be used as polymeric products [51].

27.7 Cross-linkers and chain extenders

Chain extenders or cross-linkers containing reactive and low MW -OH and amine-based mixtures have an essential part within polymer structure for PU fibers, microporous foams, and adhesives, particularly intact skin or elastomers [52]. Elastomeric characteristics of such products originate from phase separation by which soft or hard copolymer portion of the polymer is obtained similar to urethane rigid portion domains treated as cross-linkers among polyester (i.e., amorphous polyether) [53]. This separation state occurs because primarily nonpolar and less melting soft fragments are antagonistic into polar, higher melting firm fragments. The soft fragments, which are made by higher MW polyols, are reactive and usually occur in spiral fabrication; while hard fragments, created through isocyanate or chain extenders, may be rigid or nonreactive [54]. Under mechanical distortion, a piece of the soft fragment, being emphasized through uncoiling, or the rigid fragments became arranged in the force of orientation [55]. This redirection of hard fragments and subsequent strong H-bonding provides greater tensile strength, tear resistance ranges, or elongation. The selection of the chain extender may find heat and possession of chemical resistance [56]. The significant chain extenders may be 1,6-hexanediol, cyclohexane dimethanol [57]. Some of them may include glycols-based PUs Except ethylene glycol all are appropriate TPUs; [58]. Similarly, diethanolamine or triethanolamine is employed under elastic wrought foams to construct hardness and includes catalytic states [59].

27.8 Catalysts

Catalyst used in PUs could be categorized into two types: acidic and basic amine. Catalytic operation of a tertiary amine can be enhanced by nucleophilicity of the diol element. Alkyl in carboxylates is utilized as moderate Lewis acids to speed up PU production [60]. Various bases such as, conventional amine including triethylenediamine, dimethyl cyclohexylamine, bis-(2-dimethylaminoethyl) ether, dimethylethanolamine, 1,4 diazabicyclooctane), can be employed as catalysts [61]. The following procedure, extremely delicate to the quality of catalyst, is called autocatalysis. Various factors including choice of catalyst, fabrication methodology, etc. are responsible for the synthesis of a suitable catalytic system. A mixture of specific catalysts has been formulated in another study [62].

27.9 Surfactants

Surfactants play a significant role in setting and stabilizing the foams formed by urethane. Surfactants may be utilized to change properties such as foam and nonfoam PU polymers [63]. Inside foam, they are exploited to change a liquid constituent, control the cell size, strengthen the cells' constitution, and preserve underground cavities and collapse [64]. Consequently, without foaming properties, they are exploited like antifoaming agents.

27.10 Waterborne PUs

WPUs may be developed into adhesives and coatings, including little and no co-solvent, making films at optimum temperature. The as-prepared PUs exhibit better attachment to the polymeric- and glass fibers-based surfaces. Those eco-friendly polymers are nonflammable, nontoxic, or cannot pollute the air and make effluents. Overall, PUs must be insoluble underwater or hydrophobic in nature. Hence, the state of scattering it inside H_2O could be altered, that is, integrating ionic or nonionic hydrophilic portion on polymer framework. WPUs are created from

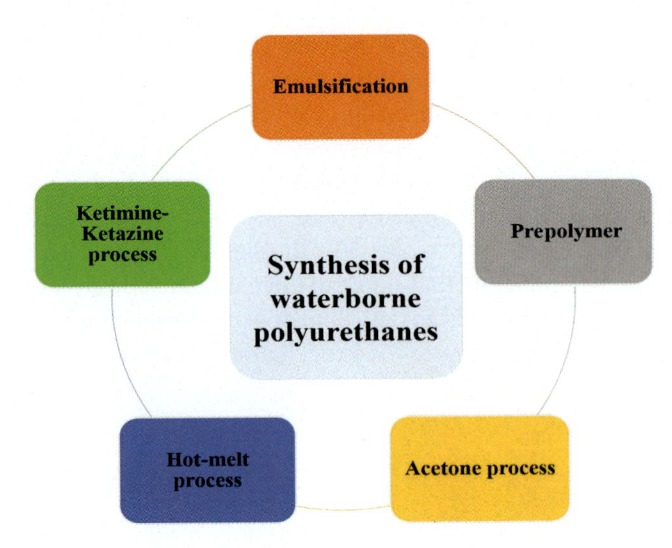

FIG 27.3 Different methods for the fabrication of waterborne polyurethanes.

two major parts: emulsifier and primary structure or backbone. This portion of the surfactant is hydrophilic in nature, and amphiphilic polymers such as grafted polymers modify scattering of the concerned PUs inside water. The hard fragments under ionic groups are hydrophilic in nature, or soft fragments play a vital part in the hydrophobic role. The emulsifier may be categorized into two main parts: external and internal parts. The internal emulsifier contains nonionic centers such as PUs oxide or ionic centers such as cationic and other kinds of surfactants [64]. Accordingly, the nonionic PU scatterings contain hydrophilic soft fragment dependent units similar to PU oxide. These scatterings are highly dispersed and stable in a wide variety of pH values.

27.11 Fabrication of WPUs

WPUs can be constructed in different ways. The procedure contains prepolymer or emulsification, hot-melt, acetone process (AP), ketimine-ketazine, and pre-polymer method or emulsification [65]. Fig. 27.3 presents different methods for the fabrication of WPUs.

27.11.1 Emulsification or prepolymer method

During the prepolymer blending process, the chain extension takes place during a heterogeneous stage, and the hydrophilic isocyanate disperses in water as a prepolymer forming agent. H_2O. The prepolymer mixing method by cycloaliphatic diisocyanates is the most utilized ordinarily due to their low reactivity of cycloaliphatic diisocyanates under water [66]. The scattering stage has to be operated for a limited period at low temperature. The crucial point of the process

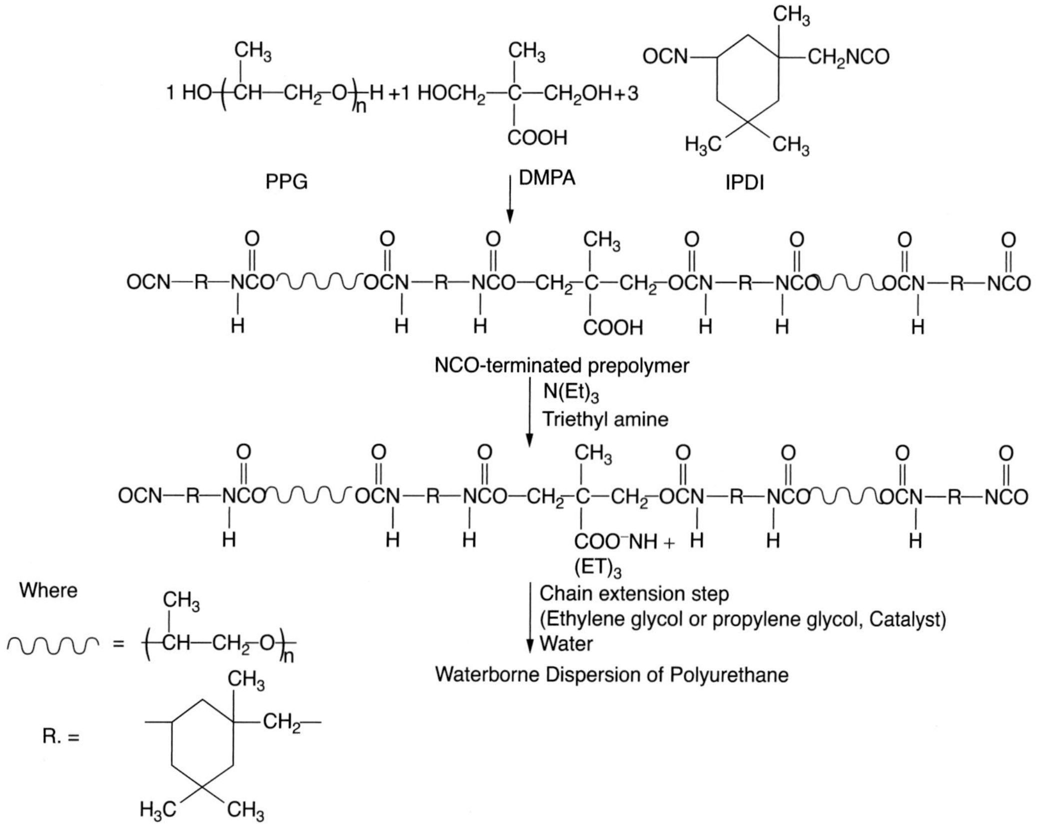

FIGURE 27.4 The preparation for water-based polyurethane dispersion via the prepolymer mixing method. Reproduced with permission from Cakic et al. [68]. Copyright © 2009 Elsevier Ltd. All rights reserved.

is the reaction of isocyanate (NCO) groups with water. For this procedure, the absolute control of the viscosity and functionality of the substance is essential.

By this mode, a moderate MW polymer or prepolymer is produced by the mixing of the right quantity of diols, containing polyester and polyether, with molar ratios of diisocyanates. Then, ionic cores are inserted into the reaction material and these cores are as long as internal surfactants [67]. Once the chain extension takes place, the stage of scattering within H_2O is executed. By this route, an amount of chitosan-based aqueous PU scatterings are fabricated, which is illustrated in Fig. 27.4 [68].

WPUs were analyzed using the alternate inquiry method depending upon 2,2-dimethoxy-2-phenyl acetophenone (DMPA) and hydroxytelechelic natural rubber (HTNR) by the prepolymer combination process. The outcomes revealed that the sample particle size decreases within an enhanced quantity of DMPA and MW of HTNR [66]. The water intake also improves within an enhanced DMPA, the hard fragment, and the epoxide elated. Yet, the increased amount of DMPA had no impact on the T_g of the WPU surface. The original WPU-founded fluorescent dyes were synthesized by connecting 4-amino-N-cyclohexyl-1,8-naphthalimide (CAN) within the PUs rings by the pre-polymer method [69]. The acquired thermal stability of PUs is enhanced. This occurrence results from the integration of naphthalimide groups inside urethane units. Moreover, the fluorescence intensity of WPU-CAN was enhanced under high temperature [70].

27.11.2 Acetone process

The AP may involve the production of a polymer through homogenous solution formation. Acetone is considered a good solvent for the production of PUs as the boiling point of the acetone is low which favors the distillation at lower temperatures compared to other solvents [71]. Acetone's particular benefit decreases the greater reactivity order from amine chain extenders within isocyanate by reversible ketamine production. Furthermore, acetone is used to manage consistency throughout this chain extension stage. Therefore, particular benefits from that procedure concisely consider [72]:

1. High reproducibility, but the chain extensions happen within the consistency of medium;
2. Better control of prepolymer or scattering of viscosity
3. The solvent-free end product is reliable.

Acetone is considered as a more appropriate and generally applied solvent through this method, yet methyl ethyl ketone may be utilized alternatively. There are still a few defects [73]. This method could be applied to form linear PUs soluble in acetone before the scattering stage [74]. The paths of the acetone method have been shown in Fig. 27.5.

Acetone is considered the most suitable solvents but methyl ethyl ketone may also be utilized instead. However, there are a few disadvantages. This operation may produce only linearly PUs, which are soluble in acetone before the dispersion process [75]. Besides the prepolymer method by forming PU scattering, this acetone procedure is the more applicable. Currently, scientists are employing an acetone method. [76].

Nanda et al. [77] evaluated impact of the ionic characteristic, the state for neutralizing, concentration of polymer, or the chain extension on aqueous PU diffusion through the acetone method. Similarly, the primary conflict among the prepolymer mixing process or the acetone method was investigated. Initially, the constant scattering was acquired within a lower concentration of DMPA (2%) as, in the prepolymer process, a large quantity (4%) was mandatory. This shows greater powerfulness of the integrated scattering units. Second, under the acetone method, the particle size was autonomous of polymer concentration. Similarly, a significant modification in viscosity was not determined under the acetone method.

27.11.3 Hot-melt process

The WPUs can be prepared by a nonsolvent-based methodology known as hot-melt process. The isocyanate capped prepolymer interacts with the urea to form a biuret unit. By conception, such a method is capping of prepolymer with urea by gaining biuret [78]. Then, such prepolymers are scattered within H_2O and treated by formaldehyde. The condensation is the final step or

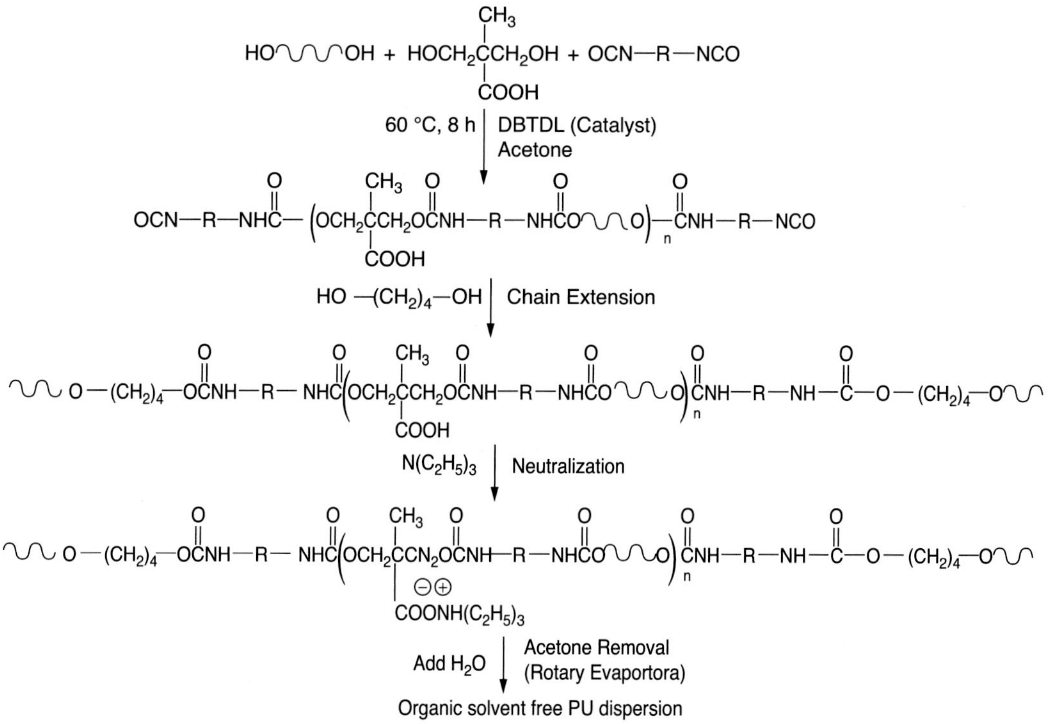

FIGURE 27.5 Elementary steps for the preparation of PU dispersions by acetone process. Reproduced with permission from Honarkar [1]. Copyright © 2018 Taylor & Francis.

the formation of the urea-based PUs. This method has been described in Fig. 27.6 [79]. Noble et al. [1] present a self-dispersing stage inside the water; such chain extensions are accomplished using methylation from biuret units through formaldehyde and reduce the pH range the polycondensation procedure. Such a method is of less importance than the formulation of surface covering. Moreover, crowned oligomers may be promptly scattered into the water in the absence of some organic (cosolvent). This procedure within urea is conducted toward graduated temperature such as >130 °C, and the subsequent oligomers are specifically scattered into lower temperature, that is, $>100°$ C to reduce the consistency. This melt-scattering procedure is applied to get WPUs across definite branching or less MW outcomes. The reaction of formaldehyde is tough to manage by using this process,. In addition, side reactions, including urethane units into PU, could happen. Then, this method provides WPUs within various polymer characteristics than the acetone method.

27.11.4 Ketimine-ketazine process

The method of ketimine-ketazine may be considered as a deviation of prepolymer mixing procedure. This method is conducted by mixing a prepolymer within a clogged amine and clogged hydrazine in the presence of air [80]. Diamine cloaked under the ketones to form keto-imines

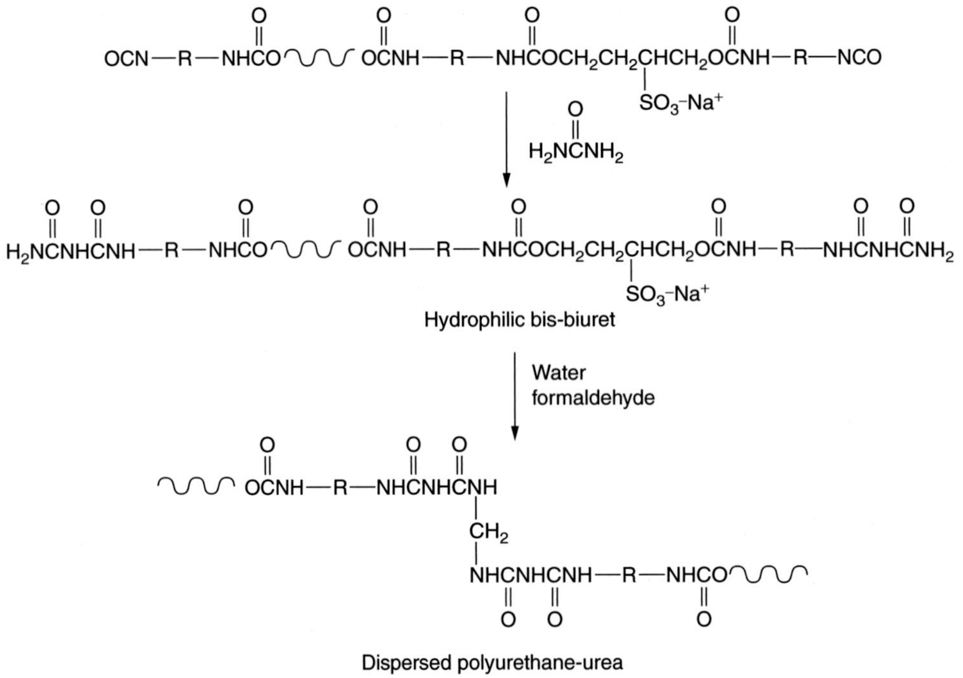

OCN—R—NHCO~~~OCNH—R—NHCOCH$_2$CH$_2$CH(SO$_3^-$Na$^+$)CH$_2$OCNH—R—NCO

H$_2$NCNH$_2$

H$_2$NCNHCNH—R—NHCO~~~OCNH—R—NHCOCH$_2$CH$_2$CH(SO$_3^-$Na$^+$)CH$_2$OCNH—R—NHCNHCNH$_2$

Hydrophilic bis-biuret

Water
formaldehyde

~~~OCNH—R—NHCNHCNH—CH$_2$—NHCNHCNH—R—NHCO~~~

Dispersed polyurethane-urea

FIGURE 27.6 Hot melt process. Reproduced with permission from Honarkar [1]. Copyright © 2018 Taylor & Francis.

may be blended to NCO prepolymer in the absence of reaction. By concerning this procedure, the substance is integrated into the water to analyze scattering. Then the hydrolysis of the keto-imine takes place to guarantee the chain extension followed by the addition of NCO-based pre-polymer. This means hydrazine that is relieved underwater creates a chain extension. The ketazines hydrolysis gradual than that of the ketimines. Fig. 27.7 illustrates the path of this method [81].

27.12 Applications of PUs

27.12.1 PU binder for LIBs

Fig. 27.8 shows different applications of PUs.

PU or Styrofoam binder is made by cross-linking a compound containing a double bond. The double bond–containing PU can scatter in the water, if we mix any reactive group such as -OH. In this way, there is no need for any organic solvent to produce an electrode. The Styrofoam or PU binder, which has high cross-linking density, gives strong elastic force and sticking power. The Styrofoam or PU is an incarnation in a modern age because, with the help of this type of double bond PU, we can store energy in different devices. PU binder is obtained by the reaction of polyol groups that contain double bonds. Polyol is a group used for cross-linking in the PU compound. The average MW of the polyol is 200–10,000, and the efficiency of the reaction is 1000–6000% [82].
~~~

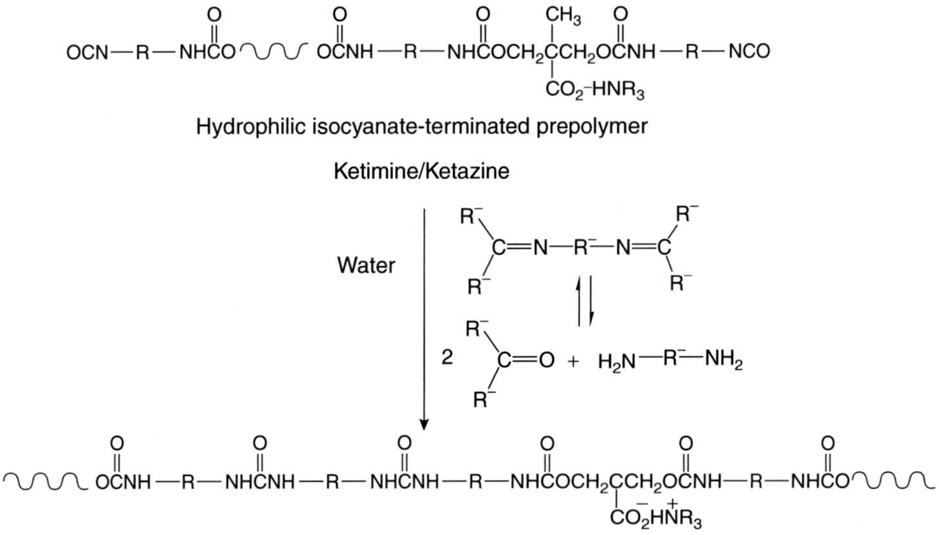

Hydrophilic isocyanate-terminated prepolymer

Ketimine/Ketazine

Water

FIGURE 27.7 Ketimine-ketazine process. Reproduced with permission from Honarkar [1]. Copyright © 2018 Taylor & Francis

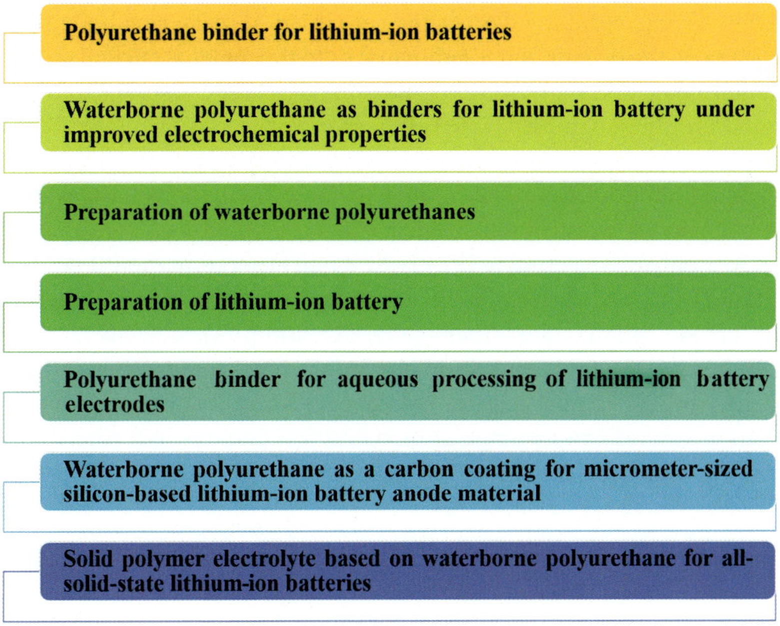

- Polyurethane binder for lithium-ion batteries
- Waterborne polyurethane as binders for lithium-ion battery under improved electrochemical properties
- Preparation of waterborne polyurethanes
- Preparation of lithium-ion battery
- Polyurethane binder for aqueous processing of lithium-ion battery electrodes
- Waterborne polyurethane as a carbon coating for micrometer-sized silicon-based lithium-ion battery anode material
- Solid polymer electrolyte based on waterborne polyurethane for all-solid-state lithium-ion batteries

FIGURE 27.8 Different applications of polyurethanes.

If the polyol's MW is low, the features of PU binder are automatically changed. If the polyol's MW is more than 10,000, then the reactivity is reduced and the PU group's viscosity is too high to handle PU. A polyol compound with a double bond can add polybutadiene with the OH group at both sides. It is also possible that a polyol with a double bond contains minimum groups of polyethylene glycol acrylate with -OH units at one side or polyethylene glycol of methacrylate with - OH group is present at the other side [83]. Cross-linking agent or water-soluble initiator used as PU binder also has high cross-linking properties, that provide enhanced elasticity to the electrode material. This is why PU binder resin is added to the electrode to retain its original volume and original stitching power against substantial mass changes occurring during charging and discharging of a battery. According to a living example of modern inventions, a full water PU binder is utilized as a PU binder on an electrode namely full of water. For such type of systems both the electrodes must be stable under aqueous environment. Also the morphologies of these components have greater effect on the performance of the system. An anode water mixture should be polished inside a current collector, similar to a copper film. An anode matrix mixture is specially made, which is directly polished onto a laminated Cu current collector to create an anode plate. Generally, when a high capacitance battery is charged, it shows a large charge and discharges of the current value. This results in the formation of an electrode having low resistivity. It may be used as a conductive agent to reduce electrodes' impedance when a basic material of anodes is C or Si, the reactivity having less conduction. Carbon black and graphite are most widely used conductive agents. For designing a cathode by employing the WPU binder, the ratio of active cathode material is calculated through a combination of active cathode materials, a conductive agent, and a WPU binder resin or a solvent. When cathode materials are directly applied onto the metallic current collector, they are dried to make a cathode plate. According to the distinct surface, the distant film is laminated onto the metallic current collectors, thereby making a cathode plate. Similarly, various metal oxides based materials are being used as cathode material for Li-ion batteries (LIBs). They may involve LiCoO , LiMnO , LiNiMnO, etc. Moreover, a cathode-based active matrix could be the mixture of Li that may have been oxidized and reduced, such as LiMnO, LiCoO, LiNiO, and LiFeC. Further, various sulfides such as VOS, TiS, or MoS along with conductive agents are used in PUs-based binders. The lithium battery would also be added in a polymer electrolyte (PE) layer by incorporating the PE ratio as an obvious composition. The PE is a mixture usually consumed in the art. The usually employed separators in Li batteries include Teflon's, glass fibers, polyester, and polypropylene. A separator can be directly coated over the electrode and dried afterward to make a film. Similarly, the separator composition is formed via a filling agent, polymer resins, and solvents. The structure of the battery is folded or wound at a rectangular or cylindrical battery case. Afterward, the organic electrolyte solution is added to the case; Thus, LIB is produced.

27.12.2 Waterborne Polyurethanes (WBPUs) as binders for LIB

LIB has amazing properties [84]. Usually, extensive explorers were centered on the active material improvements to upgrade the electrochemical characteristics of LIBs, whereas less consideration was committed for the headway at the electrically inert segments of battery cathodes, binder [85]. At the electrode, the primary impact of the binder is the bond. It maintains the active materials, increases the active material of the cathode or the conductive agent's contact, and reinforces the set liquid's active material to settle the pole piece's structure better [86]. Furthermore, the cell assembly is carried out with the help of coin cells to evaluate the electrochemical performance of the materials by using all the component required for battery formation. It maintains the stability of electrode composition [87]. Therefore, the selection of an appropriate binder is very significant for LIBs. Ordinarily, utilizing binders for LIBs mainly comprises polytetrafluoroethylene, polyvinyl alcohol, polyolefins (PE, Polypropylene, otherwise co-polymer), poly(vinylidene difluoride) (PVDF), adjusted Styrene-butadiene rubber (SBR), fluorine-based rubber. The most used binder is PVDF. However, the utilization of PVDF needs an enormous number of solvents organic in nature. WPU has a benefit of no contamination, great adhering force, or high resilience, and they are broadly utilized in textile, wrapping, and different fields.

27.12.3 Preparation of WPUs

The nonionic WPUs have been manufactured with the help of pre-polymer processing. Their synthesis was performed using a specific thermometer, automatic stirrer, and N_2 inlet or a reflux liquefier under $CaCl_2$ drying tube. During the first phase, macromolecular polydiols and isophorone diisocyanate, are assorted into the reactor or heated at 90 ± 2 °C in dry N_2 atm for 2 h. After that, it may be cooled at 50 °C, while continuously mixing a suitable quantity of 1,4-butanediol or Ac that regulates the viscosity or simultaneously, Dibutyltin dilaurate (DBTDL) and stannous octoate (0.1 wt% accordingly) are inserted using catalysts. The purpose of the remaining NCO particle is to attain a theoretical figure that is the end-point of reaction with di-butyl amine–HCl in titration. The resulting NCO terminated prepolymer is refrigerated at 40 °C or calculated deionized H_2O is inserted through stirring toward by mixing it with organic phase. Various types of WPUs such as PCDL, PTMG, BY3022, N220, and PNA are prepared as soft fragments [37].

27.12.4 Fabrication of LIBs

The LIBs are fabricated in a glove box in an inert atmosphere due to high reactivity of lithium metal anodes. For the fabrication of an LIB, lithium metal is used as anode, while cathode is any of the oxide, sulfide carbon-based materials, etc. Further the electrolyte is created through the solution of LiPF6, which includes 1 M of methyl-carbonate and ethylene carbonate in the ratio of 1:1.

Zhu et al. [88] prepared $LiFePO_4$-based LIBs via nonionic WBPUs under various delicate portions that act as a cover. Fourier Transform Infrared is utilized to describe the shapes of WPUs. Further, emulsion consistency and mechanical characteristics of the films are estimated. The charge and discharge, cycles execution, or AC of impedance spectroscopy estimation show that the main charge and discharged productivity is about 92%, the greatest discharge limit is 115 mAh/g.

27.12.5 PU binder for LIB electrodes

The significance of renewable energy resources and powerful activities is used for formulating the greater systems on small- and large-scale power reserves. Recently, LIBs are considered energy storage strategies to execute minute and chip-based energy gadgets, because of their prolonged lifetime and greater energy density. Moreover, their light weight meets the increasing requirements for the utilization of computers (PCs), smartphones, and different portable gadgets, along with electromobility resources, to make possible developments in the direction of advanced smartphone society. Li-ion and their storage gadgets consist of various materials. Such a feature exist into the permeable separator, lying among electrodes such as polyolefin while binding components within a compound electrode. These binders are considered between inactive constituents about the capability of cell phones.

On the other hand, the main role of electrode processing or its remarkable effects on electrodes' electro-chemical execution has been widely summarized. Appropriate chemical and physical characteristics for binding matters are: (1) tensile strength (cohesion), (2) thermal stability, (3) electrochemical/chemical stability, (4) flexibility, (5) viscosity (on slurries). The core function of the binder is to make stable links onto the electrode surface, such as the active components and conducting materials (i.e., cohesion). Furthermore, binders are used to assure the electrode composites' intimate contact over the current collector (adherence). The recycling of battery constituents in later life includes discharging and dissolution of the PVdF inside N-Methylpyrrolidone (NMP), creating more environmental issues. These all interests could be robustly abridged by utilizing binders that are H_2O soluble in case not removed. As a result of safer, cheaper, water-soluble binders, eco-friendly batteries have gained much importance. The processing of liquid-based electrode is much easier and cheaper in contrast to solvent typed methods such as PVdF.

Loeffler et al. [89] synthesized LIBs electrode. Their trial outcomes show the extraordinary thermal or electrochemical strength of PUs, to be used as a binder. The most significant results are PUs' capacity, which may stop current accumulator erosion that normally occurs when a fluid scattering (slurry) of lithium nickel manganese cobalt oxide NMC covers Al foils. The scanning electron microscope study demonstrates that PUs exemplify the positive dynamic particles avoiding increase in the pH values on slurry. Finally the full cell of the PUs and Carboxymethyl Cellulose (CMC) was employed to check the performance of material, which exhibit good cycling performance.

27.12.6 WBPU as a carbon coating for micrometer-sized silicon-based LIBs anode material

LIBs are an excellent device for electric vehicles due to their unique properties [84]. Graphite is essentially a negatively charged material [90]. Silicon (Si) has established enhanced interest as a new negatively charged electrode material concerning high theoretical capacity (3579 mA h g). In addition, silicon has a comparatively little voltage level, and is inexpensive, nontoxic, and extensively accessible in nature [91]. Many nanostructured silicon shapes are employed (e.g., nanoparticles, nanowire, and nanoporous structures) [35]. Such nanostructures would extraordinarily boost the cycling operation of Si extensively as they can endure profound litigation and the dilatation procedure without any cracking. Nonetheless, the nano-Si lattice is exorbitant due to its costly synthesis technique and little tapping density. However, nano-sized Si is simple to reunite as its higher surface area and surface energy, however, is hard to get through formulation at commercial-scale purpose. Micrometer-sized silicon is a better choice compared to nano-sized silicon.

Yan et al. [92] utilized WPU by carbon-covering source as micron-sized silicon. The rest of the N and O heteroatoms under pyrolysis of WPUs cooperate within the surface of oxide over silicon (Si) atoms through H-holding (Si–OHïN and Si–OHïO). Therefore, nitrogen or oxygen atoms occupied with carbon links may communicate by Li particles that help Li particle inclusion. The developed electrochemical presentation of Si@CNO cathodes may be allocated to the upgraded electrical conductivity or auxiliary dependability.

27.12.7 WBPUs-based solid PE for all-solid-state LIBs

LIBs are uncontrollably used as a fundamental vitality-stockpiling gadget for advanced cells, tablet PCs, electric vehicles, etc. The traditional LIBs utilize a lithium salt that is naturally dissolvable and the electrolyte gives reasonable ionic conductivity. In any case, the utilization of natural fluid electrolytes may cause some difficult issues such as fire, spillage, and unfortunate responses with cathodes, and flimsiness at high temperatures. All strong-state LIBs, which utilize strong state electrolytes, offer an answer for these downsides [93] and polyphosphazene. In any case, the primary issue is their low ionic conductivity and poor dimensional soundness. The PE ought to have a high ionic conductivity (surpassing 1024 S cm^2), wide electrochemical dependability window (up to 4–5 V vs Li/Li), great mechanical quality (more than 1 MPa), and fantastic warm soundness. Normally in SPEs, the ionic conductivity and mechanical quality are contrarily identified with one another. High conductivity is related to the portions' adaptability; however, the dimensional strength relies upon the polymer's inflexibility. Consequently, polymers, which have both delicate sections and hard portions, are useful.

27.13 Conclusions

This chapter presents recent progress and development to synthesize different types of waterborne PUs and their use in different fields due to a number of unique properties and wide applications. PU composites have unique properties such as excellent physical, chemical, and hydrolytic features. One of the exciting materials to incorporate with WPUs is NMs that exhibit high chemical and physical stability, high surface area, high adsorption capacity, high conductivity, making them prodigious materials for application in various fields. Due to desirable characteristics, they have been widely applied to determine and sense different compounds and materials. In conclusion, this chapter would help researchers better understand PUs and their specific properties and applications in various fields.

Acknowledgment

The authors are highly obliged to their institutes and universities for the literature services.

Conflict of interest

The authors declare that they have no known competing and financial conflicting interests.

References

[1] H. Honarkar, Waterborne polyurethanes: a review, J. Dispers. Sci. Technol. 39 (2018) 507–516.

[2] Y. Ye, Q. Zhu, The development of polyurethane, Mater. Sci. 1 (2017) 1–8.

[3] A. Cornille, R. Auvergne, O. Figovsky, B. Boutevin, S. Caillol, A perspective approach to sustainable routes for non-isocyanate polyurethanes, Eur. Polym. J. 87 (2017) 535–552.

[4] Y. Xiao, X. Fu, Y. Zhang, Z. Liu, L. Jiang, J. Lei, Preparation of waterborne polyurethanes based on the organic solvent-free process, Green Chem. 18 (2016) 412–416.

[5] V.L. Mucci, M. Hormaiztegui, M. Aranguren, Plant oil-based waterborne polyurethanes: a brief review, J. Renew. Mater. 8 (2020) 579–601.

[6] M. Dai, Y. Zhai, L. Wu, Y. Zhang, Magnetic aligned Fe_3O_4-reduced graphene oxide/waterborne polyurethane composites with controllable structure for high microwave absorption capacity, Carbon 152 (2019) 661–670.

[7] A. Rostami, M.I. Moosavi, High-performance thermoplastic polyurethane nanocomposites induced by hybrid application of functionalized graphene and carbon nanotubes, J. Appl. Polym. Sci. 137 (2020) 48520.

[8] A. Rostami, M.I. Moosavi, High-performance thermoplastic polyurethane nanocomposites induced by hybrid application of functionalized graphene and carbon nanotubes, J. Appl. Polym. Sci. 137 (2020) 48520.

[9] Q. Gao, M. Feng, E. Li, C. Liu, C. Shen, X. Liu, Mechanical, thermal, and rheological properties of $Ti_3C_2T_x$ MXene/thermoplastic polyurethane nanocomposites, Macromol. Mater. Eng. 305 (2020) 2000343.

[10] N. Nouri, M. Rezaei, R.L.M. Sofla, A. Babaie, Synthesis of reduced octadecyl isocyanate-functionalized graphene oxide nanosheets and investigation of their effect on physical, mechanical, and shape memory properties of polyurethane nanocomposites, Compos. Sci. Technol. (2020) 108170.

[11] C. Salgado, M.P. Arrieta, V. Sessini, L. Peponi, D. López, M. Fernández-García, Functional properties of photo-crosslinkable biodegradable polyurethane nanocomposites, Polym. Degrad. Stabil. 178 (2020) 109204.

[12] L. Yang, Q. Fu, H. Fu, Preparation of novel hydrophobic magnetic Fe_3O_4/waterborne polyurethane nanocomposites, J. Appl. Polym. Sci. 137 (2020) 48546.

[13] A. Amirkiai, M. Panahi-Sarmad, G.M.M. Sadeghi, M. Arjmand, M. Abrisham, P. Dehghan, H. Nazockdast, Microstructural design for enhanced mechanical and shape memory performance of polyurethane nanocomposites: Role of hybrid nanofillers of montmorillonite and halloysite nanotube, Appl. Clay Sci. 198 (2020) 105816.

[14] H.A. El-Raheem, R.Y.A. Hassan, R. Khaled, A. Farghali, I.M. El-Sherbiny, Polyurethane-doped platinum nanoparticles modified carbon paste electrode for the sensitive and selective voltammetric determination of free copper ions in biological samples, Microchem. J. 155 (2020) 104765.

[15] A. Sanati, K. Raeissi, F. Karimzadeh, A cost-effective and green-reduced graphene oxide/polyurethane foam electrode for electrochemical applications, FlatChem 20 (2020) 100162.

[16] Z. Es'haghi, F. Moeinpour, Carbon nanotube/polyurethane modified hollow fiber-pencil graphite electrode for in situ concentration and electrochemical quantification of anticancer drugs capecitabine and erlotinib, Eng. Life Sci. 19 (2019) 302–314.

[17] D. Xiang, X. Zhang, Y. Li, E. Harkin-Jones, Y. Zheng, L. Wang, C. Zhao, P. Wang, Enhanced performance of 3D printed highly elastic strain sensors of carbon nanotube/thermoplastic polyurethane nanocomposites via non-covalent interactions, Compos. B Eng. 176 (2019) 107250.

[18] Y. Yang, Z.-S. Gao, M. Yang, M.-S. Zheng, D.-R. Wang, J.-W. Zha, Y.-Q. Wen, Z.-M. Dang, Enhanced energy conversion efficiency in the surface modified BaTiO3 nanoparticles/polyurethane nanocomposites for potential dielectric elastomer generators, Nano Energy 59 (2019) 363–371.

[19] Y. Xu, X. Xie, H. Huang, Y. Wang, J. Yu, Z. Hu, Encapsulated core-sheath carbon nanotube-graphene/polyurethane composite fiber for highly stable, stretchable, and sensitive strain sensor, J. Mater. Sci. 56 (2021) 2296–2310.

[20] J.E.S. Clarindo, R.B. Viana, P. Cervini, A.B.F. Silva, E.T.G. Cavalheiro, Determination of tetracycline using a graphite-polyurethane composite electrode modified with a molecularly imprinted polymer, Anal. Lett. 53 (2020) 1932–1955.

[21] S. Guo, C. Zhang, M. Yang, Y. Zhou, C. Bi, Q. Lv, N. Ma, A facile and sensitive electrochemical sensor for non-enzymatic glucose detection based on three-dimensional flexible polyurethane sponge decorated with nickel hydroxide, Anal. Chim. Acta 1109 (2020) 130–139.

[22] Y.-I. Choi, B.-U. Hwang, M. Meeseepong, A. Hanif, S. Ramasundaram, T.Q. Trung, N.-E. Lee, Stretchable and transparent nanofiber-networked electrodes based on nanocomposites of polyurethane/reduced graphene oxide/silver nanoparticles with high dispersion and fused junctions, Nanoscale 11 (2019) 3916–3924.

[23] J. Lee, J. Kim, Y. Shin, I. Jung, Ultra-robust wide-range pressure sensor with fast response based on polyurethane foam doubly coated with conformal silicone rubber and CNT/TPU nanocomposites islands, Compos. B Eng. 177 (2019) 107364.

[24] P. Cervini, I.A. Mattioli, É.T.G. Cavalheiro, Developing a screen-printed graphite-polyurethane composite electrode modified with gold nanoparticles for the voltammetric determination of dopamine, RSC Adv. 9 (2019) 42306–42315.

[25] F. Cui, Z. Zhou, H. Feng, H.S. Zhou, Disposable polyurethane nanospiked gold electrode-based label-free electrochemical immunosensor for *Clostridium difficile*, ACS Appl. Nano Mater. 3 (2020) 357–363.

[26] I.A. Mattioli, M. Baccarin, P. Cervini, É.T.G. Cavalheiro, Electrochemical investigation of a graphite-polyurethane composite electrode modified with electrodeposited gold nanoparticles in the voltammetric determination of tryptophan, J. Electroanal. Chem. 835 (2019) 212–219.

[27] A. Wong, A.M. Santos, M. Baccarin, É.T.G. Cavalheiro, O. Fatibello-Filho, Simultaneous determination of environmental contaminants using a graphite oxide – polyurethane composite electrode modified with cyclodextrin, Mater. Sci. Eng. C 99 (2019) 1415–1423.

[28] B. Hwang, C.-H. An, S. Becker, Highly robust Ag nanowire flexible transparent electrode with UV-curable polyurethane-based overcoating layer, Mater. Des. 129 (2017) 180–185.

[29] K. Zhang, Y. Li, H. Zhou, M. Nie, Q. Wang, Z. Hua, Polyurethane/carbon fiber composite tubular electrode featuring three-dimensional interpenetrating conductive network, Carbon 139 (2018) 999–1009.

[30] P. Pedrosa, D. Machado, P. Fiedler, B. Vasconcelos, E. Alves, N.P. Barradas, N. Martin, J. Haueisen, F. Vaz, C. Fonseca, Electrochemical characterization of nanostructured Ag:TiN thin films produced by glancing angle deposition on polyurethane substrates for bio-electrode applications, J. Electroanal. Chem. 768 (2016) 110–120.

[31] A. Kemona, M. Piotrowska, Polyurethane recycling and disposal: methods and prospects, Polymers 12 (2020) 1752.

[32] M.V. Zabalov, M.A. Levina, R.P. Tiger, Polyurethanes without isocyanates and isocyanates without phosgene as a new field of green chemistry: mechanism, catalysis, and control of reactivity, Russ. J. Phys. Chem. B 13 (2019) 778–788.

[33] K. Błażek, J. Datta, Renewable natural resources as green alternative substrates to obtain bio-based non-isocyanate polyurethanes-review, Crit. Rev. Environ. Sci. Technol. 49 (2019) 173–211.

[34] A. Sienkiewicz, P. Czub, Blocked isocyanates as alternative curing agents for epoxy-polyurethane resins based on modified vegetable oils, Exp. Polym. Lett. 13 (7) (2019) 642–655.

[35] C. Zhang, H. Wang, W. Zeng, Q. Zhou, High biobased carbon content polyurethane dispersions synthesized from fatty acid-based isocyanate, Ind. Eng. Chem. Res. 58 (2019) 5195–5201.

[36] M. Ghasemlou, F. Daver, E.P. Ivanova, B. Adhikari, Polyurethanes from seed oil-based polyols: a review of synthesis, mechanical and thermal properties, Ind. Crop. Prod. 142 (2019) 111841.

[37] S. Bhoyate, M. Ionescu, D. Radojcic, P.K. Kahol, J. Chen, S.R. Mishra, R.K. Gupta, Highly flame-retardant bio-based polyurethanes using novel reactive polyols, J. Appl. Polym. Sci. 135 (2018) 46027.

[38] M. Kurańska, J.A. Pinto, K. Salach, M.F. Barreiro, A. Prociak, Synthesis of thermal insulating polyurethane foams from lignin and rapeseed based polyols: a comparative study, Ind. Crop. Prod. 143 (2020) 111882.

[39] L. Ren, X. Ma, J. Zhang, T. Qiang, Preparation of gallic acid modified waterborne polyurethane made from bio-based polyol, Polymer 194 (2020) 122370.

[40] Y. Maruoka, A. Miyata, K. Okubo, T. Izukawa, T. Hiraide, S. Matsumoto, Composition for Polyurethane Foam, Preparation for Polyurethane Foam, Polymer Polyol Preparation for Polyurethane Foam, Production Processes Therefore, and Polyurethane Foam, Google Patents US10494469B2, 2019.

[41] A. Terheiden, R. Hubel, M. Ferenz, Additive Composition Useful for Controlling the Foam Properties in the Production of Flexible Polyurethane Foams Containing Polyols Based on Renewable Raw Materials, Google Patents US9328210B2, 2016.

[42] N.M. Saad, S.A. Zubir, Palm kernel oil polyol-based polyurethane as shape memory material: effect of polyol molar ratio, J. Phys. Sci. 30 (2019) 77–89.

[43] M. Kuranska, S. Michalowski, J. Radwanska, M. Jurecka, M. Zieleniewska, L. Szczepkowski, J. Ryszkowska, A. Prociak, Bio-polyols from rapeseed oil as raw materials for polyurethane composites with natural fillers for use in cosmetics, Przem. Chem. 95 (2016) 256–262.

[44] J.A.P. da Silva, N.S.M. Cardozo, C.L. Petzhold, Enzymatic synthesis of andiroba oil based polyol for the production of flexible polyurethane foams, Ind. Crop. Prod 113 (2018) 55–63.

[45] M.E.V. Hormaiztegui, B. Daga, M.I. Aranguren, V. Mucci, Bio-based waterborne polyurethanes reinforced with cellulose nanocrystals as coating films, Prog. Org. Coat. 144 (2020) 105649.

[46] T. Wan, D. Chen, Synthesis and properties of self-healing waterborne polyurethanes containing disulfide bonds in the main chain, J. Mater. Sci. 52 (2017) 197–207.

[47] J.M. Lang, U.M. Shrestha, M. Dadmun, The effect of plant source on the properties of lignin-based polyurethanes, Front. Energy Res. 6 (2018) 4.

[48] J.M. Curtis, T.S. Omonov, E. Kharraz, Synthesis of Polyols Suitable for Castor Oil Replacement, Google Patents US10301239B2, 2019.

[49] A.S. Momodu, E.F. Aransiola, I.D. Okunade, G.O. Ogunlusi, K.N. Awokoya, I.O. Ogundari, O.T. Falope, O.W. Makinde, J.-F.K. Akinbami, Greening Nigeria's economy for industrial and environmental sustainability: polyurethane production as a test case, Nat. Resour. Forum 43 (2019) 73–81.

[50] E.B. Mubofu, Castor oil as a potential renewable resource for the production of functional materials, Sustain. Chem. Proc. 4 (2016) 1–12.

[51] H. Beneš, T. Vlček, R. Černá, J. Hromádková, Z. Walterová, R. Svitáková, Polyurethanes with bio-based and recycled components, Eur. J. Lipid Sci. Technol. 114 (2012) 71–83.

[52] V. Remya, D. Patil, V. Abitha, A.V. Rane, R.K. Mishra, Biobased materials for polyurethane dispersions, Chem. Int 2 (2016) 158–167.

[53] P. Parcheta, J. Datta, Environmental impact and industrial development of biorenewable resources for polyurethanes, Crit. Rev. Environ. Sci. Technol. 47 (2017) 1986–2016.

[54] J.O. Akindoyo, M.D.H. Beg, S. Ghazali, M.R. Islam, N. Jeyaratnam, A.R. Yuvaraj, Polyurethane types, synthesis and applications - a review, RSC Adv. 6 (2016) 114453–114482.

[55] S. Oprea, V.-O. Potolinca, V. Oprea, Synthesis and properties of new crosslinked polyurethane elastomers based on isosorbide, Eur. Polym. J. 83 (2016) 161–172.

[56] T. Gui, T. Xia, H. Wei, Z. Zhang, X. Ouyang, Investigation on effects of chain extenders and cross-linking agents of polyurethane elastomers using independent building vibration isolation sensor, Sens. Mater. 31 (12) (2019) 4069–4078.

[57] H. Sun, D. Chen, A facile co-solvent-free process for waterborne polyurethane preparation, Polym. Bull. 75 (2018) 4913–4928.

[58] X. Li, J. Ke, J. Wang, C. Liang, M. Kang, Y. Zhao, Q. Li, A new amino-alcohol originated from carbon dioxide and its application as chain extender in the preparation of polyurethane, J. CO2 Util. 26 (2018) 52–59.

[59] A. Solanki, S. Thakore, Recent studies in polyurethane-based drug delivery systems, Trends Appl. Adv. Polym. Mater. (2017) 219–244.

[60] Q. Zhang, X.-M. Hu, M.-Y. Wu, Y.-Y. Zhao, C. Yu, Effects of different catalysts on the structure and properties of polyurethane/water glass grouting materials, J. Appl. Polym. Sci. 135 (2018) 46460.

[61] A.M. Nacas, A.C. Chinellato, D.J.d. Santos, Lithium catalyst concentration influence on bio-polyols structure and polyurethane adhesives properties, Matéria (Rio de Janeiro) (2019) 12403–12411.

[62] M. Van Der Puy, D.J. Williams, H.K. Nair, D. Nalewajek, Amine Catalysts for Polyurethane Foams, Google Patents US9550854B2, 2017.

[63] M.R. Di Caprio, C. Brondi, E. Di Maio, T. Mosciatti, S. Cavalca, V. Parenti, S. Iannace, G. Mensitieri, P. Musto, Polyurethane synthesis under high-pressure CO_2, a FT-NIR study, Eur. Polym. J. 115 (2019) 364–374.

[64] S.T. Phan, S. Tamura, H. Inagaki, I. Morikawa, Polyether-Modified Silicone Composition, Surfactant, Foam Stabilizer, Polyurethane Foam Forming Composition, and Cosmetic Preparation Including Said Composition, and Method for Producing said Composition, Google Patents US20200048427A1, 2020.

[65] R. Li, Z. Shan, Asynchronous synthesis method of waterborne polyurethane with the differences of structural features and thermal conductivity, J. Polym. Res. 25 (2018) 197.

[66] K.A. Patankar, M.F. Sonnenschein, S.P. Crain, C.A. Rhoton, Polyurethane Foam from High Functionality Polyisocyanate, Google Patents US10577454B2, 2020.

[67] L. Xia, D. Cao, H. Zhang, Y. Guo, Study on the classical and rheological properties of castor oil-polyurethane pre polymer (C-PU) modified asphalt, Construct. Build. Mater. 112 (2016) 949–955.

[68] S.M. Cakic, J.V. Stamenkovic, D.M. Djordjevic, I.S. Ristic, Synthesis and degradation profile of cast films of PPG-DMPA-IPDI aqueous polyurethane dispersions based on selective catalysts, Polym. Degrad. Stabil. 94 (2009) 2015–2022.

[69] S. Nozaki, S. Masuda, K. Kamitani, K. Kojio, A. Takahara, G. Kuwamura, D. Hasegawa, K. Moorthi, K. Mita, S. Yamasaki, Superior properties of polyurethane elastomers synthesized with aliphatic diisocyanate bearing a symmetric structure, Macromolecules 50 (2017) 1008–1015.

[70] A. Kuok, C. Sipaut, M. Sundang, Synthesis and characterisation of new water-based polyurethane dispersion via solvent-free prepolymer mixing process, J. Phys. Conf. Ser. 1358 (2019) 012039.

[71] A. Tounici, J.M. Martín-Martínez, Addition of graphene oxide in different stages of the synthesis of waterborne polyurethane-urea adhesives and its influence on their structure, thermal, viscoelastic and adhesion properties, Materials 13 (2020) 2899.

[72] X.-m. Wang, Q. Li, A new method for preparing low viscosity and high solid content waterborne polyurethane—phase inversion research, Progr. Org. Coat. 131 (2019) 285–290.

[73] W.C. Stumphauzer, Self-foaming Hot Melt Adhesive Compositions and Methods of Making and Using Same, Google Patents US10279361B2, 2019.

[74] M. Serkis-Rodzeń, M. Špírková, P. Matějíček, M. Štěpánek, formation of linear and crosslinked polyurethane nanoparticles that self-assemble differently in acetone and in water, Prog. Org. Coat. 106 (2017) 119–127.

[75] M. Špírková, J. Hodan, J. Kredatusová, R. Poręba, M. Uchman, M. Serkis-Rodzeń, Functional properties of films based on novel waterborne polyurethane dispersions prepared without a chain-extension step, Prog. Org. Coat. 123 (2018) 53–62.

[76] A.P. More, S.T. Mhaske, Synthesis of polyurethane dispersion from polyesteramide polyol, Pigm. Resin Technol. 47 (2) (2018) 154–163.

[77] A.K. Nanda, D.A. Wicks, The influence of the ionic concentration, concentration of the polymer, degree of neutralization and chain extension on aqueous polyurethane dispersions prepared by the acetone process, Polymer 47 (2006) 1805–1811.

[78] M. Ruan, H. Luan, G. Wang, M. Shen, Bio-polyols synthesized from bio-based 1,3-propanediol and applications on polyurethane reactive hot melt adhesives, Ind. Crop. Prod. 128 (2019) 436–444.

[79] D. Janke, M. Cordes, K. Paschkowski, Polyurethane Hot-melt Adhesive Having a Low Content of Diisocyanate Monomers and Good Cross-linking Speed, Google Patents US9982173B2, 2018.

[80] W.-K. Liu, Y. Zhao, R. Wang, F. Luo, J.-S. Li, J.-H. Li, H. Tan, Effect of chain extender on hydrogen bond and microphase structure of biodegradable thermoplastic polyurethanes, Chin. J. Polym. Sci. 36 (2018) 514–520.

[81] M. Groenewald, T. Boekhout, C. Neuvéglise, C. Gaillardin, P.W. Van Dijck, M. Wyss, Yarrowia lipolytica: safety assessment of an oleaginous yeast with a great industrial potential, Crit. Rev. Microbiol. 40 (2014) 187–206.

[82] S.-S. Hwang, J.-H. Park, Polyurethane Binder, Electrodes Containing the Same and Lithium Battery Employing the Electrodes, Google Patents US8722230B2, 2014.

[83] M. Zheng, X. Cai, Y. Tan, W. Wang, D. Wang, H. Fei, P. Saha, G. Wang, A high-resilience and conductive composite binder for lithium-sulfur batteries, Chem. Eng. J. 389 (2020) 124404.

[84] H. Ulrich, Chemistry and Technology of Isocyanates, Wiley-Blackwell, 1996.

[85] M. Ionescu, Chemistry and Technology of Polyols for Polyurethanes, iSmithers Rapra Publishing, Shrewsbury, Shropshire, United Kingdom, 2005.

[86] D. Dieterich, E. Muller, O. Bayer, Elastomeric Polyurethanes, Google Patents US3480592A, 1969.

[87] P. Kasprzyk, E. Sadowska, J. Datta, Investigation of thermoplastic polyurethanes synthesized via two different prepolymers, J. Polym. Environ. 27 (2019) 2588–2599.

[88] C.L. Zhu, C. Tao, J.J. Bao, Y.P. Huang, G.W. Xu, Waterborne polyurethane used as binders for lithium-ion battery with improved electrochemical properties, Adv. Mater. Res. 1090 (2015) 199–204.

[89] N. Loeffler, T. Kopel, G.-T. Kim, S. Passerini, Polyurethane binder for aqueous processing of Li-ion battery electrodes, J. Electrochem. Soc. 162 (2015) A2692.

[90] A. Usman, K.M. Zia, M. Zuber, S. Tabasum, S. Rehman, F. Zia, Chitin and chitosan based polyurethanes: a review of recent advances and prospective biomedical applications, Int. J. Biol. Macromol. 86 (2016) 630–645.

[91] M.L. Listemann, K.R. Lassila, K.E. Minnich, A.C. Savoca, Hydroxyl Group-containing Blowing Catalyst Compositions for the Production of Polyurethanes, Google Patents US3480592A, 1996.

[92] C. Yan, T. Huang, X. Zheng, C. Gong, M. Wu, Waterborne polyurethane as a carbon coating for micrometre-sized silicon-based lithium-ion battery anode material, R. Soc. Open Sci. 5 (2018) 180311.

[93] J. Bao, C. Tao, R. Yu, M. Gao, Y. Huang, C. Chen, Solid polymer electrolyte based on waterborne polyurethane for all-solid-state lithium ion batteries, J. Appl. Polym. Sci. 134 (2017) 45554.

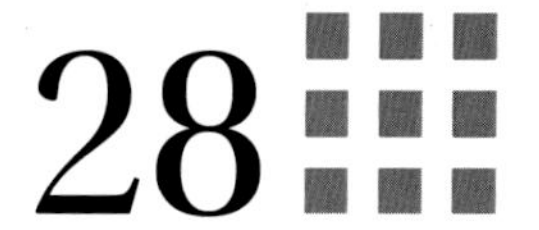

Applications of magnetic hybrid nanomaterials in Biomedicine

Andreea Cernat, Anca Florea, Iulia Rus, Florina Truta, Ana-Maria Dragan, Cecilia Cristea and Mihaela Tertis

DEPARTMENT OF ANALYTICAL CHEMISTRY, FACULTY OF PHARMACY, IULIU HATIEGANU UNIVERSITY OF MEDICINE AND PHARMACY, CLUJ-NAPOCA, ROMANIA

Chapter outline

28.1 Introduction

Magnetic nanoparticles (MNPs) represent 3D structures with small dimensions (1–100 nm) and special properties consisting of one or two parts: a core with magnetic or magnetizable properties and a coating. MNPs are superparamagnetic because of their nanoscale size, with great potentials in a variety of applications in their bare form or with a surface coating and functional groups chosen for specific uses. Ferrite nanoparticles (NPs) are the most explored MNP, which can further agglomerate into clusters to form magnetic beads. The most important property of these structures and what distinguishes them from other types of nanomaterials is magnetism, a property due to which they can be manipulated in a controlled manner by an external magnetic field. The magnetic core of MNPs is based either from iron, nickel, cobalt, chromium, manganese, gadolinium, or their oxides and other chemical compounds, while the coating represents an inorganic (gold, silica) or organic (surfactants, polymers, polysaccharides,

Biopolymeric Nanomaterials: Fundamentals and Applications. DOI: https://doi.org/10.1016/B978-0-12-824364-0.00014-9

fatty acids, etc.) functionalization layer, the later one being chosen to suit the targeted applications.

Especially when the MNPs are nano-sized, they not only have magnetic properties but also a high stability, low toxicity, high-level accumulation in the target tissue/organ, and biocompatibility, properties that qualify them even for *in-vivo* biomedical applications. Moreover, MNPs have an important advantage over other signal providers consisting in the fact that the biological samples lack magnetic content and the matrix effect is importantly diminished resulting in better analytical performances. Also, the magnetic signal can be independent avoiding unwanted interferences.

The application of MNPs in biomedical field refers to the elaboration of systems for targeted drug/gene delivery, the development of sensors and biosensors for the detection of different markers involved in severe conditions, for diagnosis and monitoring purposes, magnetic resonance imaging (MRI) as contrast enhancement agents, hyperthermia treatment, biophotonics as well as tissue engineering, bimolecular separation, and purification. Usually, the MNPs that illustrated a superparamagnetic behavior are preferred for biomedical applications, such as antibacterial, biosensors, and drug delivery. Some essential features are required for this purpose namely: the size of MNPs that should match well with the cellular level of the human body as diseases occur at cellular levels; the MNPs should be nontoxic; and should have the possibility of surface modification.

The large number of studies on this topic are continuously increasing is also a relevant indicator of the attention that the MNPs have gained especially in the biomedical field. It is fascinating to witness the development of this specific area of interest starting from the 1980s when the first applications of the MNPs were reported up until the already approved protocols.

Herein, we have presented in a critical manner data regarding the synthesis and functionalization strategies of bare MNPs. Only examples of hybrid materials applied in medical purposes such as diagnostic, monitoring, and treatment as drug delivery systems have been considered. The literature from the last 5–10 years was closely analyzed and the data presented emphasized the modifications made to improve the properties required for clinical applications that are already approved by specific organizations.

28.2 Preparation, functionalization, and coating strategies for MNPs for applications in biomedicine

There are several types of MNPs that can be applied in the biomedical field. According to their composition, the MNPs can be part of the following categories: (1) oxides or ferrites, also known as iron oxide NPs; (2) ferrites with a shell; (3) metallic NPs; (4) metallic NPs with a shell.

Ferrites represent the most known and used category of MNPs and their surface is suitable for functionalization with both organic and inorganic compounds to achieve an increased stability in aqueous media. If their diameter is less than 128 nm, they exhibit supermagnetic properties and do not present magnetism except in the presence of an external magnetic field [1,2]. This property is ilustrative as it can be exploited in many biomedical applications, including targeted drug delivery, when targeted transport can be performed without the need to include an energy source to ensure mobility at the particle level and no functionalization to ensure selectivity at the cellular targeted receptors.

Ferrites that present shells are inert and cannot be modified through functionalization based on the formation of covalent bonds. This drawback can be eliminated by the initial functionalization with silica, followed by the desired functionalization [3].

As for the *metal particles*, they only show the metallic core with high magnetic moment, being thus suitable for biomedical applications, but only if the pyrophoric property and the increased reactivity to oxidizing agents do not bother.

In the case of *metallic particles coated with a shell* obtained after the interaction with different reagents such as surfactants, polymers, or noble metals, several biomedical applications can be developed [2].

28.2.1 Synthesis strategies for MNPs applied in biomedicine

The use of MNPs in biomedical applications complicates the protocol in terms of their synthesis. Special properties are required for this type of application, so the conventional methods must be adapted to ensure optimal conditions throughout the envisaged purpose. The use of different coatings, of inorganic/organic nature, is often necessary for biomedical applications, and the synthesis strategy must be chosen according to the necessary particularities regarding the composition, shape, dimensions, and the type of functionalization chosen. In addition, the colloidal nature of MNPs complicates their synthesis and functionalization, especially when targeting *in-vivo* biomedical applications. The modification of pH and the use of special additives during their synthesis are usually adopted to make them suitable for biomedical applications. Also, the properties in the synthetic medium, such as viscosity or other fluidic properties that may influence the colloidal stability, stability in time, quality, and continuity of coatings, and behavior of MNPs are thoroughly assessed when concerning biomedical applications and sensors' design.

There are some milestones that must be considered in the synthesis of MNPs, namely establishing the optimal experimental parameters to obtain the desired characteristics for the particles (e.g., shape, size, size distribution, and surface chemistry) as well as the feasibility and reproducibility of the synthesis process, which should be as simple and as inexpensive as possible, but especially should not require complicated and laborious separation and purification steps (e.g., magnetic filtration, size-exclusion chromatography, ultracentrifugation flow field gradient, etc.).

In most approaches mentioned in the literature, the synthesis processes of MNPs are divided into two stages. The preparation stage takes place first and it is followed by the modification of the surface via specific functionalities.

There are currently two different types of preparation methods for MNPs following a physical or a chemical synthesis protocol, respectively [4]. Among the latter, co-precipitation, hydrothermal synthesis, sol-gel synthesis, thermal decomposition, sonolysis, and laser pyrolysis can be mentioned [2,5].

From the category of physical synthesis strategies, mechanical ball milling, electron beam lithography, and gas phase deposition must be mentioned. These methods require sophisticated approaches but usually do not allow control over the particle size to the nanometer scale or lower. Contrary, chemical synthesis strategies, such as wet chemical ones, are simple, economical, effective, and allow the control over size, shape, and composition of the NPs [6].

28.2.1.1 Synthesis of MNPs via physical strategies

The mechanical ball milling method refers to the crushing of larger magnetic particles into the much smaller ones, due to the impact between different particles or with the inner walls of the enclosure. Another strategy uses high energy to mechanically alloy the raw materials into nano-spinel-type ferrite. This is a simple physical method, but the resulting material has different granulations and a large amount of impurities from the synthesis process, being thus not suitable for preparing nano-sized magnetic particles with controlled morphologies [5].

MNPs with an average diameter of 10 nm were fabricated using high-energy ball milling strategy, then incorporated in nanofiber together with β-lactoglobulin, the predominant protein of the aqueous whey, by electrospinning using the assist of poly(ethylene oxide). These magnetic nanofibers showed no adverse effect on the human mesenchymal stem cell proliferation, thus confirming their suitability for biomedical applications [7]. In another study, a hybrid material based on MNPs functionalized with ionic liquid was developed using the ball milling process. A scalable production of MNPs hybrid material with controlled morphology, high surface area, and pore size of 40 nm was reported. The role of ionic liquid during this physical synthesis strategy was to guide and to form the rod-like pores containing Fe_2O_3 with good wettability and enhanced ion transfer [8].

28.2.1.2 Synthesis of MNPs via chemical strategies

Coprecipitation is one of the most popular chemical strategies applied for the synthesis of MNPs. This is an easy, reproducible and low-cost method widely used for the production of iron oxides (either Fe_3O_4 or Fe_2O_3) via nucleation and growth mechanisms, after the addition of a base to an aqueous $Fe^{2+/3+}$ salt solution in the presence of an inert gas and optimal temperature conditions [2,9].

Coprecipitation method was applied for the synthesis of magnetite superparamagnetic iron oxide NPs (SPIONs) coated with an antibiofouling polymer with special properties that can allow their use within the tumor site for *in-vivo* diagnosis of cancer [10]. Coprecipitation of MNPs was based, in this case, on a method that started from a $FeCl_3$ and $FeCl_2$ solution where NH_4OH was added in the presence of N_2 [11]. Another study was based on the coprecipitation of MNPs as a black precipitate of magnetite starting from a mixture of ferric and ferrous salts with a molar ratio of 2:1 in the presence of NH_4OH [12]. A similar strategy was applied for the synthesis of MNPs that were afterwards coated with chitosan, as a biodegradable, biocompatible, and bioactive polysaccharide polymer and which allows the use of these nanostructures as nanoheaters in hyperthermia therapy [13].

A fast and facile strategy for the synthesis of rhamnolipids-coated iron oxide NPs having 48 nm in diameter has been proposed and displayed synergistic antibacterial and antiadhesive properties against biofilms formed by *Pseudomonas aeruginosa* and *Staphylococcus aureus*. The presence of rhamnolipid shell on the NPs significantly reduced the cell adhesion by modifying the surface hydrophobicity, these NPs being a potent alternative to reduce the infection severity by inhibiting the biofilm formation in biomedical applications for antibacterial coatings and wound dressings [14]. In other study, chitosan was used as a coating material for the stabilization of iron oxide MNPs synthesized by the coprecipitation method in an alkali medium. The resulted hybrid material had an average particle diameter of about 136 nm and superparamagnetic properties, having a potential application as a contrast agent for MRI [15]. An optimized coprecipitation strategy was reported for MNPs production starting from Fe^{3+} and basic pH. The resulted NPs were then coated with poly[(methacrylic acid)-ran-(2-methacryloyloxyethyl phosphorylcholine)] via a chelating process that involved carboxylic functions, the final diameter of the hybrid stabilized materials being in the range of 10–60 nm, and having all other characteristics suitable for the elaboration of biosensors [16].

Magnetic drug targeted delivery is a strategy that can be used to improve the therapeutic efficiency on tumor cells and reduce the side effects on healthy cells, tissues, and organs. An efficient drug delivery system based on superparamagnetic NPs synthesized by coprecipitation of iron oxide followed by coating with poly citric acid, poly(ethylene glycol) for the hydroxyl end group and folic acid for the carboxyl group was developed and tested for cancer therapy. The final size of nanocarriers that combine drug targeting as well as sensing and therapy was in the range of 10–49 nm [17]. A novel nanohybrid material based on SPIONs was developed and tested as nanovector of a natural antioxidant derivate of trans-resveratrol: 4'-hydroxy-4-(3-aminopropoxy) trans-stilbene, with great potential for brain diseases treatment. The molecule was grafted onto SPIONs surface using 3-chloropropyltriethoxysilane as an organosilane coupling agent. The *in-vitro* testing of the nanohybrid material showed no influence onto the mitochondrial metabolism, but it damaged the plasma membrane having thus a potential cytotoxic effect for cancer cells [18]. Anticancer drugs doxorubicin and paclitaxel were incorporated into capsules based on the tree block poly(lactic acid)–poly(ethylene glycol)–poly(lactic acid) copolymer functionalized with oleic acid modified Fe_3O_4 MNPs obtained by coprecipitation for controlled drug release and targeted drug delivery applications. The resulted hybrid displayed a spherical morphology, diameter of 179–203 nm, no toxicity for healthy cells, but instead had a good antitumor activity in immunocompetent BABL/c mice [19].

Other popular chemical strategy for MNPs synthesis is the *hydrothermal method* that refers to the use of an aqueous solution enclosed in a specially designed and sealed reactor (autoclave operated at high pressure and temperature). The main occurring reactions are hydrolysis and oxidation, with special care for the optimal reaction conditions that influence the characteristics of the final product. A wide variety of nanostructured materials with homogenous dimensions can be synthesized in this manner, the materials being effective in applying metal oxide onto electrodes and supercapacitors [2].

Fe_3O_4-based MNPs were obtained by thermal decomposition in the presence of *N*-methyl pyrrolidone as solvent, N_2 gas for O_2 elimination, and methanol as a precipitation agent [20].

The same type of magnetic nanostructures was used for the inner functionalization of magnetic microtube-like structures together with hydrophobic polycaprolactone polymer, while the outer surface of the tubes was functionalized with an amphiphilic block copolymer composed of hydrophilic polyacrylic acid. The hybrid nanostructures showed good cell viability and cell adhesion with the HaCaT cell line and had good perspective to be used in tissue engineering and angiogenesis applications [21]. Fe_2O_3-based NPs were precipitated using a hydrothermal method then functionalized with poly(3,4-ethylenedioxythiophene) and reduced graphene oxide. The nanocomposite was further employed for the immobilization of two enzymes: acetylcholinesterase and choline oxidase, having as final goal the elaboration of a biosensor for acetylcholine detection in serum for the diagnosis of Alzheimer's disease [22]. Other Fe_3O_4-based NPs with magnetic properties were synthesized by thermal decomposition of ferric acetylacetonate in triethylene glycol and functionalized with a β-cyclodextrin core and poly(2-(dimethylamino) ethyl methacrylate cover for increased stability and biocompatibility, being a versatile promising theranostic platform [23]. Tetramethylammonium hydroxide–coated nickel ferrite ($NiFe_2O_4$) NPs with a 4.4 nm magnetic core and 15 nm hydrodynamic diameters proved to be suitable for MRI for contrast enhancement and induced heating for hyperthermia therapy [24]. Thermal decomposition of iron (III) acetylacetonate has resulted in production of magnetic nanocomposite material, allowing simultaneous cancer therapy and diagnostics after the functionalization with bovine serum albumin, polyethyleneglycol (PEG), and vascular endothelial growth factor [25].

Pyrolysis is another method for preparing uniform, crystalline, isolated MNPs based on the thermal decomposition of metallic compounds at high temperatures. These NPs also have high magnetic susceptibility and high initial magnetic susceptibility [5].

The laser pyrolysis method was applied for the synthesis of two different types of γ-Fe_2O_3-based MNPs with hydrophilic or hydrophobic behavior and high magnetization saturation. These NPs were further stabilized with L-DOPA and proved to be biocompatible, as demonstrated by a preliminary *in-vitro* study on mouse primary leukocytes and human breast carcinoma cell line MCF-7 [26].

Biosynthesis represents a green and eco-friendly method for MNPs synthesis involving the use of bacteria and other microorganisms that are responsible for reducing or oxidizing salts to generate MNPs. Different bacteria species such as *Actinobacter* sp. and *Bacillus subtilis* have been used for this purpose to synthesize MNPs with controlled features [27,28].

A green strategy of biosynthesis based on polyphenols from *Syzygium aromaticum* extract was applied and iron oxide NPs with very small dimensions with both Fe^{2+} and Fe^{3+} valences and superparamagnetic-like multiphase crystalline nature were obtained and stabilized. The organic moieties from the plant extract assured the surface functionalization with potential applications in localized hyperthermia [29]. Iron oxide–based NPs with organic functions were synthesized through a green and cost-effective strategy using a *Psidium guajava–Moringa oleifera* leaf extract. The hybrid material presented homogenous dimensions and antibacterial properties, being thus suitable for biomedical and environmental applications [30]. Another green synthesis allowed the preparation of iron oxide–based nanorods in the presence of Withania coagulans plant extract and sodium hydroxide. The resulted nanomaterial showed effective antibacterial activity against *S. aureus* and *P. aeruginosa* [31].

Microemulsion and inverse micelles is another method employed for the synthesis of MNPs (spinel ferrites) using Mn, Co, Ni, Cu, Zn, Mg, or Cd as metallic base. This synthesis strategy allowed the generation of MNPs suitable for applications in the field of electronics. The precipitation of MNPs occurs in the microemulsion environment and it is necessary to adjust the ratio between the surfactant, oil phase, water phase, and cosolvent. In this case, nucleation, growth, coalescence, and agglomeration steps must be optimized and controlled to avoid agglomeration of the synthesized NPs. Different shapes such as spheroids, oblong cross-sections, or tubes could be obtained in a controlled manner using the microemulsion method. This approach has some drawbacks, such as narrow working window for the synthesis in microemulsions, the low yield of NPs, as well as high solvent consumption, which make the synthesis very difficult to scale up [32]. Composite ferrite NPs with Ni, Zn, and Cr were prepared by a reverse micelle process combined with microemulsion involving a metal solution and ammonium hydroxide. The spinel ferrite suitable for biomedical applications was only obtained in some specific and carefully controlled pH conditions [33].

Sol-gel technique belongs to the category of chemical methods for the synthesis of MNPs, intensively used for the fabrication of metal oxides, the basic material on which the necessary functionalization is made depending on the desired application. Sol particles may interact by Van der Waals forces, hydrogen bonds, as well as by the formation polymer chains. The formation of any covalent bond must be avoided as the gelation process may be irreversible if other strong interactions are involved. A stable transparent sol system is formed in a solution, and then is concentrated into a transparent gel and dried or heat treated to prepare the 3D nanosubstructured material. The overall process involves first the synthesis of a precursor through iron alkoxides and iron salts, followed by hydrolysis and polycondensation reactions at room temperature, and further heat treatments. The conditions lead to the formation of the final product in the crystalline state with properties dependent on the experimental conditions used for the synthesis [2,5,34].

Properties such as superparamagnetic behavior and high saturation magnetization are vital for biocompatibility of MNPs. Thus, a sol-gel strategy for the synthesis of magnetite NPs functionalized with citric acid was reported. The immobilization of citric acid at the surface of MNPs occurred by adsorption on the surface of NPs, the process involving only a part of the carboxyl groups, the rest remaining free and imprinting a hydrophilic character and antiagglomeration properties. This hybrid material with magnetic properties became suitable for conjugation with specific drugs for drug delivery applications [35]. Another strategy refers to the synthesis of hematite (Fe_2O_3) NPs covered with thin films of Cr via the sol-gel spin coating method. The optimal amount of Cr in the outer layer was chosen to provide the necessary magnetic properties for the NPs, because an influence of the dopant concentration on these properties was observed [36,37]. MNPs combining Fe and Ga with sizes between 15 and 20 nm were synthesized using a polycondensation reaction by sol–gel method. These composite MNPs were found not cytotoxic being promising materials for cancer treatment by hyperthermia and drug delivery systems [38].

Sonochemical method is another important method for MNPs' synthesis in which ultrasound is applied to generate gasification bubbles in the reaction system. The process determines the simultaneous promotion of multiple reactions generating MNPs. It is a simple method and does not need special equipment or conditions, it only requires the liquid medium to perform energy transfer, and has a strong versatility for various reaction media. This simple and rapid synthesis strategy generates uniform NPs with small dimensions, suitable for biomedical applications [39].

MNPs of about 170 nm in diameter consisting of recombinant human serum albumin with incorporated hydrophilic $(NH_4)_2Ce(IV)(NO_3)_6$-γ-Fe_2O_3 were produced via a sonochemical method for MRI applications [40].

Electrodeposition is another method applied for the synthesis of MNPs in which a precursor is deposited on the surface of a substrate, generating uniform nanostructures. This procedure requires the presence of ionic dissolved precursor, namely Fe^{2+} or Fe^{3+} ions, and the adjustment of the experimental conditions (e.g., electrodeposition potential, electrolyte composition, etc.) will affect the properties of the electrochemically generated MNPs [41].

Thus, Zn^{2+} cations–doped Fe_3O_4 NPs were synthesized via a cathodic electrodeposition strategy that involved their deposition on the surface of a steel cathode. The hybrid NPs with magnetic properties were further modified with an epoxy layer to increase the biocompatibility for biomedical applications [42]. Another strategy refers to the synthesis and *in-situ* coating of magnetic iron oxide NPs with polyvinylpyrrolidone and polyethylenimine via cathodic deposition. From this electrochemical synthesis procedure, hybrid NPs with dimensions of 10–15 nm, suitable physicochemical and magnetic properties for biomedical applications resulted [43].

Polyol method represents an effective strategy for the synthesis of MNPs with homogeneous structure. The reduction and dissolution of the metal salt in the polyol solution will directly precipitate the metal, thereby forming fine metallic particles suitable for biomedical applications, including MRI and drug delivery systems due to the MNPs' excellent biocompatibility and water dispersibility [44].

The use of carbon as a coating material for MNPs may influence their biocompatibility and thus allow the use of these materials in the biomedical field. The carbon-coated ultra-small gadolinium oxide (Gd_2O_3) core-shell NPs having the average diameter of about 3 nm were synthesized using a simple polyol strategy. Based on the preliminary tests performed, this composite material proved to be suitable for biomedical applications [45].

A schematic illustration of the main strategies applied so far for the synthesis and functionalization of MNPs based on iron oxides for biomedical use is presented in Fig. 28.1 [46].

28.2.2 Functionalization and coating strategies for MNPs applied in biomedicine

Tuning the overall properties of MNPs is often required to fit targeted applications. Biomedical applications require a strict control of the MNPs interfaces for colloidal stability in a complex biological environment. Furthermore, as the MNPs are attracted to each other, their surface functionalization is important for the stabilization process. Surfactants, as sodium oleate, sodium carboxymethylcellulose, and polymers, which prevent aggregation, are some examples of such stabilizers. The modifiers for the surface of MNPs are both inorganic and polymeric (organic) materials, the last ones being both synthetic and natural. Examples of synthetic polymers used

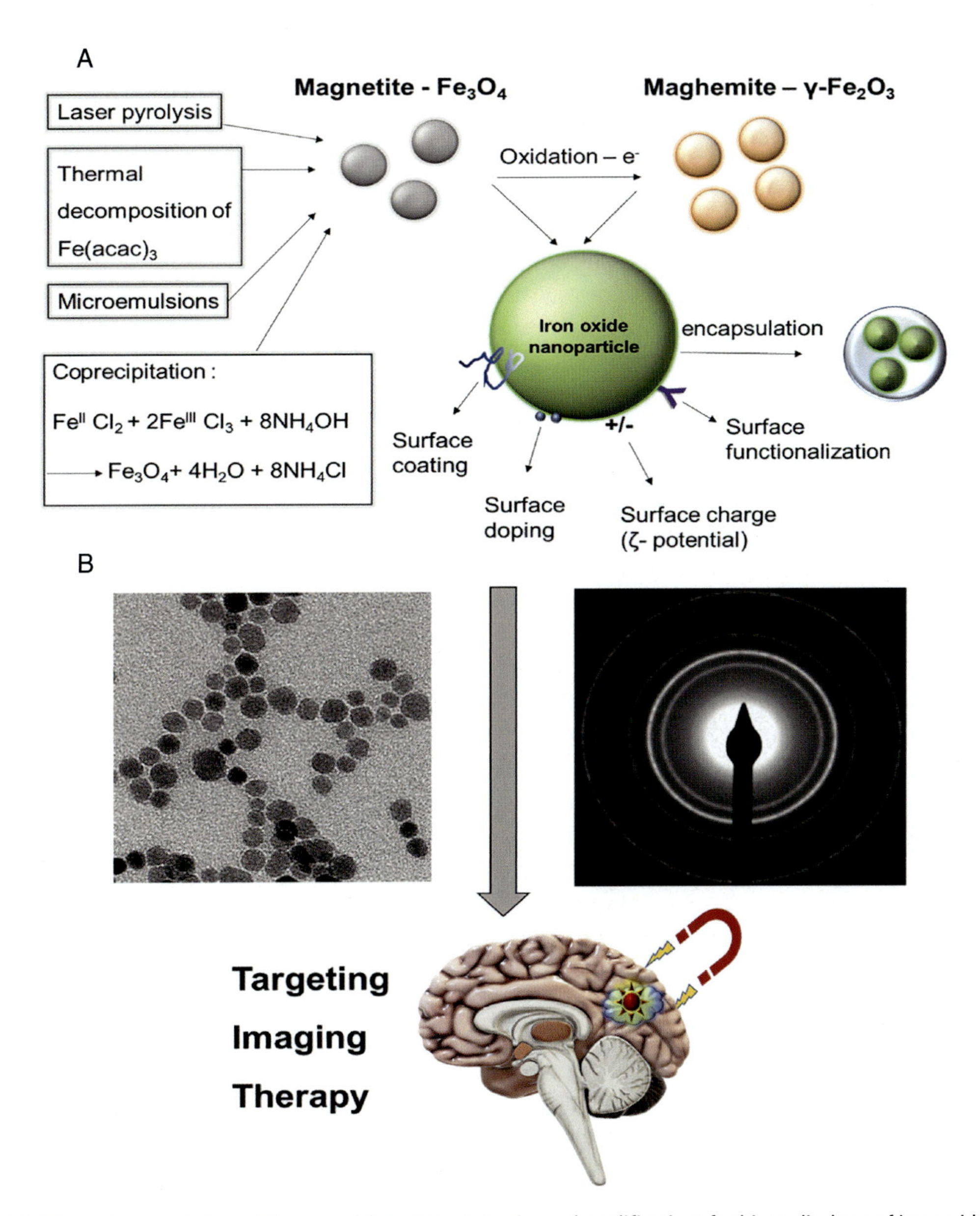

FIGURE 28.1 Representation of the main fabrication strategies and modifications for biomedical use of iron oxide nanoparticles: (A) The main methods applied for the synthesis of magnetite (Fe_3O_4, gray) and some strategies for their modification for biomedical use; (B) representative TEM image (left) and Selected Area Electron Diffraction (SAED) pattern (right) of MNPs based on magnetite toward potential biomedical applications (bottom). Reprinted under the terms of the Creative Commons CC-BY license from [46] (open access).

as coatings for MNPs' functionalization are PEG, poly(vinyl alcohol) poly(ethylene-co-vinyl acetate), poly(vinylpyrrolidone), poly(lactic-co-glycolic acid), while chitosan, gelatin, dextran are classified as natural polymers.

The stabilization of MNPs in the colloidal form is a mandatory phenomenon. The stability of these nanostructures is related to the equilibrium between repulsive and attractive forces in the medium. There are four types of forces involved in these systems, namely van der Waals forces, electrostatic repulsive forces, magnetic dipolar forces between two different particles in a suspension and steric repulsion among particles [6]. Controlling the strength of these forces is mandatory for preparation of NPs with good stability.

28.2.2.1 Functionalization of MNPs with inorganic materials

The functionalization of MNPs with inorganic materials is intensively used and involves the coating of particles with a metallic layer (e.g., Au, Pt, Ag, etc.) or with inorganic compounds, in this case, the most often used being silicon dioxide (silica), metallic oxides or sulfides (e.g., SnO_2, MnO_2, Al_2O_3, TiO_2, MgO, ZnS, Co_3S_4, Ni_3S_2, etc.).

The requirements of the medical field led to an increase in the interest in silica functionalization for researchers due to the fact that this material is not toxic and thus suitable for biomedical applications. Silica coating of MNPs occurs via the formation of cross-linking bonds when a protective layer is obtained for the magnetic core. The presence of silica at the surface of the MNPs leaves on the surface functional groups that allow for the MNPs to be applied in various fields such as catalysis, adsorption, and separation.

Fe_3O_4-based MNPs were synthesized and covered with SiO_2, for increased biocompatibility. The functionalities generated at their surface are available for the immobilization of biomolecules and in addition facilitates applications in areas such catalysis and separation [47]. A nanocomposite shell consisting in SiO_2 and hydroxyapatite was synthesized via a green strategy based on fructose as cover for Fe_3O_4-based MNPs. The hybrid material had large pores and a spherical shape and it was applied as a controlled carrier for atenolol in drug delivery system elaboration [48]. A novel green strategy was applied for the synthesis and functionalization of core-shell nanostructured MNPs covered with SiO_2. Persimmon tannin was used for the treatment of the $Fe_3O_4@SiO_2$ microspheres' surface to be used for the adsorptive recovery of Au(III) and Pd(II). The schematic representation of the synthesis and functionalization process is presented in Fig. 28.2. In addition, after the adsorption of the metallic ions, the magnetization saturation values of the MNPs were enough for an efficient magnetic separation [49].

Another simple strategy for the functionalization of MNPs with inorganic materials is to cover them with a metallic shell to avoid oxidation. The protection of the metallic layer greatly expands the application range of MNPs for applications in biomedicine and catalysis.

It is important to mention that this modification can influence the magnetization saturation of the MNPs in a different manner depending on the nature of the MNPs' core as well as the nature of the metal used for coating. For example, the functionalization results of Co, Pt, Cu, and Pd may be reversed [5].

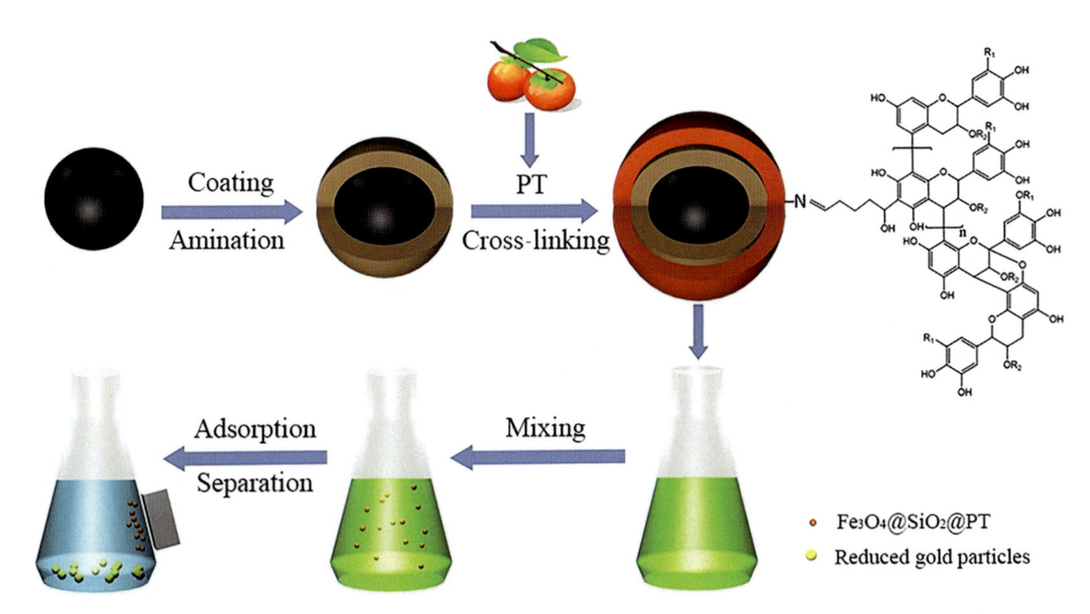

FIGURE 28.2 Functionalization of MNPs with SiO_2 by using a green strategy and plant tannins. Reprinted with permission by Elsevier from [49].

Metallic oxides based on Sn, Mn, Al, Ti, or Mg and metallic sulfides based on Zn, Co, or Ni were also widely used as covers meant to protect the MNPs. Most applications of these composite materials refer to the adsorption of pollutants and catalysis, without emphasizing biomedical applications. However, there are some studies on the use of metal oxides functionalized with metal sulfides in targeted drug delivery. For example, Fe_3O_4-ZnS spheres were prepared through a simple method and functionalized with a hollow of poly(N-isopropylacrylamide (PNIPAM). The hybrid material presented excellent magnetic, luminescence, and temperature response, and was tested for targeted drugs and drug delivery. The schematic representation of the synthesis and functionalization protocol used for the PNIPAM/Fe_3O_4–ZnS hybrid hollow spheres can be seen in Fig. 28.3A. The morphological evolution of the products during the different stages in the synthesis process is also presented in Fig. 28.3B. It can be thus clearly observed that the hollow structure has a mean diameter of 260 nm and a shell thickness of about 60 nm. The monodisperse spheres with rough surface can be observed from the scanning electron microscope (SEM) image of the product, as shown in Fig. 28.3B(g). Furthermore, the SEM observation unveils an obvious cavity in the spheres, which can directly confirm the hollow structure of these hybrid spheres Fig. 28.3B(h) [50].

Carbon-based nanomaterials such as carbon fibers, graphene, carbon nanotubes, fullerenes, nanodiamond, etc. are materials with special properties that have been intensively studied lately. Carbon fiber has excellent electrical properties, high strength, and low density, and it was employed in the iron oxide NPs composite coatings. The functionalization of carbon fiber with Fe_3O_4 NPs composites represents a good perspective for obtaining materials for applications such as absorption, batteries, and drug encapsulation systems. For example, a simple and reproducible

A

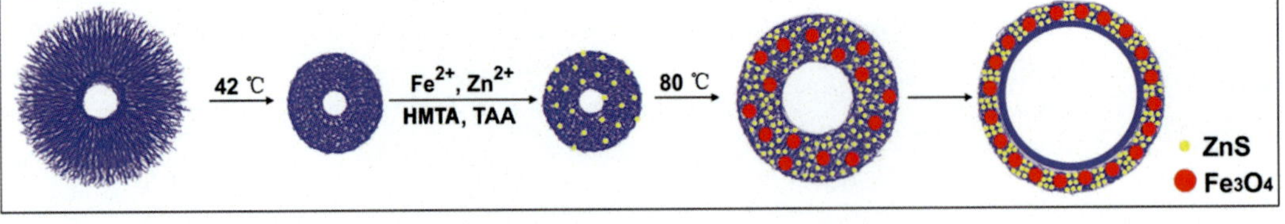

B

FIGURE 28.3 (A) Schematic representation of the PNIPAM/Fe_3O_4–ZnS hybrid hollow spheres' fabrication protocol. (B) TEM images of (a) hollow PNIPAM templates, (b) products after addition of the $Zn(Ac)_2$ solution, (c) products after addition of $FeCl_2$ solution, (d) products after reaction at 80 °C for 6 h and 12 h (e, f) SEM images of PNIPAM templates, (g and h) PNIPAM/Fe_3O_4–ZnS hybrid hollow spheres. All scale bars are 200 nm. Reprinted with permission by Elsevier from [50].

method was applied for the synthesis of microspheres of carbon and Fe_3O_4 for applications in catalysis and magnetic separation [51].

28.2.2.2 Functionalization of MNPs with graphene

Graphene represents another important carbon-based nanomaterial with 2D planar structure, large surface area, single-atom thickness, and other special features, such as excellent electrical, thermal, and mechanical ones. It also presents a delocalized π electron system between the different carbon atoms in graphene, thus it can generate a π stacking effect with benzene rings with good adsorption capacity. The use of graphene-MNPs hybrid materials is important because the presence of particles with magnetic properties makes their separation as well as controlled movement more accessible, with important perspectives on applications in targeted drug transport and separation. An innovative concept that combines MNPs based on fluorescent manganese-doped zinc sulfide functionalized in the first step with PEG and graphene oxide for increased solubility and biocompatibility and in the second step with glutathione has been used for the noncovalent incorporation of doxorubicin, an anticancer drug for drug delivery system elaboration and theranostic applications [52].

28.2.2.3 Functionalization of MNPs with organic materials

The superficial functionalization of MNPs with organic molecules is effective when biomedical applications are envisaged. The hydrophobic nature of MNPs determines their mutual interaction and leads to the formation of clusters or large-size particles. Furthermore, the functionalization of MNPs with organic compounds and biomolecules determines an increased biocompatibility and suitability for *in-vivo* application. The functionalization of MNPs with organic compounds is of great interest for many researchers and has great potential applications. MNPs functionalized with organic compounds have a core-shell structure, with a ferromagnetic core of magnetite or magnemite and an organic shell based on polymers or biomaterials. In the presence of the organic compounds, MNPs have good biocompatibility and biodegradability. Another strategy related with the modification with organic molecules is the generation of different functions on the surface of the MNPs such as aldehyde, hydroxyl, carboxyl, or amino groups, useful for further immobilization of active biological materials such as antibodies, proteins, DNA sequences, and enzymes.

The functionalization of MNPs with organic materials such as polymers, surfactants, or biomolecules is often used especially for adjusting their stability and to avoid their agglomeration in suspensions. For example, a novel green hybrid material with magnetic property was synthesized starting from Fe_3O_4 NPs coated with polydopamine and further functionalized with *n*-halogenated amines and it proved to present antimicrobial activity. The use of MNPs as templates determined the increase in surface area and allowed their facile separation and manipulation through an external magnetic field. The presence of polydopamine as a coating material for the magnetic core eliminated the risk of agglomeration and also provided active functions for further modification [53]. In another study, MNPs coated with 2,3-dimercaptosuccinic acid proved to not have toxic effects on mesenchymal cells, without significant changes in physiological parameters such as cell differentiation, proliferation, and immunomodulation [54].

Functionalization of MNPs with organic compounds also refers to small molecules such as ligands widely used when a specific function is required. Some examples in the literature are related with the modification of MNPs with organic ligands via some specific reactions to obtain a hydrophilic surface, to improve their colloidal stability, water solubility, and bioavailability, all

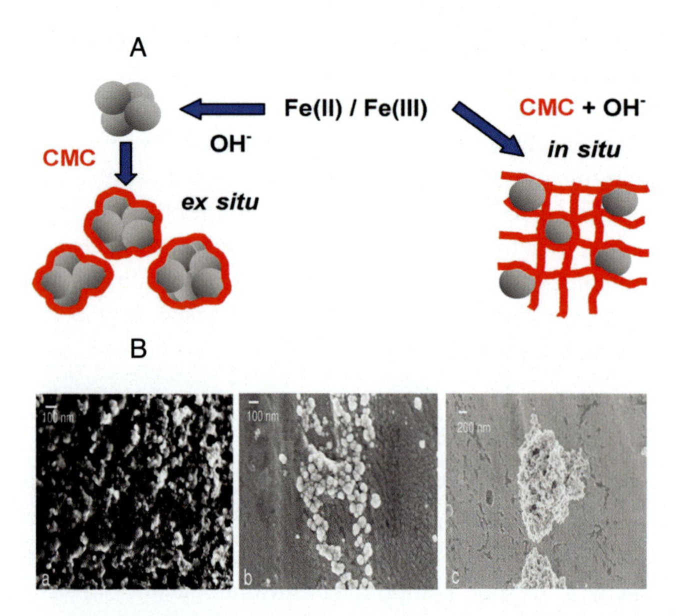

FIGURE 28.4 (A) Schematic representation of the *ex-situ/in-situ* strategies applied for the synthesis and functionalization of MNPs with carboxymethylcellulose. (B) Field Emission Scanning Electron Microscopy (FESEM) images measured for: (a) *in-situ* sample, (b) *ex-situ* sample, (c) aggregates of unmodified Fe_3O_4–based NPs. Reprinted with permission by Elsevier from [55].

these characteristics being mandatory when targeted drug delivery is envisaged as application [54]. A wide variety of biomolecules, such as proteins, peptides, antibodies, have been used to date for the functionalization of MNPs. This functionalization strategy has proven beneficial for increasing biocompatibility and in addition transforms the surface of magnetic formations into a specific or selective one for the analyte or receptor of interest with good perspectives for the biomedical field.

MNPs based on magnetite were synthesized and functionalized with carboxymethylcellulose, through an *ex-situ* and *in-situ* strategy, respectively, and the effect of the polymer on the magnetic structures was tested. In the first strategy, the magnetite NPs were synthesized then modified with the polymer, while in the second case, carboxymethylcellulose was added in the solution used for the production of MNPs. It was observed that the polymer is anchored on the surface through the carboxylate functions but does not influence the average dimension of the MNPs of about 10 nm, but the aggregation of the MNPs is significantly different for the two samples: the *in-situ* sample contains single MNPs or aggregates of few particles, whereas larger aggregates were present in the case of the sample prepared following the *ex-situ* strategy [Fig. 28.4) [55].

Furthermore, the hydrodynamic diameter in water at neutral pH and the potential measured in the same conditions explain their long-term stability, and the presence of polymer with functionalities available for further immobilization of biocompounds proves the suitability of these hybrid MNPs for biomedical applications [55].

28.3 Applications of MNPs based hybrid materials in biomedicine

28.3.1 MNPs in diagnostic

Early detection in various diseases is a cross point in the management of the therapy, with a great impact on the overall prognosis. Rapidly, low cost, but more important, effectiveness are the requirements of the sensing devices for a sustainable innovative healthcare approach. In the last decade, the detection of biomarkers has been thoroughly discussed from different viewpoints.

28.3.1.1 Applications of MNPs in medical imaging for diagnostic

Within this field of research, MNPs are featured prominently in the development of new *diagnostic and therapeutic strategies.* The capacity to generate a local magnetic field enables their use in MRI as contrast agents with already approved clinic applications in liver lesions, sentinel node detection, treatment of brain tumors, and of iron deficiency anemia in adults [56]. This technique has the advantage of the rapid acquisition of the images, high resolution, and nonionizing radiation, but its lower sensitivity is a critical point. Thus, the development of new contrasting agents has become mandatory in this field to improve the existing performances [57]. SPIONs were thoroughly studied due to their biocompatibility, capacity to form stable suspensions, hypermagnetization properties, cost-effectiveness, and the fact that they can be oriented in different parts of the human body using an external magnetic field. Fe_3O_4 NPs are used as negative contrast agents and appear dark in T_2-weighted images, which can mislead the real cause because they could be confused with the signals from bleeding, calcification, or metal deposits. Thus, the development of positive contrast agents that generate bright in T_1-weighted images became an important aspect because the unmodified NPs have poor stability in dispersions [2 3].

Food and Drug Administration (FDA) has approved SPIONs various formulations as MRI agents, but as a consequence of the continuously reevaluation of the positive/negative aspects many were withdrawn from the market. The critical point, as already highlighted, is represented by the SPIONs biocompatibility and their performance for *in-vivo* imaging. Ultra-small SPIONs were coated with a peptide layer to increase the colloidal stability and biocompatibility of the particles and to protect the magnetic core from the biological environment. The formulation is an alternative to the conventional MRI agents for patients with chronic kidney diseases at risk of nephrogenic systemic fibrosis [58]. Table 28.1 presents some representative examples regarding the use of MNPs-based hybrids in medical diagnosis.

28.3.1.2 Application of MNPs in sensor development for diagnostic and monitoring

The development of biosensing systems for these specific applications needs to meet the biomedical specific requirements such as biocompatibility and high stability. One such example is represented by the diagnosis of ovarian cancer, a disease with limited early recognizable

Table 28.1 Examples of applications of MNPs-based hybrid materials in diagnosis.

Application	Type	Target	LOD(sensors)/ dose(imaging)	Platform (hybrid)	Detection method	Real samples	Ref.
Thrombosis	Sandwich-like immunosensor	D-dimer	5 ng/mL	Superparamagnetic particles conjugated with streptavidin	GMR	Plasma	[59]
Ovarian cancer	Immunosensor	CA125 HE4 IL6	3.7 U/mL 7.4 pg/mL 7.4 pg/mL	Streptavidin-coated Fe_3O_4 MNPs	GMR	–	[60]
Infection with IAV	Sandwich-like immunosensor	nucleoprotein of the IAV	1.5×10^2 $TCID_{50}$/mL virus	Streptavidin-labeled MNPs	GMR	–	[61]
Asthma	Electrochemical biosensor	ECP	0.30 nM linear: 1–1000 mM	Heparin-modified Au@Fe_3O_4 NPs	SWV	Cell culture	[62]
Diabetes mellitus	Composite suspension	Glucose	0.15 μM linear range:0.2–20mM	Fe_3O_4@APBA	Fluorescence	Plasma	[63]
Tumor	Contrast agent for imaging	Tumor cells	–	Fe_3O_4@OA@PLA-PEG-DG	MRI	i.v. injected mice	[10]
Not mentioned	Contrast agent for imaging	Liver, kidneys, bladder	˜0.1 mmol Gd/kg	Gd_2O_3@C	MRI fluorescence	i.v. injected mice	[45]
Cancer	Contrast agent for imaging	Liver healthy cells	20 mg Fe/kg	PEG4-USPIONs	MRI	i.v. injected mice	[58]
Gastric carcinoma	Contrast agent for imaging	Cancerous gastric cells	0.05–0.5 mg/mL	GPC3@IONPs@F-ITC	MRI	Human gastric cancer cells	[64]
Cancer	Theranostic agent	Murine breast cancer cells	5 mg/kg NIFR 5 mg [Fe]/kg MRI	ICG@BSA@MCNPs	NIRF/MRI	i.v. injected mice	[65]

(continued on next page)

Table 28.1 (*continued*)

Application	Type	Target	LOD(sensors)/ dose(imaging)	Platform (hybrid)	Detection method	Real samples	Ref.
Cancer	Contrast agent for Imagining	Not mentioned	1–40 mg/L aqueous solutions	IONPs@PMSEA	MRI	Fetal bovine serum	[66]
Cancer	Contrast agent	Ascites carcinoma cells	–	IOMGNPs@PEG@DOX	MRI	i.v. injected mice	[67]
Cancer	Contrast agent	–	–	Fe_3O_4MNP@PLL@Ab	MRI	Multicellular spheroids	[68]
Cardiovascular diseases	Sandwich-like sensor	PAPP-A PCSK9 ST2	40 pg/mL for ST2 antigen	Ta/NiFe/CoFe/Cu/CoFe/IrMn/Ta MNPs@ SiO_2@streptavidin	GMR	Blood serum	[69]
Cancer	Sandwich-like sensor	CEA AFP PSA freePSA PG I PG II CYFRA21-1 NSE free-β-hCG SCC Tg CA19-9	0.5–500 ng/mL 1–1000 ng/mL 0.1–100 ng/mL 0.1–50 ng/mL 2–200 ng/mL 1–100 ng/mL 0.5–100 ng/mL 1–200 ng/mL 0.5–200 ng/mL 0.5–70 ng/mL 5–2000 ng/mL 4–800 U/mL	Al_2O_3 MNP @Si/SiO_2/Ta/PtMn/CoFe/Cu/ CoFe/NiFe @streptavidin	GMR	–	[70]
Immunodeficiency Infection Autoimmunity Hyperγ-globulinemia	Sandwich-like sensor	IgG IgM	0.07 nM 0.33 nM	EDP@streptavidin MNP @Ta/seedlayer/PtMn/CoFe/Ru/ CoFe/Cu/CoFe/Cu/Ta @SiO_2/Si_3N_4/SiO_2	GMR	–	[71]

(*continued on next page*)

Table 28.1 *(continued)*

Application	Type	Target	LOD(sensors)/ dose(imaging)	Platform (hybrid)	Detection method	Real samples	Ref.
Memory loss Alzheimer's disease	Sensor	Acetylcholine	4.0 nm/ 4.0 nM to 800 μM	Fe_2O_3NPs@PEDOT@rGO@FTO	CV	Serum	[22]
Cancer	Contrast agent for imagining	–	–	SPMNPs@Chitosan	MRI	–	[15]
Breast cancer	Sandwich-like sensor	HER2	–	Her2Ab@ Fe_3O_4 MNPs	Fluorescence microscopy	Mixtures of whole blood or mononuclear cells	[72]
Chronic liver diseases	Imaging	–	–	Fe_3O_4 NPs@GCP Fe_3O_4 NPs@PDMAEMA	MRI	Serum	[23]
MSC tracers in CT	Imaging	Human MSC	–	Au-DMSA and γ-Fe_2O_3-DMSA	CT and	Dental pulp tissues	[73]
Theranostics in breast cancer	Imaging	Breast cancer cells	–	OA- Fe_3O_4 MNPs@ PLA-PEG-PLA/DOX/PTX	MRI/ fluorescence	i.v. injected mice	[19]
Theranostics in breast cancer	Imaging	Tumor cells	–	Anti-VEGF-Ab/BSA@PEG-MNPs/DOX	MRI	i.v. injected mice	[25]
Theranostic in hyperthermia	Imaging	–	–	TMAH @$NiFe_2O_4$	MRI	–	[24]
Cancer	Imaging	Tumor cells	–	OA@SPIONPs@PLGA	MRI	i.v. injected mice	[74]
Cancer	Imaging	–	–	SPIONs@Cyanine-7	MSOT	i.v. injected mice	[75]

(continued on next page)

Table 28.1 (*continued*)

Application	Type	Target	LOD(sensors)/ dose(imaging)	Platform (hybrid)	Detection method	Real samples	Ref.
Theranostics cancer	Imaging	–	–	SPION@Chitosan	US/MRI	*In-vitro* tests	[76]
Pancreatic cancer	Imaging	t-PA ligand Gal-1 ligand	–	Fe_2O_3 MNPs@rHSA	SPECT/MRI	Serum	[77]

Applications: Chronic liver diseases: liver tumors; cirrhosis; nonalcoholic fatty liver diseases; nonalcoholic steatohepatis; *Au-DMSA*, dimercaptosuccinic acid–modified gold nanoparticles; *γ-Fe_2O_3-DMSA*, dimercaptosuccinic acid modified MNPs.

*Target biomarkers: CA*125, cancer antigen 125; *HE4*, human epididymis protein 4; *IL6*, interleukin 6; *IAV* , influenza A virus; *ECP*, eosinophil cationic protein; *PAPP-A, PCSK9 and ST2*, protein biomarkers; *IgG*, human immunoglobulin G; *IgM*, human immunoglobulin M; *HER2*, human epidermal growth factor receptor 2; *MSC*, mesenchymal stem cells; *Gal*-1, Galectin-1; *t-PA-ligands*, tissue plasminogen activator derived peptides.

Magnetic nanoparticle hybrid materials: Fe_3O_4@APBA, Fe_3O_4-based MNPs modified with phenylboronic acid; *Fe_3O_4@OA@PLA-PEG-DG*, Fe_3O_4 MNPs–modified polylactic acid-polyethylene glycol-D-glucosamine; *Gd_2O_3@C*, carbon-coated ultrasmall gadolinium oxide; *PEG4-USPIONs*, PEG4-ol-coated ultra-small supermagnetic Fe_3O_4 nanoparticles; *GPC3@IONPs@F-ITC*, Glypican-3 protein functionalized iron oxide nanoparticles loaded with fluorescein-isothiocyanate; *ICG@BSA@MCNPs*, Fe_3O_4 MNPs on carbon nanoparticles coated with bovine serum albumin and loaded with fluorescent dye indocyanine green; *IONPs@PMSEA*, Fe_3O_4 nanoparticles grafted with poly(2-(methylsulfinyl)ethyl acrylate; *IOMGNPs@PEG@DOX*, Multifunctional iron oxide MNPs with gold shell–coated and conjugated with polyethylene glycol and doxorubicin; *Fe_3O_4MNP@PLL@Ab*, Poly-L-lysine coated magnetic iron oxide nanoparticles conjugated with carbonic anhydrase specific antibody; *Ta/NiFe/CoFe/Cu/CoFe/IrMn/Ta MNPs@ SiO_2@streptavidin*, Streptavidin-labeled MNPs based on Ta/NiFe/CoFe/Cu/CoFe/IrMn/Ta coated with SiO_2; *Al_2O_3 MNP@Si/SiO_2/Ta/PtMn/CoFe/Cu/CoFe/NiFe@streptavidin*, Streptavidin MNPs based on Al_2O_3 and Si/SiO_2/Ta/PtMn/CoFe/Cu/CoFe/NiFe/Al_2O_3; *EDP@streptavidin MNP@Ta/seedlayer/PtMn/CoFe/Ru/CoFe/Cu/CoFe/Cu/Ta @SiO_2/Si_3N_4/SiO_2* , Eigen diagnosis platform based on streptavidin-coated MNPs Ta/seedlayer/PtMn/CoFe/Ru/CoFe/Cu/CoFe/Cu/Ta passivated with SiO_2/Si_3N_4/SiO_2; *Fe_2O_3NPs@PEDOT@rGO@FTO*, Iron oxide nanoparticles modified with poly(3,4-ethylenedioxythiophene and reduced graphene oxide and fluorine doped tinoxide; *SPMNPs@Chitosan*, Superparamagnetic nanoparticles based on Fe_2O_3 modified with chitosan; *Her2Ab@Fe_3O_4 MNPs*, HER2 Antibody-conjugated MNPs; *US SM Fe_3O_4 NPs*, ultra-small superparamagnetic Fe_3O_4 NPs; *SPM Fe_3O_4@ZnS NPs*, superparamagnetic Fe_3O_4@ZnS core/shell nanocomposites; *Fe_3O_4 NPs@GCP*, Fe_3O_4 NPs coated with star polymers; *Fe_3O_4 NPs@PDMAEMA*, Fe_3O_4 NPs with CD-containing star-shaped poly(2-(dimethylamino) ethyl methacrylate; *SPIONs@Cyanine-7*, SPIONs modified with cyanine-7 dye; *TMAH@$NiFe_2O_4$*, tetramethyl-ammoniumhydroxide coated nickel ferrite nanoparticles; *Fe_2O_3 MNPs@rHSA*, Maghemite MNPs modified with recombinant human serum albumin; *OA- Fe_3O_4 MNPs@ PLA-PEG-PLA/DOX/PTX*, Oleic acid–modified Fe_3O_4 MNPs incorporated into the tree block PLA-PEG-PLA copolymer for anticancer drugs doxorubicin or paclitaxel encapsulation; *Anti-VEGF-Ab/BSA@PEG-MNPs/DOX*, Anti-VEGF antibodies conjugated with bovine serum albumin–coated PEGylated MNPs for DOX delivery; *TMAH@$NiFe_2O_4$*, Tetramethylammonium hydroxide-modified $NiFe_2O_4$ MNPs; *OA@SPIONPs@PLGA*, Oleic acid–coated SPIONPs encapsulated PLGA nanospheres.

Methods: GMR, giant magnetoresistance; *NIRF/MRI*, near infrared fluorescence/magnetic resonance imaging dual imaging; *CV*, cyclic voltammetry; *LOD*, limit of detection; *SPECT-CT*, single-photon emission computed tomography-computer tomography; *US/MRI*, ultrasound/MRI; *SPECT/MRI*, single photon emission CT/MRI; *MSOT*, multispectral optoacoustic tomography.

symptoms, but with very good outcome if detected in an initial stage. A multiplexed magnetoresistive biosensor was designed to detect three biomarkers for ovarian cancer: cancer antigen 125, human epididymis protein 4, and interleukin 6 that could specifically point toward this type of cancer. It has to be underlined that the detection of a single biomarker cannot be associated with a specific cancer type due to the fact that these biomarkers could also have increased levels in other malignancies, but the detection of an entire panel has the property to directly indicate a specific type of cancer. Moreover, this portable platform was designed as flexible to be adapted for the detection of other biotargets having potential for real-case scenarios [60]. As already discussed, the assessment of tumoral biomarkers panel is extremely important either for specific and precocious diagnosis of cancer or for the monitoring of therapy efficiency. Alpha-fetoprotein, carcinoembryonic antigen, cytokeratin 19 fragment, neuron specific enolase, free β-subunit of human chorionic gonadotropin, squamous cell carcinoma, pepsinogen I and II, total prostate specific antigen, free prostate specific antigen, thyroglobulin, and carbohydrate antigen 19-9 were simultaneously detected via a giant magnetoresistive (GMR) microfluidic sensor chip. The microfluidic device based on sandwich immunoassays can be used to screen patients with lung, liver, digestive tract, and prostatic types of cancer and it has the potential to be implemented as a point-of-care device [70].

A similar approach is oriented toward the D-dimer, a degradation byproduct of cross-linked fibrin that is considered a biomarker for deep vein thrombosis and pulmonary embolism. The target was immobilized via a capture antibody and following the principle of sandwich immunoassay, a second detection biotinylated antibody was immobilized. The last step consisted in the immobilization of streptavidin functionalized with MNPs. The binding of MNPs to the GMR sensor is thus proportional to the concentration of the target which determined a change in the resistance of the sensor.

The performances of the portable microfluidic device were compared with a standardized method and the correlation studies indicated that it can satisfy the demands in clinical applications due to the high sensitivity and simplicity [59]. The same design strategy was also applied for the detection of a viral nucleoprotein, specific for the influenza A virus and the results showed that the GMR sensor had a higher sensitivity than the conventional enzyme-linked-immunosorbent serologic assay (ELISA) assays. The advantage of the GMR chip is the fact that it is composed of 64 sensors increasing the quantity of immobilized MNPs and thus the sensitivity. This approach for rapid virus monitoring with a portable device could have an important impact in public health and in the prevention of influenza epidemic/pandemic [61].

Multiplexed assay of human immunoglobulin G and M antibodies with the same type of sensor represents another application for clinical medicine. Another advantage of portable integrated platforms is that they allow the assays to be performed in remote areas or other nonclinical settings, such as home, school, and office. The clinical assays are usually performed in laboratory settings, but a new device that has the performances of the conventional methods and the simplicity of a strip test will upgrade this domain and improve the healthcare services [71]. Another positive aspect of GMR sensors is that they are cost effective and their implementation does not need supplementary financial efforts, the evidence being the wide range of studies that resulted in many proof-of-concept devices [78].

28.3.2 MNPs in treatment

MNPs can be controlled via a magnetic external field and due to their tissue permeability, could have applications into the cell delivery to different organs with impact in the clinical field. For example, mesenchymal stem cells that were labeled and manipulated magnetically by an external magnetic field to the upper hemisphere of a rodent retina, open broad perspectives toward translating cell therapies [79]. A relevant example is represented by the fact that MNPs can be forced to cross the brain-blood barrier (BBB) via an external magnetic field allowing targeted drug delivery in gliomas and metastatic tumors in the brain, that are not available for chemotherapy due to BBB. Of course, despite their magnetic properties, other features, such as size, lipophilicity and surface charge, need to be assessed because they can have a great impact in brain penetration. Magnetite (Fe_3O_4) and maghemite (γ-Fe_2O_3) can be formulated as nanoscaled particles and their surface is easily functionalized due to their atom vacancies and amphoteric OH decorations [46]. Thinking forward this approach in neuroscience will be extended to other brain pathologies such as Alzheimer's and Parkinson's maladies or to other difficult accessible internal organs such as lungs and liver.

28.3.2.1 MNPs in hyperthermia therapy

A fast forward field of applications for MNPs is represented by hyperthermia therapy, a technique used to destroy specifically cancer cells that involves temperatures between 42 and 45 °C and it can selectively heat the targeted cells from the inside. This technique has the capacity to improve the clinical outcome of conventional approaches such as radiotherapy and chemotherapy for the specific treatment of various tumor types.

MNPs can be used at the same time to deliver anticancer drugs, potentiating the antitumor activity with hyperthermia. The heating of MNPs is possible if an alternating current is applied, leading to the conversion of magnetic energy into heat [80]. There are different beliefs regarding the heating temperature. In general, the temperature considered to be efficient is between 41 and 45 °C, but some studies submit that at 50 °C or even 60–70 °C the antitumor effect is more powerful. The temperature needed may vary though, depending on the type of tumor [81]. Iron oxide-based MNPs are used for hyperthermia therapy, and some studies focused on testing cobalt ferrite MNPs that have the advantage of being more stable and prevent the aerial oxidation [80] or on ferromagnetic glass-ceramic as a thermoseed [82]. On the other hand, some studies indicate that hyperthermia is not efficient. A recent study reports that tumor cells are resistant to this type of therapy due to the nonactivation of caspase 3, a protein that activates cell apoptosis [83].

MNPs kept under an oscillating magnetic field transform the magnetic energy into heat that induces apoptotic cell death without affecting the healthy tissue, the destroyed cells being eliminated by phagocytosis. Another side of this effect is the capacity to release drugs from MNPs after delivery next to the tumor region. Of course, the size, shape, magnetic anisotropy, saturation magnetization of the NPs, and internalization within the cells are parameters that need to be optimized [80].

The use of MNPs in tumors started in 1957 when Gilchrist et al. used microrange particles for inductive heating of lymph nodes in dogs and in the early 2000s MNPs encapsulated in liposomes have been used in Europe for the treatment of glioblastoma [84]. Thus, the so-called "magnetoliposomes" are a new class of biomaterials that generate heat (killing the cancerous cells that are more sensitive to temperatures over 41 °C than healthy ones) when exposed to an oscillating magnetic field and have promising applications in hyperthermia procedures. There are two types of magnetoliposomes, the classical ones that consist in the MNPs surrounded by a bilayer of phospholipids and have approximately 20 nm and the extruded ones that have multiple MNPs inside the lipid bilayer.

28.3.2.2 MNPs in drug delivery systems

MNPs are used in treatment mostly as drug delivery systems [85]. The use of MNPs in drug delivery brings the well-known advantages, namely: the possibility of a targeted delivery with the reducing of the side effects and increasing of the efficacy and the preventing of drug degradation during transport to the site of action. What distinguishes MNPs of other drug delivery systems is their capacity of being directed to the target using an external magnetic field [86,87].

MNPs first appeared in the scientific literature as drug delivery systems in 1978, when Senyei et al. described the first drug carrier that could be directed to the site of action using an extracorporeal magnetic field. The group developed some microspheres made from an albumin matrix, Fe_3O_4 particles, and doxorubicin hydrochloride. The synthesis of the microspheres was described, as well as the equipment needed to guide them to the target [88].

MNPs used in drug delivery consist in a magnetic core functionalized by coating. Different types of magnetic cores can be used, considering the magnetic and oxidation state of the iron. The materials used in the synthesis of MNPs should assure safety and biocompatibility. From all magnetic materials, iron oxides are preferred for drug delivery because of their safety and biocompatibility. Magnetite and maghemite are the most used species of iron oxide in this purpose, the last one being considered the safest choice because of the physiological existence of Fe^{3+} into the human body. Nevertheless, the safety of MNPs in drug delivery is still under research as an eventual accumulation of the metal could be harmful. At the same time, maghemite is more stable in aqueous solution, which gives a better stability to MNPs suspension, while magnetite has a higher magnetization [86,87].

One of the major concerns in the biomedical applications is of course related to the biocompatibility issue. Even though they are considered biocompatible, the most reported source of toxicity is linked to the generation of reactive oxygen species (ROS), that induce lipid peroxidation and as a consequence the disruption of the phospholipid-bilayer membrane that ends in the cell death. *In-vivo* studies reported that SPIONs induce toxicity in the liver, kidneys, and lugs, but no effect in the brain and heart. An explanation is that negatively charged particles could disrupt the actin skeleton in the kidney and brain with no effect in the one of the heart cells [89,90].

Other features, such as size, shape, surface charge, coating and colloidal stability should also be considered in the synthesis of MNPs. The passing of the particles through endothelium from blood vessels to organs, as well as their distribution is dependent on their size. There are studies regarding the biodistribution of MNPs in the body depending on their size. While some studies found that bigger MNPs have the highest uptake in the liver [91], others concluded the opposite [92], and this suggests that other morphological characteristics also influence the distribution of MNPs in the body.

Surface charge is also important for biodistribution and cellular uptake. Some studies showed that positively charged MNPs have a better cellular uptake and better distribution in liver and spleen with faster clearance, while negatively charged MNPs get mostly in the lymph nodes and have a slower clearance [93,94]. At the same time, neutral MNPs have the highest circulation time, being less likely to be captured by the reticuloendothelial system [86,94].

Surface coating of MNPs is essential for two main reasons: to increase the biocompatibility of particles and to prevent the conglomeration. Due to their hydrophobicity and magnetic attraction, MNPs group together and form clusters, losing their specific individual properties.

There are two methods that can be used to achieve coating and colloidal stability: ligand addition and ligand exchange. In the first method, the coating material is added in the preparation process and it adheres at the surface of the particles through hydrogen bonds, electrostatic or hydrophobic interactions. Sometimes, when MNPs are stable but hydrophobic, to modify their polarity ligand exchange is used and the surface is customized with functional groups such as -OH, -SH, -COOH, $-NH_2$. This functionalization is also used when further binding of macromolecules for target delivery is wanted. A variety of natural or synthetic polymers can be used for coating, including PEG, poly(d,l-lactide), polylactic-coglycolic acid, polyvinylpyrrolidone, dextran, starch, gelatine, alginate, chitosan. Inorganic coating is also possible but less used. An important feature of the coating material is to preserve the magnetic response of MNPs to the external magnetic field [86,87].

Several studies refer to the distribution and pharmacokinetics of MNPs depending on their coating [94,95]. The molecular weights, density, as well as the binding strength of the coating material have an influence on the pharmacokinetics of MNPs. Higher density and molecular weight of the polymer used for coating is associated with longer half-life of MNPs into the body. A weak binding of the coating material at the surface of MNPs leads to its detachment, the aggregation of particles, and their accumulation in the liver. Circulation time also depends on molecular weight uniformity of the coating, a lower polydispersity index being desired for longer half-life. Usually, the functionalization of MNPs with aptamers or other molecules used for targeting decreases the circulation time of drug carriers [95].

In the majority of cases, MNPs as drug delivery systems are used to deliver antitumor drugs, a target of the tumor tissue being desired. The cancer cells have an important glycolytic activity and as a result a lower pH and high temperature than the normal ones. These particularities are known to modulate the drug release from targeted drug delivery systems. One of the drawbacks for biomedical applications regarding MNPs is defined by the aggregation of the naked NPs due to dipole–dipole interactions between them. Therefore, the association with various materials can improve their properties [13]. One such example is represented by multifunctional MNPs that contain an anticancer drug (either doxorubicin or paclitaxel), EPPT peptide and, oleic acid along with Fe_3O_4 NPs that were formulated as spherical particles with diameters approximately of 179–203 nm suitable for IV administration. The studies showed that the release of the

Table 28.2 Examples of MNPs hybrid materials used in cancer treatment as drug delivery systems.

Magnetic core	Coating/ functionalization	Antitumor agent	Dimensions (nm)	LC/EE	Release of drug	Reference
OA@Fe_3O_4	PLA-PEG-PLA	DOX PTX	179–203	EE= 49–53% EE= 57–67%	86% (DOX) 61% (PTX)	[19]
Fe_3O_4	CS-malt	5-FU	300–500	LC= 46.5–72.6 mg 5-FU/g NPs	81.14%	[99]
Fe_3O_4	APTES	CPT	12	LC= 85 mg/g	44% (pH 7.4)	[100]
γ-Fe_2O_3 @MWCNTs	PEG-PEI-FA	DOX	5	2.38±0.19 mg/g	15% (pH 7.4) 55% (pH 5.5)	[101]
γ-Fe_2O_3	SiO_2	Fe^{2+}/Toc-6-Ac	244	–	–	[102]
γ-Fe_2O_3 Fe_3O_4	CA, CP, PEG	siRNA	130	–	–	[103]
Fe_3O_4	CS	GCT	4	EE= 39%	8% (pH 7.2) 65% (pH 4.2)	[104]
Fe_3O_4	SiO_2-NH_2	ACL	32–40	EE= ˜80%	–	[96]
$CoFe_2O_4$	FA	DOX	19.6 ± 0.27	LE= 78%	90% (pH 7.4)	[80]
MNPs	PEG	Curcumin	24.33–34.24	LC= 1.9%	43.7% (pH 7.4) 53.5% (pH 5.4)	[105]
IOMGNPs	PEG	DOX	29 ± 4	LC= 10%	–	[67]
Fe_3O_4	PCA-PEG-FA	Quercetin	10–49	LC= 3.5–12.1% EE= 70.2–80.3%	83% (pH 7.4)	[17]

Magnetic core/functionalization: OA-Fe_3O_4, oleic acid-iron oxide nanoparticles; *γ-Fe_2O_3@MWCNTs*, γ-Fe_2O_3 deposited on multiwalled carbon nanotubes; *CS-Malt*, chitosan-modified with maltose; *CS*, chitosan; *APTES*, (3-aminopropyl)triethoxysilane; *PEG-PEI-FA*, polyethylene glycol-polyethyleneimine-folic acid; *PEG*, polyethylene glycol; *CA*, caffeic acid; *CP*, calcium phosphate; *FA*, folic acid; *IOMGNPs*, multifunctional iron oxide MNPs with gold shell; *PCA-PEG-FA*, polycitric acid-PEG with folic acid on the surface.
Antitumor agent: DOX, doxorubicin hydrochloride; *PTX*, paclitaxel; *5-FU*, 5-fluorouracil; *Toc-6-Ac*, acetate derivative of α-tocopherol; *siRNA*, small interfering ribonucleic acid; *CPT*, camptothecin; *GCT*, gemcitabine; *ACL*, acyclovir.
LC, loading capacity; *EE*, encapsulation efficiency.

antitumoral agents was achieved immediately at a pH value of 4.8 that is specific to tumoral cells, while at the physiological pH value of 7.4 the release was slow and continued up to 30 days. The multifunctional particles exhibited a high cytotoxicity against breast cancer cells and they also could be employed to reduce the side effects of the free drugs by modulating their release in certain conditions [19]. A similar approach this time containing star polymers that generate flexible architectures and cyclodextrins with outstanding biocompatibility and nontoxicity had the ability to suppress the toxicity of the encapsulated drugs. The association of the modifiers with the Fe_3O_4 NPs enhances the stability of the MNPs in serum samples and provides a platform for both diagnosis and treatment [23]. Plenty of MNPs with antitumor drugs alone or in combination were studied. Table 28.2 presents some examples of MNPs aimed to be used in the treatment of cancer regarding their composition, dimensions, and loading/release capacities.

Some studies describe MNPs with a double role, in drug delivery and as nanobiosensors for the detection of various molecules. For example, MNPs silica-coated for delivery of acyclovir were developed, and they can also be applied as efficient biosensors for detection of acyclovir in biomedical applications. This is possible due to the high affinity of acyclovir for the MNPs developed and the resulted adsorption of the drug molecules at the surface of the particles which brings the double application of both delivery and detection [96].

Besides antitumor drugs, MNPs are used to deliver other types of drugs such as antioxidants, antibiotic, or antimicrobial agents. Silver NPs were synthetized starting from a natural compound, 4-*N*-methylbenzoic acid, provided from medicinal plants. The antitumor, antimicrobial, and antioxidant activity of the resulted silver NPs were evaluated and the results showed an antimicrobial activity on both Gram-positive and Gram-negative bacteria, especially on *Acinetobacter baumannii*, bacteria that cause nosocomial blood, lung and urinary tract infections. Moreover, good antioxidant and antitumor activity on breast cancer cells was found [97]. Another study describes the use of MNPs to treat implant-associated infections in orthopedics which are difficult to treat because of the low concentration of antibiotics at the infection site after administration of high dose of systemic antibiotics. Thus, magnetic nanoporous silica NPs (MNPSNP) were synthetized, modified with rhodamine B isothiocyanate as fluorescence dye and PEG, and their biodistribution was followed using fluorescence microscopy on samples of blood and tissue of mice after administration of the NPs suspension [Fig. 28.5A). The NPs were found to be biocompatible, but their guidance via a magnetic field was not achieved [98].

Compared to other drug delivery systems, MNPs can influence the drug distribution using an external magnetic field. Although intense research has been made, it is difficult to describe the behavior of MNPs in blood flow under the influence of a magnetic field. The blood flow and even vessel network were tried to be simulated in various studies [106]. Some clinical trials were made over time, involving patients suffering of advanced cancer, and the results showed that MNPs concentrated at the tumor tissue and the dimensions of the tumors decreased. The limitations consisted in the fact that the magnetic field was losing strength with distance [86,107–109].

28.3.2.3 MNPs in gene therapy

Another important destination of MNPs is gene delivery for transfection, called magnetofection. A required feature of MNPs to be used in magnetofection is to be ferromagnetic. Even though Co and Mn increase the magnetism of the particles, $CoFe_2O_4$ and $MnFe_2O_4$ are less used than magnetite and maghemite because of their toxicity. Gene transfection represents the insertion of an external gene into a living cell and it is a technique used for different purposes: to produce proteins, antibodies, or viral vectors. The most difficult task in transfection is to find the proper carrier that can give specificity to the cell, efficacy to the process, and also safety. The vector (carrier) protects the fragile nucleic acid sequence from the action of nuclease. Therefore, MNPs are seen as potential vectors that can provide all of the above-mentioned requests [110].

The binding of the gene and NPs can be done by two methods. The first one involves the coating of the MNPs with a positive charged polymer which will interact electrostatically with natural negative charged genes, stabilizing them, and increasing their biocompatibility. A cationic polymer often used for this purpose is polyethylenimine due to the presence of amino groups. In the second method of binding, a complex is formed between the genes and a polycation and the product is bound to MNPs oppositely charged. Even though blood compounds can interact with

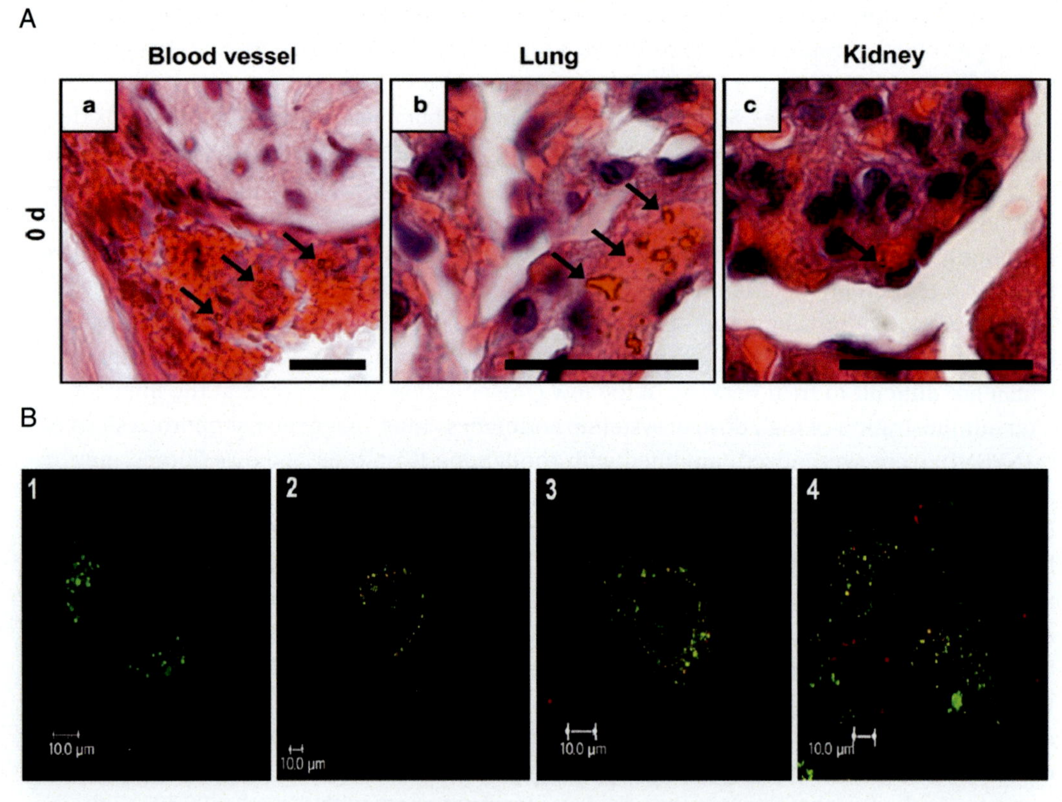

FIGURE 28.5 (A) MNPSNP cluster-detection in H.E. stained histological slices in a blood vessel (a), alveolar septa (lung, b), and glomeruli (kidney, c). All scale bars: 50 μm. Reprinted under the terms of the Creative Commons CC-BY license from [98] (open access). (B) Fluorescence confocal microscopy images showing the endosomes/lysosomes (green) and cyanine 5-labeled HER2 small interfering RNA (siHER2-Cy5) (red) in HCC1954 breast cancer cells. (1) untreated cells, (2) cells treated for 24 h with 100 nM siHER2-Cy5, (3) cells treated for 24 h with caffeic acid/calcium phosphate MNP loaded with siHER2-Cy5, (4) cells treated for 1 h with siHER2-Cy5 caffeic acid/calcium phosphate MNP under a magnetic field. Reprinted with permission by Elsevier from [103].

MNP-gene complexes and adhere to them, it was proved that under external magnetic field this does not happen and the carrier is stable in the serum over 24 h. Regarding the cell targeting and cellular uptake, it was found that MNPs are concentrated at the surface of the cell under the magnetic field, and that they shorten the cellular uptake time. The cellular uptake is achieved through phagocytosis, endocytosis, or macropinocytosis [110].

An example of transfection using MNPs is described by Cristofolini et al. [103]. SPIONs were used to deliver small interfering RNA (siRNA) for the treatment of human epidermal growth factor receptor 2 (HER2)-positive breast cancer. SPIONs were coated first with a layer of caffeic acid and after that with calcium phosphate to bind siHER2. Finally, the carrier was stabilized with PEG. In conclusion it was suggested that SPIONs could be accumulated at the tumor tissue using a magnetic field, and the cell uptake of the NPs as well as the gene silencing of HER2 were successful (Fig. 28.5B) [103].

Another approach of cancer treatment using MNPs is photodynamic therapy in which photosensitizing substances are delivered to the tumor and produce ROS under a laser irradiation of a specific wavelength. The use of photodynamic therapy is not summarized only for cancer therapy, but also for antimicrobial therapy with photoinactivation of bacteria. A recent published article describes the synthesis of a Fe_3O_4/AgNPs combined with a Zn-phthalocyanine complex and evaluates their photodynamic antimicrobial activity against *S. aureus* [111].

Last but not least, MNPs are used in regenerative medicine, a branch of medicine that is dealing with replacement or regeneration of human cells, tissues, or organs. Mesenchymal stem cells are used in this therapy due to their capacity to differentiate into various specific cells. The process monitoring is difficult though, and MNPs can help in this concern. They can be labeled and used to track mesenchymal stem cell in neurodegenerative diseases (Alzheimer's, Parkinson's, and Huntington disease), cardiovascular injury, but also to deliver progenitors in retinal or cartilage regeneration [85]. Several examples of MNPs hybrid materials used in treatment are presented in Table 28.3.

28.4 Conclusions and future trends

The continuous development of new diagnostic and therapy strategies represents one of the major challenges of the actual sanitary global context. Using different MNPs as contrast agents, many applications are enabled regarding the detection of specific markers, diagnostic procedures, drug delivery, and treatment monitored by MRI, a noninvasive imaging technique at molecular and cellular level. This insight into new customized approaches will bring a real advantage on the current clinical diagnosis techniques namely enzyme-linked immunosorbent assay, radioimmunoassay, and polymerase chain reaction that require high-cost and laboratory setting.

Requirements for biomedical applications involve several different aspects that are often hard to achieve:

- design techniques for resulting in controlled size and homogenous distribution and a high saturation magnetization;
- the activation of the surface to allow functionalization to achieve the stability at the physiological pH suitable for *in-vivo* bioapplications;
- inserting different coatings to increase the biocompatibility and diminish the toxicity for hyperthermia and drug delivery systems;
- avoiding the agglomeration of the particles by manipulating the colloidal stability via innovative formulations and functionalization techniques.

There are various strategies regarding the synthesis of bare/functionalized MNPs all having the final outcome of a new material that can be used in biomedical applications: as it is characterized by biocompatibility, nontoxicity, colloidal suspension, etc. The stability at physiological pH of the MNPs is mandatory for the bioapplications, which explains the huge amount of

Table 28.3 Examples of applications of MNPs-based hybrid materials in treatment.

Application	Target	Dose	Platform (hybrid)	Detection method	Real samples	Reference
Hyperthermia therapy in cancer	Cancer cells	10.9 mg/mL	MNPs@Chitosan	AMF	*In-vitro* study	[13]
Alzheimer's disease Parkinson's diseases Lysozyme systemic amyloidosis Diabetes mellitus type II	αLAF	0.09507 mg/mL (DC50)	MNPs@Chitosan	ThT fluorescence assay AFM	*In-vitro* study	[13]
Bacterial infection	Biofilm formed by *P. aeruginosa* and *S. aureus*	1 mg/mL	Fe_3O_4@RL	Adherent tube assay Biofilm formation CLSM	Bacterial cultures	[14]
Breast cancer	Breast cancer cells	10 mg/kg 0.3 mg/kg	OA-Fe_3O_4@PLA-PEG-PLA/DOX OA-Fe_3O_4@PLA-PEG-PLA/PTX	MRI	i.v. injected mice	[19]
Hyperthermia therapy in cancer	Cancer cells	–	$CoFe_2O_4$@FA@DOX	UV-VIS	in vitro study	[80]
Phototermal therapy in cancer	Murine breast cancer cells	1 mg/mL	MCNPs@ ICG	Photothermal imaging	Tumor-bearing mice i.v. injected via tail vein	[65]
Cancer	–	10 mg/mL	MNPs@PEG-Cur	UV-VIS	*In-vitro* study	[105]
NIR-induced chemophotothermal therapy in cancer*	Ascites carcinoma cells	–	IOMGNPs@PEG@DOX	MRI	mice i.v injected via tail vein	[67]
Hyperthermia therapy in cancer	–	2.5–11.3 mg/mL	Fe_3O_4MNP@PLL@CASAb	calorimetric analysis after AMF	–	[68]
Cancer*	–	–	$MnSiO_3$@Fe_3O_4@PEG@CDDP	MRI	–	[112]
Chemoterapy in cancer*	MDA-MB-231 HeLa	–	Fe_3O_4MNPs@PCA-PEG-FA	MRI	*In-vitro* study	[17]
Glioma	C6 rat glioma cells	–	SPIONs-CPTES-HAPtS	MTT assay FDA assay Clonogenic test	*In-vitro* study	[18]

(*continued on next page*)

Table 28.3 *(continued)*

Application	Target	Dose	Platform (hybrid)	Detection method	Real samples	Reference
Chronic liver diseases*	–	–	Fe_3O_4NPs@GCP Fe_3O_4NPs@PDMAEMA	MRI	Serum	[23]
Not mentioned	–	–	Fe_3O_4MNPs@SiO_2-NH_2@acyclovir	VSM and zetametry	–	[96]
Breast cancer*	Tumor cells	–	Anti-VEGF-Ab/BSA@PEG-MNPs/DOX	MRI	Tumor-bearing mice i.v. injected	[25]
Hyperthermia*	–	–	TMAH@$NiFe_2O_4$	MRI	–	[24]
Cancer*	Tumor cells	–	OA@SPIONPs@PLGA	MRI	Mice i.v. injected via tail vein	[74]
Cancer*	–	–	SPION@Chitosan	US/MRI	*In-vitro* tests	[76]

Applications: *theranostics (both treatment and diagnosis); chronic liver diseases: liver tumors; cirrhosis; nonalcoholic fatty liver diseases; nonalcoholic steatohepatis.
Target biomarkers: αLAF, α-lactalbumin amyloid fibrils; *MDA-MB-231*, human breast cancer cell lines; *HeLa*, human cervical cancer cells.
Magnetic nanoparticle hybrid materials: MNPs@Chitosan, chitosan-coated MNPs; *Fe_3O_4MNPs@RL,* rhamnolipid-coated iron oxide MNPs; *OA- Fe_3O_4@PLA-PEG-PLA/DOX or PTX* , Oleic acid–modified Fe_3O_4 MNPs incorporated into the tree block PLA-PEG-PLA copolymer for anticancer drugs doxorubicin or paclitaxel encapsulation; *$CoFe_2O_4$@FA@DOX,* folic acid–coated ferromagnetic cubic cobalt ferrite MNPs loaded with doxorubicin; *MCNPs@ICG,* iron oxide MNPs on carbon nanoparticles loaded with fluorescent dye indocyanine green; *MNPs@PEG-Cur,* PEGylated curcumin surface modified MNPs; *IOMGNPs@PEG@DOX,* Multifunctional iron oxide MNPs with gold shell–coated and conjugated with polyethylene glycol and doxorubicin; *Fe_3O_4MNPs@PLL@CASAb,* poly-L-lysine coated magnetic iron oxide nanoparticles conjugated with carbonic anhydrase specific antibody; *$MnSiO_3$@Fe_3O_4@PEG@CDDP,* iron oxide MNPs-decorated manganese silicate grafted with PEG and loaded with cisplatin; *Fe_3O_4MNPs@PCA-PEG-FA,* polycitric acid and PEG-coated iron oxide MNPs functionalized with PEG and incorporated with FA on the surface; *SPIONs-CPTES-HAPtS,* superparamagnetic iron oxide nanoparticles modified with 3-chlorporyltriethoxysilane and 4′-hydroxy-4-(3-aminopropoxy)-trans-stilbene; *Fe_3O_4NPs@GCP,* Fe_3O_4 NPs–coated with star polymers; *Fe_3O_4NPs@PDMAEMA,* Fe_3O_4 NPs with CD-containing star-shaped poly(2-(dimethylamino) ethyl methacrylate; *Fe_3O_4MNPs@SiO_2-NH2@Acyclovir,* acyclovir-loaded iron oxide MNPs modified with amino-functionalized silica particles; *Anti-VEGF-Ab/BSA@PEG-MNPs/DOX,* Anti-VEGF antibodies conjugated with bovine serum albumin coated PEGylated MNPs for DOX delivery; *TMAH@$NiFe_2O_4$,* Tetramethylammonium hydroxide–modified $NiFe_2O_4$ MNPs; *SPION@Chitosan*, chitosan-coated super paramagnetic iron oxide nanoparticles; *OA@SPIONPs@PLGA*, Oleic acid–coated SPIONPs-encapsulated PLGA nanospheres.
Methods: AMF, alternating magnetic field; *ThT,* thioflavin T; *AFM*, atomic force microscopy; *CLSM*, confocal laser scanning microscopy; *MRI*, magnetic resonance imaging; *MTT,* (3-(4-,5-dimethylthiazol-2-yl)-2,5-diphenyltetrazolium bromide); *FDA*, fluorescein diacetate; *VSM*, vibrating sample magnetometry; *US/MRI*, ultrasound/magnetic resonance imaging.
DC_{50}, concentration causing 50% destroying of fibril.

information in the literature on surface activation with various functionalities to ensure their circulation in the bloodstream.

Another important aspect is related to the ease of synthesis and of course the production cost to make it available at a large scale with a minimum economic impact on the medical system.

The perspective of this biomedical field is promising and the future is closer than ever in regard of the treatment of various diseases, especially in oncology, in cancer diagnosis, and treatment.

Acknowledgment

This work was supported by a grant of the Romanian Minister of Research and Innovation, CCCDI - UEFISCDI, project number PNIII-P1-1.2-PCCDI-2017-0221/59PCCDI/2018 (IMPROVE), within PNCDI III. This work has also received funding from the European Union's Horizon 2020 Research and Innovation Programme under grant agreement No. 883484/2020 (PATHOCERT).

References

[1] A. Lu, E.L. Salabas, F. Schüth, Magnetic nanoparticles: synthesis, protection, functionalization, and application, Angew. Chem. Int. Ed. 46 (2007) 1222–1244.

[2] C. Cristea, M. Tertis, R. Galatus, Magnetic nanoparticles for antibiotics detection, Nanomaterials 7 (6) (2017) 119 Available from. http://www.mdpi.com/2079-4991/7/6/119.

[3] S. Kralj, M. Rojnik, R. Romih, Effect of surface charge on the cellular uptake of fluorescent magnetic nanoparticles, J. Nanoparticle Res. 14 (2012) 1151.

[4] Y. Ma, T. Chen, M.Z. Iqbal, F. Yang, N. Hampp, A. Wu, et al., Applications of magnetic materials separation in biological nanomedicine, Electrophoresis 40 (2019) 2011–2028.

[5] S. Liu, B. Yu, S. Wang, Y. Shen, H. Cong, Preparation, surface functionalization and application of Fe_3O_4 magnetic nanoparticles, Adv. Coll. Interface. Sci. 281 (2020) 102165.

[6] S. Rahim, F.J. Iftikhar, M.I. Malik, Biomedical applications of magnetic nanoparticles, in: M.R. Shah, M. Imran, S. Ullah (Eds.), Metal Nanoparticles for Drug Delivery and Diagnostic Applications, Elsevier Inc., Amsterdam, 2020, pp. 327–354.

[7] N.A. Erfan, N.A.M. Barakat, B.J. Muller-Borer, Preparation and characterization of ß-lactoglobulin/poly(ethylene oxide)magnetic nanofibers for biomedical applications, Coll. Surf. A Physicochem. Eng. Asp. 576 (2019) 63–72.

[8] H. Kahimbi, J.M. Jeong, D.H. Kim, J.W. Kim, B.G. Choi, Facile and scalable synthesis of nanostructured Fe_2O_3 using ionic liquid-assisted ball milling for high-performance pseudocapacitors, Solid. State. Sci. 83 (2018) 201–206.

[9] I.O. Wulandari, VT. Mardila, D.J. Djoko, H. Santjojo, A. Sabarudin, Preparation and characterization of chitosan-coated Fe_3O_4 nanoparticles using ex-situ co-precipitation method and tripolyphosphate/sulphate as dual crosslinkers preparation and characterization of chitosan-coated Fe_3O_4 nanoparticles using ex-situ, IOP Conf. Sci. Mater. Ser. 299 (2018) 012064.

[10] F. Xiong, K. Hu, H. Yu, L. Zhou, L. Song, Y. Zhang, et al., A functional iron oxide nanoparticles modified with PLA-PEG-DG as tumor-targeted mri contrast agent, Pharm. Res. 34 (8) (2017) 1683–1692.

[11] H. Lee, E. Lee, D. Kim, N. Jang, Y. Jeong, S. Jon, Antibiofouling polymer-coated superparamagnetic iron oxide nanoparticles iron oxide nanoparticles as potential magnetic resonance contrast agents for in vivo cancer imaging, J. Am. Chem. Soc. (19) (2006) 7383–7389.

[12] I. Khmara, M. Kubovcikova, M. Koneracka, B. Kalska-Szostko, V. Zavisova, I. Antal, et al., Preparation and characterization of magnetic nanoparticles, Acta. Phys. Pol. A. 133 (3) (2018) 704–706.

[13] I. Khmara, M. Molcan, A. Antosova, Z. Bednarikova, V. Zavisova, M. Kubovcikova, et al., Bioactive properties of chitosan stabilized magnetic nanoparticles – Focus on hyperthermic and anti-amyloid activities, J. Magn. Magn. Mater. 513 (2020) 167056.

[14] H.F. Khalid, B. Tehseen, Y. Sarwar, S.Z. Hussain, W.S. Khan, Z.A. Raza, et al., Biosurfactant coated silver and iron oxide nanoparticles with enhanced anti-biofilm and anti-adhesive properties, J. Hazard. Mater. 364 (2019) 441–448.

[15] I. Khmara, O. Strbak, V. Zavisova, M. Koneracka, M. Kubovcikova, I. Antal, et al., Chitosan-stabilized iron oxide nanoparticles for magnetic resonance imaging, J. Magn. Magn. Mater. 474 (2019) 319–325.

[16] A.O.M. Adeoye, J.F. Kayode, B.I. Oladapo, S.O. Afolabi, Experimental analysis and optimization of synthesized magnetic nanoparticles coated with PMAMPC-MNPs for bioengineering application, St. Petersbg. Polytech. Univ. J. Phys. Math. 3 (4) (2017) 333–338.

[17] A. Mashhadi Malekzadeh, A. Ramazani, S.J. Tabatabaei Rezaei, H. Niknejad, Design and construction of multifunctional hyperbranched polymers coated magnetite nanoparticles for both targeting magnetic resonance imaging and cancer therapy, J. Colloid. Interface. Sci. 490 (2017) 64–73.

[18] F. Sallem, R. Haji, D. Vervandier-Fasseur, T. Nury, L. Maurizi, J. Boudon, et al., Elaboration of trans-resveratrol derivative-loaded superparamagnetic iron oxide nanoparticles for glioma treatment, Nanomaterials 9 (2) (2019) 287. doi:10.3390/nano9020287.

[19] A. Amani, J.M. Begdelo, H. Yaghoubi, S. Motallebinia, Multifunctional magnetic nanoparticles for controlled release of anticancer drug, breast cancer cell targeting, MRI/fluorescence imaging, and anticancer drug delivery, J. Drug. Deliv. Sci. Technol. 49 (2019) 534–546.

[20] P. Mandal, S. Panja, S. Lal, S. Kumar, S. Maji, Magnetic particle anchored reduction and pH responsive nanogel for enhanced intracellular drug delivery, Eur. Polym. J. 129 (2020) 109638.

[21] P. Mandal, S.L. Banerjee, S. Maji, S.K. Ghorai, T.K. Maiti, S. Chattopadhyay, Time-dependent self-assembly of magnetic particles tethered branched block copolymer for potential biomedical application, Appl. Surf. Sci. 527 (2020) 146649.

[22] N. Chauhan, S. Chawla, C.S. Pundir, U. Jain, An electrochemical sensor for detection of neurotransmitter-acetylcholine using metal nanoparticles, 2D material and conducting polymer modified electrode, Biosens. Bioelectron 89 (2017) 377–383.

[23] R. Cha, J. Li, Y. Liu, Y. Zhang, Q. Xie, M. Zhang, Fe_3O_4 nanoparticles modified by CD-containing star polymer for MRI and drug delivery, Coll. Surf. B. Biointerfaces 158 (2017) 213–221.

[24] E. Umut, M. Coşkun, F. Pineider, D. Berti, H. Güngüneş, Nickel ferrite nanoparticles for simultaneous use in magnetic resonance imaging and magnetic fluid hyperthermia, J. Coll. Interface. Sci. 550 (2019) 199–209.

[25] A.S. Semkina, M.A. Abakumov, A.S. Skorikov, T.O. Abakumova, P.A. Melnikov, N.F. Grinenko, et al., Multimodal doxorubicin loaded magnetic nanoparticles for VEGF targeted theranostics of breast cancer, Nanomedicine. Nanotechnology. Biol. Med. 14 (5) (2018) 1733–1742.

[26] F. Dumitrache, I. Morjan, C. Fleaca, A. Badoi, G. Manda, S. Pop, et al., Highly magnetic Fe_2O_3 nanoparticles synthesized by laser pyrolysis used for biological and heat transfer applications, Appl. Surf. Sci. 336 (2015) 297–303.

[27] A.A. Bharde, R.Y. Parikh, M. Baidakova, S. Jouen, B. Hannoyer, T. Enoki, et al., Bacteria-mediated precursor-dependent biosynthesis of superparamagnetic iron oxide and iron sulfide nanoparticles, Langmuir. 24 (16) (2008) 5787–5794.

[28] P.A. Sundaram, R. Augustine, M. Kannan, Extracellular biosynthesis of iron oxide nanoparticles by *Bacillus subtilis* strains isolated from rhizosphere soil, Biotechnol. Bioprocess. Eng. 17 (2012) 835–840.

[29] T. SJK, V. R. A, M. V., A. Muthu, Biosynthesis of multiphase iron nanoparticles using *Syzygium aromaticum* and their magnetic properties, Coll. Surf. A Physicochem. Eng. Asp. 603 (2020) 125241. https://doi.org/10.1016/j.colsurfa.2020.125241.

[30] N. Madubuonu, S.O. Aisida, A. Ali, I. Ahmad, T. Zhao, S. Botha, et al., Biosynthesis of iron oxide nanoparticles via a composite of *Psidium guavaja-Moringa oleifera* and their antibacterial and photocatalytic study, J. Photochem. Photobiol. B. Biol. 199 (2019) 111601.

[31] S. Qasim, A. Zafar, M.S. Saif, Z. Ali, M. Nazar, M. Waqas, et al., Green synthesis of iron oxide nanorods using Withania coagulans extract improved photocatalytic degradation and antimicrobial activity, J. Photochem. Photobiol. B. Biol. 204 (2020) 111784.

[32] T. Gu, Y. Zhang, S.A. Khan, T.A. Hatton, Continuous flow synthesis of superparamagnetic nanoparticles in reverse miniemulsion systems, Coll. Interface. Sci. Commun. 28 (2019) 1–4.

[33] A. Ghasemi, Real and imaginary parts of magnetic susceptibility of fine dispersed nanoparticles synthesized by reverse micelle: from superparamagnetic trend to ferrimagnetic State, J. Clust. Sci. 27 (3) (2016) 979–992.

[34] C. Sciancalepore, A.F. Gualtieri, P. Scardi, A. Flor, P. Allia, P. Tiberto, et al., Structural characterization and functional correlation of Fe_3O_4 nanocrystals obtained using 2-ethyl-1, 3-hexanediol as innovative reactive solvent in non-hydrolytic sol-gel synthesis, Mater. Chem. Phys. 207 (2018) 337–349.

[35] S. Khan, Z.H. Shah, S. Riaz, N. Ahmad, S. Islam, M.A. Raza, et al., Antimicrobial activity of citric acid functionalized iron oxide nanoparticles—superparamagnetic effect, Ceram. Int. 46 (8) (2020) 10942–10951.

[36] A. Akbar, S. Riaz, R. Ashraf, S. Naseem, Magnetic and magnetization properties of Co-doped Fe_2O_3 thin films, IEEE. Trans. Magn. 50 (8) (2014) 4–7.

[37] S. Riaz, A. Akbar, S. Naseem Ferromagnetic Effects in Cr-Doped Fe2O3 Thin Films. IEEE Trans. Magn. 2014;50(8):8–11.

[38] J. Sánchez, D.A. Cortés-Hernández, J.C. Escobedo-Bocardo, R.A. Jasso-Terán, A. Zugasti-Cruz, Bioactive magnetic nanoparticles of Fe–Ga synthesized by sol–gel for their potential use in hyperthermia treatment, J. Mater. Sci. Mater. Med. 25 (10) (2014) 2237–2242.

[39] G. Marchegiani, P. Imperatori, A. Mari, L. Pilloni, A. Chiolerio, P. Allia, et al., Ultrasonics sonochemistry sonochemical synthesis of versatile hydrophilic magnetite nanoparticles, Ultrason. Sonochem. 19 (4) (2012) 877–882.

[40] I. Rosenberger, C. Schmithals, J. Vandooren, S. Bianchessi, P. Milani, E. Locatelli, et al., Physico-chemical and toxicological characterization of iron-containing albumin nanoparticles as platforms for medical imaging, J. Control. Release 194 (2014) 130–137.

[41] R. Wang, Y. Wan, F. He, Y. Qi, W. You, H. Luo, The synthesis of a new kind of magnetic coating on carbon fibers by electrodeposition, Appl. Surf. Sci. 258 (2012) 3007–3011.

[42] M. Jouyandeh, J.A. Ali, M. Aghazadeh, K. Formela, M.R. Saeb, Z. Ranjbar, et al., Curing epoxy with electrochemically synthesized $ZnxFe_{3-x}O_4$ magnetic nanoparticles, Prog. Org. Coat. 136 (2019) 105246.

[43] I. Karimzadeh, M. Aghazadeh, M.R. Ganjali, T. Doroudi, P.H. Kolivand, Preparation and characterization of iron oxide (Fe_3O_4) nanoparticles coated with polyvinylpyrrolidone/polyethylenimine through a facile one-pot deposition route, J. Magn. Magn. Mater. 433 (2017) 148–154.

[44] S. Majidi, F.Z. Sehrig, S.M. Farkhani, M.S. Goloujeh, A. Akbarzadeh, Current methods for synthesis of magnetic nanoparticles., Artif. Cells Nanomed. Biotechnol. 44 (2) (2016) 722–734.

[45] H. Yue, S. Marasini, M.Y. Ahmad, S.L. Ho, H. Cha, S. Liu, et al., Carbon-coated ultrasmall gadolinium oxide (Gd_2O_3@C) nanoparticles: application to magnetic resonance imaging and fluorescence properties, Coll. Surf. A. Physicochem. Eng. Asp. 586 (2020) 124261.

[46] L.L. Israel, A. Galstyan, E. Holler, J.Y. Ljubimova, Magnetic iron oxide nanoparticles for imaging, targeting and treatment of primary and metastatic tumors of the brain, J. Control. Release 320 (2020) 45–62.

[47] T. Kieu, H. Ta, M. Trinh, N.V. Long, T. Thanh, M. Nguyen, et al., Synthesis and surface functionalization of Fe_3O_4-SiO_2 core-shell nanoparticles with 3-glycidoxypropyltrimethoxysilane and 1,1-carbonyldiimidazole for bio-applications, Coll. Surf. A. Physicochem. Eng. Asp. 504 (2016) 376–383.

[48] S. Mortazavi-derazkola, M. Salavati-niasari, H. Khojasteh, Green synthesis of magnetic Fe_3O_4/SiO_2/HAp nanocomposite for atenolol delivery and in vivo toxicity study, J. Clean. Prod. 168 (2017) 39–50.

[49] R. Fan, H. Min, X. Hong, Q. Yi, W. Liu, Q. Zhang, et al., Plant tannin immobilized Fe_3O_4@SiO_2 microspheres: A novel and green magnetic bio-sorbent with superior adsorption capacities for gold and palladium, J. Hazard. Mater. 364 (1) (2019) 780–790.

[50] G. Liu, D. Hu, M. Chen, C. Wang, L. Wu, Multifunctional PNIPAM/Fe_3O_4-ZnS hybrid hollow spheres: synthesis, characterization, and properties, J. Coll. Interface Sci. 397 (2013) 73–79.

[51] Q. An, M. Yu, Y. Zhang, W. Ma, J. Guo, C. Wang, Fe_3O_4@carbon microsphere supported Ag-Au bimetallic nanocrystals with the enhanced catalytic activity and selectivity for the reduction of nitroaromatic compounds, J. Phys. Chem. C. 116 (42) (2012) 22432–22440.

[52] S. Dinda, M. Kakran, J. Zeng, T. Sudhaharan, S. Ahmed, D. Das, et al., Grafting of ZnS:Mn-doped nanocrystals and an anticancer drug onto graphene oxide for delivery and cell labeling, Chempluschem 81 (1) (2016) 100–107.

[53] N. Akter, L. Chowdhury, J. Uddin, A.K.M.A. Ullah, M.H. Shariare, M.S. Azam, N-halamine functionalization of polydopamine coated Fe_3O_4 nanoparticles for recyclable and magnetically separable antimicrobial materials, Mater. Res. Express 5 (11) (2018) 115007.

[54] S. Liu, B. Yu, S. Wang, Y. Shen, H. Cong, Preparation, surface functionalization and application of Fe_3O_4 magnetic nanoparticles, Adv. Coll. Interface Sci. 281 (2020) 102165.

[55] M. Maccarini, A. Atrei, C. Innocenti, R. Barbucci, Interactions at the CMC/magnetite interface : implications for the stability of aqueous dispersions and the magnetic properties of magnetite nanoparticles, Coll. Surf. A Physicochem. Eng. Asp. 462 (2014) 107–114.

[56] A. Van de Walle, J.E. Perez, A. Abou-Hassan, M. Hémadi, N. Luciani, C. Wilhelm, Magnetic nanoparticles in regenerative medicine: what of their fate and impact in stem cells? Mater. Today Nano. (2020) 11. doi:10.1016/j.mtnano.2020.100084.

[57] M. Abd Elkodous, G.S. El-Sayyad, I.Y. Abdelrahman, H.S. El-Bastawisy, A.E. Mohamed, F.M. Mosallam, et al., Therapeutic and diagnostic potential of nanomaterials for enhanced biomedical applications, Coll. Surf. B. Biointerfaces 180 (2019) 411–428.

[58] H.L. Chee, C.R.R. Gan, M. Ng, L. Low, D.G. Fernig, K.K. Bhakoo, et al., Biocompatible peptide-coated ultrasmall superparamagnetic iron oxide nanoparticles for in vivo contrast-enhanced magnetic resonance imaging, ACS Nano. 12 (7) (2018) 6480–6491.

[59] Y.Z. Gao, L. Zhang, W.S. Huo, S. Shi, J. Lian, Y.H. Gao, An integrated giant magnetoresistance microfluidic immuno-sensor for rapid detection and quantification of D-dimer, Chin. J. Anal. Chem. 43 (6) (2015) 802–807.

[60] T. Klein, W. Wang, L. Yu, K. Wu, K.L.M. Boylan, R.I. Vogel, et al., Development of a multiplexed giant magnetoresistive biosensor array prototype to quantify ovarian cancer biomarkers, Biosens. Bioelectron. 126 (2019) 301–307.

[61] V.D. Krishna, K. Wu, A.M. Perez, J.P. Wang, Giant magnetoresistance-based biosensor for detection of influenza A virus, Front. Microbiol. 7 (2016) 1–8.

[62] C.Y. Lee, L.P. Wu, T.T. Chou, Y.Z. Hs ieh, Functional magnetic nanoparticles-assisted electrochemical biosensor for eosinophil cationic protein in cell culture, Sens. Actuators B. Chem. 257 (2018) 672–677.

[63] J. Li, X. Li, R. Weng, T. Qiang, X. Wang, Glucose assay based on a fluorescent multi-hydroxyl carbon dots reversible assembly with phenylboronic acid brush grafted magnetic nanoparticles, Sens. Actuators. B. Chem. (2020) 304. doi:10.1016/j.snb.2019.127349.

[64] C. Weizhi, Y. Zhongheng, Human gastric carcinoma cells targeting peptide-functionalized iron oxide nanoparticles delivery for magnetic resonance imaging, J. Neurol. Sci. (2020) 116544.

[65] S. Song, H. Shen, T. Yang, L. Wang, H. Fu, H. Chen, et al., Indocyanine green loaded magnetic carbon nanoparticles for near infrared fluorescence/magnetic resonance dual-modal imaging and photothermal therapy of tumor, ACS Appl. Mater. Interfaces 9 (11) (2017) 9484–9495.

[66] J. Yan, S. Li, F. Cartieri, Z. Wang, T.K. Hitchens, J. Leonardo, et al., Iron oxide nanoparticles with grafted polymeric analogue of dimethyl sulfoxide as potential magnetic resonance imaging contrast agents, ACS Appl. Mater. Interfaces 10 (26) (2018) 21901–21908.

[67] N.S. Elbialy, M.M. Fathy, R. AL-Wafi, R. Darwesh, U.A. Abdel-dayem, M. Aldhahri, et al., Multifunctional magnetic-gold nanoparticles for efficient combined targeted drug delivery and interstitial photothermal therapy, Int. J. Pharm 554 (2019) 256–263.

[68] M. Kubovcikova, M. Koneracka, O. Strbak, M. Molcan, V. Zavisova, I. Antal, et al., Poly-L-lysine designed magnetic nanoparticles for combined hyperthermia, magnetic resonance imaging and cancer cell detection, J. Magn. Magn. Mater 475 (2019) 316–326.

[69] Y. Wang, W. Wang, L. Yu, L. Tu, Y. Feng, T. Klein, et al., Giant magnetoresistive-based biosensing probe station system for multiplex protein assays, Biosens. Bioelectron 70 (2015) 61–68.

[70] Y. Gao, W. Huo, L. Zhang, J. Lian, W. Tao, C. Song, et al., Multiplex measurement of twelve tumor markers using a GMR multi-biomarker immunoassay biosensor, Biosens. Bioelectron 123 (2019) 204–210.

[71] J. Choi, A.W. Gani, D.J.B. Bechstein, J.R. Lee, P.J. Utz, S.X. Wang, Portable, one-step, and rapid GMR biosensor platform with smartphone interface, Biosens. Bioelectron 85 (2016) 1–7.

[72] A.H. Haghighi, M.T. Khorasani, Z. Faghih, F. Farjadian, Effects of different quantities of antibody conjugated with magnetic nanoparticles on cell separation efficiency, Heliyon 6 (4) (2020) e03677.

[73] L.H.A. Silva, J.R. Silva, G.A. Ferreira, R.C. Silva, E.C.D. Lima, R.B. Azevedo, et al., Labeling mesenchymal cells with DMSA-coated gold and iron oxide nanoparticles: assessment of biocompatibility and potential applications, J. Nanobiotechnol 14 (1) (2016) 1–15.

[74] J. Mosafer, K. Abnous, M. Tafaghodi, H. Jafarzadeh, M. Ramezani, Preparation and characterization of uniform-sized PLGA nanospheres encapsulated with oleic acid-coated magnetic-Fe_3O_4 nanoparticles for simultaneous diagnostic and therapeutic applications, Coll. Surf. A Physicochem. Eng. Asp. 514 (2017) 146–154.

[75] T. Anani, A. Brannen, P. Panizzi, E.C. Duin, A.E. David, Quantitative, real-time in vivo tracking of magnetic nanoparticles using multispectral optoacoustic tomography (MSOT) imaging, J. Pharm. Biomed. Anal. 178 (2020) 112951.

[76] J.Z. Sun, Y.C. Sun, L. Sun, Synthesis of surface modified Fe_3O_4 super paramagnetic nanoparticles for ultra sound examination and magnetic resonance imaging for cancer treatment, J. Photochem. Photobiol. B. Biol. 197 (2019) 111547.

[77] I. Rosenberger, A. Strauss, S. Dobiasch, C. Weis, S. Szanyi, L. Gil-Iceta, et al., Targeted diagnostic magnetic nanoparticles for medical imaging of pancreatic cancer, J. Control. Release 214 (2015) 76–84.

[78] C. Ren, Q. Bayin, S. Feng, Y. Fu, X. Ma, J. Guo, Biomarkers detection with magnetoresistance-based sensors, Biosens. Bioelectron 165 (2020) 112340.

[79] A. Yanai, U.O. Häfeli, A.L. Metcalfe, P. Soema, L. Addo, C.Y. Gregory-Evans, et al., Focused magnetic stem cell targeting to the retina using superparamagnetic iron oxide nanoparticles, Cell Transplant 21 (6) (2012) 1137–1148.

[80] C. Dey, K. Baishya, A. Ghosh, M.M. Goswami, A. Ghosh, K. Mandal, Improvement of drug delivery by hyperthermia treatment using magnetic cubic cobalt ferrite nanoparticles, J. Magn. Magn. Mater 427 (2017) 168–174.

[81] A.S. Garanina, V.A. Naumenko, A.A. Nikitin, E. Myrovali, A.Y. Petukhova, S.V. Klimyuk, et al., Temperature-controlled magnetic nanoparticles hyperthermia inhibits primary tumor growth and metastases dissemination, Nanomedicine. Nanotechnology. Biol. Med. 25 (2020) 102171.

[82] D. Bizari, A. Yazdanpanah, F. Moztarzadeh, BaO–Fe_2O_3 containing bioactive glasses: a potential candidate for cancer hyperthermia, Mater. Chem. Phys. 241 (2020) 122439.

[83] X. Tang, F. Cao, W. Ma, Y. Tang, B. Aljahdali, M. Alasir, et al., Cancer cells resist hyperthermia due to its obstructed activation of caspase, Reports. Pract. Oncol. Radiother. 25 (3) (2020) 323–326.

[84] B. Thiesen, A. Jordan, Clinical applications of magnetic nanoparticles for hyperthermia, Int. J. Hyperth. 24 (6) (2008) 467–474.

[85] P.D.C. Carvalho De Jesus, D.S. Pellosi, A.C. Tedesco, Magnetic nanoparticles: applications in biomedical processes as synergic drug-delivery systems, in: A-M Holban, A.M. Grumezescu (Eds.), Materials for Biomedical Engineering: Nanomaterials-based Drug Delivery, Elsevier Inc., Amsterdam, 2019, pp. 371–396.

[86] M.A. Agotegaray, V.L. Lassalle, Magnetic nanoparticles as drug delivery devices, in: M.A. Agotegaray, V.L. Lassalle (Eds.), Silica-coated Magnetic Nanoparticles: An Insight into Targeted Drug Delivery and Toxicology, Springer International Publishing,, New York, NY, 2017, pp. 9–26.

[87] S. Mirza, M.S. Ahmad, M.I.A. Shah, M. AteeqM.R. Shah, M. Imran, S. Ullah (Eds.), Magnetic nanoparticles: drug delivery and bioimaging applications, Metal Nanoparticles for Drug Delivery and Diagnostic Applications (2020) 189–213.

[88] A. Senyei, K. Widder, G. Czerlinski, Magnetic guidance of drug-carrying microspheres, J. Appl. Phys. 49 (6) (1978) 3578–3583.

[89] J. Gavard, S. Hanini, C. Kacem, Ammar, et al., Evaluation of iron oxide nanoparticle biocompatibility, Int. J. Nanomed. 6 (2011) 787.

[90] M. Nedyalkova, B. Donkova, J. Romanova, G. Tzvetkov, S. Madurga, V. Simeonov, Iron oxide nanoparticles - In vivo/in vitro biomedical applications and in silico studies, Adv. Colloid. Interface. Sci. 249 (2017) 192–212.

[91] M. Agotegaray, A. Campelo, R. Zysler, F. Gumilar, C. Bras, A. Minetti, et al., Influence of chitosan coating on magnetic nanoparticles in endothelial cells and acute tissue biodistribution, J. Biomater. Sci. Polym. Ed. 27 (11) (2016) 1069–1085.

[92] L. Yang, H. Kuang, W. Zhang, Z.P. Aguilar, Y. Xiong, W. Lai, et al., Size dependent biodistribution and toxicokinetics of iron oxide magnetic nanoparticles in mice, Nanoscale 7 (2) (2015) 625–636.

[93] M.P. Calatayud, B. Sanz, V. Raffa, C. Riggio, M.R. Ibarra, G.F. Goya, The effect of surface charge of functionalized Fe_3O_4 nanoparticles on protein adsorption and cell uptake, Biomaterials 35 (24) (2014) 6389–6399.

[94] M. Salimi, S. Sarkar, S. Fathi, A.M. Alizadeh, R. Saber, F. Moradi, et al., Biodistribution, pharmacokinetics, and toxicity of dendrimer-coated iron oxide nanoparticles in BALB/c mice, Int. J. Nanomed. 13 (2018) 1483–1493.

[95] H. Arami, A. Khandhar, D. Liggitt, K.M. Krishnan, In vivo delivery, pharmacokinetics, biodistribution and toxicity of iron oxide nanoparticles, Chem. Soc. Rev. 44 (23) (2015) 8576–8607.

[96] X. Xie, L. Zhang, W. Zhang, R. Tayebee, A. Hoseininasr, H.H. Vatanpour, et al., Fabrication of temperature and pH sensitive decorated magnetic nanoparticles as effective biosensors for targeted delivery of acyclovir anti-cancer drug, J. Mol. Liq. 309 (2020) 113024.

[97] M.S. AlSalhi, K. Elangovan, A.J.A. Ranjitsingh, P. Murali, S. Devanesan, Synthesis of silver nanoparticles using plant derived 4-N-methyl benzoic acid and evaluation of antimicrobial, antioxidant and antitumor activity, Saudi. J. Biol. Sci. 26 (5) (2019) 970–978.

[98] H.C. Janßen, N. Angrisani, S. Kalies, F. Hansmann, M. Kietzmann, D.P. Warwas, et al., Biodistribution, biocompatibility and targeted accumulation of magnetic nanoporous silica nanoparticles as drug carrier in orthopedics, J. Nanobiotechnol 18 (1) (2020) 1–18.

[99] L. Alupei, C.A. Peptu, A.M. Lungan, J. Desbrieres, O. Chiscan, S. Radji, et al., New hybrid magnetic nanoparticles based on chitosan-maltose derivative for antitumor drug delivery, Int. J. Biol. Macromol. 92 (2016) 561–572.

[100] A.R. Patil, M.S. Nimbalkar, P.S. Patil, A.D. Chougale, P.B. Patil, Controlled release of poorly water soluble anticancerous drug camptothecin from magnetic nanoparticles, Mater. Today Proc. 23 (2020) 437–443.

[101] X. Ge, M. Fu, X. Niu, X. Kong, Atomic layer deposition of γ-Fe2O3 nanoparticles on multi-wall carbon nanotubes for magnetic drug delivery and liver cancer treatment, Ceram. Int. (2020) 1–7.

[102] B.A. Zasońska, V.I. Pustovyy, A.V. Babinskiy, O.M. Palyvoda, V.F. Chekhun, I. Todor, et al., Combined antitumor effect of surface-modified superparamagnetic maghemite nanoparticles and a vitamin E derivative on experimental Walker-256 mammary gland carcinosarcoma, J. Magn. Magn. Mater. 471 (2019) 381–387.

[103] T. Cristofolini, M. Dalmina, J.A. Sierra, A.H. Silva, A.A. Pasa, F. Pittella, et al., Multifunctional hybrid nanoparticles as magnetic delivery systems for siRNA targeting the HER2 gene in breast cancer cells, Mater. Sci. Eng. C. 109 (2020) 110555.

[104] M. Parsian, G. Unsoy, P. Mutlu, S. Yalcin, A. Tezcaner, U. Gunduz, Loading of gemcitabine on chitosan magnetic nanoparticles increases the anti-cancer efficacy of the drug, Eur. J. Pharmacol. 784 (2016) 121–128.

[105] M. Ayubi, M. Karimi, S. Abdpour, K. Rostamizadeh, M. Parsa, M. Zamani, et al., Magnetic nanoparticles decorated with PEGylated curcumin as dual targeted drug delivery: synthesis, toxicity and biocompatibility study, Mater. Sci. Eng. C. 104 (2019) 109810.

[106] N. Hedayati, A. Ramiar, M.M. Larimi, Investigating the effect of external uniform magnetic field and temperature gradient on the uniformity of nanoparticles in drug delivery applications, J. Mol. Liq. 272 (2018) 301–312.

[107] M. Johannsen, U. Gneveckow, L. Eckelt, A. Feussner, N. Waldöfner, R. Scholz, et al., Clinical hyperthermia of prostate cancer using magnetic nanoparticles: presentation of a new interstitial technique, Int. J. Hyperth. 21 (7) (2005) 637–647.

[108] M.W. Wilson, R.K. Kerlan, N.A. Fidelman, A.P. Venook, J.M. LaBerge, J. Koda, et al., Hepatocellular carcinoma: regional therapy with a magnetic targeted carrier bound to doxorubicin in a dual MR imaging/conventional angiography suite-initial experience with four patients, Radiology 230 (1) (2004) 287–293.

[109] P. Münster, D. Marchion, E. Bicaku, M. Schmitt, H.L. Ji, R. DeConti, et al., Phase I trial of histone deacetylase inhibition by valproic acid followed by the topoisomerase II inhibitor epirubicin in advanced solid tumors: a clinical and translational study, J. Clin. Oncol. 25 (15) (2007) 1979–1985.

[110] Q. Bi, X. Song, A. Hu, T. Luo, R. Jin, H. Ai, et al., Magnetofection: magic magnetic nanoparticles for efficient gene delivery, Chin. Chem. Lett. 31 (12) (2020) 3041–3046. https://doi.org/10.1016/j.cclet.2020.07.030.

[111] S. Mapukata, N. Nwahara, T. Nyokong, The photodynamic antimicrobial chemotherapy of *Staphylococcus aureus* using an asymmetrical zinc phthalocyanine conjugated to silver and iron oxide based nanoparticles, J. Photochem. Photobiol. A. Chem. 402 (2020) 112813.

[112] X. Sun, G. Zhang, R. Du, R. Xu, D. Zhu, J. Qian, et al., A biodegradable $MnSiO_3@Fe_3O_4$ nanoplatform for dual-mode magnetic resonance imaging guided combinatorial cancer therapy, Biomaterials 194 (2019) 151–160.

Natural biopolymeric nanomaterials for tissue engineering: overview and recent advances

Vaishali Pawar, Sneha Ravi and Rohit Srivastava

DEPARTMENT OF BIOSCIENCES AND BIOENGINEERING, INDIAN INSTITUTE OF TECHNOLOGY, BOMBAY, MAHARASHTRA, INDIA

Chapter outline

29.1 Introduction

Every year, millions of people across the globe suffer from tissue loss or organ failure. The human body has limited capacity to regenerate most of the tissues and organs; however, it fails to retain its integrity and functions in the event of a serious damage. Thus, there exists a big gap between the ever-increasing patient waitlist and the available organs for donation. Faced with these dire issues, research seeks to facilitate and expedite this regenerative process by stimulating the patient's own restorative potential or by making use of a substitute for the lost/damaged

Biopolymeric Nanomaterials: Fundamentals and Applications. DOI: https://doi.org/10.1016/B978-0-12-824364-0.00029-0

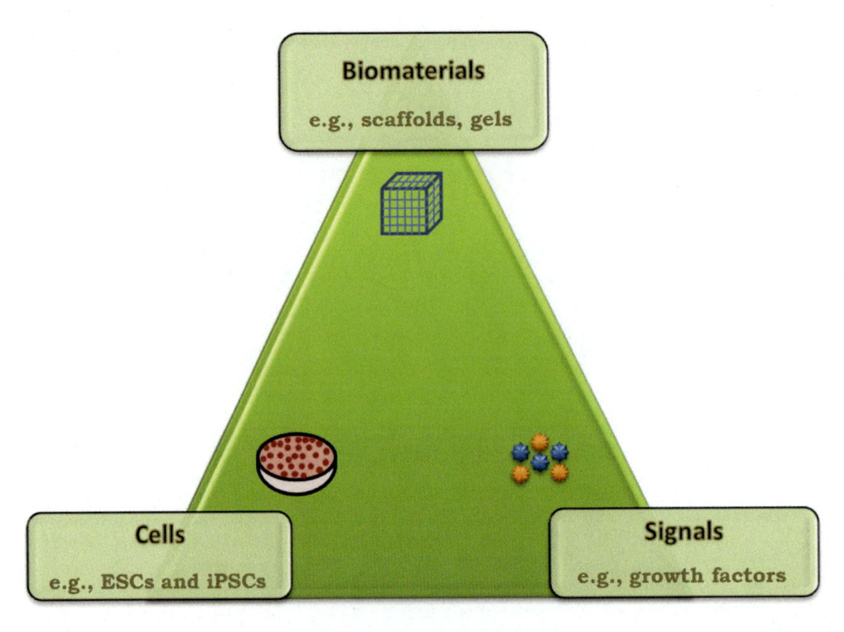

FIGURE 29.1 Key components of tissue engineering Triad [3] (ESCs: embryonic stem cells, iPSCs: induced pluripotent stem cells).

tissue or organ [1]. Tissue engineering is an interdisciplinary field that combines the principles of biomaterials, cell transplantation, and engineering to develop functional tissue substitutes and/or promote endogenous regeneration. Tissue engineering uses scaffolds, cells, biologically active molecules, growth factors, etc. to design the tissue substitutes for damaged tissues [2]. The production of an engineered tissue requires a construct with a three-dimensional environment, similar to that of a porous scaffold. The addition of appropriate cells and growth factors to the construct forms the tissue engineering "triad" as shown in Fig. 29.1.

The porous scaffold provides support and a suitable environment for the cells to proliferate and synthesize new tissue while the specific growth factors such as transforming growth factor-β (TGF-β), epidermal growth factor (EGF), insulin-like growth factor (IGF-I), fibroblast growth factor (FGF), vascular endothelial growth factor (VEGF), etc. facilitate the growth and division of specific cells. It is essential to modify the elements of the triad so as engineer the tissue with regard to its application. For effective tissue engineering, it is crucial to understand the mechanism and interaction between each component of the triad [3]. Thus, the ultimate aim of tissue engineering is to engineer constructs that restore, maintain, or regenerate damaged tissues.

Tissue engineering, being a multidisciplinary field, has widespread applications and thus, can provide solutions to a large number of problems. 3D cultures and models bridge the gap between 2D cultures and animal models. Tissue engineering can overcome the problems associated with organ donation, transplantation and rejection, tissue damage or loss, mechanical devices that fail to perform all the functions associated with the tissues, and surgical reconstruction [4].

The repair and regeneration of tissues, such as bone, cartilage, tendons, and ligaments, pave way for the entry of tissue engineering into the field of orthopedics [5]. Autografts, allografts, and synthetic grafts made from polymers can be replaced with high-quality constructs made from cells. The injection of these cells into the injured area, (e.g., ligaments) can expedite the repair process that includes assembly of collagen fibers and other extracellular matrix molecules, release of specific growth factors, and the initiation of associated immune responses [6]. Breast tissue engineering is another developing field that uses adipose tissue engineering strategies. Post mastectomy reconstruction involves substituting for lost skin and soft tissue volume to reconstruct the appearance of the breast. Currently used silicone breast implants and expanders for abdominal flap reconstruction using autologous tissues are subject to complications, such as silicone or saline leak and infections, and muscle atrophy overtime, respectively. Breast tissue engineering offers a great potential in eliminating the drawbacks with soft tissue augmentation in future [7]. Another important application is skin tissue engineering in plastic and reconstructive surgery. It can obliterate donor site morbidity related with skin grafts, as well as device *in-vitro* models to study malignant melanoma, psoriasis, atopic eczema, vitiligo, and its associated pathophysiological mechanisms [8]. Tissue engineering also has its branches spread out into the fields of dentistry, heart, pancreas, vascular system, lungs, stem cell therapy, and can thus emerge as a valuable tool for the next-generation biomedical research and engineering [3,4,9].

Irrespective of the tissue engineering strategy used, the proper selection of cells and biomaterials is of utmost importance. The desired properties of the tissue to be engineered drive this selection process. Cells such as keratinocytes, fibroblasts, embryonic stem cells, induced pluripotent stem cells, mesenchymal stem cells (MSCs), hematopoietic stem cells, etc. can be chosen depending on its application [8]. Biomaterials used in tissue engineering can be broadly classified into natural and synthetic, based on its origin. Thus, a biomaterial can be defined as a natural or synthetic substance designed to interact or mimic biological systems for medical treatment. Natural biomaterials can be further classified into protein-based, polysaccharide-based, decellularized tissue–derived biomaterials, etc. while synthetic biomaterials can be divided into aliphatic esters, conducting polymers, cell interactive polymer, etc. [10]. Both natural and synthetic biopolymers have been extensively used for treatment purposes. Certain advantages of natural biopolymers over synthetic are as follows: biocompatibility, biodegradability, and remodeling. Synthetic biopolymers have structural dissimilarity with their native tissue/organs and possess variations in their composition. They have low biocompatibility and low tissue remodeling ability. Naturally derived biopolymers, due to their advantages, play an important role in replacing and restoring the function of damaged tissues and organs. They aid in cell adhesion, migration, proliferation, and differentiation, all of which are key steps in the repair process [10].

Polymeric biomaterials in the form of scaffolds, hydrogels, sponges, membranes, etc. have many applications in tissue engineering. Elastin, a major protein component of the tissues, is a biopolymer containing amino acid chains with elasticity. It has applications in skin tissue engineering, as elastin is known to induce migration and differentiation of keratinocytes, which is essential to repair and reestablish the epidermis [11,12]. Polysaccharide-based biopolymers find their application as a neural tissue substitute that enhances attachment and proliferation of

PC12 cells, in dermal tissue engineering which aids adhesion, growth, and proliferation of 3T3 mouse fibroblasts and soft tissue repair [11,13]. Polyhydroxyalkanoates, another class of natural biopolymers derived from bacterial sources, are used in valves to aid tissue repair, cardiovascular patches, articular cartilage repair scaffolds, bone graft substitutes, and nerve guides [14]. Thus, biopolymers find their application in different aspects of tissue engineering, which include treatment of bone, soft tissues (cartilage, tendons, ligaments, etc.), skin, skeletal muscle, neural tissues, and vascular tissues.

29.2 Overview of biopolymeric nanomaterials in tissue engineering

Over the past few decades, there has been a strong interest in the implementation of nanomaterials in tissue engineering. Nanotechnology, a versatile field with numerous applications, has made significant contributions to the progress of tissue engineering and biomedical devices, along with therapies in cancer treatment, infections, diabetes, etc. The introduction of nanomaterials (nanoparticles, nanosheets, nanocrystals, etc.) into biopolymeric matrices (scaffolds, hydrogels, films, etc.) followed by 3D printing opens new avenues by creating 3D structures with enhanced properties and multifunctionality [15].

Certain properties that biopolymers and nanomaterials should possess are as follows:

1. *Biocompatibility*: The material should be nontoxic and allow scaffold replacement by the proteins synthesized and secreted by the host cells [4].
2. *Biodegradability*: The material should allow the host macrophages to infiltrate and clear the cellular debris.
3. *Mechanical support*: The material should provide mechanical support until the newly formed tissues mature and become functional [6].
4. *Antimicrobial*: The material should contain antimicrobial properties or should be loaded with certain antimicrobial agents to prevent the development of infections.
5. *Bioinertness*: The material should be inert or minimize the inflammatory and immunological reactions to prevent further tissue damage [8].
6. *Porous*: Ideal pore diameter of the material is important to facilitate cell migration, proliferation, and movement of growth factors across the material and enhance the regeneration process [6].
7. *Durability*: The material should be durable and sustain throughout the regeneration process without early degradation.
8. *Ductility*: The material should be able to bear the load without getting deformed.
9. *Clinical compliance*: The material should comply with all the guidelines under the Good Manufacturing Practice [4].

For bioprinting purposes, the biomaterial should be "printable", that is, the material should possess two main properties in addition to those mentioned above; rheology (shear thinning and viscosity) and cross-linking ability (covalent, ionic, stereocomplex, etc.) [8].

Table 29.1 Nanomaterials used for tissue engineering and their host polymers.

Nanomaterials	Polymer matrix	Applications	References
PLGA nanofibers	Polycaprolactone	Cell therapy and tissue regeneration	[15,19]
Carbon nanotubes	Silicon substrate	Tissue regeneration	[18]
PLLA/PCL nanofibers	Collagen	Blood vessel regeneration	[16]
nHA and PLGA nanospheres	Polyethylene glycol	Osteochondral tissue regeneration	[15,20]
SWNTs	Gelatin	Cardiac tissue engineering	[21,22]
MWCNTs	PEGDA	Nerve regeneration	[23]
Hydroxyapatite nanocrystals	Gelatin	Osteoblast regeneration	[18]
MWCNTs	Polycaprolactone	Cardiac tissue engineering	[24]
Graphene oxide nanosheets	GelMA	Myocardial vasculogenesis	[21,25]
PCL nanofibers	Gelatin	Tooth regeneration	[26,27]
Nanosilicates	GelMA	Bone tissue engineering	[28]
nHA nanocomposite	Collagen	Bone regeneration	[16,29]
SWCNTs	Agarose	3D organ printing	[30]

PLGA, poly(lactic-co-glycolic acid); *PLLA*, poly (l-lactide); *PCL*, poly(ε-caprolactone); *nHA*, nanohydroxyapatite; *SWNTs*, single-walled carbon nanotubes, *MWNTs*, multiwalled carbon nanotubes; *PEGDA*, poly(ethylene glycol) diacrylate; *GelMA*, gelatin methacryloyl

Generally, tissue constructs for engineering hard and soft tissue are developed using natural and synthetic biopolymers. However, the use of biopolymers alone has drawbacks such as low thermal stability and poor mechanical and barrier properties. Thus, to overcome these challenges, approaches to incorporate nanomaterials in the matrices of biopolymers are being studied extensively [15]. These nanomaterials can confer additional properties to the constructs such as targeted drug delivery, controlled drug release, and nanoscale topography that resemble the natural extracellular matrix (ECM) [16]. Current nanomaterials used in tissue constructs along with biopolymers are in the form of nanoparticles, nanosheets, nanocomposite films, nanotubes, nanohybrid membranes, etc. Polymeric nanoparticles can be further divided into protein nanoparticles (albumin, keratin, silk, and gelatin) and polysaccharide nanoparticles (alginate and chitosan). The first biopolymeric nanoparticle was developed using albumin (protein nanoparticle) and nonbiodegradable synthetic polymers [17]. The toxicity due to the nonbiodegradable polymer led to the use of biodegradable polymers such as poly (lactic-co-glycolic acid) (PLGA) and polyanhydride. Thus, nanoparticles such as liposomes, niosomes, virus-like-particle, proteins, etc. were used in the preparation of tissue constructs. Hydroxyapatite is a ceramic biomaterial that has resemblance with bone mineral. Nano-sized hydroxyapatite (nHA) has been widely used for hard tissue repair in orthopedics and dentistry [18]. Rheological properties such as shear thinning were greatly enhanced with the incorporation of nanofibrillated cellulose into the matrix of alginate that was used for printing cartilage tissues of meniscus and ears [15]. In recent studies, a variety of nanocomposite materials made of poly (propylene fumarate) and single-walled carbon nanotubes (SWCNT) have been prospected for its promising use in bone-tissue engineering [18]. Similarly, a list of few nanomaterials, the polymer matrix, and their applications in tissue engineering is given in Table 29.1.

29.3 Biopolymeric nanomaterial used in tissue engineering

So far, we have seen the properties and applications of biopolymers and nanomaterials in tissue engineering. Now, we will discuss in detail few of the many biopolymeric nanomaterials used in tissue engineering.

29.3.1 Cellulose-based nanomaterials

Cellulose is the most abundant biopolymer found in plants (e.g., birch, *Pinus radiata, Syzygium cumini,* etc.), few algal (*Cladophora, Cystoseira myrica,* etc.), bacterial (*Acetobacter, Pseudomonas, Achromobacter, Agrobacterium, Rhizobium,* etc.), and fungal (*Trichoderma, Aspergillus)* species [31]. It is a linear chain polysaccharide composed of repeating units of a disaccharide called cellobiose (β (1→4) linked D-glucose units) [32]. Cellulose produced by plant and bacteria is identical in structure; however, bacterial cellulose does not contain lignin and hemicelluloses and hence, does not require extensive purification [33]. It possesses desirable properties such as biodegradability, biocompatibility, good mechanical strength, high water-absorbing capacity, etc. and thus, finds its relevance in soft and hard tissue engineering [34]. Nanocellulose can be described as cellulose nanostructures with at least one dimension in the nanorange (1–100 nm). Therefore, depending on the shape and morphology, nanocellulose can be categorized into nanorods, nanofibrils, nanowhiskers, nanocrystals, etc. [31].

Cellulose has been widely used for the engineering of bone, cartilage, blood vessels, skin, neural tissues, adipose tissues, and many more. A study by Muller et al. was dedicated toward neural tissue engineering [35]. The study involved the growth of PC12 neuronal cells *in vitro* on polypyrrole-coated bacterial nanocellulose composite scaffolds. This resulted in the development of a composite scaffold with topographic features and electric activity, especially electrical conductivity that is a prerequisite for nerve tissue regeneration. Torgbo and Sukyai [36] and Hickey and Pelling [37] provided a detailed review on various bacterial and plant cellulose-based scaffolds, their properties, molecular mechanisms, methods for preparation, and applications in tissue engineering. Gao et al. [38] constructed cellulose nanofibers/quaternized chitin/organic rectorite composites for enhanced wound healing. *In-vivo* assessment on Male Sprague Dawley rats revealed quick hemostasis function, increased collagen synthesis, and neovascularization, thereby accelerating wound healing 3 days earlier compared to that with the traditional gauze. Further, porous cellulose scaffold activated in calcium hydroxide and coated with calcium phosphate was designed for cartilage tissue engineering [34], the use of nanocellulose as bioink for 3D printing tissues [39], particularly skin tissue, has been discussed [40], the development of cellulose nanofibers/hydroxyapatite composite has been studied for bone tissue engineering [41,42], the development of *in-vitro* adipose tissue model by 3D culturing for adipocytes in bacterial nanocellulose scaffolds for adipose tissue engineering [43], and the use of bacterial cellulose scaffold for growth of human dural fibroblasts for the reconstruction of dura mater using a dural patches following a cerebrospinal fluid leak [44].

29.3.2 Gelatin-based nanomaterials

Gelatin is a natural biopolymer obtained by partial or complete acid hydrolysis or thermal denaturation of collagen. As gelatin is a denatured product of collagen, it has an added advantage of being able to prevent immunogenicity and probable pathogen transmission as seen in the case of collagen scaffolds [45]. Gelatin is a polyampholyte, which is 13% positively charged, 12% negatively charged, and 11% hydrophobic. The general structural representation is $(Gly\text{-}X\text{-}Pro)_n$, where "X" is largely arginine, alanine, methionine, aspartic acid, lysine, or valine [46]. Other amino acids such as leucine, threonine, phenylalanine, histidine, tyrosine, serine, and cysteine make up a smaller percentage of gelatin. Unlike collagen that has a fixed structure and molecular weight, gelatin has a heterogeneous structure, which depends on cleavage of the collagen fiber during the denaturation process of gelatin [47]. Properties of gelatin include biocompatibility, biodegradability, flexible rheology, that is, viscosity and Young's modulus (gelatin can behave like a Newtonian fluid or a non-Newtonian fluid based on the pH, temperature, and concentration), cross-linking, low antigenicity, chemical modification potential, and high absorbing capacity [46,47].

The above-stated properties make gelatin, a desirable scaffold or nanomaterial in the fields of tissue engineering, such as cardiac, nerve, bone, vascular, corneal, skin, hepatic tissue engineering, etc. Shin et al. created a carbon nanotube embedded GelMA hydrogel for cardiac tissue engineering that showed spontaneous and synchronous beating of the heart in neonatal rat cardiomyocytes [48]. Liu et al. constructed a biomimetic nanofibrous gelatin/apatite composite scaffold for bone tissue engineering [45]. *In-vitro* experiments demonstrated that the nanofibrous gelatin scaffolds enhanced cell adhesion and proliferation compared to Gelfoam. Further incorporation of bone-like-apatite onto the surface of the nanofibrous-gelatin scaffold enhanced osteoblastic cell differentiation and the mechanical strength of the scaffold. Ghasemi-Mobarakeh et al. [49] designed an electrospun polycaprolactone (PCL)/gelatin nanofibrous scaffold to study and understand the nanofiber alignment and use of gelatin for nerve tissue engineering. In this study, they reported that PCL/gelatin scaffold in the ratio 70:30 enhanced nerve differentiation and proliferation compared to the PCL nanofibrous scaffold. Additionally aligned nanofibers supported and enhanced nerve cell neurite growth and differentiation. Further, Nikkhah et al. [47] discussed various application of gelatin nanomaterials and scaffolds in corneal tissue engineering and ophthalmology [50,51], skin tissue engineering [52,53], and other miscellaneous fields.

29.3.3 Silk protein and its nanocomposites

For decades, silk has been well known to the textile industries and to physicians for sutures [54]. Silks are natural fibrous proteins present in glands of silk-producing arthropods (such as silkworms, spiders, scorpions, mites and bees) and are spun into fibers during their metamorphosis [55,56]. The spun silk fibers are composed of two proteins: a central protein called fibroin (70%–80%), which is embedded, and bonded in a glue-like coating made up of sericin proteins

(20%–30%) [54,57]. Silk proteins generally comprise four different structural components: (1) elastic β-spirals, (2) crystalline β-sheets containing Gly-X (X being Ala, Ser, Thr, Val), (3) tight amino acid repeats forming α-helices, and (4) spacer regions [55,56]. Silk fibroin proteins have been extensively used for the development of composite materials such as fibers, membranes, hydrogels, films, sponges, etc. and have been implemented in the engineering and regeneration of bones, cartilage, tendons, etc. [57]. Properties of silk protein include good biocompatibility, water-based processing, biodegradability, chemical modification of the functional groups, and mechanical properties (elasticity, strain hardening behavior, tensile strength, etc.) [55].

In recent years, silk protein-based composites have been extensively investigated as one of the promising candidates for application in various fields, such as structural applications, biosensors and biomedical applications, tissue engineering, and drug delivery systems. Patra et al. [58] studied the ability of nonmulberry silk protein fibroin obtained from *Antheraea mylitta* as a scaffold for cardiac tissue engineering. The *in-vitro* study demonstrated enhanced cell adhesion, cellular metabolic activity, response to extracellular stimuli, cell-to-cell communication, and contractility of 3-day postnatal rat cardiomyocytes on silk fibroin. They also reported that *A. mylitta* silk fibroin exhibits properties similar to fibronectin, which is a natural component of cardiomyocytes and hence, demonstrate that *A. mylitta* silk fibroin 3D scaffolds are suitable for the engineering of cardiac patches. Correia et al. [59] investigated a variety of silk protein scaffolds for bone tissue engineering using human adipose-derived stem cells (hASCs) to develop autologous bone grafts. The scaffold were seeded with hASCs and cultured in osteogenic medium containing a decellularized trabecular bone. Of the different scaffolds fabricated, silk scaffold fabricated using hexafluoro-2-propanol displayed improved bone tissue formation by increased bone protein production (osteopontin, collagen type I, bone sialoprotein), enhanced calcium deposition, and total bone volume. Harkin et al. [60] explored the use of silk fibroin in ocular tissue reconstruction. They reviewed the advantages and disadvantages of using fibroin in the cellular reconstruction of corneoscleral limbus, corneal stroma, corneal endothelium, and outer blood-retinal barrier (Ruysch's complex) and concluded that fibroin's strength, structural versatility, and potential for modification make it a worthy candidate for further exploration. Further, use of silk fibroin/hydroxyapatite composites for bone tissue engineering [61,62] and scaffolds [63] for other applications has been discussed in a number a studies. Other applications, such as the use of silk for nerve regeneration, are yet to be fully investigated.

29.3.4 Chitin- and chitosan-based nanomaterials

Chitin is the second most abundant polymer in nature after cellulose and is derived from porifera (sponges), shells of marine crustaceans, exoskeleton of arthropods, and fungi [64]. Chitin, is a long chain, linear, semicrystalline polysaccharide composed of covalently bonded β (1→4)-2-(acetylamino)-2-deoxy-D-glucose (N-acetyl D-glucosamine) units [65]. Chitosan is a derivative of chitin that is obtained by deacetylation of chitin [120] using chemical hydrolysis under severe alkaline conditions or enzymatic hydrolysis using chitinase deacetylase. Chitosan is composed of β(1→4)-2-(acetylamino)-2-deoxy-D-glucose (N-acetyl D-glucosamine) and β-(1→4) -2-amino-2-deoxy-D-glucose (D-glucosamine) units. The distribution of these units can be random or blocked depending on the method of deacetylation. Although chitin and chitosan have structural similarities, they possess differences, such as chitin is insoluble in most organic solvents while chitosan is soluble in dilute organic solvents such as acetic acid and formic acid [66]. Chitin and chitosan are biodegradable, biocompatible, nonantigenic, nontoxic, hemostatic, antibacterial, mucoadhesive, and possess high water absorbing capacity [67,68]. Their particular structural similarity to glycosaminoglycans, which forms a majority of the extracellular matrix of the bone, has made it a favorable polymer in tissue engineering [66].

The fields in tissue engineering that make use of chitin- and chitosan-based scaffolds and nanomaterial majorly include bone, cardiac, nerve and skin tissue engineering. Chen et al. [69] synthesized chitosan-MWCNTs/hydroxyapatite nanocomposites by a novel *in-situ* precipitation method for bone tissue engineering. *In-vitro* studies showed that these nanocomposite scaffolds were capable of promoting the attachment and proliferation of preosteoblasts. Further, chitosan films improved the outgrowth of neurites when compared to chitin films, thus indicating the ability to modify nerve cell affinity based on the amine content in the polysaccharide [70]. Zhang et al. [71] developed a novel method to fabricate chitosan-derived sandwich tubular scaffold for blood vessel tissue engineering. The scaffold was found to possess good swelling property, a burst strength of 4000 mmHg, and high suture-retention strength. The scaffold also supported the growth and proliferation of vascular smooth muscle cells.

29.3.5 Pectin-based nanomaterials

Pectin is a complex, natural heteropolysaccharide and is the most abundant, multifunctional component, which is normally produced during the initial stages of growth of primary cell wall of most plants [72]. It also forms the main components of the peel and pulp of several fruits such as orange, lemon, grape fruit, lime, and apple pomace [73]. Pectin is composed of [1,4] linked d-galacturonic acid (GaIA) residues and galacturonic acid methyl ester units and contains a variety of neutral sugars, such as arabinose, galactose, rhamnose, and other sugars. Pectin has been widely used to prepare scaffolds for bone tissue engineering owing to the ionic and/or polar interactions of pectin and its ability to modulate cell behavior and enhance the adhesion and proliferation of osteoblasts [72]. Properties of pectin that make it suitable for the development of nanoparticles, scaffolds, etc. in tissue engineering are as follows: natural, biocompatible, biodegradable, water-soluble, nontoxic, and easily modifiable.

Vedhanayagam et al. [74] worked on the synthesis of collagen-silver-pectin nanoparticle-based scaffold for wound healing and tissue engineering applications. The *in-vitro* biocompatibility assay revealed that the collagen-silver-pectin nanoparticle-based scaffold provided higher antibacterial activity against Gram-positive and Gram-negative bacteria and also enhanced the growth and viability of keratinocytes. Archana et al. [75] investigated the biocompatibility, antimicrobial, and *in-vivo* wound healing properties of a titanium dioxide nanoparticle-loaded chitosan-pectin ternary nanodressing. The wound dressing supported the growth of NIH3T3 and L929 fibroblast cells. *In-vivo* open excision-type wound healing efficiency of the prepared nanodressing was examined in adult male albino rats and it was found that the dressing enhanced wound healing with high wound closure rate. In addition the nanodressing for wound also possessed good antibacterial activity, high swelling properties, high water vapor transmission rate, and excellent hydrophilic nature, which suggest potential application for tissue engineering along with wound healing. Munarin et al. [76] used pectin modified with an arginylglycylaspartic acid (RGD)-containing oligopeptide as an extracellular matrix alternative to immobilize cells for bone tissue regeneration. The grafting of the RGD peptide on pectin backbone improved MC3T3-E1 preosteoblasts adhesion and proliferation within the microspheres. Furthermore, the cells spread out from the microspheres and organized themselves into 3D structures producing a mineralized extracellular matrix.

29.3.6 Zein-based nanoparticles

Zein is a natural amphiphilic polymer and the principal protein in maize that comprises about 50% of the maize protein. Pure zein is odorless and tasteless solid, which is insoluble in water, soluble in alcohol and edible [77]. It has a molecular weight of about 40 kDa and belongs to the class of prolamins. It can be processed into a nontoxic biopolymer that undergoes a naturally controlled degradation process [78]. Structurally, zein contains a hydrophobic core and a hydrophilic surface. The hydrophobic internal core serves as an excellent water barrier and is therefore used as a coating material. Zein is available commercially in two forms; yellow zein (contains high amounts of xanthophyll and has a purity of about 88%–90%) and white zein (contains negligible xanthophyll and has a purity of about 96%) [79]. Zein is biocompatible, biodegradable, edible, antioxidant, antibacterial, can be chemically modified, and contains good thermoplastic properties and adhesive properties and hence has been used extensively for the development of films, fibers, coatings, composites, nanoparticles, and microspheres along with other natural polymers [79,80].

Fereshteh et al. [81] developed PCL/zein-coated 45S5 bioactive glass scaffolds and studied the mechanical properties, *in-vitro* bioactivity, and drug release for bone tissue engineering. The coated scaffolds showed good mechanical properties with sustained drug release. In addition, a dense bone-like apatite layer formed on the surface of PCL/zein-coated scaffolds immersed for 14 days in simulated body fluid. Thus, the developed scaffolds exhibit attractive properties for application in bone tissue engineering. Babitha and Korrapati [78] fabricated a biodegradable zein–polydopamine polymeric scaffold impregnated with titanium dioxide nanoparticles for skin tissue engineering. They investigated the thermal stability of the scaffold and its potential as a suitable wound dressing material. Further, owing to its nanotopographic structure, *in-vitro* studies demonstrated enhanced cell adhesion, proliferation, and migration using immortalized human keratinocyte (HaCaT) and mouse embryonic fibroblast (3T6-Swiss albino) cell lines. In addition, the *in-vivo* excisional wound healing experiment corroborated with the results of the *in-vitro* studies. Thus, the nanofibrous scaffold is anticipated to be an alternative, cost-effective biomaterial for skin tissue engineering applications. Hadavi et al. [82] synthesized zein nanoparticles as a carrier system for BMP-6 derived peptide for enhanced osteogenic differentiation of C2C12 cells. The results demonstrated that the BMP-6 peptide activated the expression of RUNX2, a transcription factor, which in turn regulated the gene expression of *SPP1* and *BGLAP*, the osteogenic marker genes. Therefore, the peptide-loaded zein nanoparticles act as an osteoinductive material, which may be used to repair small area of bone defects. Few other studies that focused on bone tissue engineering include the preparation and evaluation of electrospun zein/hydroxyapatite fibers by Zhang et al. [83], the fabrication of a biomimetic Zein polydopamine nanofibrous scaffold impregnated with BMP-2 peptide conjugated titanium dioxide nanoparticle by Babitha et al. [84], and vancomycin-loaded poly (sodium 4-styrene sulfonate)-modified hydroxyapatite nanoparticles in zein-based scaffold for bone defect treatment by Babaei et al. [80]. With the major application of zein in bone tissue engineering, zein has also been investigated for cartilage tissue engineering [85] and periodontal tissue engineering [86].

29.3.7 Keratin-based nanoparticles

Keratin is a tough, insoluble, cysteine-rich intermediate filament protein and can be used in a variety of biomedical applications [87,88]. Keratin fibers consist of an inner cortex region and an outer cuticle layer. The protein structure of keratin is composed of two classes: type I (acidic) keratin and type II (basic) keratin. They are found as hard keratins (5% sulfur) in hair, horns, nails, fur, feathers, and are classified as types Ia (acidic-hard) and IIa (basic-hard), whereas they are found as soft keratins (1% sulfur) in skin (cytoskeleton of the epithelial cells, stratum corneum) are classified as Ib (acidic-soft) and IIb (basic-soft). The classification depends on amino acid composition, distribution, and functions [89]. Further, on the basis of their secondary structure, they can be subdivided into α- (50%–60%), β-, and γ- (20%–30%) keratins. Keratin contains RGD (Arg-Gly-Asp) and LDV (Leu-Asp-Val) sequences, which are cell adhesion sequences, making it ideal for tissue engineering applications [87]. Other properties of keratin that make it suitable for tissue engineering are biocompatibility, biodegradability, nonimmunogenic, potential to accelerate the growth of fibroblasts for extracellular matrix production, and the presence of cellular interaction sites [87,89]. Recent studies involve the fabrication of keratin coatings, hydrogels, nanoparticles, scaffolds, films, etc.

Silva et al. [90] developed a novel keratin/alginate hybrid hydrogels for applications in soft tissue engineering. The ability of keratin to mimic the extracellular matrix was combined with the excellent chemical and mechanical stability of alginate to produce 2D and 3D hybrid hydrogels. Human umbilical vein endothelial cells were used to assess the cell interactions and it was found that the hybrid hydrogels supported cell attachment, spreading, and proliferation. Xu et al. [91] developed water-stable scaffolds composed of ultrafine keratin fibers oriented randomly and evenly in three dimensions for cartilage tissue engineering. The scaffold was formed from the highly cross-linked keratin extracted from chicken feathers, which was de-cross-linked and disentangled into linear and aligned molecules with a preserved molecular weight. The solution was then electrospun into scaffolds with ultrafine keratin fibers oriented randomly in three dimensions. Due to their highly cross-linked molecular structures, keratin scaffolds showed intrinsic water stability. Adipose-derived MSCs were able to penetrate deeper, proliferate, and chondrogenically differentiate, thus showing potential for cartilage tissue engineering [91]. Further, application of keratin in periodontal tissue engineering has been studied by Lee et al. [92], novel wound dressing based on nanofibrous poly(hydroxybutyrate-co-hydroxyvalerate)-keratin mats have been developed by Yuan et al. [88], and PCL/keratin-based composite nanofibers for biomedical applications have also been investigated [93].

29.3.8 Alginate-based nanomaterial

Alginate is a naturally occurring anionic, linear unbranched, chain-forming, hydrophilic, colloidal heteropolysaccharide that is found in the cell walls of brown algae and exopolysaccharide of some bacteria, such as *Pseudomonas* and *Azotobacter* [62,94,95]. Structurally, it is randomly arranged in blocks of α-L-guluronic acid (G block) and β-D-mannuronic acid (M block) residues and linked through 1,4- glycosidic linkages [96]. These blocks can be arranged consecutively (repeating units of GGGG or MMMM blocks), alternately (GMGMGMGM), or randomly, and the ratio and distribution varies depending on the species, source, and seasonal conditions. The free hydroxyls (−OH) and carboxyl (−COOH) groups enable alginates to form intramolecular hydrogen bonds and enhance tunability. Commercial production of alginate is from the *Laminaria* spp. (*Laminaria hyperborea, Laminaria digitata, and Laminaria japonica*), *Macrocystis pyrifera, Ascophyllum nodosum, Durvillaea antarctica,* and *Sargassum* spp. It is produced using a wide range of methods, including controlled gelification using Ca^{2+} ions, ionotropic gelation via intermolecular interactions, spray drying followed by cross-linking, alginate nanoaggregates through self-assembly, electrospinning and electrospraying, thermally induced phase separation, etc. Alginate has structural similarity with the ECM, which supports cell adhesion and can undergo chemical modifications [95,97]. In addition, alginate polymers are biodegradable, biocompatible, mucoadhesive, nonantigenic, hemocompatible, nontoxic, and possess chelating ability [98,99]. Thus, alginate-containing nanomaterials, such as scaffolds and hydrogels, are widely investigated for tissue engineering of bone, cartilage, and skin.

Coluccino et al. [100] developed bioactive TGF-β1/hydroxyapatite alginate-based scaffolds using freeze-drying technique for osteochondral tissue repair. The scaffolds were fabricated from a mixture of calcium cross-linked alginates followed by addition of TGF-β1 in the chondral layer and hydroxyapatite (HA) granules in the bony layer (as a bioactive signal) to create an osteoinductive surface for the cells. The scaffold promoted MSCs adhesion, proliferation, and demonstrated *in-vivo* biocompatibility. Thus, the functional and biomechanical properties make this scaffold a promising candidate for osteochondral tissue-engineering applications. Duruel et al. [101] synthesized sequential IGF-1 and BMP-6 releasing chitosan/alginate/PLGA hybrid scaffolds for periodontal tissue engineering to repair or regenerate the destructed or lost periodontium by improving functions of cells in the remaining tissue. *In-vitro* cell culture studies demonstrated the ability of the chitosan/alginate/PLGA hybrid scaffolds to induce proliferation and osteoblastic differentiation of cementoblasts when compared to the IGF-1 and BMP-6 free chitosan scaffold. Esfandiariy et al. [102] compared the aggrecan synthesis potential of natural articular chondrocytes and hASCs encapsulated in alginate microparticles and cultured in chondrogenic medium with and without TGF-β3 for 3 weeks. Their findings indicated that differentiated chondrocytes (DCs) with and without TGF-β3 synthesized more aggrecan than natural chondrocytes on day 14, while on day 21, DCs without TGF-β3 had a higher production than the other groups. Thus, application of TGF-β3 resulted in an increase in the amount of aggrecan in DCs on day 14 but a decrease on day 21. As aggrecan is an important chondrogenic marker, it was concluded that hASCs could be a reliable alternative cell source for cartilage tissue engineering in future. Further, Lauritano et al. [103] and Farokhi et al. [97] discussed the potential of alginate-based scaffolds in periodontal and cartilage tissue engineering, respectively. Zia et al. [96] review numerous alginate-based bionanocomposites, De Silva et al. [104] fabricated magnesium oxide nanoparticles reinforced electrospun alginate-based nanofibrous scaffolds, and Goodarzi et al. [105] investigated efficacy of the developed alginate-based hydrogel containing taurine-loaded chitosan nanoparticles for tissue engineering and biomedical applications.

29.3.9 Miscellaneous

Lignins possess an array of properties that make them advantageous for tissue engineering. Lignins possess properties such as biocompatibility, biodegradability, valorization potential, high mechanical strength, antimicrobial, and antioxidant capacity [106]. Particularly, lignin's potential as a reducing agent in the synthesis and manufacture of various high-value compounds has made it a favorable biopolymer in the field of nanomaterials [107]. Quraishi et al. [108] described a novel method to develop hybrid alginate-lignin aerogels for tissue engineering and regenerative medicine. They utilized pressurized carbon dioxide for gelation, and the characterization studies and cell studies revealed that the gel was noncytotoxic and supported the adhesion and proliferation of mouse L929 fibroblast cells. Wang et al. [109] engineered PCL/lignin nanofibers as an antioxidant scaffold for nerve tissue engineering. The results showed that the nanofibers supported the growth and proliferation of Schwann cells and dorsal root ganglion neurons by enhancing myelin basic protein expressions of Schwann cells and neurite outgrowth of dorsal root ganglion neurons. Wang et al. [110] developed lignin/PCL nanofibers for hydroxyapatite biomineralization in hard-tissue engineering. The study demonstrated that incubation of the electrospun lignin/PCL matrix in a simulated body fluid facilitated the growth of hydroxyapatite crystals, which were found to be mechanically and structurally similar to the native bone. In addition, the mineralized lignin/PCL nanofibrous films promoted efficient adhesion and proliferation of osteoblasts by directing filopodial extension.

Heparin contains high negative charge (approximately -75) due to the prevalence of sulfate and carboxylate groups, which mediates its electrostatic interactions with many proteins, such as growth factors, proteases, and chemokines. These interactions bring about the stabilization of growth factors, such as FGF and VEGF, and have thus been employed in the design of materials including scaffolds for tissue regeneration [111]. Qi Tan et al. [112] created a heparin/chitosan nanoparticle-immobilized decellularized bovine jugular vein scaffold to increase the loading capacity and allow for controlled release of VEGF for regeneration of decellularized tissue-engineered scaffolds. The scaffolds immobilized with heparin/chitosan nanoparticles stimulated endothelial cell proliferation and exhibited highly effective localization and sustained release of VEGF for several weeks *in vitro*. In addition, *in-vivo* experiments on mouse subcutaneous implantation model demonstrated the localization of VEGF due to the utilization of heparin/chitosan nanoparticles significantly increased fibroblast infiltration, extracellular matrix production, and accelerated vascularization [113]. Hyaluronic acid possesses unique properties such as biocompatibility, biodegradability, chemical modifications, gel-forming properties, viscoelastic properties, and structural flexibility [114,115]. Thus, hyaluronic acid–based nanomaterials (micelles, polymersomes, and inorganic nanoparticle formulations), scaffolds, hydrogels, etc. are extensively used for tissue engineering and regeneration purposes. Domingues et al. [32] developed a new class of injectable hydrogels composed of adipic acid dihydrazide-modified hyaluronic acid and aldehyde-modified hyaluronic acid reinforced with varying amounts of aldehyde-modified cellulose nanocrystals (a-CNCs). The study demonstrated that hyaluronic acid–CNCs nanocomposite hydrogels supported the growth and proliferation of hASCs under *in-vitro* culture conditions due to higher structural integrity and potential interaction of microenvironmental cues with CNC's sulfate groups. They also reported that hASCs encapsulated in hyaluronic acid–CNCs hydrogels demonstrated the ability to spread within the volume of gels and exhibited pronounced proliferative activity.

29.4 Clinical status update

Although tissue engineering has attracted extensive attention as a new therapeutic alternative for replacing or repairing damaged tissues and organ transplantation, its clinical translation has been a very slow moving process with limited clinical trials. This is because the biggest obstacle impeding the effective translation is safety concerns in humans. However, tremendous research is being carried out to expedite this process [116]. Of the several engineered tissues, cartilage, bone, skin, bladder, vascular grafts, cardiac tissues, etc. have entered into clinical trials, while the complex tissues, such as liver, lung, kidney, and heart, have been constructed in the laboratories and undergone *in-vivo* testing; however, these complex tissues face a plethora of challenges that hinders their clinical translation [117]. Nevertheless, there have been several successful trials resulting in tissue-engineered commercial products approved by the Food and Drug Administration (FDA) and European Medicines Agency (EMA), such as Dermagraft (repair of diabetic foot ulcers), TransCyte (treatment of mid-partial thickness burns), Durepair (repair of dura mater), Restore (soft tissue reinforcement), Osteomesh (craniofacial repair), Apligraf (venous leg ulcers repair), and OrCel (treatment of burn wounds) [118], and therapies, such as Holoclar (limbal stem cell therapy to repair the corneal injury) [119].

29.5 Conclusion and future perspectives

Tissue engineering and regenerative medicine is an interdisciplinary field encompassing many disciplines including engineering, medicine, and science. The field has gained an enormous recognition owing to its potential to replace damaged human tissues and organs to restore normal function. Although the early concept of tissue engineering stemmed from cell culture techniques, recent advances over the past few decades combine multiple innovative technologies to accelerate the translation of clinical therapies. In the bid to do so, few successful clinical translations include implantation of tissue engineered tubularized urethras using autologous smooth muscle and epithelial cells into urethral defect sites, full-grown trachea, small arteries, and skin grafts. While livers, lungs, and hearts have been successfully grown in labs, they are still a long way from being ready for implantation. 3D printing, an additive manufacturing technology, is one of the most explored scaffold fabrication technologies. Current approaches in biomaterials research aim to develop smart biomaterials that provide bioactive functions and have been investigated with the objective of facilitating and enhancing tissue regeneration by controlling the biological activities of cells and releasing regulatory factors to deliver outcomes more successfully than conventional therapies. With the rapid advancements made in tissue engineering, we anticipate the elimination of organ donation and organ rejection and thus the overall burden on health care services in future. Although the field currently faces a plethora of difficulties and challenges, coordinated efforts among researchers from various disciplines can expedite clinical applicability in the future and present a viable therapeutic option with life-extending benefits of tissue replacement or repair.

References

[1] F.M. Chen, X. Liu, Advancing biomaterials of human origin for tissue engineering, Prog. Polym. Sci. 53 (2016) 86–168.

[2] www.nibib.nih.gov. NIBIB Tissue Engineering and Regenerative Medicine [Internet]. 2019. Available from: www.nibib.nih.gov. (Accessed on June 2020).

[3] F. Akter, Principles of Tissue Engineering, in: F Akter (Ed.), Tissue Engineering Made Easy, University of Cambridge, Cambridge, United Kingdom: Academic Press, Elsevier, 2016, pp. 3–16.

[4] C. Castells-Sala, M. Alemany-Ribes, T. Fernández-Muiños, L. Recha-Sancho, P. López-Chicón, C.A.-. Re-verté, Javier Caballero-Camino AM-G and CES, Current Applications of Tissue Engineering in Biomedicine, J. Biochip. Tissue. chip. 2 (2013).

[5] A.C.-V. Suryavanshi, V. Borse, Vaishali Pawar SK and RS, Material advancements in bone-soft tissue fixation devices, Sci. Adv. Today 2 (2016) 1–12.

[6] E.W. Yates, A. Rupani, G.T. Foley, W.S. Khan, S. Cartmell, SJ. Anand, Ligament tissue engineering and its potential role in anterior cruciate ligament reconstruction, Stem. Cells. Int. 2012 (2012) 1–6.

[7] CW. Patrick, Breast tissue engineering, Annu. Rev. Biomed. Eng. 6 (2004) 109–130.

[8] S.P. Tarassoli, Z.M. Jessop, A. Al-Sabah, N. Gao, S. Whitaker, S. Doak, et al., Skin tissue engineering using 3D bioprinting: an evolving research field, J. Plast. Reconstr. Aesthetic. Surg. 71 (5) (2018) 615–623.

[9] D. Kaigler, D. Mooney, Tissue engineering's impact on dentistry, J. Dent. Educ. 65 (5) (2001) 456–462.

[10] BHaT Le, M. T, D. Nguyen, D. Minh, Naturally derived biomaterials: preparation and application, in: T Le Bao Ha, T Minh, D Nguyen, D Minh (Eds.), Regenerative Medicine and Tissue Engineering, InTechOpen, London, UK, 2013, pp. 247–274.

[11] LA. Loureiro dos Santos, Natural polymeric biomaterials: processing and properties, Ref. Modul. Mater. Sci. Mater. Eng. 2001 (2017) 609–615.

[12] J.F. Almine, D.V. Bax, S.M. Mithieux, L.N. Smith, J. Rnjak, A. Waterhouse, et al., Elastin-based materials, Chem. Soc. Rev. 39 (9) (2010) 3371–3379.

[13] S. Tiwari, R. Patil, P. Bahadur, Polysaccharide based scaffolds for soft tissue engineering applications, Polymers. (Basel) 11 (1) (2018) 1–23.

[14] Q. Wu, Y. Wang, GQ. Chen, Medical application of microbial biopolyesters polyhydroxyalkanoates, Artif. Cells Blood Subs. Biotechnol. 37 (1) (2009) 1–12.

[15] M. Hassan, K. Dave, R. Chandrawati, F. Dehghani, VG. Gomes, 3D printing of biopolymer nanocomposites for tissue engineering: nanomaterials, processing and structure-function relation, Eur. Polym. J. 121 (2019) 109340.

[16] M. Goldberg, R. Langer, X. Jia, Nanostructured materials for applications in drug delivery and tissue engineering, J. Biomater. Sci. Polym. Ed. 18 (3) (2007) 241–268.

[17] S. Sundar, J. Kundu, SC. Kundu, Biopolymeric nanoparticles, Sci. Technol. Adv. Mater. 11 (1) (2010) 014104.

[18] I. Armentano, M. Dottori, E. Fortunati, S. Mattioli, JM. Kenny, Biodegradable polymer matrix nanocomposites for tissue engineering: a review, Polym. Degrad. Stab. 95 (11) (2010) 2126–2146.

[19] N. Maurmann, D.P. Pereira, D. Burguez, F. Pereira, P.I. Neto, R.A. Rezende, et al., Mesenchymal stem cells cultivated on scaffolds formed by 3D printed PCL matrices, coated with PLGA electrospun nanofibers for use in tissue engineering, Biomed. Phys. Eng. Exp. 3 (4) (2017) 045005.

[20] N.J. Castro, J. O'Brien, LG. Zhang, Integrating biologically inspired nanomaterials and table-top stereolithography for 3D printed biomimetic osteochondral scaffolds, Nanoscale 7 (33) (2015) 14010–14022.

[21] R. Amezcua, A. Shirolkar, C. Fraze, DA. Stout, Nanomaterials for cardiac myocyte tissue engineering, Nanomaterials 6 (7) (2016) 133.

[22] J. Zhou, J. Chen, H. Sun, X. Qiu, Y. Mou, Z. Liu, et al., Engineering the heart: evaluation of conductive nanomaterials for improving implant integration and cardiac function, Sci. Rep. 4 (2014) Article number: 3733.

[23] S.J. Lee, W. Zhu, M. Nowicki, G. Lee, D.N. Heo, J. Kim, et al., 3D printing nano conductive multi-walled carbon nanotube scaffolds for nerve regeneration, J. Neural. Eng. 15 (1) (2018) 016018.

[24] C.M.B. Ho, A. Mishra, P.T.P. Lin, S.H. Ng, W.Y. Yeong, Y.J. Kim, et al., 3D printed polycaprolactone carbon nanotube composite scaffolds for cardiac tissue engineering, Macromol. Biosci. 17 (4) (2017) 1600250.

[25] A. Paul, A. Hasan, K.H. Al, A.K. Gaharwar, V.T.S. Rao, M. Nikkhah, et al., Injectable graphene oxide/hydrogel-based angiogenic gene delivery system for vasculogenesis and cardiac repair, ACS Nano. 8 (8) (2014) 8050–8062.

[26] X. Yang, F. Yang, X.F. Walboomers, Z. Bian, M. Fan, JA. Jansen, The performance of dental pulp stem cells on nanofibrous PCL/gelatin/nHA scaffolds, J. Biomed. Mater. Res. A 93 (1) (2010) 247–257.

[27] M. Chieruzzi, S. Pagano, S. Moretti, R. Pinna, E. Milia, L. Torre, et al., Nanomaterials for tissue engineering in dentistry, Nanomaterials 6 (7) (2016) 1–21.

[28] J.R. Xavier, T. Thakur, P. Desai, M.K. Jaiswal, N. Sears, E. Cosgriff-Hernandez, et al., Bioactive nanoengineered hydrogels for bone tissue engineering: a growth-factor-free approach, ACS Nano. 9 (3) (2015) 3109–3118.

[29] R. Murugan, S. Ramakrishna, Development of nanocomposites for bone grafting, Compos. Sci. Technol. 65 (2005) 2385–2406.

[30] A. Nadernezhad, N. Khani, G.A. Skvortsov, B. Toprakhisar, E. Bakirci, Y. Menceloglu, et al., Multifunctional 3D printing of heterogeneous hydrogel structures, Sci. Rep. 6, (2016) 33178.

[31] L. Bacakova, J. Pajorova, M. Bacakova, A. Skogberg, P. Kallio, K. Kolarova, et al., Versatile application of nanocellulose: from industry to skin tissue engineering and wound healing, Nanomaterials 9 (2) (2019).

[32] R.M.A. Domingues, M. Silva, P. Gershovich, S. Betta, P. Babo, S.G. Caridade, et al., Development of injectable hyaluronic acid/cellulose nanocrystals bionanocomposite hydrogels for tissue engineering applications, Bioconjug. Chem. 26 (8) (2015) 1571–1581.

[33] K. Novotna, P. Havelka, T. Sopuch, K. Kolarova, V. Vosmanska, V. Lisa, et al., Cellulose-based materials as scaffolds for tissue engineering, Cellulose 20 (5) (2013) 2263–2278.

[34] F.A. Müller, L. Müller, I. Hofmann, P. Greil, M.M. Wenzel, R. Staudenmaier, Cellulose-based scaffold materials for cartilage tissue engineering, Biomaterials 27 (21) (2006) 3955–3963.

[35] D. Muller, J.P. Silva, C.R. Rambo, G.M.O. Barra, F. Dourado, FM. Gama, Neuronal cells behavior on polypyrrole coated bacterial nanocellulose three-dimensional (3D) scaffolds, J. Biomater. Sci. Polym. Ed. 24 (11) (2013) 1368–1377.

[36] S. Torgbo, P. Sukyai, Bacterial cellulose-based scaffold materials for bone tissue engineering, Appl. Mater. Today 11 (2018) 34–49.

[37] R.J. Hickey, AE. Pelling, Cellulose biomaterials for tissue engineering, Front. Bioeng. Biotechnol. 7 (2019) 45.

[38] H. Gao, Z. Zhong, H. Xia, Q. Hu, Q. Ye, Y. Wang, et al., Construction of cellulose nanofibers/quaternized chitin/organic rectorite composites and their application as wound dressing materials, Biomater. Sci. 7 (6) (2019) 2571–2581.

[39] C.C. Piras, S. Fernández-Prieto, WM. De Borggraeve, Nanocellulosic materials as bioinks for 3D bioprinting, Biomater. Sci. 5 (10) (2017) 1988–1992.

[40] A. Rees, L.C. Powell, G. Chinga-Carrasco, D.T. Gethin, K. Syverud, K.E. Hill, et al., 3D bioprinting of carboxymethylated-periodate oxidized nanocellulose constructs for wound dressing applications, Biomed. Res. Int. 2015 (2015) 1–7.

[41] Y. Huang, J. Wang, F. Yang, Y. Shao, X. Zhang, K. Dai, Modification and evaluation of micro-nano structured porous bacterial cellulose scaffold for bone tissue engineering, Mater. Sci. Eng. C 75 (2017) 1034–1041.

[42] S. Eftekhari, I. El Sawi, Z.S. Bagheri, G. Turcotte, H. Bougherara, Fabrication and characterization of novel biomimetic PLLA/cellulose/hydroxyapatite nanocomposite for bone repair applications, Mater. Sci. Eng. C 39 (1) (2014) 120–125.

[43] P. Krontiras, P. Gatenholm, DA. Hagg, Adipogenic differentiation of stem cells in three-dimensional porous bacterial nanocellulose scaffolds, J. Biomed. Mater. Res. B. Appl. Biomater 103 (1) (2015) 195–203.

[44] E. Goldschmidt, M. Cacicedo, S. Kornfeld, M. Valinoti, M. Ielpi, P.M. Ajler, et al., Construction and in vitro testing of a cellulose dura mater graft, Neurol. Res. 38 (2016) 25–31.

[45] X. Liu, L.A. Smith, J. Hu, PX. Ma, Biomimetic nanofibrous gelatin/apatite composite scaffolds for bone tissue engineering, Biomaterials 30 (12) (2009) 2252–2258.

[46] AO. Elzoghby, Gelatin-based nanoparticles as drug and gene delivery systems: reviewing three decades of research, J. Control. Release 172 (3) (2013) 1075–1091.

[47] M. Nikkhah, M. Akbari, A. Paul, A. Memic, A. Dolatshahi-Pirouz, A. Khademhosseini, Gelatin-based biomaterials for tissue engineering and stem cell bioengineering, Biomaterials from Nature for Advanced Devices and Therapies. Editor(s): Nuno M. Neves, Rui L. Reis, Arizona State University: Wiley-Blackwell, Arizona, United States, 2016, pp. 37–62.

[48] S.R. Shin, S.M. Jung, M. Zalabany, K. Kim, P. Zorlutuna, S.B. Kim, et al., Carbon-nanotube-embedded hydrogel sheets for engineering cardiac constructs and bioactuators, ACS Nano. 7 (3) (2013) 2369–2380.

[49] L. Ghasemi-Mobarakeh, M.P. Prabhakaran, M. Morshed, M.H. Nasr-Esfahani, S. Ramakrishna, Electrospun poly(ε-caprolactone)/gelatin nanofibrous scaffolds for nerve tissue engineering, Biomaterials 29 (34) (2008) 4532–4539.

[50] T. Mimura, S. Amano, S. Yokoo, S. Uchida, S. Yamagami, T. Usui, et al., Tissue engineering of corneal stroma with rabbit fibroblast precursors and gelatin hydrogels, Mol. Vis. 14 (2008) 1819–1828.

[51] M.M. Jumblatt, D.M. Maurice, BD. Schwartz, A gelatin membrane substrate for the transplantation of tissue cultured cells, Transplantation 29 (6) (1980) 498–499.

[52] C.K. Perng, C.L. Kao, Y.P. Yang, H.T. Lin, L.W. Bin, Y.R. Chu, et al., Culturing adult human bone marrow stem cells on gelatin scaffold with pNIPAAm as transplanted grafts for skin regeneration, J. Biomed. Mater. Res. A 84 (3) (2008) 622–630.

[53] J. Mao, L. Zhao, K. De Yao, Q. Shang, G. Yang, Y. Cao, Study of novel chitosan-gelatin artificial skin in vitro, J. Biomed. Mater. Res. A 64 (2) (2003) 301–308.

[54] N. Kasoju, U. Bora, Silk fibroin in tissue engineering, Adv. Healthc. Mater. 1 (4) (2012) 393–412.

[55] B. Kundu, R. Rajkhowa, S.C. Kundu, X. Wang, Silk fibroin biomaterials for tissue regenerations, Adv. Drug. Deliv. Rev. 65 (4) (2013) 457–470.

[56] V. Kearns, A.C. Macintosh, A. Crawford, V.P. Hatton, Silk-Based Biomaterials For Tissue Engineering, in: N. Ashammakhi, R. Reis, F. Chiellni (Eds.), Topics in Tissue Engineering, University of Sheffield, Claremont Crescent, Sheffield, S10 2TA, UK: Biomaterials and Tissue Engineering, 2008.

[57] Z.H. Li, S.C. Ji, Y.Z. Wang, X.C. Shen, H. Liang, Silk fibroin-based scaffolds for tissue engineering, Front. Mater. Sci. 7 (2013) 237–247.

[58] C. Patra, S. Talukdar, T. Novoyatleva, S.R. Velagala, C. Mühlfeld, B. Kundu, et al., Silk protein fibroin from *Antheraea mylitta* for cardiac tissue engineering, Biomaterials 33 (9) (2012) 2673–2680.

[59] C. Correia, S. Bhumiratana, L.P. Yan, A.L. Oliveira, J.M. Gimble, D. Rockwood, et al., Development of silk-based scaffolds for tissue engineering of bone from human adipose-derived stem cells, Acta Biomater 8 (7) (2012) 2483–2492.

[60] D.G. Harkin, K.A. George, P.W. Madden, I.R. Schwab, D.W. Hutmacher, T.V. Chirila, Silk fibroin in ocular tissue reconstruction, Biomaterials 32 (10) (2011) 2445–2458.

[61] H.J. Kim, U.J. Kim, H.S. Kim, C. Li, M. Wada, G.G. Leisk, et al., Bone tissue engineering with premineralized silk scaffolds, Bone 42 (6) (2008) 1226–1234.

[62] M. Farokhi, F. Mottaghitalab, S. Samani, M.A. Shokrgozar, S.C. Kundu, R.L. Reis, et al., Silk fibroin/hydroxyapatite composites for bone tissue engineering, Biotechnol. Adv. 36 (1) (2018) 68–91.

[63] P. Bhattacharjee, B. Kundu, D. Naskar, H.W. Kim, T.K. Maiti, D. Bhattacharya, et al., Silk scaffolds in bone tissue engineering: an overview, Acta Biomater 63 (2017) 1–17.

[64] V. Pawar, M.C. Bavya, K.V. Rohan, R. Srivastava, Advances in polysaccharide-based antimicrobial delivery vehicles, in: B. Li, T. Moriarty, T.X.M. Webster (Eds.), Racing for the Surface. 1st edn, Switzerland: Springer, Cham, 2020, pp. 267–295.

[65] F. Croisier, C. Jérôme, Chitosan-based biomaterials for tissue engineering, Eur. Polym. J. 49 (4) (2013) 780–792.

[66] S. Deepthi, J. Venkatesan, S.K. Kim, J.D. Bumgardner, R. Jayakumar, An overview of chitin or chitosan/nano ceramic composite scaffolds for bone tissue engineering, Int. J. Biol. Macromol. 93 (2016) 1338–1353.

[67] V. Pawar, R. Srivastava, Chitosan-polycaprolactone blend sponges for management of chronic osteomyelitis: a preliminary characterization and in vitro evaluation, Int. J. Pharm. 568 (2019) 776–787.

[68] V. Pawar, M. Dhanka, R. Srivastava, Cefuroxime conjugated chitosan hydrogel for treatment of wound infections, Coll. Surf. B Biointerfaces 173 (2018) 776–787 Available from:. https://doi.org/10.1016/j.colsurfb.2018.10.034 .

[69] L. Chen, J. Hu, X. Shen, H. Tong, Synthesis and characterization of chitosan-multiwalled carbon nanotubes/hydroxyapatite nanocomposites for bone tissue engineering, J. Mater. Sci. Mater. Med. 24 (2013) 1843–1851.

[70] TL. Yang, Chitin-based materials in tissue engineering: Applications in soft tissue and epithelial organ, Int. J. Mol. Sci. 12 (3) (2011) 1936–1963.

[71] L. Zhang, Q. Ao, A. Wang, G. Lu, L. Kong, Y. Gong, et al., A sandwich tubular scaffold derived from chitosan for blood vessel tissue engineering, J. Biomed. Mater. Res. A 77 (2) (2006).

[72] L. Zhao, J. Li, L. Zhang, Y. Wang, J. Wang, B. Gu, et al., Preparation and characterization of calcium phosphate/pectin scaffolds for bone tissue engineering, RSC. Adv. 6 (2016) 62071–62082.

[73] A. Noreen, Z.I.H. Nazli, J. Akram, I. Rasul, A. Mansha, N. Yaqoob, et al., Pectins functionalized biomaterials; a new viable approach for biomedical applications: a review, Int. J. Biol. Macromol. 101 (2017) 254–272.

[74] M. Vedhanayagam, M. Nidhin, N. Duraipandy, N.D. Naresh, G. Jaganathan, M. Ranganathan, et al., Role of nanoparticle size in self-assemble processes of collagen for tissue engineering application, Int. J. Biol. Macromol. 99 (2017) 655–664.

[75] D. Archana, J. Dutta, PK. Dutta, Evaluation of chitosan nano dressing for wound healing: characterization, in vitro and in vivo studies, Int. J. Biol. Macromol. 57 (2013) 193–203.

[76] F. Munarin, S.G. Guerreiro, M.A. Grellier, M.C. Tanzi, M.A. Barbosa, P. Petrini, et al., Pectin-based injectable biomaterials for bone tissue engineering, Biomacromolecules 12 (3) (2011) 568–577.

[77] G. Labib, Overview on zein protein: a promising pharmaceutical excipient in drug delivery systems and tissue engineering, Expert. Opin. Drug. Deliv. 15 (1) (2018) 65–75.

[78] S. Babitha, PS. Korrapati, Biodegradable zein-polydopamine polymeric scaffold impregnated with TiO_2 nanoparticles for skin tissue engineering, Biomed. Mater. 12 (5) (2017) 055008.

[79] R. Paliwal, S. Palakurthi, Zein in controlled drug delivery and tissue engineering, J. Control. Release 189 (2014) 108–122.

[80] M. Babaei, A. Ghaee, J. Nourmohammadi, Poly (sodium 4-styrene sulfonate)-modified hydroxyapatite nanoparticles in zein-based scaffold as a drug carrier for vancomycin, Mater. Sci. Eng. C 100 (2019) 874–885.

[81] Z. Fereshteh, P. Nooeaid, M. Fathi, A. Bagri, AR. Boccaccini, Mechanical properties and drug release behavior of PCL/zein coated 45S5 bioactive glass scaffolds for bone tissue engineering application, Data Br. 4 (2015) 524–528.

[82] M. Hadavi, S. Hasannia, S. Faghihi, F. Mashayekhi, H. Homazadeh, SB. Mostofi, Zein nanoparticle as a novel BMP6 derived peptide carrier for enhanced osteogenic differentiation of C2C12 cells, Artif. Cells, Nanomed. Biotechnol. 46 (Suppl 1) (2018) 559–567.

[83] M. Zhang, Y. Liu, Y. Jia, H. Han, D. Sun, Preparation and evaluation of electrospun Zein/HA fibers based on two methods of adding HA nanoparticles, J. Bionic. Eng. 11 (1) (2014) 115–124.

[84] S. Babitha, M. Annamalai, M.M. Dykas, S. Saha, K. Poddar, J.R. Venugopal, et al., Fabrication of a biomimetic ZeinPDA nanofibrous scaffold impregnated with BMP-2 peptide conjugated TiO_2 nanoparticle for bone tissue engineering, J. Tissue. Eng. Regen. Med. 12 (4) (2018) 991–1001.

[85] Y.X. Lin, Z.Y. Ding, Z.X. Bin, S.T. Li, D.M. Xie, Z.Z. Li, et al., In vitro and in vivo evaluation of the developed PLGA/HAp/Zein scaffolds for bone-cartilage interface regeneration, Biomed. Environ. Sci. 28 (1) (2015) 1–12.

[86] X. Yan-Zhi, W. Jing-Jing, Y.P. Chen, J. Liu, N. Li, F.Y. Yang, The use of zein and Shuanghuangbu for periodontal tissue engineering, Int. J. Oral Sci. 2 (3) (2010) 142–148.

[87] N. Bhardwaj, W.T. Sow, D. Devi, K.W. Ng, B.B. Mandal, NJ. Cho, Silk fibroin-keratin based 3D scaffolds as a dermal substitute for skin tissue engineering, Integr. Biol. (United Kingdom) 7 (1) (2015) 53–63.

[88] J. Yuan, J. Geng, Z. Xing, K.J. Shim, I. Han, J.C. Kim, et al., Novel wound dressing based on nanofibrous PHBV-keratin mats, J. Tissue. Eng. Regen. Med. 9 (9) (2015) 1027–1035.

[89] F. Costa, R. Silva, AR. Boccaccini, Fibrous protein-based biomaterials (silk, keratin, elastin, and resilin proteins) for tissue regeneration and repair, in: M.A. Barbosa, M.C.L. Martins (Eds.), Peptides and Proteins as Biomaterials for Tissue Regeneration and Repair, Woodhead Publishing, Cambridge, 2018, pp. 175–204.

[90] R. Silva, R. Singh, B. Sarker, D.G. Papageorgiou, J.A. Juhasz, J.A. Roether, et al., Hybrid hydrogels based on keratin and alginate for tissue engineering, J. Mater. Chem. B 2 (33) (2014) 5441–5451.

[91] H. Xu, S. Cai, L. Xu, Y. Yang, Water-stable three-dimensional ultrafine fibrous scaffolds from keratin for cartilage tissue engineering, Langmuir 30 (28) (2014) 8461–8470.

[92] H. Lee, K. Noh, S.C. Lee, I.K. Kwon, D.W. Han, I.S. Lee, et al., Human hair keratin and its-based biomaterials for biomedical applications, Tissue. Eng. Regen. Med. 11 (4) (2014) 255–265.

[93] A. Edwards, D. Jarvis, T. Hopkins, S. Pixley, N. Bhattarai, Poly(e-caprolactone)/keratin-based composite nanofibers for biomedical applications, J. Biomed. Mater. Res. B. Appl. Biomater 103 (1) (2015) 21–30.

[94] V. Pawar, H. Topkar, R. Srivastava, Chitosan nanoparticles and povidone iodine containing alginate gel for prevention and treatment of orthopedic implant associated infections, Int. J. Biol. Macromol. 115 (2018) 1131–1141.

[95] I.P.S. Fernando, W.W. Lee, E.J. Han, G. Ahn, Alginate-based nanomaterials: fabrication techniques, properties, and applications, Chem. Eng. J. 391 (2020) 123823.

[96] F. Zia, M. Salman, M. Ali, R. Iqbal, A. Rasul, M. Najam-ul-Haq, et al., Alginate-based bionanocomposites, in: K Mahmood Zia, F Jabeen, MN Anjum, S Ikram (Eds.), Bionanocomposites [Internet], Elsevier, 2020, pp. 173–205. Available from:. http://www.sciencedirect.com/science/article/pii/B9780128167519000088.

[97] M. Farokhi, F. Jonidi Shariatzadeh, A. Solouk, H. Mirzadeh, Alginate based scaffolds for cartilage tissue engineering: a review, Int. J. Polym. Mater. Polym. Biomater 69 (4) (2020) 230–247.

[98] J. Sun, H. Tan, Alginate-based biomaterials for regenerative medicine applications, Materials. (Basel) 6 (4) (2013) 1285–1309.

[99] V. Pawar, V. Borse, R. Thakkar, R. Srivastava, Dual-purpose injectable doxorubicin conjugated alginate gel containing polycaprolactone microparticles for anti-cancer and anti-inflammatory therapy, Curr. Drug. Deliv. 15 (5) (2018) 716–726.

[100] L. Coluccino, P. Stagnaro, M. Vassalli, S. Scaglione, Bioactive TGF-β1/HA alginate-based scaffolds for osteochondral tissue repair: design, realization and multilevel characterization, J. Appl. Biomater. Funct. Mater. 14 (1) (2016) e42–e52.

[101] T. Duruel, A.S. Çakmak, A. Akman, R.M. Nohutcu, M. Gümüşderelioğlu, Sequential IGF-1 and BMP-6 releasing chitosan/alginate/PLGA hybrid scaffolds for periodontal regeneration, Int. J. Biol. Macromol. 104 (2017) 232–241.

[102] E. Esfandiariy, B. Hashemibeni, M. Hatef, M. Ansar, M. Mardani, S. Zarkesh-Esfahani, et al., A comparative study of aggrecan synthesis between natural articular chondrocytes and differentiated chondrocytes from adipose derived stem cells in 3D culture, Adv. Biomed. Res. 1 (1) (2012) 24.

[103] D. Lauritano, L. Limongelli, G. Moreo, G. Favia, F. Carinci, Nanomaterials for periodontal tissue engineering: chitosan-based scaffolds. A systematic review, Nanomaterials 10 (4) (2020) 605.

[104] R.T. De Silva, M. Mantilaka, K.L. Goh, S.P. Ratnayake, G.A.J. Amaratunga, KMN. De Silva, Magnesium oxide nanoparticles reinforced electrospun alginate-based nanofibrous scaffolds with improved physical properties, Int. J. Biomater 2017 (2017) 1391298.

[105] A. Goodarzi, M. Khanmohammadi, S. Ebrahimi-Barough, M. Azami, A. Amani, A. Baradaran Rafii, et al., Alginate-Based Hydrogel Containing Taurine-Loaded Chitosan Nanoparticles in Biomedical Application, Arch. Neurosci. 6 (2) (2019) e86349.

[106] M.H. Sipponen, H. Lange, C. Crestini, A. Henn, M. Österberg, Lignin for nano- and microscaled carrier systems: applications, trends, and challenges, ChemSusChem 12 (10) (2019) 2039–2054.

[107] C. Thulluri, S.R. Pinnamaneni, P.R. Shetty, U. Addepally, Synthesis of lignin-based nanomaterials/nanocomposites: recent trends and future perspectives, Ind. Biotechnol. 12 (3) (2016) 153–160.

[108] S. Quraishi, M. Martins, A.A. Barros, P. Gurikov, S.P. Raman, I. Smirnova, et al., Novel non-cytotoxic alginate–lignin hybrid aerogels as scaffolds for tissue engineering, J. Supercrit. Fluids 105 (2015) 1–8.

[109] J. Wang, L. Tian, B. Luo, S. Ramakrishna, D. Kai, X.J. Loh, et al., Engineering PCL/lignin nanofibers as an antioxidant scaffold for the growth of neuron and Schwann cell, Coll. Surf. B Biointerfaces 169 (2018) 356–365.

[110] D. Wang, J. Jang, K. Kim, J. Kim, CB. Park, Tree to bone": lignin/polycaprolactone nanofibers for hydroxyapatite biomineralization, Biomacromolecules 20 (7) (2019) 2684–2693.

[111] Y. Liang, KL. Kiick, Heparin-functionalized polymeric biomaterials in tissue engineering and drug delivery applications, Acta Biomater 10 (4) (2014) 1588–1600.

[112] Q. Tan, H. Tang, J. Hu, Y. Hu, X. Zhou, Y. Tao, et al., Controlled release of chitosan/heparin nanoparticle-delivered VEGF enhances regeneration of decellularized tissue-engineered scaffolds, Int. J. Nanomed. 6 (2011) 929–942.

[113] M. Fathi-Achachelouei, H. Knopf-Marques, C.E. Ribeiro da Silva, J. Barthès, E. Bat, A. Tezcaner, et al., Use of nanoparticles in tissue engineering and regenerative medicine, Front. Bioeng. Biotechnol. 7 (2019) 113.

[114] M. Hemshekhar, R.M. Thushara, S. Chandranayaka, L.S. Sherman, K. Kemparaju, KS. Girish, Emerging roles of hyaluronic acid bioscaffolds in tissue engineering and regenerative medicine, Int. J. Biol. Macromol. 86 (2016) 917–928.

[115] M. Farid, Hyaluronic acid and derivatives for tissue engineering, J. Biotechnol. Biomater s3 (2013) , 001.

[116] P. Sharma, P. Kumar, R. Sharma, V.D. Bhatt, PS. Dhot, Tissue engineering; current status & futuristic scope, J. Med. Life 12 (2019) 225–229.

[117] P.K. Chandra, S. Soker, A. Atala, Tissue engineering: current status and future perspectives, in: R Lanza, R Langer, A Atala (Eds.), Principles of Tissue Engineering, Elsevier, Cambridge, MA, United States, 2020, pp. 1–35.

[118] Y.S. Wong, C.Y. Tay, S.S.V. Feng Wen, LPT. Engineered, Polymeric biomaterials for tissue engineering, Curr. Tissue. Eng. 1 (2012) 41–53.

[119] B.T. O'Donnell, C.J. Ives, O.A. Mohiuddin, BA. Bunnell, Beyond the present constraints that prevent a wide spread of tissue engineering and regenerative medicine approaches, Front. Bioeng. Biotechnol. 7 (2019) 95.

[120] V. Pawar, U. Bulbake, W. Khan, R. Srivastava, et al., Chitosan sponges as a sustained release carrier system for the prophylaxis of orthopedic implant-associated infections, Int. J. Biol. Macromol. 134 (2019) 100–112. https://doi.org/10.1016/j.ijbiomac.2019.04.190.

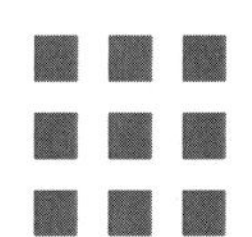

Index

Page numbers followed by "*f*" and "*t*" indicate, figures and tables respectively.

M

T